ACCESSING AutoCAD ARCHITECTURAL DESKTOP™ RELEASE 2

WILLIAM WYATT

ACCESSING AutoCAD
ARCHITECTURAL
DESKTOP™ RELEASE 2

Autodesk
Press

Thomson Learning™

Africa • Australia • Canada • Denmark • Japan • Mexico • New Zealand • Philipines
• Puerto Rico • Singapore • Spain • United Kingdom • United States

NOTICE TO THE READER

Publisher does not warrant or guarantee any of the products described herein or perform any independent analysis in connection with any of the product information contained herein. Publisher does not assume, and expressly disclaims, any obligation to obtain and include information other than that provided to it by the manufacturer.

The reader is expressly warned to consider and adopt all safety precautions that might be indicated by the activities herein and to avoid all potential hazards. By following the instructions contained herein, the reader willingly assumes all risks in connection with such instructions.

The publisher makes no representation or warranties of any kind, including but not limited to, the warranties of fitness for particular purpose or merchantability, nor are any such representations implied with respect to the material set forth herein, and the publisher takes no responsibility with respect to such material. The publisher shall not be liable for any special, consequential, or exemplary damages resulting, in whole or part, from the readers' use of, or reliance upon, this material. Autodesk does not guarantee the performance of the software and Autodesk assumes no responsibility or liability for the performance of the software or for errors in this manual.

Trademarks

Autodesk, AutoCAD and the AutoCAD logo are registered trademarks of Autodesk, Inc. Thomson Learning uses "Autodesk Press" with permission from Autodesk, Inc. for certain purposes. Windows is a trademark of the Microsoft Corporation. All other product names are acknowledged as trademarks of their respective owners.

Autodesk Press Staff
Executive Director: Alar Elken
Executive Editor: Sandy Clark
Acquisitions Editor: Michael Kopf
Developmental Editor: John Fisher
Executive Marketing Manager: Maura Theriault
Executive Production Manager: Mary Ellen Black
Production Manager: Larry Main
Art and Design Coordinator: Mary Beth Vought
Marketing Coordinator: Paula Collins
Technology Project Manager: Tom Smith

Cover illustration by Scott Keidong

COPYRIGHT © 2001 Thomson Learning™.

Printed in Canada
1 2 3 4 5 6 7 8 9 10 XXX 05 04 03 02 01 00
For more information, contact
Autodesk Press
3 Columbia Circle, Box 15-015
Albany, New York USA 12212-15015;
or find us on the World Wide Web at http://www.autodeskpress.com

All rights reserved Thomson Learning 2000. The text of this publication, or any part thereof, may not be reproduced or transmitted in any form or by any means, electronic or mechanical, including photocopying, recording, storage in an information retrieval system, or otherwise, without prior permission of the publisher.

You can request permission to use material from this text through the following phone and fax numbers.
Phone: 1-800-730-2214; Fax: 1-800-730-2215; or visit our Web site at www.thomsonrights.com

Library of Congress Cataloging-in-Publication Data

Wyatt, William
   Accessing AutoCAD Architectural Desktop, release 2 / William Wyatt.
     p. cm.
   ISBN 0-7668-1262-6
Architectural drawing—Computer-assisted instruction. 2. AutoCAD. I. Title

   NA2728 .W96 2000
   720'.28'402855369—dc21

00-026874

CONTENTS

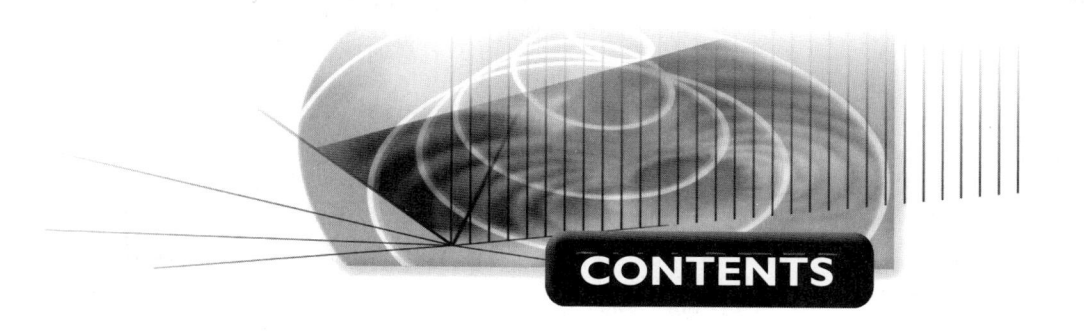

PREFACE
FEATURES OF THIS EDITION xviii
STYLE CONVENTIONS xix
HOW TO USE THIS BOOK xix
    Command Access Tables xix
ONLINE COMPANION xxi
WE WANT TO HEAR FROM YOU xxi
ABOUT THE AUTHOR xxi
    Dedication xxi
ACKNOWLEDGEMENTS xxii

CHAPTER 1 INTRODUCTION TO ARCHITECTURAL DESKTOP
OVERVIEW 1
    Advantages of Object Technology 1
ARCHITECTURAL DESKTOP RELEASE 2 2
STARTING ARCHITECTURAL DESKTOP 3
ARCHITECTURAL DESKTOP SCREEN LAYOUT 6
    Shortcut Menus 8
    Architectural Desktop Toolbars 10
    Steps to Creating a Profile 13
TUTORIAL 1.1 CREATING AND USING A PROFILE 13
    Architectural Desktop Resources 16
    Templates 16
    Defining Display Control for Objects and Viewports 25
    Accessing the Viewer 31
TUTORIAL 1.2 VIEWING OBJECTS IN LAYOUTS 34
SUMMARY 39
REVIEW QUESTIONS 39

CHAPTER 2 CREATING FLOOR PLANS
INTRODUCTION 41
OBJECTIVES 41
DRAWING WALLS 42
    Creating Walls with the Add Wall Command 42
TUTORIAL 2.1 DRAWING WALLS USING POLYLINE CLOSE 50
TUTORIAL 2.2 USING ORTHO CLOSE 53
TUTORIAL 2.3 CREATING WALLS 57

SETTING UP FOR DRAWING AND EDITING WALLS 59
    Using the Node Object Snap...60
    Grips of the Wall..60
    Setting Justification 61
    Setting the Cleanup Radius 62
STEPS TO SETTING THE CLEANUP RADIUS 63
STEPS TO EDITING THE CLEANUP RADIUS OF EXISTING WALLS 63
    Managing Wall Error Markers 64
STEPS FOR TURNING ON GRAPH DISPLAY REPRESENTATION OF
WALLS 66
    Using Temporary Tracking Point to Draw Interior Walls 69
TUTORIAL 2.4 USING TRACKING TO DRAW INTERIOR WALLS AND
DISPLAYING CLEANUP RADII 70
    Using AutoCAD Offset to Draw Additional Interior Walls....................76
    Using Grips to Edit Walls..78
TUTORIAL 2.5 CREATING INTERIOR PARTITIONS WITH GRIPS AND
COPY-STRETCH 79
TUTORIAL 2.6 USING GRIPS TO CHANGE WALL LENGTH....................81
    Editing Walls with AutoCAD Editing Commands.....................................83
TUTORIAL 2.7 DRAWING EXTERIOR WALLS OF THE FLOOR PLAN 84
    Editing Walls using Modify Wall 88
TUTORIAL 2.8 USING THE MATCH OPTION TO CHANGE WALLS............92
SETTING WALL PROPERTIES...94
    Changing Wall Direction with Reverse Wall 99
TUTORIAL 2.9 USING THE CONVERT TO WALL COMMAND TO
CONVERT A SKETCH TO WALLS 100
    Creating Wall Dimensions...102
TUTORIAL 2.10 USING THE CREATE WALL DIMENSIONS
COMMAND TO PLACE DIMENSIONS 103
    Setting Dimension Properties 105
SUMMARY 108
REVIEW QUESTIONS 109
    Project 110
EXERCISE 2.11 DRAWING THE WALLS OF A FLOOR PLAN 110

CHAPTER 3 ADVANCED WALL FEATURES

INTRODUCTION..111
OBJECTIVES..111
ACCESSING WALL STYLES 111
STEPS TO SETTING A WALL STYLE CURRENT 112
CREATING WALL STYLES..113
    Creating a New Wall Style Name..114
STEPS TO CREATING A NEW WALL STYLE...115
    Editing a Wall Style 116

STEPS TO OVERRIDING THE SYSTEM DEFAULT DISPLAY
  REPRESENTATION FOR A WALL STYLE 130
TUTORIAL 3.1 CREATING A WALL STYLE 133
    Enhancing Wall Styles with Display Properties 137
    Changing a Wall Style with Wall Modify 141
TUTORIAL 3.2 CONTROLLING DISPLAY OF WALL COMPONENTS 142
    Exporting and Importing Wall Styles 148
STEPS TO EXPORTING A WALL STYLE 152
STEPS TO IMPORTING A WALL STYLE 152
    Purging Wall Styles 152
TUTORIAL 3.3 IMPORTING AND EXPORTING WALL STYLES 153
CREATING ENDCAPS 154
STEPS TO ASSIGNING AN ENDCAP TO A WALL STYLE 158
    Modifying Endcaps 160
OVERRIDING ENDCAPS AND PRIORITY OF WALL STYLES 161
    Editing Endcaps using Wall Style Overrides 163
    Using Priority Overrides 165
TUTORIAL 3.4 CREATING AND EDITING AN ENDCAP 166
CREATING WALL MODIFIER STYLES 174
    Attaching a Wall Modifier Using Wall Properties 176
    Changing Wall Modifiers 179
TUTORIAL 3.5 CREATING WALL MODIFIERS 181
CREATING CLEANUP GROUPS 188
TUTORIAL 3.6 CREATING AND USING A WALL GROUP 191
CREATING ADDITIONAL FLOORS 193
TUTORIAL 3.7 CREATING THE BASEMENT FLOOR PLAN 195
CREATING A FOUNDATION PLAN 205
TUTORIAL 3.8 CREATING A FOUNDATION PLAN 206
SUMMARY 213
REVIEW QUESTIONS 214
PROJECT 215
EXERCISE 3.1 CREATING A WALL STYLE 215

CHAPTER 4 PLACING DOORS AND WINDOWS

INTRODUCTION 217
OBJECTIVES 217
PLACING DOORS IN WALLS 218
    Setting the Swing 221
    Locating the Insertion Point of the Door 222
    Defining Door Properties 225
CHANGING DOORS WITH MODIFY DOOR 228
    Changing the Angle of Door Swing 229
TUTORIAL 4.1 INSERTING DOORS 230
CREATING DOOR STYLES 239
    Creating a New Door Style 242

STEPS TO CREATING A NEW DOOR STYLE..242
EDITING A DOOR STYLE..242
TUTORIAL 4.2   CREATING AND EDITING A DOOR STYLE.......................251
EDITING A DOOR WITH DOOR PROPERTIES...260
   Shifting a Door Within the Wall 263
TUTORIAL 4.3   REPOSITIONING A DOOR WITHIN THE WALL.................264
TUTORIAL 4.4   USING REPOSITION WITHIN WALL TO OFFSET A
  DOOR A SPECIFIC DISTANCE...266
   Shifting the Door Along a Wall 268
PLACING WINDOWS IN WALLS 269
   Defining Window Properties 273
   Using Modify Window 276
   Editing a Window with Grips..277
STEPS TO USING GRIPS TO COPY A WINDOW.......................................279
TUTORIAL 4.5   INSERTING WINDOWS IN A DRAWING 280
CREATING A WINDOW STYLE 283
   Defining Window Style Properties...285
STEPS TO CREATING A NEW WINDOW STYLE..286
   Editing a Window Style..286
TUTORIAL 4.6   CREATING A WINDOW STYLE 292
CREATING OPENINGS 298
STEPS TO ADDING AN OPENING TO A WALL 300
   Modifying an Opening with Modify Opening..301
CREATING DOORS, WINDOWS AND OPENINGS USING AEC
  PROFILES..305
   Creating an Aec Profile...305
TUTORIAL 4.7   CREATING AN AEC PROFILE AND USING THE
  PROFILE FOR A DOOR STYLE..307
IMPORTING AND EXPORTING DOOR AND WINDOW STYLES 309
STEPS TO IMPORTING DOOR STYLES 310
TUTORIAL 4.8 EXPORTING DOOR STYLES TO A NEW FILE.....................313
SUMMARY 315
REVIEW QUESTIONS 315
PROJECTS...316
EXERCISE 4.1   INSERTING DOORS IN A FLOOR PLAN............................316
EXERCISE 4.2   INSERTING WINDOWS IN A FLOOR PLAN......................316
EXERCISE 4.3   INSERTING AND REPOSITIONING DOORS AND
  WINDOWS IN A FLOOR PLAN..317

CHAPTER 5 CREATING ROOFS

INTRODUCTION..319
OBJECTIVES...319
CREATING A ROOF WITH ADD ROOF...319
   Creating a Hip Roof...325
TUTORIAL 5.1   CREATING A HIP ROOF 325

Contents

DEFINING ROOF PROPERTIES..330
    Editing Roof Properties of an Existing Roof 333
    Using Modify Roof to Change Roofs..336
CREATING A GABLE ROOF 337
TUTORIAL 5.2 CREATING A GABLE ROOF 337
CREATING A SHED ROOF...342
TUTORIAL 5.3 CREATING A SHED ROOF...342
EDITING AN EXISTING ROOF TO CREATE GABLES 345
STEPS TO CREATING A GABLE BY EDITING ROOF PROPERTIES......346
    Using Edit Roof Edges/Faces to Edit Roof Planes 346
    Using Grips to Edit a Roof...348
STEPS TO CREATING A GABLE USING GRIPS 349
CREATING A GAMBREL ROOF USING DOUBLE SLOPED ROOF........350
TUTORIAL 5.4 CREATING A GAMBREL ROOF.................................350
CREATING A ROOF USING CONVERT TO ROOF................................358
TUTORIAL 5.5 CREATING A ROOF USING ROOF CONVERT 358
USING MASS GROUPS TO DETERMINE ROOF INTERSECTIONS 360
TUTORIAL 5.6 USING MASS GROUPS TO IDENTIFY ROOF
  INTERSECTIONS...363
CREATING DORMERS..366
TUTORIAL 5.7 CREATING ROOFS FOR DORMERS 368
EXTENDING WALLS TO THE ROOF 379
TUTORIAL 5.8 EXTENDING WALLS TO THE ROOF 384
CREATING A FLAT ROOF..386
TUTORIAL 5.9 CREATING A FLAT ROOF...386
SUMMARY 396
REVIEW QUESTIONS 396
PROJECT 397
EXERCIS 5.1 CREATING A SHED ROOF 397

CHAPTER 6 STAIRS AND RAILINGS

INTRODUCTION..399
OBJECTIVES..399
USING THE ADD STAIR COMMAND 400
    Creating a Stair..400
    Designing Stairs with Stair Properties 404
STEPS TO CREATING A STAIR WITH A FIXED RISE AND NUMBER
  OF RISERS 411
STEPS TO CREATING A STAIR WITH A FIXED RUN AND FIXED RISE........414
    Stair Default Settings 416
CREATING STRAIGHT STAIRS..416
TUTORIAL 6.1 CREATING A STRAIGHT STAIR 419
CREATING U SHAPED STAIRS..423
STEPS TO CREATING A U SHAPED STAIR 425
TUTORIAL 6.2 CREATING A U SHAPED STAIR 427

CREATING MULTI-LANDING STAIRS 433
STEPS TO CREATING A MULTI-LANDING STAIR............................433
CREATING SPIRAL STAIRS 435
TUTORIAL 6.3 CREATING A SPIRAL STAIR.....................................436
MODIFYING EXISTING STAIRS WITH MODIFY STAIR 440
    Editing a Stair Using Grips 442
TUTORIAL 6.4 USING GRIPS TO MIRROR THE STAIR 443
STAIR STYLES 446
    Creating Stair Styles 446
STEPS TO CREATING A NEW STAIR STYLE.....................................448
    Defining the Attributes of a Stair Style 449
TUTORIAL 6.5 CREATING A STAIR STYLE AND CONTROLLING
  DISPLAY..455
    Exporting Stair Styles..463
STEPS TO EXPORTING A STAIR STYLE 464
STEPS TO IMPORTING A STAIR STYLE 464
    Purging Stair Styles 465
CREATING RAILINGS..466
    Default Properties of a Railing..469
DESIGNING THE RAILING WITH RAILING PROPERTIES..................469
    Changing the Railing with Modify Railing 478
DISPLAYING THE STAIR IN MULTIPLE LEVELS 479
TUTORIAL 6.6 PLACING A RAILING ON A STAIR 480
SUMMARY 496
REVIEW QUESTIONS 497
PROJECT 498
EXERCISE 6.1 CREATING A MULTI-LANDING STAIR WITH RAILINGS 498

CHAPTER 7 USING AND CREATING SYMBOLS

INTRODUCTION..499
OBJECTIVES...499
SETTING THE SCALE FOR SYMBOLS AND ANNOTATION...............500
    Changing Units in the Middle of a Drawing.............................504
USING THE DESIGNCENTER 507
    Defining the Default Symbol Menu 509
    Displaying Design Content Toolbars...511
INSERTING SYMBOLS 512
STEPS TO DRAGGING SYMBOLS INTO A DRAWING 513
TUTORIAL 7.1 INSERTING SYMBOLS USING DRAG AND DROP......515
INSERTING AND MODIFYING MULTI-VIEW BLOCKS......................516
    Using Multi-View Block Properties to Enhance Placement.........520
    Editing a Multi-View Block 523
TUTORIAL 7.2 INSERTING MULTI-VIEW BLOCKS WITH PRECISION 528
CREATING A MULTI-VIEW BLOCK...541
    Defining the Properties of a New Multi-View Block 543

TUTORIAL 7.3 CREATING AND MODIFYING MULTI-VIEW BLOCKS........546
IMPORTING AND EXPORTING MULTI-VIEW BLOCKS...............................554
STEPS TO EXPORTING MULTI-VIEW BLOCKS 554
INSERTING MULTIPLE FIXTURES 554
CREATING MASKING BLOCKS 558
    Defining Mask Block Properties 561
STEPS TO CREATING A MASK BLOCK...564
INSERTING MASKING BLOCKS..565
CHANGING MASKING BLOCKS...567
ATTACHING OBJECTS TO MASKING BLOCKS...571
CREATING SYMBOLS FOR THE DESIGNCENTER....................................573
    Content Type...575
    Insert Options 575
    Display Options 577
TUTORIAL 7.4 CREATING AND MODIFYING MASK BLOCKS.............579
SUMMARY 585
REVIEW QUESTIONS 586
PROJECT 587
EXERCISE 7.1 INSERTING SYMBOLS 587

CHAPTER 8 ANNOTATING THE DRAWING

INTRODUCTION 589
OBJECTIVES 589
PLACING ANNOTATION ON A DRAWING 590
CREATING BREAK MARKS..591
    Placing Straight Cut Lines 595
PLACING DETAIL MARKS..595
PLACING ELEVATION SYMBOLS...599
PLACING LEADERS IN THE DRAWING 603
TUTORIAL 8.1 INSERTING BREAK MARKS, DETAIL MARKS,
   ELEVATION MARKS, AND LEADERS 606
MISCELLANEOUS..614
    Creating Dimensions 614
STEPS TO CHANGING THE FRACTION FORMAT OF THE
   AEC_ARCH_I STYLE..616
    Fire Rating Lines 618
    Match Lines...620
    North Arrows 623
    Datum Elevation 625
    Revision Clouds 626
    Section Marks...629
    Title Marks..632
TUTORIAL 8.2 PLACING DIMENSIONS..636
CREATING TAGS AND SCHEDULES FOR OBJECTS................................647
    Door and Window Tags...648

STEPS TO PLACING A WINDOW TAG	649
Room and Finish Tags	651
TUTORIAL 8.3 PLACING TAGS FOR DOORS, WINDOWS AND ROOMS	654
Object Tags	660
Wall Tags	663
EDITING TAGS AND SCHEDULE DATA	664
USING SCHEDULE TABLE STYLES	666
ADDING A SCHEDULE TABLE	670
TUTORIAL 8.4 STEPS FOR CREATING A WINDOW SCHEDULE	672
RENUMBERING TAGS	674
UPDATING A SCHEDULE	675
EDITING THE CELLS OF A SCHEDULE	677
SUMMARY	679
REVIEW QUESTIONS	679
PROJECTS	680
EXERCISE 8.1 CREATING DOOR AND WINDOW SCHEDULES	680

CHAPTER 9 CREATING ELEVATIONS AND SECTIONS

INTRODUCTION	681
OBJECTIVES	681
CREATING THE MODEL FOR ELEVATIONS AND SECTIONS	682
TUTORIAL 9.1 CREATING THE MODEL FOR SECTIONS AND ELEVATIONS	683
CREATING A BUILDING ELEVATION LINE	688
CREATING AN ELEVATION	690
CONTROLLING THE VIEW OF THE ELEVATION	692
TUTORIAL 9.2 CREATING THE ELEVATION	693
Properties of the Section/Elevation Mark	698
Entity Display of the Elevation or Section	703
TUTORIAL 9.3 REFINING THE DISPLAY OF THE ELEVATION	706
REVISING THE ELEVATION WITH UPDATE ELEVATION	713
Creating 2D Elevation Drawings	714
TUTORIAL 9.4 CREATING ELEVATION DRAWINGS USING HIDDEN LINE PROJECTION	717
USING THE CREATE SECTION COMMAND	722
Placing a Section Mark	722
Using Existing Polylines as Section Lines	724
Creating the Section	725
Properties of the Section Mark	727
STEPS TO DEFINING THE LINETYPE AND WIDTH OF THE SECTION LINE	728
Revising the Section with the Update Section Command	730
Properties of Sections and Elevations	731

TUTORIAL 9.5   CREATING A SECTION USING HIDDEN LINE
  PROJECTION 732
SUMMARY 739
REVIEW QUESTIONS 739
PROJECT 740
EXERCISE 9.1   CREATING 3D ELEVATION AND 2D ELEVATION
  DRAWINGS 740

CHAPTER 10   CREATING MASS MODELS

INTRODUCTION 741
OBJECTIVES 741
CREATING MASS MODELS 742
    Inserting Mass Elements 742
STEPS TO INSERTING A MASS ELEMENT 754
TUTORIAL 10.1   CREATING MASS ELEMENTS 755
    Modifying Mass Elements 759
    Modifying Mass Elements with Grips 762
TUTORIAL 10.2   MODIFYING MASS ELEMENTS 764
    Creating Mass Elements Using Extrusion and Revolution 771
TUTORIAL 10.3   CREATING MASS ELEMENTS BY EXTRUDING
  PROFILES 771
    Creating New Profiles 773
STEPS TO CREATING A NEW PROFILE 777
STEPS TO EXPORTING A PROFILE 779
STEPS TO IMPORTING A PROFILE 779
TUTORIAL 10.4   CREATING NEW PROFILES 780
    Creating New Profiles from Existing Profiles 782
    Importing Profiles 783
TUTORIAL 10.5   IMPORTING AND EXPORTING PROFILES 783
CREATING MASS ELEMENTS TO REPRESENT BUILDING
  COMPONENTS 789
TUTORIAL 10.6   CREATING A MASS ELEMENT FOR A FLOOR SYSTEM 789
SUMMARY 797
REVIEW QUESTIONS 797
PROJECT 798
EXERCISE 10.1   CREATING A FLOOR SYSTEM USING MASS ELEMENTS ...798

CHAPTER 11   CREATING COMPLEX MODELS

INTRODUCTION 799
OBJECTIVES 799
CREATING GROUPS FOR MASS ELEMENTS 800
ADDING MASS ELEMENTS TO A GROUP 802
    Using Attach Elements to Add Mass Elements to a Group 803
    Detaching Elements from a Mass Group 805

TUTORIAL 11.1 CREATING MASS ELEMENTS WITH GROUPS.................806
    Boolean operations with Mass Elements..................812
    Using the Model Explorer..................816
    Mouse Operations in the Model Explorer 820
    Using the Tree View Shortcut Menus..................824
    Creating Mass Elements with the Model Explorer..................824
TUTORIAL 11.2 BUILDING MODELS USING THE EXPLORER..................826
    Referencing Mass Models..................830
TUTORIAL 11.3 CREATING REFERENCE MASS ELEMENTS..................836
    Creating Mass Elements Above the XY Plane 840
    Creating Floorplate Slices and Boundaries 841
    Converting the Slice to a Polyline 846
TUTORIAL 11.4 CREATING A FLOORPLATE SLICE AND BOUNDARY......847
SUMMARY 854
REVIEW QUESTIONS 855
PROJECT 856
EXERCISE 11.1 CREATING A MASS GROUP AND SLICING THE
  MASS MODEL 856

CHAPTER 12 CREATING SPACES AND BOUNDARIES

INTRODUCTION..................857
OBJECTIVES..................857
CREATING SPACES FOR SPACE PLANNING 858
    Editing Space Properties..................862
    Positioning and Sizing of Spaces 867
TUTORIAL 12.1 INSERTING SPACES 871
    Using Space Styles..................877
    Creating a Space Style..................879
    Modifying Spaces..................882
    Converting Polylines and Slices to a Space..................884
    Converting a Slice to a Space..................885
    Dividing and Combining Spaces 886
TUTORIAL 12.2 CREATING A SPACE STYLE..................890
USING SPACE BOUNDARIES..................899
    Adding Space Boundaries..................899
    Specifying the Properties of the Space Boundary 902
TUTORIAL 12.3 CREATING SPACE BOUNDARIES..................907
    Modifying Space Boundaries..................909
TUTORIAL 12.4 WORKING WITH BOUNDARIES..................915
    Creating Spaces Boundaries from Spaces, Slices, and Sketches..................919
STEPS TO CREATING SPACE BOUNDARY FROM A SLICE..................923
    Adding Boundary Edges..................924
    Editing Boundary Edges 925
    Deleting a Boundary Using Remove Boundary Edges..................926
    Using the Anchor to Boundary Command..................928

TUTORIAL 12.5 CONVERTING SPACES TO SPACE BOUNDARIES..............930
USING SPACE INQUIRY TO ADJUST SPACES..940
        Changing Spaces with SPACESWAP..942
        Generating Room Labels 943
        Generating Walls from Space Boundaries 943
TUTORIAL 12.6 ADJUSTING SPACES 944
SUMMARY 948
REVIEW QUESTIONS 948
PROJECT 949
EX12.1 POSITIONING SPACES TO CREATE SPACE BOUNDARIES AND
  WALLS 949

CHAPTER 13 DRAWING COMMERCIAL STRUCTURES

INTRODUCTION..953
OBJECTIVES..953
CREATING STRUCTURAL GRIDS..954
        Specifying the Size of the Column Grid on Screen 958
STEPS TO DYNAMICALLY SIZING A RECTANGULAR COLUMN GRID.......959
STEPS TO DYNAMICALLY SIZING A RADIAL COLUMN GRID.....................960
        Defining the Column for the Grid...961
REFINING THE COLUMN GRID WITH PROPERTIES 963
TUTORIAL 13.1 CREATING A COLUMN GRID 968
MODIFYING THE COLUMN GRID...971
        Adding and Removing Column Grid Lines 972
        Editing the Column Mass Elements...975
        Labeling the Column Grid...976
        Dimensioning the Column Grid 978
ADDING A COLUMN WITHOUT A GRID 979
TUTORIAL 13.2 LABELING A COLUMN GRID 981
CREATING CEILING GRIDS 985
STEPS FOR PLACING A CEILING GRID 988
ADJUSTING THE PROPERTIES OF THE CEILING GRID.................................990
TUTORIAL 13.3 CREATING A CEILING GRID 993
CHANGING THE CEILING GRID 996
        Adding Boundaries and Holes..997
        Adding and Removing Ceiling Grid Lines..1000
TUTORIAL 13.4 EDITING THE CEILING GRID 1004
USING LAYOUT CURVES TO PLACE COLUMNS 1007
        Refining the Intersection of Columns and Walls 1009
TUTORIAL 13.5 USING LAYOUT CURVES...1011
VIEWING THE MODEL...1015
        Creating a Walkthrough..1017
TUTORIAL 13.6 CREATING A WALKTHROUGH 1019
SUMMARY 1021

REVIEW QUESTIONS 1022
PROJECTS 1022
EXERCISE 13.1   CREATING COLUMNS AND COLUMN GRIDS 1022
EXERCISE 13.2   CREATING A CEILING GRID 1023

PREFACE

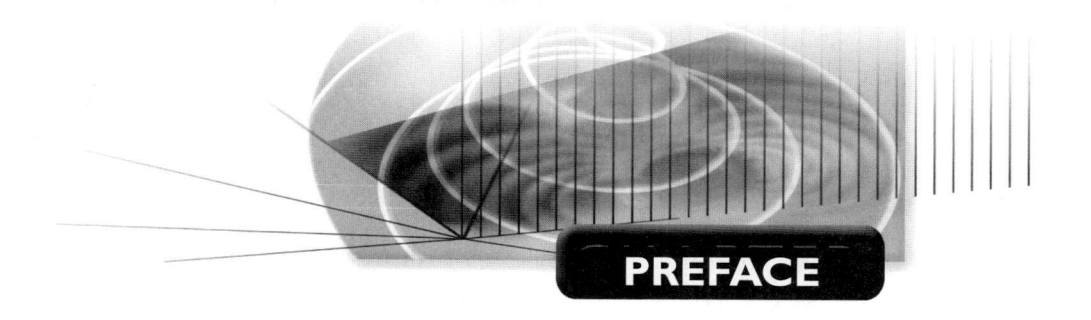

AutoCAD Architectural Desktop is the state-of-the-art software for architectural design. Recently, programs of study in architectural technology and architecture have integrated computer-aided design within the curriculum. This is the first text to integrate the tools of AutoCAD Architectural Desktop in order to create architectural working drawings. AutoCAD Architectural Desktop provides these programs with the software to create architectural working drawings or mass models. It provides 3D design using Object Technology, which allows the user to create instant perspectives, dynamic sections, elevations, and building components.

AutoCAD Architectural Desktop evolved from the Softdesk8 products, which enhanced the productivity of technicians and architects. The first release of AutoCAD Architectural Desktop was based upon AutoCAD Release 14 in 1998. This text is based upon AutoCAD Architectural Desktop Release 2, which utilizes AutoCAD 2000. These products increase the productivity of professionals and students when the proper methods for using the software are utilized.

The intent of the software may not be realized if the software is used prematurely without systematic instruction. Buildings should not be built without drawings, and software should not be used without instruction.

With this in mind, Accessing AutoCAD Architectural Desktop has been written in order to provide detailed and systematic explanations of the tools of AutoCAD Architectural Desktop for creating architectural working drawings. The book includes techniques and tutorials that apply the software to the creation of drawings for residential and commercial buildings.

The first nine chapters of the book provide the necessary instruction regarding the use of AutoCAD Architectural Desktop for the creation of architectural working drawings. The tutorials include the procedures for creating architectural floor plans, the basics of creating walls, wall styles, wall modifier styles, and object display control. This format allows the reader to immediately begin drawing floor plans. The remaining chapters are intended to assist in creating mass models. Tutorials include instruction in creating mass models and the development of the mass model, for space planning, space boundaries, and generation of walls. The final chapter, Drawing Commercial Structures, will be of interest to those who create drawings for commercial structures because it includes structural grids, ceiling grids, and walkthroughs using the camera.

FEATURES OF THIS EDITION

Included in Accessing AutoCAD Architectural Desktop are tutorials, which provide step by step instruction in the use of AutoCAD Architectural Desktop to develop architectural working drawings for residential and commercial buildings. The tutorials, command access tables, and the contents of the CD together help a student identify the appropriate commands and techniques to develop the working drawings. The features of this edition are summarized below:

- Command access tables are provided for each new command introduced. The text includes 173 command access tables, which describe in table format how to access the commands from the menu bar, command line, toolbar, or shortcut.
- Architectural Desktop™ R2 from the Autodesk Student Portfolio is located on the CD in the back of the book. Included on the CD is the ADT Commands Excel file that lists all the commands of command access tables in the text.
- Tutorials are included to introduce and reinforce the options of the commands. The 75 tutorials of the text include applications of residential and commercial buildings. Throughout the tutorials, two buildings are developed and the techniques to create traditional architectural working drawings are emphasized.
- The CD shipped with the book includes the drawings and solutions for the tutorials. Students can repeat the tutorials to reinforce their knowledge of the commands.
- The Instructor Guide includes additional drawing exercises for each chapter. The drawings for the assignments and the solutions are located in the *ADT Tutor\Supplement* directory on the CD.
- In each chapter display representation and display configurations are described when appropriate to correctly display objects to create architectural working drawings.
- Development of complex roofs, roof intersections, and dormers for residential and commercial buildings are included.
- Development of schedules and the exporting of schedule data in the MicroSoft Excel file format is included to enhance the control of data related to the building. The properties and property sets used in the schedule tables are included on the CD. The CD also includes Autodesk White Papers, which provide an overview of AutoCAD Architectural Desktop R2, layering, and migration from Softdesk.
- The appendices and an advanced section entitled "Enhancing Schedules and Schedule Data" from chapter 8 are located in PDF format on the CD.
- Custom views and walkthroughs that can be created from the model are included in the tutorials to enhance the presentation of the design.

- A completed model of residence is included on the CD. This model can be reconstructed with the external reference files included on the CD.

STYLE CONVENTIONS

Throughout the book you are requested to select commands and respond to the command in the command line or on screen with the mouse. The text style conventions are used systematically to enhance the understanding and recall of the commands. The style conventions of the text are as follows:

Text	Style Example
Commands	**WALLADD**
Menu and toolbar names	**Design>Walls>Add Wall**
Dialog box elements	The **Edit** button
Command line prompts	`Elevation line start point:`
Keyboard input	Press ENTER to end the command.
User Input	Type **OFFICE_CAD** in the field.
File and directory names	The *Aec arch (metric).dwt* template

HOW TO USE THIS BOOK

The design of each chapter of the text is to introduce the commands of Architectural Desktop to allow you to gain the skills and understanding of the software to create architectural working drawings. Each chapter includes an Introduction and Objectives. Read carefully the objectives of the chapter to determine the commands and types of drawings created in the chapter. The commands and the screen captures of the related dialog boxes should be reviewed in detail prior to performing the tutorial. The purpose of the tutorial is not to finish quickly but to gain a hands-on experience in using the commands to create drawings. You may find it most helpful to repeat the tutorial to gain recognition of the commands on the toolbar and the format of the dialog boxes. When most commands are selected, dialog boxes open, allowing you to set the properties of an object. These dialog boxes usually provide access to one or more other dialog boxes to define additional properties of an object. The book includes screen captures of these dialog boxes to allow you to study the commands when you might not have access to the software. However, access to the software as you read the book greatly enhances the learning process. Summary and review questions are provided at the end of each chapter. Additional project exercises are included in most chapters.

COMMAND ACCESS TABLES

The commands included in the text are summarized and presented in command access tables. Included in each table is a list of methods to access the command. Command access tables within the chapter include the methods to access the command from the menu bar, command prompt, toolbar, and shortcut menus.

Command access tables are provided when a new command is introduced, and the commands are presented in the same order as needed to develop a set of architectural working drawings. Most commands can be accessed in several ways, and the tutorials do not require you to access the command using all possible options. Shortcut menu options should be experienced so that you can determine your preference.

The command access table for the **Add Door** command (**DOORADD**) is shown below as an example.

Menu bar	Design>Doors>Add Door
Command prompt	DOORADD
Toolbar	Select Add Door from the Doors/Windows/Openings toolbar shown in Figure 4.2
Shortcut	Right-click over drawing area; choose Design>Doors>Add

Table 4.1 *Accessing the Add Door command*

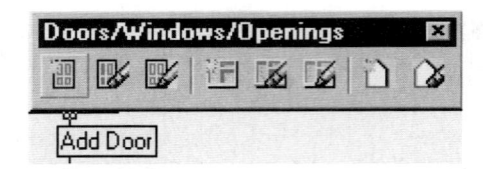

Figure 4.2 *Add Door command on the Doors/Windows/Openings toolbar*

Because most of these commands are long and difficult to type, the tutorials will emphasize the use of the toolbars and dialog boxes. A comprehensive list of the commands in the command access tables is included in the *ADT Commands.xls* Excel file on the CD. It is recommended that after reading each chapter and performing the tutorials, you open this file and review how the commands of the chapter are accessed. The *ADT Commands.xls* file includes the feature, chapter number, toolbar, tool tip, shortcut, and command. It is organized alphabetically by feature. You can elect to sort the *ADT Command.xls* file to format the command list alphabetically or by chapter number, according to your preferences.

Drawing files for all tutorials are located on the CD shipped with the book. The tutorials will develop floor plans, elevations, and other plans, which include the necessary annotation and schedules for architectural working drawings. The drawing files are located in the chapter directories of the ADT Tutor directory of the CD. Copy the ADT Tutor directory from the CD to your computer. Create a student directory in the ADT Tutor directory on your computer. When you open a drawing file located in the ADT Tutor directory, save the drawing to your student directory. It is recommended that the tutorials be opened and then saved to your student directory prior to your performing the operations of the tutorial. Therefore, the original tutorial will remain unchanged, allowing you to repeat the tutorial if necessary.

The drawing files and the solutions for the project exercises located at the end of each chapter are included on the CD. Perform the project exercises at the end of the chapter to test your knowledge of the commands in the chapter.

ONLINE COMPANION

If you have access to the Internet, you can access the Online Companion. Additional resources and links to other sites are available. Access the Online Companion at: http://www.autodeskpress.com/onlinecompanion.html. Accessing this address allows you to view other resources available from Autodesk Press.

WE WANT TO HEAR FROM YOU

We welcome your comments and suggestions regarding the contents of this text. Your input will result in the improvement of future publications. Please forward your comments and questions to:

The CADD Team
C/O Autodesk Press
3 Columbia Circle
P.O. Box 15015
Albany, N.Y. 12212-5015

ABOUT THE AUTHOR

William G. Wyatt, Sr. is an instructor at John Tyler Community College in Chester, Virginia. He has taught architectural drafting and related technical courses in the Architectural Engineering Technology program since 1972. He earned his Doctor of Education from Virginia Tech and his Masters of Science and Bachelor of Science degrees in Industrial Technology from Eastern Kentucky University. He earned his Associate in Applied Science in Architectural Technology from John Tyler Community College. He is a certified Architectural and Building Construction Technician.

DEDICATION

This book is dedicated to several people. First, to the memory of my parents, Leslie and Catherine Wyatt, whose love has been a cornerstone of stability in my life. Next I would like to thank my father-in-law and mother-in-law, Kermit and Helen Hedahl, for their continuous family support. In addition, a special thanks to our children: Leslie, Sarah, and William Jr., for their acceptance and understanding of the time needed to complete this project. Lastly, to my wife, Bevin Hedahl Wyatt, who has provided continuous encouragement. Her love and support has helped make this project possible.

ACKNOWLEDGEMENTS

The author would like to thank and acknowledge the team of reviewers who reviewed the chapters and provided guidance and direction to the work. A special thanks to the following reviewers, who reviewed the chapters in detail:

> Paul Adams, Denver Technical College, Denver, Colorado
> Paul N. Champigny, New England Institute of Technology, Warwick, Rhode Island
> Lynn A. Gurnett, York County Technical College, Wells, Maine
> Christopher LeBlanc, Porter & Chester Institute, Chicopee, Massachusetts
> Joseph M. Liston, Westark College, Fort Smith, Arizona
> Jeff Porter, Porter & Chester Institute, Watertown, Connecticut
> Charles T. Walling, Silicon Valley College, Walnut Creek, California

A special thanks to Bill Chisholm, John Tyler Community College and Lark Ziegler Rilling, Hutchins, Texas, who reviewed the preliminary manuscript and provided comments for the improvement of the text. In addition, I greatly appreciate the assistance and encouragement from Bill Proctor of CADD Microsystems, Inc. Bill carefully and thoroughly edited the manuscript providing comments to enhance the text.

I would like to thank Ronald A. Williams, LTD, Autodesk Education Representative of Virginia, for their support and encouragement to this project.

The author would like to acknowledge and thank the following staff from Delmar Publishers:

> Publisher: Alar Elken
> Executive Editor: Sandy Clark
> Acquisitions Editor: Mike Kopf
> Developmental Editor: John Fisher
> Production Manager: Larry Main
> Art and Design Coordinator: Mary Beth Vought
> Editorial Assistant: Jasmine Hartman
> Marketing Manager: Maura Theriault
> Marketing Coordinator: Paula Collins

The author would like to acknowledge and thank the following people:

> Copyediting: Gail Taylor
> Composition: Vince Potenza of SoundLightMind Media Design and Development

CHAPTER 1

Introduction to Architectural Desktop

OVERVIEW

AutoCAD® Architectural Desktop™ is intended to assist the designer in developing architectural working drawings as well as preliminary schematics and computer models of a building. The first nine chapters of this book include the tools available in Architectural Desktop to develop architectural working drawings. These chapters include the development and editing of walls, doors, windows, stairs, and roofs. The remainder of the book explains the special tools available for computer modeling of a commercial building and the development of associated working drawings.

Architectural Desktop creates components of a building as objects. The objects consist of walls, doors, windows, stairs, and roofs. These objects interact with each other to reduce the need for editing. Door and window objects are anchored to the wall to enhance the placement of the unit. The object approach to design decreases editing of individual building components and can be used to create a three-dimensional model of a building. In addition, Architectural Desktop includes tools for creating schedules that allow you to extract a comprehensive list of doors, windows, furniture, rooms, and wall types from the drawing.

ADVANTAGES OF OBJECT TECHNOLOGY

Using Architectural Desktop requires a basic knowledge of AutoCAD because AutoCAD is integrated into the Architectural Desktop. Doors and windows as well as the walls are inserted in a drawing through ObjectARX™ technology. These objects consist of several entities that have associative interaction with other objects when inserted. Doors and windows are easily placed in the wall, and when placed, they behave as magnets within the wall, and the wall is correctly edited for the window or door. Using ObjectARX technology reduces the need to edit the drawing. Architectural Desktop objects are three-dimensional and consist of several components. For instance, a door object consists of a door panel, frame, stop and swing components.

Another advantage of object technology is the use of the Object Viewer, which is a separate window designed for viewing the object. The object can be viewed in orthographic, isometric, and perspective views. The navigator of the Object Viewer allows the object to be viewed dynamically from any vantage point. All objects placed in a drawing can be viewed in the Object Viewer. The Object Viewer can be used to view interior spaces or individual object components in perspective as shaded or rendered images. Rendered views of objects can be saved as bitmap image files and exported to other documents.

ARCHITECTURAL DESKTOP RELEASE 2

AutoCAD Architectural Desktop Release 2 includes the new features of AutoCAD 2000 with the enhancements of the first AutoCAD Architectural Desktop release. The Multiple Document Environment of AutoCAD 2000 allows you to work on more than one drawing at a time. Choosing Window>Tile Vertically from the menu bar allows you to view and work on the two drawings, as shown in the figure below. This feature would allow you to work on a detail or section while viewing the floor plan of the same project. In addition, symbols and objects can be copied and pasted across drawings through Windows operations.

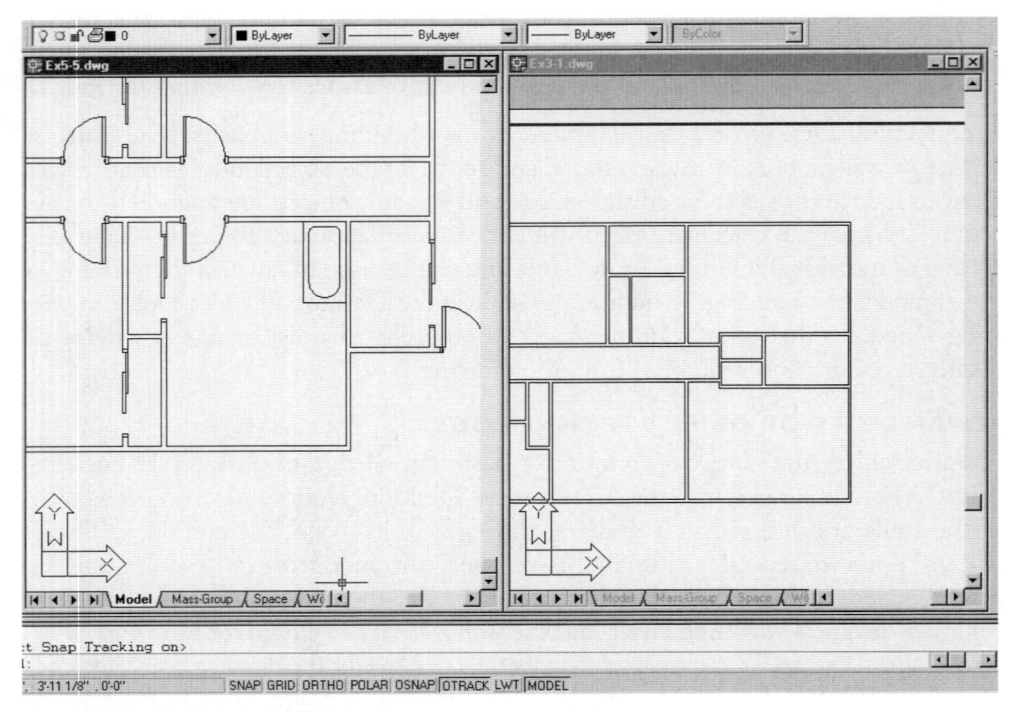

Figure 1.1 *Multiple AutoCAD drawings open at one time*

The DesignCenter™ of AutoCAD Architectural Desktop provides access to three Design Content submenus: Architectural Desktop Imperial Content, CSI Imperial Content, and Metric Content. The Custom view of the DesignCenter displays an image of each symbol in the design menu. It allows you to preview the block prior to inserting it in the drawing. Shown below is the DesignCenter display of the Electric group of the Imperial Design directory.

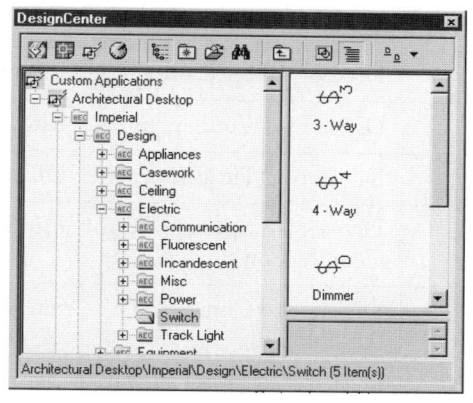

Figure 1.2 *Architectural symbols of the DesignCenter*

The symbol groups of the Content directories are listed in Table 1.1

These extensive libraries of symbols allow you to easily place symbols in a drawing from the DesignCenter. The Design CSI directory consists of symbols grouped according to the CSI Masterformat. Therefore, a plumbing fixture would be selected from the 15400- Plumbing Fixtures group or Division 15.

Additional door and window styles have been added to the Architectural Desktop. New additions include French doors, glass doors, revolving doors, garage doors, shutters, jalousie windows, bay windows, and bow windows. The Windows Custom Imperial Style resource file also provides five window styles including muntins, as shown in Figure 1.3.

STARTING ARCHITECTURAL DESKTOP

Start Architectural Desktop by choosing the **Start** button of Windows NT4 or Windows 98 and selecting the **Programs>AutoCAD Architectural Desktop> AutoCAD Architectural Desktop**. Once AutoCAD Architectural Desktop is launched, the **Startup** dialog box will be displayed, similar to the **Startup** dialog box of AutoCAD, as shown in Figure 1.4.

The four buttons across the top of the **Startup** dialog box allow you to **Open a Drawing, Start from Scratch, Use a Template,** or **Use a Wizard**. Begin a new

Content>Imperial> Design	Content>Imperial> Design CSI	Content>Metric> Design
Appliances	Division 1 General Requirements	Bathroom Fittings
Casework	Division 2 Site Construction	Domestic Furniture
Ceiling Fixtures	Division 10 Specialties	Electrical Services
Electrical Fixtures	Division 11 Equipment	Kitchen Fittings
Equipment	Division 12 Furnishings	Office Furniture
Furniture	Division 13 Special Construction	Pipe and Ducts
Plumbing	Division 14 Conveying Systems	Site
Sites	Division 15 Mechanical	
	Division 16 Electrical	

Table 1.1 *Symbol groups of the Content directories*

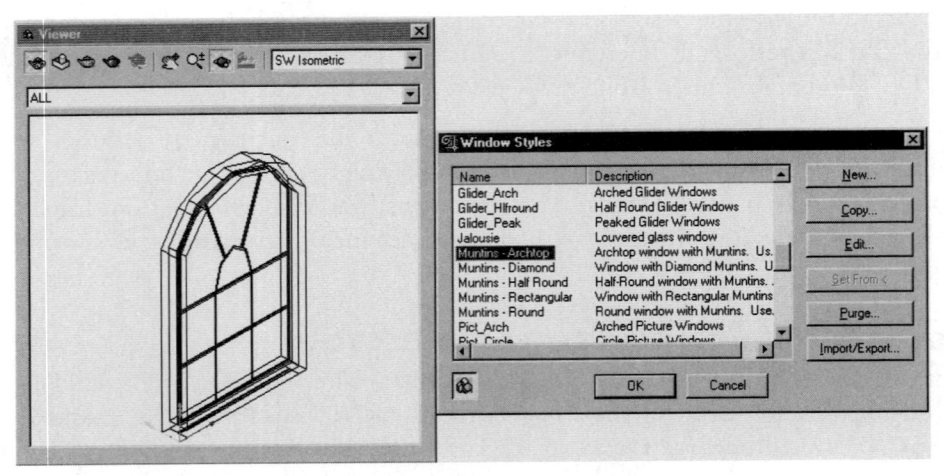

Figure 1.3 *Window styles of the Windows Custom Imperial Style file*

drawing by selecting the **Use a Template** button. New drawings in Architectural Desktop should utilize the Aec arch (imperial) or the Aec arch (metric) template. These

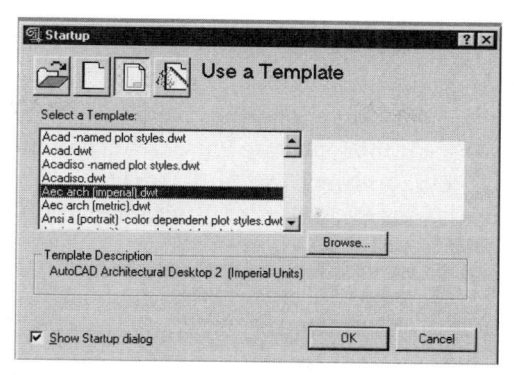

Figure 1.4 *Startup dialog box*

templates are listed in the template window of the **Startup** dialog box. The Aec arch (imperial) template is designed for using imperial units, whereas the Aec arch (metric) template uses metric units.

Start a new drawing using imperial units by selecting the **Use a Template** button at the top of the **Startup** dialog box and selecting the Aec arch (imperial).dwt template. There are no graphic entities in the Aec arch (imperial) or the Aec arch (metric) template. Therefore, the preview box to the right of the template list will remain blank when either of the templates is selected.

When you create a new drawing using a template, the default drawing name is Drawing1 and it has not been saved to a directory. To save a drawing with a specific name, choose **SaveAs** from the **File** menu and insert a new drawing name in the file name edit box, as shown below:

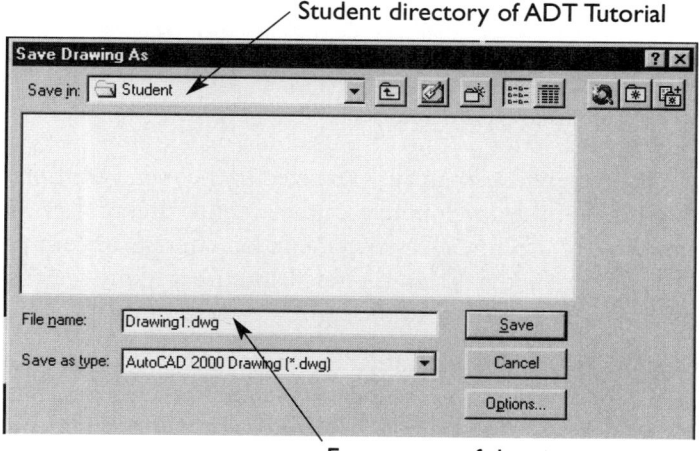

Figure 1.5 *Save Drawing As dialog box*

ARCHITECTURAL DESKTOP SCREEN LAYOUT

The Architectural Desktop screen includes a menu bar at the top of the screen. The left portion of the menu bar includes the standard AutoCAD menu bar. The right portion of the menu bar consists of the following four Architectural Desktop menu headings: **Concept, Design, Documentation** and **Desktop**. The remainder of the AutoCAD Architectural Desktop screen is identical to that of AutoCAD.

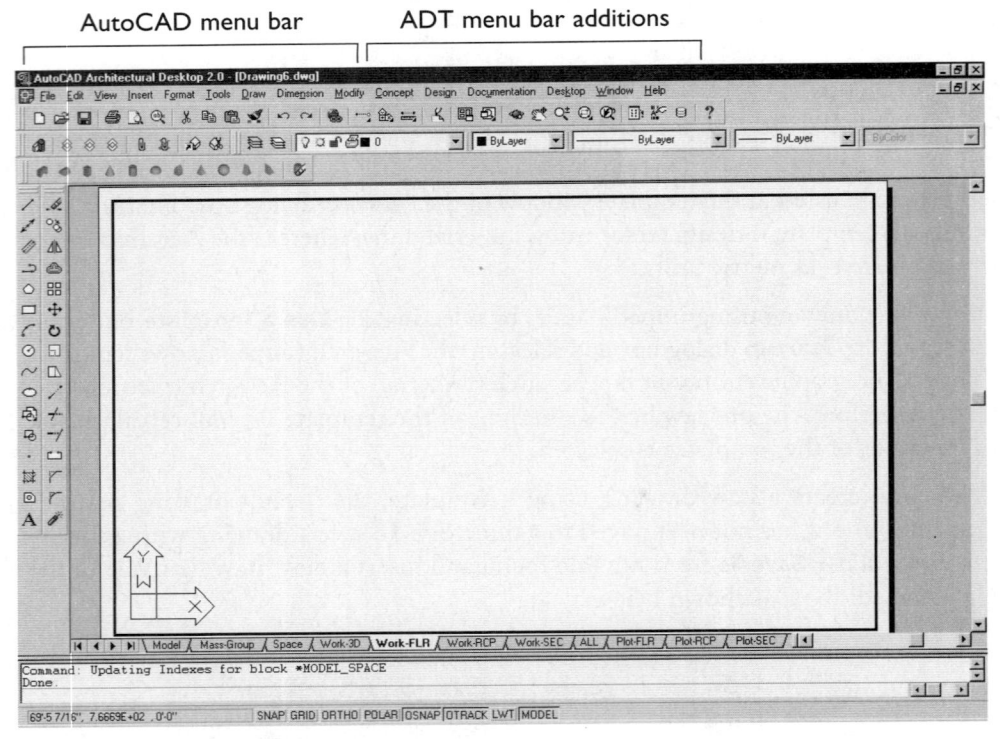

Figure 1.6 *Menu bar of Architectural Desktop*

Commands of Architectural Desktop are organized in the menu bar from left to right parallel to the progress of the development of the architectural project. The **Concept** menu portion of the Architectural Desktop menu bar consists of commands used in the early stages of design. As the design is refined and the working drawings are developed, the menu bar options further to the right are used.

The **Concept** menu includes those commands for creating the components of the initial design phase—creating mass models, space planning, slicing floorplates and converting boundaries to walls. The content of the Concept pull down menu is shown in Figure 1.7.

The **Design** menu consists of commands for inserting symbols from the design part library, inserting doors, windows, and generating roofs, elevations, and sections. The content of the **Design** menu is shown Figure 1.8.

Commands necessary for annotating and developing the final construction documents are located on the **Documentation** menu. The content of the **Documentation** menu is shown Figure 1.9.

The **Desktop** menu includes commands for drawing setup, layer management and display control. The content of the **Desktop** menu is shown Figure 1.10.

Choose options from the menu bar by clicking with the mouse on the menu heading and then moving the mouse pointer down to the desired menu heading and clicking

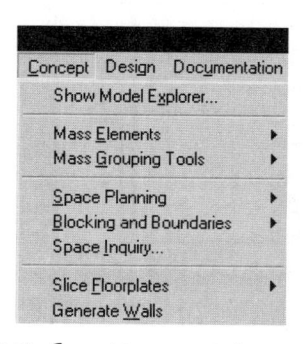

Figure 1.7 Concept menu options on the menu bar

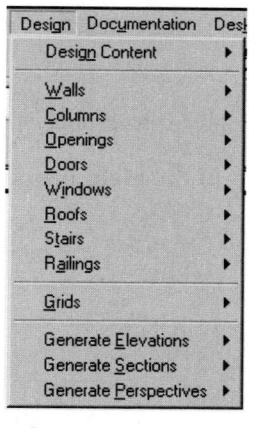

Figure 1.8 Design menu options on the menu bar

Figure 1.9 Documentation menu options on the menu bar

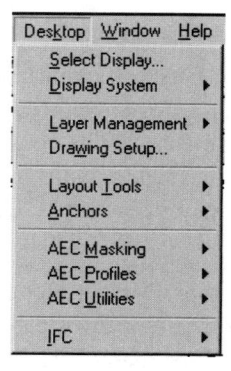

Figure 1.10 Desktop menu options on the menu bar

on that option. Clicking on the menu bar expands the menu to the first level of the menu. Menu selections with an arrow to the right will expand to lower levels of the menu. Once the menu is cascaded, you can make the menu choice by clicking with the mouse when the pointer is positioned over the desired menu choice.

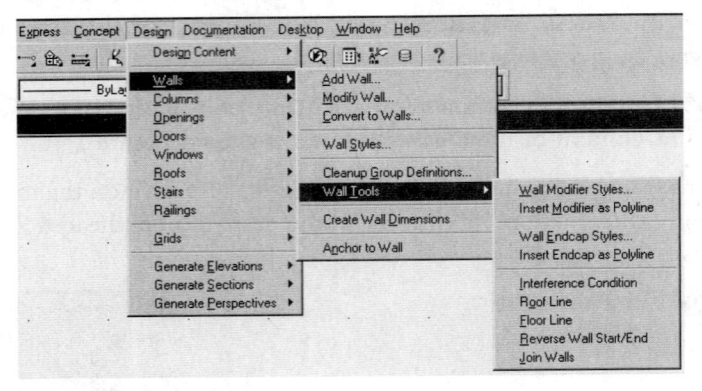

Figure 1.11 *Walls cascaded menu*

The content of the **Help** menu is shown below:

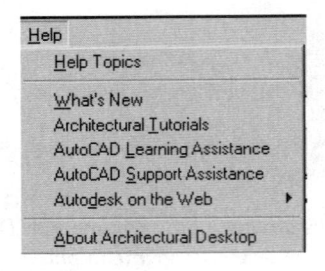

Figure 1.12 *Help menu options on the menu bar*

SHORTCUT MENUS

Architectural Desktop commands can also be selected from shortcut menus. Shortcut menus, displayed when you right-click the mouse, provide quick access to Architectural Desktop commands. The content of the shortcut menu varies according to where the cursor is located when you right-click. If you right-click when the cursor is positioned over the drawing area and no object has been selected, the shortcut menu is displayed in Figure 1.13.

Included in this shortcut menu are **Utilities** and **Design** menu options. The **Utilities** and **Design** menu options provide access to Architectural Desktop commands. The cascaded **Utilities** menu options are shown above. The **Design** cascaded menu is shown in Figure 1.14.

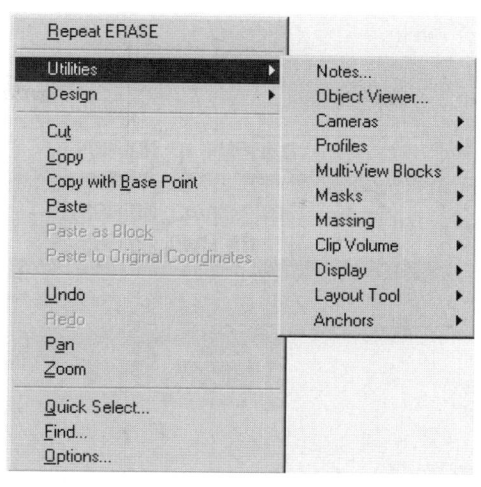

Figure 1.13 Shortcut menu of the drawing area when no object is selected

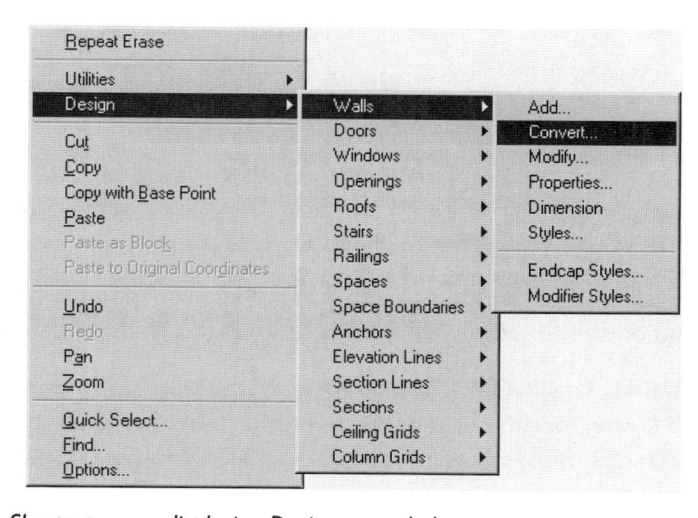

Figure 1.14 Shortcut menu displaying Design cascaded menu

The **Design** shortcut menu includes options for most of the commands included on the **Design** menu. Also included on this shortcut menu are commands for placing and editing walls and other basic editing commands such as **CUT**, **COPY**, and **PASTE**. **ZOOM** and **PAN** commands are also included on the shortcut menu.

If you right-click when an Architectural Desktop object has been selected, the shortcut menu includes the menu to edit the selected object. As shown below, a door was selected and the shortcut menu includes the following editing menu options: **Edit Door Style**, **Modify Door**, and **Door Properties**.

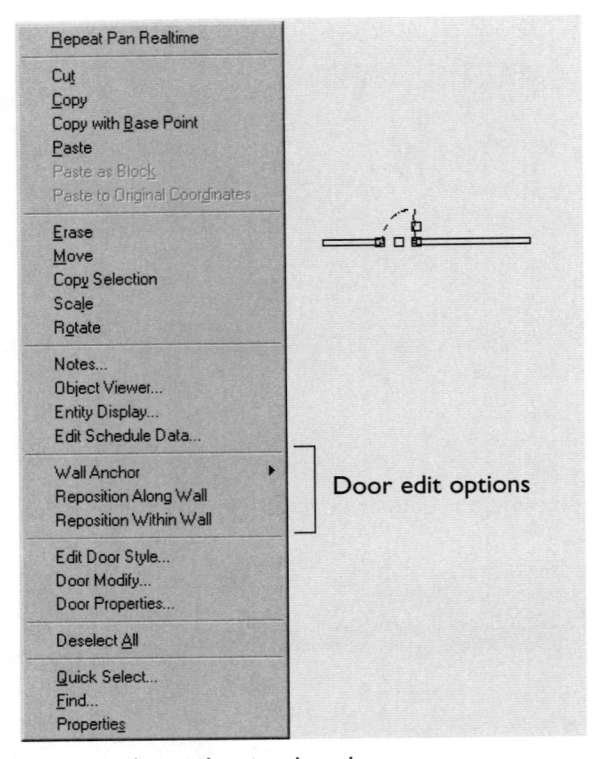

Figure 1.15 *Shortcut menu when a door is selected*

Therefore the quick method of editing an object is to select it and then right-click and choose the menu option for editing that object.

ARCHITECTURAL DESKTOP TOOLBARS

The AutoCAD **Draw, Modify, Standard,** and **Object Properties** toolbars are displayed when Architectural Desktop opens. Use the **TOOLBAR** command to display the Architectural Desktop toolbars. You can execute this command by typing **TOOLBAR** in the command line or right-clicking with the mouse over any existing toolbar and choosing **Customize** from the shortcut menu. The **TOOLBAR** command opens the **Toolbars** dialog box as shown in Figure 1.16.

The **Toolbars** dialog box includes a toolbar list and a menu group list. There are three menu groups included with Architectural Desktop: ACAD, AECARCHX, and EXPRESS. Selecting the AECARCHX menu group will change the toolbar list to the Architectural Desktop toolbars.

The toolbar names displayed in the toolbar list box vary according to the menu group selected. Selecting the ACAD menu group will display the names of the AutoCAD toolbars such as the **Draw** and **Modify** toolbars. The Architectural Desktop

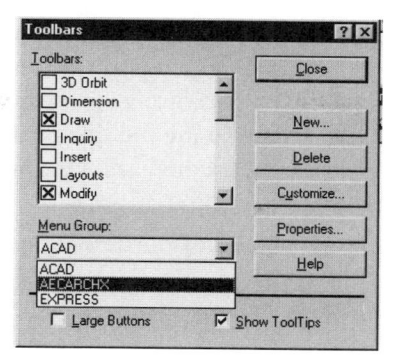

Figure 1.16 *Toolbars dialog box*

toolbars are included in the AECARCHX menu group. If a check is in the box to the left of the toolbar name, that toolbar will be displayed. The toolbars of the AECARCHX are shown in Figure 1.17.

Obviously, not all Architectural Desktop toolbars are needed during any phase of design. During the conceptual phase of design, the **Mass Elements, Mass Tools,**

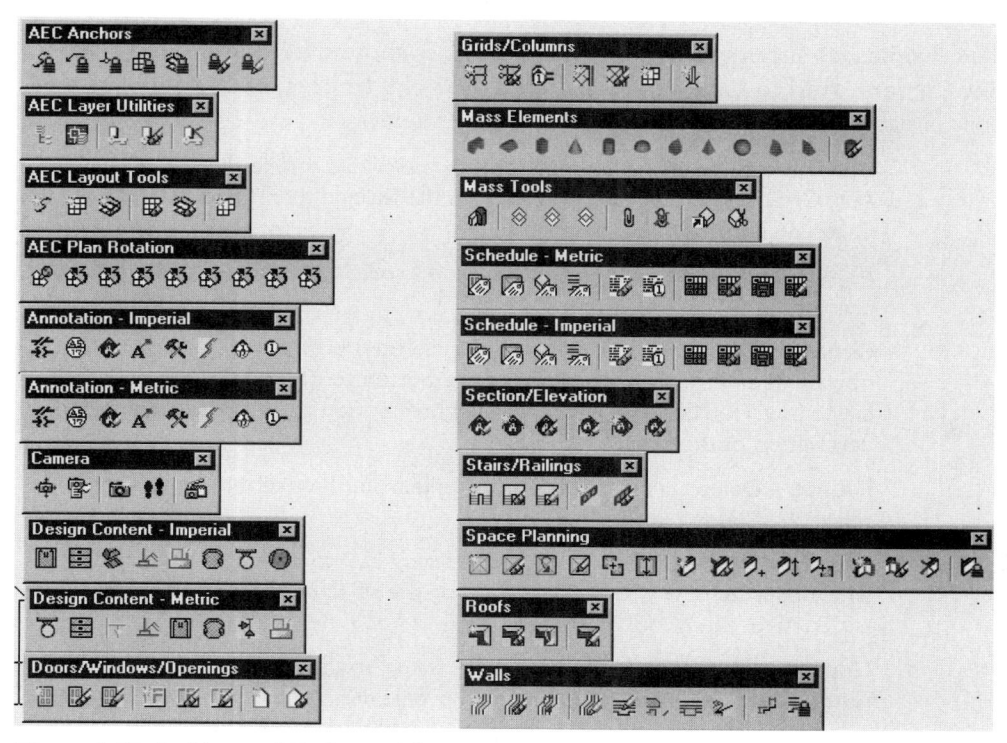

Figure 1.17 *Architectural Desktop toolbars*

and **Space Planning** toolbars are used, whereas during design development, **Walls**, **Doors**, and **Windows** toolbars are used. To retain the display of desired toolbars, you can create a profile, which saves the current display of toolbars as a named profile. To create a profile, type OPTIONS in the command line or right-click with the mouse over the command window and select **Options**. The **Options** dialog box includes the **Profiles** tab, as shown in Figure 1.18.

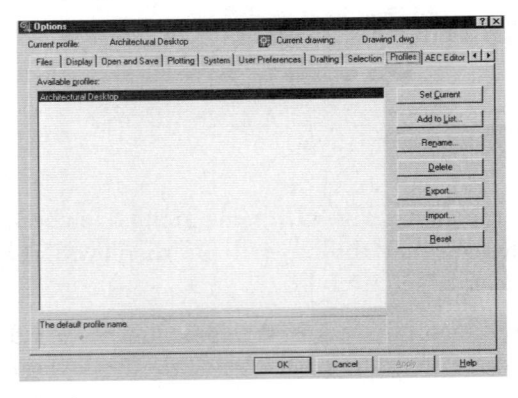

Figure 1.18 *Options dialog box*

The **Profiles** tab includes a list of available profiles and the following seven buttons: **Set Current, Add to List, Rename, Delete, Export, Import,** and **Reset**. The purpose of each of the seven buttons is summarized below:

> **Set Current** – Applies a profile from the available profiles list to the current AutoCAD session. Select a profile from the list and then select the **Set Current** button to apply a profile.
>
> **Add to List** – Creates a new name for a profile through the **Add Profile** dialog box.
>
> **Rename** – Renames an existing profile. Select a profile from the available list of profiles and then select the **Rename** button to open the **Change Profile** dialog box. The **Change Profile** dialog box is used to change the name and description of the profile.
>
> **Delete** – Deletes a profile. Select a profile from the available list and then select the **Delete** button.
>
> **Export** – Opens the **Export Profile** dialog box to create an external file, which consists of the toolbar settings for the selected profile. The saved file has an ARG extension.
>
> **Import** – Opens the **Import Profile** dialog box to bring into a session of AutoCAD the settings of an exported profile file.
>
> **Reset** – Restores the settings to the system default.

STEPS TO CREATING A PROFILE

1. Open toolbars desired for the named profile in Architectural Desktop.
2. Enter **OPTIONS** in the command line.
3. Select the Profiles tab.
4. Select the Add to List button.
5. Enter a name and description for the new profile as shown below:

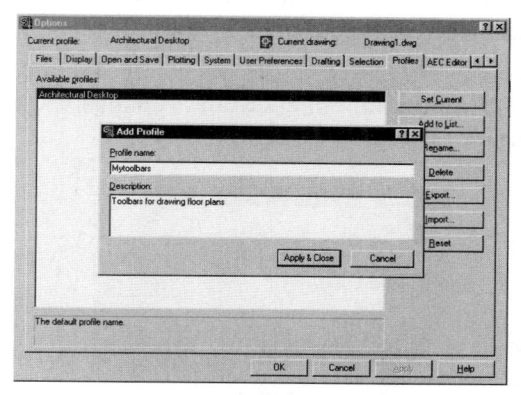

Figure 1.19 The Add Profile dialog box and the Profiles tab of the Options dialog box

6. Select the Apply & Close button.
7. Select the desired profile
8. Select the Set Current button.
9. Select OK to close the Options dialog box.

The Architectural Desktop profile is the default profile created upon installation. It includes the Architectural Desktop pull-down menu and the **Draw** and **Modify** toolbars. **Mass Elements** and **Mass Tools** toolbars of the Architectural Desktop are displayed in the initial installation. You can display additional toolbars using the **TOOLBAR** command and you can save them to a profile using the **Profile** tab of the Options dialog box. Profiles can be created that consist of toolbars for each design phase and recalled as needed. In the following tutorial, you will create a named profile that displays the **Doors/Windows/Openings**, **Stairs/Railings**, and **Walls** toolbars.

TUTORIAL 1.1 CREATING AND USING A PROFILE

1. Open Architectural Desktop by clicking **Start** on the Windows menu bar and select **Programs>AutoCAD Architectural Desktop 2.0**.
2. Select the **Use a Template** button of the **Startup** dialog box, select the *Aec arch (imperial).dwt* template, and select **OK** to dismiss the dialog box.

3. Choose **File>SaveAs** from the menu bar.
4. Select the down arrow of the **Save In** list and set the directory to your student directory in the *ADT_Tutor* directory.
5. Type **Ex1-1.dwg** in the **File Name** edit field of the **Save Drawing As** dialog box.
6. Select the **Save** button to save the file and dismiss the dialog box.
7. Right-click over any toolbar with the mouse and choose **Customize** from the shortcut menu to execute the **TOOLBAR** command.
8. Edit the Menu Group to AECARCHX in the **Toolbars** dialog box and check the **Doors/Windows/Openings**, **Stairs/Railings**, and **Walls** toolbars as shown in Figure 1.20.

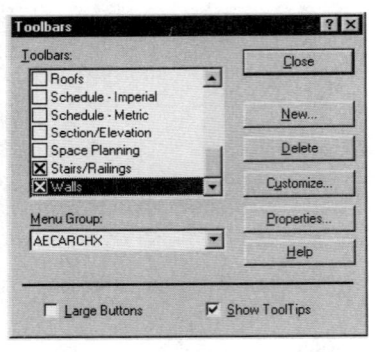

Figure 1.20 *Toolbars dialog box displaying Architectural Desktop toolbars*

9. Select the **Close** button to close the **Toolbars** dialog box.
10. Right-click over the command line with the mouse and choose **Options** on the shortcut menu.
11. Select the **Profiles** tab of the **Options** dialog box as shown in Figure 1.21.

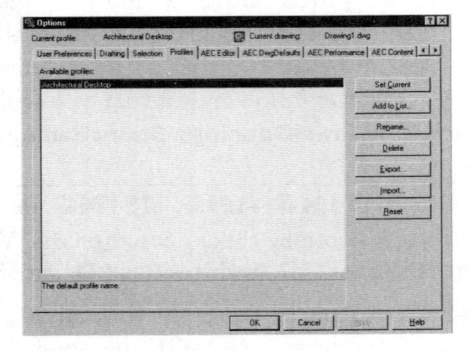

Figure 1.21 *Profiles tab of the Options dialog box*

12. Select the **Add to List** button and type **Floorplan** as the profile name.
13. Press TAB and type **toolbars for floor plans** in the **Description** field as shown in Figure 1.22.

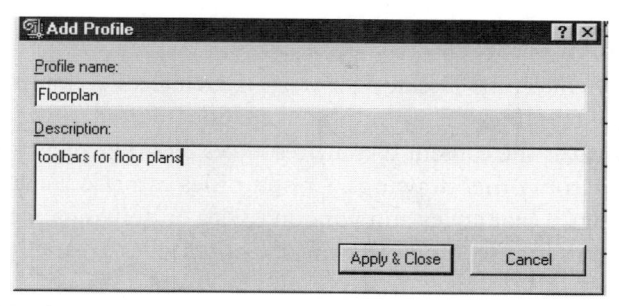

Figure 1.22 *Add Profile dialog box for creating a profile*

14. Select the **Apply & Close** button to save the new profile.
15. Select the Floorplan profile and select the **Set Current** button.
16. Select **OK** to dismiss the **Options** dialog box.
17. The Architectural Desktop menu and toolbars should now display the following toolbars: **Walls**, **Doors/Windows/Openings**, and **Stair/Railings** as shown in Figure 1.23.
18. Save the drawing.

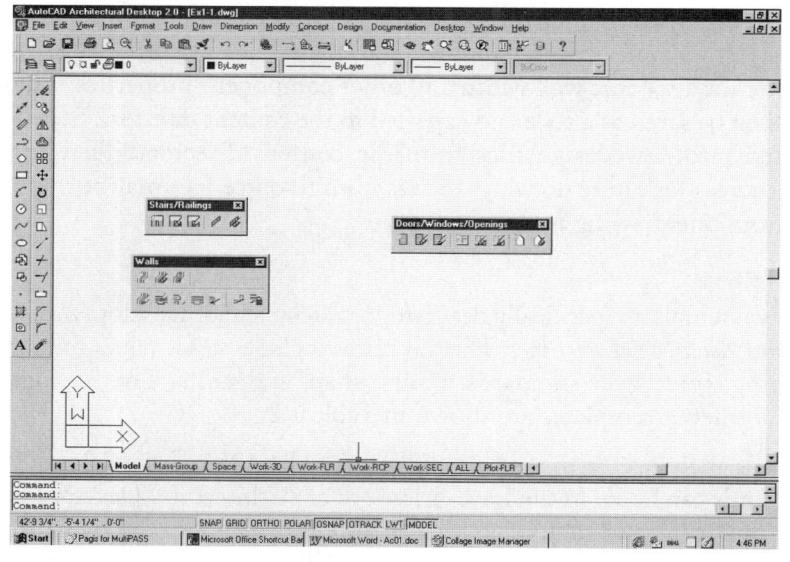

Figure 1.23 *Toolbars displayed with the Floorplan profile*

Note that profiles are not retained across drawings; rather the profile of the last user will be displayed when Architectural Desktop is reopened. Therefore, you can use the **OPTIONS** command to set your profile current to display the toolbars that you prefer.

ARCHITECTURAL DESKTOP RESOURCES

The resources of Architectural Desktop for creating architectural drawings include template files and content files. The templates include custom viewport and display configurations, whereas the content files are resources files that allow you to access settings and symbols from other drawings. Content files can also transfer plot settings from other drawings. Door styles and window styles of drawings can be exported to the Content directory for use in new drawings. Content files of Architectural Desktop include

> **Layer standards** – including AecLayerStd and AIA layer standards
>
> **Schedule formats** – including schedules for door, window, wall and casework
>
> **Symbol libraries** – including symbols for electrical, plumbing, appliances, casework, and furniture
>
> **Plot configuration**s – including plot settings
>
> **Profile shapes** – including custom shapes used to develop mass models
>
> **Styles** – including door, window, space, stair, and wall styles

The content files are located in the *Program Files\AutoCAD Architectural 2\Content* directory. Building components represented in architectural drawing files usually consist of styles for walls, doors, windows, and spaces. The use of styles allows you to create a style for a specific wall width and other component properties. These properties can then be saved as a style and exported to the content directory. Styles can then be imported into new design files from the content directory. Therefore, styles become resources for future drawings. Styles from resource files may be imported and exported from one drawing to another.

TEMPLATES

There are two templates specifically designed for Architectural Desktop: *Aec arch (imperial).dwt* and *Aec arch (metric).dwt*. The templates include preset values for annotation, dimensioning, grid, layer standards, limits, snap, and units. The settings for the imperial and metric templates are shown in Table 1.2.

The tutorials included in this book utilize the AIA (256 Colors) layer standard. Layer standards can be set in the Layer Manager of Architectural Desktop. The Layer Manager can also be used to develop layer snapshots, which save the layer display controls. The Layer Manager is discussed in Appendix C.

Setting	Imperial	Metric
Annotation Plot Size	1/8"	3 mm
Dimension Styles	Aec_Arch_I	Aec_Arch_M & ISO25
Dimension Style for Stair	Aec_Stair_I	Aec_Stair_M
Grid	10'	1000 mm
Layer Standards	AIA Long Format	BS1192 Descriptive
Limits	0,0 to 100',60'	-1.49,-1.48 to 81.49,48.89
Scale Factor	96	50
Snap	1'	100 mm
Units	Architectural	Millimeters
Units Precision	1/16 Precision	0.0 Precision

Table 1.2 *Settings for the imperial and metric templates*

Using Layouts

Each of the templates includes layout tabs specifically designed for the display of Architectural Desktop objects. The layout tabs at the bottom of the drawing area, as shown below, provide quick access to model space and paper space viewports of the template. Layout tabs are a new feature included in Architectural Desktop Release 2 and AutoCAD 2000. AutoCAD entities are created in either model space or paper space. In previous versions, TILEMODE was turned ON (**TILEMODE=1**) to work solely in model space or turned OFF to access paper space. When TILEMODE is ON, viewports are created with the VPORTS command. These viewports allow you to view the design from various orientations to create a top, front, side or isometric view. Architectural Desktop display representations control the display of objects in floating viewports when **TILEMODE** is **OFF** and you are drawing in layouts provided in Architectural Desktop.

In order for you to fully utilize the display control features of Architectural Desktop, TILEMODE should be turned OFF (**TILEMODE=0**). When TILEMODE mode is turned OFF, floating viewports can be created. (The viewports displayed when you select any of the layout tabs other than Model are floating viewports.) Therefore selecting the MASS-GROUP layout tab turns OFF TILEMODE and displays two floating viewports as shown below. You can open the *Ex1-3.dwg* of the *ADT Tutor* directory and view the layout tabs shown below:

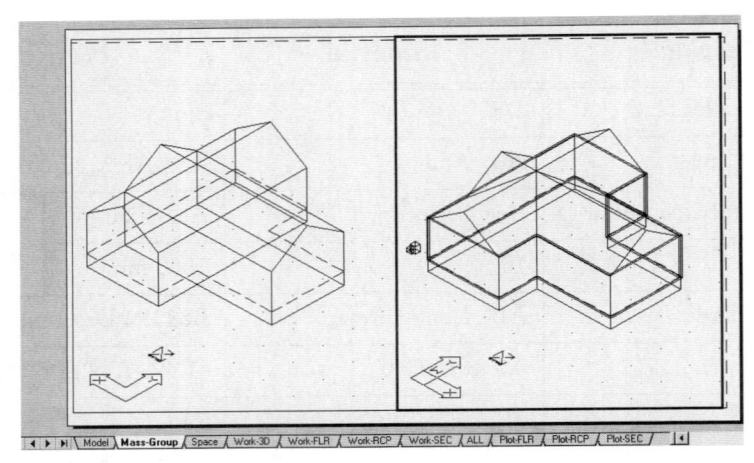

Figure 1.24 *Mass-Group Layout tab*

The white area of the screen represents a sheet of paper. This area varies according to the size of the printer configured for printing in AutoCAD. The borders inside the white area are edges of the floating viewports. Entities drawn inside these viewports are model space entities.

Selecting any of the layout tabs other than Model allows you to toggle between model space and paper space using the Model/Paper toggle in the Status Bar. When these layout tabs are used, the Model toggle is selected to create objects in model space. To create a title block or drawing border, toggle to paper space by selecting the Paper toggle in the Status Bar.

Architectural Desktop Layouts
The Aec_arch (imperial) and the Aec_arch (metric) templates include layout tabs, which allow you to quickly toggle to the predefined paper space viewports. The layout tabs of the Aec arch (metric) template are as follows: Model, Mass-Group, Space, Work-3D, Work-FLR, Work-RCP, Work-SEC, Plot-FLR, Plot-RCP, Plot-SEC. The Aec_arch (imperial) template includes the same layouts as the metric template with the addition of the All tab. The All tab allows you to view all the floating viewports in paper space, as shown in Figure 1.25.

There are nine viewports displayed by the All tab. Each viewport has a customized display control configuration assigned to it. The display control configuration turns on or off the display of certain Architectural Desktop objects in that viewport. The display configuration for the viewport will be discussed later in the chapter.

The viewports shown in Figure 1.25 are numbered in the upper right corner of the viewport. Viewports 1 through 3 are designed for creating conceptual design, including mass modeling and space planning. Viewports 4, 5, 7, and 8 are designed for design

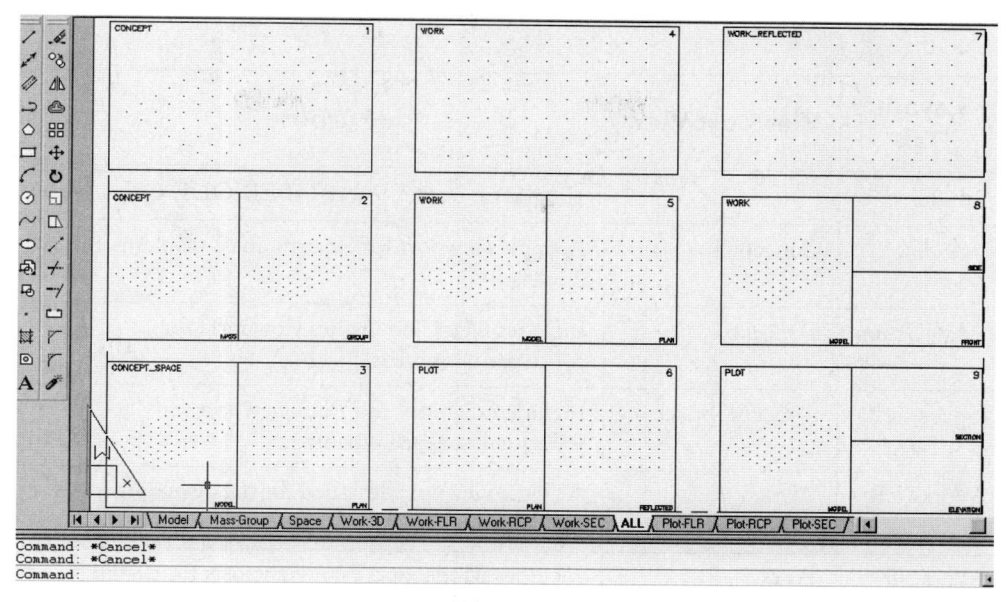

Figure 1.25 *Viewports of the Aec Arch (imperial) template*

development, including placing walls, doors, windows, and roofs of a building. Viewports 6 and 9 are designed for plotting the building.

The layout tabs at the bottom of the graphics screen allow you to quickly toggle to each of the nine viewports and three additional viewports that are locked. The Plot-FLR, Plot-RCP, and Plot-SEC are locked viewports designed for plotting the plan, elevation or section, and reflected ceiling plan drawings. The Plot-FLR view is locked to the top view and the design may not be viewed as a pictorial because the viewport is designed for plotting. The viewports are locked to provide the 1/8"=1'-0" scale. You can unlock the viewport using the **–VPORTS** command. The layouts and viewports of the Aec_arch (imperial) template are summarized in Table 1.3.

In addition to the Layout tabs, the templates include saved views, which allow you to quickly view one of the nine viewports and to activate a viewport to work on the design in model space. To access a saved view, select the All tab then enter **DDVIEW** or **V** at the command line. The **View** dialog box will be displayed as shown in Figure 1.26.

The nine saved views are named VP1 through VP9, to correspond with each of the nine floating viewports of the Aec_arch (imperial) template. To display the Concept 2 viewport, you select VP2 from the list of saved views, select the **Set Current** button and select the **OK** button. Saved views provide a quick method of displaying certain viewports of the Architectural Desktop templates without using the layout tabs.

Layout Tab	Viewport View	Purpose
Model		Toggles on model space; TILEMODE ON.
	Concept 1	A single viewport for creating mass elements for models.
MassGroup	Concept 2	Includes Mass and Group viewports for constructing models.
Space	Concept_Space 3	Includes a Model and Plan viewport for constructing space planning.
WorkFLR	Work 4	A single viewport designed for the development of the design.
Work3D	Work 5	Includes Model and Plan viewports for design development
	Plot 6	Includes Plan and Reflected viewports for plotting design development models.
Work RCP	Work_Reflected 7	Includes a single viewport for design development of reflected ceiling plans.
	Work 8	Includes Model, Side and Front viewports used for constructing models of design development.
	Plot 9	Includes Model, Section, and Elevation viewports for the plot of design development models.
All		Displays all paper space viewports.
Plot-FLR		Plotting of floor plans scaled view of the model at 1/8"=1'-0".
Plot-RCP		Plotting of reflected ceiling plans scaled view at 1/8"=1'-0".
Plot-SEC		Plotting of sections and elevations scaled view at 1/8"=1'-0".
Work-SEC		Includes pictorial and elevation views for sections or elevations.

Table 1.3 *Layouts and viewports of the Aec_arch (imperial) template*

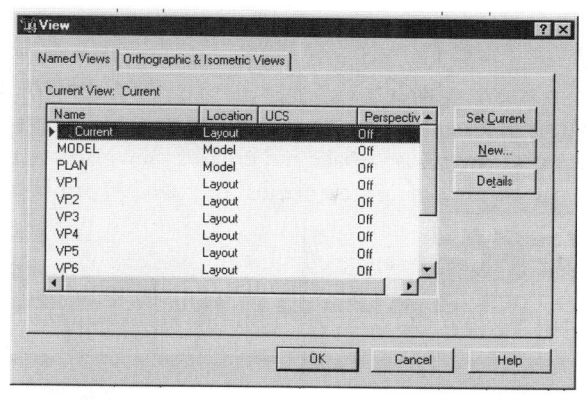

Figure 1.26 *Saved views of the Aec_Arch (imperial) template*

Two additional saved views included are the Plan and Model saved views. If you are currently in paper space and select the Model or Plan saved view, you will be prompted to select a viewport window to toggle ON model space. The Model saved view will toggle ON model space and display the design using the SW Isometric view. The Plan saved view will switch to model space and display the top view of the design. If model space is current, the Model and Plan saved views switch the view to pictorial and top views respectively.

Layout Selection

Selecting the appropriate layout to begin a project depends upon the project requirements. The Mass Group tab should be selected to create a mass model of a building. Shown below are the layout tabs of the *ADT Tutor\Ex 1-3.dwg*. The Mass-Group tab displays two viewports using the SW Isometric View in each viewport. The mass elements created to model the building are shown below in the left viewport and the mass group created from the mass elements is shown in the right viewport. The right viewport displays the mass elements combined as a mass group.

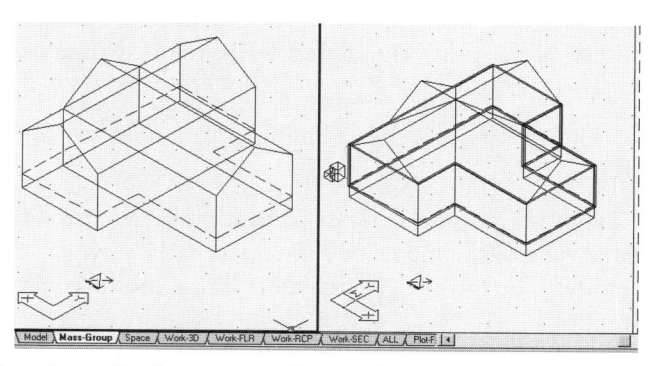

Figure 1.27 *Mass-Group floating viewport display of mass elements and a mass group*

The Space layout tab is designed to assist in space planning. It consists of two viewport windows. In the figure below, a space and space boundary were created from the mass elements shown above. The left viewport is a southwest view of a space boundary and the right viewport displays a plan view of the space. The space entity is not displayed in the pictorial view. The Space tab creates viewports with display configurations to control the display of space entities for space planning.

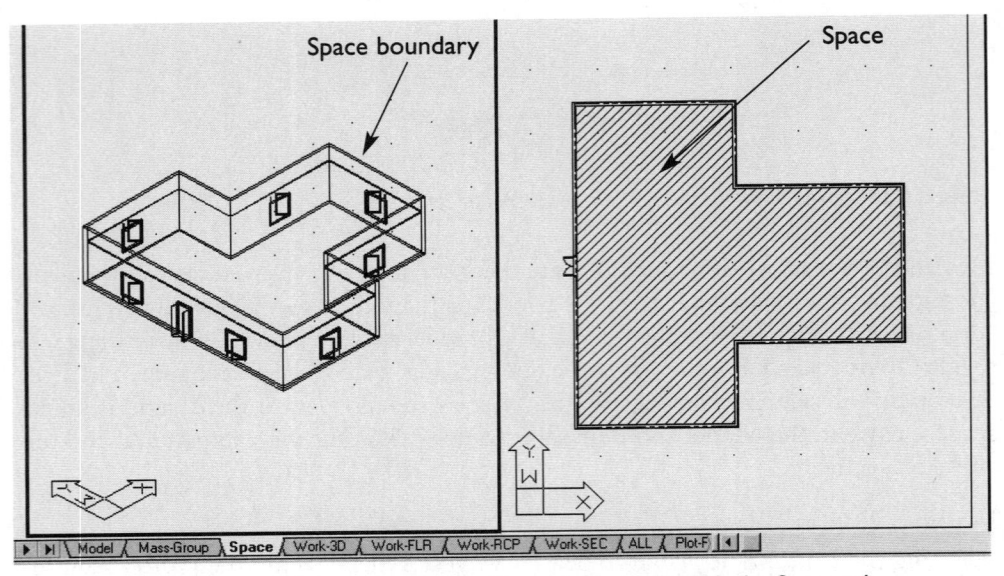

Figure 1.28 *The space and space boundary objects displayed through the Space tab*

The Work_3D tab (Figure 1.29) displays two viewports of a model. The left viewport displays walls, space boundaries, and mass elements in a southwest view. The right viewport displays the walls, mass elements, spaces and space boundaries in plan view. This viewport provides a pictorial view of the model as the design progresses.

The Work-FLR layout tab creates one viewport providing a full screen to view the design. Figure 1.30 displays space boundaries, walls, doors, windows, and mass elements.

The Work-RCP layout tab displays the ceiling grid for a reflected ceiling plan drawing as shown in Figure 1.31.

The Work-SEC layout tab (Figure 1.32) displays three viewports. The left viewport consists of a pictorial view of the model and the upper right viewport consist of a section view of the model. The lower right viewport of the model consists of an elevation view of the model.

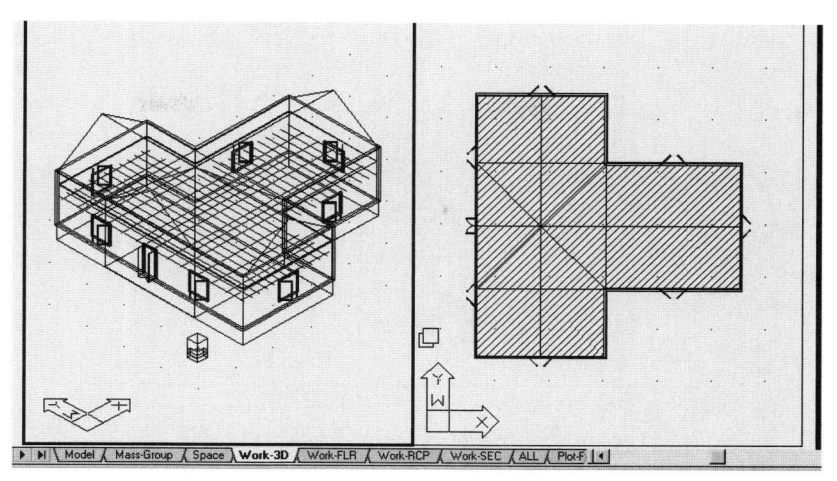

Figure 1.29 *Work-3D viewport displays mass model, walls, doors, windows, and ceiling*

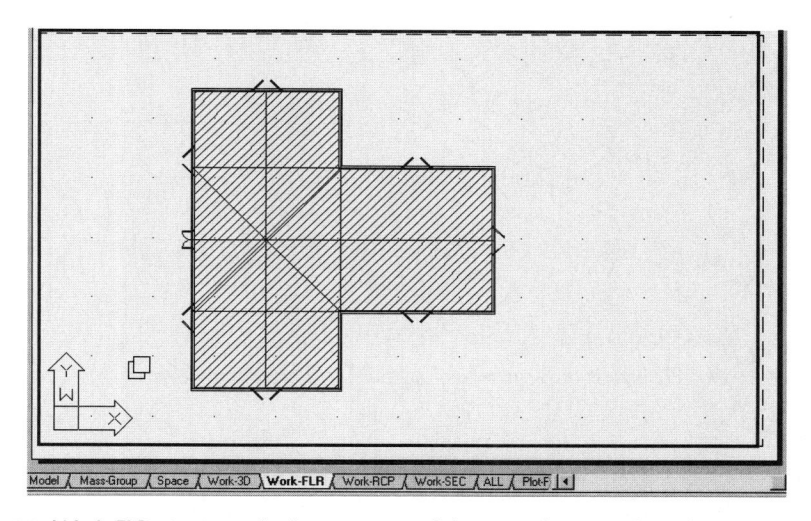

Figure 1.30 *Work-FLR viewport displays mass model, space, doors, and windows*

The Plot-FLR layout tab displays a single viewport. The viewport creates a plan view of the design specifically for plotting the floor plan. The Plot-FLR layout tab is shown in Figure 1.33.

The Plot-RCP layout tab (Figure 1.34) displays a single viewport; however, the object's display is controlled to produce a reflected ceiling plan

The Plot-SEC layout tab (Figure 1.35) consists of two viewports for plotting elevations and sections of a design.

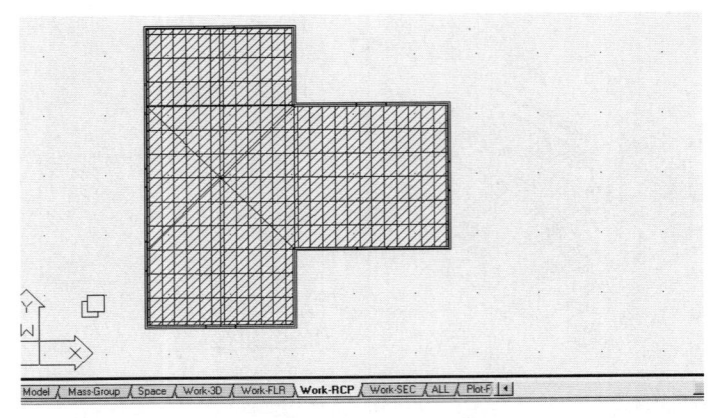

Figure 1.31 Work-RCP displays mass model, space, and reflected ceiling

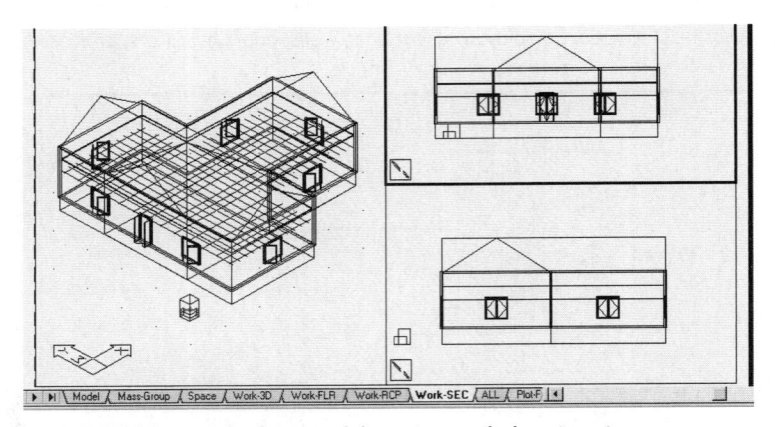

Figure 1.32 Work-SEC layout displays model, section, and elevation views

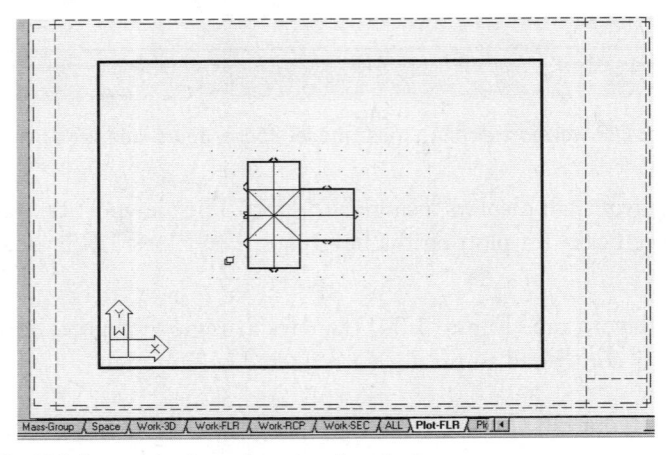

Figure 1.33 Plot-FLR layout displays plan view for plotting

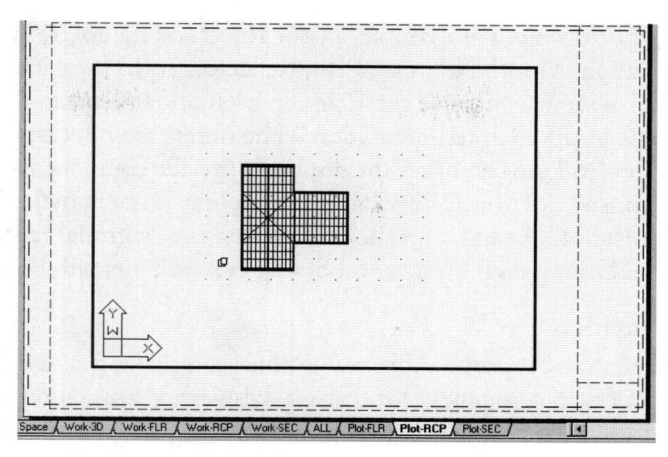

Figure 1.34 *Plot-RCP layout displays plan view of model and reflected ceiling*

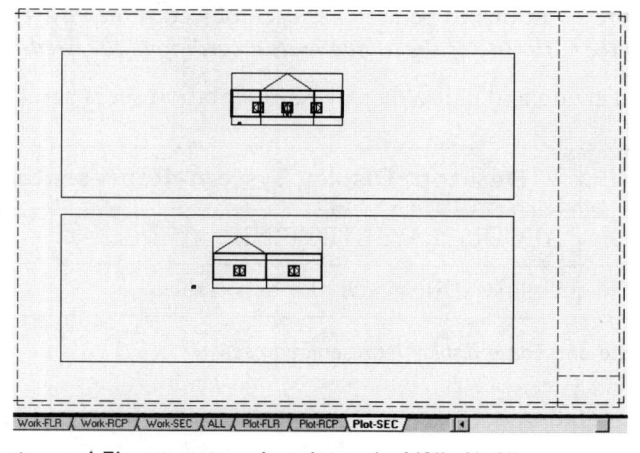

Figure 1.35 *Section and Elevation viewed at the scale 1/8"=1'–0"*

DEFINING DISPLAY CONTROL FOR OBJECTS AND VIEWPORTS

Each viewport described above has an assigned display configuration for the viewport. The Architectural Desktop viewports are named according to the assigned display configuration, which controls the display of objects in a viewport. Display control in Architectural Desktop consists of three levels: display representations, display representation sets, and display configuration.

Display Representations

Display representations control how individual objects are displayed. The display of each component of an object can be controlled for visibility, layer, color, and linetype. The display representation for an object varies according to the desired representation of that object in the drawing. For example, a door object has the following display representations: Elevation, Model, Nominal, Plan, Reflected, and Threshold Plan. A door when shown in plan view would use the Plan representation. Certain components of the door are visible in the Plan representation while others are not displayed. Each of these representations will turn on or off the display of specific components of the object. A door object consists of the following components: Door panel, Frame, Stop, Swing, Glass, Threshold A, and Threshold B. When the Nominal representation is assigned, only the Door panel, Frame, and Swing components will be displayed.

Display Representation Sets

A display representation set groups objects and their components as a defined set and turns on visibility of each component for that set. Display representation sets allow display representations to be grouped together in a set to obtain the appropriate object representation for a drawing. There are nineteen display representation sets defined in the Aec arch (imperial) template: each consists of a specific list of objects and a defined set of representations to be used for each object. *A display representation set therefore controls the visibility of the components according to the needs of the drawing.*

The command to create and edit a display representation set is accessed as follows:

Menu bar	Desktop>Display System>Representation Sets
Command prompt	AECDISPLAYSETDEFINE
Shortcut menu	Utilities>Display>Display Sets>Table

Table 1.4 *To create and edit a display representation set*

To view a representation set, choose **Desktop>Display System>Representation Sets** from the menu bar and the **Display Representation Sets** dialog box will be displayed, as shown in Figure 1.36.

The **Display Representation Sets** dialog box lists the name and description of the display representation sets. Display the object visibility defined for a selected set by selecting the display representation set from the list and then selecting the **Edit** button to open the **Display Representation Set** dialog box, as shown in Figure 1.37.

The object display is defined by object and by set. Selecting the **Display Control (by Object)** tab will open the **Display Control (by Object)** dialog box as shown in Figure 1.38.

The left section of the dialog box lists the Architectural Desktop objects to be displayed and the upper right section of the dialog box lists the various representations for the object. If the representations of the object are checked, those representations of the

Introduction to Architectural Desktop 27

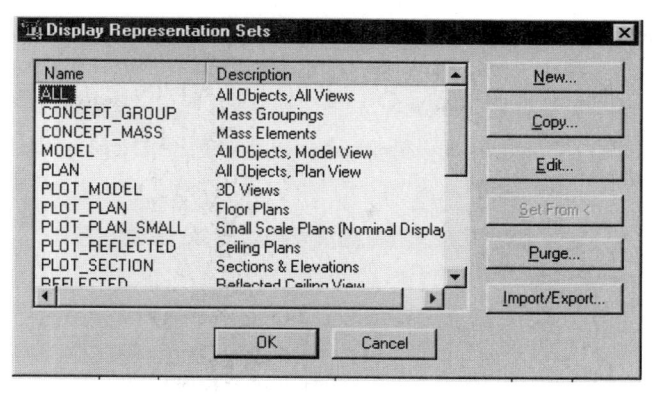

Figure 1.36 *Display Representation Sets dialog box*

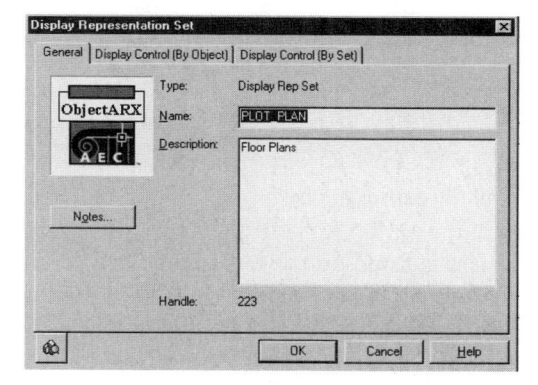

Figure 1.37 *Determining object display defined in the Display Representation Set dialog box*

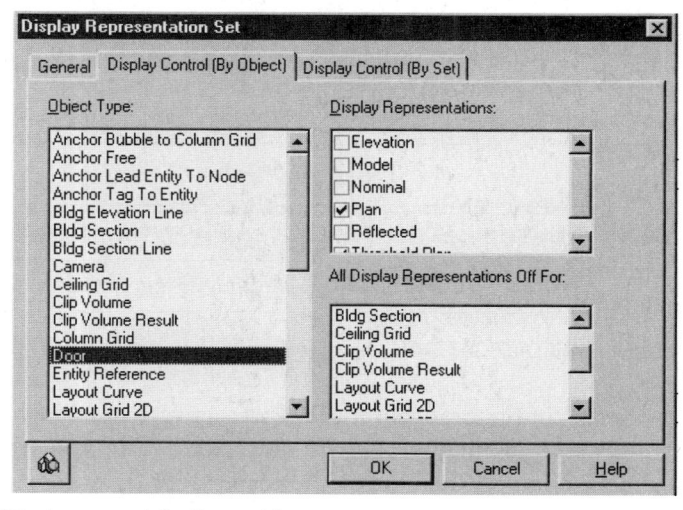

Figure 1.38 *Display control for Door object*

object are applied. In Figure 1.38 the Door object will be represented through the Plan representation. The objects that are not included in the representation set are listed in the **All Display Representations Off For:** section of the dialog box.

Selecting the **Display Control (by Set)** tab will open the **Display Control (By Set)** dialog box, as shown in Figure 1.39.

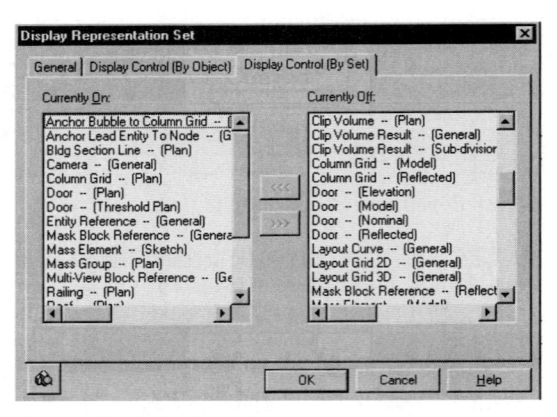

Figure 1.39 *Display Control (By Set) dialog box*

Notice that the Door object is listed on the left to be displayed through the Plan and Threshold Plan representations. Listed on the right as turned off are the Elevation, Model, Nominal, and Reflected representations of the Door object.

Display Configurations

The third level of display control uses display configurations to apply one or more display representation sets to a viewport. The display configuration of a viewport can be identified or changed by the **AECDISPLAYCONFIGATTACH** command, which is accessed as follows:

Menu bar	Desktop>Select Display
Command prompt	AECDISPLAYCONFIGATTACH ENTER
Shortcut menu	Utilities>Display>Display Configurations>Attach to Viewport

Table 1.5 *The AECDISPLAYCONFIGATTACH command*

To identify the display configuration of a viewport, activate the viewport and then choose **Desktop>Select Display** from the menu bar. The **Viewport Display Configuration** dialog box will open with the display configuration of the active viewport highlighted as shown in Figure 1.40.

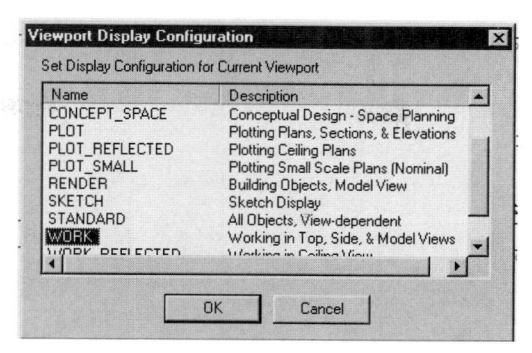

Figure 1.40 *Viewport Display Configuration for the active viewport*

Selecting another display configuration and selecting **OK** to dismiss the dialog box will change the display configuration assigned to an active viewport. Therefore, this command allows you to identify the current display configuration or change the display configuration assigned to a viewport. Display configurations are defined by the **AECDISPLAYCONFIGDEFINE** command, which is accessed as follows (Table 1.6):

Menu bar	Desktop>Display Systems>Display Configurations
Command prompt	DISPLAYCONFIGDEFINE
Shortcut	Right-click over drawing area and choose Utilities>Display>Display Configuration>Table

Table 1.6 *The AECDISPLAYCONFIGDEFINE command*

To edit a display configuration, choose **Desktop>Display Systems>Display Configurations** from the menu bar. The **Display Configurations** dialog box will open as shown in Figure 1.41.

The display configuration that is current will be highlighted when the **Display Configurations** dialog box opens. In Figure 1.41, the Work display configuration is current. To edit a display configuration, select the display configuration from the display list and select the **Edit** button. Selecting the **Edit** button opens the **Display Configuration** dialog box, which consists of two tabs: **General** and **Configuration**, as shown in Figure 1.42.

The purpose of each of the two tabs is described below.

>**General** – Allows you to edit the name and description of the display configuration.

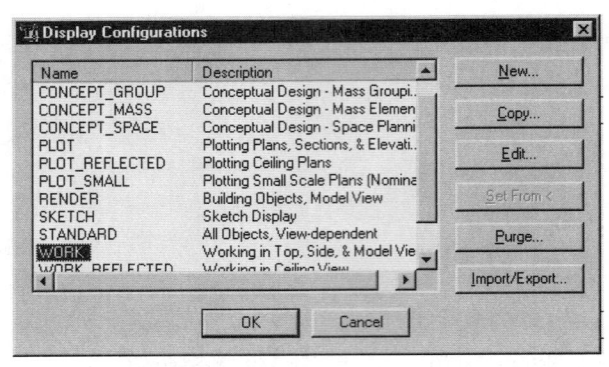

Figure 1.41 *Edit the Display Configurations dialog box to specify object display for viewport*

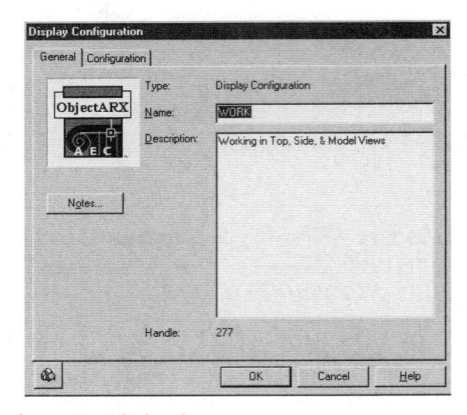

Figure 1.42 *Display Configuration dialog box*

Configuration – Allows you to assign display representation sets to define the display configuration. In Figure 1.43 the display configuration will apply the Work_Plan display representation set when the design is viewed from the top, and the Work_Section display representation set will be applied when the design is viewed from the left, right or front.

This display configuration is therefore View Directional Dependent. A View Directional Dependent display configuration will control the display of an object dependent upon the view.

Although display configurations can be edited, the display configurations available in the template are usually adequate for creating architectural working drawings. In essence, the beginner does not need to edit the display configurations if the Aec arch (imperial) or the Aec arch (metric) template is used. Display configurations will cause objects to be visible in one viewport but not necessarily in all viewports. In future tutorials, such display settings will become obvious, and future chapters will discuss

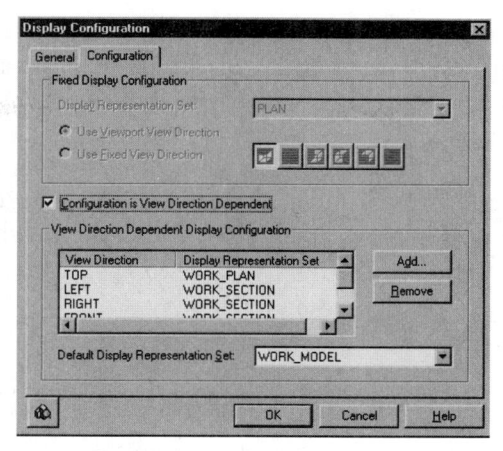

Figure 1.43 View Direction set for display configuration

how display configurations can be used to create the correct display of objects for working drawings.

ACCESSING THE VIEWER

When Architectural Desktop objects are inserted in the drawing, a dialog box opens allowing you to specify the size and properties of the object. If a door is created, the **Add Doors** dialog box is edited to set the width, height and other properties. The **Add Doors** dialog box is shown below:

Floating viewer button

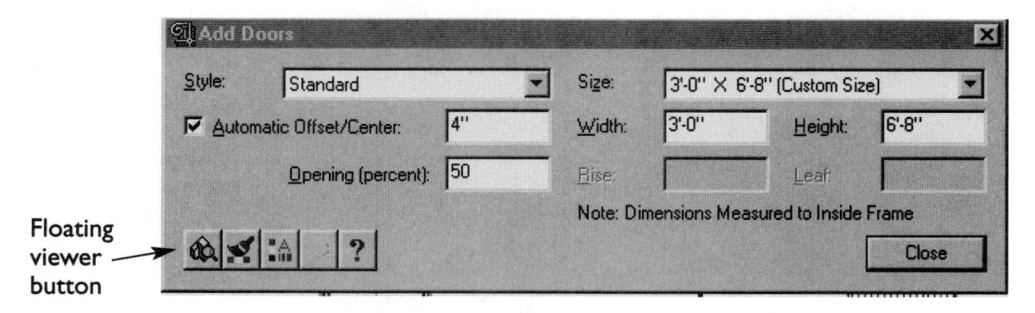

Figure 1.44 Settings for a Standard door in open position displayed in the Viewer

Each of the dialog boxes for placing the object includes a **Floating Viewer** button in the lower left corner. Selecting this button will open the Viewer, which displays the current object. In Figure 1.45, the Viewer displays the door according to the properties defined in the active dialog box. This allows you to edit in the **Add Doors** dialog box and view the results in the Viewer prior to placing the door in the drawing. The Viewer is also available from the shortcut menu of the drawing area (**Utilities>Object Viewer**). The Viewer allows you to edit the type of view displayed by selecting one of the buttons in the upper left corner. The flyout in the upper right corner allows you to change the view to one of the following views: Top, Bottom, Front, Back, Left, Right, Southwest, Southeast, Northeast, Northwest. The flyout below the view buttons allows you to select a display representation for the object, as shown in Figure 1.45.

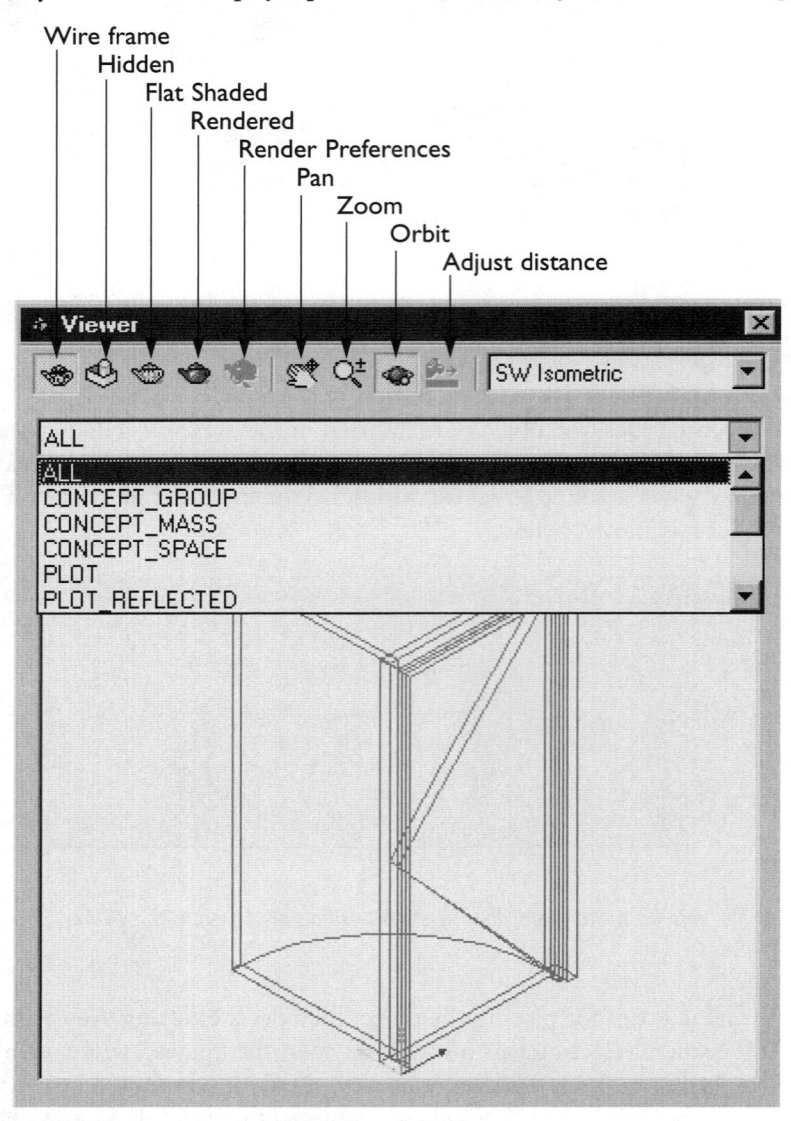

Figure 1.45 *Display representation for object of the Viewer*

You can view the object using **Pan, Zoom,** or **Orbit** when you move your mouse over the Viewer. The view option can be selected from the Viewer toolbar. The **Orbit** option is the default viewing method when the Viewer opens. If the **Orbit** view is selected, you can view the object from virtually any rotation and tilt angle. If you move your mouse over the view and right-click, a shortcut menu allows you to select other viewing options, as shown in Figure 1.46.

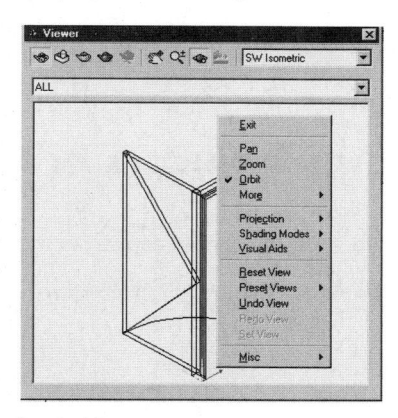

Figure 1.46 *Shortcut menu for the Viewer*

The menu options of this shortcut menu are summarized below:

> **Exit** – Closes the Viewer.
>
> **Pan** – Executes Realtime Pan on the image in the Viewer.
>
> **Zoom** – Executes Realtime Zoom on the image in the Viewer.
>
> **Orbit** –Allows you to click with the mouse to rotate and tilt the object. The Orbit mode of viewing the object is toggled ON by default.
>
> **More** – Flyout allows you to select **Zoom Window, Zoom Extents, Zoom Center, Zoom In, Zoom Out.**
>
> **Projection** – Flyout allows you to view the object in orthographic views (parallel) or in perspective.
>
> **Shading Mode** – Flyout includes the following display methods: **Wireframe, Hidden, Flat Shaded, Rendered.**
>
> **Visual Aids** – Flyout allows you to select a **Compass** or **Grid** to aid in visualizing the object.
>
> **Reset View** – Returns view of the object to the original view.
>
> **Preset Views** – Flyout allows you to select one of the following views of the object: Top, Bottom, Left, Right, Front, Back, Southwest, Southeast, Northeast, Northwest.

Undo View – Performs an Undo to the view changes of the Viewer.

Redo View – Returns to view prior to executing the Undo.

Set View – This option is not active for the Floating Viewer. If an object in the drawing is selected, then the Object Viewer is opened and the Set View option allows the view in the drawing area to be equal to that displayed in the Object Viewer.

Misc – Flyout allows you to copy the image to the clipboard, save the file as a bitmap, or open the **Render Preference**s dialog box when **Render is activated.**

TUTORIAL 1.2 VIEWING OBJECTS IN LAYOUTS

1. Open Ex1-2.dwg
2. Select the All tab; your screen should appear as shown in Figure 1.47.

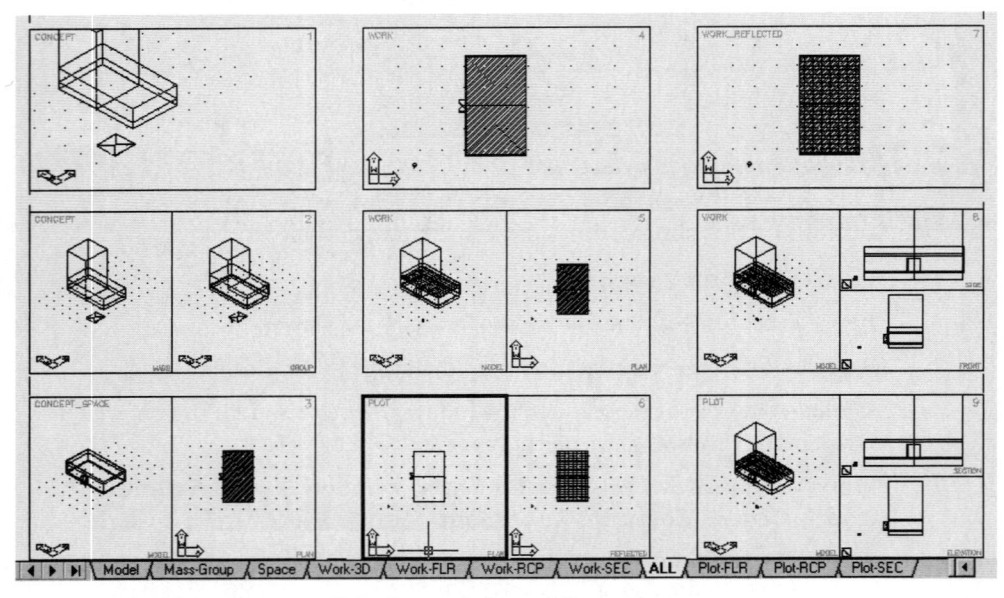

Figure 1.47 *Nine viewports of the Aec arch (imperial) template*

3. Select the Mass-Group layout tab. The mass elements and groups should be displayed as shown in Figure 1.48.
4. Select the Space layout tab. The display representation for the space object has turned off the display of the hatching in the pictorial view as shown in the left viewport in Figure 1.49.
5. Select the Work-RCP tab to display the Work Reflected Ceiling Viewport as shown in Figure 1.50.

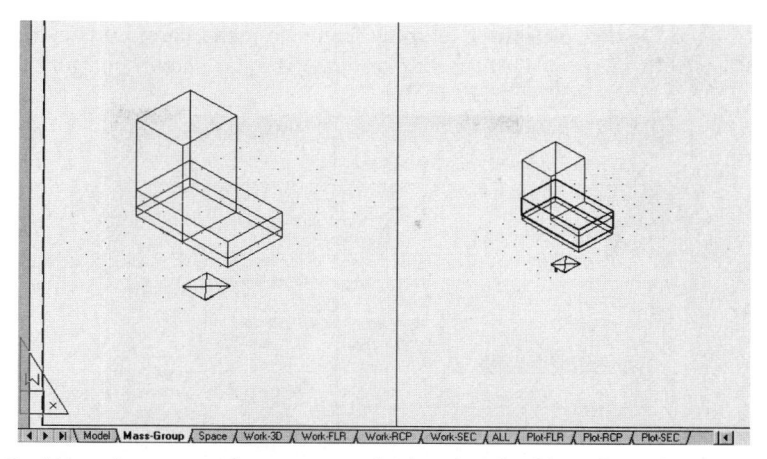

Figure 1.48 *Mass elements and mass group displayed in the Mass-Group layout tab*

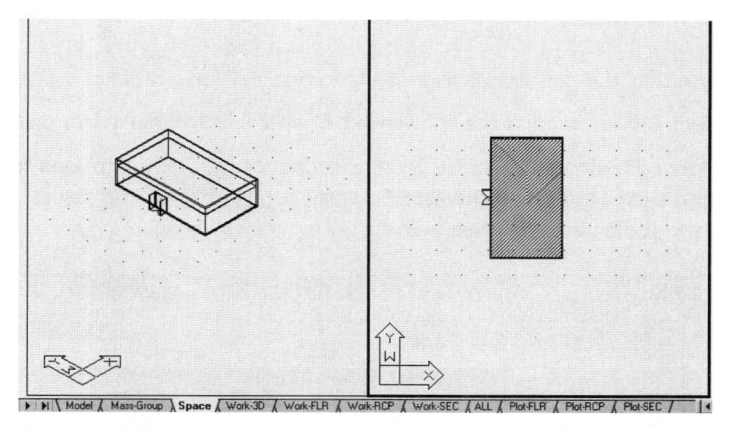

Figure 1.49 *Space and Space Boundary displayed in the Concept-Space viewport*

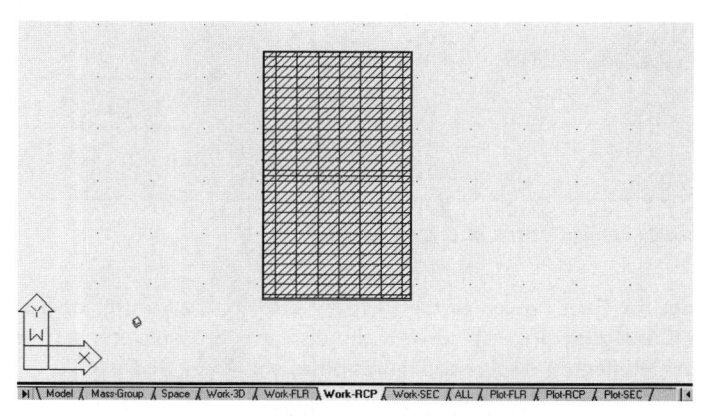

Figure 1.50 *Reflected ceiling plan displayed in the Work-RCP layout tab*

6. Choose **Desktop>Select Display** from the menu bar to display the **Viewport Display Configuration** dialog box as shown in Figure 1.51.

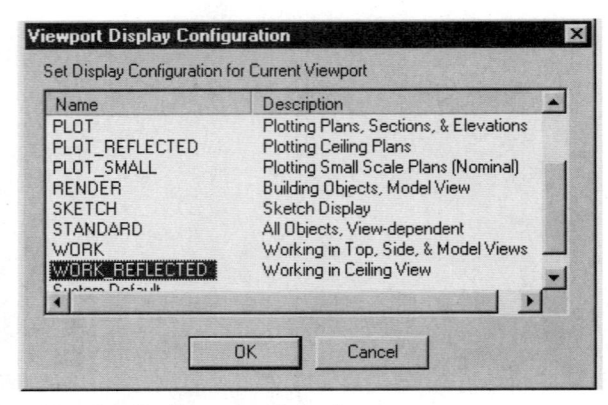

Figure 1.51 *Viewport Display Configuration dialog box*

7. Note that the current display configuration is Work_Reflected.
8. Select **OK** to dismiss the **Viewport Display Configuration** dialog box.
9. Choose **Desktop>Display System>Display Configurations** from the menu bar. Scroll down the list of display configurations and select Work_Reflected as shown below:

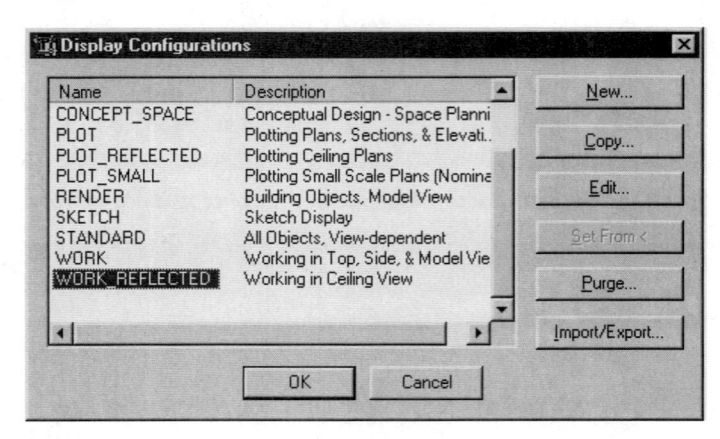

Figure 1.52 *Display configurations of current viewport*

10. Select the **Edit** button of the **Display Configuration** dialog box; then select the **Configuration** tab to view the display representation sets used for the Work_Reflected display configuration as shown in Figure 1.53.
11. Note that the Work_Model display representation set is the default representation set and that the Reflected representation set is applied when the design

Introduction to Architectural Desktop **37**

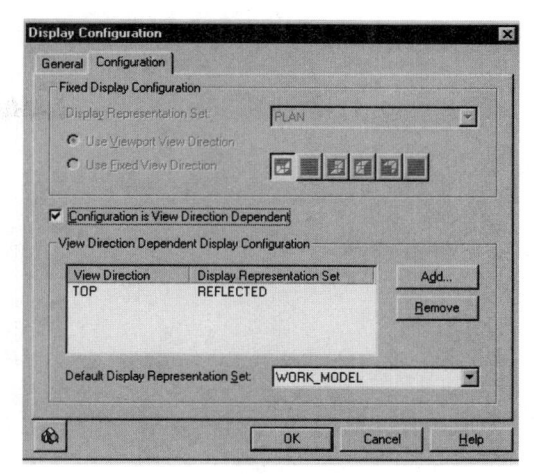

Figure 1.53 *View direction defined in the Display Configuration viewport*

is viewed from the top. Therefore, when the view is set to any view other than top, the Work_Model display representation set is applied.

12. Select **OK** to dismiss the **Display Configuration** dialog box.
13. Select **OK** to dismiss the **Display Configurations** dialog box.
14. To view the definition of the Reflected representation set, choose **Desktop>Display Systems>Representation Sets** from the menu bar.
15. Select the Reflected display representation set from the list as shown below:

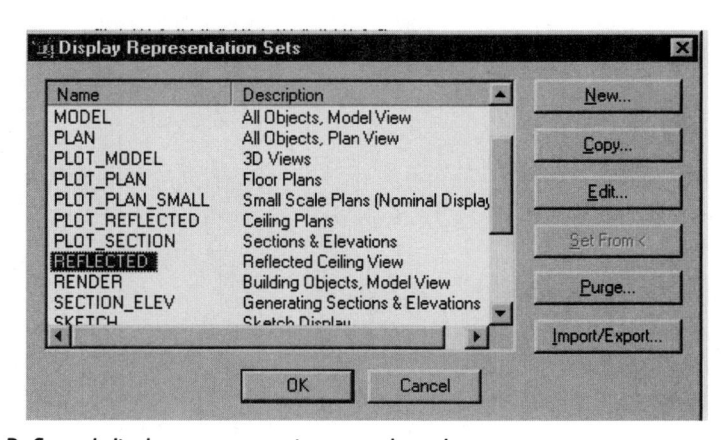

Figure 1.54 *Reflected display representation set selected*

16. Select the **Edit** button to open the **Display Representation Set** dialog box. Select the **Display Control (By Object)** tab and then select the Ceiling Grid object from the **Object Type** list as shown below:

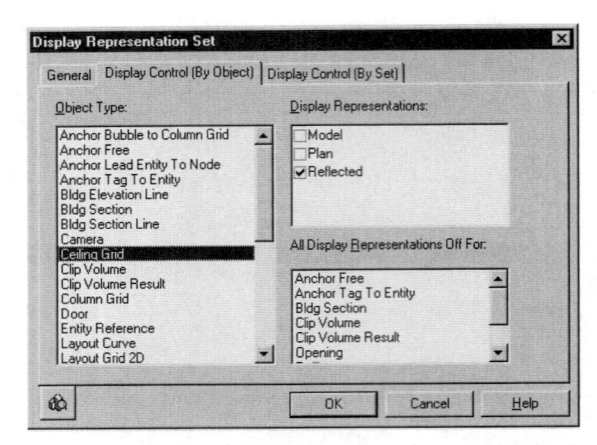

Figure 1.55 *Ceiling Grid display of the Display Representation Set dialog box*

17. Note that the Reflected representation set is checked for the Ceiling Grid, whereas Plan and Model representation sets do not include the display of the Ceiling Grid.
18. Select **OK** to dismiss the **Display Representation Set** dialog box.
19. Select **OK** to dismiss the **Display Representation Sets** dialog box.
20. Select the Plot-FLR tab, and then select **Zoom Extents** from the **View** flyout of the **Standard** toolbar. The display of the Plot-FLR layout is appropriate for plotting a floor plan. Note that the reflected ceiling plan is not displayed in this view.
21. Save the file.

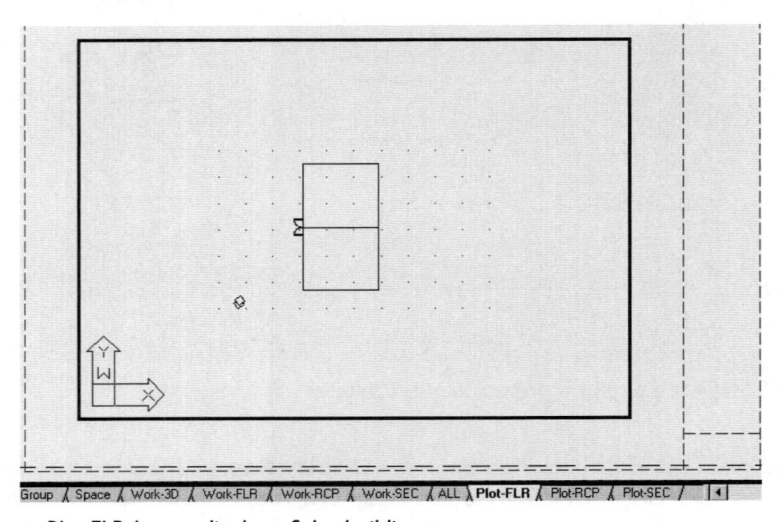

Figure 1.56 *Plot-FLR layout display of the building*

SUMMARY

1. Doors, windows, and stairs are inserted in the drawing as objects.
2. Objects inserted in a drawing interact with other objects to reduce the need to edit the wall.
3. The menu bar to the right of the AutoCAD menu bar includes the Architectural Desktop menu options.
4. Toolbars for the Architectural Desktop are loaded from the Aecarchx Menu Group through the **TOOLBAR** command.
5. Profiles can be created to retain the display of selected toolbars.
6. The resource files for Architectural Desktop are located in the *Program Files\AutoCAD Architectural 2\Content* directory.
7. Templates for beginning an Architectural Desktop file are Aec arch (imperial).dwt and Aec arch (metric).dwt.
8. Seven floating viewports are used to develop the design and two viewports are designed for plotting in each template.
9. A display representation set identifies which objects are displayed in a viewport.
10. Display configurations identify one or more display representation sets to govern the display of objects in a viewport.
11. Display configurations can be defined according to the view of the drawing.
12. The Object Viewer is included in each object dialog box, allowing you to view an object prior to placing it in the drawing.

REVIEW QUESTIONS

1. Doors and windows are inserted in a drawing as _____.
2. The menu group that includes the Architectural Desktop toolbars is _____.
3. The resource files included in Architectural Desktop include files for _____, _____, _____.
4. Resource files are located in the _____ directory.
5. The layout tab selected for developing mass models of a building would be _____.
6. The layout tab designed for plotting the finished floor plan is _____.
7. An object's display is controlled in the _____ _____ _____.
8. A display configuration for a viewport defines _____.
9. Determine the display configuration for a layout tab or viewport by selecting _____.

10. Obtain a perspective view or one or more objects by selecting _____.
11. The template that includes the resource files for Architectural Desktop is _____.
12. Architectural Desktop can be installed to provide access to the Imperial _____ or the _____ symbol library.
13. What procedure can be used to modify a door or window using shortcut menus?
14. Profiles are created and set current through the _____ command.
15. The default layer standard of the imperial template is _____.
16. The display of each component of an object is controlled in the object's _____ _____.
17. The command used to edit a display representation set is the _____.
18. List ten of the layouts included in drawings created from the Aec_arch (imperial) template.
19. The limits of the Aec_arch (imperial) template are set to _____.
20. The default scale factor for symbols and dimensions of the Aec_arch (imperial) template is _____.

CHAPTER 2

Creating Floor Plans

INTRODUCTION

This chapter introduces the commands to draw and modify walls using the Standard wall style. The focus of the chapter is the placement of the Standard wall and the tools available for precise placement of walls for the creation of a floor plan. Options of the **Add Wall** command (**AECADDWALL**) and the **Modify Wall** command (**AEC-MODIFYWALL**) will be explained. The tools to place walls with precision such as GRID, OBJECT SNAP, and grips, will also be explained. Finally, the conversion of AutoCAD geometry to walls using the **Convert to Walls** command (**AECWALL-CONVERT**) will be explained.

OBJECTIVES

After completing this chapter, you will be able to

- Draw straight or curved walls using the **Add Wall** command (**AECAD-DWALL**)
- Set justification, offset, width, and height properties of walls
- Use the **Polyline Close** and **Ortho Close** options to close a polygon shape
- Use the **Offset** option to shift the handle for the wall relative to the justification line
- Use the **Modify Wall** command (**AECMODIFYWALL**) to edit wall properties
- Use grips to identify the justification of a wall
- Use object snaps to place walls with precision
- Set the cleanup radius to avoid wall error markers
- Use tracking and AutoCAD modify commands to draw and edit walls
- Use the **Modify Wall** command (**AECMODIFYWALL**) to edit and identify wall properties

Convert existing AutoCAD geometry to walls with the **Convert to Walls** command

DRAWING WALLS

Drawing walls is the beginning point for the creation of floor plans for residential or commercial structures. Walls created in Architectural Desktop are objects that have three-dimensional properties of width, height, and length. Therefore the wall can be viewed in plan to create a floor plan or viewed as an isometric to create a pictorial drawing of the building. Walls placed in a drawing have attributes defined by their style. The simplest style of wall is the Standard wall style that consists of only two wall lines. Additional styles of walls can be created that consist of several wall lines to represent more complex walls. The use of Wall Styles to create more complex walls will be discussed in the next chapter.

Because walls are inserted as three-dimensional objects, Architectural Desktop will automatically clean up the intersection of adjoining walls. Object technology controls the placement of door and window objects to eliminate the need for editing the wall when a window or door is placed.

CREATING WALLS WITH THE ADD WALL COMMAND

The commands to insert and modify walls are shown on the **Walls** toolbar. To display the **Walls** toolbar, enter the **TOOLBAR** command in the command line or position the mouse pointer over any toolbar button; then right-click and choose **Customize** from the shortcut menu. When the **TOOLBAR** command is executed, the **Toolbars** dialog box will open as shown in Figure 2.1.

Edit the **Toolbars** dialog box, setting the Menu Group to AECARCHX, and scroll down the list of toolbars to Walls. Check the Walls toolbar by clicking in the **Walls** check box and the Walls toolbar will be displayed. The commands on the **Walls** toolbar are shown in Figure 2.2.

Access the **Add Wall** command as follows:

Menu bar	Design>Walls>Add Wall
Command prompt	AECWALLADD
Toolbar	Select Add Wall from the Walls toolbar shown in Figure 2.3
Shortcut	Right-click over drawing area and choose Design>Walls>Add

Table 2.1 *Accessing the Add Wall command*

Creating Floor Plans   43

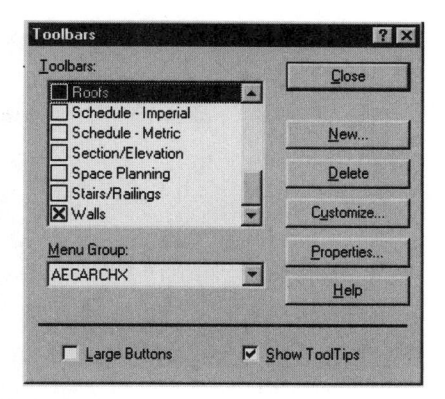

Figure 2.1 *Toolbars dialog box*

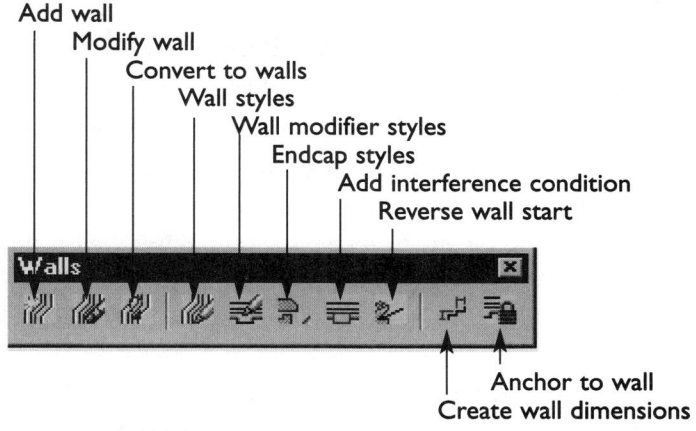

Figure 2.2 *Commands on the Walls toolbar*

Figure 2.3 *Selecting the Add Wall command from the Walls toolbar*

When the **Add Wall** command is executed, you are prompted for a start point to begin the wall. The **Add Walls** dialog box will open as shown in Figure 2.4.

The **Add Walls** dialog box consists of **Style** and **Group** drop-down lists and **Width, Height, Justify,** and **Offset** edit fields. The buttons located in the lower left corner will

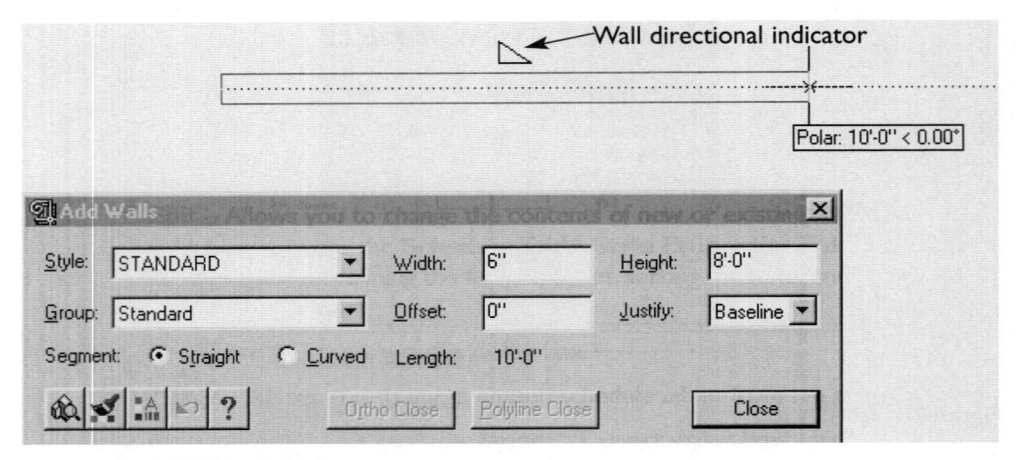

Figure 2.4 Add Walls dialog box

execute the **Floating Viewer, Match, Wall Properties, Undo** and **Help** options of the command. The options of the **Add Walls** dialog box are as follows:

Style – List of all wall styles imported in the drawing.

Group – List of wall cleanup groups.

Width – Allows you to set the width of the wall.

Offset – Allows you to set a distance to offset the handle for placing the wall.

Height – Allows you to set the height of the wall.

Justify – List of justifications for the wall, which include Baseline, Center, Left, and Right.

Straight – Toggles wall type to straight wall segments.

Curved – Toggles wall type to curved wall segments.

Ortho Close – Closes a polygon with wall segment perpendicular to beginning wall segment.

Polyline Close – Closes a polygon with a wall segment back to the beginning wall segment.

Floating Viewer – Opens the Viewer, allowing you to view the current wall style.

Match – Prompts you to select an existing wall style to set current for a new wall.

Properties – Opens the **Wall Properties** dialog box, allowing you to view the properties of the current wall.

Undo – Performs an Undo of the last wall segment.

? – Executes Help for placing walls.

Close – Closes the Add Walls dialog box.

The **Add Walls** dialog box can be edited prior to or after the start point for a wall is selected. If the wall start point has been set, you can still change the default wall properties by editing the **Add Walls** dialog box or selecting the options from the command line. The sequence of the **Add Wall** command is as follows:

```
Command:
AECWALLADD
Start point or
[STyle/Group/WIdth/Height/OFfset/Justify/Match/Arc]:
```
(Specify the start point.)
```
End point or
[STyle/Group/WIdth/Height/OFfset/Justify/Match/Arc]:
```
(Specify the end point.)

You can select options for this command by entering the characters capitalized in the options list at the command prompt. Therefore, to select the **OFfset** option you would enter **OF**. The first command prompt allows you to specify the location for beginning the wall. The next prompt allows you to enter the location for the end of the wall segment. The length of the wall is specified just as you would specify the length of an AutoCAD line by using absolute coordinates, direct distance entry, and relative coordinates or by clicking with the mouse. Once the wall is begun, the distance from the start point to the current pointer location is dynamically displayed on the tracking pointer when the **Auto Track** tooltip is toggled on in the **Drafting** tab of the **Options** dialog box. The Length of the wall is dynamically displayed in the **Add Walls** dialog box, as shown in Figure 2.4. Other edit fields of the **Add Walls** dialog box allow you to set the width and height of the wall.

> **NOTE:** when you edit a dialog box, the graphic screen becomes inactive. To reactivate the drawing area after editing a dialog box, you need to click in the drawing area prior to selecting coordinate locations on the drawing area. When you click in the drawing area, the title bar of the **Add Walls** dialog box will become dim, indicating it is inactive.

The wall shown in Figure 2.4 is 10' long, created from the Standard wall style with a width of 6" and height of 8'. The lines created to represent this wall are automatically placed on the layer A-Wall. When an object is created by Architectural Desktop, it is automatically placed on the correct layer. The default color for A-Wall layer is color 50 (yellow).

Setting Justification for Precision Drawing

The wall in Figure 2.4 was drawn from left to right. A wall directional indicator marker is displayed above the wall as the wall segment is being drawn. The directional indicator marker shown above the wall points away from the origin of the wall segment. The wall directional indicator is helpful in understanding the justification of a wall. The justification of a wall sets the location of the handle for placing the wall relative

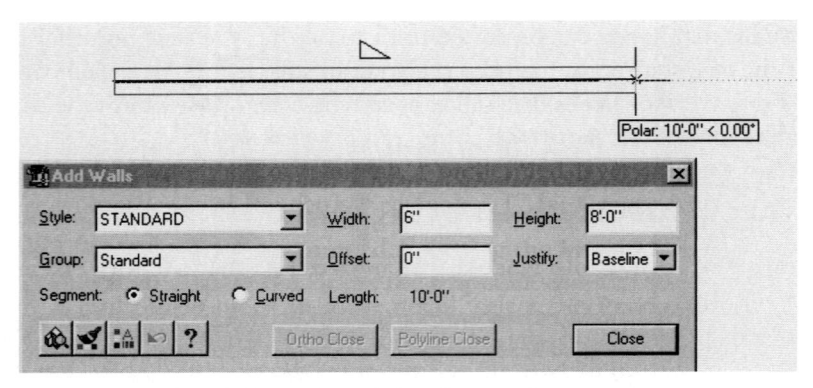

Figure 2.5 Wall drawn from left to right with Baseline justification

Figure 2.6 Wall drawn from left to right with Left justification

Figure 2.7 Wall drawn from left to right with Center justification

to the wall width. A wall drawn with right justification has its handle on the right when you orient yourself at the start point and face toward the end of the wall.

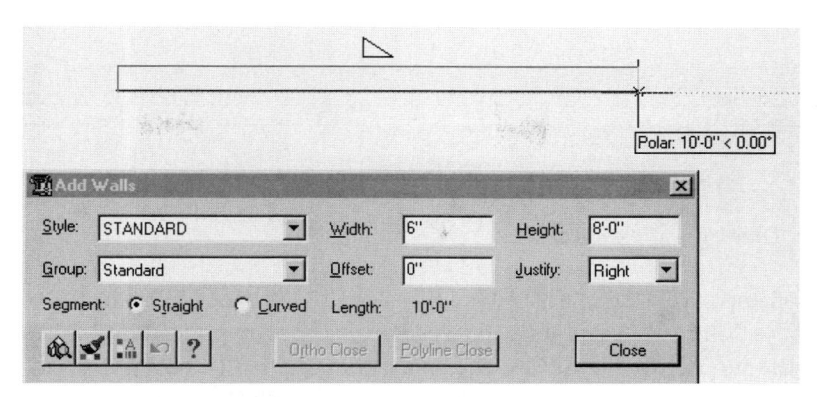

Figure 2.8 *Wall drawn from left to right with Right justification*

Types of justifications available are Right, Left, Center, and Baseline. Figures 2.5 to 2.8 illustrate each type of justification and its handle position.

The Standard wall style consists of only two lines, which are offset one-half of the value entered for the wall width on each side of the Baseline. The value entered for the wall width is balanced one-half on each side of the zero Baseline. Complex walls created through the use of wall styles can consist of wall lines offset a specific distance from the Baseline. However, the Standard wall style is unique because both its Baseline and Center justification result in locating the wall handle in the center position.

Creating Straight Walls

The **Segment** toggles of the **Add Walls** dialog box include **Straight** or **Curved** toggles. The **Straight** toggle is the default wall type that is retained after curved walls are drawn. Therefore you will be prompted to create straight walls when the **Add Wall** command is executed. To draw a straight wall, select the **Add Wall** command, specify the width or height in the **Add Walls** dialog box, and then specify the beginning and ending points of the wall. The following command sequence demonstrates how to draw the straight wall shown in Figure 2.9.

```
(Select Add Wall from the Walls toolbar)
Command: _(AECWALLADD)
Start point or
[STyle/Group/WIdth/Height/OFfset/Justify/Match/Arc]:
5',5' ENTER (P1)
End point or
[STyle/Group/WIdth/Height/OFfset/Justify/Match/Arc]:
@20',0 ENTER (P2)
End point or
[STyle/Group/WIdth/Height/OFfset/Justify/Match/Arc/
Undo]: ENTER
```

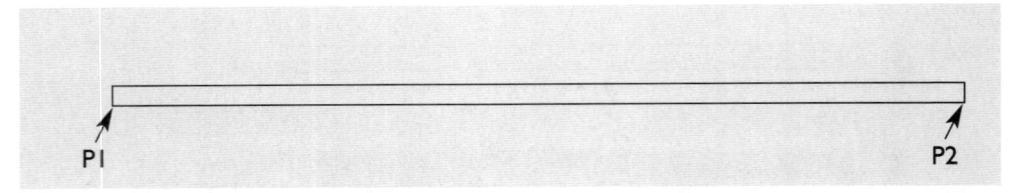

Figure 2.9 *Drawing a straight wall*

The first prompt, Start point or [STyle/Group/WIdth/Height/OFfset/ Justify/Match/Arc]: allows you to establish the start point (P1) for the wall by a click of the mouse. In Figure 2.9, the start point is established by absolute coordinates. The second prompt: End point or [STyle/Group/WIdth/Height/ OFfset/Justify/Match/Arc]: allows you to establish the endpoint of the wall. The endpoint can be established by any of the following four coordinate entry methods: absolute coordinate, relative coordinate, polar coordinate or direct distance entry. In the example in Figure 2.9, relative coordinates were used to draw the wall 20' long.

Drawing walls in Architectural Desktop is very similar to drawing a line in AutoCAD. Notice in the command sequence above that the **Undo** option is added to the command prompt once the endpoint of the wall is established. The **Undo** button is also activated in the **Add Walls** dialog box after one wall segment is drawn. Selecting the **Undo** option allows you to change the location of the endpoint without exiting the **Add Wall** command. To exit the command, press ENTER; the null response will cause the **Add Wall** command to be terminated.

Designing Curved Walls

You can draw curved walls by toggling ON the **Curved** radio button of the **Add Walls** dialog box. You create a curved wall by establishing three points: the beginning point, a point along the curve, and the end point of the curved wall. To draw a curved wall, toggle ON **Curved** in the **Add Walls** dialog box as shown in Figure 2.10.

Figure 2.10 *Toggling ON Curved to create curved walls*

The following command sequence illustrates the command line prompts and entries to create the curved wall shown in Figure 2 11.

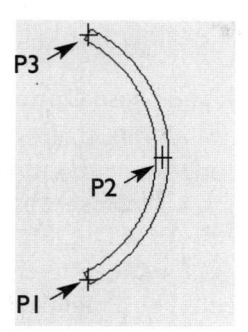

Figure 2.11 *Defining points for a curved wall*

(Select Add Wall from the Walls toolbar and select the Curved toggle)
```
Command: _AecWallAdd
Start point or
[STyle/Group/WIdth/Height/OFfset/Justify/Match/Arc]:
DBOX
Start point or
[STyle/Group/WIdth/Height/OFfset/Justify/Match/Line]:
2',2' (P1)
Mid point or
[STyle/Group/WIdth/Height/OFfset/Justify/Match/Line]:
@3',5' (P2)
End point or
[STyle/Group/WIdth/Height/OFfset/Justify/Match/Line]:
@0,10' (P3)
Mid point or
[STyle/Group/WIdth/Height/OFfset/Justify/Match/Line/
Undo]: ENTER
```
(Ends the Add Wall command)

In the command sequence above, the letters DBOX occur in the command line when the **Curved** toggle is set ON in the dialog box. These letters indicate that an option or toggle has been set in the **Add Walls** dialog box. The first prompt allows you to identify the start point (P1), the beginning point of the curved wall. In the example above, the start point is established by absolute coordinates. Note that included in the options is the Line option, which allows you to toggle back to drawing straight wall segments.

The Mid point prompt allows you to establish a point (P2) for the curved wall to pass through. This point and all succeeding points can be established with any of the

AutoCAD coordinate entry methods. Once the mid point is established, additional points can be set to end the curved wall or additional mid point locations can be added to continue drawing additional curved wall segments. Drawing a curved wall is similar to drawing an arc through three points with the **ARC** command of AutoCAD.

Closing Polygon Shapes with Polyline Close Option

The **Polyline Close** button of the **Add Walls** dialog box is used to create a final wall segment of a polygon to close the polygon without specifying the coordinates of the endpoint. This option is similar to the **Close** option when AutoCAD lines are drawn. The **Close** option of the AutoCAD **LINE** command allows you to draw several line segments and enter a C for **Close** to close the last line segment to the beginning of the first line segment. The **Polyline Close** button will draw a final wall segment from the last wall segment endpoint back to the beginning of the first wall segment. This button becomes active after two wall segments have been drawn. The following tutorial provides an exercise for drawing walls and using the **Polyline Close** button.

TUTORIAL 2.1 DRAWING WALLS USING POLYLINE CLOSE

1. Select **New** from the **Standard** toolbar.
2. Select the **Use a Template** button in the **Create New Drawing** dialog box.
3. Select from the **Select a Template** list the Aec arch (imperial) template and select **OK** to dismiss the **Create New Drawing** dialog box.
4. Choose **File>SaveAs** from the menu bar and save the drawing as *Ex2-1.dwg* in your student directory.
5. Select the All layout tab; enter **V** to begin the **VIEW** command.
6. Select VP5 saved view and select the **Set Current** button and then the **OK** button to dismiss the **View** dialog box. The Work 5 viewport should now be displayed.
7. Toggle on model space by clicking the Paper/Model toggle of the Status Bar.
8. Select the right viewport named Plan to activate the viewport.
9. Display the **Walls** toolbar using the **TOOLBAR** command. To execute the **TOOLBAR** command, move the mouse pointer over any toolbar button and right-click. Choose **Customize** from the shortcut menu to display the **Toolbars** dialog box. Edit the **Toolbars** dialog box, setting the Menu Group to AECARCHX. Scroll down the list of toolbars to Walls and select the check box for the Walls toolbar. Select the **Close** button to close the **Toolbars** dialog box.

10. Verify that Ortho is ON (Ortho in the Status Bar should appear pushed in).
11. Begin drawing walls by selecting the **Add Wall** command (**AECADDWALL**) from the **Walls** toolbar shown in Figure 2.12.

Figure 2.12 *Add Wall command on the Walls toolbar*

12. Edit the **Add Walls** dialog box as shown in Figure 2.13 to Width=8", Height=8', Offset=0; Justify = Right; Style= Standard; Group=Standard (Figure 2.13).

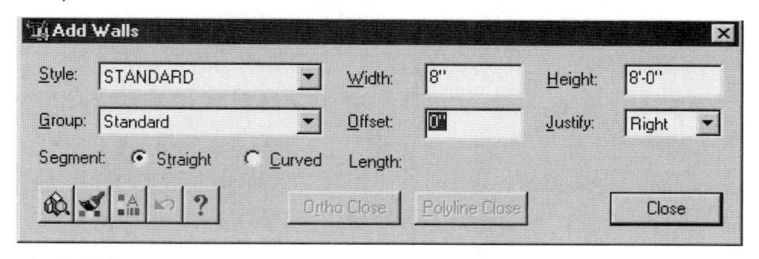

Figure 2.13 *Add Walls dialog box for drawing the first wall segment*

13. Begin drawing the wall by clicking with the mouse in the PLAN viewport to activate the drawing area and obtain a pointer.
14. Respond to the following command line prompts, referring to Figure 2.14.

```
Command: _(AECWALLADD)
Start point or
[STyle/Group/WIdth/Height/OFfset/Justify/Match/Arc]:
2',2' ENTER (p1)
```
(Move mouse to right)
```
End point or
[STyle/Group/WIdth/Height/OFfset/Justify/Match/Arc]:
50' ENTER (p2)
```
(Move mouse up)
```
End point or
[STyle/Group/WIdth/Height/OFfset/Justify/Match/Arc/
Undo]: 28' ENTER (p3)
```
(Move mouse to left)
```
End point or
[STyle/Group/WIdth/Height/OFfset/Justify/Match/Arc/
Undo/Close/ORtho]: 20' ENTER (p4)
```

```
(Move mouse up)
End point or
[STyle/Group/WIdth/Height/OFfset/Justify/Match/Arc/
Undo/Close/ORtho]: 16' ENTER (p5)
(Move mouse to left)
End point or
[STyle/Group/WIdth/Height/OFfset/Justify/Match/Arc/
Undo/Close/ORtho]: 30' ENTER (p6)
End point or
[STyle/Group/WIdth/Height/OFfset/Justify/Match/Arc/Un
do/Close/ORtho]: DBOX (select the Polyline Close
button of the Add Walls dialog box.)
```

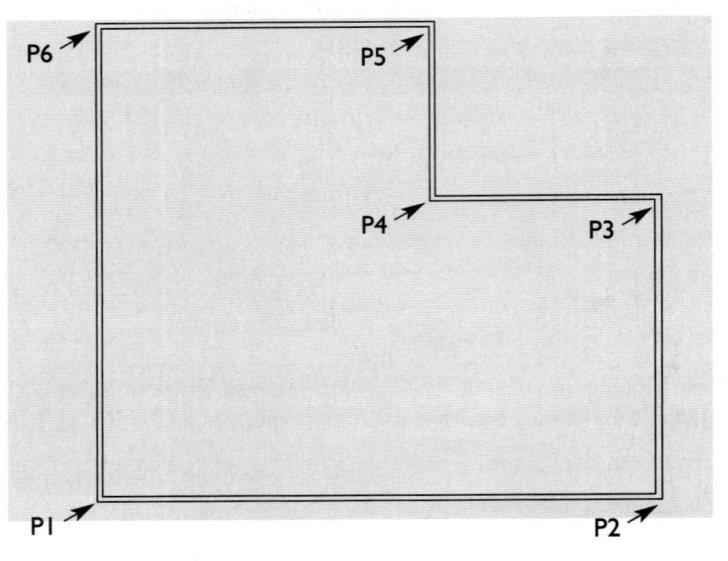

Figure 2.14 *Polyline Close option used to draw the last wall segment*

15. Save the drawing.

Using the Ortho Close Option

The **Ortho Close** option draws the final wall segment of a polygon to connect at an angle of 90 degrees to the beginning of the first wall segment. This option allows you to quickly draw a rectangle after establishing the length and width distances by the first two wall segments. Execute this option by selecting the **Ortho Close** button in the **Add Walls** dialog box or entering OR in the command line.

The following tutorial uses the **Ortho Close** option to draw a rectangle 68'-3" by 28'-0" as shown in Figure 2.15.

Creating Floor Plans 53

Figure 2.15 *Creating walls for a rectangular shape using Ortho Close*

TUTORIAL 2.2 USING ORTHO CLOSE

1. Select **New** from the **Standard** toolbar.
2. Select the **Use a Template** button in the **Create New Drawing** dialog box.
3. Select the Aec arch (imperial) and select **OK** to dismiss the **Create New Drawing** dialog box.
4. Choose **File>SaveAs** from the menu bar and save the drawing as *Ex2-2.dwg* in your directory.
5. Select the Work-FLR layout tab.
6. Verify that the model/paper space toggle is set to Model in the Status Bar.
7. Begin drawing walls by selecting the **Add Wall** command (**AECADDWALL**) from the **Walls** toolbar.
8. Edit the **Add Walls** dialog box as shown in Figure 2.16 to Width=3 1/2", Height=8', Offset=0; Justify = Right; Style= Standard; Group=Standard.

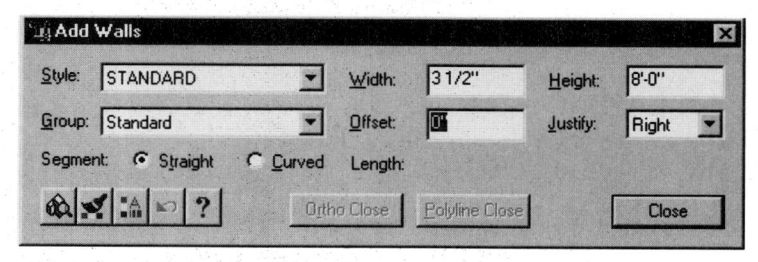

Figure 2.16 *Add Walls dialog box for drawing the first wall segment*

9. Select in the drawing area to activate the drawing area and obtain a pointer.
10. Verify that Ortho is ON (Ortho button in the Status bar appears pushed in).

11. Respond to the following command line prompts, referring to Figure 2.17.

    (Select Add Wall from the Walls toolbar)
    ```
    Command:_(AECWALLADD)
    Start point or
    [STyle/Group/WIdth/Height/OFfset/Justify/Match/Arc]:
    2',2' (p1 shown in Figure 2.17)
    ```
    (Move mouse to right)
    ```
    End point or
    [STyle/Group/WIdth/Height/OFfset/Justify/Match/Arc]:
    68'3 (p2 shown in Figure 2.17)
    ```
    (Move mouse up)
    ```
    End point or
    [STyle/Group/WIidth/Height/OFfset/Justify/Match/Arc]:
    28' (p3 shown in Figure 2.17)
    ```
    (Select the Ortho Close button of the Add Walls dialog box)
    ```
    End point or
    [STyle/Group/WIdth/Height/OFfset/Justify/Match/Arch/
    Undo/Close/Ortho]: DBOX
    Point on wall in direction of close:
    ```
    (Click in the drawing area to activate the drawing area and then select a location near point p4 shown in Figure 2.17.)

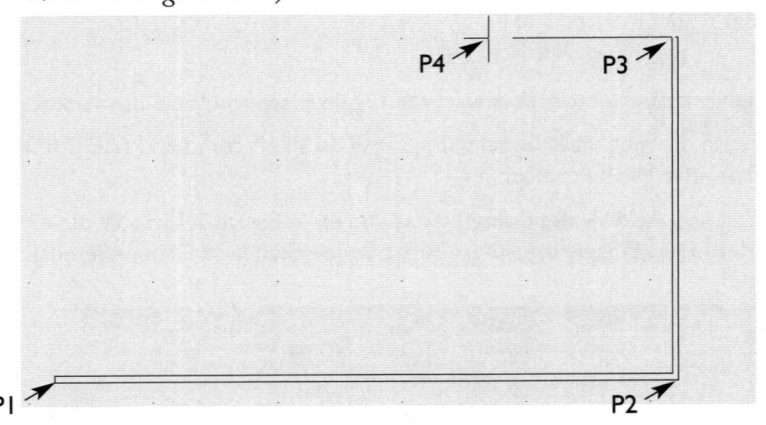

Figure 2.17 *Identifying a point for Ortho Close to complete the rectangle*

12. Save the drawing.

The **Ortho Close** button of the **Add Walls** dialog is not active until two wall segments are drawn. In the command sequence above, the **Ortho Close** button is selected after you draw the wall through points P1, P2, and P3. When the **Ortho Close** button is selected, you are prompted Point on wall in direction of close: In response to this prompt, set ON AutoCAD ORTHO (F8 toggle) and select a point on the drawing area perpendicular to the last wall segment and in the direction of the beginning of the first wall segment. Since Ortho is on, the next wall segment will be drawn perpendicular to the last and **Ortho Close** will close the polygon to form a rectangle.

If the first two wall segments were not perpendicular to each other when **Ortho Close** was used, the final wall segment will still close to the first wall segment at an angle of 90 degrees as shown in Figure 2.18. In each example, points P1, P2, and P3 were established by walls drawn through specific locations. Point P4 was selected in response to the prompt `Point on wall in direction of close:` to close the polygon.

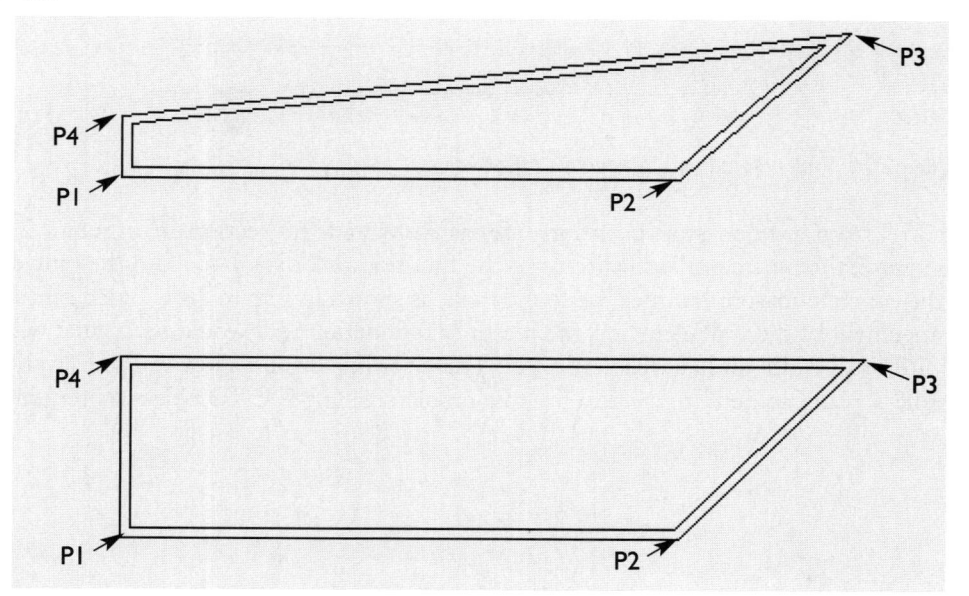

Figure 2.18 *Ortho Close used to complete the walls of a polygon*

Using the Offset Option to Move the Wall Handle

The **Offset** edit field of the **Add Walls** dialog box allows you to enter a distance to offset the handle for the wall relative to the justification of the wall. If you enter an offset of 2" using a right wall justification, the handle for the wall will be located 2" beyond the right wall face, as shown in Figure 2.19. The wall in the figure is 8" wide and is being drawn from left to right.

If a negative value is entered for the offset distance, the handle for the same wall as shown in Figure 2.19 will move up 2" from the right face, as shown in Figure 2.20.

The ability to shift the handle relative to the justification of a wall is useful in creating a foundation plan. In residential construction drawings, the masonry walls are off-

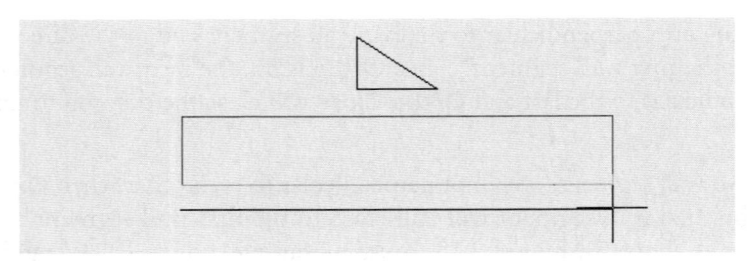

Figure 2.19 *Wall created with 2" offset*

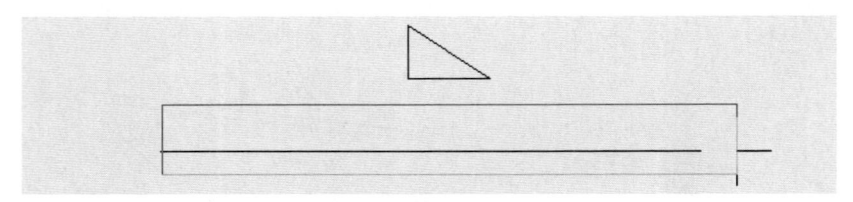

Figure 2.20 *Wall created with a negative offset value*

set 3/4" from a wood stud partition line, as shown in the section in Figure 2.21. Therefore, the outside of stud surface of the first floor is offset 3/4" from the outside of the concrete masonry unit of the foundation, as shown in Figure 2.21. Based on this arrangement of the wall elements, the masonry foundation wall would be created with a 7 5/8" wall width and an offset of –3/4". These settings would allow you to trace the first floor plan geometry to create a foundation plan using the END object snap mode.

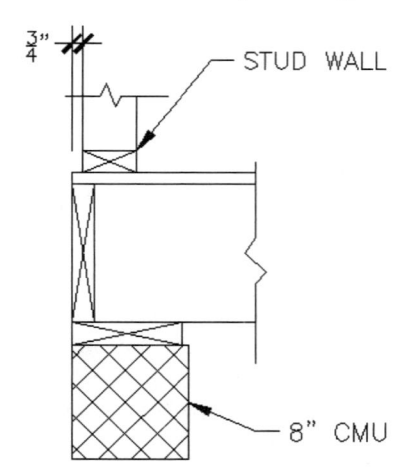

Figure 2.21 *Wall detail of wood wall and masonry foundation construction*

In commercial construction the offset distance could be set according to the distance a wall is offset from the column centerline. This would allow you to trace the column line to draw a wall offset the desired distance between the column line and the wall.

TUTORIAL 2.3 CREATING WALLS

1. Select **New** from the **Standard** toolbar.
2. Select the **Use a Template** button in the **Create New Drawing** dialog box.
3. Select the Aec arch (imperial) template and select **OK** to dismiss the **Create New Drawing** dialog box.
4. Choose **File>SaveAs** from the menu bar and save the drawing as *Ex2-3.dwg* in your student directory.
5. Select the All layout tab; type **V** to begin the **VIEW** command.

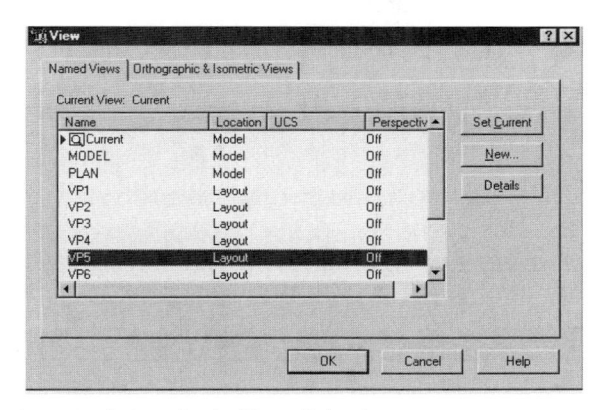

Figure 2.22 *Selecting saved views in the View dialog box*

6. Select VP5 saved view and select the **Set Current** button and then the **OK** button to dismiss the View dialog box. The Work 5 viewport should now be displayed.
7. Toggle on model space by clicking the Paper/Model toggle of the Status Bar.
8. Click in the right viewport named Plan to activate the drawing area.
9. Begin drawing walls by choosing the **Add Wall** command (**AECADDWALL**) from the **Walls** toolbar shown in Figure 2.23.
10. Edit the **Add Walls** dialog box as shown in Figure 2.24 to Width=8", Height=8', Offset=0; Justify = Right, Style= Standard and Group = Standard.

Figure 2.23 *Add Wall command on the Walls toolbar*

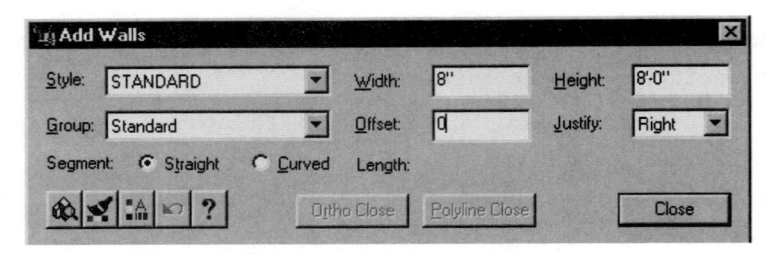

Figure 2.24 *Add Walls dialog for drawing the first wall segment*

11. Begin drawing the wall by clicking in the PLAN viewport to activate the drawing area and obtain a pointer.

12. Respond to the following command line prompts, referring to Figure 2.26:

Command: _(AECWALLADD)

(Select F8 to turn ORTHO ON.)

```
Start point or
[STyle/Group/WIdth/Height/OFfset/Justify/Match/Arc]:
5',5' ENTER (p1)
```
(Move the mouse to the right and notice that the length of the wall is dynamically displayed in the Add Walls dialog box.)
```
End point or
[STyle/Group/WIdth/Height/OFfset/Justify/Match/Arc]:
40' ENTER (p2)
```
(Toggle on Curved walls in the Add Walls dialog as shown in Figure 2.25.)
(Click in the Plan viewport to reactivate the drawing area.)
```
Mid point or
[STyle/Group/WIdth/Height/OFfset/Justify/Match/Line/
Undo]: @3',12' (p3)
```
(Note that the curved wall will begin to rubber-band through this point.)
```
End point or
[STyle/Group/WIdth/Height/OFfset/Justify/Match/Line]:
@0,24' (p4)
```
(Note that this location establishes the endpoint of the curved wall.)
```
Mid point or
[STyle/Group/WIdth/Height/OFfset/Justify/Match/Line/
Undo/Close/ORtho]: (Select the Straight toggle of
the Add Walls dialog.)
```
(Click in the Plan viewport to reactivate the drawing area and move the mouse to the left.)
```
End point or
[STyle/Group/WIdth/Height/OFfset/Justify/Match/Arc/
Undo/Close/ORtho]: 40' ENTER (p5)
End point or
```

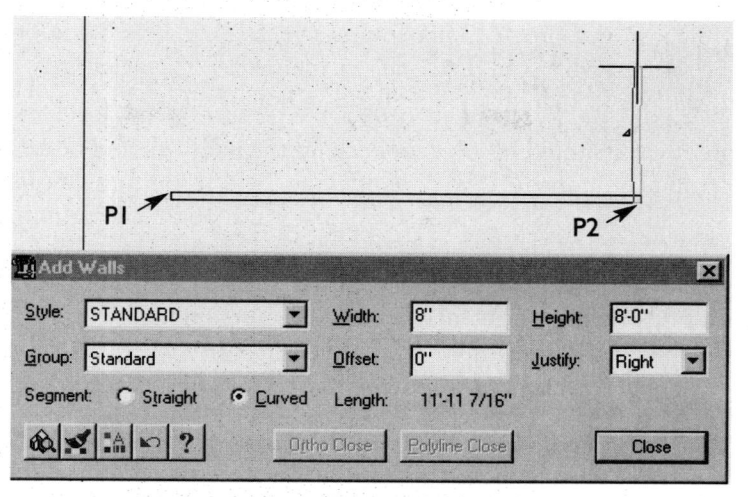

Figure 2.25 *Toggling ON Curved in the Add Walls dialog box to create a curved wall*

[STyle/Group/WIdth/Height/OFfset/Justify/Match/Arc/
Undo/Close/ORtho]: *(Select the Polyline Close button of the Add Walls dialog box.)* DBOX
The last wall segment will be added as shown in Figure 2.26.

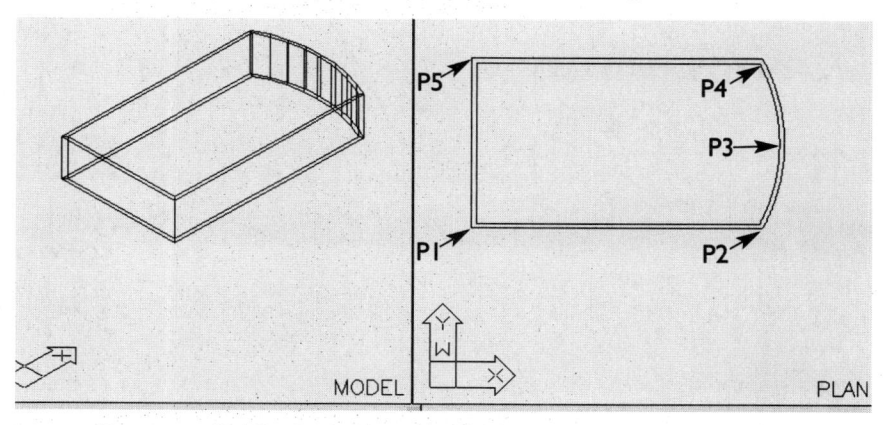

Figure 2.26 *Exterior walls of a floor plan created with curved and straight walls*

13. Notice that the walls appear in pictorial view in the left viewport.
14. Save the drawing.

SETTING UP FOR DRAWING AND EDITING WALLS

When a new file is created to draw a floor plan, the Aec arch (imperial) template is used. The Work-FLR layout tab is best suited to drawing the floor plan because it consists of one viewport and provides maximum screen area. The Node object snap mode should be selected when you draw walls or manipulate mass elements. The Node object snap allows you to snap to the justification line of walls.

Representation of walls should be kept simple, and the width should be set to that of the wall's structural component. If you are drawing 2 x 4 wood frame walls, the wall width for the wall should be set to 3 1/2", the actual width of a 2 x 4. Setting the width to the actual size of the structural component allows you to snap to the lines representing the wall for specific location dimensions. More complex walls such as brick veneer frame or brick veneer with concrete masonry units should be drawn using wall styles. Those walls consist of multiple wall surfaces for the structural components.

USING THE NODE OBJECT SNAP

The Node object snap can be used to snap to the endpoint and midpoint of the wall along its justification line. The points for the Node object snap are located on both the bottom and top of the wall along the justification line. Therefore, the Node mode can be used to connect existing walls or when placing other objects such as windows and doors are placed. A pictorial view of the Node object snap locations for the Standard Wall Style is shown in Figure 2.27.

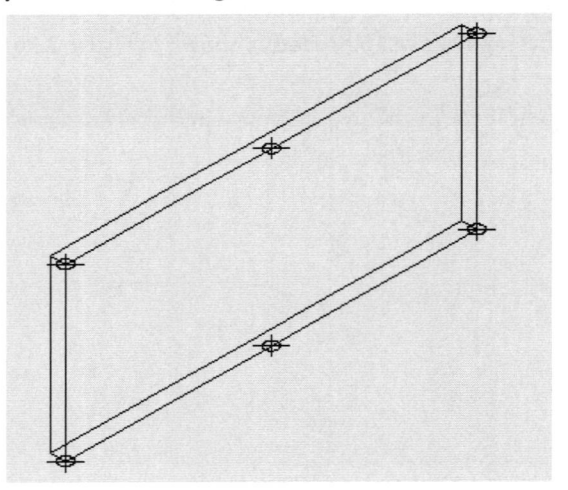

Figure 2.27 *Node object snap locations for a wall*

GRIPS OF THE WALL

In addition to using the Node mode of Object Snap, you can use AutoCAD grips located on the wall justification line. The grips are located at the ends and midpoint of the wall justification line. The grips are only on the bottom plane of the wall, as shown in Figure 2.28. These grip points can be used to edit the wall just as grips are used to edit AutoCAD lines. Displaying the grips of a wall is a quick way to see the justification of a wall. The figure displays the grips for each of the following wall justifications: Left, Center, Right, and Baseline. All walls were drawn from left to right.

SETTING JUSTIFICATION

Prior to beginning the floor plan, identify which wall justification best suits your needs. If you are working from a sketch that identifies the overall dimensions to the outer wall surface, the left or right wall justification should be used. The justification of the wall sets the location of the handle for placing the wall. This handle is located at the bottom of the wall or at the zero Z coordinate. Therefore, as the wall is drawn the bottom of the wall is placed at Z=0 for all wall vertices. If a wall is drawn with center justification, the handle for the wall is located at the bottom of the wall and is centered between the sides of the wall.

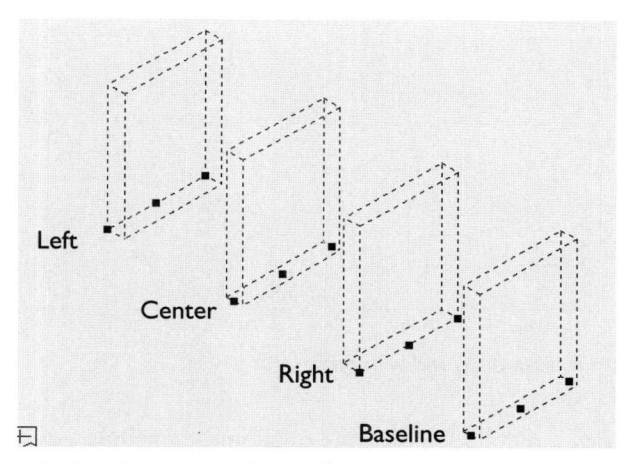

Figure 2.28 *Grips displayed on the justification lines*

When drawing a floor plan, you often start with a basic shape such as a rectangle. The overall dimensions of the basic shape are set by the outer wall surfaces. To draw walls dimensioned to the outer surfaces, set the justification to right and draw the walls from left to right or in a counterclockwise direction. When the justification is set to Right and the polygon is drawn in a counterclockwise direction, the handles for the wall are on the outer wall lines. The same effect is obtained if the walls are drawn with left justification and drawn in a clockwise direction.

If you are drawing a wood frame structure, the overall dimensions of the structure should usually be a multiple of 16" for framing efficiency. Since studs and joists are usually placed 16" o.c., the building length should be a multiple of 16". Therefore, the overall dimensions of the building should not be a random distance but rather a distance that is a specific multiple of 16". Setting the SNAP and GRID distance to 16" will aid in controlling wall lengths to multiples of 16". With SNAP, GRID, and the appropriate wall justification, the outer wall surfaces will be snapped to dimensions that are multiples of 16 inches as the mouse pointer is moved. In Figure 2.29, the GRID and SNAP were set to 16". The right justification of the walls is shown by the location of the grips. The exterior walls were drawn counterclockwise and the interior walls were drawn from bottom to top, or in a counterclockwise direction. Notice that all the dimensions are multiples of 16" and are placed relative to the justification line.

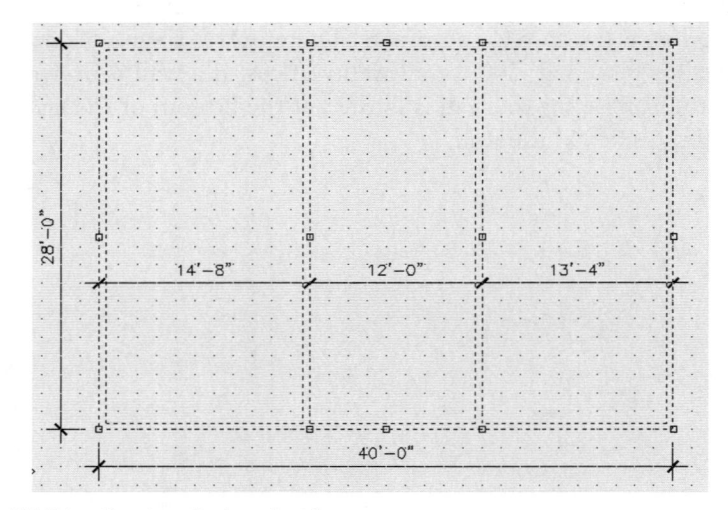

Figure 2.29 *Wall justification displayed with grips*

In summary, to draw a rectangle given the outside dimensions use

- Right Justified walls and draw in a counterclockwise direction

  Or

- Left Justified walls and draw in a clockwise direction

SETTING THE CLEANUP RADIUS

When you draw walls in Architectural Desktop, the intersections of the walls are cleaned up according to the radius dimension set by the cleanup radius. In addition, the cleanup radius controls how close you have to be to an existing wall when you begin drawing it near another existing wall in order for it to be joined to the wall. The endpoint of a wall is joined to an existing wall if that wall is within the range of the cleanup radius. If the cleanup radius is set to 12", a new wall will be trimmed and joined to other walls within a 12" radius of its endpoints. This allows you to select points to start and end the wall without using object snaps to join walls.

STEPS TO SETTING THE CLEANUP RADIUS

1. Right-click over the drawing area and choose **Options** from the shortcut menu.
2. Select the **Aec DwgDefaults** tab of the **Options** dialog box.
3. Edit the **Wall Cleanup Radius** as shown in Figure 2.30.

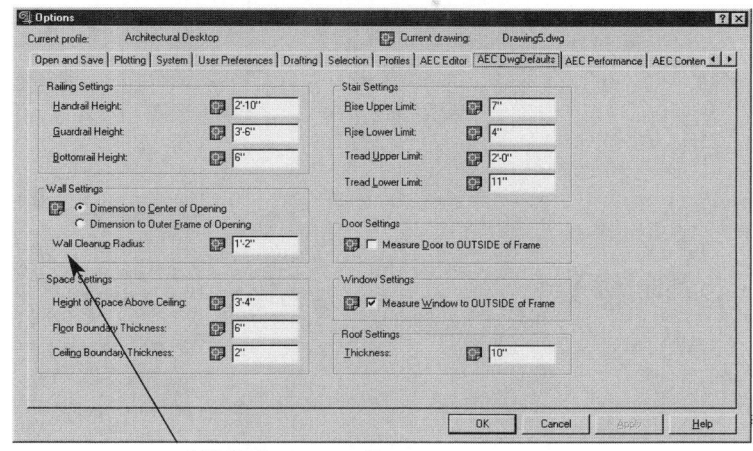

Wall cleanup radius

Figure 2.30 *Editing Wall Cleanup Radius in the AEC DwgDefaults tab of the Options dialog box*

4. Select **OK** to dismiss the Options dialog box.

Setting the **Wall Cleanup Radius** in the **Desktop Preferences** will control the joining of all future walls with existing walls. The default value for the cleanup radius is 1'-2" when the Aec arch (imperial) template is used. Depending on the wall width and density of walls in the drawing, the cleanup radius may need to be adjusted. If the cleanup radius is set equal to the wall thickness, the walls will usually clean up and blend with others without errors. Editing the cleanup radius in the **Wall Properties** dialog box of selected walls will change the cleanup radius of a previously drawn wall.

STEPS TO EDITING THE CLEANUP RADIUS OF EXISTING WALLS

1. Click on a wall to highlight and select the wall.
2. Right-click to display the shortcut menu shown in Figure 2.31.
3. Choose **Wall Properties** from the shortcut menu to display the **Wall Properties** dialog box shown in Figure 2.32.
4. Select the **Dimensions** tab and edit the cleanup radius as shown in Figure 2.33.

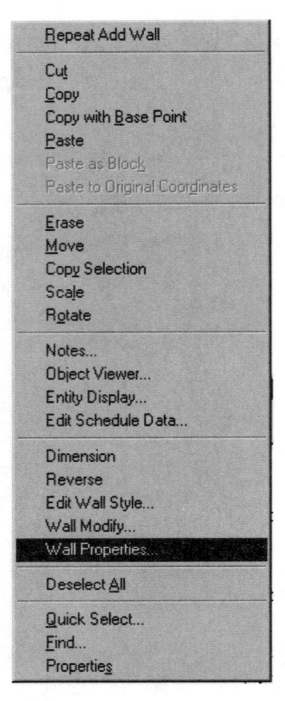

Figure 2.31 *Choosing Wall Properties from the shortcut menu*

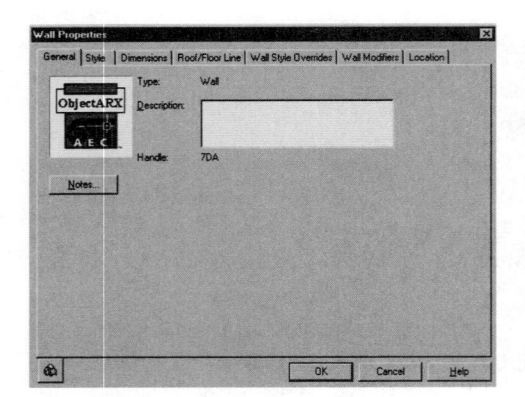

Figure 2.32 *General tab of the Wall Properties dialog box*

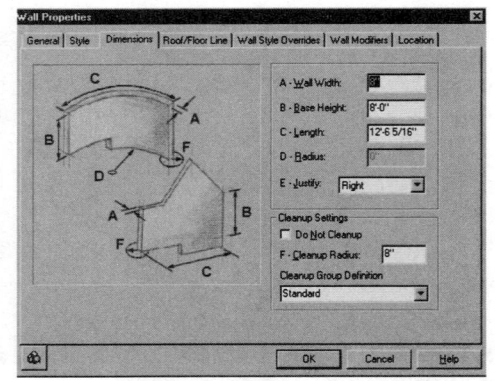

Figure 2.33 *Dimensions tab of the Wall Properties dialog box*

MANAGING WALL ERROR MARKERS

A wall error marker, as shown on the left in Figure 2.34, will be displayed in substitution for a wall being drawn if a wall is too short or coincident with another, or it incorrectly intersects other walls within the cleanup radius. Wall error markers are displayed at the midpoint of the wall and are colored red to alert you to the problem. The cleanup radius distance controls the definition of when a wall is too short. If the cleanup radius is set to 12", walls must be drawn longer than 12" to avoid the display of the wall error marker. The wall error marker shown in Figure 2.34 on the left resulted from a wall 6" long being drawn with a 12" cleanup radius setting. The wall error marker is the Defect Warning component of the wall object and is placed on the DefPoints layer. Therefore, the wall error marker will not print because the entities of the DefPoints layer are not plotted.

You can display the circle representing the cleanup radius by applying the Graph display representation for wall objects of the current display configuration. The Graph display representation is the only representation that displays the cleanup radius. The walls on the right in Figure 2.34 are displayed using the Graph display representation.

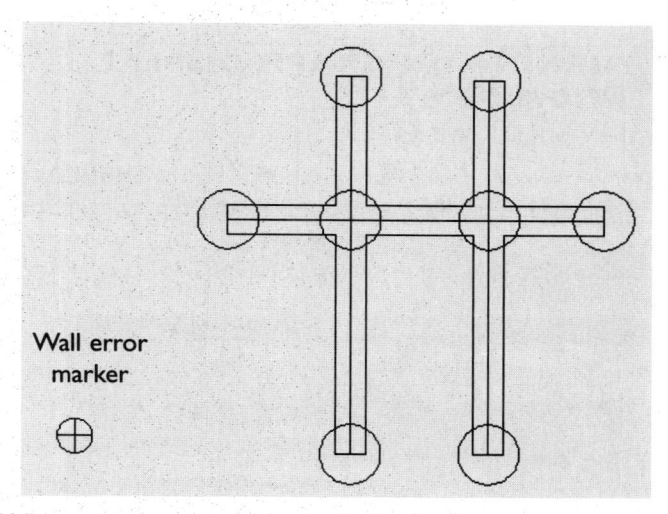

Figure 2.34 *Wall error marker and cleanup radii of wall objects*

Turn ON the Graph display representation for walls by editing the current display representation set. To determine the current display representation set, you must first determine the display configuration being used for the viewport. The display configuration for a viewport is determined by the **DISPLAYCONFIGATTACH** command (right-click over the drawing area and choose **Utilities>Display>Display Configurations>Attach to Viewport** from the shortcut menu). One or more display representation sets can be assigned for a display configuration. If the floor plan is developed with the Work-FLR layout tab, the Work display configuration is active for this viewport. It is a View Direction Dependent configuration and includes the following view directions and display representation sets:

View	Display Representation Set
Top	Work-Plan
Left	Work-Section
Right	Work-Section
Front	Work-Section
Back	Work-Section
All other views	Work-Model

Therefore, when the wall is viewed from the top or plan view, the Work-Plan display representation set is applied. Editing this wall display representation set to include Graph for walls will display the cleanup radii.

STEPS FOR TURNING ON GRAPH DISPLAY REPRESENTATION OF WALLS

1. Select the Work-FLR layout tab.
2. Determine viewport display configuration using the **DISPLAYCONFIGATTACH** command (choose **Desktop>Select Display** from the menu bar). The **Viewport Display Configuration** dialog box opens with current display configuration selected as shown in Figure 2.35.

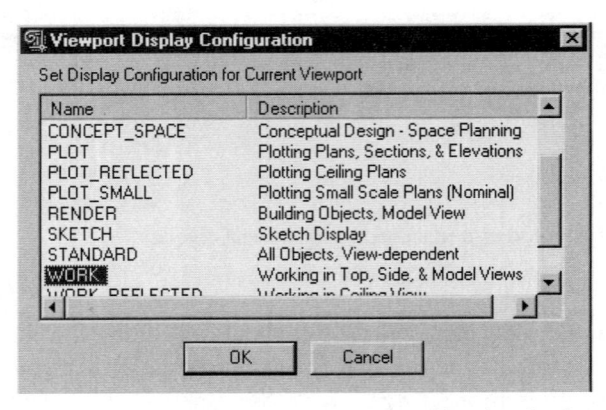

Figure 2.35 *Current viewport display shown in the Viewport Display Configuration dialog box*

3. Select **OK** to dismiss the **Viewport Configuration** dialog box.
4. To determine the display representation sets assigned to the display configuration, open the **Display Configurations** dialog box using the **DISPLAYCONFIGDEFINE** command (right-click over the drawing area and choose **Utilities>Display>Display Configurations>Table**) as shown in Figure 2.36.

Creating Floor Plans 67

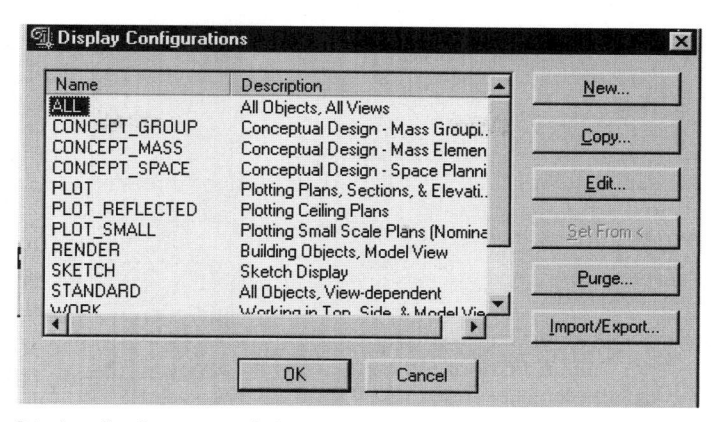

Figure 2.36 *Display Configurations dialog box*

5. Select the Work display configuration and then select the **Edit** button of the **Display Configurations** dialog box.
6. Select the **Configuration** tab of the **Display Configuration** dialog box to determine which display representation is used for the Top view as shown in Figure 2.37.
7. Select **OK** to dismiss the **Display Configuration** dialog box and then select **OK** to dismiss the **Display Configurations** dialog box (Figure 2.38).

Display representation for TOP (Plan View)

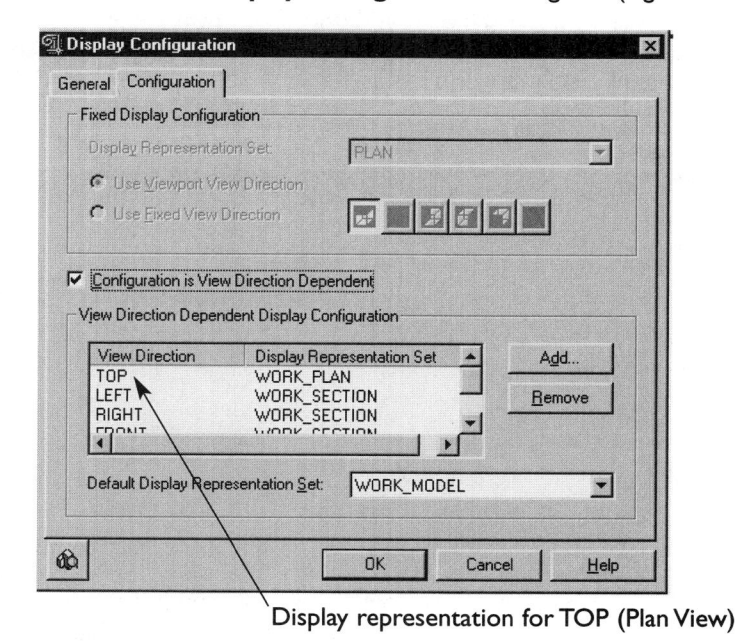

Figure 2.37 *Display representations in the Display Configuration dialog box*

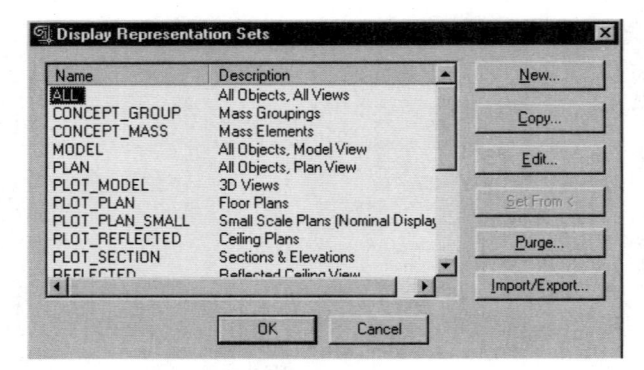

Figure 2.38 *Display Representation Sets dialog box*

8. Open the **Display Representation Sets** dialog box to edit the Work-Plan display representation set using the **DISPLAYSETDEFINE** command (right-click over drawing area and choose **Utilities>Display>Display Sets>Table**).
9. Select the Work-Plan display representation set in the **Name** list, and then select the **Edit** button.
10. Select the **Display Control (By Object)** tab in the **Display Representation Set** dialog box.
11. Select Wall from the **Object Type** list box on right and check the Graph display representation in the **Display Representation** list box as shown in Figure 2.39.
12. Select **OK** to dismiss all dialog boxes.

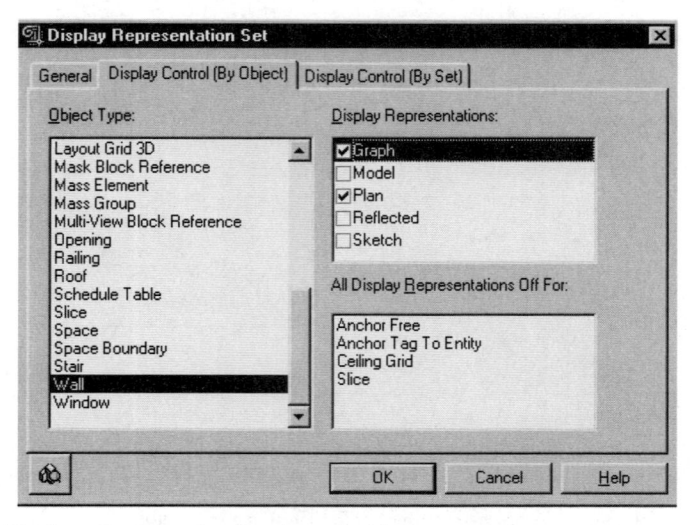

Figure 2.39 *Display representation for the wall object*

13. Type **RE** to regenerate the viewport.

The Graph display representation displays the connecting lines of the wall and the cleanup radii. Display of these two wall components allows you to edit the wall and determine which wall is causing the display of the wall error marker. Selecting the wall error marker will display its grips on the justification line. Selecting an endpoint grip, making it hot, allows you to stretch the wall out of the range of the cleanup radius that has caused the display of the wall error marker.

If a wall error marker is displayed to represent a short wall and the design requires such a short wall, the cleanup radius should be reduced to remove the wall error marker. Reduce the cleanup radius to a value less than the length of the wall and the wall error marker will be removed and the wall displayed. To edit the cleanup radius, select the wall error marker and then right-click to display the shortcut menu. Choose **Wall Properties** from the shortcut menu and edit the **Cleanup Radius** value located on the **Dimensions** tab of the **Wall Properties** dialog box. Wall error markers can also occur when walls are drawn with coincident locations. You can remove the wall error marker by erasing it; when you erase the wall error marker the wall it represents is also erased.

USING TEMPORARY TRACKING POINT TO DRAW INTERIOR WALLS

Once the exterior walls are drawn, interior walls can be added using tracking for precise placement of the interior walls. Placing walls using **Temporary Tracking Point** provides an easy method of placing walls relative to the existing geometry of other walls. Using **Temporary Tracking Point** allows you to track from a point without anchoring the beginning of the wall. In essence, tracking allows you to measure from existing geometry in a direction set by the movement of the mouse. Tracking is most effective if ORTHO is set ON because it limits the mouse pointer movement to horizontal and vertical directions. Once the direction is set, you can enter a distance using Direct Distance Entry.

Access **Temporary Tracking Point** as follows:

Menu bar	Not available
Command prompt	TT
Toolbar	Select Temporary Tracking Point from the Object Snap toolbar shown in Figure 2.40
Shortcut	Hold down Shift; right-click over drawing area; choose Temporary Tracking Point

Table 2.2 *Accessing Temporary Tracking Point*

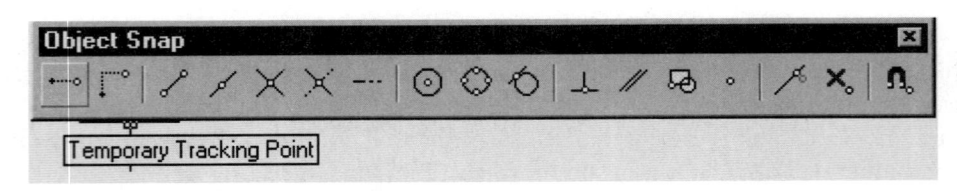

Figure 2.40 *Temporary Tracking Point option on the Object Snap toolbar*

TUTORIAL 2.4 USING TRACKING TO DRAW INTERIOR WALLS AND DISPLAYING CLEANUP RADII

1. Open *Ex2-2* from the *ADT Tutor\Ch2* directory.
2. Choose **File>SaveAs** from the menu bar and save the drawing as *Ex2-4.dwg* to your student directory.
3. Select the Work-FLR layout tab. The Work-FLR viewport should now be displayed.
4. Right-click over the OSNAP button of the Status Bar and choose **Settings** from the shortcut menu. Select the Node mode of Object Snap. Check Object Snap On (F3) and verify that the Midpoint and Endpoint modes of Object Snap are deselected.
5. Select **OK** to dismiss the **Drafting Settings** dialog box. Verify that Object Snap Tracking is ON by pressing F11.
6. Verify that Polar Tracking is ON by pressing F10.
7. Begin the **Add Wall** command by right-clicking with the mouse over the drawing area and choosing **Design>Walls>Add** from the shortcut menu.
8. Edit the **Add Walls** dialog box to Width=3 1/2", Height=8', Offset=0; Justify = Right, Style= Standard, and Group=Standard. Draw the new wall by referring to Figure 2.41 and the following command sequence:

(Select in the drawing area to activate the viewport.)
Command: _(AECWALLADD)
Start point or
[STyle/Group/WIdth/Height/OFfset/Justify/Match/Arc]:
_tt *(Select Temporary Tracking Point option from the Object Snap flyout on the Standard toolbar.)*
Specify temporary OTRACK point:
(Move the pointer to the outer line of the exterior wall (p1) as shown in Figure 2.42 to display the Node object snap marker; then click.)
(Move the mouse to the right, referring to Figure 2.42)
Start point or
[STyle/Group/WIdth/Height/OFfset/Justify/Match/Arc]:
31'2-3/4" ENTER (p2)

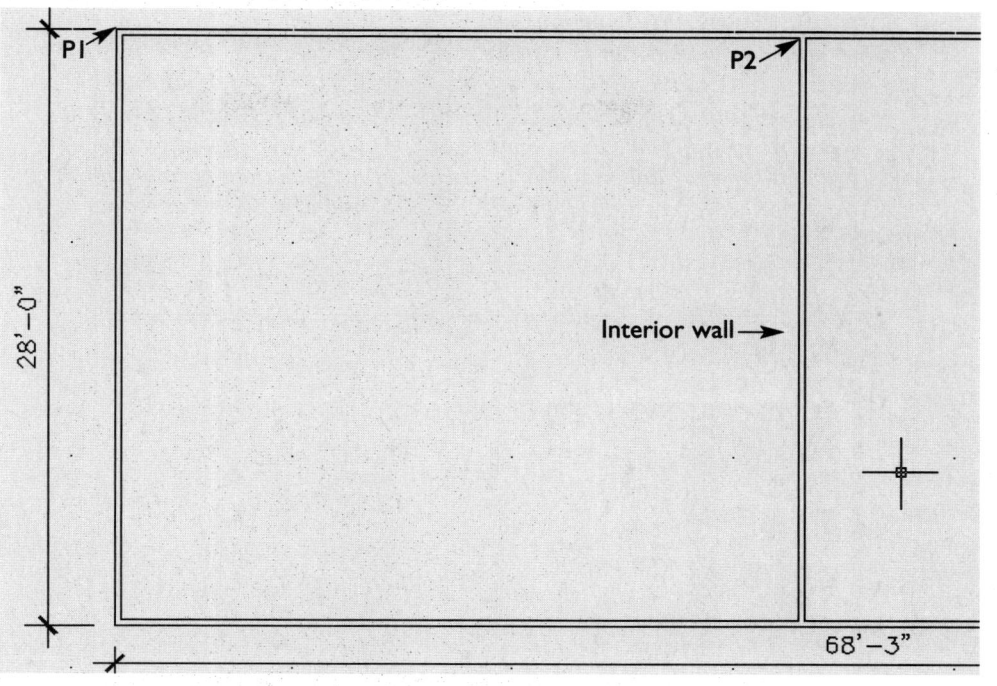

Figure 2.41 *Using Temporary Tracking Point to create interior walls*

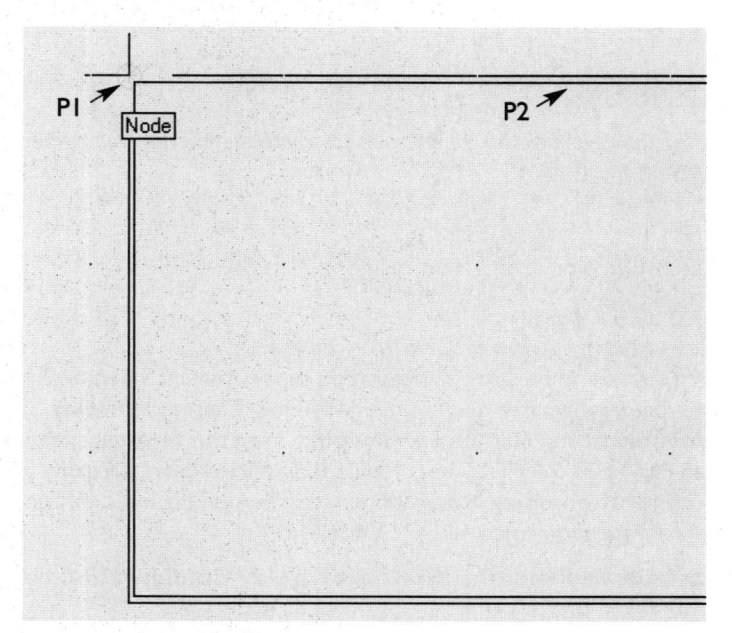

Figure 2.42 *Tracking from the Node object snap of the exterior wall*

(Note: This exits tracking and begins the wall.)
(Move the mouse down in a vertical direction.)
```
End point or
[STyle/Group/WIdth/Height/OFfset/Justify/Match/Arc]:
```
(Select Perpendicular Object Snap from the Object Snap flyout on the Standard toolbar and move the mouse down to the bottom horizontal line; click when the Perpendicular marker is displayed at point p3 as shown in Figure 2.43.)

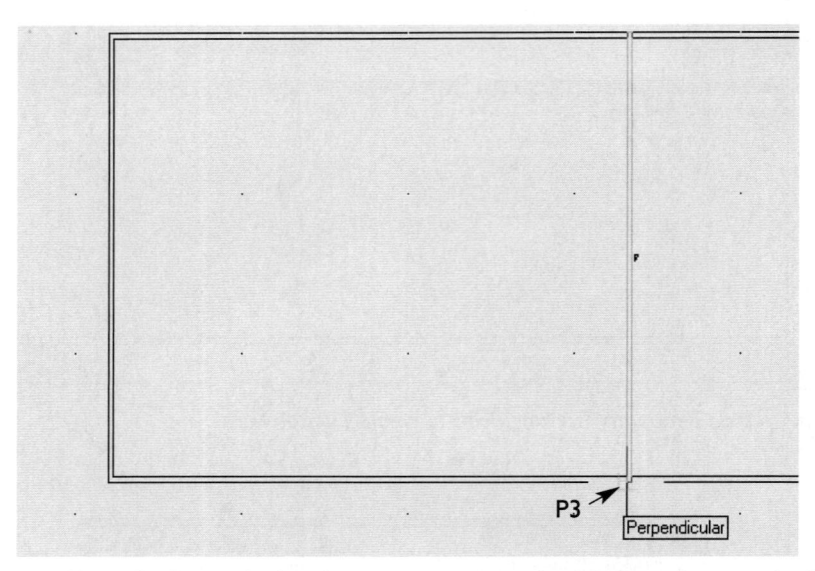

Figure 2.43 Using the Perpendicular object snap to create the interior wall perpendicular to the exterior wall

```
End point or
[STyle/Group/WIdth/Height/OFfset/Justify/Match/Arc/
Undo]: ENTER
```
(Note: Pressing Enter will exit the command.)

9. To determine the display configuration of the current viewport, right-click over the drawing area and choose **Utilities>Display>Display Configurations>Attach to Viewport** from the shortcut menu (**DISPLAYCONFIGATTACH** command). The **Viewport Display Configuration** dialog box will open as shown in Figure 2.44. The Work display configuration is current for the viewport.

10. Select **OK** to dismiss the **Viewport Display Configuration** dialog box.

11. To view the display representation sets of the display configuration, right-click over the drawing area and choose **Utilities>Display>Display**

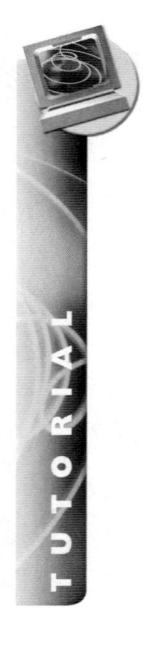

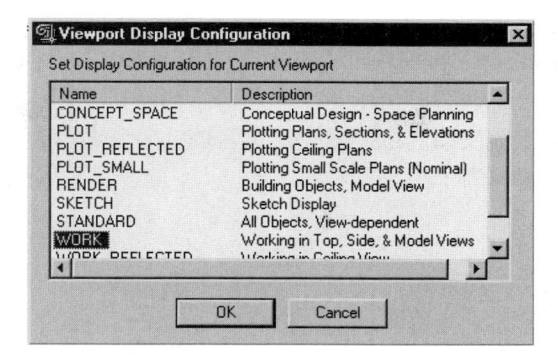

Creating Floor Plans 73

Figure 2.44 *Viewport Display Configuration dialog box*

Configurations>Table (**DISPLAYCONFIGDEFINE** command). The **Display Configuration** dialog box opens as shown in Figure 2.45.

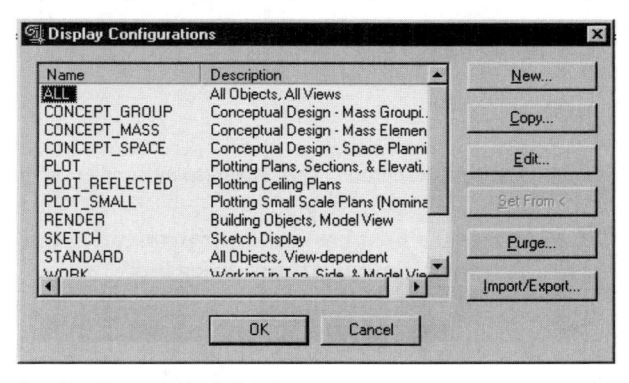

Figure 2.45 *Display Configurations dialog box*

12. Select the Work display configuration and select the **Edit** button to open the **Display Configuration** dialog box as shown in Figure 2.46.

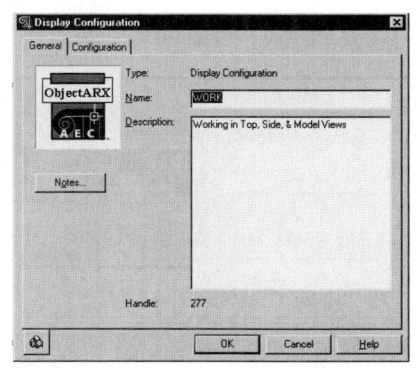

Figure 2.46 *Display Configuration dialog box*

13. Select the **Configuration** tab to view the display representation sets assigned to the display configuration as shown in Figure 2.47. Note that the Top view direction is assigned the Work-Plan display representation set.

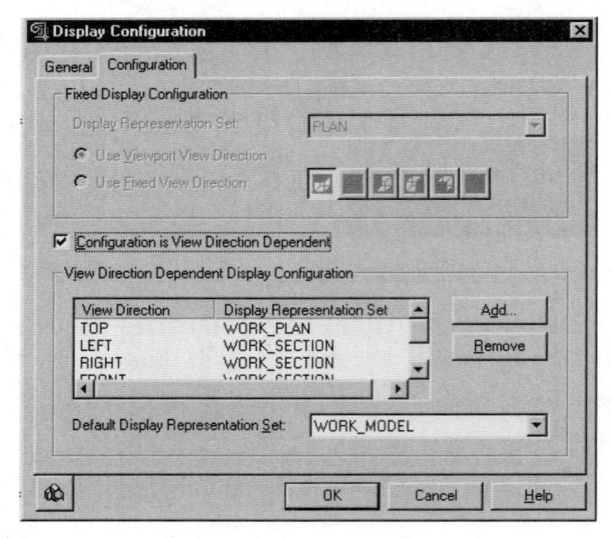

Figure 2.47 *Determining display representation sets of a display configuration*

14. Select **OK** to dismiss the **Display Configuration** dialog box and select **OK** to dismiss the **Display Configurations** dialog box.

15. To edit the Work-Plan display representation set, right-click over the drawing area and choose **Utilities>Display>Display Sets>Table (DISPLAYSET-DEFINE** command). The **Display Representation Sets** dialog box opens as shown in Figure 2.48.

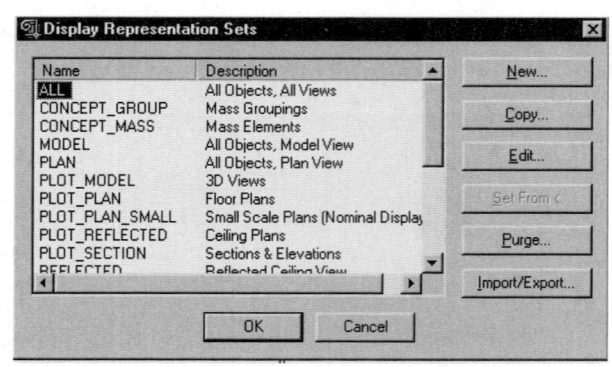

Figure 2.48 *Display Representation Sets dialog box*

16. Select Work-Plan from the list of display representation sets and select the **Edit** button.

17. Select the **Display Control (By Object)** tab.
18. Select Wall from the **Object Type** list as shown in Figure 2.49.

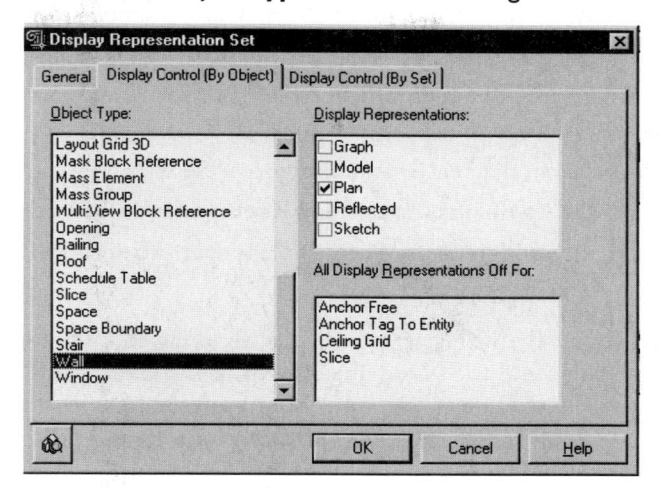

Figure 2.49 *Display Control (By Object) tab of Display Representation Set dialog box*

19. Check Graph in the **Display Representation** list box.
20. Select **OK** to dismiss the **Display Representation Set** dialog box.
21. Select **OK** to dismiss the **Display Representation Sets** dialog box.
22. Type **REGENALL** in the command line and press ENTER (regenerates the display of the drawing in all viewports). Cleanup radii should be displayed as shown in Figure 2.50.

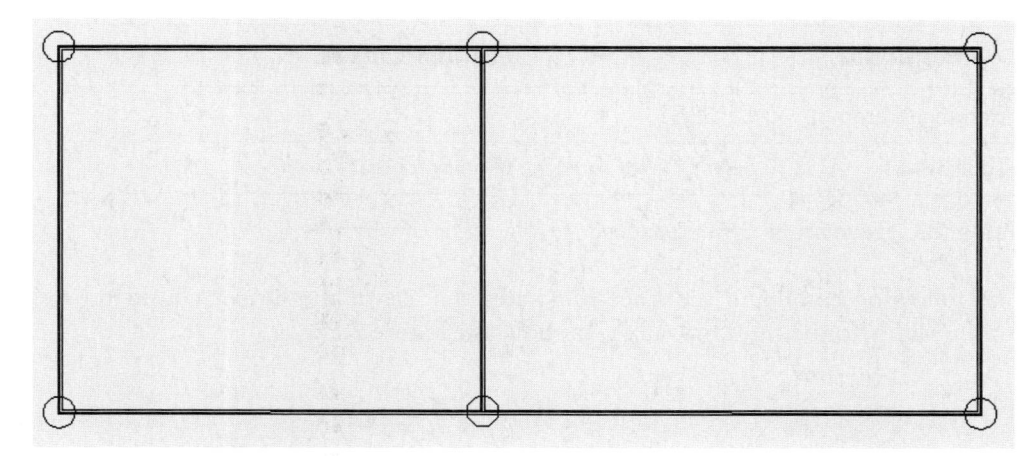

Figure 2.50 *Display of cleanup radii*

23. Turn off the Graph display representation. Right-click over the drawing area and choose **Utilities>Display>Display Sets>Table**.
24. Select Work-Plan for the list of display representation sets and select the **Edit** button.
25. Select the **Display Control (By Object)** tab.
26. Select Wall from the **(Object Type)** list.
27. Clear the check from **Graph** display representation.
28. Select **OK** to dismiss the **Display Representation Set** dialog box.
29. Select **OK** to dismiss the **Display Representation Sets** dialog box.
30. Save the drawing.

In the tutorial above, the wall start point (p2) was delayed by the selection of a temporary tracking point and the Node object snap of the exterior wall (p1) was used to set the temporary point. Direct distance entry was used to shift the temporary point to the right 31'-2 3/4". In essence, tracking allows you to defer the start point for the wall and measure from existing geometry to set the beginning of a wall. Other object snaps in addition to the Node object snap can be used to track about existing wall entities to place new walls.

USING AUTOCAD OFFSET TO DRAW ADDITIONAL INTERIOR WALLS

The **Offset** command can also be used to copy walls. This allows the walls to be copied parallel to their present position according to the distance and direction specified. Because the walls are objects, both lines of the wall are copied and their intersection with other walls is automatically cleaned up.

Execute the **Offset** command as follows:

Menu bar	Modify>Offset
Command prompt	OFFSET
Toolbar	Select Offset from the Modify toolbar

Table 2.3 *Executing the Offset command*

The following command sequence illustrates the use of the **Offset** command to copy a wall 6' to the right of wall A, as shown in Figure 2.52.

```
Command: _offset
Specify offset distance or [Through]: 6'
Select object to offset or <exit>: (Select wall A.)
Specify point on side to offset: (Select a point to the right
of A near point p1.)
```

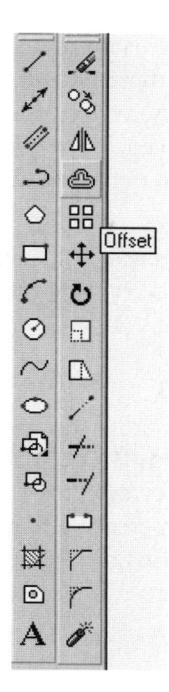

Figure 2.51 *Selecting Offset from the Modify toolbar*

```
Select object to offset or <exit>: ENTER
```
(Note: Wall B is created as a copy of A.)

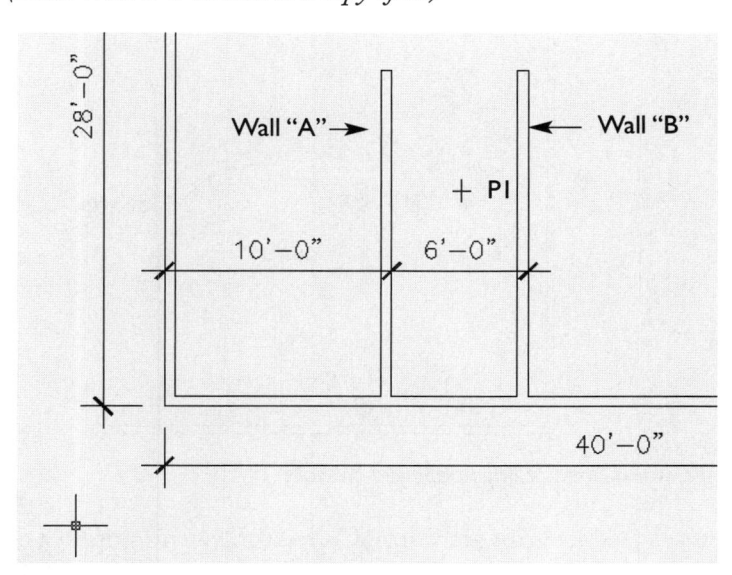

Figure 2.52 *Copying Wall "A" using the Offset command*

In this example, the new wall was created 6' to the right of the original. The 6' offset distance copies the wall 6' to the right of the original wall. When the offset command is used to create specific room sizes, the thickness of the wall must be added to the offset distance. To create a tub enclosure using 3 1/2" walls, which has a 5' inside clear dimension between partitions, make the offset distance 5'-3 1/2".

> **Note:** if you are creating partitions with specific inside room dimensions, the wall thickness must be added to the desired room size when you use the Offset command.

USING GRIPS TO EDIT WALLS

The grips of a wall can be used to modify the length of a wall or to copy the wall. All grip editing operations of **MOVE**, **MIRROR**, **ROTATE**, **SCALE**, and **STRETCH** are executable on the grips of a wall. However, the **MOVE** and **STRETCH** options when used with **COPY** are most useful in creating floor plans. Each wall has grips located at each endpoint and midpoint. Grips displayed on the wall are cold, warm, or hot, as shown from left to right in Figure 2.53.

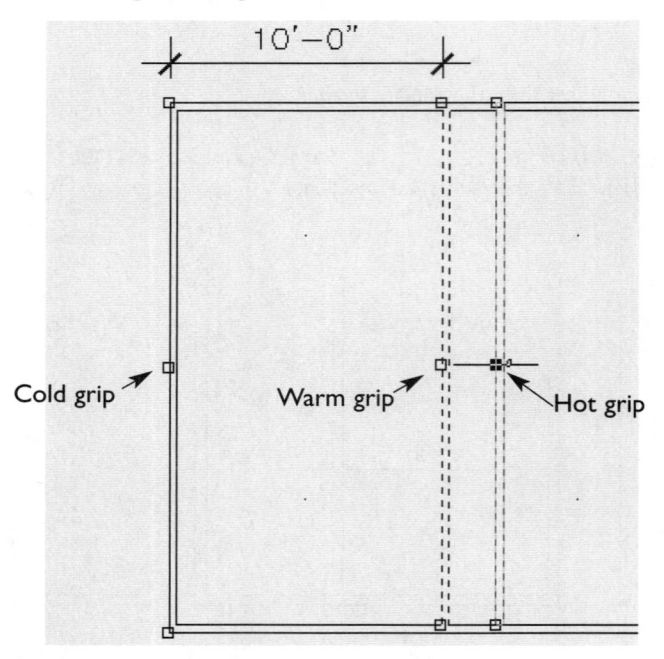

Figure 2.53 *Cold, warm, and hot grips displayed on walls*

When a wall is selected its grips are warm. Convert the warm grip to a cold grip by pressing ESC or make the warm grip hot by selecting it.

Warm Grips – Displayed for a wall when you click on the wall. This highlights the wall, indicating that it is selected for editing, and displays the grips.

Cold Grips – Warm grips are made cold when you press ESC, which dismisses the wall as a selected object and removes the highlight from the wall, although the grips remain displayed. Press ESC to dismiss cold grips.

Hot Grips – When warm grips are selected, they become hot, causing the grip box to fill. Once grips are hot, the grip point can be edited with the operations of **MOVE, MIRROR, ROTATE, SCALE,** and **STRETCH**. Once a grip is hot, convert it to a warm grip by pressing ESC.

Shift to Add – To make more than one grip hot, hold SHIFT down prior to clicking on the first grip.

You can copy a wall once the wall is selected and warm grips are displayed by making the midpoint grip of the wall hot. After making the grip hot, use the Copy-Stretch operation to copy the wall. The use of **COPY-STRETCH** grip operations is shown in Figure 2.54.

TUTORIAL 2.5 CREATING INTERIOR PARTITIONS WITH GRIPS AND COPY-STRETCH

1. Open *Ex2-4.dwg* from the *ADT Tutor\Ch2* directory.
2. Choose **File>Save As** from the menu bar and save the drawing as *Ex2-5.dwg* in your student directory.
3. Verify that ORTHO is ON.
4. Refer to Figure 2.54 and click on wall "A" to display its grips and select the wall.

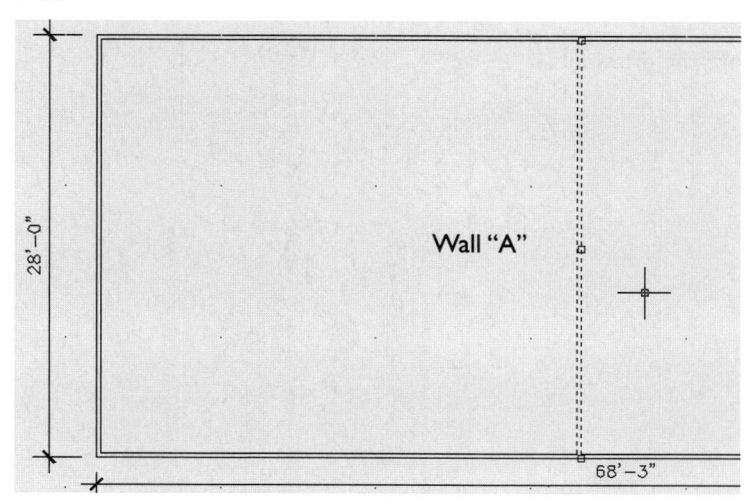

Figure 2.54 *Clicking on the middle grip to make it hot*

5. Refer to Figure 2.55 and click on the warm grip to make it hot as shown in the figure.

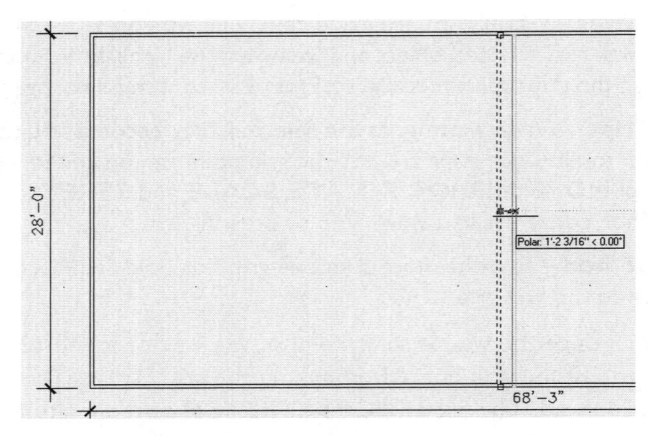

Figure 2.55 *Copying a wall using Copy-Stretch grip operations*

6. Right-click to display the **Grips** shortcut menu.
7. Choose **Copy** from the shortcut menu.
8. Move the mouse to the right and type 13'2-1/2" and press ENTER in the command line.
9. Press ESC to end grip editing.
10. Press ESC to deselect the wall.
11. Press ESC to clear the cold grips. The wall is COPY-STRETCHED to the right 13'-2 1/2" as shown in Figure 2.56.
12. Save the drawing.

Figure 2.56 *Wall copied using COPY-STRETCH with grips*

In addition to using grips to copy a wall, you can use the grips to lengthen or shorten a wall. In Tutorials 2.4 and 2.5, walls were placed using tracking and grip editing. You can extend or shorten a wall by selecting its grip at the end of the wall. When walls cross each other in Architectural Desktop, the walls are not cut into two separate components. The grips of all the walls are displayed in Figure 2.57 to illustrate the beginning and endpoints of the walls. The vertical interior partitions do not change the grip location of the exterior walls. You can create rooms for the floor plan by editing the grips of interior partitions. In the following tutorial, the grips of a vertical partition will be stretched to shorten the wall.

TUTORIAL 2.6 USING GRIPS TO CHANGE WALL LENGTH

1. Open *Ex 2-5.dwg* from the *ADT Tutor\Ch2* directory.
2. Choose **File>SaveAs** from the menu bar and save the drawing as *Ex2-6* in your student directory.
3. Refer to Figure 2.57 and select all walls by creating a selection window with the mouse.
4. Make the bottom grip of Wall B interior partition hot by clicking on the grip as shown in Figure 2.58.

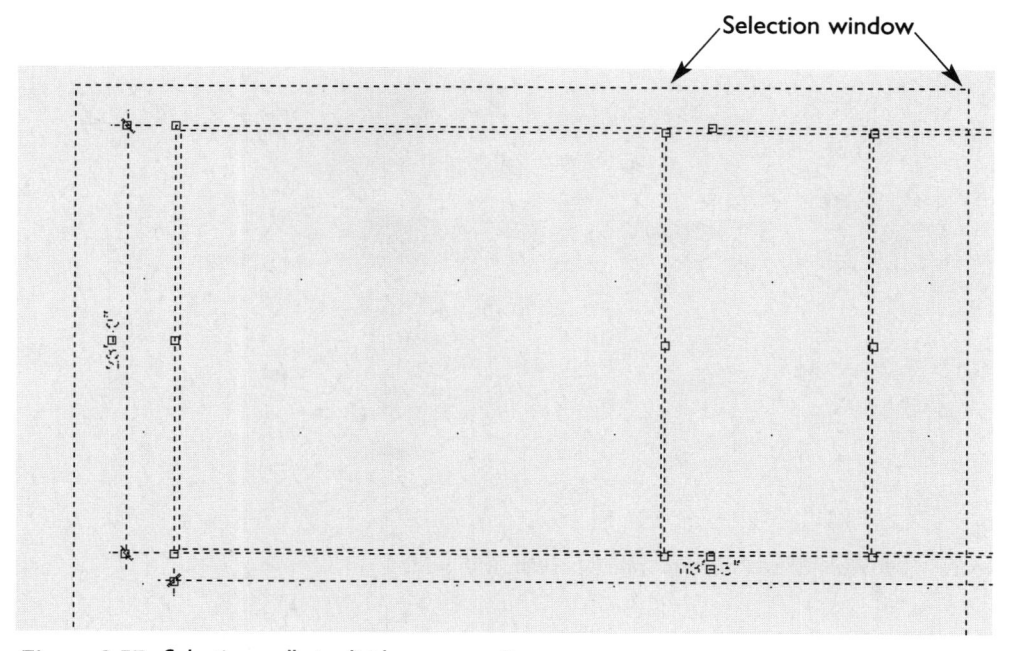

Figure 2.57 *Selecting walls to display warm grips*

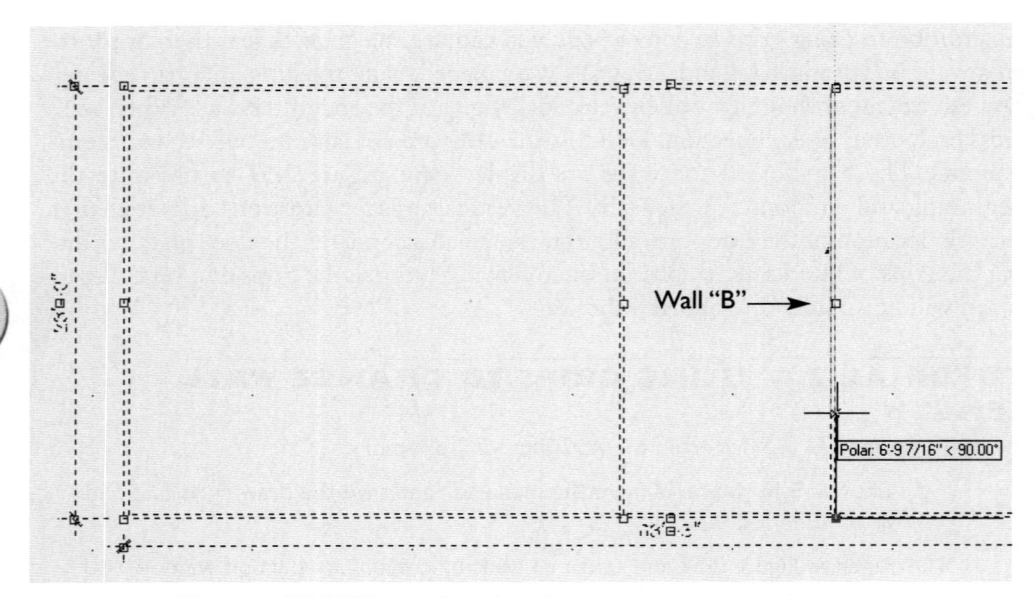

Figure 2.58 *Shortening Wall "B" using Stretch with grips*

    5. Turn Ortho ON and then move the mouse pointer up and type **12'** ENTER in the command line to stretch the wall up as shown in Figure 2.59.

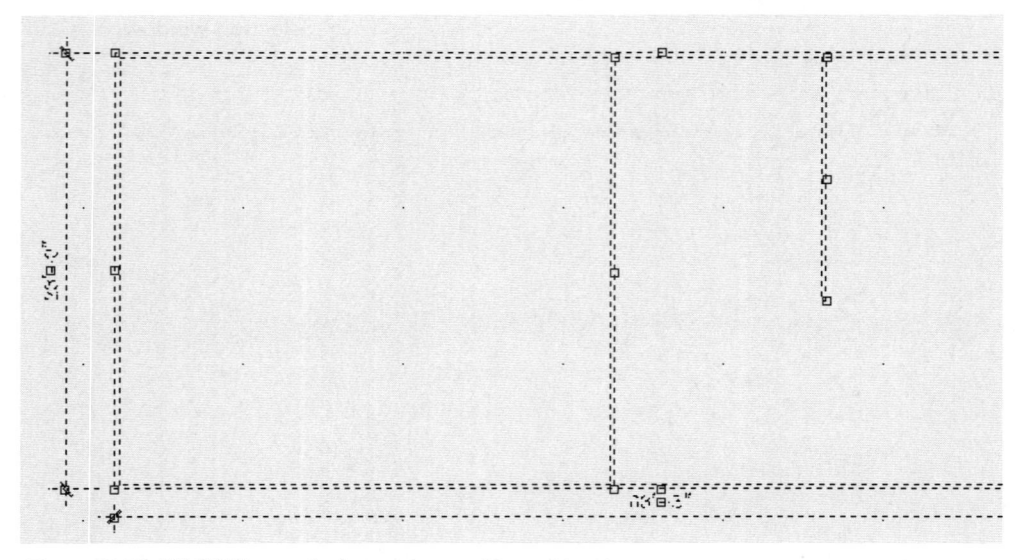

Figure 2.59 *Wall "B" stretched to a new position with grips*

    6. Press ESC to exit the command, then press ESC to deselect the wall.

7. Press ESC to clear the cold grips; the walls should now appear as shown in Figure 2.60.

Figure 2.60 *Wall "B" in the new position*

8. Save the drawing.

EDITING WALLS WITH AUTOCAD EDITING COMMANDS

Most all other AutoCAD commands can be used to edit the walls of a drawing. Walls can be edited with the following AutoCAD commands on the **Modify** toolbar:

- **Erase**
- **Copy**
- **Mirror**
- **Offset**
- **Array**
- **Move**
- **Rotate**
- **Scale**
- **Stretch**
- **Trim**
- **Extend**
- **Break**
- **Align**

The **FILLET**, **CHAMFER** and **LENGTHEN** commands cannot be used to edit walls. The **EXPLODE** command should not be used on Architectural Desktop objects because the object definition is lost when the object is exploded.

In summary, you can create the floor plan using the **Add Wall** command on the **Walls** toolbar by

- Using the viewport Work 4 or Work-FLR layout tab of the A-design-model template
- Setting width, height, and justification for the wall
- Drawing the exterior walls with right justification in a counterclockwise direction or drawing the exterior walls with left justification in a clockwise direction
- Using tracking and Offset to draw the interior walls
- Using grips and other AutoCAD editing commands to edit the interior walls to create the desired rooms

TUTORIAL 2.7 DRAWING EXTERIOR WALLS OF THE FLOOR PLAN

1. Open *Ex2-6* from the *ADT Tutor\Ch2* directory.
2. Choose **File>SaveAs** from the menu bar and save the file as *Ex2-7.dwg* in your student directory.
3. Verify that Ortho and Object Snap buttons in the Status Bar are pushed in. Right-click over the Object Snap button of the Status bar and verify that Node Object Snap is selected.
4. Select the **Add Wall** button on the **Walls** toolbar.
5. Edit the **Add Walls** dialog box to the values shown in Figure 2.61: Width= 3 1/2", Offset= 0, Height= 8', Justify = Right, Style= Standard, and Group = Standard.

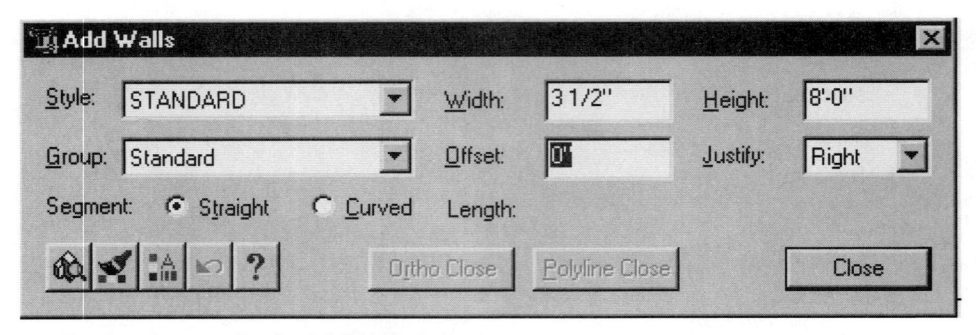

Figure 2.61 *Settings for the Add Walls dialog box*

6. Refer to Figure 2.62 and the following command sequence to place the interior partition shown using Temporary Tracking Point.

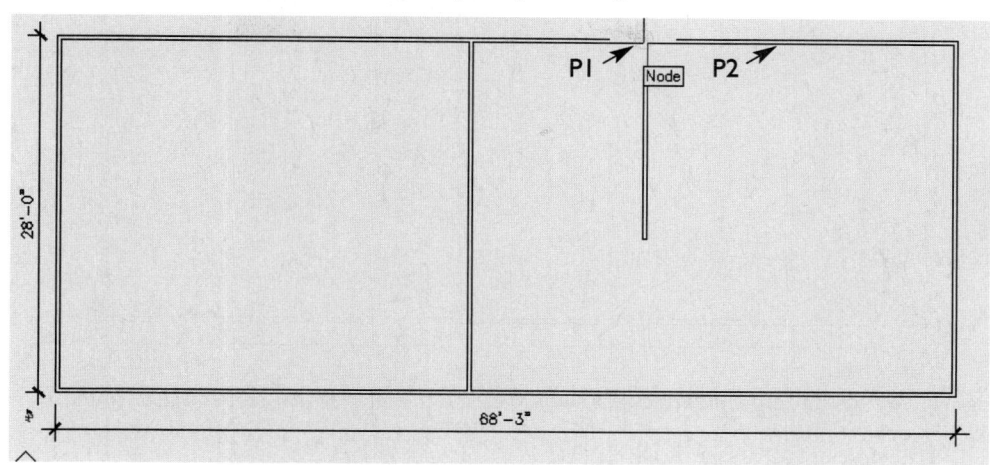

Figure 2.62 *Selecting the node of the wall for the temporary tracking point*

```
Command: _(AECWALLADD)
Start point or
[STyle/Group/WIdth/Height/OFfset/Justify/Match/Arc]:
```
(Select Temporary Tracking Point from the Object Snap flyout on the Standard toolbar.) _tt
```
Specify temporary OTRACK point:
```
(Move the mouse to point (p1) and click when the node object snap marker is displayed as shown in Figure 2.62.)

(Toggle off the running OBJECT SNAP by pressing F3; move the mouse to the right near p2 of Figure 2.62.)
```
Start point or
[STyle/Group/WIdth/Height/OFfset/Justify/Match/Arc]:
7'10-1/2 ENTER
```
(Select the Perpendicular OBJECT SNAP from the Standard toolbar.)
```
End point or
[STyle/Group/WIdth/Height/OFfset/Justify/Match/Arc]:
```
(Move the mouse down to lower horizontal wall; click when the Perpendicular object snap marker appears as shown in Figure 2.63.)
```
End point or
[STyle/Group/WIdth/Height/OFfset/Justify/Match/Arc/
Undo]: ENTER
```
(Note: Pressing ENTER *will end the Add Wall command and your drawing should appear as shown in Figure 2.64.)*

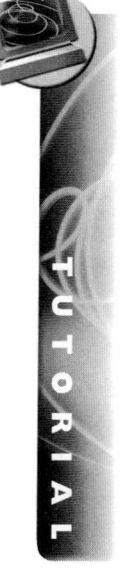

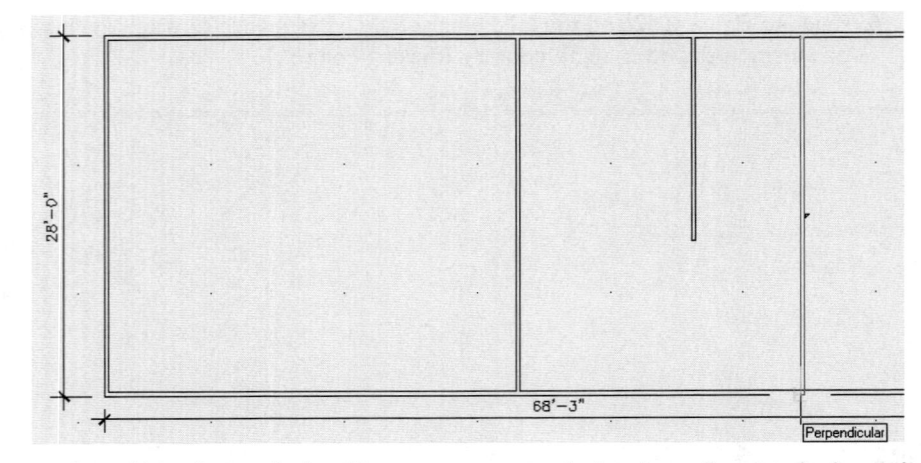

Figure 2.63 *Using Perpendicular object snap to create the interior wall perpendicular to the exterior wall*

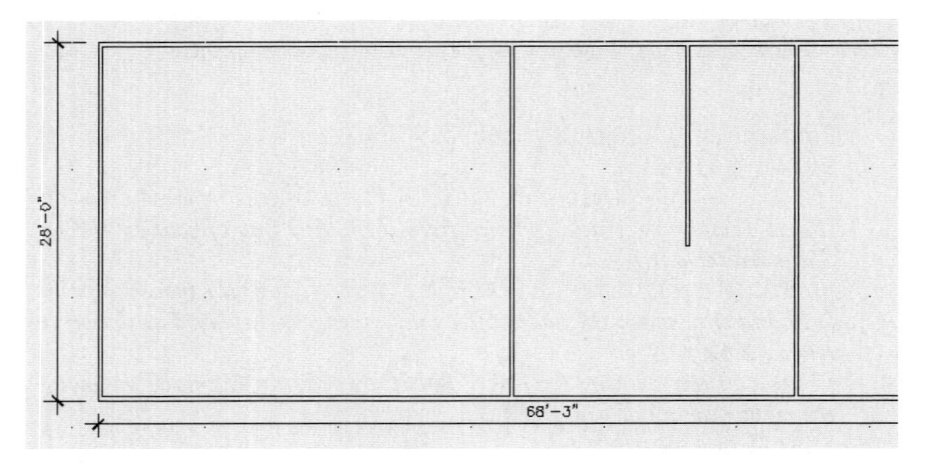

Figure 2.64 *Interior wall created with Temporary Tracking Point object snap*

7. Use the **Offset** command to copy interior walls. Select **Offset** from the AutoCAD **Modify** toolbar and refer to Figure 2.65 and the following command sequence:

```
Command: _offset
Specify offset distance or [Through] <Through>: 12'5
ENTER
Select object to offset or <exit>: (Select the wall at p1.)
Specify point on side to offset: (Select below the wall at point p2.)
Select object to offset or <exit>: ENTER
(Pressing ENTER ends the OFFSET command.)
```

```
Command: ENTER
(Pressing ENTER repeats the last command.)
Command: _offset
Specify offset distance or [Through] <12'5">: 3'3-
1/2 ENTER
Select object to offset or <exit>: (Select the wall at p3.)
Specify point on side to offset: (Select a point below, p4.)

Select object to offset or <exit>: (Select the last wall at
p5.)
Specify point on side to offset: (Select to the right of the
last wall at p6.)
Select object to offset or <exit>: ENTER
(Pressing ENTER ends the Offset command.)
Command:
```

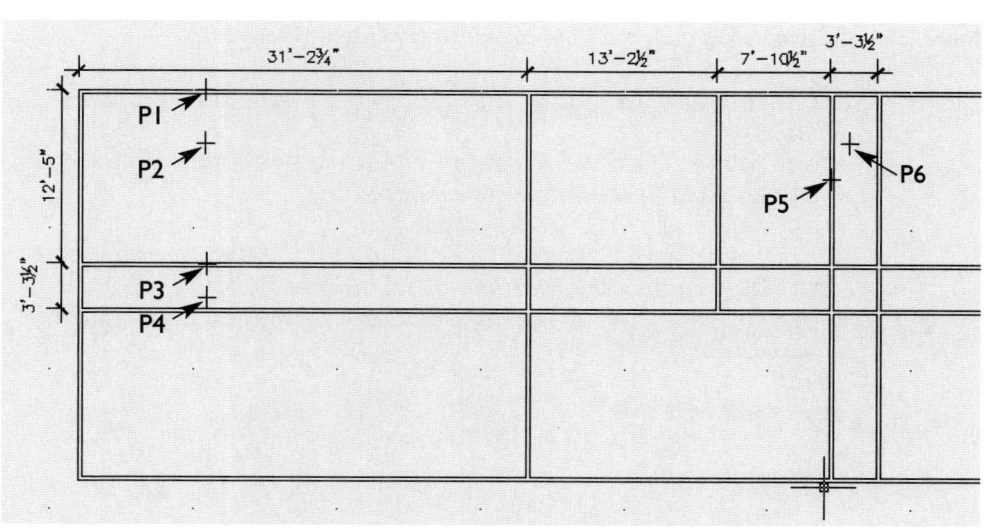

Figure 2.65 *Creating additional interior walls with the Offset command*

8. Use the **TRIM** command to remove walls and create rooms. Refer to Figure 2.66 and the following command sequence:

```
(Select TRIM from the Modify toolbar.)
Command: _trim
Current settings: Projection=UCS Edge=None
Select cutting edges ...
Select objects: ENTER (Selects all objects of the drawing as a cutting
edge.)
Select object to trim or [Project/Edge/Undo]: (Select
wall at p1.)
```

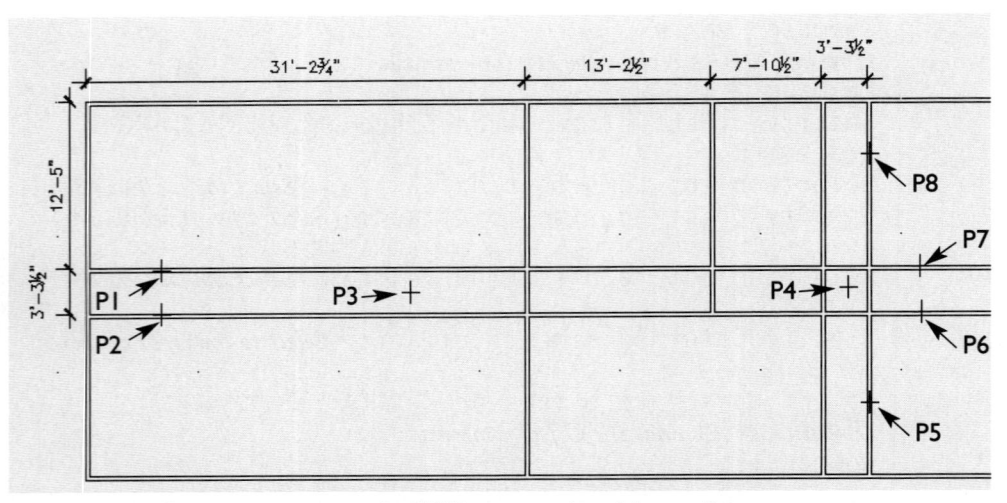

Figure 2.66 *Creating rooms using the TRIM command to delete wall segments*

```
Select object to trim or [Project/Edge/Undo]: (Select
wall at p2.)
Select object to trim or [Project/Edge/Undo]: F
```
ENTER *(sets Fence as the object selection method.)*
```
First fence point: (Select at point p3.)
Specify endpoint of line or [Undo]: (Select near point p4.)
```
ENTER *(Walls crossing this fence line are selected.)*
```
Select object to trim or [Project/Edge/Undo]: (Select
wall at point p5.)
Select object to trim or [Project/Edge/Undo]: (Select
wall at point p6.)
Select object to trim or [Project/Edge/Undo]: (Select
wall at point p7.)
Select object to trim or [Project/Edge/Undo]: (Select
wall at point p8.)
Select object to trim or [Project/Edge/Undo]: ENTER
```
(Pressng ENTER *ends the Trim command).*
(The drawing should now appear as shown in Figure 2.67.)

9. Save the drawing.

EDITING WALLS USING MODIFY WALL

The **Modify Wall** command opens a dialog box for quickly editing the width, height, offset, justification and style of a wall. After you select the **Modify Wall** command and select the wall to be edited, the **Modify Wall** dialog box opens, which consists of the edit fields describing the current properties of the wall. This dialog box is nearly identical to the **Add Wall** dialog box, except that it includes an **Apply** button.

Figure 2.67 Rooms created after editing

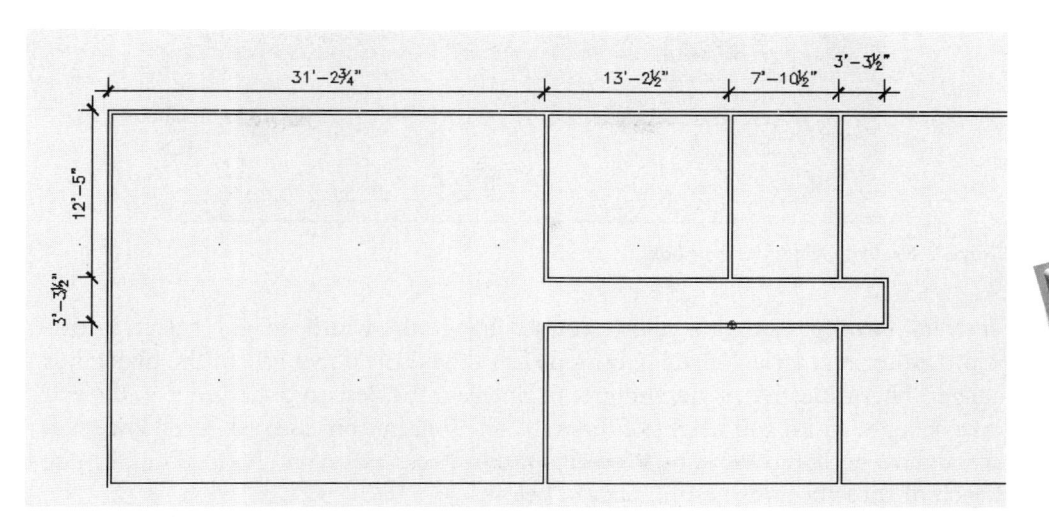

Access the **Modify Wall** command as follows:

Menu bar	Design>Walls>Modify Wall
Command prompt	AECWALLMODIFY
Toolbar	Select Modify Wall from the Walls toolbar shown in Figure 2.68
Shortcut	Right-click over drawing area and choose Design>Walls>Wall Modify
Shortcut	Select wall; then right-click and choose Wall Modify

Table 2.4 Modify Wall commands

Figure 2.68 Modify Wall command on the Walls toolbar

When the **Modify Wall** command is executed, you are prompted to select walls. You can select one or several walls for editing. After you select a wall for editing, the **Wall Modify** dialog box will open as shown in Figure 2.69.

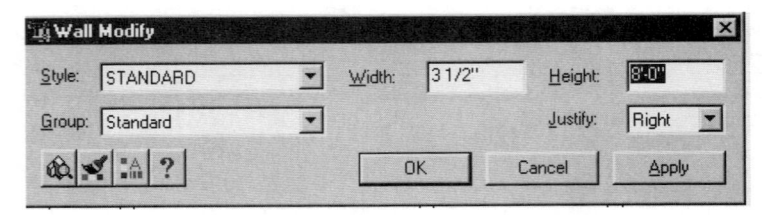

Figure 2.69 *Wall Modify dialog box*

The **Wall Modify** dialog box allows you to change the width, height, offset, justification, group, and style. Selecting options from the drop-down lists of the dialog box changes the justification, style, and group. Entering the desired dimensions in the edit boxes changes width and height of the wall. The four buttons located in the lower left corner allow access to **Floating Viewer**, **Match**, **Properties**, and **Help**. The purpose of each of these buttons is summarized below.

> **Floating Viewer** – Opens the Viewer and allows you to view the wall object.
>
> **Match** – Prompts you to select another wall in the drawing to match its properties.
>
> **Properties** – Executes the **Wall Prop** command, which allows you to specify additional properties of the wall.
>
> **?** – Executes the HELP command for the **Modify Wall** command.

After you edit the **Wall Modify** dialog box, the **Apply** button is active and the changes will be applied if either the **OK** or **Apply** button is selected. However, selecting the **Apply** button allows you to see the change in the wall prior to dismissing the dialog box. Selecting **OK** will change the wall and dismiss the dialog box.

When the **Wall Modify** command is executed and a wall selected, the current properties of the selected wall are displayed in the **Wall Modify** dialog box. The **Apply** button of the **Wall Modify** dialog box is inactive until the dialog box is edited as shown in Figure 2.70.

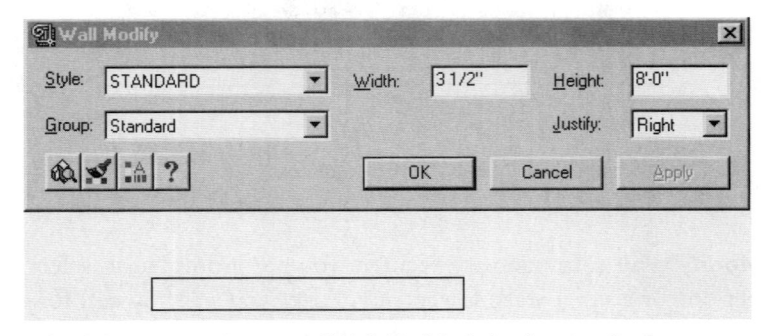

Figure 2.70 *Apply button is inactive until Wall Modify dialog box is edited*

The options of the **Modify Wall** command can also be selected from the command line prompts. To select options of the command, enter the capitalized letter of the options listed: Style/Group/Width/Height/Justify/Match. The following command sequence illustrates how the wall justification, height, and style would be changed in the command line.

```
Command: _AecWallModify
Select walls: (Select a wall.) 1 found
Select walls: ENTER (Null response ends selection)
Wall modify [Style/Group/Width/Height/Justify/Match]: J
Justification [Left/Center/Right/Baseline] <Right>: L
Wall modify [Style/Group/Width/Height/Justify/Match]: H
Height <8'-0">: 10'
Wall modify [Style/Group/Width/Height/Justify/Match]: S
Style name or [?] <STANDARD>: GYP_06
Wall modify [Style/Group/Width/Height/Justify/Match]: ENTER
```
(Pressing ENTER *ends the command)*

In the command sequence above, the wall was edited without the use of the dialog box. Entering information in the command line requires more typing and therefore increases the amount of time required for editing. The specific name of a wall style had to be correctly entered in the command line for changes to be applied to the wall style. Therefore, editing a wall using the **Wall Modify** dialog box is quicker and more efficient.

Select a wall, right-click, and then choose **Wall Modify** from the shortcut menu to access the **Modify Wall** command. The shortcut menu provides quicker access to the command because you do not have to locate the **Walls** toolbar and cycle through the prompts. To display the shortcut menu, click on a wall and then right-click to display the shortcut menu as shown in Figure 2.71.

Choosing **Wall Modify** will execute the **Modify Wall** command and display the **Wall Modify** dialog box. This approach provides faster access to the **Modify Wall** command, and the grips are displayed, allowing you to view the wall justification prior to editing the wall.

Setting Wall Properties with the Match Option

The **Match** button of the **Modify Wall** command allows you to select another wall in the drawing and match its properties. When this option is executed, you can match all the properties of another wall or select any one of the following properties: Style,

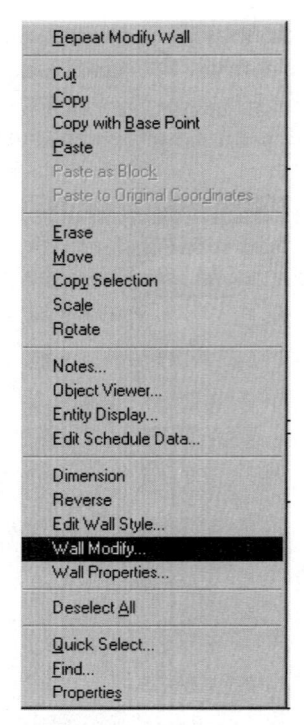

Figure 2.71 *Choosing Wall Modify from the shortcut menu of a selected wall*

Group, Width, Height, and Justification. To select only a certain property, enter the option in the command line. The command line is used to select the options of **Match**. The following tutorial illustrates how the **Match** option is used in modifying the wall shown in Figure 2.72. The horizontal wall is a CMU_Cavity_10 wall, whereas the vertical wall is a CONC_08 wall.

TUTORIAL 2.8 USING THE MATCH OPTION TO CHANGE WALLS

1. Open *Ex 2-8* from the *ADT Tutor\Ch2* directory.
2. Choose **SaveAs** from the menu bar and save the file as *Match* in your student directory.
3. Select **Modify Wall** from the **Walls** toolbar.
4. Respond to the command line prompts as shown in the following command sequence and Figure 2.72.

    ```
    Command: _AecWallModify
    Select walls: 1 found
    ```
    (Select the vertical wall to edit it as shown in Figure 2.72.)

```
Select walls: ENTER (Press ENTER to end wall selection.)
Wall modify [Style/Group/Width/Height/Justify/Match]:
DBOX (Select the Match button of the Wall Modify dialog box.)
Select a wall to match: (Select the horizontal wall to copy its
```
properties as shown in Figure 2.72.)
```
Match [Style/Group/Width/Height/Justify] <All>: ENTER
```
(Press ENTER to accept all properties.)
```
Wall modify [Style/Group/Width/Height/Justify/Match]:
ENTER (Press ENTER to end the command.)
Command:
```
(The Match option changes the walls as shown in Figure 2.73.)

Figure 2.72 Using Match to change wall properties

5. Save the drawing.

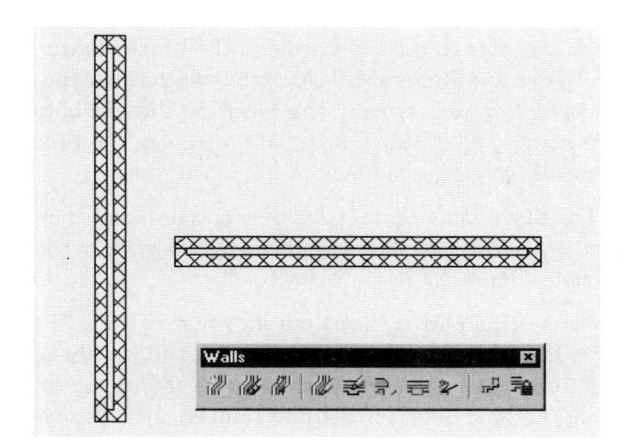

Figure 2.73 Walls edited to match each other with the Match button

In the command sequence above, the vertical wall was changed to match the horizontal wall. The prompt: `Match [Style/Group/Width/Height/Justify] <All>:` allows you to identify which properties of the horizontal wall you want to assign the vertical wall. For instance, this option would allow you to match the style or width of a wall but not its justification.

SETTING WALL PROPERTIES

The Properties button of the **Wall Modify** dialog box permits you to view the properties of a wall in a series of seven dialog boxes. This button executes the **Wall Properties** command from within the **Modify Walls** command. Access the **WALLPROPS** command as follows:

Menu bar	Design>Walls>Modify Wall; Properties button
Command prompt	WALLPROPS
Toolbar	Select Modify Wall from the Walls toolbar; select Properties button
Shortcut	Right-click over the drawing area and choose Design>Walls>Wall Properties
Shortcut	Select a wall; then right-click and choose Wall Properties

Table 2.5 *Accessing the WALLPROPS command*

Editing each of seven tabs modifies the properties of a wall. The seven tabs are summarized below:

General – The **General** tab shown in Figure 2.74 allows you to add a description or notes to the wall segment. The **Notes** button allows you to access the **Text** and **Reference Doc** tabs, which allow you to add text information regarding a wall segment. The **Floating Viewer** button located in the lower left corner will display the wall in the Viewer. The **Floating Viewer** button is available from all seven tabs.

Style – The **Style** tab includes a list of wall styles loaded in the drawing. You can select any of the styles listed and change the style of the wall to the selected style (Figure 2.75).

Dimensions – The **Dimensions** tab shown in Figure 2.76 allows you to change the **Width, Height, Length, Radius**, and **Justify** settings, and **Cleanup Settings**. Editing the **Length** value allows you to change the length of the wall. The start point for the wall remains anchored while the second point location moves. The angle of the wall remains the same when the length is edited. The start and end points remain anchored as the radius is changed.

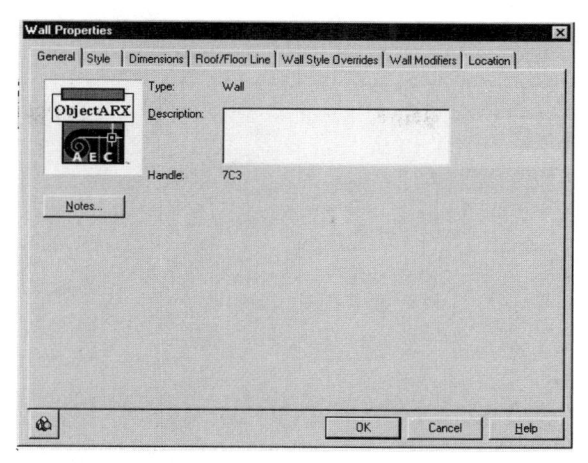

Figure 2.74 General tab of the Wall Properties dialog box

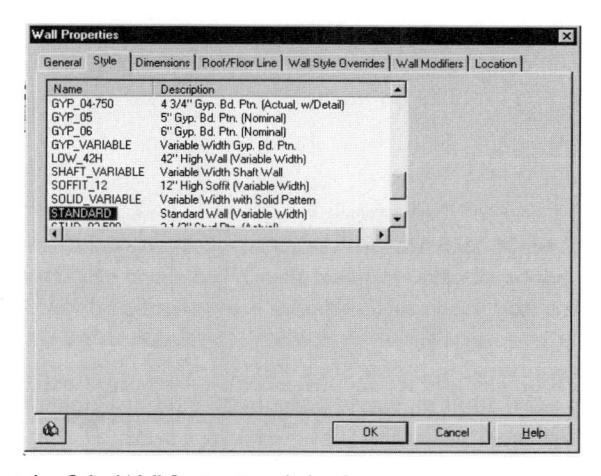

Figure 2.75 Style tab of the Wall Properties dialog box

> **Note:** When a curved wall is placed, the options do not allow you to define the wall radius; however, the Wall Properties Dimensions tab allows you to edit the curve to a specific radius.

In this tab the radius of the curved wall is displayed but not the length of the curve.

Roof/Floor Line – The **Roof/Floor Line** tab allows you to edit how the top and bottom of the wall meet the floor or roof as shown in Figure 2.77. The **Add Gable** and **Vertex** buttons allow you to extend a wall up to the roof as necessary when a gable roof is used. The **Insert Vertex** and **Edit Vertex** buttons allow you to step the bottom or top of a wall. The properties of the wall shown in Figure 2.77 include wall gable.

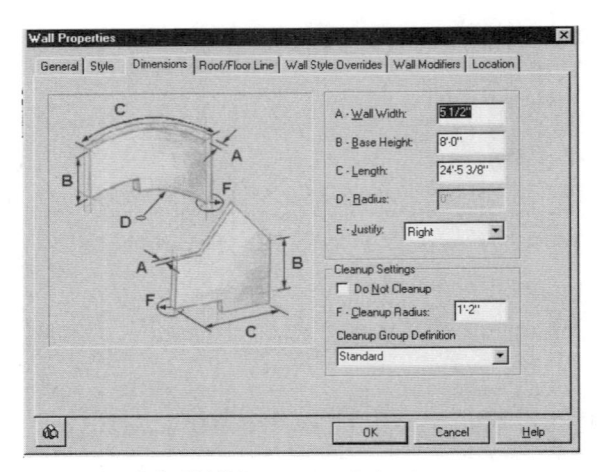

Figure 2.76 *Dimensions tab of the Wall Properties dialog box*

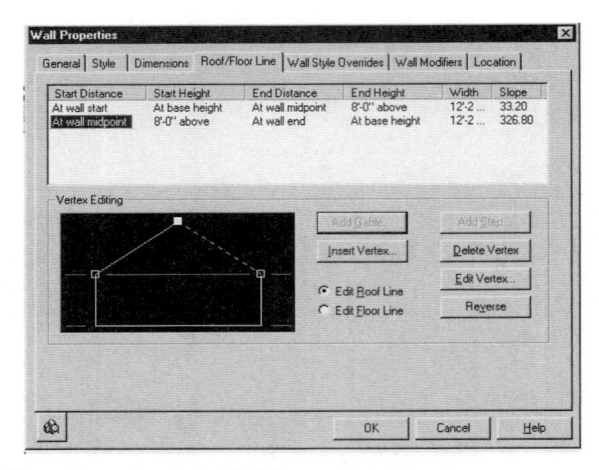

Figure 2.77 *Roof/Floor Line tab of the Wall Properties dialog box.*

Wall Style Overrides – The **Wall Style Overrides** tab allows you to override the endcap and wall component priority defined in a wall style. Editing this dialog box will create wall properties for a wall segment that are an exception to other wall segments of the same wall style. The **Wall Style Overrides** dialog box shown in Figure 2.78 includes three wall components assigned a priority not common to the wall style. The **Priority** number shown for each component determines how the components clean up at an intersection. Priorities vary from 1 to 99. A wall component with a priority of 1 is considered high and will clean up and override other components.

Wall Modifiers – The **Wall Modifiers** tab allows you to assign a wall modifier to a wall. Wall modifiers are representations of materials used to finish a

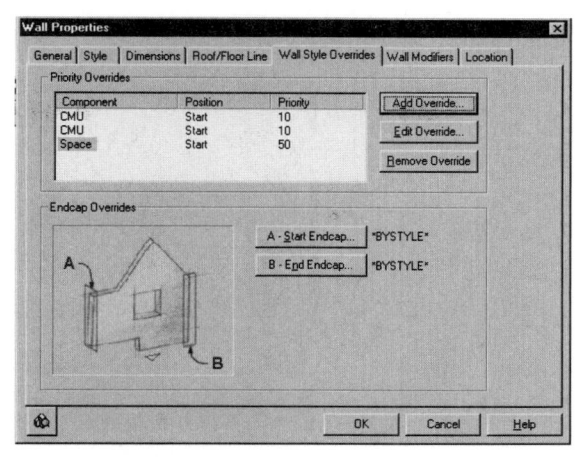

Figure 2.78 *Wall Style Overrides tab of the Wall Properties dialog box*

wall, such as brick pilasters or other projections not common throughout the length of the wall. The projection shown in Figure 2.79 is an example of a wall modifier applied to a wall according to the values shown in the **Wall Modifiers** tab of the dialog box.

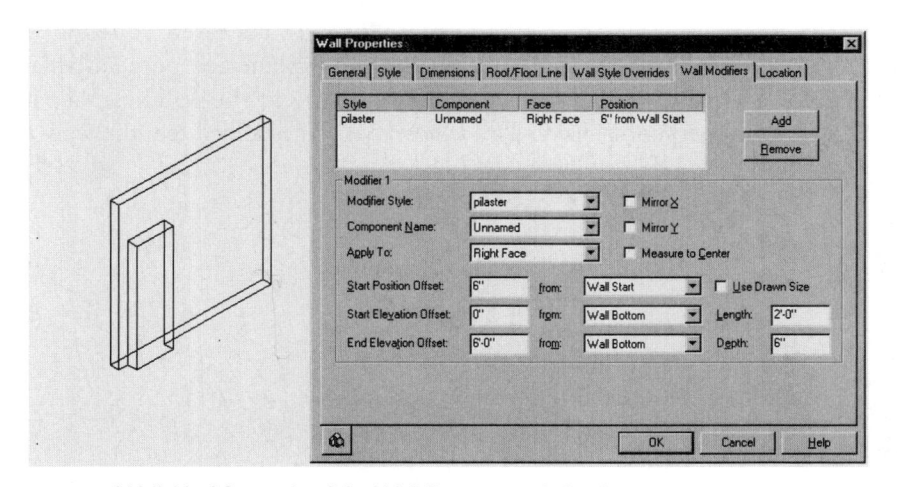

Figure 2.79 *Wall Modifiers tab of the Wall Properties dialog box*

Location – The **Location** tab describes the location of the wall in three-dimensional space relative to the World Coordinate System or the User Coordinate System. As shown in Figure 2.80, if a wall is rotated 45 degrees from the World Coordinate System, the angle will be indicated in the **Rotation** edit field.

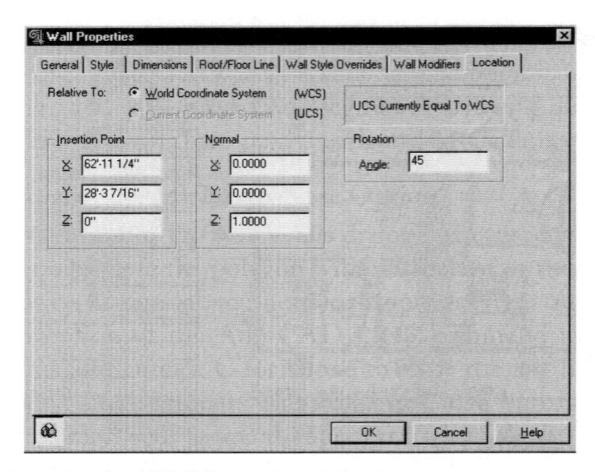

Figure 2.80 *Location tab of the Wall Properties dialog box*

The **Insertion Point** is the location of the wall based on the World Coordinate System. The **Normal** value is the orientation of the wall relative to the World Coordinate System. If **Normal** values are Z=1, X=0, Y=0, the wall is normal parallel to the Z plane and perpendicular to XY. This setting is the most typical because walls are drawn as extrusions in the Z direction. If **Normal** values were edited to X=1, Z=0, Y=0, the wall would flip to normal parallel to the X plane and perpendicular to the YZ plane. A Y=1, X=0, Z=0, **Normal** value would orient the wall normal parallel to the Y plane and perpendicular to XZ. If **Normal** values are edited, the walls will flip as shown in Figure 2.81.

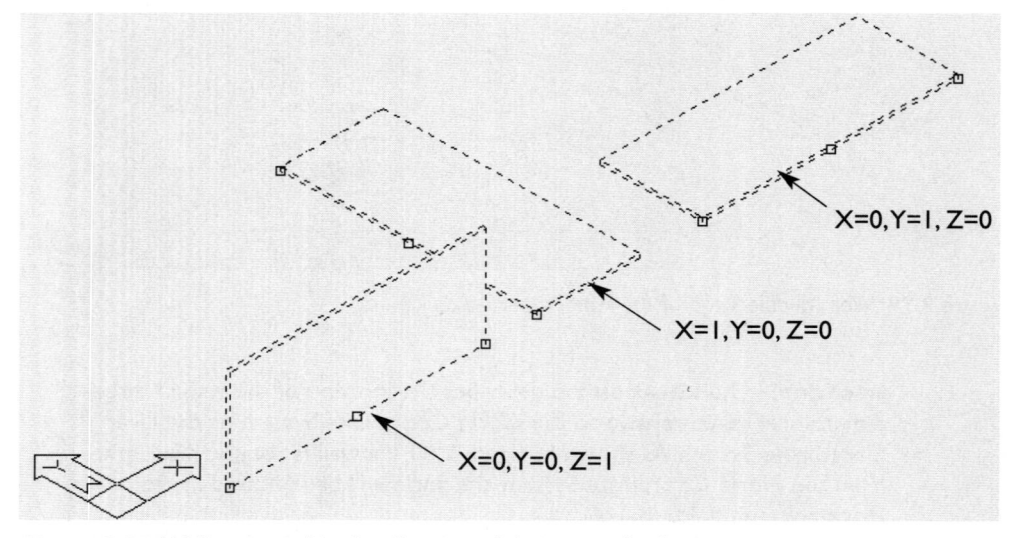

Figure 2.81 *Walls extruded in the direction of their normal values*

CHANGING WALL DIRECTION WITH REVERSE WALL

The **Reverse Wall** command allows you to convert the direction in which a wall is drawn. This command is most useful in flipping the thickness of a wall to the other side of the justification line.

Access the **Reverse Wall Start/End** command as follows:

Menu bar	Design>Wall>Wall Tools>Reverse Wall Start/End
Command prompt	WALLREVERSE
Toolbar	Select Reverse Wall Start/End from the Walls toolbar as shown in Figure 2.82
Shortcut	Select a Wall and right-click; choose Reverse.

Table 2.6 *Accessing the Reverse Wall command*

Figure 2.82 *Reverse Wall Start/End command on the Walls toolbar*

When this command is executed, you are prompted to select a wall. The direction the wall is drawn in is reversed when the wall is selected. If the wall is drawn with left or right justification, the thickness of the wall will flip to the other side. The **Reverse Wall** command does not move the justification line; it simply reverses the direction in which the wall is drawn. In Figure 2.83, the two walls drawn on the left with left justification were copied to the right. The **Reverse Wall Start/End** command was applied to the top right wall, which flipped the wall thickness to the other side of the justification line. Notice that the wall directional indicator is reversed for the top wall.

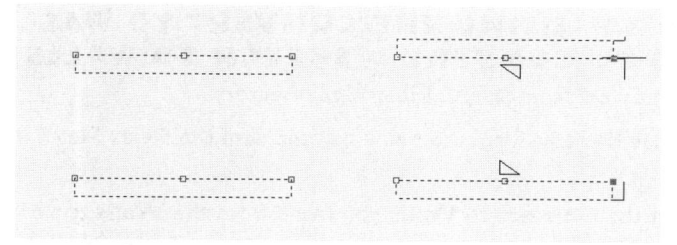

Figure 2.83 *Effects of the Reverse Wall Start/End command on existing walls*

CONVERTING GEOMETRY TO WALLS

The **Convert to Walls** command (**AECWALLCONVERT**) on the **Walls** toolbar allows you to convert lines, arcs, circles and polylines to wall objects. This command allows you to create a sketch of the floor plan and convert the sketch geometry to walls. The sketch can consist of lines, arcs, circles or polylines. Since the **POLYGON** and **RECTANG** commands create rectangles and regular polygon shapes using polylines, these shapes can be converted to wall objects.

Access **Convert to Walls** command as follows:

Menu bar	Design>Walls>Convert to Walls
Command prompt	WALLCONVERT
Toolbar	Select Convert to Walls on the Walls toolbar shown in Figure 2.84
Shortcut	Right-click over the drawing area and choose Design>Walls>Convert

Table 2.7 *Accessing the Convert to Walls command*

Figure 2.84 *Convert to Walls command on the Walls toolbar*

When the **Convert to Walls** command is executed, you are prompted to select the geometry to convert to walls. After you select the geometry, the **Wall Properties** dialog box opens, allowing you to define the properties of the new walls. In the following tutorial, the geometry shown in Figure 2.85 is converted to walls.

TUTORIAL 2.9 USING THE CONVERT TO WALL COMMAND TO CONVERT A SKETCH TO WALLS

1. Open *Ex 2-9* from the *ADT Tutor\Ch2* directory.
2. Choose **SaveAs** from the menu bar and save the file as *Sketch* in your student directory.
3. Select the **Convert to Walls** command from the **Walls** toolbar, and then respond to the command line prompts as shown in the following command sequence and Figure 2.85.

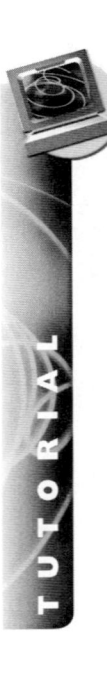

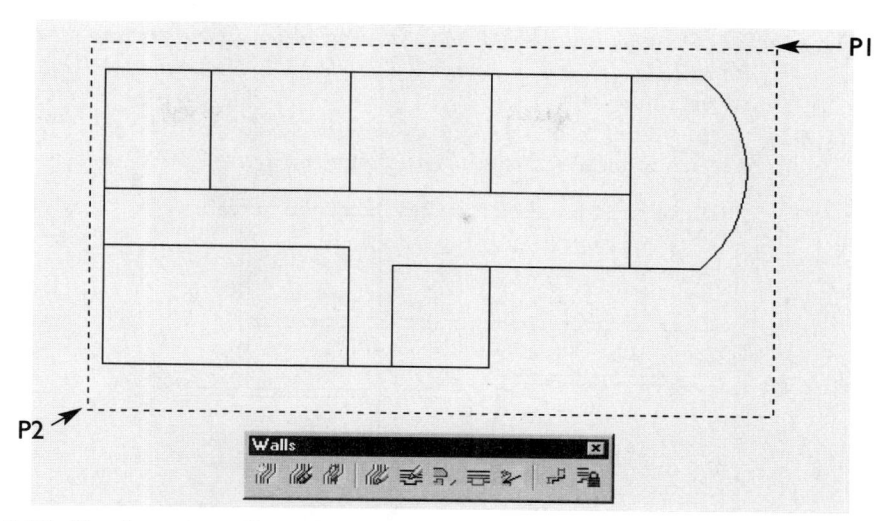

Figure 2.85 *Sketch geometry for creating walls*

```
Command: _AecWallConvert
Select lines, arcs, circles, or polylines to convert
into walls: 
```
(Select point P1 to specify a corner of the window as shown in Figure 2.85.)
`Other corner:` *(Select a point near P2 to select all the geometry.)*
`16 found`
`Select lines, arcs, circles, or polylines to convert into walls:` ENTER *(Press ENTER to accept objects selected)*
`Erase layout geometry? <Y>:` ENTER *(Press ENTER to erase existing geometry.)*
(After you select the geometry the Wall Properties dialog box opens as shown in Figure 2.86.)

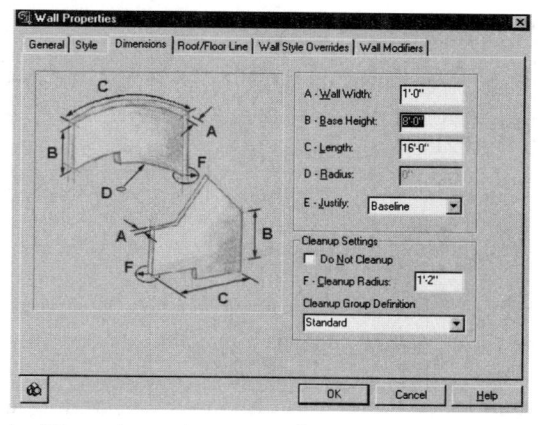

Figure 2.86 *Editing the Dimensions tab to set wall properties for new walls*

(Select the Dimensions tab of the Wall Properties dialog box and set the Width to 12" as shown in Figure 2.86; select OK to dismiss the Wall Properties dialog box.)
```
16 new wall(s) created
```
(The sketch is converted to walls as shown in Figure 2.87.)

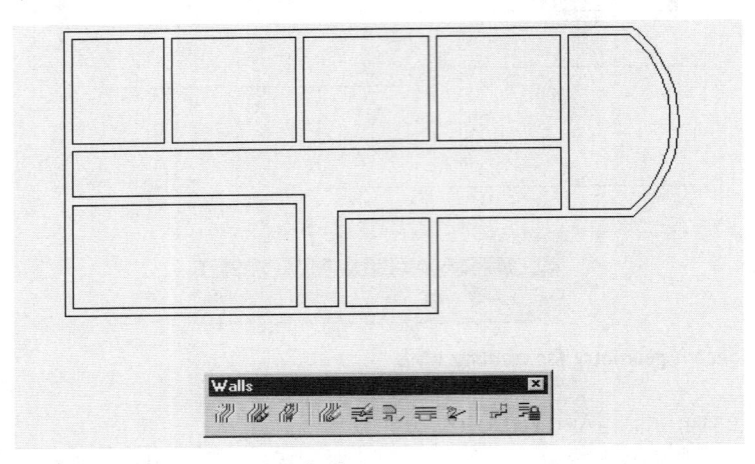

Figure 2.87 *Walls converted from sketched geometry*

4. Save the drawing.

CREATING WALL DIMENSIONS

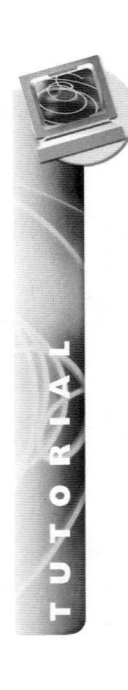

The **Create Wall Dimensions** command (**AECWALLDIM**) on the **Walls** toolbar will allow you to quickly place preliminary dimensions for one or more wall objects of a drawing. This command prompts you to select the walls and to locate the dimension line.

Access the **Create Wall Dimensions** command as follows:

Menu bar	Design>Walls>Create Wall Dimensions
Command prompt	WALLDIM
Toolbar	Select Create Wall Dimension from the Walls toolbar
Shortcut	Right-click over the drawing area and then choose Design>Walls>Dimension

Table 2.8 *Accessing the Create Wall Dimensions command*

When the **Create Wall Dimensions** command is selected, you are prompted to select walls for dimensioning. You can select one wall or several walls. After selecting

Figure 2.88 *Create Wall Dimensions command on the Walls toolbar*

the walls, you terminate the selection process by pressing ENTER and select a side to place the dimension line. The point selected for locating the dimension becomes the point that the dimension line is drawn through. Drafting conventions for architectural drawings permit the dimension line to be placed a random distance away from the walls to enhance the clarity of the dimensions.

The next prompt is for a `Second point or [Perp]`. This prompt establishes the angle for the dimensioning. Entering **P** for perpendicular will place the dimensions with extension lines perpendicular to the selected wall. Selecting a point at an angle to the first point will create aligned dimensions at an angle specified by the selected points. Turning on ORTHO aids in placing the dimensions parallel to the walls. You can erase any dimensions created with this command that are inappropriate without erasing the entire chain. If interior partitions are included in the walls selected, the location dimensions of the exterior and interior partitions include dimensions placed to the center of the interior partitions.

The following tutorial illustrates the use of the **Create Wall Dimensions** command for dimensioning walls.

TUTORIAL 2.10 USING THE CREATE WALL DIMENSIONS COMMAND TO PLACE DIMENSIONS

1. Open *Ex 2-10* from the *ADT Tutor\Ch2* directory.
2. Choose **SaveAs** from the menu bar and save the file as *Dimensions* in your student directory.
3. Verify that Ortho is turned ON in the Status Bar.
4. Choose **Create Wall Dimensions** from the **Walls** toolbar and then respond to the following command prompts as shown in the command sequence and figure in Figure 2.89.

   ```
   Command: _AecWallDim
   Select walls: (Select a point near P1 as shown in Figure 2.89.)
   Specify opposite corner: (Select a point near P2 as shown in Figure 2.89.)
   7 found
   Select walls: ENTER
   Pick side to dimension: (Select a point near P3 to locate the
   ```
   dimension line as shown in Figure 2.89.)

Second point or [Perp] *(Select a point near P4 to set the angle of the dimension line as shown in Figure 2.89.)*
(Linear dimensions created for the walls as shown in Figure 2.90.)

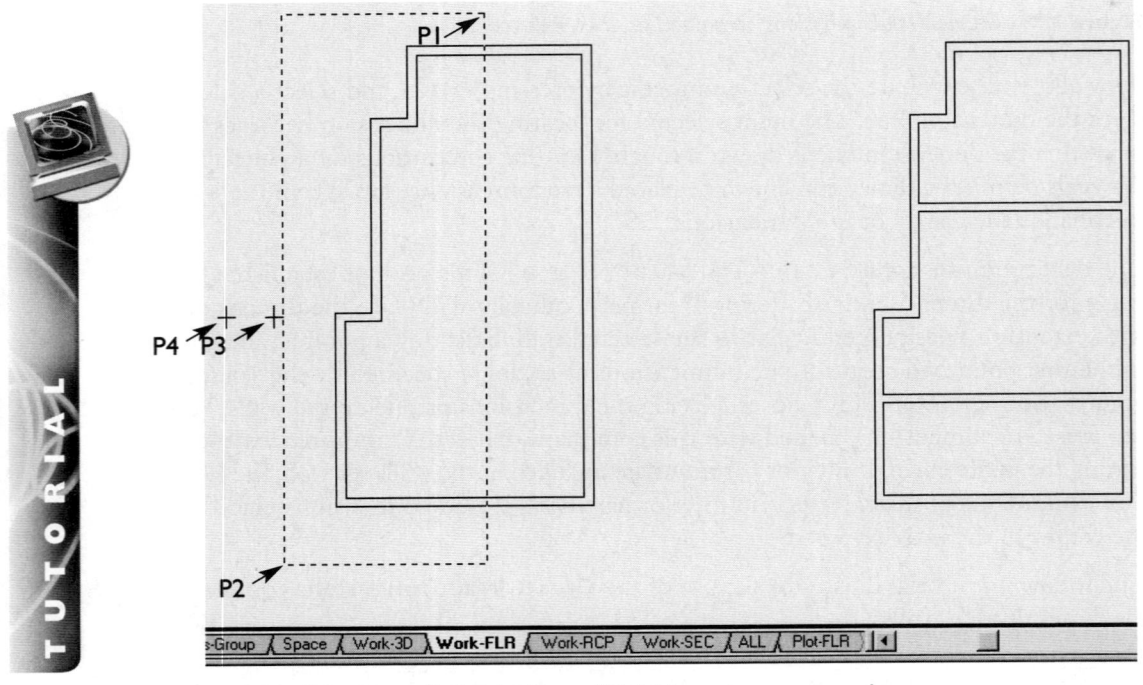

Figure 2.89 *Selecting walls for the Create Wall Dimensions command*

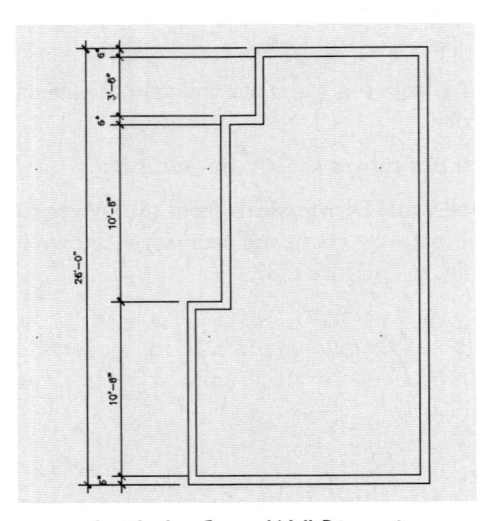

Figure 2.90 *Dimensions created with the Create Wall Dimensions command*

5. Choose **Create Wall Dimensions** from the **Walls** toolbar and then respond to the following command prompts as shown in the command sequence and figure in Figure 2.91.

   ```
   Command: _AecWallDim
   Select walls: (Select a point near P5 as shown in Figure 2.91.)
   Specify opposite corner: (Select a point near P6 as shown in Figure 2.91.)
   ```

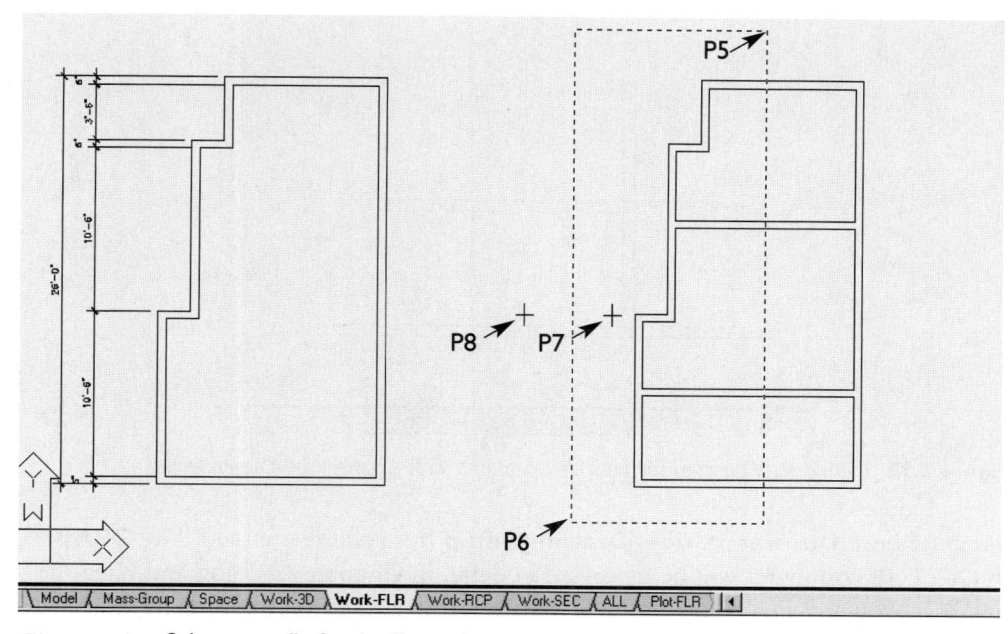

Figure 2.91 Selecting walls for the Create Wall Dimensions command

```
9 found
Select walls: ENTER
Pick side to dimension: (Select a point near P7 to locate the dimension line as shown in Figure 2.91.)
Second point or [Perp] (Select a point near P8 to set the angle of the dimension line as shown in Figure 2.91.)
(Linear dimensions created for the walls as shown in Figure 2.92.)
```

6. Save the drawing.

SETTING DIMENSION PROPERTIES

Dimensions are placed with the Aec_Arch_I dimension style, which is included in the Aec_arch (imperial) template. The dimensioning style includes an architectural tick arrowhead and is scaled to fit the drawing as specified by the **DRAWINGSETUP** com-

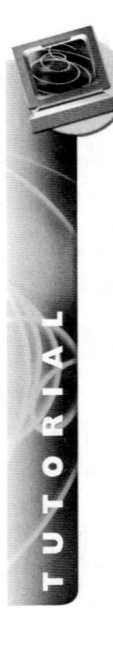

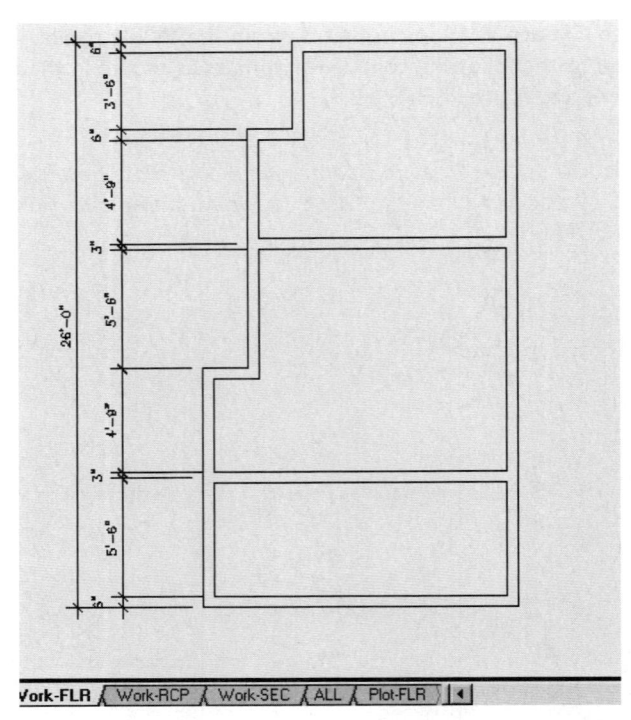

Figure 2.92 *Dimensions created for interior partitions with Create Wall Dimensions*

mand (choose **Documentation>Drawing Setup** from the menu bar). The **DRAW-INGSETUP** command will be discussed in detail in Chapter 7, Using and Creating Symbols. You can view and edit the attributes of the dimensioning style by selecting the **Dimension Style** button on the **Dimension** toolbar as shown in Figure 2.93.

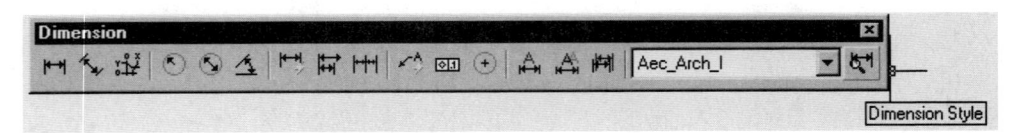

Figure 2.93 *Selecting Dimension Style button on the Dimension toolbar*

Selecting **Dimension Styles** on the **Dimension** toolbar will open the **Dimension Style Manager** dialog box as shown in Figure 2.94.

You can view and modify the properties of the dimension style by selecting the **Modify** button for the dimension style. The **Modify Dimension Style** dialog box for the Aec_Arch_I dimension style is shown in Figure 2.95.

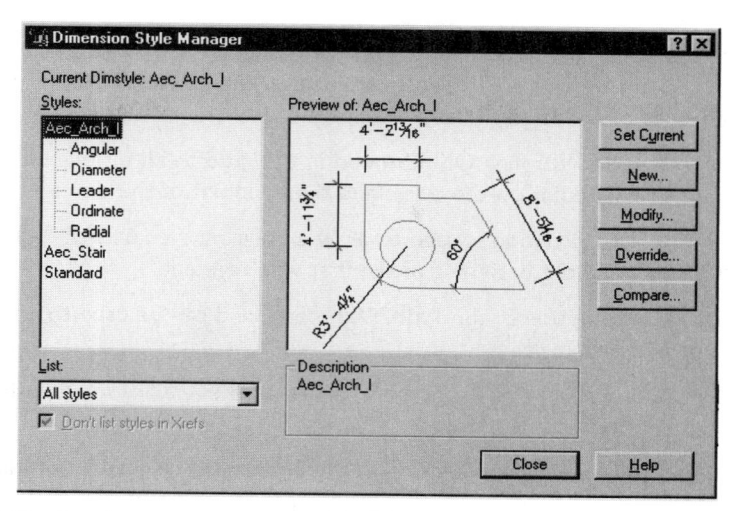

Figure 2.94 *Viewing dimension style properties using the Dimension Style Manager dialog box*

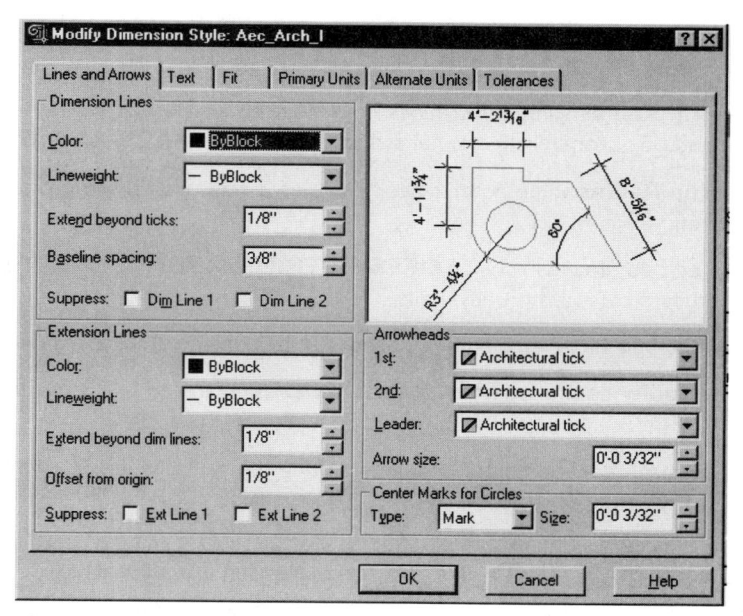

Figure 2.95 *Properties of a dimension style displayed in the Modify Dimension Style dialog box*

Additional information regarding dimensioning of walls will be included in Chapter 8, Annotating the Drawing. The **Create Wall Dimensions** command can be useful in placing preliminary dimensions on a floor plan as the plan is developed.

SUMMARY

1. Walls are created with the **Add Wall** command (**AECWALLADD**) as objects with properties of width, height, and justification.

2. Draw curved walls by toggling ON **Curved** in the **Add Wall** dialog box and identifying a start point, a point on the curve and the endpoint of the curve.

3. If the **Polyline Close** option is used to draw a polygon, the last wall segment is drawn to connect to the beginning of the first wall segment.

4. The **Ortho Close** option of the **Add Wall** command can be used to create walls in the shape of a rectangle.

5. The **Offset** option shifts the justification handle off the justification line the distance specified as the **Offset** value.

6. The handle for placing a wall is located at the bottom of the justification line.

7. Object Snap Nodes are located at the beginning, middle, and end of the justification line along the top and bottom.

8. Grips of a wall are located at the beginning, middle, and end on the bottom of the justification line.

9. The **Cleanup Radius** value determines how close to an existing wall a new wall can begin and be automatically connected to the wall.

10. The **Cleanup Radius** value determines how long a wall must be to avoid the wall error marker.

11. AutoCAD edit commands such as **OFFSET**, **TRIM**, **EXTEND**, and **STRETCH** can be used to create and modify walls.

12. The grips of a wall allow easy editing of the wall to copy, stretch, and move walls.

13. The **Wall Modify** command (**AECWALLMODIFY**) is used to change the properties of a wall.

14. You can perform detail editing of a wall by editing the wall properties of a wall using the **WALLPROPS** command.

15. Wall modifiers can be added to a wall to form a projection such as a pilaster with the **WALLPROPS** command.

16. Polylines can be converted to walls with the **Convert to Walls** command (**AECWALLCONVERT**).

Creating Floor Plans 109

REVIEW QUESTIONS

1. The option of the **Add Wall** command that allows you to move the wall handle a specified distance from the wall is _____.

2. The single object snap mode that allows you to snap to the midpoint and endpoints of the wall justification line is _____.

3. The Z coordinate of the wall justification line is _____.

4. The grips of a wall are located at _____, _____ and _____.

5. The nodes of a wall are located at the _____.

6. The width of a wall is changed by the _____ command.

7. The length of a wall can be changed by the _____.

8. The handle of the wall is located on the _____ of the wall.

9. List three methods for selecting the **Add Wall** command.

10. What is the purpose of the wall directional indicator?

11. Wall modifiers are defined for a wall with the _____ command.

12. The priority number assigned to a wall component can vary from _____ to control how the wall component will clean up with other walls.

13. Wall components with low priority numbers will _____ the display of wall components with high priority numbers.

14. Create a curved wall by establishing three points: _____, _____, _____.

15. The _____ Close option draws the final wall segment of a polygon to connect at an angle of 90 degrees to the beginning of the first wall segment.

16. The option of the **Wall Properties** command that controls how close you have to be to an existing wall for it to merge with the current wall is _____.

17. Often you can remove wall error markers by editing the _____ of the wall.

18. The following commands on the **Modify** toolbar cannot be used to modify walls _____, _____, _____.

19. The _____ command allows you to change the direction in which a wall has been drawn.

20. The **Convert to Walls** command (**AECWALLCONVERT**) on the **Walls** toolbar allows you to convert _____, _____, _____ and _____ to wall objects.

PROJECT

EXERCISE 2.11 DRAWING THE WALLS OF A FLOOR PLAN

Open *EX2-7* and insert additional interior walls as shown in Figure 2.96. Save the file as *Proj 2-11* in your student directory. You can verify your wall locations by placing temporary dimensions.

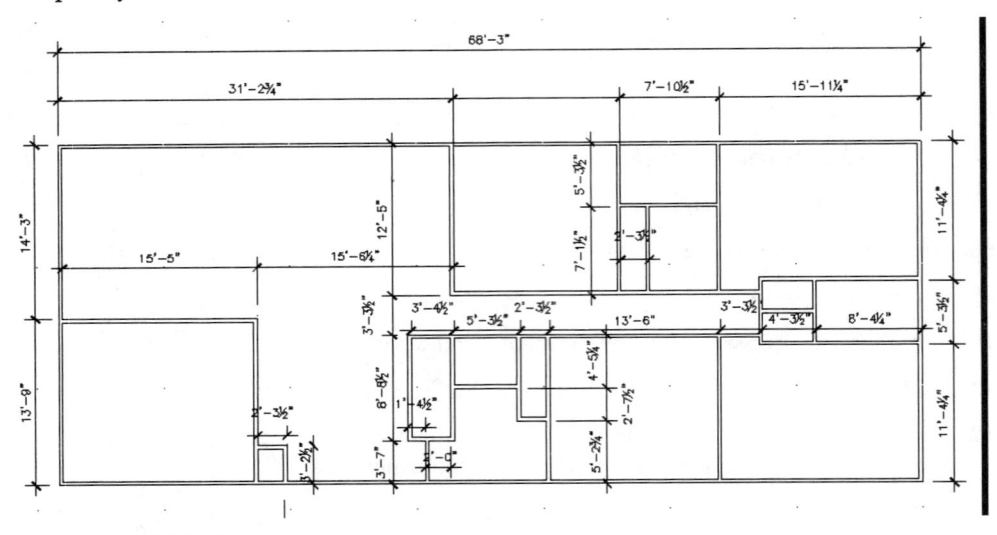

Figure 2.96 *Wall placement dimensions*

CHAPTER 3

Advanced Wall Features

INTRODUCTION

Architectural Desktop provides additional tools to enhance the creation of walls that consist of multiple wall components. The use of wall styles, wall groups, wall caps, and wall surface modifiers increases the speed of design and the detail of wall representation.

OBJECTIVES

After completing this chapter, you will be able to

- Create a wall style
- Change the wall style of a wall
- Control the display of wall components by editing display properties of wall styles
- Create wall cleanup groups and use cleanup groups to control wall intersections
- Create wall endcap styles and wall modifiers to modify the properties of walls

ACCESSING WALL STYLES

Although the Standard wall style may be adequate in many design applications, it consists of only two wall lines. Other wall styles can be used to create wall representations for complex walls that consist of several wall components. Wall styles allow the designer to create a wall style for each wall type. The wall style can then be selected from the **Style** drop-down list of the **Add Walls** dialog box. The Aec arch (imperial) template includes 38 wall styles. The Aec arch (metric) template includes only the Standard wall style.

Any of the included wall styles can be copied and edited to create new wall styles. Wall styles can also be imported and exported across design files. A list of the wall styles loaded in the drawing is displayed in the **Style** drop-down list of the **Add Walls** dialog box. If an existing style has a predefined width, the **Width** edit field of the **Add Walls** dialog will be inactive.

To use a wall style other than Standard, select another style from the **Style** drop-down list in the **Add Walls** dialog box. The steps to selecting a wall style are shown below:

STEPS TO SETTING A WALL STYLE CURRENT

1. Select **New** from the **Standard** toolbar.
2. Select the **Use a Template** button in the **Create New Drawing** dialog box.
3. Select the Aec arch (imperial) template and select **OK** to dismiss the **Create New Drawing** dialog box.
4. Select the Work-FLR layout.
5. Select **Add Wall** on the **Walls** toolbar, as shown in Figure 3.1.

Figure 3.1 *Add Wall command on the Walls toolbar*

6. Select the down arrow in the **Style** drop-down list of the **Add Walls** dialog box, as shown in Figure 3.2.

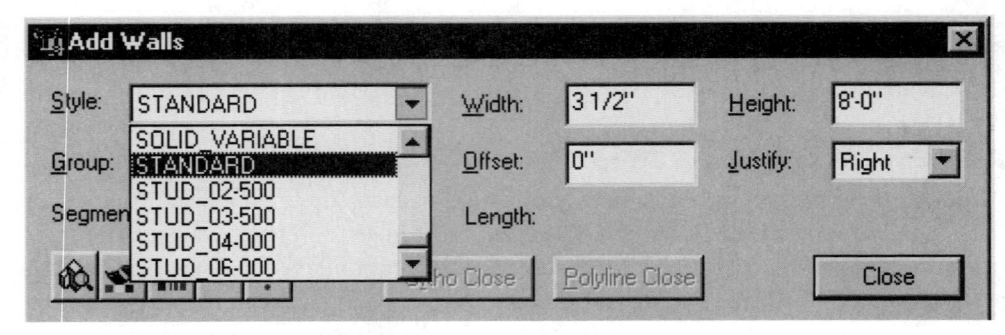

Figure 3.2 *Style list of the Add Walls dialog box*

7. Scroll down the list of styles and select the desired style.
8. Click in a model space viewport to activate the drawing area.
9. Click in a model space viewport to specify the start point of the wall using the selected wall style.

CREATING WALL STYLES

The **Wall Styles** command (**AECWALLSTYLE**) is used to create and edit the definition of a wall style. To view and edit the existing wall styles loaded in a drawing, access **Wall Styles** as follows:

Menu bar	Design>Walls>Wall Styles
Command prompt	AECWALLSTYLE
Toolbar	Select Wall Styles on the Walls toolbar as shown in Figure 3.3
Shortcut	Right-click over drawing area; choose Design>Walls>Style
Shortcut	Select a wall; right-click and choose Edit Wall Style

Table 3.1 *Accessing Wall Styles*

Figure 3.3 *Wall Styles command on the Walls toolbar*

When the **Wall Styles** command is executed, the **Wall Styles** dialog box opens, as shown in Figure 3.4. It includes a list of all wall styles loaded in a drawing.

This dialog box is used to create, copy, edit, import, and export walls styles and purge them from a drawing. The Aec arch (imperial) template includes 38 different wall styles; they are described in Appendix A. The simplest is the Standard wall style, which consists of only two wall lines. Since the Standard wall style is simple, it is often used as the base for new wall styles. The buttons to the right of the wall style list perform the following functions:

>**New** – Creates a new name for a new wall style. This button is not used to set the description or properties of the wall style.

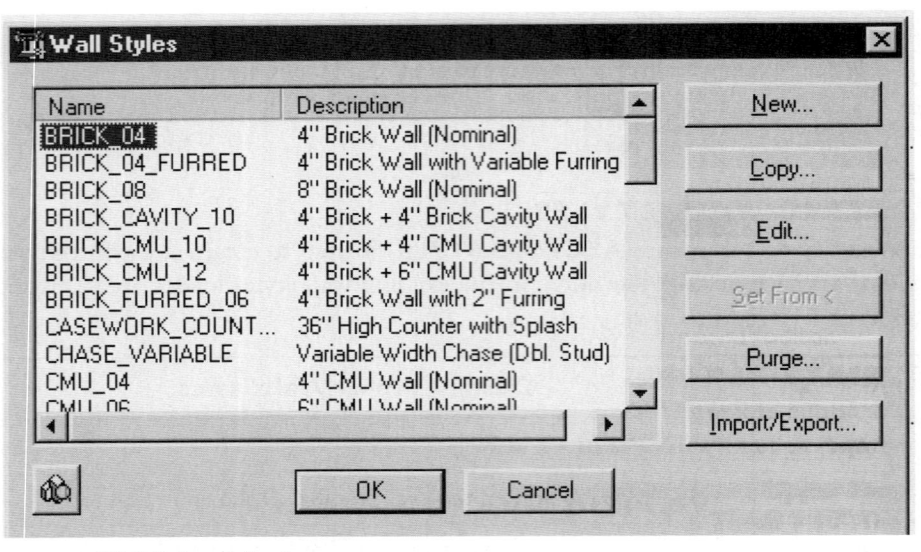

Figure 3.4 *Wall Styles dialog box*

Copy – Copies an existing wall style so as to create a new named wall style with the same properties as the selected wall style.

Edit – Creates a description for a wall style or to review or define new wall style properties.

Set from – Inactive button.

Purge – Deletes wall styles from the current list.

Import/Export – Transfers defined wall styles to a resource file or imports wall styles from a resource file into a drawing file.

The **New** button of the **Wall Styles** dialog box allows you to add a new name to the list of wall styles. Once a new name has been created, the **Edit** button of the **Wall Styles** dialog box is used to define specific properties of the new style. Editing the wall style allows you to define the number and size of wall components, endcap styles, hatching, and display representation of the wall components.

CREATING A NEW WALL STYLE NAME

The first step in creating a new wall style is to establish a new name. Names for wall styles should be kept short and descriptive. Spaces are permitted in the wall style name; however, the (") inch symbol is not accepted. Once a name has been created, it will be added to the names listed in the **Style** list of the **Add Walls** dialog box.

STEPS TO CREATING A NEW WALL STYLE

1. Select **New** from the **Standard** toolbar.
2. Select the **Use a Template** button in the **Create New Drawing** dialog box.
3. Select the Aec arch (imperial) template, and then select **OK** to dismiss the **Create New Drawing** dialog box.
4. Select the Work-FLR layout.
5. Select **Wall Styles** on the **Walls** toolbar, as shown in Figure 3.5.

Figure 3.5 *Selecting Wall Styles from the Walls toolbar*

6. Select the **New** button of the **Wall Styles** dialog box to open the **Name** dialog box as shown in Figure 3.6.

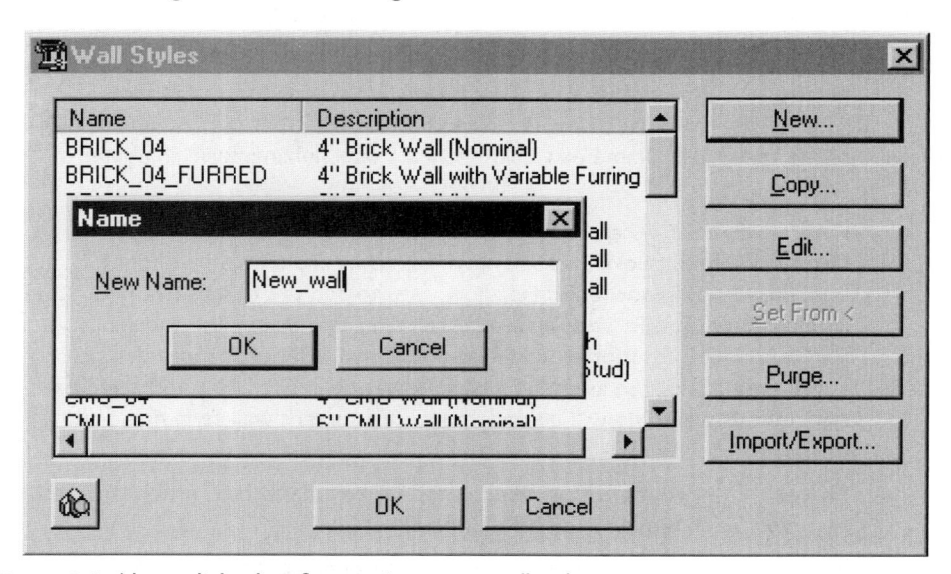

Figure 3.6 *Name dialog box for entering a new wall style name*

7. Type a new name in the **Name** dialog box.
8. Select the **OK** button to dismiss the **Name** dialog box.

This procedure does not allow you to enter a description or edit specific properties for the new wall style; it only creates a new name. To set specific properties for a wall style, you must select the **Edit** button of the **Wall Styles** dialog box.

EDITING A WALL STYLE

Once a name has been created for a new wall style, editing the wall style sets the properties defined for the wall style. Editing a wall style can occur prior to or during the use of the wall style. The current definition for the wall style is applied to all instances of the wall in the drawing. Therefore, if a wall style is defined initially to consist of three wall lines with a total width of 12" and is later edited to a width of 10", all walls placed in the drawing using this style will automatically be updated to 10"-wide walls. Using wall styles allows you to globally edit all the walls of an entire building by changing the original wall style definition.

To define a wall style, select the **Edit** button of the **Wall Styles** dialog box. Selecting the **Edit** button opens the **Wall Style Properties** dialog box, which consists of the following four tabs: **General**, **Endcaps**, **Components**, and **Display Properties**. Each tab is used to define the characteristics of a wall style.

General

The **General** tab for a new wall style named BRICK_CMU_12 is shown in Figure 3.7.

Name – You can edit the name by clicking on the text in the **Name** edit field. Editing the name in essence renames the wall style. You cannot delete all of the text in the Name edit field—at least one alphanumeric character must remain.

Description – You can edit the description by entering text in the **Description** edit field. Because new wall styles are created without a description, the **Description** edit field allows you to add the descriptive information. It can include specific sizes and material types that distinguish the wall style. The text in the description edit field is optional. If the name and description of a wall style are edited after the wall style has been used, all previous insertions of this wall style will be updated to the current wall style name and description.

Notes – The **Notes** button displays the **Notes** dialog box, which includes two tabs, **Text Notes** and **Reference Docs**.

Text Notes – To add text to the **Text Notes** tab, click in the edit field and begin typing, as shown in Figure 3.8. Information regarding construction notes or costs for a given wall style can be entered in this tab. The **Select All** but-

Advanced Wall Features | 117

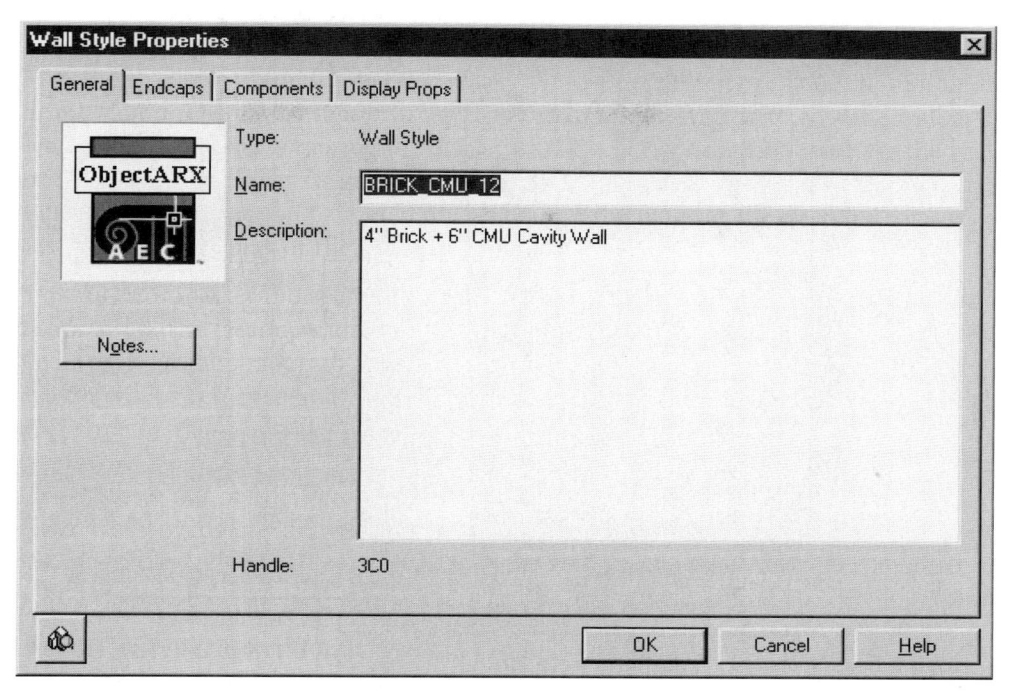

Figure 3.7 *General tab of the Wall Style Properties dialog box*

ton in the lower right corner of the dialog box allows you to select all the text in the **Notes** dialog box. Once the text is selected, a right-click of the mouse displays a shortcut menu that includes **Undo**, **Cut**, **Copy**, **Paste**, and **Delete** operations. Choosing these shortcut menu options allows the selected text to be manipulated through Windows operations and the Clipboard.

Reference Docs – The **Reference Docs** tab shown in Figure 3.9 allows you to list other files that can contain details of the construction or text files that describe the wall. Selecting the **Add** button shown in the lower left corner opens the **Select Reference Document** file dialog box, which allows you to select a file and directory path for reference files. Selecting the **Edit** button allows you to edit the path shown for the reference document file or the description of the file. The **Description** field would be useful for describing the content of the reference document. The **Delete** button is used to delete a file from the list. Using a reference file list allows you to cross-reference drawings while working on a design file.

Viewer – The **Floating Viewer** icon located in the lower left corner of the **General** tab will open the **Viewer** and allow you to view sample graphics of the current wall

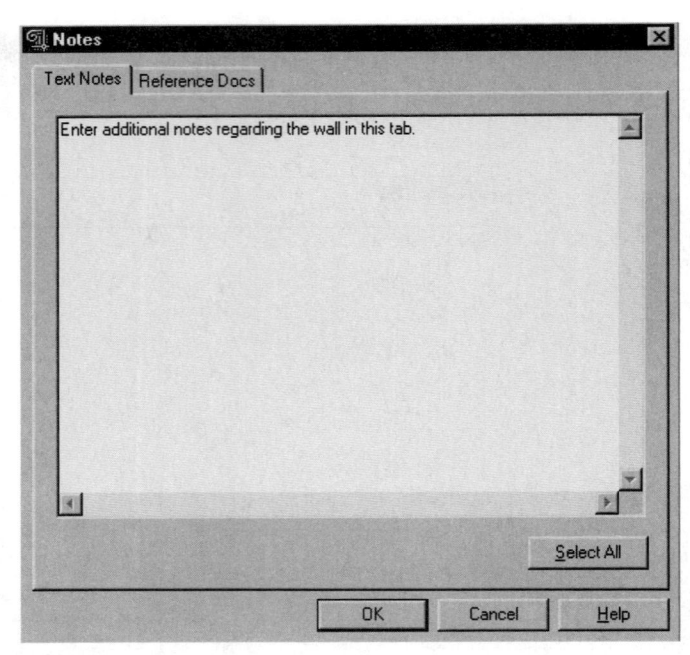

Figure 3.8 *Text Notes tab of the Notes dialog box*

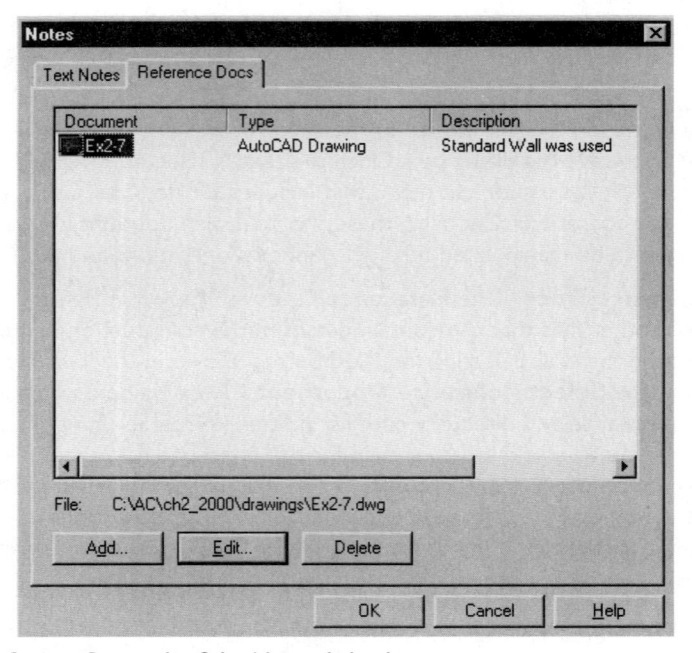

Figure 3.9 *Reference Docs tab of the Notes dialog box*

style. A view of the wall can be either an orthographic or a selected pictorial view. The view shown in Figure 3.10 is the top view of the BRICK_CMU_12 wall style, which consists of several wall lines and hatching as shown in the **Viewer**.

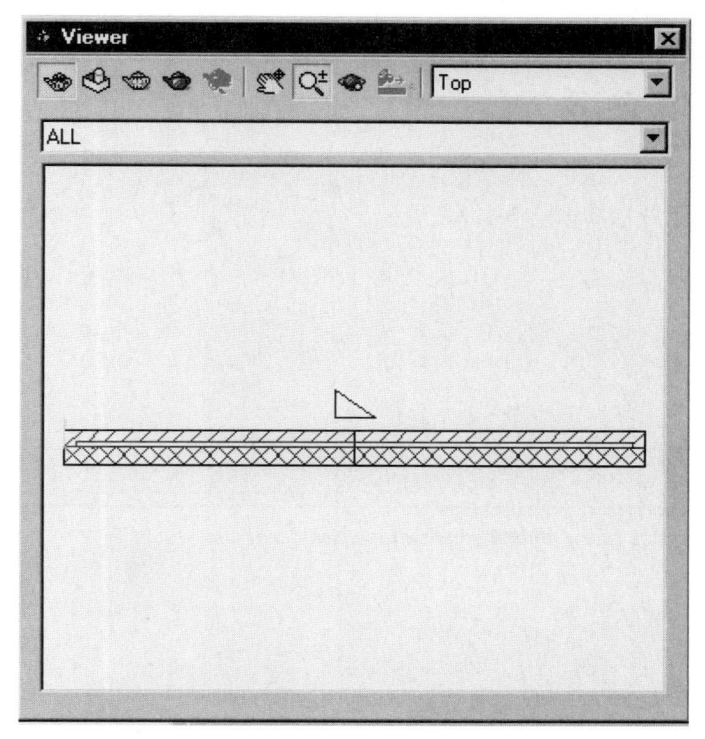

Figure 3.10 *Wall displayed in the Viewer*

Endcaps

The **Endcaps** tab, shown in Figure 3.11, allows you to define an endcap style to be used for the ends of walls or openings in a wall around windows or doors. In the figure, the Cavity_12 endcap is assigned to the **Wall Endcap Style**, and the Standard Endcap Style is assigned to the **Opening Endcap Style**.

Endcap styles allow you to specify how the wall components terminate at the end of a wall. In Figure 3.12, the Cavity_12 endcap style is assigned to cap the end of the wall. Selecting the **Wall Endcap Style** or **Opening Endcap Style** button allows you to select from a list of saved endcap styles for the drawing.

> **Wall Endcap Style** – The **Wall Endcap Style** button is used to assign an endcap style for the ends of a wall. When the **Wall Endcap Style** button is

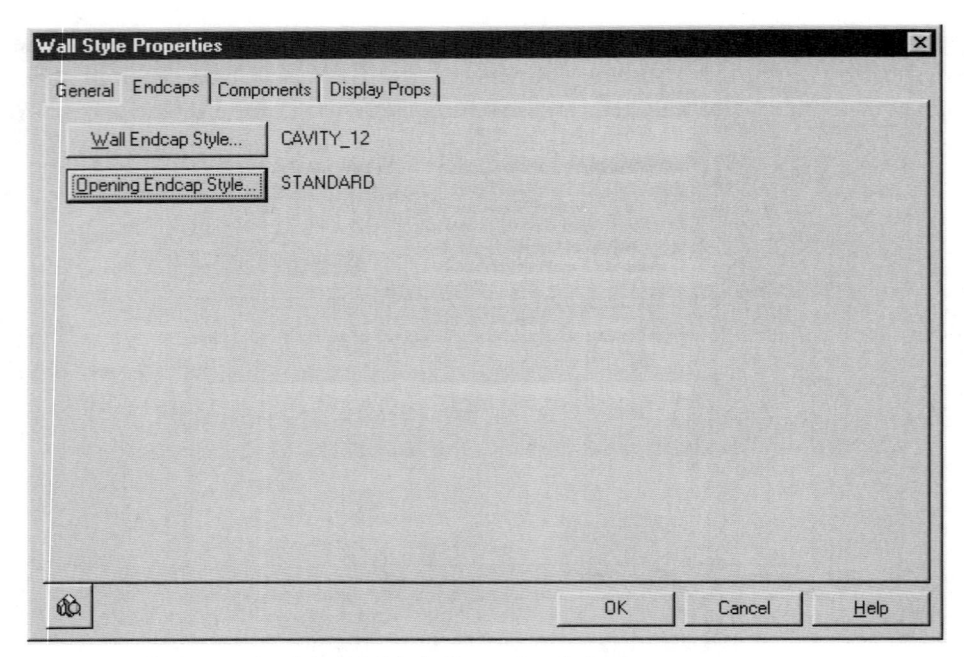

Figure 3.11 Endcaps tab of the Wall Style Properties dialog box

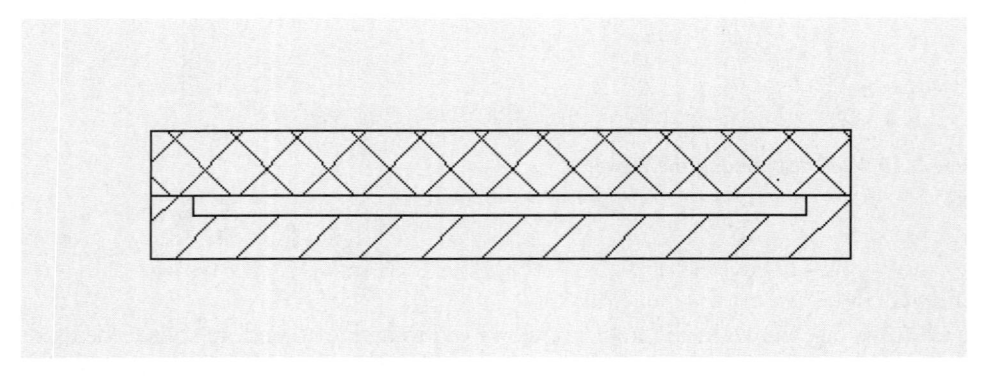

Figure 3.12 Wall with Cavity_12 endcap style

selected, a dialog box allows you to select from a list of saved endcap styles, as shown in Figure 3.13.

Opening Endcap Style – The **Opening Endcap Style** button is used to display and select the endcap styles for openings created for doors and windows. Endcap styles included in the Aec arch (imperial) template are Standard, Cavity_10, Cavity_12, CMU_FND_8, CMU_FND_12.

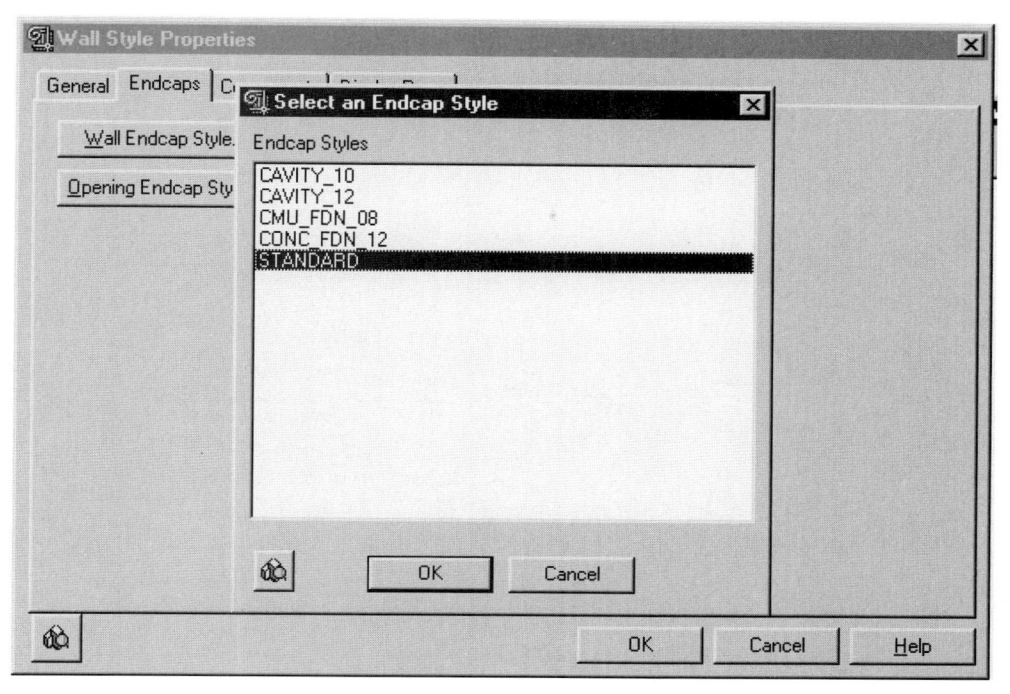

Figure 3.13 *Selecting an endcap style from Select an Endcap Style dialog box*

Components

The top section of the **Components** tab, shown in Figure 3.14, consists of a list of wall components. The **Width**, **Priority**, and **Edge Offset** assigned to each component are displayed in the component list.

> **Add** – The **Add** button is used to create new wall components. Selecting the **Add** button will create a new component identical to the component highlighted in the component list.
>
> **Remove** – The **Remove** button will remove a wall component from the wall component list. Remove components by first selecting the **Index** number of the wall component and then selecting the **Remove** button.
>
> **Component list** – Each component is identified by a number in the **Index** column. The components of the BRICK_CMU_12 wall style are shown above. The BRICK_CMU_12 wall consists of Brick, Space, and CMU wall components. The names assigned to the components assist the designer in understanding which wall lines represent a building component. You can set properties of each component by selecting the **Index** number for the component at the top of the dialog box and then modifying the edit fields located in the lower portion of the dialog box.

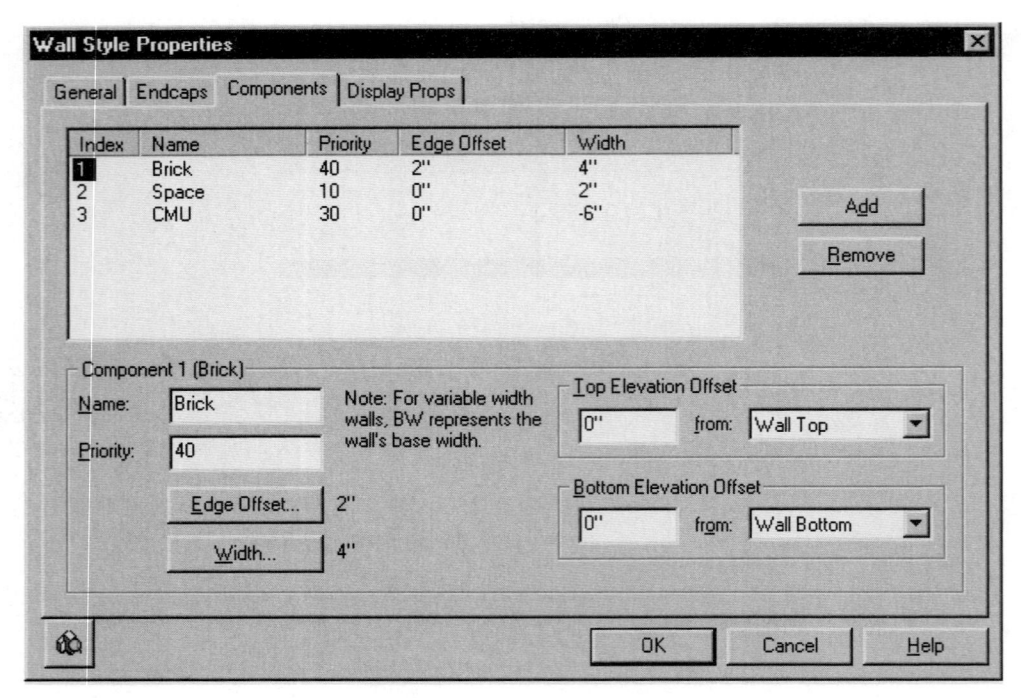

Figure 3.14 *Components tab of the Wall Style Properties dialog box*

The lower section of the **Components** tab includes edit fields for defining the attributes of each wall component as described below:

Name – The **Name** edit field allows you to enter the name for the wall component. Spaces are permitted in the name; however, the name should be kept short and descriptive.

Priority – The **Priority** edit field allows you to assign a priority number from 1 to 99 to control how the wall component will clean up with other walls. Wall components with low priority numbers will override the display of wall components with high priority numbers. The walls shown in Figure 3.15 were drawn with different priority numbers. The horizontal wall and the left wall have component priority set to 10. The right wall has its component priority set to 80. The walls with priority of 10 override the display of wall components with priority of 80.

Assigning priority numbers to wall components would be useful for controlling the display of components of fire-rated walls, which are displayed without penetrations from other non-rated walls. Therefore, in the example shown above, the right vertical wall could correctly represent a non-rated wall that

Advanced Wall Features 123

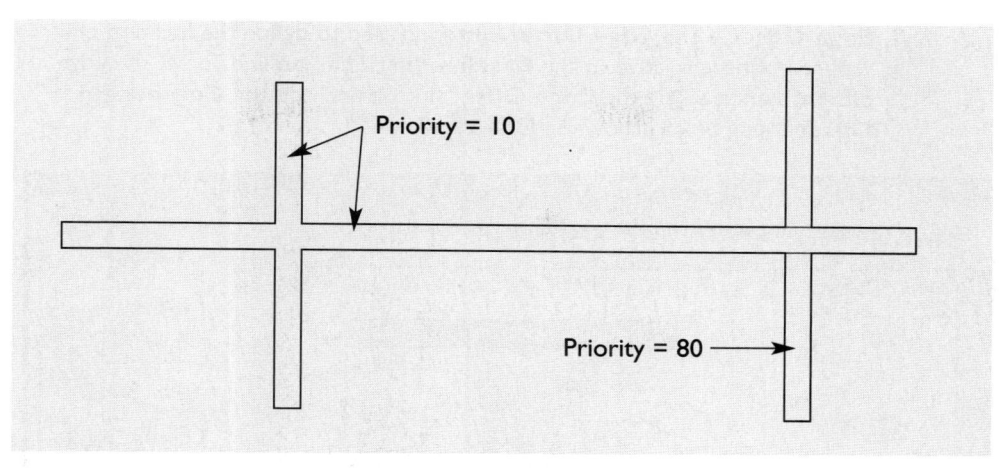

Figure 3.15 *Walls with low priority numbers overriding walls with high priority numbers*

intersects the horizontal fire-rated wall. Assigning priority to wall components controls the cleanup behavior of a wall when wall components intersect other walls.

Note that the Space component of the BRICK_CMU_12 wall shown in Figure 3.14 is assigned the lowest priority number. Therefore, the lines that represent the Space will override the lines that represent the brick and CMU materials as shown in Figure 3.16. The CMU priority number is set to 30 and the Brick to 40; this sets the display of the CMU to override that of the Brick when the walls intersect.

Figure 3.16 *The Brick_CMU_12 wall style*

Edge Offset – The **Edge Offset** button is used to define the location of a wall component relative to the **Baseline Justification** line. To set the edge offset distance, select the **Edge Offset** button and edit the **Component Offset** dialog box as shown in Figure 3.17.

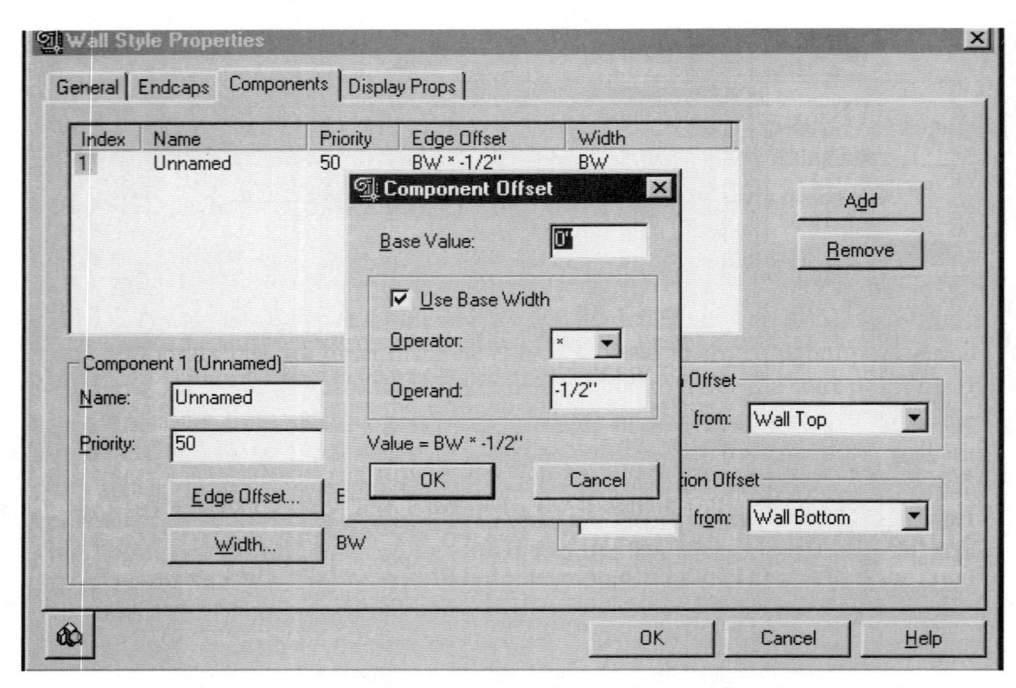

Figure 3.17 *Variable width edge offset defined in the Component Offset dialog box of the Standard wall*

The **Component Offset** dialog box includes options to set the edge offset distance as a fixed width or variable width. Select variable width by checking the **Use Base Width** check box. This method of specifying the edge offset distance is used in the Standard wall style shown above. When the **Use Base Width** box is checked, the edge offset is determined as a function of the wall width. The Base Width value is the value entered in the **Width** edit field of the **Add Walls** dialog box. The settings of the dialog box shown in Figure 3.17 set the **Operator** to multiply and the **Operand** to -1/2. The **Width** from the **Add Walls** dialog box is then multiplied by -1/2 to determine the edge offset distance. Other **Operator** options include addition, subtraction, and division. The **Operand** can be any numeric value. Therefore, if you were drawing the Standard wall style and set the Width to 6", the offset distance would be -3 (6 x − 1/2 − 3).

The variable width method can also be used with a value entered for the **Base Value** edit field. Any value entered in the **Base Value** edit box when the **Use Base Width** check box is selected will be added to arithmetic operations of the **Base Width**. This method adds the **Base Value** to the offset distance determined by arithmetic operations defined with the **Base Width**.

The edge offset can be set to a fixed width. The fixed width method will set the edge offset distance as specified in the **Base Value** edit field when **Use Base Width** is deselected. Therefore the value entered in the **Base Value** edit field will be used as the edge offset distance. This is the method used for the components in the BRICK_CMU_12 wall style as shown in Figure 3.14.

Width – The width of a wall component is determined in a manner similar to the edge offset. Create a fixed width by specifying a width in the **Base Value** edit field and deselecting **Use Base Width**. You can specify a variable width by checking the **Use Base Width** check box. When the **Use Base Width** check box is deselected for the **Width** and **Edge Offset**, the control of the component is set solely in the style definition, and the **Width** edit field of the **Add Walls** dialog box is disabled. The fixed width method is used to set the width of the BRICK_CMU_12 brick component, as shown in Figure 3.18.

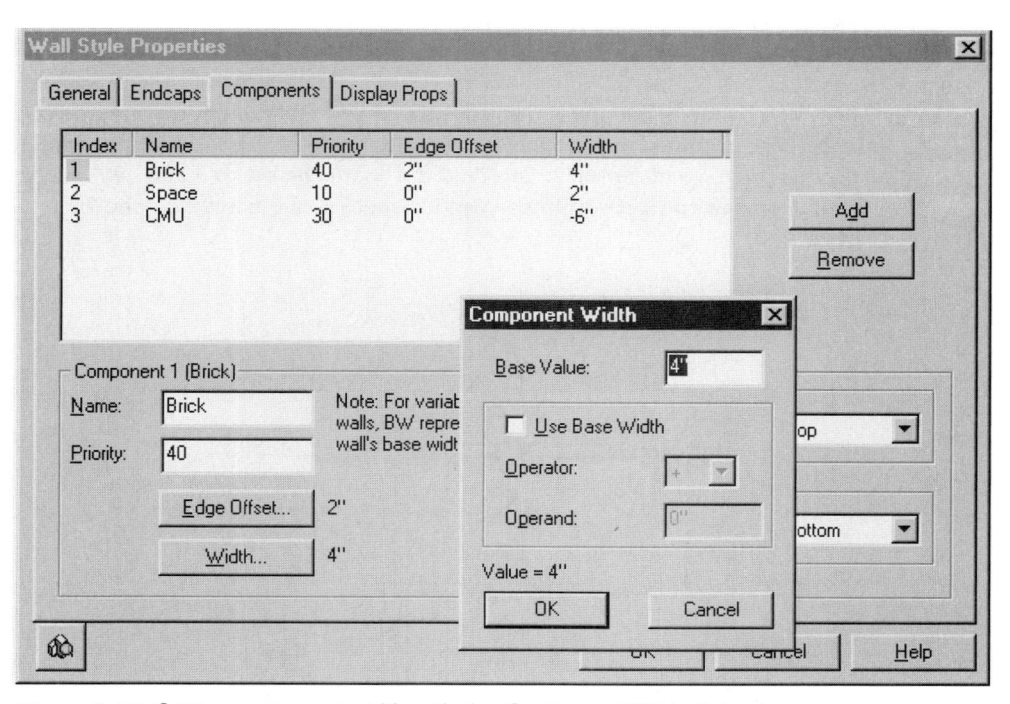

Figure 3.18 *Setting component width with the Component Width dialog box*

Positive edge offset and width dimensions are generated above the baseline when a wall is drawn from left to right, as shown in Figure 3.19. If negative values are used for edge offset and width dimensions, the distances are generated below baseline for a wall drawn from left to right.

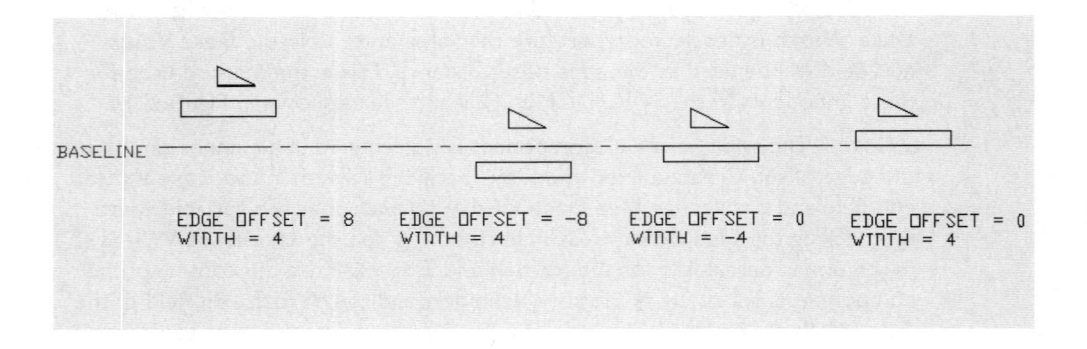

Figure 3.19 *Effects of Edge Offset and Width settings on walls*

Therefore, to represent a wall that consists of a 4" component on one side of the baseline and an 8" component on the other side of the baseline, the edge offset distance could remain at zero and the widths be set to 4 and –8. A wall drawn from left to right with these widths would result in a wall as shown in Figure 3.20.

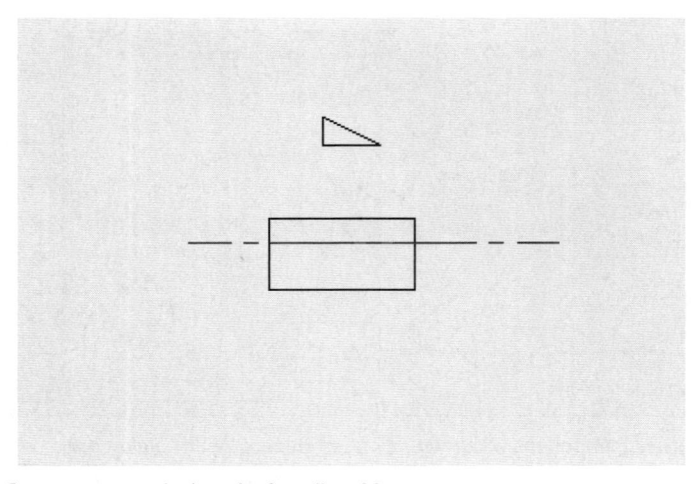

Figure 3.20 *Components with 4 and –8 wall widths*

Top Elevation Offset – The **Top Elevation Offset** section allows you to set the top elevation of a wall component. The distance is defined from the Wall Top, Base Height, Baseline or Wall Bottom. This allows a wall component to have a top elevation different from other components. In Figure 3.21, the 4" wall component has a top elevation of 4' above the wall bottom. This 4" wall component could represent brick veneer placed in front of a masonry wall component.

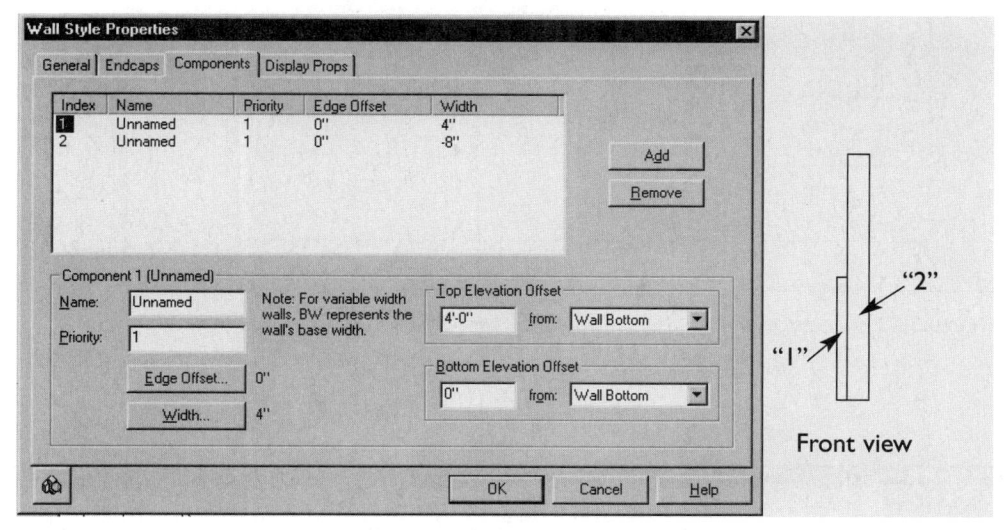

Figure 3.21 *Effects of one component's Top Elevation Offset 4'-0"*

In addition, a wall could be drawn eight feet high with the height of one component restricted down three feet below the top of another wall component. Wall components can be assigned a negative elevation offset distance to reduce the height of a wall component. Wall components cannot extend above the Base Height of the wall. The Base Height of the wall is specified in the **Height** edit field of the **Add Walls** dialog box. Therefore, no component can project above this basic height. Figure 3.22 shows an 8'-high wall. The Top Elevation of the CMU wall component is defined -3' from the Base Height as shown in the **Wall Style Properties** dialog box. The -3' Top Elevation Offset reduces the height of the CMU component relative to the Base Height.

Bottom Elevation Offset – The **Bottom Elevation Offset** section allows you to set the elevation of the bottom of each component. The distance entered can be measured from the Wall Top, Base Height, Baseline, or Wall Bottom. Setting the Bottom Elevation above the wall bottom allows you to create a cavity wall, as shown in Figure 3.23.

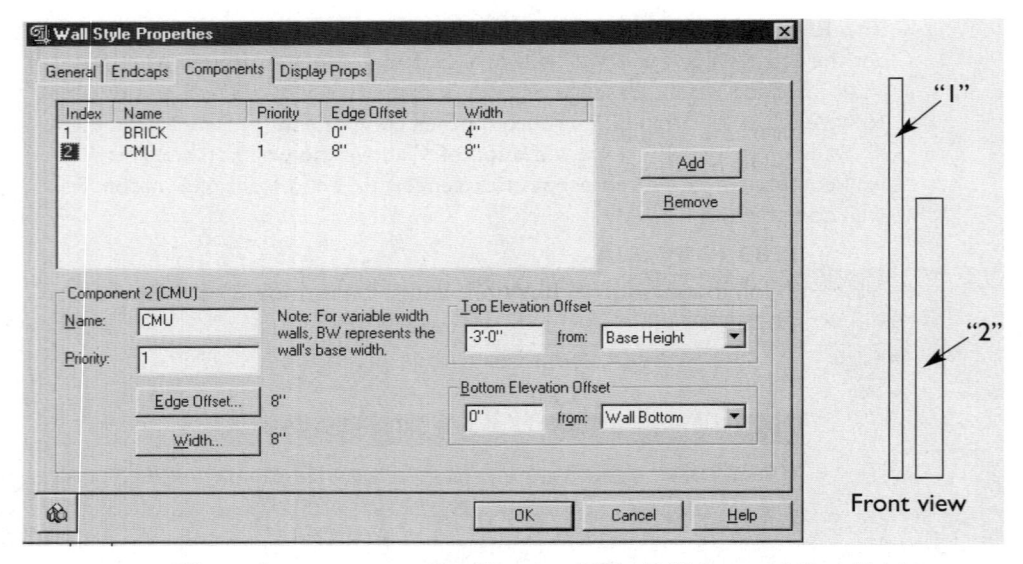

Figure 3.22 *Effects of one component Top Elevation Offset 3'-0" from wall Base Height*

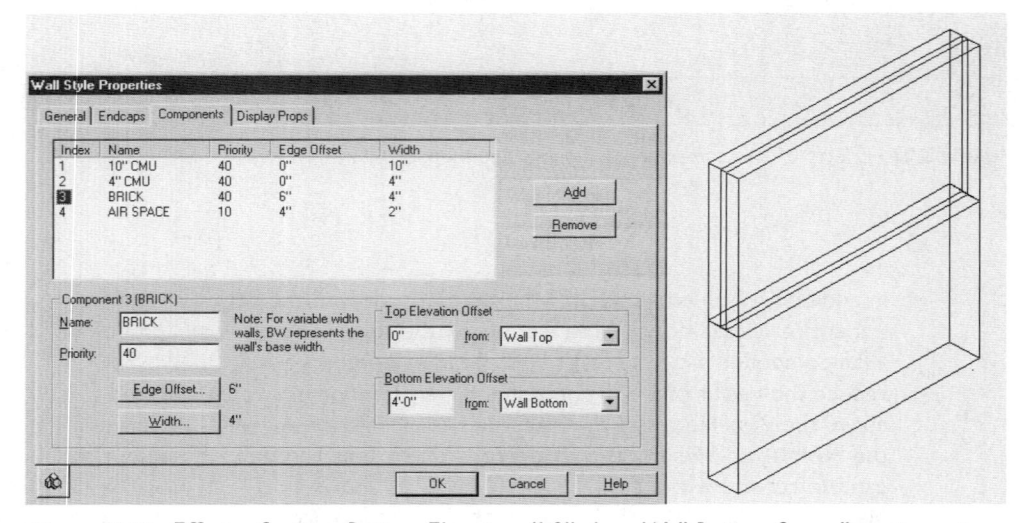

Figure 3.23 *Effects of setting Bottom Elevation 4'-0" above Wall Bottom for wall components*

In the figure shown above, the 10" CMU wall has a **Bottom Elevation Offset** of zero and **Top Elevation Offset** of 4'. The brick, space and 4" CMU components have a **Bottom Elevation Offset** of 4' and a **Top Elevation Offset** of zero. Controlling the elevation of wall components allows the wall style to simulate actual construction.

Display Props

The Display Props tab allows you to set the display of the components of the wall. The **Attach Override, Remove Override,** and **Edit Display Props** buttons allow you to change how an object is displayed in a viewport.

> **Display Representation** – The **Display Representation** drop-down list at the top of the **Display Props** tab lists display representation sets available for an object. The display representation marked by an asterisk is the current display representation set used in the active viewport.
>
> **Property Source** – The **Property Source** column lists object display categories for the current viewport. The **Wall Style** category controls the display according to the definition in the wall style. The **System Default** category displays the wall using global display settings for the current wall.

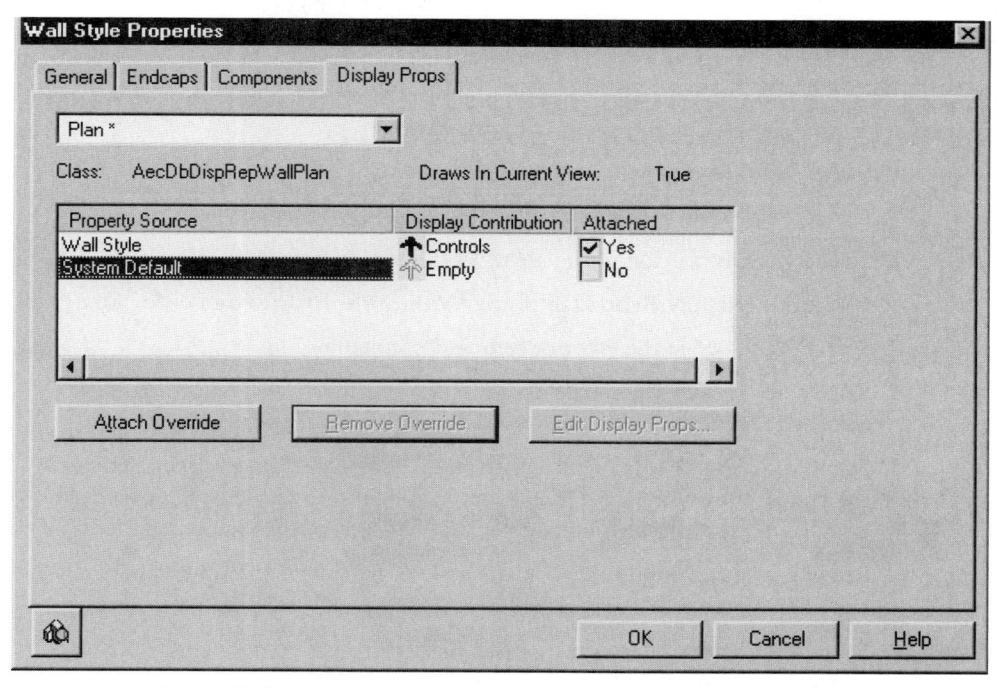

Figure 3.24 *System Default settings for display properties of wall components*

> **Display Contribution** – The **Display Contribution** column indicates the status of the current state of object display. The options of the display contribution are Controls, Empty, and Overridden. Controls indicates that the display of the object is being controlled by that display category. The wall style is controlling the display of the wall object, as shown in Figure 3.24. An Empty

Display Contribution indicates that the display category has not been overridden. Overridden indicates that exceptions to the original display settings have been created.

Attached – The **Attached** column has Yes and No options that indicate whether the display category has exceptions attached as overrides.

If the **System Default** settings are controlling display of the wall style, it will be listed as Controls and the **Attached** column checked as Yes. This condition does not allow you to set the components of a wall unique to that style.

When **System Default** settings are overridden for a wall style, the **Controls Display Contribution** column will be marked with a red X to indicate that the **System Default** is overridden. When **System Default** is overridden, you can set the display properties of a wall style unique to the style. To set display properties of a wall style unique to a wall style, perform the following steps.

STEPS TO OVERRIDING THE SYSTEM DEFAULT DISPLAY REPRESENTATION FOR A WALL STYLE

1. Select **Wall Styles** from the **Walls** toolbar.
2. Select a wall style such as Standard from the **Name** list.
3. Select the **Edit** button of the **Wall Styles** dialog box.
4. Select the **Display Props** tab of the **Wall Style Properties** dialog box.
5. Select Wall Style in the **Property Source** column.
6. Select the **Attach Override** button. (Note: a Red X will indicate that the **System Default** display is overridden, as shown in Figure 3.25.)
7. Select the **Edit Display Props** button to display the **Entity Properties** dialog box as shown in Figure 3.26. Note: Editing the **Entity Properties** dialog box allows you to set visibility, color, and hatching of the wall components.
8. Select the **Hatching** tab to edit the style of hatching for a wall component as shown in Figure 3.27.
9. Select the **Other** tab to edit specific exceptions for component display as shown in Figure 3.28.
10. Select the **OK** button of the **Entity Properties** dialog box to dismiss the dialog box.
11. Select the **OK** button of the **Wall Style Properties** dialog box to dismiss the dialog box.
12. Select the **OK** button of the **Wall Styles** dialog box to dismiss the dialog box.

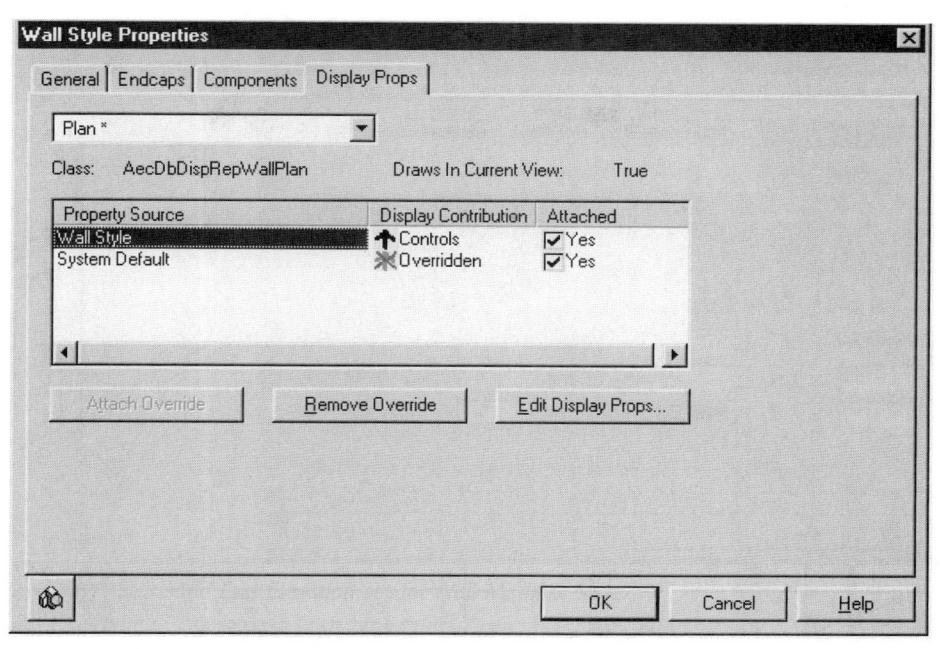

Figure 3.25 *Overriding System Default display properties of wall components*

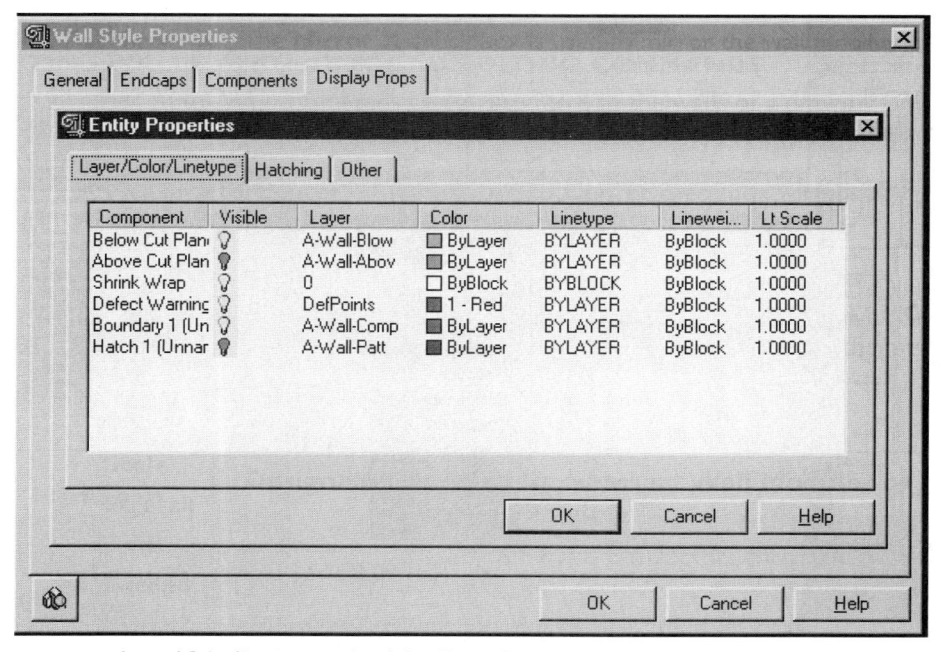

Figure 3.26 *Layer/Color/Linetype tab of the Entity Properties dialog box*

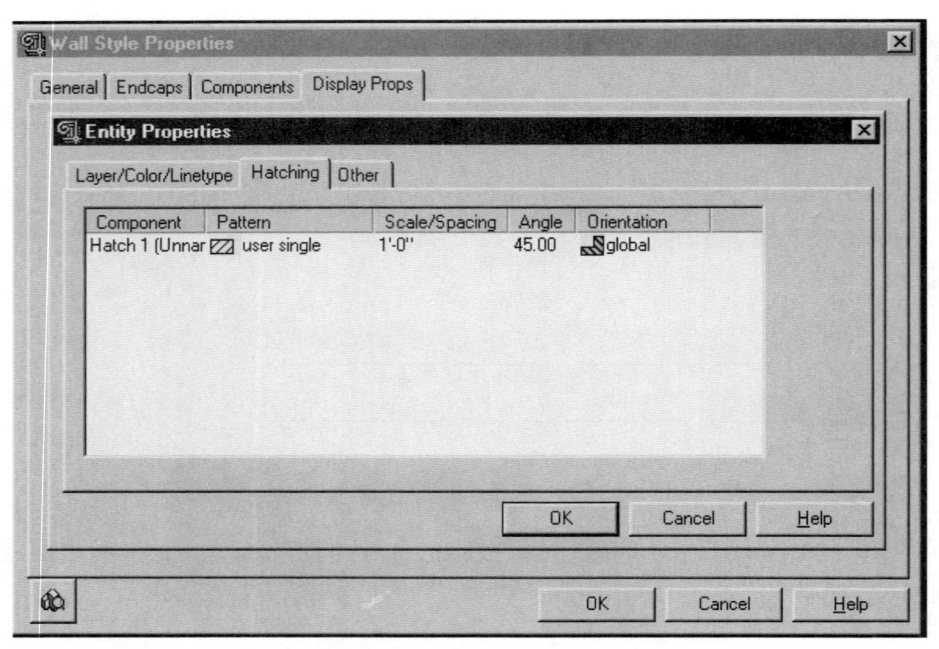

Figure 3.27 *Hatching tab of the Entity Properties dialog box*

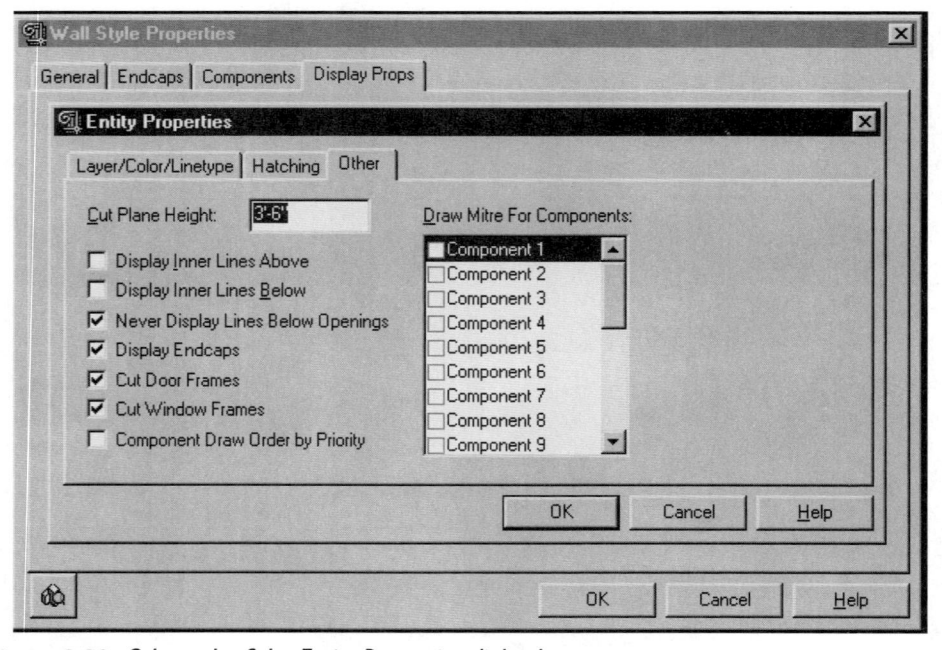

Figure 3.28 *Other tab of the Entity Properties dialog box*

In summary, the **Edit** button of the **Wall Style Properties** dialog box allows you to define endcap styles, wall components, and the display properties of each component of a wall style.

TUTORIAL 3.1 CREATING A WALL STYLE

1. Open *Ex3-0* from the *ADT Tutor\Ch3* directory.
2. Choose **File>SaveAs** from the menu bar and save the drawing as *Ex3-1* in your directory.
3. Select the Work-FLR layout tab.
4. Select **Wall Styles** from the **Walls** toolbar, as shown in Figure 3.29.
5. Select the **New** button of the **Wall Styles** dialog box.

Figure 3.29 *Wall Styles command on the Walls toolbar*

6. Type **Frame_Brick** as the name of the new wall style in the **Name** dialog box.
7. Select the **OK** button to dismiss the **Name** dialog box.
8. Select Frame_Brick as a style name and select the **Edit** button, as shown in Figure 3.30.
9. Type the description **Wood frame brick veneer** in the **Description** edit field as shown in Figure 3.31.

In the following steps you will create two components as described below:

Index	Name	Priority	Edge-Offset	Width
1	Wood	40	0	3.5
2	Brick	40	-5.125	3.625

Table 3.2 *Steps to create two components*

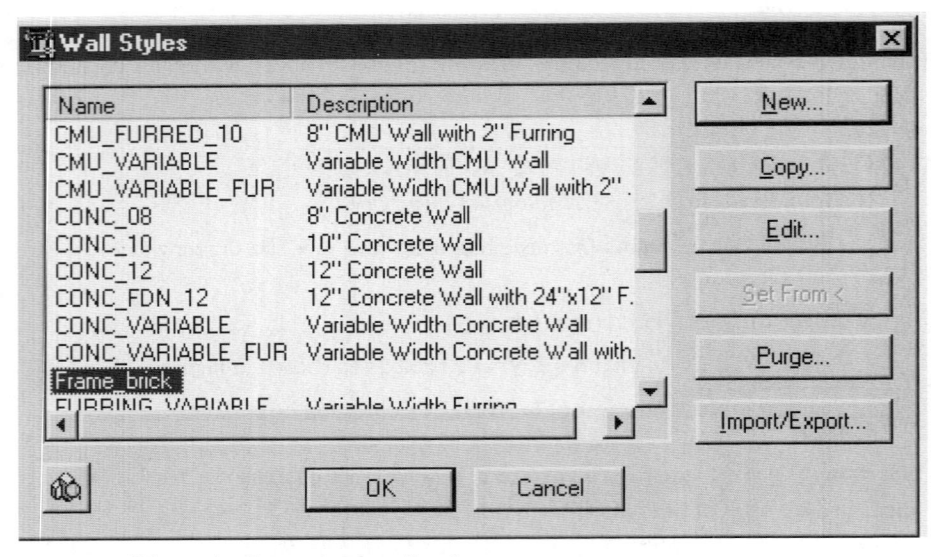

Figure 3.30 *Editing the Frame_brick wall style*

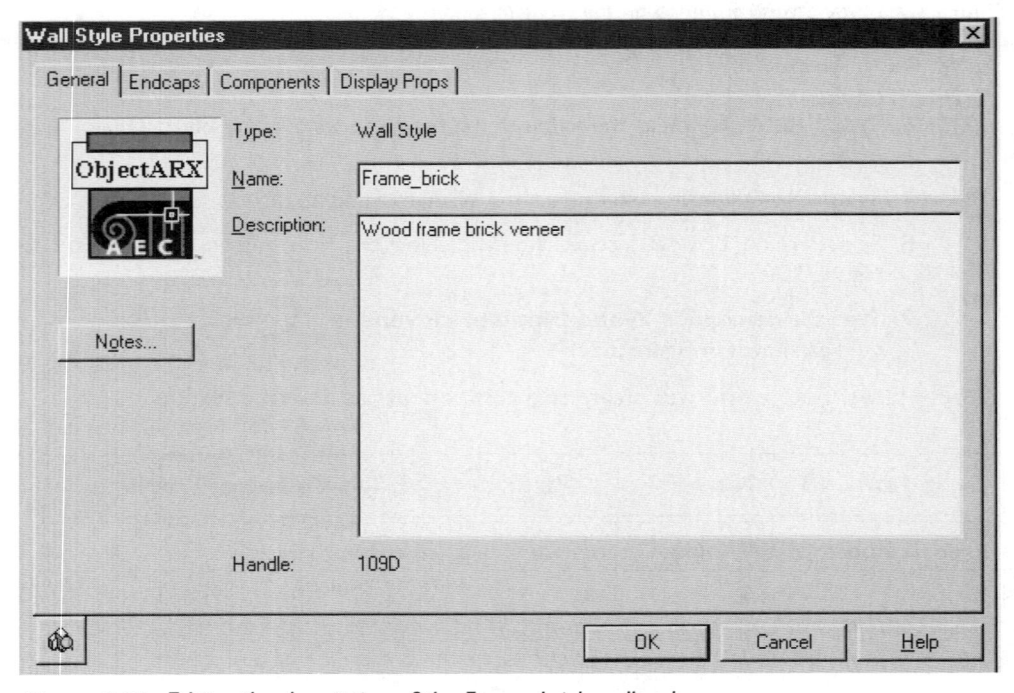

Figure 3.31 *Editing the description of the Frame_brick wall style*

10. To create the Wood component, select the **Components** tab and select **Index 1**. Select Unnamed in the **Name** edit field and type **Wood**, as shown in Figure 3.32.

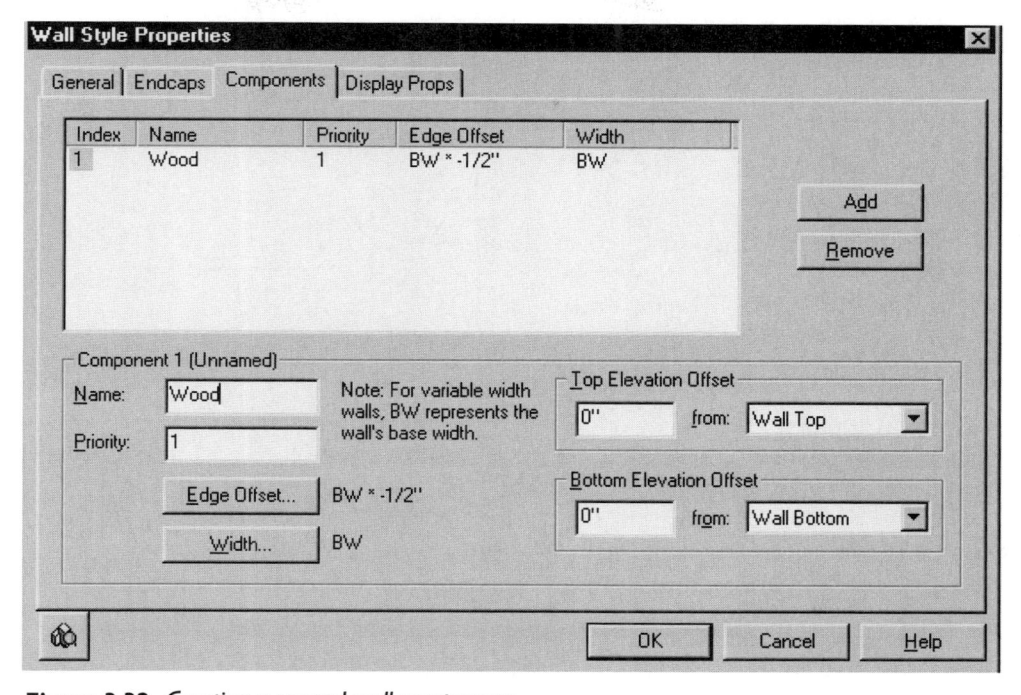

Figure 3.32 *Creating a named wall component*

11. Press the TAB to edit the **Priority** to **40**.
12. Select the **Edge Offset** button, clear the **Use Base Width** check box, and verify that the **Base Value** is set to 0 as shown in Figure 3.33. Select **OK** to dismiss the **Component Offset** dialog box.
13. Select the **Width** button, clear the **Use Base Width** check box, and edit the **Base Value** to 3.5 as shown in Figure 3.34. Select **OK** to dismiss the **Component Width** dialog box.

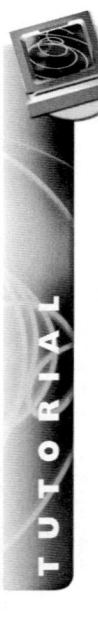

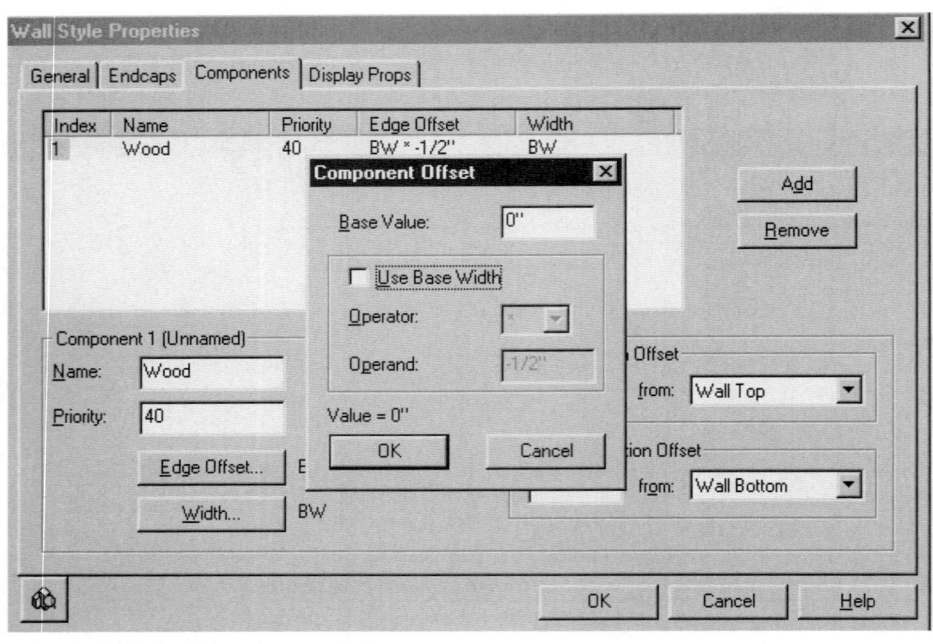

Figure 3.33 *Editing the Component Offset dialog box.*

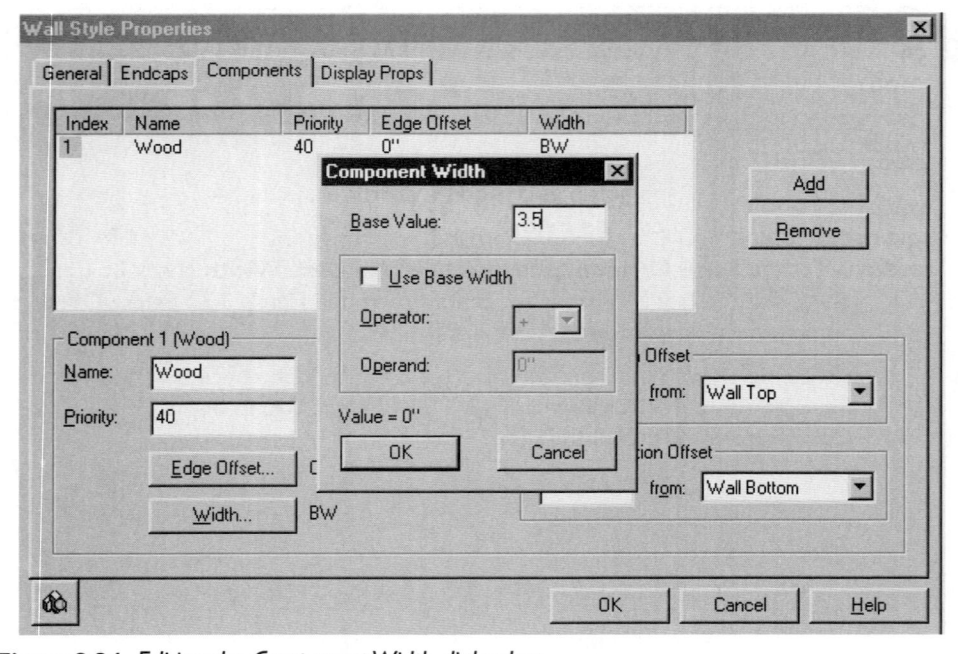

Figure 3.34 *Editing the Component Width dialog box*

14. Select the **Add** button to add another component. Edit the values for **Name**, **Priority**, **Edge Offset** and **Width** as follows.

Index	Name	Priority	Edge-Offset	Width
2	Brick	40	-5.125	3.625

Table 3.3 *Adding another component*

15. Verify that the **Name**, **Priority**, **Edge Offset** and **Width** values are as shown in Figure 3.35.

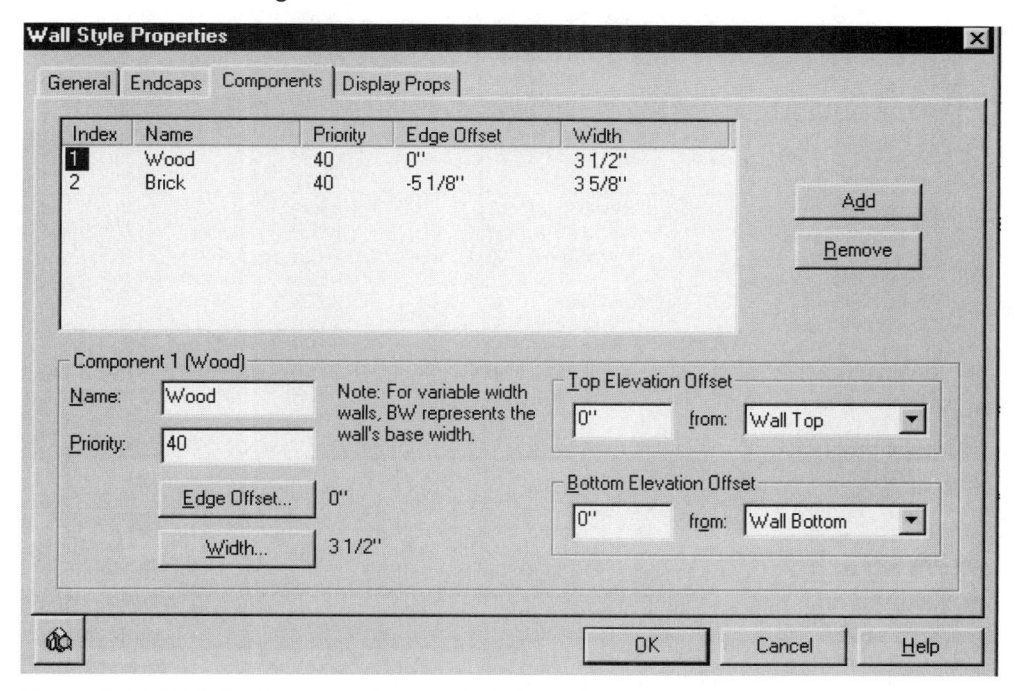

Figure 3.35 *Wall Component settings for the Brick wall component*

16. Select the **OK** button to dismiss the **Wall Style Properties** dialog box.
17. Select the **OK** button to dismiss the **Wall Styles** dialog box.
18. Save the drawing as *Ex3-1.dwg*.

ENHANCING WALL STYLES WITH DISPLAY PROPERTIES

The display properties of wall objects in Architectural Desktop can be used to enhance the appearance of a wall. Hatching can be added or color and linetype altered by the display controls of the wall style. When brick is used as a wall component, the width of the brick is usually hatched with ANSI31 hatch pattern. You can turn hatching on or off by controlling the display properties associated with the wall style. Editing display properties can turn off the display of an entire wall component.

To edit the display of a wall style, you must override the normal properties of the wall style. When you edit the display properties of a wall style, the **Entity Properties** dialog box opens. The **Entity Properties** dialog box for the Frame_Brick wall style created in Tutorial 3.1 is shown in Figure 3.36.

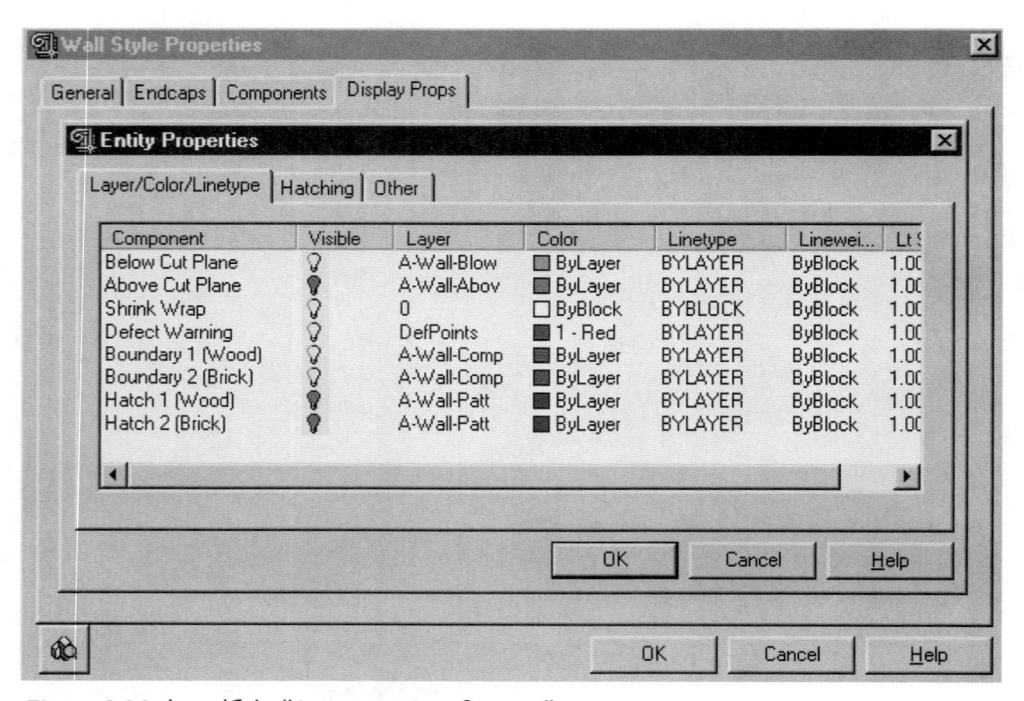

Figure 3.36 *Layer/Color/Linetype settings for a wall component*

The **Entity Properties** dialog box shown in Figure 3.36 is for a wall that consists of only two wall components: wood and brick.

Layer/Color/Linetype

The **Layer/Color/Linetype** tab allows control of visibility, layer, color, linetype, line weight, and linetype scale for each component of the wall object. Components of the wall object are described below.

Below Cut Plane – The Below Cut Plane component consists of the entity representation of the wall located at an elevation below the cutting plane. The default cutting plane elevation is 3'-6". You can alter the height of the cutting plane by editing the **Other** tab of **Entity Properties** dialog box. Turning on visibility turns on entities to represent objects below the cutting plane.

Above Cut Plane – The Above Cut Plane component consist of entities representing the object at an elevation above the cutting plane. In the dialog box shown above, the visibility of components above the cutting plane is turned off.

Shrink Wrap – The Shrink Wrap component is displayed when the **Wall Interference** command is applied to a wall that intersects with other objects. Shrink-wrap entities add emphasis to the intersection of objects.

Defect Warning – The Defect Warning component is the wall error marker displayed when walls are drawn either too short or coincident with other walls.

Boundary – The Boundary is the wall components that represent the width of the wall. In the figure above, Boundary 1 (Wood) and Boundary 2 (Brick) are the wall components created as Index 1 and Index 2. You can turn the display of each component off by turning off the visibility of the component. Wall styles can consist of 20 boundaries.

Hatch – The Hatch component is the hatch pattern for each wall component. Hatch 1 (Wood) and Hatch 2 (Brick) are the hatch patterns applied to each of the wall components. To display hatching of a component, turn on the light bulb in the visibility column.

Hatching

The **Hatching** tab allows you to define the style of hatching used for each hatch boundary. It is displayed for Plan and Reflected display representations. The **Hatching** tab of the Frame_Brick wall style, shown in Figure 3.37, includes the wood and brick components.

The type of hatch pattern, scale, angle, and orientation for the component can be set by the **Hatching** tab. You can edit the hatch pattern by selecting the **Pattern** column for the desired hatch. The default pattern is User Single, which will fill the hatch area with parallel lines. You can select other hatch patterns by selecting the User Single option of the **Pattern** column to display the **Hatch Pattern** dialog box as shown in Figure 3.38.

The **Hatch Pattern** dialog box allows you to select the pattern source from the **Pattern** drop-down list: Predefined, User-Defined, Custom, or Solid Fill. The Predefined option allows access to the AutoCAD hatch patterns. The User-Defined option, the default pattern of Architectural Desktop, will apply a single or double line hatch. The Custom option allows you to select custom hatch patterns from other hatch files, and the Solid Fill option will fill the hatch area with solid shading.

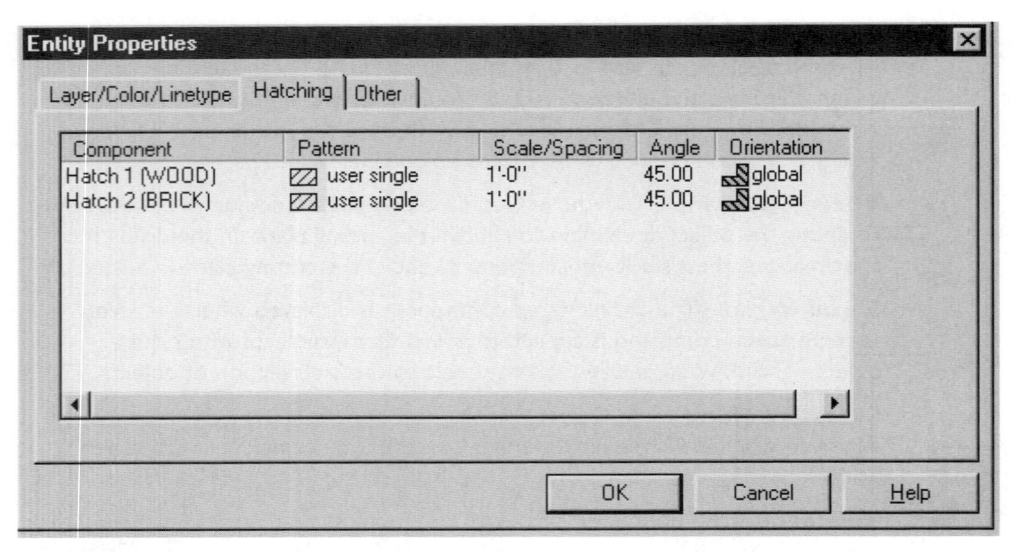

Figure 3.37 *Hatching tab for wall components*

Figure 3.38 *Setting hatch pattern for wall components*

The hatch scale, angle, and orientation can be edited directly in the **Hatching** tab of the dialog box. To edit the scale of the hatch pattern, select the current scale; the scale value will be highlighted and you can overtype the desired value for the scale. The orientation of the hatch can be set to either Object or Global. The Global option will apply the angle of the hatch to all wall segments as if the wall were one continuous area, as shown in Figure 3.39 on the right. The Object orientation of hatch will orient each hatch according to the orientation of the wall. The Object orientation is shown in Figure 3.39 on the left.

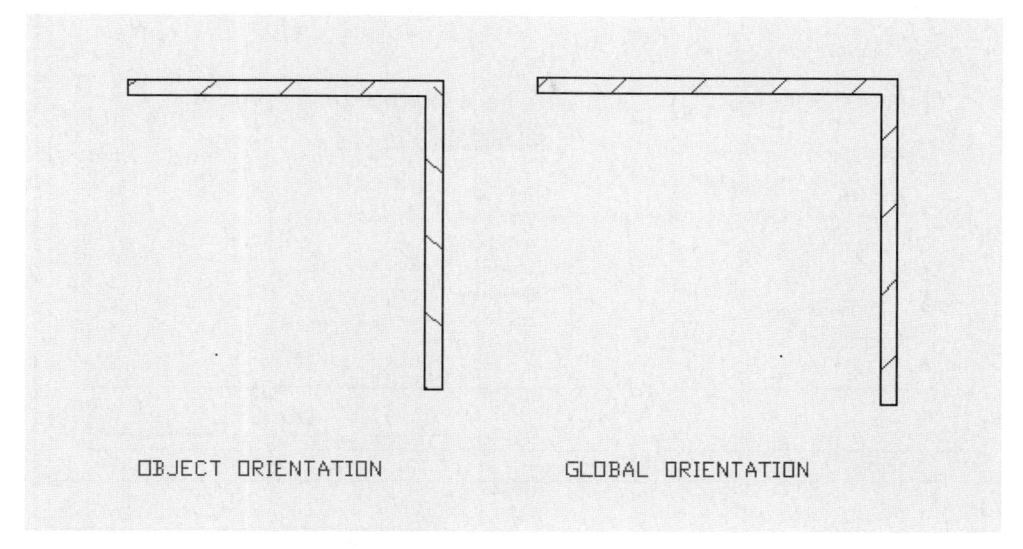

Figure 3.39 *Object and Global hatch orientation*

Other

The **Other** tab, shown in Figure 3.40, allows you to control the elevation of the cutting plane. The elevation of the cutting plane controls the visibility of lines when controlled according to their elevation. Walls below the cutting plane elevation are not displayed in the Work-RCP layout tab. Therefore, setting the cutting plane elevation can turn on or off the visibility of wall lines depending upon their elevation.

The lower portion of the **Other** tab includes check boxes for turning on the visibility of openings, door frames, window frames, and inner wall lines that are cut by the cutting plane. Miter lines can also be checked for each of the wall components.

CHANGING A WALL STYLE WITH WALL MODIFY

The style of a wall can be changed by the **Wall Modify** command (**AECWALLMODIFY**). To change the style of a wall, select the wall; then right-click to display the shortcut menu and choose **Modify Wall**. The **Wall Modify** dialog box will be displayed, allowing you to select another wall style from the style list. The selected wall will be converted to the new wall style. Therefore the **Wall Modify** command allows you to quickly change the style of one or more walls.

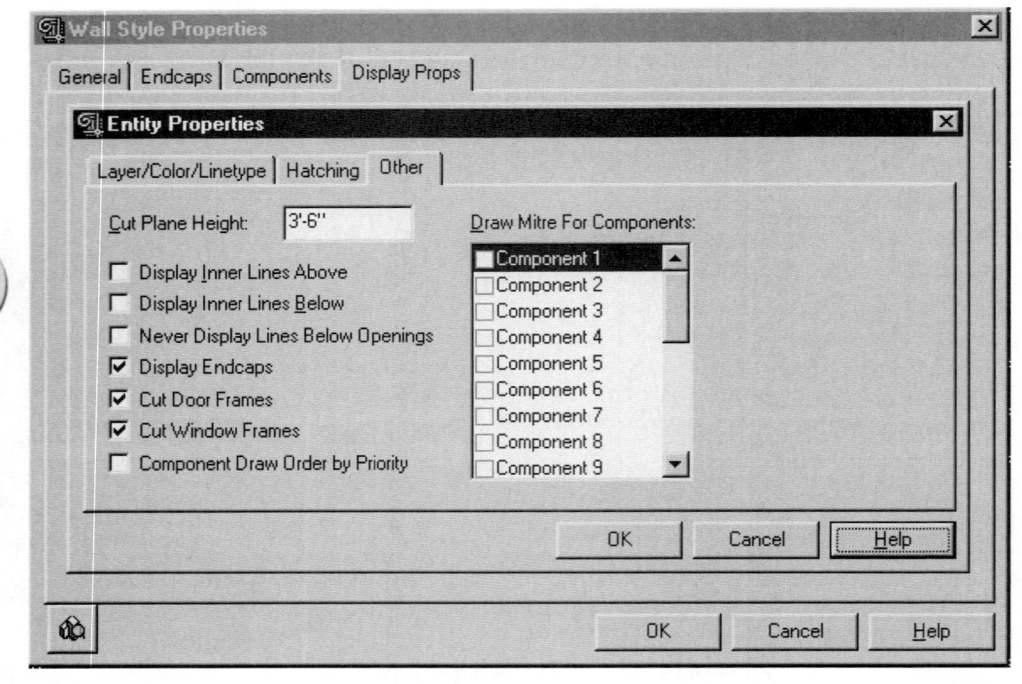

Figure 3.40 *Other tab of Entity Properties dialog box*

TUTORIAL 3.2 CONTROLLING DISPLAY OF WALL COMPONENTS

1. Open *Ex3-1.dwg* from the *ADT_Tutor\Ch3* directory and save the file as *Ex3-2* in your student directory.

2. Select the four exterior walls and then right-click to display the shortcut menu as shown in Figure 3.41.

3. Choose **Wall Modify** from the shortcut menu and select Frame_Brick wall style from the **Style** list; set **Justify** to Baseline; select **OK** to dismiss the **Wall Modify** dialog box.

   (Note: setting wall justification to Baseline will retain overall dimensions located to the outside of the stud.)

   The exterior wall should be displayed with the Frame_Brick wall style as shown in Figure 3.42.

4. Zoom a window from 12,12 to 60,60 to display the front right corner of the building as shown in Figure 3.43.

Advanced Wall Features | 143

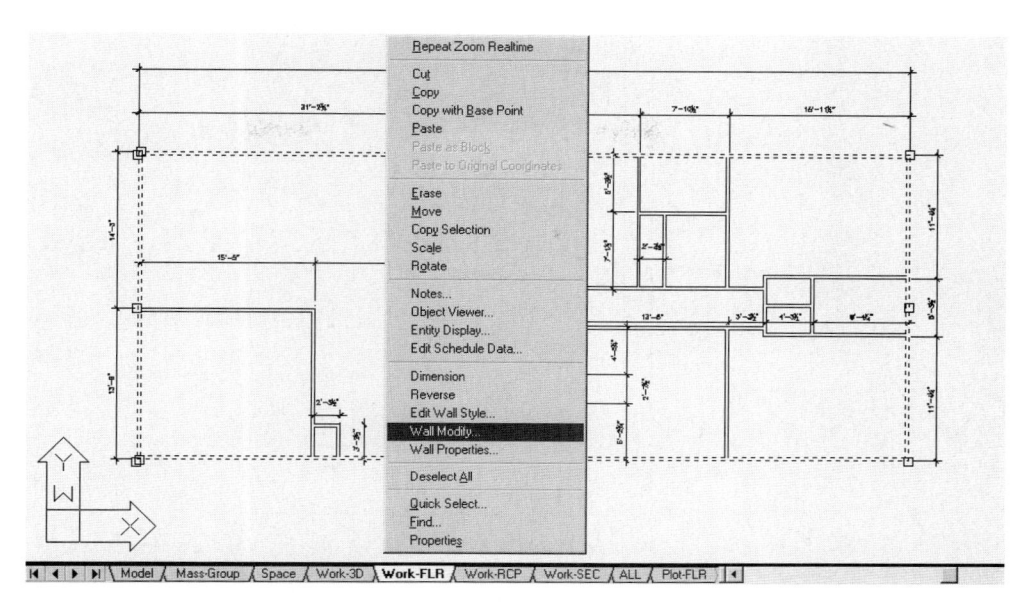

Figure 3.41 *Selecting Wall Modify from the shortcut menu*

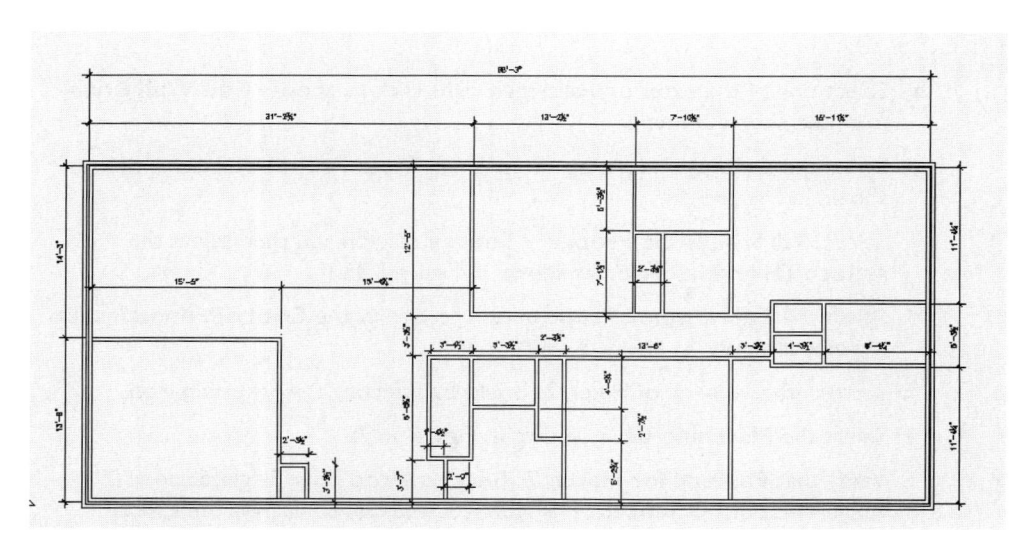

Figure 3.42 *Exterior wall converted to Frame_Brick veneer wall style*

Figure 3.43 *Frame_Brick veneer without hatch*

5. Select one of the exterior walls; then right-click to choose **Edit Wall Style}** from the shortcut menu.
6. Select the **Display Props** tab of the **Wall Style Properties** dialog box, as shown in Figure 3.44.
7. Select Wall Style in the **Property Source** column and then select the **Attach Override** button as shown in Figure 3.45.
8. Select the **Edit Display Props** button to display the **Entity Properties** dialog box as shown in Figure 3.46.
9. Turn on the visibility of Hatch 2 (Brick) by selecting the light bulb icon.
10. Select the **Hatching** tab as shown in Figure 3.47:
11. Verify that **Pattern for Hatch 2** (Brick) is set to User Single, Scale 1'-0", Angle = 45, and **Orientation** is global. Select **OK** to dismiss the **Entity Properties** dialog box.
12. Verify that the **System Default** is Overridden as shown in Figure 3.48.

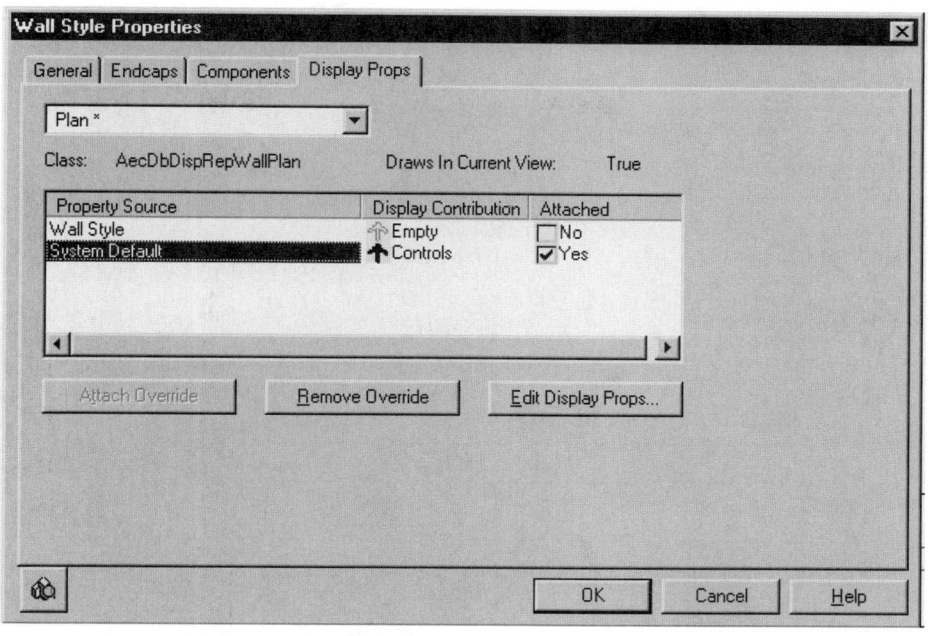

Figure 3.44 *Display Props tab of the Wall Style Properties dialog box*

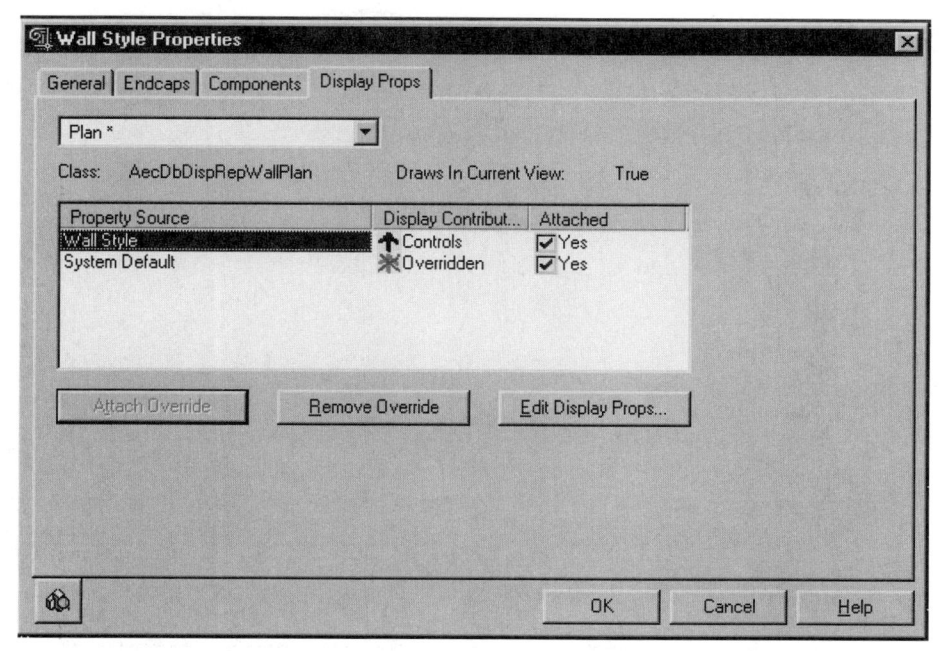

Figure 3.45 *Selecting the Wall Style Property Source for Display Props tab*

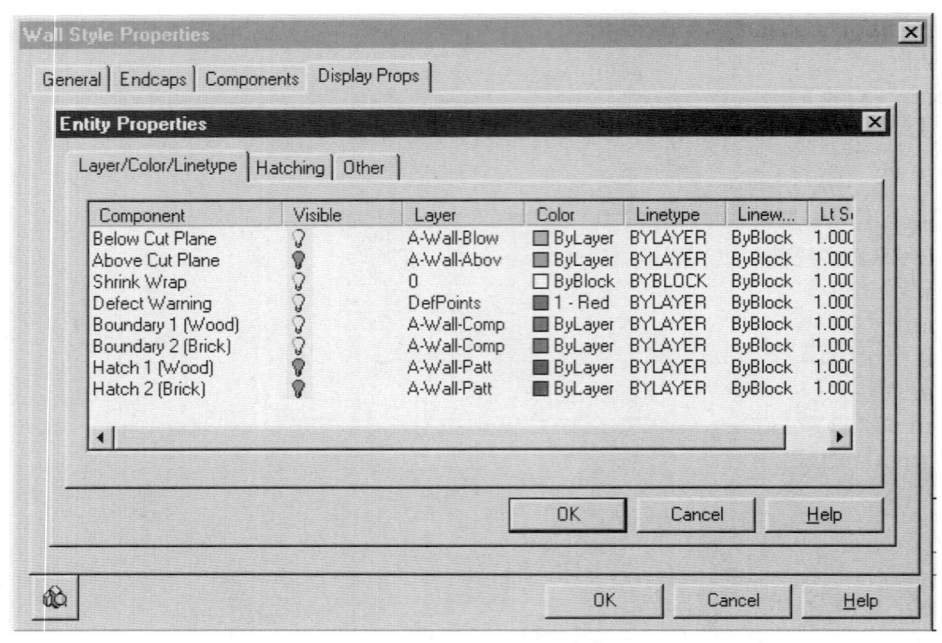

Figure 3.46 *Layer/Color/Linetype tab of Entity Properties dialog box*

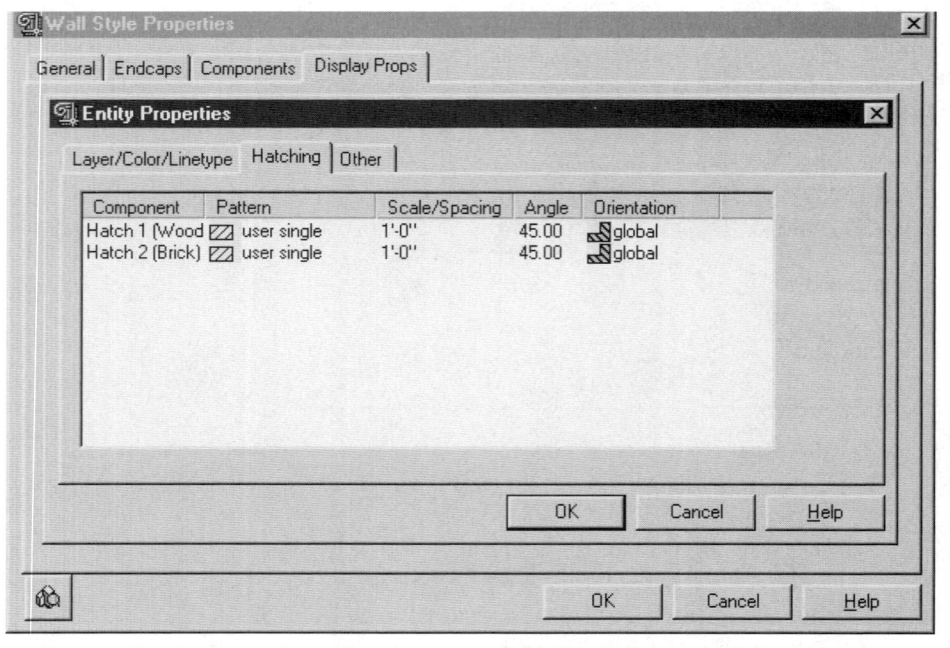

Figure 3.47 *Hatch settings for wall components of the Entity Properties dialog box*

Advanced Wall Features 147

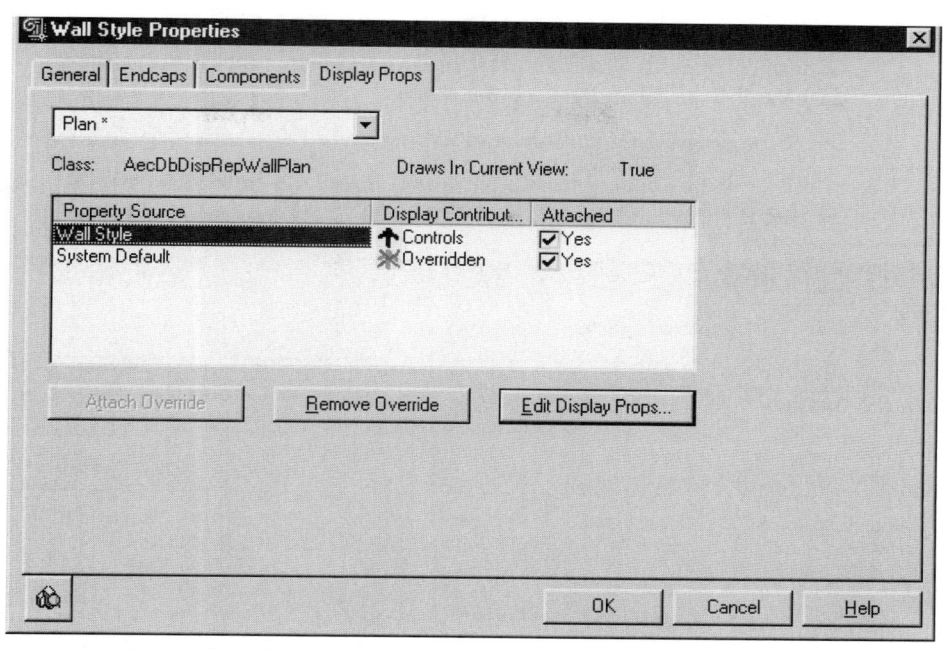

Figure 3.48 *System Default overriden to set wall style display properties*

13. Select **OK** to dismiss the **Wall Style Properties** dialog box. The exterior wall should now display hatching for the brick wall component as shown in Figure 3.49.

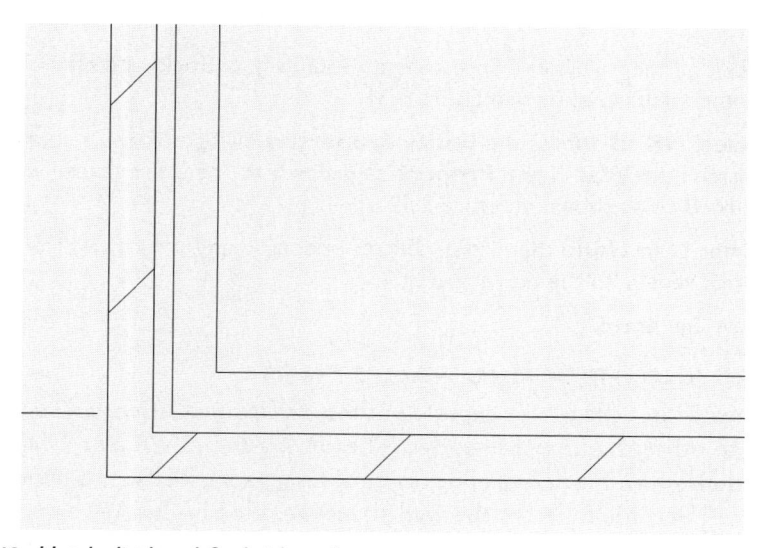

Figure 3.49 *Hatch displayed for brick wall components*

14. Turn off the display of the hatching for the brick veneer by editing the **Display Props** tab. Select an exterior wall and then right-click and choose **Edit Wall Style** from the shortcut menu.
15. Select the **Display Props** tab of the **Wall Style Properties** dialog box.
16. Select the **Edit Display Props** button to display the **Entity Properties** dialog box as shown in Figure 3.50.

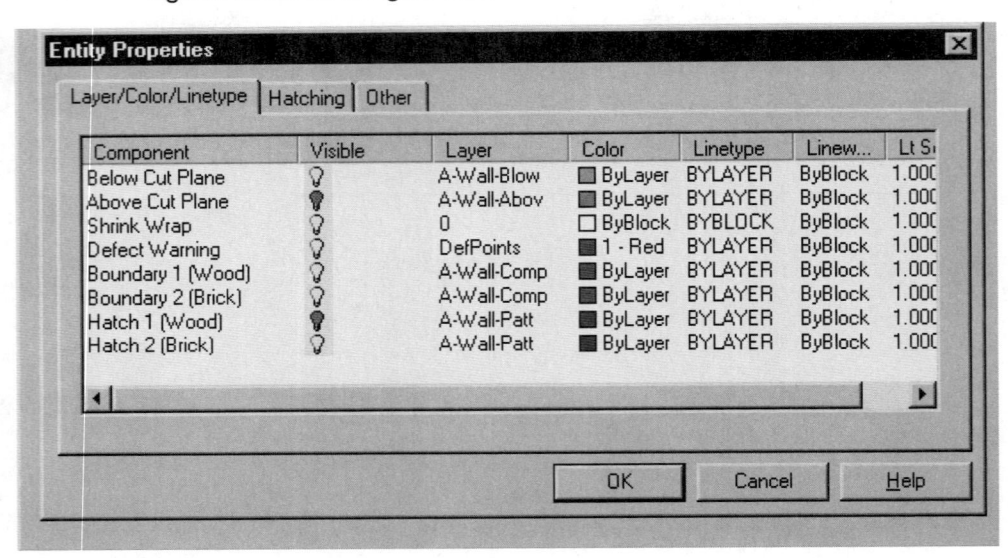

Figure 3.50 Brick wall component hatch turned ON

17. Turn off the visibility of Shrink Wrap, Boundary 2 (Brick) and Hatch 2 (Brick) components as shown in Figure 3.51.
18. Select **OK** to dismiss the **Entity Properties** dialog box and select **OK** to dismiss the **Wall Style Properties** dialog box. The brick veneer visibility is turned off as shown in Figure 3.52.
19. Enter **U** to **Undo** the display control operation and return visibility of the brick veneer to the drawing area.
20. Save the drawing.

EXPORTING AND IMPORTING WALL STYLES

Wall styles created in a file can be exported to the *Architectural2\Content\Imperial\Styles* directory or to other resource directories. The *Architectural2\Content\Imperial\Styles* directory includes files for doors, endcaps, profiles, spaces, walls, and windows. The *Walls (imperial).dwg* file includes the wall styles included in the Aec arch (imperial) template. New wall styles created in a drawing can be saved to the *Walls (imperial).dwg*

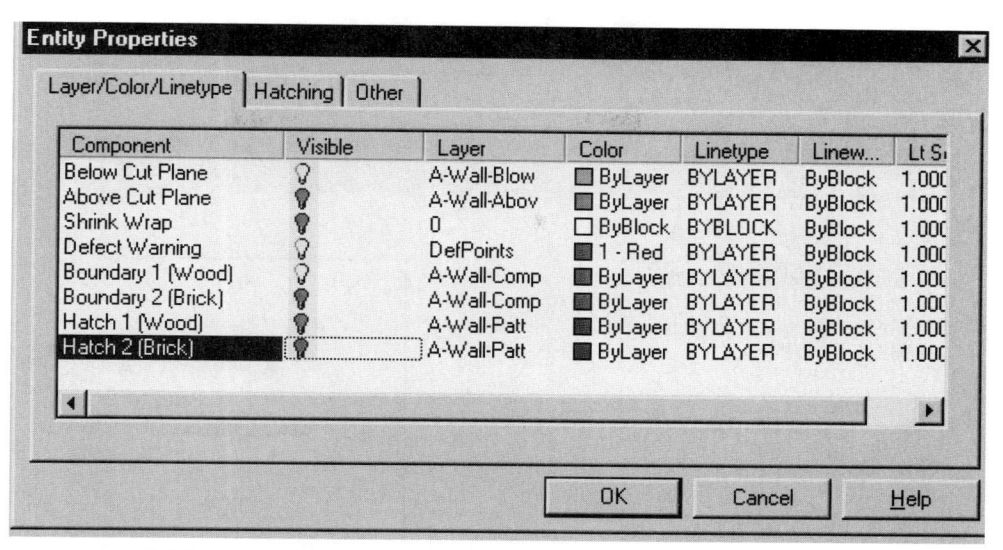

Figure 3.51 Wall component Boundary 2 and Hatch 2 turned OFF

Figure 3.52 Brick veneer wall component turned OFF

file or saved to another new file. The wall styles of the *Walls (imperial).dwg* can be imported to other files.

Export wall styles from a drawing by selecting the **Import/Export** button of the **Wall Styles** dialog box. Selecting this button opens the **Import/Export** dialog box as shown in Figure 3.53.

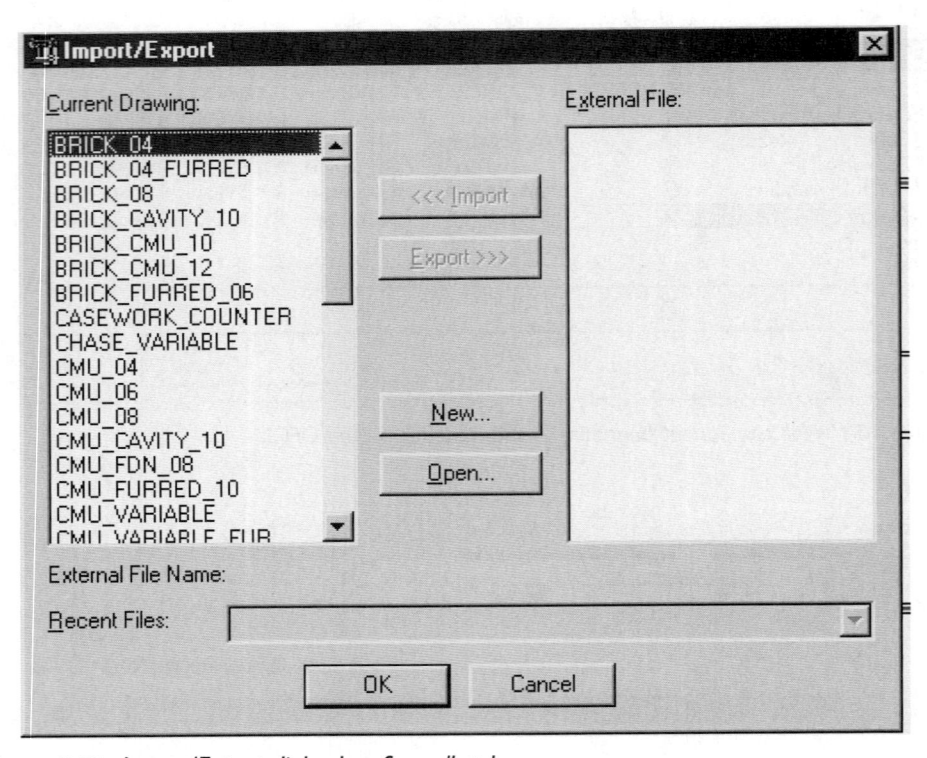

Figure 3.53 *Import/Export dialog box for wall styles.*

The options of this dialog box are as follows:

> **Import** – The **Import** button allows you to copy selected wall styles from the external file to the current file.
>
> **Export** – The **Export** button allows you to copy selected wall styles from the current file to an external resource file.
>
> **New** – The **New** button opens the **New Drawing File** dialog box to create a new file for wall styles.
>
> **Open** – The **Open** button opens the **File to Import From** dialog box, allowing you to open a file for import/export.

Current Drawing – The **Current Drawing** list includes wall styles that exist in the current drawing.

External File – The **External File** list includes the wall styles that exist in the file opened by the **File to Import From** dialog box.

Once a file is opened or created, wall styles can be selected for import or export. If a file that is opened contains wall styles, they will be listed in the **External File** list. If you select wall styles from the **Current Drawing** list, the **Export** button becomes active and allows you to export the selected style to the external file. If a style from the **External File** list is selected, the **Import** button becomes active and you can import the wall style from the external file to the current file. To select all styles from the list, select the first style; hold down SHIFT and select the last desired wall style. To select more than one wall style from the list of wall styles, hold CTRL down and select the desired wall styles.

If wall styles of a drawing are selected for export to a file that contains the same named wall styles, the **Import/Export - Duplicate Names Found** dialog box will open, displaying the duplicate names. There are three options: **Leave Existing**, **Overwrite Existing**, and **Rename to Unique**, as shown in Figure 3.54.

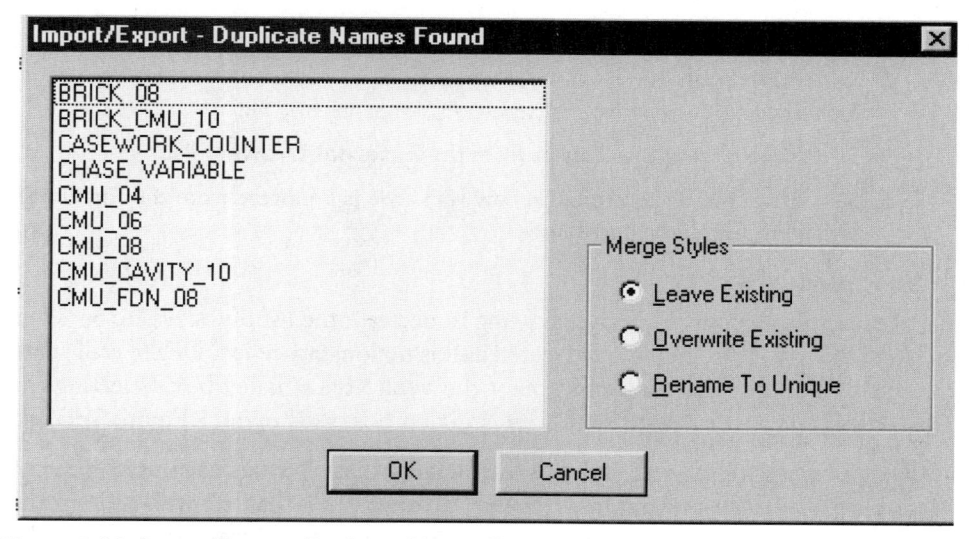

Figure 3.54 *Import/Export - Duplicate Names Found dialog box*

The **Leave Existing** option will not replace the wall style of the destination drawing with that of the source file. The **Overwrite Existing** option revises the wall style definition of the destination file with that of the wall style of the source file. The **Rename to Unique** option will append a number such as 2, 3, or 4 to the wall style name from the source file to distinguish it from the wall style of the destination drawing.

In summary, the steps to exporting and importing walls styles are as follows:

STEPS TO EXPORTING A WALL STYLE

1. Select the **Wall Style** command (**AECWALLSTYLE**) from the **Walls** toolbar.
2. Select the **Import/Export** button of the **Wall Styles** dialog box.
3. Select the **Open** button and open the *Architectural2\Content\Imperial\Styles\Walls(Imperial).dwg* file.
4. Select one or more wall styles from the **Current Drawing** list.
5. Select the **Export** button. The new wall style is exported from the current drawing to the external file.

STEPS TO IMPORTING A WALL STYLE

1. Select the **Wall Styles** command (**AECWALLSTYLE**) from the **Walls** toolbar.
2. Select the **Import/Export** button of the **Wall Styles** dialog box.
3. Select the **Open** button and open the *Architectural2\Content\Imperial\Styles\Walls(Imperial).dwg* file.
4. Select one or more wall styles from the **External Drawing** list.
5. Select the **Import** button. The new wall style is imported from the **External Drawing** list to the current file.

PURGING WALL STYLES

Wall styles can also be purged from a drawing to decrease the list of wall styles or delete a wall style that has been incorrectly created and is no longer needed. Delete wall styles from a file by selecting the **Purge** button of the **Wall Styles** dialog box. Selecting the **Purge** button opens the **Purge Wall Styles** dialog box, as shown in Figure 3.55.

This dialog box lists all the wall styles of the drawing that have not been used. You can deselect a wall style for purging by clearing the check box for that wall style in the **Purge Wall Styles** dialog box. If you want to delete all unused wall styles, select the **OK** button, and all unused wall styles will be deleted from the drawing.

Advanced Wall Features | 153

Figure 3.55 *Wall styles listed in the Purge Wall Styles dialog box*

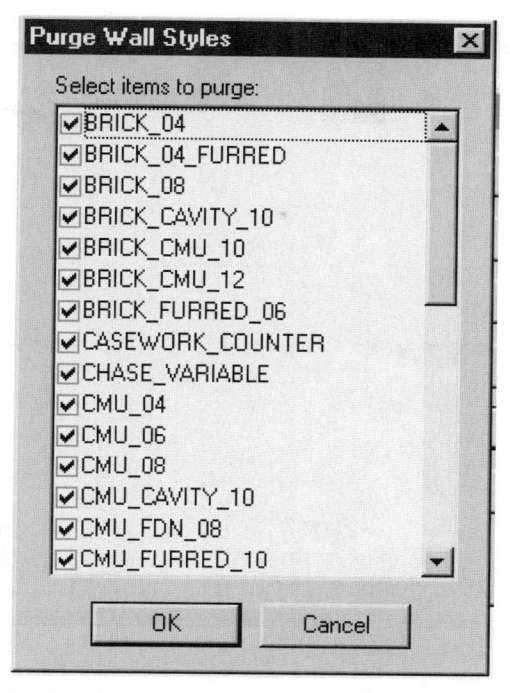

TUTORIAL 3.3 IMPORTING AND EXPORTING WALL STYLES

1. Open *Ex3-2.dwg* from the *ADT_Tutor\Ch3* directory and save the drawing as *Ex3-3.dwg* in your student directory.
2. Select the **Wall Styles** command from the **Walls** toolbar.
3. Select the **Import/Export** button in the **Wall Styles** dialog box.
4. Select the **New** button to open the **New Drawing File** dialog box and create a *Wall Residential.dwg* file in the AutoCAD Architectural2\Content\Imperial\Styles directory.
5. Select the **Save** button to dismiss the **New Drawing File** dialog box.
6. Select the Frame_Brick wall style from the **Current Drawing** list as shown in Figure 3.56.

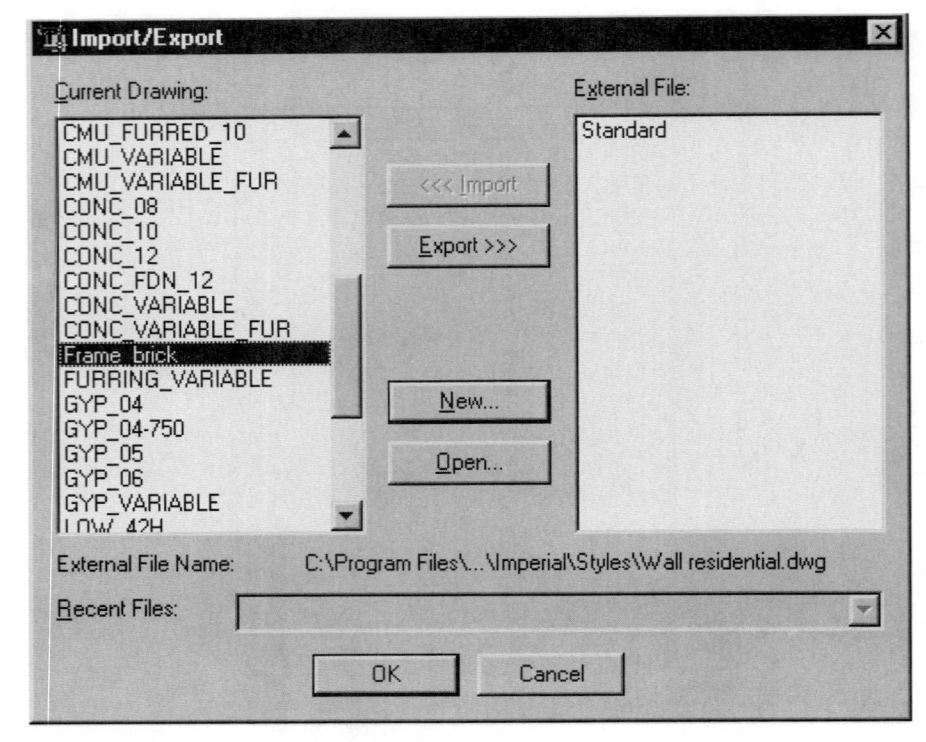

Figure 3.56 Frame_Brick wall style selected for Export

7. Select the **Export** button. and the Frame_Brick style will appear in the **External File** list on the left as shown in Figure 3.57.
8. Select the **OK** button to dismiss the **Import/Export** dialog box. The AutoCAD warning box will appear as shown in Figure 3.58 to confirm whether you want to save the changes made in the new file; select Yes to save the changes.
9. Save the drawing.

CREATING ENDCAPS

When openings are created in a wall or when a wall terminates, the treatment of the ends is controlled through endcap styles. The Standard endcap style terminates the wall with a single line perpendicular to the wall. Other endcap styles available in the Aec arch (imperial) template include Cavity_10, Cavity_12, CMU_FDN_08, and CMU_FDN_12. The Cavity endcaps are designed to cap the wall with brick veneer or other cavity wall construction. The cavity endcaps styles are used on the Cavity wall

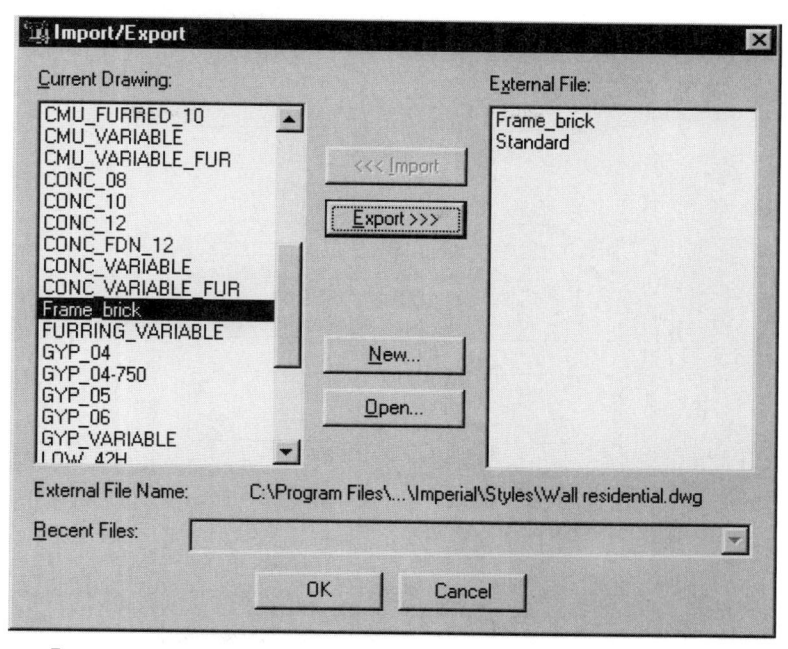

Figure 3.57 Frame_Brick veneer wall style exported to the Wall Residential drawing

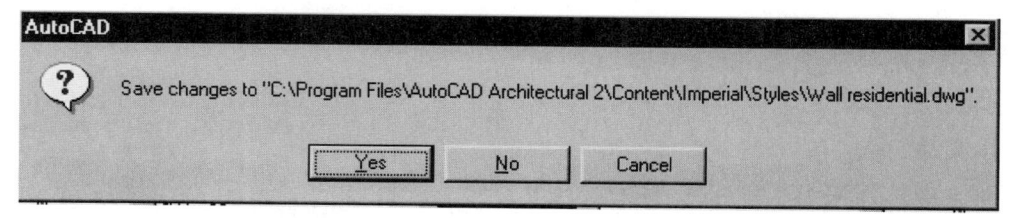

Figure 3.58 AutoCAD warning message box prior to saving wall style

styles. The CMU_FDN_08 includes an endcap that forms the footing with the wall.

The walls shown in Figure 3.59 are wall styles and endcaps included in the Aec arch (imperial) template. The first wall is created from the Standard wall style using the Standard endcap style, which terminates the wall with a plane perpendicular to the length of the wall.

The second wall is the Brick_Cavity_10 wall style with the Cavity_10 wall endcap style. This endcap style closes the cavity of the 4" brick wall component. The third wall consists of the Conc_FDN_12 wall style with the CONC_FDN_12 endcap style. The endcap closes the wall with a chamfer. The wall consists of a 12" concrete wall with

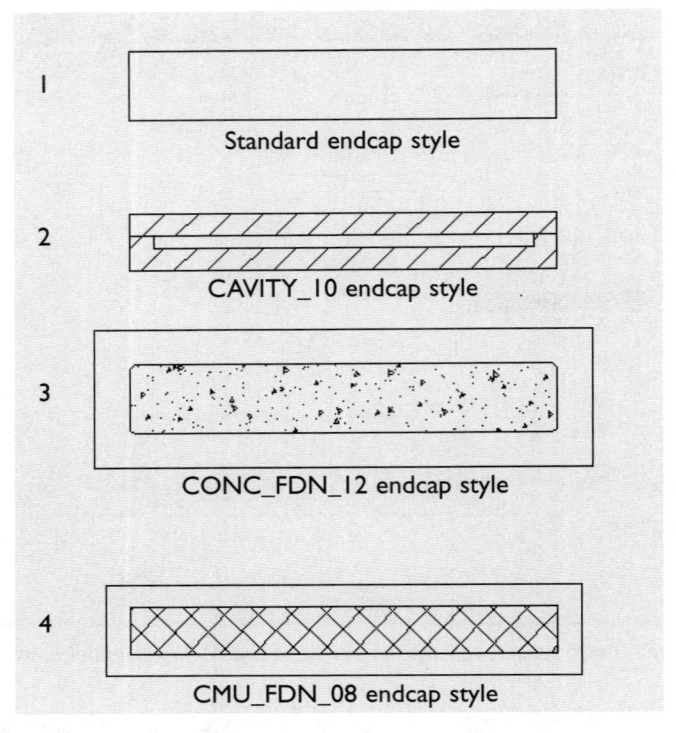

Figure 3.59 Endcap styles included in the Aec arch (imperial) template

a 24" x 12" footing. The fourth wall consists of the CMU_FDN_08 wall style and the CMU_FDN_08 endcap style. This endcap style creates an extension of the footing beyond the wall. The footing is created as a separate wall component, and the endcap design extends the footing wall beyond the masonry wall.

These four examples are applications of the endcap style to enhance the wall design. Endcaps are created from polylines and assigned to the components of a wall. An endcap can be used with various wall styles; however, walls with multiple components should be capped with endcaps linked to each wall component.

Access **Endcap Styles** as shown in Table 3.4.

Menu bar	Design>Walls>Wall Tools>Wall Endcap Styles
Command prompt	WALLENDCAPSTYLE
Toolbar	Select Endcap Styles on the Walls toolbar shown in Figure 3.60
Shortcut	Right-click over the drawing area and choose Design>Walls>Endcap Styles

Table 3.4 *Accessing Endcap styles*

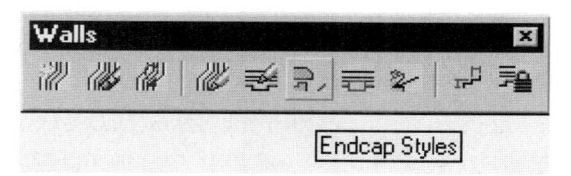

Figure 3.60 *Endcap Styles command on the Walls toolbar*

When the command is executed, the **Endcap Styles** dialog box will open as shown in Figure 3.61.

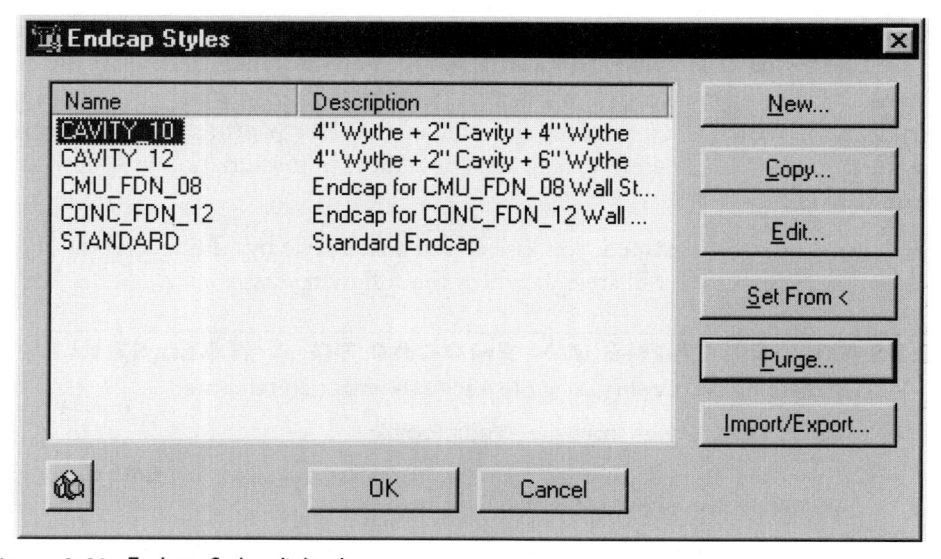

Figure 3.61 *Endcap Styles dialog box*

The **Endcap Styles** dialog box is similar to the **Wall Styles** dialog box and consists of the following buttons:

New – Used to create a new name for an endcap style.

Copy – Used to copy an endcap style definition and assign a name to the new endcap style.

Edit – Used to edit the name and description of an endcap style.

Set From< – Enables you to select polyline geometry to define the endcap style.

Purge – Used to remove unused endcap styles from the drawing.

Import/Export – Used to open the **Import/Export** dialog box, allowing you to copy endcap styles to a resource file or to copy endcap styles from a resource file to the current drawing file.

The resource files for endcaps are located in the *\AutoCAD \Architectural2\Content\ Imperial\Styles\Endcaps.dwg*. Endcap styles are imported to and exported from files in a procedure similar to that for wall styles.

To create a new endcap, first draw a polyline in the shape of the endcap. Use the **Set From<** button of the **Endcap Styles** dialog box to select the polyline and assign it to a wall component. The polyline does not have to be drawn the actual size of the wall. The shape of the polyline can be applied to walls of different widths. However, the ends of the polyline should align vertically because the shape is applied to the end of the wall. Shown in Figure 3.62 are two polylines that were used to form endcaps. Adjacent to the endcaps are examples of walls created that use the endcap style. The polyline that capped the wall with unequal length polylines in essence bent the end of the wall when the endcap style was attached. Note that geometries of the endcaps differ in size; however, when attached to a wall of the same width, the endcap geometry adjusts to the wall. Therefore, endcap styles adjust to the variation in wall widths to close the wall.

Once the endcap style is created, you assign it to a wall style by editing the wall style. To assign an endcap to a wall style, perform the following steps:

STEPS TO ASSIGNING AN ENDCAP TO A WALL STYLE

1. Create a new drawing using the Aec Arch (imperial) template.
2. Select **Wall Styles** from the **Walls** toolbar.
3. Select the desired wall style from the Name list and select the **Edit** button of the **Wall Styles** dialog box.
4. Select the **Endcaps** tab as shown in Figure 3.63.

Advanced Wall Features   159

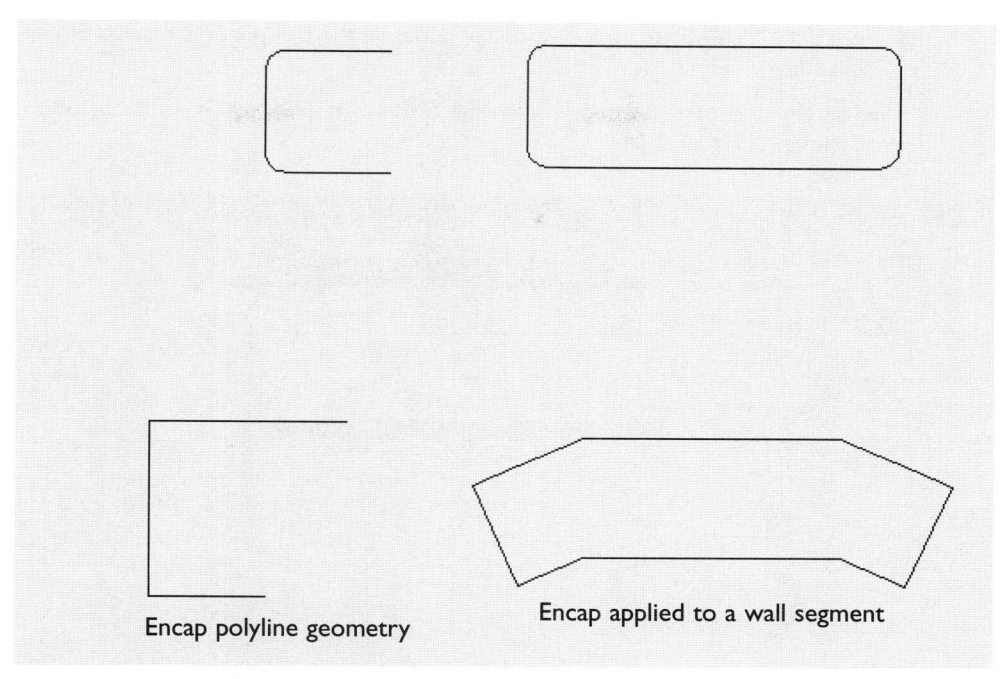

Encap polyline geometry   Encap applied to a wall segment

Figure 3.62 *Applications of endcaps to walls*

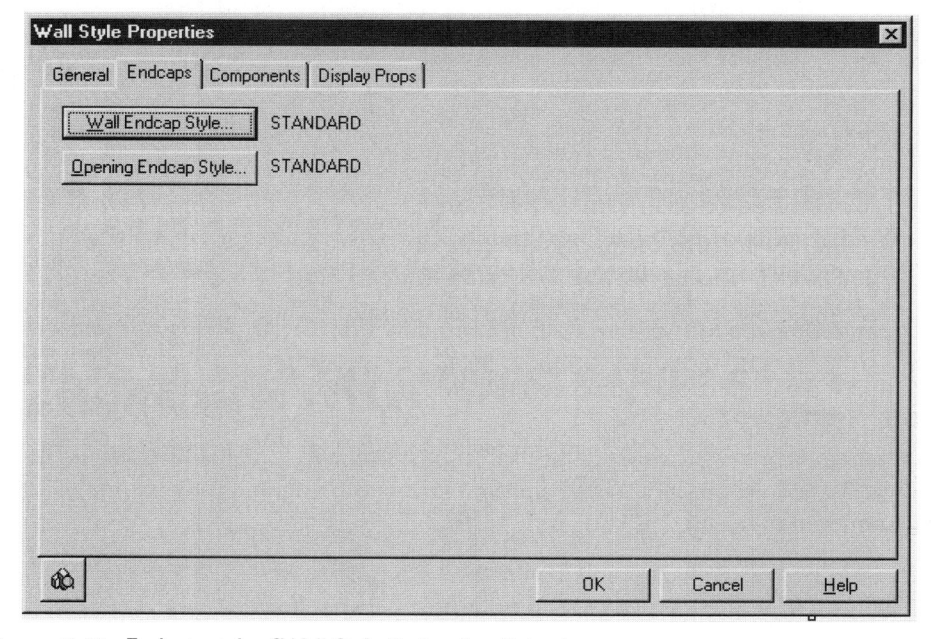

Figure 3.63 *Endcaps tab of Wall Style Properties dialog box*

5. Select the **Wall Endcap Style** button to open the **Select an Endcap Style** dialog box.
6. Select the desired endcap style from the **Select an Endcap Style** dialog box as shown in Figure 3.64.

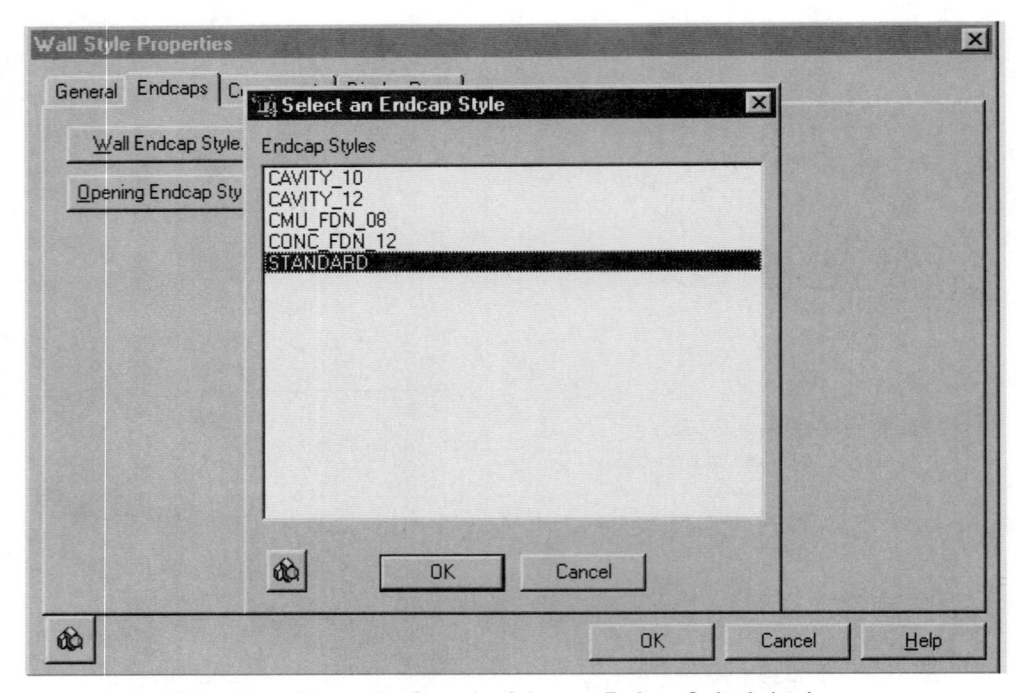

Figure 3.64 *Selecting endcap styles from the Select an Endcap Style dialog box*

7. Select **OK** to dismiss the **Select an Endcap Style** dialog box.
8. Select **OK** to dismiss the **Wall Style Properties** dialog box.
9. Select **OK** to dismiss the **Wall Styles** dialog box.

MODIFYING ENDCAPS

If an existing endcap style exists and needs to be modified, the **WALLENDCAP** command can be used to insert the endcap style in the drawing as a polyline for editing. This command is selected as shown in Table 3.5.

Menu bar	Design>Walls>Wall Tools>Insert Endcaps as Polyline
Command Prompt	WALLENDCAP

Table 3.5 *Selecting the WALLENDCAP command*

NOTE: If you are accessing the command through the menu bar, the **Endcap Styles** dialog box will open. The desired endcap style can then be selected and you will be prompted to enter the insertion point of the Pline.

If this command is selected from the command line, the command has two options: **Define** and **as Pline**. You must select the **as Pline** option to insert the endcap in the drawing. The **Define** option can be used to create an endcap style, whereas the **as Pline** option will insert the selected endcap as polyline geometry in the drawing. When the endcap is inserted as a polyline, the geometry can be edited and saved as an endcap. You can save the endcap using a new name. However, if the revised geometry is saved with the same endcap name as that of the original endcap, all endcaps with the same name will be updated to the new geometry definition.

The fourth wall shown in Figure 3.59 is the CMU_FND_08 wall style. This wall style assigns CMU_FND_08 endcap to the wall. This endcap is shown in Figure 3.65 inserted in the drawing as a polyline.

This endcap consist of the two components shown on the right. This endcap is attached to the CMU_FND_08 wall. The components of the CMU_FND_08 wall are shown in the **Components** tab of the **Wall Style Properties** dialog box in Figure 3.66.

The Index 1 component named CMU is capped by the geometry shown in the lower portion of Figure 3.65. The Index 2 component named Footing is capped by the larger "C" shaped geometry shown on the right in the upper portion of Figure 3.65. The CMU_FND_08 endcap attaches both polyline shapes to the end of the wall.

You can modify the CMU_FND_08 endcap by editing the original polyline geometry and saving the endcap using the same name, CMU_FND_08. This technique will automatically update all walls that used the CMU_FND_08 endcap.

OVERRIDING ENDCAPS AND PRIORITY OF WALL STYLES

Endcap styles and priority of wall components assigned in a wall style are applied to all walls placed in the drawing. However, the **Wall Properties** command (**WALL-**

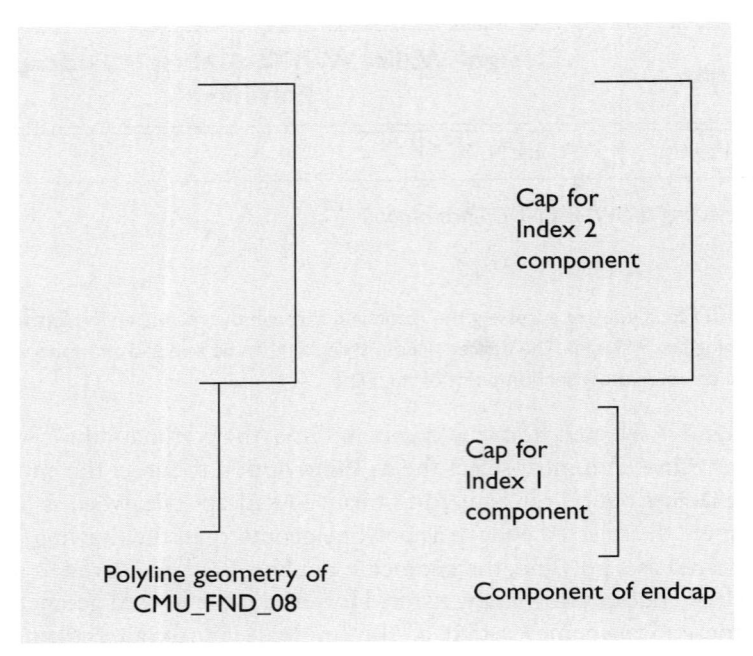

Figure 3.65 *Components of a wall endcap*

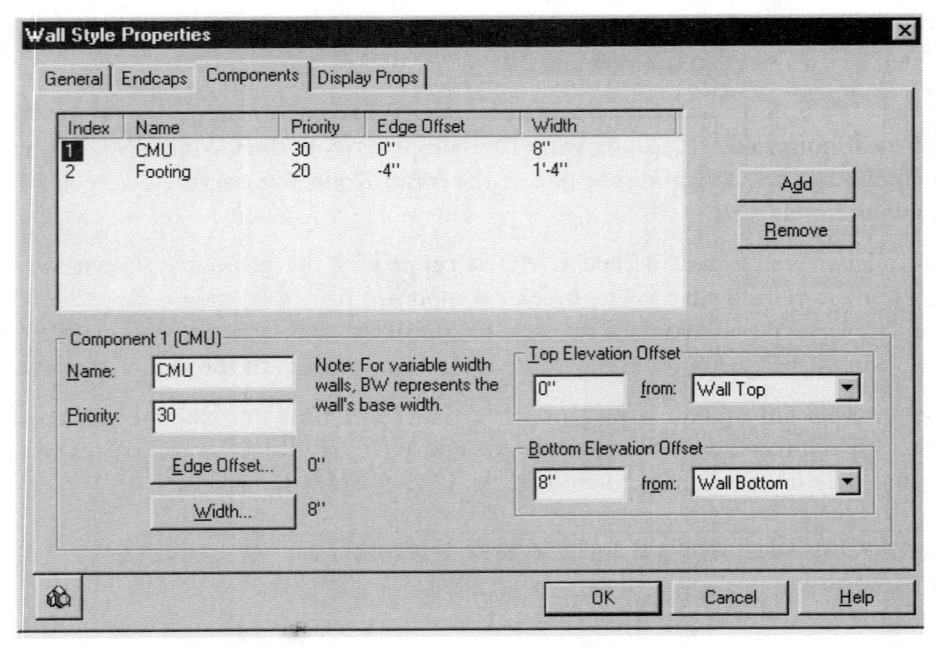

Figure 3.66 *Components tab of the Wall Style Properties dialog box*

PROPS) allows you to create exceptions for selected walls. In Figure 3.67, all walls are the same style; however, the left vertical wall has been edited with **Wall Style Overrides** of the **WALLPROPS** command. The priority number of the top end of the left wall was decreased to remove the merger of the walls. The endcap of the lower end of the left wall was changed to the bull nose endcap style.

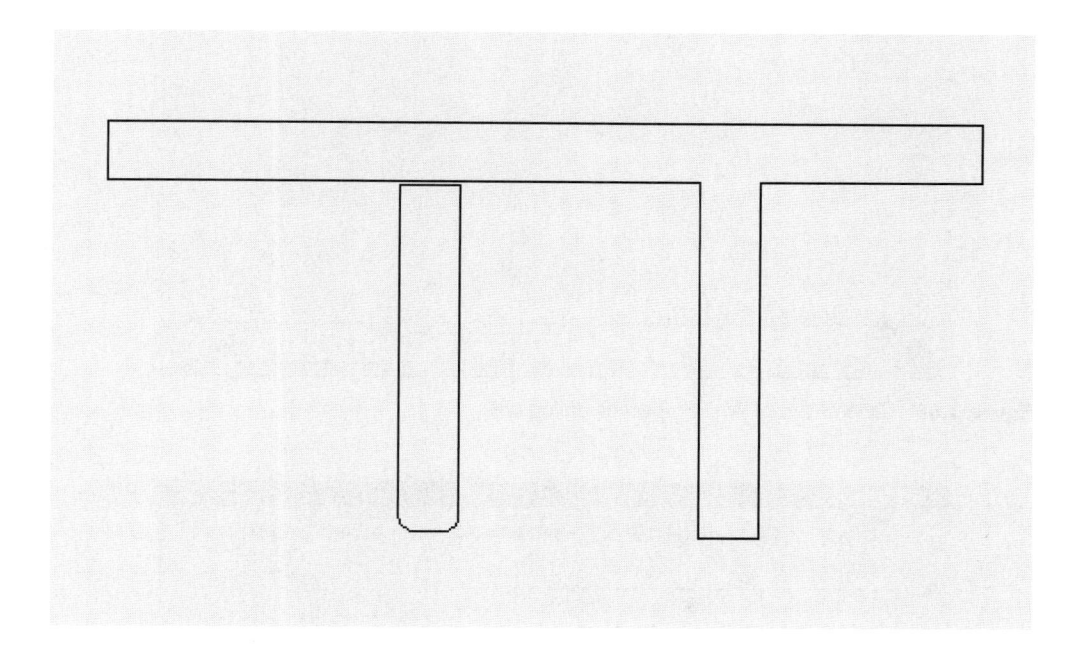

Figure 3.67 *Effects of overriding endcap and priority of walls*

To apply different endcaps and priority to selected walls, execute the **Wall Properties** command (**WALLPROPS**), which opens the **Wall Properties** dialog box as shown in Figure 3.68.

Select the **Wall Style Overrides** tab as shown in Figure 3.69.

EDITING ENDCAPS USING WALL STYLE OVERRIDES

The Wall Style Overrides tab consists of a **Priority Overrides** section at the top and **Endcap Overrides** at the bottom, which includes two buttons: **A-Start Endcap** and **B-End Endcap**. The purpose of each of these buttons is described below:

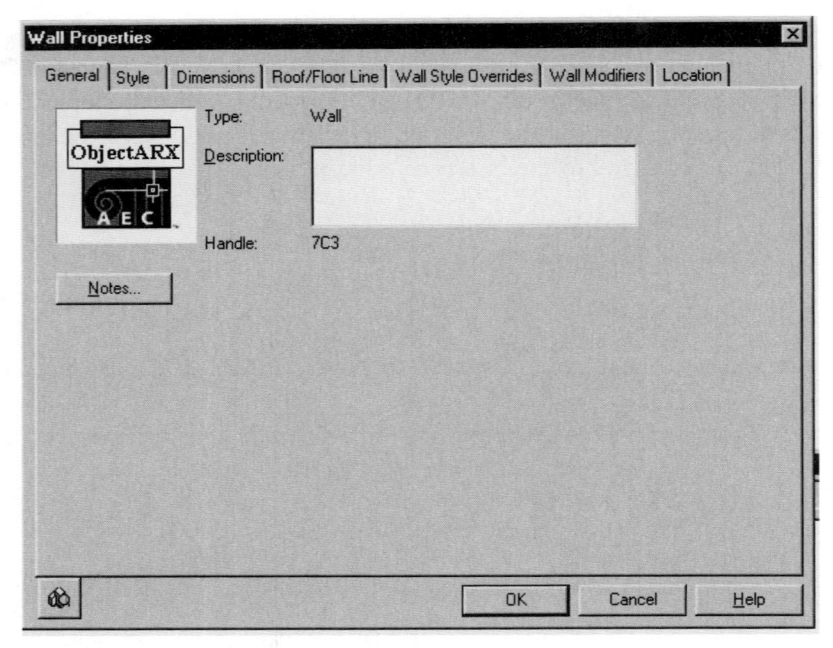

Figure 3.68 *General tab of Wall Properties dialog box*

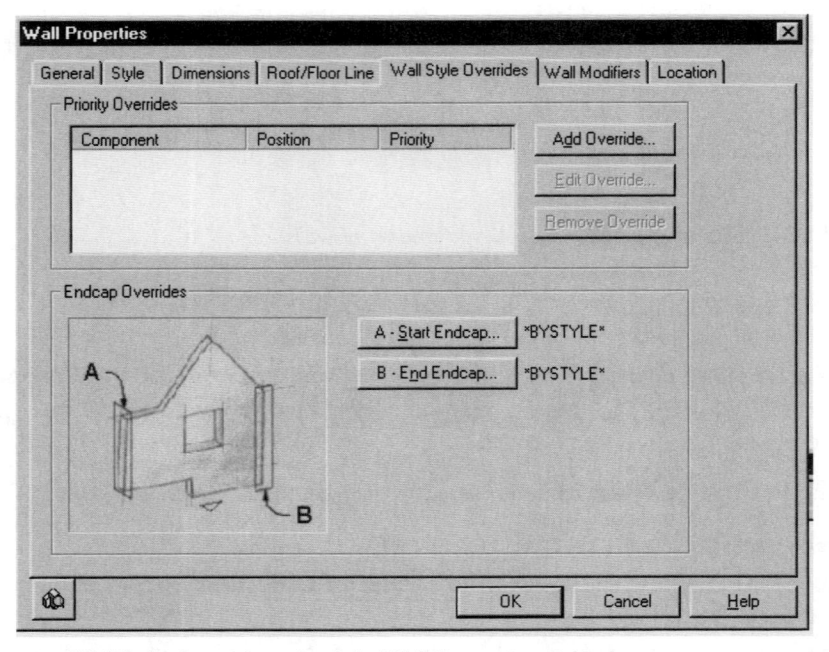

Figure 3.69 *Wall Style Overrides tab of the Wall Properties dialog box*

A-Start Endcap – Changes the endcap style used at the beginning of the wall. The default endcap style assigned to the start of the wall is listed to the right of the **A-Start Endcap** button.

B-End Endcap – Changes the endcap style used at the end of the wall. The default endcap style assigned to the end of the wall is listed to the right of the **B-End Endcap** button.

The BYSTYLE Endcap style is listed adjacent to the **A-Start Endcap** and **B-End Endcap** buttons if no override has been assigned. Selecting either button will open the **Select an Endcap Style** dialog box, as shown in Figure 3.70.

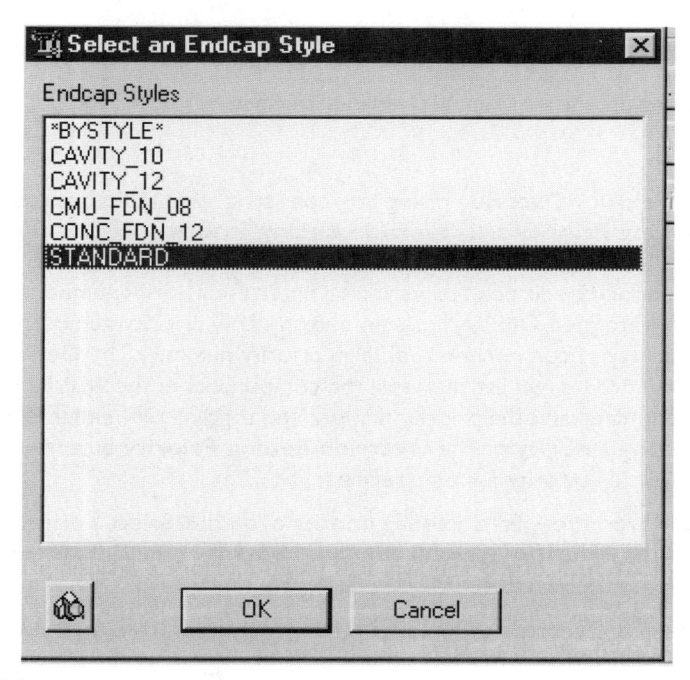

Figure 3.70 *Select an Endcap Style dialog box*

Selecting an endcap style from the dialog box will override the global endcap style defined in the wall style definition.

USING PRIORITY OVERRIDES

The top section of the **Wall Style Overrides** tab allows you to set an override priority for the end of the wall that is different from other walls of the same wall style. This section consists of a **Priority Overrides** list box and three buttons to create, edit and remove the overrides. The purpose of each of the buttons is described below:

Add Override – Opens the **Priority Override** dialog box as shown in Figure 3.71.

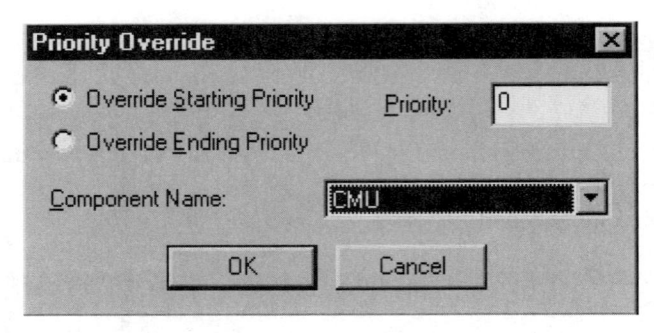

Figure 3.71 *Editing priority with Priority Override dialog box*

The **Priority Override** dialog box consists of two radio buttons: **Override Starting Priority** and **Override Ending Priority**. Toggling either radio button ON will set an override for the start of the wall or the end of the wall. The **Priority** edit box allows you to enter the priority number for the end of the wall toggled ON by the radio buttons. Low priority numbers will override the display of components with high priority numbers. The **Component Name** drop-down list shows all the components of the wall. Editing the component name and the priority number and toggling ON either the **Override Starting Priority** or the **Override Ending Priority** button will establish an override priority for the specified wall.

Edit Override – Edits existing priority overrides. Select a priority override from the **Priority Overrides** list and then select the **Edit Override** button to open the **Priority Override** dialog box.

Remove Override – Removes priority overrides from the **Priority Overrides** list.

TUTORIAL 3.4 CREATING AND EDITING AN ENDCAP

1. Open the \ADT_Tutor\Ch3\Ex3-4.dwg and save the drawing as Endcap.dwg in your student directory.
2. Zoom the extents of the drawing.
3. Select the **Wall Styles** command from the **Walls** toolbar.
4. Select the 10_bull_nose wall style and select the **Edit** button of the **Wall Style Properties** dialog box.
5. Select the **Components** tab and note that there are two components, as shown in Figure 3.72.

Advanced Wall Features 167

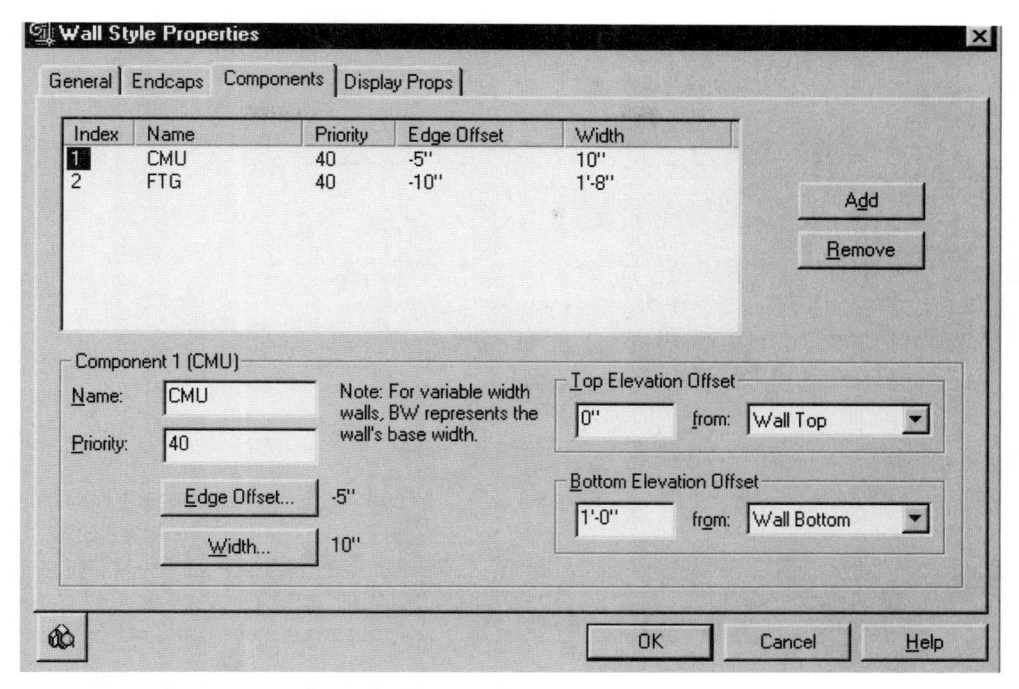

Figure 3.72 *Components of a wall to have endcaps applied*

6. Select **OK** to dismiss the **Wall Style Properties** dialog box.
7. Select **OK** to dismiss the **Wall Styles** dialog box.
8. Select **Endcap Style** from the **Walls** toolbar.
9. Select the **New** button.
10. Enter the name BULL_NOSE in the **Name** dialog box; select **OK** to dismiss the dialog box.
11. Select the **Set From<** button and respond to the command sequence as shown below.

   ```
   Command: _AecWallEndCapStyle
   Select a polyline: (Select the polyline on the right at P1 as shown in Figure 3.73.)
   Enter component index for this segment <1>: ENTER
   (Pressing ENTER assigns this polyline to wall component Index 1, the CMU.)
   Add another component? [Yes/No] <N>: Y ENTER
   Select a polyline: (Select the polyline on the left at P2 as shown in Figure 3.73.)
   ```

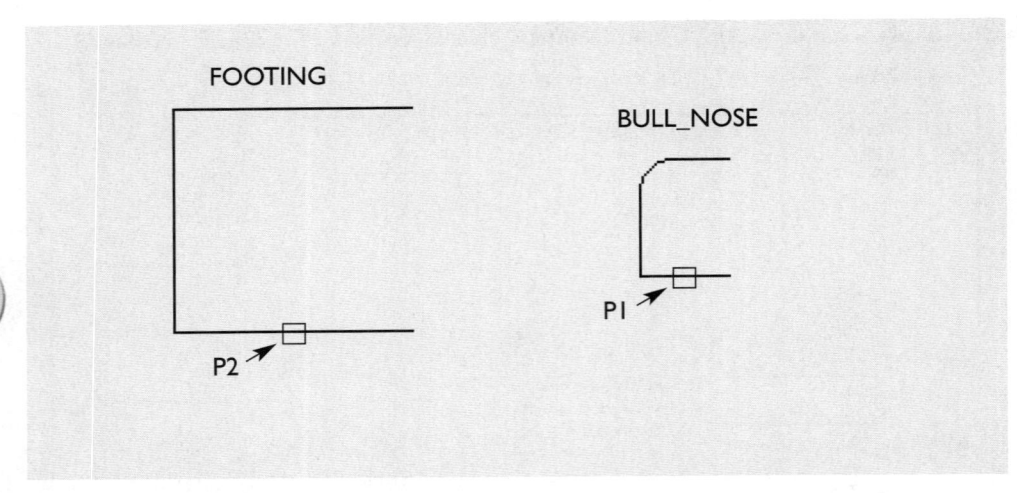

Figure 3.73 *Polyline geometry for endcaps*

```
Enter component index for this segment <2>: ENTER
```
(Pressing ENTER *assigns this polyline to wall component Index 2, the footing.)*
```
Add another component? [Yes/No] <N>: ENTER
Enter return offset <0">: ENTER
```
(Pressing ENTER *attaches the endcap to the wall without extending the wall by the endcap.)*

12. Select the **OK** button to dismiss the **Endcap Styles** dialog box.
13. Select the **Wall Styles** command from the **Walls** toolbar.
14. Select the 10_bull_nose wall style and select the **Edit** button.
15. Select the **Endcaps** tab of the **Wall Style Properties** dialog box.
16. Select the **Wall Endcap Style** button to display the **Select an Endcap Style** dialog box as shown in Figure 3.74.
17. Select the BULL_NOSE endcap style. Select **OK** to dismiss the **Select an Endcap Style** dialog box.
18. Select **OK** to dismiss the **Wall Style Properties** dialog box.
19. Select **OK** to dismiss the **Wall Styles** dialog box.
20. Select **Add Walls** from the **Walls** toolbar.
21. Select 10_bull_nose from the **Style** list as shown in Figure 3.75.
22. Draw a wall 8' long.

Advanced Wall Features   169

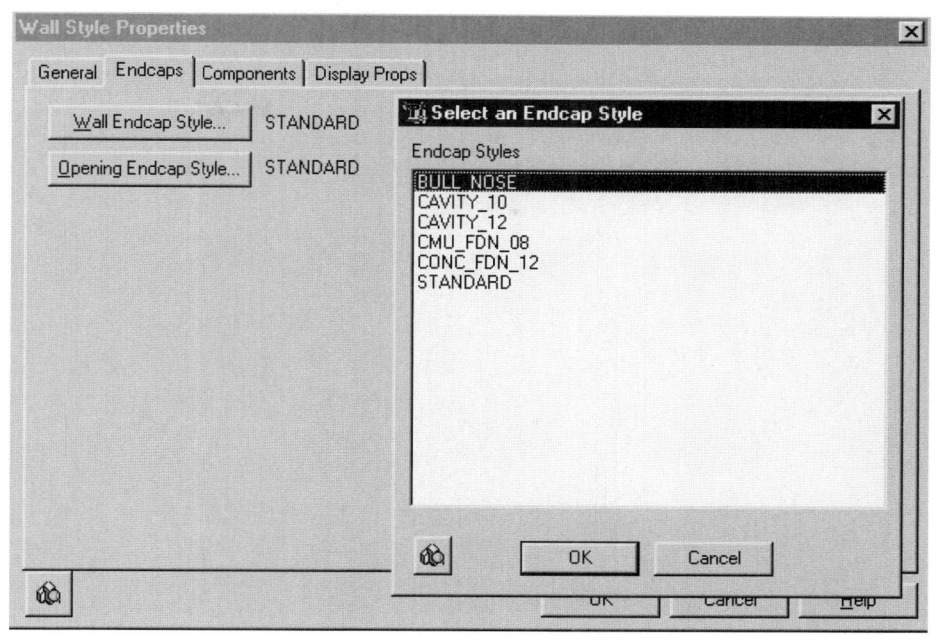

Figure 3.74 *Select an Endcap Style dialog box*

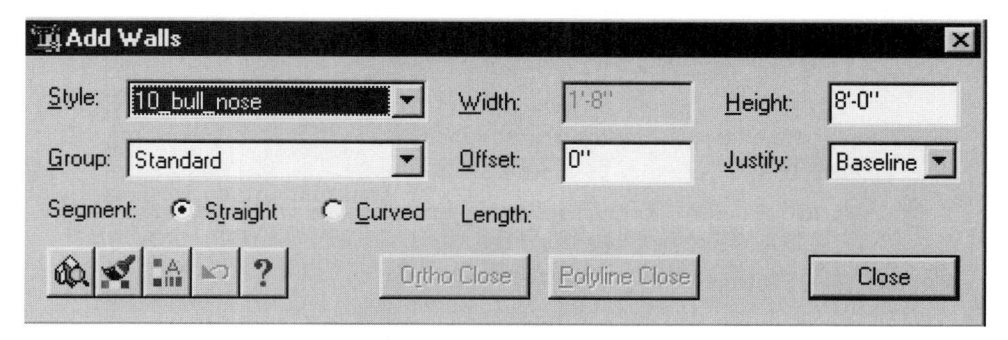

Figure 3.75 *Creating a wall with the new endcap style*

23. Select the southwest isometric view option from the **View** flyout on the **Standard** toolbar. The wall should appear as shown in Figure 3.76.
24. Type **SHADE** and press ENTER in the command line.
25. To modify the wall endcap style definition, choose **Design> Walls>Wall Tools>Insert Endcap as Polyline** from the menu bar.
26. Select the BULL_NOSE endcap from the **Endcap Styles** dialog box and select **OK** to dismiss the **Endcap Styles** dialog box.

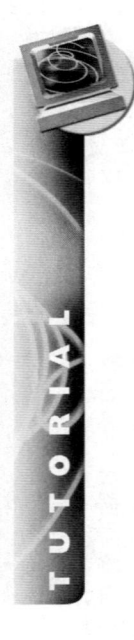

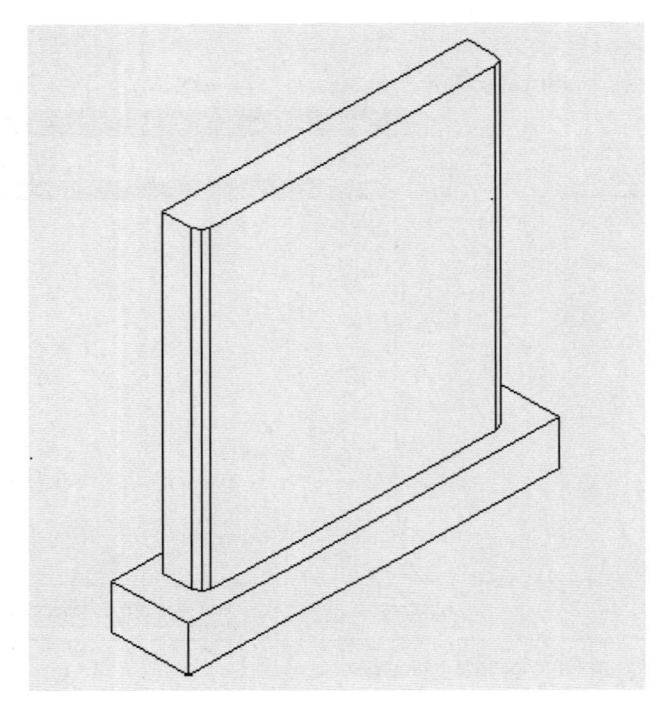

Figure 3.76 *Southwest isometric view of wall with endcaps*

27. Insert the endcap geometry by responding to the command line prompts as shown below:

    Select insertion point: *(Select a location on the graphics screen.)*
       [2] AcDbPolyline(s) created
    Wall endcap [Define/as Pline]:
    (After you select the insertion point, the command ends itself and inserts the endcap as a polyline as shown in Figure 3.77.)
28. Select the **Fillet** command from the **Modify** toolbar.
29. Refer to Figure 3.77 and respond to the command line prompts as shown below:

    Command:
    FILLET
    Current settings: Mode = TRIM, Radius = 0'-0 "
    Select first object or [Polyline/Radius/Trim]: r ENTER
    Specify fillet radius <0'-0 ">: 2 ENTER
    (Press ENTER *to repeat the command)*

Advanced Wall Features | 171

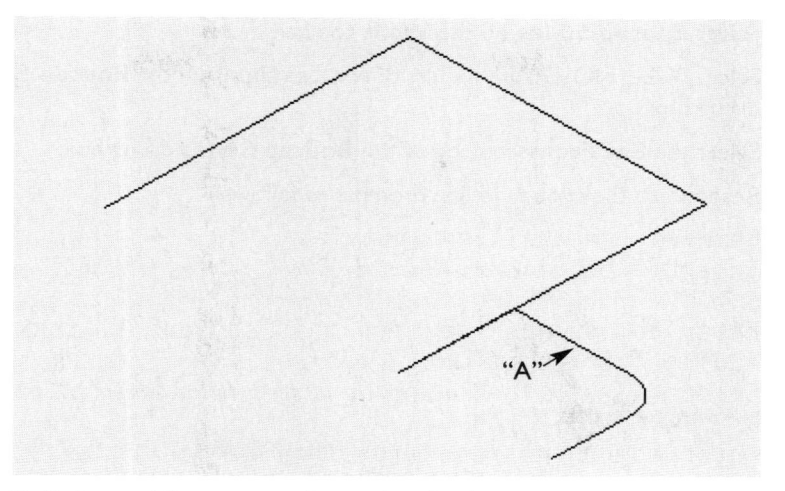

Figure 3.77 *Endcap polyline geometry inserted in the drawing*

```
Command:
FILLET
Current settings: Mode = TRIM, Radius = 0'-2"
Select first object or [Polyline/Radius/Trim]: p
```
(selects the Polyline option.)
```
Select 2D polyline:
```
(Select the filleted endcap labeled "A"; both corners of the endcap will now be filleted as shown in Figure 3.78.)
```
2 lines were filleted.
```

Figure 3.78 *Components of wall endcap geometry*

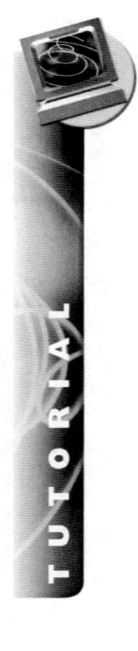

30. Select **Endcap Styles** on the **Walls** toolbar.
31. Select BULL_NOSE from the list of endcaps shown in the **Endcap Styles** dialog box.
32. Select the **Set From<** button of the **Endcap Styles** dialog box.
33. Respond to the command line prompts as follows:

    ```
    Command: _AecWallEndCapStyle
    Select a polyline: (Select the filleted polyline labeled "A" in Figure 3.78.)
    Enter component index for this segment <1>: ENTER
    Add another component? [Yes/No] <N>: Y ENTER
    Select a polyline: (Select the larger polyline labeled "B" in Figure 3.78.)
    Enter component index for this segment <2>: ENTER
    Add another component? [Yes/No] <N>: ENTER
    Enter return offset <0">: ENTER
    ```
34. Select **OK** to dismiss the **Endcap Styles** dialog box.
35. Zoom to display the 8' wall created in Step 22.
36. Type **HIDE** and press ENTER in the command line to hide hidden lines. Note that the revised endcap shape is applied to the wall as shown in Figure 3.79.

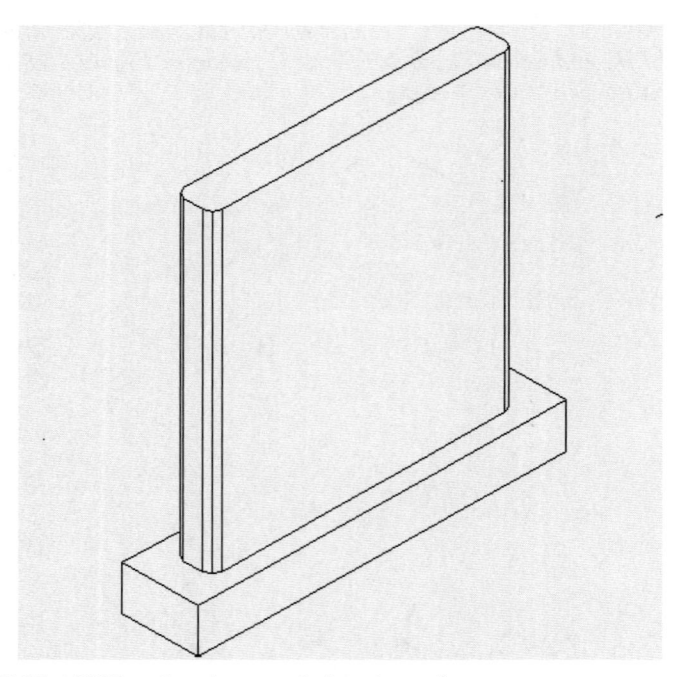

Figure 3.79 BULL_NOSE wall endcap applied to the wall

37. Select the **Top View** from the **View** flyout on the **Standard** toolbar.
38. Draw a vertical wall 10' long down from the left end of the 8' wall. Use the 10_bull_nose wall style as shown in Figure 3.80.

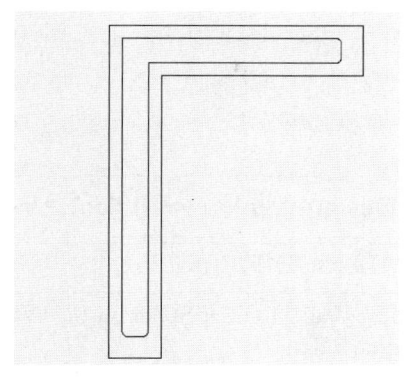

Figure 3.80 *Plan view of wall with the 10_bull_nose wall style*

39. Click on the vertical wall; right-click and choose Wall Properties from the shortcut menu.
40. Select the **Wall Style Overrides** tab of the **Wall Properties** dialog box.
41. Select the B-End Endcap button.
42. Select the Standard endcap style; select **OK** to dismiss the **Select an Endcap Style** dialog box. Select **OK** to dismiss the **Wall Properties** dialog box. The lower end of the wall should change as shown in Figure 3.81.
43. Save the drawing.

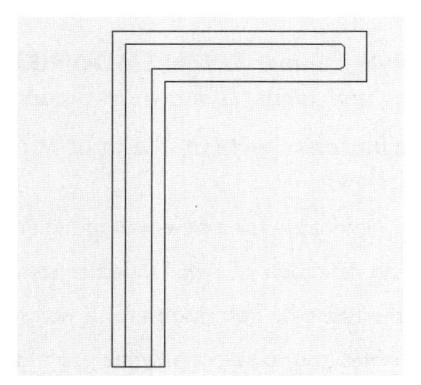

Figure 3.81 *Wall endcap changed for the lower end of the wall*

CREATING WALL MODIFIER STYLES

Wall modifiers can be created to add projections to a wall at the beginning, end, or at a specified distance from the beginning, end, or midpoint of a wall. Wall modifiers can consist of rectangular projections or polyline shapes saved as wall modifier styles. You can create wall pilasters or other decorative projections of the wall using wall modifier styles.

Access **Wall Modifier Style** as follows:

Menu bar	Design>Walls>Wall Tools>Wall Modifier Styles
Command prompt	WALLMODIFIERSTYLE
Toolbar	Select Wall Modifier Styles on the Walls toolbar as shown in Figure 3.82
Shortcut	Right-click over the drawing area and choose Design>Walls>Modifier Styles

Table 3.6 Acessing the Wall Modifier Style

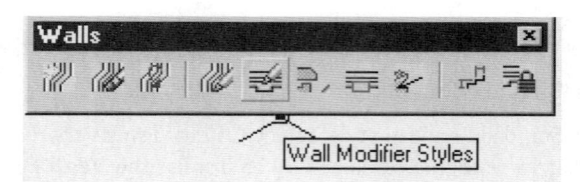

Figure 3.82 Wall Modifier Styles command on the Walls toolbar

When the **Wall Modifier Style** command (**WALLMODIFIERSTYLE**) is executed, the **Wall Modifier Styles** dialog box opens, as shown in Figure 3.83.

The purpose of each of the buttons shown on the right of the **Wall Modifier Styles** dialog box is summarized below:

New – Creates a new name for a new wall modifier style.

Copy – Copies the definition of a wall modifier style to a new name.

Edit – Changes the name or description for a wall modifier style.

Set From< – Enables you to select polyline geometry to define a wall modifier style.

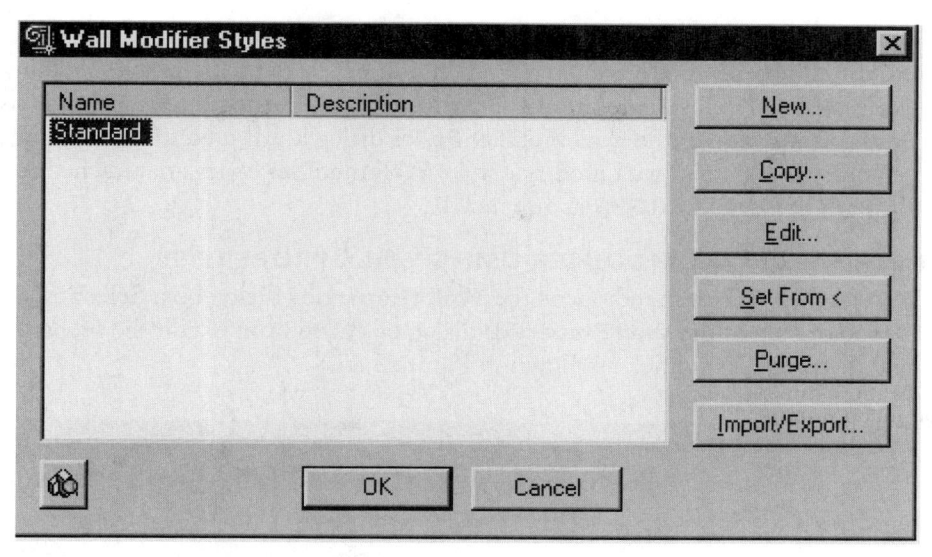

Figure 3.83 Wall Modifier Styles dialog box

> **Purge** – Removes unused wall modifiers from the wall modifier style list.
>
> **Import/Export** – Saves wall modifier styles to a resource file or imports wall modifier styles into a drawing from a resource file.

You can create rectangular projections along a wall as a wall modifier using the Standard wall modifier style. Projections other than rectangular require the creation of a new wall modifier style. To create a wall modifier style, draw the profile of the wall projection as an open polyline. Examples of polylines used as wall modifiers are shown in Figure 3.84.

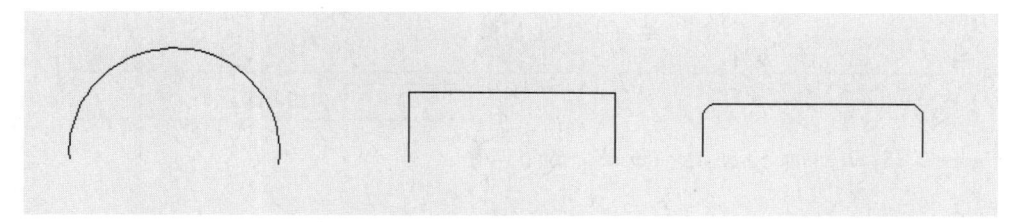

Figure 3.84 Sample geometry for wall modifier style

The profile can be drawn the actual size of the desired projection or drawn as a sketch and applied with specified length and depth dimensions. If a wall modifier style is created from geometry drawn actual size, it can be applied to a wall using the actual dimensions of the original geometry. Wall modifier styles created from sketched polyline geometry are inserted with the length and depth of the projection specified. The **Set From<** option of the **Wall Modifier Styles** dialog box is used to save the polyline geometry as a named wall modifier style. Wall modifier styles are attached to a wall through the **WALLPROPS** command.

ATTACHING A WALL MODIFIER USING WALL PROPERTIES

The **WALLPROPS** command opens the **Wall Properties** dialog box. Selecting the **Wall Modifiers** tab of the **Wall Properties** dialog box allows you to edit the properties assigned to the wall modifier, as shown in Figure 3.85.

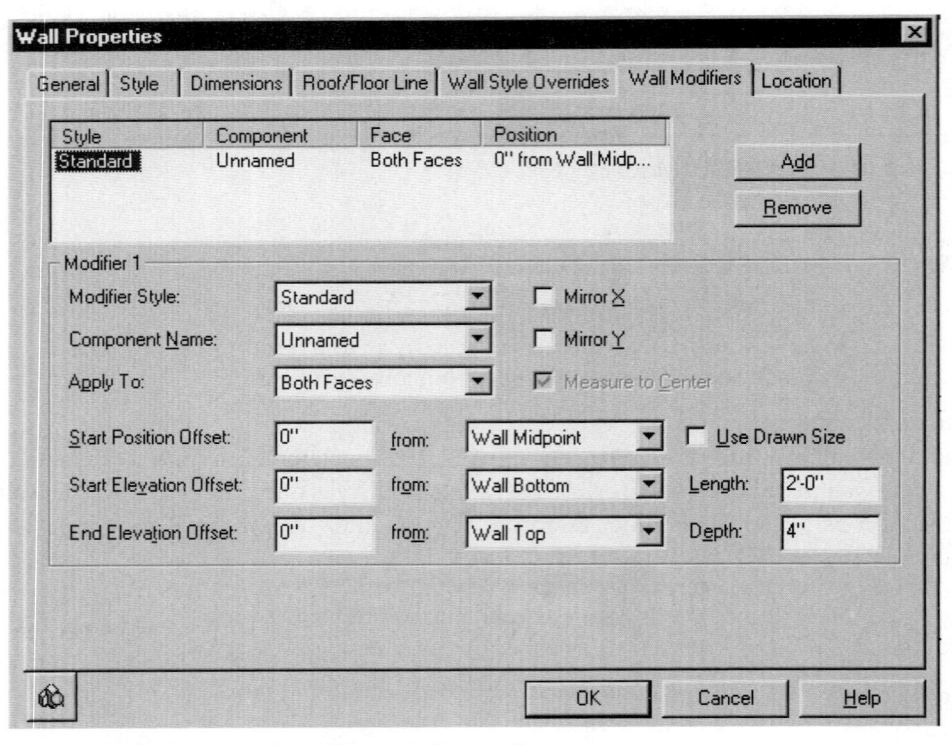

Figure 3.85 *Defining a wall modifier style for a wall*

The top portion of the **Wall Modifiers** tab lists the wall modifiers that are currently attached to the selected wall. The **Add** and **Remove** buttons as described below can be used to add additional wall modifiers or remove a modifier.

Add – Activates the dialog box allowing you to assign a wall modifier style to the selected wall. The Standard wall modifier style will be used as the default unless other wall modifier styles are selected.

Remove – Detach the wall modifier from the wall. Remove a wall modifier style from the wall by selecting the wall modifier style from the wall modifier style list and then selecting the **Remove** button.

The lower portion of the dialog box is used to edit properties and conditions of attachment to the wall. The purpose of each of the edit fields of the lower portion of the Wall Modifiers tab is described below.

Modifier Style – The **Modifier Style** drop-down list shows defined wall modifier styles that have been created in the drawing. Selecting a wall modifier style from the list will apply the predefined modifier style to the wall.

Component Name – The **Component Name** drop-down list shows the wall components of the selected wall. Wall modifier styles can be applied to each wall component of a wall style.

Apply To – The **Apply To** drop-down list allows the wall modifier style to be applied to the Left, Right or Both Faces of the wall.

Start Position Offset – The **Start Position Offset** edit box allows you to set the position horizontally along the wall to locate the wall modifier. The distance entered can be a positive or negative number. The **from** list allows the start position to be established relative to the wall start, wall end, or wall midpoint. Setting the distance to zero and measuring the distance relative to the wall midpoint will locate the projection in the middle of the wall.

Start Elevation Offset – The **Start Elevation Offset** edit box establishes the elevation of the bottom of the projection. The bottom of the projection can be defined to start at the bottom of the wall or at some distance from the bottom. The **Start Elevation Offset** distance and the **from** list allow the distance to be defined from the wall top, wall base height, wall baseline, or wall bottom. The wall top and wall bottom are the top and bottom surfaces of the wall component being modified. Because wall components can be defined with different elevations in the wall style definition, the wall baseline and base height options are provided. Wall baseline is the bottom of the baseline justification line, and the base height is the top of the baseline justification line. Establish the distance between the baseline and the wall base height by setting the height in the **Add Walls** dialog box.

End Elevation Offset – The **End Elevation Offset** section establishes the elevation of the top of the projection. The **End Elevation Offset** distance and the **from** list define the distance from the wall top, wall base height, wall baseline, or wall bottom for the top of the projection. The top of the projection is defined in the same manner as the bottom. Note that if the start elevation offset and end elevation offset are set to the same distance and reference surface, the modifier will become invisible because the projection is starting and ending at the same elevation.

Setting a distance in the start elevation offset from the wall bottom will lift the projection up from the bottom. Editing the start position offset, start elevation offset, and end elevation offset would allow the creation of the wall modifiers as shown in Figure 3.86.

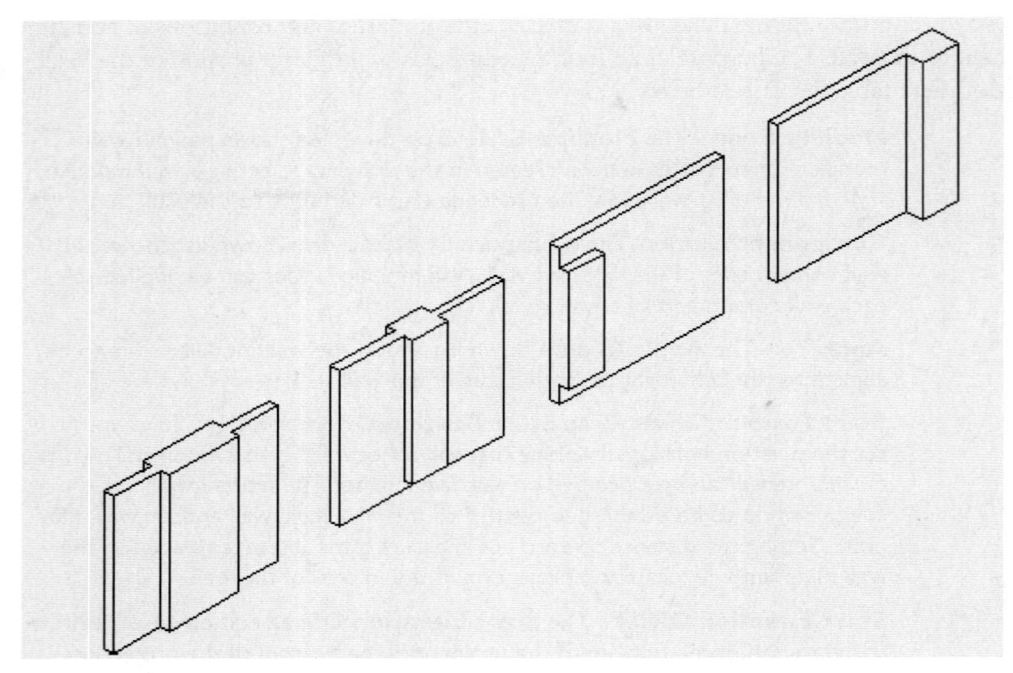

Figure 3.86 *Examples of wall modifier styles*

Notice in Figure 3.86 that the projections are positioned at the beginning, midpoint, and end of the wall. In Figure 3.87, the start position offset is set to zero relative to the wall end. This projection has a start elevation of 1'-0" from the wall bottom and an end elevation offset of a negative 1'-0".

Use Drawn Size – The **Use Drawn Size** check box turns OFF the size established in the **Length** and **Depth** edit fields. If this box is checked, the wall modifier will be created instead according to the actual size of the polyline geometry used to create the wall modifier style.

Length – The **Length** edit field allows you to establish the length of the wall modifier. The distance entered in the **Length** edit box can be the same as the wall length.

Advanced Wall Features   179

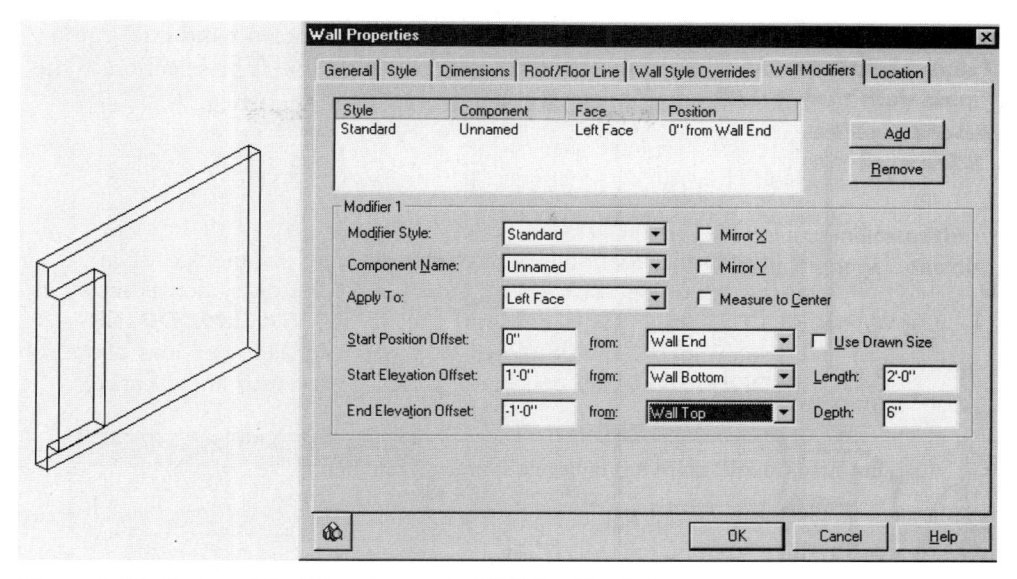

Figure 3.87 *Start and End Elevation set for Wall Modifier Style*

Depth – The **Depth** edit field sets the distance the wall modifier will project from the wall.

Mirror X – The **Mirror X** check box is used to mirror the wall modifier in the X direction.

Mirror Y – The **Mirror Y** check box is used to mirror the wall modifier in the Y direction.

Measure to Center – The **Measure to Center** check box will position the wall modifier relative to its center along the wall.

CHANGING WALL MODIFIERS

If a wall modifier style is created incorrectly, it can be changed with the **WALLMODIFIER** command. This command can be used to create or change a wall modifier style.

Access the **WALLMODIFIER** command as follows:

Menu bar	Design>Walls>Wall Tools>Insert Wall Modifier as Polyline
Command: prompt	WALLMODIFIER

Table 3.7 *Acessing the WALL MODIFIER command*

Selecting the command by entering **WALLMODIFIER** in the command line requires you to enter the **Define** or **as Pline** option in the command line. The selection of the **as Pline** option is shown below, in the following command sequence:

```
Command: wallmodifier
Wall modifier [Define/as Pline]: P ENTER
```

The purpose of each option is defined below.

>   **Define** – The **Define** option is used to create a new wall modifier style. Selecting this option opens the **Wall Modifier Styles** dialog box as shown in Figure 3.88. This dialog box is identical to that of the **WALLMODIFIERSTYLE** command. This option allows you to perform all the options of the **WALLMODIFIERSTYLE** command discussed previously in this chapter.

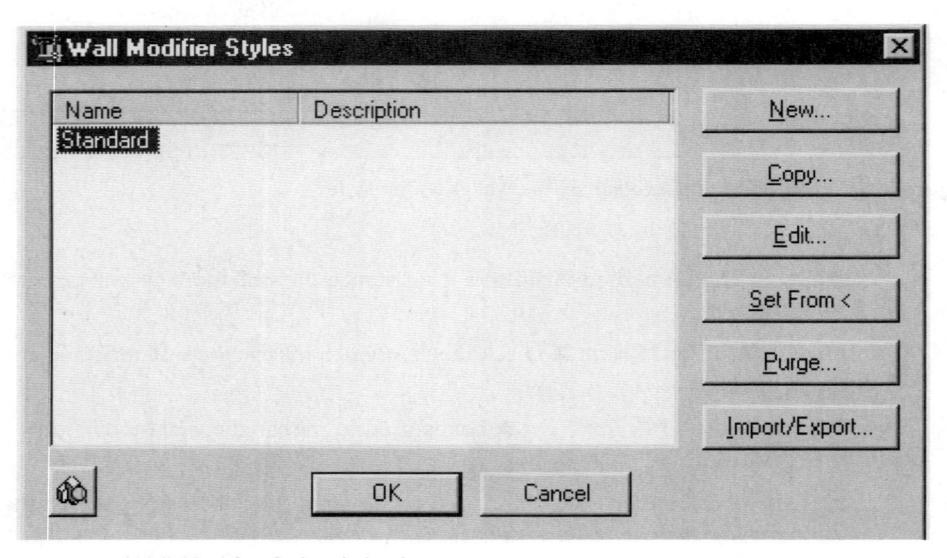

Figure 3.88 *Wall Modifier Styles dialog box*

>   **as Pline** – Selecting the **as Pline** option will open the **Modifier Styles** dialog box as shown in Figure 3.89.
>
>   **Note:** This command requires input from the command line after a wall modifier style is selected in the dialog box.
>
>   This command option allows you to insert the wall modifier style in the current drawing as a polyline. After it is inserted, the polyline can be edited and redefined as a wall modifier style. After selecting a wall modifier style you are prompted for the insertion point, as shown in the following command sequence:

Advanced Wall Features | 181

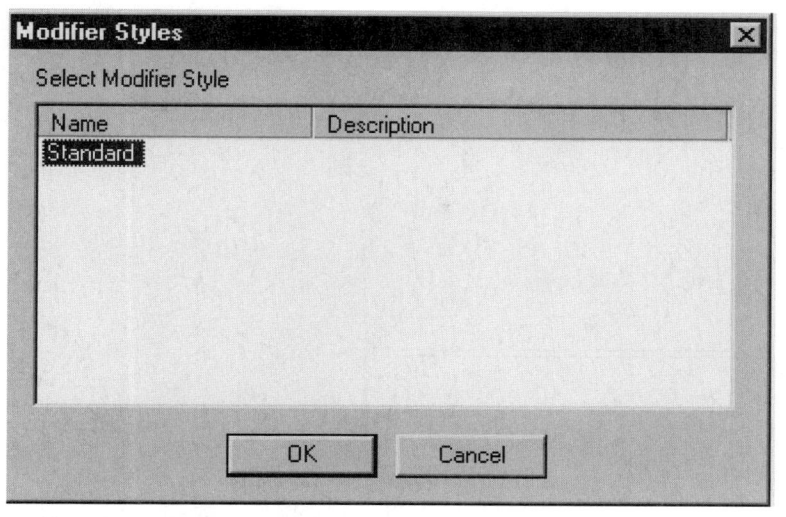

Figure 3.89 *Modifier styles listed in Modifier Styles dialog box*

```
Wall modifier [Define/as Pline]: p
Select insertion point: (Select a location for the insertion of the
    polyline.)
```
Once the polyline is inserted in the drawing, it can be modified and saved as a wall modifier style. A new wall modifier style can be created from the geometry or the same name can be used for the wall modifier style. If the revised geometry is saved with the original wall modifier style name, all wall modifier styles with that same name will be updated to the new geometry definition. Using the original wall modifier name allows you to quickly change the profile of the wall modifiers.

TUTORIAL 3.5 CREATING WALL MODIFIERS

1. Open *Ex3-3.dwg* from the *ADT_Tutor\Ch3* directory and choose **File>SaveAs** from the menu bar to save the file as *Ex3-5* in your student directory.

2. Select **Layers** from the **Object Propertie**s toolbar and then select A-Anno-Dims to freeze the layer of the dimensions.

3. Draw an arc with a 4" radius with an included angle of 180 degrees. Draw the arc in the room shown in Figure 3.90.

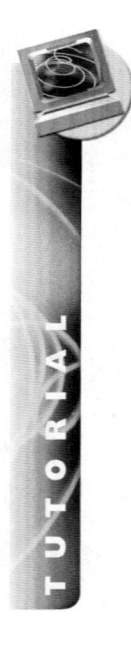

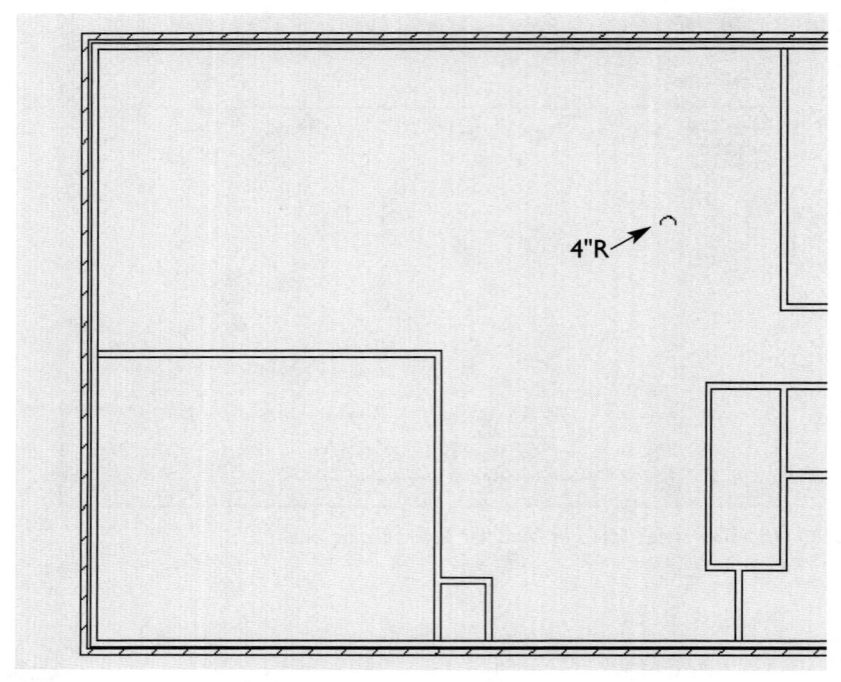

Figure 3.90 *Arc geometry for wall modifier style*

4. Change the arc to a polyline by entering **PEDIT** in the command line and responding to the command prompts as follows:

   ```
   Command: pedit
   Select polyline: (Select the arc as shown in Figure 3.90.)
   Object selected is not a polyline
   Do you want to turn it into one? <Y> ENTER
   ```
   (Converts the arc to a polyline)
   ```
   Enter an option [Close/Join/Width/Edit vertex/Fit/Spline/Decurve/Ltype gen/Undo]:   ENTER
   ```
   (Select ENTER to end the PEDIT command.)

5. Right-click over the drawing area and choose **Design>Walls>Modifier Styles** from the shortcut menu.

6. Select the **New** button of the **Modifier Styles** dialog box to open the **Name** dialog box, as shown in Figure 3.91.

Advanced Wall Features 183

Figure 3.91 *Name dialog box for creating a new wall modifier style name*

7. Type the name **Bar_support** and select **OK** to dismiss the **Name** dialog box.
8. Select the **Set From<** button in the **Wall Modifier Styles** dialog box.
9. Select the arc drawn in step 3.
10. Select the **Edit** button and enter the following description: **36" bar support**.
11. Select **OK** to dismiss the **Wall Modifier Style Properties** dialog box.
12. Select **OK** to dismiss the **Wall Modifier Styles** dialog box.
13. Right-click over the drawing area and choose **Design>Walls>Styles** from the shortcut menu.
14. Select the **New** button of the **Wall Styles** dialog box.
15. Enter the name **Bar** in the **Name** dialog box.
16. Select **OK** to dismiss the **Name** dialog box.
17. Select the **Edit** button of the **Wall Styles** dialog box.
18. Select the **Components** tab and edit the name for Index 1 to Wood and the Priority to 40 as shown in Figure 3.92.

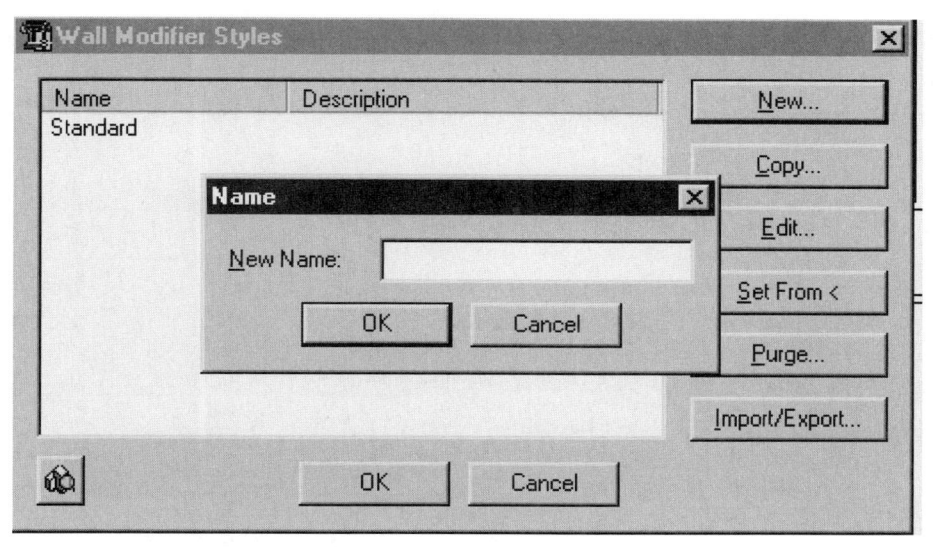

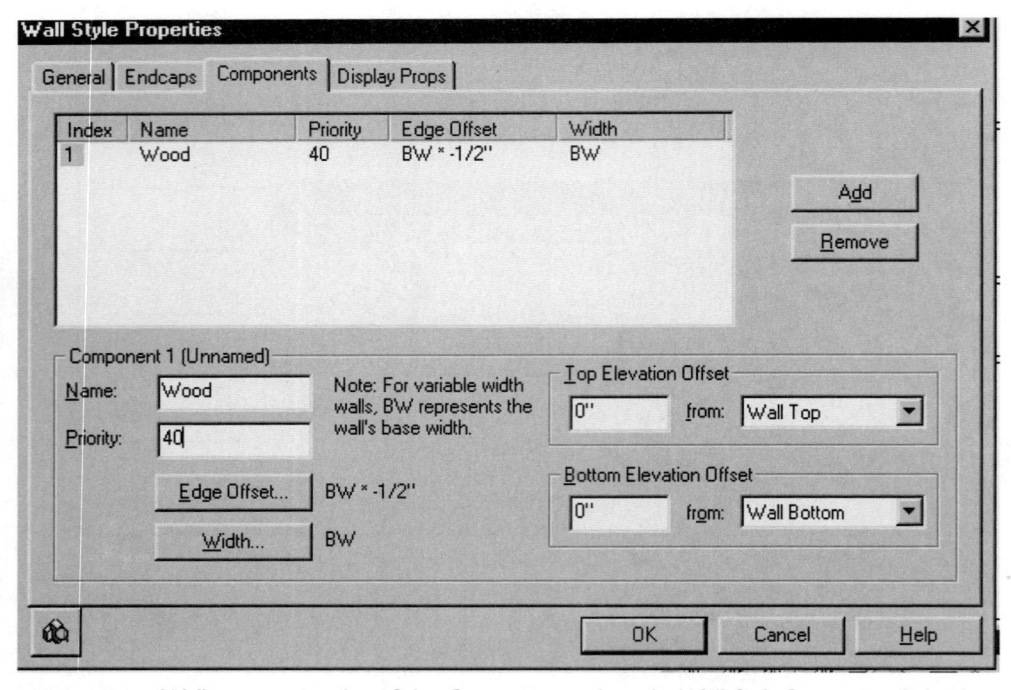

Figure 3.92 *Wall components list of the Components tab in the Wall Style Properties dialog box*

19. Select the **Display Props** tab; select Wall Style as the **Property Source**.
20. Select the **Attach Override** button; then select the **Edit Display Props** button to display the **Entity Properties** dialog box.
21. Select the **Other** tab of the **Entity Properties dialog box**, and then type **3'-0"** in the Cutting Plane Height edit field to lower the Cutting Plane from 3'-6", as shown in Figure 3.93.
22. Select **OK** to dismiss the **Entity Properties** dialog box.
23. Select **OK** to dismiss the **Wall Style Properties** dialog box.
24. Select **OK** to dismiss the **Wall Styles** dialog box.
25. Select the **Add Wall** button on the **Walls** toolbar. Select the Bar wall style and draw a left justified wall, 3'-0" high and 9' long from the corner of the existing wall, as shown in Figure 3.94.
26. Select the wall drawn in step 25; right-click and choose **Wall Properties** from the shortcut menu.
27. Select the **Wall Modifiers** tab of the **Wall Properties** dialog box.

Advanced Wall Features 185

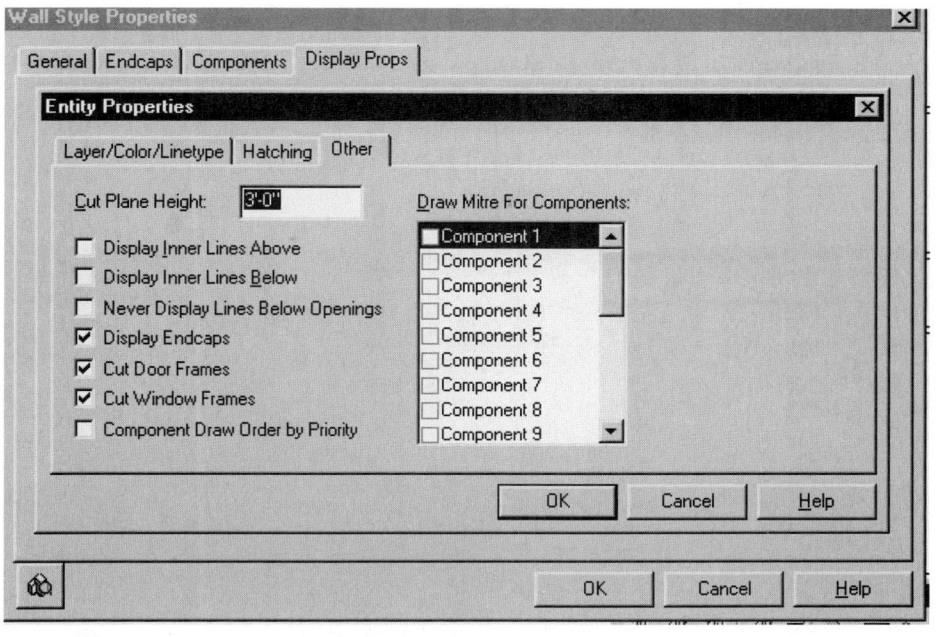

Figure 3.93 *Editing the cutting plane elevation in the Other tab of the Entity Properties dialog box*

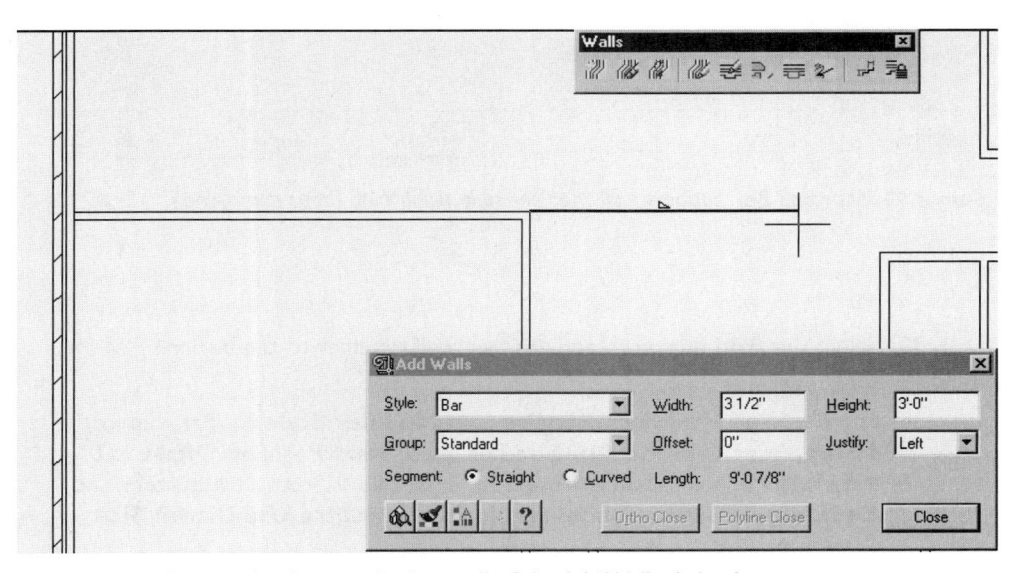

Figure 3.94 *Settings for drawing the bar wall of the Add Walls dialog box*

28. Select the **Add** button to add a wall modifier.
29. Edit the **Wall Modifiers** dialog box: set **Modifier Style}** to Bar_support, **Component** to Wood, **Apply** to Left Face, **Start Position Offset** to 0" from Wall Start, **Start Elevation Offset** to 0" from Wall Bottom, **End Elevation Offset** to 0" from Wall Top, and select the **Use Drawn Size** check box, as shown in Figure 3.95.

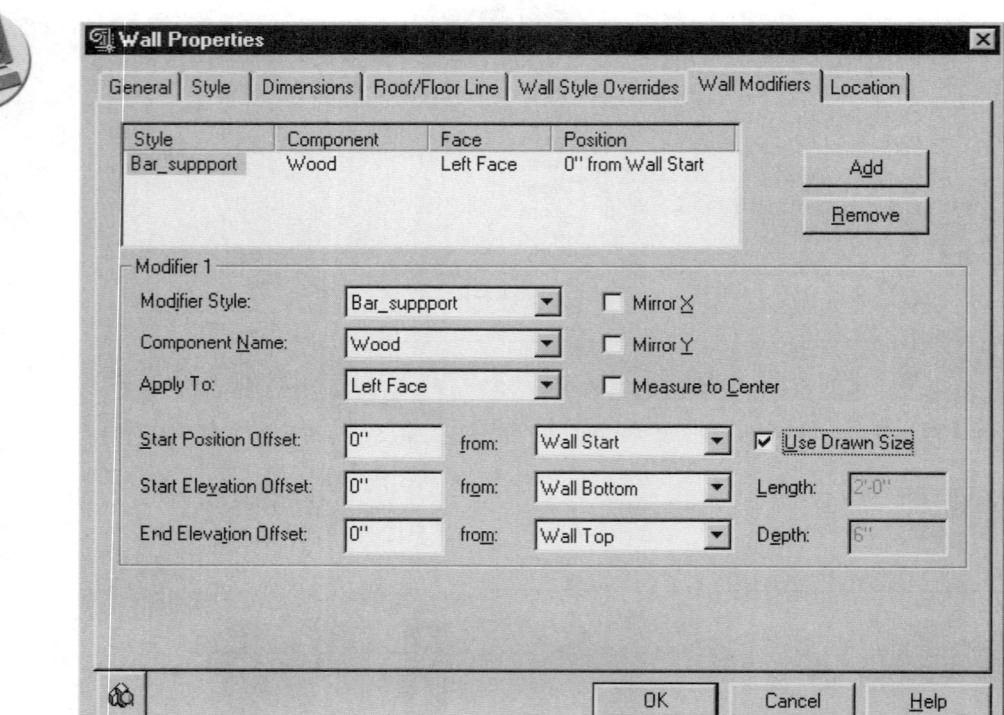

Figure 3.95 *Assigning Bar_support wall modifier style using Wall Props command*

30. Select the **Add** button to add another wall modifier to the beginning of the wall.
31. Edit the **Wall Modifiers** dialog box: set **Modifier Style** to "Bar_support", **Component** to Wood, **Apply** to Left Face, **Start Position Offset** to 0" from Wall Midpoint, **Start Elevation Offset** to 0" from Wall Bottom, **End Elevation Offset** to 0" from Wall Top, and select the **Use Drawn Size** check box.
32. Select the **Add** button to add another wall modifier.

Advanced Wall Features 187

33. Edit the **Wall Modifiers** dialog box: set **Modifier Style** to Bar_support, **Component** to Wood, **Apply** to Left Face, **Start Position Offset** to 0" from Wall End, **Start Elevation Offset** to 0" from Wall Bottom, **End Elevation Offset** to 0" from Wall Top, and select the **Use Drawn Size** check box. The modifier settings for the three components should be as shown in Figure 3.96.

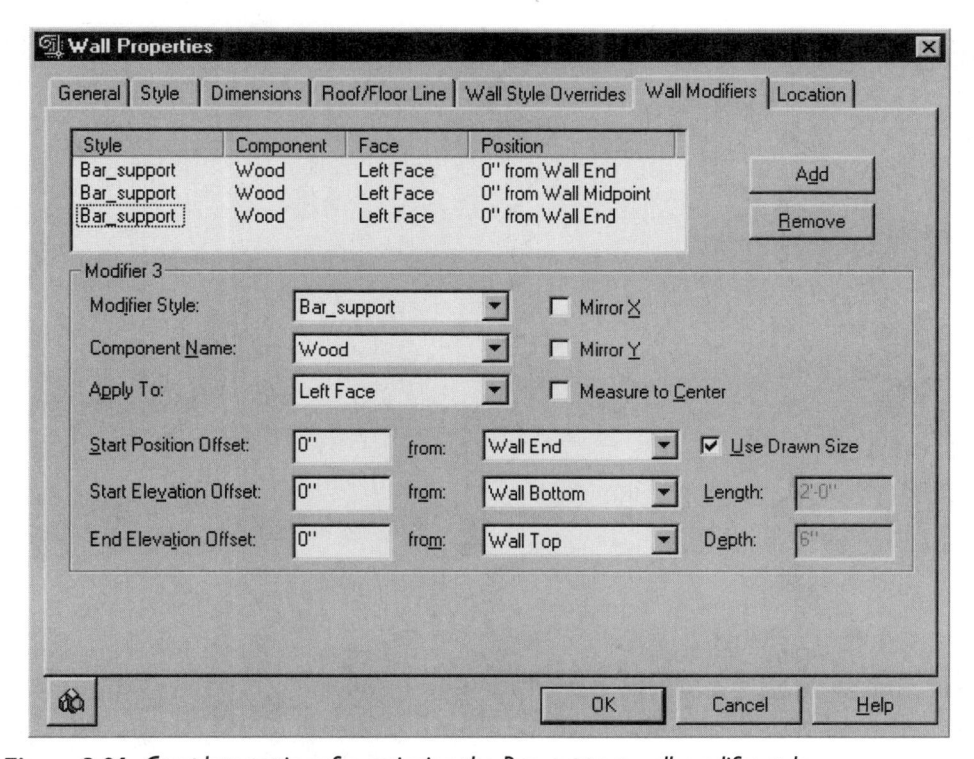

Figure 3.96 *Complete settings for assigning the Bar_support wall modifer style*

34. Select **OK** to dismiss the **Wall Properties** dialog box. The Arc should be added as a wall modifier to each end and to the middle of the wall, as shown in Figure 3.97.
35. Save the drawing.

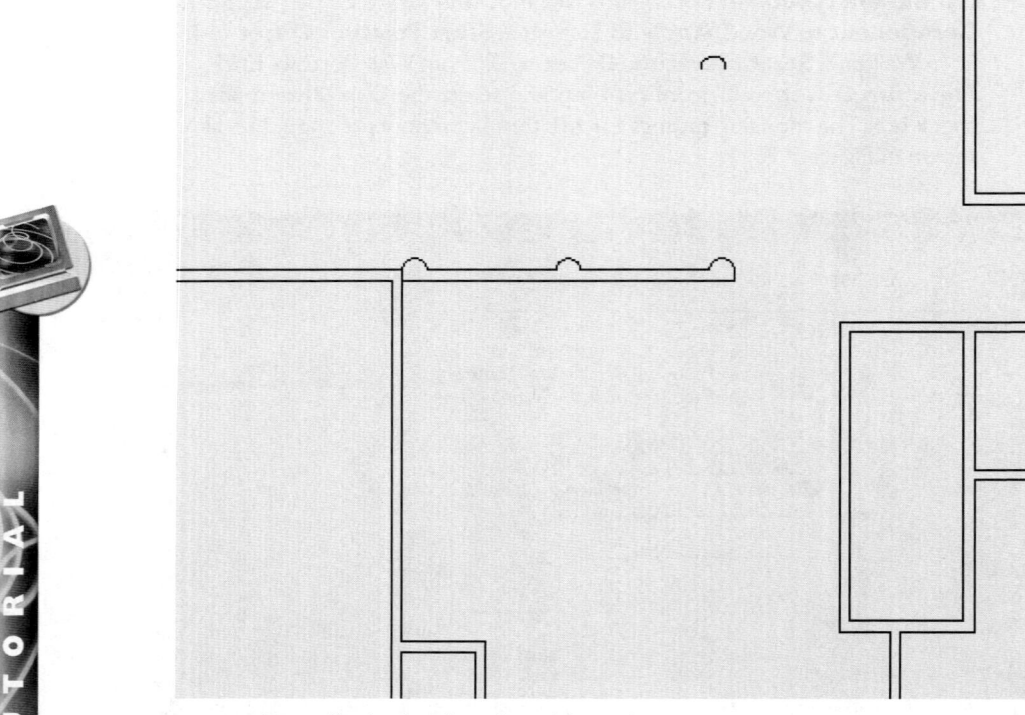

Figure 3.97 *Bar wall edited with wall modifier style*

CREATING CLEANUP GROUPS

Cleanup groups are created to control the merger of walls during automatic cleanup. Regardless of style name, walls merge with other walls and the intersections are cleaned up according to wall component priority number. The Priority number of wall components controls how the walls clean up. Walls with components of equal priority numbers merge together if they share the same group.

Walls of different groups will not blend together regardless of wall component priority number. When cleanup groups are assigned to walls, the walls that share the same group will merge based upon the priority number of the wall component. Walls that do not share the same cleanup group will not merge regardless of the wall component priority number.

Access **Cleanup Group** as follows:

Menu bar	Design>Walls>Cleanup Group Definition
Command prompt	WALL (Select CLeanup option)

Table 3.8 *Acessing the Cleanup Group*

When the **Cleanup Group Definition** command is executed, the **Cleanup Group Definitions** dialog box will open as shown in Figure 3.98.

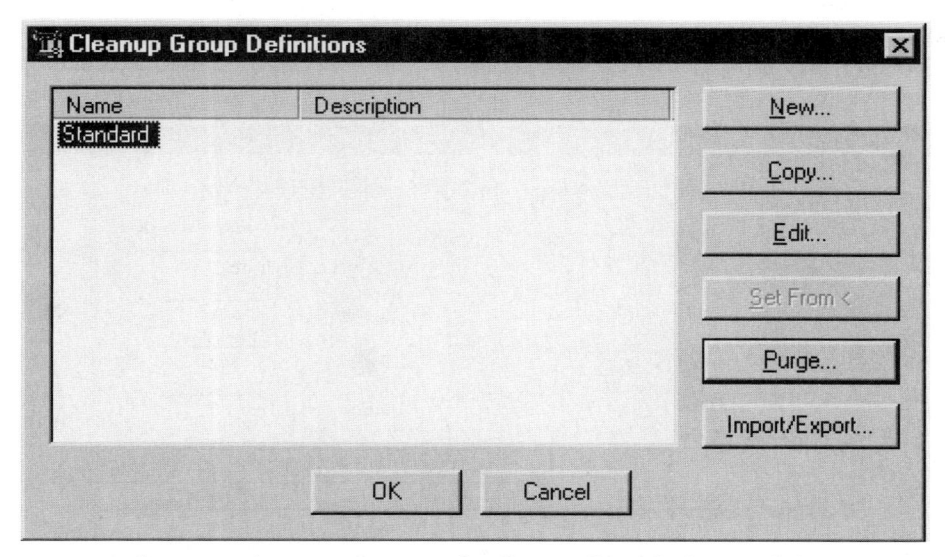

Figure 3.98 *Creating a cleanup group using the Cleanup Group Definitions dialog box*

The **Cleanup Group Definitions** dialog box includes a list of the cleanup groups defined for the drawing. The purpose of each of the buttons on the right of the dialog box is summarized below:

    New – Creates a new name for a cleanup group.

    Copy – Copies the definition of a cleanup group and assign a new name.

    Edit – Changes the name or description for a cleanup group.

    Set From< – Inactive in this command.

    Purge – Removes unused cleanup groups from the list.

Import/Export – Saves the cleanup group definitions to a resource file or imports cleanup groups into a drawing from a resource file.

The name of the cleanup group does not matter; however, walls with the same cleanup group name will merge together. Walls with different cleanup group names will not blend together. Select cleanup groups for a wall when it is created by editing the **Group** list of the **Add Walls** dialog box. You can assign existing walls a different cleanup group by editing the **Group** list of the **Wall Modify** dialog box when the **Wall Modify** command (**AECWALLMODIFY**) is executed.

The cleanup group of existing walls can also be changed by the **Wall Props** command (**WALLPROPS**). The **Dimensions** tab of the **Wall Properties** dialog box includes a **Cleanup Settings** section that allows you to select the cleanup group, as shown in Figure 3.99.

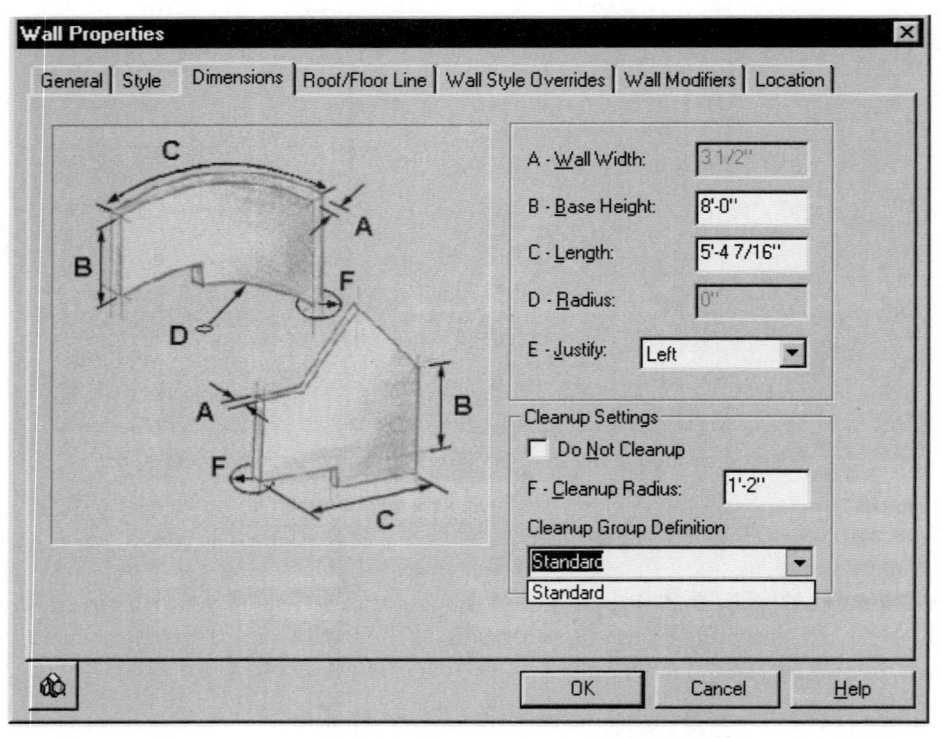

Figure 3.99 *Assigning a cleanup group in the Dimensions tab of the Wall Properties dialog box*

TUTORIAL 3.6 CREATING AND USING A WALL GROUP

1. Open *Ex3-6* from the \ADT Tutor\Ch3 directory and choose **File>SaveAs** from the menu bar; save the file as *Groups.dwg* in your student directory.

   The stud walls and footing walls were created with equal priority.

2. Choose **Design>Walls>Cleanup Group Definitions** from the menu bar to display the **Cleanup Group Definition**s dialog box as shown in Figure 3.100.

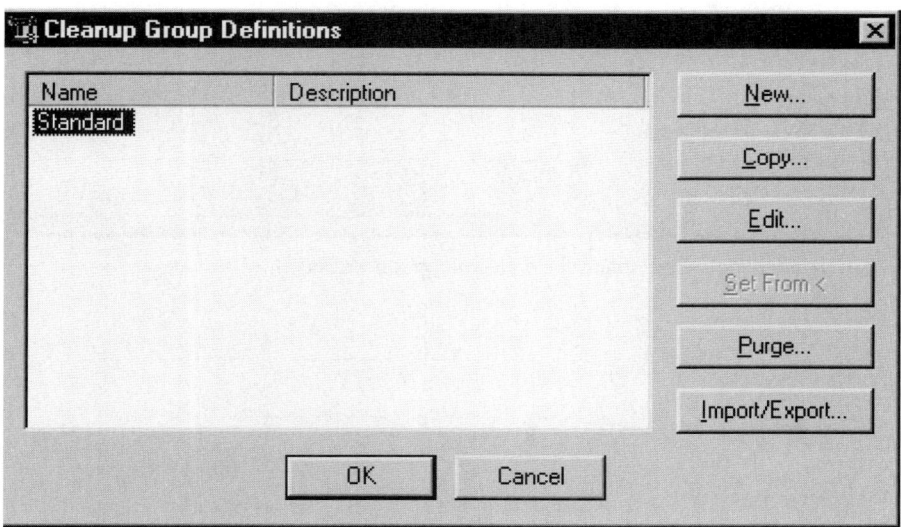

Figure 3.100 *Cleanup Group Definitions dialog box*

3. Select the **New** button and type the name **Wood** in the **Name** dialog box.
4. Select **OK** to dismiss the **Name** dialog box.
5. Select the **Edit** button in the **Cleanup Group Definition**dialog box and insert the description **Wood** walls as shown in Figure 3.101.
6. Select the **OK** button to dismiss the **Cleanup Group Definition Properties** dialog box.
7. Select the **OK** button to dismiss the **Cleanup Group Definitions** dialog box.
8. Select the 3 1/2" stud walls as shown in Figure 3.102.
9. Right-click to display the shortcut menu and choose **Wall Modify**.

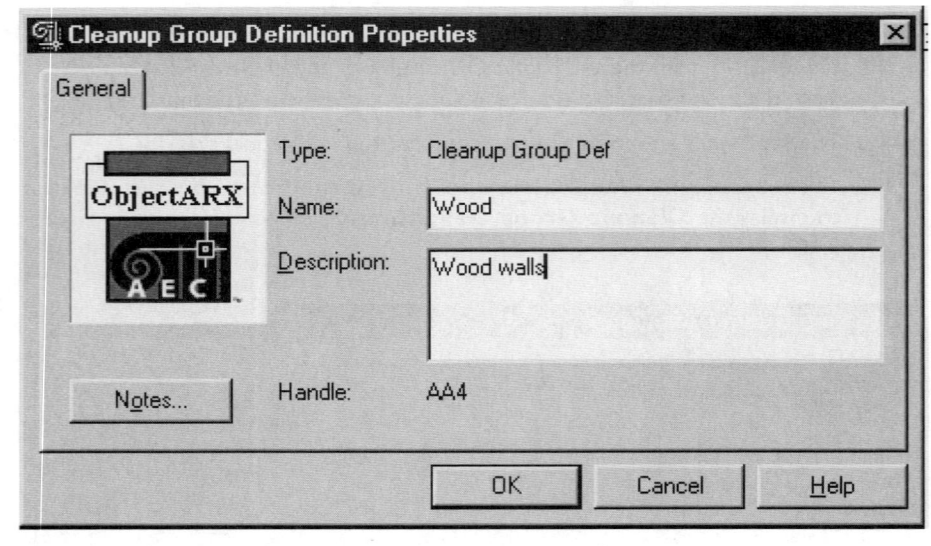

Figure 3.101 *Creating a description for a cleanup group definition*

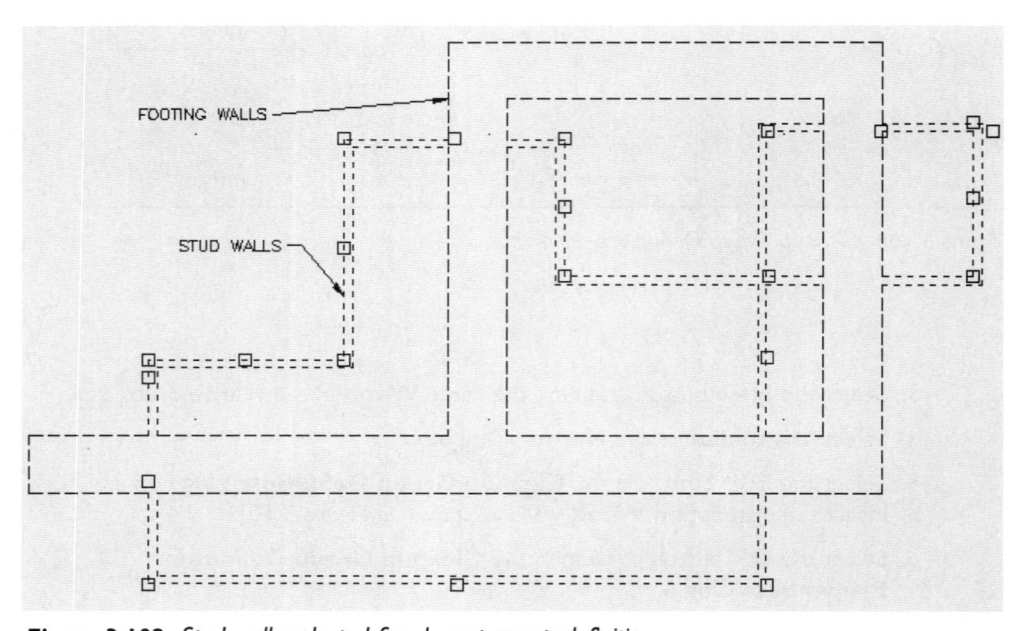

Figure 3.102 *Stud walls selected for cleanup group definition*

10. Edit the Group to **Wood** in the **Wall Modify** dialog box. Then select the **OK** button to dismiss the **Wall Modify** dialog box.

The walls should no longer merge; therefore the wood walls will override the display of the footing wall as shown in Figure 3.103.

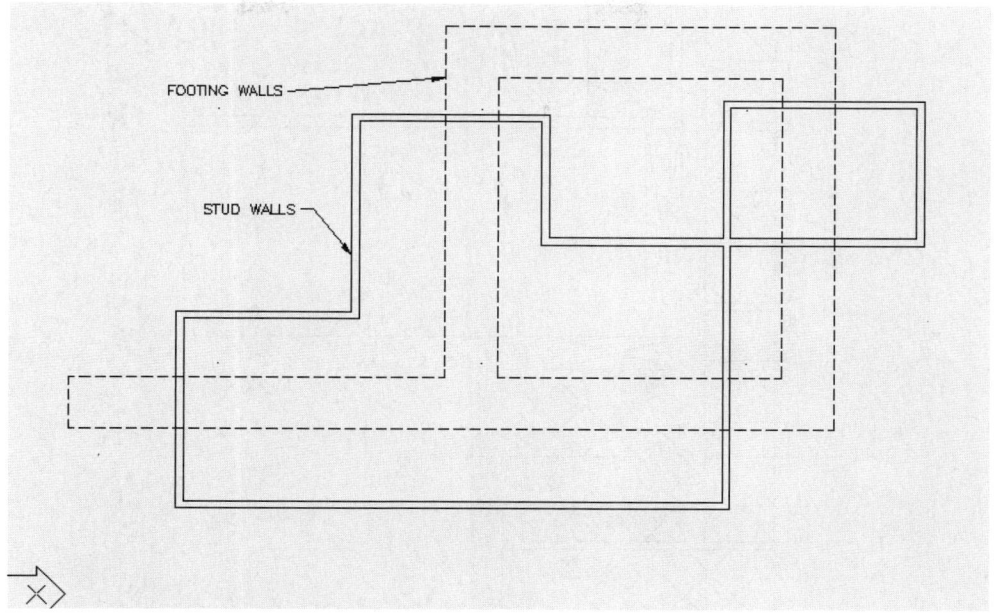

Figure 3.103 *Walls after cleanup group definition applied*

11. Save the drawing.

CREATING ADDITIONAL FLOORS

After you develop a floor plan, additional floors can be developed based upon the shape of the current floor. To ensure vertical alignment between floors, save the first floor plan as the second floor plan. Therefore, the second floor plan becomes a copy of the first floor plan. You can then edit the second floor plan by deleting non–load bearing walls and creating the rooms of the second floor.

If the exterior walls of one floor differ in construction from those of another floor, the wall styles of each floor can be created to permit the vertical alignment of the walls. The exterior walls of a basement in residential construction usually differ from those of the first floor. The wall section shown in Figure 3.104 illustrates the alignment of the wall components of each level.

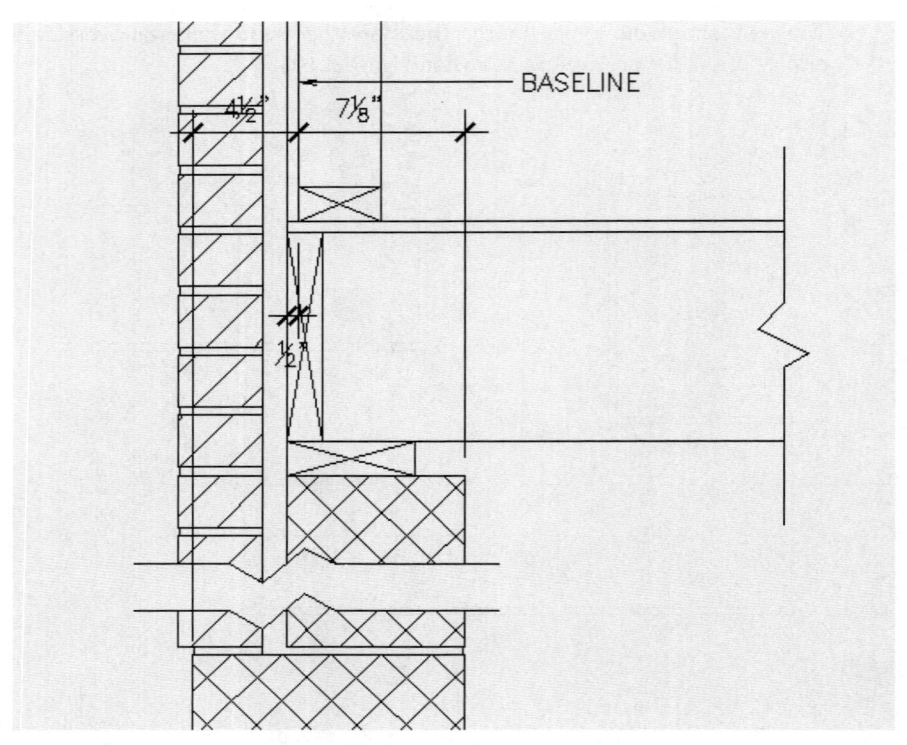

Figure 3.104 Wall section for basement construction

The baselines of each floor can be used to align the wall components of each floor. The baseline location of the walls for each floor will then create the control points for the alignment of floors.

Figure 3.105 shows the Frame_Brick wall style created in Tutorial 3.1 for the first floor plan and a wall style created for the basement level. The baselines of each wall are shown by centerlines. The wall style for the basement plan includes the brick and block components positioned relative to the baseline.

The components of the basement wall style shown above consist of:

Name	Priority	Edge Offset	Width
Block	40	−1/2	7-5/8
Brick	40	−4-1/2	3-5/8

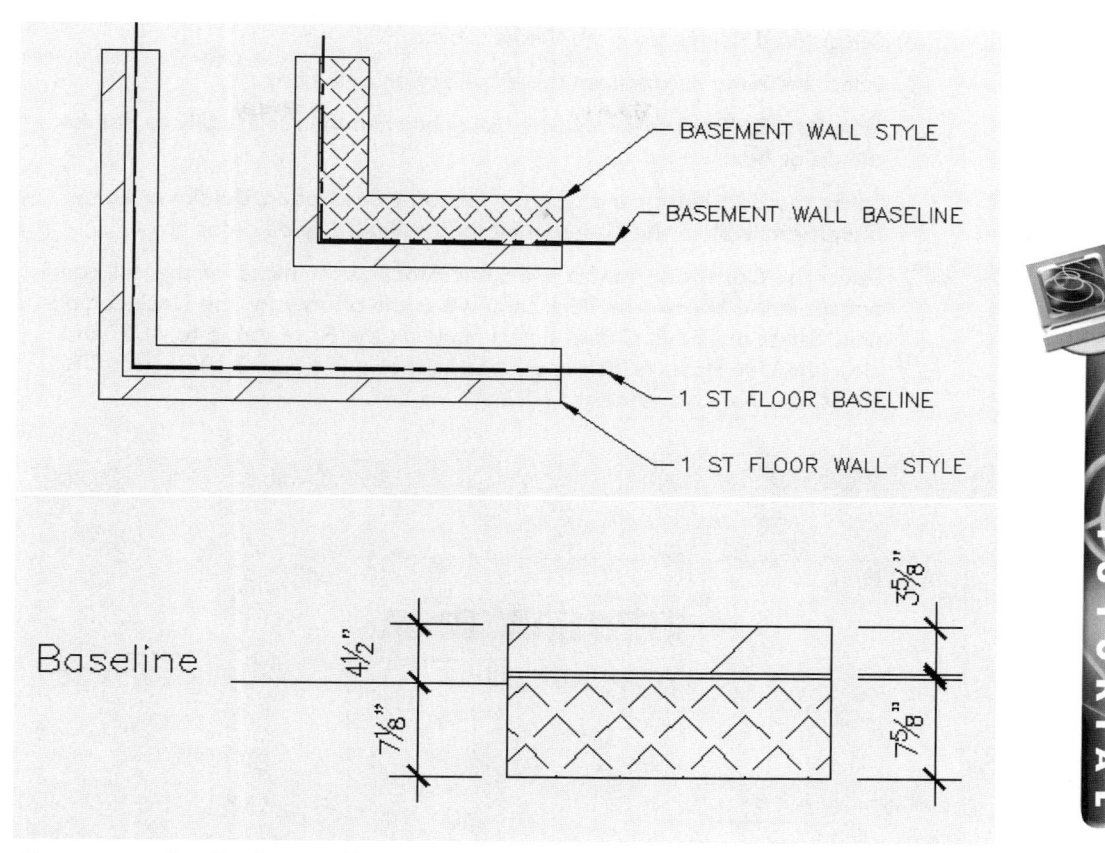

Figure 3.105 *Baseline locations for wall styles*

Therefore, if the first floor plan is copied to create the basement level, you can use the **Modify Wall** command to change the wall style of the basement exterior walls to a Basement wall style to align the walls vertically.

> **Note:** You can check the vertical alignment of levels by attaching floor plan 1 as a reference file to the basement floor plan. The reference file should be attached at an elevation higher than the basement wall height to avoid wall error markers if the walls interfere.

TUTORIAL 3.7 CREATING THE BASEMENT FLOOR PLAN

1. Open \ADT_Tutor\Ch3\Ex3-5.dwg
2. Choose **File>SaveAs** from the menu bar and save the drawing as *Ex3-7.dwg* in your student directory.

3. Select **Wall Styles** from the **Walls** toolbar.
4. Select the **New** button from the **Wall Styles** dialog box.
5. Type **BASEMENT** in the **Name** dialog box and then select **OK** to dismiss the dialog box.
6. Select the **Edit** button and type the following description: **Brick veneer basement wall** in the **General** tab of the **Wall Styles** dialog box.
7. Select the **Components** tab and enter **Block** as the name for the wall component in the **Name** edit field. Enter **40** as the priority for the Block component. Select the **Edge Offset** button and edit the **Base Value** to –1/2" and clear the **Use Base Width** check box as shown in Figure 3.106. Select **OK** to dismiss the **Component Offset** dialog box.

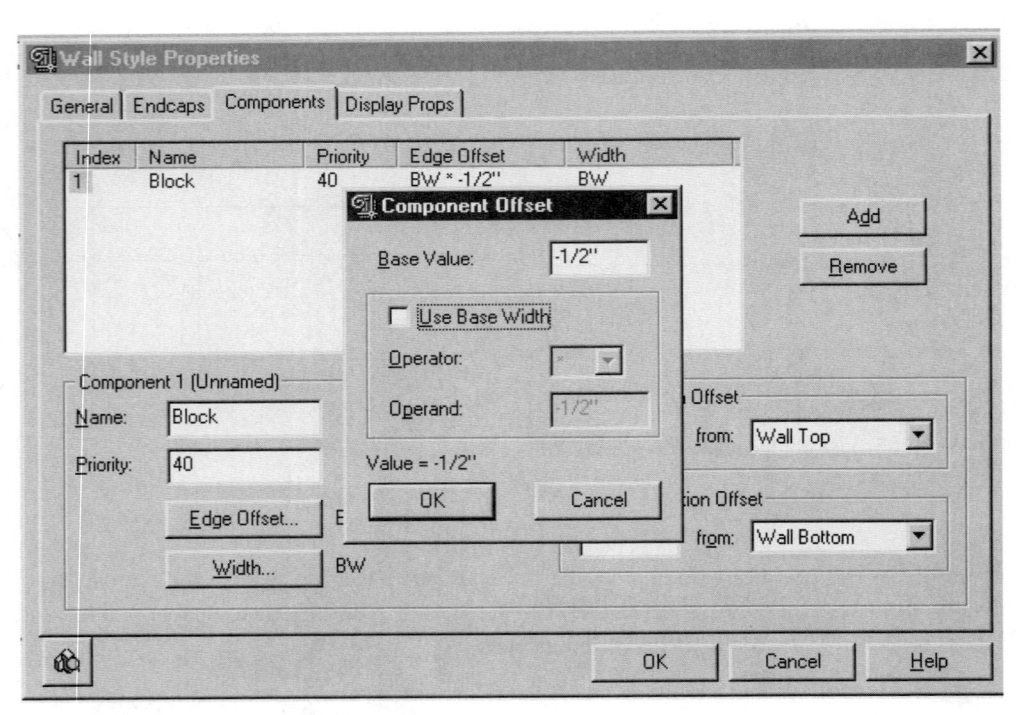

Figure 3.106 *Component Offset dialog box*

8. Select the **Width** button of the **Wall Styles** dialog box; then edit the **Base Value** to 7-5/8" and clear the **Use Base Width** check box. Select **OK** to dismiss the **Component Width** dialog box. Verify that the component properties are set as shown in Figure 3.107.

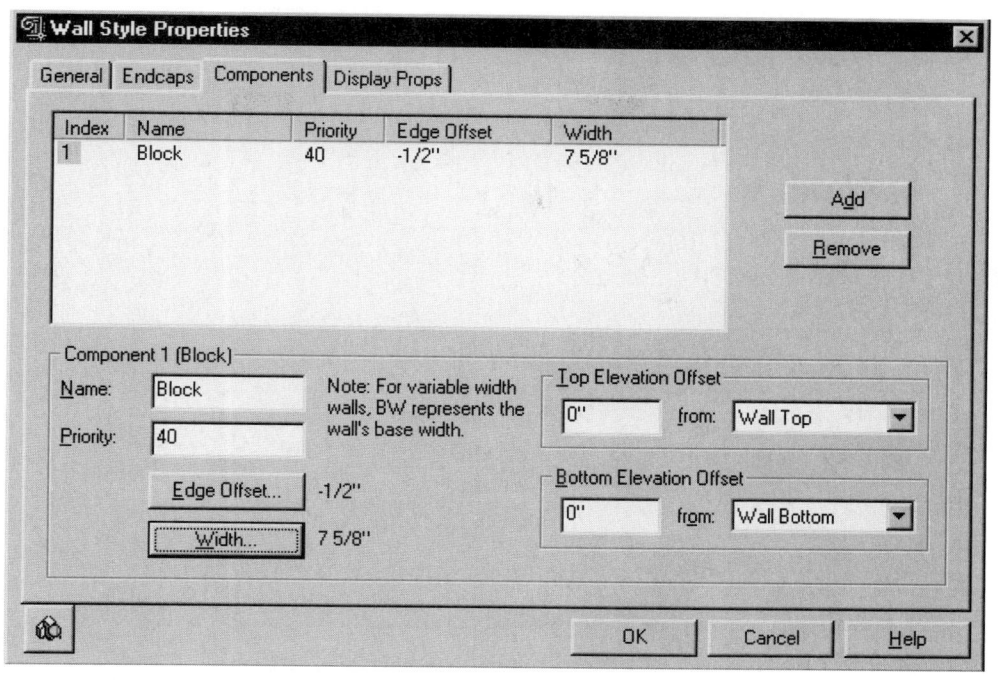

Figure 3.107 *Block component of the Basement wall style*

9. Select the **Add** button to add another wall component. Edit the name of the new component to Brick and set its priority to 40.

10. Select the **Edge Offset** button and edit the **Base Value** to –4-1/2". Verify that the **Use Base Width** check box is cleared. Select **OK** to dismiss the **Component Offset** dialog box.

11. Select the **Width** button and edit the **Base Value** to 3-5/8". Verify that the **Use Base Width** check box is cleared. Select **OK** to dismiss the **Component Width** dialog box. The component width, offset, and priorities should be set as shown in Figure 3.108.

12. Select the **Display Props** tab.

13. Select the **Wall Style Property Source** and then select the **Attach Override** button. Select the **Edit Display Properties** button to open the **Entity Properties** dialog box as shown in Figure 3.109.

14. Select the light bulb to turn ON the visibility of Hatch 1 (Block) and Hatch 2 (Brick). Select the **Hatching** tab and then select the **Pattern** column for Hatch 1 (Block) to open the **Hatch Pattern** dialog box, as shown in Figure 3.110.

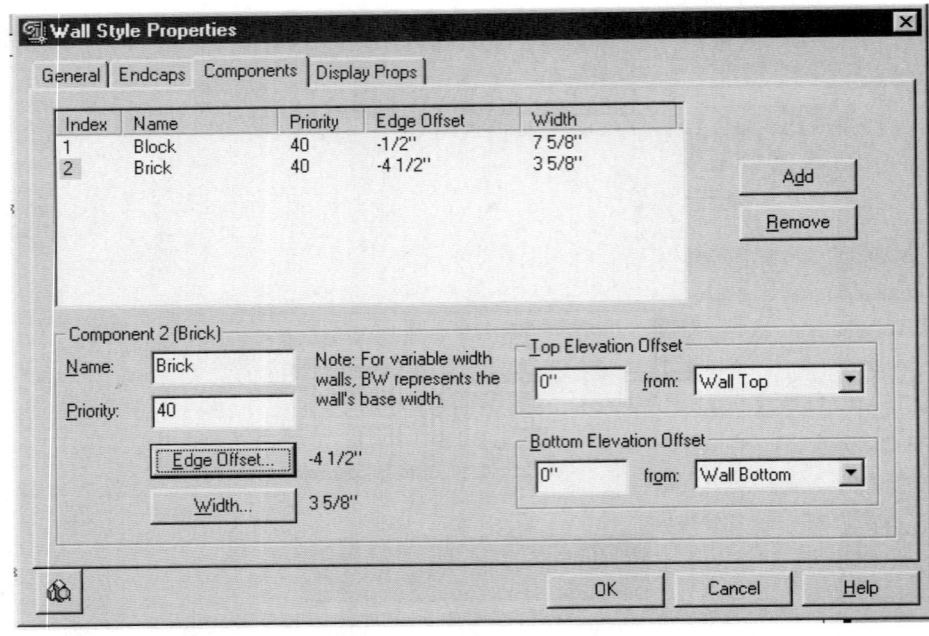

Figure 3.108 *Brick component settings of the Basement wall style*

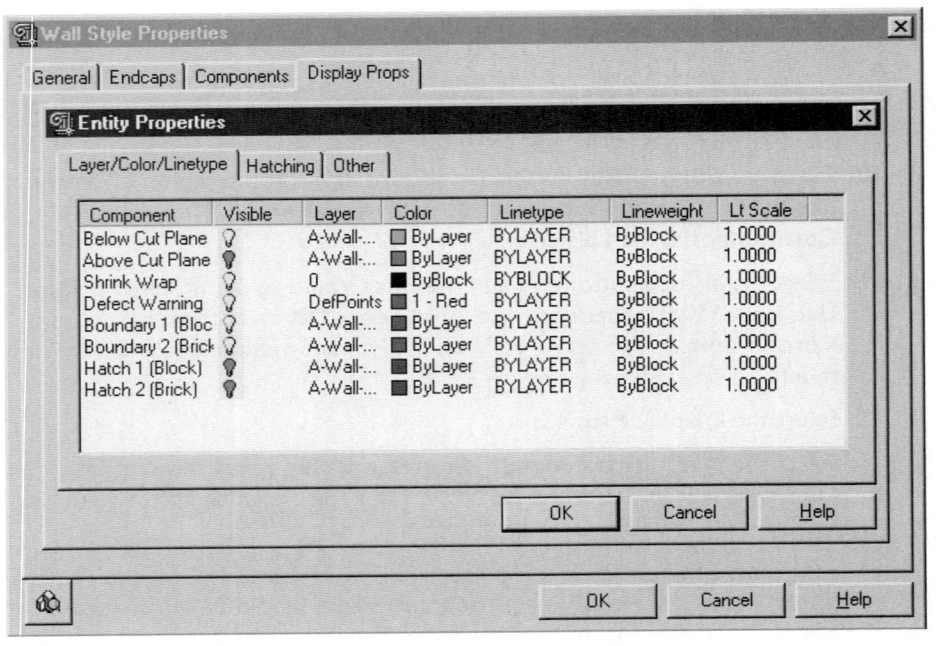

Figure 3.109 *Entity Properties of the Basement wall style*

Advanced Wall Features 199

Figure 3.110 *Hatch Pattern dialog box for setting hatch properties*

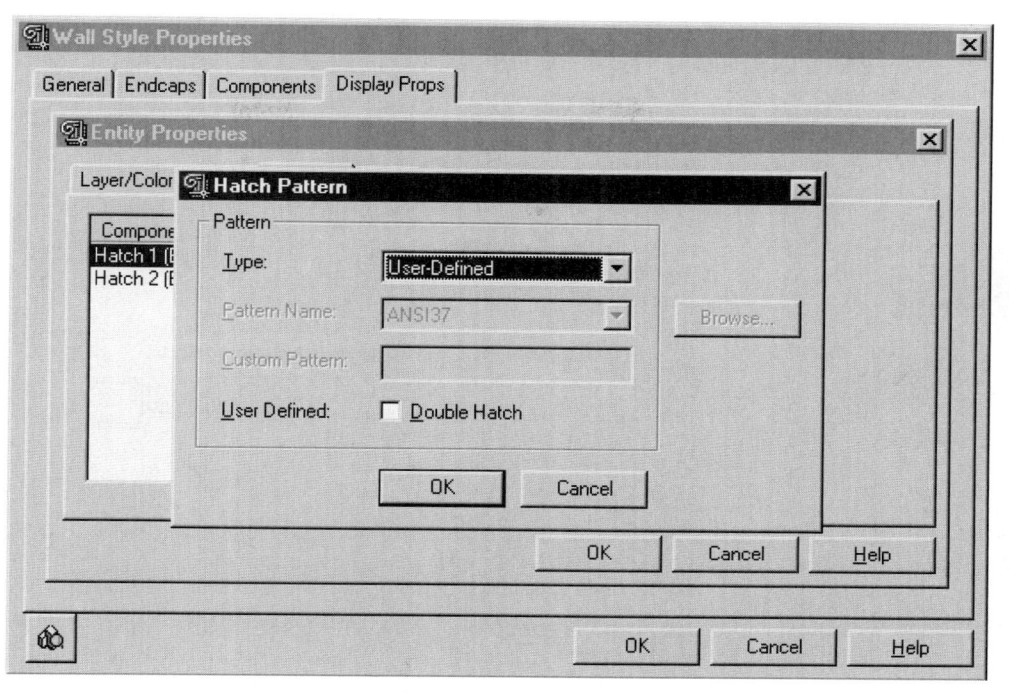

15. Edit the **Type** by selecting Predefined from the list and then select ANSI37 from the **Pattern** name list. Select **OK** to dismiss the **Hatch Pattern** dialog box. Edit the **Angle** column to 90 and edit the **Scale/Spacing** to 24 for the Hatch 1 (Block) component. Select **OK** to dismiss the **Entity Properties** dialog box. Select **OK** to dismiss the **Wall Style Properties** dialog box.

16. Select **OK** to dismiss the **Wall Styles** dialog box.

17. Select **Extend** from the **Modify** toolbar and refer to Figure 3.111 and the following command sequence.

    ```
    Command: _extend
    Current settings: Projection=UCS Edge=None
    Select boundary edges ... (Select the exterior wall at point P1.)
    Select objects: 1 found ENTER

    Select object to extend or[Project/Edge/Undo]: (Select
    ```
    the interior wall at point P2.)
    ```
    Select object to extend or [Project/Edge/Undo]:
    ```
    ENTER

Command:
Wall is extended as shown in Figure 3.112.

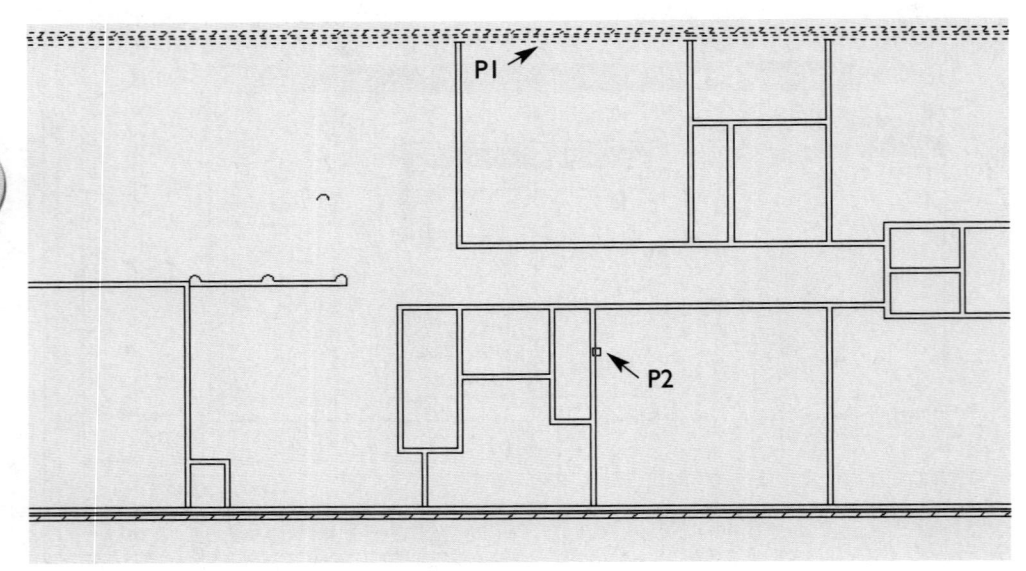

Figure 3.111 *Boundary selected for the EXTEND command*

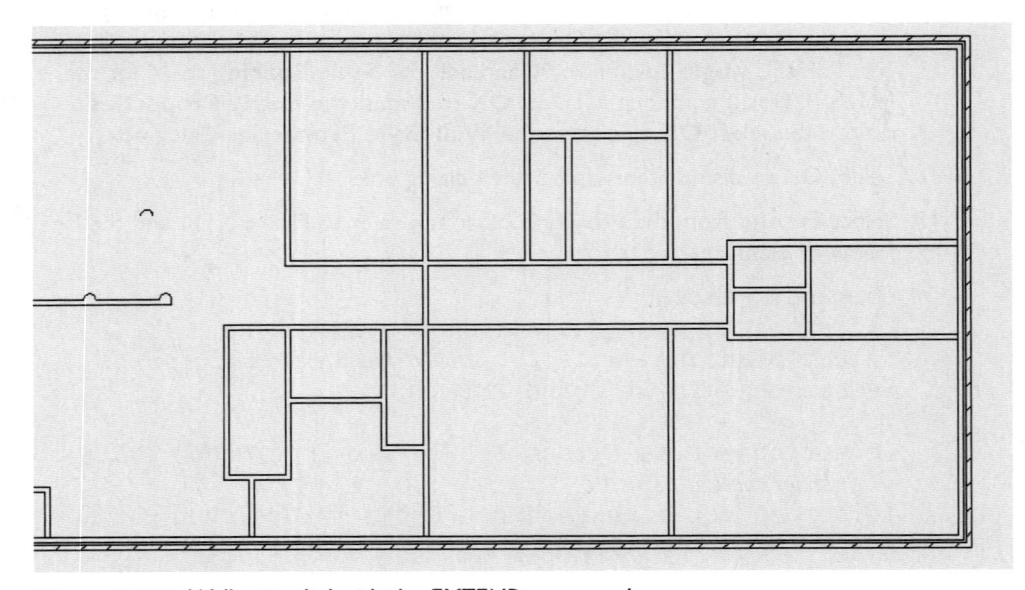

Figure 3.112 *Wall extended with the EXTEND command*

18. Select the **Trim** command from the **Modify** toolbar and refer to Figure 3.113 and the following command sequence.

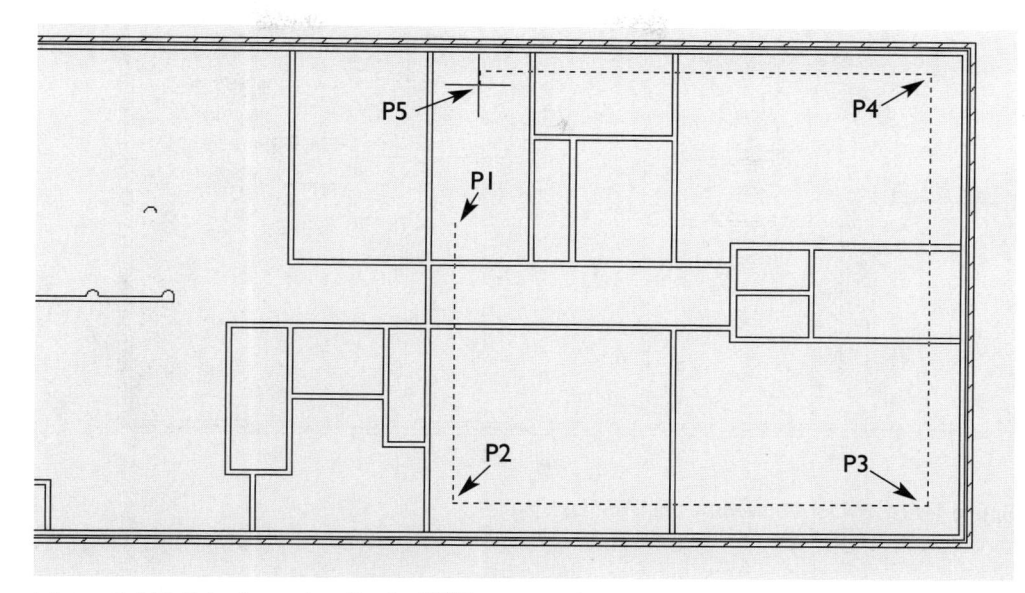

Figure 3.113 *Selection points for the TRIM command*

```
Command: _trim
Current settings: Projection=UCS Edge=None
Select cutting edges ...
Select objects: ENTER    (Select ENTER to select all objects as cutting edges.)
Select object to trim or [Project/Edge/Undo]: F  (Type F to select Fence selection method.)
First fence point: (Select point P1.)
Specify endpoint of line or [Undo]: (Select point P2.)
Specify endpoint of line or [Undo]: (Select point P3.)
Specify endpoint of line or [Undo]: (Select point P4.)
Specify endpoint of line or [Undo]: (Select point P5.)
No intersections in the interior of the curve.
Select object to trim or [Project/Edge/Undo]:  ENTER
(Select ENTER to end selection.)
Command:
```

Interior walls are trimmed as shown in Figure 3.114.

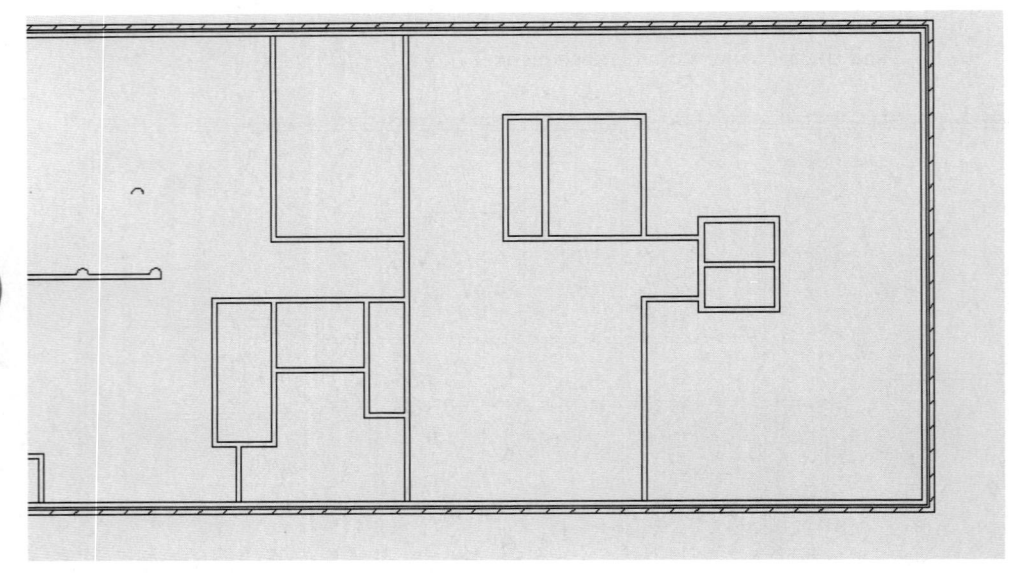

Figure 3.114 *Walls removed by the TRIM command*

19. Select the **Erase** command from the **Modify** toolbar and erase the walls shown in Figure 3.115 and in the following command sequence.

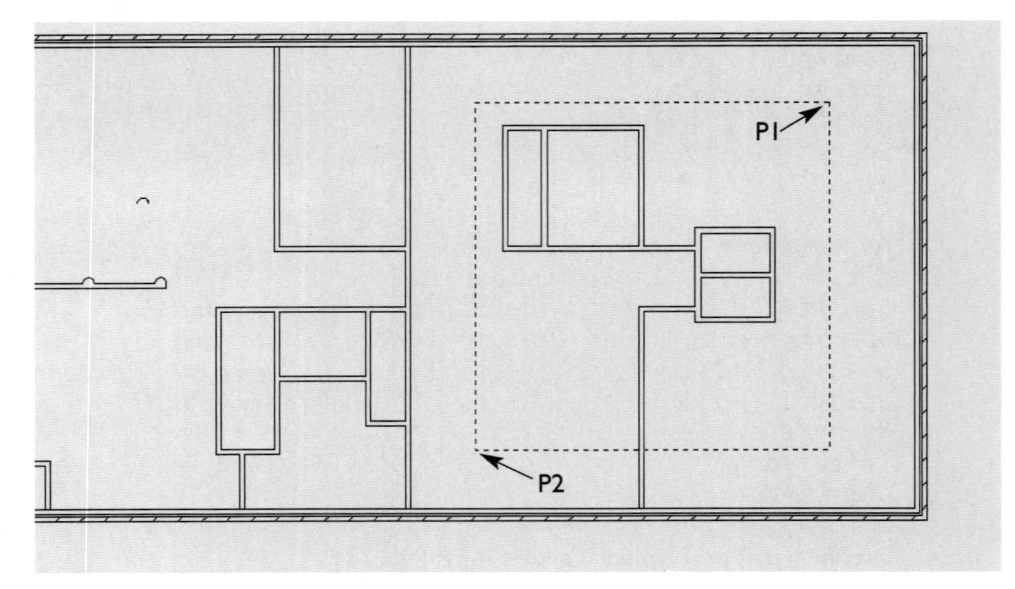

Figure 3.115 *Crossing selection window for erasing walls*

```
Command:  _erase
Select objects: (Select points P1 and P2 Crossing selection window.)
Specify opposite corner: 12 found
Select objects:   ENTER
Command:
```
20. Select the **Erase** command from the **Modify** toolbar and erase the walls shown in Figure 3.116 and the following command sequence.

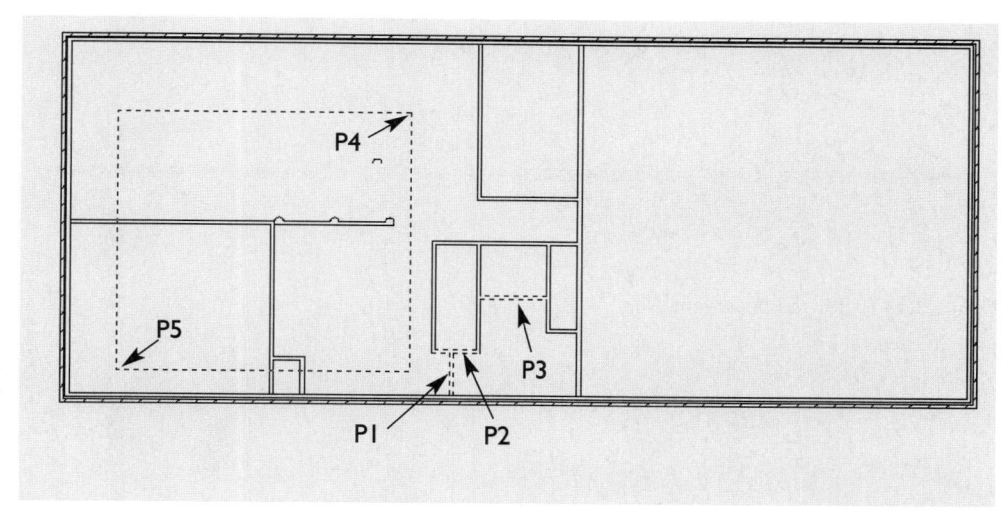

Figure 3.116 *Wall selection for the ERASE command*

```
Command: e ERASE
Select objects: (Select wall at point P1.) 1 found
Select objects: (Select wall at point P2.) 1 found, 2 total
Select objects: (Select wall at point P3.) 1 found, 3 total
Select objects: (Select walls by crossing selection window from P4 to
P5.) Specify opposite corner: 6 found, 9 total
Select objects:   ENTER (Pressing ENTER ends selection)
```
The floor plan should now appear as shown in Figure 3.117.

21. Select the **Extend** command from the **Modify** toolbar and extend the wall shown in Figure 3.118 and in the following command sequence.

```
Command:   EXTEND
Current settings: Projection=UCS Edge=None
Select boundary edges ...
Select objects: (Select exterior wall at P1 in Figure 3.118.) 1
found
```

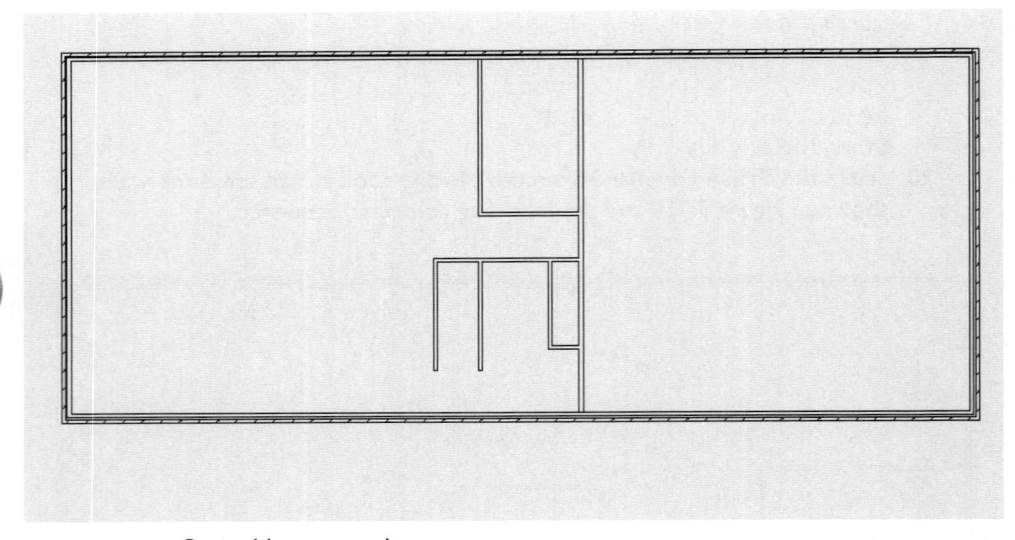

Figure 3.117 Revised basement plan

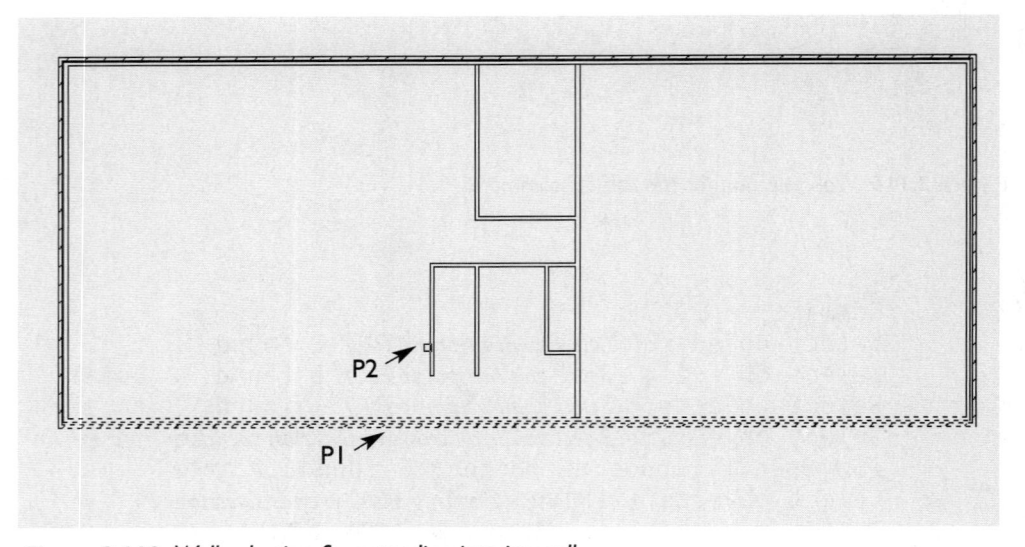

Figure 3.118 Wall selection for extending interior wall

```
Select objects:   ENTER   (Press ENTER to end boundary selection)
Select object to extend or [Project/Edge/Undo]: (Select
wall at point P2 in Figure 3.118.)
Select object to extend or [Project/Edge/Undo]:
ENTER  (Select ENTER to end object selection)
```

22. Select the four exterior walls, and then right-click and choose **Wall Modify** from the shortcut menu.
23. Select the Basement wall style from the **Style** list and verify the justification is set to Baseline in the **Wall Modify** dialog box. Select **OK** to dismiss the **Wall Modify** dialog box. The exterior walls should appear as shown in Figure 3.119.

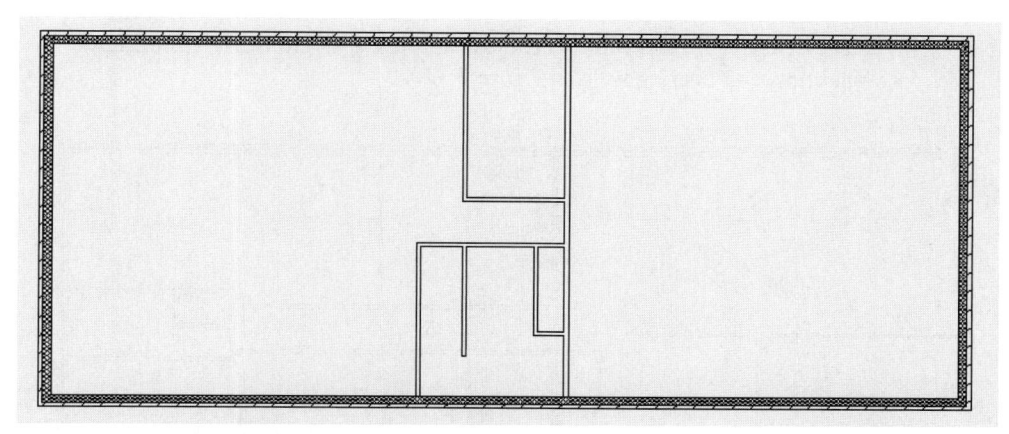

Figure 3.119 *Revised Basement plan*

24. Save the drawing.

CREATING A FOUNDATION PLAN

A foundation plan for a house that does not have a basement is created in a manner similar to the basement plan described above. The vertical alignment of the structural components is established through controlling the baseline locations of each floor. The wall height is reduced to the anticipated height of the foundation wall. Piers can be inserted as columns with layout curves. Layout curves allow you to equally space the piers along a centerline; they will be discussed in Chapter 13. You can create the concrete footing for the foundation wall and the pier footing by customizing the foundation wall style.

> **Note:** To represent double joists and girders, create a DOUBLE_JOIST wall style which has a component with zero width, zero offset and center linetype. Turn off the display of other components in the wall style. Change the wall style of interior walls that require double joists or girders to this DOUBLE_JOIST joist wall style.

TUTORIAL 3.8 CREATING A FOUNDATION PLAN

1. Open *Ex3-8* from the *ADT_Tutor\Ch3* directory.
2. Choose **File>SaveAs** from the menu bar and save the file as *Found.dwg* in your student directory.
3. Select **Layers** from the **Object Properties** toolbar and freeze the A-Anno-Dims layer. Select **OK** to dismiss the **Layer Properties Manager**.
4. Select **Erase** from the **Modify** toolbar and erase all interior walls except walls shown as bearing walls in Figure 3.120.

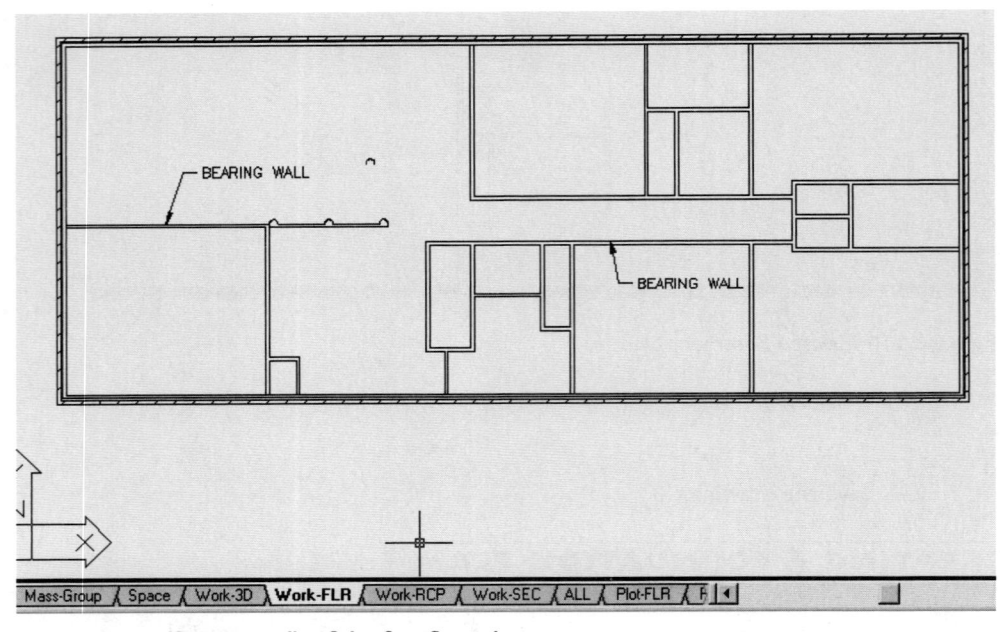

Figure 3.120 *Bearing walls of the first floor plan*

The interior partitions are retained to identify the centerline for the piers as shown in Figure 3.121.

5. Select **Wall Styles** from the **Walls** toolbar.
6. Select the **Import/Export** button from the **Wall Styles** dialog box to open the **Import/Export** dialog box.
7. Select the **Open** button of the **Import/Export** dialog box to open the **File to Import From** dialog box.

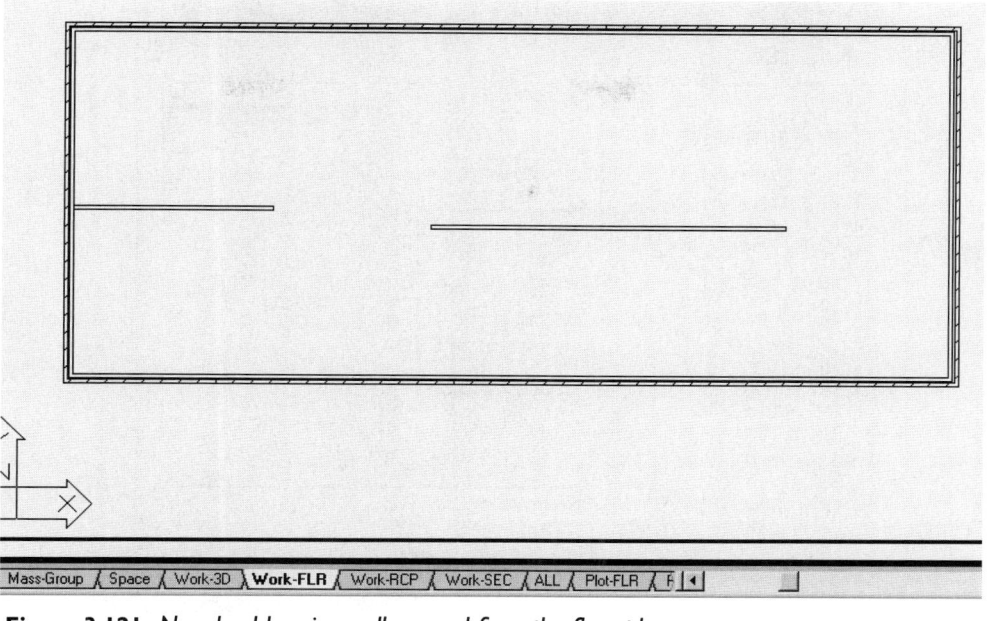

Figure 3.121 *Non–load bearing walls erased from the floor plan*

8. Select the *Ex3-7.dwg* from your student directory and then select the **Open** button to open the file and close the **File to Import From** dialog box.
9. Select Basement from the **External File** list and then select the **Import** button as shown in Figure 3.122.
10. Select **OK** to dismiss the **Import/Export** dialog box.
11. Select the Basement wall style and then select the **Copy** button of the **Wall Styles** dialog box to open the **Name** dialog box.
12. Type **Found12** in the **Name** dialog box and then select **OK** to dismiss the **Name** dialog box.
13. Select the **Edit** button of the **Wall Styles** dialog box to open the **Wall Style Properties** dialog box.
14. Type **Brick veneer foundation wall 12** in the **Description** field.
15. Select the **Components** tab and select the Brick component from the components list. Edit the **Bottom Elevation Offset** to 12" from Wall Bottom for the Brick component.

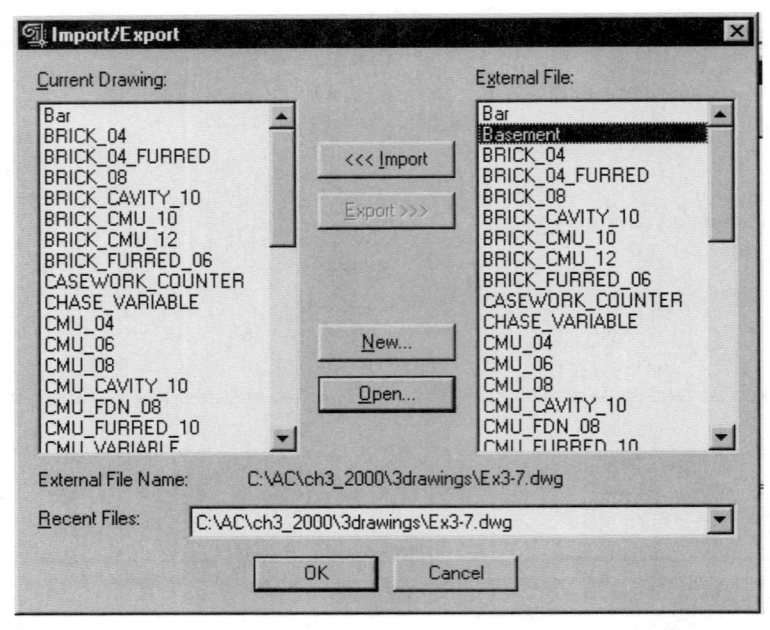

Figure 3.122 *Importing the Basement wall style*

16. Select the Block component and edit the **Bottom Elevation Offset** to 12" from Wall Bottom as shown in Figure 3.123.
17. Select the **Add** button to add a new component.
18. Edit the new component as follows: set **Name** to Footing, **Priority** to 40, **Edge Offset** to –10.5, **Width** to 2'-0", **Top Offset Elevation** to 1'-0" from Wall Bottom, and **Bottom Elevation Offset** from Wall Bottom to 0, as shown in Figure 3.124.
19. Select the **Display Props** tab and then select Wall Style in the **Property Source** list and select the **Edit Display Props** button to open the **Entity Properties** dialog box.
20. Select the **Below Cut Plane** component and then select the **Linetype** column to open the **Select Linetype** dialog box. Select Dashed from the linetype list and select **OK** to dismiss the **Select Linetype** dialog box as shown in Figure 3.125.

Advanced Wall Features 209

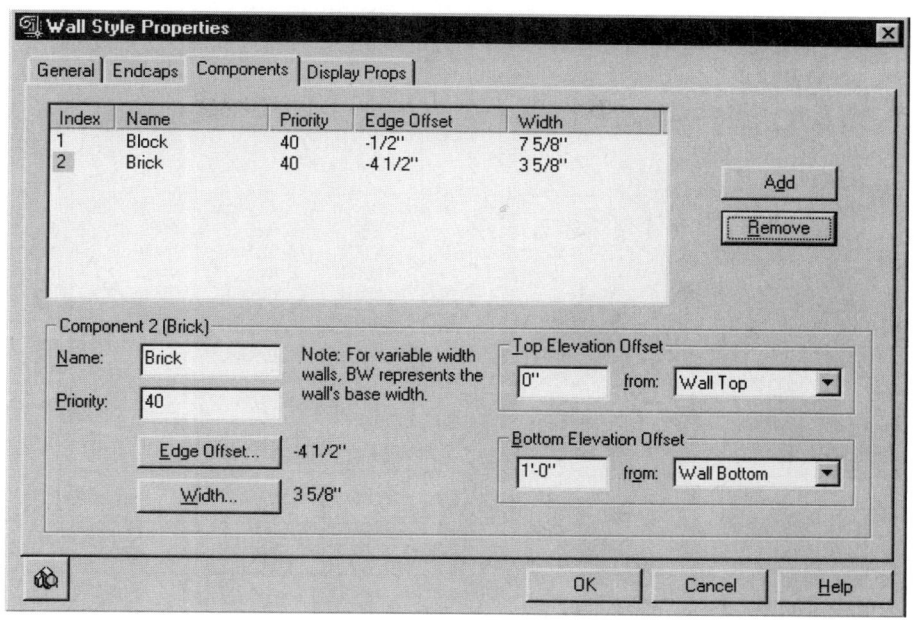

Figure 3.123 *Bottom elevation of brick and block component raised to top of footing*

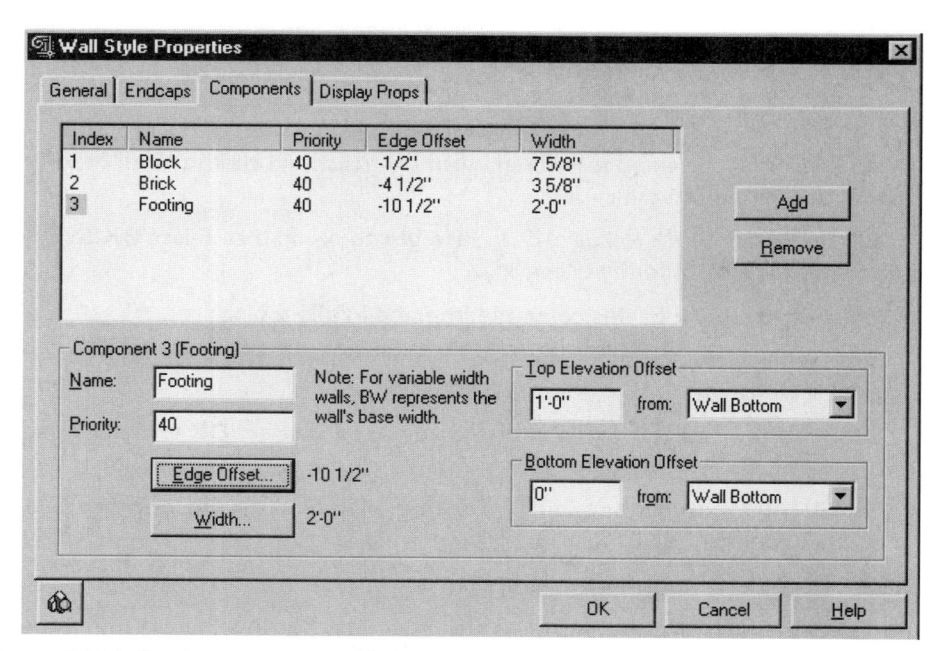

Figure 3.124 *Footing component added to wall style*

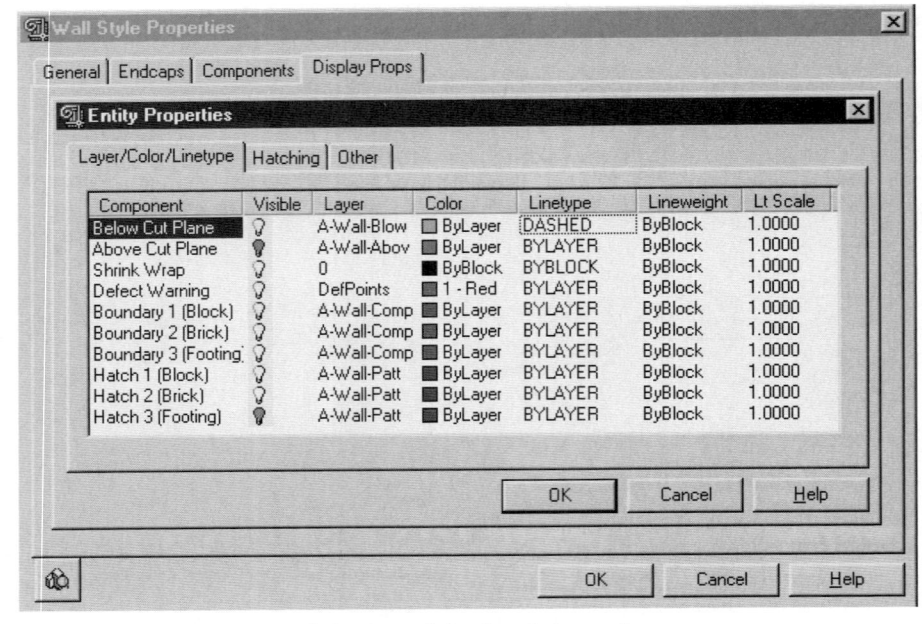

[Figure: Wall Style Properties dialog box showing Display Props tab and Entity Properties sub-dialog with Layer/Color/Linetype tab displaying components including Below Cut Plane, Above Cut Plane, Shrink Wrap, Defect Warning, Boundary 1 (Block), Boundary 2 (Brick), Boundary 3 (Footing), Hatch 1 (Block), Hatch 2 (Brick), Hatch 3 (Footing)]

Figure 3.125 *Entity Properties dialog box of the foundation wall*

21. Select **OK** to dismiss the **Entity Properties** dialog box.
22. Select **OK** to dismiss the **Wall Style Properties** dialog box.
23. Select **OK** to dismiss the **Wall Styles** dialog box.
24. Select the four exterior walls and then right-click and choose **Wall Modify** from the shortcut menu.
25. Change the **Style** to Found12 and **Height** to 4' and then select **OK** to dismiss the **Wall Modify** dialog box.
26. Select the **Wall Styles** command from the **Walls** toolbar.
27. Select Found12 from the wall style list and select the **Copy** button to open the **Name** dialog box.
28. Type **Pier_Footing** in the **Name** dialog box and select **OK** to dismiss the **Name** dialog box.
29. Select the **Edit** button and type **Footing 24 x 12** in the **Description** field of the **General** tab.
30. Select the **Components** tab.
31. Select the Brick component and then select the **Remove** button to remove that component. Repeat this step for the Block component.

32. Select the Footing component and change the **Edge Offset** to -12" and select **OK** to dismiss the **Component Offset** dialog box. Edit the **Top Elevation Offset** to 0 from Wall Top. The revised **Components** tab should now appear as shown in Figure 3.126.

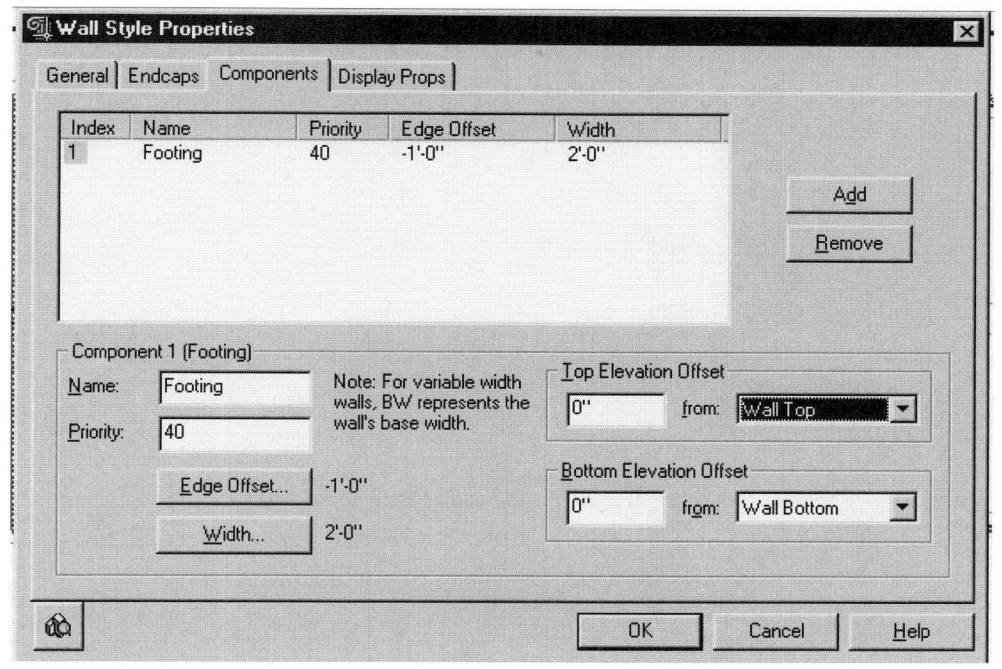

Figure 3.126 Revised footing component for the pier footing

33. Select **OK** to dismiss the **Wall Style Properties** dialog box.
34. Select **OK** to dismiss the **Wall Styles** dialog box.
35. Select the **Add Wall** command from the **Walls** toolbar. Edit the **Add Walls** dialog box as follows: set the **Style** to Pier_Footing, **Height** to 12" and **Justification** to Baseline.
36. Respond to the following command line prompts to place the pier footing as shown in the command sequence and in Figures 3.127 and 3.128.

```
Command: _AecWallAdd
Start point or
[STyle/Group/WIdth/Height/OFfset/Justify/Match/Arc]:
node    ENTER
```

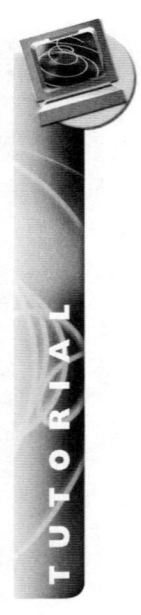

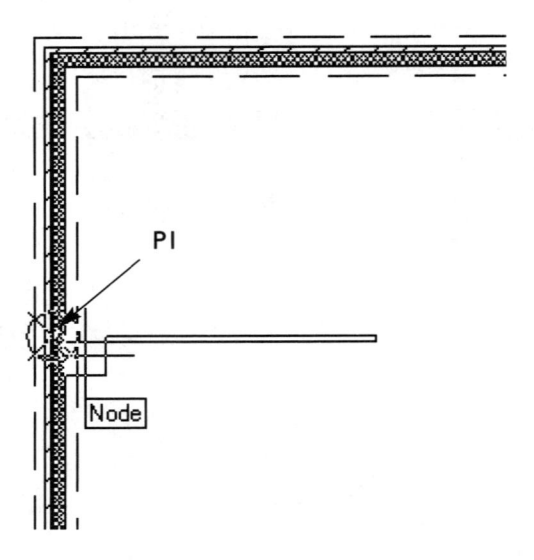

Figure 3.127 *Drawing the footing wall from the node of the interior wall*

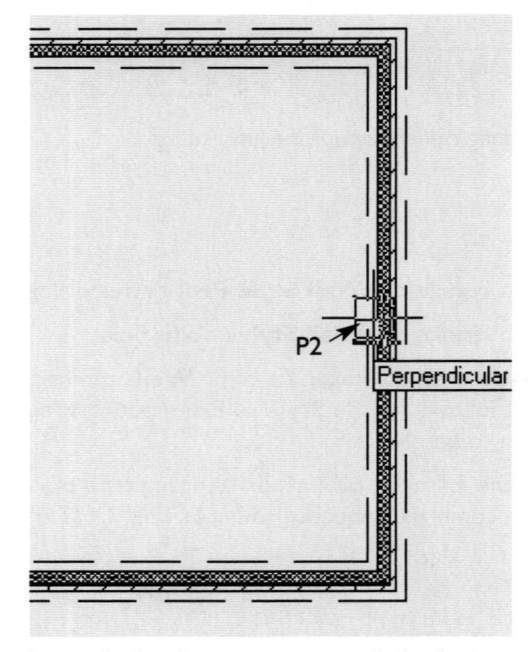

Figure 3.128 *Use the Perpendicular object snap to specify the footing wall endpoint*

```
Of (Select a point near P1 as shown in Figure 3.127.)
End point or
[STyle/Group/WIdth/Height/OFfset/Justify/Match/Arc]:
perp   ENTER
To (Select a point near P2 as shown in Figure 3.128.)
End point or
[STyle/Group/WIdth/Height/OFfset/Justify/Match/Arc/Un
do]:   ENTER   (Select ENTER to end the command.)
```

37. Select the Erase command from the Modify toolbar and erase the interior walls as shown at P1 and P2 in Figure 3.129.

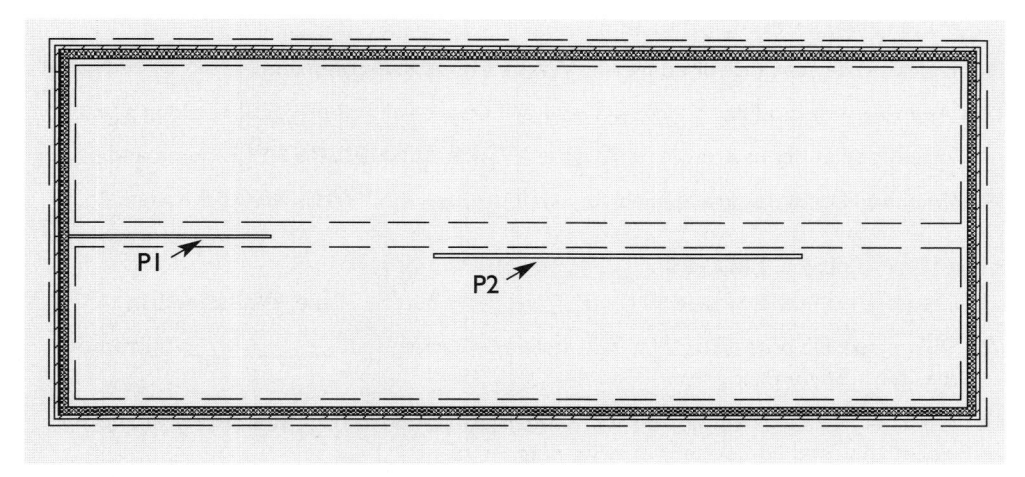

Figure 3.129 *Erase interior bearing walls*

38. Save the drawing.

SUMMARY

1. Wall styles are created through the **Wall Styles** command (**AECWALLSTYLE**).

2. Wall components are created by the **Wall Styles** command; they represent sub-assemblies of a wall and can be defined with unique widths and elevations.

3. The edge offset distance of a wall component positions one edge of the wall component from the baseline.

4. The width of a wall component establishes the distance between the surfaces of a wall component.

5. The wall component priority number controls the cleanup of wall components when they intersect with other walls.

6. Wall endcap styles are applied to a wall component to close the end of a wall component at wall openings or wall ends.
7. Wall styles include display properties control settings that control layer, visibility, color, and linetype of wall entities.
8. Display control of a wall style can include the display of hatching or wall components.
9. Wall styles can be exported to resource files and imported from resource files to other files.
10. Endcap styles for wall components are created by the **Wall Endcap Style** command (**WALLENDCAPSTYLE**).
11. Endcap styles can be edited by the **WALLENDCAP** command.
12. Applying wall modifier styles to a wall will create vertical projections along a wall.
13. Wall modifier styles are created by the **WALLMODIFIERSTYLE** command.
14. Wall modifier styles are applied to a wall through the **WALLPROPS** command.

REVIEW QUESTIONS

1. Create a wall style by selecting the _____ button of the **Walls** toolbar.
2. Define specific properties of a wall style by selecting the _____ button of the **Wall Style** dialog box.
3. A wall component with a priority number of 10 will override the display of another wall with a wall component priority number of _____.
4. Toggling _____ **Use Base Width** in the **Component Offset** dialog box will set the **Edge Offset** to variable.
5. If the **Edge Offset** of a wall component is set to −8 and the wall is drawn from left to right, the component will be drawn (above, below) the justification line.
6. If the **Width** of a wall component is to −8 and the wall is drawn from left to right, the component will be drawn (above, below) the justification line.
7. The top elevation offset of a wall component can be specified relative to _____, _____, _____, and _____.
8. Hatching a wall component is controlled by _____.
9. The geometry used to create an endcap must be a _____.
10. Wall modifier styles are created from _____.
11. Wall modifiers are attached to a wall by the _____ command.
12. Create wall cleanup groups by selecting the _____ command.

13. The Index number of the endcap component determines the _____ of the wall component capped by the endcap style.
14. The _____ command is used to create an endcap style.
15. Endcap styles are assigned to a wall with the _____ command.
16. Wall modifiers can be placed along the length of the wall at the following locations _____, _____, and _____.
17. Wall styles are transferred to other drawings with the _____ command.
18. Describe the procedure to change the style of a wall using the shortcut menu.
19. Describe the procedure of changing the definition of the wall style for an existing wall.
20. The entity display can be controlled for up to _____ boundaries.

PROJECT

EXERCISE 3.1 CREATING A WALL STYLE

Create a new drawing in your student directory named *Ex3-9*. Create a wall style named Basement6 to represent the basement wall as shown in the section in Figure 3.130.

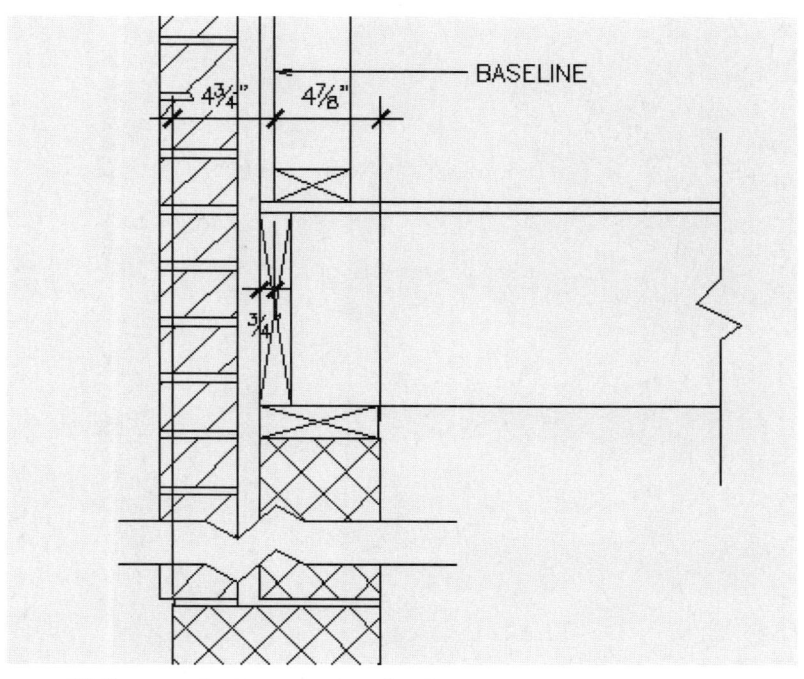

Figure 3.130 *Wall section for Basement6 wall style*

The wall should consist of two components, named brick and CMU. Edit the entity display properties of the wall to create an ANSI 31 hatch pattern for the brick component and ANSI 37 hatch pattern for the CMU wall component. The completed Basement wall style should appear as shown in Figure 3.131.

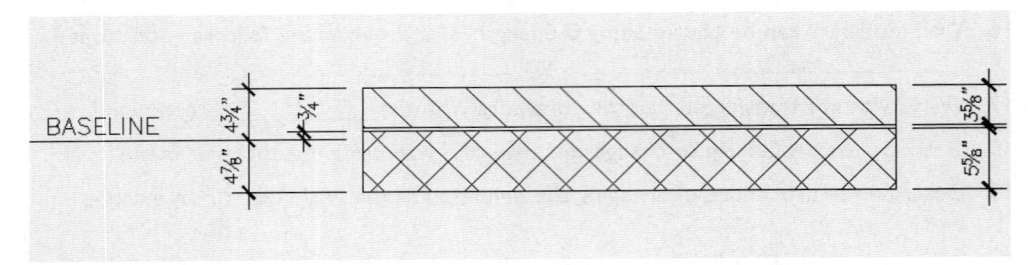

Figure 3.131 *Basement wall style*

CHAPTER 4

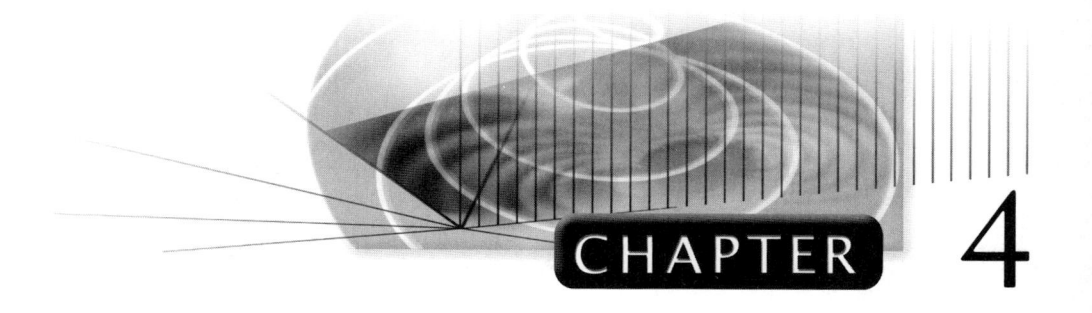

Placing Doors and Windows

INTRODUCTION

Doors and Windows are placed in the drawing as objects with three-dimensional properties. When doors, windows, and openings are placed in a wall, the wall is automatically trimmed and edited. The swing, hinge location, and position within and along the wall can be edited in Architectural Desktop. Door and window styles can be created, imported, and exported from resource files to improve design.

OBJECTIVES

After completing this chapter, you will be able to

- Place doors with precision, specifying swing, insertion point, and properties using **Add Door**

- Edit swing, opening percent, and hinge location of doors and windows

- Change size, location, and properties of doors using **Modify Door**

- Create door styles with specific dimensions, type, shape, and display properties

- Use **Door Properties** to edit door style, dimensions, display properties, and anchor of a door

- Shift doors and windows within and along walls using the **REPOSITIONWITHIN** and **REPOSITIONALONG** commands

- Place windows in walls with precision using **Add Window** (**WINDOWADD**)

- Edit the style and dimensions of a window using **Window Properties**

- Edit the size and style of a window using **Modify Window** command

Edit the location and swing of doors and windows using grips

Create and modify window styles using the **Window Styles** command

Place and modify openings in walls using **Add Opening** and **Modify Opening**

Create an AEC Profile and apply the AEC Profile shape to create a custom door

Import and Export door and window styles

The commands to place and modify doors, windows and openings are on the **Doors/Windows/Openings** toolbar as shown in Figure 4.1.

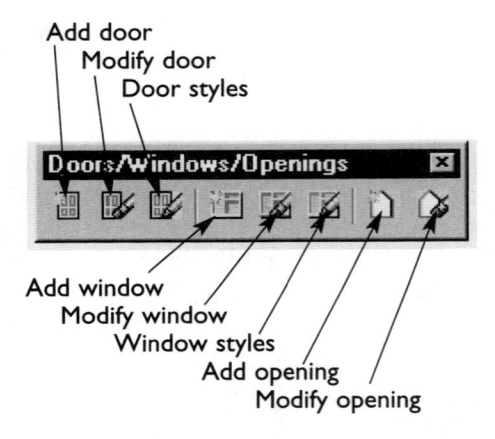

Figure 4.1 *Doors/Windows/Openings toolbar*

PLACING DOORS IN WALLS

The **Add Door** command is used to place doors in a drawing. See Table 4.1 for command access.

Menu bar	Design>Doors>Add Door
Command prompt	DOORADD
Toolbar	Select Add Door from the Doors/Windows/Openings toolbar shown in Figure 4.2
Shortcut	Right-click over drawing area; choose Design>Doors>Add

Table 4.1 *Add Door access*

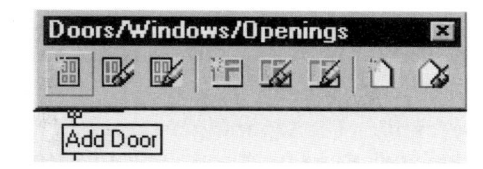

Figure 4.2 *Add Door command on the Doors/Windows/Openings toolbar*

When the **Add Door** command (**DOORADD**) is executed, the **Add Doors** dialog box opens, allowing you to edit the properties of the door, as shown in Figure 4.3.

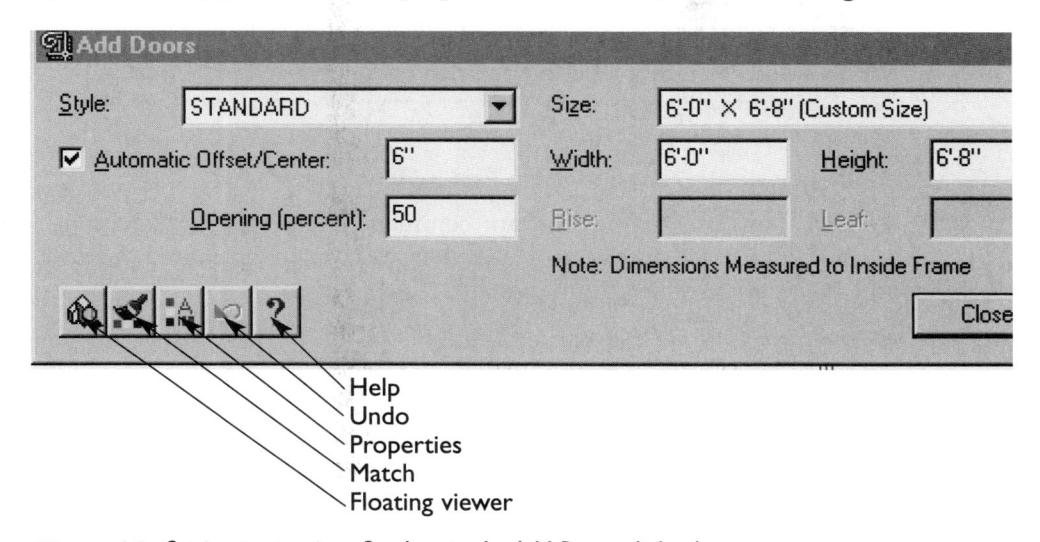

Figure 4.3 *Setting properties of a door in the Add Doors dialog box*

The **Add Doors** dialog box consists of the following edit fields and buttons:

Style – Lists door styles included in the drawing.

Size – Lists the preset sizes for doors.

Width – Allows you to enter the desired width of the door prior to placement.

Height – Allows you to enter the desired height of the door prior to placement.

Automatic Offset/Center – When selected, the **Automatic Offset/Center** check box sets placement to the center of the wall or an offset distance from the ends of the wall. If selected, the distance to offset from the ends of the wall for door placement is placed in the edit field to the right of the check box.

Opening (percent) – The **Opening** percent edit field allows you to set the angle of the door swing as a percentage of a 180-degree angle. A 50 percent swing will place the door open at an angle of 90 degrees.

Floating Viewer – Opens the **Viewer** and allows you to view the door prior to placement.

Match – Allows you to set the size and style of a door by selecting an existing door in the drawing.

Properties – Opens the **Door Properties** dialog box, which can be used to select and edit the door style, dimensions, endcap, and anchor properties.

Undo – Becomes active after the insertion point is defined for a door. Selecting **Undo** will remove the insertion point definition, allowing you to select a different insertion point for the door.

Help – Opens the Help file for the **Add a Door** topic.

To place a door in a wall, select the **Add Door** command. After you select the command, the mouse pointer is converted to a select box and you are prompted to select a wall or space boundary as shown in Figure 4.4.

```
Command: _AecDoorAdd
Select wall, space boundary or RETURN: (Select a wall.)
Insert point or
[STyle/WIdth/HEight/OPening/Auto/Match]: (Select an insert point for the door.)
Insert point or
[STyle/WIdth/HEight/OPening/Auto/Match/Undo]:    ENTER
```

The first prompt of the command is to select a wall or space boundary or to press ENTER (RETURN). Selecting a wall or space boundary will lock the door anchor to a location along the wall or space boundary. However, you can place a freestanding door by pressing ENTER rather than selecting a wall or space boundary. If you select a wall or space boundary, the door object will rubber-band along the wall or space boundary as you move the mouse. You anchor the door to a location along the wall or space boundary by selecting a point with the mouse when prompted for the insert point in the command line (Figure 4.4).

Prior to anchoring the door in the wall, you can edit the **Add Doors** dialog box by clicking in the dialog box, shifting the focus from the drawing area to the dialog box. You can then set the properties of the door in the dialog box. To return the focus back to the drawing area, click anywhere in the drawing area and then select a location for the door on the wall or space boundary. You can continue to toggle between the draw-

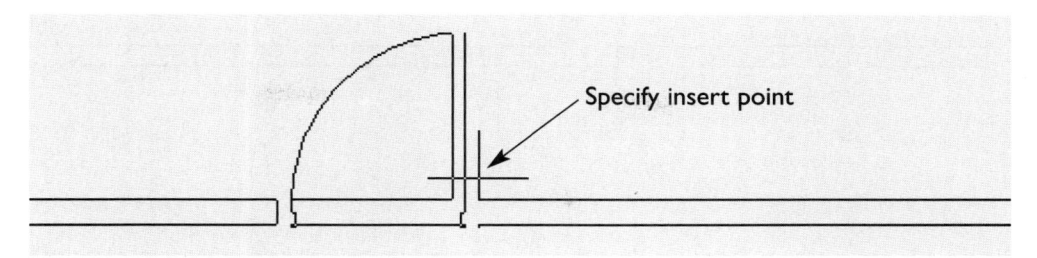

Figure 4.4 *Wall selected for placing a door*

ing area and the **Add Doors** dialog box until the size and style of the door are satisfactory. Each time the **Add Doors** dialog box is edited, the phantom image of the door is revised when the drawing area is activated. Toggling between the drawing area and the **Add Doors** dialog box allows you to edit the properties of a door prior to placing it.

You can set door sizes and styles solely by using the **Add Doors** dialog box or you can define them by using the options of the command line. Use command line options of the **Add Door** command by entering the capitalized letters of the command options as shown below. In the following command sequence, typing **WI** in the command line sets the width of the door.

```
Command: _AecDoorAdd
Select wall, space boundary or RETURN: (Wall selected.)
Insert point or
[STyle/WIdth/HEight/OPening/Auto/Match]: WI ENTER
Width <2'-0">: 3'
Insert point or
[STyle/WIdth/HEight/OPening/Auto/Match]: (Insertion point
selected.)
```

Using the command line input to set the properties of a door is slower because it requires more typing. Changing the style of door using command line options is difficult because it requires recalling and typing the name of a style in the command line. Using the **Add Doors** dialog is a more efficient method for setting the options of the **Add Door** command.

SETTING THE SWING

After selecting the wall, establish the door swing by placing the pointer above, below, left, or right relative to the insert point, which you then select. Prior to selecting the insert point, the movement of the mouse shifts the door swing. The position of the door swing when the insert point is selected finalizes the position for the door swing. In Figure 4.4 the mouse pointer was moved above the wall to set the door swing above the wall. In Figure 4.5 the mouse pointer was positioned below the wall to set the door swing below the wall.

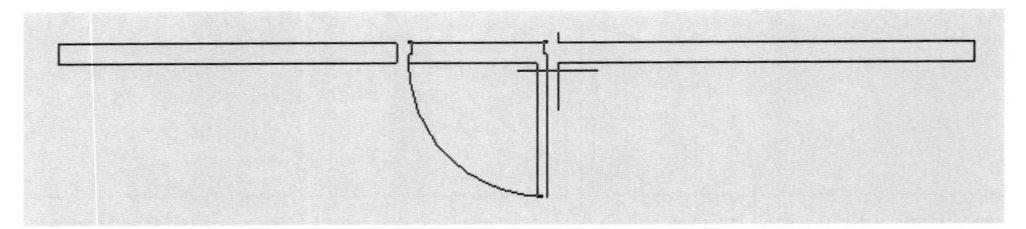

Figure 4.5 *Setting the swing of the door by moving the mouse pointer below the insertion point*

Moving the mouse pointer to the left of the door will flip the door to the left side, as shown in Figure 4.6.

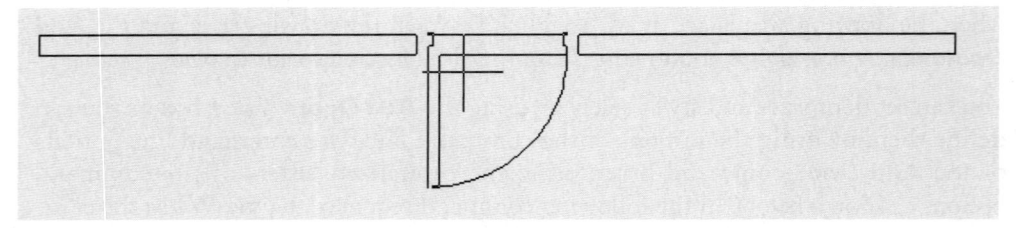

Figure 4.6 *Moving the hinge point of a door by moving the mouse pointer to the left of the insert point*

Moving the mouse pointer to the left and above the wall will flip the door swing above and to the left as shown in Figure 4.7.

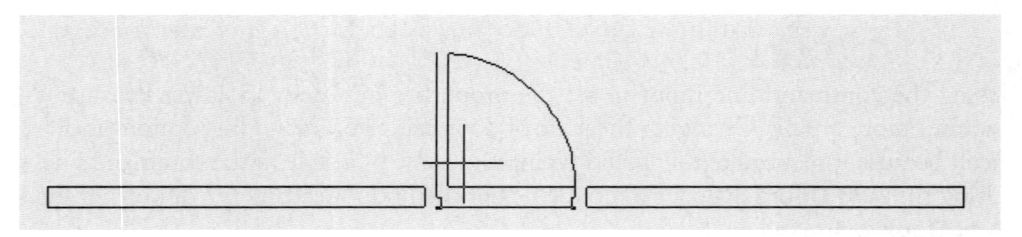

Figure 4.7 *Moving the hinge point of a door by moving the mouse to the left and above the insert point*

LOCATING THE INSERTION POINT OF THE DOOR

The handle or the insertion point for the doors in Figures 4.4 and 4.5 is on the right. The handle for a door is always on the side of the door nearest the start point of the wall. The wall shown in each figure above was drawn from right to left; therefore the handle for the door is on the right.

The grips of a door can be used to reverse the handle and swing of the door once a door is inserted. They are not used to change the size of the door, only to change the location and swing of the door. The three grips displayed along the wall in Figure 4.8 can be used to stretch the door to a new position within the wall. The grip near the leaf of the door is used to flip the swing of the door. In Figure 4.8, grips are displayed, and the grip near the leaf of the door is hot. The hot grip has been selected and stretched below the wall to move the door below the wall.

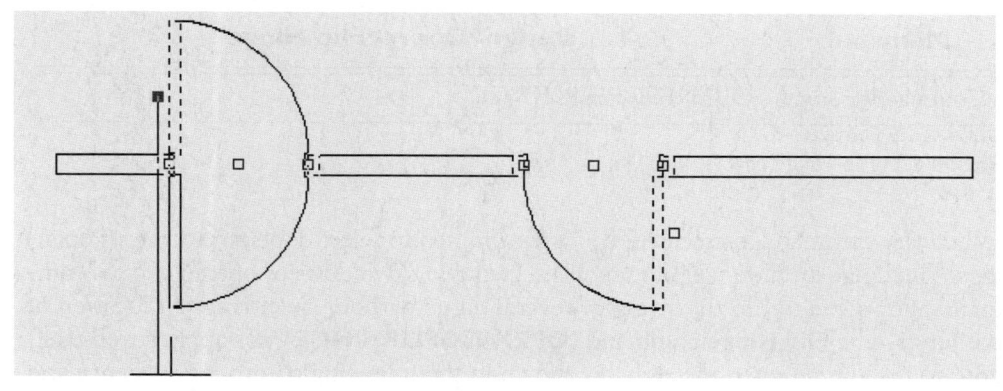

Figure 4.8 Editing the swing of a door with grips

If the hot grip is stretched to the right, the door swing will flip to the right as shown in Figure 4.9.

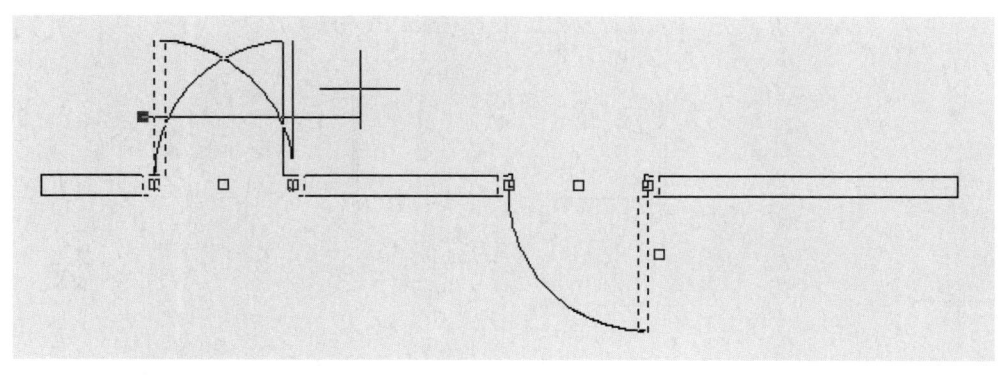

Figure 4.9 Editing the hinge location with grips

Therefore, using grips allows you to quickly change the swing of the door to new positions. The insertion point for the door is always located on the side of the unit nearest the door leaf. If you use grips to flip the door, the insertion point for the door also flips. A door can be placed with precision with the Temporary Tracking Point object snap and tracking from existing geometry to specify the insert point of a door.

The hinge location can also be changed with the **OPENINGFLIPHINGE** command, which flips the location of the hinge. See Table 4.2 for command access.

Menu bar	Design>Doors>Flip Hinge
Command prompt	OPENINGFLIPHINGE

Table 4.2 *Flip Hinge access*

When this command is executed, you are prompted to select doors, windows, or openings. The hinge location will flip for all selected doors, windows or openings. This command allows you to flip the swing of several doors without specifying the location of the hinge. The **Flip Hinge** command (**OPENINGFLIPHINGE**) was applied to the left door to flip the hinge to the right, as shown in the following command sequence and in Figure 4.10.

```
Command: _AecOpeningFlipHinge
Select doors, windows, or openings: (Select the door near P1
as shown in Figure 4.10.) 1 found
Select doors, windows, or openings: ENTER   (Press ENTER
to end the selection.)
(Hinge of door flipped as shown at right in Figure 4.10.)
```

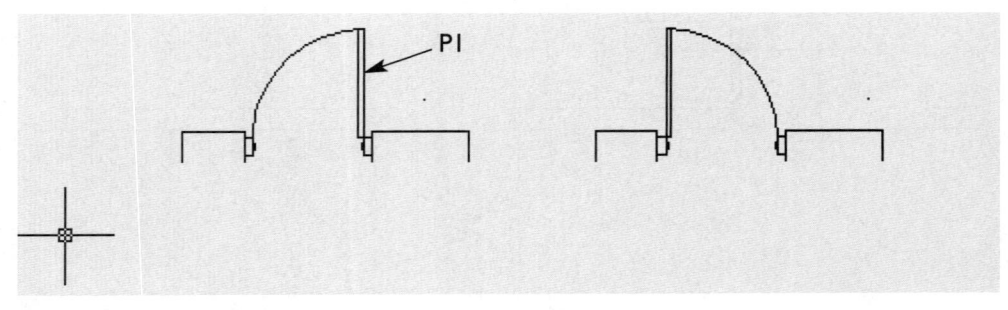

Figure 4.10 *Editing the hinge location with the Flip Hinge command*

The swing location can also be changed by the **OPENINGFLIPSWING** command, which flips the swing of one or more doors. See Table 4.3 for command access.

Menu bar	Design>Doors>Flip Swing
Command prompt	OPENINGFLIPSWING

Table 4.3 *Flip Swing access*

When this command is executed, you are prompted to select doors, windows, or openings. The swing location will flip for all selected doors, windows, and openings. The **Flip Swing** command (**OPENINGFLIPSWING**) was applied to the door on the left to flip the swing as shown in the following command sequence and in Figure 4.11.

```
Command: _AecOpeningFlipSwing
Select doors or windows:  (Select the door at P1 as shown on the
left in Figure 4.11.) 1 found
Select doors or windows:  (ENTER  (Press ENTER to end door
selection.)
(Door swing flipped down as shown in Figure 4.11.)
```

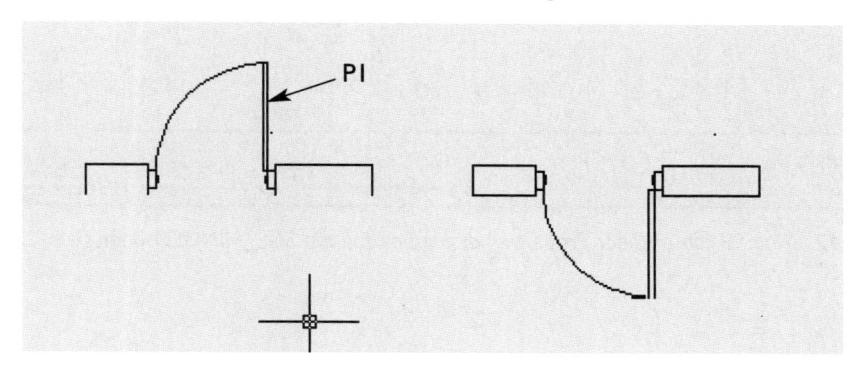

Figure 4.11 *Editing door swing with the Flip Swing command*

> **Note: Flip Hinge** and **Flip Swing** can be applied to windows and are located on the **Design>Windows** menu.

DEFINING DOOR PROPERTIES

You can view the properties of a door being inserted by selecting the **Properties** button of the **Add Doors** dialog box. Selecting the **Properties** button opens the **Door Properties** dialog box, which includes three tabs: **General**, **Style**, and **Dimensions**. Each of these tabs is described below.

General

The **General** tab allows you to edit the description of the selected door style. Figure 4.12 shows the **General** tab for the SGL_HINGED door style.

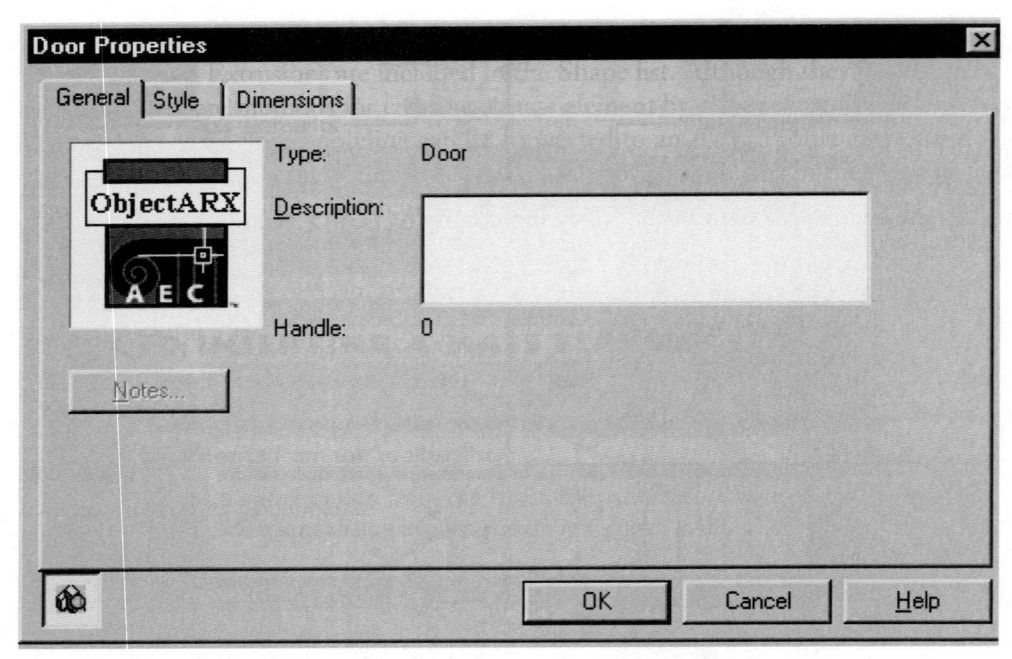

Figure 4.12 *General tab of Door Properties dialog box for the SGL_HINGED door style*

Style

The **Style** tab of the dialog box lists the available door styles. Figure 4.13 shows the **Style** tab for the SGL_HINGED door. Selecting the **Object Viewer** button while this dialog is opened allows you to view the graphic representation of each style. You can toggle to other styles by selecting a style from the style list, and the **Viewer** will display each door style as shown in Figure 4.13. The **Viewer** can be used to view the style and size of the door prior to the door being inserted.

Dimensions

The Dimensions tab includes two sections: **Standard Sizes** and **Current Size**. The standard sizes are the door sizes saved with the style definition. The current sizes are set in the **Add Doors** dialog box. Door height, width, and percent of opening can be set in the **Dimensions** tab. Figure 4.14 shows the **Dimensions** tab for the SGL_HINGED style.

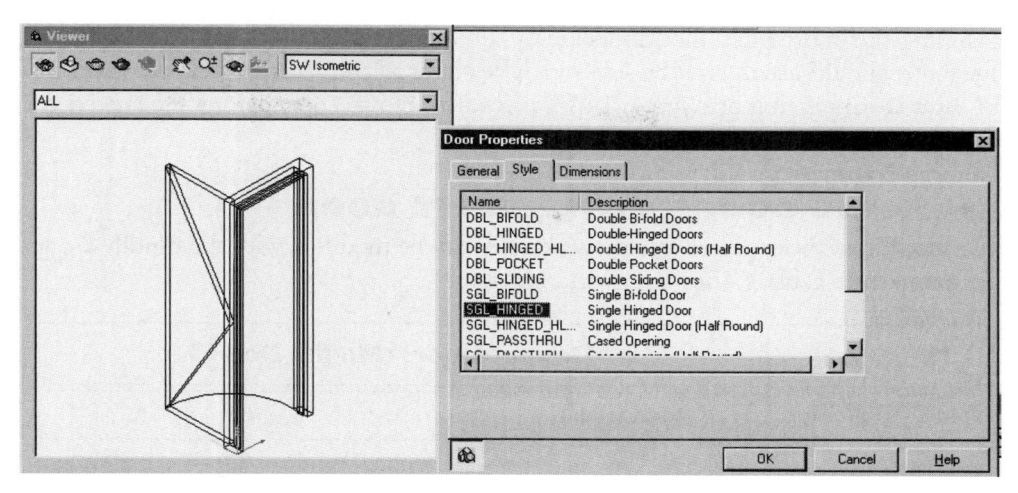

Figure 4.13 *Style tab of the Door Properties dialog box with Viewer open*

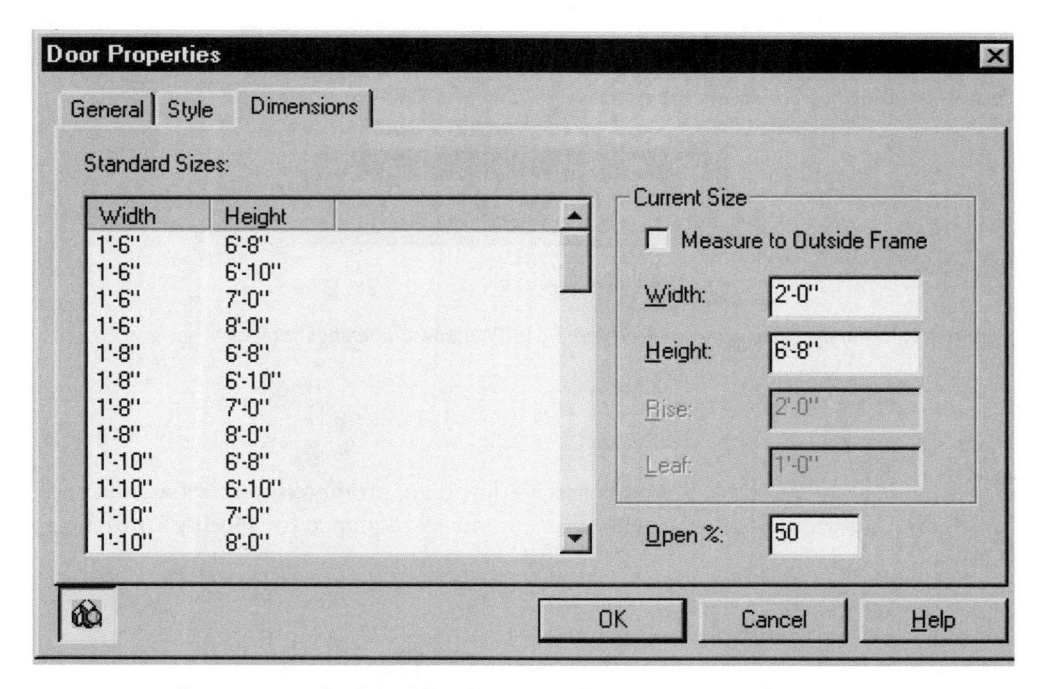

Figure 4.14 *Dimensions tab of the Door Properties dialog box for the SGL_HINGED style*

When the **Measure to Outside of Frame** check box is selected, the width and height dimensions of the door are applied to the outside of the doorjamb. When this option is not selected, the width and height dimensions are applied only to the door, excluding the doorjamb. Since doors are usually sized according to the door size, this check box should usually not be selected. The status of this check box is displayed in the **Add Doors** dialog box shown in Figure 4.3 as **Note: Dimensions Measured to Inside Frame**.

CHANGING DOORS WITH MODIFY DOOR

After doors have been placed in a drawing, they can be modified with the **Modify Door** command. See Table 4.4 for command access.

Menu bar	Design>Doors>Modify Door
Command prompt	DOORMODIFY
Toolbar	Select Modify Door from the Doors/Windows/Openings toolbar shown in Figure 4.15
Shortcut	Right-click over drawing area; choose Design>Doors>Modify Door

Table 4.4 *Modify Door access table*

Figure 4.15 *Modify Door command on the Doors/Windows/Openings toolbar*

When the **Modify Door** command is executed, you are prompted to select a door, and the **Modify Doors** dialog box opens. The command sequence for **Modify Door** is as follows:

```
Command: _AecDoorModify
Select doors: (Select a door for editing) 1 found
Select doors: ENTER (Pressing ENTER ends selection)
Door modify [Style/Width/HEight/Opening/Match]: DBOX
(Modify Doors dialog box opens allowing you to edit the dialog box.)
```

The **Modify Doors** dialog box shown in Figure 4.16 is similar to the **Add Doors** dialog box.

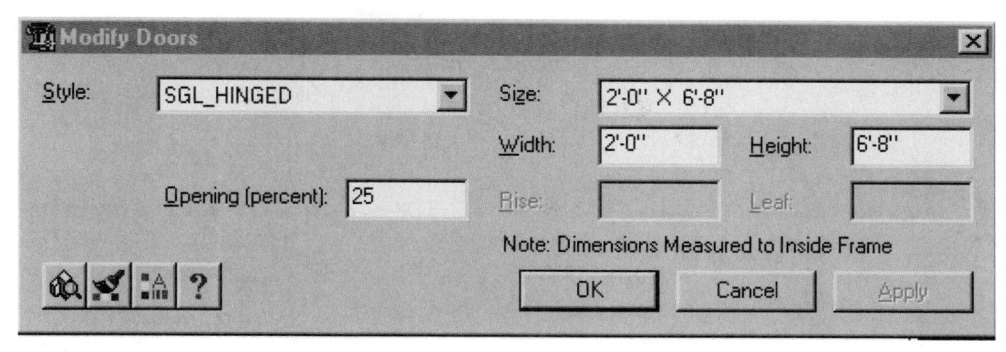

Figure 4.16 Modify Doors dialog box

Editing the **Modify Doors** dialog box will change the door style, size and percent of opening of a door. After editing the dialog box, select the **Apply** button to change the door and retain the dialog box. Selecting **OK** applies the changes to the door and dismisses the **Modify Doors** dialog box. You can also edit the door by selecting the options in the command line as shown in the command sequence above.

CHANGING THE ANGLE OF DOOR SWING

Although the **Modify Door** command can be used to change the angle of the door swing, **Opening Percent** (**OPENINGPERCENT**) is specifically designed to change the percent of opening of windows and doors. The **Opening Percent** command allows you to edit the angle of the door swing without using a dialog box. See Table 4.5 for command access.

Menu bar	Design>Doors>Change Swing Opening
Command prompt	OPENINGPERCENT

Table 4.5 Opening Percent access

When this command is executed, you are prompted to select doors, windows, or openings. The opening percent will change for all selected doors or windows. The **Opening Percent** command (**OPENINGPERCENT**) was applied to the door on the left in Figure 4.17 to decrease the opening, as shown at the right.

> **Note:** The **Change Swing Opening** command can be applied to windows and is also located on the **Design>Windows** menu.

The following command sequence demonstrates the use of **Opening Percent** to change the angle of the door swing, as shown in Figure 4-17.

```
Command: _AecOpeningPercent
Select doors or windows: (Select door on left at P1 as shown in
Figure 4.17.) 1 found
Select doors or windows: ENTER  (Press ENTER to end door
selection.)
Opening percent [Toggle] <50>: 25 ENTER  (Door swing
changes as shown on right.)
```

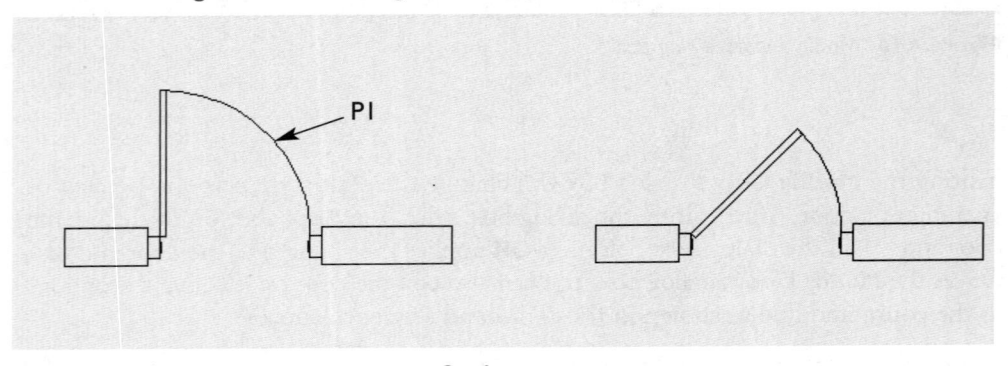

Figure 4.17 *Changing opening percent of a door*

TUTORIAL 4.1 INSERTING DOORS

1. Open \ADT Tutor\Ch4\Ex4-1.dwg.

2. Choose **File>SaveAs** from the menu bar and save the drawing as *Door1.dwg* in your student directory.

3. Verify that Running Object Snap is turned OFF in the Status Bar.

4. Select **Zoom Window** from the **Standard** toolbar and window the area shown in Figure 4.18.

5. Move the mouse pointer over the **Walls** toolbar and right-click; choose the **Doors/Windows/Openings** toolbar from the shortcut menu.

6. Select **Add Door** from the **Doors/Windows/Openings** toolbar.

7. Edit the **Add Doors** dialog box as follows: select the **Automatic Offset/Center** check box, offset distance = 4", **Opening (percent)** = 50, **Style** = SGL_HINGED, **Width** = 2'-0", and **Height** = 6'-8" as shown in Figure 4.19.

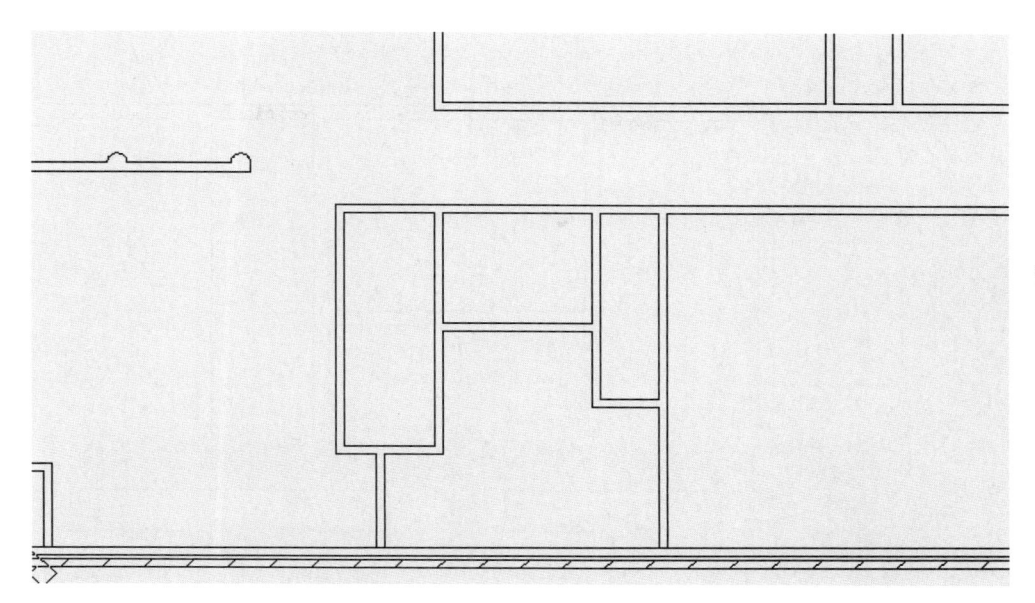

Figure 4.18 *Area of floor plan for inserting doors*

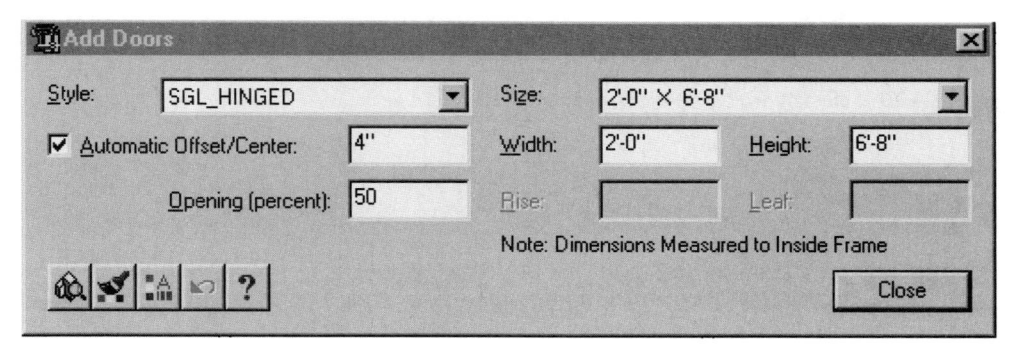

Figure 4.19 *Settings for Add Doors dialog box*

8. Verify that **Note: Dimensions Measured to Inside Frame** is set in the **Add Doors** dialog box. If necessary, edit this setting by selecting the **Properties** button of the **Add Doors** dialog box. Select the **Dimensions** tab of the **Door Properties** dialog box and clear the **Measure to Outside of Frame** check box.

9. Place the door by responding to the AutoCAD prompts as follows:
   ```
   Command: _AecDoorAdd
   Select wall, space boundary or RETURN: 
   ```
   (Select the wall as shown in the Figure 4.20 to place the 2'-0" door.)

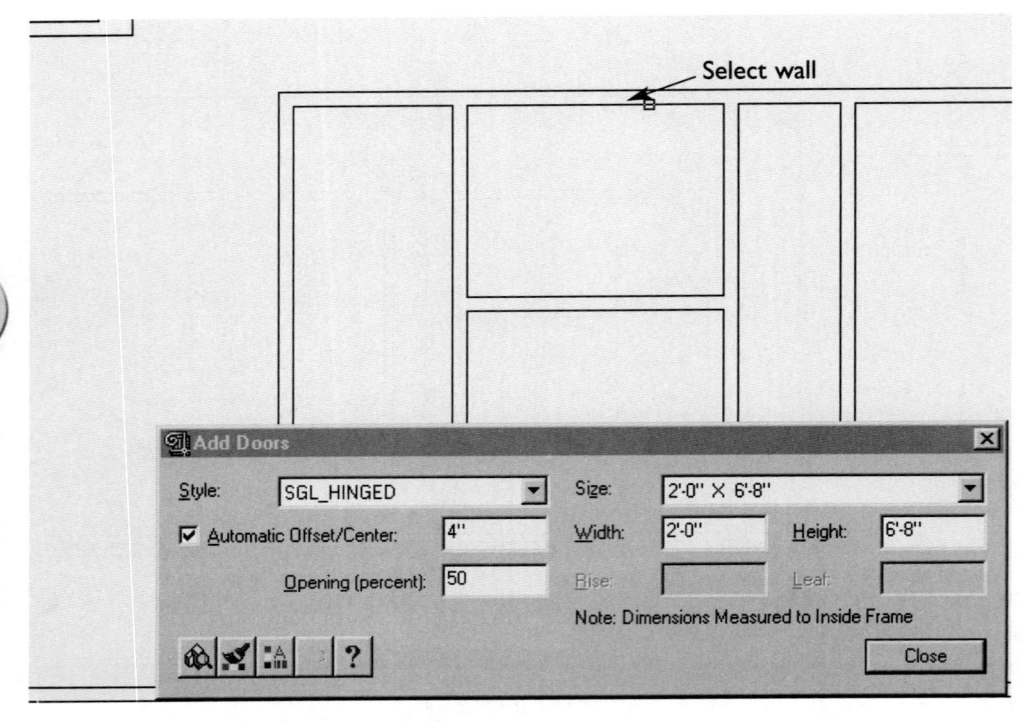

Figure 4.20 *Selecting wall for placing door*

```
Insert point or
[STyle/WIdth/HEight/OPening/Auto/OFfset/Match]: (Select
a point below the selected wall to specify the swing as shown in Figure 4.21.)
Insert point or
[STyle/WIdth/HEight/OPening/Auto/OFfset/Match/Undo]:
ENTER
```
10. Remain in the **Add Door** command and move the pointer to the right as shown in Figure 4.22.
11. Select the **Add Doors** dialog box to activate the dialog box.
12. Edit the **Width** field to 2'-6".
13. Select the graphics screen to change focus from the toolbar to the drawing area.
14. Select a point near point P1 to specify the swing of the door, as shown in Figure 4.23.
15. Press ENTER to end the command.

Placing Doors and Windows 233

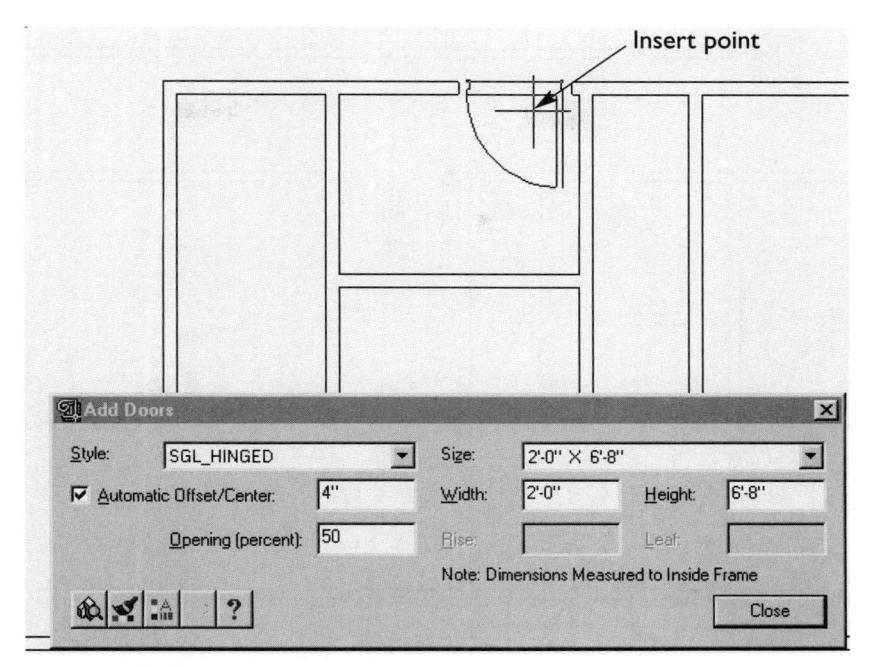

Figure 4.21 *Determining door swing during door placement*

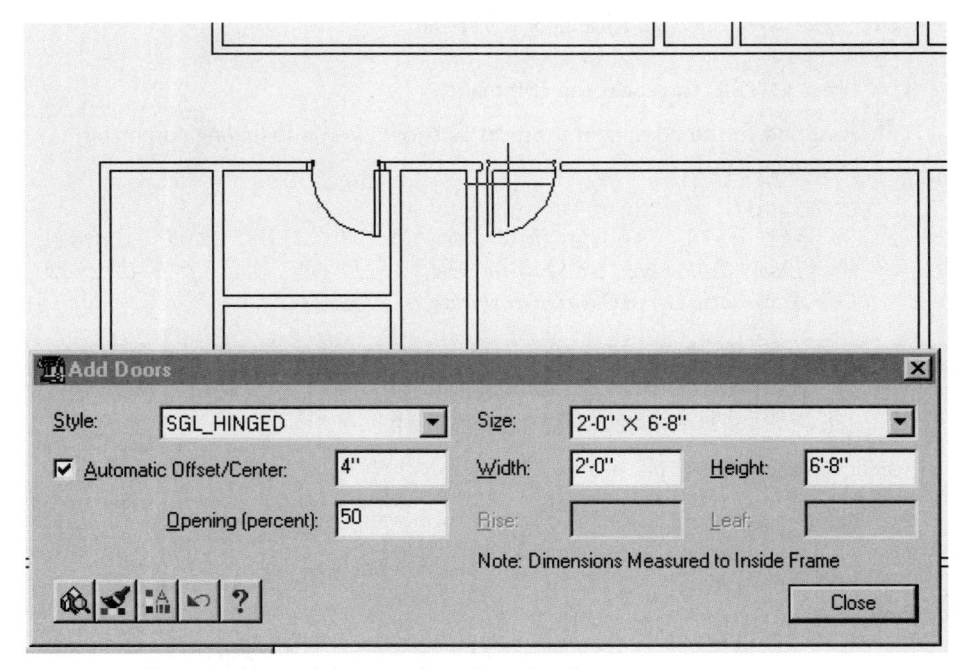

Figure 4.22 *Placing additional doors in the selected wall*

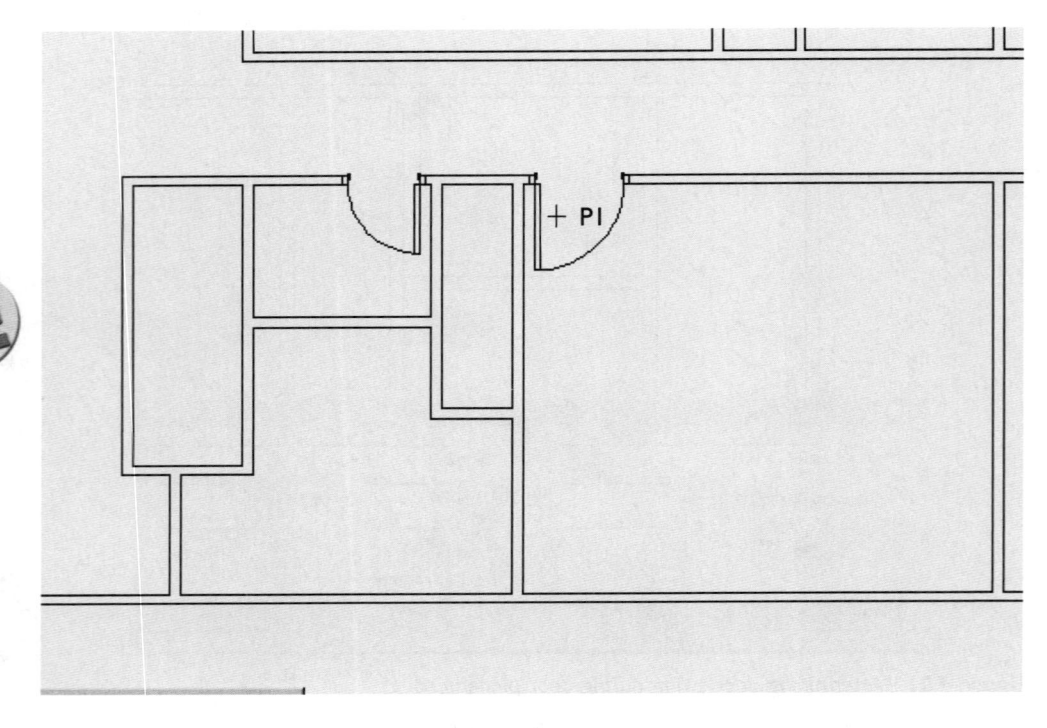

Figure 4.23 *Door swing specified while door is inserted*

16. Press ENTER to repeat the command.
17. Respond to the command prompts as shown in the following command sequence.

    ```
    Command: _AecDoorAdd
    Select wall, space boundary or RETURN: DBOX
    ```
    (Edit the Add Doors dialog box; set Style to DBL_SLIDING, Width to 6'-0"; check ON Automatic Offset/Center as shown in Figure 4.24.)

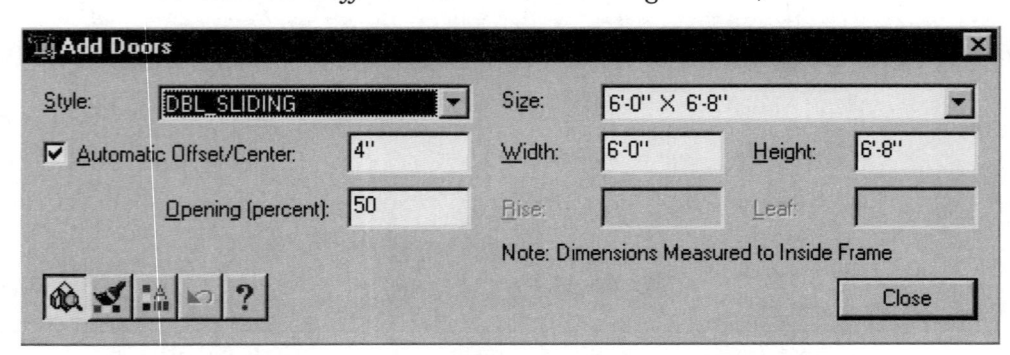

Figure 4.24 *Editing the Add Doors dialog box for placing the double sliding door*

```
Select wall, space boundary or RETURN: (Select the wall as
shown in Figure 4.25.)
```

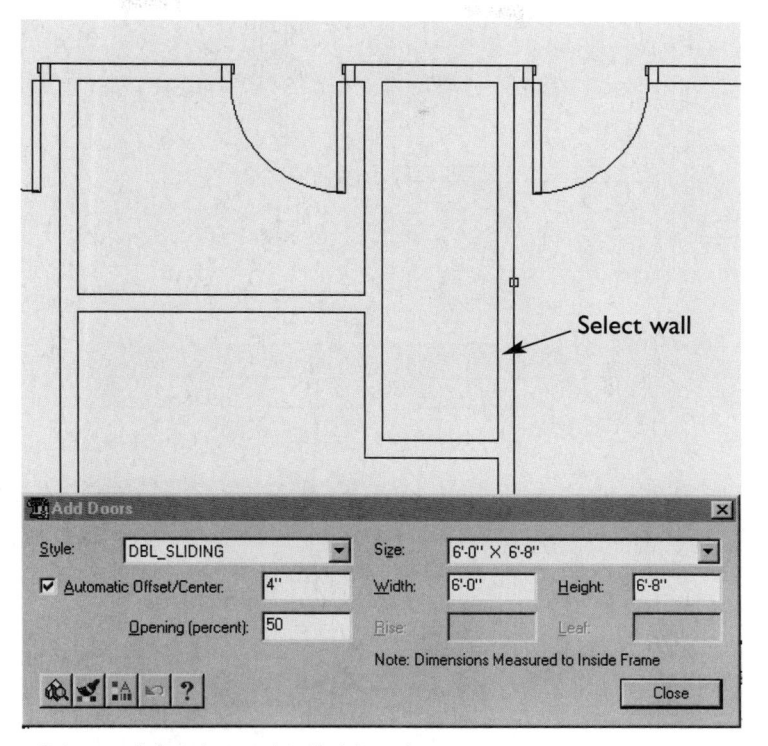

Figure 4.25 *Select wall for placing the 6' sliding door*

```
Insert point or
[STyle/WIdth/HEight/OPening/Auto/OFfset/Match]: (Select
the insertion point within the wall as shown in Figure 4.26.)
Insert point or
[STyle/WIdth/HEight/OPening/Auto/OFfset/Match/Undo]:
ENTER   (Press ENTER to end the command.)
```
The sliding door should snap to the center of the wall as shown in Figure 4.27.

18. Zoom a window around the upper left corner of the building as shown in Figure 4.28.

19. Select the exterior wall; then right-click to choose **Edit Wall Style** from the shortcut menu.

20. Select the **Display Props** tab of the **Wall Style Properties** dialog box.

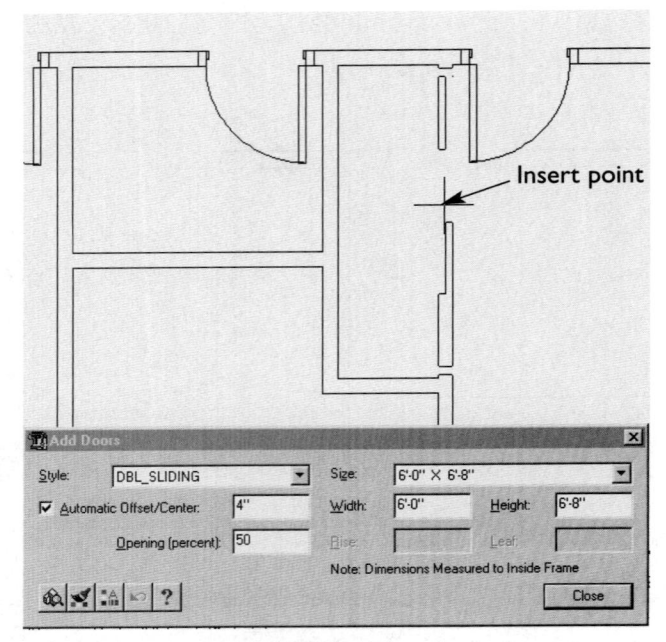

Figure 4.26 *Specifying insertion point for double sliding door*

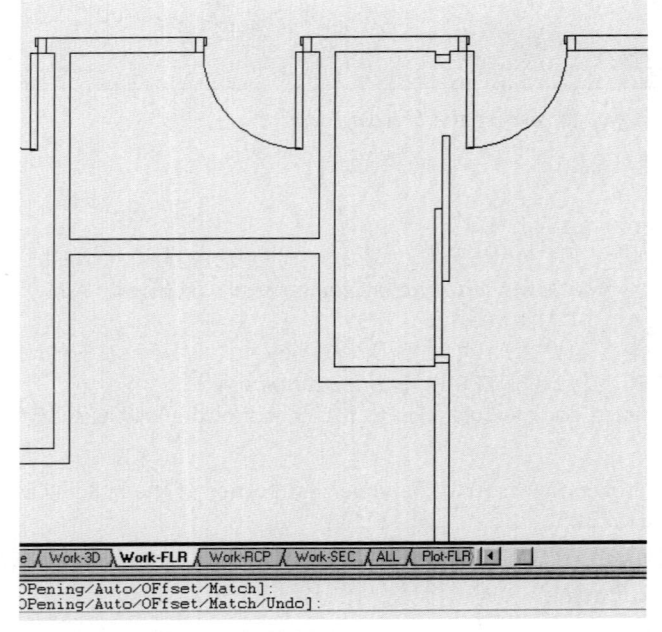

Figure 4.27 *Double sliding door placed in wall*

Placing Doors and Windows 237

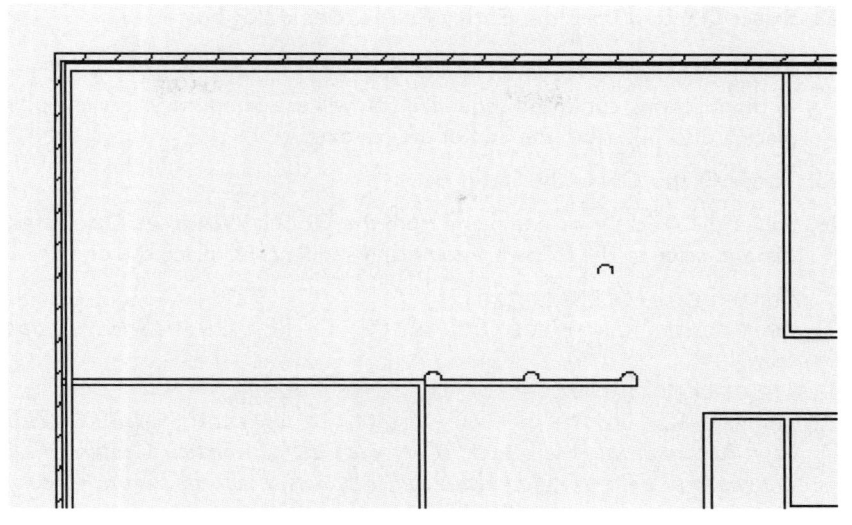

Figure 4.28 *Zoom window for placing exterior door*

21. Select the **Edit Display Props** button to display the **Entity Properties** dialog box.
22. Select the visibility (light bulb icon) to turn OFF Boundary 2 (Brick) and Hatch 2 (Brick) as shown in Figure 4.29.

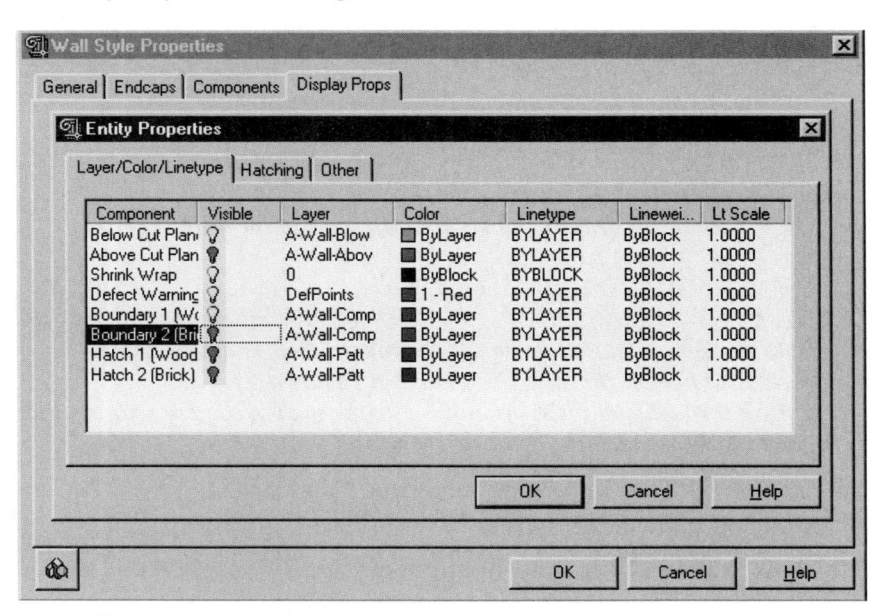

Figure 4.29 *Entity Properties dialog box*

23. Select **OK** to dismiss the **Entity Properties** dialog box.
24. Select **OK** to dismiss the **Wall Style Properties** dialog box.

    In the following command sequence you will use Temporary Tracking points to place a door 17' from the end of the building.

25. Toggle Ortho ON in the Status Bar.
26. Select the **Add Door** command from the **Doors/Windows/Openings** toolbar and refer to the following command sequence to place the door.

    ```
    Command: _AecDoorAdd
    Select wall, space boundary or RETURN: (Select the upper wall.)
    Insert point or
    [STyle/WIdth/HEight/OPening/Auto/Match]: DBOX (Edit Style to Standard, Width to 3'-0", toggle OFF Automatic Center/Offset, and Height to 6'-8" in the Add Walls dialog box as shown in Figure 4.30.)
    ```

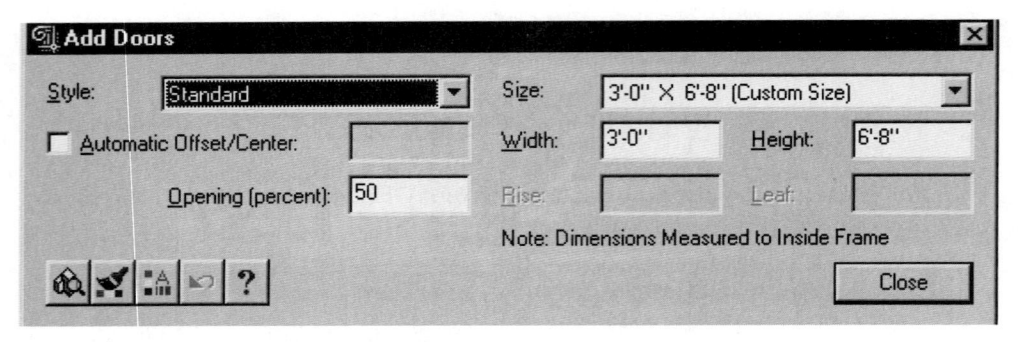

Figure 4.30 Settings in the Add Doors dialog box

```
(Left-click in the drawing area to return focus to the drawing area.)
Insert point or
[STyle/WIdth/HEight/OPening/Auto/Match]: track ENTER
First tracking point: _endp of (Select Endpoint object snap from the Object Snap flyout on the Standard toolbar and select the end of outer stud line on the corner as shown in Figure 4.31.)
(Move the mouse to the right and below the wall selected so that the door swing will be inside the room.)
Next point: 17' ENTER
Next point: (Press ENTER to end tracking.) ENTER
Insert point or
[STyle/WIdth/HEight/OPening/Auto/Match/Undo]: ENTER
(Pressing ENTER ends the Add Door command.)
```

Figure 4.31 *Tracking from endpoint of exterior wall*

27. Select the exterior wall; then right-click to choose **Edit Wall Style** from the shortcut menu.
28. Select the **Display Props** tab of the **Wall Style Properties** dialog box.
29. Select the **Edit Display Props** button to display the **Entity Properties** dialog box.
30. Toggle ON the visibility of Boundary 2 (Brick) and Hatch 2 (Brick) components.
31. Select **OK** to dismiss the **Entity Properties** dialog box.
32. Select **OK** to dismiss the **Wall Style Properties** dialog box.
33. Select the **Modify Door** command on the **Doors/Windows/Openings** toolbar.
34. Select the rear door placed in step 26 and then press ENTER (pressing ENTER ends the selection set).
35. Edit the **Modify Doors** dialog box to **Opening (percent)** = 25 as shown in Figure 4.32.
36. Select **OK** to execute the change and dismiss the **Modify Doors** dialog box. The door should now appear as shown in Figure 4.33.
37. Save the drawing.

CREATING DOOR STYLES

When the **Add Door** command (**DOORADD**) is used, the **Style** drop-down list includes door styles that are preset according to shape, type, and dimensions in the drawing. These styles are included as part of the Aec_Arch_Imperial template. The

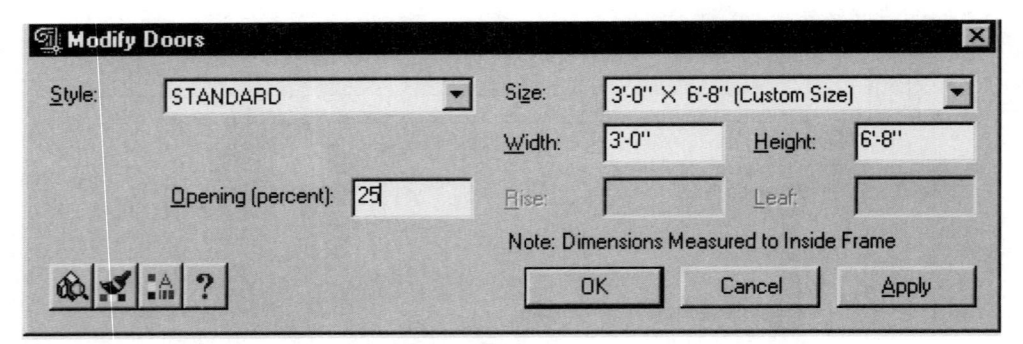

Figure 4.32 *Editing opening percentage using Modify Doors dialog box*

Figure 4.33 *Door opening modified to 25 percent*

door styles can be edited and additional styles created through the **Door Styles** command (**DOORSTYLE**). See Table 4.6 for command access.

Menu bar	Design>Doors>Door Styles
Command prompt	DOORSTYLE
Toolbar	Select Door Styles from Doors/Windows/Openings toolbar shown in Figure 4.34
Shortcut	Right-click over drawing area; choose Design>Doors>Door Styles

Table 4.6 *Door Styles access*

Figure 4.34 *Door Styles command on the Doors/Windows/Openings toolbar*

When the **Door Styles** command is executed, the **Door Styles** dialog box opens as shown in Figure 4.35. It lists door styles of the current drawing and the buttons to create, copy, edit, import, and export door styles.

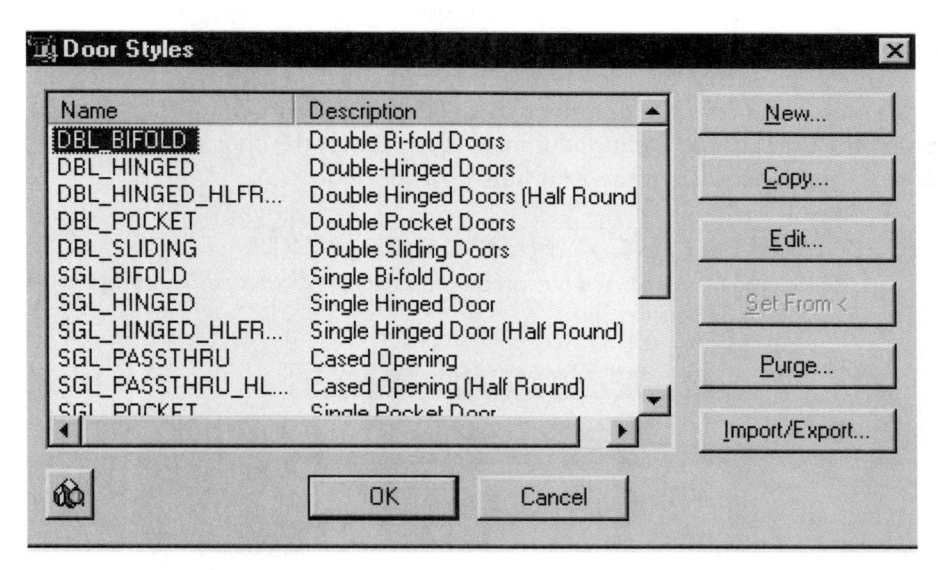

Figure 4.35 *Door Styles dialog box*

Door styles of a drawing are listed in the dialog box in a similar manner to other style dialog boxes. The purpose of each of the six buttons on the right is summarized below:

> **New** – Creates new names for door styles.
>
> **Copy** – Copies an existing style to a new name.
>
> **Edit** – Edits the description and properties of a door style.
>
> **Set from<** – Inactive unless a new style is created from an Aec profile. This button is used to apply an Aec profile to create a new door shape.
>
> **Purge** – Deletes saved styles from the drawing.

Import/Export – Transfers saved door styles to other files or copies saved door styles from a file to the current drawing.

Note: Additional door and window styles are in the \AutoCAD Architectural 2\Content\Imperial\Styles directory.

CREATING A NEW DOOR STYLE

Create a new door style by selecting **Door Styles** from the **Doors/Windows/Openings** toolbar and then selecting the **New** button in the **Door Styles** dialog box. The initial properties of the new door style are inherited from the door style that is highlighted in the door style list. Because the Standard door style is a simple door without any saved sizes, it is a good style to use as the base for creating other door styles. The **New** button of the **Door Styles** dialog box will open the **Name** dialog box, which allows you to assign a name for the new door style. Add the description of the new style by selecting the **Edit** button. Define additional properties of the door style by editing the dialog box that opens when the Edit button is selected.

STEPS TO CREATING A NEW DOOR STYLE

1. Select the **Door Styles** button on the **Doors/Windows/Openings** toolbar as shown in Figure 4.36.

Figure 4.36 *Door Styles command on the Doors/Windows/Openings toolbar*

2. Select the Standard door style from the door style list.
3. Select the **New** button.
4. Type a name in the **Name** dialog box as shown in Figure 4.37.
5. Select **OK** to dismiss the **Name** dialog box.

EDITING A DOOR STYLE

Define the properties of a door style by editing the dialog box that opens when the **Edit** button of the **Door Styles** dialog box is selected. As shown in Figure 4.38, selecting the **Edit** button opens the **Door Style Properties** dialog box, which includes the following tabs: **General**, **Dimensions**, **Design Rules**, **Standard Sizes**, and **Display Props**. The purpose and function of each tab follow.

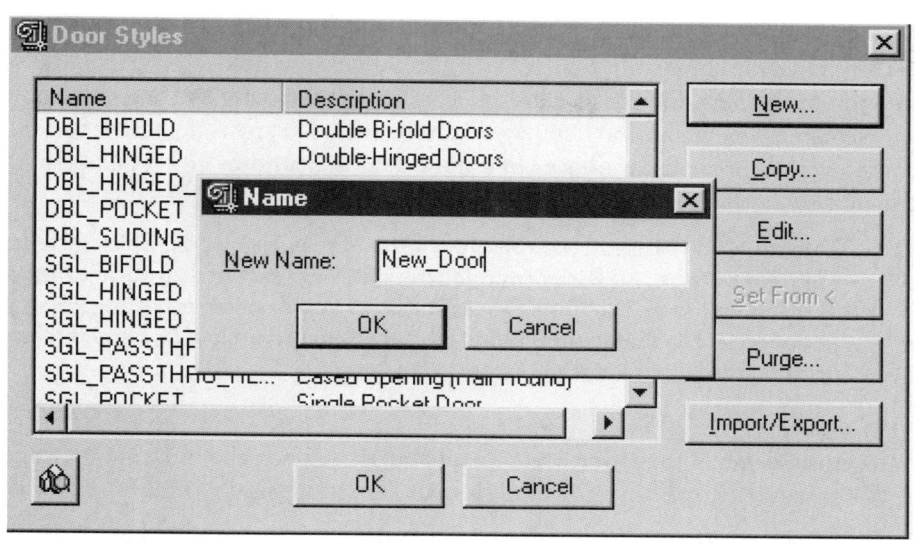

Figure 4.37 Name dialog box for creating a new door style

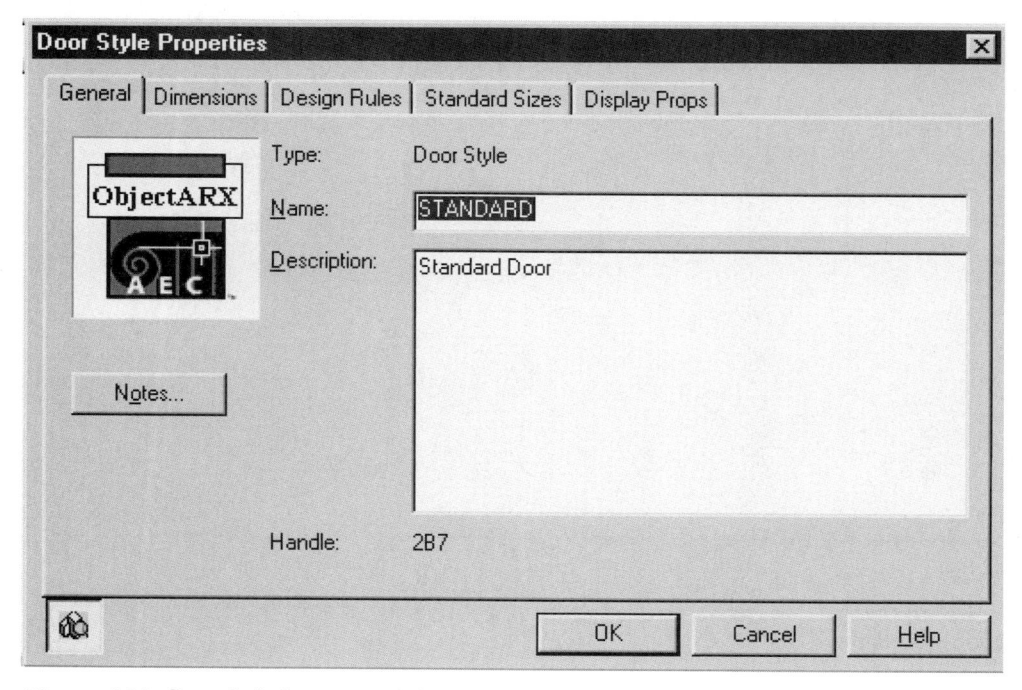

Figure 4.38 Door Style Properties dialog box

General

The **General** tab, shown in Figure 4.38, includes edit fields for the name and description of the new door style. This tab is useful in correcting spelling errors in the door style name. The name and description can be edited at any time during the drawing. However, all previous insertions of a door style will automatically be updated to the current name and description.

> **Notes** – The **Notes** button will open the **Notes** dialog box, which includes the **Text Docs** and **Reference Docs** tabs. These tabs are identical to those for the wall style. The **Text Docs** tab allows you to enter notes regarding a door style. The **Reference Docs** tab allows you to list the files and directories that include related information.

Dimensions

The **Dimensions** tab of the **Door Style Properties** dialog box allows you to set dimensions of the frame, stop, and door thickness, as shown in Figure 4.39.

> **Frame** – The **Frame** section of the **Dimensions** tab includes edit fields for setting the **A-Width** and **B-Depth** of a door frame. The **Auto-Adjust to Width of Wall** check box determines whether the depth is fixed or varies

Door stop dimensions

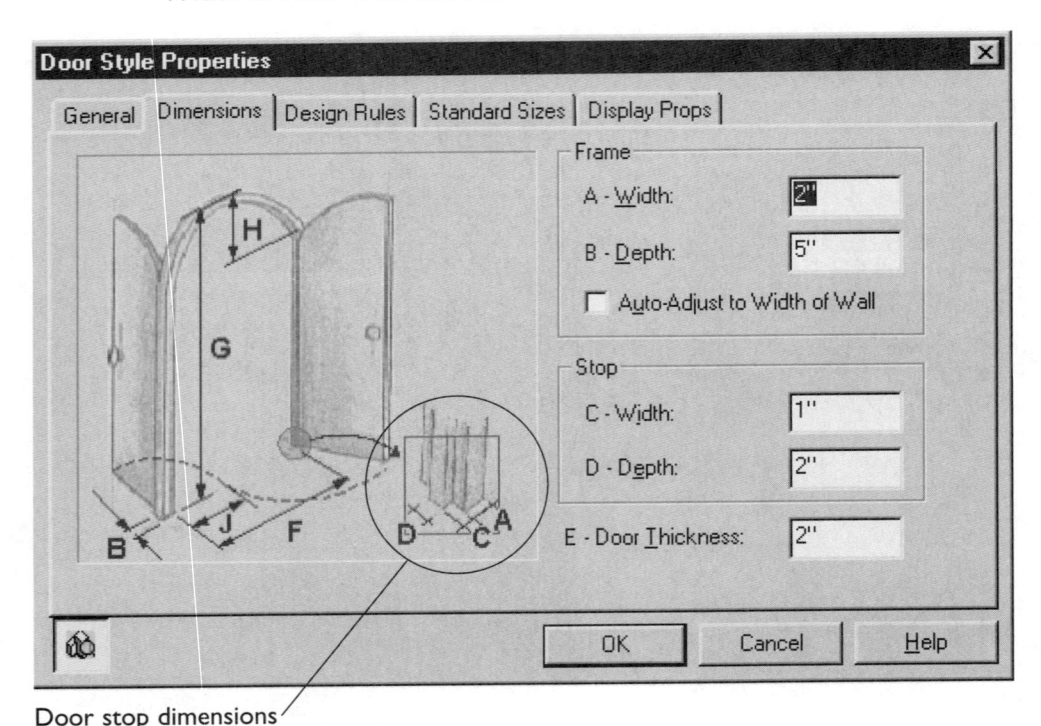

Figure 4.39 *Dimensions tab of the Door Style Properties dialog box*

according to the width of the wall. When this check box is selected, the frame depth adjusts to the wall width, as shown on the left in Figure 4.40. If the **Auto-Adjust to Width of Wall** is not checked, the frame depth is fixed to the depth listed in the **B-Depth** edit field. The door on the right in Figure 4.40 has a fixed door frame depth.

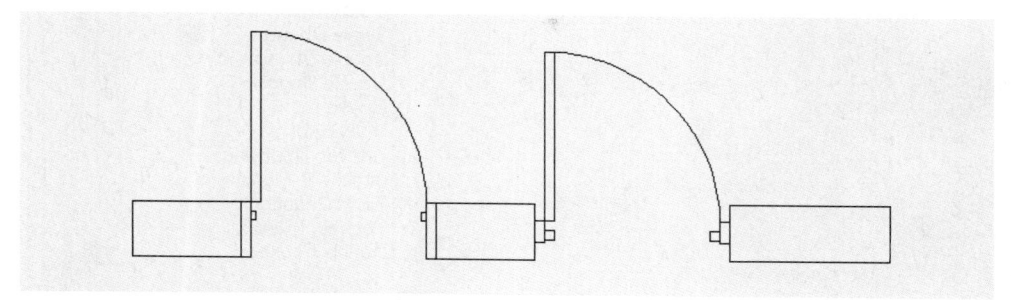

Figure 4.40 *Auto-adjusted and fixed frames*

The **Width** dimension for a frame in residential construction should be set equal to the total of the doorjamb and brick molding or exterior trim thickness. Therefore, the **Width** dimension of door styles for residential construction consists of the total unit thickness exclusive of the interior trim. The **Width** dimension of the frame in commercial construction is set according to dimensions of the steel frame.

> **Stop** – The size of the stop for a door frame shown in Figure 4.39 is specified by dimensions **C-Width** and **D-Depth**. Attaching a wood stop to the frame usually creates the stop in residential construction. The stop size and style therefore varies. The stop in commercial construction is formed with the steel frame and varies according to the manufacturer of the steel frame.
>
> **Door thickness** – The **E-Door Thickness** dimension is the thickness of the door. Doors used in residential construction are typically 1 3/8" or 1 3/4". Doors used in commercial construction are usually 2" or more in thickness.

Design Rules

The **Design Rules** tab, shown in Figure 4.41, defines the shape and type of door used for the style.

The **Design Rules** tab includes two sections: **Shape** and **Door Type**.

> **Shape** – The **Shape** section allows you to choose from predefined shapes or custom shapes. The predefined shapes include rectangular, half-round, quarter-round, arch, gothic, and peak pentagon. This list is extensive and satisfactory for most design applications. However, you can use other shapes by toggling on the **Custom** option. The shapes in the **Custom** list are Aec Profiles, saved profile shapes that can be applied to objects in Architectural Desktop.

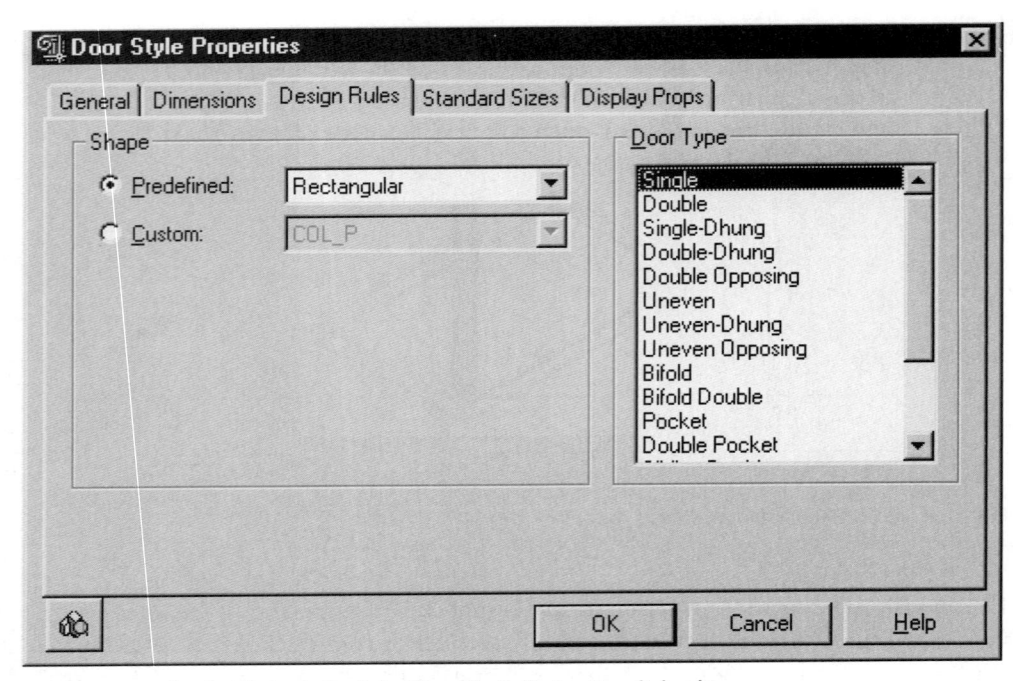

Figure 4.41 *Design Rules tab of the Door Style Properties dialog box*

Door Type –The types of doors that can be created are shown in Figures 4.42 and 4.43. The name assigned to each type is shown below the figure. Note that the garage door is named Overhead and is shown in Figure 4.43.

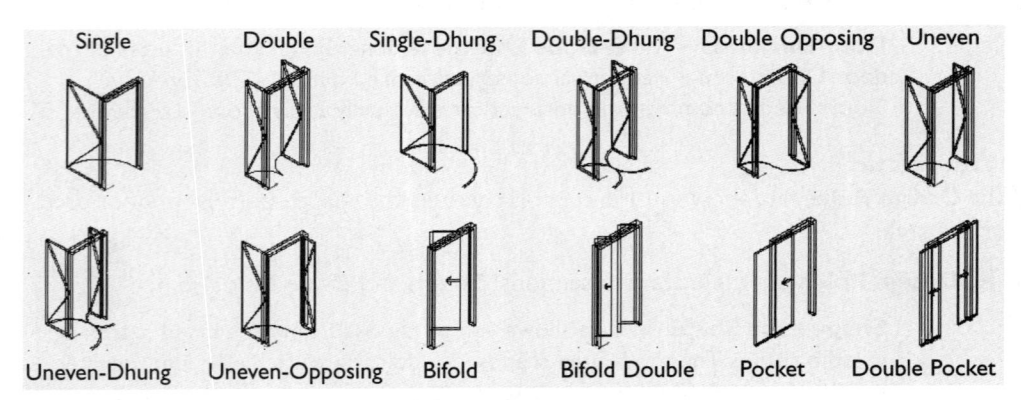

Figure 4.42 *Door types available in Architectural Desktop*

Placing Doors and Windows 247

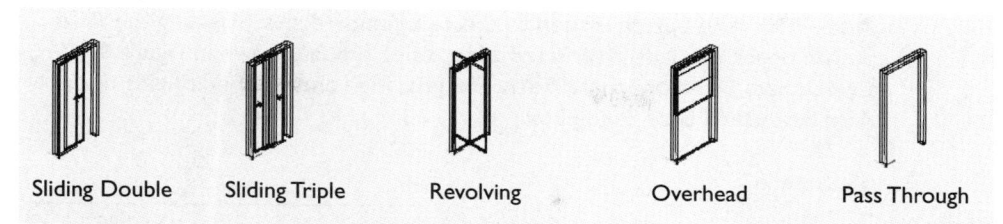

Figure 4.43 *Door types including the Overhead garage door*

Standard Sizes

The **Standard Sizes** tab is used to create predefined door sizes associated with a door style. The Standard door style does not have predefined sizes, as shown in the **Standard Sizes** tab of Figure 4.44.

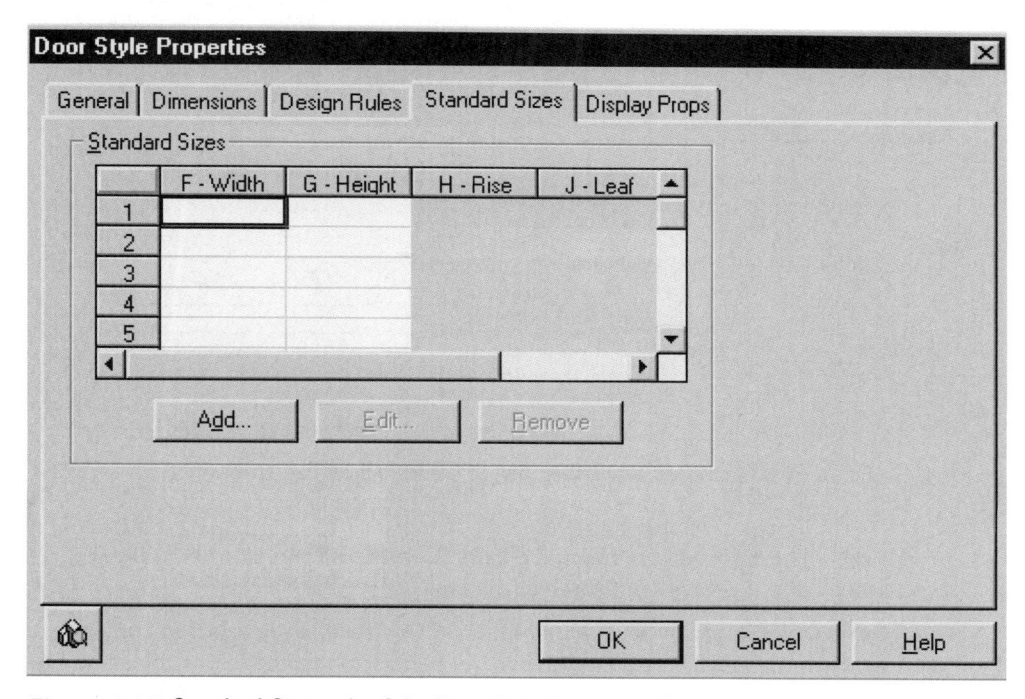

Figure 4.44 *Standard Sizes tab of the Door Style Properties dialog box*

>**Sizes** – The sizes listed in the **Standard Sizes** table include **F-Width**, **G-Height**, **H-Rise**, and **J-Leaf**. These dimensions are illustrated on the **Dimensions** tab in Figure 4.39. The **H-Rise** is inactive unless the door shape is Gothic, Arch, or Peak Pentagon. The **J-Leaf** column is inactive unless the door type is set to Uneven.

Add – The **Add** button is used to create a standard size. Selecting the **Add** button opens the **Add Standard Size** dialog box, as shown in Figure 4.45. Create sizes by editing the **Width**, **Height**, **Rise** and **Leaf** edit fields of the **Add Standard Size** dialog box.

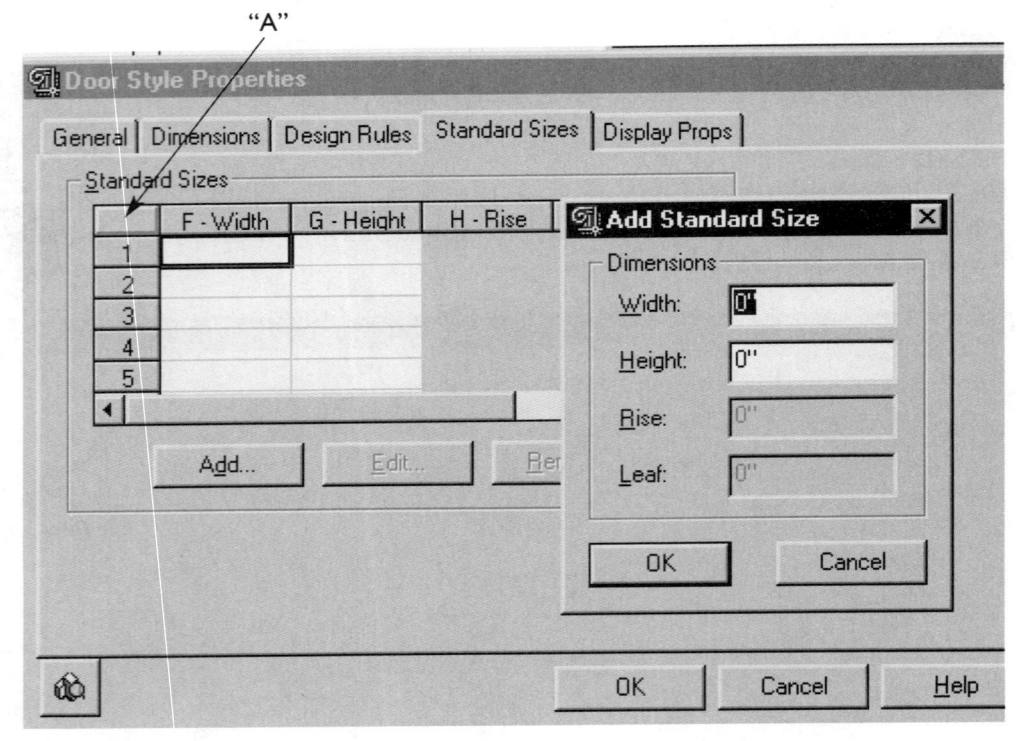

Figure 4.45 *Edit the Add Standard Size dialog box to create a standard size*

Edit – The **Edit** button opens the **Edit Standard Size** dialog box, allowing you to edit the sizes highlighted in the table of standard sizes.

Remove – Selecting the **Remove** button will remove the selected components of the table. Selecting the row number button to the left of the standard size list will select the entire row, allowing you to remove the row. Whereas if you select the header button as shown at "A" in Figure 4.45, the entire table is selected and can be removed.

Display Props

The **Display Props** tab allows you to control the display of a door. The display representations for the door style are listed in the display representations list at the top of the dialog box. Figure 4.46 shows the display representations for a door style.

Placing Doors and Windows 249

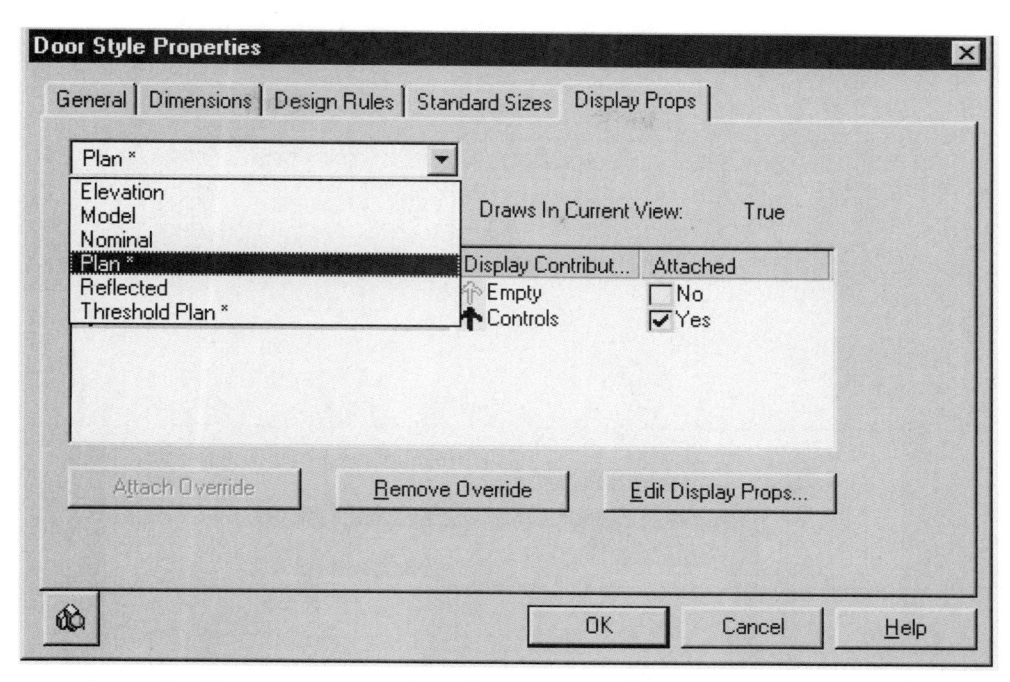

Figure 4.46 *Display representations of the door style*

The display representations marked with an asterisk are the current display representations for the viewport.

The **Property Source** includes a **Door Style** and **System Default**. The buttons below the property source table—**Attach Override**, **Remove Override**, and **Edit Display Props**—are used to set the display of the door style.

>**Attach Override** – Marks the selected property source as overridden.
>
>**Remove Override** – Removes an override that has been attached to the select property source.
>
>**Edit Display Props** – Used to edit the display of the door when an Override has been attached. Selecting the **Edit Display Props** button opens the **Entity Properties** dialog box to control the visibility, layer, color, linetype, lineweight, and linetype scale of the components of the door style. Figure 4.47 shows the **Entity Properties** dialog box for the Plan display representation, which includes the door panel, frame, stop, and swing. Toggle the visibility of each of these components off or on by selecting the light bulb icon.

You can add thresholds to the display of a door by editing the Threshold display representation. The Threshold and Plan display representations are applied to the Work-

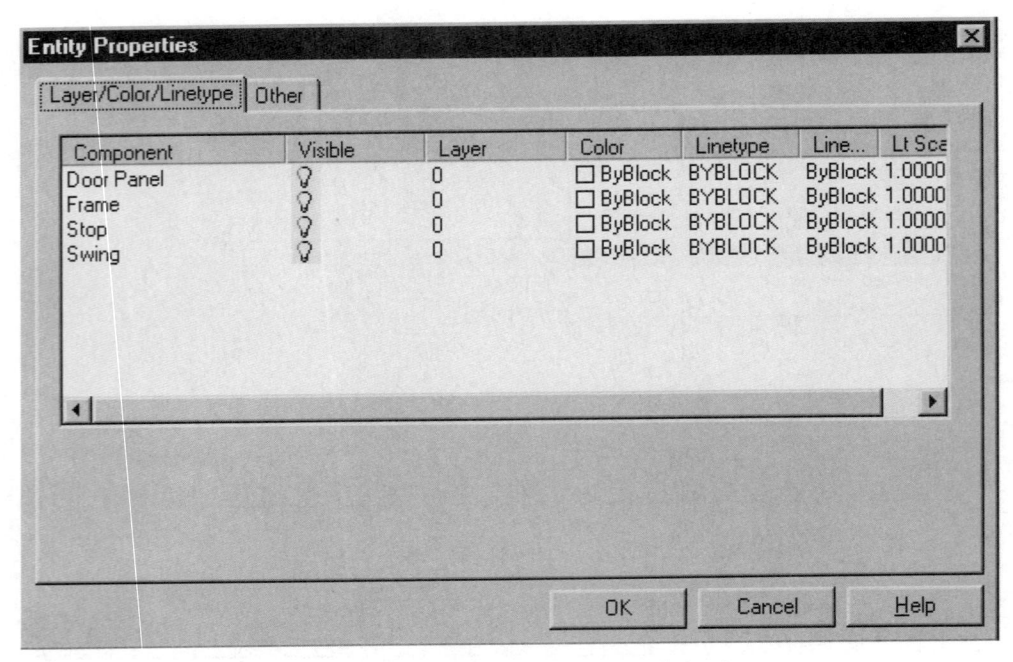

Figure 4.47 Layer/Color/Linetype tab of the Entity Properties dialog box

FLR and Plot-FLR viewports. The Threshold display representation **Entity Properties** dialog box is shown in Figure 4.48.

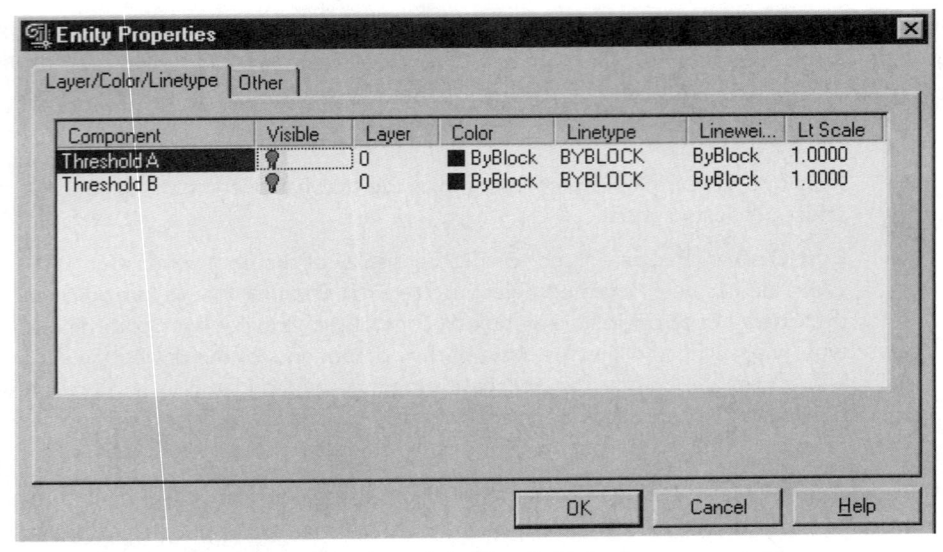

Figure 4.48 Entity Properties dialog box of the Threshold display representation

Toggling on the light bulb icon will display Threshold A, the inner threshold, or Threshold B, the outer threshold.

TUTORIAL 4.2 CREATING AND EDITING A DOOR STYLE

1. Open \ADT Tutor\Ch4\Ex4-2.dwg.

2. Choose **File>SaveAs** from the menu bar and save the drawing as *Door2.dwg* in your student directory.

3. Select **Zoom Window** from the **View** flyout and then respond to command prompts as shown below.

   ```
   Command: '_zoom
   Specify corner of window, enter a scale factor (nX
   or nXP), or
   [All/Center/Dynamic/Extents/Previous/Scale/Window]
   <real time>: _w
   Specify first corner: 20',0 ENTER
   Specify opposite corner: 50',25' ENTER
   ```

4. Select the **Door Styles** command from the **Doors/Windows/Openings** toolbar.

5. Select the Standard style from the door style list.

6. Select the **New** button of the **Door Styles** dialog box.

7. Type **SINGLE_INT_RES** in the **Name** dialog box as shown in Figure 4.49.

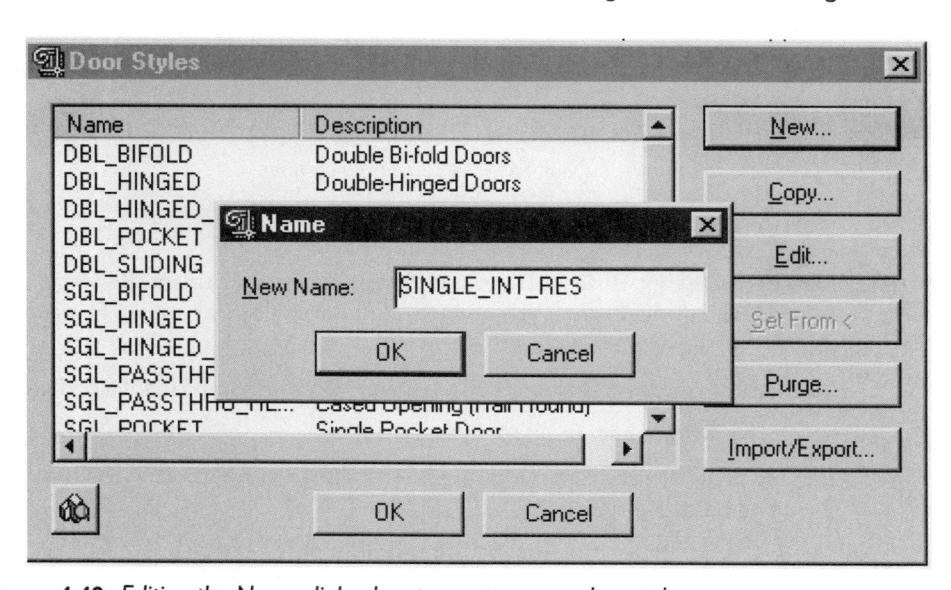

Figure 4.49 *Editing the Name dialog box to create a new door style*

8. Select **OK** to dismiss the **Name** dialog box.
9. Select the **Edit** button of the **Door Styles** dialog box.
10. Select the **General** tab and type **Single hinged interior doors** in the description edit field.
11. Select the **Dimensions** tab and edit the following fields: Frame **A-Width** = 2 1/4", Frame **B-Depth** = 4 9/16"; Stop **C-Width** = 1/2", **D-Depth** = 2"; **E-Door Thickness** = 1 3/8" as shown in Figure 4.50.

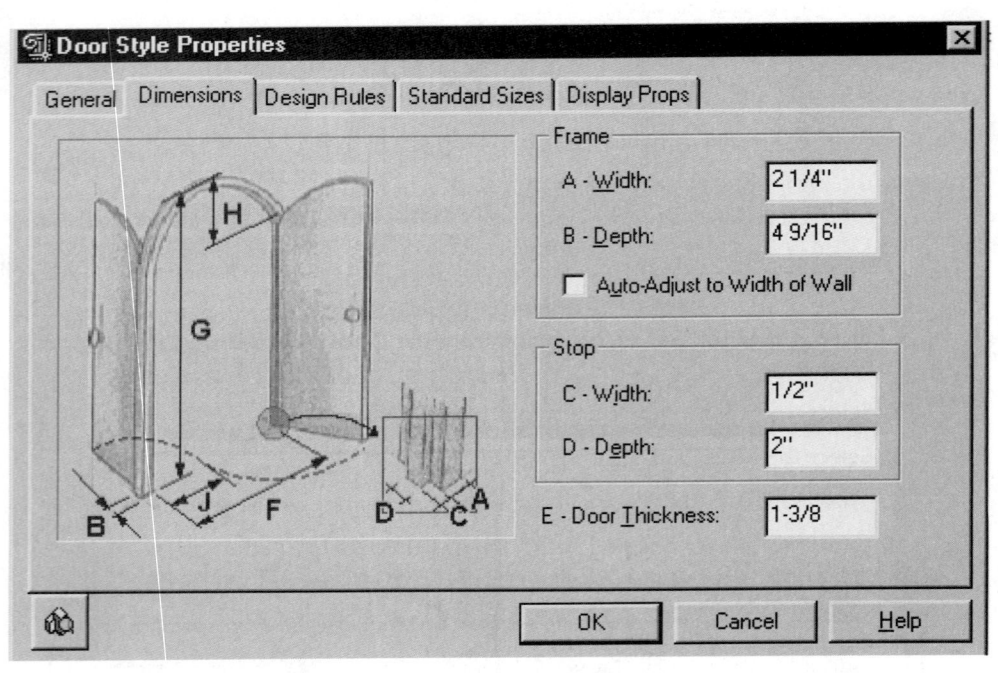

Figure 4.50 *Dimensions tab of the Door Style Properties dialog box*

12. Select the **Design Rules** tab and edit **Shape** to Predefined and Rectangular. Select Single in the **Door Type** section of the dialog box as shown in Figure 4.51.
13. Select the **Standard Sizes** tab of the **Door Style Properties** dialog box.
14. Select the **Add** button to add a size.
15. Edit the **Add Standard Size** dialog box as shown in Figure 4.52. Set width to 2'-6" and height to 6'-8".
16. Select **OK** to dismiss the **Add Standard Size** dialog box.

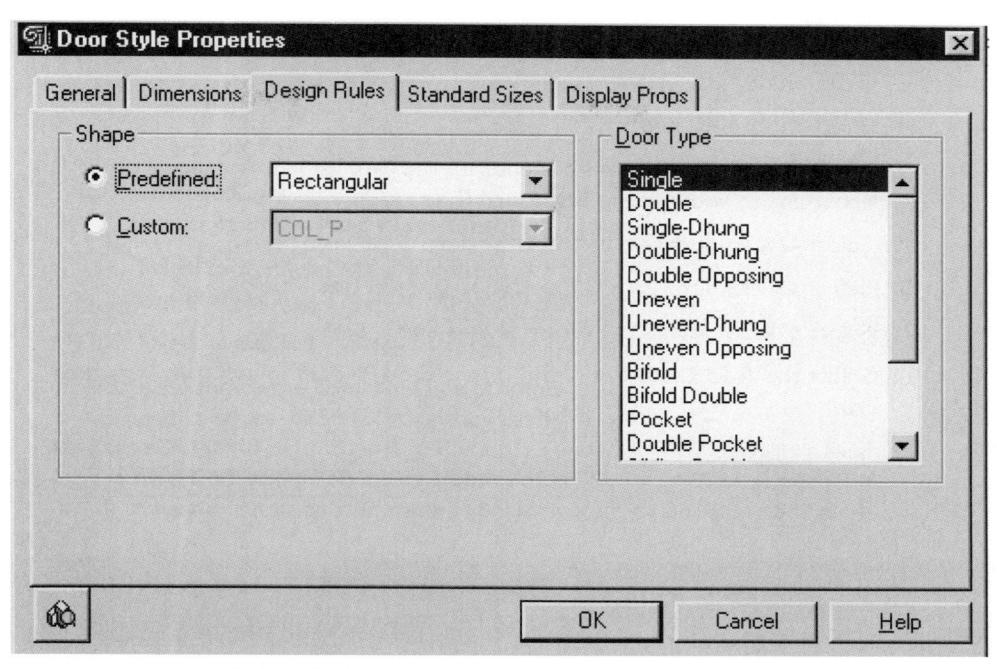

Figure 4.51 *Design Rules tab of the Door Style Properties dialog box*

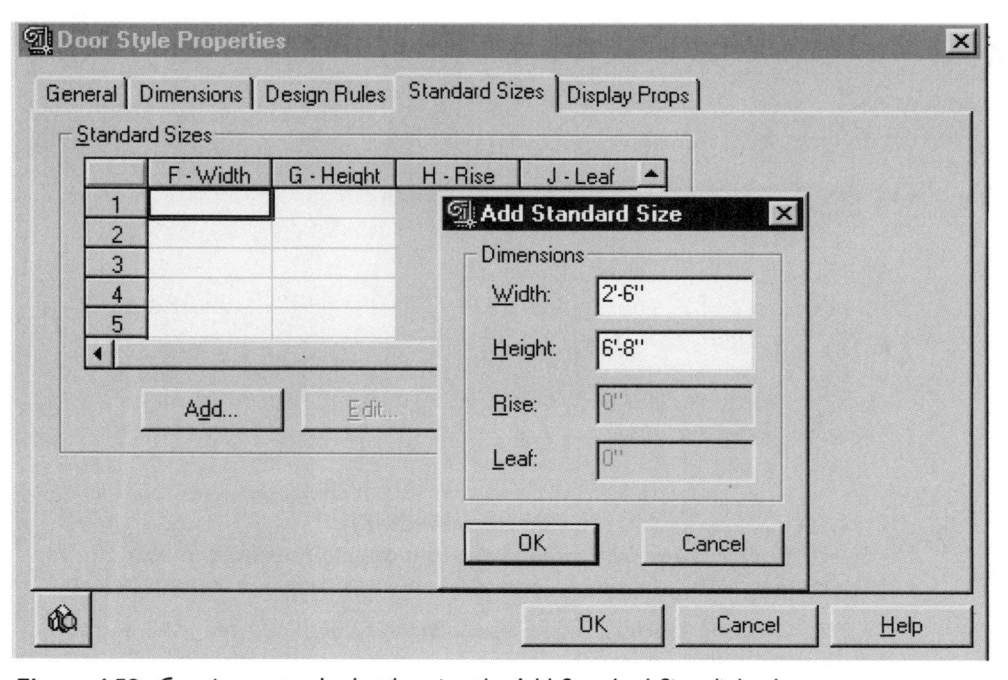

Figure 4.52 *Creating a standard style using the Add Standard Size dialog box*

17. Select the **Add** button and create the following new sizes:

    2'-0" width, 6'-8" height

    3'-0" width, 6'-8" height

18. Note that as the new sizes are entered, they are listed in ascending order according to width in the **Standard Sizes** table.
19. Select **OK** to dismiss the **Add Standard Size** dialog box.
20. Select **OK** to dismiss the **Door Style Properties** dialog box.
21. Select **OK** to dismiss the **Door Styles** dialog box.
22. Select the **Add Door** command on the **Doors/Windows/Openings** toolbar.
23. Select SINGLE_INT_RES door style, select **Automatic Offset/Center** and set the size to 2'-6" x 6'-8" in the **Add Doors** dialog box, as shown in Figure 4-53. Then respond to the command line prompts as shown below.

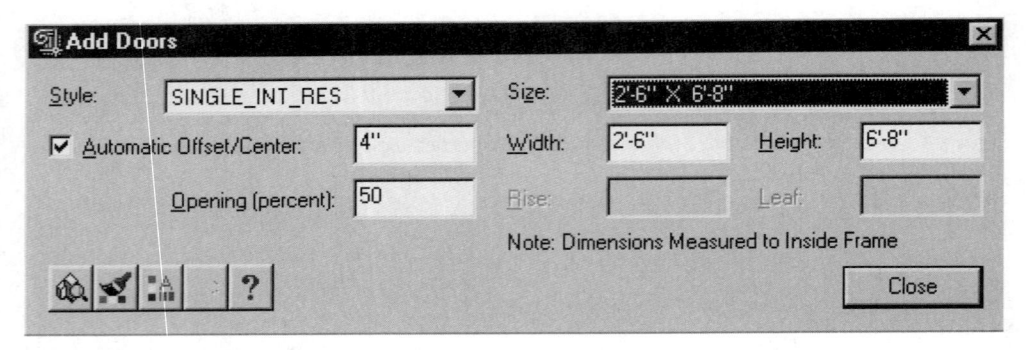

Figure 4.53 *Add Doors dialog box*

```
Command: _AecDoorAdd
Select wall, space boundary or RETURN: (Select wall "B" as
shown in Figure 4.54.)
Insert point or
[STyle/WIdth/HEight/OPening/Auto/OFfset/Match]: (Select
a point near P1 as shown in Figure 4-54.)
Insert point or
[STyle/WIdth/HEight/OPening/Auto/OFfset/Match/Undo]:
ENTER  (Pressing ENTER ends the command.)
```

24. Select **Modify Door** from the **Doors/Windows/Openings** toolbar and select door "A" as shown in Figure 4-54 above, and then press ENTER.
25. Edit the **Modify Doors** dialog box: set **Style** to SINGLE_INT_RES, and select the **Apply** button to edit the door.

Placing Doors and Windows 255

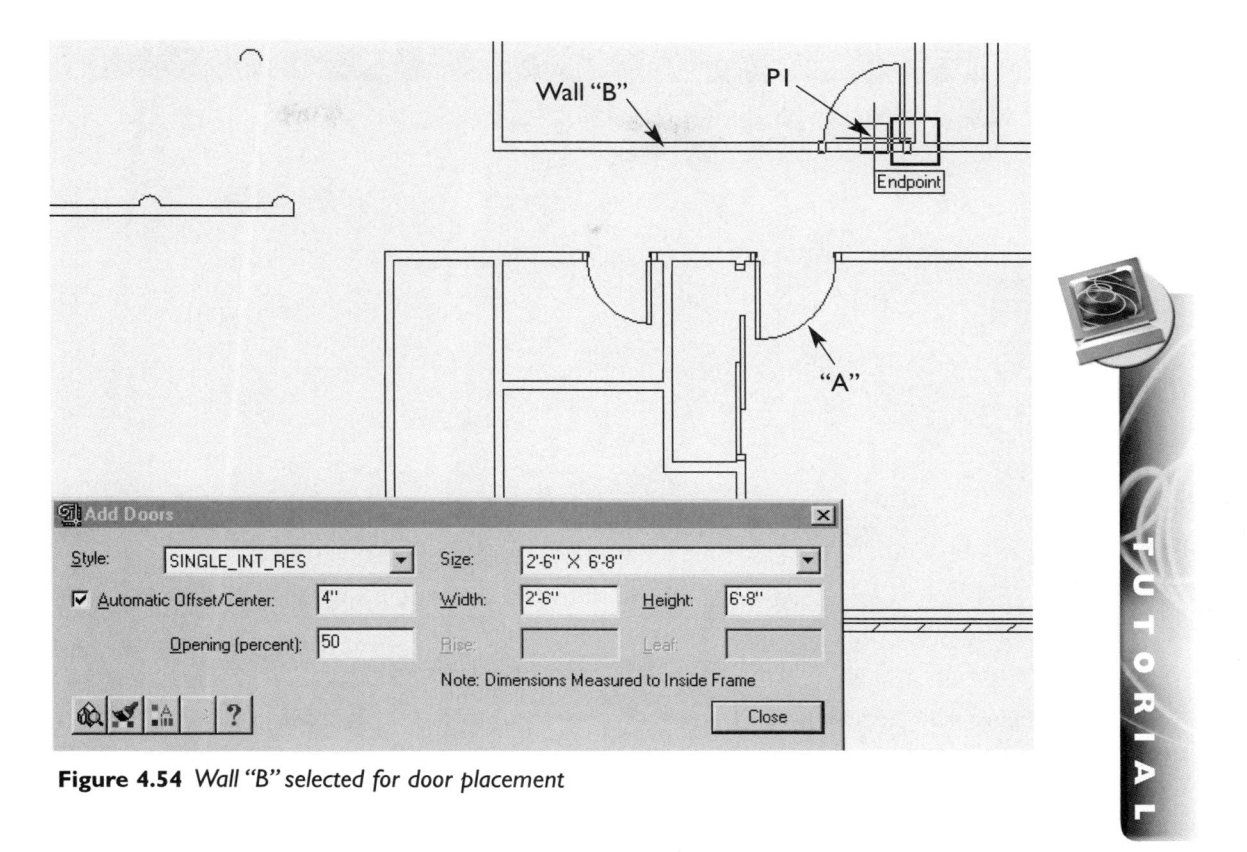

Figure 4.54 *Wall "B" selected for door placement*

26. Select **OK** to dismiss the **Modify Doors** dialog box.
27. Select the **Door Styles** command from the **Doors/Windows/Openings** toolbar.
28. Select the SINGLE_INT_RES door style from the door style list, and then select the **New** button of the **Door Styles** dialog box.
29. Type **SINGLE_EXT_RES** in the **Name** dialog box.
30. Select **OK** to dismiss the **Name** dialog box.
31. Select the **Edit** button of the **Door Styles** dialog box.
32. Select the **General** tab and type *Exterior hinged door* in the **Description** edit field.
33. Select the **Dimensions** tab and edit the following fields: Frame **A-Width** = 2-1/2", Frame **B-Depth** = 5-9/16", clear the **Auto-Adjust to Width of Wall** check box, Stop **C-Width** = 1/2", Stop **D- Depth** = 2", and **E-Door Thickness** = 1-3/4" as shown in Figure 4.55.

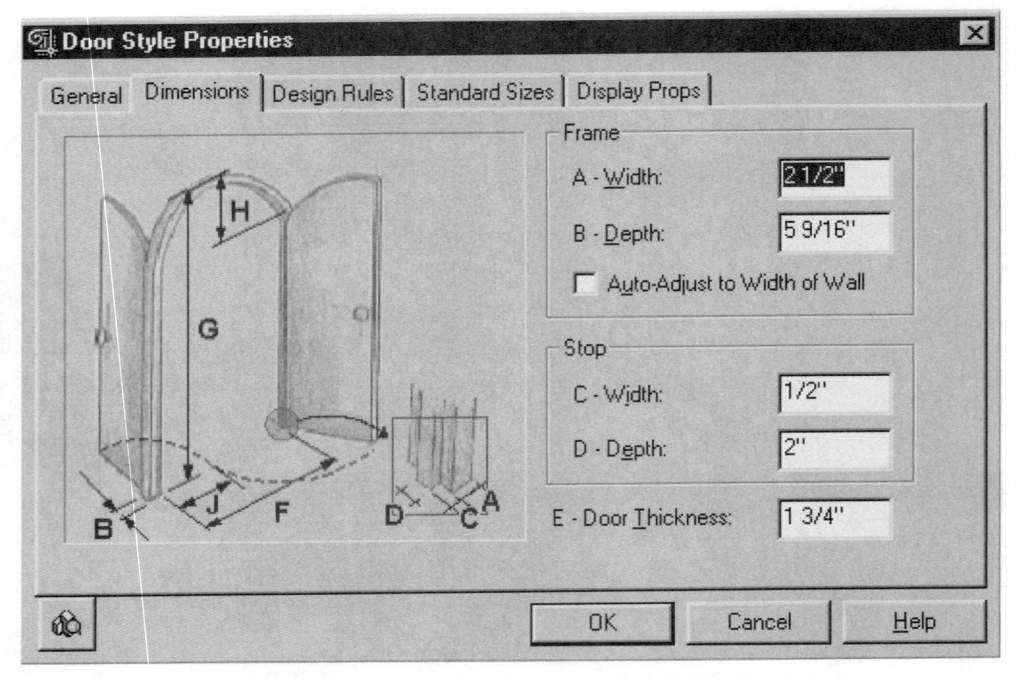

Figure 4.55 *Settings for the Dimensions tab of the Door Style Properties dialog box*

34. Select the **Design Rules** tab and edit the **Shape** to Predefined and Rectangular. Select Single in the **Door Type** section of the dialog box as shown in Figure 4.56.
35. Select the **Standard Sizes** tab of the **Door Style Properties** dialog box.
36. Select the **Add** button to add a size.
37. Edit the **Add Standard Size** dialog box as shown in Figure 4.57. Set width to 3'–0" and height to 6'–8".
38. Select **OK** to dismiss the **Add Standard Size** dialog box.
39. Select the **Display Props** tab and edit the display representation to Threshold Plan.
40. Select Door Style **Property Source** in the **Door StyleProperties** dialog box.
41. Select the **Attach Override** button in the **Door Style Properties** dialog box.
42. Select the **Edit Display Props** button to open the **Layer/Color/Linetype** tab of the **Entity Properties** dialog box.

Placing Doors and Windows 257

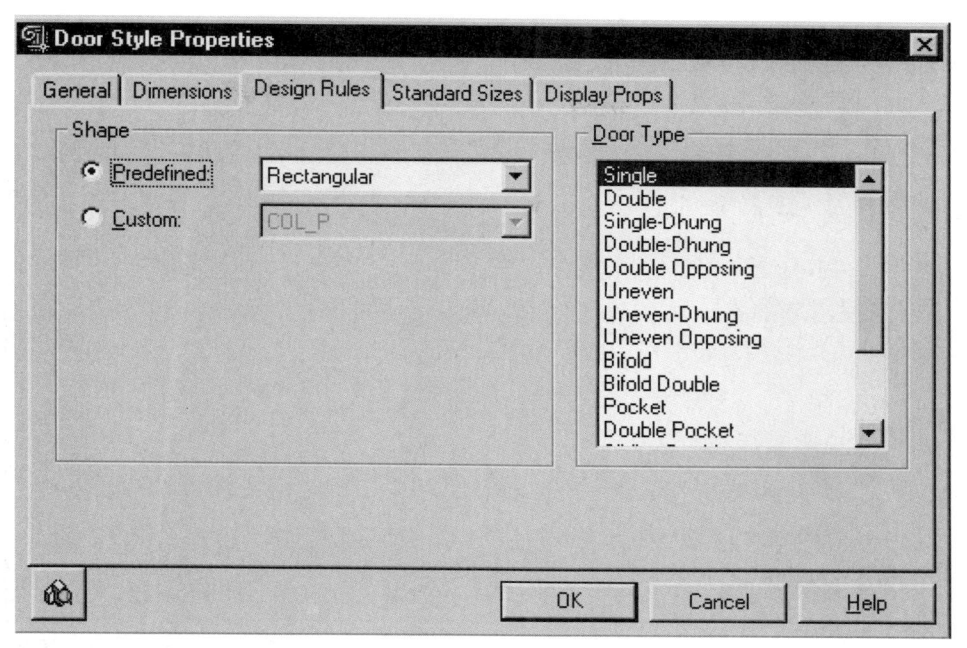

Figure 4.56 Design Rules tab of the Door Style Properties dialog box

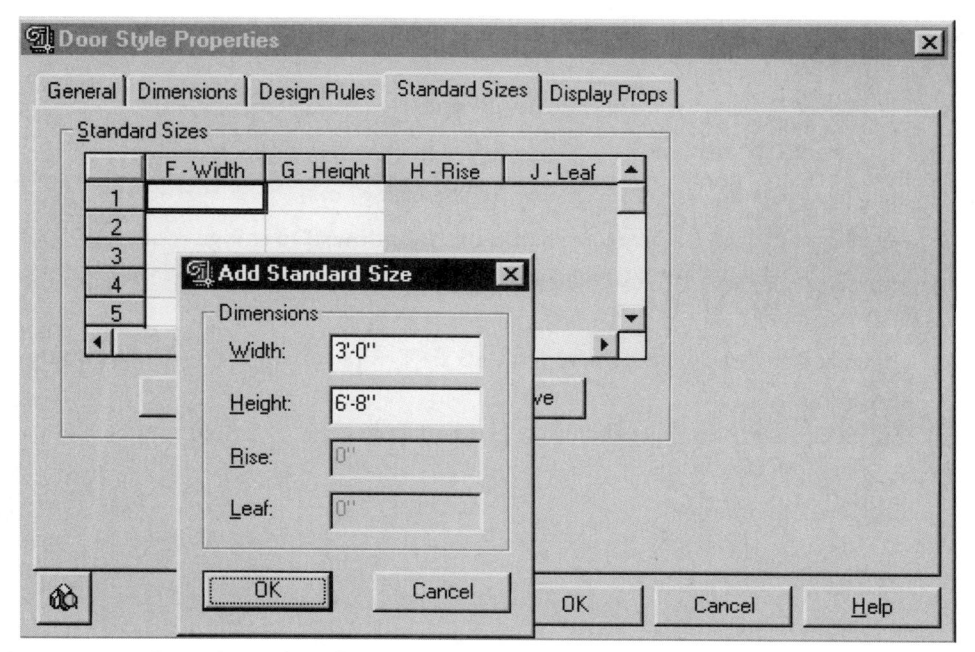

Figure 4.57 Creating a style with a standard size

43. Turn on the visibility (toggle light bulb icon) of Threshold B in the **Layer/Color/Linetype** tab of the **Entity Properties** dialog box, as shown in Figure 4.58.

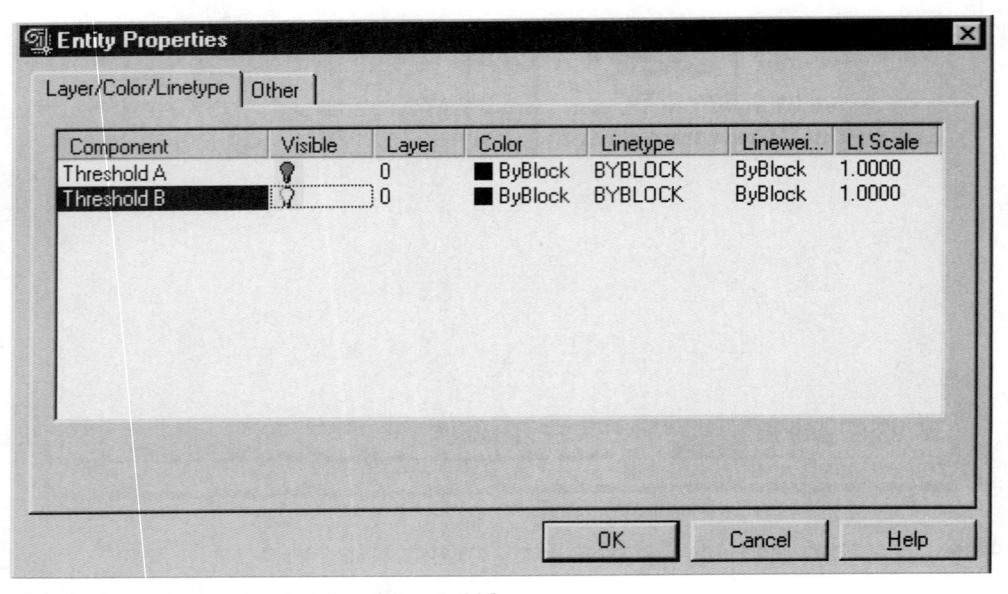

Figure 4.58 *Turn on the visibility of Threshold B*

44. Select **OK** to dismiss the **Entity Properties** dialog box.
45. Select **OK** to dismiss the **Door Style Properties** dialog box.
46. Select **OK** to dismiss the **Door Styles** dialog box.
47. Select **Modify Door** from the **Doors/Windows/Openings** toolbar.
48. Select the rear door as shown in Figure 4.59. Then press ENTER to finish the selection.
49. Edit the style of the door to SINGLE_EXT_RES in the **Modify Doors** dialog box.
50. Select **OK** to dismiss the dialog box and display the threshold as shown in Figure 4.60.
51. Save the drawing.

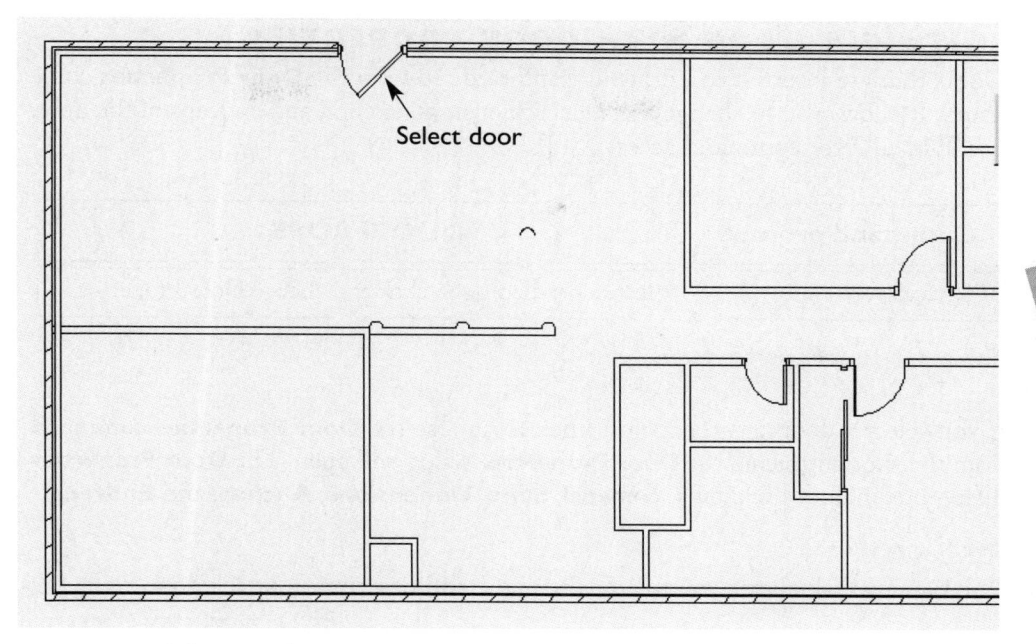

Figure 4.59 *Select rear door*

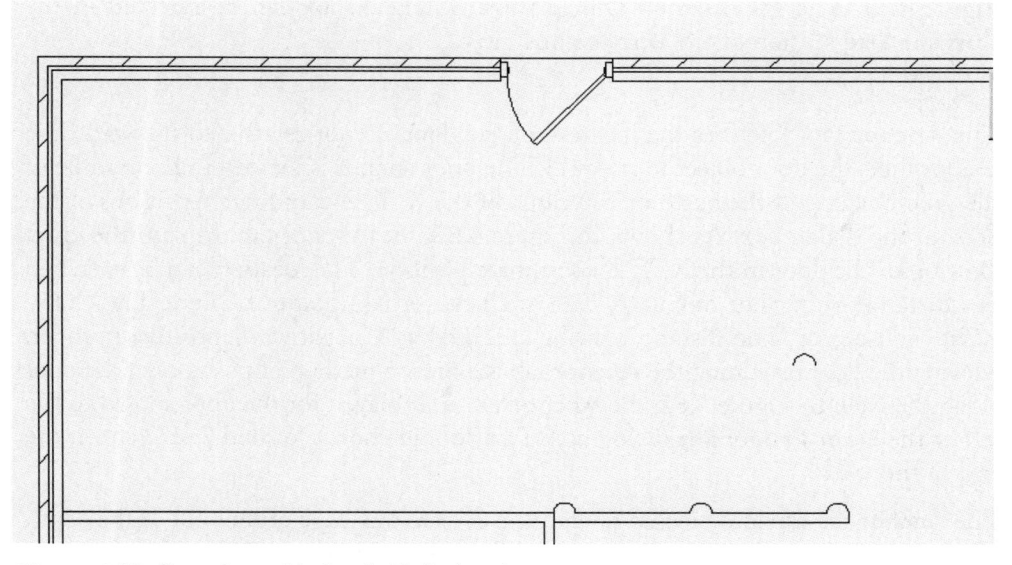

Figure 4.60 *Rear door with threshold displayed*

EDITING A DOOR WITH DOOR PROPERTIES

Doors that are inserted in a drawing can be edited with the **Door Properties** command. It allows you to change the style, dimensions, anchor, and endcaps of the door. See Table 4.7 for command access.

Command prompt	**DOORPROPS**
Shortcut	Select door; then right-click and choose Door Properties

Table 4.7 *Door Properties access*

If you select a door and then right-click to choose the **Door Properties** command from the shortcut menu, the **Door Properties** dialog will open. The **Door Properties** dialog box includes five tabs: **General**, **Style**, **Dimensions**, **Anchor**, and **Endcaps**.

General
The **General** tab allows you to edit the name and description of the door.

Style
The **Style** tab allows you to select another door style.

Dimensions
The **Dimensions** tab allows you to edit the width and height properties as shown in Figure 4.61. The **Measure to Outside Frame** check box can be selected in the **Current Size** section of the **Dimensions** tab.

Anchor
The **Anchor** tab describes the location of the door anchor relative to the wall. The anchor links the door object to the wall. The door anchor is located in the middle of the door opening at the bottom. You can edit the X, Y, Z coordinate positions of the door in the dialog box. As shown in Figure 4.62, the Anchor tab displays the exact location of the door in the X, Y, Z coordinate position. The location can be specified as a distance from Start of Curve, End of Curve, or Midpoint of Curve. The Curve is the wall object. The distance can be specified as a negative or positive number. Moving the door by editing the **Anchor** tab is a precise method of positioning the door along the wall. In Figure 4.62, the **Anchor** tab is displayed for the door shown to the left of the **Door Properties** dialog box. The door anchor is located 7'-1/2" from the end of the wall.

The **Anchor** tab can also be used to shift the door to one edge of the wall. If the depth of the door frame is fixed, the door frame will be placed in the wall centered about the

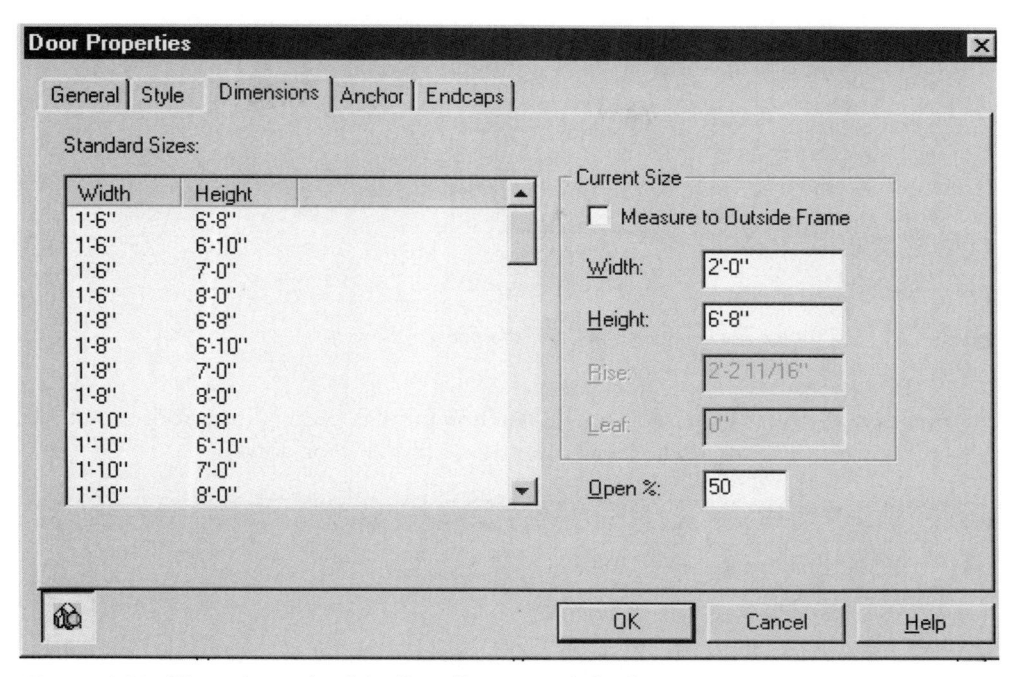

Figure 4.61 Dimensions tab of the Door Properties dialog box

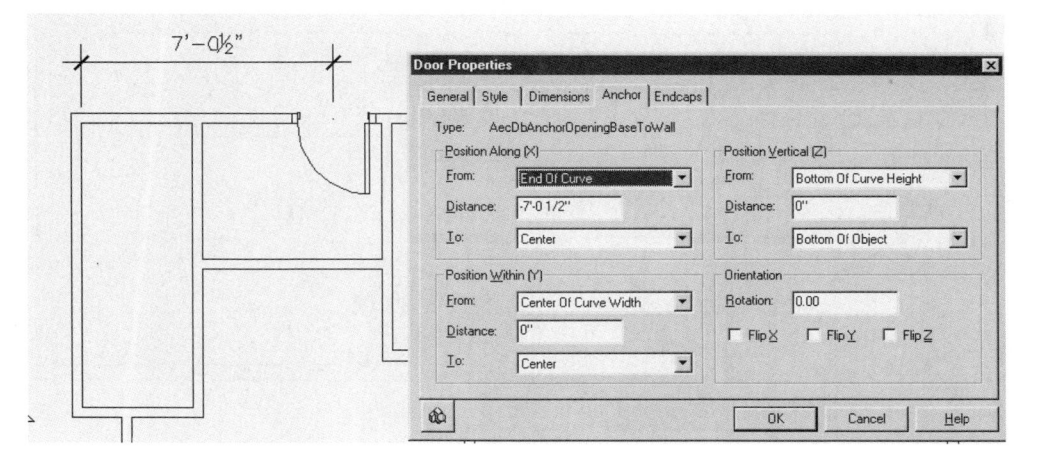

Figure 4.62 Anchor tab of the Door Properties dialog box

width of the wall. In Figure 4.63, the left door is centered about the thickness of the wall. The door on the right was edited to set it flush with the edge of the wall.

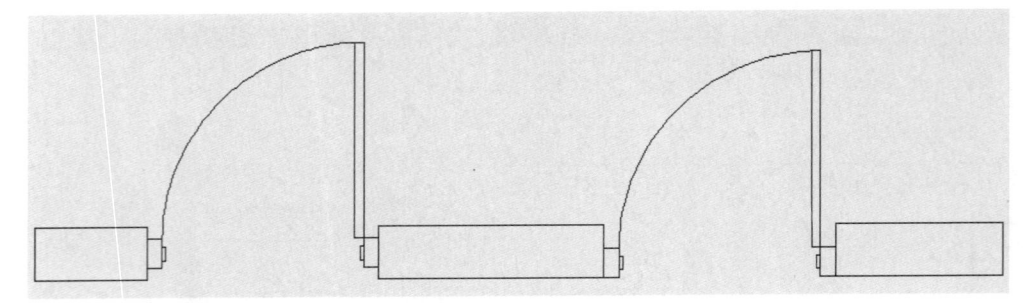

Figure 4.63 *Editing the Anchor tab to position the door in the wall*

The **Position Within (Y)** section of the **Anchor** tab has been edited from zero to 1 3/4", as shown in Figure 4.64, to shift the frame of the door down.

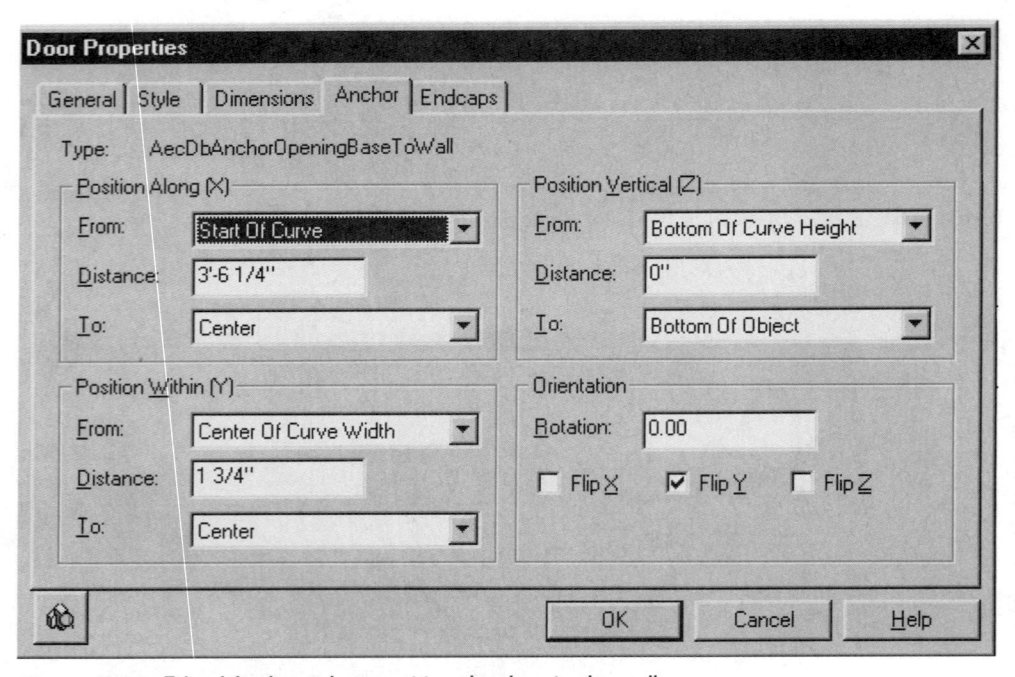

Figure 4.64 *Edited Anchor tab to position the door in the wall*

Endcaps

The **Endcaps** tab allows you to edit the wall endcap used on each end of the door. The **Endcap** drop-down lists shown in Figure 4.65 allow you to assign a different wall endcap for each side of the door.

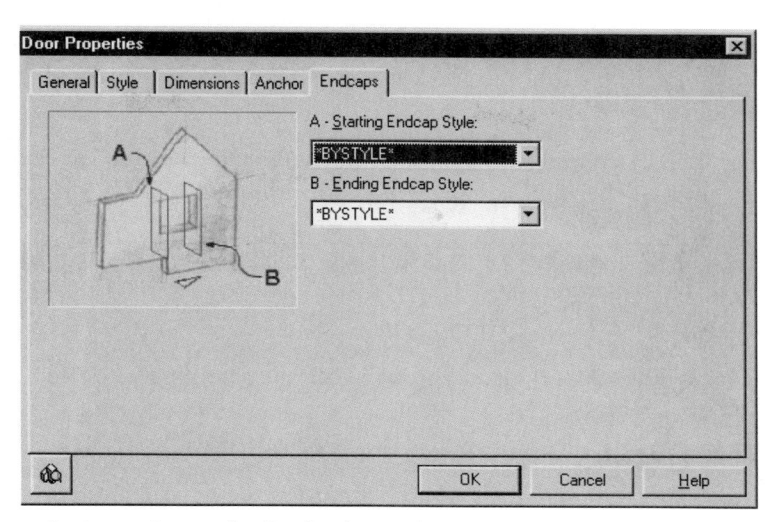

Figure 4.65 Setting endcap styles for the door style

SHIFTING A DOOR WITHIN THE WALL

A door frame can be shifted within the wall, flush to a wall surface, or centered about the wall width with the **Reposition Within** command (**REPOSITIONWITHIN**). This command allows you to shift the door frame relative to the surface of the wall. Door frames that are set in masonry walls are often positioned to an edge of the wall rather than centered in the wall. The **Reposition Within Wall** command of Architectural Desktop allows you to automatically set the door frame to a selected edge of the wall. The command can also be used to shift the frame back to the center of the wall.

Access **Reposition Within Wall** as follows:

Menu bar	Design>Doors>Reposition>Within Wall
Command prompt	REPOSITIONWITHIN
Shortcut	Select door, right click and select Reposition Within Wall

Table 4.8 Reposition Within Wall command access table

When the **Reposition Within Wall** command (**REPOSITIONWITHIN**) is selected, you are prompted to select doors, windows, or openings to adjust. The command has two options: **Offset** and **Center**. The **Offset** option allows you to position the frame to one of the two outer surfaces of the wall. An offset distance can be specified to position the frame a specified distance from the wall surface. The **Center** option will center the frame between the two surfaces of the wall width. If an offset distance is entered, the frame will shift the specified distance relative to the center or offset position. The default offset distance is zero.

The following tutorial demonstrates the use of **Reposition Within Wall** to center a door frame to the center of the wall width, as shown in Figure 4.66.

TUTORIAL 4.3 REPOSITIONING A DOOR WITHIN THE WALL

1. Open *Ex4-3* from the *ADT Tutor\Ch4* directory.
2. Choose **File>SaveAs** from the menu bar and save the file as *Door3* in your student directory.
3. Choose **Design>Doors>Reposition>Within Wall** from the menu bar.
4. Respond as shown in the following command sequence and referring to Figure 4.66 to reposition the door in the wall.

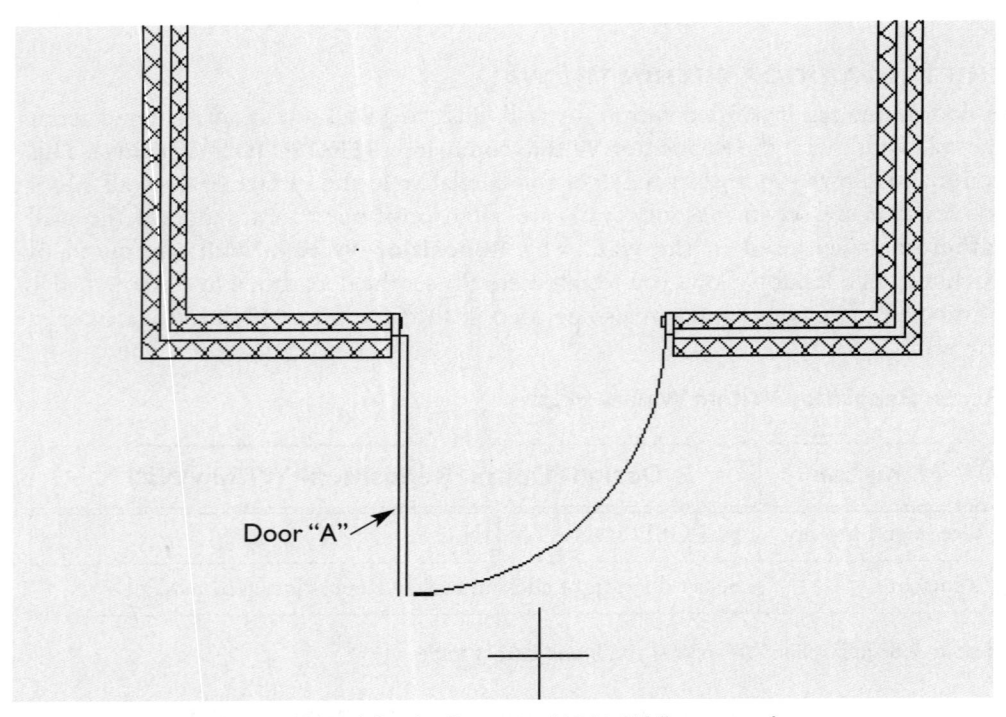

Figure 4.66 *Door frame selected for the Reposition Within Wall command*

```
_AecRepositionWithin
Select doors, windows, or openings:
```
(Select door "A" shown in Figure 4.66.) `1 found`

```
Select doors, windows, or openings: ENTER  (Press ENTER
to end selection)
Current offset is 0".  (Zero is the default offset distance.)
Select side to offset or [Offset/Center]: C ENTER
[1] anchor(s) were modified.
```
(Door is shifted to the center of the wall width as shown in Figure 4.67)

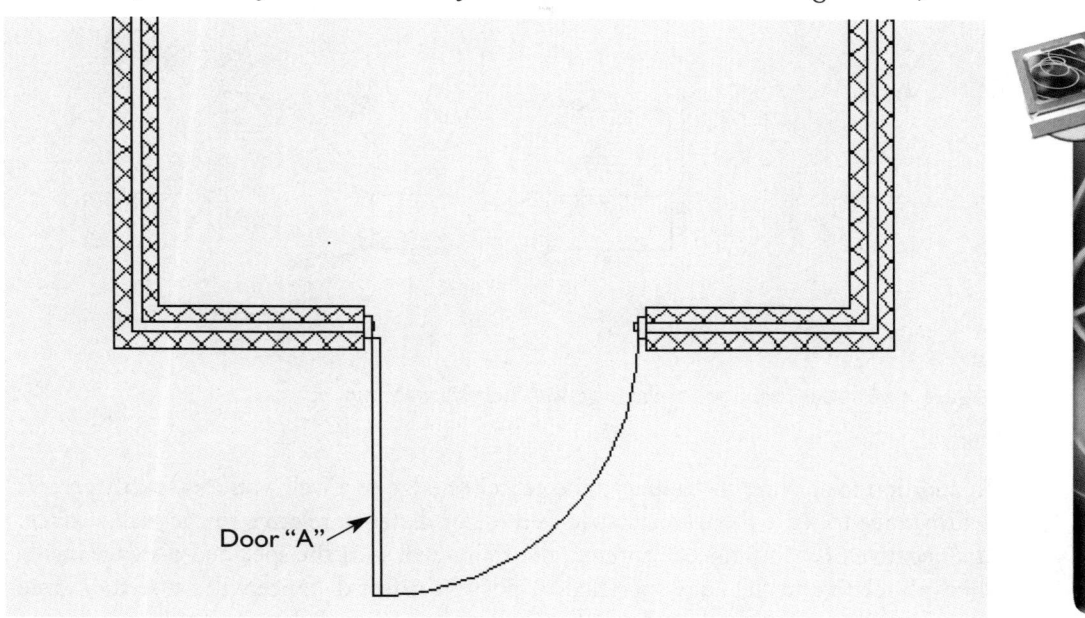

Figure 4.67 Door frame shifted to center with Reposition Within Wall command

5. Save the drawing.

The **Reposition Within Wall** command can also be used to shift the door frame to an edge of the wall width. The following command sequence demonstrates the use of **Reposition Within Wall** to offset the door frame to the edge of the wall as shown in Figure 4.68.

```
(Choose Design>Doors>Reposition>Within from the menu
bar.)
_AecRepositionWithin
Select doors, windows, or openings:  (Select door "A" shown
at left in Figure 4.68.) 1 found
Select doors, windows, or openings: ENTER  (Press
{KEY}Enter to end selection)
Current offset is 0".  (Zero is the default offset distance)
```

```
Select side to offset or [Offset/Center]: (Select a point
near P1 below the frame.)
[1] anchor(s) were modified.
```
(The frame is shifted to the lower wall edge as shown at right in Figure 4.68.)

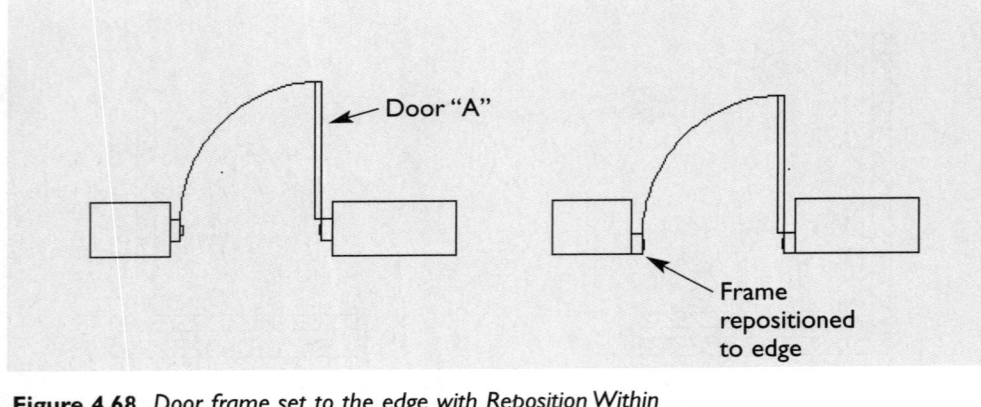

Figure 4.68 *Door frame set to the edge with Reposition Within*

In addition to shifting the frame to the center or edge of a wall, you can specify an offset distance to shift the frame a specified offset distance relative to the wall surface. If a negative offset distance is entered, the frame will shift the specified distance inside the wall relative to the edge specified. A positive offset distance will cause the frame to project from the wall surface. The following tutorial demonstrates the use of **Reposition Within Wall** to offset the door frame to the edge of the wall with a -1" offset distance.

TUTORIAL 4.4 USING REPOSITION WITHIN WALL TO OFFSET A DOOR A SPECIFIC DISTANCE

1. Open \ADT Tutor\Ch4\Ex4-4.dwg and save the file to your student directory as Door4.

2. Select the door, and then right-click and choose **Reposition Within Wall** from the shortcut menu. Respond as shown in the following command sequence to reposition the door in the wall.

```
_AecRepositionWithin
Current offset is 0". (Default offset distance)
Select side to offset or [Offset/Center]: O ENTER
Enter offset <0>: -1 ENTER
Select side to offset or [Offset/Center]: (Select a point
above the door near P1 as shown in Figure 4.69.)
```

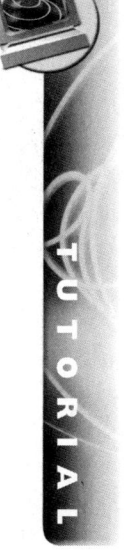

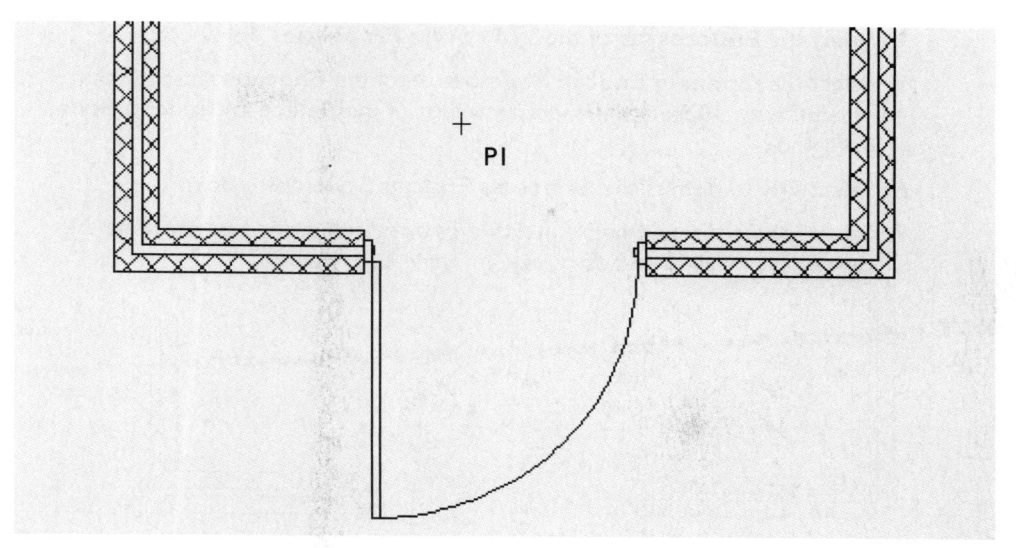

Figure 4.69 Selecting the direction of offset for the Reposition Within Wall command

```
[1] anchor(s) were modified.
```
(The frame is shifted to within 1" of the upper wall surface as shown in Figure 4.70.)

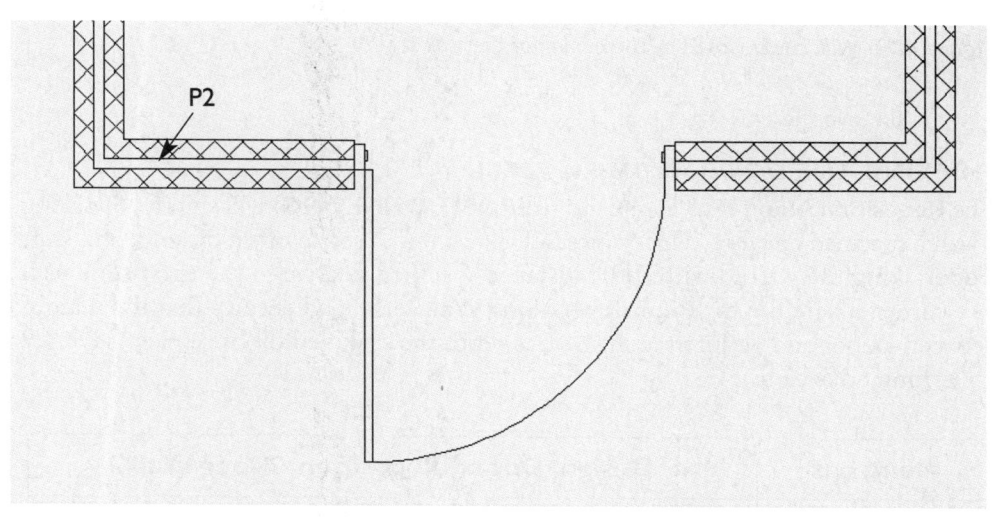

Figure 4.70 Shifting the door frame a specified distance with the Reposition Within Wall command

3. Select the wall at P2 as shown in Figure 4.70, and then right-click and choose **Edit Wall Style** from the shortcut menu.

4. Select the **Endcaps** tab of the **Wall Style Properties** dialog box.
5. Select the **Opening Endcap Style** button of the **Endcaps** tab, and then select Cavity_10 from the endcap style list of the **Select an Endcap Style** dialog box.
6. Select **OK** to dismiss the **Select an Endcap Style** dialog box.
7. Select **OK** to dismiss the **Wall Style Properties** dialog box and view the cavity of the wall closed, as shown in Figure 4.71.

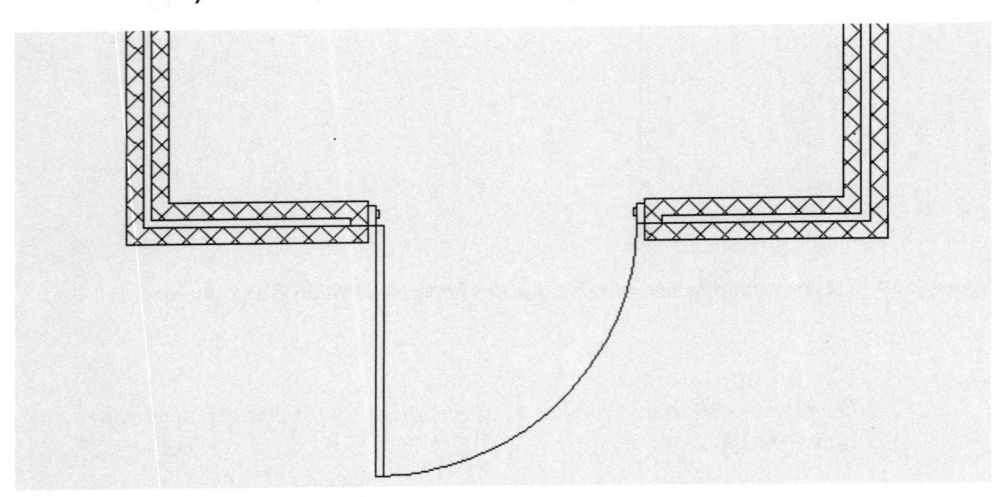

Figure 4.71 *Wall cavity closed by the editing of the wall style*

8. Save the drawing.

SHIFTING THE DOOR ALONG A WALL

The **Reposition Along Wall** command (**REPOSITIONALONG**) will shift a door along a wall a specified distance. This command allows you to set an offset distance and shift a door along the wall to within the distance specified relative to the end of the wall or a corner. Therefore the **Reposition Along Wall** command ensures that the margin between a door and an intersecting wall is set to the specified distance. See Table 4.9 for command access.

Menu bar	Design>Doors>Reposition>Along Wall
Command prompt	REPOSITIONALONG
Shortcut	Select door, right click and select Reposition Along Wall

Table 4.9 *Reposition Along Wall access table*

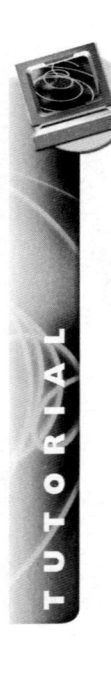

The **Reposition Along Wall** command can be used to edit doors inserted with the wrong **Automatic Offset/Center** value. The offset value of the **Reposition Along Wall** command positions an existing door in a manner similar to the way as a new door is positioned with the **Automatic Offset/Center** option of the **Add Door** command.

The following command sequence and Figure 4.72 demonstrate the use of **Reposition Along Wall** (**REPOSITIONALONG**) to shift a door frame along a wall to within 12" of a corner.

```
Command: _AecRepositionAlong
Select object: (Select door A as shown in Figure 4-72.)
Current offset is 0", Ignore openings: No.
Select point near wall or [Offset/Ignore openings]:
O ENTER   (Selects Offset option.)
Enter offset<0">: 12 ENTER   (Specifies offset distance)
Select point near wall or [Offset/Ignore openings]:
(Select a point near the corner P1 as shown in Figure 4.72.)
(Door moves to within 1'-0" of the corner of the walls as shown on the
example to the right.)
Select object: ENTER   (Pressing ENTER ends the command.)
```

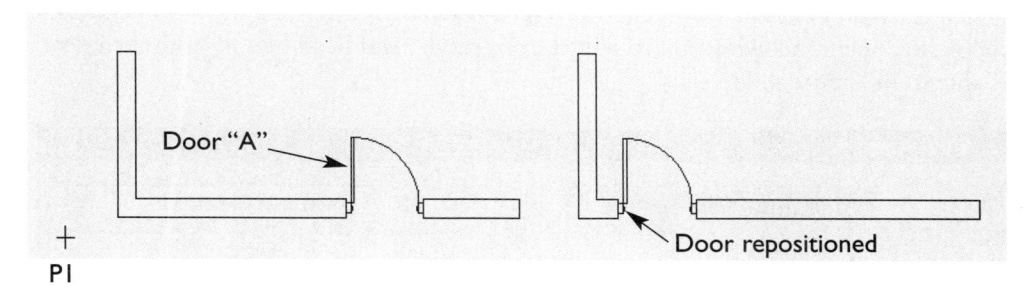

Figure 4.72 *Door adjusted with Reposition Along Wall command*

Note: The **Reposition Within Wall** and **Reposition Along Wall** commands can also be used to shift windows in a wall.

PLACING WINDOWS IN WALLS

The **Add Window** command is used to place Windows in a drawing. See Table 4.9 for command access.

Menu bar	Design>Windows>Add Window
Command prompt	WINDOWADD
Toolbar	Select Add Window from the Doors/Windows/Openings toolbar shown in Figure 4.73
Shortcut	Right-click over drawing area; choose Design>Windows>Add

Table 4.10 *Add Window access*

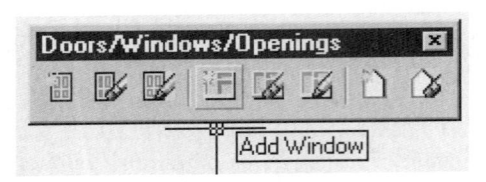

Figure 4.73 *The Add Window command on the Doors/Windows/Openings toolbar*

When the **Add Window** command (**WINDOWADD**) is executed, the **Add Windows** dialog box opens, allowing you to edit the size, style, and head height of the window, as shown in Figure 4.74.

Figure 4.74 *Add Windows dialog box*

The **Add Windows** dialog box consists of the following edit fields and buttons:

Style – The **Style** drop-down list displays list of window styles.

Size – The **Size** list displays the preset sizes for windows of the selected style.

Width – The **Width** edit box allows you to enter the desired width of the window prior to placement.

Height – The **Height** edit box allows you to enter the desired height of the window prior to placement.

Automatic Offset/Center – The **Automatic Offset/Center** check box sets placement to the center of the wall or an offset distance from the ends of the wall. If this check box is selected, the distance to offset from the ends of the wall for door placement is specified in the edit field to the right of the check box.

Head Height – The head height is the distance from the zero Z coordinate that the header of the window is placed.

Floating Viewer – The **Floating Viewer** button opens the **Viewer** and allows you to view the window prior to placement.

Match – The **Match** button allows you to set the size and style of a window by selecting an existing window in the drawing.

Properties – The **Properties** button opens the **Window Properties** dialog box, which allows you to select window styles and size properties of the window.

Undo – The **Undo** button becomes active when a window is placed; selecting **Undo** will remove the last insertion of a window. After selecting the **Undo** button, you remain in the **Add Window** command and can specify a location for the window.

Help – Selecting the **Help** button opens the Help file to the Add a Window topic.

To place a window in a wall, select the **Add Window** command from the **Doors/Windows/Openings** toolbar. The mouse pointer is converted to a select box after you select the command, and you are prompted to select a wall or space boundary as follows:

```
Command: _AecWindowAdd
Select wall, space boundary or RETURN: (Select a wall or
space boundary.)
Insert point or [STyle/WIdth/HEight/HEAd
height/Auto/Match]: (Select a location for the window in the wall.)
Insert point or [STyle/WIdth/HEight/HEAd
height/Auto/Match/Undo]: (Select additional locations for windows
in the same wall or space boundary.)
```

After you select the wall, the window will dynamically slide along the wall until an insertion point is defined by a left-click of the mouse. The **Add Windows** dialog box can be edited before or after an insertion point for the window is specified. You can place additional windows in the same wall by moving the mouse along the wall and selecting additional insertion points or entering relative coordinates.

The handle or insertion base point for a window is located on the end of a window toward the beginning of the wall. In Figure 4.75, the wall was drawn from left to right; therefore the insertion point for the window will be on the left side of the window.

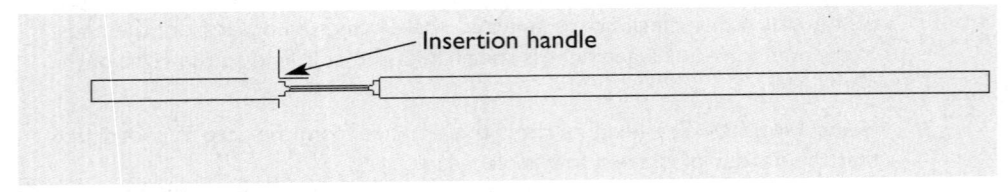

Figure 4.75 *Insertion handle of window drawn from left to right*

The insertion point of the window cannot be toggled to the center of the window. However, the **Automatic Offset/Center** option of the **Add Windows** dialog box can be used to center a window within the length of the wall. When this check box is selected, the window will be centered along the length of the wall. If a window is placed near a corner, the window will be offset from the corner according to the distance specified in the distance edit field. In Figure 4.76, the window on the left was placed with **Automatic Offset/Center** selected and the distance set to 3'-0". The insertion point was selected near the left corner to position the window within 3'–0" of the corner. The window on the right was placed with the same settings except that the insertion point was selected near point P1 on the wall.

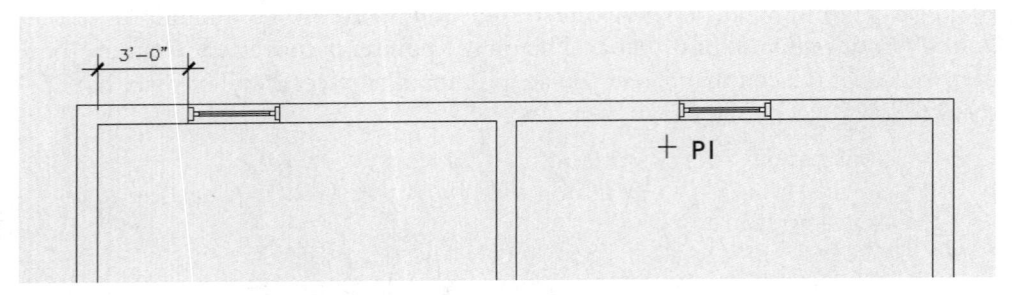

Figure 4.76 *Using the Automatic Offset/Center check box to place a window.*

You can also place windows using the **ARRAY** command; it can be used to place a series of windows in a wall a specified distance apart. In Figure 4.77, the window at the far left was placed first and copied to the right with a rectangular array. The windows are 3'-0" wide and were arrayed using 1 row, 5 columns with the distance between columns set to 4'.

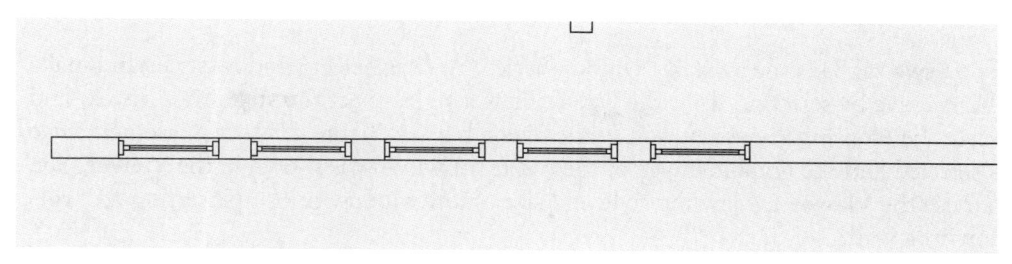

Figure 4.77 *Creating an array of windows*

DEFINING WINDOW PROPERTIES

You can set the properties of a window prior to inserting a window in a wall. Selecting the **Window Properties** button in the **Add Windows** dialog box will open the **Window Properties** dialog box, which consists of three tabs: **General**, **Style**, and **Dimensions**. These tabs are described below.

General

The **General** tab, shown in Figure 4.78, allows you to edit the description of the selected window style.

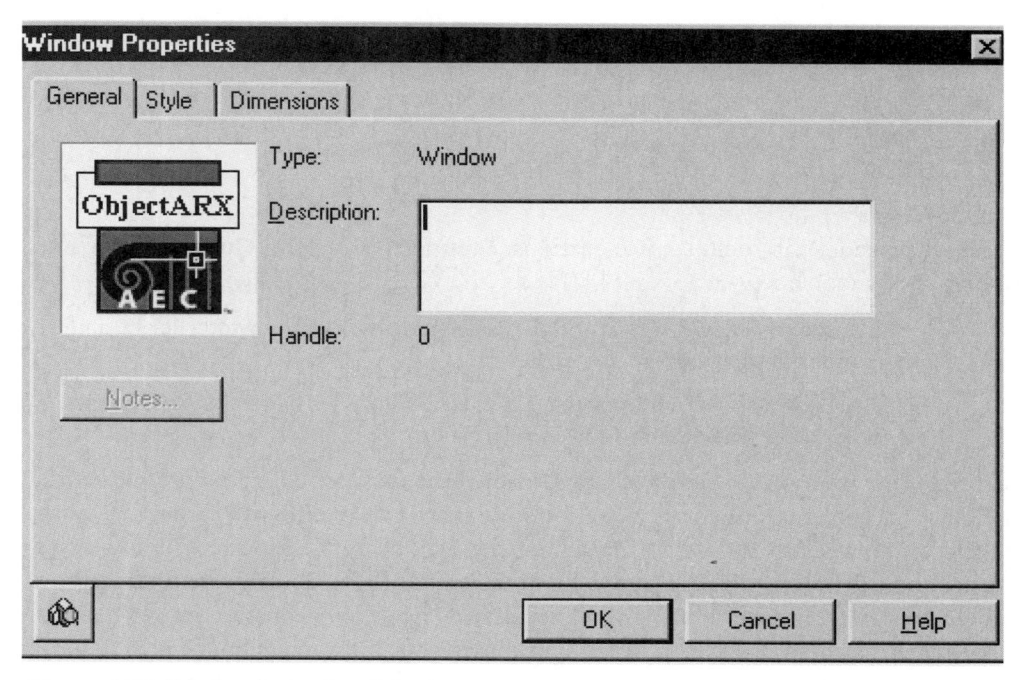

Figure 4.78 *Window Properties dialog box*

Style

The **Style t**ab lists the available window styles. You can select window styles and make them active by selecting from the list. To view a style, select the style from the list and select the **Floating Viewer** button in the dialog box. In Figure 4.79, the Casement style is selected and the graphic image of the casement window is shown in the **Viewer**. You can use the **Viewer** to view the style and size of any window prior to inserting the window in a wall.

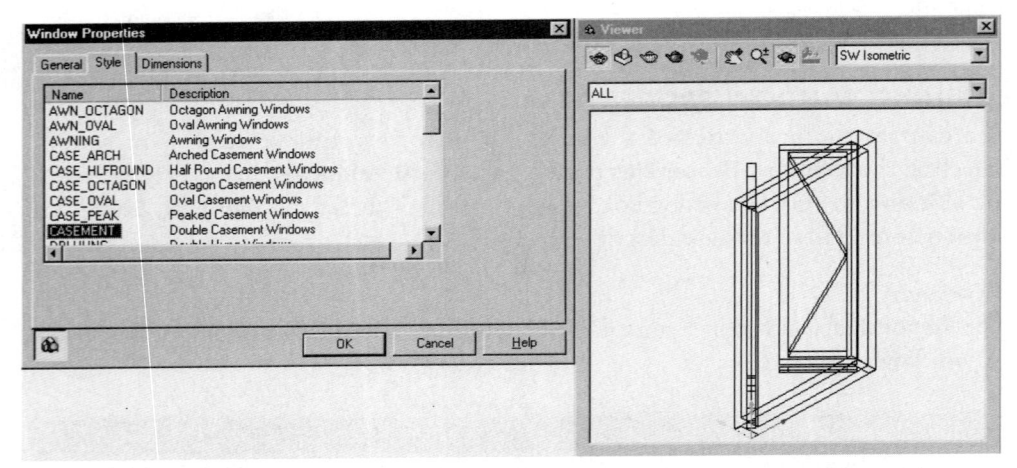

Figure 4.79 *Using the Viewer to view a style in the Window Properties dialog box*

Dimensions

The Dimensions tab includes two sections: **Standard Sizes** and **Current Size**. The **Dimensions** tab shown in Figure 4.80 is for the CASE_ARCH window style.

> **Standard Sizes** – The **Standard Sizes** section lists the width, height, and rise of the selected window style.
>
> **Current Size** – The **Current Size** section includes edit fields for setting other sizes while the **Add Window** dialog box is open.

Editing the appropriate fields of the **Dimensions** tab can change window width, height, and percent of opening. When the **Measure to Outside of Frame** check box is selected, the width and height measurements will be applied to the window frame rather than the sash opening size. The **Measure to Outside of Frame** dimension would be appropriate if the window were inserted in masonry construction. When the check box is cleared, the width and height dimensions are applied to specify the window size as its sash opening size. When this check box is cleared, the **Dimensions Measured to Inside Frame** system is used. This system allows you to specify the sash

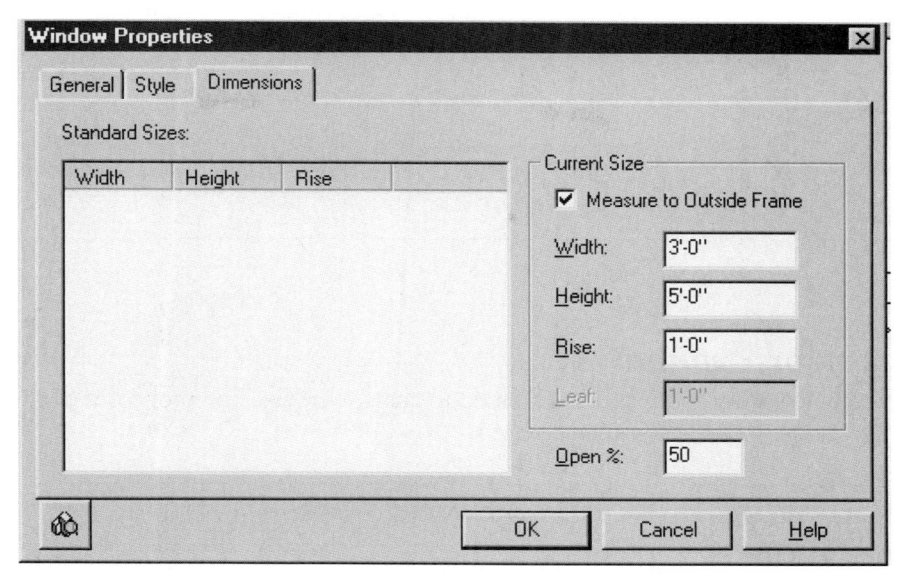

Figure 4.80 *Dimensions tab of the Window Properties dialog box*

opening size as the window dimension and adjust the window frame in the window style definition to obtain the rough opening of the unit. The width of frame dimension specified in the window style is added to the width and height of the sash opening size to determine the overall Frame Opening Height and Frame Opening Width properties when the window schedule is generated.

> **NOTE:** Use the **Measure to Outside of Frame** system to specify the overall window unit dimension for masonry wall construction. The width and height of the window are converted to the Frame Opening Height and Frame Opening Width properties of the window schedule.
>
> Use the **Dimensions Measured to Inside Frame** system to specify window sash opening size and determine overall rough opening by adjusting the Frame A-Width dimension in the window style. Note that the Frame A-Width dimension is added to the height and width of the sash opening to determine the Frame Opening Width and Frame Opening Height dimensions for the window schedule data.

The **Open %** edit field controls the amount the window is displayed in the open position. The percentage is applied to 90 degrees; therefore 50 percent will result in a casement window open at an angle of 45 degrees. The **Rise** edit field will be active on window styles created from the arch, gothic, peak pentagon, and trapezoid shapes. The rise is the distance from the top of the window in which the arch begins. A 1' rise was used to create the arch, gothic, peak pentagon and trapezoid windows shown in Figure 4.81.

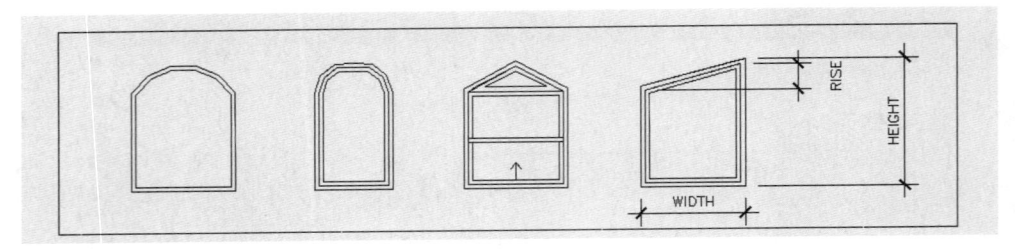

Figure 4.81 Rise of windows

USING MODIFY WINDOW

The **Modify Window** command can be used to change the size and properties of a window. See Table 4.11 for command access.

Menu bar	Design>Windows>Modify Window
Command prompt	WINDOWMODIFY
Toolbar	Select Modify Window from Doors/Windows/Openings toolbar shown in Figure 4.82
Shortcut	Right-click over drawing area; choose Design>Windows>Modify
Shortcut	Select window, right-click and select Window Modify

Table 4.11 Modify Window access

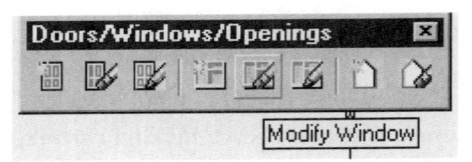

Figure 4.82 Modify Window command on the Doors/Windows/Openings toolbar

When the **Modify Window** command is selected, you are prompted to select a window. The command sequence for **Modify Window** is as follows:

```
Command: _AecWindowModify
Select windows: (Select a window for editing.) 1 found
Select windows: ENTER (Press ENTER to end selection.)
Window modify [Style/Width/HEight/HEAd
height/Match]: DBOX (Edit the Modify Windows dialog box.)
```

After selecting a window, you can edit its properties using the **Modify Windows** dialog box or entering the command options in the command window. The **Modify Windows** dialog box shown in Figure 4.83 allows you to change the size, style and head height of the window.

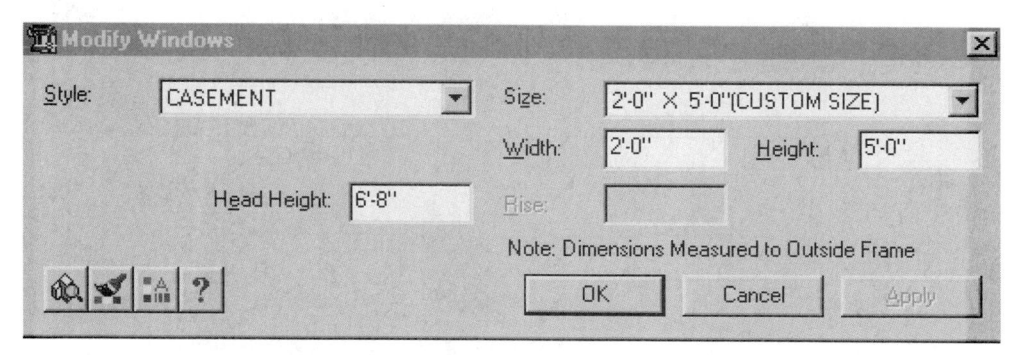

Figure 4.83 *Modify Windows dialog box.*

The **Properties** button can also be selected to edit window properties from within the **Modify Windows** dialog box. When the **Properties** button is selected, the **Window Properties** dialog box opens; it includes the following five tabs: **General**, **Style**, **Dimensions**, **Anchor**, and **Endcaps**. The **General**, **Style,** and **Dimensions** tabs are identical to those of the **Window Properties** dialog box that opens when the **Properties** button in the **Add Windows** dialog box is selected. The **Anchor** and **Endcap** tabs are added to the **Window Properties** dialog box when you edit a window that has been inserted. The **Anchor** tab shown in Figure 4.84 allows you to edit the position and orientation of the window.

The **Endcaps** tab, shown in Figure 4.85, allows you to define the endcap style for the wall on each side of the window. The BYSTYLE endcap shown in the dialog box is the endcap style defined by the style of the wall. Therefore, the starting and ending endcaps allow you to define endcaps as exceptions to those defined by the wall style.

EDITING A WINDOW WITH GRIPS

The grips of a window can be used to edit the swing of the window or move the window in the wall. The grips of a casement window are shown in Figure 4.86. The three grips along the center of the wall can be used to shift the window in the wall or copy the window to other locations in the wall. The grip near the sash of the window can be used to edit the direction of the swing. In Figure 4.86, the grip near the casement sash has been selected and moved above the wall. Selecting a stretch point above the wall will flip the swing of the casement window to the opposite side of the wall.

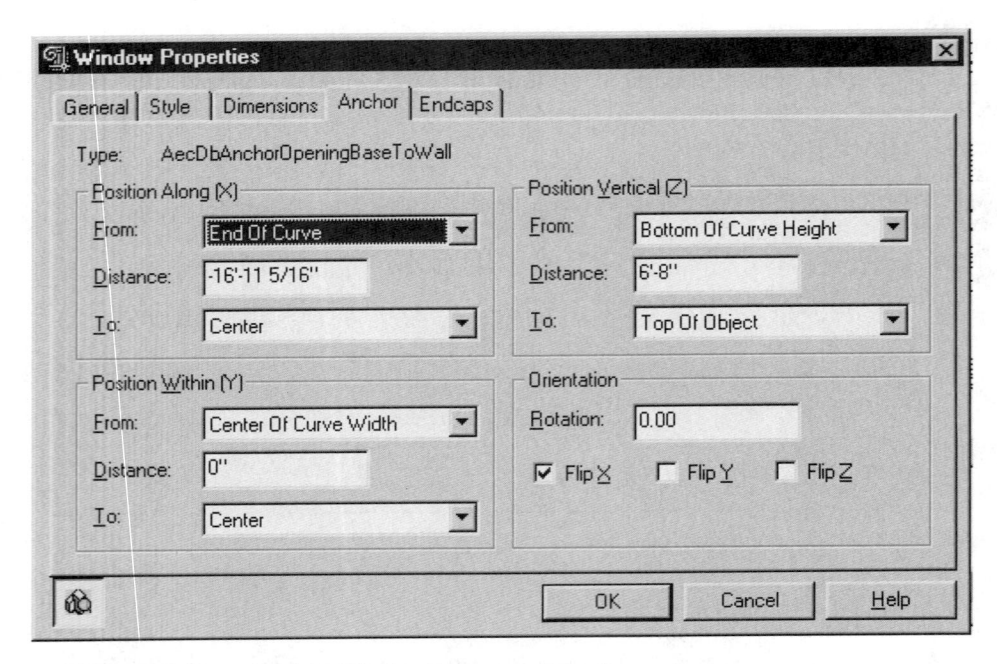

Figure 4.84 *Anchor tab of the Window Properties dialog box*

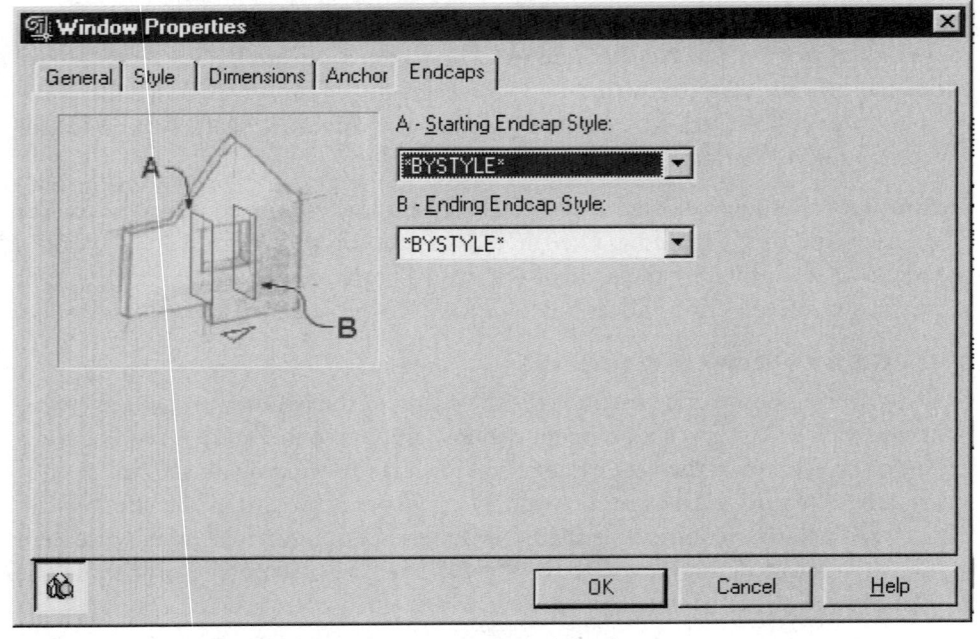

Figure 4.85 *Endcaps tab of the Window Properties dialog box*

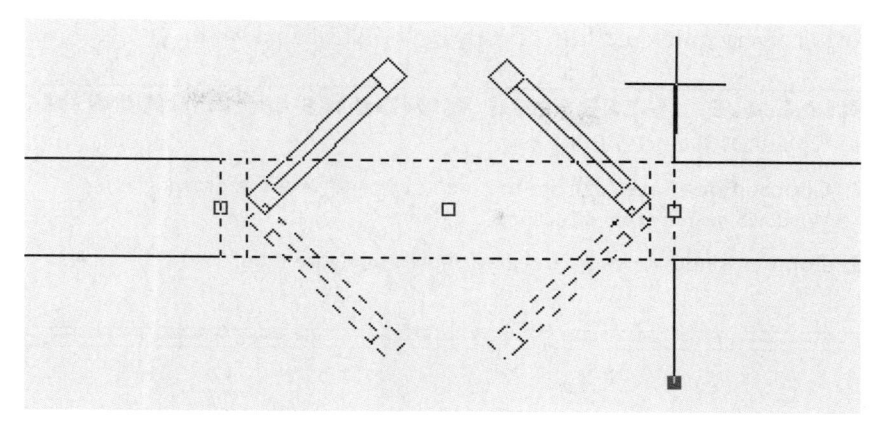

Figure 4.86 *Using grips to edit window swing*

Grips can also be used to copy a window to another location along the wall. If a design specifies the center to center dimension between windows, you can insert one window and use the center grip of the window to **COPY_STRETCH** the window the given distance.

STEPS TO USING GRIPS TO COPY A WINDOW

1. Select an existing window to display its grips.
2. Select the middle grip to make it hot, as shown in Figure 4.87.

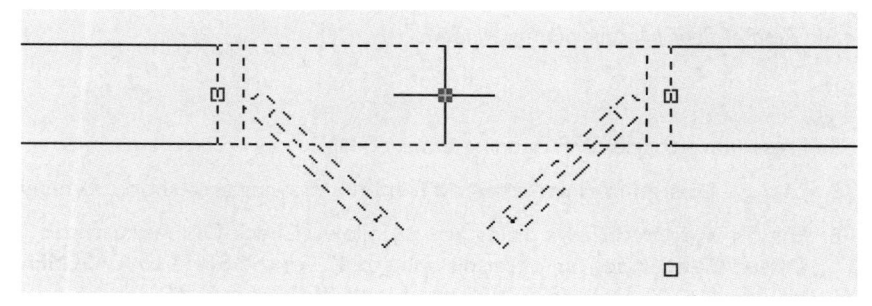

Figure 4.87 *Selecting center grip to copy a window*

3. Right-click to display the shortcut menu and choose **Copy**.
4. Move the mouse in the direction to copy the window.
5. Enter the distance to copy the window and press ENTER.
6. Press ESC to turn off the warm grips.
7. Press ESC to turn off the cold grips.

Grips provide a very quick method of copying windows in a wall.

TUTORIAL 4.5 INSERTING WINDOWS IN A DRAWING

1. Open *ADT Tutor\Ch5\Ex4-5.dwg*.
2. Choose **File>SaveAs** from the menu bar and save the drawing as *Window5.dwg* in your directory.
3. Zoom a window around the area shown in Figure 4.88.

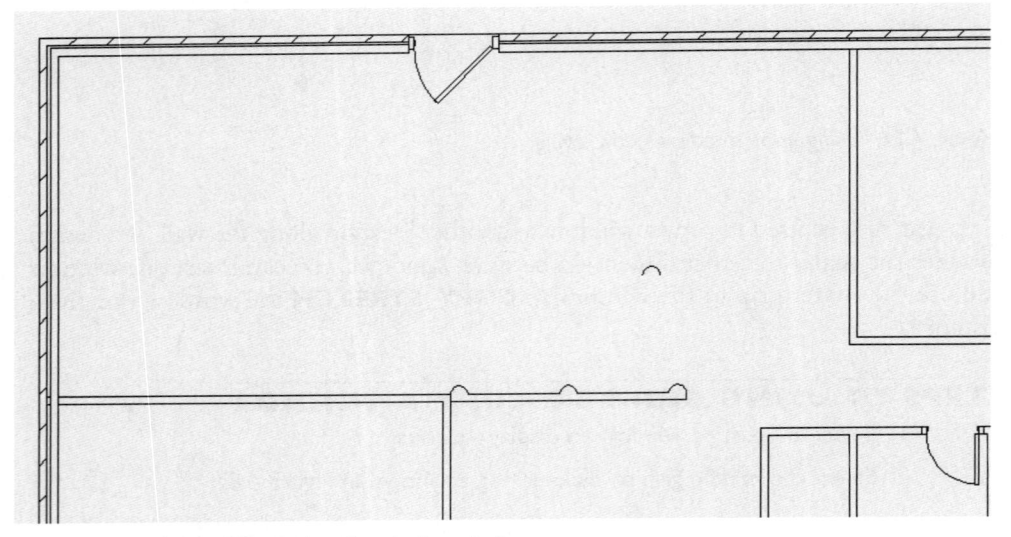

Figure 4.88 *Area of floor plan for placing windows*

4. Press F3 to toggle OFF Running Object Snap.
5. Choose **Design>Windows>Add** from the drawing area shortcut menu.
6. Edit the **Add Windows** dialog box as follows: Check ON **Automatic Offset/Center**, set the offset distance to 4", set the **Style** to CASEMENT, **Width** = 4'-0", **Height** = 3'-0", and **Head Height** = 6'-8".
7. Select the **Properties** button in the **Add Windows** dialog box and select the **Dimensions** tab. Clear the **Measure to Outside of Frame** check box and edit the **Open %** to 50 as shown in Figure 4.89.
8. Select **OK** to dismiss the **Window Properties** dialog box.
9. Verify the settings of the **Add Windows** dialog box as shown in Figure 4.90.
10. Refer to Figure 4.91 and place the window by responding to the AutoCAD prompts as follows:

Placing Doors and Windows 281

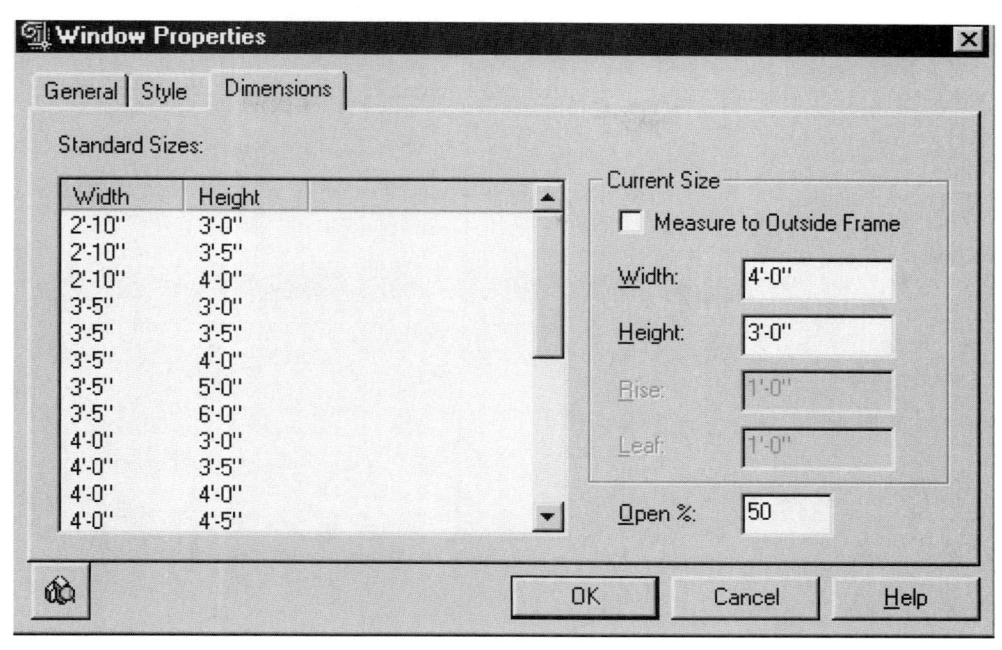

Figure 4.89 *Editing Window Properties using the Dimensions tab*

Figure 4.90 *Add Windows dialog box settings for placing a window*

```
Command: _AecWindowAdd
Select wall, space boundary or RETURN:
```
(Select the wall at P1 in Figure 4.91.)
```
Insert point or [STyle/WIdth/HEight/HEAd
height/Auto/OFfset/Match]:
```
(Move the pointer above the wall to set the window swing outside of the building and then select near P2 to insert the window as shown in Figure 4.91.)

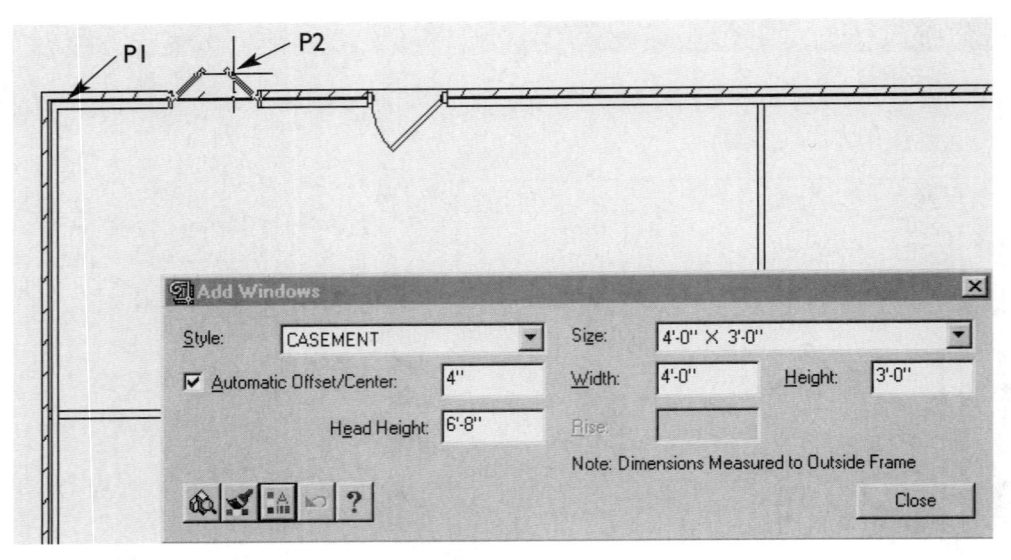

Figure 4.91 *Specifying the swing of the window*

```
Insert point or [STyle/WIdth/HEight/HEAd
height/Auto/OFfset/Match/Undo]:
```
(Move the pointer to the right and above the outside wall and then select near point P3 to insert the second window as shown in Figure 4.92.)

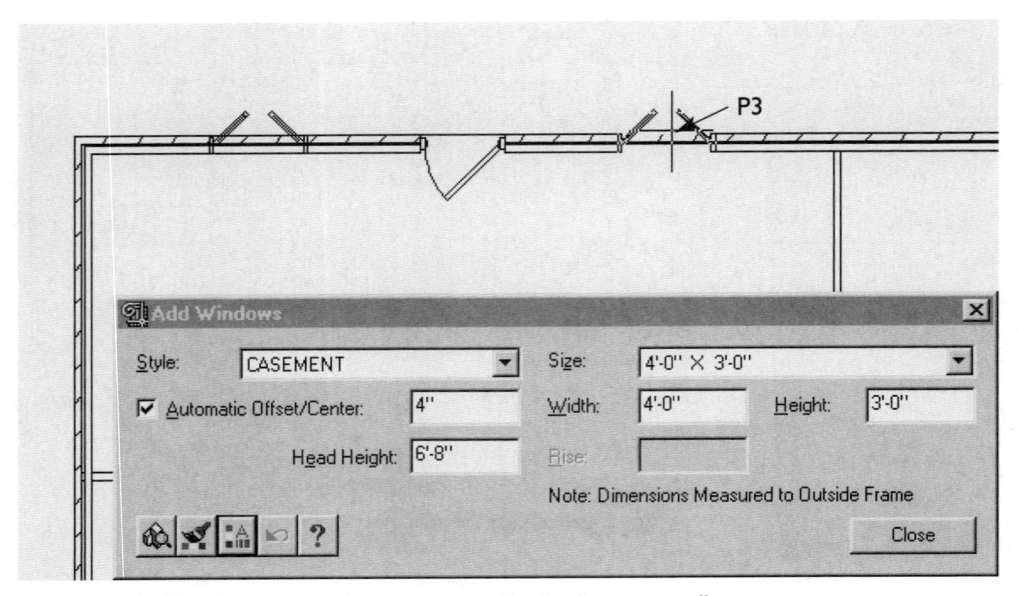

Figure 4.92 *Placing the second casement window in the same wall.*

```
Insert point or [STyle/WIdth/HEight/HEAd
height/Auto/OFfset/Match/Undo]:  ENTER
```
(Pressing ENTER *ends the command.)*

The two windows are centered in each respective wall as shown in Figure 4.93.

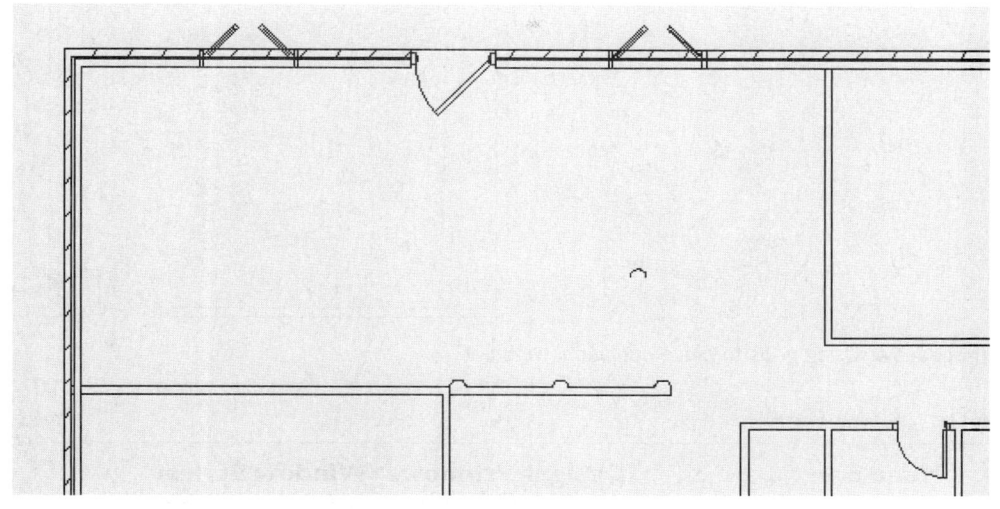

Figure 4.93 *Two windows placed in the center of the wall segments*

11. Select the right window to display its grips.
12. Select the middle grip of the window as shown in Figure 4.94.
13. Right-click with the mouse and choose **Copy** from the shortcut menu.
14. Move the mouse to the right and type **18'** and press ENTER in the command line.
15. Press ESC to clear the hot grip.
16. Press ESC to clear the cold grip.
17. Save the drawing.

CREATING A WINDOW STYLE

Window styles can be created that include a saved list of sizes, frame, and sash parameters. Using styles of windows improves uniformity in placing windows. There are thirty-five window styles included in the Aec_arch_Imperial template, all of which can be used to create other styles. The **Add Windows** dialog box **Style** list allows you to select the saved window styles. You can edit and create window styles using the **Window Styles** command (**WINDOWSTYLE**). See Table 4.12 for command access.

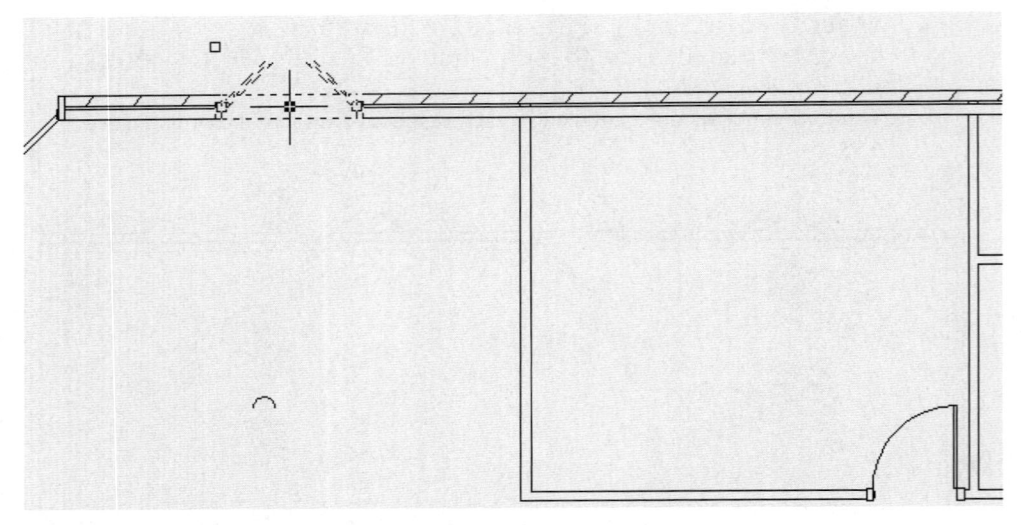

Figure 4.94 *Using grips to place additional windows*

Menu bar	Design>Windows>Window Styles
Command prompt	WINDOWSTYLE
Toolbar	Select Window Styles from Doors/Windows/Openings toolbar shown in Figure 4.95
Shortcut	Right-click over drawing area; choose Design>Windows>Styles

Table 4.12 *Window Styles access*

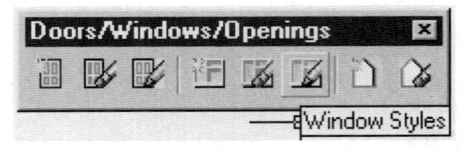

Figure 4.95 *Window Styles command on the Doors/Windows/Openings toolbar*

Selecting the **Window Styles** command opens the **Window Styles** dialog box shown in Figure 4.96. It lists the existing window styles and the buttons to create, copy, edit, set from, import, and export window styles.

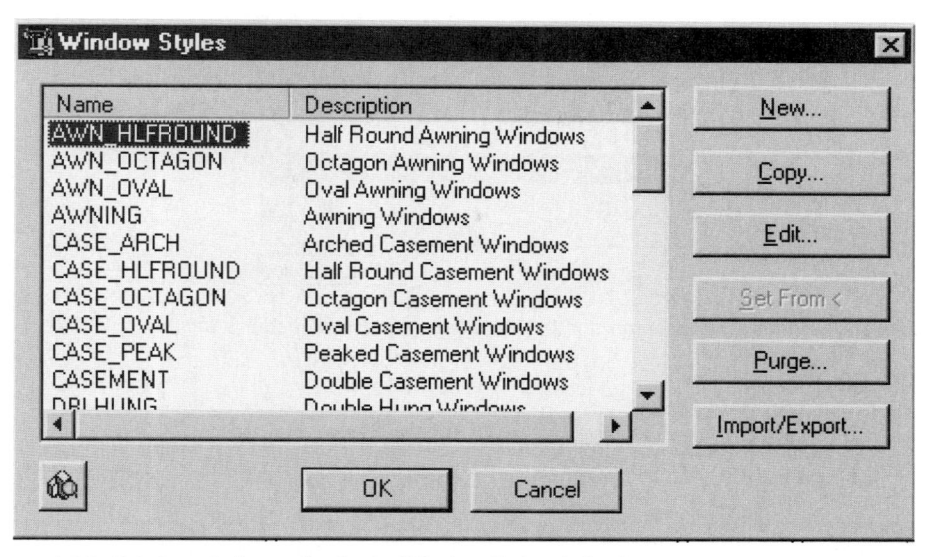

Figure 4.96 Existing window styles in the Window Styles dialog box

The window styles of the drawing are listed in the dialog box in a similar manner to the door styles list of the Door Styles dialog box. The purpose of each of the six buttons on the right is summarized below:

> **New** – Creates new names for window styles.
>
> **Copy**– Copies an existing style to a new name.
>
> **Edit** – Edits the description and properties of a window style.
>
> **Set from** – Inactive button.
>
> **Purge** – Deletes saved styles from a drawing.
>
> **Import/Export** – Transfers saved window styles to resource files or copies saved window styles from a resource file to the current drawing.

DEFINING WINDOW STYLE PROPERTIES

You create a window style by selecting the **Window Styles** command on the **Doors/Windows/Openings** toolbar and selecting the **New** button in the **Window Styles** dialog box. Selecting the **New** button will open the **Name** dialog box, which allows you to assign a new name for the window style. You can add the description of the new style by selecting the **Edit** button. Define additional properties of the window style by editing the dialog box that opens when the **Edit** button is selected.

STEPS TO CREATING A NEW WINDOW STYLE

1. Select **Window Styles** on the **Doors/Windows/Openings** toolbar as shown in Figure 4.97.

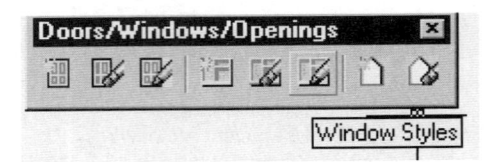

Figure 4.97 *Window Styles command on the Doors/Windows/Openings toolbar*

2. Select the **New** button in the **Window Styles** dialog box.
3. Type the new style name in the **Name** dialog box as shown in Figure 4.98.

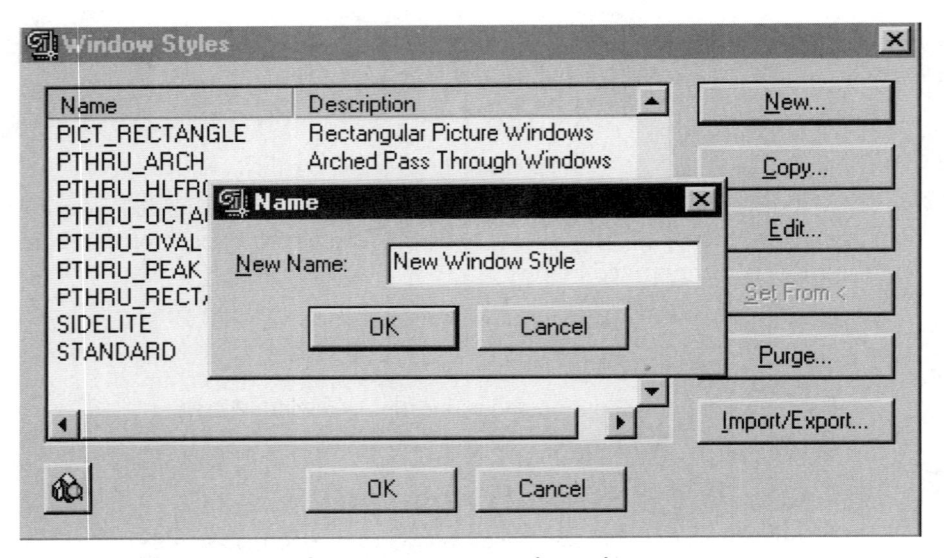

Figure 4.98 *Name dialog box for creating a new window style name*

4. Select **OK** to dismiss the **Name** dialog box.

EDITING A WINDOW STYLE

There are thirty-five window styles included in the Aec_Arch_Imperial template. A style, which includes sizes available from a window manufacturer, can be created. Each style consists of a predefined shape and a type of window. Edit a window style by selecting the **Edit** button in the **Window Styles** dialog box. The **Window Style**

Properties dialog box shown in Figure 4.99 includes the following tabs: **General, Dimensions, Design Rules, Standard Sizes,** and **Display Props**.

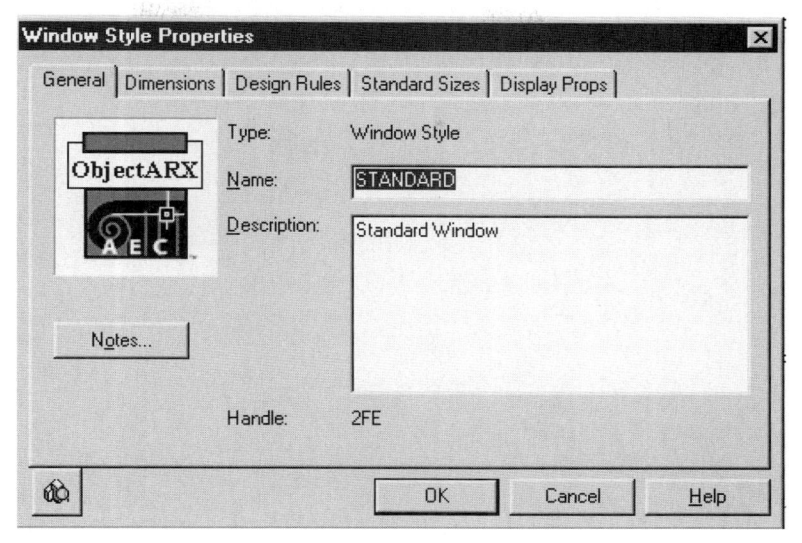

Figure 4.99 Window Style Properties dialog box

Each tab of the **Window Style Properties** dialog box is described below:

General
The **General** tab allows you to edit the name and description of a saved window style. This tab is useful in correcting spelling errors in a named style or in renaming an existing style. The **Notes** button provides access to a **Notes** dialog box, which includes the **Text Docs** and **Reference Docs** tabs. This **General** tab is similar to the **General** tab of the **Door Style Properties** dialog box.

Dimensions
The **Dimensions** tab allows you to set dimensions of the frame, sash, and glass, as shown in Figure 4.100.

> **Frame** – The **Frame** section includes edit fields for setting the **A-Width** and **B-Depth** of the window frame. The **A-Width** dimension is the distance between the sash or glass frame and the wall stud or masonry component. It can be used to determine the rough opening or overall unit dimension if the **Measure to Inside of Frame** option is selected in the **Window Properties** dialog box. The **B-Depth** of the frame can be controlled by the **Auto-Adjust to Width of Wall** check box or set to a specific width, which is retained regardless of wall width. When the **Auto-Adjust to Width of Wall** check box is selected, the frame depth adjusts to the wall width.

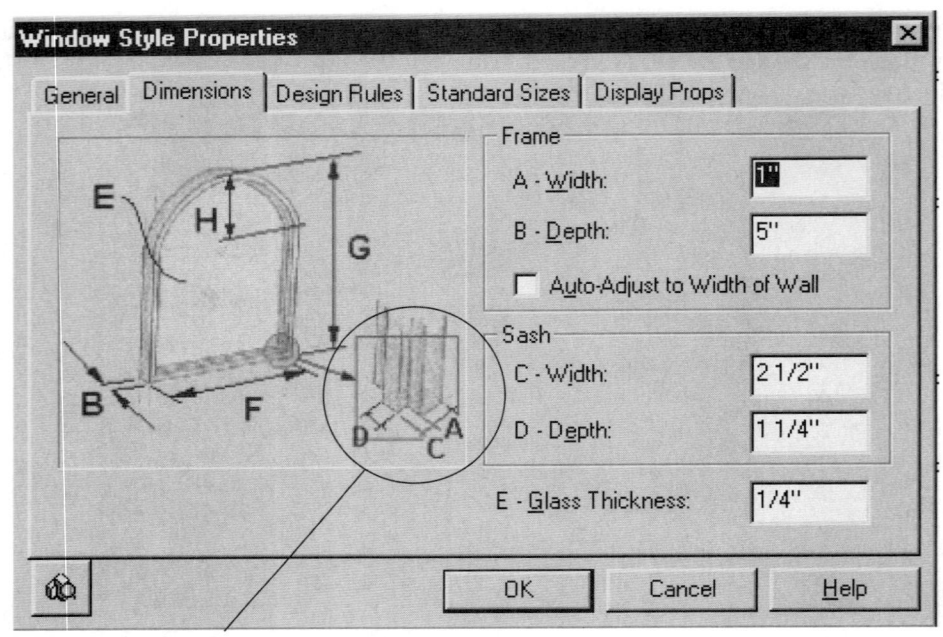

Sash component dimensions

Figure 4.100 *Dimensions tab of the Window Style Properties dialog box*

Sash – The **Sash** section includes edit boxes for setting the **C-Width** and **D-Depth** of the sash. These dimensions vary according to the window manufacturer. The **C-Width** dimension is the distance the sash projects from the frame. The **D-Depth** dimension is the thickness of the sash as shown graphically in Figure 4.100.

Glass – The glass thickness can be entered in the **Glass Thickness** field.

Design Rules

The **Design Rules** tab includes **Shape** and **Window Type** sections, as shown in Figure 4.101.

Shape – The **Shape** section includes radio buttons to select predefined shapes or custom Aec_Profiles. The predefined shapes include thirteen shapes, including rectangle, round, half round, quarter round, oval, arch, gothic, Isosceles triangle, right triangle, peak pentagon, octagon, hexagon, and trapezoid. Each of these shapes can be applied to a window type to create a window style.

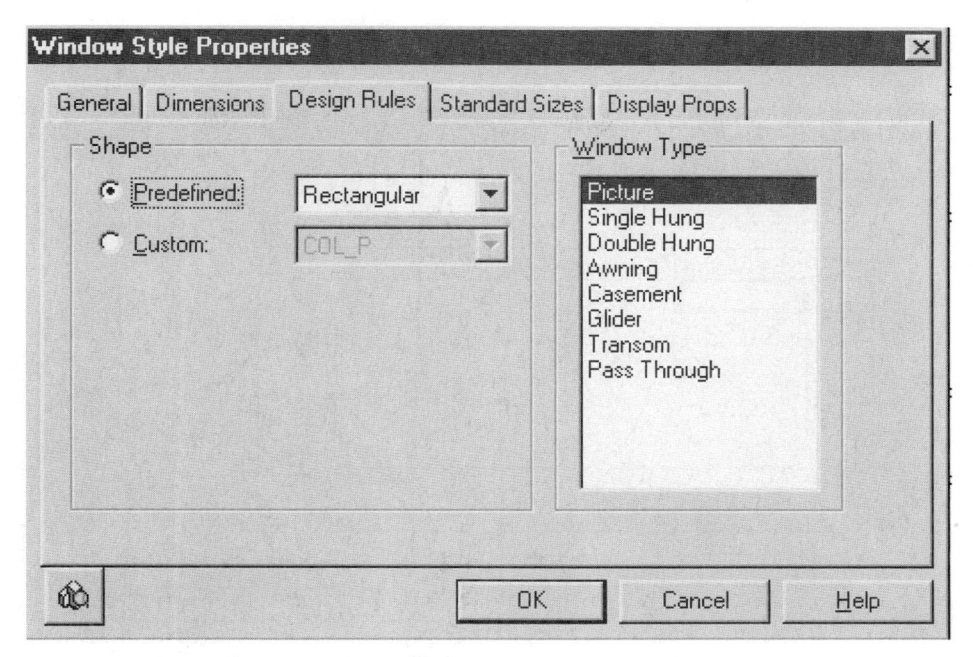

Figure 4.101 Design Rules tab of the Window Style Properties dialog box

The **Custom** radio button allows access to shapes created as an Aec_Profile. These shapes include geometry created from AutoCAD entities and transformed to Aec_Profiles. Use of Aec_Profiles will be discussed later in the chapter.

Window Type – The **Window Type** list box includes various types of windows that are manufactured, such as double hung, casement, picture and glider

Standard Sizes

The **Standard Sizes** tab provides a table of sizes for a window style. The Standard window style does not include any preset sizes, as shown in Figure 4.102.

Standard Sizes – The sizes are listed in a size table that includes **F-Width**, **G-Height**, **H-Rise**, and **J-Leaf**. These dimensions are illustrated on the **Dimensions** tab of the **Window Style Properties** dialog box. The **H-Rise** is inactive unless the window shape is Arch, Gothic, Peak, or Trapezoid. The **J-Leaf** column is inactive.

Add – The **Add** button located below the size table is used to create a standard size. Selecting the **Add** button opens the **Add Standard Size** dialog box as shown in Figure 4.103.

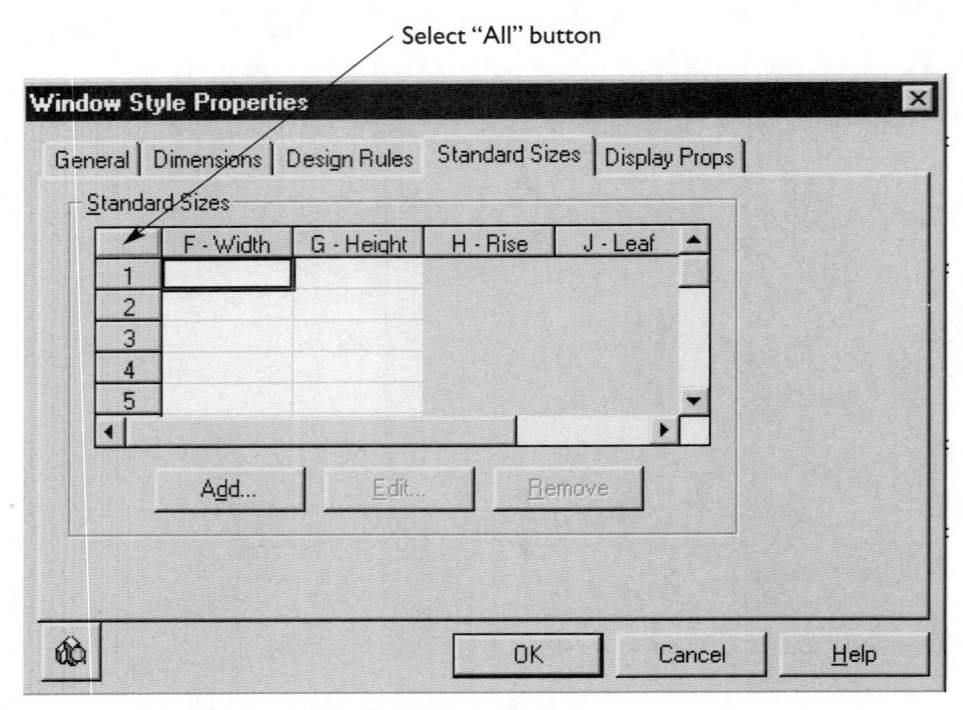

Figure 4.102 *Standard Sizes of the Standard window style*

Figure 4.103 *Creating standard sizes by editing Add Standard Size dialog box*

Edit – The **Edit** button located below the size table opens the **Edit Standard Size** dialog box, which is used to edit the sizes that are highlighted in the table of standard sizes.

Remove – The **Remove** button allows you to remove part or all of the standard sizes. If you select a size from the table, the **Remove** button is activated and you can select it to delete the selected size. If you select the header button at the top of the column, as shown in Figure 4.99, the entire schedule will be selected, and then if you select the **Remove** button, the entire table is deleted.

Display Props

The **Display Props** tab allows you to control the display of a window. The display representations for current viewport of the window style are listed in the display representation list at the top of the dialog box. The current display representation is suffixed with an asterisk. The **Display Props tab** for a window style using the Plan display representation is shown in Figure 4.104.

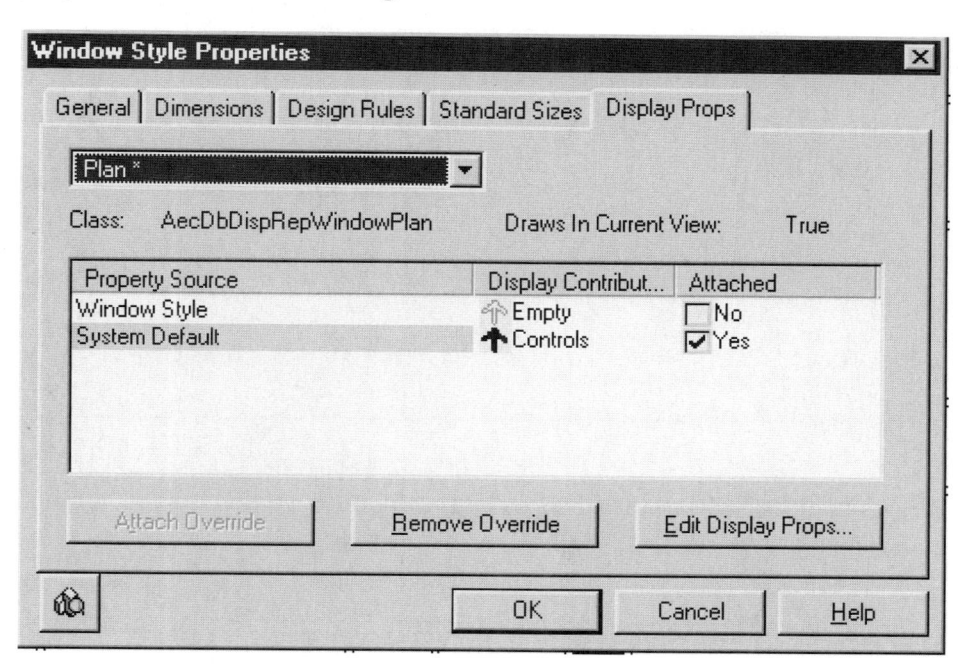

Figure 4.104 *Display Props tab of the Window Style Properties dialog box*

The property source table includes a Window Style and System Default. The buttons below it are **Attach Override, Remove Override,** and **Edit Display Props**. These buttons are used to set the display of the window style.

Attach Override – Marks the selected property source with an Override marker and allows override display properties to be set.

Remove Override –Removes an override that has been attached to the selected property source.

Edit Display Props– Used when an override has been attached to edit the display of the window style. Selecting **Edit Display Props** opens the **Entity Properties** dialog box to control the visibility, layer, color, linetype, lineweight, and linetype scale of the components of the window style, as shown in Figure 4.105.

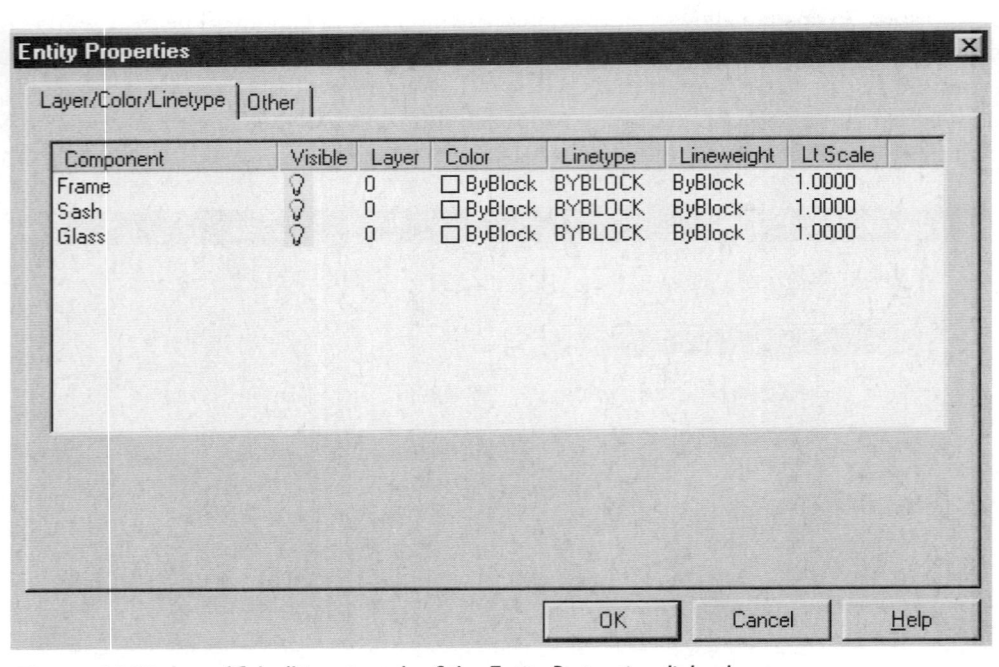

Figure 4.105 *Layer/Color/Linetype tab of the Entity Properties dialog box*

TUTORIAL 4.6 CREATING A WINDOW STYLE

1. Open *ADT Tutor\Ch6\Ex4-6.dwg*.
2. Choose **File>SaveAs** from the menu bar and save the drawing as *Window6.dwg* in your directory.
3. Select **Window Styles** from the **Doors/Windows/Openings** toolbar.
4. Select Casement from the **Window Styles** list and then select the **Copy** button in the **Window Styles** dialog box.
5. Enter the name ANDERSON_CASEMENT in the **Name** dialog box as shown in Figure 4.106.

Figure 4.106 *Creating a new window style name*

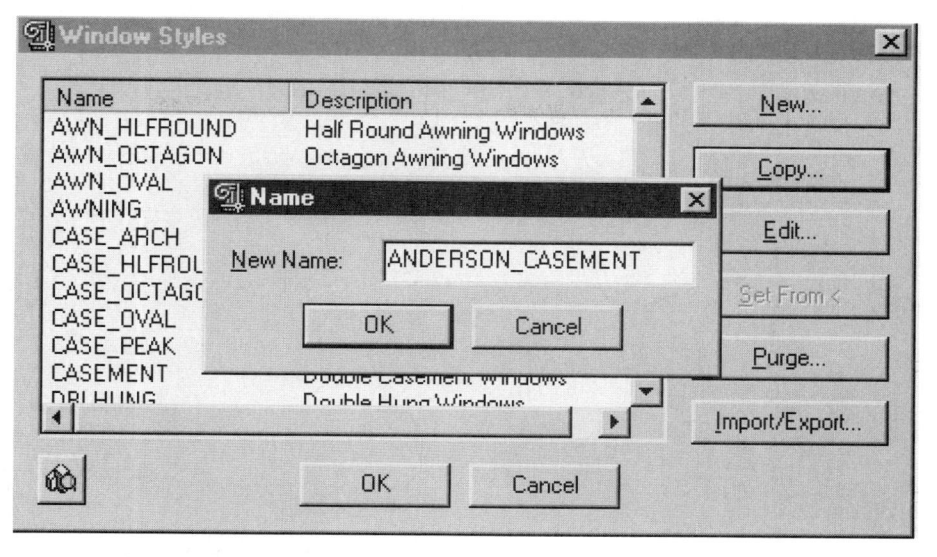

6. Select **OK** to dismiss the **Name** dialog box.
7. Select the **Edit** button in the **Window Styles** dialog box.
8. Select the **General** tab and type **Double Casement Windows sample sizes** in the **Description** edit field.
9. Select the **Dimensions** tab and edit the **Frame** section: clear the **Auto-Adjust to Width of Wall** check box, set Frame **A-Width** to 2", **B-Depth** to 5-9/16", Sash **C-Width** to 2", Sash **D-Depth** to 2", and **E-Glass Thickness** to _" as shown in Figure 4.107.
10. Select the **Design Rules** tab; verify that **Shape** is set to Predefined and Rectangular and **Window Type** to Casement as shown in Figure 4.108.
11. Select the **Standard Sizes** tab and remove all existing sizes by selecting the header button above row 1 shown in Figure 4.109; then select the Remove button.
12. Select the **Add** button in the **Window Style Properties** dialog box.
13. Enter the width of 2'-4" and height of 3'-4-13/16" in the **Add Standard Size** dialog box as shown in Figure 4.110.
14. Select **OK** to dismiss the **Add Standard Size** dialog box.
15. Select the **Add** button of the **Window Style Properties** dialog box and enter the width of 4'-8 1/2" and height of 3'-4 13/16".

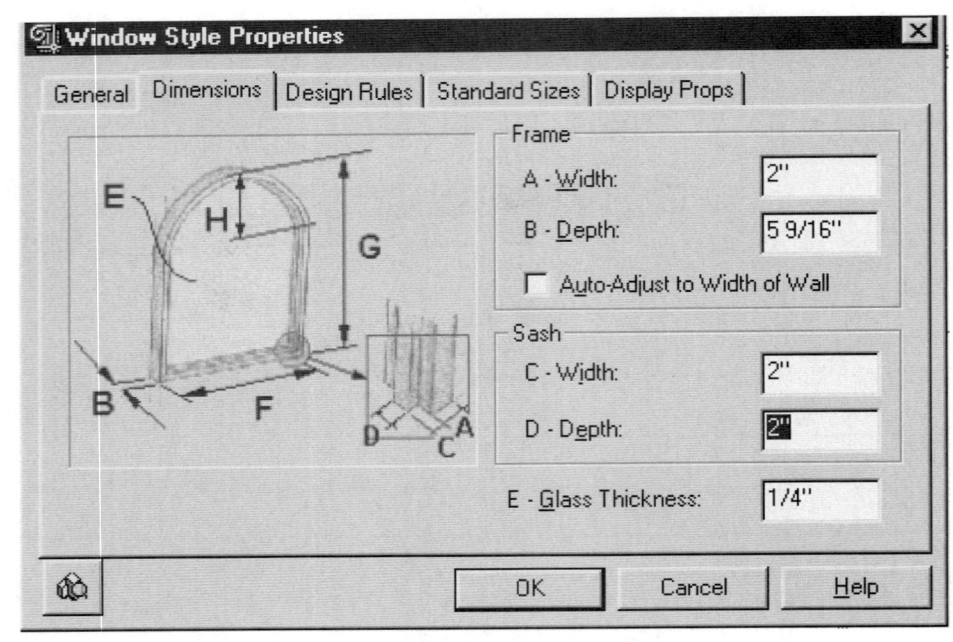

Figure 4.107 *Dimensions tab of the Window Properties dialog box*

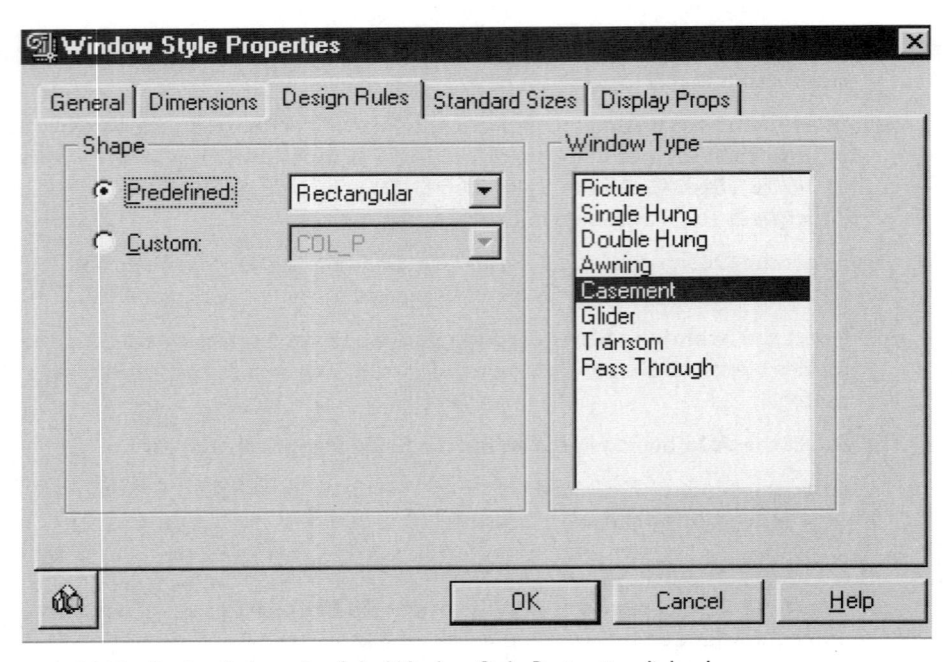

Figure 4.108 *Design Rules tab of the Window Style Properties dialog box*

Placing Doors and Windows 295

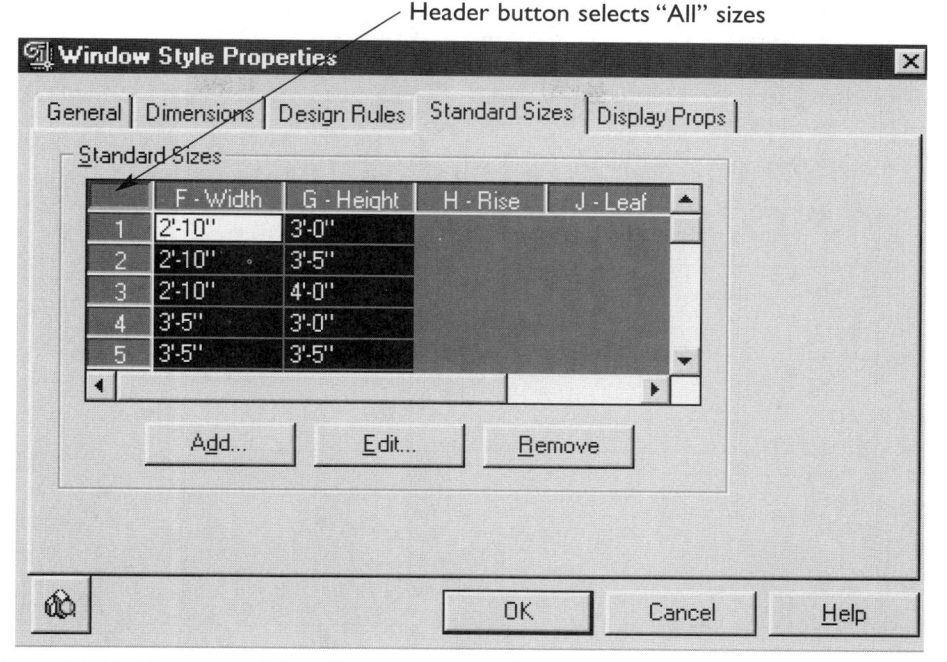

Figure 4.109 *Removing all sizes from a window style*

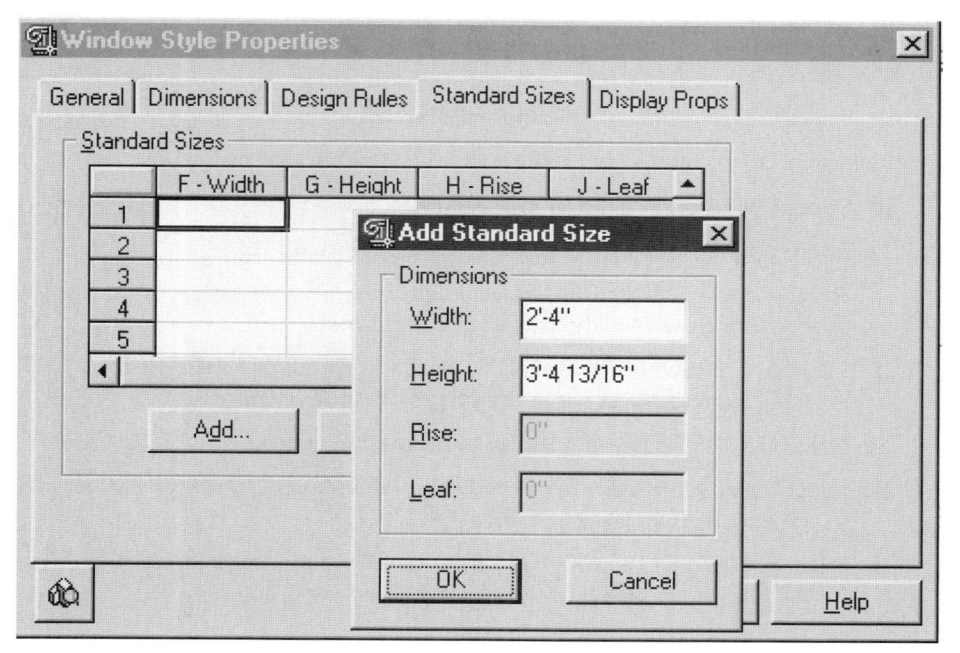

Figure 4.110 *Edited Add Standard Size dialog box*

16. Select **OK** to dismiss the **Add Standard Size** dialog box. The standard sizes should be displayed as shown in Figure 4.111.

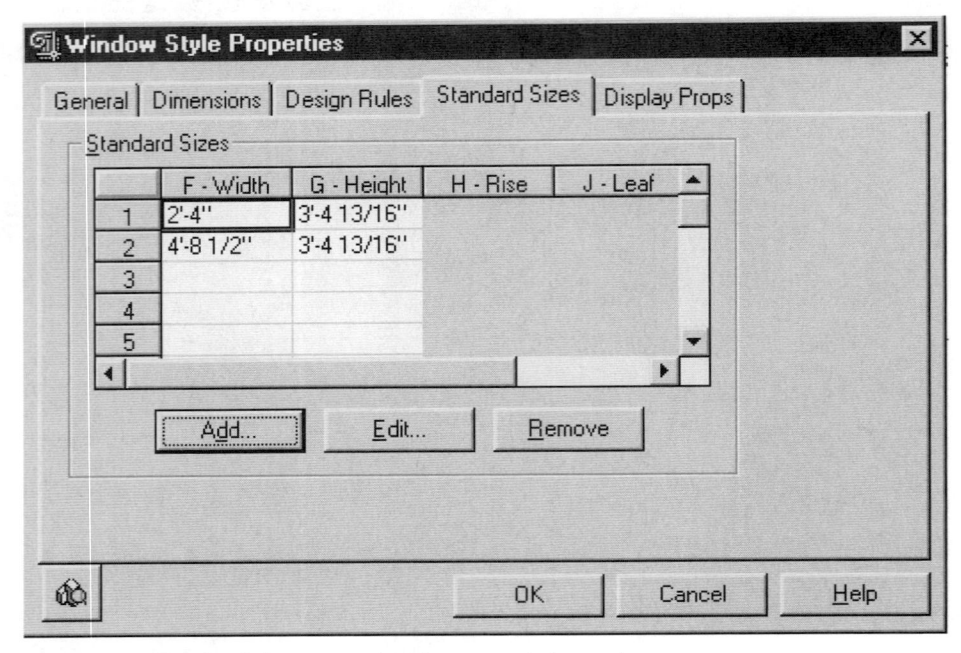

Figure 4.111 *Standard sizes created with a new window style*

17. Select **OK** to dismiss the **Standard Sizes** dialog box.
18. Select **OK** to dismiss the **Window Style Properties** dialog box.
19. Select **OK** to dismiss the **Window Styles** dialog box.
20. Zoom a window around the area shown in Figure 4.112.
21. Select windows "A" and "B" shown in Figure 4.113; then right-click and choose **Window Modify** from the shortcut menu.
22. Select ANDERSON_CASEMENT from the **Style** list and select 4'-8 1/2" x 3'-4-13/16" from the **Size** list in the **Modify Window** dialog box.
23. Select **OK** to dismiss the **Modify Window** dialog box.
24. Select **Modify Window** from the **Doors/Windows/Openings** toolbar and select window "C" shown in Figure 4.113.
25. Select ANDERSON_CASEMENT from the **Style** list and select 2'-4" x 3"-4-13/16" from the **Size** list in the **Modify Window** dialog box.
26. Select **OK** to dismiss the **Modify Window** dialog box.

Placing Doors and Windows **297**

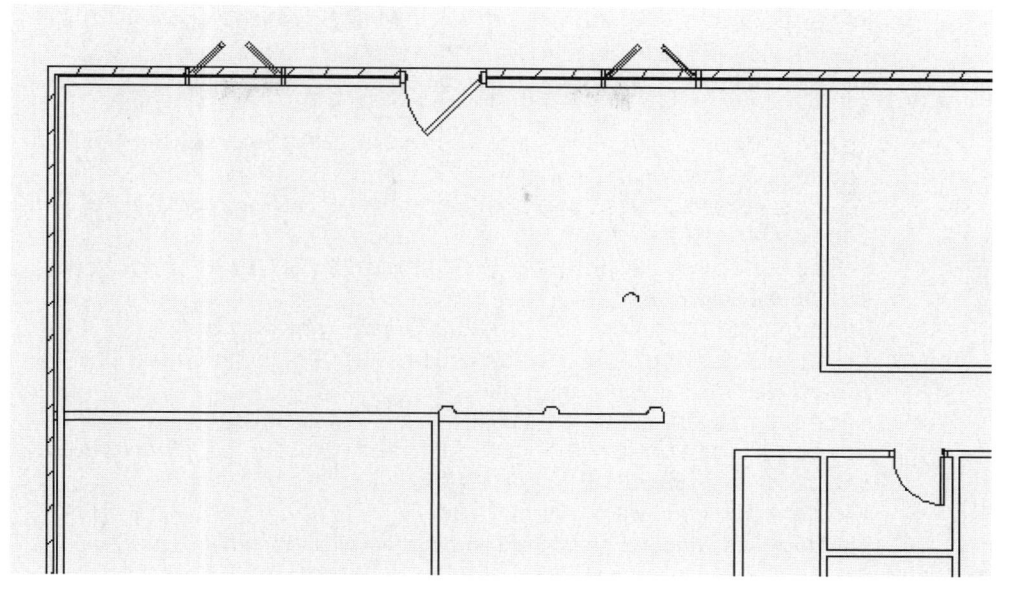

Figure 4.112 *Area of building for changing window styles*

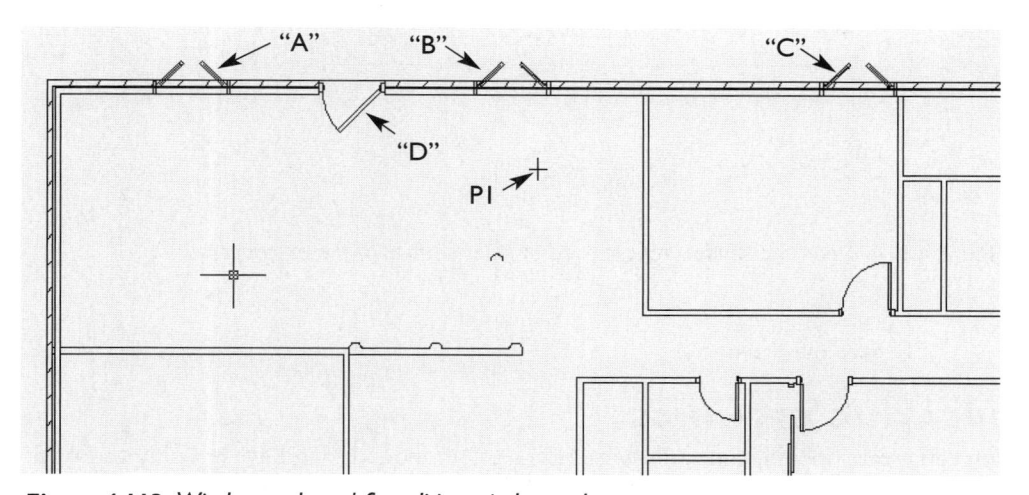

Figure 4.113 *Windows selected for editing window style*

27. Choose **Design>Windows>Reposition>Within Wall** from the menu bar.
28. Refer to Figure 4.113 and the following command sequence to reposition the windows and door.

    ```
    Command: _AecRepositionWithin
    ```

```
Select doors, windows, or openings: (Select window "A".) 1
found
Select doors, windows, or openings: (Select door "D".) 1
found, 2 total
Select doors, windows, or openings: (Select window "B".) 1
found, 3 total
Select doors, windows, or openings: (Select window "C".) 1
found, 4 total
Select doors, windows, or openings: ENTER (Pressing
ENTER ends selection)
Current offset is -1". (Note default value may vary.)
Select side to offset or [Offset/Center]: O ENTER
(Selects the Offset option)
Enter offset<-1">: 0 ENTER
Select side to offset or [Offset/Center]: (Select a point
near point P1 as shown in Figure 4.113.)
[4] anchor(s) were modified.
```

The door and windows are shifted within the wall as shown in Figure 4.114.

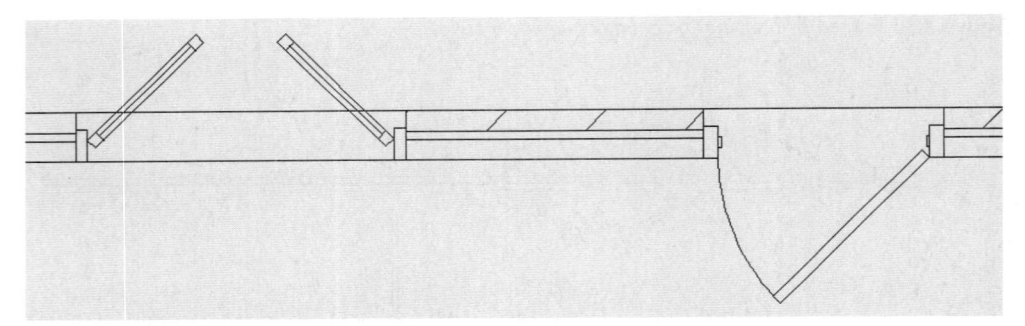

Figure 4.114 *Door and window repositioned to inner surface of the exterior wall*

29. Save the drawing.

CREATING OPENINGS

You can create openings in walls without inserting doors or windows by using the **Add Opening** command (**OPENINGADD**). See Table 4.13 for command access.

Selecting the **Add Opening** command (**OPENINGADD**) will open the **Add Opening** dialog box, as shown in Figure 4.116.

You place an opening in a wall or space boundary by setting the shape, width, and height. The opening does not include a frame. The shape of the opening can be a predefined or custom shape. The predefined shapes include rectangular, round, half

Menu bar	Design>Openings>Add Opening
Command prompt	OPENINGADD
Toolbar	Select Add Opening from the Doors/Windows/Openings shown in Figure 4.115
Shortcut	Right-click over drawing area; choose Design>Openings>Add

Table 4.13 Add Opening access

Figure 4.115 Add Opening command on the Doors/Windows/Openings toolbar

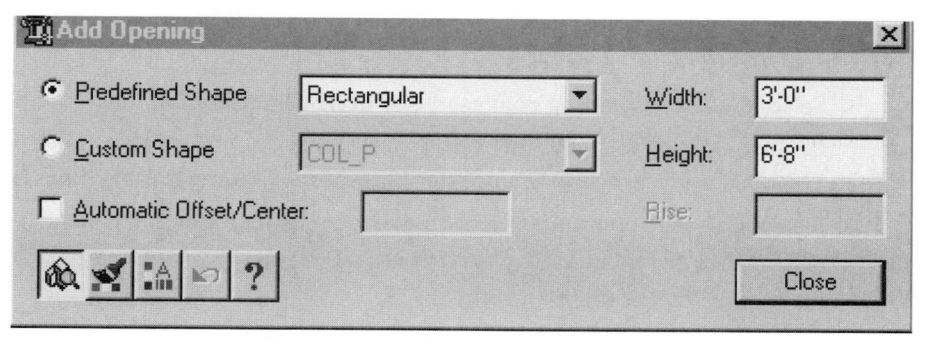

Figure 4.116 Add Opening dialog box

round, quarter round, oval, circle, arched, trapezoid, gothic, or isosceles triangle. Custom shapes are shapes created from Aec_Profiles, which can be created from any closed polyline shape. The **Automatic Offset/Center** option is also available for placing the opening. The **Properties** button of the **Add Opening** dialog box will open the **Opening Properties** dialog box, which includes the **General** tab, used to edit the description of an opening, and the **Dimensions** tab, shown in Figure 4.117, which allows you to change the shape and size of the opening.

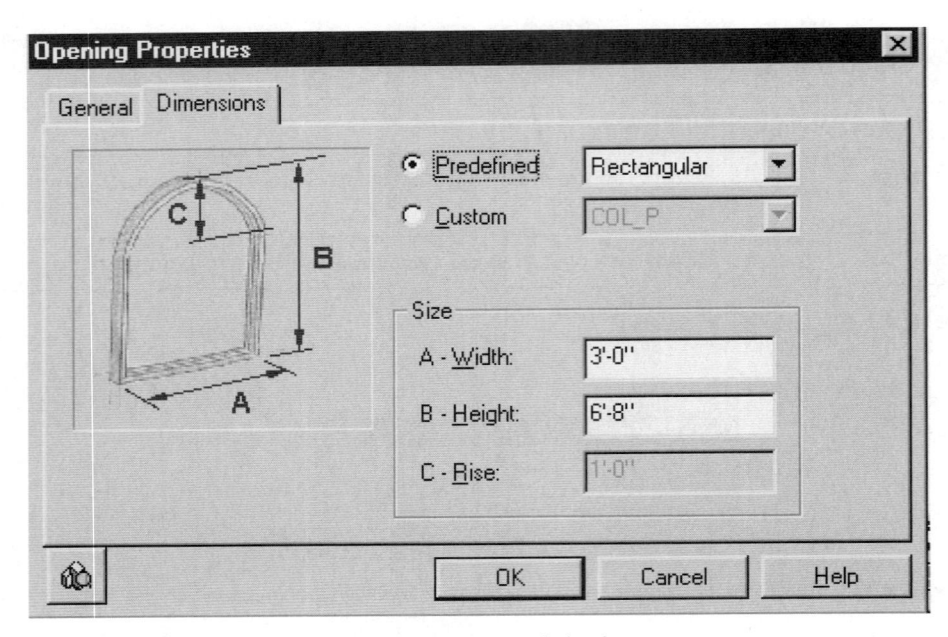

Figure 4.117 *Dimensions tab of Opening Properties dialog box*

STEPS TO ADDING AN OPENING TO A WALL

1. Select the **Add Opening** command (**OPENINGADD**) from the **Doors/Windows/Openings** toolbar.
2. Edit the shape and size of the opening in the **Add Opening** dialog box.
3. Respond to the command prompt as shown in the following command sequence.

   ```
   Command: _AecOpeningAdd
   Select wall, space boundary or RETURN:
   ```
   (Select a wall for the opening.)
   ```
   Insert point or
   [SHape/Custom/WIdth/Height/Auto/Match]:
   ```
   (Select the insertion point to specify the location of the opening.)
   ```
   Insert point or
   [SHape/Custom/WIdth/Height/Auto/Match/Undo]: ENTER
   ```
   (Pressing ENTER *ends the command.)*

4. Select **SE Isometric** from the **View** flyout on the **Standard** toolbar.
5. Enter **HIDE** in the command line to view the opening, as shown in Figure 4.118.

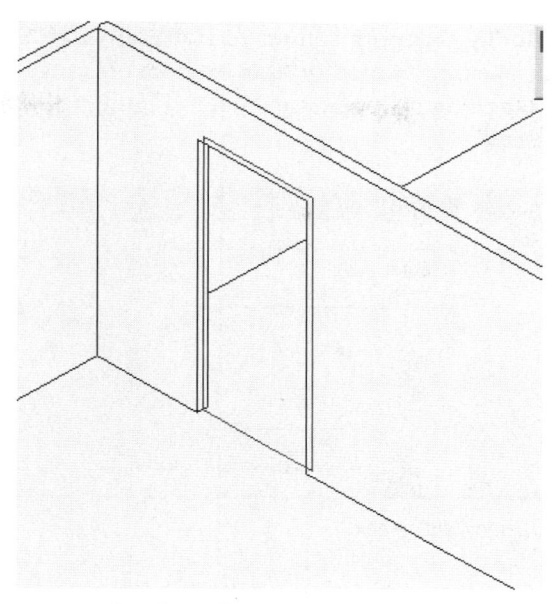

Figure 4.118 *Opening inserted in the wall*

MODIFYING AN OPENING WITH MODIFY OPENING

After an opening has been placed in a wall or space boundary, it can be edited for size and shape by the **Modify Opening** command (**OPENINGMODIFY**). See Table 4.14 for command access.

Menu bar	Design>Opening>Modify
Command prompt	OPENINGMODIFY
Toolbar:	Select Modify Opening from the Doors/Windows/Openings toolbar shown in Figure 4.119
Shortcut	Right-click over drawing area; choose Design>Openings>Modify

Table 4.14 *Modify Opening access*

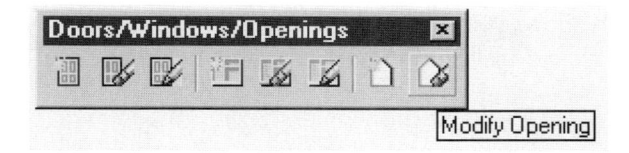

Figure 4.119 *Modify Opening command on the Doors/Windows/Openings toolbar*

After selecting the **Modify Opening** command (**OPENINGMODIFY**), you will be prompted to select an opening to modify. One or more openings can be selected for editing. The **Modify Opening** dialog box, shown in Figure 4.120, will open after an opening has been selected.

Figure 4.120 *Modify Opening dialog box*

You can edit the **Modify Opening** dialog box to change the shape and size of an opening. Selecting the **Properties** button of the **Modify Opening** dialog box will open the Opening Properties dialog box. This dialog box includes four tabs: **General**, **Dimensions**, **Anchor**, and **Endcaps**.

General
The **General** tab, shown in Figure 4.121, allows you to edit the description of the opening.

Dimensions
The **Dimensions** tab, shown in Figure 4.122, allows you to edit the shape and size of the opening.

Anchor
The **Anchor** tab, shown in Figure 4.123, allows you to shift the opening in the X, Y, Z directions. Openings are anchored in a manner similar to that for anchoring door and window objects.

Endcaps
The **Endcaps** tab allows you to change the endcap style assigned to each end of the wall opening, as shown in Figure 4.124.

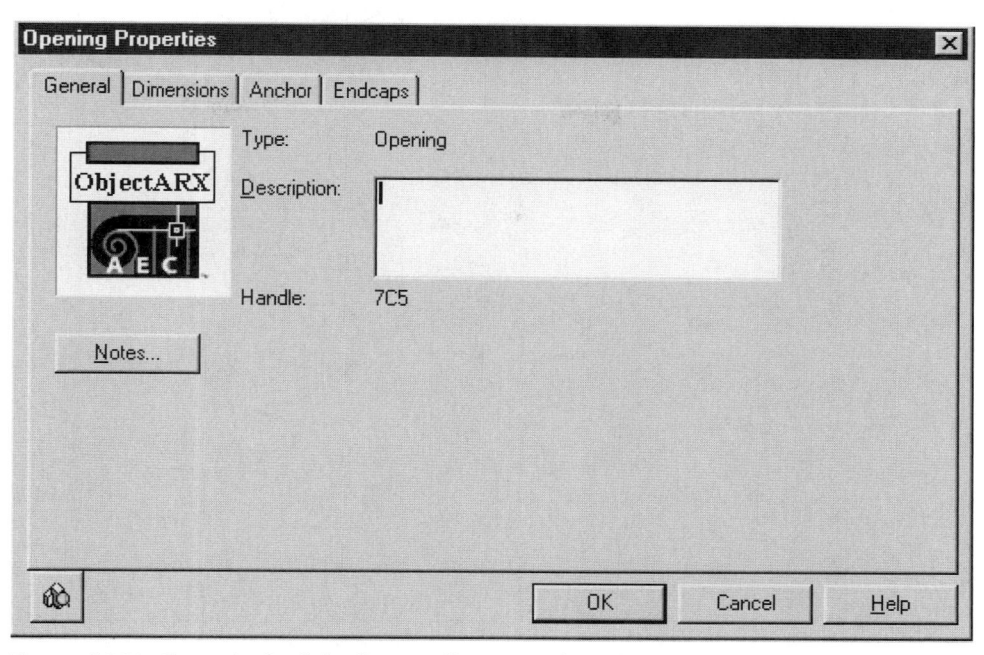

Figure 4.121 *General tab of the Opening Properties dialog box*

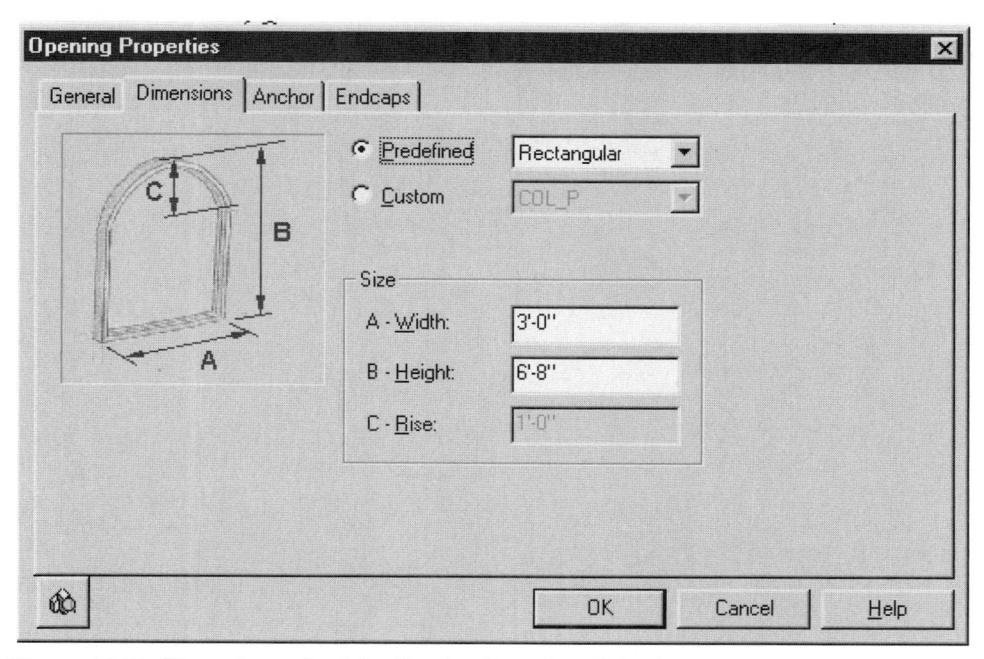

Figure 4.122 *Dimensions tab of the Opening Properites dialog box*

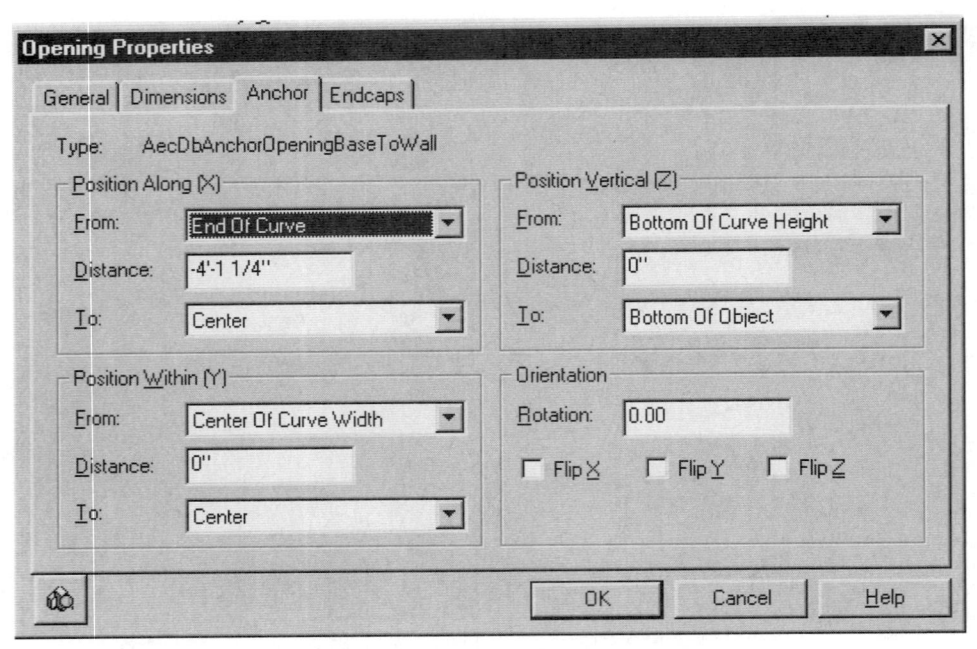

Figure 4.123 *Anchor tab of the Opening Properties dialog box*

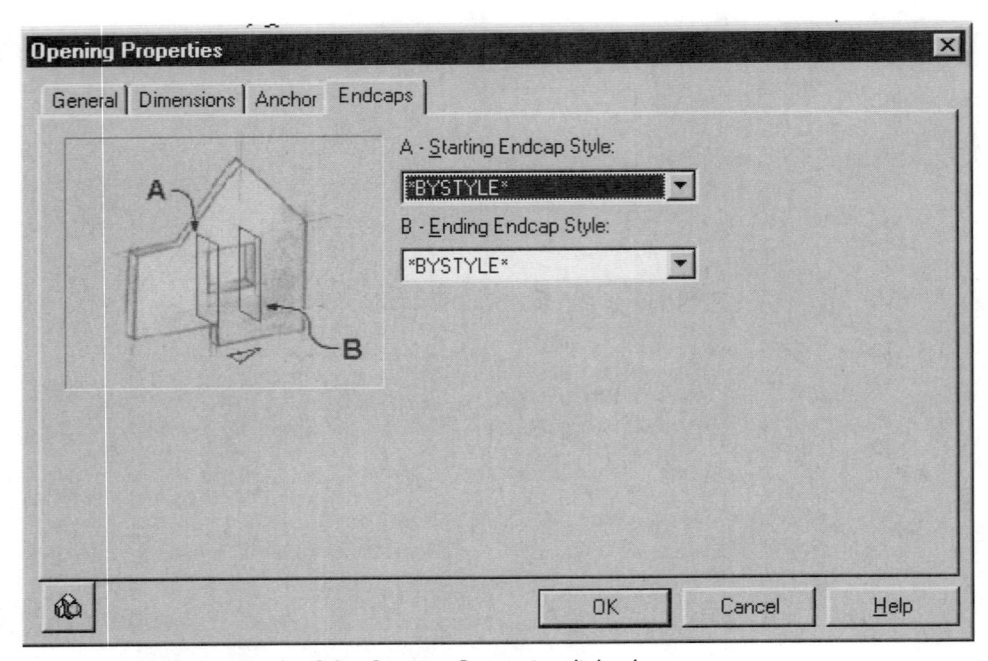

Figure 4.124 *Endcaps tab of the Opening Properties dialog box*

CREATING DOORS, WINDOWS AND OPENINGS USING AEC PROFILES

Doors, windows and openings can be created in the shape of an Aec Profile. The Aec Profile is a shape created in Architectural Desktop that can be applied to create the shape for a door, window, or opening. An Aec_Profile is created from a closed polyline and converted to an Aec_Profile through the **AECPROFILEDEFINE** command. After the profile is created, it can be assigned as a shape for a door, window, or opening.

CREATING AN AEC PROFILE

Table 4.15 shows the command access for the **Define AEC Profile** command (**AECPROFILEDEFINE**).

Menu bar	Desktop>AEC Profile>Define AEC Profile
Command prompt	AECPROFILEDEFINE
Shortcut	Right-click over drawing area; choose Utilities>Profiles>Table

Table 4.15 *Define AEC Profile access*

When the **Define AEC Profile** command (**PROFILEDEFINE**) is executed, the **Profiles** dialog box opens as shown in Figure 4.125.

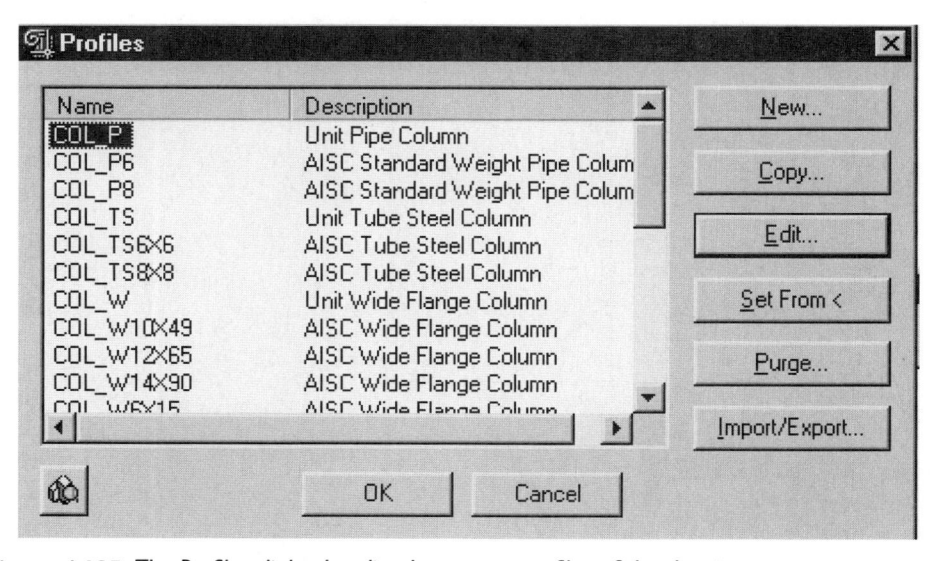

Figure 4.125 *The Profiles dialog box list the current profiles of the drawing*

The buttons on the right of the **Profiles** dialog box are described below.

> **New** – Opens the **Name** dialog box and creates a new name for a profile.
>
> **Copy**– Copies a profile with its description to new named profile.
>
> **Edit** – Edits the description of the profile.
>
> **Set From<** – Defines the polyline geometry for the profile highlighted in the **Profiles** dialog box.
>
> **Purge** – Opens the **Purge Profiles** dialog box. Profiles in the **Purge Profiles** dialog box can be selected and deleted from the drawing.
>
> **Import/Export** – Transfers saved profiles to other files or copies profiles from other drawings to the current drawing.

Profiles are created from a closed polyline. One or more polylines can be selected to create the profile. If additional polylines are selected, they can be used to create a void in the previous shape.

In the following tutorial, an Aec Profile is created from polyline shapes shown in Figure 4.126. The new profile will be used to create a custom shaped door.

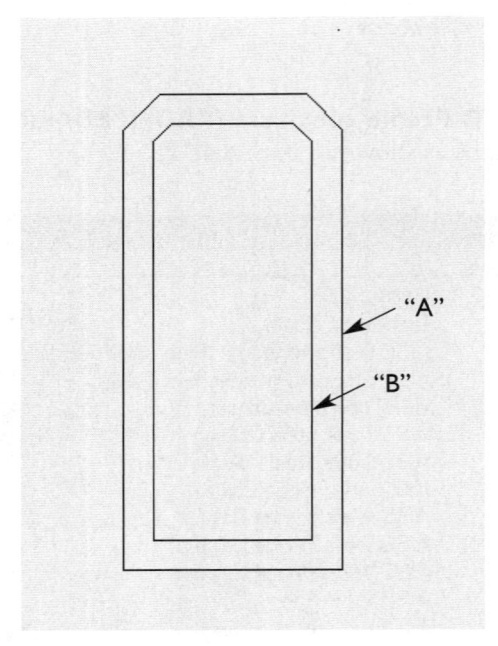

Figure 4.126 *Polylines used to create AEC Profile*

TUTORIAL 4.7 CREATING AN AEC PROFILE AND USING THE PROFILE FOR A DOOR STYLE

1. Open *Ex 4-7* from the *\ADT Tutor\Ch4* directory

2. Choose **File>SaveAs** from the menu bar and save the drawing as *Profile7* in your student directory.

3. Choose **Desktop>Aec Profile>Define AEC Profile** from the menu bar to open the **Profiles** dialog box.

4. Select the **New** button to open the **Name** dialog box.

5. Type **Chamfer** in the **Name** dialog box as shown in Figure 4.127.

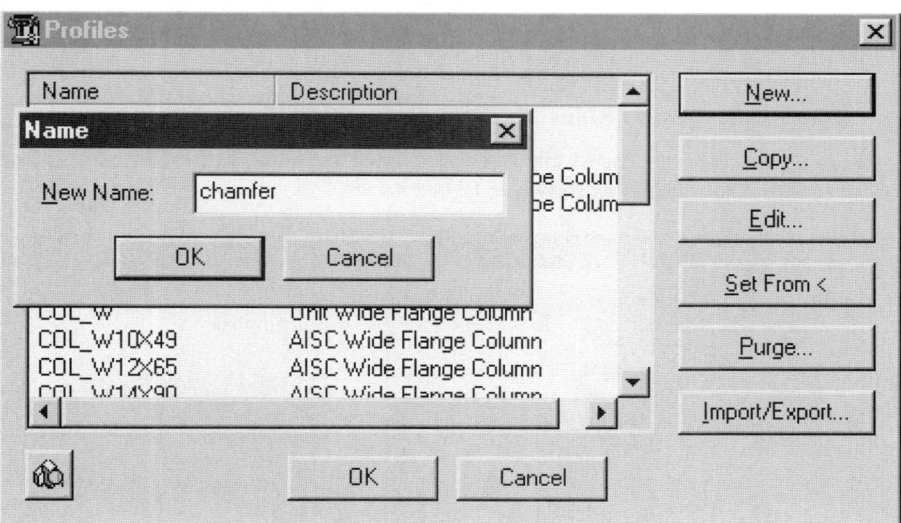

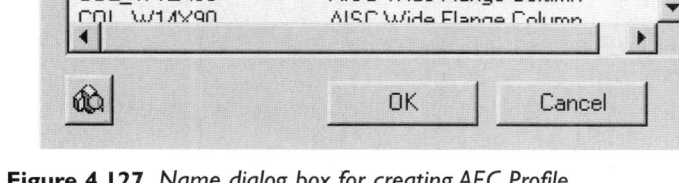

Figure 4.127 *Name dialog box for creating AEC Profile*

6. Select **OK** to dismiss the **Name** dialog box.

7. Select the **Set From<** button in the **Profiles** dialog box.

8. Respond to the following command line prompts:

    ```
    Command: _AecProfileDefine
    Select a closed polyline: (Select geometry "A" shown in Figure 4.126.)
    Add another ring? [Yes/No] <N>: y ENTER (creates a second ring)
    Select a closed polyline: (Select geometry "B" shown in Figure 4.126.)
    ```

```
Ring is a void area? [Yes/No] <Y>: ENTER
```
(Sets geometry "B" as a void.)
```
Add another ring? [Yes/No] <N>: ENTER
```
(Pressing ENTER ends the selection.)

9. Select **OK** to dismiss the **Profiles** dialog box. A new profile is created.
10. Select **Door Styles** from the **Doors/Windows/Openings** toolbar.
11. Select the **New** button of the **Door Styles** dialog box.
12. Type **CHAMFER** in the **Name** dialog box.
13. Select **OK** to dismiss the **Name** dialog box.
14. Select Chamfer from the **Style** list; select the **Edit** button in the **Door Styles** dialog box.
15. Edit the description to **Chamfer Single Hinged Door** in the **General** tab.
16. Select the **Design Rules** tab of the **Door Style Properties** dialog box.
17. Edit the **Design Rules** tab: select the **Custom** radio button and select Single as the **Door Type** as shown in Figure 4.128.

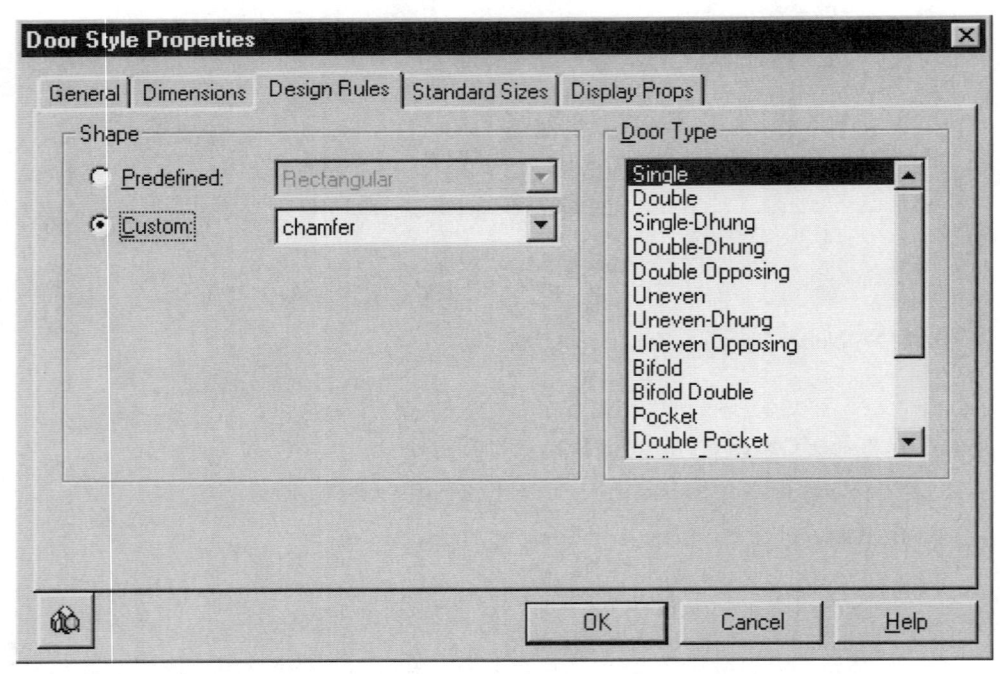

Figure 4.128 *Setting shape to custom to assign an AEC Profile to the door style*

18. Select **OK** to dismiss the **Door Style Properties** dialog box.

19. Select **OK** to dismiss the **Door Styles** dialog box.
20. Select **Add Door** from the **Doors/Windows/Openings** toolbar.
21. Edit the **Add Doors** dialog box; set the style to CHAMFER, width to 3', and height to 6'-8". Place the door in the middle of the wall shown at the right of the profile.
22. Select **SW Isometric** from the **View** flyout on the **Standard** toolbar.
23. Type **HI** (**HIDE** command) in the command line to hide lines in the view. The door should appear as shown in Figure 4.129.

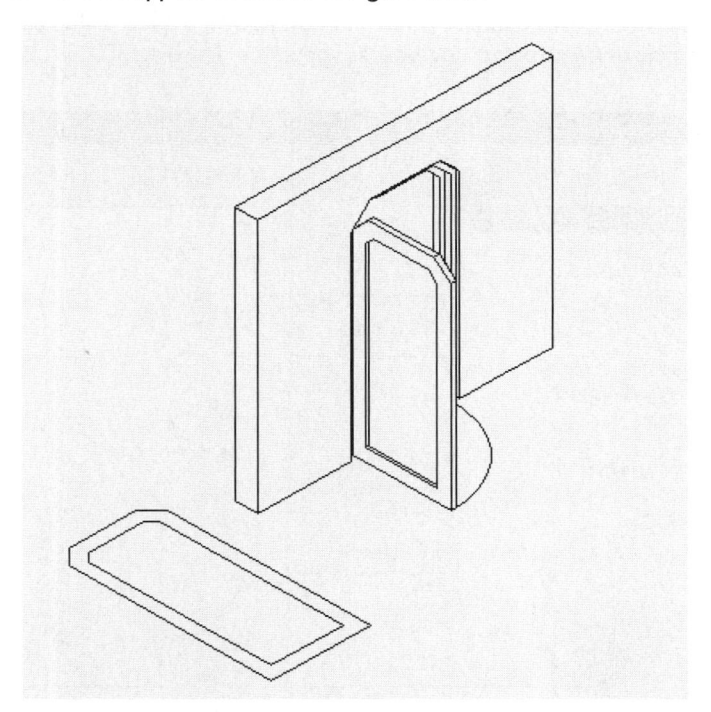

Figure 4.129 *Door created from the CHAMFER profile*

24. Save the drawing.

IMPORTING AND EXPORTING DOOR AND WINDOW STYLES

Door and window styles that are created in a drawing can be exported to resource files for future use in other drawings. The resource files are located in the \AutoCAD\Architectural2\Content\Imperial\Styles directory. Copies of door and window styles of the Aec Arch (imperial) template are in this directory in the Doors (Imperial) and Windows (Imperial) files. The Doors-Custom (Imperial) and Windows-Custom

(Imperial) files include additional door and window styles. Door Styles are imported to a drawing by the **Door Styles** command.

STEPS TO IMPORTING DOOR STYLES

1. Select **New** from the **Standard** toolbar.
2. Select **Use a Template** and select the Aec arch (imperial) template; select **OK** to dismiss the **Create New Drawing** dialog box.
3. Select **Door Styles** from the **Doors/Windows/Openings** toolbar.
4. Select the **Import/Export** button in the **Door Styles** dialog box to open the **Import/Export** dialog box, as shown in Figure 4.130.

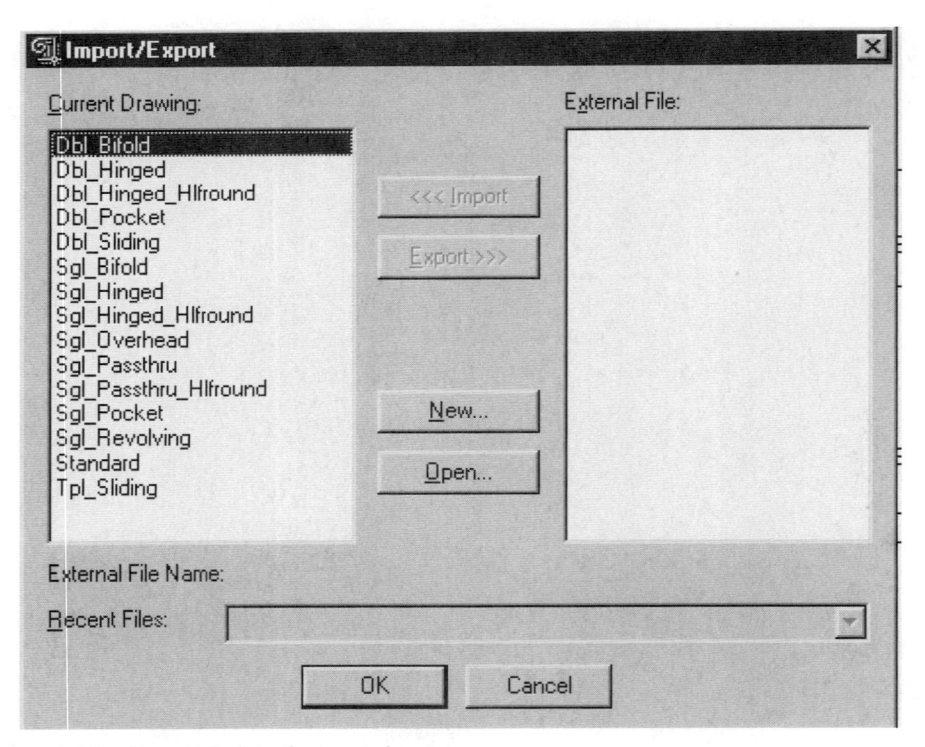

Figure 4.130 *Door style list of current drawing*

5. Select the **Open** button.
6. Select the *AutoCAD Architectural2\Content\Imperial\Styles\Doors-Custom (Imperial).dwg* file from the **File to Import From** dialog box, as shown in Figure 4.131.

Placing Doors and Windows 311

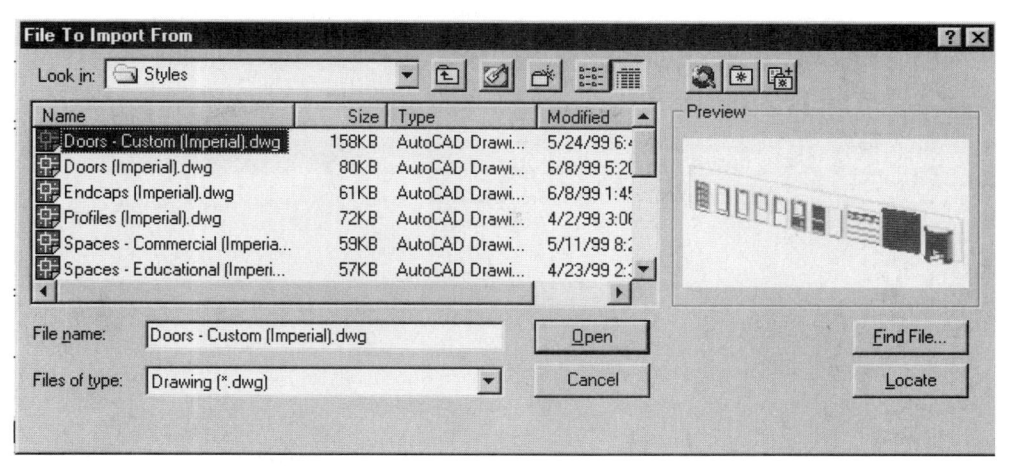

Figure 4.131 *Importing styles from resouce files*

7. Select the **Open** button in the **File to Import From** dialog box.
8. Hold down SHIFT and select the first and last style listed in the **External File** list on the right, as shown in Figure 4.132.

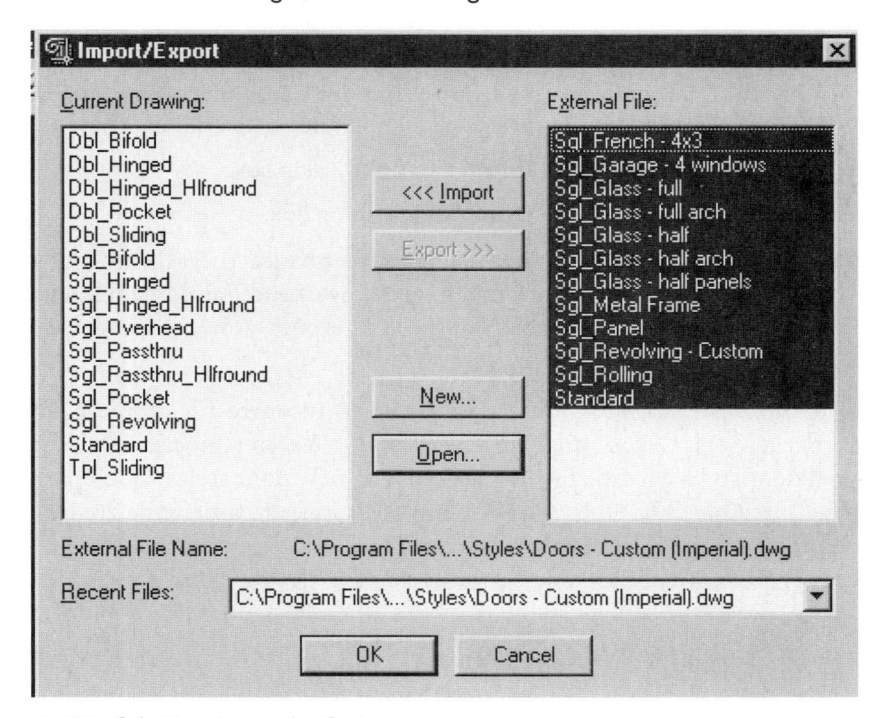

Figure 4.132 *Selecting door styles for import*

9. Select the **Import** button in the center of the **Import/Export** dialog box. The **Import/Export - Duplicate Names Found** dialog box will open as shown in Figure 4.133.

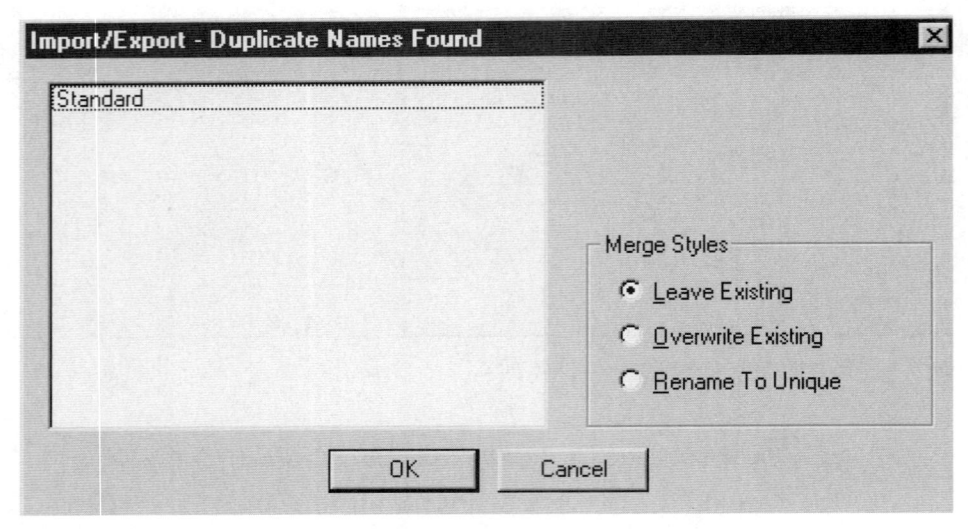

Figure 4.133 *Import/Export - Duplicate Names Found dialog box*

10. Select the **OK** button in the **Import/Export - Duplicate Names Found** dialog box to leave the existing definition of the Standard door style.
11. Select **OK** to dismiss the **Import/Export** dialog box.
12. Select **OK** to dismiss the **Door Styles** dialog box.

The doors styles imported from this file greatly enhance the selection of doors. Included in the imported styles are French, glass, overhead, raised panel, revolving, rolling, and metal frame. The overhead and rolling doors allow you to place garage doors in a design.

Door styles can be exported from a design file to resource files for use in other drawings. Because the Doors (Imperial) and Doors-Custom (Imperial) styles include the original door styles provided with the software, new door styles should be placed in a new resource file. The steps to exporting door styles to a new file are described in the following tutorial. The door styles created in Door2 drawing are included in the *Ex4-8* drawing file, which is exported to the *MY DOOR STYLES* resource file.

TUTORIAL 4-8 EXPORTING DOOR STYLES TO A NEW FILE

1. Open \ADT Tutor\Ch4\Ex4-8.
2. Choose **File>SaveAs** from the menu bar and save the drawing as Export8.dwg in your student directory.
3. Select **Door Styles** from the **Doors/Windows/Openings** toolbar.
4. Select the **Import/Export** button in the **Door Styles** dialog box.
5. Select the **New** button to open the **New Drawing File** dialog box as shown in Figure 4.134. Select the **Save In** drop-down list and set the file path to \ADT Tutor\Student.

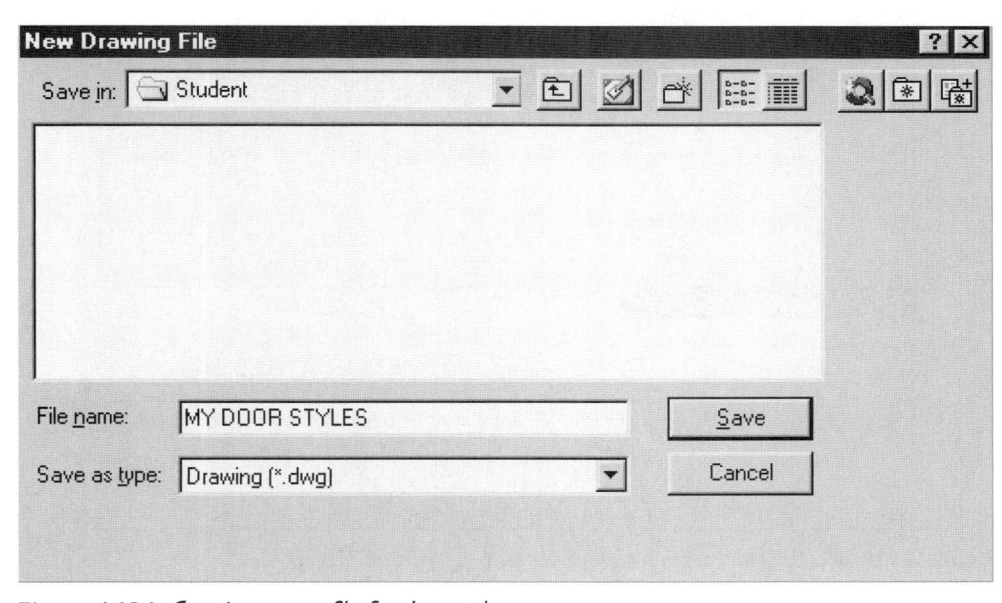

Figure 4.134 Creating a new file for door styles

6. Type **MY DOOR STYLES** in the **File name** edit field of the **New Drawing File** dialog box, as shown in Figure 4.134.
7. Select the **Save** button to save the file name and dismiss the **New Drawing File** dialog box.
8. Select all the door styles from the **Current** drawing list, and then select the **Export** button.
9. The **Import/Export - Duplicate Names Found** dialog box opens, which lists the Standard door style as duplicate; select **OK** to accept the Leave

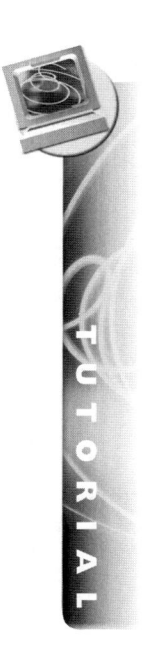

Existing definition and dismiss the **Import/Export - Duplicate Names Found** dialog box.

10. Select **OK** to dismiss the **Import/Export** dialog box.
11. The AutoCAD warning dialog box opens as shown in Figure 4.135; select **Yes** to accept the change.

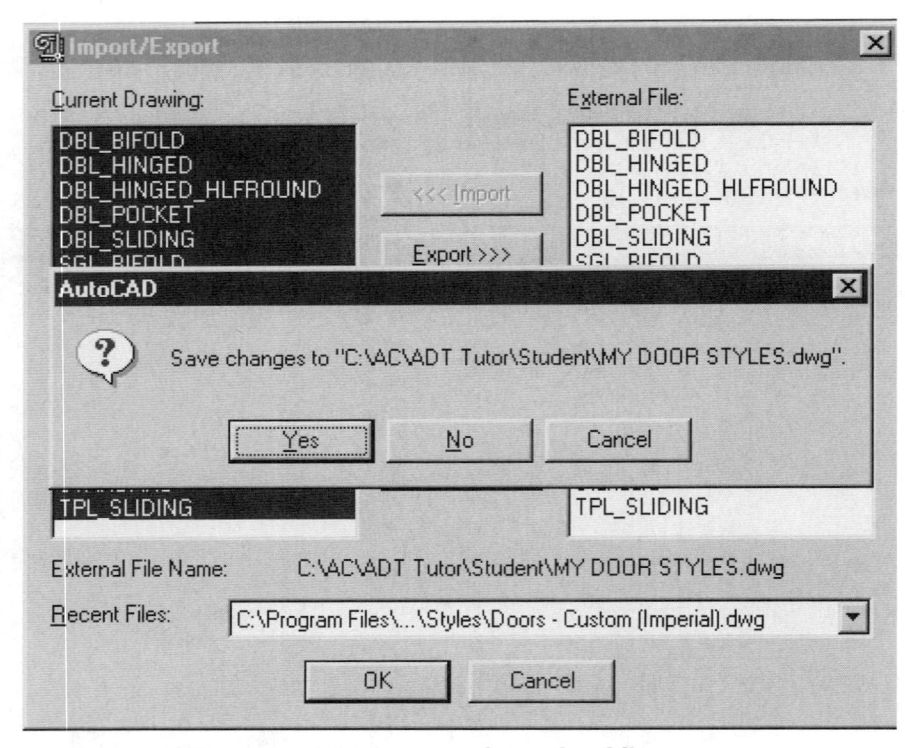

Figure 4.135 *AutoCAD warning dialog box to verify transfer of files*

12. Select **OK** to dismiss the **Door Styles** dialog box.
13. Save the drawing.

Window styles are imported and exported in a similar manner. The window styles of the Aec Arch (Imperial) template are included in the *Windows (Imperial)* file of the *\AutoCAD Architectural 2\Content\Imperial\Styles* directory. Additional window styles provided in the *Windows-Custom (Imperial)* file of the *Styles* directory are bay, bow, jalousie, muntins, 2d frames, and windows with shutters. The additional styles of windows can be imported to a drawing to enhance the display of windows.

SUMMARY

1. Doors are inserted in a drawing with the **Add Door** command (**DOORADD**).
2. Door styles are created with the **Door Styles** command (**DOORSTYLE**).
3. The insertion handle for doors and windows is located on the end of the unit near the start of the wall.
4. **Flip Hinge**, **Flip Swing**, and grips can be used to change the location and swing of a door or window.
5. Openings can be created with a specified width and height.
6. Doors, windows, and openings can be created in the shape of an AEC Profile.
7. Door and window styles can be imported from and exported to resource files.

REVIEW QUESTIONS

1. The style of a door is defined with the _____ _____ command.
2. A single hinged door will display at an angle of 90 degrees if the opening percent is set to _____.
3. The **Measure to Outside of Frame** check box is selected in the _____.
4. The fastest method of changing the swing of a door is to _____.
5. To shift a frame to the outer surface of a wall, the _____ _____ command should be used.
6. The insertion handle for a door or window is _____.
7. Door styles are imported into a drawing by the _____ command.
8. The overhead garage door style is located in the _____ file.
9. The endcaps of the wall at each end of the door unit are controlled by _____.
10. Wall openings do not have defined _____, unlike doors and windows.
11. Describe the procedure to automatically center a door or window in a wall.
12. Door sizes can be preset to a style in the _____ _____ _____ dialog box.
13. Describe the procedure to move a door along a wall using grips.
14. The _____ of an Arch window defines its radius.
15. You can create custom shapes of doors and windows by specifying an _____ for the shape.

PROJECTS

EXERCISE 4.1 INSERTING DOORS IN A FLOOR PLAN

Open *EX4-9* and insert the following doors, as shown in Table 4.16 and in Figure 4.136. Save the file as *Proj9*. Use the **Automatic Offset/Center** to place the doors.

DOOR MARK	SIZE	STYLE
A	2'-6" x 6'-8"	SINGLE_INT_RES
B	2'-0" x 6'-8"	SINGLE_INT_RES
C	3'-0" x 6'-8"	SINGLE_EXT_RES
D	3'-0" x 6'-8"	DOUBLE SLIDING
E	4'-0" x 6'-8"	DOUBLE SLIDING
F	2'-6" x 6'-8"	Sgl_French 4 x 3

Table 4.16 *Door sizes and styles for Exercise 4.1*

EXERCISE 4.2 INSERTING WINDOWS IN A FLOOR PLAN

Open *EX4-10* and insert the following windows, as shown in Table 4.17 and Figure 4.136. Save the file as *Proj10*. Use the **Automatic Offset/Center** when placing windows. Reposition all exterior windows and doors frames to the interior surface of the wall.

WINDOW MARK	SIZE	STYLE
1	2'-4" X 3'-4 13/16"	ANDERSON_CASEMENT
2	4'-8 1/2" X 3'-4 13/16"	ANDERSON_CASEMENT

Table 4.17 *Window sizes and styles for Exercise 4.2*

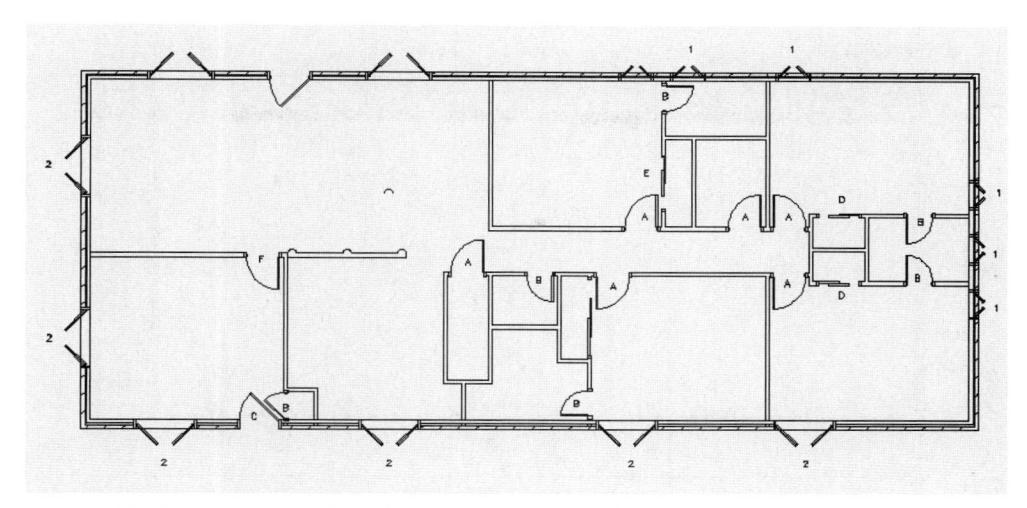

Figure 4.136 *Door and window placement in the floor plan*

EXERCISE 4.3 INSERTING AND REPOSITIONING DOORS AND WINDOWS IN A FLOOR PLAN

Open *Ex4-11* and insert the following windows and doors, as shown in Table 4.18 and Figure 4.137. Reposition all exterior windows and door frames to the interior surface of the wall. Deselect **Auto Adjust to Width of Wall** in the **Window Style Properties** dialog box and set width to 8" for the awning window style. Save the file as *Proj11*.

WINDOW MARK	SIZE	STYLE
A	3'–6" X 2'–4"	AWNING
DOOR MARK	**SIZE**	**STYLE**
1	16' X 7'	GARAGE DOOR
2	3'–0" X 6'–8"	SGL_HINGED
3	4'–0" X 6'–8"	DBL_BIFOLD
4	2'–6" X 6'–8"	SGL_HINGED

Table 4.18 *Sizes and styles for Exercise 4.3*

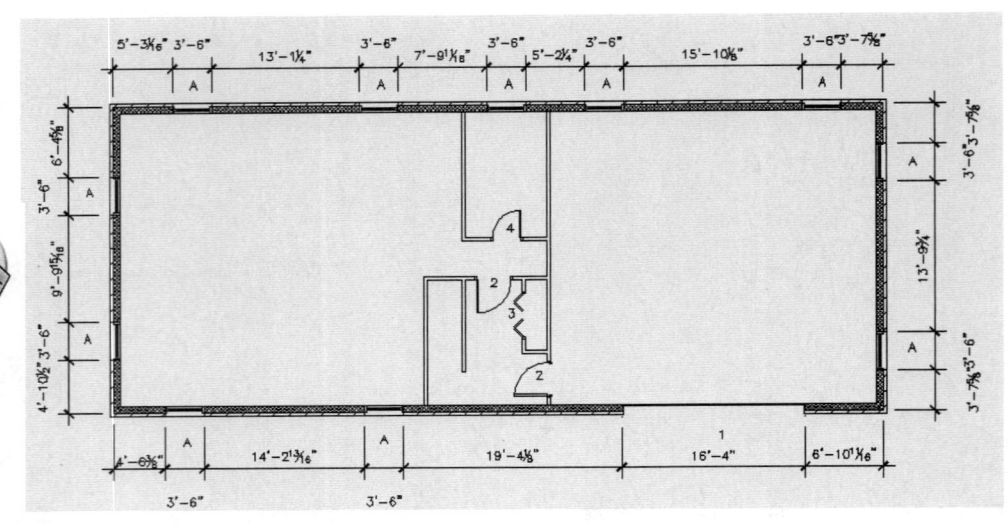

Figure 4.137 *Door and window placement in the basement floor plan*

CHAPTER 5

Creating Roofs

INTRODUCTION

Simple or complex roofs can be placed on a building through the **Add Roof** command. The roof can be applied to the walls of the building or created from a polyline. This chapter includes tutorials that develop the following roofs: shed, gable, gambrel, hip and dormer. Multiple roofs can be constructed and merged together through mass groups.

OBJECTIVES

After completing this chapter, you will be able to

- Use **Add Roof** (**ROOFADD**) to create hip, gable, shed, and gambrel roofs
- Edit roof slope, thickness, and fascia angle using **Roof Properties**
- Modify roof plate height, slope, and overhang using **Modfiy Roof**
- Use grips to change a hip roof to a gable roof
- Use **Convert to Roof** to create a roof from a closed polyline
- Merge roofs to determine the intersection points using mass groups
- Create dormer roofs that merge with the main roof

CREATING A ROOF WITH ADD ROOF

The commands for creating and editing roofs are located on the **Roofs** toolbar shown in Figure 5.1. The **Add Roof** command (**ROOFADD**) is used to place a roof. The edges of a roof can be modified by grips, **Edit Roof Edges/Faces**, or **Modify Roof**. The **Convert to Roof** command converts a closed polyline or closed walls to develop a roof.

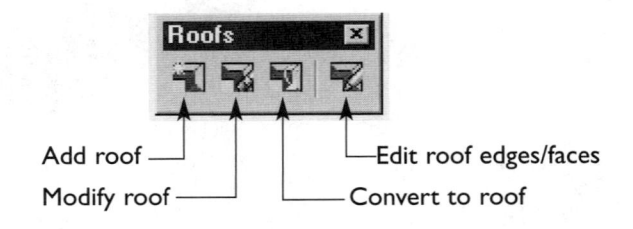

Figure 5.1 *Roofs toolbar*

Use the **Add Roof** command (**ROOFADD**) to create roofs by selecting the walls at the corners of the roof. See Table 5.1 for command access.

Menu bar	Design>Roofs>Add Roof
Command prompt	ROOFADD
Toolbar	Select Add Roof from the Roofs toolbar as shown in Figure 5.2
Shortcut	Right-click over drawing area; then choose Design>Roofs>Add

Table 5.1 *Add Roof access*

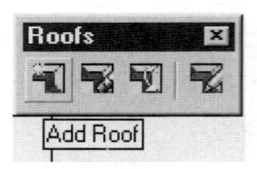

Figure 5.2 *Add Roof command on the Roofs toolbar*

Selecting the **Add Roof** command (**ROOFADD**) will open the **Add Roof** dialog box shown in Figure 5.3.

This dialog box consists of a **Shape** drop-down list, **Gable** and **Overhang** check boxes, and edit boxes for setting slope and overhang. Buttons for the **Viewer**, **Match**, **Properties**, **Undo**, and **Help** are located in the lower left corner of the dialog box. The creation of the roof is based upon a specified slope for the roof and the height of the plate. Some of the options of the dialog box are inactive, depending on the type of shape specified.

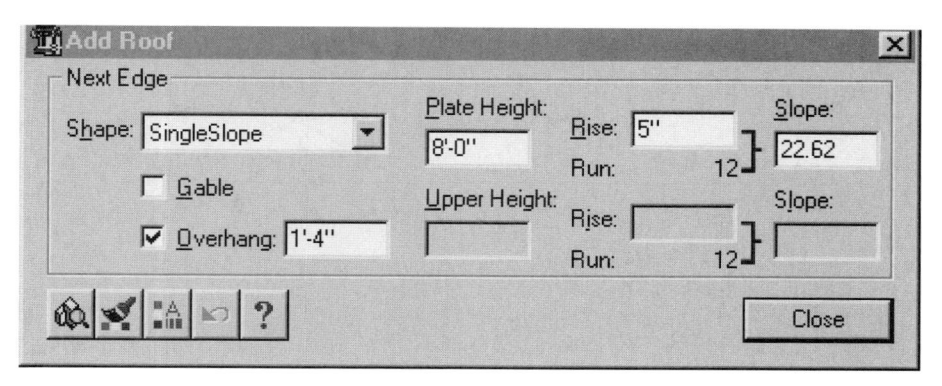

Figure 5.3 Add Roof dialog box for creating a roof

The options of the command can be selected from the dialog box or the command line. The options listed in the command line are

```
Roof point or
[Shape/Gable/Overhang/overhangValue/PHeight/PRise/PSl
ope/UHeight/URise/USlope/Ma
tch]:
```

To select the options, type in the command line the letters capitalized in each of the options. Editing the dialog box to select an option is a faster method of defining roof properties. The options of the **Add Roof** dialog box are described below:

> **Shape** – There are two **Shape** options: SingleSlope and DoubleSlope (see Figure 5.4). A single slope roof extends a roof plane at an angle from the **Plate Height**. A double slope roof includes a single slope and adds another slope, which begins at the intersection of the first slope, and the height specified for the second slope. Select the **Shape** option in the command line by typing **S**.
>
> **Gable** – The **Gable** check box, when selected, turns off the slope of the roof plane. To create a gable edge of the roof, select **Gable** prior to identifying the first corner of the gable end. Select the **Gable** option in the command line by typing **G**.
>
> **Plate Height** – The **Plate Height** edit field allows you to specify the height of the top plate from which the roof plane is projected. The height is relative to the XY plane, which has a Z coordinate of zero. Type **PH** in the command line to select the **PH**eight option and set the plate height.
>
> **Rise** – The **Rise** is the slope value entered to obtain the slope angle of a single slope roof. The **Rise** value is used to set the angle of the roof based on a run value of 12. As shown in Figure 5.3, a rise of 5 creates a 5/12 roof, which forms a slope angle of 22.62 degrees. The angle formed is the arc tangent of

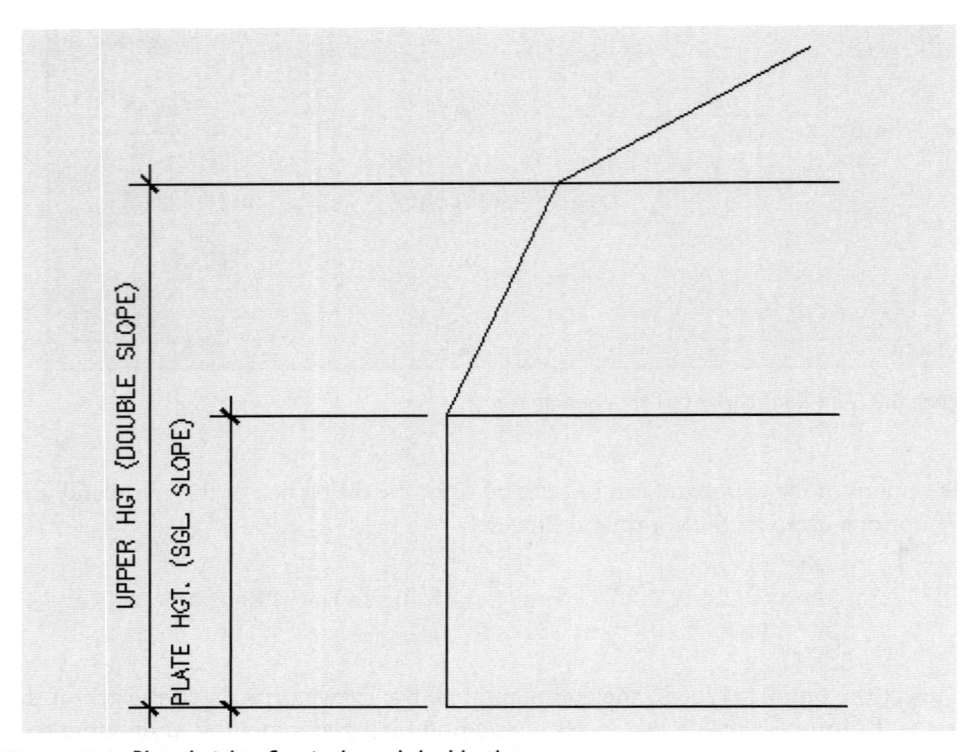

Figure 5.4 *Plate heights for single and double slopes*

the rise value divided by 12. Type **PR** to select the **PRise** option in the command line.

Slope – The slope is the angle of the roof plane from horizontal. Slope angles can be entered directly or derived from the rise value. If slope angles are entered, the equivalent rise value is displayed in the **Rise** edit field. Type **PS** to select the **PSlope** option in the command line.

Upper Height – The **Upper Height** edit field is the height defined to begin the second slope of a double slope roof. If **Shape** is toggled to DoubleSlope, the **Upper Height** and upper **Rise** edit fields become active. The second slope begins at the **Upper Height** distance defined from the XY plane. Type **UH** to select the **Uheight** option in the command line.

Rise (upper) – The **Rise** value determines the slope angle of the second slope. The **Rise** value for the upper slope is based upon a run of 12. Type **UR** to select the **URise** option in the command line.

Slope (upper) – The **Slope** angle value defines the slope angle for the upper slope. The slope (upper) can be entered as an angle or derived from the upper rise value. Type **US** to select the **USlope** option in the command line.

Overhang toggle – The **Overhang** check box turns on or off the development of a roof overhang. The overhang extends the roof down from the plate height to create the overhang. Type **O** to select the **Overhang** option in the command line.

Overhang – The **Overhang** edit field is the horizontal distance defined for the overhang. Type **V** to select the **overhangValue** option in the command line.

Floating Viewer button – The **Floating Viewer** button opens the **Viewer** window, which displays the graphics of the roof.

Match button – The **Match** button prompts you to select an existing roof to match its properties.

Properties button – The **Properties** button opens the **Roof Properties** dialog box.

Undo button – The **Undo** button will undo the last roof operation during the creation of a roof.

? – The question mark opens the Roof Overview help file.

Close button – The **Close** button closes the **Add Roof** dialog box.

When the **Add Roof** command (**ROOFADD**) is executed, the **Add Roof** dialog box opens and you are prompted to select roof points for each corner of the roof. The points selected for the roof edge are located at the bottom of the wall with a Z coordinate of zero. Selecting the points using the Intersection object snap will locate the edge of the roof plane on either the inner or outer wall surface of a corner. The Node object snap locates the roof plane on the justification line of the wall. Selecting the outer wall surface for the roof points will generate a roof construction similar to the truss roof construction as shown in Figure 5.5. Note that the lower plane of the roof intersects the outer wall surface.

If the inner wall surface is selected to determine the roof plane, the construction is similar to that of conventional framed rafters. Using the Intersection object snap mode to select the inner roof plane will result in the lower plane of the roof being generated from the top of the inner wall, as shown in Figure 5.6.

As points are selected, the roof is dynamically constructed. If errors are made in selecting points, the **Undo** button of the **Add Roof** dialog box can be used to remove the point selected. Roof planes will be created at the slope angle specified in the dialog box from the edge specified by each pair of points. Because a sloped roof plane is generated from each edge, a hip roof is created by default.

The plate height for a roof should be set equal to the wall height. The plate height is measured relative to the XY plane with a zero Z coordinate.

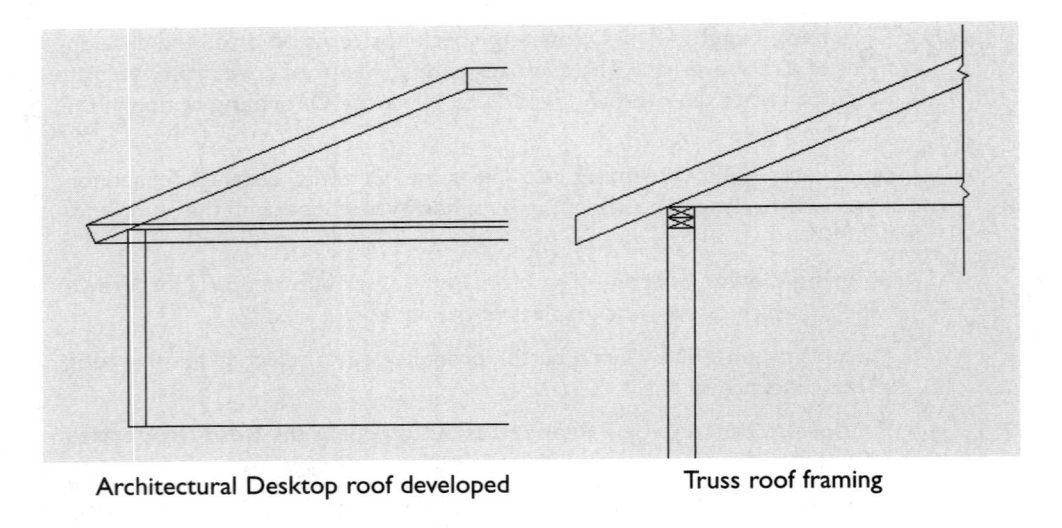

Architectural Desktop roof developed   Truss roof framing

Figure 5.5 *Roof developed by the selection of outer lines of the wall to define the roof*

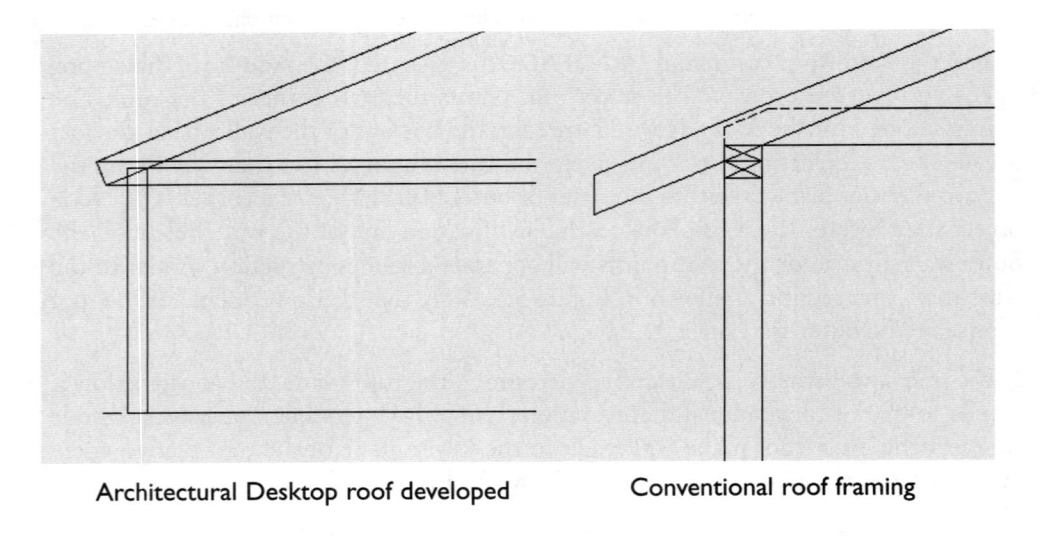

Architectural Desktop roof developed   Conventional roof framing

Figure 5.6 *Roof developed by the selection of inner lines of the walls to define the roof*

Note: You can select **Zoom Window** from the **Standard** toolbar to zoom a window around desired roof points while creating a roof. It is a transparent zoom and does not cancel the active command. Selecting transparent **Zoom** and **Pan** commands from the **Standard** toolbar while creating a roof can increase precision in selecting roof points.

CREATING A HIP ROOF

A hip roof is created by default when the SingleSlope **Shape** option is selected. To create a hip roof, set the **Shape** to **SingleSlope** and set the **Rise** for the roof planes. Selecting the corners of the building identifies each edge of the roof. The slope of each edge can be set independently of other roof edges. The slope of an edge should be defined prior to selecting the second point that defines that edge of the roof. The roof is created dynamically as the corners of the building are selected; therefore the **Undo** button can be selected to deselect a corner of the roof as it is developed. The roof object is created on the layer A-Roof, which has the color 20 (red). Freezing the A-Roof layer turns off the display of the roof.

TUTORIAL 5.1 CREATING A HIP ROOF

1. Open *ADT Tutor\Ch5\EX5-1.dwg*.
2. Choose **File>SaveAs** from the menu bar and save the drawing as *HIP1.dwg* in your student directory.
3. Move the mouse pointer to the OSNAP toggle of the status bar and right-click and then choose **Settings** from the shortcut menu. Edit the **Object Snap** tab of the **Drafting Settings** dialog box as shown in Figure 5.7. Verify

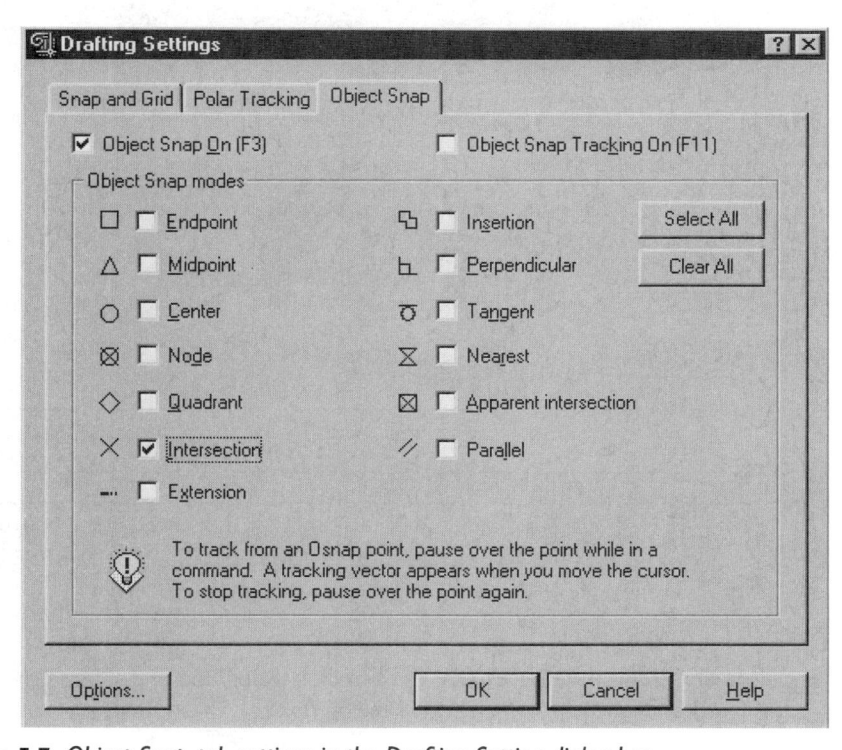

Figure 5.7 *Object Snap tab settings in the Drafting Setting dialog box*

that **Object Snap** is checked ON and check ON the Intersection mode. Clear all other object snap modes. Select **OK** to dismiss the **Drafting Standards** dialog box.

4. Select the **Layers** button on the **Object Properties** toolbar. Select the layers named A_door and A_Glaz, and then select the **Freeze** buttons for each layer. Select **OK** to dismiss the **Layer Properties Manager**.

5. Turn off the display of brick hatching and the Brick wall component by selecting the exterior wall; right-click and choose **Edit Wall Style** from the shortcut menu.

6. Select the **Display Props** tab of the **Wall Style Properties** dialog box.

7. Select the **Edit Display Props** button to open the **Entity Properties** dialog box.

8. Select the Boundary 2 (Brick) and Hatch 2 (Brick) components and turn off the visibility by selecting the light bulb icon as shown in Figure 5.8.

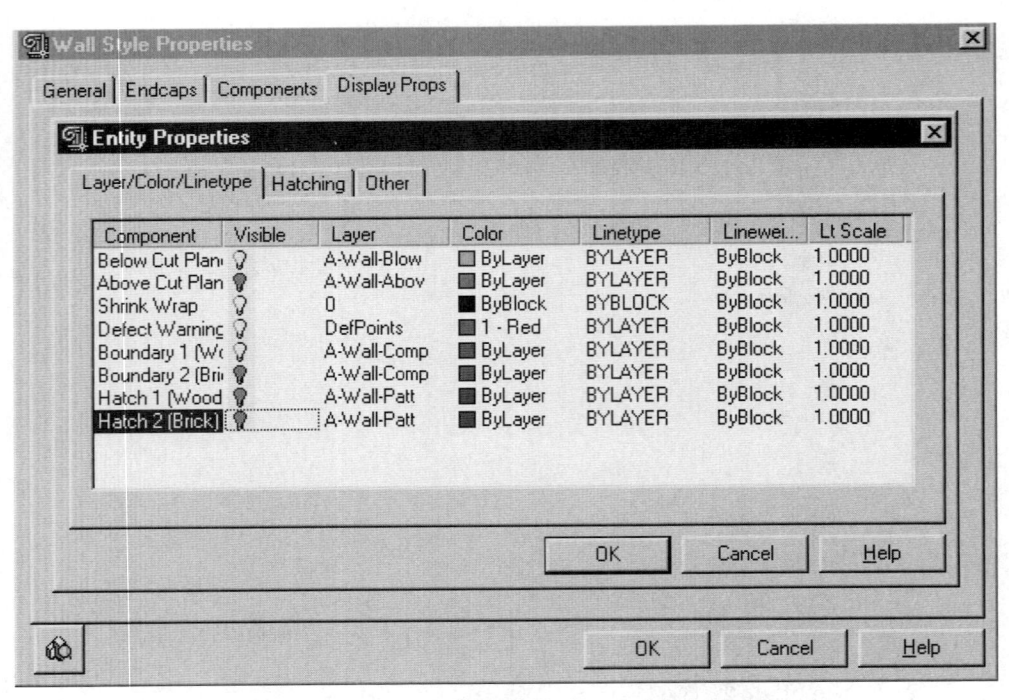

Figure 5.8 *Layer/Color/Linetype settings of the Entity Properties dialog box*

9. Select **OK** to dismiss the **Entity Properties** dialog box and select **OK** to dismiss the **Wall Style Properties** dialog box.

10. Select **Zoom Window** from the **Standard** toolbar and window the lower left corner of the building as shown in Figure 5.10.

11. Select the **Add Roof** command from the **Roofs** toolbar and set the **Shape** to SingleSlope, **Plate Height** to 8', **Rise** = 5, **Overhang** check box selected and overhang distance equal to 1'-8", as shown in Figure 5.9.

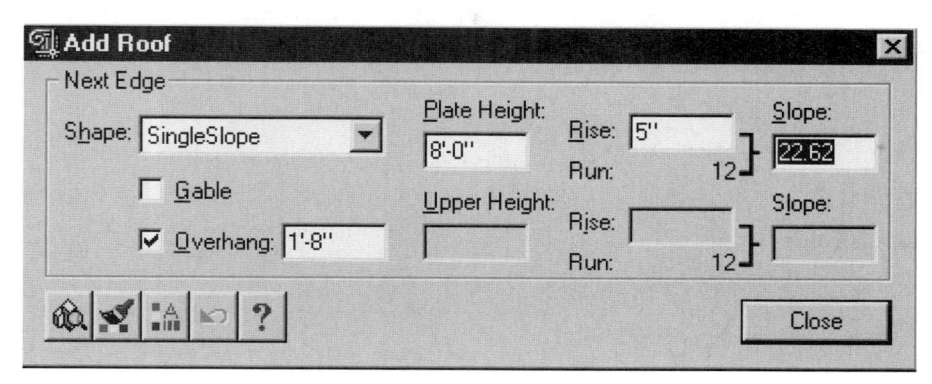

Figure 5.9 *Settings for the Add Roof dialog box*

12. Respond to the following command line prompts by selecting roof points as specified below.

```
Command: _AecRoofAdd
Roof point or
[Shape/Gable/Overhang/overhangValue/PHeight/PRise/PSl
ope/UHeight/URise/USlope/Ma
tch]:
```
(Select the intersection of lines representing the outside of stud, point P1 in Figure 5.10.)

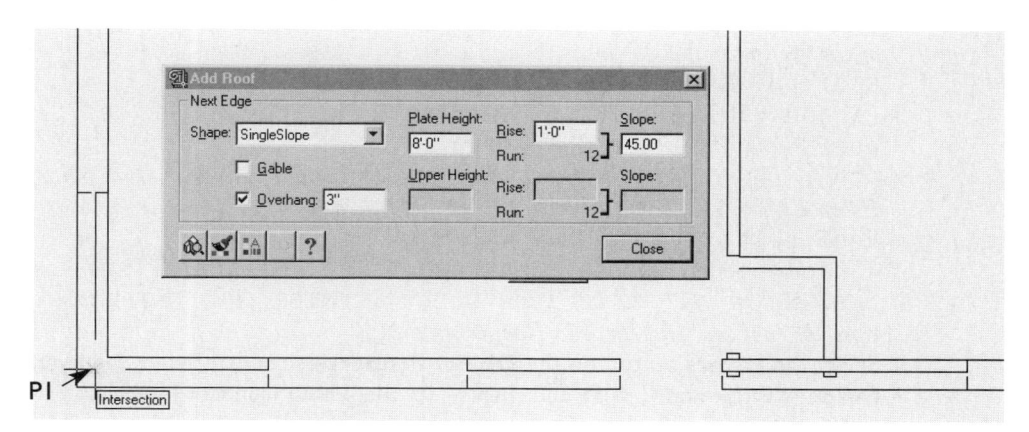

Figure 5.10 *First point for creating the roof plane*

```
Roof point or
[Shape/Gable/Overhang/overhangValue/PHeight/PRise/PSl
ope/UHeight/URise/USlope/Ma
tch]:
```
(Select the intersection of lines representing the outside of stud, point P2 in Figure 5.11.)

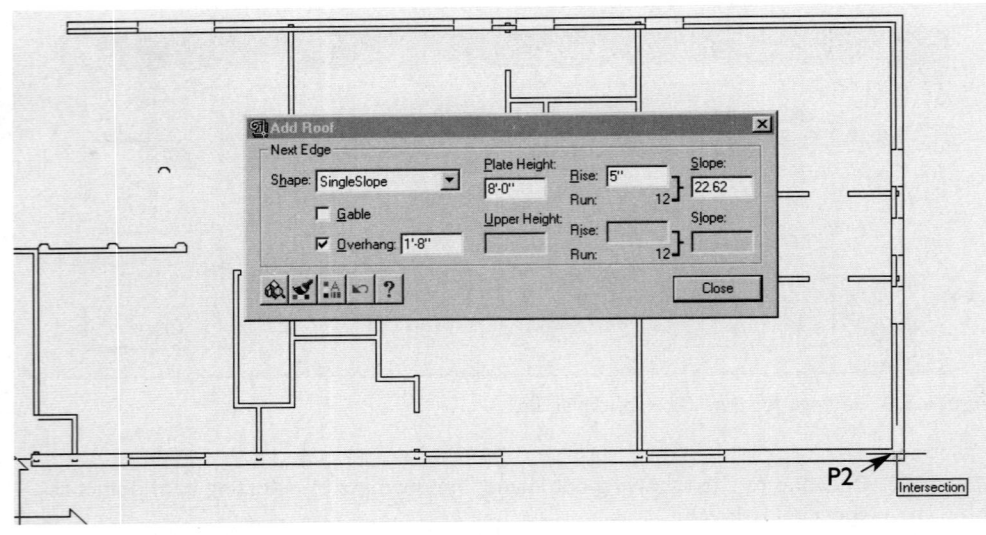

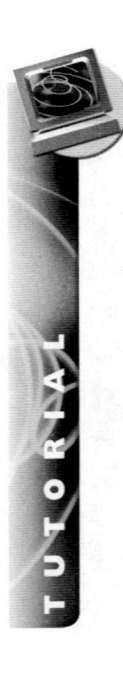

Figure 5.11 *Second point for creating the first roof plane*

```
Roof point or
[Shape/Gable/Overhang/overhangValue/PHeight/PRise/PSl
ope/UHeight/URise/USlope/Ma
tch]:
```
(Select the intersection of lines representing the outside of stud, point P3 in Figure 5.12. Note that selecting point P3 creates the dynamic display of the roof plane.)
```
Roof point or
[Shape/Gable/Overhang/overhangValue/PHeight/PRise/PSl
ope/UHeight/URise/USlope/Ma
tch]:
```
(Select the intersection of lines representing the outside of stud, point P4 in Figure 5.13.)
```
Roof point or
[Shape/Gable/Overhang/overhangValue/PHeight/PRise/PSl
ope/UHeight/URise/USlope/Match]: ENTER
```
(Press ENTER to close the roof and end the Add Roof command.)

13. Select the **Layers** button on the **Object Properties** toolbar. Select the layer named A_Door and A_Glaz, and then select the **Thaw** (sun icon) for each layer.

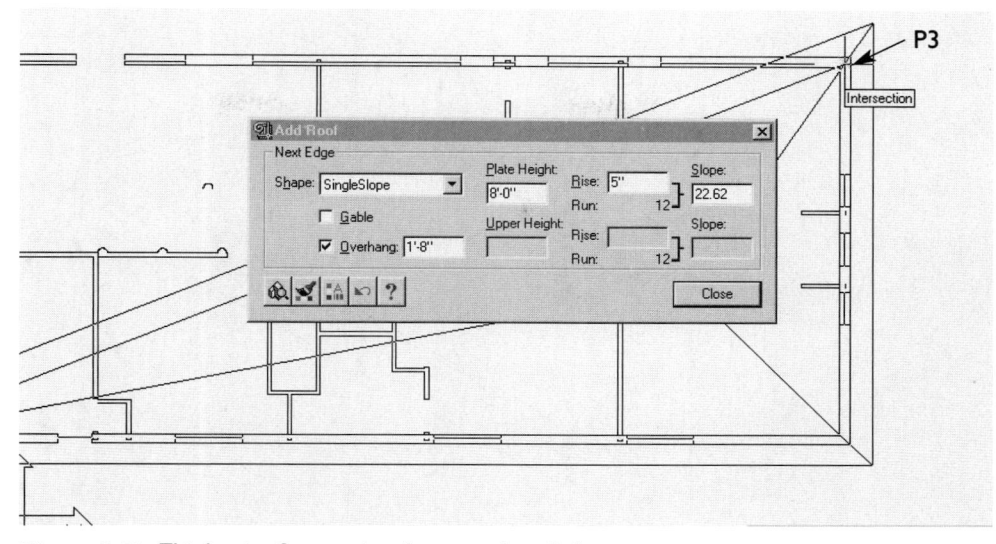

Figure 5.12 *Third point for creating the second roof plane*

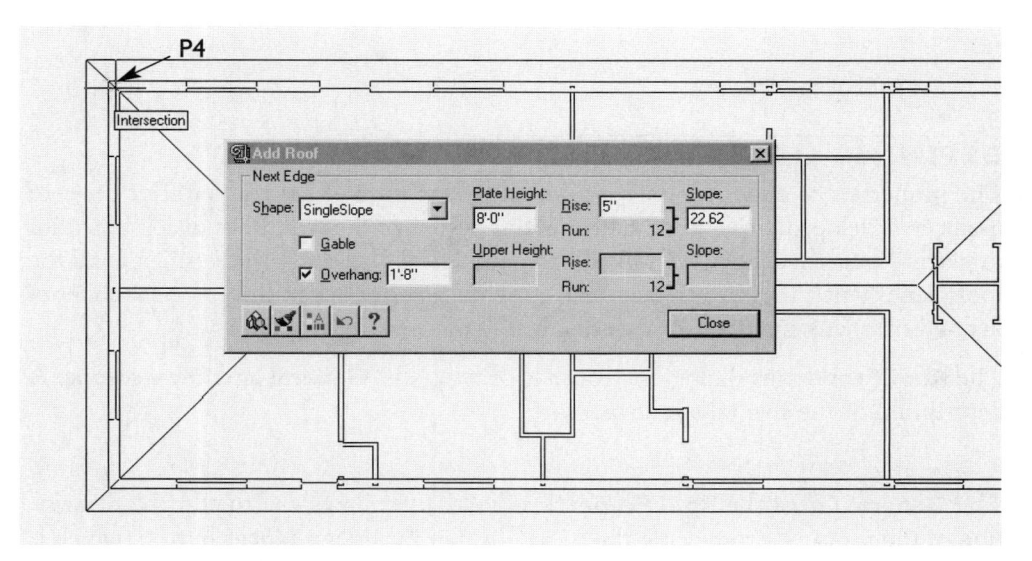

Figure 5.13 *Fourth point for creating the third roof plane*

14. Select **OK** to dismiss the **Layer Properties Manager**.
15. Select **SW Isometric** from the **View** flyout on the **Standard** toolbar.
16. Type **HI** and press ENTER in the command line to execute the **HIDE** command. The roof should appear as shown in Figure 5.14.

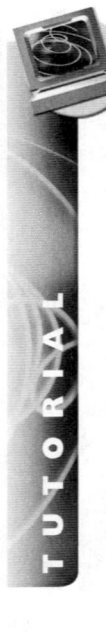

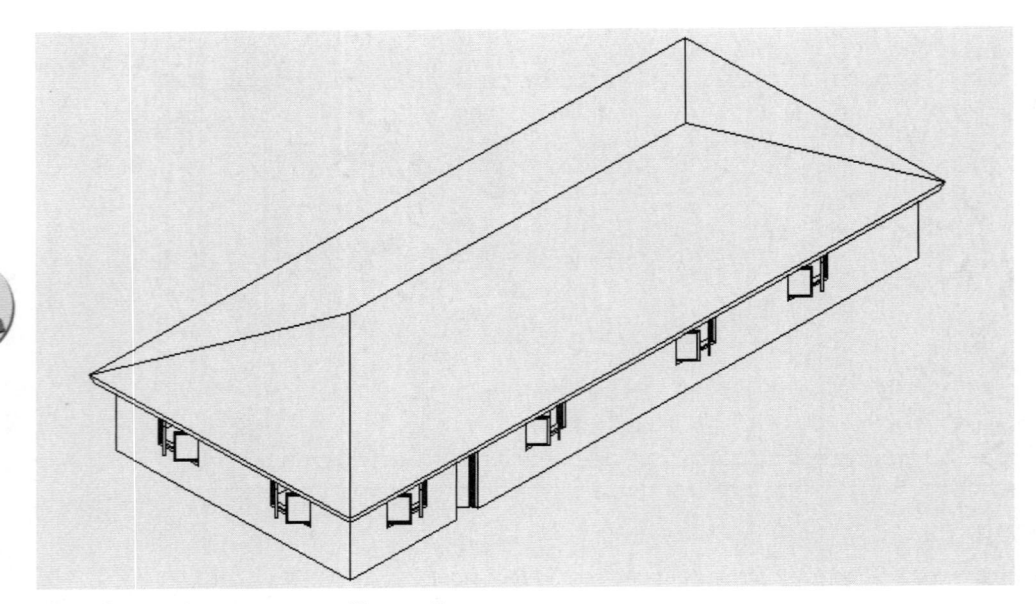

Figure 5.14 *Isometric view of hip roof*

17. Save the drawing.

DEFINING ROOF PROPERTIES

The properties of a roof can be set prior to the creation of the roof or after the roof has been developed. Selecting the **Properties** button of the **Add Roof** dialog box prior to placing a roof will allow you to edit the properties and apply the settings to all the roof planes when they are created. Selecting the **Properties** button in the **Add Roof** dialog box opens the **Roof Properties** dialog box shown in Figure 5.15.

The **Roof Properties** dialog box consists of two tabs: **General** and **Dimensions**. A description of the two tabs follows.

General

The **General** tab of the **Roof Properties** dialog box allows you to edit the description of the roof prior to placing the roof. This tab includes a **Notes** button, which is inactive until a roof is created. If an existing roof is selected, the **Notes** button can be selected to open the **Notes** dialog box, consisting of two tabs: the **Text Notes** tab allows you enter text information regarding the roof and the **Reference Docs** tab allows you to list other files that include relevant information regarding the roof.

Dimensions

The **Dimensions** tab of the **Roof Properties** dialog box consists of two sections, **Selected Roof Edges** and **Roof Faces (by Edge)**, as shown in Figure 5.16.

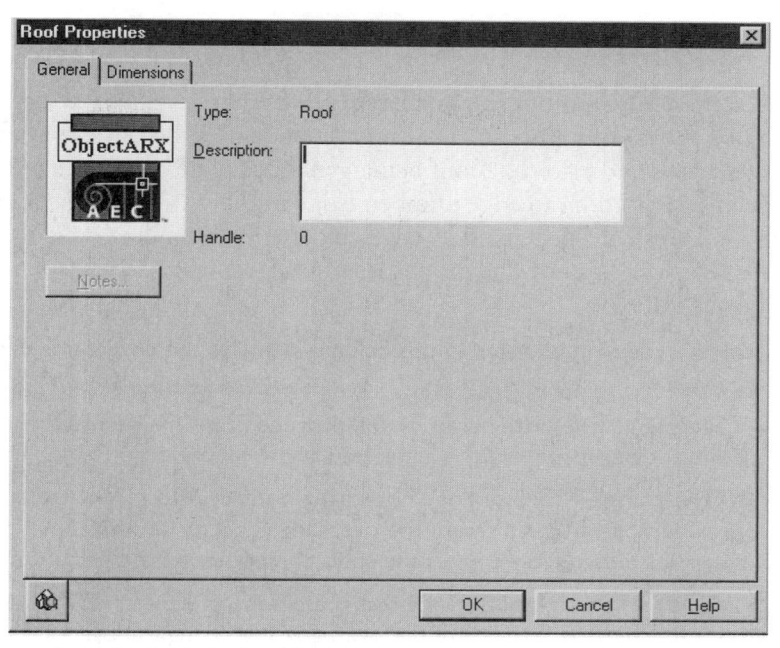

Figure 5.15 *General tab of the Roof Properties dialog box*

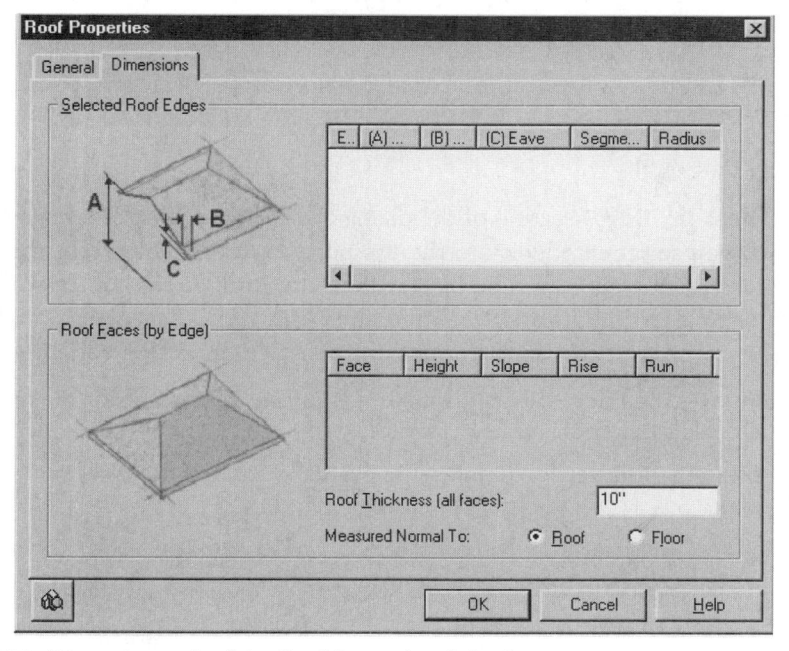

Figure 5.16 *Dimensions tab of the Roof Properties dialog box*

Selected Roof Edges

The **Selected Roof Edge** list box shows the roof planes that have been created by edge. The dialog box shown in Figure 5.16 does not include a list because the **Properties** button has been selected prior to a roof being generated. The properties of each roof edge are specified according to edge number. Edge number 0 is the first edge selected when a roof is created. The remaining edges are listed in ascending order in the same sequence as they were created. The properties of each edge are described in columns of the list box as follows:

Edge – The number listed in this column identifies the edge number of the roof plane.

A(Height) – The plate height or the distance from the XY plane that the roof plane begins. The roof is generated at the specified slope from this point.

B(Overhang) – The horizontal overhang distance. The roof extends at the specified slope angle to create the overhang equal to the horizontal overhang distance. The overhang creates the soffit component of the roof.

C(Eave) – The vertical distance that the overhang projects down from the top plate over the distance of the overhang. Increasing the overhang and roof slope will cause the **C (Eave)** dimension to increase.

Segments – The number of divisions created to cover a curved roof edge.

Radius – The radius of a curved roof generated to fit a curved wall.

In Figure 5.17, Edge 0 covers the curved portion of the wall and consists of six segments.

Roof Faces (by Edge)

The **Roof Faces (by Edge)** section of the dialog box includes a list of faces created for each edge. If you select an edge from the top portion of the dialog box, the properties of the face for that edge will be displayed in the **Roof Faces (by Edge)** list box. The height, slope, rise, and run of each face are described. The slope of the face can be edited; however, the remaining dimensions of a face cannot be edited.

Face – The **Face** column identifies the number of the face for the roof edge. Double slope roofs include two faces.

Height – The vertical distance the overhang projects down from the top plate.

Slope – The angle of the roof measured in degrees. The slope value can be edited to specify the angle of the roof.

Rise – The angle of the roof as the vertical change in the roof over a horizontal distance of one foot.

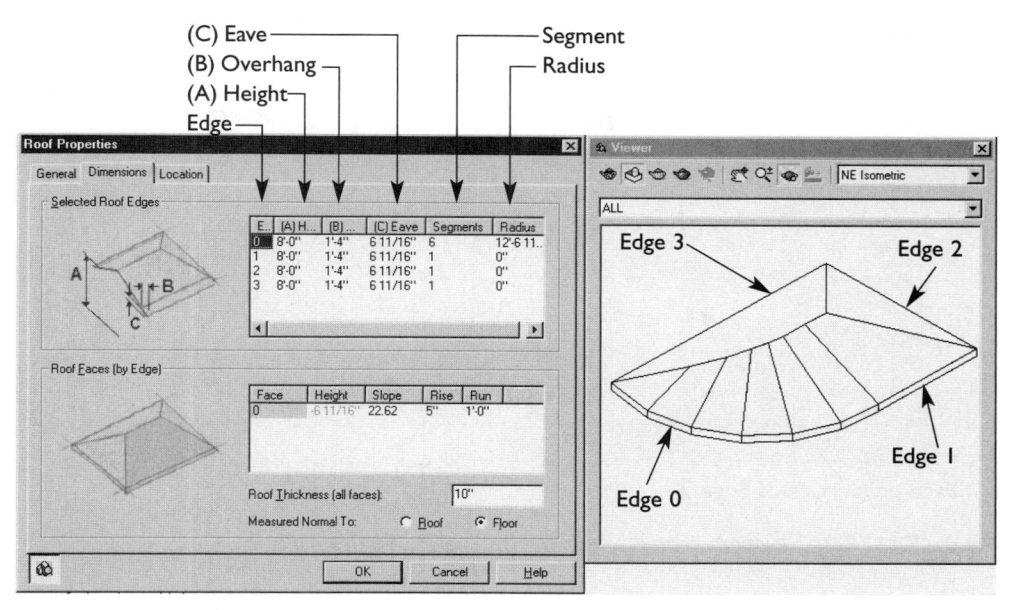

Figure 5.17 The Dimensions tab of the Roof Properties dialog box for a curved roof

Run – The horizontal distance of the roof used to define the rise of the slope.

Roof Thickness (all Faces) – The thickness of the roof plane. The **Roof Thickness** default value is 10". Roof thickness should be set according to the dimensions of structural members.

Measured Normal to: Roof or Floor – When the **Roof** radio button is toggled ON, the roof fascia is generated perpendicular to the face of the roof. When the **Floor** radio button is toggled ON, the roof fascia is generated perpendicular to the face of the floor.

The elevation of the roof face shown in Figure 5.18 is dimensioned according to its **Roof Properties** dialog box. The fascia is vertical because **Floor** is toggled ON.

EDITING ROOF PROPERTIES OF AN EXISTING ROOF

The **Roof Properties** command (**ROOFPROPS**) is used to edit the properties of an existing roof. See Table 5.2 for command access.

When the **Roof Properties** command is executed, you are prompted to select a roof. The properties of the selected roof are displayed in a **Roof Properties** dialog box as shown in Figure 5.19.

The **General** and **Dimensions** tabs of the **Roof Properties** dialog box are identical in content to the **Roof Properties** dialog box opened when you select the **Properties** button from within the **Add Roof** command. The **Dimensions** tab of the **Roof**

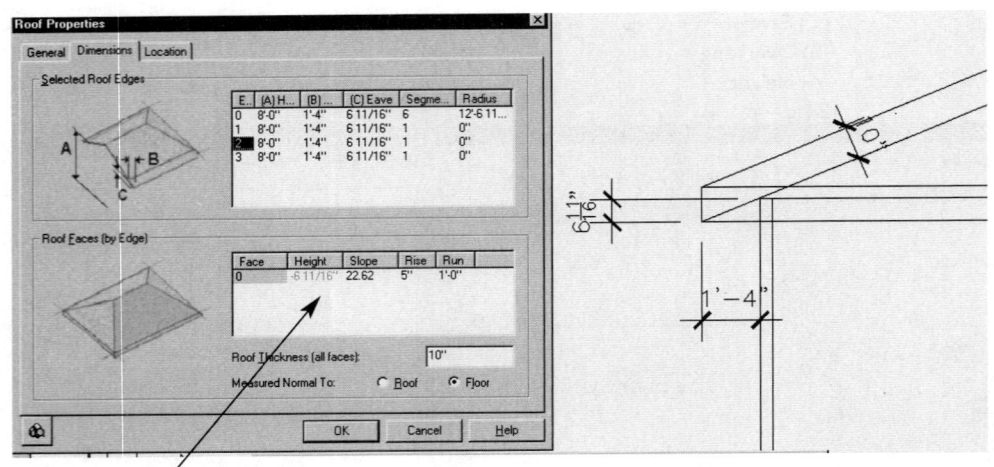

Face properties of Edge 2

Figure 5.18 *Roof elevation created by editing the Roof Properties dialog box*

Command: prompt	ROOFPROPS
Shortcut	Right-click over the drawing area; choose Design>Roofs>Properties
Shortcut	Select roof; right-click and choose Roof Properties

Table 5.2 *Roof Properties access*

Properties dialog box for the roof developed in Tutorial 5.1 is shown in Figure 5.20.

The first roof plane developed in Tutorial 5.1 is listed as Edge 0 in the **Selected Roof Edges** section of the dialog box. The roof edges are listed in the same order as they were developed. You can edit the roof face properties by selecting the roof edge in the **Selected Roof Edges** section and then editing the face properties in the **Roof Faces (By Edge)** section.

The **Location** tab is added to the **Roof Properties** dialog box when the **Roof Properties** command is applied to an existing roof. As shown in Figure 5.21, the tab identifies the location of the roof relative to the World Coordinate System using X, Y, Z coordinate values. The **Rotation** edit field allows the roof to be rotated about its insertion point.

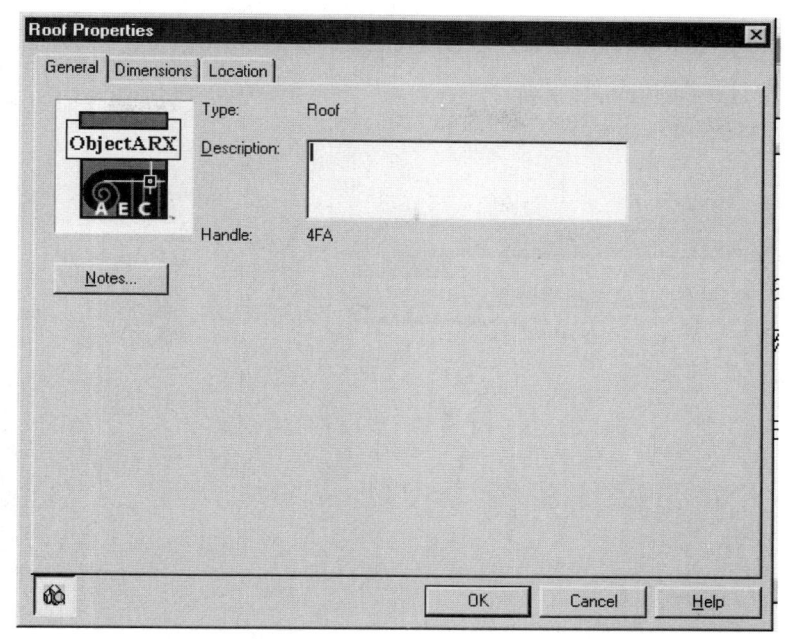

Figure 5.19 Roof Properties dialog box

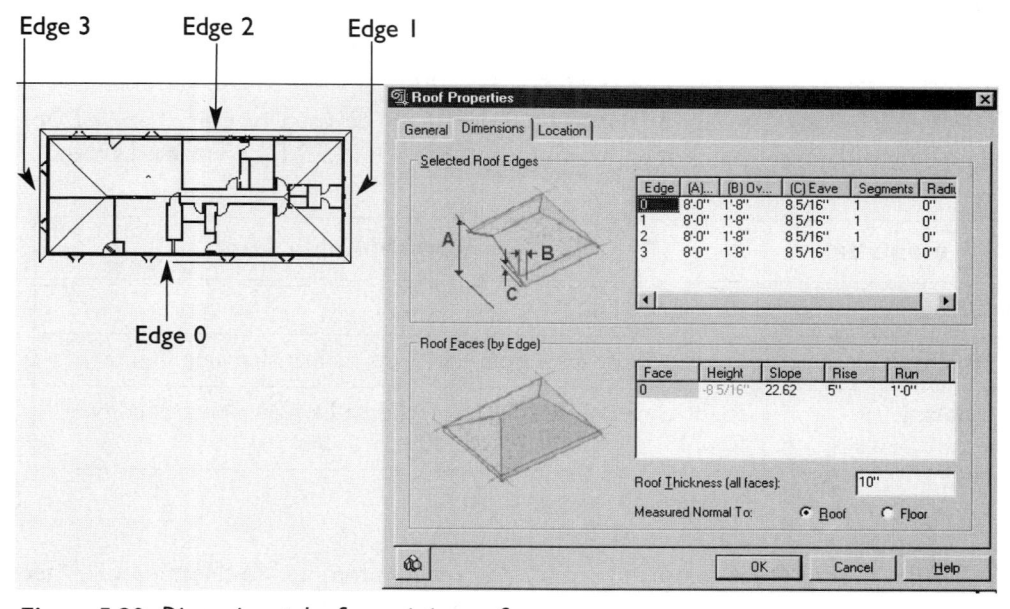

Figure 5.20 Dimensions tab of an existing roof

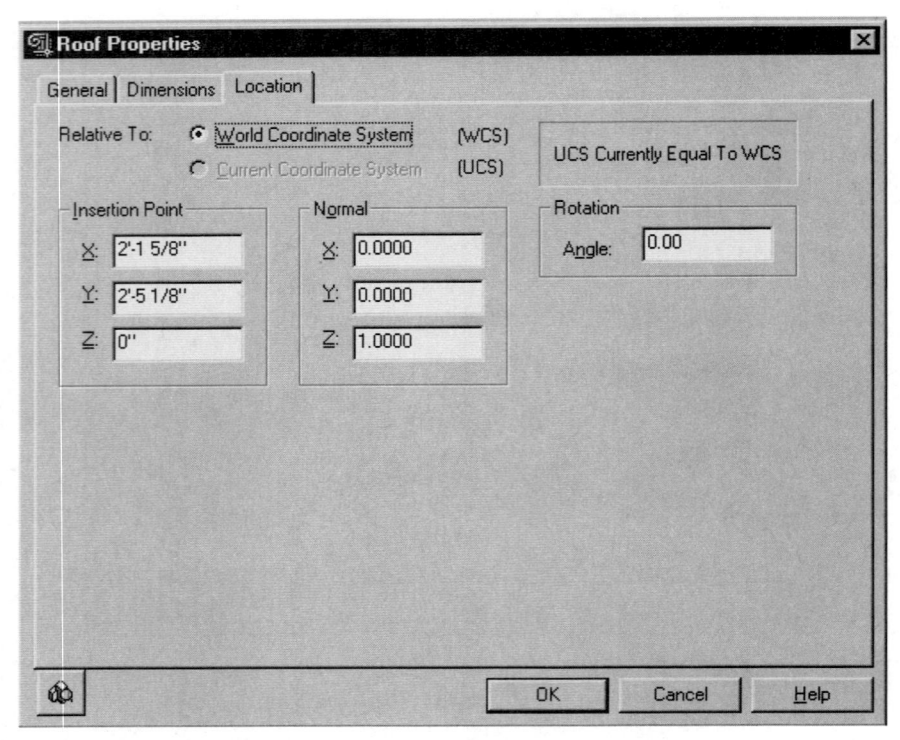

Figure 5.21 *Location tab of the Roof Properties dialog box*

USING MODIFY ROOF TO CHANGE ROOFS

The **Modify Roof** command (**ROOFMODIFY**) can be used to edit shape, slope, plate height, and overhang of an existing roof. See Table 5.3 for command access.

Menu bar	Design>Roofs>Modify Roof
Command prompt	ROOFMODIFY
Toolbar	Select Modify Roof from the Roofs toolbar shown in Figure 5.22
Shortcut	Right-click over drawing area; choose Design>Roofs>Modify Roof

Table 5.3 *Modify Roof access*

When the **Modify Roof** command is selected, you are prompted to select a roof. After you select a roof, the **Modify Roof** dialog box will open, as shown in Figure 5.23.

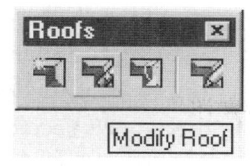

Figure 5.22 *Modify Roof command on the Roofs toolbar.*

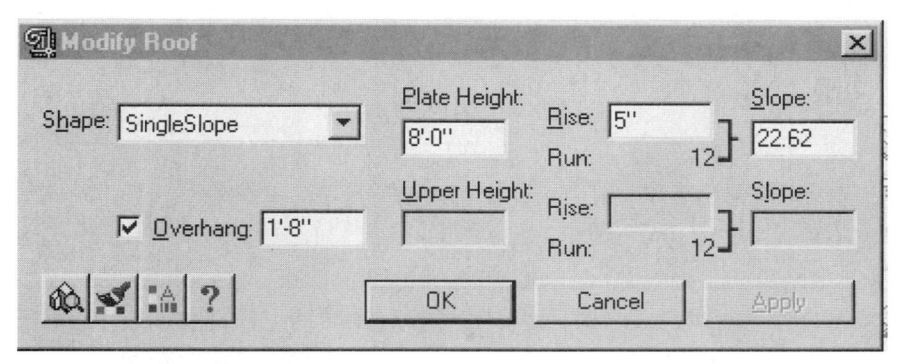

Figure 5.23 *Modify Roof dialog box*

Changing the edit fields in the **Modify Roof** dialog box and selecting the **Apply** button will change the roof. The **Modify Roof** dialog box allows you to change the parameters of the roof in a dialog box similar to the **Add Roof** dialog box. You can edit each edge of the roof by selecting the **Properties** button to access the **Roof Properties** dialog box.

CREATING A GABLE ROOF

A gable roof is created in a procedure similar to that for creating the hip roof. Define the gable ends of the roof by selecting **Gable** in the **Add Roof** dialog box when the locations of the roof plane are defined. The **Gable** check box sets the slope to 90 degrees for the roof edges defining the gable. When a gable roof is created, the overhang on the gable ends is usually reduced. The **Overhang** check box can be selected and adjusted as each roof plane is created.

TUTORIAL 5.2 CREATING A GABLE ROOF

1. Open *ADT Tutor\Ch5\Ex5-2.dwg* and save the drawing as *Gable2.dwg* in your student directory.

2. Move the mouse pointer to the OSNAP toggle of the status bar and right-click; then choose **Settings** from the shortcut menu. Check ON the Intersection mode of object snap and clear all other modes of object snap.

Check ON **Object Snap On(F3)** at the top of the **Object Snap** tab and select **OK** to dismiss the **Drafting Standards** dialog box.

3. Select the **Add Roof** command and set the **Shape** to SingleSlope, **Plate Height** to 8', **Overhang** check box selected and overhang distance equal to 1'-4", and **Rise** to 5, as shown in Figure 5.24.

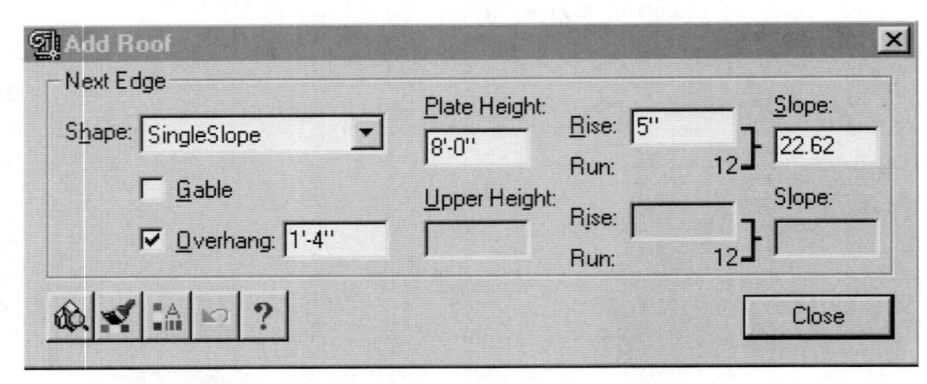

Figure 5.24 *Add Roof dialog box settings for first roof plane*

4. Respond to the following command line prompts by selecting points to define the roof:

```
Command: _AecRoofAdd
Roof point or
[Shape/Gable/Overhang/overhangValue/PHeight/PRise/PSl
ope/UHeight/URise/USlope/Match]: DBOX
```
(Edit dialog box: set the Overhang to 1'-4", Plate Height to 8' and Rise to 5)

(Select the drawing area to activate the drawing area)

```
Roof point or
[Shape/Gable/Overhang/overhangValue/PHeight/PRise/PSl
ope/UHeight/URise/USlope/Match]:
```
(Select the intersection of the inner lines of the corner at point P1 in Figure 5.25.)

```
Roof point or
[Shape/Gable/Overhang/overhangValue/PHeight/PRise/PSl
ope/UHeight/URise/USlope/Match]:
```
(Select the intersection of the inner lines of the corner at point P2 in Figure 5.26.)

```
Roof point or
[Shape/Gable/Overhang/overhangValue/PHeight/PRise/PSl
ope/UHeight/URise/USlope/Match]: DBOX
```
(Edit the Add Roofs dialog box: select Gable, edit Overhang to 6".)

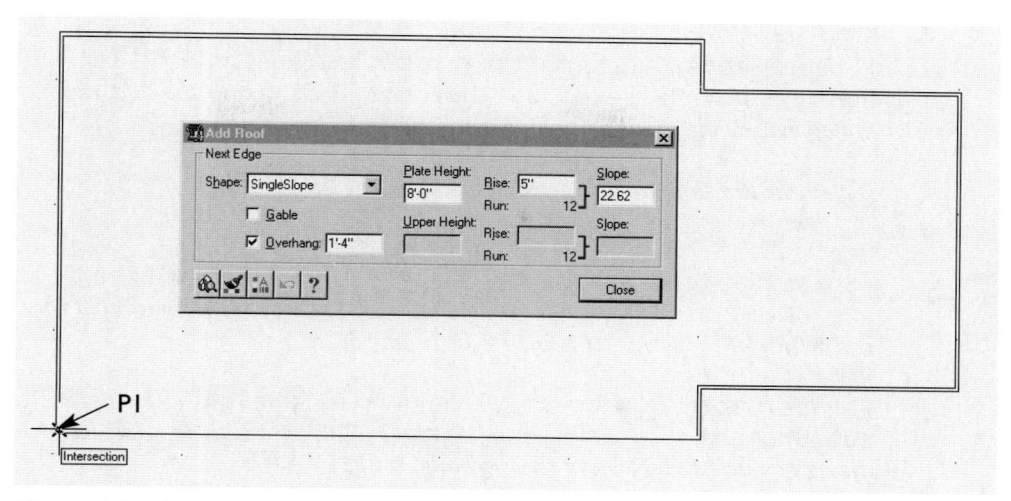

Figure 5.25 *First point to define the first roof plane*

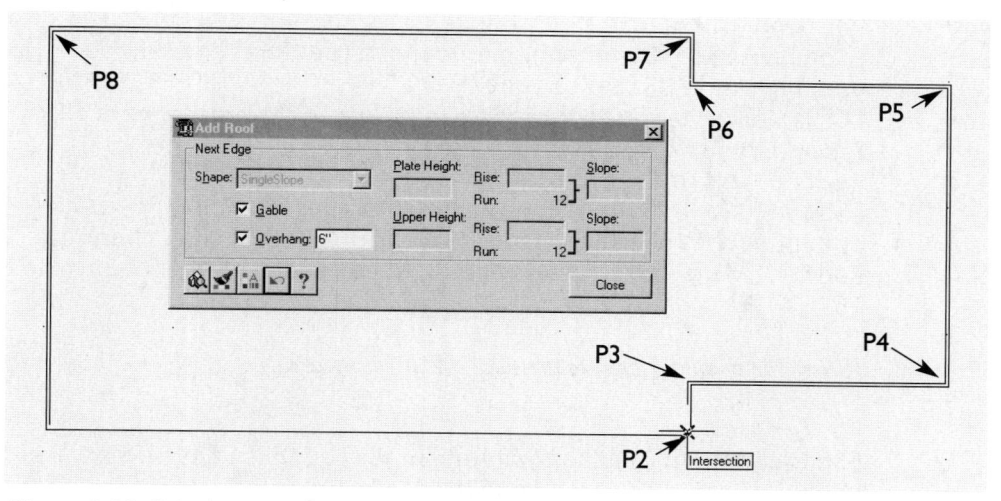

Figure 5.26 *Point locations for creating the gable roof*

```
Roof point or
[Shape/Gable/Overhang/overhangValue/PHeight/PRise/PSl
ope/UHeight/URise/USlope/Match]:
```
(Select the intersection of the inner walls at point P3 in Figure 5.26.)

Roof point or
[Shape/Gable/Overhang/overhangValue/PHeight/PRise/PSl
ope/UHeight/URise/USlope/Match]: DBOX *(Edit the Add Roof dialog box: deselect Gable, and set overhang to 16".)*

(Select the drawing area to activate the drawing area)

Roof point or
[Shape/Gable/Overhang/overhangValue/PHeight/PRise/PSl
ope/UHeight/URise/USlope/Match]: *(Select the intersection of the inner lines of the corner at point P4 in Figure 5.26.)*
Roof point or
[Shape/Gable/Overhang/overhangValue/PHeight/PRise/PSl
ope/UHeight/URise/USlope/Match]: DBOX *(Edit the Add Roof dialog box: select Gable, and set overhang to 6".)*

(Select the drawing area to activate the drawing area)

Roof point or
[Shape/Gable/Overhang/overhangValue/PHeight/PRise/PSl
ope/UHeight/URise/USlope/Ma
tch]: *(Select the intersection of the inner lines of the corner at point P5 in Figure 5.26.)*
Roof point or
[Shape/Gable/Overhang/overhangValue/PHeight/PRise/PSl
ope/UHeight/URise/USlope/Match]: DBOX (Edit the Add Roof dialog box: deselect Gable, and set overhang to 16".)

(Select the drawing area to activate the drawing area.)

Roof point or
[Shape/Gable/Overhang/overhangValue/PHeight/PRise/PSl
ope/UHeight/URise/USlope/Match]: *(Select the intersection of the inner lines of the corner at point P6 in Figure 5.26.)*
Roof point or
[Shape/Gable/Overhang/overhangValue/PHeight/PRise/PSl
ope/UHeight/URise/USlope/Match]: DBOX *(Edit the Add Roof dialog box, select Gable, and set overhang to 6".)*

(Select the drawing area to activate the drawing area.)

Roof point or
[Shape/Gable/Overhang/overhangValue/PHeight/PRise/PSl
ope/UHeight/URise/USlope/Match]: *(Select the intersection of the inner lines of corner at point P7 in Figure 5.26.)*

Roof point or
[Shape/Gable/Overhang/overhangValue/PHeight/PRise/PSl
ope/UHeight/URise/USlope/Match]: DBOX *(Edit the Add Roof dialog box, deselect Gable, and set overhang to 16".)*

(Select the drawing area to activate the drawing area.)

Roof point or
[Shape/Gable/Overhang/overhangValue/PHeight/PRise/PSl
ope/UHeight/URise/USlope/Match]: *(Select the intersection of the inner lines of the corner at point P8 in Figure 5.26.)*
Roof point or
[Shape/Gable/Overhang/overhangValue/PHeight/PRise/PSl
ope/UHeight/URise/USlope/Match]: ENTER *(Press ENTER to close the roof to P1 and end the command.)*

5. Select **SE Isometric** from the **View** flyout on the **Standard** toolbar.

6. Type **HI** and press ENTER in the command line to execute the **HIDE** command and hide lines. The final roof should appear as shown in Figure 5.27.

7. Save the drawing.

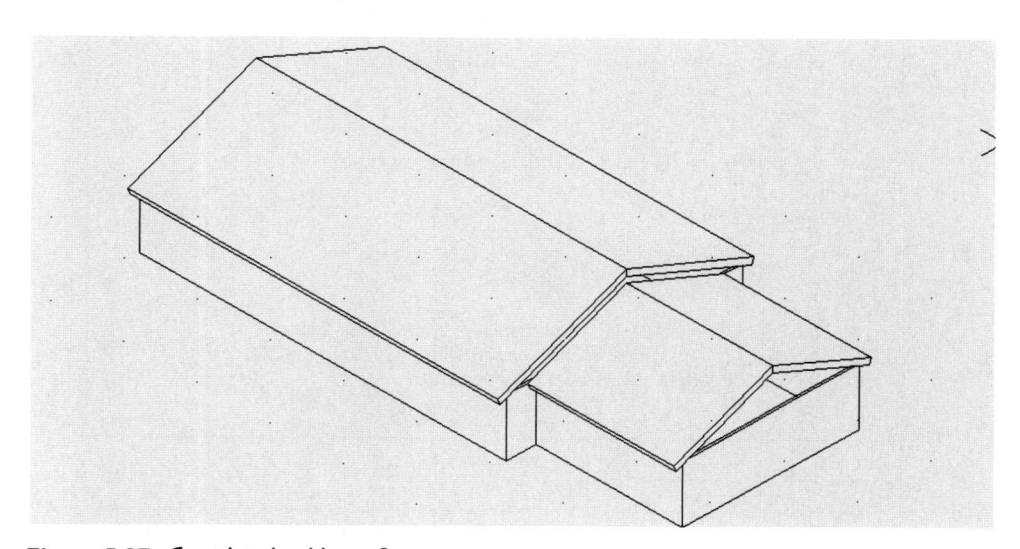

Figure 5.27 *Completed gable roof*

CREATING A SHED ROOF

Create a shed roof by selecting the **Gable** check box for three of the four sides of a roof. The slope of the non–gable side of the roof creates the shed. The **Gable** check box cannot be selected for more than three sides. Each time it is used, the roof for that edge of the building has a vertical slope. Shed roofs are often used in combination with other roof types.

TUTORIAL 5.3 CREATING A SHED ROOF

1. Open *ADT Tutor\Ch5\Ex5-3.dwg* and save the drawing as *Shed3.dwg* in your student directory.

2. To set the object snap, move the mouse pointer to the OSNAP toggle of the status bar and right-click; choose **Settings** from the shortcut menu. Check ON the Intersection mode and clear all other object snap modes. Select the **Object Snap On(F3)** check box to turn ON object snaps, as shown in Figure 5.28. Select **OK** to dismiss the **Drafting Settins** dialog box.

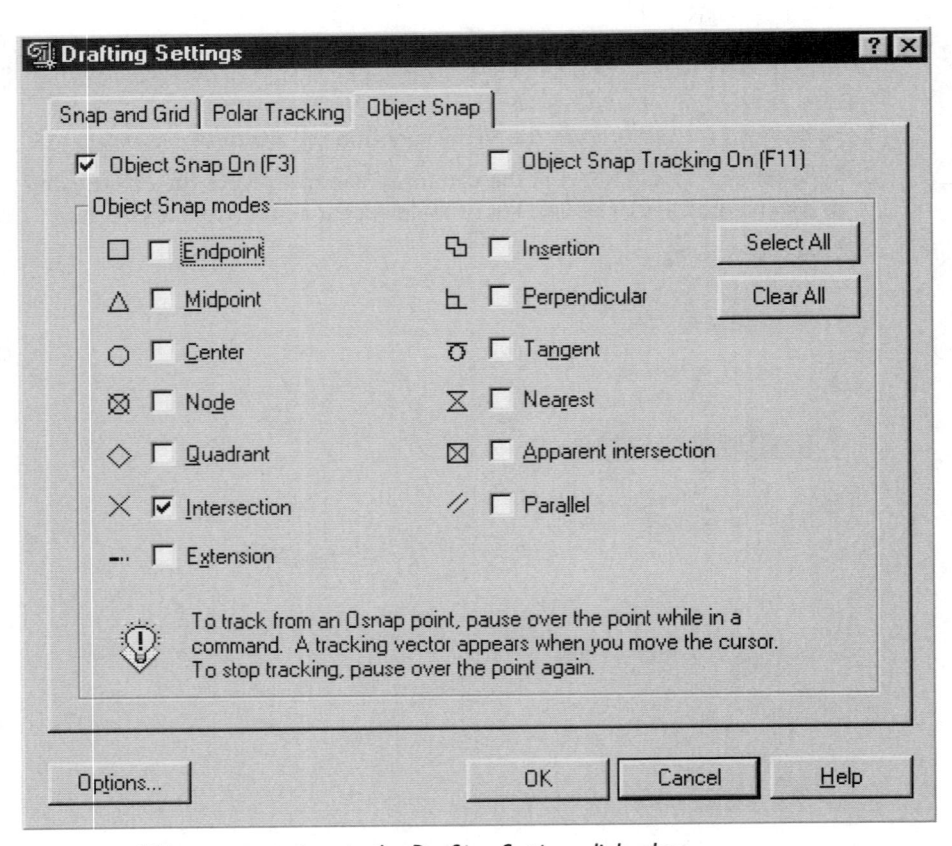

Figure 5.28 *Object snap settings in the Drafting Settings dialog box*

3. Zoom a window around the building from P1 to P2 as shown in Figure 5.29.

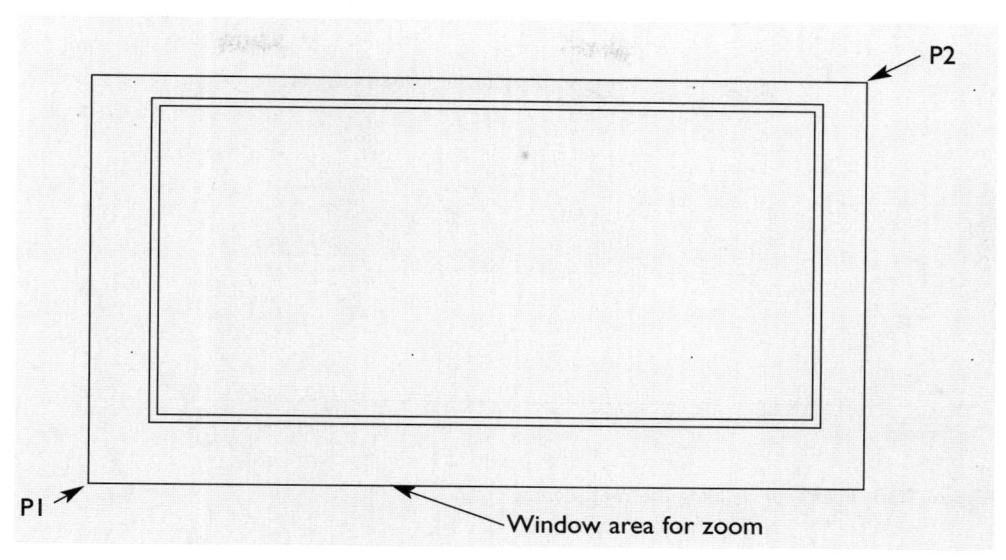

Figure 5.29 *Zoom window for viewing the building*

4. Select the **Add Roof** command and set the **Shape** to SingleSlope, **Plate Height** to 8', **Rise** to 5, select the **Overhang** check box and set the overhang distance to 1'-4" as shown in Figure 5.30. Respond to the following command line prompts by selecting roof points as specified below:

```
Roof point or
[Shape/Gable/Overhang/overhangValue/PHeight/PRise/PSl
ope/UHeight/URise/USlope/Match]: DBOX
```
(Edit the Add Roof dialog box: deselect Gable, and set overhang to 16".)

(Select the drawing area to activate the drawing area.)

```
Roof point or
[Shape/Gable/Overhang/overhangValue/PHeight/PRise/PSl
ope/UHeight/URise/USlope/Match]:
```
(Select the intersection of the inner lines of the corner at point P1 in Figure 5.30.)

```
Roof point or
[Shape/Gable/Overhang/overhangValue/PHeight/PRise/PSl
ope/UHeight/URise/USlope/Match]:
```
(Select the intersection of the inner lines of the corner at point P2 in Figure 5.30.)

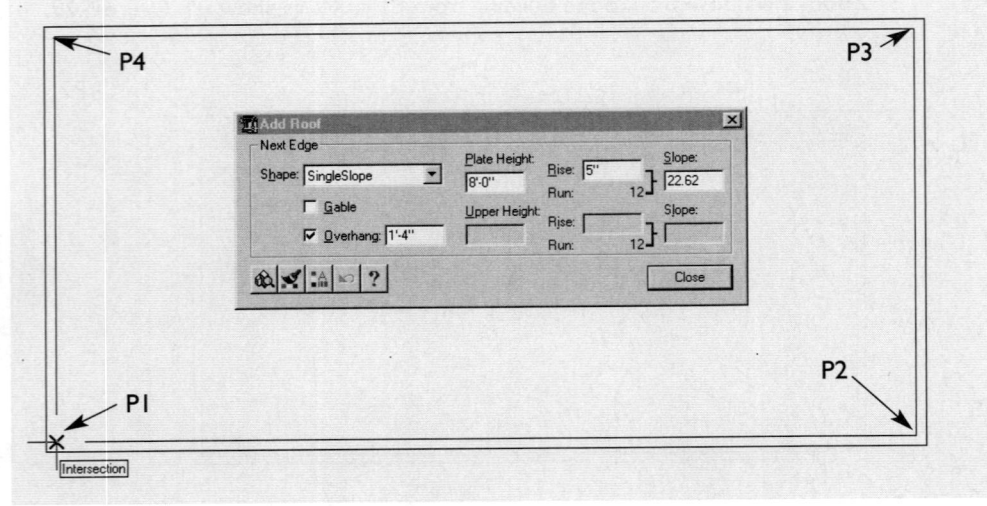

Figure 5.30 *Points for defining the roof planes*

```
Roof point or
[Shape/Gable/Overhang/overhangValue/PHeight/PRise/PSl
ope/UHeight/URise/USlope/Match]: DBOX
```
(Edit the Add Roof dialog box: select Gable, and set the overhang to 6".)

(Select the drawing area to activate the drawing area.)

```
Roof point or
[Shape/Gable/Overhang/overhangValue/PHeight/PRise/PSl
ope/UHeight/URise/USlope/Match]:
```
(Select the intersection of the inner lines of the corner at point P3 in Figure 5.30.)

(Verify that Gable check box is selected and overhang is set to 6".)

```
Roof point or
[Shape/Gable/Overhang/overhangValue/PHeight/PRise/PSl
ope/UHeight/URise/USlope/Match]:
```
(Select the intersection of the inner lines at the corner at point P4 in Figure 5.30.)

```
Roof point or
[Shape/Gable/Overhang/overhangValue/PHeight/PRise/PSl
ope/UHeight/URise/USlope/Match]: ENTER
```
(Pressing ENTER closes the roof to P1 and ends the command.)

5. Select **SW Isometric** from the **View** flyout to view the shed roof as shown in Figure 5.31.

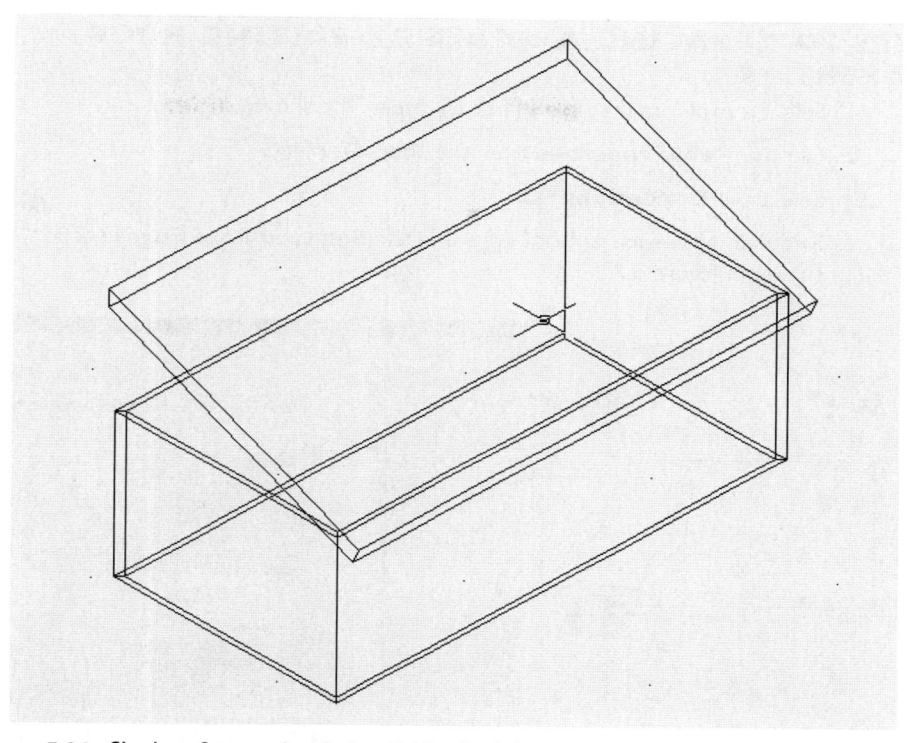

Figure 5.31 Shed roof created with the Gable check box

6. Save the drawing.

EDITING AN EXISTING ROOF TO CREATE GABLES

It is often easier to create roof gables by developing a roof for the entire structure and then editing selected roof planes to create the gable. A roof for an entire structure can be developed with a single slope. The single slope roof creates a hip roof. The **Roof Properties** command can then be used to edit each plane of roof. The **Roof Properties** command opens the **Roof Propertie**s dialog box, which lists each edge of the roof. The slope and overhang of each edge can then be edited. The roof edges are listed in the **Roof Properties** dialog box in the same order that the edges were created when the roof was developed.

In Figure 5.32, roof edge 3 can be edited to change it to a gable end by editing the slope value shown in the **Roof Faces (By Edge)** portion of the **Dimensions** tab. When the slope is changed, the adjoining roof planes extend or adjust to the different slope. The steps to creating a gable roof using the **Roof Properties** command follow.

STEPS TO CREATING A GABLE BY EDITING ROOF PROPERTIES

1. Select a roof and then right-click to display the shortcut menu.
2. Choose **Roof Properties** from the shortcut menu.
3. Select the **Dimensions** tab.
4. Select an edge such as Roof Edge 3 in the **Selected Roof Edges** section as shown in Figure 5.32.

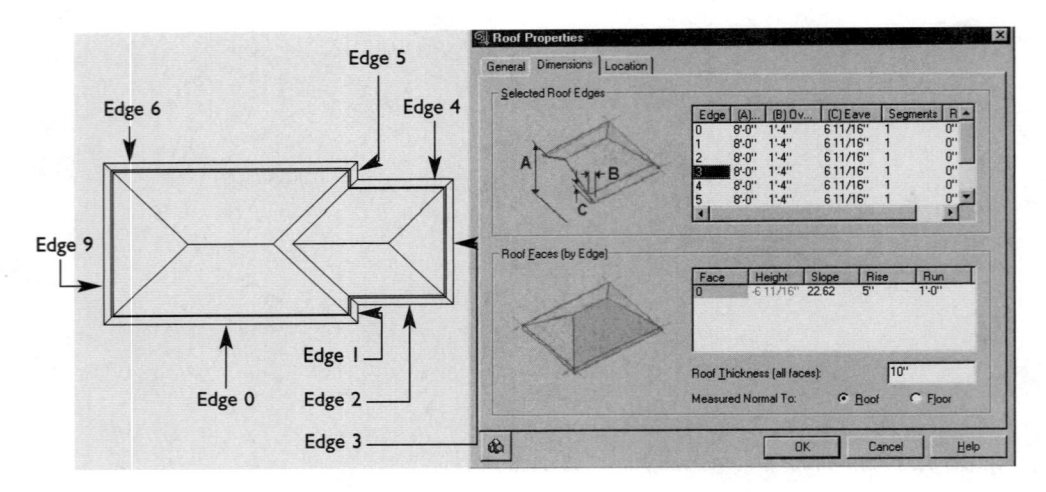

Figure 5.32 *Properties of the roof edges of a roof*

5. Edit the **Slope** value from 22.62 to 90 in the **Roof Faces (by Edge)** section as shown in Figure 5.33.

The roof plane is edited to form a gable as shown in Figure 5.34.

USING EDIT ROOF EDGES/FACES TO EDIT ROOF PLANES

Each plane of a roof can be edited with the **Roof Edit Edges/Faces** command (**ROOFEDITEDGES**). This command will allow you to select a plane of the roof and edit its properties. See Table 5.4 for command access.

When the **Edit Roof Edges/Faces** command (**ROOFEDITEDGES**) is executed, you are prompted to select an edge as shown below.

```
_AecRoofEditEdges
Select a roofedge:
```
(Select a roof edge and press ENTER *)*

Selecting an edge opens the **Edit Roof Edges** dialog box, as shown in Figure 5.36.

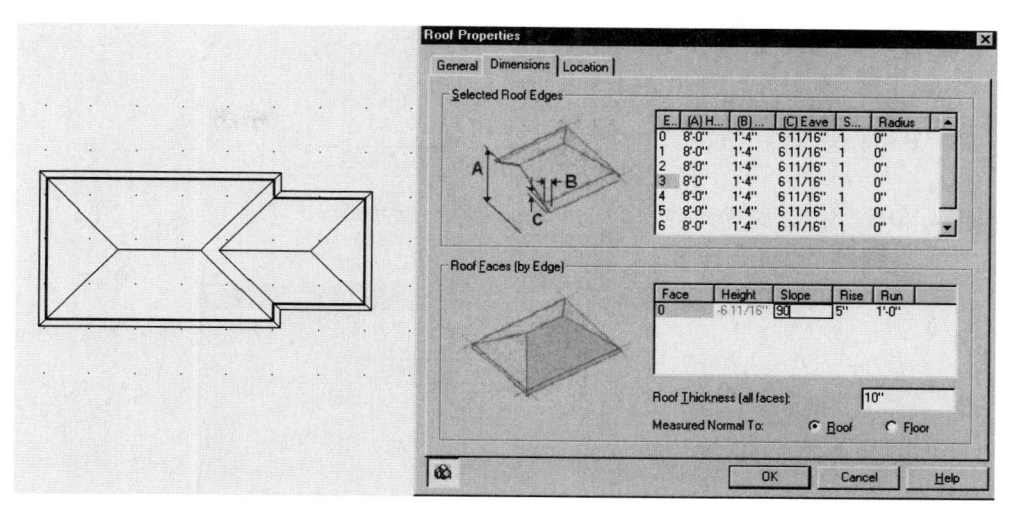

Figure 5.33 *Editing the slope of a roof edge*

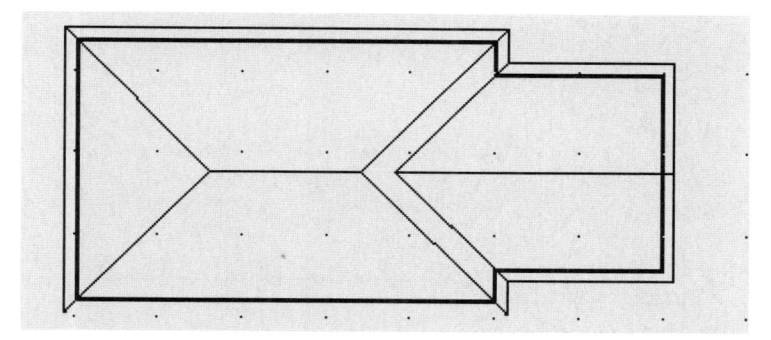

Figure 5.34 *Selected roof edge converted to a gable using Roof Properties*

Menu bar	Design>Roofs>Edit Roof Edges/Faces
Command prompt	ROOFEDITEDGES
Toolbar	Select Edit Roof Edges/Faces from the Roofs toolbar shown in Figure 5.35
Shortcut	Right-click over drawing area; choose Design>Roofs>Edit Edges

Table 5.4 *Edit Roof Edges/Faces access*

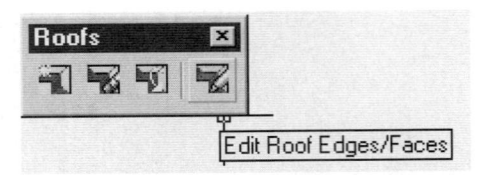

Figure 5.35 *Edit Roof Edges/Faces command on the Roofs toolbar*

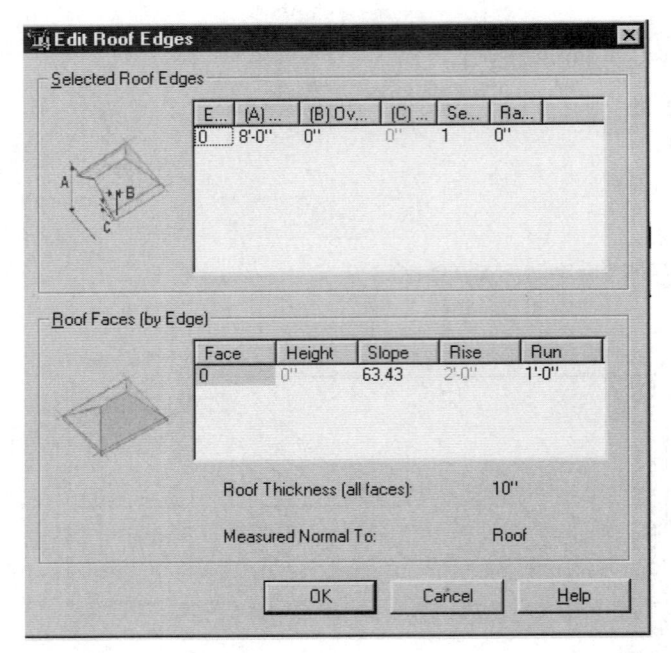

Figure 5.36 *Edit Roof Edges dialog box.*

The **Edit Roof Edges** dialog box is similar to the **Dimensions** tab of the **Roof Properties** dialog box. It allows you to edit the roof slope, overhang, and height. In contrast to the **Roof Properties** command, which displays properties for all roof edges, this dialog box displays only information regarding the plane selected. Therefore, if you select only one plane, the **Edit Roof Edges** dialog box will display information regarding only that one plane.

USING GRIPS TO EDIT A ROOF

The roof planes that are created have grips on each edge of the roof plane. Selecting a roof will display the grips. Grips are located on the midpoint and endpoints of the roof edge at the point of intersection between the roof and the wall at plate height. Grips are also located at the top plane of the roof where each plane intersects with the adjoining roof plane. Each grip can be selected and stretched to a new position. The grips of the ridge can be stretched to create a gable roof. Editing the grips on the ridge of a roof is a quick method of converting a hip plane to a gable.

STEPS TO CREATING A GABLE USING GRIPS

1. Select a roof to display its grips.
2. Select the ridge grip as shown in Figure 5.37.

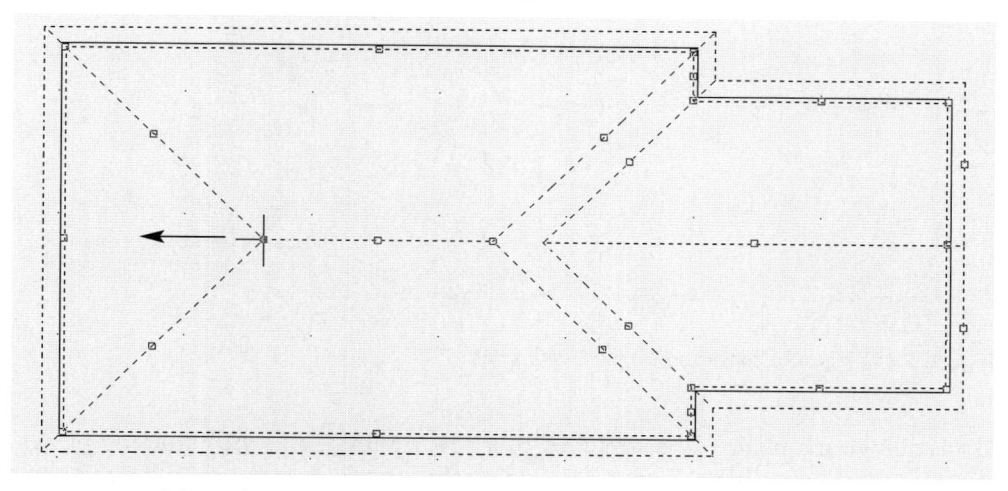

Figure 5.37 *Editing the grip on the ridge of the roof*

3. Select the hot grip and stretch it to the left; specify the new location by selecting a point to the left near P1 as shown in Figure 5.38.

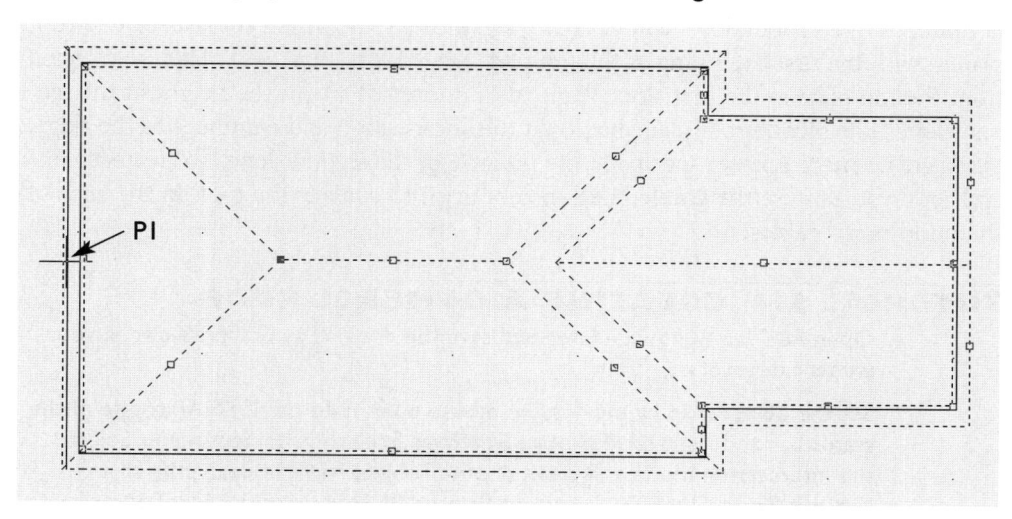

Figure 5.38 *Stretching the hot grip on the ridge*

4. Press ESC twice to remove the display of grips. The hip roof plane is converted to a gable as shown in Figure 5.39.

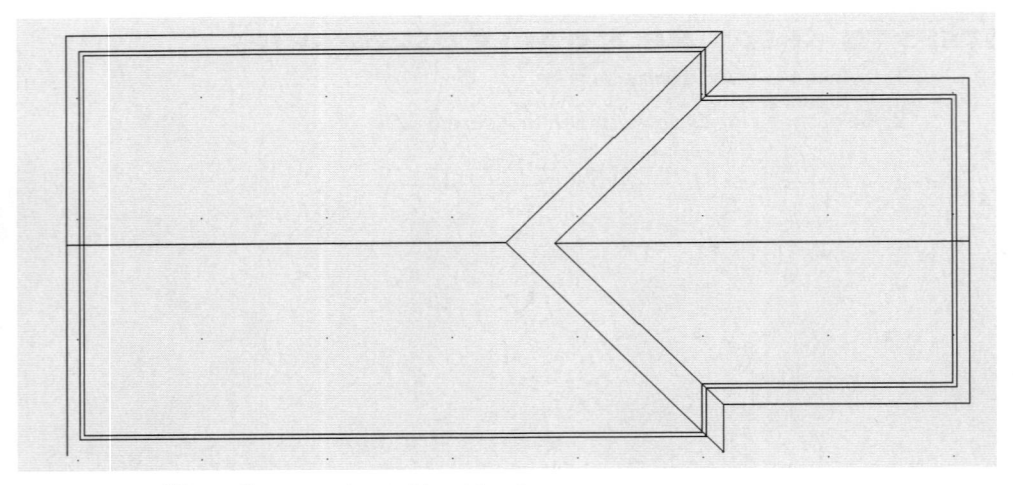

Figure 5.39 *Hip roof converted to gable with grips*

Grips allow each plane to be stretched to a new position, and the adjoining planes adjust to the new position. Each ridge grip can be stretched to any point near or outside of the building to convert the hip roof to a gable.

CREATING A GAMBREL ROOF USING DOUBLE SLOPED ROOF

A double sloped roof allows you to create a gambrel roof, which consists of two roof planes, with the first beginning at the top plate height. This first roof plane is projected from the top plate at the first slope angle until it intersects with the height of the second slope. The second roof plane begins at this intersection and continues to the ridge. Walls are not necessary for specifying the position of the second slope. To create a gambrel roof, you can use the **Gable** check box or grips to convert the roof on the ends of the building to gables.

TUTORIAL 5.4 CREATING A GAMBREL ROOF

1. Open *ADT Tutor\Ch5\Ex5-4.dwg* and save the drawing as *Gambrel4.dwg* in your student directory.

2. Set the object snap by moving the mouse pointer to the OSNAP toggle of the status bar and right-click; choose **Settings** from the shortcut menu. Select the Intersection mode and clear all other object snap modes. Verify that **Object Snap On(F3)** is checked. Select **OK** to dismiss the **Drafting Settings** dialog box.

3. Select the **Add Roof** command and respond to the command line prompts specifying roof points, as shown in the following command sequence:

   ```
   Roof point or
   [Shape/Gable/Overhang/overhangValue/PHeight/PRise/PSl
   ope/UHeight/URise/USlope/Match]: DBOX
   ```
   (Edit the Add Roof dialog box: set the Shape to DoubleSlope, Plate Height to 8', Rise to 24" for the first slope, Upper Height to 16', Upper Rise to 5 and Overhang to 8" as shown in Figure 5.40.)

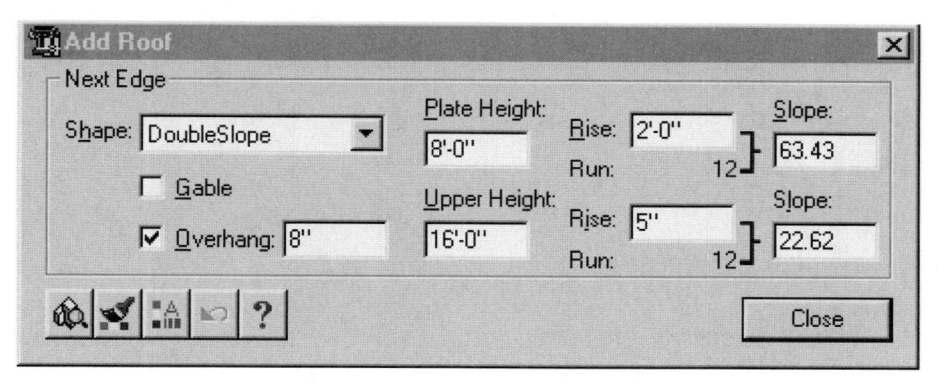

Figure 5.40 *Add Roof dialog box for creating the first edge of a gambrel roof*

(Select the drawing area to activate the drawing area.)

```
Roof point or
[Shape/Gable/Overhang/overhangValue/PHeight/PRise/PSl
ope/UHeight/URise/USlope/Ma
tch]:
```
(Select the intersection of the inner lines of the corner at point P1 in Figure 5.41.)

```
Roof point or
[Shape/Gable/Overhang/overhangValue/PHeight/PRise/PSl
ope/UHeight/URise/USlope/Match]:
```
(Select the intersection of the inner lines of the corner at point P2 in Figure 5.41.)

```
Roof point or
[Shape/Gable/Overhang/overhangValue/PHeight/PRise/PSl
ope/UHeight/URise/USlope/Match]:
```
(Select the intersection of the inner lines of the corner at point P3 in Figure 5.41.)
```
Roof point or
[Shape/Gable/Overhang/overhangValue/PHeight/PRise/PSl
ope/UHeight/URise/USlope/Match]:
```
(Select the intersection of the inner lines of the corner at point P4 in Figure 5.41.)

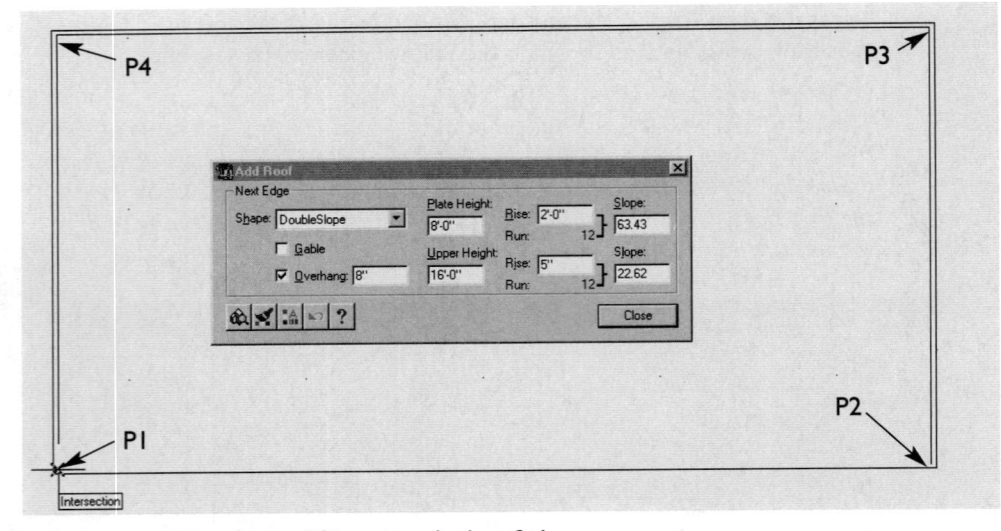

Figure 5.41 *Points for specifying a gambrel roof plane*

```
Roof point or
[Shape/Gable/Overhang/overhangValue/PHeight/PRise/PSl
ope/UHeight/URise/USlope/Match]: ENTER
```
(Pressing ENTER *closes the roof to P1 and ends the command.)*

4. Create a gable using grips by selecting all roof planes and selecting the ridge grip as shown in Figure 5.42.

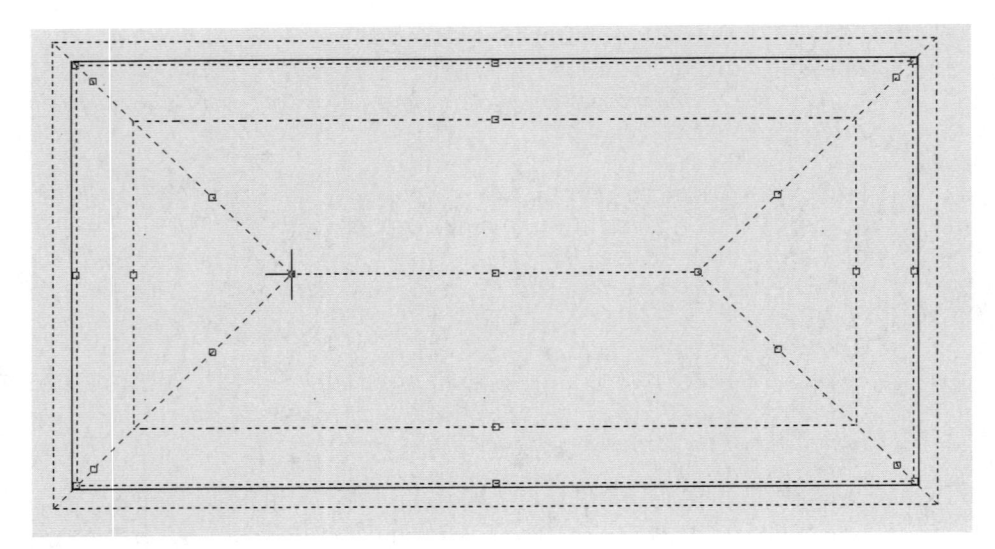

Figure 5.42 *Selecting grip at ridge*

5. Stretch the hot grip to the left as shown in Figure 5.43.

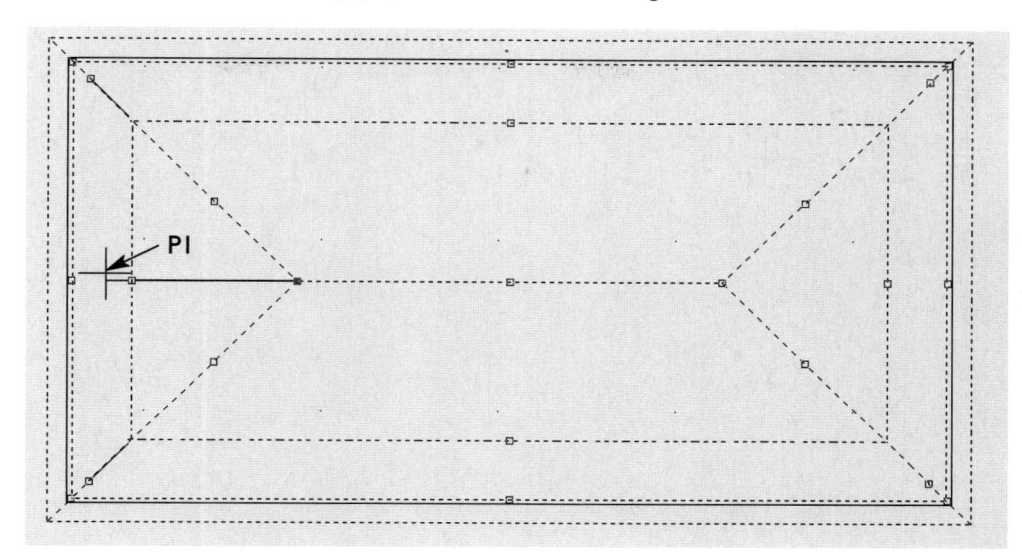

Figure 5.43 *Stretching ridge grip to the left*

6. Left-click on the ridge of the second slope to make it hot, as shown in Figure 5.44.

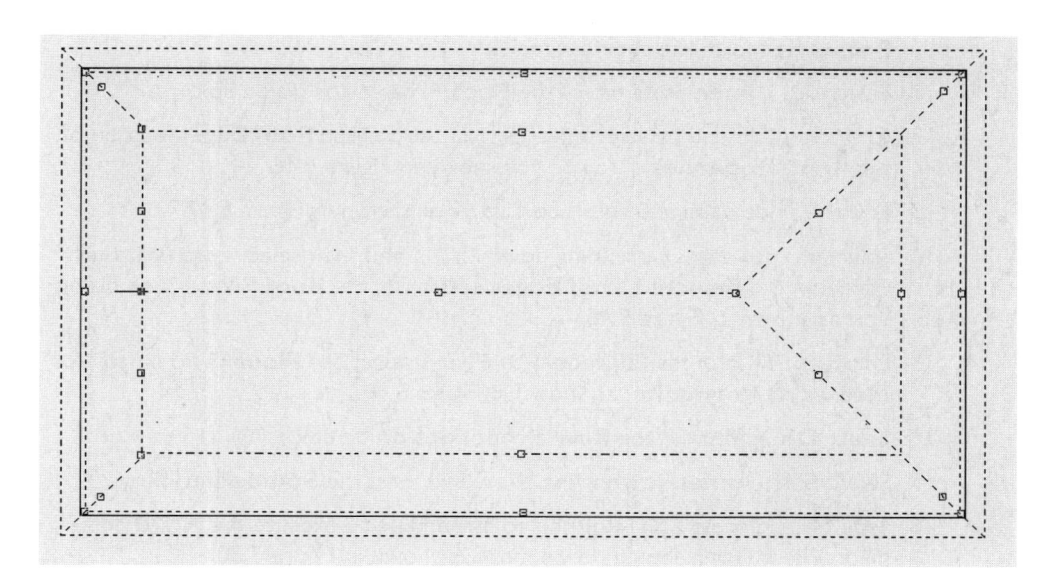

Figure 5.44 *Selecting ridge grip of second slope*

7. Stretch the hot grip to the left as shown in Figure 5.45.

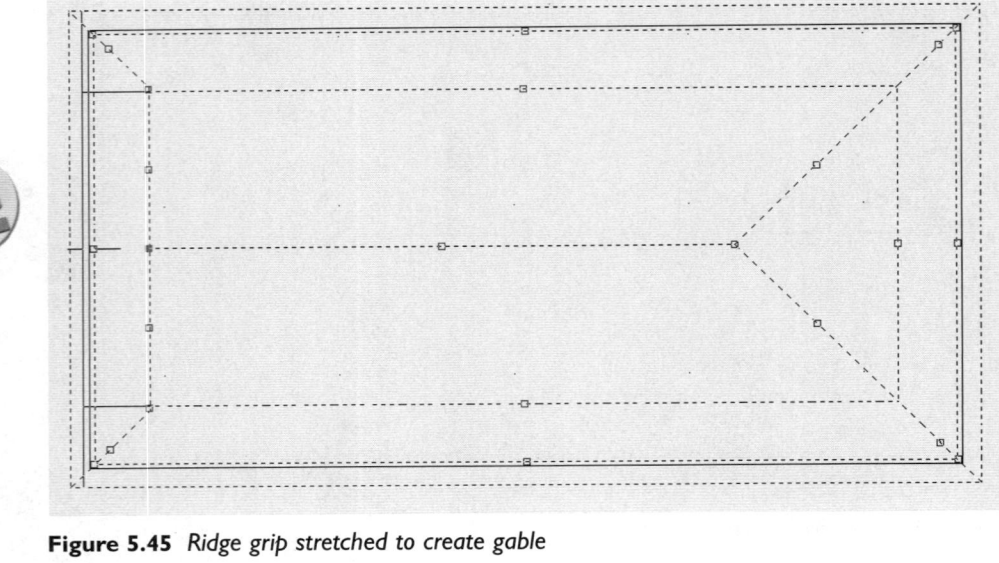

Figure 5.45 *Ridge grip stretched to create gable*

8. Press ESC to clear the selection of the roof.
9. Press ESC to clear the grips.
10. Select the roof by left-clicking on the roof; right-click and choose **Roof Properties** from the shortcut menu.
11. Select the **Dimensions** tab of the **Roof Properties** dialog box.
12. Create a gable end by selecting edge 1 in the **Select Roof Edges** section of the **Roof Properties** dialog box as shown in Figure 5.46.
13. Edit the slope of Face 0 and Face 1 to 90 as shown in Figure 5.47.
14. Select all roof edges by holding down SHIFT and then select Edge 0 through Edge 4 in the **Selected Roof Edges** section of the **Roof Properties** dialog box, as shown in Figure 5.48.
15. Edit **Roof Thickness (all faces)** to 6" and select the **Floor** radio button for **Measured Normal To:** as shown in Figure 5.49.
16. Select **OK** to dismiss the **Roof Properties** dialog box.
17. Select **SW Isometric** from the **View** flyout on the **Standard** toolbar.
18. Type **HI** and press ENTER in the command line to execute the **HIDE** command. The building should now appear as shown in Figure 5.50.
19. Save the drawing.

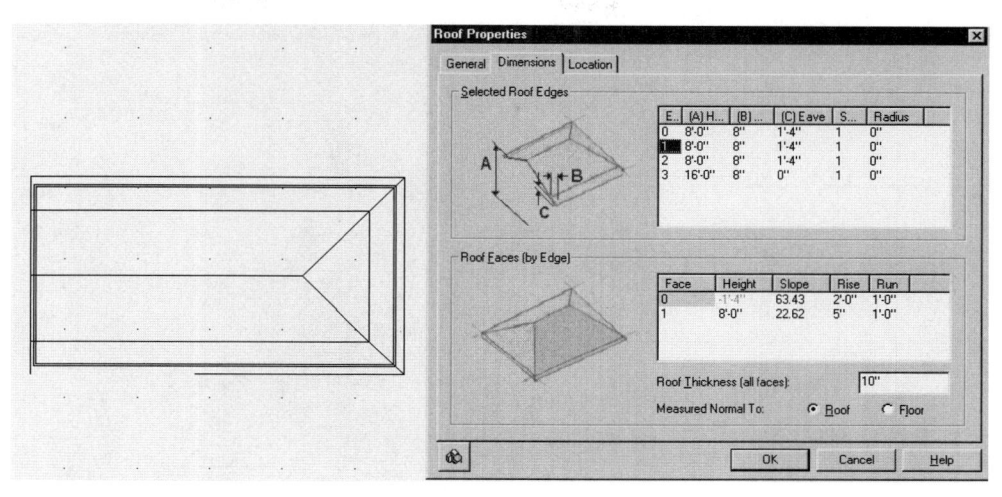

Figure 5.46 *Creating gables using Roof Properties*

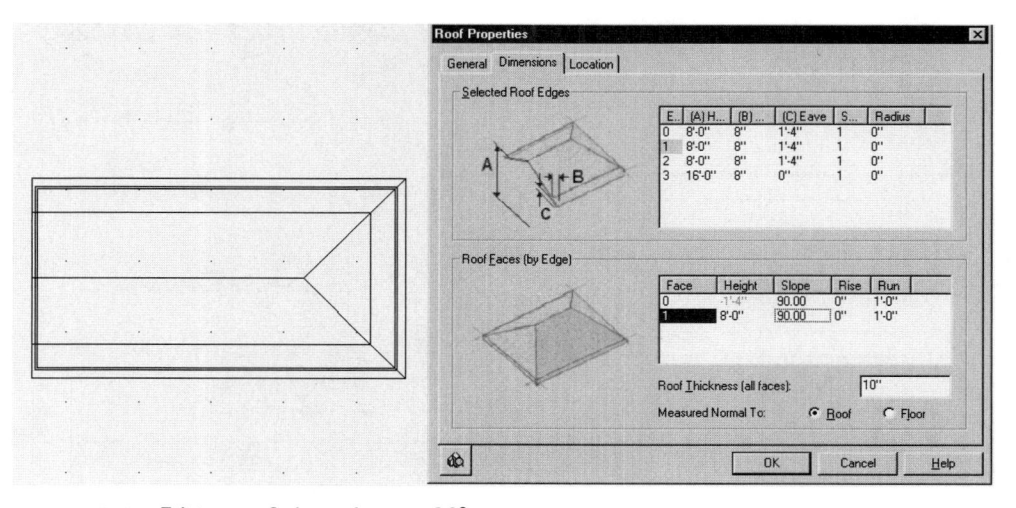

Figure 5.47 *Editing roof plane slope to 90°*

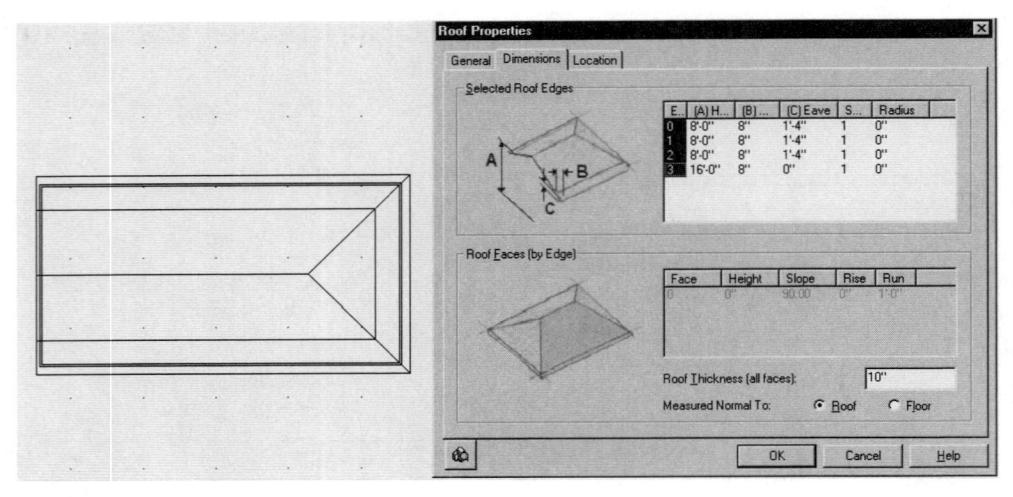

Figure 5.48 *Editing roof edges thickness*

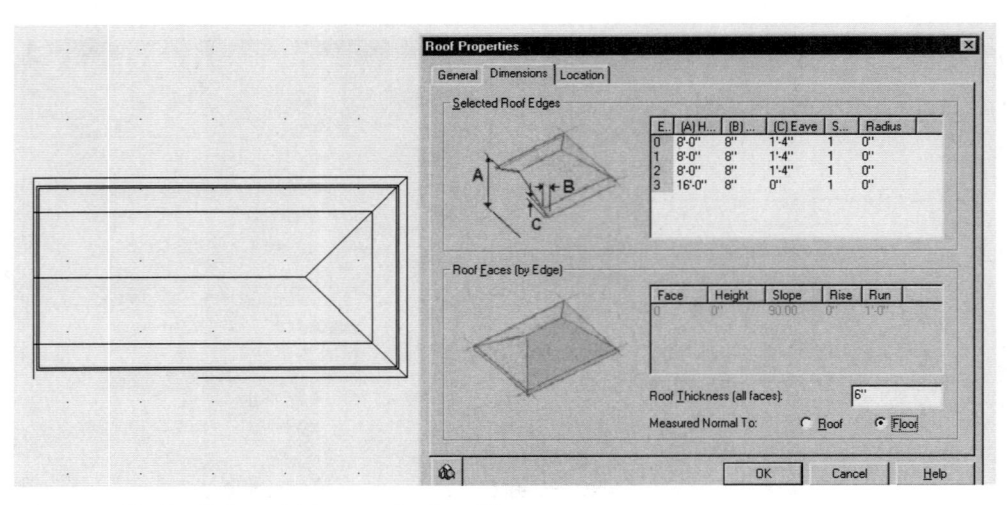

Figure 5.49 *Roof plane thickness edited to 6"*

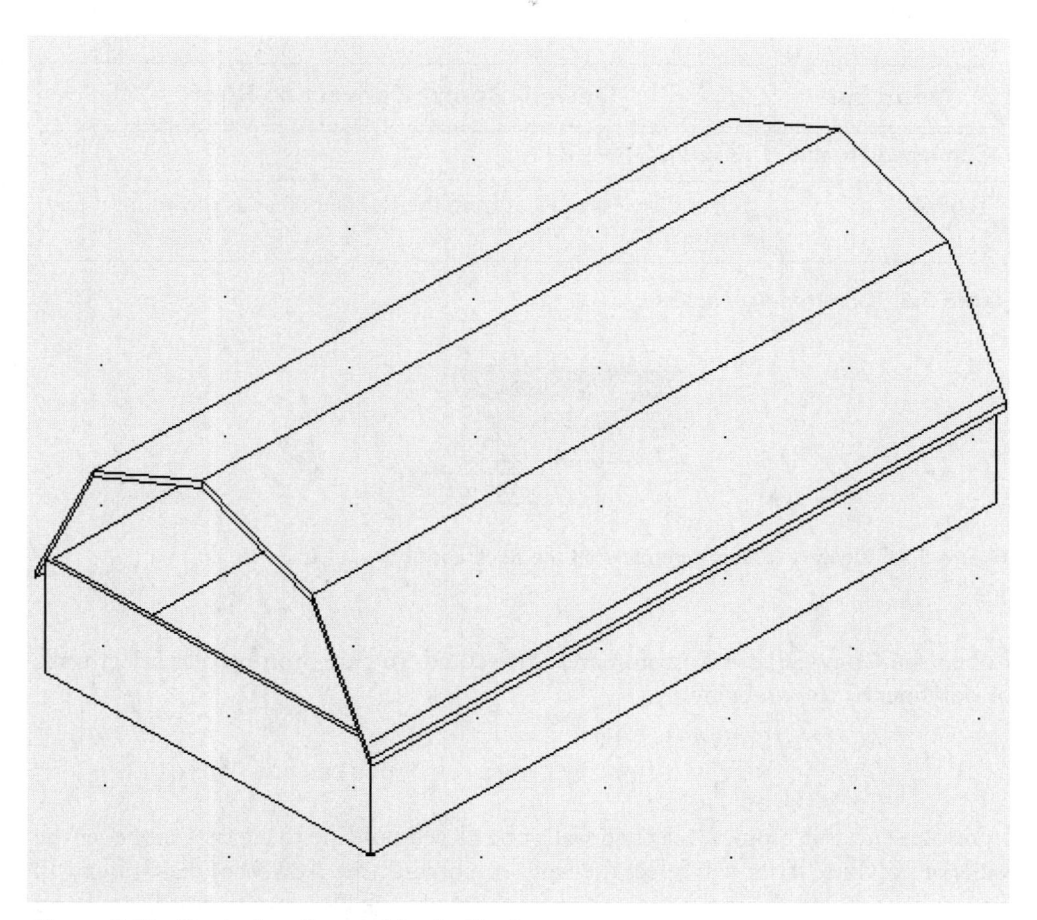

Figure 5.50 *Gambrel roof created by double slope*

CREATING A ROOF USING CONVERT TO ROOF

The **Convert to Roof** command (**ROOFCONVERT**) will create a roof for the selected walls without specifying each corner of the roof. **Convert to Roof** creates a roof from a closed series of walls or a closed polyline. The selected walls must be closed to convert the walls to a roof. The walls can be selected using a crossing or window selection method or they can be selected individually. See Table 5.5 for command access.

Menu bar	Design>Roofs>Convert to Roof
Command prompt	ROOFCONVERT
Toolbar	Select Convert to Roof from the Roofs toolbar as shown in Figure 5.51

Table 5.5 *Convert to Roof access*

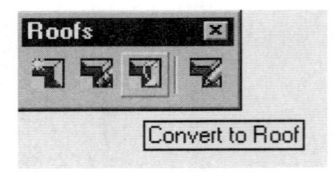

Figure 5.51 *Convert to Roof command on the Roofs toolbar*

When the **Convert to Roof** command is executed, you are prompted to select walls or polylines as shown below:

```
AecRoofConvert
Choose walls or polylines to create roof profile:
Select objects:
```

In response to this prompt, select all walls or a closed polyline to create a roof over the walls or polyline. After you select the walls or polyline, the **Add Roof** dialog box will open, allowing you to set the properties for the roof. The **Convert to Roof** command is a quick method of creating a roof over curved walls.

TUTORIAL 5.5 CREATING A ROOF USING ROOF CONVERT

1. Open *ADT Tutor\Ch5\Ex5-5.dwg*.
2. Choose **File>SaveAs** from the menu bar and save the file as *Convert5.dwg* in your student directory.

Creating Roofs   359

3. Select the **Convert to Roof** command from the **Roofs** toolbar and respond to the command line prompts as shown in the following command sequence.

```
Command: _AecRoofConvert
Choose walls or polylines to create roof profile:
Select objects: All ENTER
130 found
Select objects: ENTER
Erase layout geometry? [Yes/No] <Y>: N ENTER
Roof modify
[Shape/Match/overhangValue/Overhang/PHeight/PRise/PSl
ope/UHeight/URise/USlope]:
DBOX (Edit the Modify Roof dialog box as shown in Figure 5.52. Set
Overhang to 1'–4" and select OK to dismiss the Modify Roof dialog box.)
```

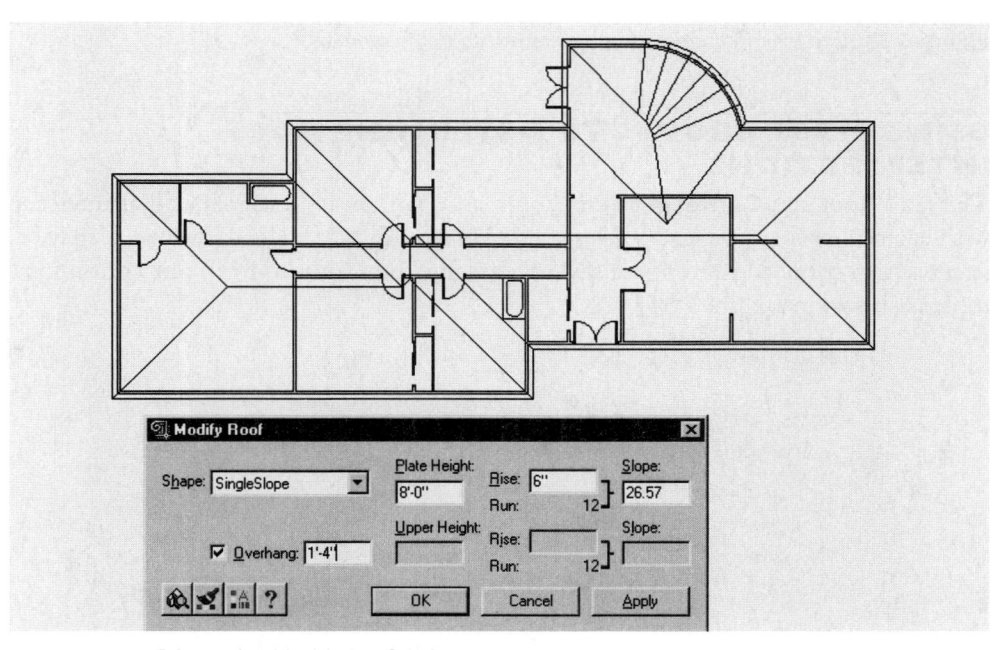

Figure 5.52 *Editing the Modify Roof dialog box*

The complex roof with a 1'–4" overhang is developed as shown in Figure 5.53.

4. Save the drawing.

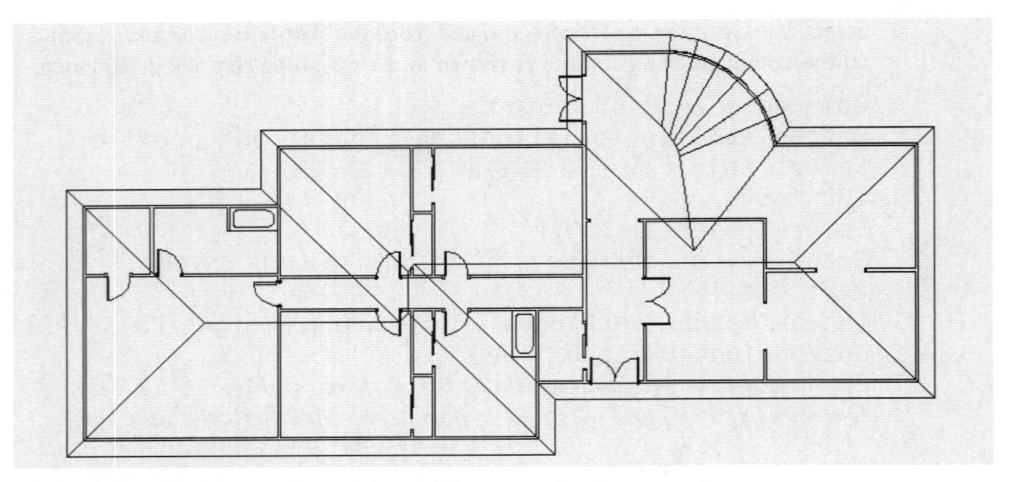

Figure 5.53 *Complex roof created using Convert to Roof command*

USING MASS GROUPS TO DETERMINE ROOF INTERSECTIONS

The **Add Roof** and **Convert to Roof** commands will create roof planes that intersect with each other for complex roofs created from continuous walls. However, it may be necessary to connect the roofs of several detached buildings to form one continuous roof, as shown in Figure 5.54.

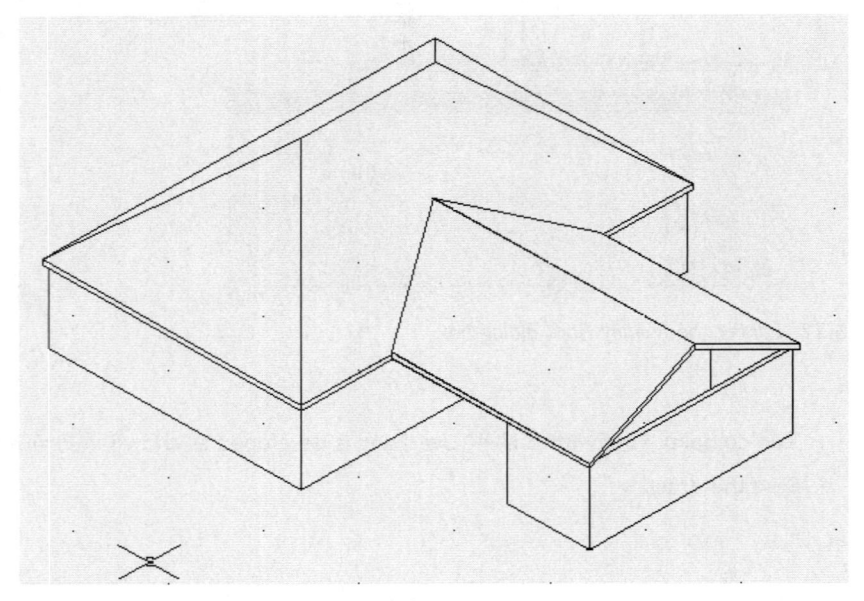

Figure 5.54 *Roof connecting to another building*

To accomplish this objective, create a roof for each building and then stretch the roof of one building to intersect with the other. When the two roof objects are stretched to intersect with each other, the valley intersections of the roofs are not displayed. To determine the valley intersections of the roof planes, create a mass group. The mass group allows the roof planes to be merged together, which defines the valley intersections of the roofs.

To create a mass group, use the **Add Mass Group** command (**MASSGROUPADD**). The purpose of this command is to group together two or more mass elements to form a mass model of a building. The development of mass models is discussed in detail in Chapter 11. However, this command can also be used to unite roof objects of the drawing. See Table 5.6 for command access.

Menu bar	Concept>Mass Grouping Tools>Add Mass Group
Command prompt	MASSGROUPADD
Toolbar	Select Add Mass Group from the Mass Tools toolbar shown in Figure 5.55
Shortcut	Right-click over the drawing area; choose Utilities>Massing>Mass Group>Add

Table 5.6 *Add Mass Group access*

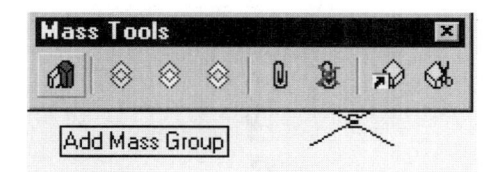

Figure 5.55 *Add Mass Group command on the Mass Tools toolbar*

When the **Add Mass Group** command is executed, you are prompted to set the location and rotation of a mass group marker, as shown in the following command sequence.

```
Command: _AecMassGroupAdd
Location:
Rotation angle <0.00>:
```

The location and rotation of the mass group marker can be at any position or rotation within the drawing. This marker is simply a graphical representation of the mass group. When you select the mass group marker, all elements attached to the mass group marker and included in the mass group are selected. The marker and all elements attached to the mass group are placed on the A_Mass_Grps layer. This layer has the color 130, which is clearly distinguished from the roof object, which has the color 20. Freezing the A_Mass_Grps layer will turn off the display of the mass group.

The second step in determining roof intersections is to attach two or more roof objects to the mass group. Attach roof objects to the mass group using the **Attach Elements** command (**MASSGROUPATTACH**). See Table 5.7 for command access.

Menu bar	Concept>Mass Group Tools>Attach Elements
Command prompt	MASSGROUPATTACH
Toolbar	Select Attach Elements from the Mass Tools toolbar shown in Figure 5.56
Shortcut	Right-click over the drawing area; choose Utilities>Massing>Mass Group>Attach Elements

Table 5.7 *Attach Elements access*

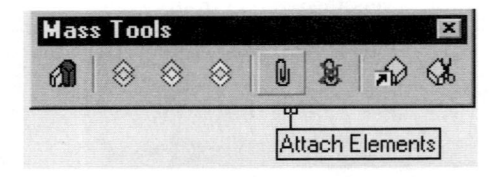

Figure 5.56 *Attach Elements command on the Mass Tools toolbar*

When you select the **Attach Elements** command (**MASSGROUPATTACH**), you are prompted to select a mass group and then select the elements to attach to the group, as shown in the following command sequence:

```
Command: _AecMassGroupAttach
Select a mass group :
Select elements to attach:
```

Select the mass group by selecting the mass group marker. After selecting the mass group marker, you are prompted to Select elements to attach. The roof planes of two or more buildings are selected as the elements. The roof objects selected for the group will be merged together through Boolean addition to become one object. The roof objects, when added together, form a mass group attached to the mass group marker. If the mass group marker is erased, the mass group is deleted and only the original roof objects are displayed. The original roof objects remain separate objects and can be edited. The properties of the roof can be edited at any time, and the graphics of the mass group will adjust to the changes.

TUTORIAL 5.6 USING MASS GROUPS TO IDENTIFY ROOF INTERSECTIONS

1. Open *ADT Tutor\Ch5\Ex5-6.dwg*.
2. Choose **File>SaveAs** from the menu bar and save the drawing as *Mgroup6.dwg* in your student directory.
3. Select **Top** from the **View** flyout on the **Standard** toolbar.
4. Select the roof of the smaller building to display its grips.
5. Select the top ridge grip to make it hot and stretch the ridge up 20' as shown in the following command sequence and in Figure 5.57.

```
Command:
** STRETCH **
Specify stretch point or [Base
point/Copy/Undo/eXit]: 20' ENTER
```
(Specify distance for direct distance entry.)

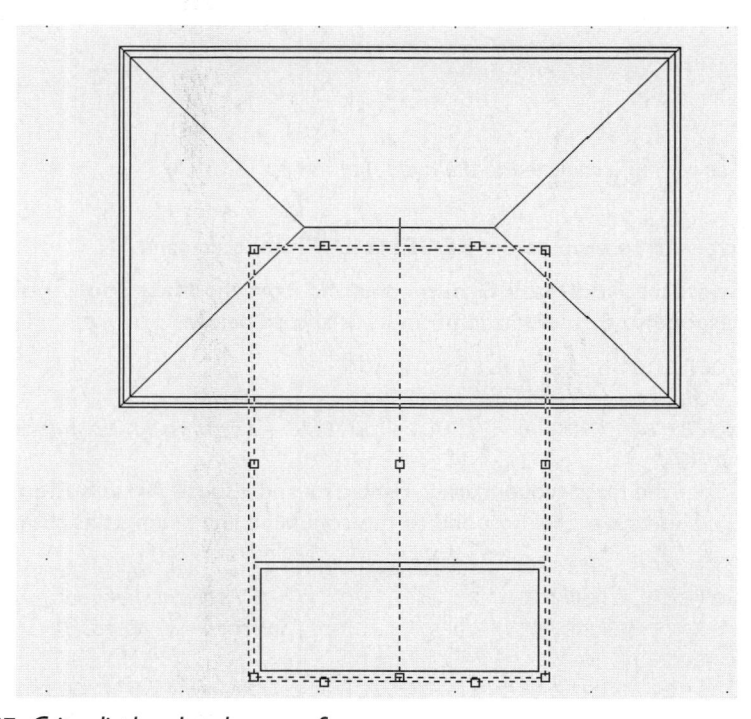

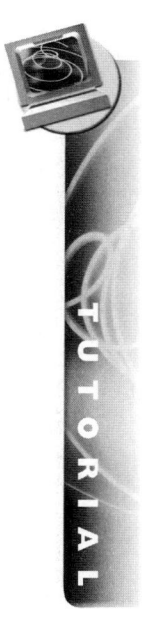

Figure 5.57 *Grips displayed on lower roof*

6. Press ESC twice to clear the grips of the roof.
7. Select **SW Isometric** from the **View** flyout on the **Standard** toolbar.
8. Type **HI** and press ENTER in the command line to execute the **HIDE** command, and the two roofs should appear as shown in Figure 5.58.

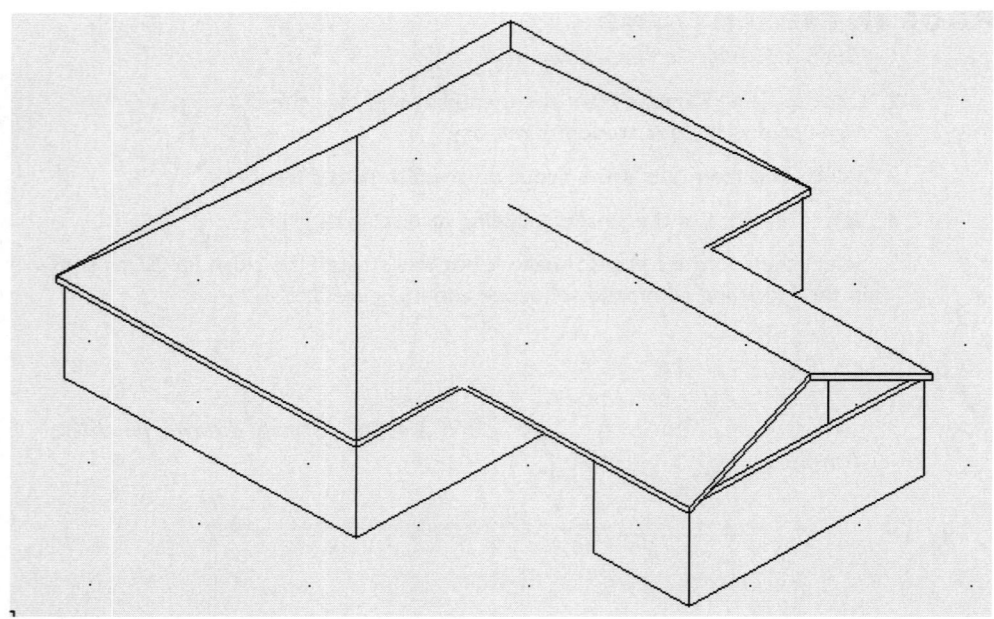

Figure 5.58 *Lower roof connected to the main roof*

9. Type **RE** to execute the **REGENERATION** command.
10. Select the **Add Mass Group** command from the **Mass Tools** toolbar and respond to the command prompts as shown below.

    ```
    Command: _AecMassGroupAdd
    Location: 20',30' ENTER
    Rotation angle <0.00>: ENTER
    ```
    (The mass group marker is placed.)
11. Select the mass group marker, right-click and choose **Attach Elements** from the shortcut menu. Respond to the command line prompts as shown below.

    ```
    Command: MassGroupAttach
    Select elements to attach:
    ```
    (Select the smaller roof.) 1 found
    ```
    Select elements to attach:
    ```
    (Select the larger roof.) 1 found, 2 total
    ```
    Select elements to attach: ENTER
    ```
    (Press ENTER *to end the selection.)*

```
Attaching elements...
2 element(s) attached
```
12. Type **HI** and press ENTER in the command line to execute the **HIDE** command. The intersection of the roofs should appear as shown in Figure 5.59.

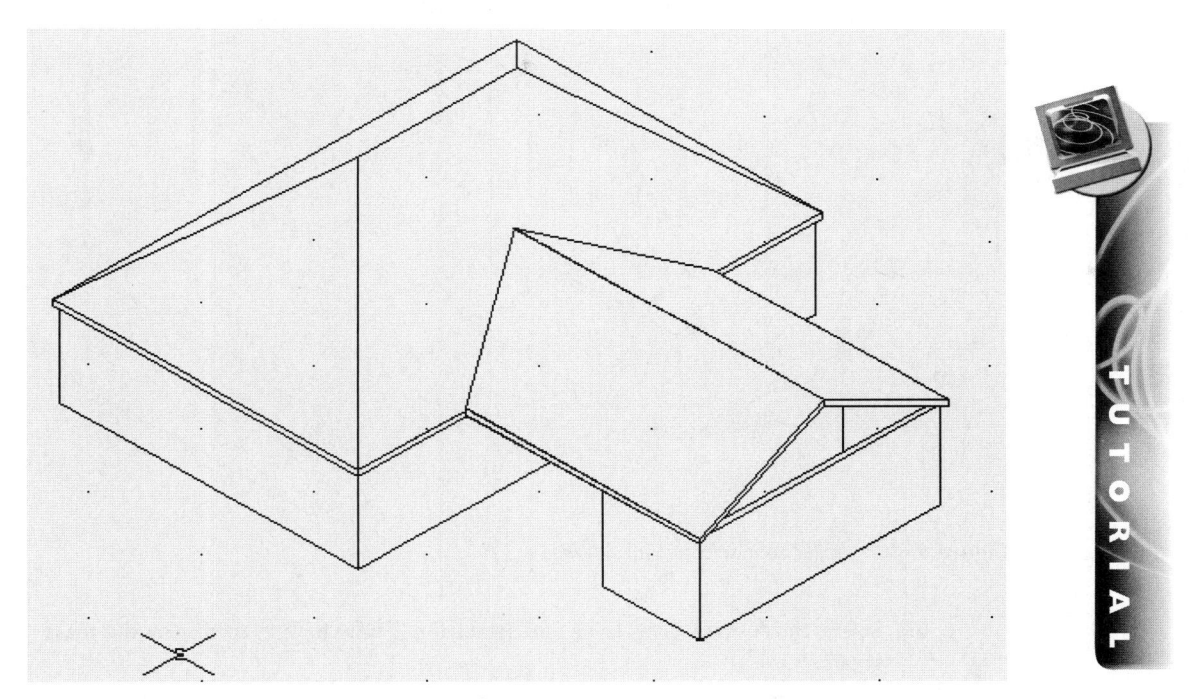

Figure 5.59 *Mass group displaying the roof intersection*

13. Select the **Layers** button on the **Object Properties** toolbar to display the **Layer Properties Manager** dialog box.
14. Select the A-Mass-Grps layer and select the **Freeze** button to freeze the layer.
15. Select **OK** to dismiss the **Layer Properties Manager**.
16. Left-click on the smaller front roof as shown in Figure 5.60.
17. Right-click to display the shortcut menu and choose **Roof Modify**.
18. Edit the **Rise** to 8 in the **Modify Roof** dialog box and then select **OK** to dismiss the dialog box.
19. Select the **Layers** button on the **Object Properties** toolbar to display the **Layer Properties Manager** dialog box.

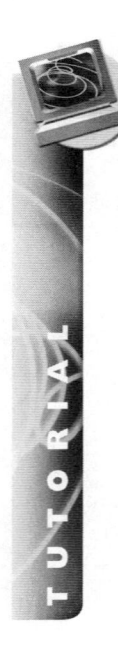

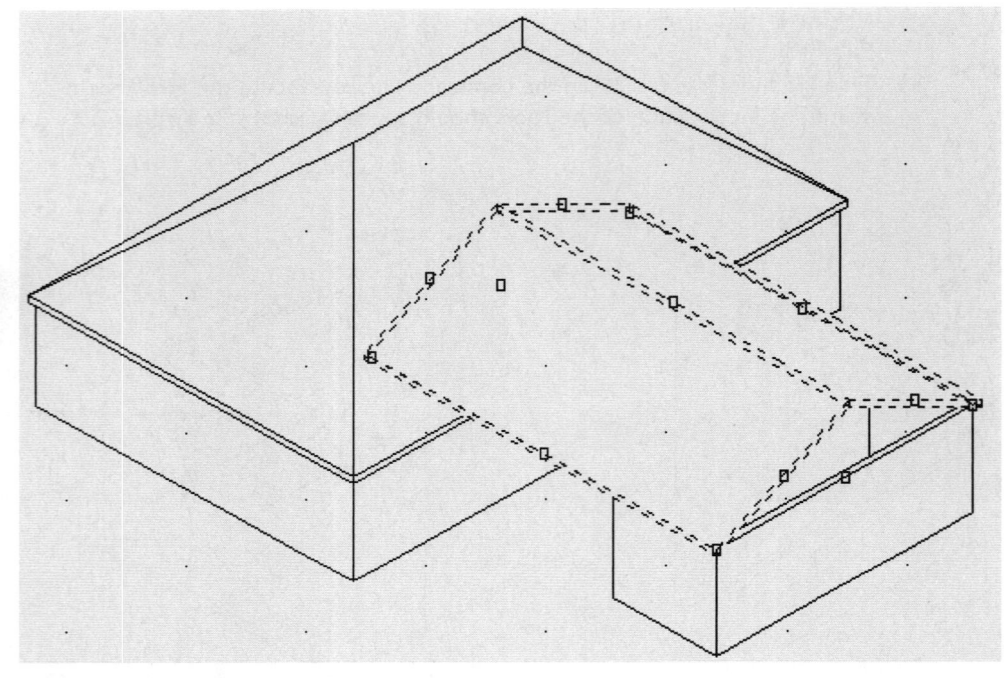

Figure 5.60 *Smaller roof selected for editing*

20. Select the A-Mass-Grps layer and select the **Thaw** button to display the mass group.
21. Select **OK** to dismiss the **Layer Properties Manager**.
22. Type **HI** and press ENTER in the command line to execute the **HIDE** command. The revised roof intersection should appear as shown in Figure 5.61.
23. Save the drawing.

CREATING DORMERS

Dormers are created as projections from a roof for the placement of windows in the second floor of a Cape Code style house. The walls for the dormers and other walls for the second floor are drawn first. The second floor is designed based upon the layout of the first floor. The first floor could be attached as a reference file for the development of the second floor. The perimeter of the first floor should be outlined on the second floor drawing to orient the second floor to the first floor. The walls of the second floor and a polyline outlining the walls of the first floor are shown in Figure 5.62.

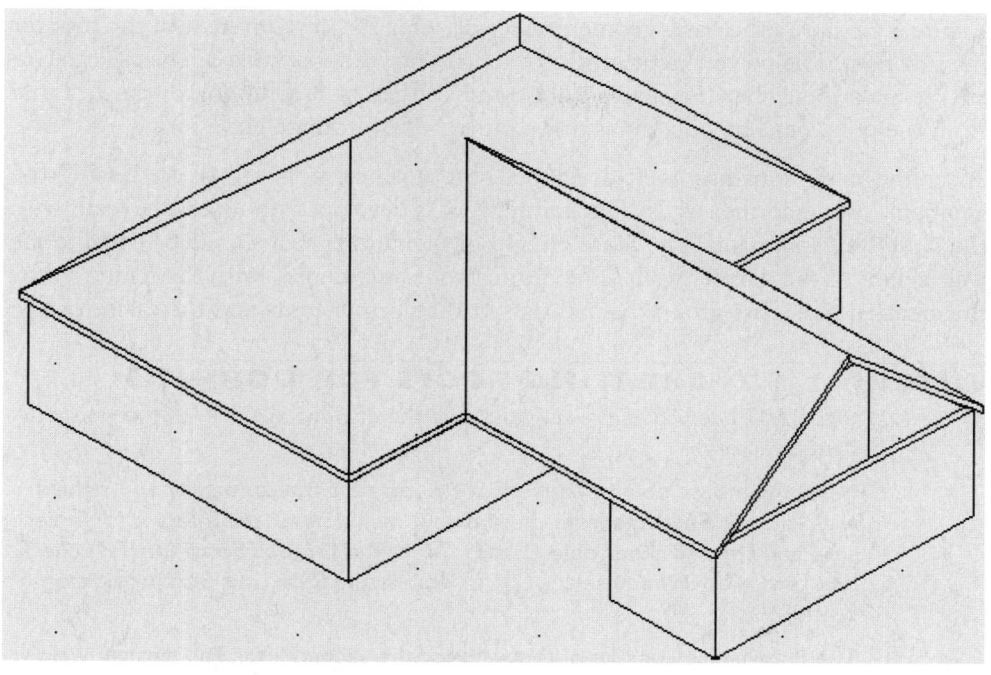

Figure 5.61 Mass group displayed for small roof after editing

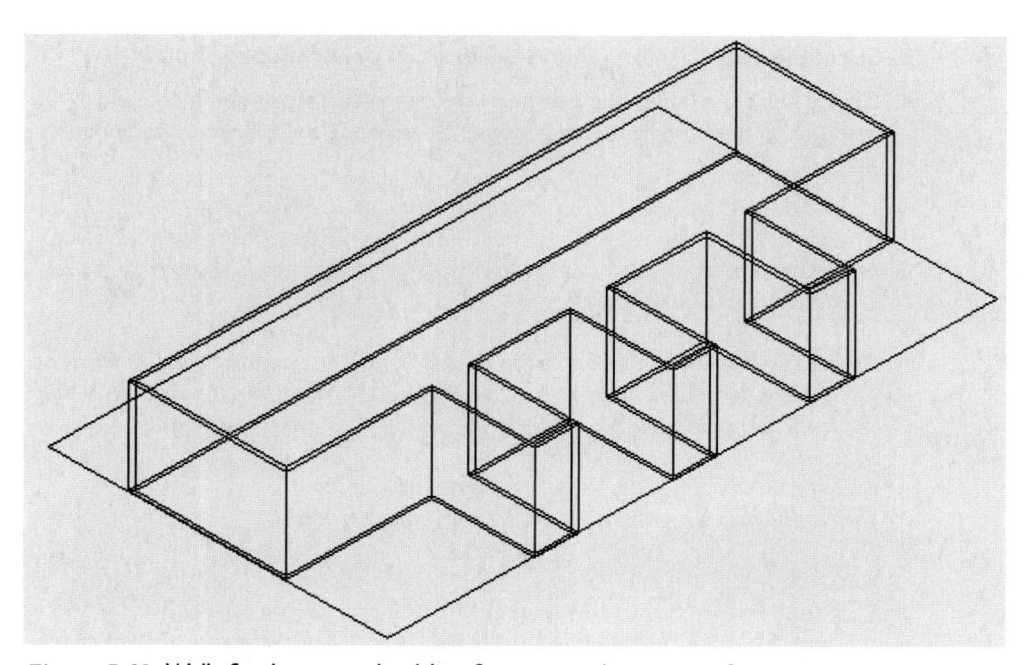

Figure 5.62 Walls for dormer and polyline for creating the main roof

Figure 5.62 shows the basic geometry necessary for the development of the roof for the first floor. The polyline representing the first floor exterior walls can be used to create the main roof. The **Convert to Roof** command can be used to quickly create a roof from the polyline. The polyline is drawn at a Z=0 coordinate elevation.

After the main roof has been developed, the dormer roofs are created using the geometry of the dormer walls. The dormer roof is developed on top of the walls with the **Add Roof** command. The **Plate Height** of the dormer roof should be equal to the wall height of the dormer walls. The dormer roofs are merged with the main roof by the creation of a mass group that consists of the dormer roofs and the main roof.

TUTORIAL 5.7 CREATING ROOFS FOR DORMERS

1. Open *ADT Tutor\Ch5\Ex5-7.dwg* and save the drawing as *Cape7.dwg* in your student directory.

2. Move the mouse pointer to the OSNAP toggle of the status bar and right-click; choose **Settings** from the shortcut menu. Select the Intersection object snap and clear all other object snaps. Select the **Object Snap On(F3)** check box to turn it ON and select **OK** to dismiss the **Drafting Settings** dialog box.

   The drawing file consists of the walls for the second floor. The existing walls shown are 8' high. A polyline is included in the drawing, which is a trace of the first floor outer wall line.

3. Select the **Top** view from the **View** flyout on the **Standard** Toolbar.

4. Create a gable roof for the dormer; select the **Add Roof** command and respond to the command line prompts by selecting the following wall points.

   ```
   Roof point or
   [Shape/Gable/Overhang/overhangValue/PHeight/PRise/PSl
   ope/UHeight/URise/USlope/Match]: DBOX
   ```
   (Edit the Add Roof dialog box setting: Shape to SingleSlope, Plate Height to 8', Rise to 6", select Gable, deselect Overhang.)

   (Select the Properties button of the Add Roof dialog box; edit the Dimensions tab, setting Roof Thickness to 6", and toggle ON Floor for Measured Normal To as shown in Figure 5.63.)

   (Select OK to dismiss the Roof Properties dialog box.)
   (Select the drawing area to activate the drawing area.)

   ```
   Roof point or
   [Shape/Gable/Overhang/overhangValue/PHeight/PRise/PSl
   ope/UHeight/URise/USlope/Match]:
   ```
   (Select the intersection of the outer lines of the corner at point P1 in Figure 5.64.)

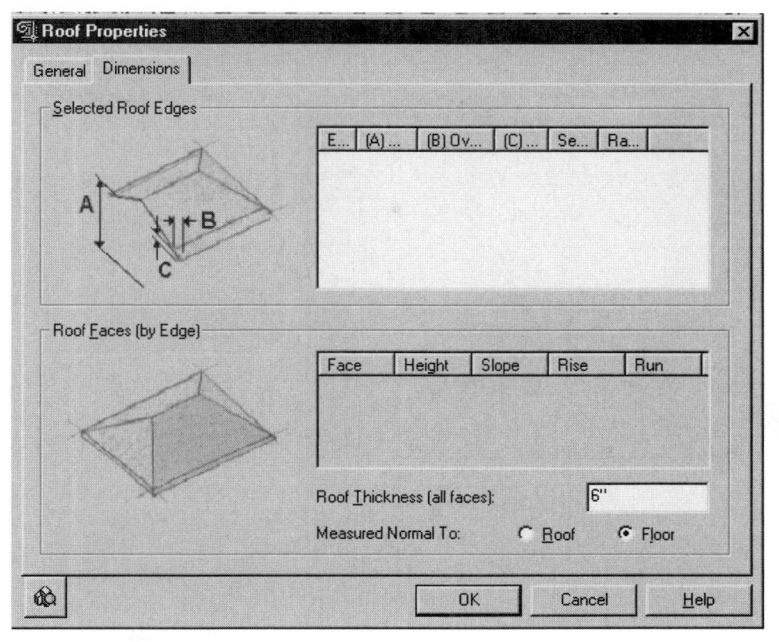

Figure 5.63 Roof Properties dialog box edited

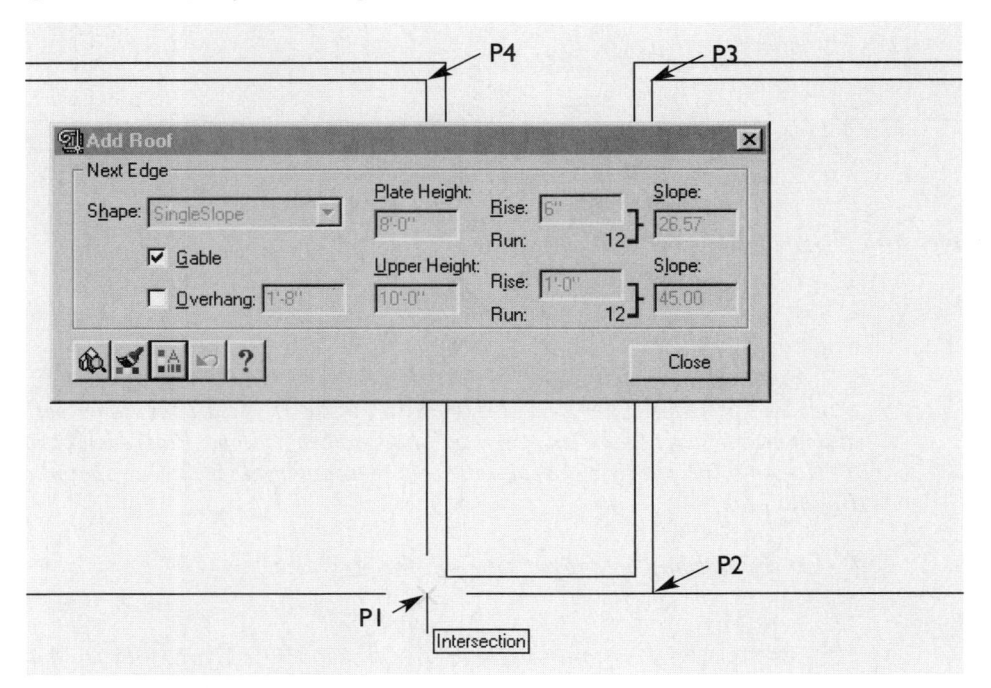

Figure 5.64 First point to define the roof

```
Roof point or
[Shape/Gable/Overhang/overhangValue/PHeight/PRise/PSl
ope/UHeight/URise/USlope/Match]:
```
(Select the intersection of the outer lines of the corner at point P2 in Figure 5.65.)

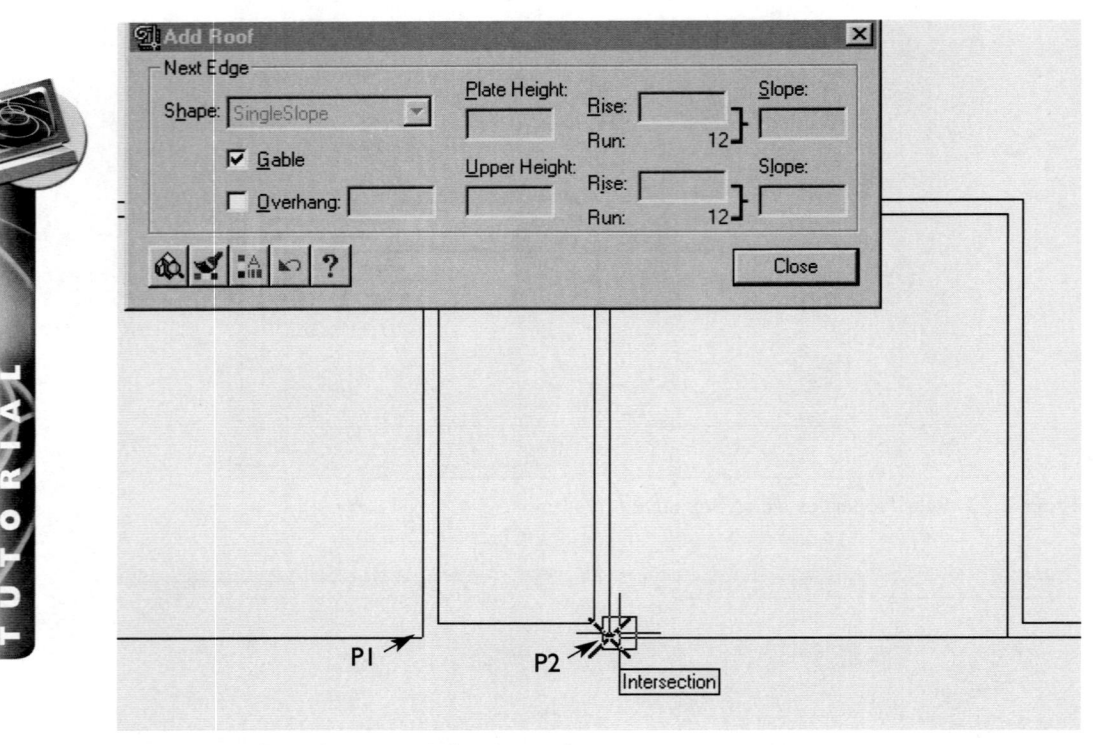

Figure 5.65 *Second point to define the roof*

```
Roof point or
[Shape/Gable/Overhang/overhangValue/PHeight/PRise/PSl
ope/UHeight/URise/USlope/Match]: DBOX
```
(Edit the Add Roof dialog box: deselect Gable and then set Shape to SingleSlope, Plate Height to 8', Rise to 6" for the first slope, and select Overhang and set to 4" as shown in Figure 5.66.)

(Select the drawing area to activate the drawing area.)

```
Roof point or
[Shape/Gable/Overhang/overhangValue/PHeight/PRise/PSl
ope/UHeight/URise/USlope/Match]:
```
(Select the intersection of the wall lines of the corner at point P3 in Figure 5.66.)

Figure 5.66 *Third point to define the roof*

```
Roof point or
[Shape/Gable/Overhang/overhangValue/PHeight/PRise/PSl
ope/UHeight/URise/USlope/Match]: DBOX
```
(Edit the Add Roof dialog box: select Gable and then set the Shape to SingleSlope, Plate Height to 8', Rise to 6", deselect Overhang as shown in Figure 5.67.)

(Select the drawing area to activate the drawing area.)

```
Roof point or
[Shape/Gable/Overhang/overhangValue/PHeight/PRise/PSl
ope/UHeight/URise/USlope/Match]:
```
(Select the intersection of the wall lines of the corner at point P4 in Figure 5.67.)

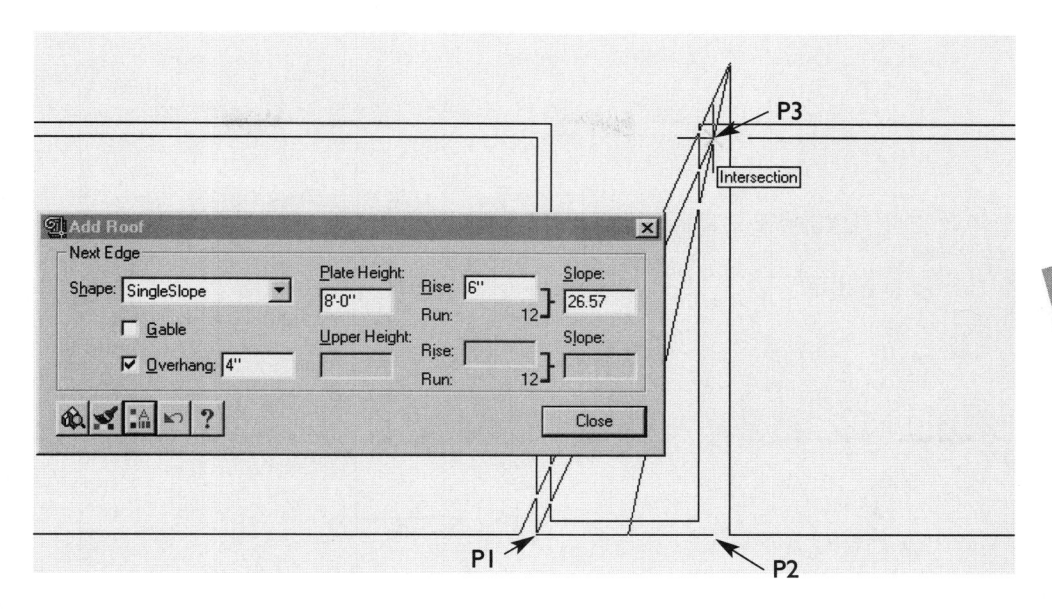

```
Roof point or
[Shape/Gable/Overhang/overhangValue/PHeight/PRise/PSl
ope/UHeight/URise/USlope/Match]: DBOX
```
(Edit the Add Roof dialog box: deselect Gable and then set Shape to SingleSlope, Plate Height to 8', Rise to 6" for the first slope, and select Overhang and set to 4" as shown in Figure 5.68.)

(Select the drawing area to activate the drawing area.)

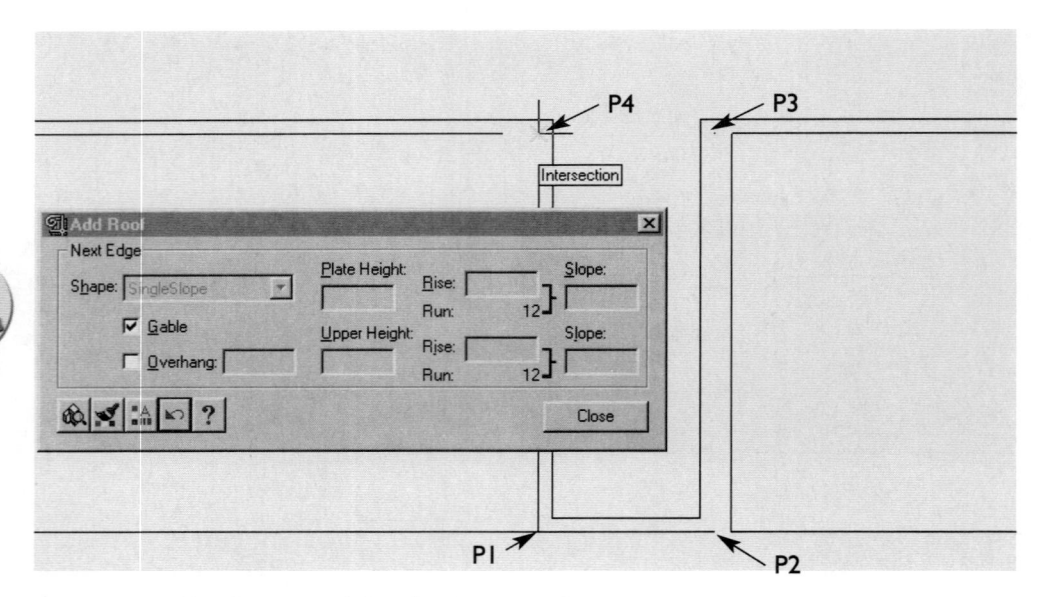

Figure 5.67 *Fourth point to define the dormer roof*

```
Roof point or
[Shape/Gable/Overhang/overhangValue/PHeight/PRise/PSl
ope/UHeight/URise/USlope/Match]: (Select the intersection of the
outer wall lines at point P1 shown in Figure 5.68.)
[Shape/Gable/Overhang/overhangValue/PHeight/PRise/PSl
ope/UHeight/URise/USlope/Match]:  ENTER  (Press ENTER to end
the command.)
```

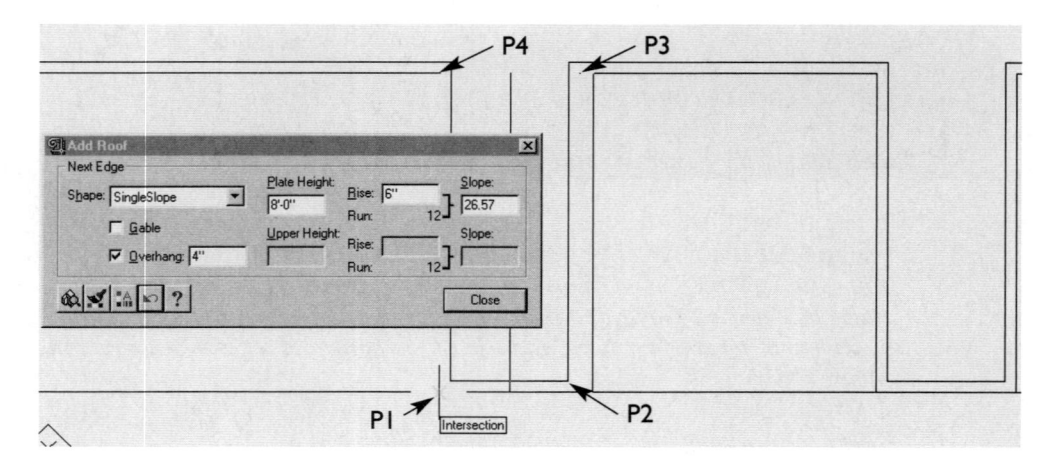

Figure 5.68 *Point P1 reselected to complete the dormer roof*

5. Select the **Add Roof** command and create a roof on each of the remaining dormer walls using the same procedure as shown in step 4 above.

6. To create a gable roof for the main portion of the house, choose **Design>Roof>Convert to Roof** from the menu bar and respond to the command line prompts as follows:

```
Command: _AecRoofConvert
Choose walls or polylines to create roof profile:
```
(Select the polyline at P1 in Figure 5.69.)
```
Select objects: 1 found
```

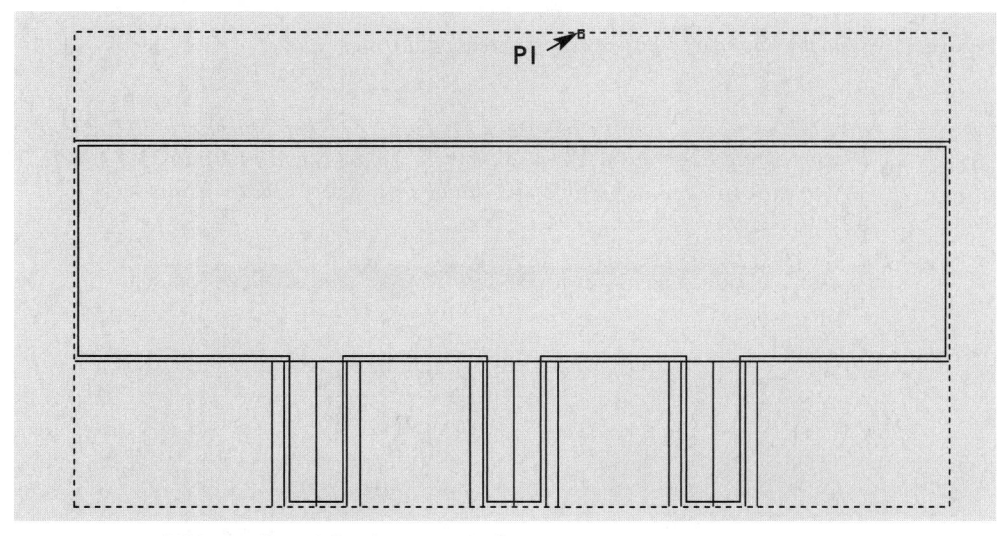

Figure 5.69 *Polyline selected for the main roof*

```
Select objects: ENTER
```
(Press ENTER to end selection of walls or polyline)
```
Erase layout geometry? [Yes/No] <Y>: n ENTER
```
(Type N to retain geometry.)

```
Roof modify
[Shape/Match/overhangValue/Overhang/PHeight/PRise/PSlope/UHeight/URise/USlope]:
DBOX
```
(Edit the Modify Roof dialog box: set Shape to SingleSlope, Plate Height = 0, Rise = 14, and select Overhang and set overhang distance equal to 12", as shown in Figure 5.70.)

(Select the Properties button of the Modify Roof dialog box and select the Dimensions tab. Select all edges by holding down ENTER *and select the first*

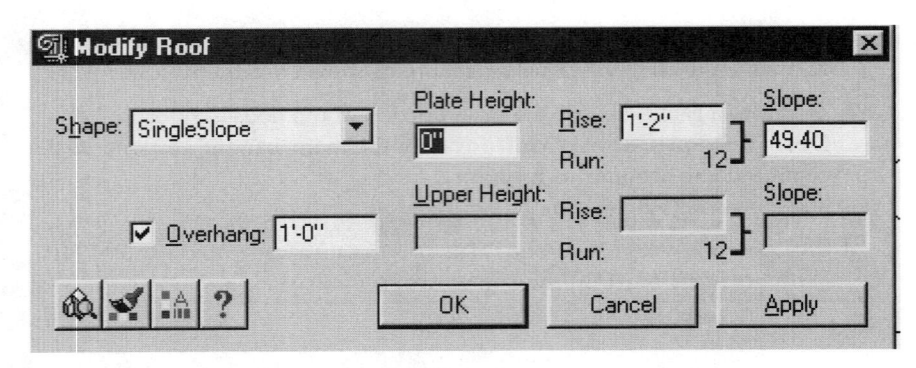

Figure 5.70 *Modify Roof dialog box setting for creating the main roof*

and last edge as shown in Figure 5.71. Edit the Roof Thickness (all faces) to 6" and toggle Floor for Measured Normal To as shown in Figure 5.71. Select OK to dismiss the Roof Properties dialog box. Select OK to dismiss the Modify Roof command.)

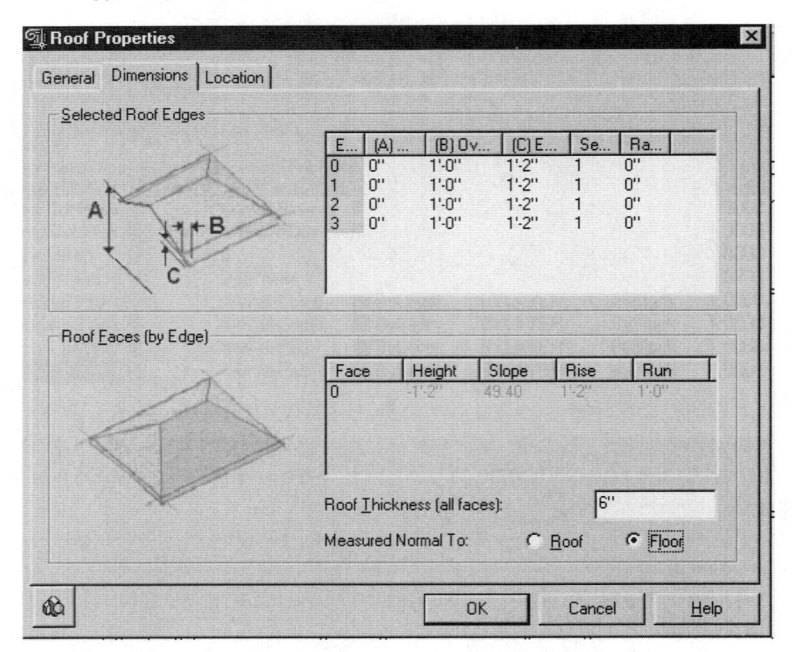

Figure 5.71 *Roof Properties dialog box edited for the main roof*

7. Select **Edit Roof Edges/Faces** from the **Roofs** toolbar and select the left end of the building at P1 as shown in Figure 5.72 and then press ENTER to end selection.

Figure 5.72 *Roof edge selected for editing*

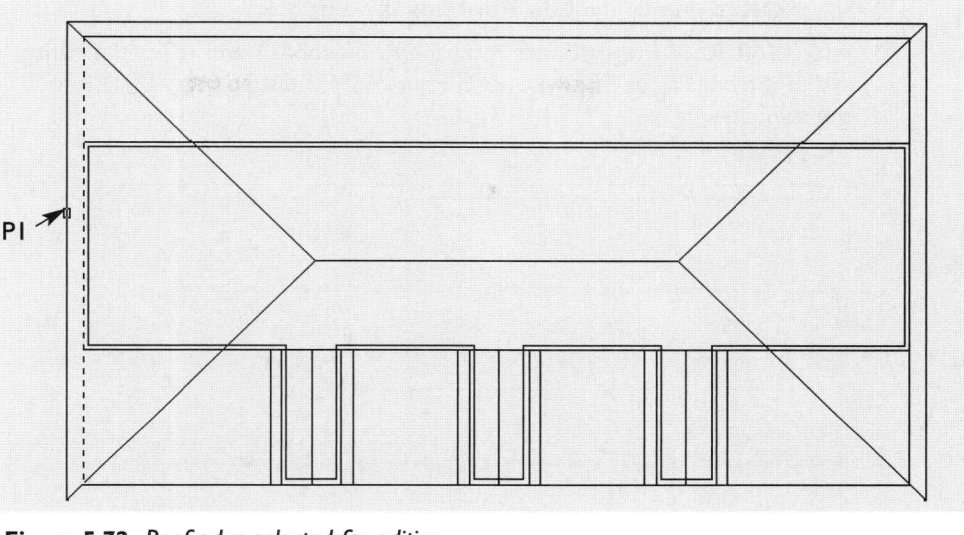

8. Edit the **Edit Roof Edges** dialog box, setting **Slope** to 90 in the **Roof Faces (by Edge)** section as shown in Figure 5.73.

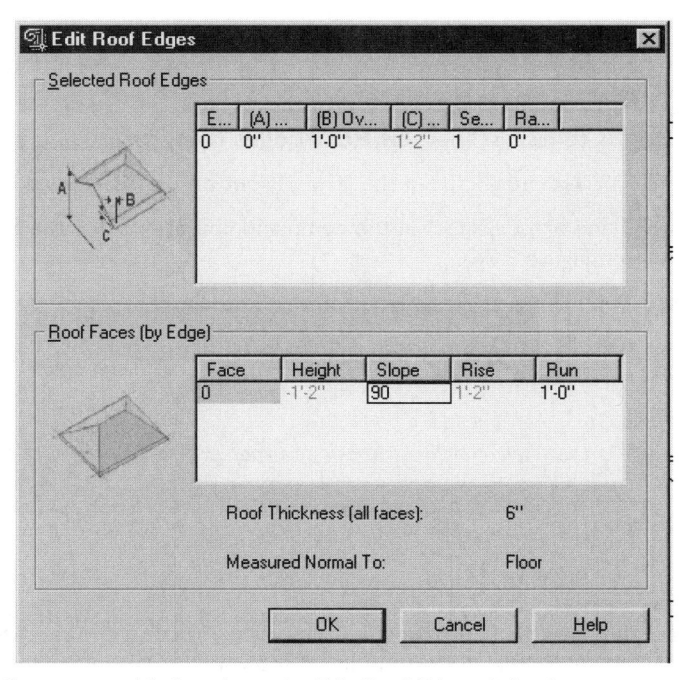

Figure 5.73 *Creating a gable by editing the Edit Roof Edges dialog box*

9. Select **OK** to dismiss the **Edit Roof Edges** dialog box.
10. Select **Edit Roof Edges/Faces** from the Roofs toolbar and select the right end of the building at P1 as shown in Figure 5.74 and then press ENTER to end selection.

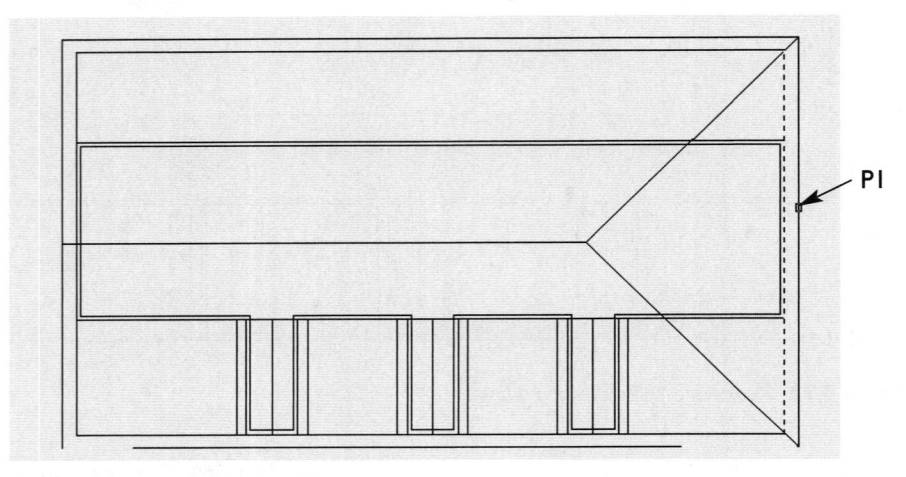

Figure 5.74 *Right roof edge selected*

11. Edit the **Edit Roof Edges** dialog box, setting **Slope** to 90 in the **Roof Faces (by Edge)** section as shown in Figure 5.75.
12. Select **OK** to dismiss the **Edit Roof Edges** dialog box.
13. Select **SW Isometric** from the **View** flyout on the **Standard** toolbar.
14. Type **HI** and press ENTER in the command line to execute the **HIDE** command (see Figure 5.76).
15. Select **Top** from the **View** flyout on the **Standard** toolbar.
16. Select **Add Mass Group** from the **Mass Tools** toolbar and respond to the command prompt as follows:

    ```
    Command: _AecMassGroupAdd
    Location: (Select a point above the roof near P1 in Figure 5.77.)

    Rotation angle <0.00>: ENTER (Press ENTER key to accept zero rotation.)
    ```

17. Select the mass group marker and then right-click and choose **Attach Elements** from the shortcut menu. Respond to the command prompts as follows:

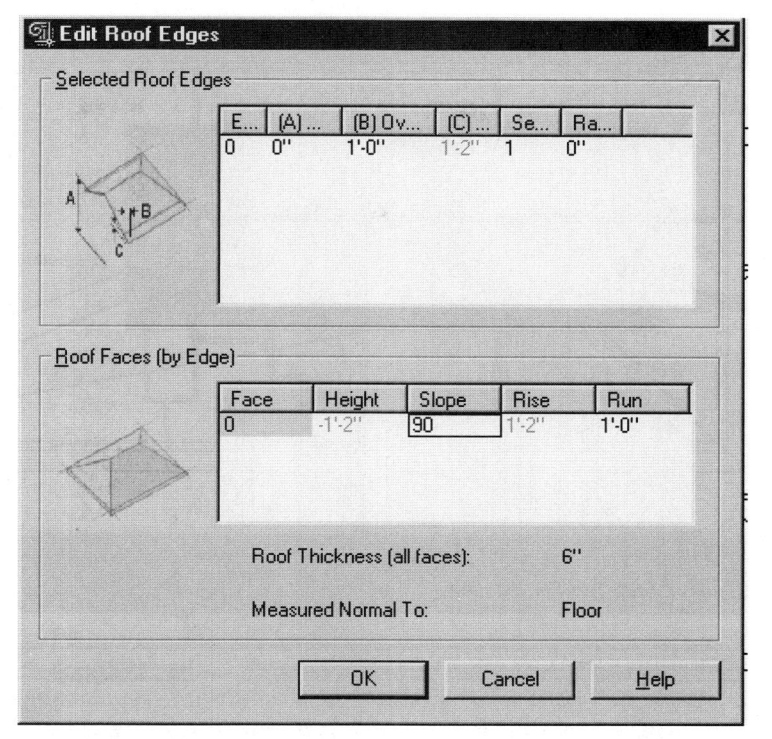

Figure 5.75 *Creating a gable end using the Edit Roof Edges dialog box*

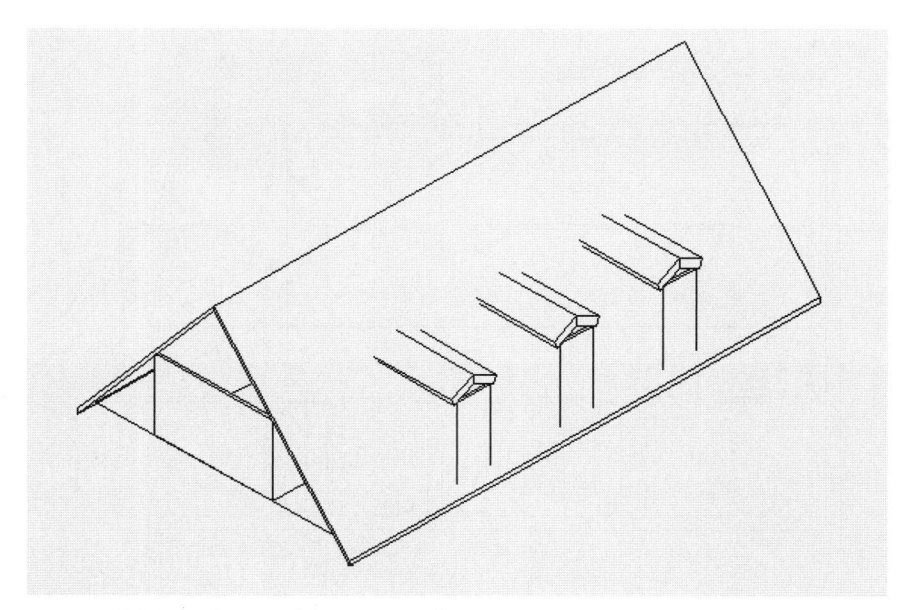

Figure 5.76 *Gable roof created for main roof*

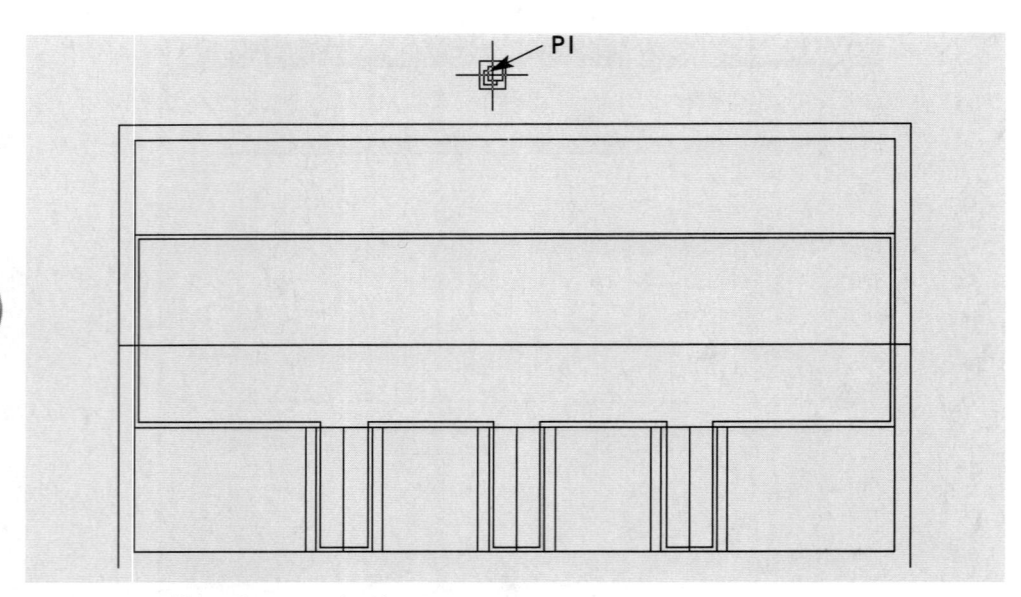

Figure 5.77 *Mass Group marker location*

```
Command: MassGroupAttach
Select elements to attach: 
```
(Use a crossing window by selecting from point P1 to point P2 as shown in Figure 5.78.) `Specify opposite corner: 14 found`

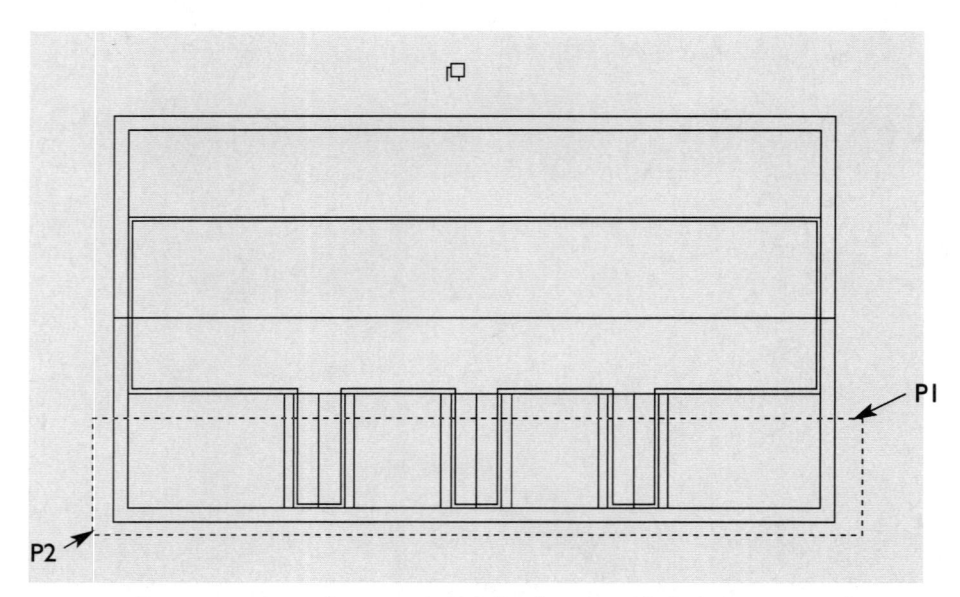

Figure 5.78 *Crossing window selection to select the dormer roofs and the main roof.*

```
Select elements to attach: ENTER (Press ENTER to end selec-
tion.)
Attaching elements...
14 element(s) attached.
Command:
```
18. Select **SW Isometric** from the **View** flyout on the **Standard** toolbar.
19. Type **HI** and press ENTER in the command line to execute the **HIDE** command. The dormer roofs are combined with the main roof as shown in Figure 5.79.

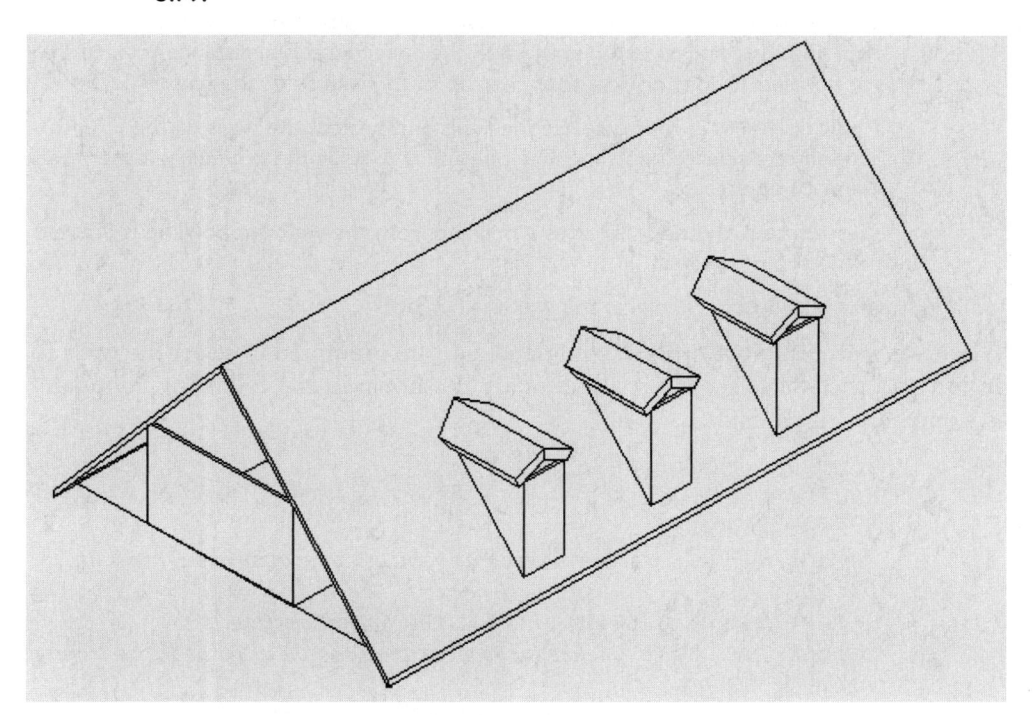

Figure 5.79 Display of roofs attached to the mass group

20. Save the drawing.

EXTENDING WALLS TO THE ROOF

Gable ends and shed roofs create a roof surface, which is projected from a wall that does not intersect with the wall. When the gable roof is created for a gable or shed roof, the wall and roofs do not intersect. You can modify the height of the wall according to the roofline using the **ROOFLINE** command. See Table 5.8 for command access.

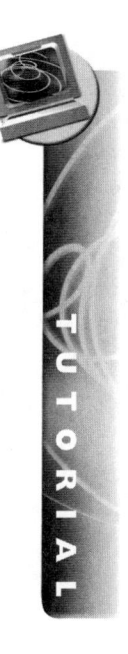

Menu bar	Walls>Wall Tools>Roof Line
Command prompt	ROOFLINE

Table 5.8 *ROOFLINE access*

The **ROOFLINE** command allows you to project the height of the walls to a roof or a polyline. The options of the command are described below.

Offset – Allows you to specify the distance the wall is projected toward the roof line. This option extends the wall the specified offset distance toward the roof profile. It does not contour the top of the wall to fit the roof.

Project – Allows you to select the walls and project the walls to the intersection of a selected polyline. The polyline is used as the boundary rather than the roof object.

Generate polyline – Creates a polyline from the wall; the polyline is created at the top of the wall.

Auto project – Projects the wall to the roof.

When you select the **ROOFLINE** command, you are prompted to select the walls to project and then to select the roof boundary, as shown in the following command sequence and Figure 5.80.

```
Command: roofline
RoofLine [Offset/Project/Generate polyline/Auto project]: A ENTER
Select walls: (Select the wall at point P1 as shown in Figure 5.80.)
1 found
Select walls: ENTER (Press ENTER to end selection.)
Select entites: (Select the roof entity as shown at point P2 in Figure 5.80.) 1 found
Select entites: ENTER (Press ENTER to end roof selection.)
[1] Wall cut line(s) converted.
(Wall extended to roof as shown in Figure 5.81.)
Select entites: ENTER (Press ENTER to end the command.)
```

The base of the wall can be edited to conform to the profile of the roof. Dormer walls supported by the roof or other vertical projections that meet the roof from above can be trimmed by the **FLOORLINE** command. See Table 5.9 for command access.

The options of the command are described below.

Offset – Allows you to specify the distance the bottom of the wall is projected toward the entity selected. The offset option extends the wall the specified offset distance from the floor. In Figure 5.82, three wall segments were drawn

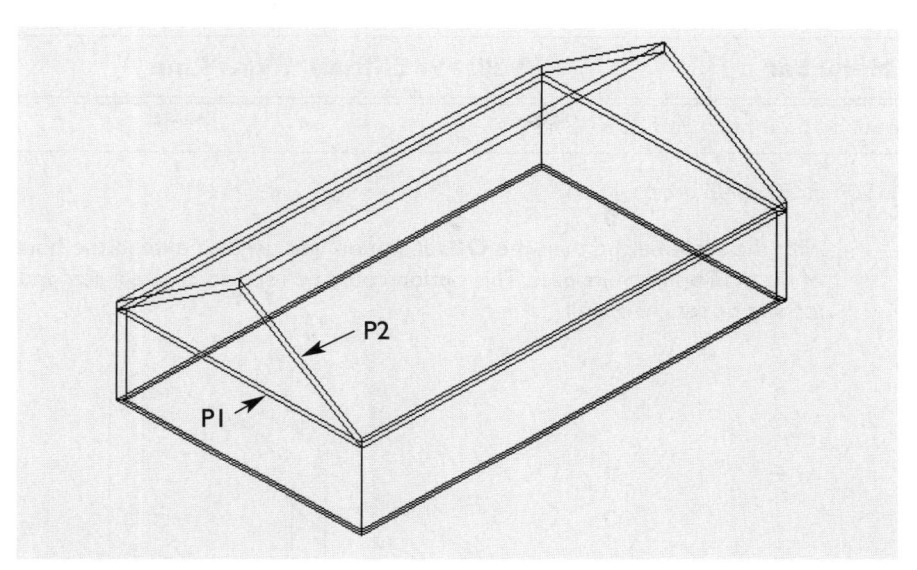

Figure 5.80 Roof and wall selections for the ROOFLINE command

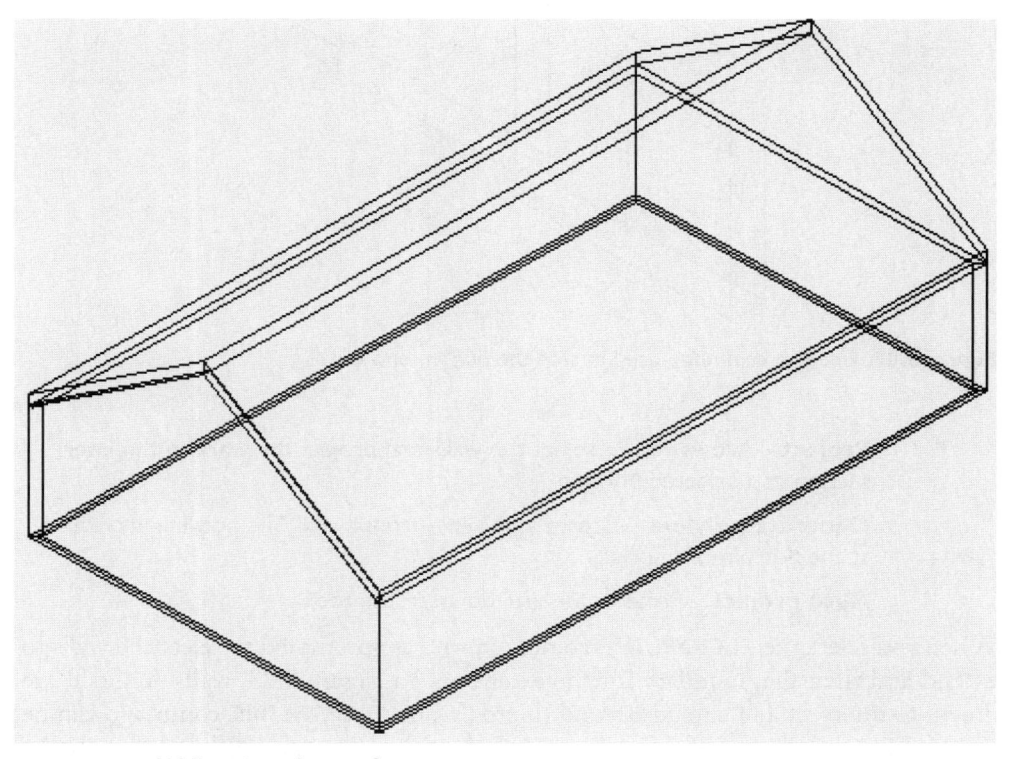

Figure 5.81 Wall projected to roof

Menu bar	Walls>Wall Tools>Floor Line
Command prompt	FLOORLINE

Table 5.9 *FLOORLINE access*

with the same height; then the **Offset** option was used to extend the bottom of the wall up from its base. This option could be used to create a stepped footing or retaining wall.

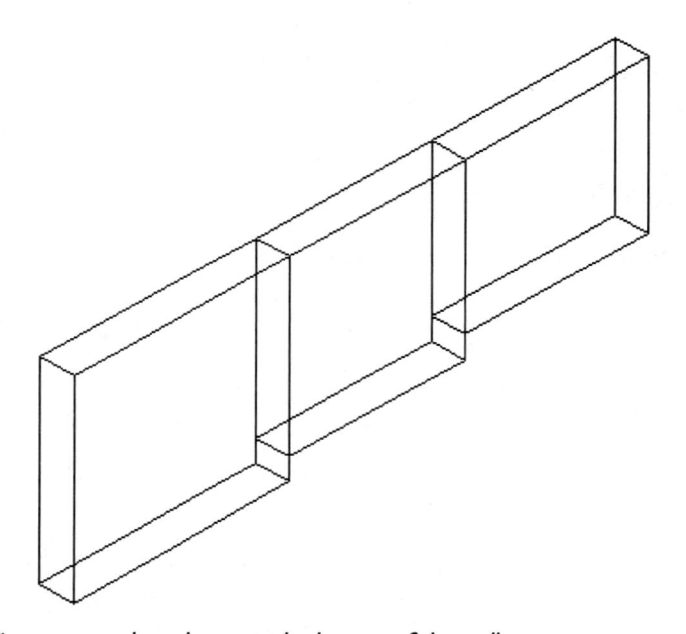

Figure 5.82 *Floorline command used to step the bottom of the wall*

Project – Allows you to select the walls and project the walls to the intersection of a selected polyline.

Generate polyline – Create a polylines from a wall. The polyline is created at the bottom of the wall.

Auto project – Projects the wall down to the roof.

When you select the **FLOORLINE** command, you are prompted to select the walls to extend and then the boundary for the extension. In Figure 5.83, walls A and B are drawn to intersect and extend beyond the roof. The **FLOORLINE** command can be used to trim the bottom of the walls according to the boundary of the roof as shown in the following command sequence.

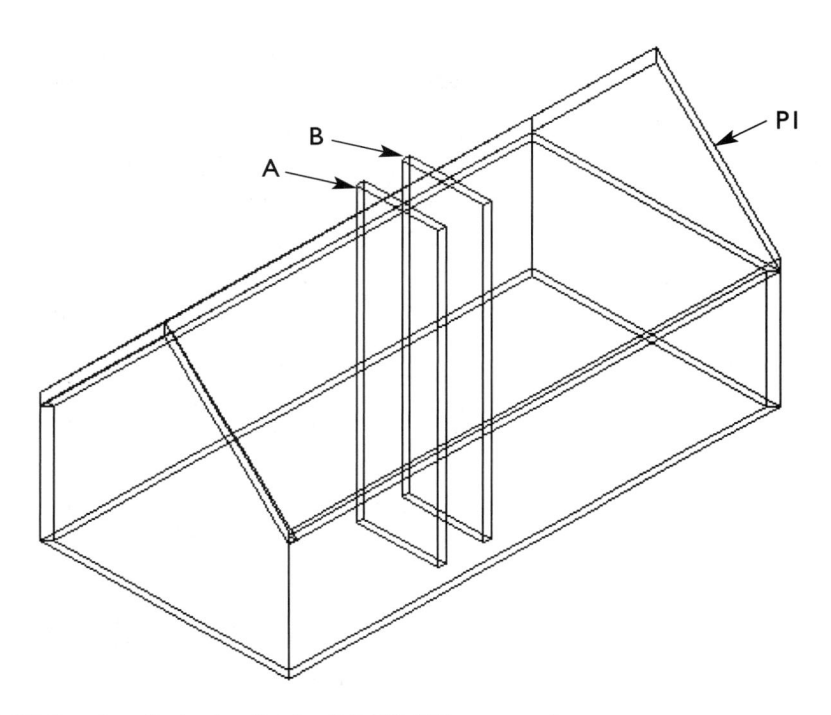

Figure 5.83 *Wall and roof selection for the FLOORLINE command*

```
Command: floorline
FloorLine [Offset/Project/Generate polyline/Auto
project]: a ENTER
```
Select walls: *(Select wall "A" in Figure 5.83.)* 1 found
Select walls: *(Select wall "B" in Figure 5.83.)* 1 found, 2 total
Select walls: ENTER *(Press ENTER to end wall selection.)*
Select entites: (Select the roof at point P1 in Figure 5.83.) 1 found
Select entites: ENTER *(Press {ENTER to end roof selection.)*
[2] Wall cut line(s) converted.
FloorLine [Offset/Project/Generate polyline/Auto project]: ENTER *(Press ENTER to end the command.)*
Command: hi ENTER *(Select the HIDE command to display the walls intersecting the roof as shown in Figure 5.84.)*

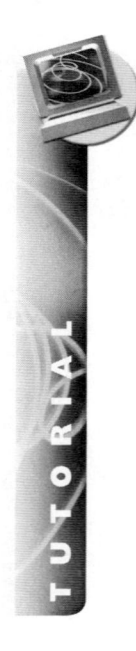

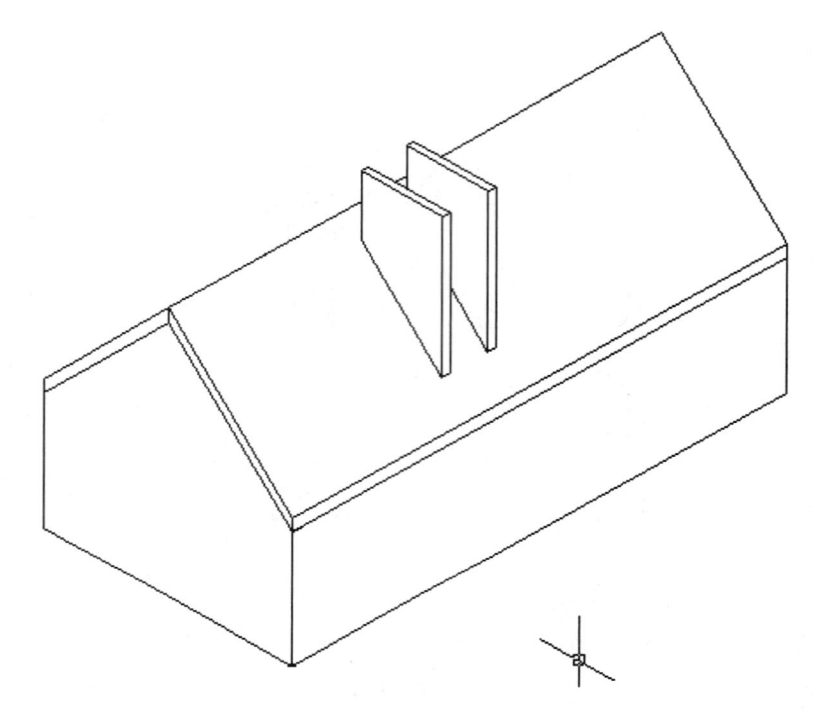

Figure 5.84 *Walls trimmed by the roof using the FLOORLINE command*

TUTORIAL 5.8 EXTENDING WALLS TO THE ROOF

1. Open *ADT Tutor\Ch5\Ex5-8.dwg* and save the drawing as *Roofline8.dwg* in your student directory.

2. Move the mouse pointer to the OSNAP toggle of the status bar and right-click; choose **Settings** from the shortcut menu. Select the Endpoint object snap and clear all other object snaps. Select the **Object Snap On(F3)** check box to turn it ON and select **OK** to dismiss the **Drafting Settings** dialog box.

   ```
   Command: _AecRoofLine
   RoofLine [Offset/Project/Generate polyline/Auto project]: A ENTER
   Select walls: F ENTER
   First fence point:
   Specify endpoint of line or [Undo]: (Select near point P1 in Figure 5.85.)
   Specify endpoint of line or [Undo]: (Select near point P2 in Figure 5.85.)
   ```

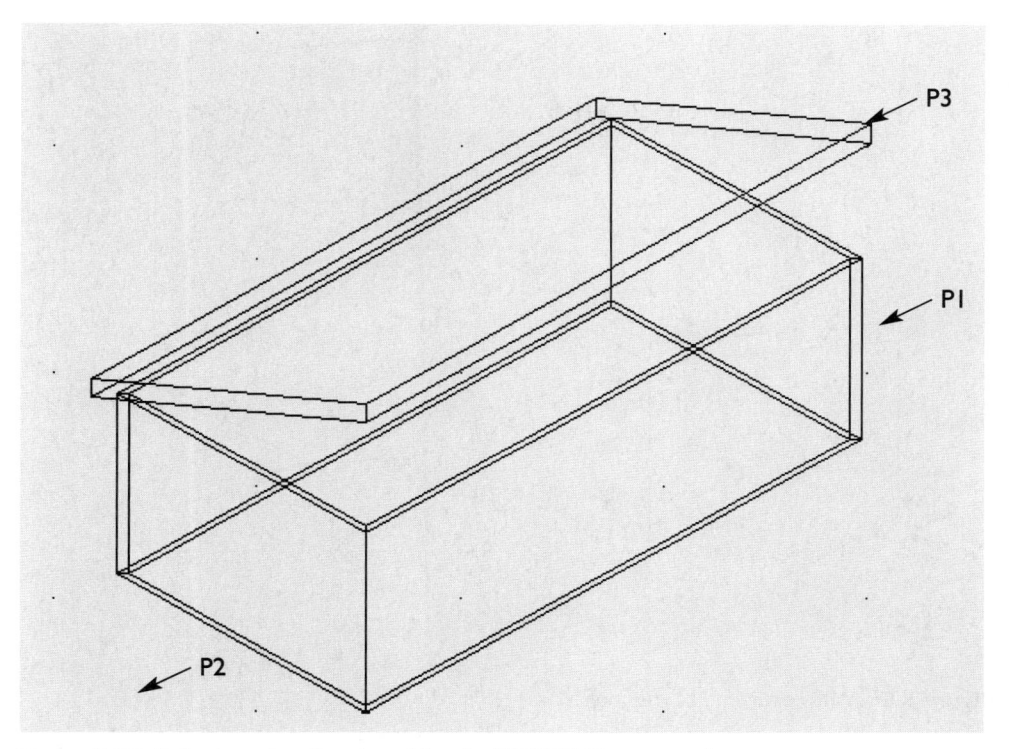

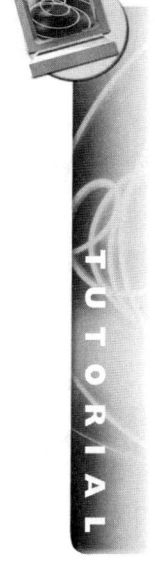

Figure 5.85 *Selection of wall and roof for the ROOFLINE command*

```
3 found
Select entites: 1 found
```
(Select the roof near point P3 in Figure 5.85.)
```
Select entites: ENTER
```
(Press ENTER to end roof selection.)
```
[3] Wall cut line(s) converted.
RoofLine [Offset/Project/Generate polyline/Auto project]: ENTER
```
(Press ENTER to end the command.)
```
Command: HI ENTER
```
(Walls extended to the roof as shown in Figure 5.86.)

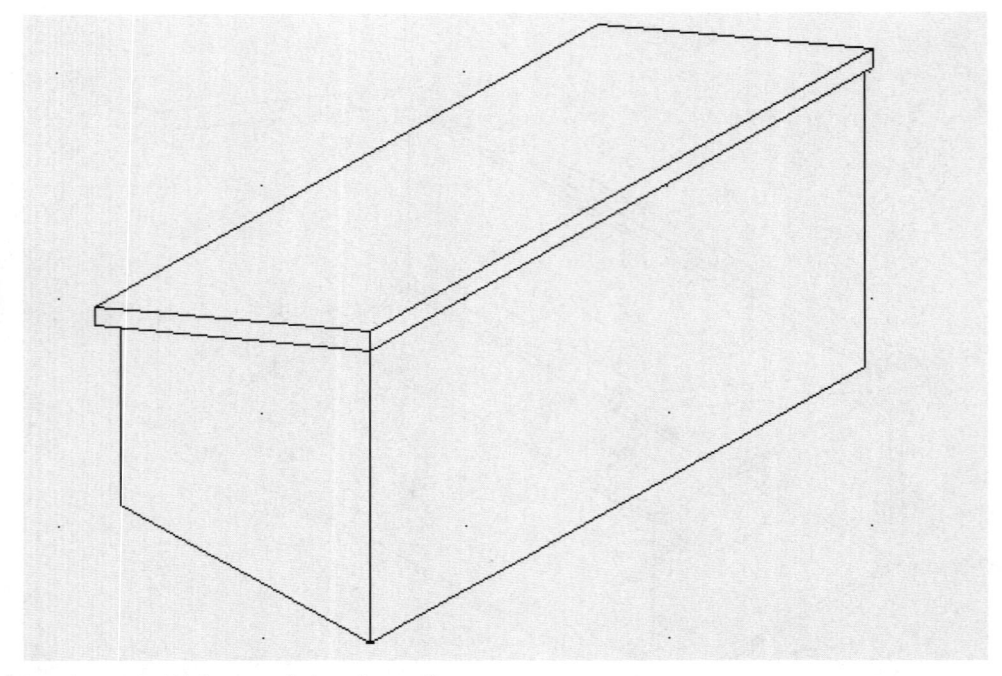

Figure 5.86 *Walls extended to the roof.*

3. Save the drawing.

CREATING A FLAT ROOF

A flat roof is often used in commercial construction. Create a flat roof using the **Add Roof** command and setting the slope to zero. The walls of a building with a flat roof usually extend beyond the roof to hide the roof and any roof-mounted equipment. Therefore, a wall style with multiple components should be used to represent this construction. Figure 5.87 shows the **Components** tab of the **Wall Style Properties** dialog box for a parapet wall.

The flat roof is created inside the exterior walls with zero slope and the plate height is set to the roof elevation at the bottom of the roof.

TUTORIAL 5.9 CREATING A FLAT ROOF

1. Open *ADT Tutor\Ch5\Ex5-9.dwg* and save the drawing as *Flat9.dwg* in your student directory.
2. Select the Work-FLR layout tab.
3. Select the top 8" CMU wall, and then right-click and choose **Edit Wall Style** from the shortcut menu.

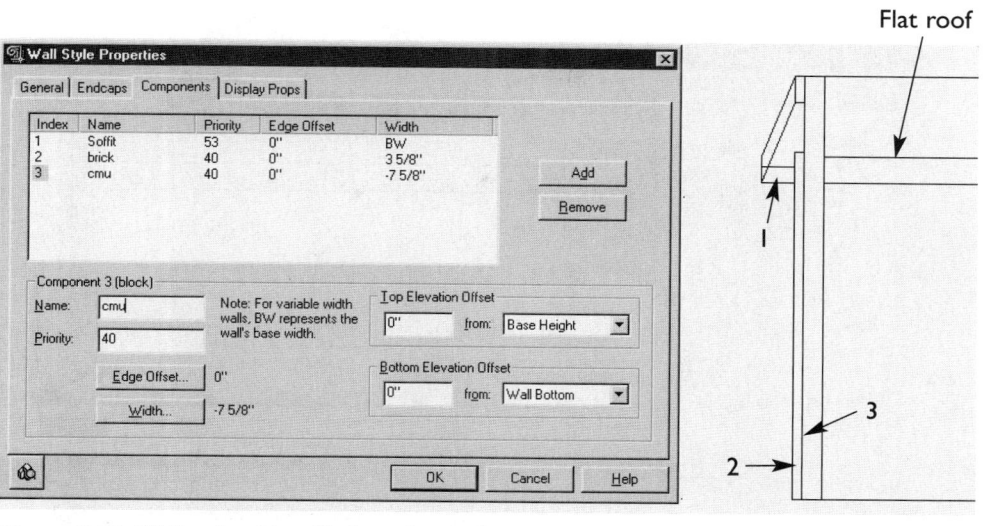

Figure 5.87 Wall style with soffit for a flat roof

4. Select the **Display Props** tab, and then select the **Edit Display Props** button to open the **Entity Properties** dialog box.
5. Select the light bulb to toggle OFF the Hatch I component.
6. Select **OK** to dismiss the **Entity Properties** dialog box.
7. Select **OK** to dismiss the **Wall Style Properties** dialog box.
8. Move the mouse pointer to the OSNAP toggle of the status bar and right-click; choose **Settings** from the shortcut menu. Select the Endpoint object snap and clear all other object snaps. Select the **Object Snap On(F3)** check box to toggle it ON and select **OK** to dismiss the **Drafting Settings** dialog box.

The drawing file consists of a soffit wall style, as shown in Figure 5.87.

9. To create a Flat roof, select the **Add Roof** command and respond to the command line prompts by selecting the following wall points.

```
Roof point or
[Shape/Gable/Overhang/overhangValue/PHeight/PRise/PSl
ope/UHeight/URise/USlope/Match]: DBOX
```
(Edit the Add Roof dialog box: set Shape to SingleSlope, Plate Height to 10'-6", Rise to 0 , select Gable, deselect Overhang.)

(Select the Properties button of the Add Roof dialog box, edit the Dimensions tab setting Roof Thickness to 10", and select the Floor for Measured Normal To as shown in Figure 5.88.)

(Select OK to dismiss Roof Properties dialog box.)

Figure 5.88 *Roof Properties dialog box edited*

```
_AecRoofAdd
Roof point or
[Shape/Gable/Overhang/overhangValue/PHeight/PRise/PSl
ope/UHeight/URise/USlope/Ma
tch]: DBOX
```

(Select Zoom Window from the Standard toolbar and respond to the following command prompts.)

```
Roof point or
[Shape/Gable/Overhang/overhangValue/PHeight/PRise/PSl
ope/UHeight/URise/USlope/Ma
tch]: '_zoom
>>Specify corner of window, enter a scale factor (nX
or nXP), or
[All/Center/Dynamic/Extents/Previous/Scale/Window]
```

```
<real time>: _w
>>Specify first corner: (Select point P1 as shown in Figure
5.89.)
  >>Specify opposite corner: (Select point P2 as shown in Figure
5.89.)
```

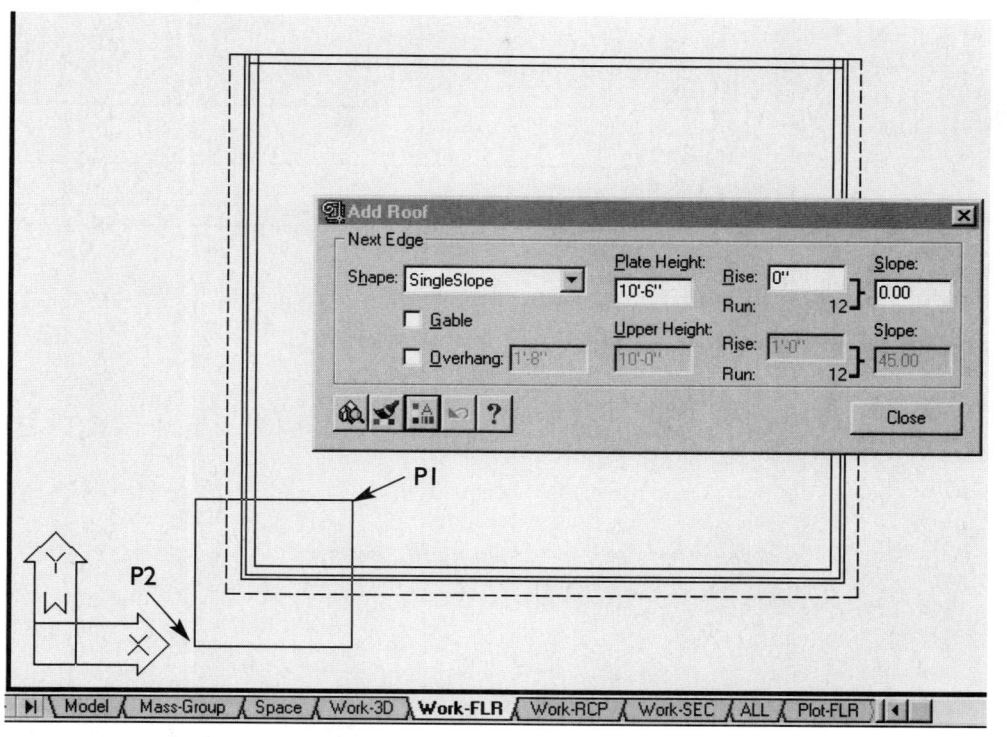

Figure 5.89 *Zoom window specified*

```
Resuming AECROOFADD command.
Roof point or
[Shape/Gable/Overhang/overhangValue/PHeight/PRise/PSl
ope/UHeight/URise/USlope/Ma
tch]: (Select the endpoint shown at P1 in Figure 5.90.)
```

(Select Zoom Previous from the Standard toolbar.)

```
Roof point or
[Shape/Gable/Overhang/overhangValue/PHeight/PRise/PSl
ope/UHeight/URise/USlope/Match]: '_zoom
>>Specify corner of window, enter a scale factor (nX
or nXP), or
```

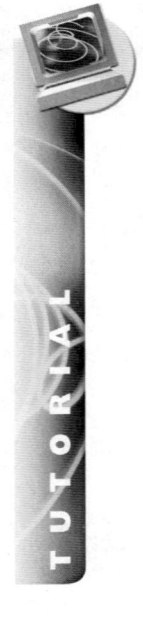

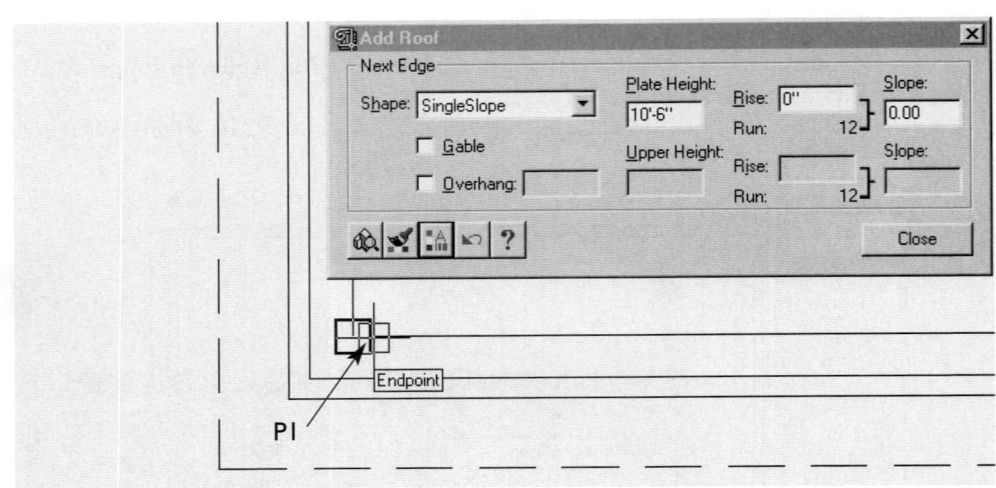

Figure 5.90 *Beginning of roof specified*

```
[All/Center/Dynamic/Extents/Previous/Scale/Window]
<real time>: _p
Resuming AECROOFADD command.
```

(Select Zoom Window from the Standard toolbar.)

```
Roof point or
[Shape/Gable/Overhang/overhangValue/PHeight/PRise/PSl
ope/UHeight/URise/USlope/Ma
tch]: '_zoom
>>Specify corner of window, enter a scale factor (nX
or nXP), or
[All/Center/Dynamic/Extents/Previous/Scale/Window]
<real time>: _w
>>Specify first corner:
```
(Select point P1 as shown in Figure 5.91.)
```
>>Specify opposite corner:
```
(Select point P2 as shown in Figure 5.91.)
```
Resuming AECROOFADD command.

Roof point or
[Shape/Gable/Overhang/overhangValue/PHeight/PRise/PSl
ope/UHeight/URise/USlope/Match]:
```
(Select the endpoint as shown at P2 in Figure 5.92.)
(Select Zoom Previous from the Standard toolbar.)

Creating Roofs 391

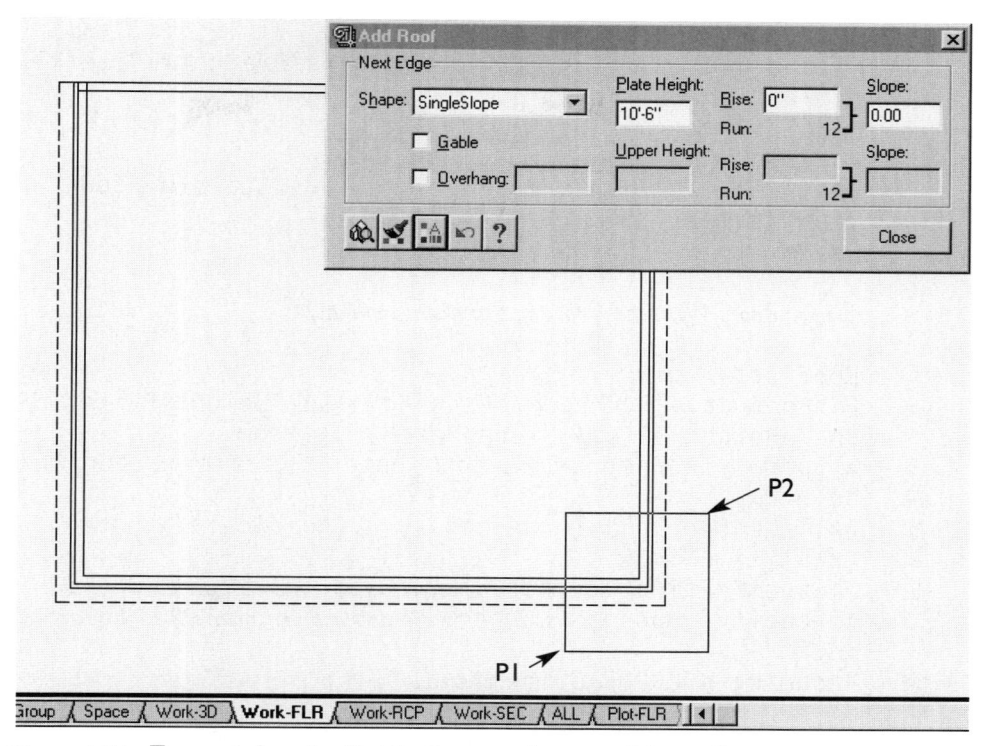

Figure 5.91 Zoom window specified for the second corner of the roof

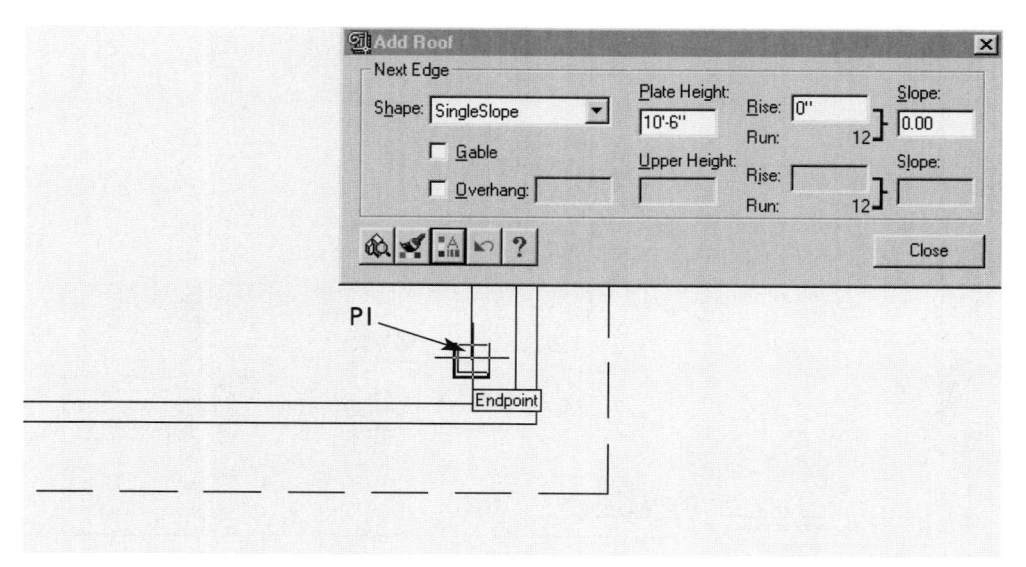

Figure 5.92 Roof corner for the second point specified

```
Roof point or
[Shape/Gable/Overhang/overhangValue/PHeight/PRise/PSl
ope/UHeight/URise/USlope/Match]: '_zoom
>>Specify corner of window, enter a scale factor (nX
or nXP), or
[All/Center/Dynamic/Extents/Previous/Scale/Window]
<real time>: _p
Resuming AECROOFADD command.
```

(Select Zoom Window from the Standard toolbar.)

```
Roof point or
[Shape/Gable/Overhang/overhangValue/PHeight/PRise/PSl
ope/UHeight/URise/USlope/Match]: '_zoom
>>Specify corner of window, enter a scale factor (nX
or nXP), or
[All/Center/Dynamic/Extents/Previous/Scale/Window]
<real time>: _w
>>Specify first corner: (Select a point near P1 in Figure 5.93.)
>>Specify opposite corner: (Select a point near P2 in Figure
5.93.)
Resuming AECROOFADD command.
```

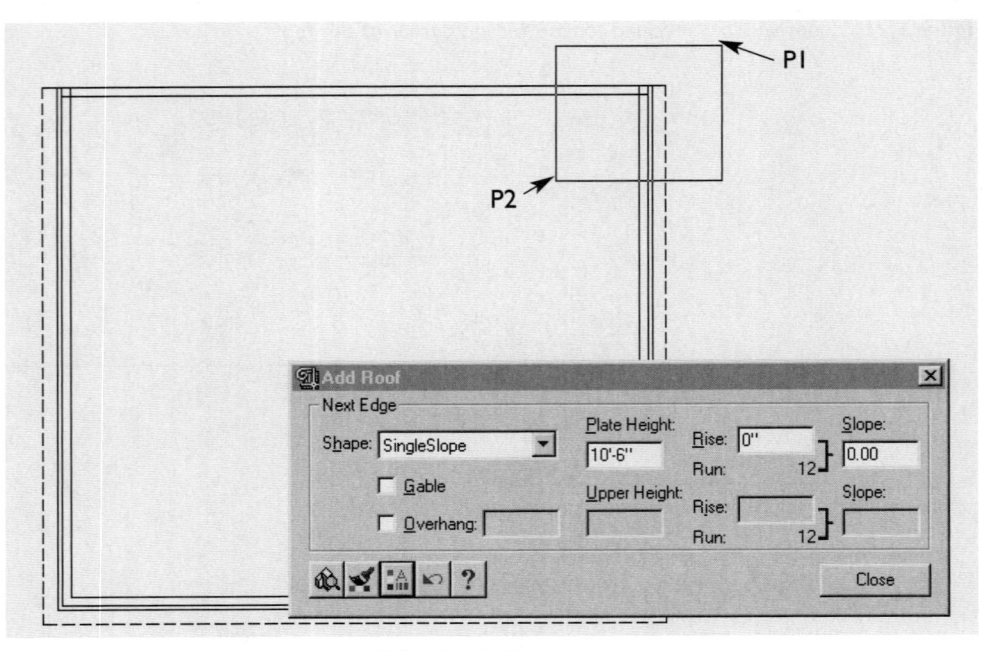

Figure 5.93 *Zoom window specified for the third point*

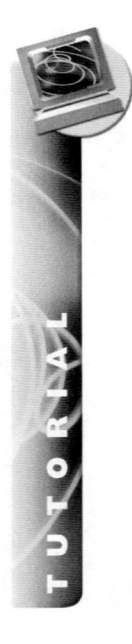

```
Roof point or
[Shape/Gable/Overhang/overhangValue/PHeight/PRise/PSl
ope/UHeight/URise/USlope/Ma
tch]:
```
(Select the endpoint as shown at P3 in Figure 5.94.)

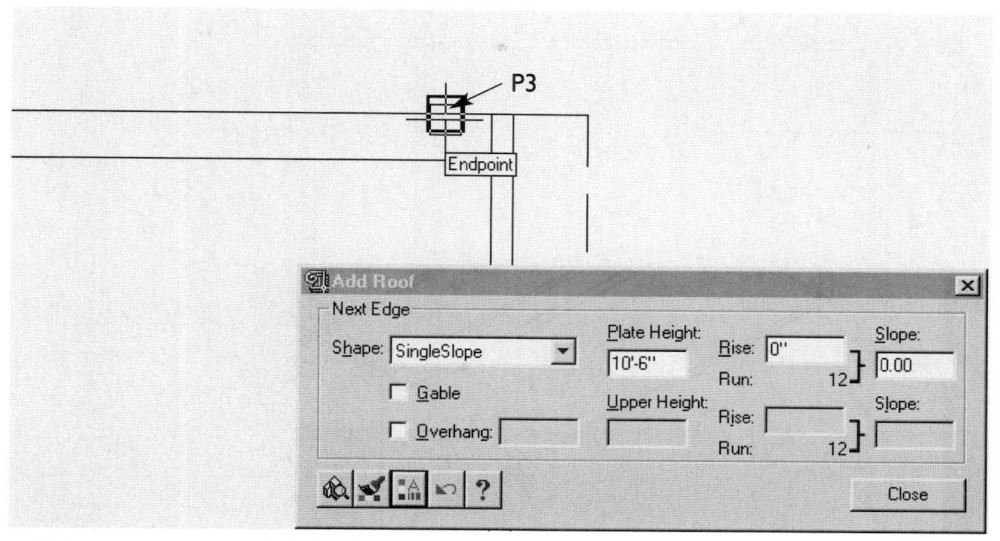

Figure 5.94 *Third roof point specified*

(Select Zoom Previous from the Standard toolbar.)
```
Roof point or
[Shape/Gable/Overhang/overhangValue/PHeight/PRise/PSl
ope/UHeight/URise/USlope/Match]: '_zoom
>>Specify corner of window, enter a scale factor (nX
or nXP), or
[All/Center/Dynamic/Extents/Previous/Scale/Window]
<real time>: _p
Resuming AECROOFADD command.
```

(Select Zoom Window from the Standard toolbar.)
```
Roof point or
[Shape/Gable/Overhang/overhangValue/PHeight/PRise/PSl
ope/UHeight/URise/USlope/Match]: '_zoom
>>Specify corner of window, enter a scale factor (nX
or nXP), or
[All/Center/Dynamic/Extents/Previous/Scale/Window]
<real time>: _w
```

```
>>Specify first corner: (Select a point near P1 in Figure 5.95.)
>>Specify opposite corner: (Select a point near P2 in Figure
5.95.)
```

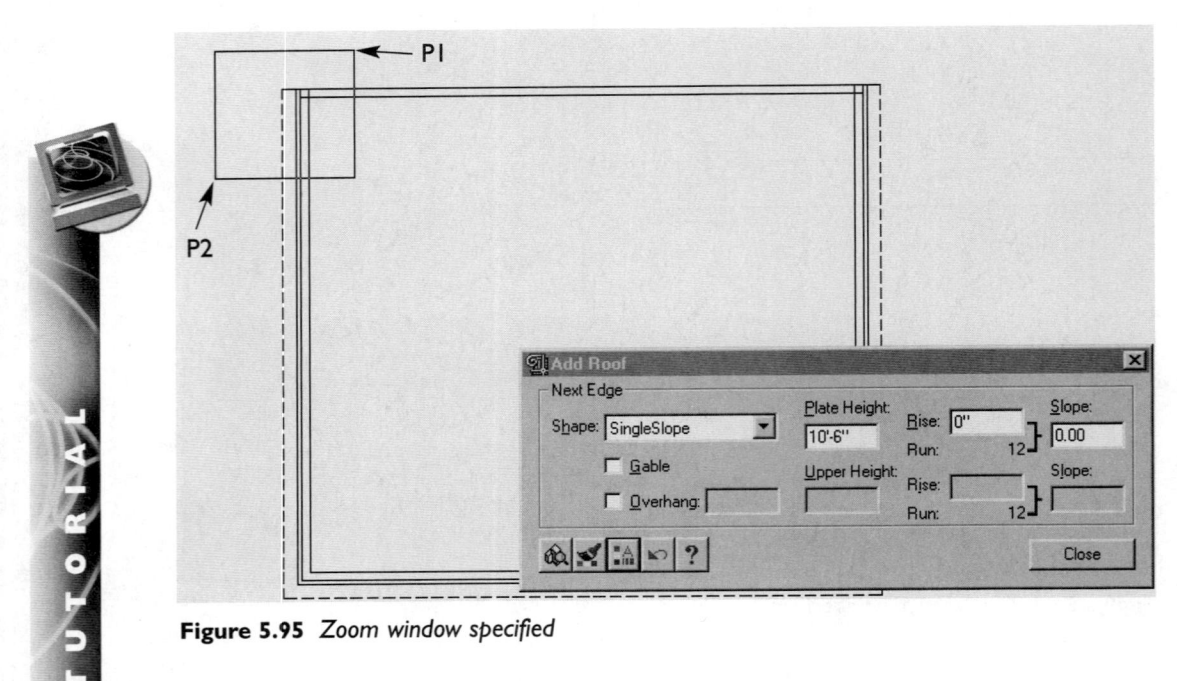

Figure 5.95 *Zoom window specified*

```
Resuming AECROOFADD command.
Roof point or
[Shape/Gable/Overhang/overhangValue/PHeight/PRise/PSl
ope/UHeight/URise/USlope/Ma
tch]: (Select the endpoint of the wall as shown at point P4 in Figure
5.96.)
Roof point or
[Shape/Gable/Overhang/overhangValue/PHeight/PRise/PSl
ope/UHeight/URise/USlope/Ma
tch]: ENTER (Pressing ENTER closes the roof to P1 and ends the Add
Roof command.)
```
10. Select **SW Isometric** from the **View** flyout on the **Standard** toolbar.

11. Type **HI** and press ENTER in the command line to view the roof as shown in Figure 5.97.

12. Save the drawing.

Creating Roofs **395**

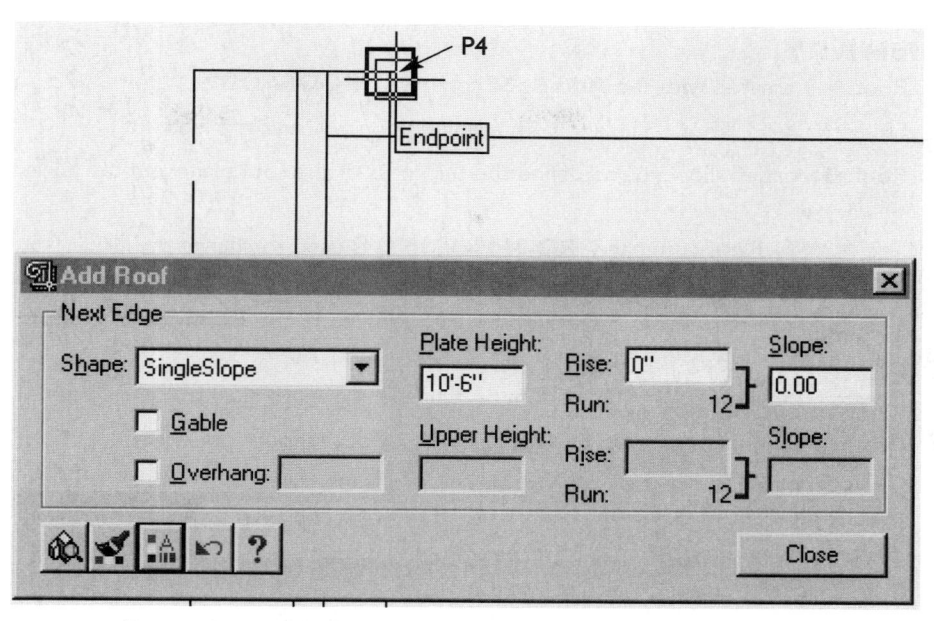

Figure 5.96 *Four roof point specified*

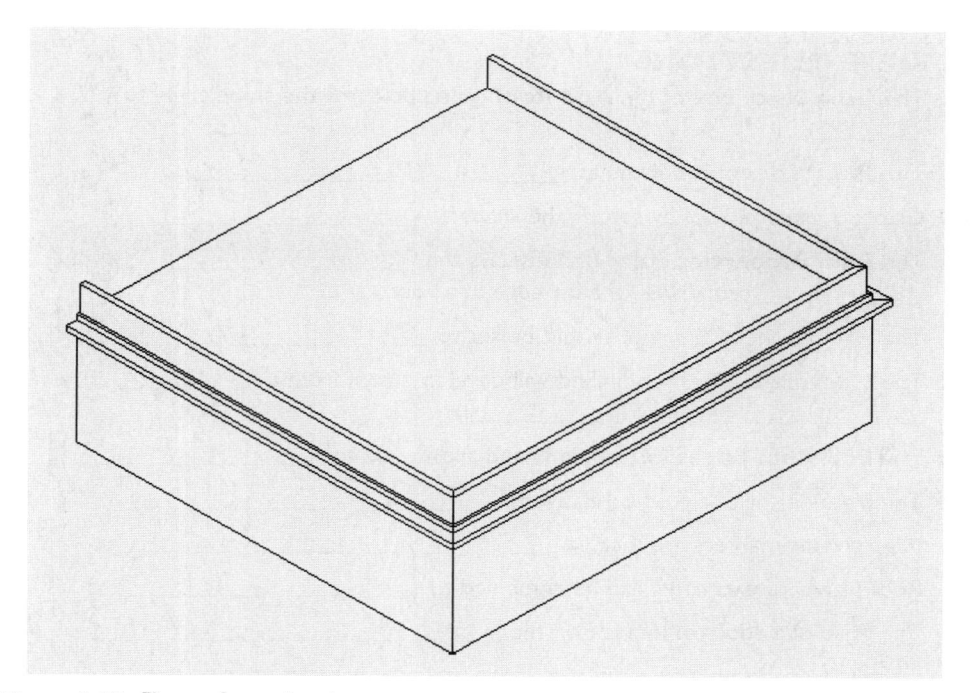

Figure 5.97 *Flat roof completed*

SUMMARY

1. Roofs are created with the **Add Roof** command (**ROOFADD**).
2. Using the **Add Roof** command with a single slope will create a hip roof.
3. Roof properties allow you to define the thickness of the roof plane and the angle of the fascia.
4. The **Modify Roof** command (**ROOFMODIFY**) is used to change the slope, plate height, shape, and overhang of an existing roof.
5. The angle of the roof can be defined by the ratio of the rise to run or by a slope angle.
6. Create a gable end of a gable roof by setting the roof edge slope to 90 degrees.
7. Create a shed roof by selecting **Gable** for three of the four roof edges.
8. Create the gambrel roof by applying the double slope shape to two edges of the roof.
9. The **Convert to Roof** command (**ROOFCONVERT**) is used to create a roof from a closed polyline or closed wall segments.
10. Attaching roof planes to a mass group combines roof planes to determine points of intersection.

REVIEW QUESTIONS

1. The Gable check box of the **Add Roof** dialog box sets the slope angle to _____.
2. The angle of a roof is defined as _____ _____.
3. Create a gambrel roof by setting the shape to _____.
4. The **Roof Properties** dialog box displays the _____, _____, and _____ properties of each edge of a roof.
5. The plate height of the roof should be set to _____.
6. The **Convert to Roof** command will develop a roof from a _____ or _____.
7. The **Edit Roof Edges/Faces** command allows you to _____.
8. The plate height of a roof is relative to _____.
9. Mass groups markers are located _____.
10. Roof planes of two roofs can be combined by _____.
11. The grips of a roof are located on the _____, _____.
12. Walls can be extend to the gable roof plane with the _____ command.

13. Create flat roofs by setting the slope angle to _____.
14. Represent rafter size and the decking thickness by adjusting the _____ in the _____ dialog box.
15. Describe the procedure for creating a shed roof.

PROJECT

EX5.1 CREATING A SHED ROOF

1. Open *ADT Tutor\Ch5\Ex5-10.dwg*
2. Save the drawing to your student directory as *Proj5-10*.
3. Create a shed roof for the overhang through points 1 through 4 as shown in Figure 5.98.

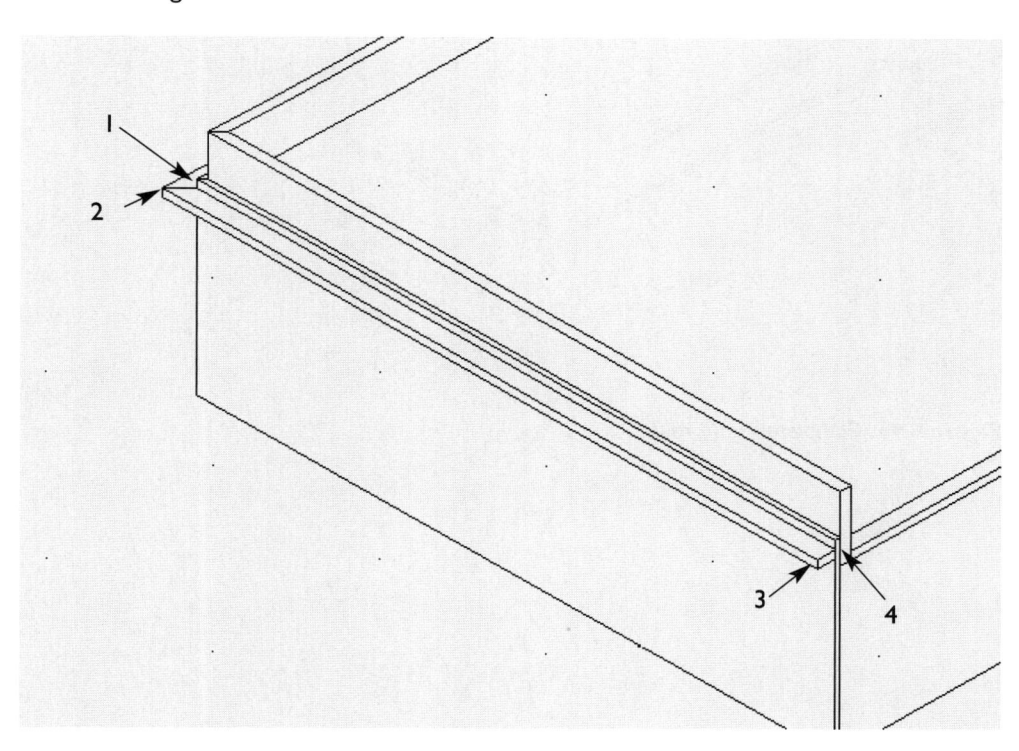

Figure 5.98 *Roof points for creating the soffit roof*

4. Roof segments: all roof segments will have no overhang, fascia normal to the floor, thickness of 10", and the plate height equal to 10'-6". Segment 2-3 is sloped 2'-7" rise whereas segments 1-2, 3-4, and 4-1 are gable.

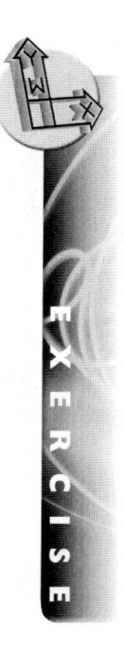

5. Repeat the procedure stated above to place a shed roof on the remaining soffits.

The completed roof is shown in Figure 5.99.

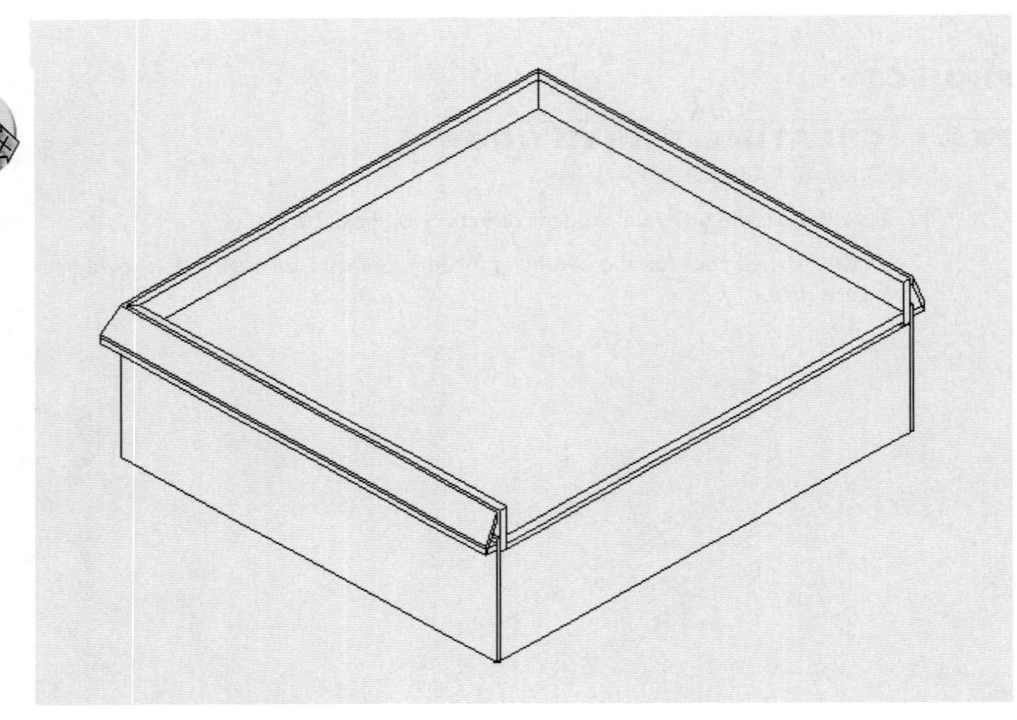

Figure 5.99 *Completed soffit roof*

CHAPTER 6

Stairs and Railings

INTRODUCTION

During the development of the floor plan, stairs must be inserted to determine if the space allocated for the stair well is sufficient. Railings and exterior steps must ultimately be added to the plan. Architectural Desktop provides the commands for placing stairs and adding railings.

OBJECTIVES

After completing this chapter, you will be able to

- Use the **Add Stair** command (**STAIRADD**) to create Straight, Multi-landing, U Shaped, U Shaped Winder, and Spiral stairs

- Change the number of risers, tread size and run of a stair using **Stair Properties** (**STAIRPROPS**)

- Access the **Options** dialog box to edit default stair settings

- Change the size and properties of stairs using the **Modify Stair** (**STAIRMODIFY**) and grips

- Create, edit, import, and export stair styles using the **Stair Styles** command (**STAIRSTYLE**)

- Turn off the visibility of the lower or upper portions of stairs using the display properties of a stair style

- Use the **Add Railing** command (**RAILINGADD**) to create railings for freestanding railings and railings attached to stairs

- Change the components and dimensions of railings using **Modify Railing** command (**RAILINGMODIFY**)

USING THE ADD STAIR COMMAND

Stairs are placed in a drawing with the **Add Stair** command. The available shapes of stairs include U Shaped, U Shaped Winder, Multi-landing, and Spiral. The available space for the stair is a major consideration for the selection of the shape. The **Stair Properties** dialog box allows you to apply code rules for a stair to determine the necessary space required prior to inserting the stair. The code rules specify the upper and lower limit dimensions for the risers and treads.

Stair styles available include Cantilever, Concrete, Standard, Steel Open, Steel-Pan, Wood-Housed, and Wood Saddled. Each of these styles changes the appearance of the stair and the means of support for the stair. Prior to placing the stair, you should first review the stair properties.

The commands for placing and editing stairs are located on the **Stairs/Railings** toolbar, shown in Figure 6.1.

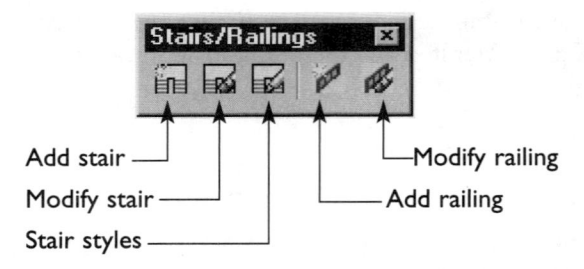

Figure 6.1 *Commands on the Stairs/Railings toolbar*

The commands on the **Stairs/Railings** toolbar allow you to create and modify a stair. The **Stair Styles** command allows you to edit the properties of a stair style. After the stair is placed, handrails, balusters, and guardrails can be added through the **Add Railing** command. The **Modify Railing** command allows you to change existing railings.

CREATING A STAIR

The **Add Stair** command (**STAIRADD**) is used to create a stair. See Table 6.1 for command access.

The **Add Stairs** dialog box opens when the **Add Stair** command is executed, allowing you to set the shape, style, dimensions, and landing of the stair, as shown in Figure 6.2.

Menu bar	Design>Stairs>Add Stair
Command prompt	AECSTAIRADD
Toolbar	Select Add Stair from the Stairs/Railings toolbar as shown in Figure 6.1
Shortcut	Right-click over the drawing area; then choose Design>Stairs>Add

Table 6.1 *Stair access*

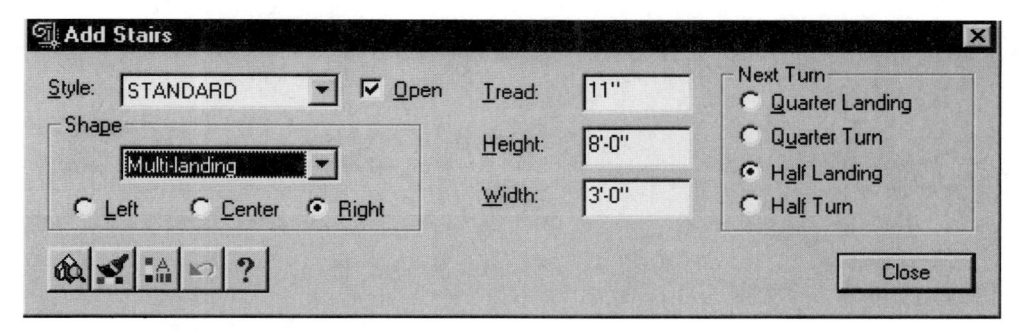

Figure 6.2 *Add Stairs dialog box*

Set the options of the command by editing the dialog box or selecting options from the command line. The command line prompt for the **Add Stair** command includes the options as shown below.

```
Command: _AecStairAdd
Flight start point or
[SHape/STyle/Tread/Height/Width/Justify/Open/Next/Match/Undo]:
```

The purpose of each option is as follows.

> **Style** – The **Style** drop-down list includes the following stair construction styles: Cantilever, Concrete, Standard, Steel Open, Steel Pan, Wood Housed, or Wood Saddle.
>
> **Open** – The **Open** check box, if selected, will create stairs with open risers.
>
> **Shape** – The shapes of stairs can be selected from the list that includes U Shaped, U Shaped Winder, Multi-landing, or Spiral. Each shape is defined below and illustrated in Figure 6.3.
>
> **U Shaped** – Includes two runs of equal length with a half landing.

U Shaped Winder – Includes two runs of equal length with a quarter landing. The treads continue around the turn of the stair to create the winders.

Multi-landing – Can include stair runs of different lengths and the turns can include any of the following: quarter landing, quarter turn, half landing, and half turn.

Note: The Multi-landing stair shape is used to create straight stairs and stairs with landings.

Spiral – Develops a curved stair.

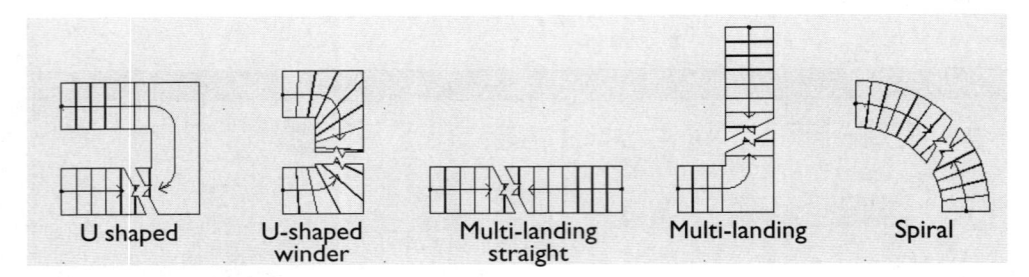

Figure 6.3 *Stair shapes*

Justification – The justification options for the stair are **Left**, **Center**, and **Right**. The justification option specifies the location of the handles for placing the stair. If you choose a left justified stair, the start and end points, identified when the stair is placed, will be located on the left of the stair.

Tread – The **Tread** edit box allows you to preset the dimension of the tread. The tread dimension can be locked or unlocked in the **Design Rules** tab of the **Stair Properties** dialog box. If it is locked in that dialog box, the **Tread** edit box here will be inactive.

Height – The **Height** edit field allows you to enter the finish floor to finish floor design dimension for the stair.

Width – The **Width** edit field allows you to enter the width of the stair. The width dimension is the total width of the stair unit. Therefore, the clear width can be reduced when stringers are housed. Figure 6.4 includes a 3'–0" stair with housed stringers on the left and saddled stringers on the right. Regardless of the construction style, the 3'–0" dimension is inclusive of all stair elements.

Next Turn – The Next Turn section includes the following radio buttons: **Quarter Landing**, **Quarter Turn**, **Half Landing**, and **Half Turn**. Each of these toggles can be used to determine the type of landing. When you create a multi-landing stair, a **Next Turn** radio button is used to define the style of

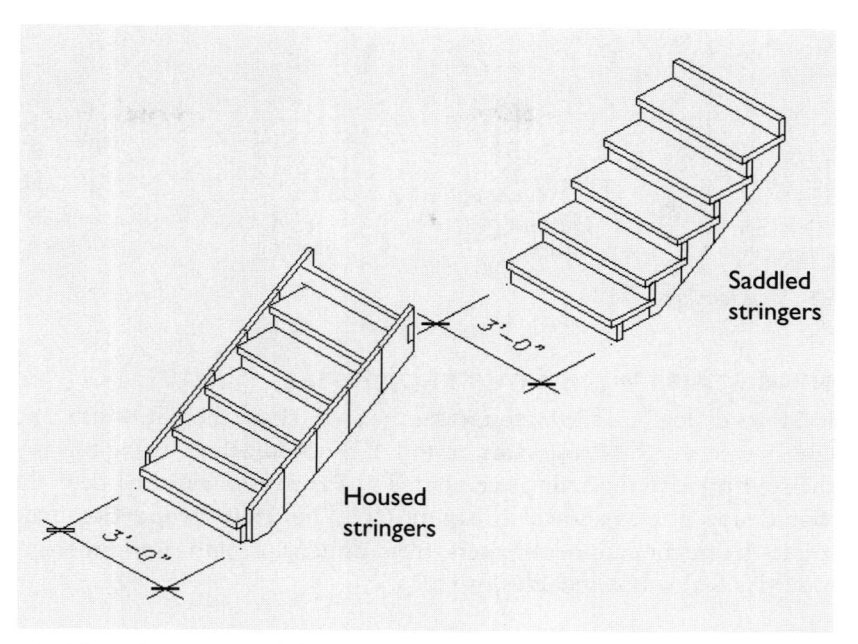

Figure 6.4 Housed stringers on left and saddled stringers on right

the landing. Examples of each style of landing are defined below and shown in Figure 6.5.

Quarter Landing – Turns a right angle and includes a flat landing.

Quarter Turn – Turns a right angle and the treads turn in the landing.

Half Landing – Reverses the direction of the stair and includes a flat landing.

Half Turn – Reverses the direction of the stair and the treads turn in the landing.

Floating Viewer – The **Floating Viewer** button allows you to preview the stair as currently defined in the **Add Stairs** dialog box prior to placement.

Match – The **Match** button allows you to select an existing stair to match its properties. It allows you match the style, width, height, tread, and open riser options from an existing stair.

Properties – The **Properties** button opens the **Stair Properties** dialog box, which includes tabs for setting the dimensions and code limitations of the stair.

Undo – Select the **Undo** button to remove the last action taken in the creation of the stair.

? – Select the **?** button to open the Help topic on Stairs.

Quarter landing   Quarter turn   Half landing   Half turn

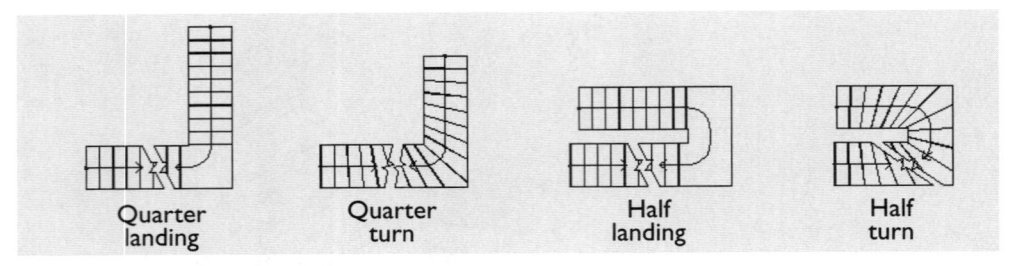

Figure 6.5 *Stair landings*

DESIGNING STAIRS WITH STAIR PROPERTIES

The **Add Stairs** dialog box allows you to set the style, shape, height, width, and landing design. In essence, the **Properties** button of the **Add Stairs** dialog box is used to design the stair prior to the stair placement. The **Properties** button opens the **Stair Properties** dialog box as shown in Figure 6.6. The **Stair Properties** dialog box allows you to define the number of risers, tread dimension, and riser dimension. The Stair Properties dialog box includes five tabs.

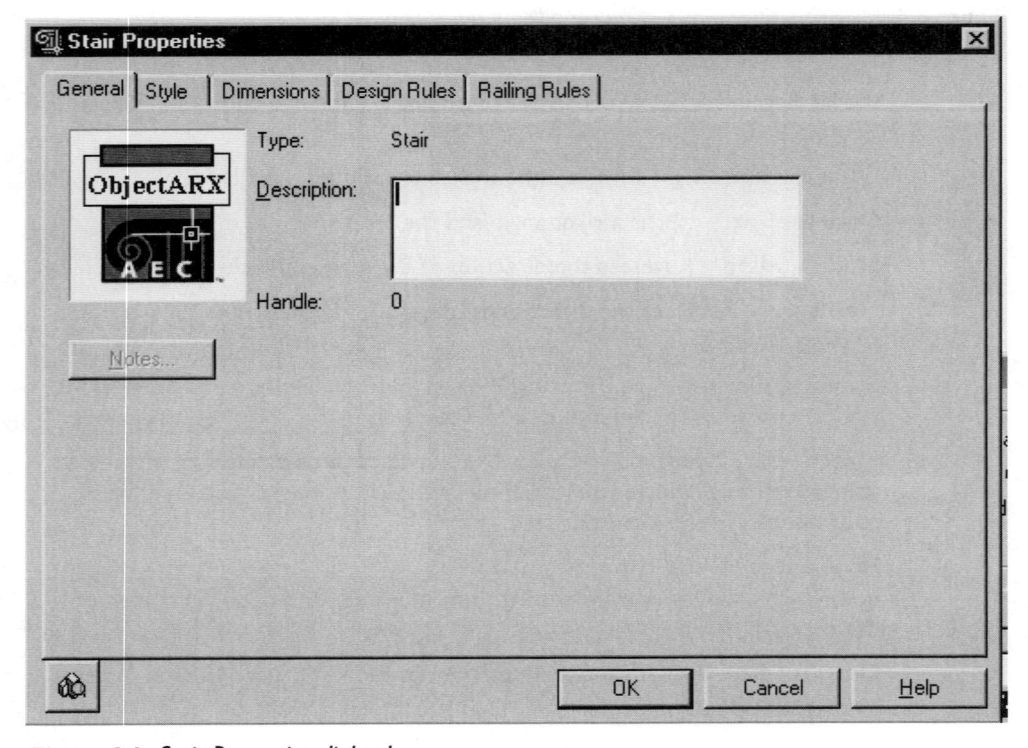

Figure 6.6 *Stair Properties dialog box*

The **Dimensions, Design Rules,** and **Railing Rules** tabs of the **Stair Properties** dialog box allow you to set the parameters for the development of the stair. The **Design Rules** and **Dimensions** tabs include edit fields that calculate the horizontal run, riser height and tread dimensions based upon the height of the stair and specified code rules. When creating a stair, you are prompted for the start point and end points, and then the stair is developed based upon the parameters specified in the tabs of the **Stair Properties** dialog box.

The **Stair Properties** dialog box for an existing stair is displayed in the dialog box of the **STAIRPROPS** command. It includes an additional tab, **Location**. See Table 6.2 for command access.

Command prompt	STAIRPROPS
Toolbar	Select Modify Stair; select the Properties button of the Modify Stair dialog box.
Shortcut	Right-click over drawing area; choose Design>Stairs>Properties
Shortcut	Select the stair; then right-click and choose Stair Properties

Table 6.2 STAIRPROPS access

General

The **General** tab of the **Stair Properties** dialog box allows you to enter a description for the stair. The description edit field includes an unlimited text field for placing a description regarding the location or specifications of the stair. The **Notes** button becomes active for an existing stair. The **Notes** button opens the **Notes** dialog box, containing two tabs: **Text Notes**, which allows you to add notes regarding the stair, and **Reference File**, used to list files related to the design project.

Style

The **Style** tab includes a list of the stair construction styles. These styles are also listed in the **Style** list of the **Add Stairs** dialog box. You can select a stair construction style from the list as shown in Figure 6.7.

Dimensions

The **Dimensions** tab includes edit fields for defining the width, length, rise, number of risers and other specific dimensions for the stair. The dimensions shown in the sketch on the **Dimensions** tab define stair values, as shown in Figure 6.8.

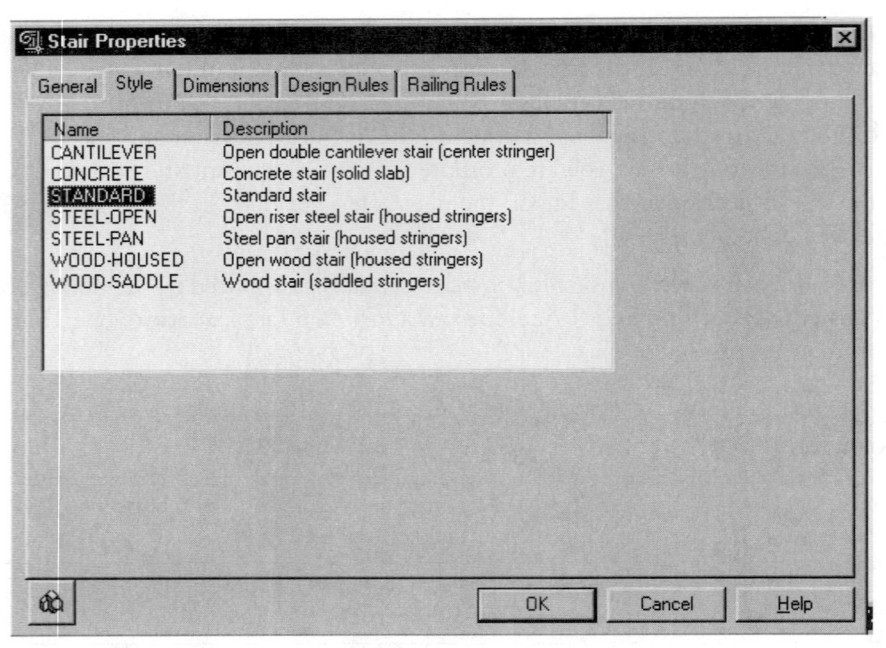

Figure 6.7 *Style tab of the Stair Properties dialog box*

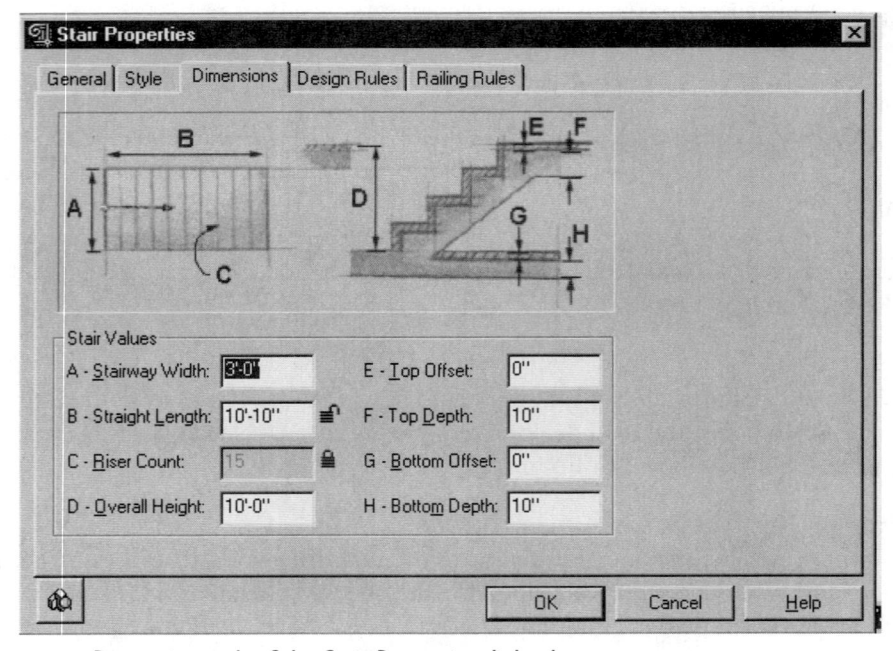

Figure 6.8 *Dimensions tab of the Stair Properties dialog box*

The dimensions in the **Stair Values** section are described below:

A-Stairway Width – Defines the width of the stair. Therefore the tread dimension is reduced when housed stringers are used. The stairway width dimension develops a stair to fit between existing walls.

B-Straight Length – The horizontal length or run of the stair. This length can be locked or unlocked. Selecting the lock to the right of edit field will lock the length to a specific value. If a straight length is specified and locked, the tread dimension in the **Design Rules** tab is automatically unlocked and adjusted according to the required number of treads for the run. The **Design Rules** tab includes upper and lower limit dimensions for the tread and riser, which control the number of risers and treads required.

C-Riser Count – The total number of risers required. The number of risers can be locked to a specific number, which causes the straight length to adjust according to the code rules of the **Design Rules** tab. If the riser count is unlocked, the quantity of risers can be specified within the range of riser and tread limit dimensions of the **Code Limits** section of the **Design Rules** tab.

D-Overall Height – The rise or vertical distance from finish floor to finish floor.

E-Top Offset – The thickness of the finish floor applied to the top tread.

F-Top Depth – The thickness of the supporting stringer at the top of the stair.

G-Bottom Offset – The thickness of the lower floor for the finish floor material at the bottom of the stair.

H-Bottom Depth – The thickness of the bottom floor.

Design Rules

The **Design Rules** tab, shown in Figure 6.9, consists of two sections: **Component Relationships**, which defines the dimensions of the risers, treads, and nosing, and **Code Limits**, which allows you to specify the range of dimensions acceptable for the riser and tread.

Component Relationships

J-Tr. Depth – The thickness of the tread when applied to the stringer.

K-R.Depth – The thickness of the riser as applied to the stringer.

L-Nosing – The distance the tread overhangs the riser. The nosing distance is defined in the stair style definition; therefore, it is inactive in the **Design Rules** tab.

M-Riser – The height of the riser or the vertical distance between adjacent treads. The riser height is calculated by the division of the total height by the number of risers.

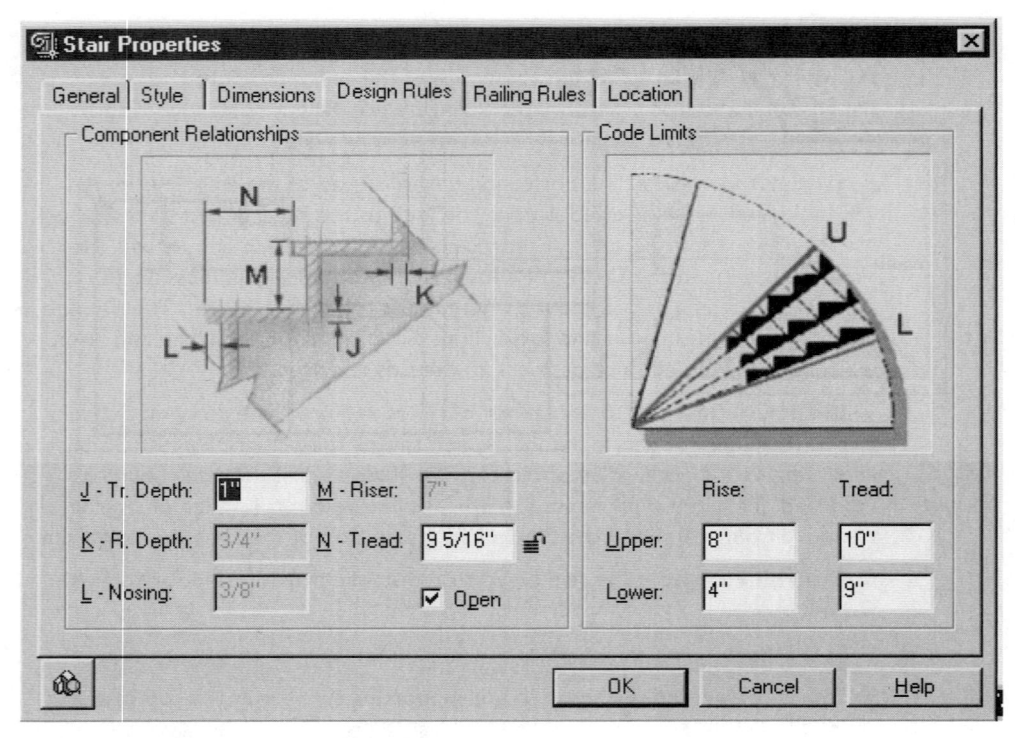

Figure 6.9 *Design Rules tab of the Stair Properties dialog box*

N-Tread– The **Tread** distance is the horizontal distance between nosings of adjacent treads. The tread distance can be locked to a specific dimension or unlocked. If it is unlocked, it is calculated by the division of the total straight length by the number of treads.

Open–If selected, creates a stair with open risers.

Code Limits

The **Code Limits** section of the **Design Rules** tab specifies the upper and lower limits of the riser and tread sizes as required by building code requirements.

Railing Rules

The **Railing Rules tab** includes two edit fields to define the distance to extend the handrail at the top and bottom landings. As shown in Figure 6.10, the extension of the handrail can be an absolute distance or, if the **Add Tread Depth** check box is selected, an absolute distance can be added to the tread depth.

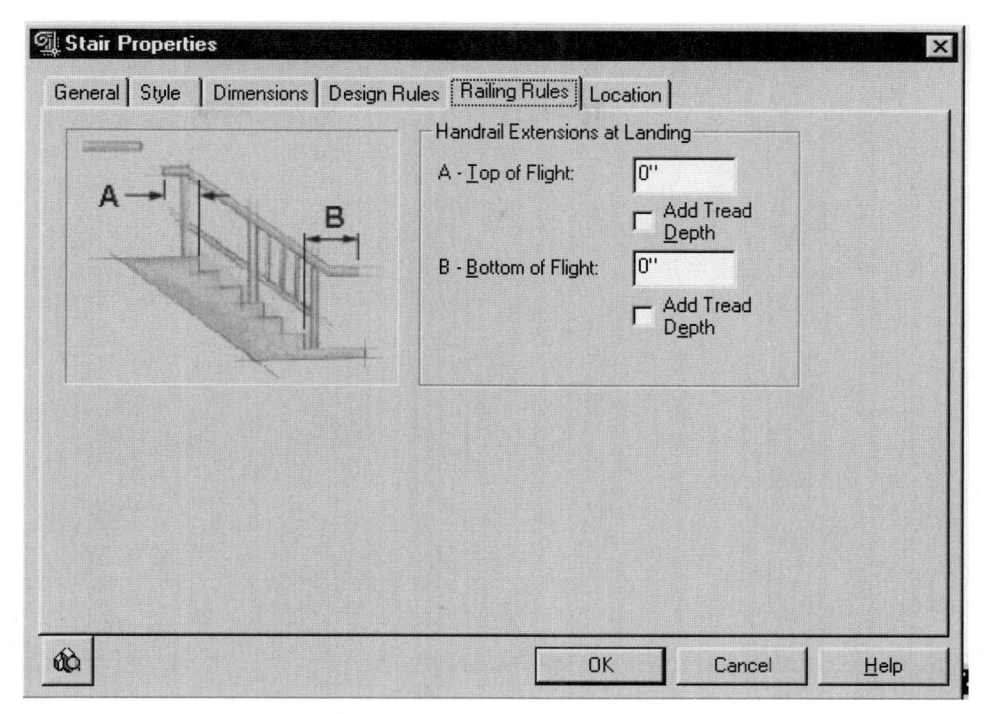

Figure 6.10 *Railing Rules tab of the Stair Properties dialog box*

Location

The **Location** tab of the **Stair Properties** dialog box is displayed if an existing stair is selected through the **STAIRPROPS** command. The **Location** tab defines the coordinates of the insertion point of the stair and its rotation. The Z coordinate can be edited to move the stair below the current floor level. The **Normal** edit fields allows the stair to be rotated perpendicular to a plane. In Figure 6.11, the number 1.0000 listed in the **Normal Z** column indicates that the stair is developed perpendicular to the XY plane. The **Location** tab is not included in the **Stair Properties** dialog box unless a stair has been placed.

You can accomplished the design of the stairs by setting the overall height, tread, straight length and riser count variables of the Dimensions and Design Rules tabs. The tread, straight length, and riser count variables can be locked. Two of these variables can be locked, allowing the third variable to change and create the stair design. If the riser count, tread dimension, or straight length is edited to conflict with the Code Limits of the Design Rules tab, Figure 6.12 shows the warning box that will be displayed.

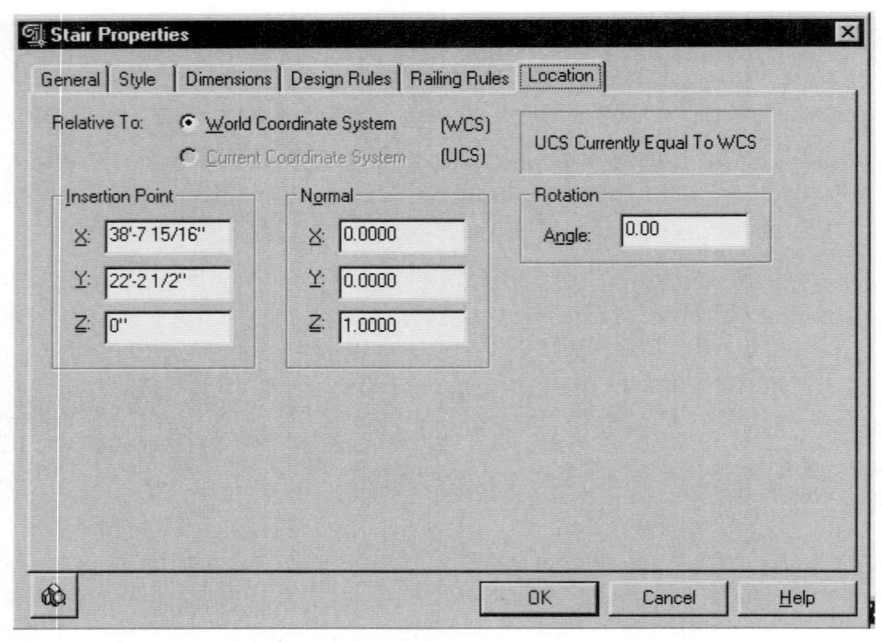

Figure 6.11 *Location tab of the Stair Properties dialog box*

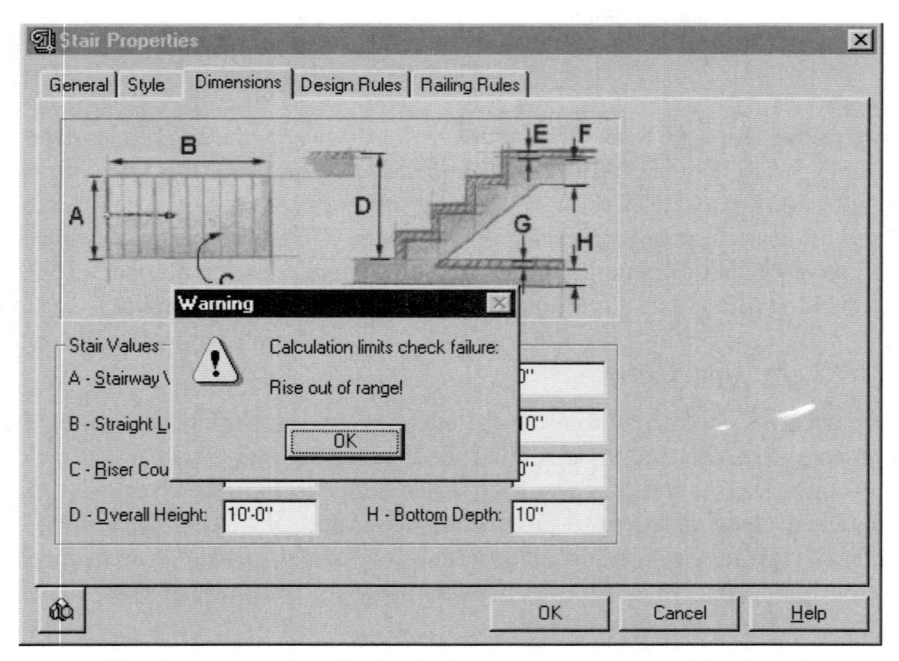

Figure 6.12 *Warning message box displayed when components exceed the Code Limits*

The overall height of a stair is fixed based upon the construction and finish floor to finish floor dimension. The number of risers is adjusted to establish a riser height within the range of the code limits. The number of risers determines the number of treads, which is always equal to the total number of risers minus one. Therefore the straight length is established by the multiplication of the tread dimension by one less than the total number of risers. The tread dimension can then be adjusted to reduce the overall straight length without the number of risers or treads changing.

The straight length can be locked to a specific dimension and the number of risers locked. When the straight length is locked, the tread size is adjusted to fit within the total straight length. Therefore, if the number of risers and the total straight length are locked, the tread dimension is adjusted accordingly.

If the tread dimension and the number of risers are both locked, the straight length is adjusted to equal the product of the tread dimension multiplied by one less than the total number of risers.

> **Note:** The default values of the **Add Stairs** and **Stair Properties** dialog boxes are not saved with the drawing. Therefore, the default values are those set by previous users and may require you to edit the **Dimensions** and **Design Rules** tabs to clear the **Warning: Calculation Limits Check Failure** message box.

STEPS TO CREATING A STAIR WITH A FIXED RISE AND NUMBER OF RISERS

The design of a stair is based upon the vertical rise or height for the stair and the establishment of the number of risers. The rise of the stair is determined from a wall section that includes specific dimensions regarding the floor system and ceiling height. Once the finish floor to finish floor dimension is determined, it is entered in the **Height** edit field of the **Add Stairs** dialog box. The **Dimensions** and **Design Rules** tabs of the **Stair Properties** dialog box can then be used to restrain the design within the code limitations. Given a fixed rise and fixed number of risers, the riser and tread dimensions are adjusted to comply with the code restrictions. If you select a number beyond the code limitation, a warning box will be displayed. Select **OK** to dismiss the warning box and increase the number of risers. The following steps illustrate the design of a stair for a residence.

1. Determine the rise or the vertical dimension between finish floor to finish floor as shown in Figure 6.13.
2. Select the **Add Stair** command on the **Stairs/Railings** toolbar.
3. Enter the rise of 8'-11" in the **Height** edit field of the **Add Stairs** dialog box.
4. Select the **Properties** button of the **Add Stairs** dialog box.

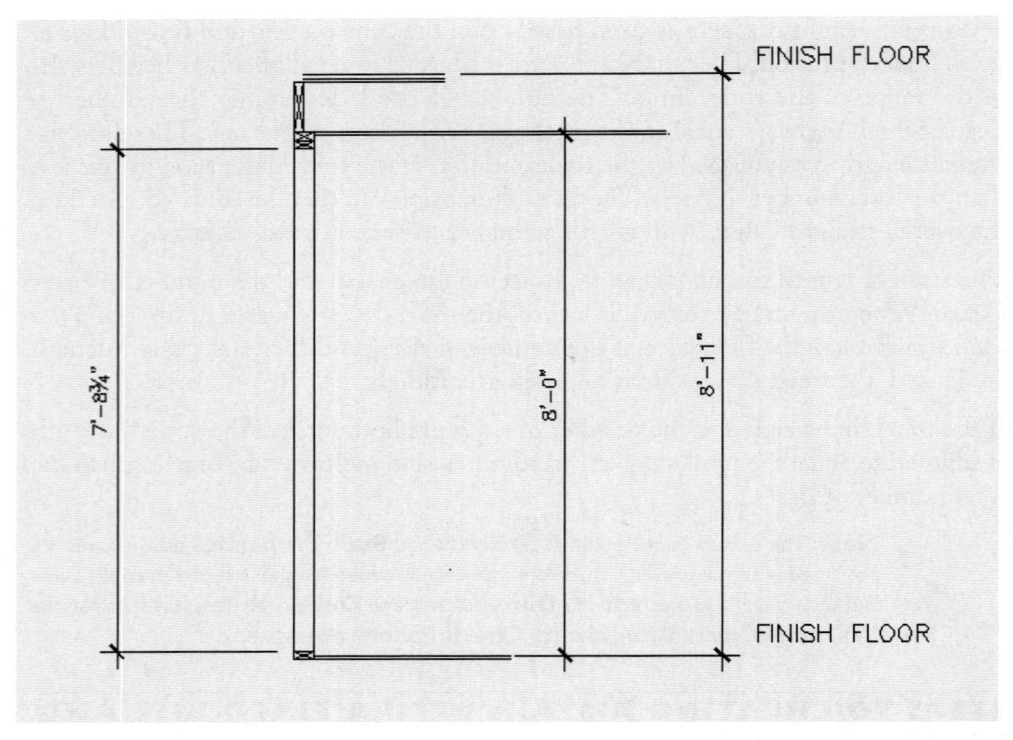

Figure 6.13 Rise of stair

5. Select the **Design Rules** tab of the **Stair Properties** dialog box.
6. Edit the **Code Limits** to **Rise Upper** = 7-3/4", **Rise Lower** = 4", **Tread Upper** = 1'–0", and **Tread Lower** = 10".
7. Select the lock to unlock **N-Tread** and set the **N-Tread** to the minimum tread value of 10" as shown in Figure 6.14.
8. Select the **Dimensions** tab, unlock the **C-Riser Count** and edit the **C-Riser Count** to 14, as shown in Figure 6.15.
9. Note that the **B-Straight Length** is adjusted to 10'–10". Select the **Design Rules** tab and note that **M-Riser** is 7-5/8" and **N-Tread** is 10".
10. Check an alternative solution by selecting the **Dimensions** tab and editing the **C-Riser Count** to 15.
11. Note that the **B-Straight Length** is adjusted to 11'–8"; select the **Design Rules** tab and note that **M-Riser** is 7-1/8" and **N-Tread** is 10".
12. Therefore several solutions are available that comply with the code limitations.

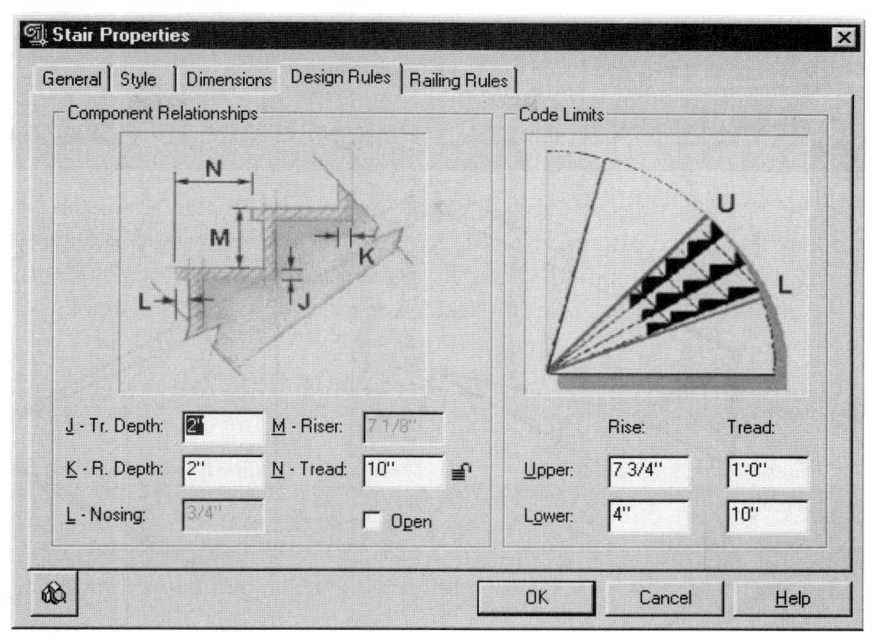

Figure 6.14 *Code Limits of Design Rules tab*

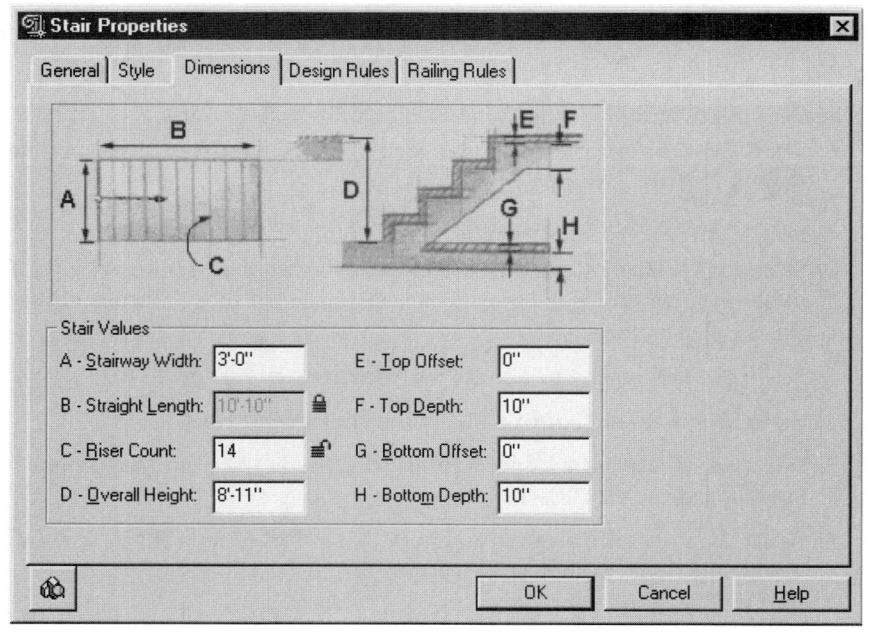

Figure 6.15 *Dimensions tab of the Stair Properties dialog box*

13. Select **OK** to dismiss the **Stair Properties** dialog box.
14. Select start and end locations for the stair flight.

STEPS TO CREATING A STAIR WITH A FIXED RUN AND FIXED RISE

If the rise and run of a stair are fixed, the **Dimensions** and **Design Rules** tabs can be used to determine the minimum straight length with the minimum tread size. After determining the minimum straight length, you can then set the straight length to a fixed value greater than the minimum and determine the associated tread and riser dimensions.

The following steps illustrate the use of the **Stair Properties** dialog box to design a stair with a fixed rise and fixed run for a residence.

1. Determine the rise or the vertical dimension from finish floor to finish floor as shown in Figure 6.13.
2. Select the **Add Stair** command on the **Stairs/Railings** toolbar.
3. Enter the rise of 8'–11" in the **Height** edit field in the **Add Stairs** dialog box.
4. Select the **Properties** button of the **Add Stairs** dialog box.
5. Select the **Design Rules** tab of the **Stair Properties** dialog box.
6. Edit the **Code Limits** for the upper and lower values for the riser and tread as shown in Figure 6.14.
7. Set the **N-Tread** to 10", the minimum acceptable tread size.
8. Select the **Dimensions** tab and unlock the **C-Riser Count** lock, and then set the **C-Riser Count** to 14.
9. Note that the **B-Straight Length** is equal to 10'–10".
10. Select the **C-Riser Count** edit field and decrease the riser count to 13. The **Warning** box shown in Figure 6.16 is displayed because this value exceeds the code limitations.
11. Select the **OK** button to dismiss the **Warning** dialog box.
12. Select the **C-Riser Count** and edit the riser count to 14.
13. Select the **B-Straight Length** to unlock the edit field, and then enter 11'–0" (a value greater than 10'–10").
14. Select the **Design Rules** tab and note that the **M-Rise** is 7-5/8" and the **N-Tread** is 10-1/8" as shown in Figure 6.17.

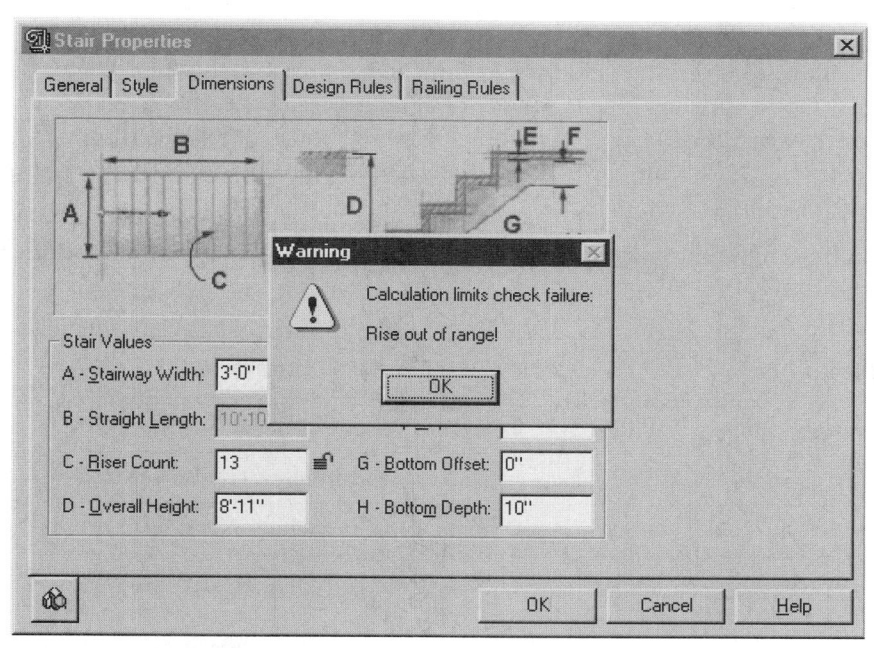

Figure 6.16 *Warning box indicating design exceeds the code limitations*

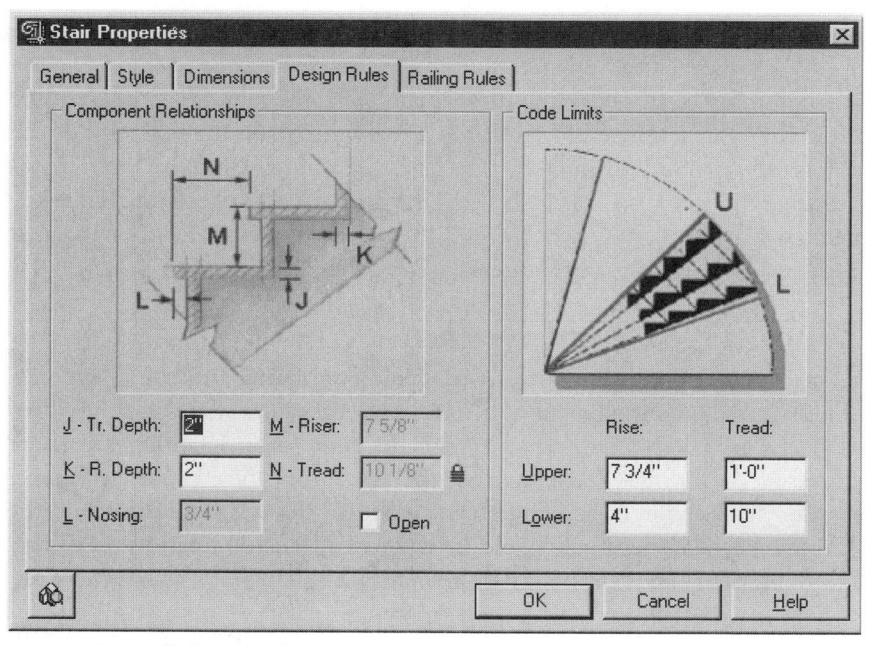

Figure 6.17 *Design Rules tab of the Stair Properties dialog box*

STAIR DEFAULT SETTINGS

The default settings for the **Add Stair** command are preset in the **AEC DwgDefaults** tab of the **Options** dialog box. If you are designing a residence, the initial default settings for the stair should be set in that dialog box. If you set the default settings prior to selecting the **Add Stair** command, fewer **Warning** dialog boxes will be displayed when the stair properties are set.

See Table 6.3 for command access to the **AEC DwgDefaults** tab of the **Options** dialog box.

Menu bar	Tools>Options
Command prompt	OPTIONS
Shortcut	Right-click over the command window; choose Options
Shortcut	Right-click over the drawing area; choose Options

Table 6.3 *Options access*

Selecting the **OPTIONS** command opens the **Drawing Defaults** tab of the **Options** dialog box, as shown in Figure 6.18.

The four tabs on the right of the **Options** dialog box are unique to Architectural Desktop: **AEC Editor**, **AEC DwgDefaults**, **AEC Performance**, and **AEC Content**. (Due to the number of tabs of the **Options** dialog box, the tabs are wider than the dialog box and therefore you need to select the arrow buttons on the top right of the dialog box to scroll right and display the tabs.) The **AEC DwgDefaults** tab includes the **Stair Settings** section (the upper and lower limits for the riser and tread) and the **Railing Settings** section, which will preset the upper and lower limits settings of the **Design Rules** tab of the **Stair Properties** dialog box.

CREATING STRAIGHT STAIRS

Straight stairs are created with the **Add Stair** command and the Multi-landing shape. The Multi-landing shape is also used to create stairs with landings. The stair object is placed on the A-Flor-Strs layer, which has the color 12. Stairs are created up from the zero Z coordinate. After setting the stair properties, you select a point for the beginning of the stair. Specifying an endpoint for the stair that is beyond the end of the stair creates a straight stair; because the point selected for the end of the stair is beyond the end of the stair, no landing is created.

To place a straight stair, select the **Add Stair** command and edit the **Add Stairs** dialog box to establish the height, width and stair properties. An example of the points

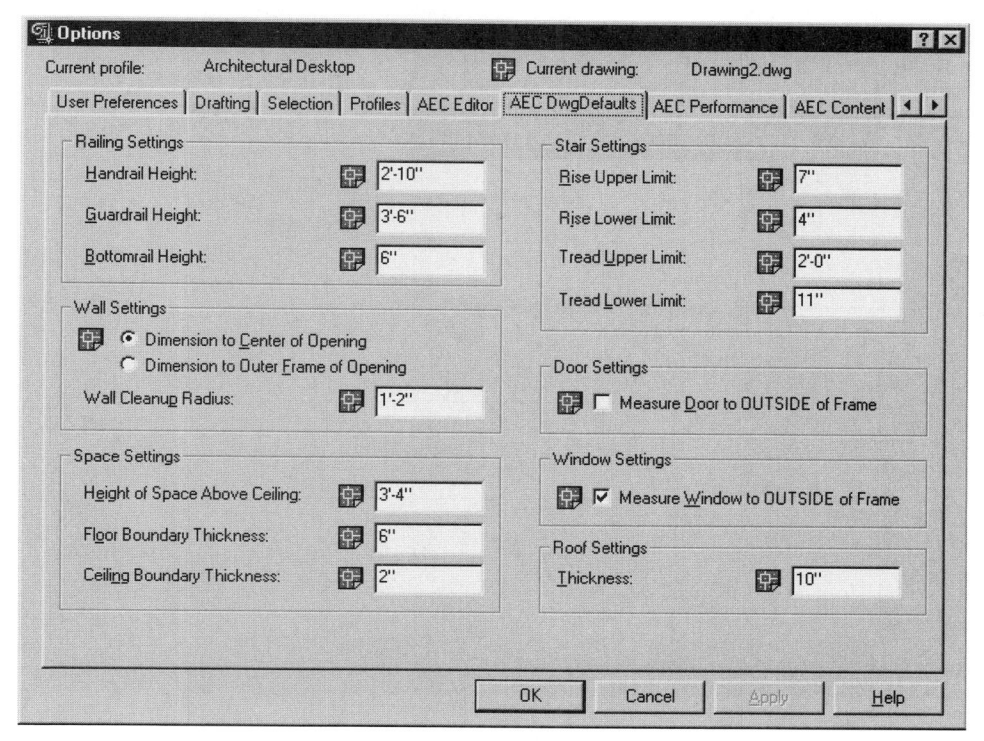

Figure 6.18 *The AEC DwgDefaults tab of the Options dialog box*

selected to create a straight stair is shown in the following command sequence and Figure 6.19.

```
Command: _AecStairAdd
Flight start point or
[SHape/STyle/Tread/Height/Width/Justify/Open/Next/Match/Undo]: (Select P1 at the endpoint of the wall in Figure 6.19.)
Flight end point or
[SHape/STyle/Tread/Height/Width/Justify/Open/Next/Match/Undo]: (Select point P2 in Figure 6.19.)
Flight start point or
[SHape/STyle/Tread/Height/Width/Justify/Open/Next/Match/Undo]: ENTER
```

The straight stair created by the selection of points P1 and P2 is shown in Figure 6.20.

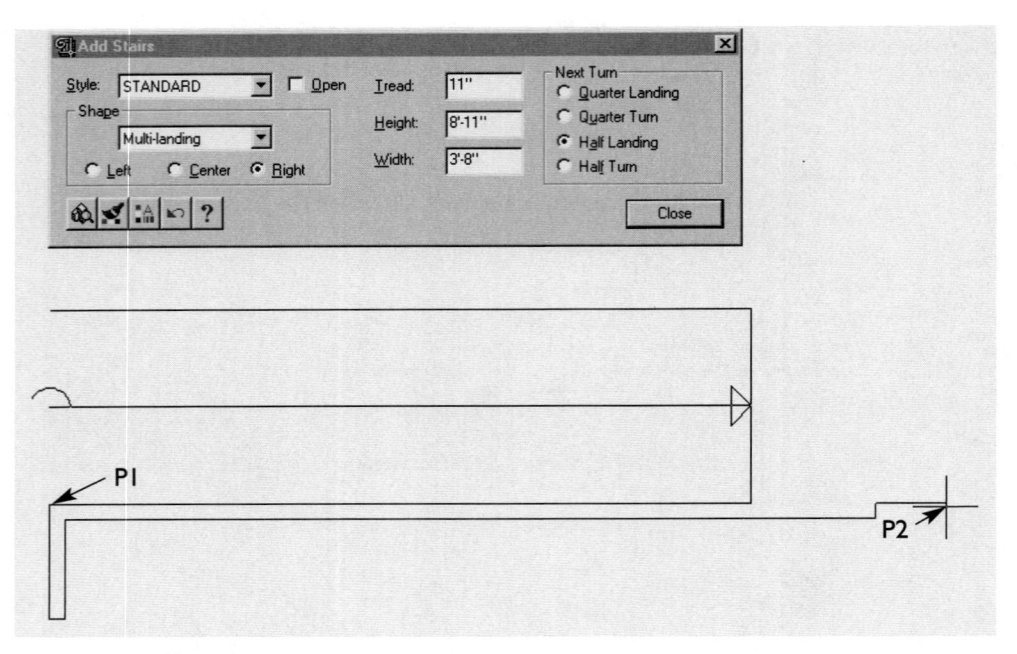

Figure 6.19 Start and end points specified for a straight stair

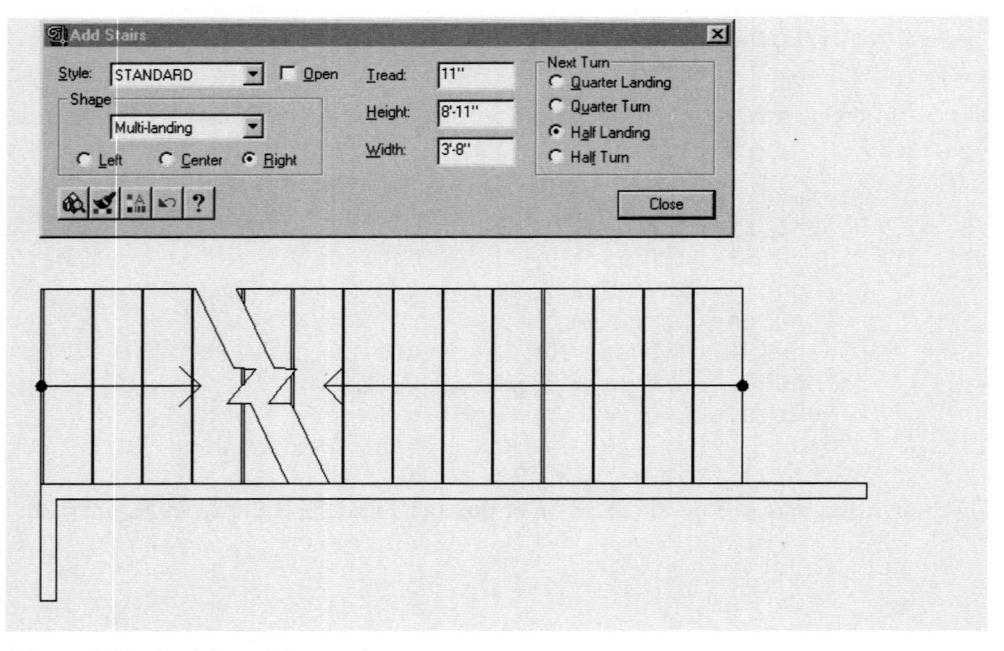

Figure 6.20 Straight stair inserted

TUTORIAL 6.1 CREATING A STRAIGHT STAIR

1. Open *ADT Tutor\Ch6\EX6-1.dwg*
2. Choose **File>SaveAs** from the menu bar and save the drawing as *Straight1.dwg* in your directory.
3. Right-click over the command window and choose **Options**.
4. Select the **AEC DwgDefaults** tab of the **Options** dialog box.
5. Refer to Figure 6.21 and edit the **Stair Settings** section as follows: **Rise Upper Limit** = 7-3/4", **Rise Lower Limit** = 6", **Tread Upper Limit** = 12", **Tread Lower Limit** = 10".

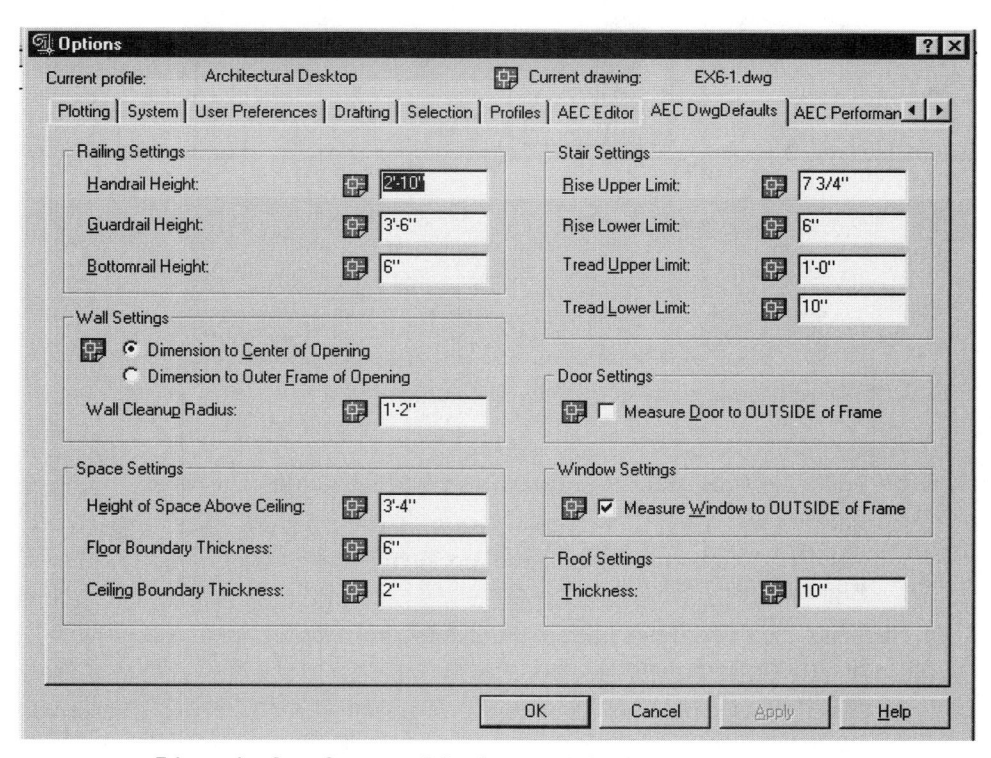

Figure 6.21 *Editing the Stair Settings of the Options dialog box*

6. Select **OK** to dismiss the **Options** dialog box.
7. Select the Section layout tab and verify that the rise of the stair equals 8'–10 1/2" as shown in Figure 6.22.
8. Select the Work-FLR layout tab.

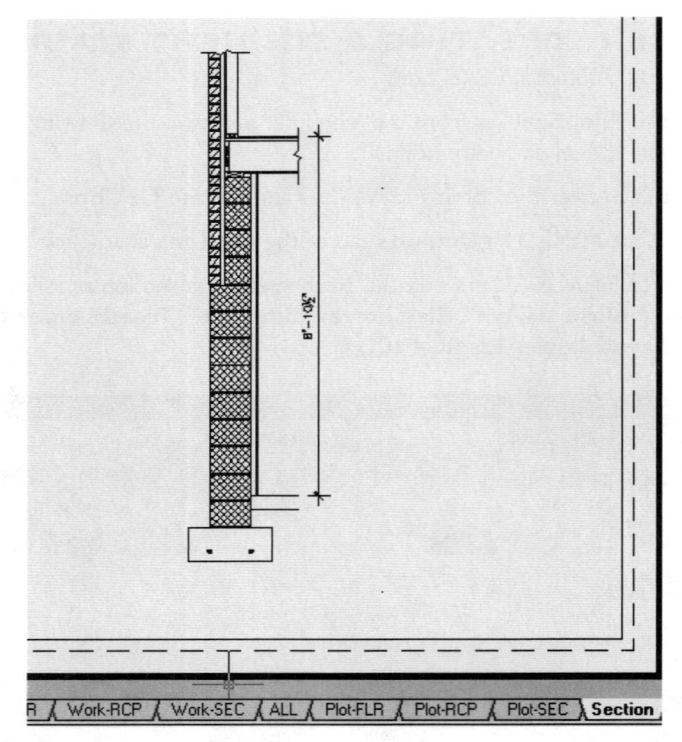

Figure 6.22 Determine the stair rise from the wall section

9. Select the **Add Stair** command from the **Stairs/Railings** toolbar.
10. Refer to Figure 6.23 and edit the **Add Stairs** dialog box as follows: **Style** = Standard, **Shape** = Multi-landing, **Next Turn** = Half Landing, and select the **Right** radio button for right justification.
11. Select the **Properties** button of the **Add Stairs** dialog box to open the **Stair Properties** dialog box.
12. Select the **Dimensions** tab and set the **A-Stairway Width** to 3'–0". Unlock the **B-Straight Length** lock and set it to 11'–8-1/2". Edit the **Overall Height** to 8'10-1/2" as shown in Figure 6.24.
13. Select the **Design Rules** tab and note that the **M-Riser** dimension is 7-1/8" and the **N-Tread** dimension is 10-1/16 as shown in Figure 6.25.
14. Select **OK** to dismiss the **Stair Properties** dialog box.
15. Refer to the following command sequence and specify the location of the stair as shown in Figures 6.26, 6.27, and 6.28.

(Select the drawing area to activate the drawing area.)

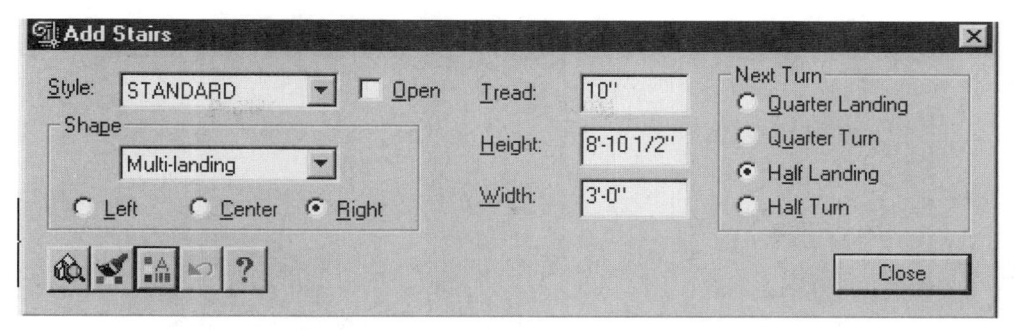

Figure 6.23 Editing Add Stairs dialog box

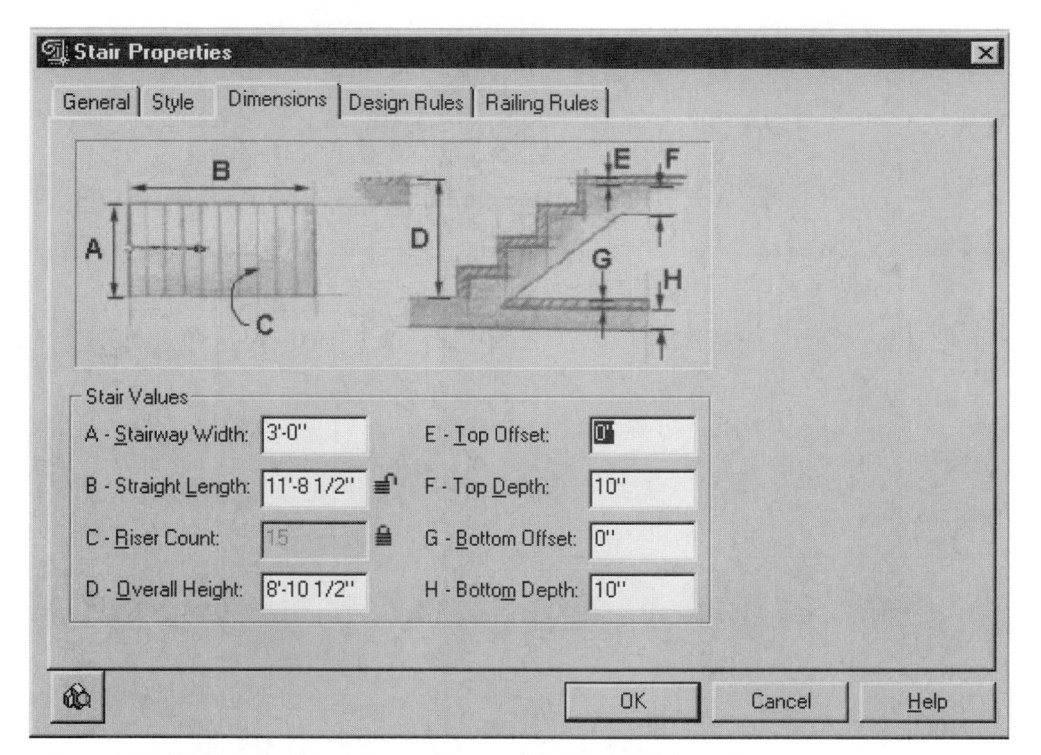

Figure 6.24 Editing the Dimensions tab to specify the run of the stair

```
Command: _AecStairAdd
Flight start point or
[SHape/STyle/Tread/Height/Width/Justify/Open/Next/Match/Undo]: (Select point P1 in Figure 6.26.)
```

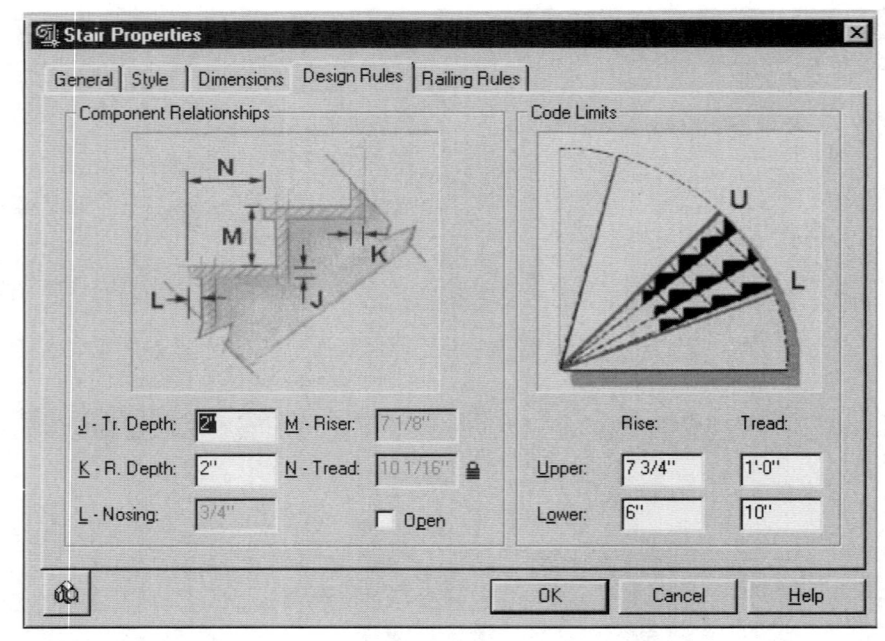

Figure 6.25 Riser and tread sizes displayed in the Design Rules tab

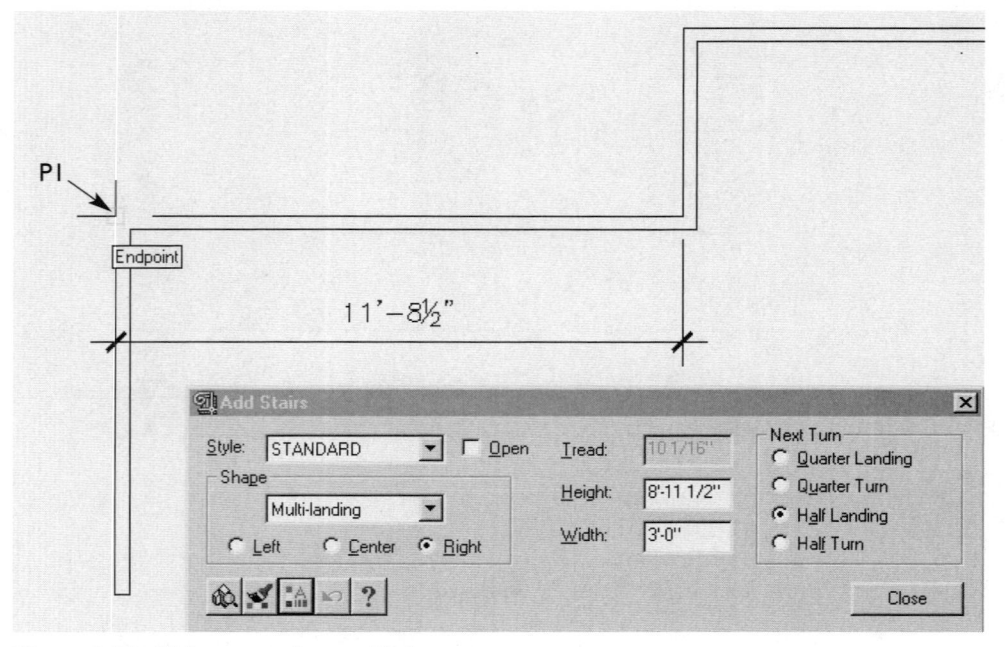

Figure 6.26 Flight start point specified

```
Flight end point or
[SHape/STyle/Tread/Height/Width/Justify/Open/Next/
Match/Undo]: (Select point P2 beyond the end of the wall as shown in
Figure 6.27.)
```

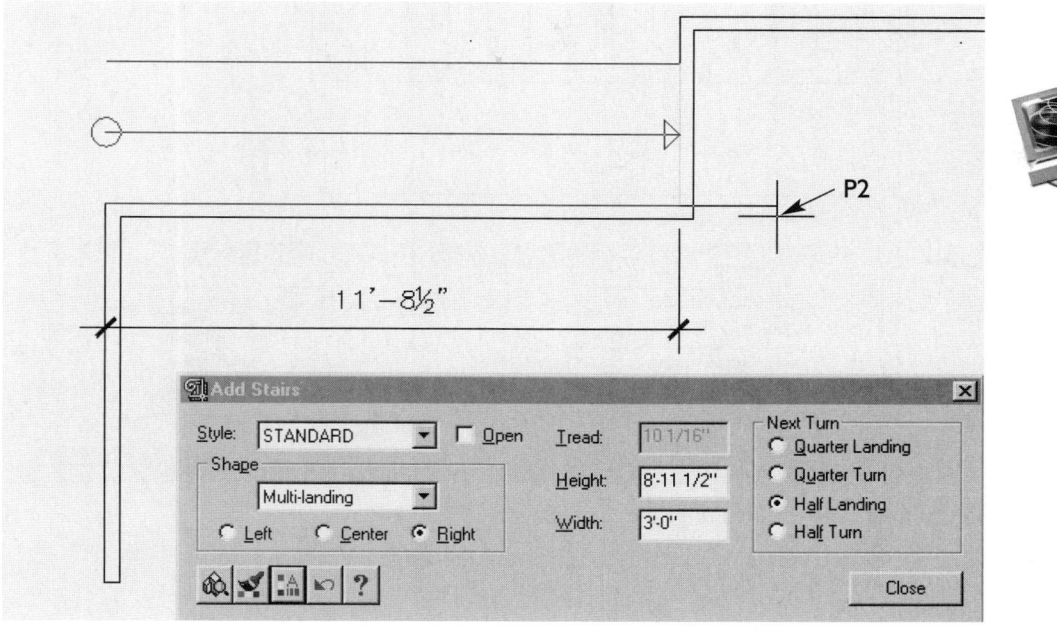

Figure 6.27 *End of flight specified*

```
Flight start point or
[SHape/STyle/Tread/Height/Width/Justify/Open/Next/
Match/Undo]: ENTER
(Stair is created as shown in Figure 6.28.)
```
16. Save the drawing.

CREATING U SHAPED STAIRS

The U Shape and U Shape Winder stair shapes are used to create U Shaped stairs. U Shaped stairs are divided into two equal runs. They can include a flat landing or a winder landing between the two runs. A flat landing is created with the U Shape stair, while the U Shape Winder continues the treads in the landing space using the Half Turn landing, as shown in Figure 6.29.

The stair is created based upon the height and width specified in the **Add Stairs** dialog box. Two equal stair runs are calculated based upon the total rise and width of the stair. Selecting the U Shaped stair shape turns off all landing styles in the **Next Turn**

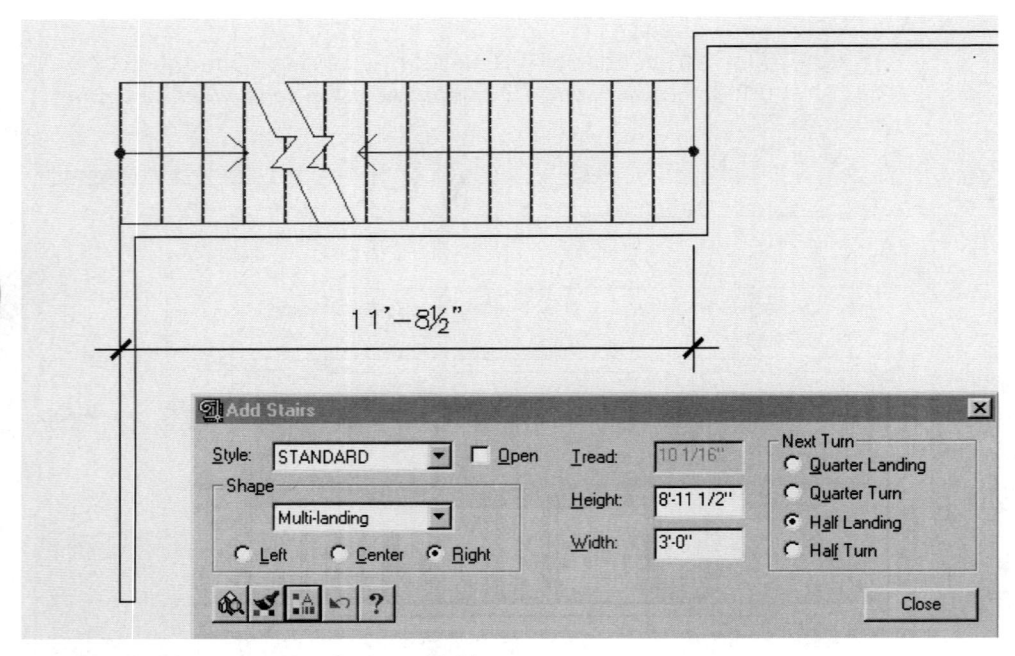

Figure 6.28 *Stair created to fit the available run*

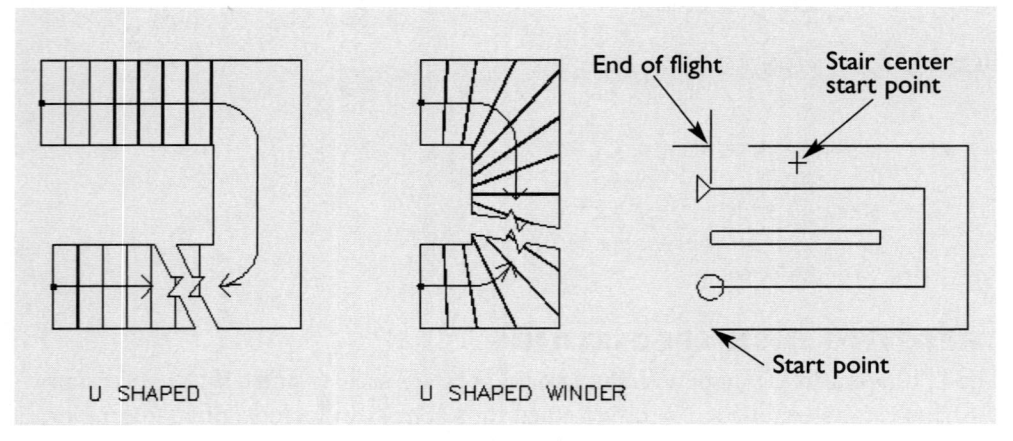

Figure 6.29 *U Shaped and U Shaped Winder stair shapes*

section of the **Add Stairs** dialog box except the **Half Landing** style. Therefore the U Shape stair is created with a flat landing.

After selecting the U Shaped style of stair, you are prompted for the beginning and end of the stair as shown in Figure 6.29. The beginning and end of the stair specify the total width of the stair, and the length of the stair is calculated in the **Design Rules** tab of the **Stair Properties** dialog box. After establishing the beginning and end points, you are prompted for the stair center, as shown in Figure 6.29, to locate which side of the start and end points to place the stair. The stair center can be specified anywhere in the intended area of the stair. You are not prompted for a specific landing location because the two runs are equal. After the stair is created, you can move it using the **MOVE** command to fit existing walls.

STEPS TO CREATING A U SHAPED STAIR

1. Select **Add Stair** from the **Stairs/Railings** toolbar.
2. Edit the **Add Stairs** dialog box specifying the dimensions of the tread, height, and width of the stair. Edit the **Shape** to U-shaped and set the justification to **Right** as shown in Figure 6.30.

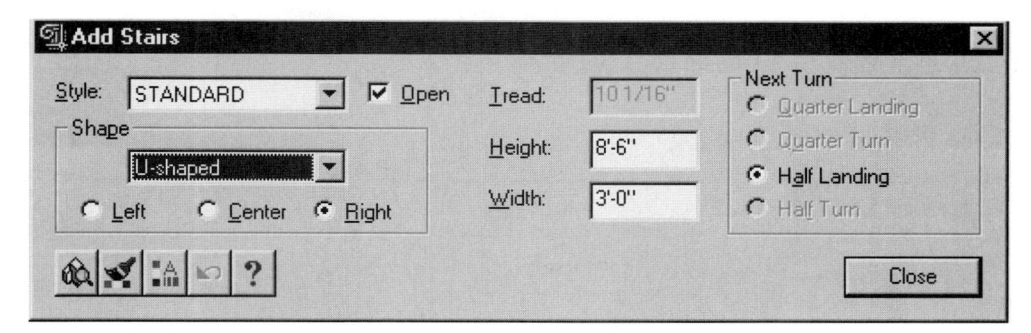

Figure 6.30 *Editing Add Stairs dialog box to create U Shaped stairs*

3. Respond to the command sequence as shown below.

```
Command: AECSTAIRADD
Flight start point or
[SHape/STyle/Tread/Height/Width/Justify/Open/Next/Mat
ch/Undo]: DBOX
```
(Edit Shape to U-shaped and justification to Right.) (Select the drawing area to activate the drawing area)
```
Flight start point or
[SHape/STyle/Tread/Height/Width/Justify/Open/Next/
Match/Undo]:
```
(Select point P1 in Figure 6.31.)
```
Flight end point or
[SHape/STyle/Tread/Height/Width/Justify/Open/Next/
Match/Undo]:
```
(Select point P2 in Figure 6.31.)
```
U-Stair center or
[SHape/STyle/Tread/Height/Width/Justify/Open/Next/
Match/Undo]:
```
(Select point P3 in Figure 6.32 to specify the location of the stair.)

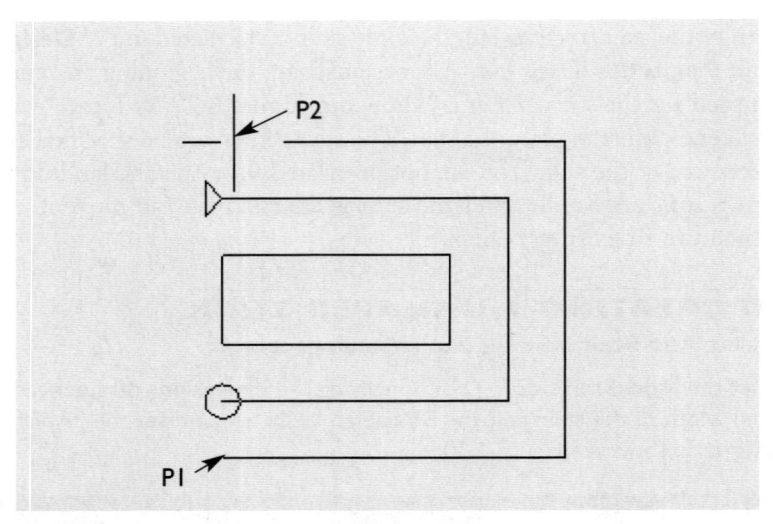

Figure 6.31 *Selecting start and end points of the stair*

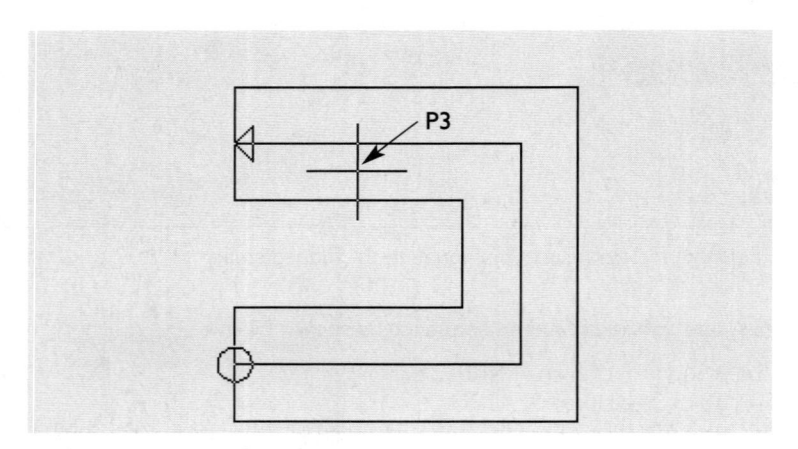

Figure 6.32 *Selecting center of a stair*

```
Flight start point or
[SHape/STyle/Tread/Height/Width/Justify/Open/Next/
Match/Undo]: ENTER
```
The U Shaped stair is created as shown in Figure 6.33.

If the U Shaped Winder shape is used for the creation of the stairs, the **Half Turn** landing is used. Selecting U-shaped Winder in the **Add Stairs** dialog box turns off all landing styles in the **Next Turn** section except the **Half Turn** landing, as shown in Figure 6.34.

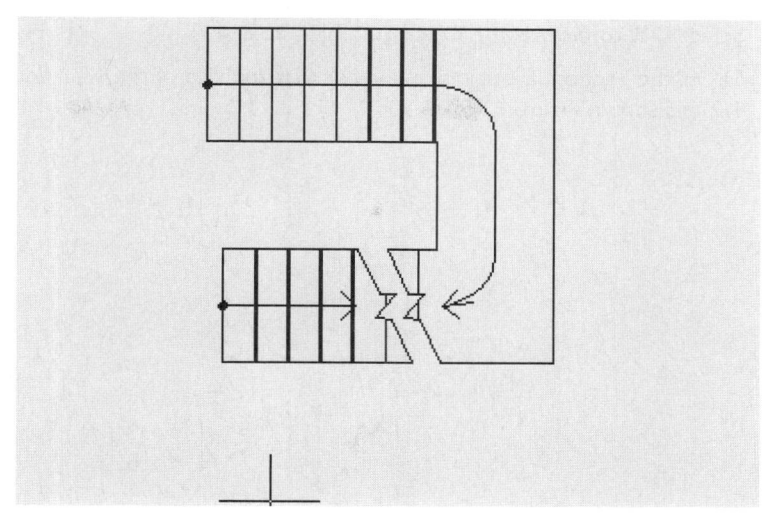

Figure 6.33 U Shaped stair

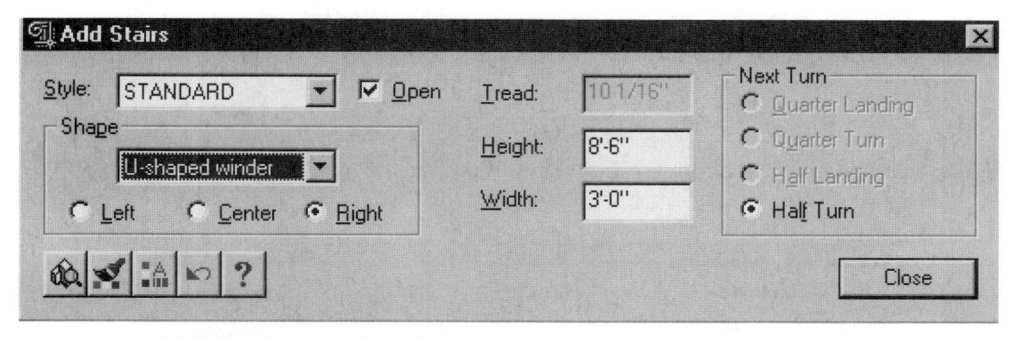

Figure 6.34 Half Turn landing of the U Shaped Winder stair shape

TUTORIAL 6.2 CREATING A U SHAPED STAIR

1. Open *ADT Tutor\Ch6\EX6-2.dwg*.

2. Choose **File>SaveAs** from the menu bar and save the drawing as *Ushaped2.dwg* in your student directory.

3. Right-click over the drawing area and choose **Options** from the shortcut menu.

4. Select the **AEC DwgDefaults** tab in the **Options** dialog box.

5. Edit the **Stair Settings** section of the **AEC DwgDefaults** tab as follows: **Rise Upper Limit** = 7-3/4", **Rise Lower Limit** = 6", **Tread Upper Limit** = 12", **Tread Lower Limit** = 10".

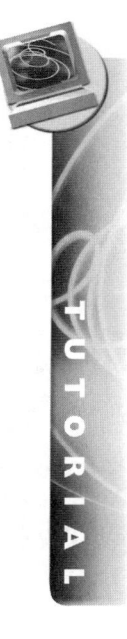

6. Select **OK** to dismiss the **Options** dialog box.
7. Select the Section layout tab and verify that the rise of the stair equals 9'–0 1/2", as shown Figure 6.35.

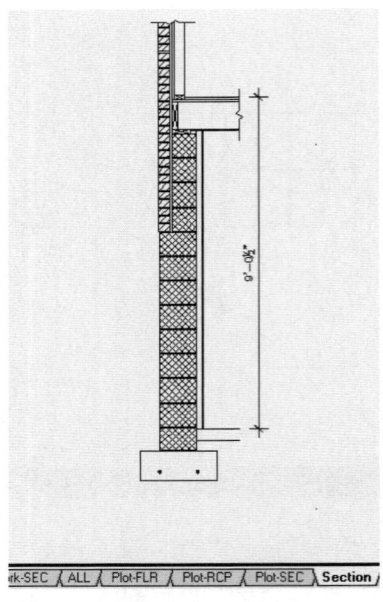

Figure 6.35 *Determine the stair rise from the wall section*

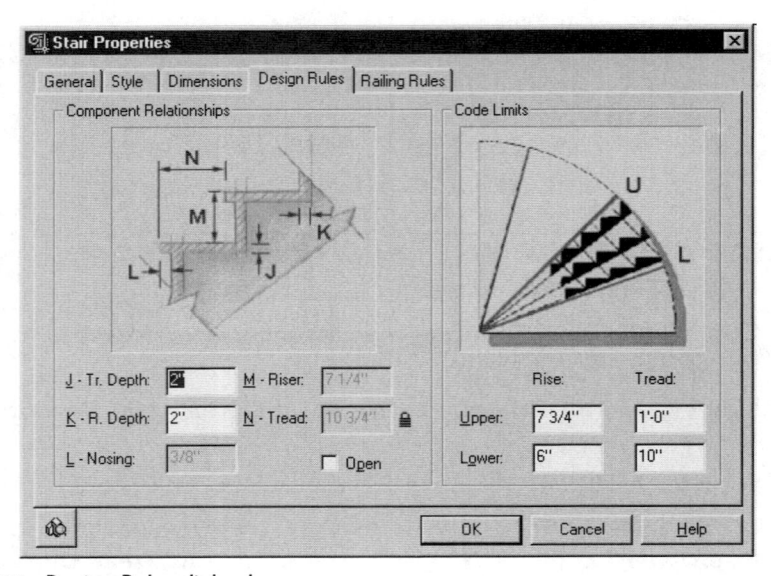

Figure 6.36 *Design Rules dialog box*

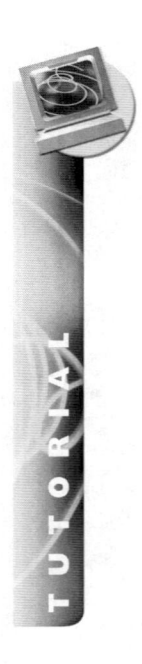

8. Select the Work-FLR layout tab.
9. Select F3 to turn off Running Object Snaps.
10. Select the **Add Stair** command from the **Stairs/Railings** toolbar.
11. Select the **Properties** button to open the **Stair Properties** dialog box.
12. Select the **Dimensions** tab and edit the **C-Riser Count** to 15 and **D-Overall Height** to 9'–0-1/2".
13. Select the **Design Rules** tab and note that the **M-Riser** dimension is 7-1/4 and the **N-Tread** dimension is 10-3/4, as shown in Figure 6.36.
14. Select **OK** to dismiss the **Stair Properties** dialog box.
15. Refer to Figure 6.37 and edit the **Add Stairs** dialog box as follows: **Style** = Standard, **Shape** = U-shaped, **Width** = 3'–0", **Next Turn** = Half Landing, and select the **Left** justification radio button.
16. Refer to the following command sequence and specify the location of the stair as shown in Figures 6.38, 6.39, and 6.40.

    (Select the drawing area to activate the drawing area)
    ```
    Command: _AecStairAdd
    Flight start point or
    [SHape/STyle/Tread/Height/Width/Justify/Open/Next/
    Match/Undo]:endp ENTER
    of
    ```
    (Select point P1 in Figure 6.38.)

    ```
    Flight end point or
    [SHape/STyle/Tread/Height/Width/Justify/Open/Next/
    Match/Undo]: perp ENTER
    to
    ```
    (Select point P2 in Figure 6.39.)

    ```
    U-Stair center or
    [SHape/STyle/Tread/Height/Width/Justify/Open/Next/
    Match/Undo]:
    ```
    (Select a point near point P3 in Figure 6.40.)
    ```
    Flight start point or
    [SHape/STyle/Tread/Height/Width/Justify/Open/Next/
    Match/Undo]: ENTER
    ```

17. Select **NW Isometric** from the **Standard** toolbar to display the stair as shown in Figure 6.41.
18. Select **Top View** from the **Standard** toolbar.
19. Select **Move** from the **Modify** toolbar and move the stair as specified in the following command sequence.

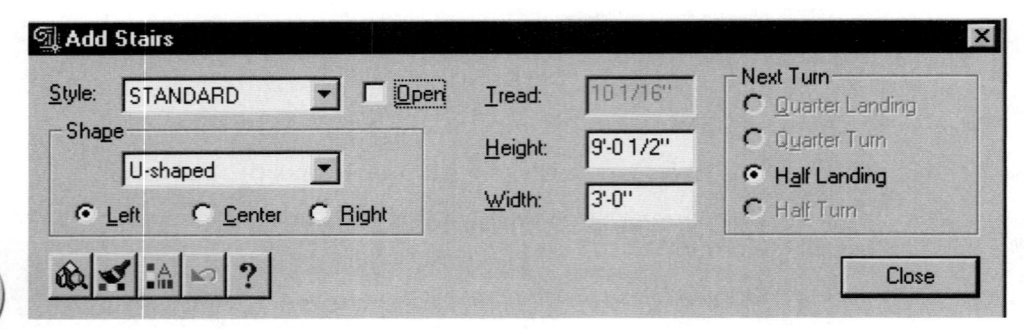

Figure 6.37 *Add Stairs dialog box*

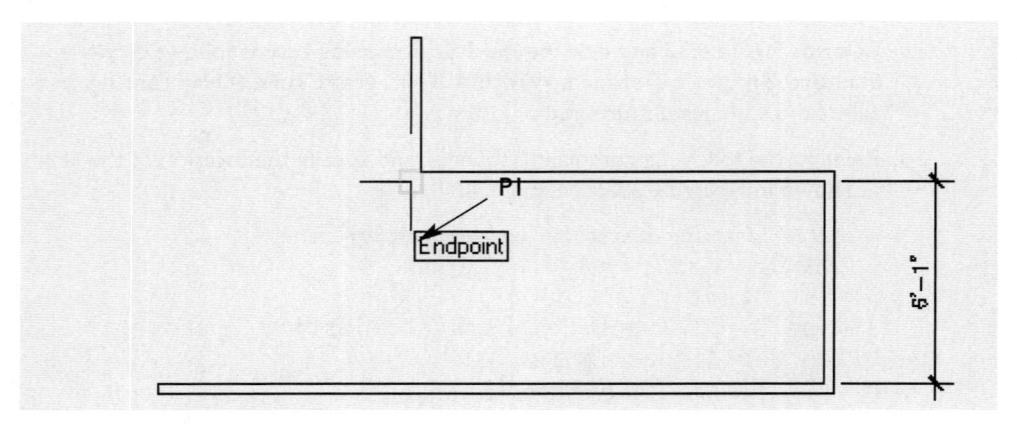

Figure 6.38 *Flight start point specified*

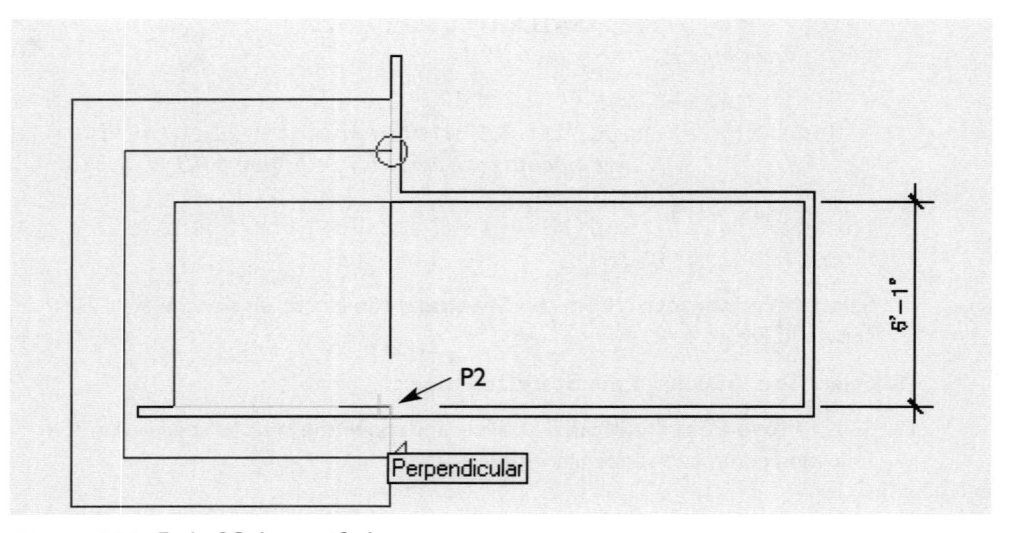

Figure 6.39 *End of flight specified*

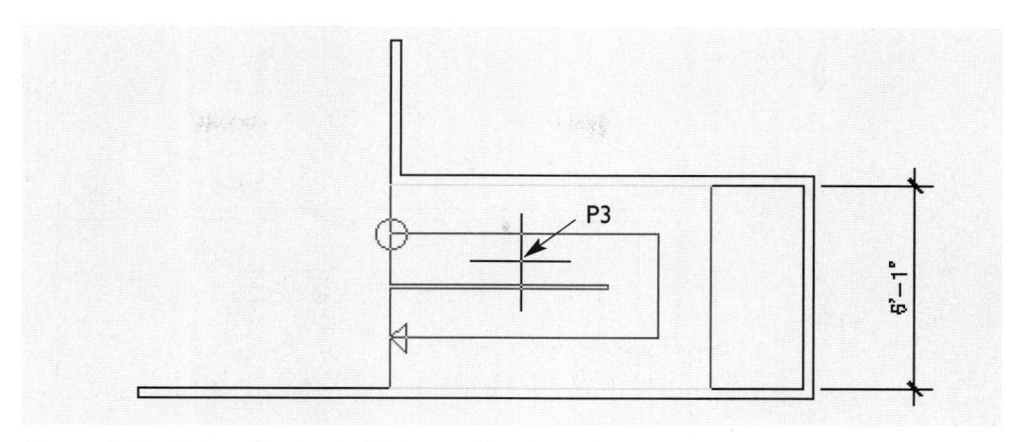

Figure 6.40 *Center of stair specified to position the stair*

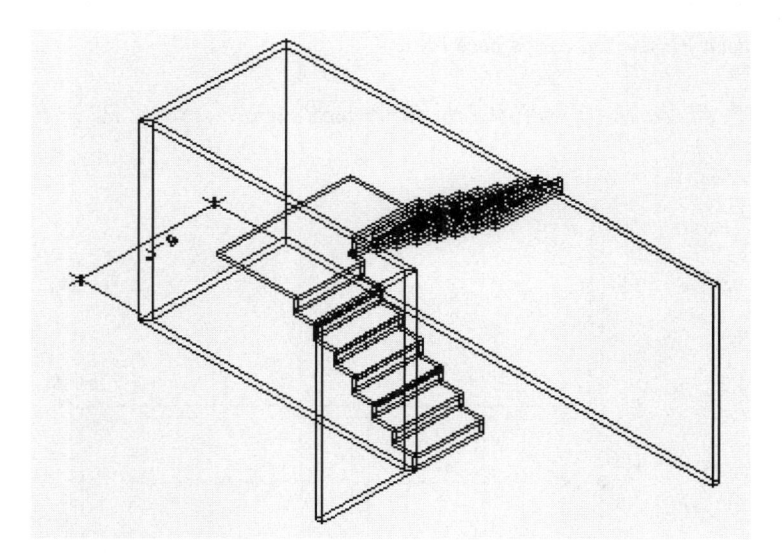

Figure 6.41 *North West Isometric view of the U Shaped stair*

```
Command: _move
Select objects: (Select the stair) 1 found

Select objects: ENTER
Specify base point or displacement: end ENTER
of (Select the end of the landing at point P1 in Figure 6.42.)

Specify second point of displacement or <use first
point as displacement>: end ENTER
```

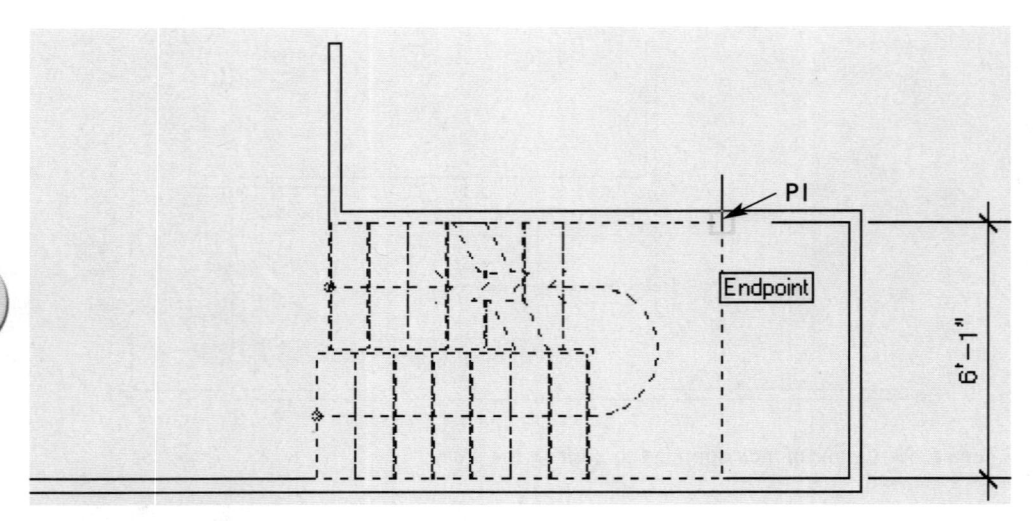

Figure 6.42 Point P1 selected as the base point

Of *(Select the endpoint of the inner wall surface at point P2 in Figure 6.43.)*
Command:

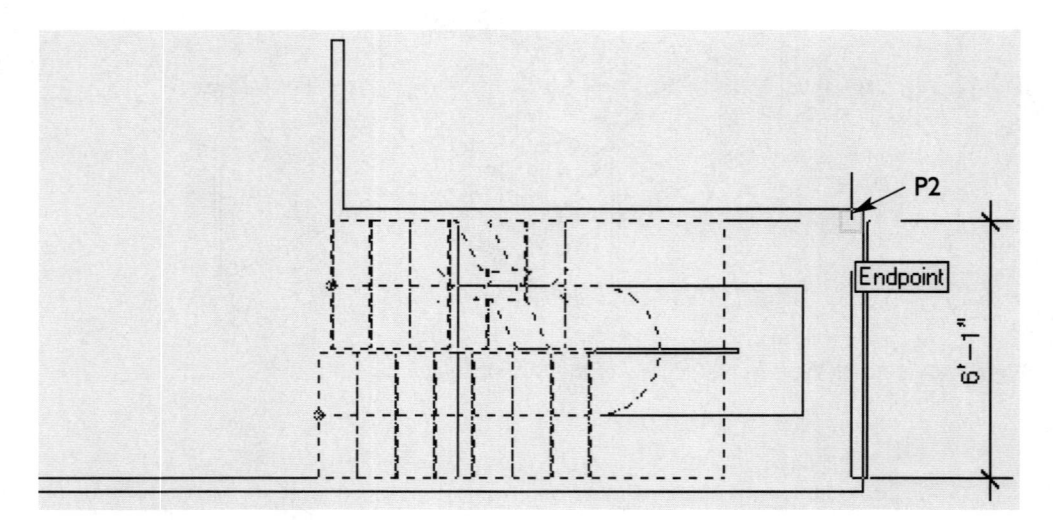

Figure 6.43 Point P2 selected as the second point of displacement

Stair is positioned with the **MOVE** command, as shown in Figure 6.44.

20. Save the drawing.

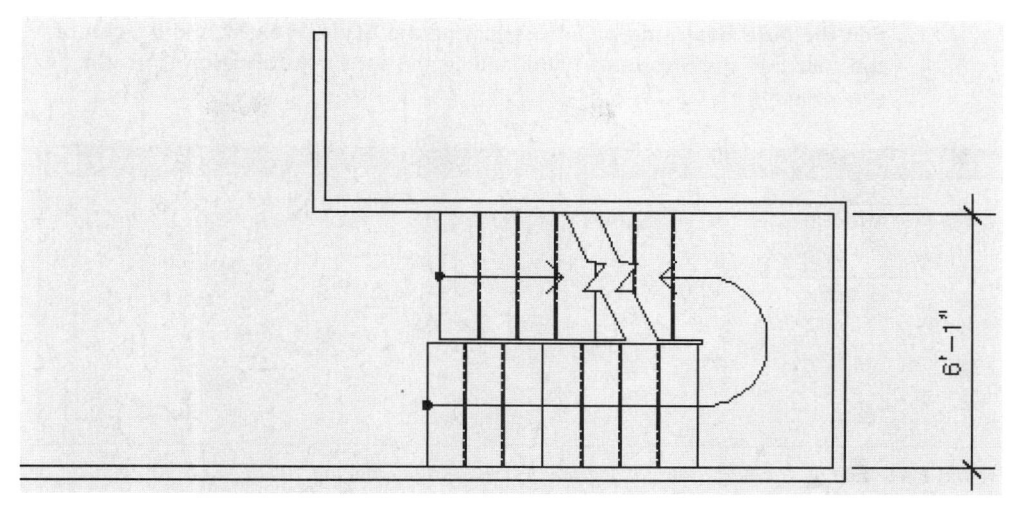

Figure 6.44 Stair postioned with the MOVE command

CREATING MULTI-LANDING STAIRS

Multi-landing stairs allow you to create one or more landings of the stair; they are created with the Multi-landing shape. The type of landing is specified in the **Next Turn** section of the **Add Stairs** dialog box. The Multi-landing shape allows the following four types of landings: **Quarter Landing**, **Quarter Turn**, **Half Landing**, and **Half Turn**. The **Quarter Landing** and **Quarter Turn** will create a landing for the stair to turn 90 degrees. The **Half Landing** and **Half Turn** create a landing that allows you to reverse the direction of stair, as shown in Figure 6.45.

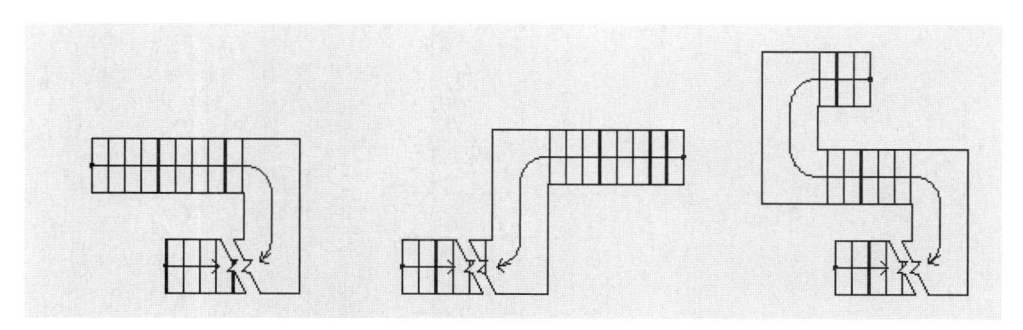

Figure 6.45 Half landings of the multi-landing stairs

STEPS TO CREATING A MULTI-LANDING STAIR

1. Select **Add Stair** from the **Stairs/Railings** toolbar.

2. Edit the **Add Stairs** dialog box, specifying the **Style**, **Height**, and **Width** of the stair. Edit the **Shape** to Multi-landing and set the justification to **Right**, as shown in Figure 6.46.

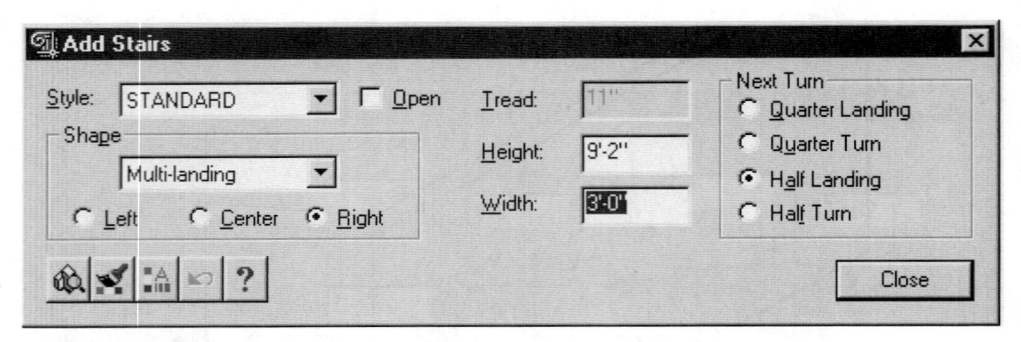

Figure 6.46 *Editing the Add Stairs dialog box to create multi-landing stairs*

3. Respond to the command sequence as shown below.

```
Command: AECSTAIRADD
Flight start point or
[SHape/STyle/Tread/Height/Width/Justify/Open/Next/
Match/Undo]: DBOX
```
(Edit Shape to Multi-landing and select Right justification.)
(Select the drawing area to activate the drawing area)
```
Flight start point or
[SHape/STyle/Tread/Height/Width/Justify/Open/Next/
Match/Undo]:
```
(Select point P1 in Figure 6.47.)

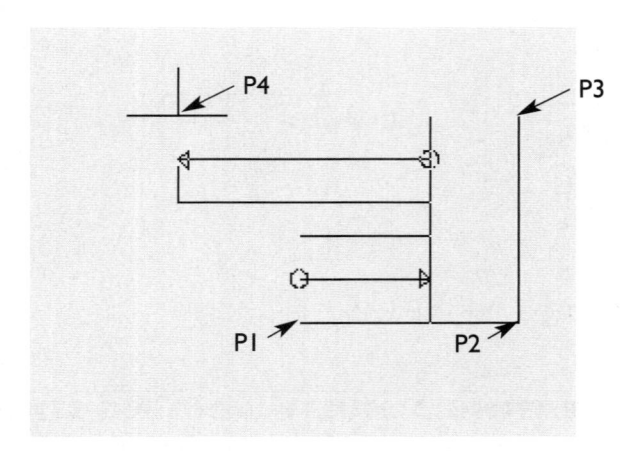

Figure 6.47 *Selecting flight start and end points*

```
Flight end point or
[SHape/STyle/Tread/Height/Width/Justify/Open/Next/
Match/Undo]:
```
(Select point P2 in Figure 6.47 to specify the start of the landing.)
```
Flight end point or
[SHape/STyle/Tread/Height/Width/Justify/Open/Next/
Match/Undo]:
```
(Select point P3 in Figure 6.47 to specify the end of the landing.)
```
Flight end point or
[SHape/STyle/Tread/Height/Width/Justify/Open/Next/Mat
ch/Undo]:
```
(Select point P4 in Figure 6.47 to specify the end of the stair.)
```
Flight end point or
[SHape/STyle/Tread/Height/Width/Justify/Open/Next/
Match/Undo]: ENTER
```
The Multi-landing stair is created as shown in Figure 6.48.

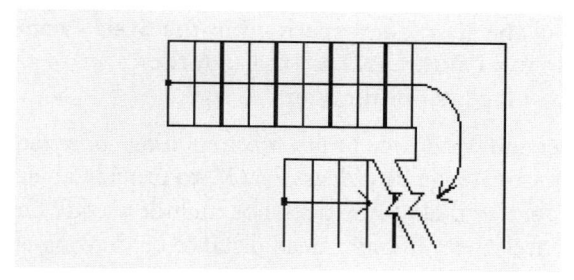

Figure 6.48 *Multi-landing stair*

CREATING SPIRAL STAIRS

The Spiral shape is used to create spiral stairs. You specify the center and radius of the spiral stair prior to specifying the beginning and end of the stair flight. It is difficult to determine the radius of the stair that will fit in an existing space. Therefore, after placing the stair, you can change the location of the center and the radius by editing the grips of the stair, shown in Figure 6.49.

The grips of the spiral stair are located at the center of the arc of the spiral stair and at the beginning of the stair. Editing the grips allows you change the radius and location of the start point. In Figure 6.49, the grip at the center of the spiral stair is being stretched to decrease the radius; therefore, the start point for the flight remains stationary. After editing the curve of the stair, you can move the stair to position it in alignment with existing walls.

The **Stair Properties** dialog box can be used to design the stair riser height. The **B-Straight Length** of a spiral stair is the length of the centerline curving with the spi-

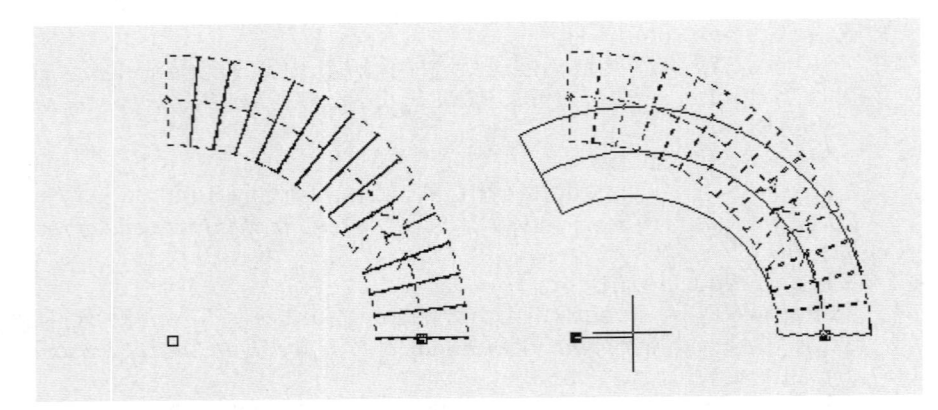

Figure 6.49 *Grips of a spiral stair*

ral stair. The radius of the stair is not specified in the **Stair Properties** dialog box. However, you can draw an AutoCAD arc through the curve of the spiral stair and determine the radius and center of the stair.

In addition to the riser and tread code limits, most building codes require that the tread depth 12" from the narrow edge be at least 7 1/2", to provide an adequate tread surface. The **Stair Properties** dialog box does not include a code check for this tread dimension. You can check the minimum tread distance by drawing an arc with a radius equal to the spiral stair, offsetting the arc 12" and then measuring the tread at this location. Measuring the tread along the arc that is 12" from the narrow edge will allow you to check the spiral stair for code compliance, as shown in Figure 6.50.

TUTORIAL 6.3 CREATING A SPIRAL STAIR

1. Open *ADT Tutor\Ch6\EX6-3.dwg*.
2. Choose **File>SaveAs** from the menu bar and save the drawing as *Spiral3.dwg* in your student directory.
3. Select the Section layout tab and verify that the rise of the stair equals 9'–2 1/2" as shown in Figure 6.51.
4. Select the Work-FLR layout tab.
5. Select F3 to turn off Running Object Snaps.
6. Select the **Center Mark** from the **Dimension** toolbar and select the curved wall. The center of the wall will be marked with a center mark as shown in Figure 6.52.
7. Select the **Add Stair** command from the **Stairs/Railings** toolbar.
8. Select the **Properties** button to open the **Stair Properties** dialog box.

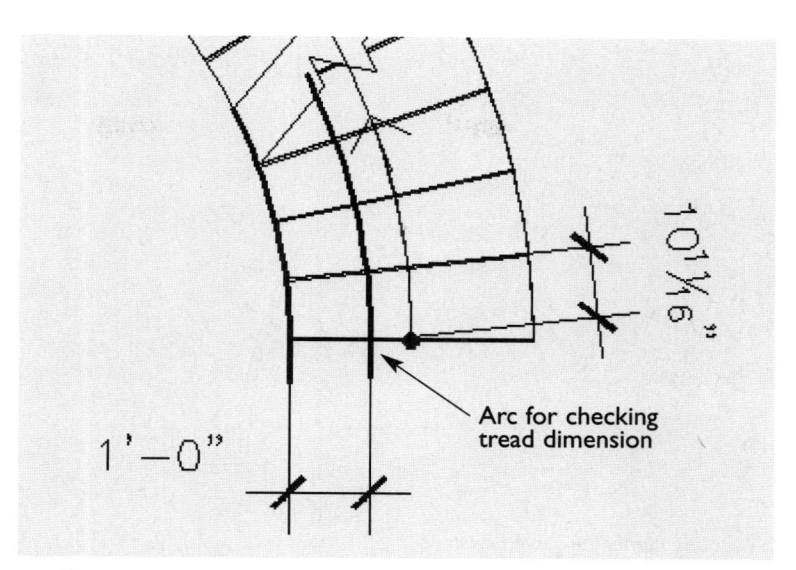

Figure 6.50 Checking the minimum tread 12" from the narrow edge of the spiral stair

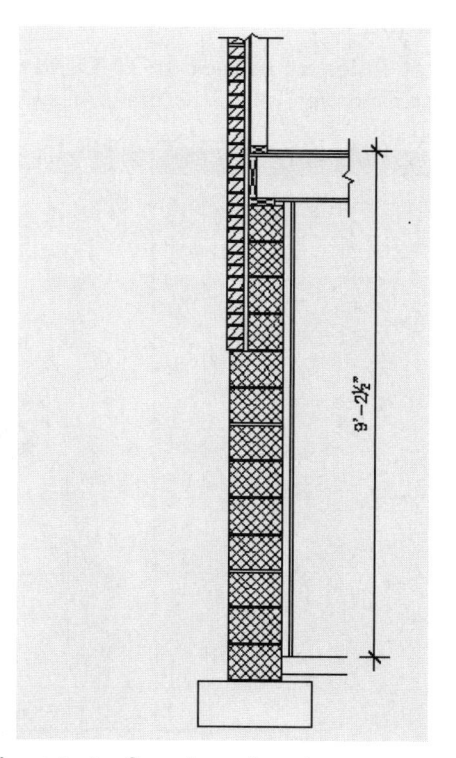

Figure 6.51 Determine the stair rise from the wall section

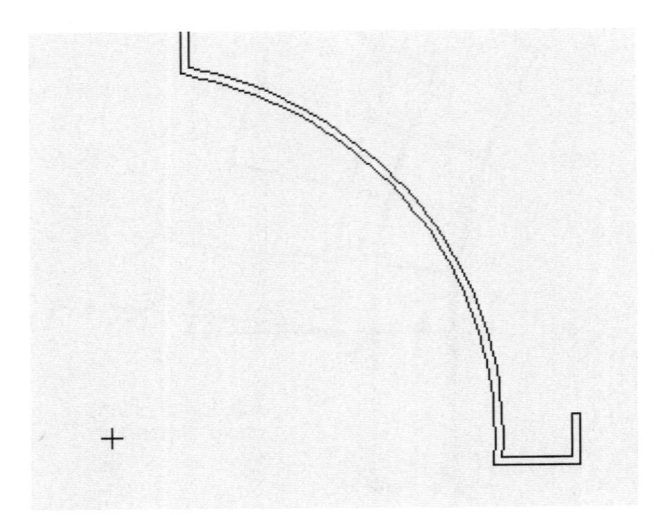

Figure 6.52 *Locating center of curved wall*

9. Select the **Dimensions** tab and edit the **C-Riser Count** to 16 and **D-Overall Height** to 9'–2-1/2.
10. Select the **Design Rules** tab and edit the **N-Tread** dimension to 10". Note that the **M-Riser** dimension is 6-15/16, as shown in Figure 6.53.

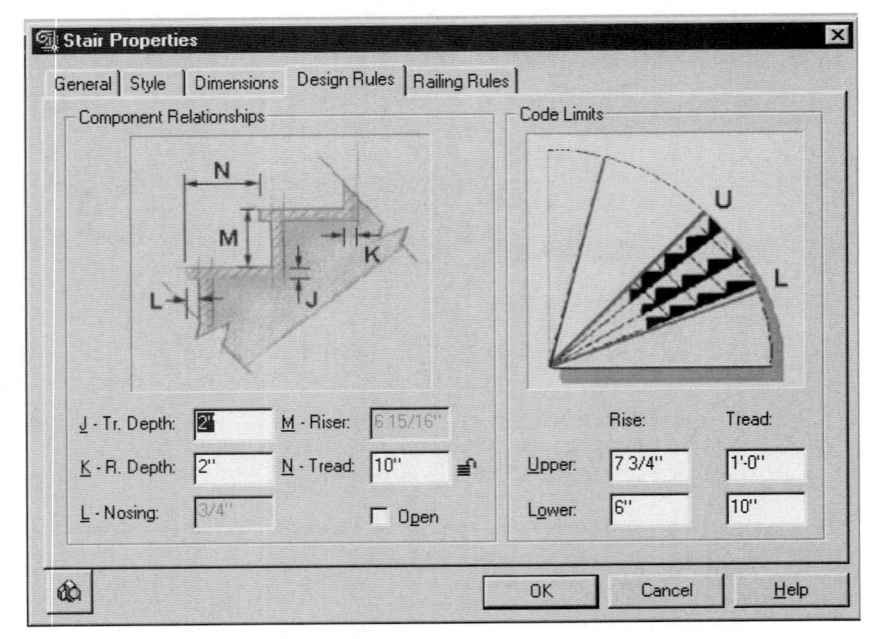

Figure 6.53 *Add Stairs dialog box for the spiral stair*

11. Select **OK** to dismiss the **Stair Properties** dialog box.
12. Edit the **Add Stairs** dialog box as follows: **Style** = Standard, **Shape** = Spiral, **Width** = 3'–0", and select the **Right** justification radio button.
13. Refer to the following command sequence and specify the location of the stair as shown in Figures 6.54 and 6.55.

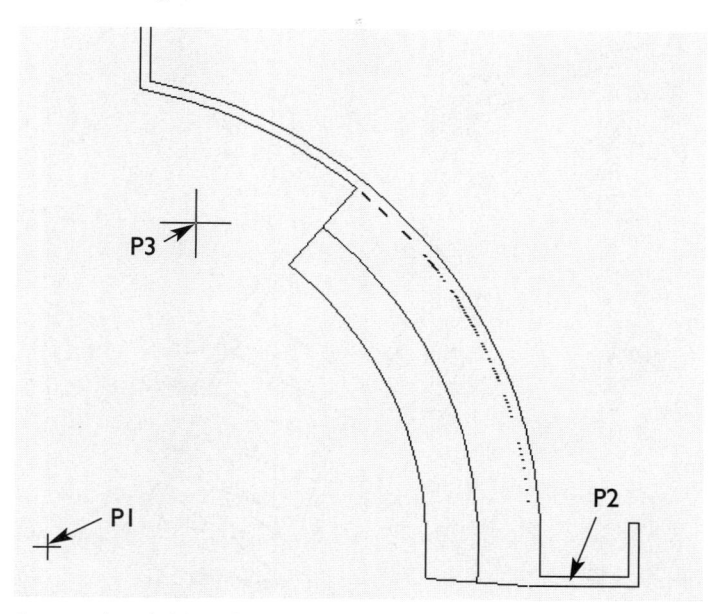

Figure 6.54 *Start and end points for the spiral stair*

```
Command: _AecStairAdd
Center of spiral stair or
[SHape/STyle/Tread/Height/Width/Justify/Open/Next/
Match/Undo]: DBOX
```
(Select the drawing area to activate the drawing area)
```
Center of spiral stair or
[SHape/STyle/Tread/Height/Width/Justify/Open/Next/
Match/Undo]: int ENTER
of
```
(Select point P1 in Figure 6.54.)
```
Start point or
[SHape/STyle/Tread/Height/Width/Justify/Open/Next/
Match/Undo]: endp ENTER
of
```
(Select the end of the wall at point P2 in Figure 6.54.)
```
Flight direction or
[SHape/STyle/Tread/Height/Width/Justify/Open/Next/
Match/Undo]:
```
(Select a point near P3 in Figure 6.54.)

```
Center of spiral stair or
[SHape/STyle/Tread/Height/Width/Justify/Open/Next/
Match/Undo]: ENTER
```
(Spiral stair created as shown in Figure 6.55.)

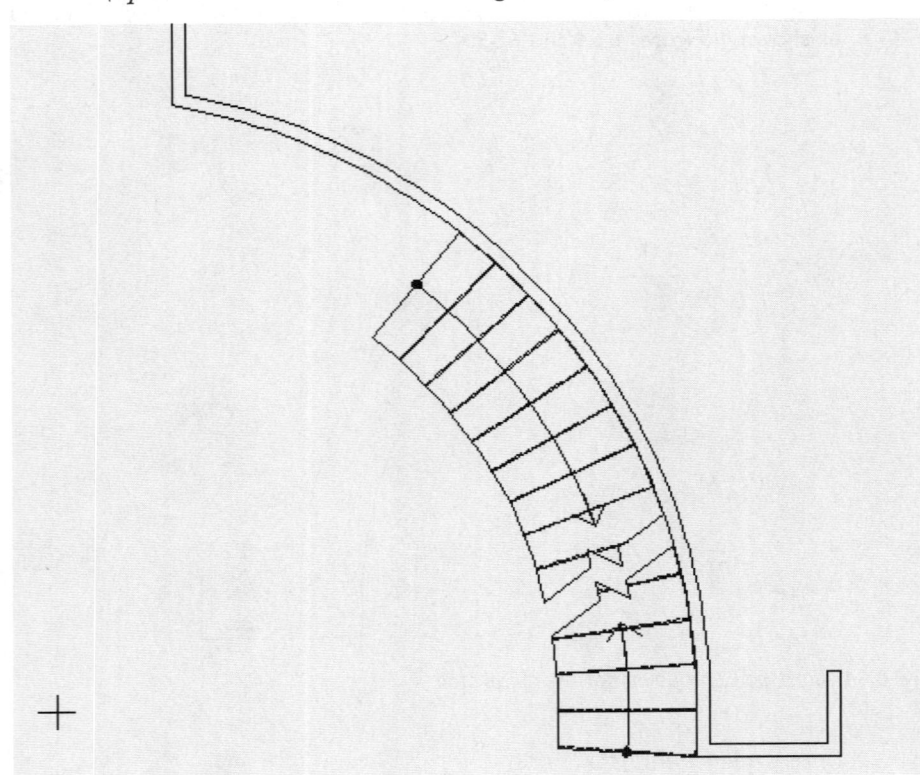

Figure 6.55 *Spiral stair*

14. Save the drawing.

MODIFYING EXISTING STAIRS WITH MODIFY STAIR

Existing stairs can be edited by the **Modify Stair** command. The **Style, Open Tread, Tread size, Height,** and **Width** can be changed with the command. See Table 6.4 for command access.

When you select the **Modify Stair** command, you are prompted to select a stair, and then the **Modify Stair** dialog box opens as shown in Figure 6.57.

The **Modify Stair** dialog box displays the dimensions and style of the selected stair. It includes all of the options of the **Add Stair** dialog box except the **Shape, Justification**

Stairs and Railings **441**

Menu bar	Design>Stairs>Modify Stair
Command prompt	STAIRMODIFY
Toolbar	Select Modify Stair from the Stairs/Railings toolbar as shown Figure 6.56
Shortcut	Right-click over the drawing area and choose Design>Stairs>Modify Stair
Shortcut	Select stair; right-click and choose Modify Stair

Table 6.4 *Modify Stair access*

Figure 6.56 *The Modify Stair command on the Stairs/Railings toolbar*

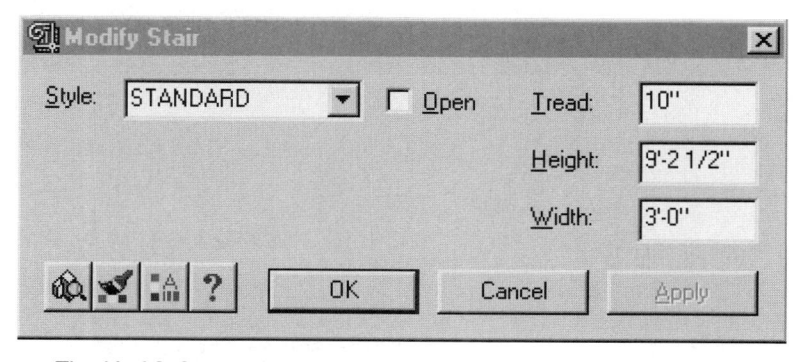

Figure 6.57 *The Modify Stair dialog box*

and landing style. To edit a stair, select the **Properties** button or edit any of the following fields: **Style, Open Risers, Tread, Height,** and **Width**. After editing the **Modify Stair** dialog box, select the **Apply** button to execute the change in the stair. If the width of a stair is increased or decreased, the change occurs equally on each side of the centerline of the stair. Changes to the stair properties that change the **B-Straight Length** are applied to the end of the stair. Therefore, the beginning of the stair remains stationary.

EDITING A STAIR USING GRIPS

Each stair shape can be edited by the grips of the stair. Stretching the grips of the stair allows you to quickly adjust the start and end points of the stair flight. If a multi-landing stair is edited by grips, it is shortened or lengthened a distance equal to the tread size. Figure 6.58 shows the grips of each stair shape.

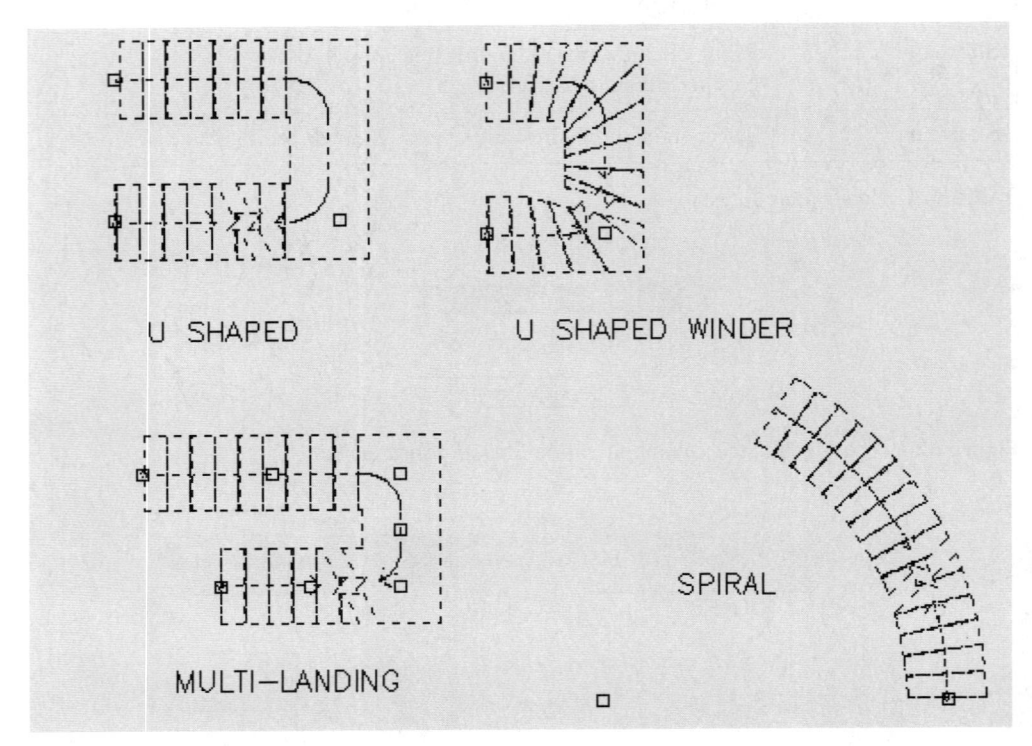

Figure 6.58 *Grips of stair shapes*

The grips of the U Shaped and U Shaped Winder stairs will not stretch to extend the run of the stairs. The grips of these stairs can be used to move or rotate the entire stair to a new position. The grips of the Multi-landing stairs can be stretched to alter the beginning or end of the stair. The radius and start point of the spiral stair can be edited with grips.

Reversing the Direction of Stairs with Grips

All stairs are created in the up direction from the start point. Therefore, a stair is developed up from the start point. You can use the mirror command or use grips to reverse the stair to go in the down direction. The three grips of the stair are located at the Z=0 coordinate. If you mirror the stair about the midpoint grip, the stair will remain in about the same location. If you mirror the stair about an endpoint grip, the stair will flip in the mirror operation the length of the stair.

If you are drawing a floor plan that includes a stair going down below the current level, the stair can be moved down in the Z direction a distance equal to the distance between levels. The **MOVE** command can be used to move the stair in the Z direction.

TUTORIAL 6.4 USING GRIPS TO MIRROR THE STAIR

1. Open *ADT Tutor\Ch6\Ex 6-4.dwg*.
2. Choose **File>SaveAs** from the menu bar and save the drawing as *Mirror4.dwg* in your student directory.
3. Select the **Right View** from the **View** flyout on the **Standard** toolbar to view the stair as shown in Figure 6.59.

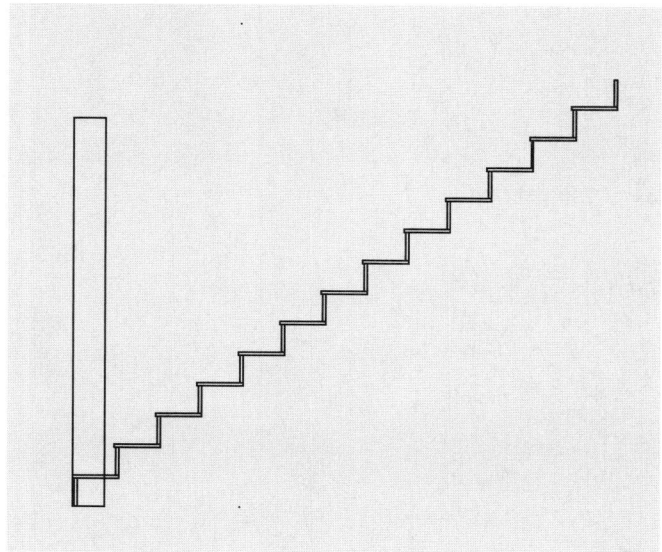

Figure 6.59 *Right side view of stair*

4. Select the stair to display the grips of the stair.
5. Select the midpoint grip as shown in Figure 6.60.
6. Right-click and choose **Mirror** from the shortcut menu.
7. Move the mouse above the middle grip and left-click.
8. Press ESC to deselect the grip.
9. Press ESC to deselect the stair. The stair direction is reversed, as shown in Figure 6.61

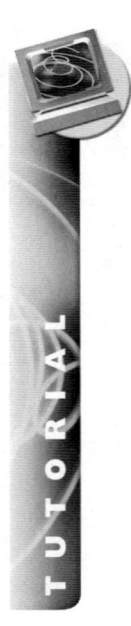

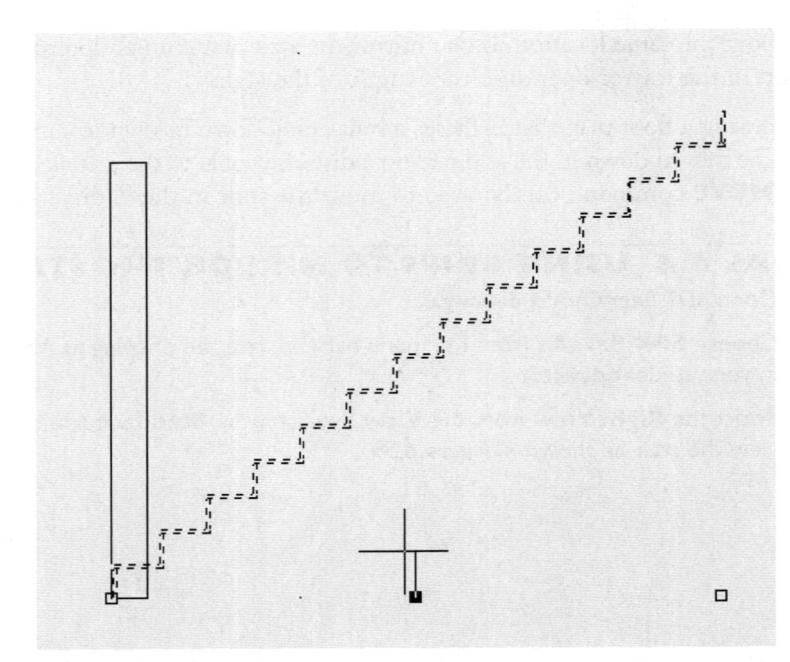

Figure 6.60 *Midpoint grip selected*

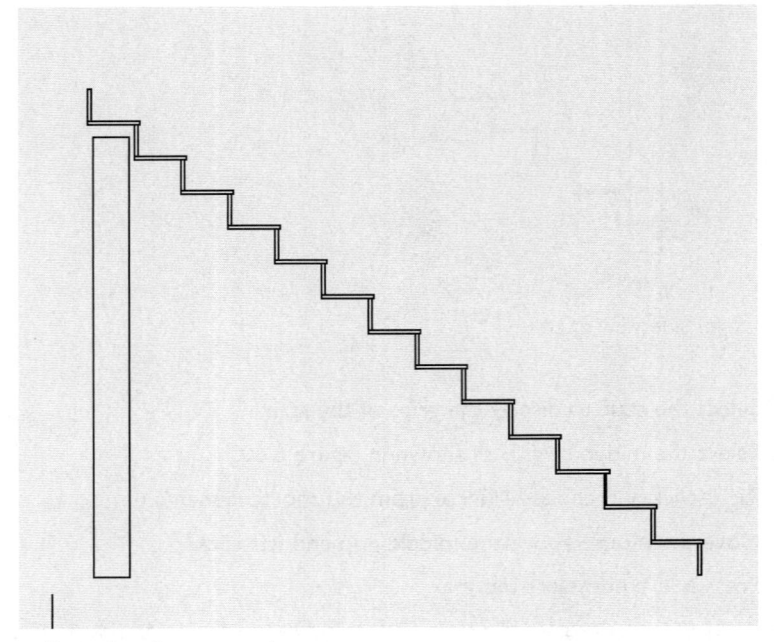

Figure 6.61 *Stair direction reversed*

10. Select the **Top View** from the **View** flyout on the **Standard** toolbar.
11. Select the stair; then right-click and choose **Move** from the shortcut menu.
12. Respond to the command sequence as shown below.

    ```
    Command: _move 1 found
    Specify base point or displacement: (Select a point in the
    drawing area.)
    Specify second point of displacement or
    <use first point as displacement>: @0,0,-8'11.5
    ENTER (Specify the displacement equal to the height of the stair.)
    ```
13. Select the **Right View** from the **Standard** toolbar to view the stair as shown in Figure 6.62.

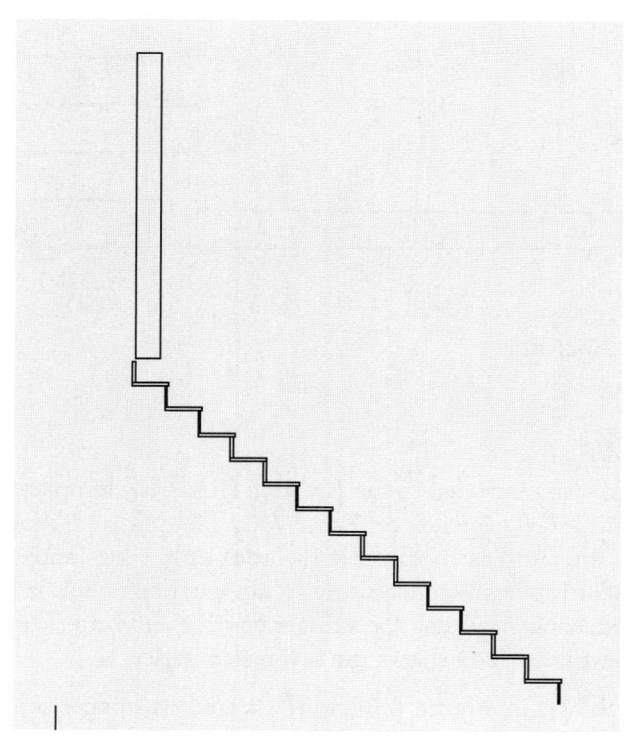

Figure 6.62 *Stair moved down in the negative Z direction*

14. Select the **Top View** from the **Standard** toolbar to view the plan of the stair as shown in Figure 6.63.
15. Save the drawing.

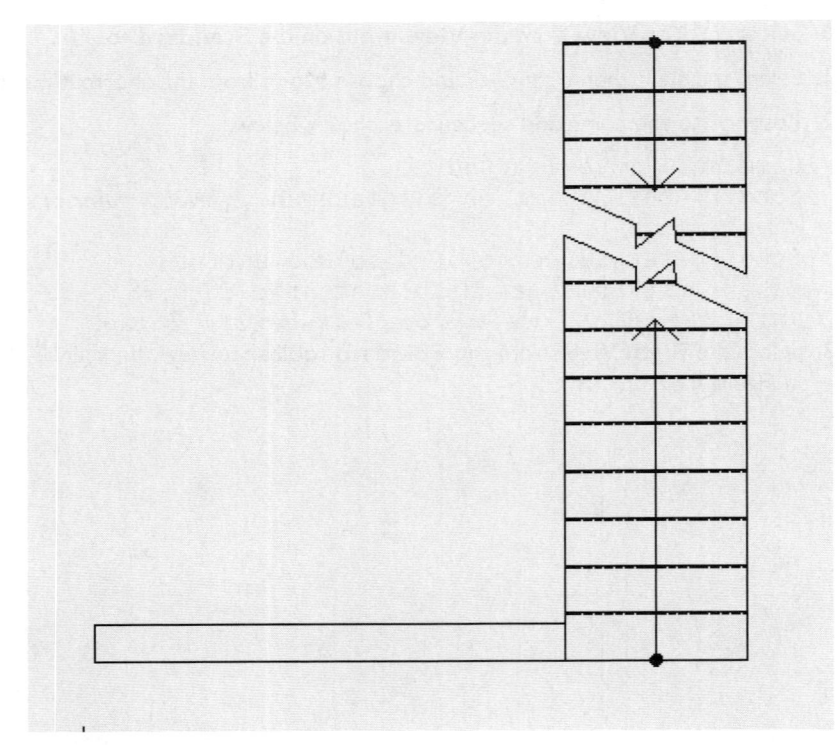

Figure 6.63 *Plan view of stair*

STAIR STYLES

The seven styles of stairs included in the Aec arch (Imperial) template are Cantilever, Concrete, Standard, Steel Open, Steel Pan, Wood Housed, and Wood Saddle. The Standard style is the simplest because it includes only risers and treads, with no stringer. The Standard style provides a simple representation of a stair. However, other stair styles add more detail regarding the stringer position and better represent the complexity of the stair. Figure 6.64 shows the seven stair styles.

The stair styles shown in Figure 6.64 were created as open riser to display the stringer. The Concrete style is a monolithic concrete stair and therefore the riser cannot be open. The Cantilever style includes only one stringer in the center of the stair. The Wood Housed style places the treads between the stringers.

CREATING STAIR STYLES

Stair styles can be created, edited, imported, and exported by the **Stair Styles** command. The specific location and size of the stringer can be established through the **Stair Styles** command. In addition, the visibility of the upper or lower half of the stairs

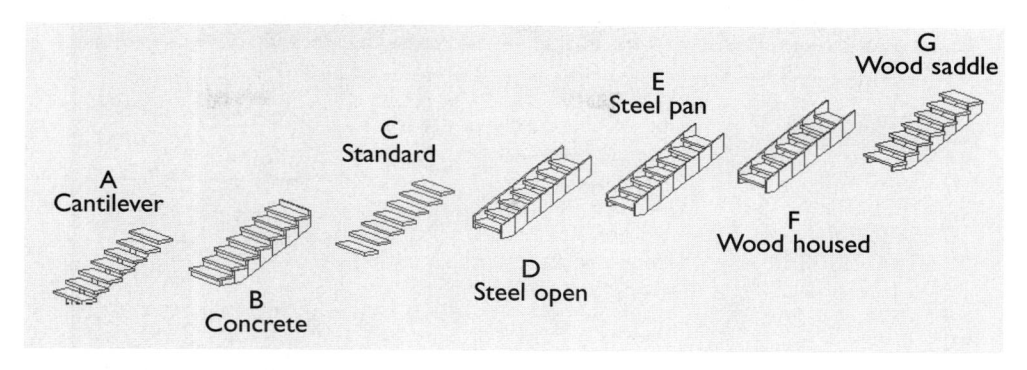

Figure 6.64 Stair styles

is controlled in the display properties of the stair style. See Table 6.5 for command access.

Menu bar	Design>Stairs>Stair Styles
Command prompt	STAIRSTYLE
Toolbar	Select Stair Styles from the Stairs/Railings toolbar as shown in Figure 6.65
Shortcut	Right-click over the drawing area; choose Design>Stairs>Styles
Shortcut	Select a stair; right-click and choose Edit Stair Style

Table 6.5 Stair Styles access

Figure 6.65 Stair Styles command on the Stairs/Railings toolbar

When the **Stair Styles** command is selected from the **Stairs/Railings** toolbar, the **Stair Styles** dialog box opens as shown in Figure 6.66.

The stair styles listed in the dialog box include all styles loaded in the drawing. Additional styles can be imported from other drawings and the styles of the current

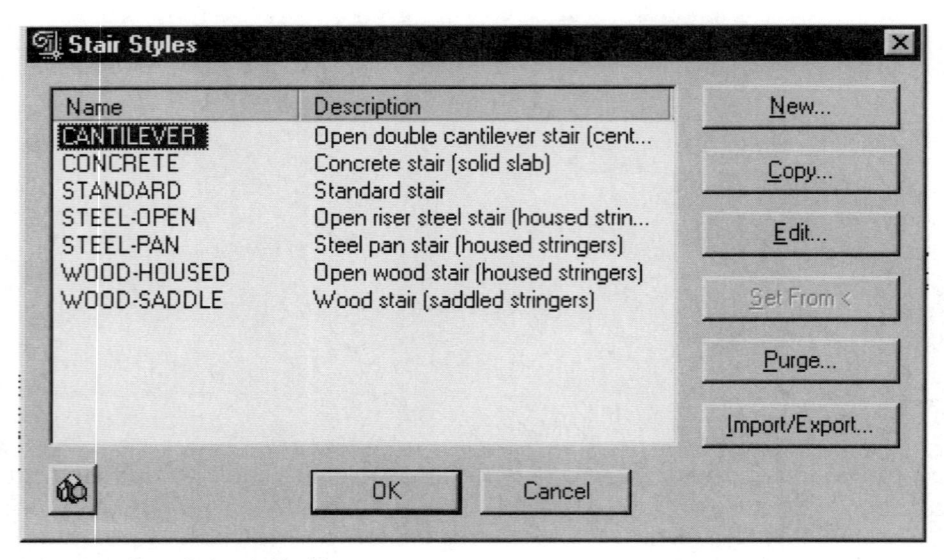

Figure 6.66 Stair Styles dialog box

drawing can be exported to common resource files. The purpose of each of the buttons in the dialog box is described below:

> **New** – Creates a new name for a style.
>
> **Copy** – Copies an existing stair style so as to create a new named stair style that includes the properties of the original stair style.
>
> **Edit** – Creates the description of the stair style and defines specific properties of the new style.
>
> **Set from>** – Inactive for stair styles.
>
> **Purge** – Deletes unused stair styles from the stair style list.
>
> **Import/Export** – Transfers stair styles to common resource files or imports stair styles from resource files to the current drawing.

Create a new stair style by selecting the **New** button in the **Stair Styles** dialog box. The **New** button opens the **Name** dialog box, which allows you to enter the name for the new style. Creating the name is the first step; however, the stair style has no specific stringer definition. The **Edit** button of the **Stair Styles** dialog box allows you to add a description to the stair style and add specific stringer properties.

STEPS TO CREATING A NEW STAIR STYLE

1. Select **Stair Styles** from the **Stairs/Railings** toolbar to open the **Stair Styles** dialog box as shown in Figure 6.66.

2. Select the **New** button in the **Stair Styles** dialog box to open the **Name** dialog box as shown in Figure 6.67.

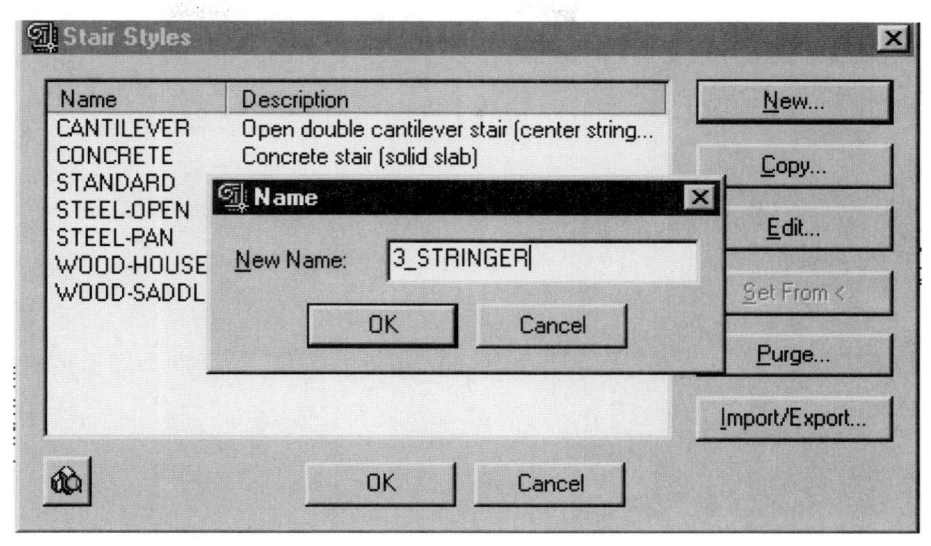

Figure 6.67 Name dialog box

3. Type the new name **3_Stringer** in the **Name** dialog box.
4. Select **OK** to dismiss the **Name** dialog box.
5. Select the **Edit** button of the **Stair Styles** dialog box and add a description to the new style as shown in Figure 6.68.

DEFINING THE ATTRIBUTES OF A STAIR STYLE

The **Edit** button of the **Stair Styles** dialog box includes three tabs: **General**, **Dimensions**, and **Display Props**.

General

The **General** tab, shown in Figure 6.68, allows you to add a description to the stair style and edit the name of the style. The **Notes** button opens the **Notes** dialog box, consisting of **Text Notes** tab, which allows you to add notes regarding the style, and the **Reference Docs** tab, which includes provisions for creating a file list. The file list can include files that are details of the stair or text documents regarding the specifications for the stair. The **Floating Viewer** button located in the lower left corner of the **Stair Styles** dialog box opens the **Viewer** window, which allows you to view the stair style prior to inserting the stair in the drawing.

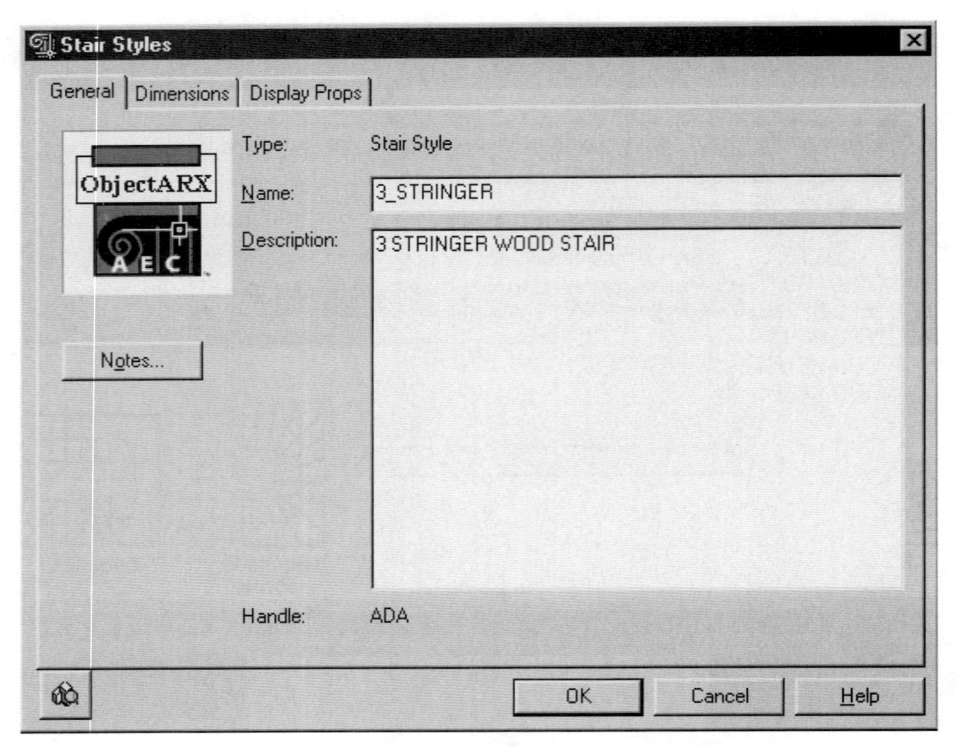

Figure 6.68 *Adding a description to the General tab of the Stair Styles dialog box*

Dimensions

The **Dimensions** tab of the **Stair Styles** dialog box, shown in Figure 6.69, allows you to specify the stringer size, position, type and alignment.

The **Dimensions** tab includes a **Stringer Values** section at the top that list the stringers of the stair. The lower portion of the dialog box is used to define the dimensions and position of each of the stringers. Each of the options of the **Dimensions** tab is described below.

> **A-Stringer End Offset** – Locates the stringer from the edge of the stair the specified distance.
>
> **B-Stringer Width** – The actual width of the stringer.
>
> **C-Stringer Waist/Slab Depth** – The minimum width between the bottom surface of the stringer and tread location on the stringer. This distance specifies the slab thickness of the concrete slab for the stair.
>
> **D-Stringer Total Depth** – The dimension of the stringer inclusive of the tread.
>
> **E-Nosing Depth** – The dimension of the nosing of the stair.

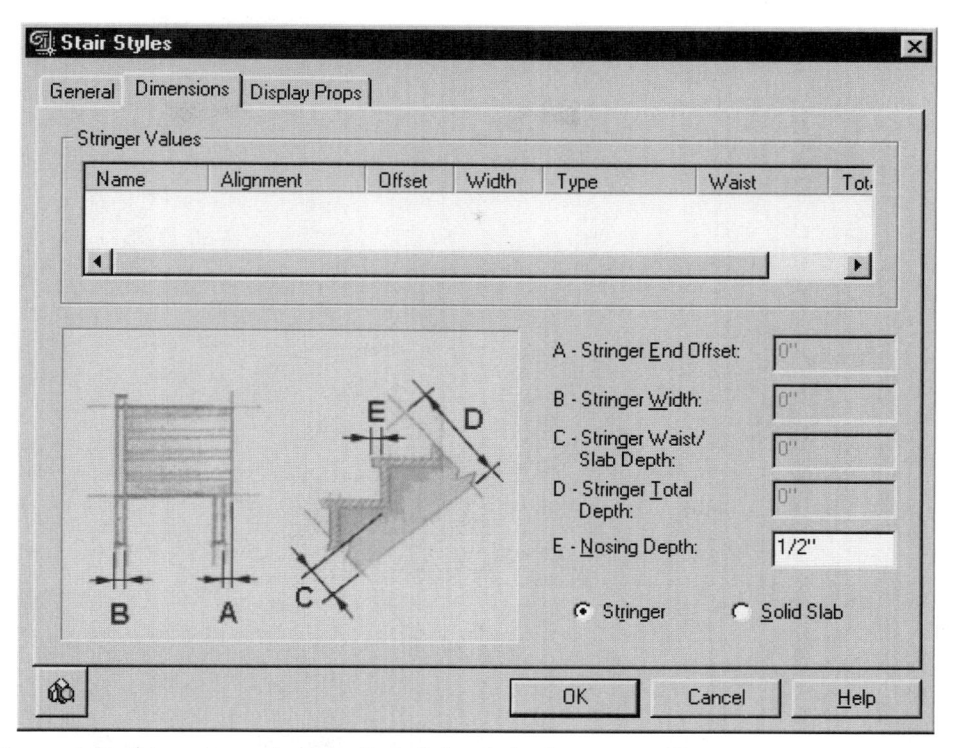

Figure 6.69 *Dimensions tab of the Stair Styles dialog box*

> **Stringer** – If selected, creates a stair that includes stringers.
>
> **Solid Slab** – If selected, creates a stair without stringers.

Selecting the **Name** column in the **Stringer Values** section creates an edit field for entering a name. When you press ENTER after entering the name, the stringer values will become active, as shown in Figure 6.70.

Each time you select this **Name** column, a new stringer is added. The edit fields in the lower portion of the dialog box can be used to edit the stringer selected in the upper **Stringer Values** section. You can also select the dimensions in the **Stringer Values** section and overtype new values directly in that section. Selecting the **Alignment** option opens a dialog box to change the alignment to Left, Center or Right. Selecting the **Type** option opens a dialog box to toggle the type to Saddled or Housed.

Display Props

The **Display Props** tab allows you to set the display properties of the stair style. The **Display Props** tab is similar to that of other style dialog boxes, as shown in Figure 6.71.

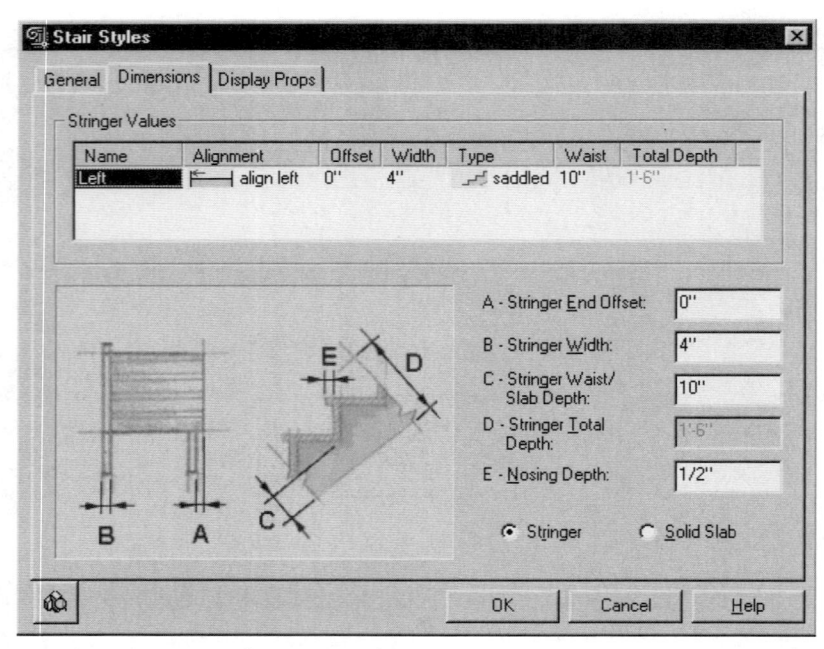

Figure 6.70 *A new stringer value is created*

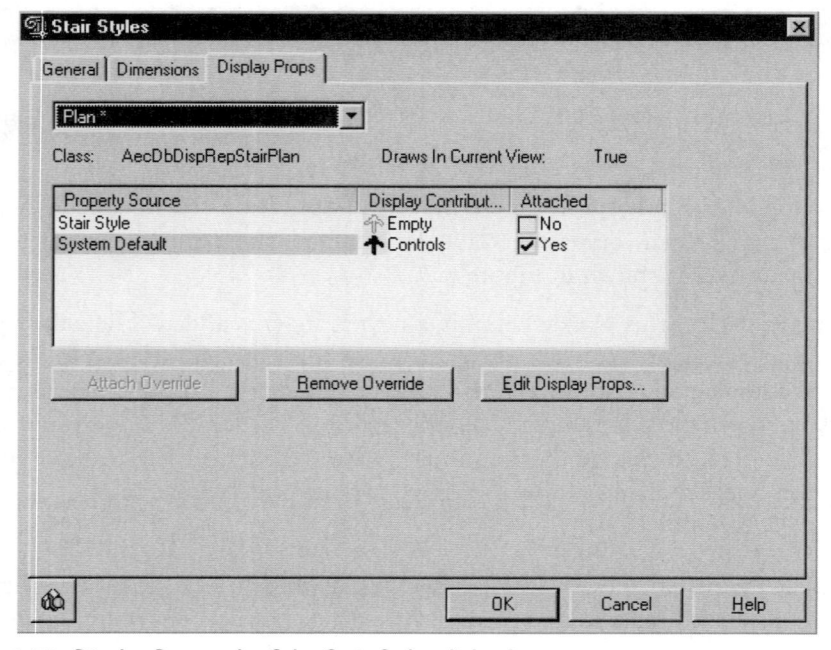

Figure 6.71 *Display Props tab of the Stair Styles dialog box*

The **Display Props** tab includes the following options:

Display Representation – The **Display Representation** drop-down list shows the display representation available for stair objects. In Figure 6.71, the Plan display representation is marked by an asterisk and it is the current display representation being used in the active viewport. In addition to the Plan display representation, the Model and Reflected display representations are available for stairs.

Property Source – The **Property Source** lists object display categories for the current viewport. Therefore, the display properties can be controlled by the Stair Style or by the System Default category.

Display Contribution – The **Display Contribution** column indicates the current state of the object category. Display contribution can be Controls, Empty, or Overridden. In Figure 6.71, the System Default category is controlling the display of the stair. An Empty display configuration indicates that no overrides have been set regarding the display of the object. In the figure, the stair style is shown as empty because the System Default is controlling the display. The Overridden condition exists when the System Default is overridden and the stair style is displayed with unique display options.

Attached – The **Attached** column includes Yes and No options, which indicate whether overrides or exceptions have been defined for the display of the object.

Attach Override – The **Attach Override** button allows you create an override of the display properties for a stair style or system default.

Remove Override – If a **Property Source** has an override defined, you can remove it by selecting the property source and then selecting the **Remove Override** button.

Edit Display Props – The **Edit Display Props** button opens the **Entity Properties** dialog box, which allows you to control the display of each component of the stair. The **Entity Properties** dialog box consists of the **Layer/Color/Linetype** and **Other** tabs.

Layer/Color/Linetype

Selecting the **Edit Display Props** button on the **Display Props** tab opens the **Entity Properties** dialog box. The **Layer/Color/Linetype** tab of the **Entity Properties** dialog box is shown in Figure 6.72.

The **Entity Properties** dialog box allows you to turn off the display of the objects of the upper half or lower half of the stair. This option allows you to turn off the upper stair representation if the plan you are creating represents the lower floor of a two-story building. Color, linetype and linetype scale can also be controlled for each component of the stair.

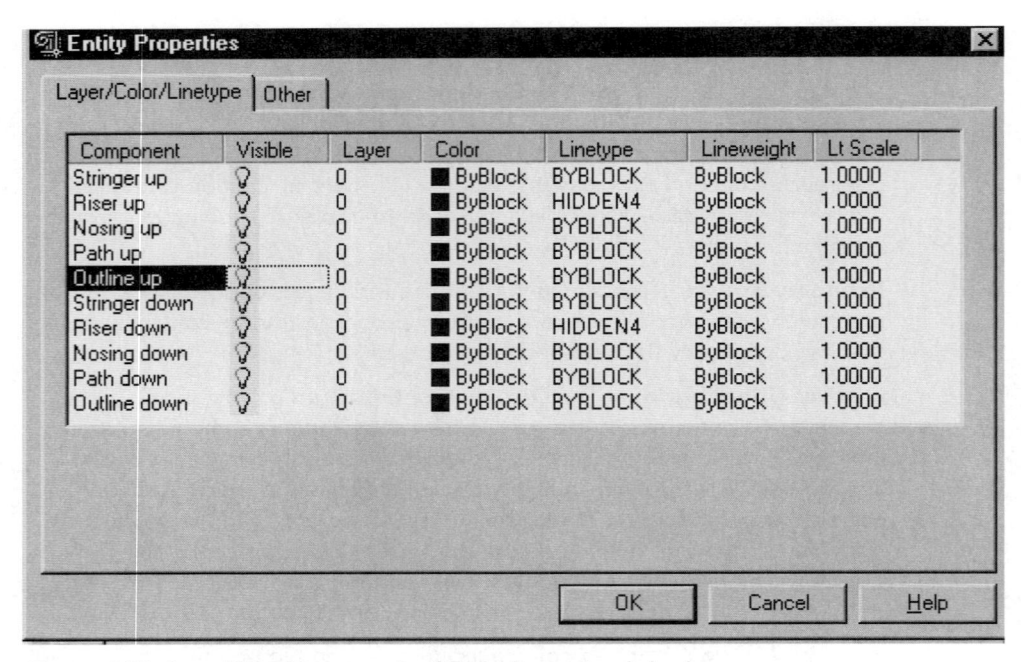

Figure 6.72 *Layer/Color/Linetype tab of Entity Properties dialog box*

Other

The **Other** tab of the **Entity Properties** dialog box consists of the **Cut Plane**, **Arrow**, and **Break Mark** sections, as shown in Figure 6.73.

Cut Plane

> **Elevation** – The vertical distance from the zero Z coordinate of the cutting plane for creating the location of the lower half and the upper half of the stairs. The elevation of the cutting plane determines the location of the break line for the stair.
>
> **Distance** – The horizontal distance separating the break lines of the upper and lower stair sections.
>
> **Angle** – The angle of the break lines of each half of the stairs.

Arrow

> **Size** – Allows you to define the dimension of the directional arrow for the stair.
>
> **Offset** – The distance from the break line to the head of the directional arrow.
>
> **Dim Style** – The dimensioning style used to define the style of the directional arrow.

Stairs and Railings 455

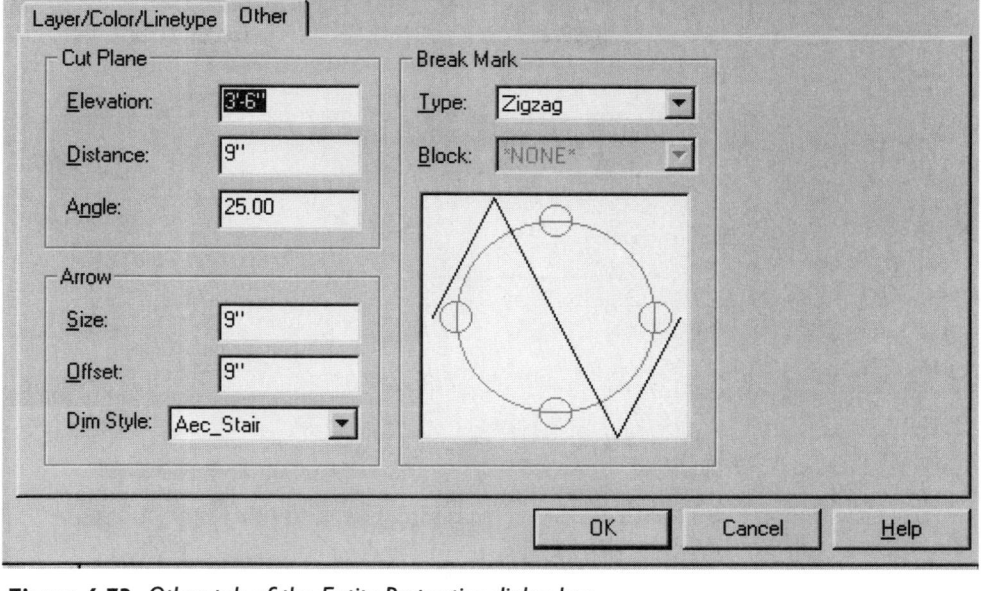

Figure 6.73 *Other tab of the Entity Properties dialog box*

Break Mark

Type – Lists the break marks: None, Curved, Zigzag, Custom shapes. If Custom is selected, the blocks of the drawing can be selected from the **Block** list. The break mark selected is displayed in the window below the **Block** edit field.

Block – Lists blocks of the drawing that can be used for a custom break mark.

TUTORIAL 6.5 CREATING A STAIR STYLE AND CONTROLLING DISPLAY

1. Open *ADT Tutor\Ch6\EX6-5.dwg*.
2. Choose **File>SaveAs** from the menu bar and save the drawing as *Style5.dwg* in your student directory.
3. Select **Stair Styles** on the **Stairs/Railings** toolbar to open the **Stair Styles** dialog box.
4. Select Standard from the stair style list in the **Stair Styles** dialog box.
5. Select the **New** button of the **Stair Styles** dialog box.
6. Enter the name **3 Stringer** in the **Name** dialog box as shown in Figure 6.74.

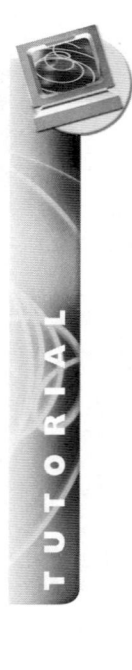

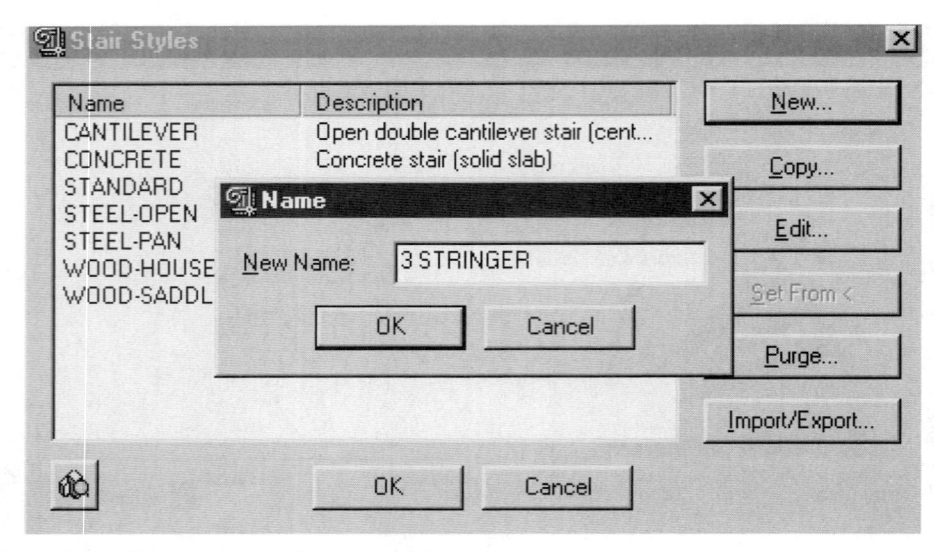

Figure 6.74 *Enter the name for a new style*

7. Select **OK** to dismiss the **Name** dialog box.
8. Select the **Edit** button of the **Stair Styles** dialog box.
9. Type **Three stringer stair** in the **Description** edit field.
10. Select the **Dimensions** tab of the **Stair Styles** dialog box.
11. Select in the **Name** column to create a stringer definition, as shown in Figure 6.75.
12. Type **left** as the stringer name as shown in Figure 6.76.
13. Verify that the **Alignment** is set to align left and edit the following: **A-Stringer End Offset** to 2", **B-Stringer Width** to 1–1/2", **C-Stringer Waist/Slab Depth** to 3", **Stringer Total Depth** to 11–1/4", **E-Nosing Depth** to 3/4", and select **Stringer**, as shown in Figure 6.77.
14. Select the **Name** column and type **Center** as the name of the new stringer.
15. Select the **Alignment** column and select Center in the **Stringer Alignment** dialog box as shown in Figure 6.78.
16. Select **OK** to dismiss the **Stringer Alignment** dialog box.
17. Edit the **Offset** to zero for the center stringer.
18. Select the **Name** column of the **Stringer Values** and type the name **Right**.
19. Select the **Alignment** column of the right stringer to open the **Stringer Alignment** dialog box.

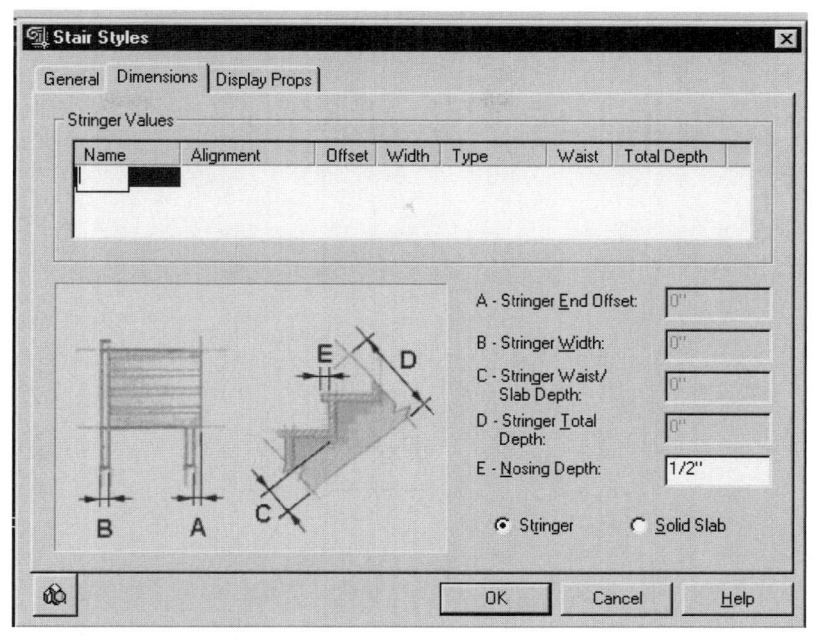

Figure 6.75 *Creating the left stringer*

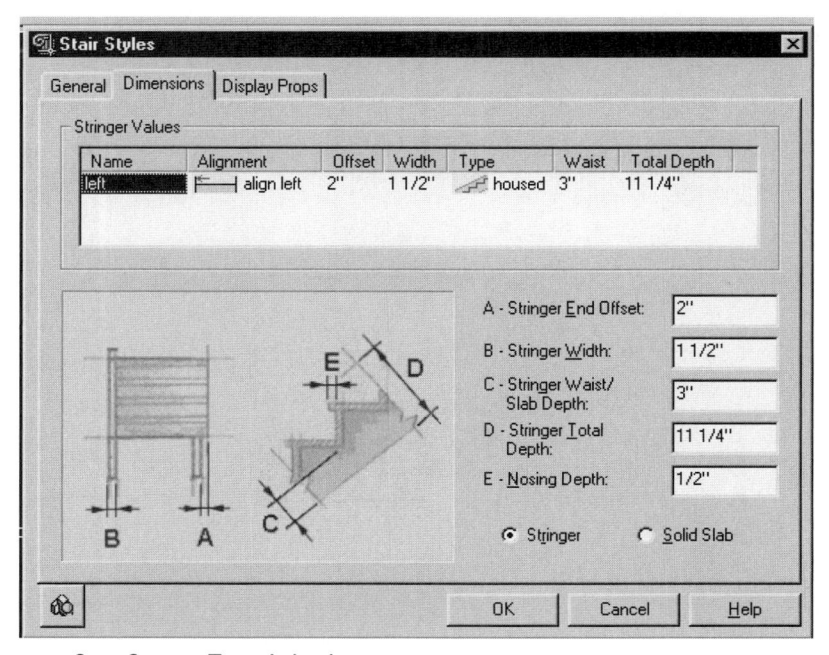

Figure 6.76 *Stair Stringer Type dialog box*

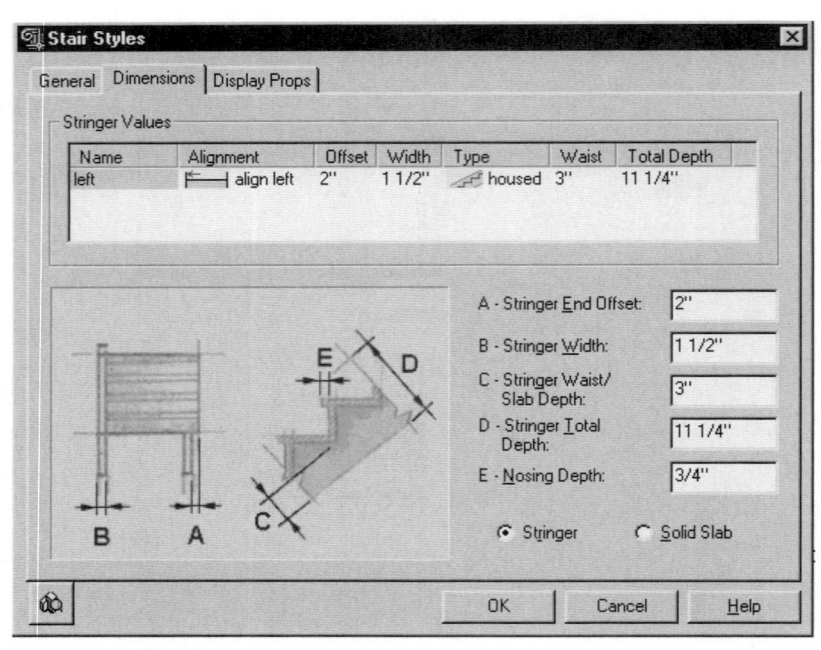

Figure 6.77 *Editing Dimensions tab to define a style*

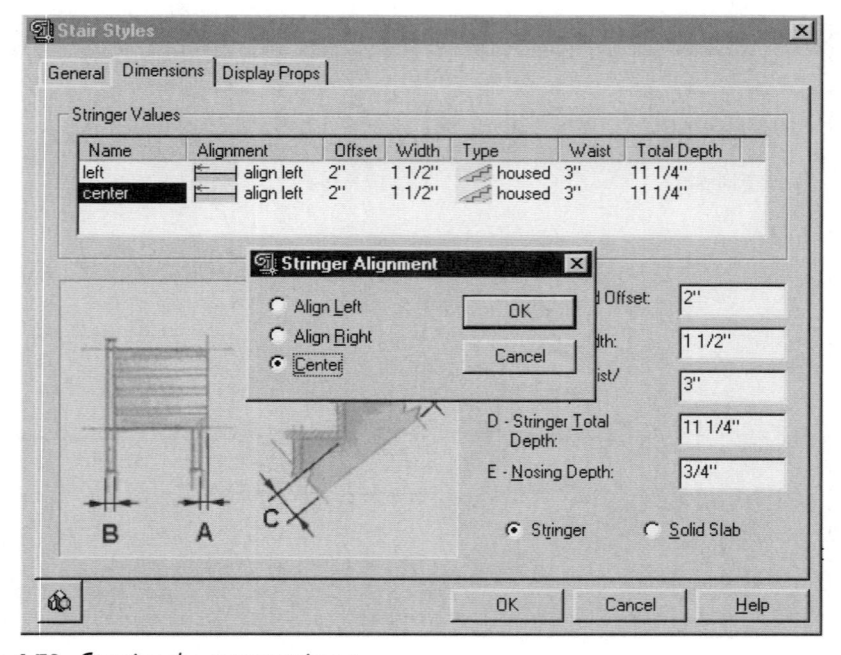

Figure 6.78 *Creating the center stringer*

20. Select the **Align Right** radio button and then select **OK** to dismiss the **Stringer Alignment** dialog box.
21. Select the **Offset** column and edit the offset to 2" as shown in Figure 6.79.

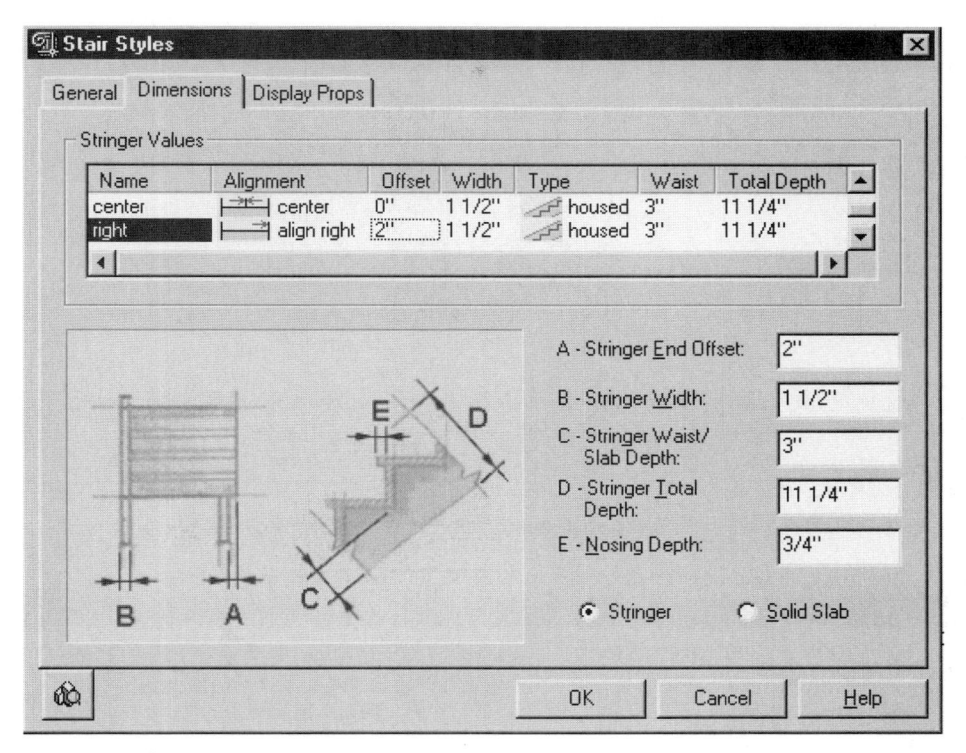

Figure 6.79 *Creating the right stringer*

22. Select the **Display Props** tab of the **Stair Styles** dialog box.
23. Select the System Default property source and select the **Edit Display Props** button.
24. In the **Entity Properties** dialog box, select the light bulb icon to turn OFF the display of the following stair components: **Stringer down**, **Riser down**, **Nosing down**, **Path down**, and **Outline down**, as shown in Figure 6.80.
25. Select the **Other** tab of the **Entity Properties** dialog box.
26. Edit the **Cut Plane Elevation** to 4'-0" as shown in Figure 6.81.
27. Select **OK** to dismiss the **Entity Properties** dialog box.
28. Select **OK** to dismiss the **Stair Styles** dialog box.

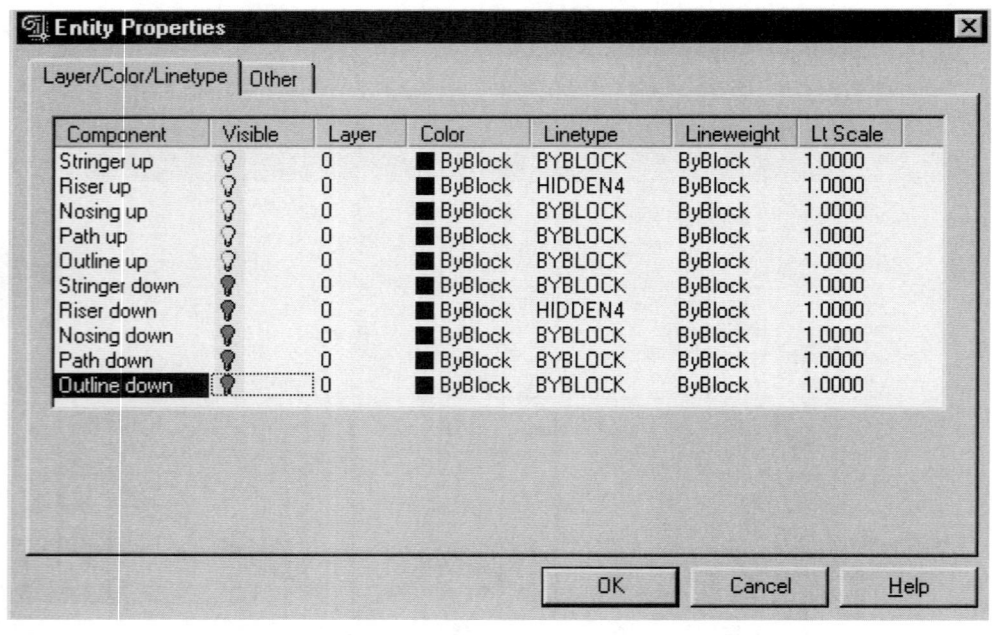

Figure 6.80 *Editing Layer/Color/Linetype tab of the Entity Properties dialog box*

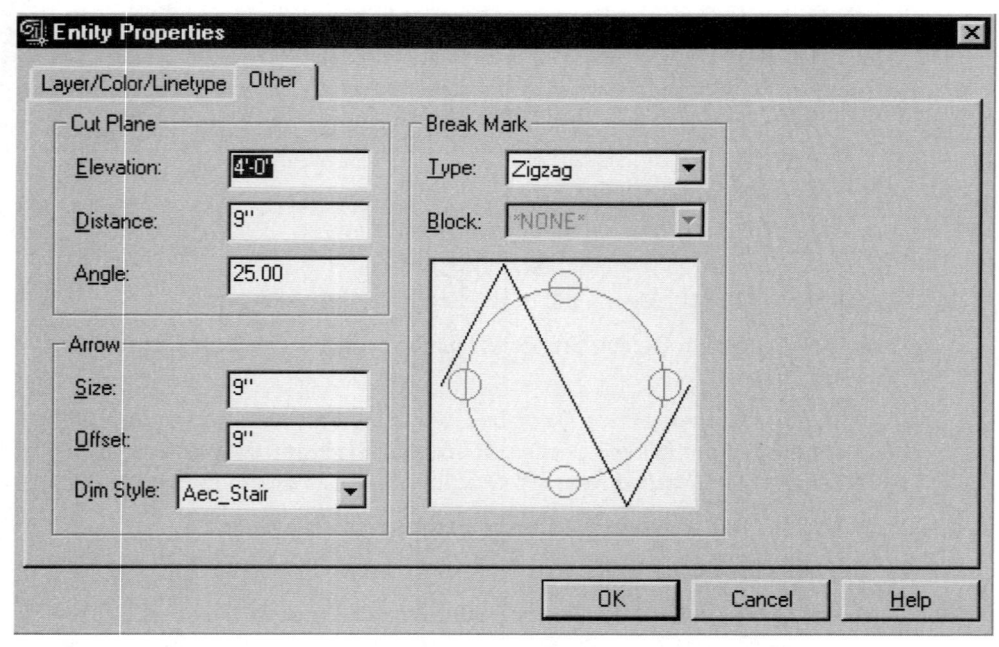

Figure 6.81 *Editing the cutting plane height in the Other tab of the Entity Properties dialog box*

29. Select the existing stair of *Ex6-5*, and then right-click and choose **Stairs Modify** from the shortcut menu.
30. Select the 3 Stringer style from the **Style** list as shown in Figure 6.82.

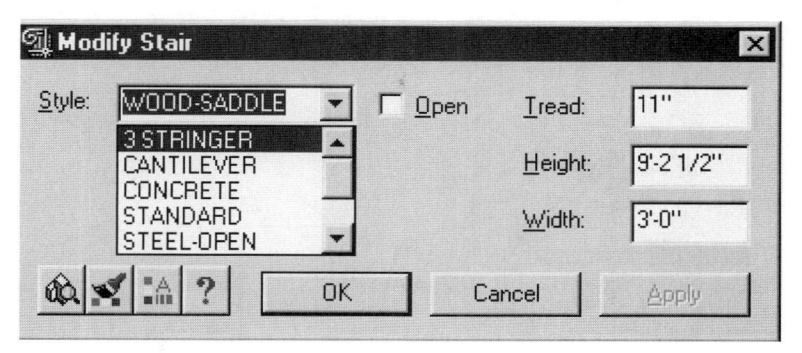

Figure 6.82 *Editing the stair with Modify Stair*

31. Select the **Apply** button of the **Modify Stair** dialog box.
32. Select the **Properties** button of the **Modify Stair** dialog box.
33. Select the **Dimensions** tab of the **Stair Properties** dialog box.
34. Edit the **C-Riser Count** to 16 as shown in Figure 6.83.
35. Select the **Design Rules** tab to verify that the **M-Riser** is 6-15/16" and lock the **N-Tread** value to 11".
36. Select **OK** to dismiss the **Stair Properties** dialog box.
37. Select **OK** to dismiss the **Modify Stair** dialog box. The stair is displayed with an error marker as shown Figure 6.84.
38. Select the stair, and then right-click and choose **Stairs Modify** from the shortcut menu.
39. Select the **Properties** button in the **Modify Stair** dialog box.
40. Select the **Design Rules** tab of the **Stair Properties** dialog box.
41. Unlock the **N-Tread** variable and enter 10-1/2" as the tread dimension.
42. Select **OK** to dismiss the **Stair Properties** dialog box.
43. Select **OK** to dismiss the **Modify Stair** dialog box. The stair error marker should be cleared.
44. Save the drawing.

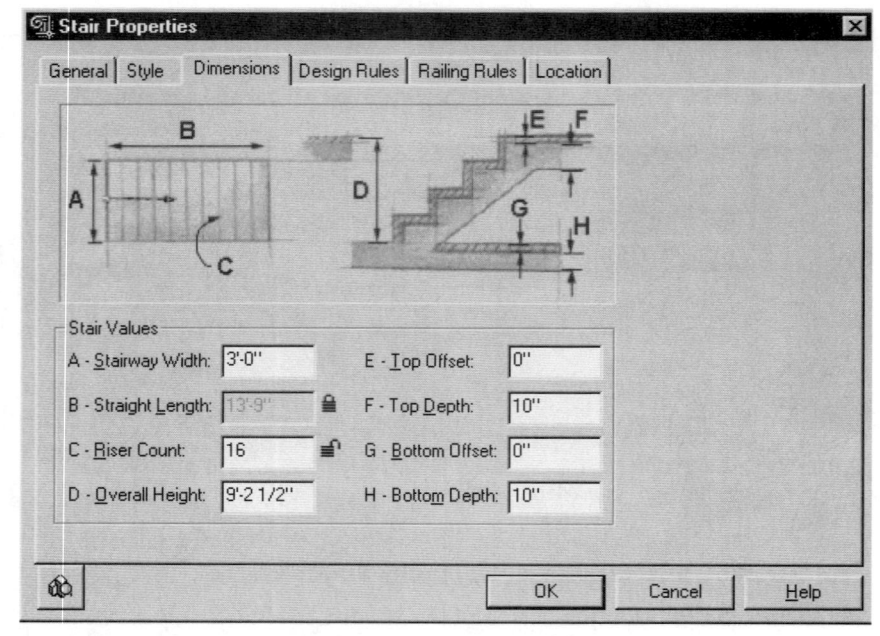

Figure 6.83 Dimensions tab of the Stair Properties dialog box

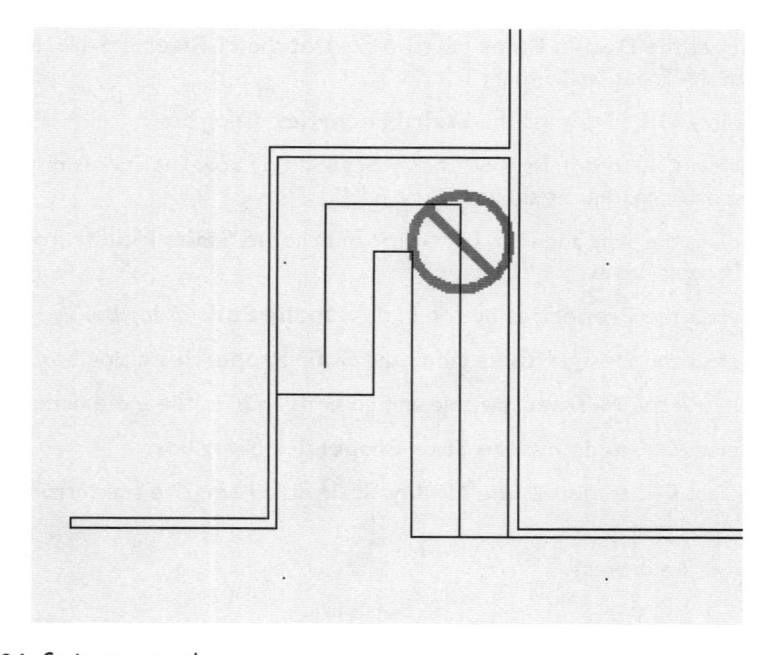

Figure 6.84 Stair error marker

EXPORTING STAIR STYLES

Stair styles developed in a drawing can be exported to other drawings or to a common resource file. Stair styles are similar to wall styles and door styles, since they can be transferred to other drawings. The stair styles included in the Aec arch (Imperial) template are also included in the *\AutoCAD Architectural 2\Content\Imperial\Styles\Stairs (Imperial).dwg*. Therefore, if the stair styles of a drawing are deleted you can reload them by importing the stair styles from the *\AutoCAD Architectural 2\Content\Imperial\Styles\Stairs (Imperial).dwg*. A new stair style can be exported to the *AutoCAD Architectural 2\Content\Imperial\Styles\Stairs (Imperial).dwg* drawing, which will allow you to import the stair style into new drawing files.

Stair styles are imported and exported through the **Import/Export** button of the **Stair Styles** dialog box. Therefore, to import or export a stair style, select the **Stair Styles** button on the **Stairs/Railings** toolbar, and then select the **Import/Export** button to open the **Import/Export** dialog box as shown in Figure 6.85.

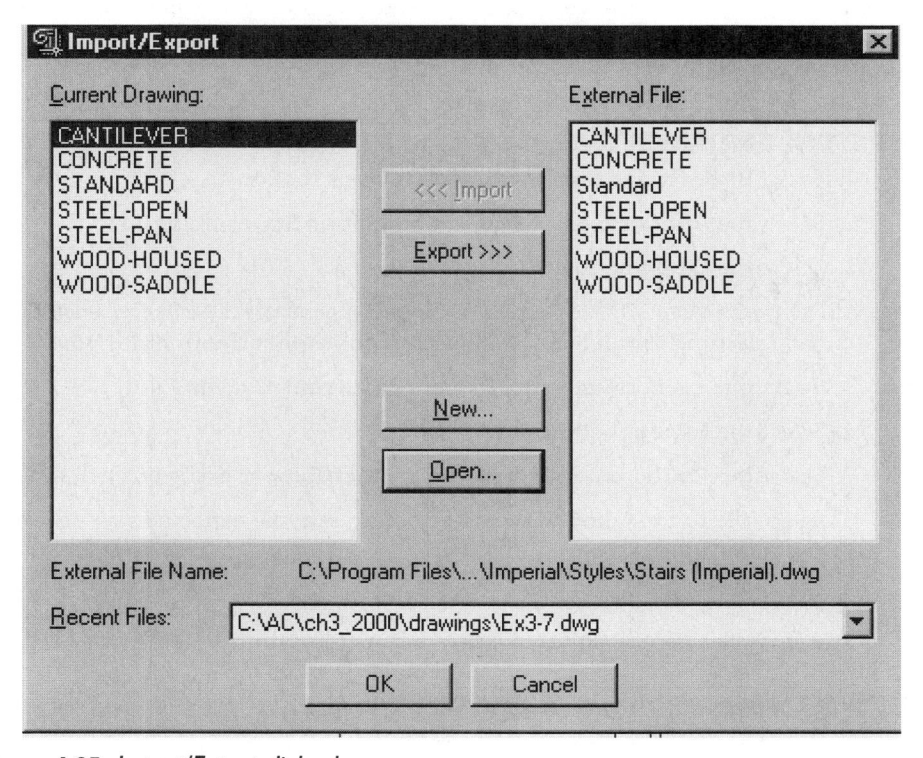

Figure 6.85 *Import/Export dialog box*

The options of this dialog box are as follows:

Import – The **Import** button allows you to copy a stair style from an external file into the current drawing.

Export – The **Export** button allows you to copy a stair style from the current drawing to other drawings or to a common external resource file.

New – The **New** button opens the **New Drawing File** dialog box, which allows you to create a new file for Stair styles.

Open – The **Open** button opens the **File to Import From** dialog box, which allows you to select a file for import or export of the stair style.

Current Drawing – The **Current Drawing** list box shows the stair styles that exists in the current drawing.

External File – The **External File** list box includes a list of the stair styles that exist in the file opened by the **File to Import From** dialog box.

You can begin to import or export stair styles by selecting the **Open** button. Selecting the **Open** button opens the **File to Import From** dialog box, which allows you to select a file for importing or exporting stair styles. Once the file is opened, you can export or import stair styles.

STEPS TO EXPORTING A STAIR STYLE

1. Select **Stair Styles** from the **Stairs/Railings** toolbar.
2. Select the **Import/Export** button of the **Stair Styles** dialog box.
3. Select the **Open** button.
4. Select the *AutoCAD Architectural 2\Content\Imperial\Styles\Stairs (Imperial).dwg* and select the **Open** button in the **Files to Import From** dialog box.
5. Select one or more stair styles from the **Current Drawing** list.
6. Select the **Export** button.
7. Select the **OK** button to dismiss the **Import/Export** dialog box.
8. Select YES to the AutoCAD message box to save the exported stair style to the *Stairs (Imperial).Dwg*.
9. Select **OK** to dismiss the **Stair Styles** dialog box.

    A new stair style is exported from the current drawing to an external file.

STEPS TO IMPORTING A STAIR STYLE

1. Select **Stair Styles** from the **Stairs/Railing**s toolbar.
2. Select the **Import/Export** button of the **Stair Styles** dialog box.
3. Select the **Open** button and open the *AutoCAD Architectural 2\Content\Imperial\Styles\Stairs (Imperial).dwg*.

4. Select one or more stair styles from the **External** list box.
5. Select the **Import** button.
6. Select the **OK** button to dismiss the **Import/Export** dialog box.
7. Select **OK** to dismiss the **Stair Styles** dialog box.

   A new style is imported from the external file to the current file.

PURGING STAIR STYLES

Stair styles that exist in a drawing but that have not been used can be deleted from the drawing. Selecting the **Purge** button of the **Stair Styles** dialog box opens the **Purge Stair Styles** dialog box, as shown in Figure 6.86.

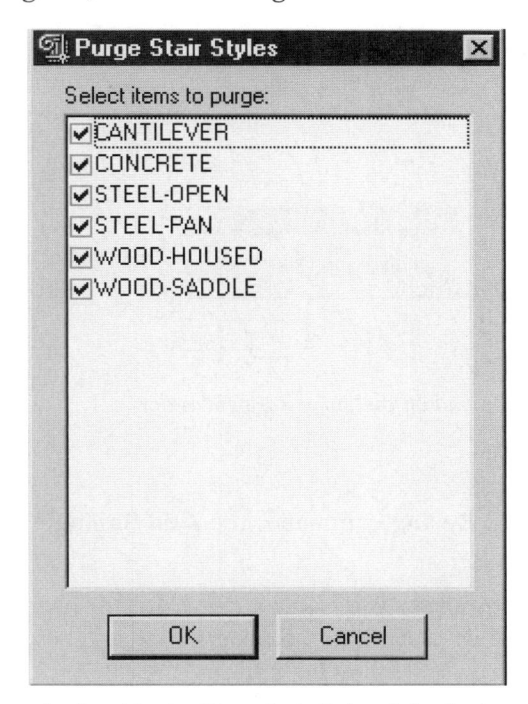

Figure 6.86 *Stair styles listed in the Purge Stair Styles dialog box*

The stair styles listed in the **Purge Stair Styles** dialog box include all existing styles that have not been used in the drawing. (The Standard stair style is not included in the list regardless of its use; it always remains in the drawing and is never included in the purge operation.) To deselect a stair style listed in the **Purge Stair Styles** dialog box, click in the check box to remove the check mark. Selecting the **OK** button will delete all stair styles checked in the **Purge Stair Styles** dialog box.

CREATING RAILINGS

The **Add Railing** command is used to add a railing to one or both sides of the stair. Railings can consist of post, balusters, guardrail, bottom rail and handrail. Railings can be added to a stair or be freestanding. See Table 6.6 for command access.

Menu bar	Design>Railings>Add Railing
Command prompt	RAILINGADD or RAILING
Toolbar	Select Add Railing from the Stairs/Railings toolbar as shown in Figure 6.87
Shortcut	Right-click over the drawing area; choose Design>Railings>Add

Table 6.6 Add Railing access

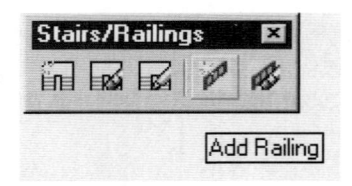

Figure 6.87 Add Railing command on the Stairs/Railings toolbar

When you select the **Add Railing** command, the **Add Railing** dialog box opens as shown in Figure 6.88.

The **Add Railing** dialog box allows you to place a guardrail or handrail. The handrail and guardrail can be placed with or without balusters, post, and bottom rails. The **Add Railing** dialog box consists of the following options.

> **Handrail** – When the **Handrail** check box is selected, a handrail will be generated with the railing.
>
> **Guardrail** – When **Guardrail** is checked, a guardrail is created between posts.
>
> **Posts** – When **Posts** is checked, the railing will include newel posts at each end of the stair.
>
> **Balusters** – When **Balusters** is checked, balusters are placed below the handrail or guardrail.

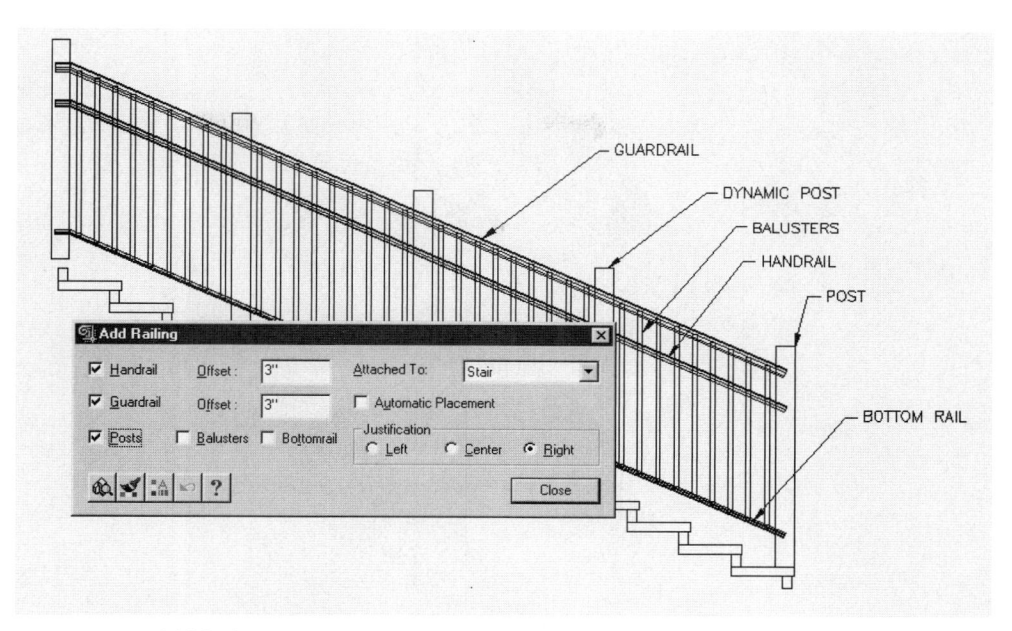

Figure 6.88 *Add Railing dialog box*

> **Bottomrail** – When **Bottomrail** is checked, the balusters terminate at a bottomrail. When **Bottomrail** is not checked, the balusters terminate at the floor or stair tread.
>
> **Attached to** – The **Attached to** list includes the Stair and None options. If **Attached to** is set to Stair, the railing is created as an attachment to a stair. The None option allows you to place a freestanding railing or guardrail. When the None option is used, the **Automatic Placement** and **Justification** options are inactive.
>
> **Automatic Placement** – The **Automatic Placement** check box, when selected, will create a railing on the side of the stair selected. The railing is automatically placed continuously from the beginning to the end of the stair. If **Automatic Placement** is not selected, the **Left, Center**, and **Right** justifications become active and the start and end point of the stair can be specified.
>
> **Justification** – The **Justification** section includes **Left, Center** and **Right** options. These options are available only if **Automatic Placement** is not selected. Each of the justification options allows you to define the beginning and end locations for the railing. If the **Left** justification is toggled ON, the railing will be created on the left edge of the stair when oriented at the start point of the stair. The **Center** justification will allow the railing to be placed in the center of the width of the stair. The **Right** justification allows you to place the rail on the right of the stair. Figure 6.89 shows each of the justifications.

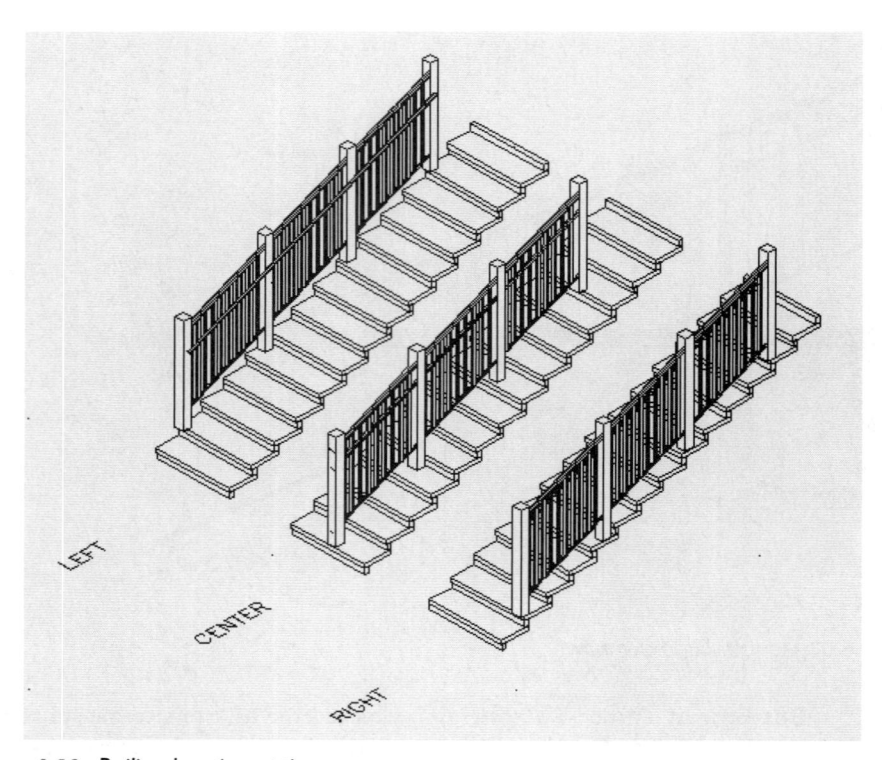

Figure 6.89 *Railing location options*

Floating Viewer – The **Floating Viewer** button opens the **Viewer,** which allows you to view the railing object prior to inserting the railing. Figure 6.90 shows a railing with bottom rail, handrail, guardrail, and balusters.

Match – The **Match** button allows you to select an existing railing and copy one or more of its settings for the current railing. After selecting the **Match** button, you are prompted to select an existing railing, as shown in the following command sequence.

```
Command:
AECRAILINGADD
Railing start point or
[Attach/Guardrail/Handrail/boTtomrail/Balusters/Posts
/Match/Undo]: DBOX
```
(Select the Match button of the Add Railing dialog box.)
```
Select a railing to match:
```
(Select an existing railing.)
```
Match [Guardrail/Handrail/Lower
rails/Balusters/Posts] <All>:
```
(Type the option to match or press ENTER *to select all options.)*

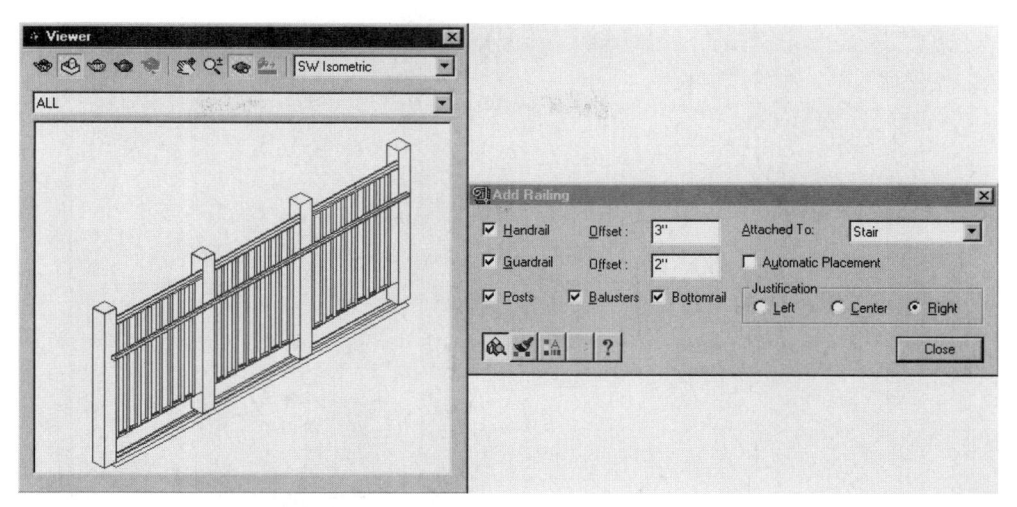

Figure 6.90 Stair displayed in the Viewer

Pressing ENTER selects the **All** default option and all properties of the railing are matched. Therefore, the **Add Railing** dialog box is changed according to that of the selected railing. However, you can elect to match only a specific property by entering one of the following options in the command line: **Guardrail**, **Handrail**, **Lower rails**, **Balusters**, or **Posts**.

Properties – The **Properties** button opens the **Railing Properties** dialog box, which consists of three tabs: **General, Dimensions**, and **Design Rules** The **Railing Properties** dialog box allows you to set the profile, size, and location of the railing components.

Undo – The **Undo** button allows you to undo the points specified for the railing location.

? – The **?** opens the Help file for Railings.

DEFAULT PROPERTIES OF A RAILING

The default location of railing components are set in the **Aec DwgDefaults** tab of the **Options** dialog box. The **Options** dialog box shown in Figure 6.91 includes a **Railing Settings** section. The default height of the **Handrail**, **Guardrail**, and **Bottomrail** are defined above the zero Z coordinate. Editing each of these heights will decrease the amount of editing in the **Railing Properties** dialog box.

DESIGNING THE RAILING WITH RAILING PROPERTIES

The **Add Railing** dialog box allows you to specify which components will be included in the railing. The **Properties** button of the **Add Railing** dialog box allows you to specify size and profile for each individual component of the railing prior to placing

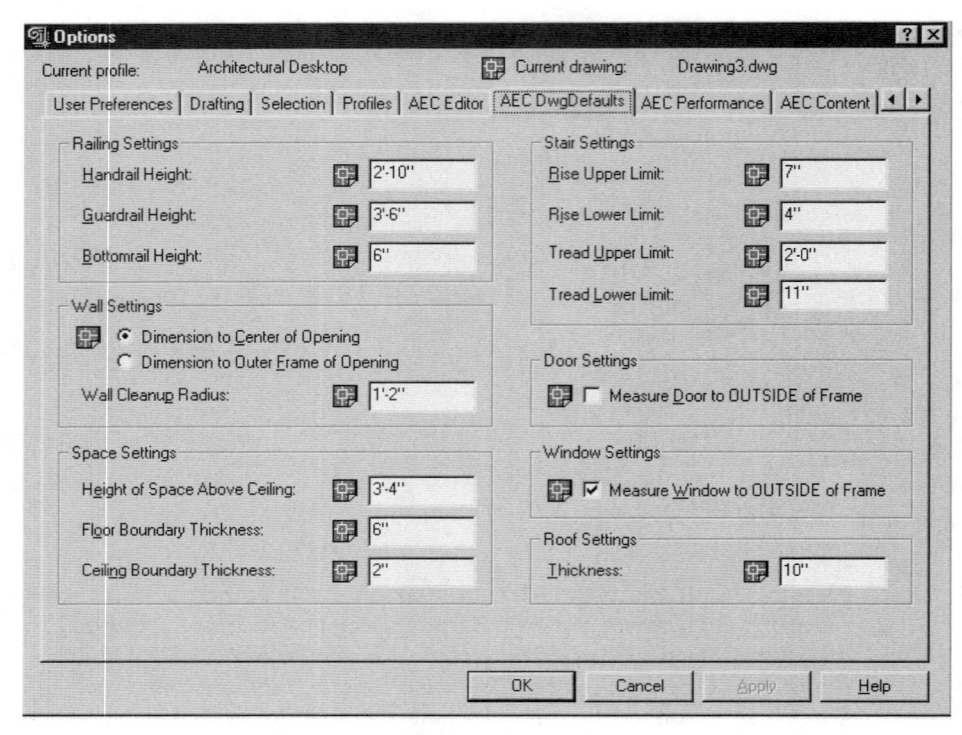

Figure 6.91 *Railing Settings of the AEC DwgDefaults tab*

the railing in the drawing. The properties of an existing railing can be edited by the **RAILINGPROPS** command. See Table 6.7 for command access.

Command prompt	RAILINGPROPS
Toolbar	Select Add Railing from the Stairs/Railings toolbar; select Properties button
Toolbar	Select Modify Railing from the Stairs/Railings toolbar; select Properties button
Shortcut	Select Railing; right-click and choose Railing Properties
Shortcut	Right-click over the drawing area and choose Design>Railing>Properties

Table 6.7 *RAILINGPROPS access*

When the **RAILINGPROPS** command is selected, you are prompted in the command line to select a railing. The selection of an existing railing opens the **Railing Properties** dialog box, as shown in Figure 6.92.

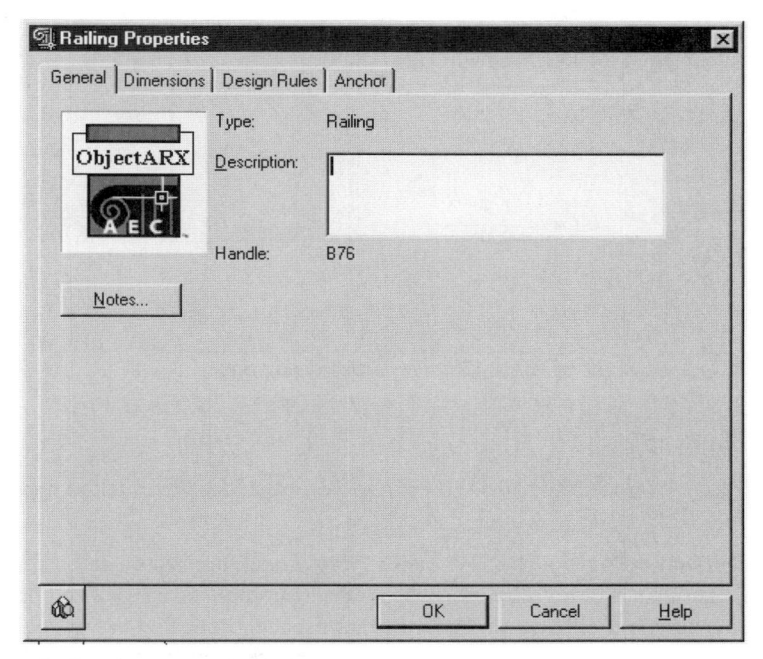

Figure 6.92 *Railing Properties dialog box*

The **Railing Properties** dialog box for an existing railing consists of the following tabs: **General**, **Dimensions**, **Design Rules**, and **Location**, which is only included if an existing stair is selected. The tabs of the **Railing Properties** dialog box allow you to design the size and dimensions of the railing components.

General
The **General** tab, shown in Figure 6.92, allows you to add a description to the railing. The **Notes** button is inactive unless you are editing the properties of an existing railing. If an existing stair is selected, the **Notes** button opens the **Notes** dialog box, which consists of the **Text Notes** and **Reference Docs** tabs. These tabs are similar to those of other Architectural Desktop objects because they allow you to enter text and a list of files related to the stair.

Dimensions
The **Dimensions** tab includes the **Vertical Components** and **Horizontal Components** sections, as shown in Figure 6.93.

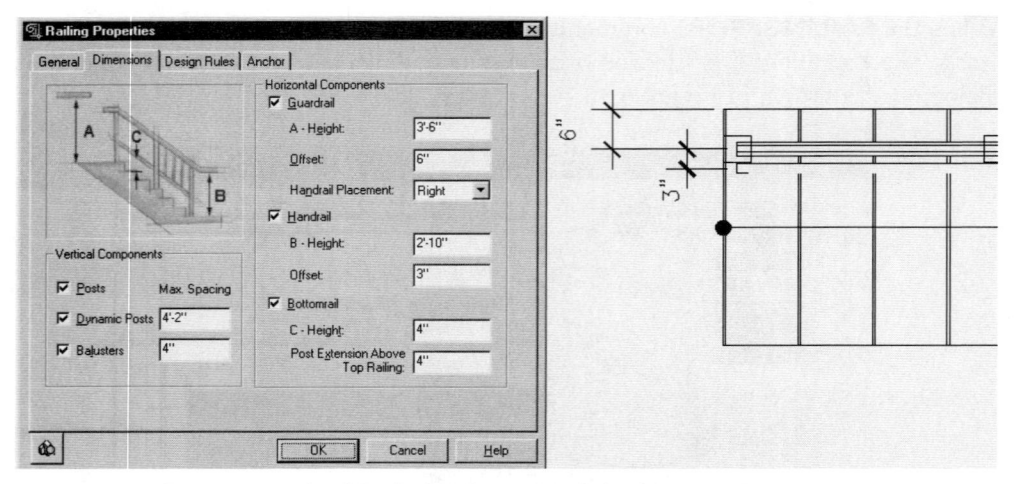

Figure 6.93 *Dimensions tab of the Railing Properties dialog box*

Vertical Components

Posts – The **Posts** check box allows you to select posts at the ends of the stair.

Dynamic Posts – The **Dynamic Posts** check box, when selected, will place intermediate posts along the railing. The maximum spacing of the posts is specified in the edit field. The maximum spacing specified in Figure 6.93 is 4'–2". Therefore, when the railing is created, dynamic posts are inserted along the railing length at least every 4'–2". If **Dynamic Posts** is not selected, no intermediate posts will be placed along the railing.

Balusters – The **Balusters** check box selects balusters for the railing. The horizontal distance between balusters is specified in the edit field. The balusters are uniformly centered between the end posts and the dynamic posts.

Horizontal Components

Guardrail – The **Guardrail** check box allows you to select a guardrail. The guardrail is the top rail of the railing.

A-Height – The height of the guardrail above the tread or floor if the guardrail not attached to a stair. The default height of the guardrail is set in the **Aec DwgDefaults** tab of the **Options** dialog box.

Offset – Specifies the distance from the edge of the stair to the center of the guardrail. The 6" Offset distance shown in Figure 6.93 positions the guardrail 6" from the edge of the stair. The **Offset** distance is also displayed in the **Guardrail Offset** edit field of the **Add Railing** dialog box. The **Offset** distance allows you to precisely position the railing relative to the stair.

Handrail Placement – You can place the handrail to the left, right or center of the guardrail by selecting Left, Right, or Center in the **Handrail Placement** list.

Handrail – The **Handrail** check box selects the handrail.

B-Height – The height of the handrail above the tread of the stair or floor. The default handrail height is set in the **Aec DwgDefaults** tab of the **Options** dialog box.

Offset – Specifies the distance from the center of the handrail to the center of the guardrail.

Bottomrail – The **Bottomrail** check box selects the bottomrail. If the **Bottomrail** is not selected, the balusters terminate at the floor.

C-Height – The height of the post above the top railing.

Post Extrusion Above Top Railing – Specifies the height of the post above the railing. The extension value can be set to zero to create a continuous top railing.

Design Rules

The **Design Rules** tab shown in Figure 6.94 allows you to specify the dimension and shape of the railing components. The top portion of the **Design Rules** tab is a list of components and the **Dimensions** section below is used to edit each railing component.

The components of the railing include **Guardrail, Handrail, Bottomrail, Post, Dynamic Post**, and **Baluster**. To edit the profile and size of a component, select the component name in the component list; then edit the **Profile Name, Width, Depth** or **Rotation** in the **Dimensions** section of the **Design Rules** tab.

The shape of each component is developed from the Aec Profile listed in the **Profile Name** column in the **List of Components** section of the dialog box. The components listed in Figure 6.94 are developed from the circular and rectangular Aec Profiles. Therefore, extruding the circular shape, which has a diameter of 2", creates the guardrail. When the rectangular Aec Profile is used, the width and depth can be specified to create a rectangular component.

Profile Name – The **Profile Name** drop-down list shows the Aec Profiles loaded in the drawing. Display a list and description of the Aec Profiles of the drawing in the **Profiles** dialog box by choosing **Desktop>Aec Profiles>Define Aec Profile** from the menu bar. The **Profiles** dialog box is shown in Figure 6.95.

There are five AEC Profiles specifically designed for handrails shown in the drawing in Figure 6.95. The profiles HR_CAP1, HR_CAP2, and HR_CAP3 are designed to receive the balusters in the guardrail.

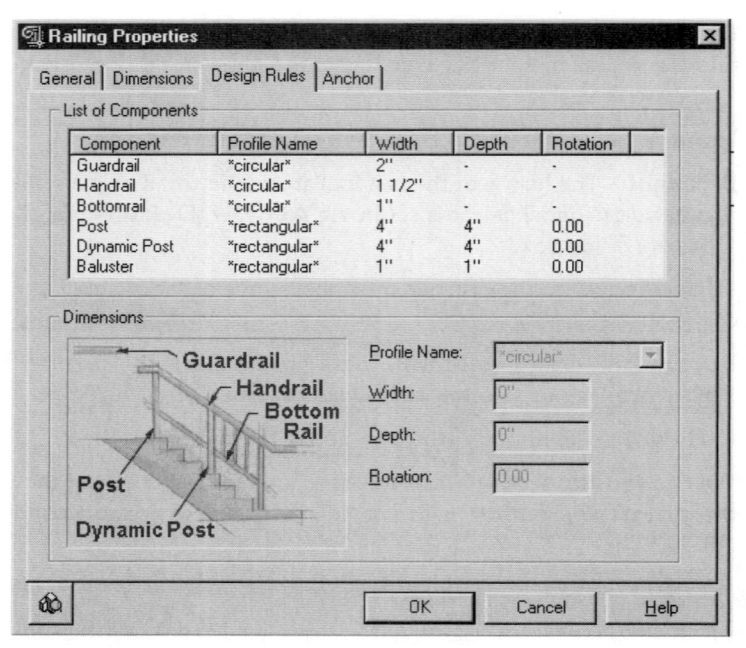

Figure 6.94 *Design Rules tab of the Railing Properties dialog box*

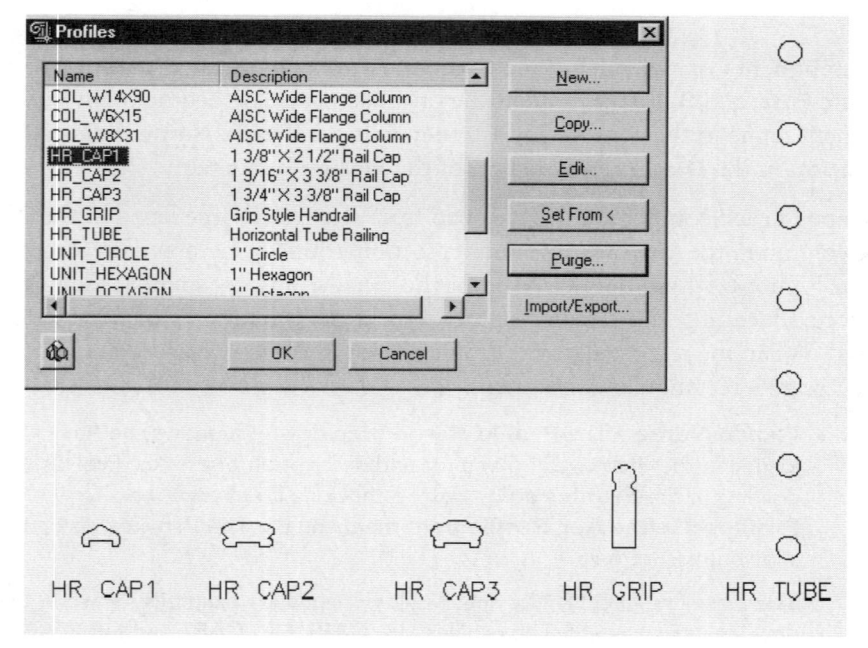

Figure 6.95 *AEC Profiles used as handrail shapes*

> **Width** – Allows you to enter the width.
>
> **Depth** – Allows you to enter the depth of the component.
>
> **Rotation** – Allows you to rotate the component.

Figure 6.96 shows the **Design Rules** tab and a railing developed using various Aec Profiles. The width and depth dimensions have been edited to create end posts that are larger than the dynamic posts. The HR_GRIP Aec Profile was used to create the handrail. The HR_CAP1 Aec Profile was used to create the guardrail.

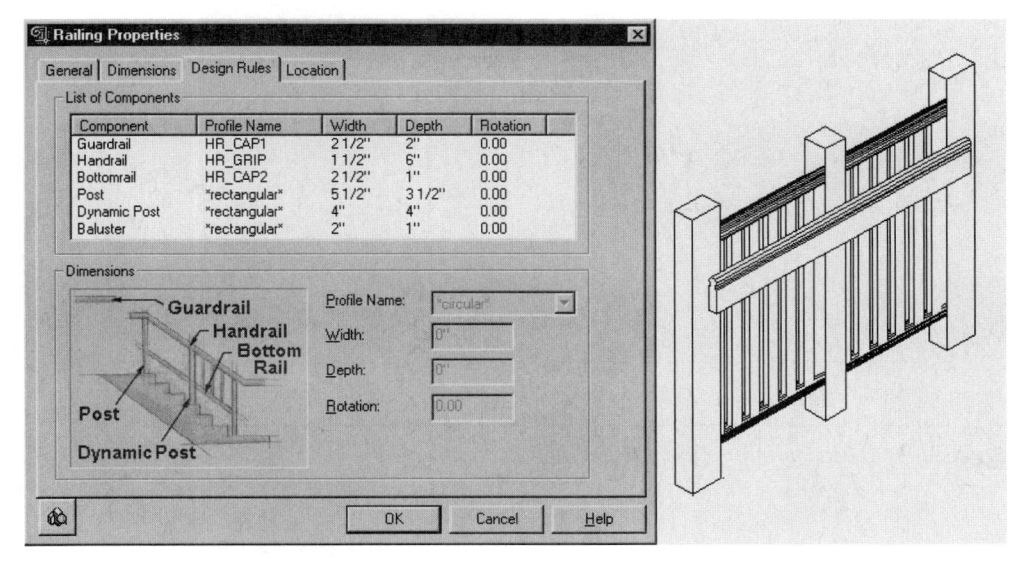

Figure 6.96 *Railing created with various Aec Profiles*

Location

The **Railing Properties** dialog box for railings not attached to a stair has a **Location** tab, shown in Figure 6.97. It includes **Insertion Point**, **Normal** and **Rotation** sections. Each of these sections displays the position of an existing railing. The **Location** tab is not included in the **Railing Properties** dialog box if the **RAILINGPROPS** command is executed prior to a railing being placed.

A description of each option of the Location tab follows.

> **Relative to-World Coordinate System** – Selected to display the location using the world coordinate system.
>
> **Relative to-Current Coordinate System** – Selected to display the location using the User Coordinate System.

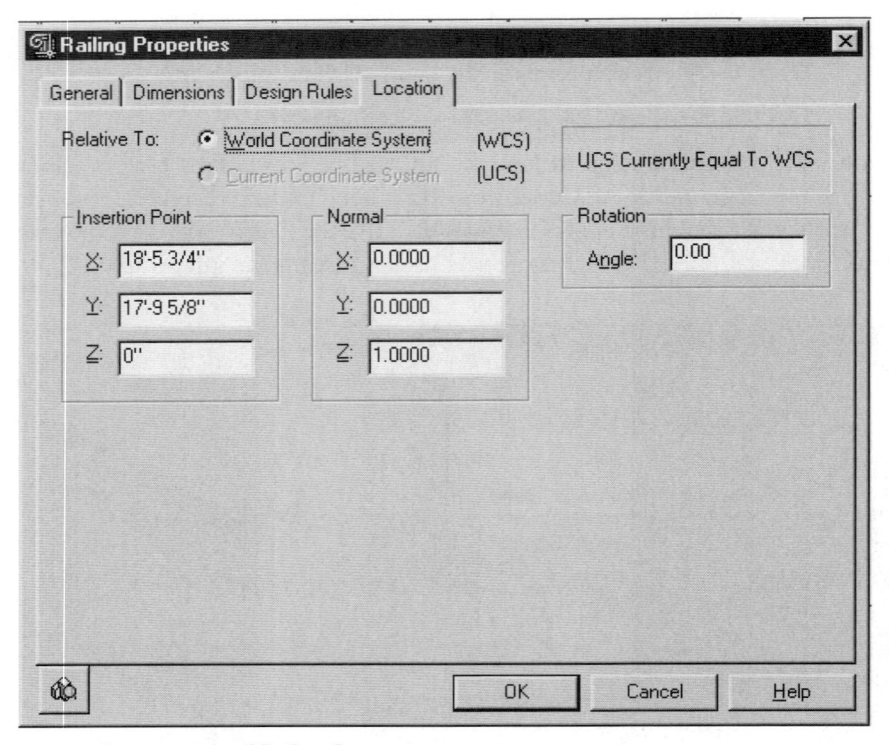

Figure 6.97 *Location tab of Railing Properties*

Insertion Point – The **Insertion Point** edit fields displays the X, Y, Z coordinate position of the railing. The insertion point of a railing is the location of the start point for the railing.

Normal – The **Normal** edit fields include X, Y, and Z. In Figure 6.97, the Z value is set to 1.0000; this extrudes the posts in the positive Z direction and develops the railing in the XY plane. If the X edit field is set to 1, the railing will be developed in the YZ plane. The railing will be developed in the XZ plane if 1 is entered in the Y edit field and 0 in the X and Z edit fields.

Rotation – Allows you to rotate the railing in its plane.

Anchor

The **Railing Properties** dialog box for railings attached to a stair has an **Anchor** tab rather than a **Location** tab. The **Anchor** tab displays the properties of the railing relative to the stair geometry. It includes edit fields for justification and railing offset from the beginning and end of the stair, as shown in Figure 6.98.

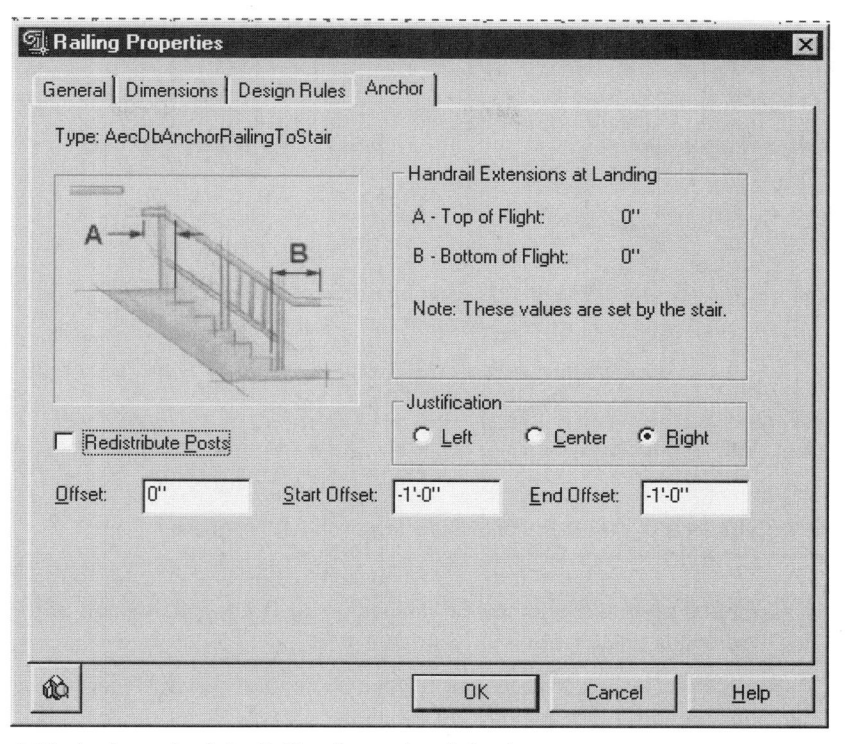

Figure 6.98 *Anchor tab of the Railing Properties dialog box*

The **Anchor** tab includes the following options.

> **Handrail Extensions at Landing** – The **Handrail Extensions at Landing** section displays the amount of handrail extension as defined by the stair properties. The **Railing Rules** tab of the **Stair Properties** dialog box allows you to set the amount of handrail extension at the beginning and end of the stair. The distances set in the **Railing Rules** tab are also displayed in the **Handrail Extensions at Landing** section.
>
> **Redistribute Posts** – The **Redistribute Posts** check box, if selected, will adjust the post along the railing after the start and end offset of the railing have been adjusted. The **Redistribute Posts** check box is selected for the stairs shown on the right in Figure 6.99. The **Redistribute Posts** check box shifts a post to the beginning of the stair.
>
> **Offset** – Allows you to change the distance the handrail is placed from the edge of the stair. Negative offset values will shift the railing beyond the bounds of the stair.
>
> **Start Offset** – Allows you specify where the railing starts relative to the stair. A negative offset value extends the railing beyond the stair. The railing is

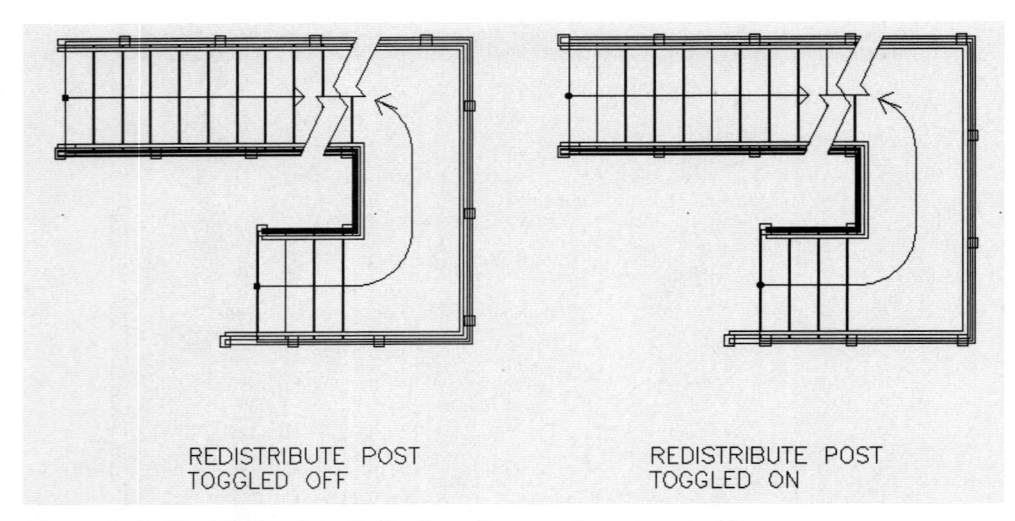

Figure 6.99 The effects of the Redistribute Posts toggle on post positions

extended beyond the extension defined in the **Railing Rules** tab of the **Stair Properties** ddialog box.

End Offset – Extends the stair beyond the end of the stair. Negative offset values extend the railing beyond the stair.

Justification–The **Justification** section allows you to select the railing placement to the **Left, Center**, or **Right** on the stair.

CHANGING THE RAILING WITH MODIFY RAILING

Existing railings can also be changed by the Modify Railing command. It provides access to the Railing Properties button and allows you to quickly edit the components of the railing. See Table 6.8 for command access.

Menu bar	Design>Railings>Modify Railing
Command prompt	RAILINGMODIFY
Toolbar	Select Modify Railing from the Stairs/Railings toolbar as shown in Figure 6.100
Shortcut	Right-click over the drawing area and choose Design>Railings>Modify
Shortcut	Select railing; then right-click and choose Railing Modify

Table 6.8 Modify Railing access

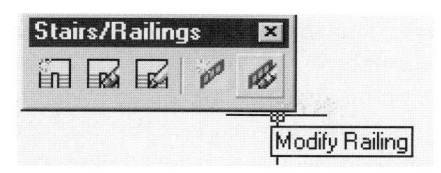

Figure 6.100 *The Modify Railing command on the Stairs/Railings toolbar*

If you select a railing and then right-click and choose **Modify Railing**, the **Modify Railing** dialog box will open, as shown in Figure 6.101.

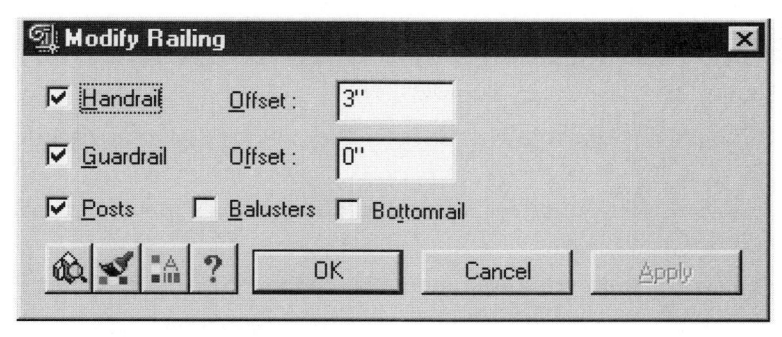

Figure 6.101 *Modify Railing dialog box*

The **Modify Railing** dialog box consists of the following check boxes and edit fields.

> **Handrail** – Allows you to select a handrail. The **Handrail** check box includes an **Offset** edit field that allows you to define the distance to offset the handrail from the guardrail.
>
> **Guardrail** – Allows you to select a guardrail. The **Guardrail** check box includes an **Offset** edit field that allows you specify the distance to offset the guardrail from the stair.
>
> **Posts** – Allows you to select the end posts and dynamic posts.
>
> **Balusters** – Allows you to select balusters.
>
> **Bottomrail** – Allows you to select a bottom rail. If the bottom rail is not selected, the balusters terminate at the floor.

DISPLAYING THE STAIR IN MULTIPLE LEVELS

Stairs are displayed on each level of a floor plan. The upper half of the stair is displayed on the plan of the upper level and the lower half of the stair is displayed on the plan of the lower level. Because the stair is created from the lower level up, the stair is part of the lower level floor plan, and the upper level plan does not include a stair. To display a stair in the upper level, you should create a WBLOCK of the stair from the lower level and insert it in the upper level floor plan. When you create the WBLOCK of the stair, specify the base point equal to 0,0,0, which will ensure the vertical alignment of the stair. However, this procedure requires that one floor level be created from the geometry of the other so that the absolute coordinates of walls remain identical. Therefore, when the stair block is then inserted in the upper level floor plan, it is inserted at the same X,Y location but at an elevation below the upper level equal to the stair rise. If the stair rise is 9'–2", the stair block would be inserted at 0,0,–9'-2". You can then explode the stair and edit the entity display properties of the stair to freeze the lower level of the stair in the upper floor plan.

TUTORIAL 6.6 PLACING A RAILING ON A STAIR

1. Open *ADT Tutor\Ch6\EX6-6.dwg*.
2. Choose **File>SaveAs** from the menu bar and save the drawing as *Railing6.dwg* in your student directory.
3. Select **Stair Styles** from the **Stairs/Railings** toolbar to open the **Stair Styles** dialog box.
4. Select the **Import/Export** button in the **Stair Styles** dialog box.
5. Select the **Open** button in the **Import/Export** dialog box, and then edit the **Look in** drop-down list to *\ADT_Tutor\Ch6\Style5.dwg*. Select the **Open** button in the **File to Import From** dialog box.
6. Select the 3 Stringer stair style from the **External File** list; then select the **Import** button as shown in Figure 6.102.
7. Select **OK** to dismiss the **Import/Export** dialog box.
8. Select **OK** to dismiss the **Stair Styles** dialog box.
9. Right-click over the drawing area and choose **Options** from the shortcut menu.
10. Edit the **AEC DwgDefaults** tab as follows: **Riser Upper Limit** = 7-3/4", **Riser Lower Limit** = 6, **Tread Upper Limit** = 12", and **Tread Lower Limit** = 10. Select **OK** to dismiss the **Options** dialog box.
11. Select the right wall of the stairwell and refer to the following command line sequence to stretch the wall 6" up as shown in Figure 6.103.

    ```
    Command: (Select the right wall; then select the lower grip and move the
    mouse up as shown in Figure 6.103.)
    ** STRETCH **
    Specify stretch point or [Base
    point/Copy/Undo/eXit]: 6   ENTER
    Command: *Cancel*  ESC
    Command: *Cancel*  ESC
    ```

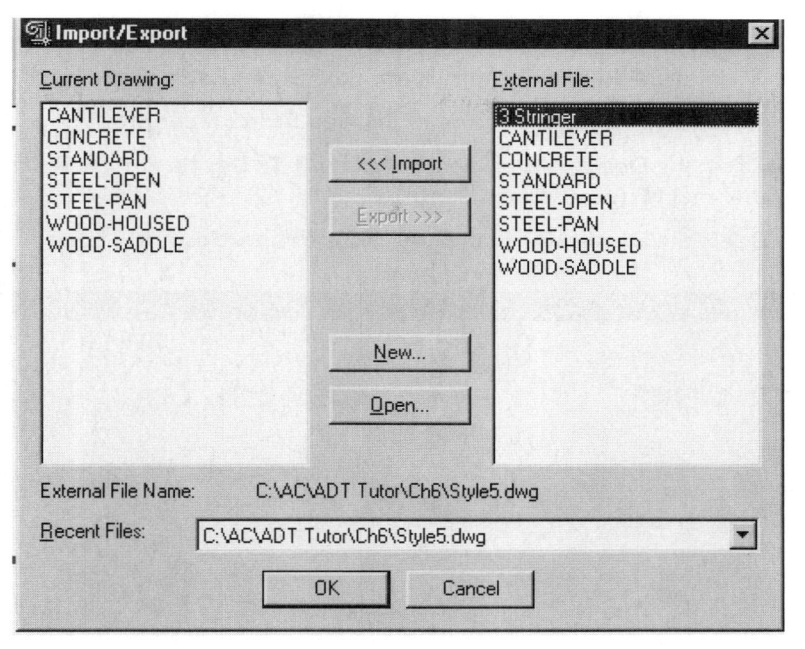

Figure 6.102 Import/Export dialog box

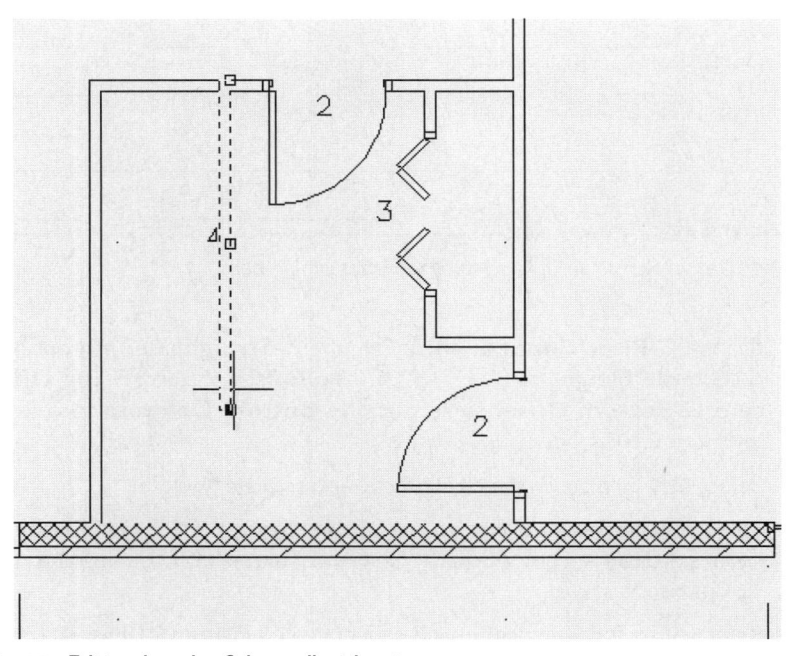

Figure 6.103 Editing length of the wall with grips

The wall is stretched 6" up.

12. Select the **Add Stair** command from the **Stairs/Railings** toolbar.
13. Select the **Properties** button of the **Add Stairs** dialog box.
14. Select the **Design Rules** tab and edit the **J-Tr Depth** to 1", **K-R Depth** to 3/4" and **N-Tread** dimension to 10".
15. Select the **Dimensions** tab of the **Stair Properties** dialog box.

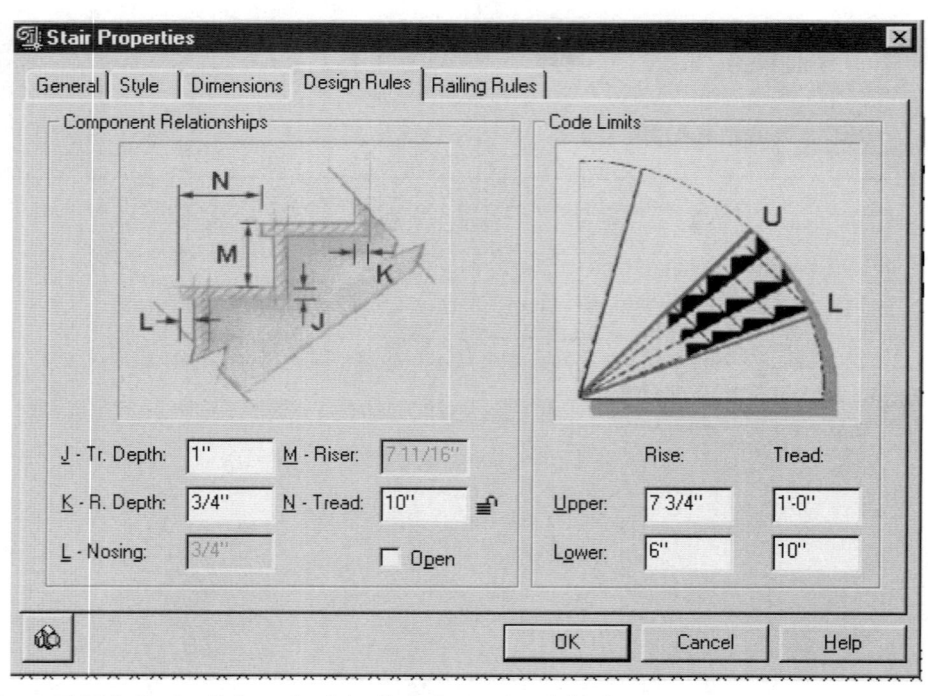

Figure 6.104 *Design Rules tab of the Stair Properties dialog box*

16. Set the **C-Riser Count** to 14, unlock the **B-Straight Length** option, set the **D-Overall Height** to 8'–11-1/2", **E-Top Offset** to zero, **F-Top Offset** to zero, **G-Bottom Offset** to zero, and **H-Bottom Depth** to zero, as shown in Figure 6.105.
17. Select **OK** to dismiss the **Stair Properties** dialog box.
18. Edit the **Add Stairs** dialog box as follows: **Style** = 3 Stringer, **Shape** = Multi-landing, **Justify** = Left, **Width** = 37, **Next Turn** = Half Landing and deselect **Open** risers.
19. Refer to the following command sequence and Figures 6.106, 6.107, and 6.108 to place the stair.

Stairs and Railings 483

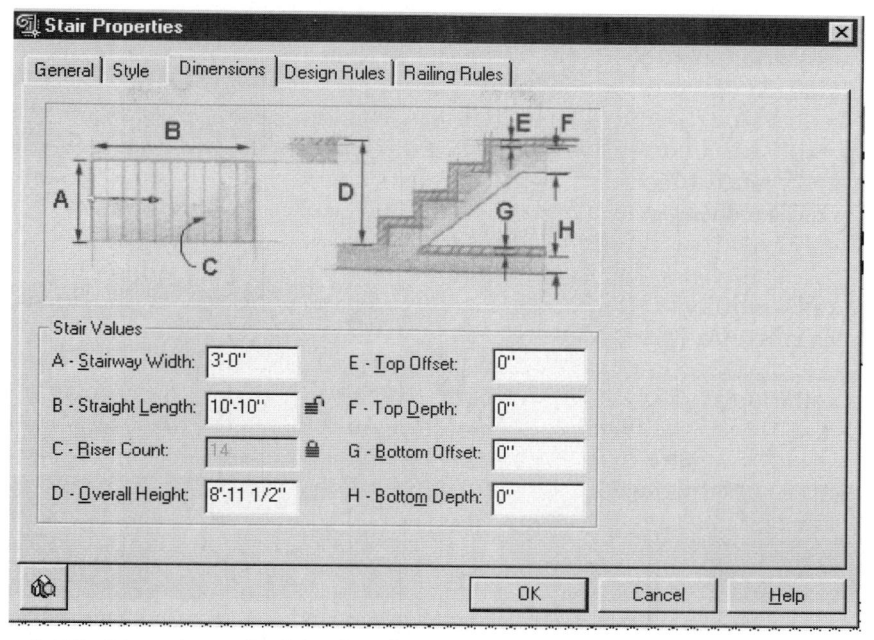

Figure 6.105 *Dimensions tab of the Stair Properties dialog box*

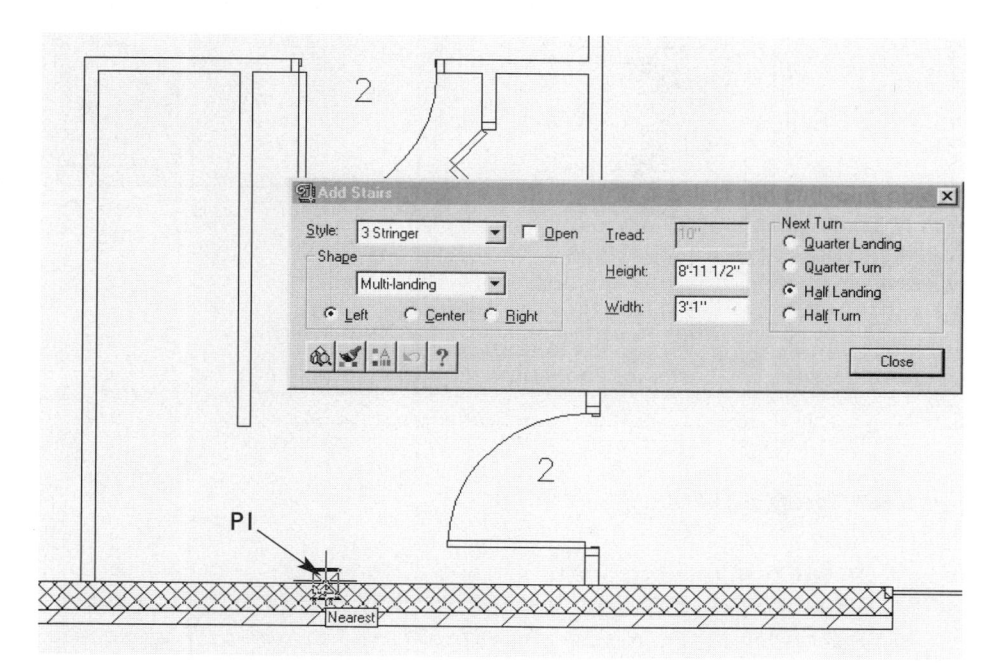

Figure 6.106 *Specify start point of stairs*

(Select in the drawing area to activate the drawing area.)
```
Command: _AecStairAdd
Flight start point or
[SHape/STyle/Tread/Height/Width/Justify/Open/Next/
Match/Undo]: nea ENTER to
```
(Select the wall near point P1 in Figure 6.106.)
(Move the mouse to the left of P1.)
```
Flight end point or
[SHape/STyle/Tread/Height/Width/Justify/Open/Next/
Match/Undo]: 22 ENTER
```
(Specifies end of flight.)
```
Flight end point or
[SHape/STyle/Tread/Height/Width/Justify/Open/Next/
Match/Undo]: int ENTER
to
```
(Refer to Figure 6.107 and select point P2 to specify the end of the landing.)

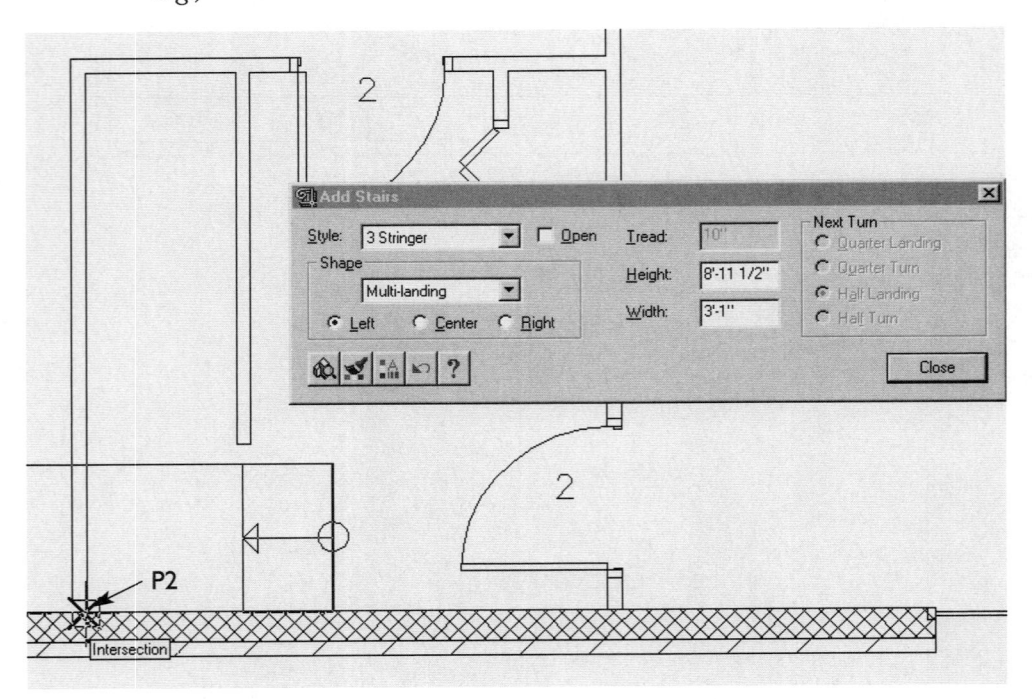

Figure 6.107 *Specify end of landing*

```
Flight end point or
[SHape/STyle/Tread/Height/Width/Justify/Open/Next/
Match/Undo]:
```
(Refer to Figure 6.108 and select point P3 beyond the end of the wall to specify the end of the stair flight.)

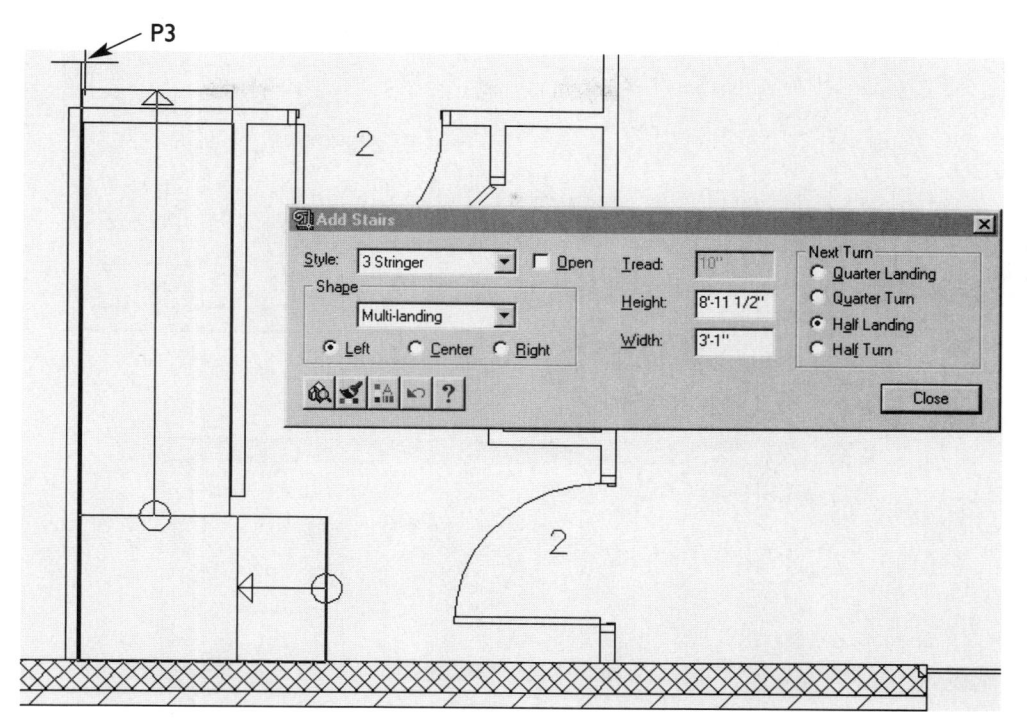

Figure 6.108 *Specify end of stair flight*

```
Flight start point or
[SHape/STyle/Tread/Height/Width/Justify/Open/Next/
Match/Undo]: ENTER (Press ENTER to end the command.)
```

20. Select the **Add Railing** command from the **Stairs/Railings** toolbar.

21. Edit the **Add Railing** dialog box as follows: deselect the **Handrail,** select the **Guardrail,** select **Post**s, select **Balusters**, deselect **Bottomrail**, set **Offset** to 0, **Attach to** to Stair, deselect **Automatic Placement**, and select **Right** justification, as shown in Figure 6.109.

22. Select the **Properties** button of the **Add Railing** dialog box.

23. Select the **Dimensions** tab of the **Railing Properties** dialog box: select **Posts**, deselect **Dynamic Posts**, select **Balusters**, **Max Spacing** to 4", select **Guardrail** and set the **A-Height** to 37", **Offset** to 2, **Handrail Placement** to Right and **Post Extension Above Top Railing** to 6". Verify that **Handrail** and **Bottomrail** are deselected, as shown in Figure 6.110.

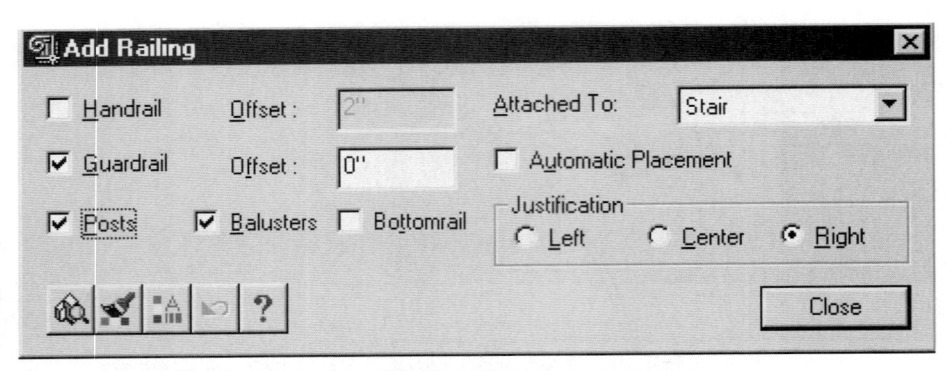

Figure 6.109 Add Railing dialog box defining railing components

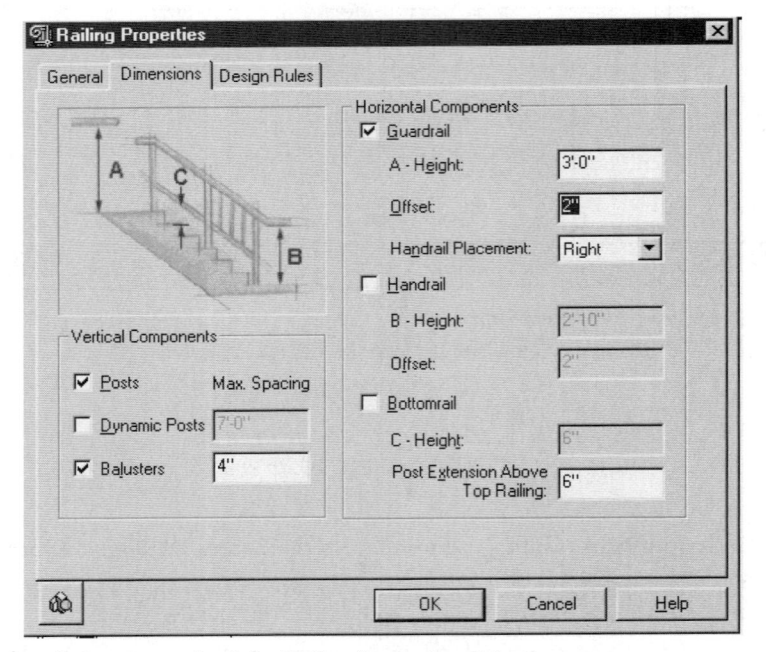

Figure 6.110 Dimensions tab of the Railing Properties dialog box

24. Select the **Design Rules** tab and select the Handrail component. Select HR_CAP1 from the **Profile Name** list. Edit the **Width** and **Dept**h to 1-1/2" as shown in Figure 6.111.

25. Select **OK** to dismiss the **Railing Properties** dialog box.

26. Refer to the following command sequence and Figures 6.112 and 6.113 to place the railing.

Stairs and Railings **487**

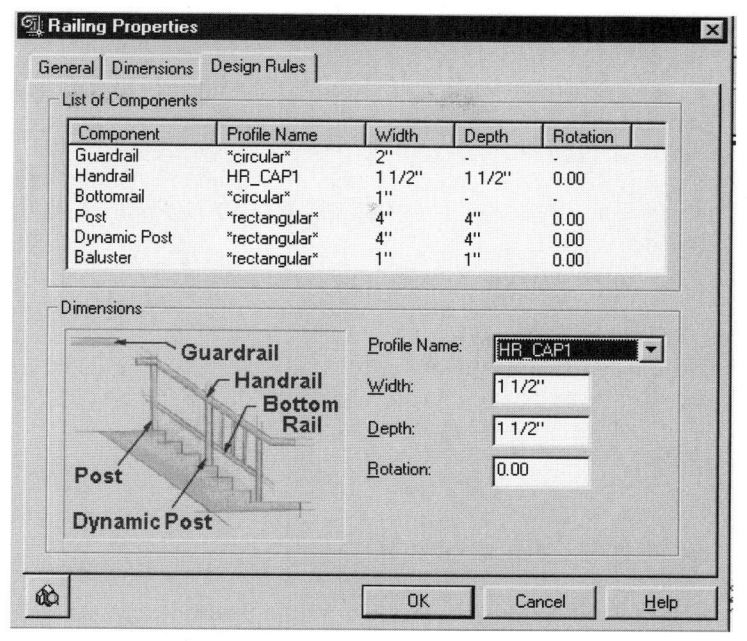

Figure 6.111 Design Rules tab of the Railing Properties dialog box

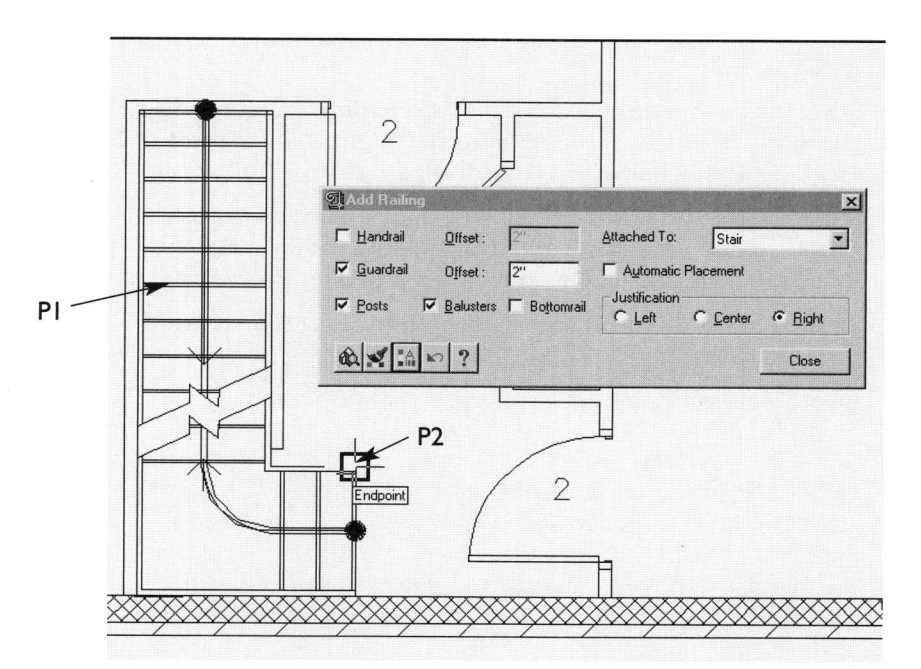

Figure 6.112 Specify start point of railing

```
Command: AECRAILINGADD
Select a stair or
[Attach/Guardrail/Handrail/boTtomrail/Balusters/Posts
/Match/Undo]: DBOX
Select a stair or
[Attach/Guardrail/Handrail/boTtomrail/Balusters/Posts
/Match/Undo]:
```
(Select the stair at point P1 in Figure 6.112.)
```
Railing start point or
[Attach/Guardrail/Handrail/boTtomrail/Balusters/Posts
/Match/Undo]: end ENTER of
```
(Select the stair at point P2 to begin the railing as shown in Figure 6.112.)
```
Railing end point or
[Attach/Guardrail/Handrail/boTtomrail/Balusters/Posts
/Match/Undo]: end ENTER of
```
(Select the stair at point P3 in Figure 6.113.)
```
Select a stair or
[Attach/Guardrail/Handrail/boTtomrail/Balusters/Posts
/Match/Undo]: ENTER
Command:
```

27. Select the **Add Railing** command from the **Stairs/Railings** toolbar.

28. Edit the **Add Railing** dialog box as follows: select **Handrail**, deselect **Guardrail,** deselect **Posts**, deselect **Balusters**, deselect **Bottomrail**, set **Attach to** to Stair, deselect **Automatic Placement**, and set **Justification** to **Right**.

29. Select the **Properties** button of the **Add Railing** dialog box. Select the **Dimensions** tab and edit the **Handrail Height** to 36", **Offset** to 2" and verify that **Posts**, **Balusters**, **Guardrail,** and **Bottomrail** are deselected.

30. Select the **Design Rules** tab and the Handrail component and select the Circular profile from the **Profile Name** list, as shown in Figure 6.114. Set the **Width** to 1-1/2". Select **OK** to dismiss the **Railing Properties** dialog box.

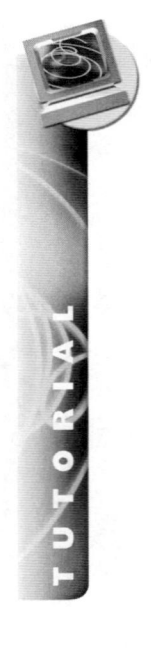

31. Refer to the following command sequence and Figures 6.115 and 6.116 to place the railing.

    (Select the drawing area to activate the drawing area.)
    ```
    Command: _AecRailingAdd
    Select a stair or
    [Attach/Guardrail/Handrail/boTtomrail/Balusters/Posts
    /Match/Undo]:
    ```
    (Select stair at point P1 in Figure 6.115.)
    ```
    Railing start point or
    [Attach/Guardrail/Handrail/boTtomrail/Balusters/Posts
    /Match/Undo]: end ENTER
    ```
    (Select start point for the railing at point P2 in Figure 6.115.)
    ```
    Railing end point or
    ```

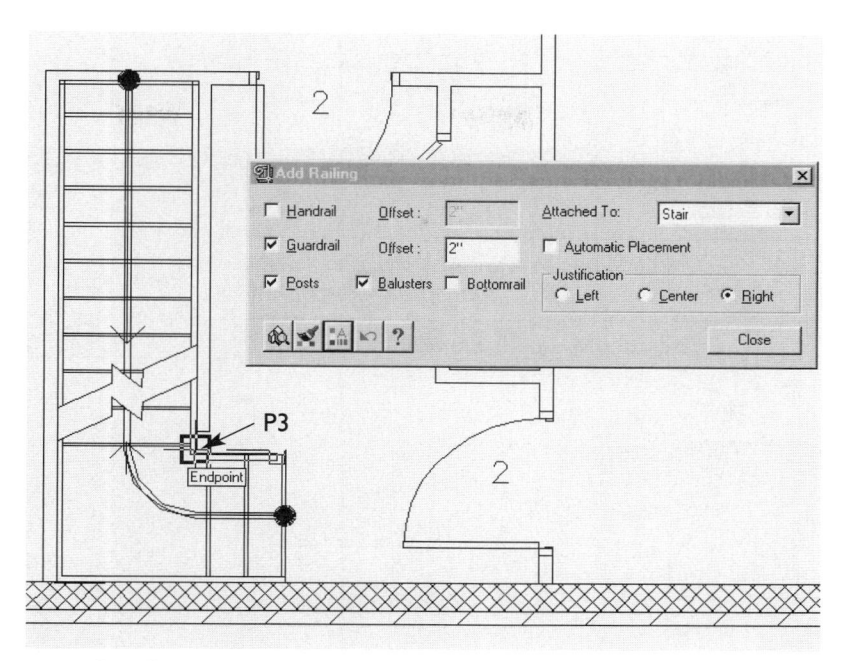

Figure 6.113 Specify end of railing

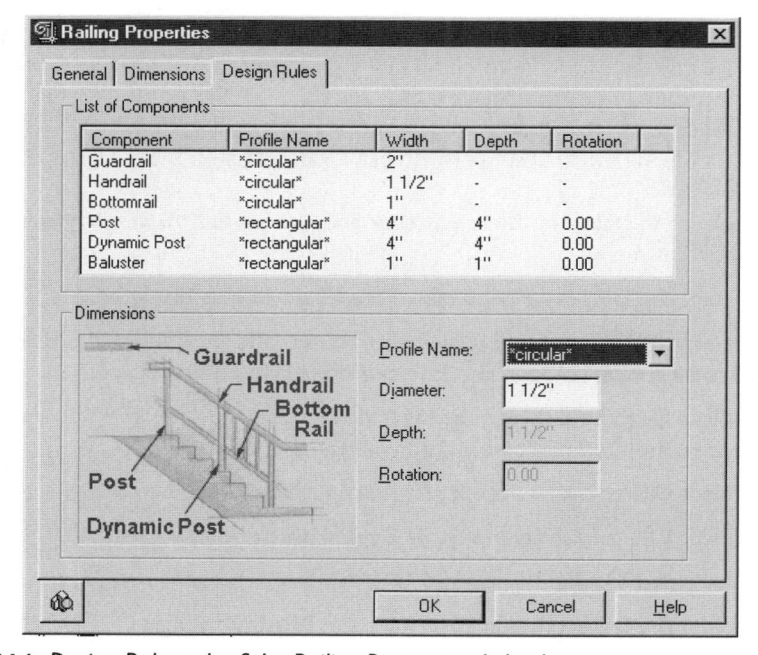

Figure 6.114 Design Rules tab of the Railing Properties dialog box

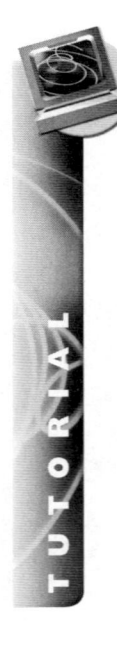

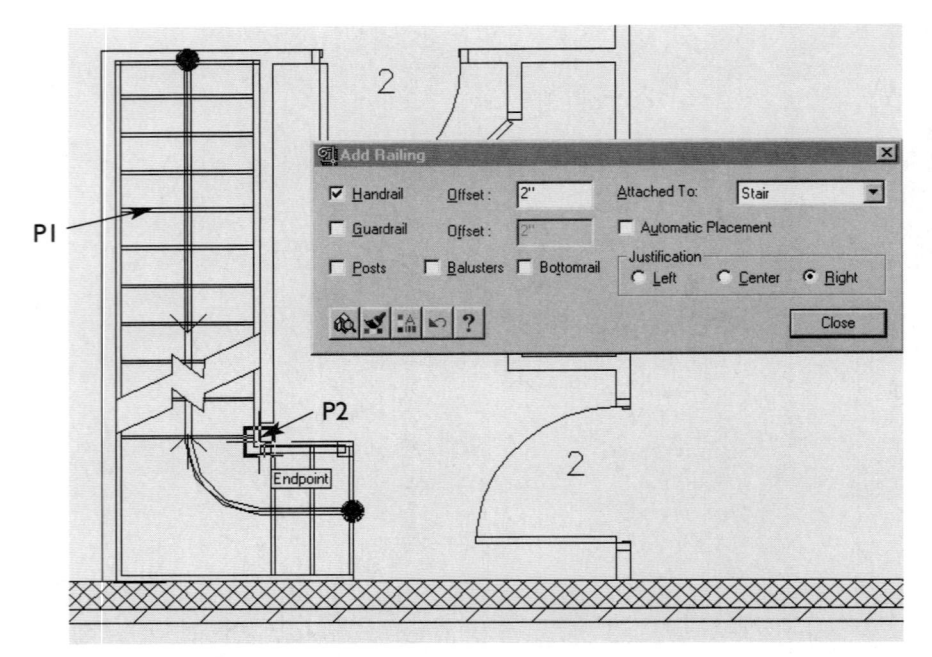

Figure 6.115 *Specify start point of railing*

```
[Attach/Guardrail/Handrail/boTtomrail/Balusters/Posts
/Match/Undo]: end ENTER of
```
(Select end of stair at point P3 as shown in Figure 6.116.)
```
Select a stair or
[Attach/Guardrail/Handrail/boTtomrail/Balusters/Posts
/Match/Undo]: ENTER
```

32. Select the stair, and then right-click and choose **Edit Stair Style** from the shortcut menu.
33. Select the **Display Props** tab of the **Stair Styles** dialog box.
34. Select the System Default property source, and then select the **Edit Display Props** button to open the **Entity Properties** dialog box.
35. Toggle OFF the display of Stringer down, Riser down, Nosing down, Path down, and Outline down, as shown Figure 6.117.
36. Select **OK** to dismiss the **Entity Properties** dialog box.
37. Select **OK** to dismiss the **Stair Styles** dialog box.
38. Verify that ORTHO is toggled ON.
39. Select the **Stretch** command from the **Modify** toolbar.

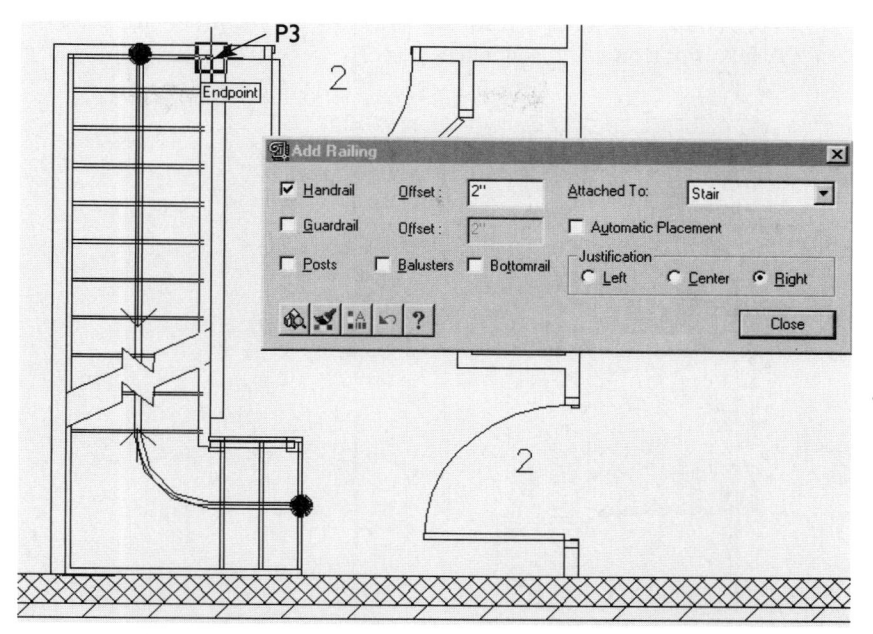

Figure 6.116 Specify end of railing

Figure 6.117 Turn off the display of the down stair components

40. To provide a 3'–0" corridor, stretch the closet as shown in Figure 6.118 and the following command sequence.

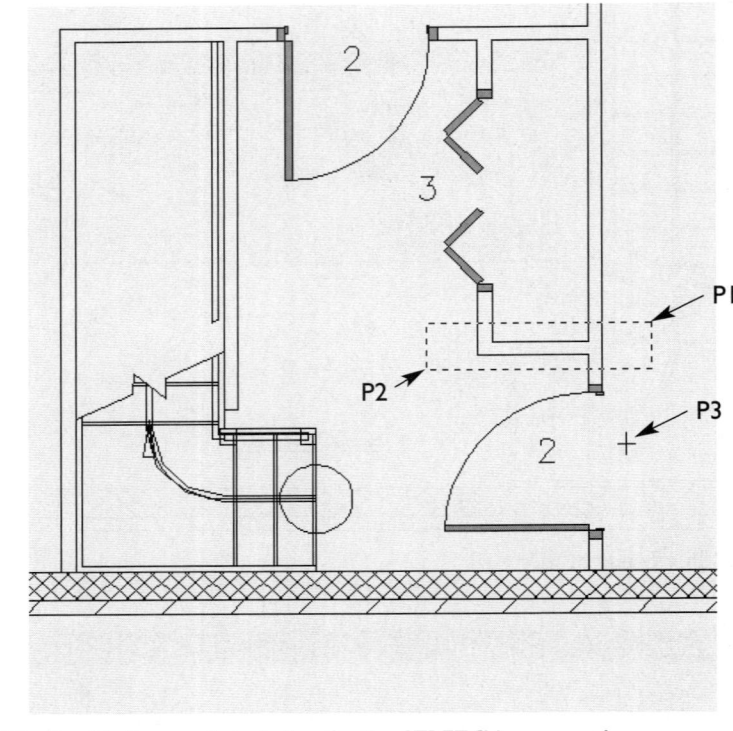

Figure 6.118 *Specify the crossing-window for the STRETCH command*

```
Command: _stretch
Select objects to stretch by crossing-window or
crossing-polygon...
Select objects: (Specify the crossing-window by selecting point P1,
and then point P2 as shown in Figure 6.118.) Specify opposite
corner: 3 found
Select objects: ENTER
Specify base point or displacement: (Select point P3 in the
drawing area and move the mouse up.)
Specify second point of displacement: 12 ENTER
```
The closet wall is moved up as shown in Figure 6.119.

41. Type **WBLOCK** in the command line and then press ENTER to open the Write Block dialog box.

42. Edit the **Write Block** dialog box as shown in Figure 6.120: select **Objects** as the **Source,** set the **Base point** to 0,0,0, **File name** to stair.dwg, **Location**

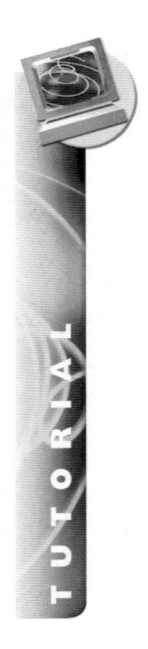

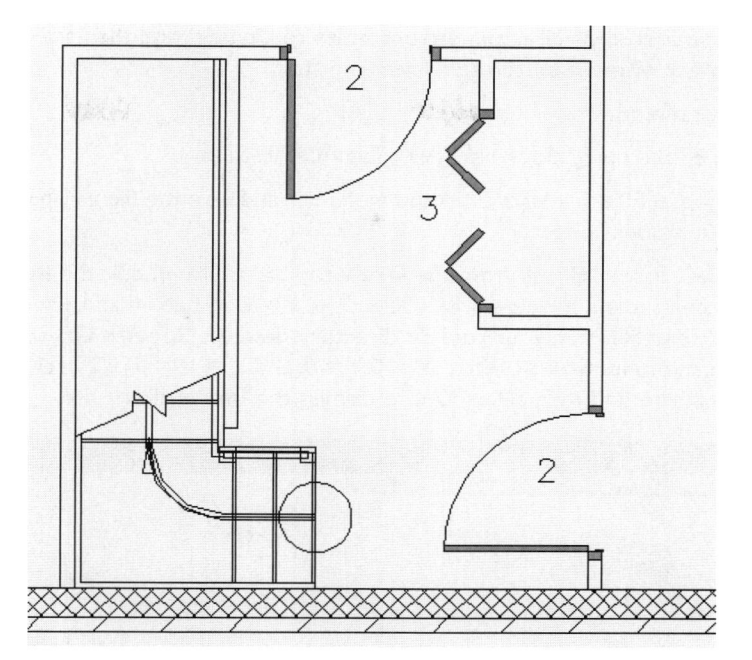

Figure 6.119 Wall stretched 12"

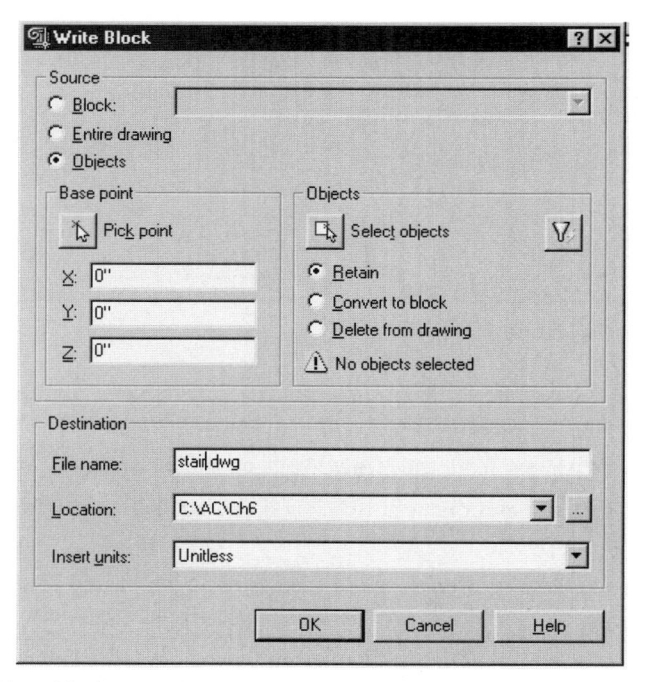

Figure 6.120 Write Block dialog box

to your student directory, **Insert units** to Unitless, and then select the **Select objects** button and select the stair.

43 Save the drawing.

44 Open the upper floor plan \ADT_Tutor\Ch5\Hip1.dwg.

45 Choose **File>SaveAs** from the menu bar and save the file as *UpStair.dwg* in your student directory.

46 Select **Insert Block** from the **Draw** toolbar and then edit the **Insert** dialog box as shown in Figure 6.121. Select the **Browse** button and select the *Stair.dwg* file from your student directory, deselect **Specify On-screen** and set the **Insertion point** to X = 0, Y = 0, and Z = –8'-11-1/2", select **Explode**, and then select **OK** to dismiss the **Insert** dialog box.

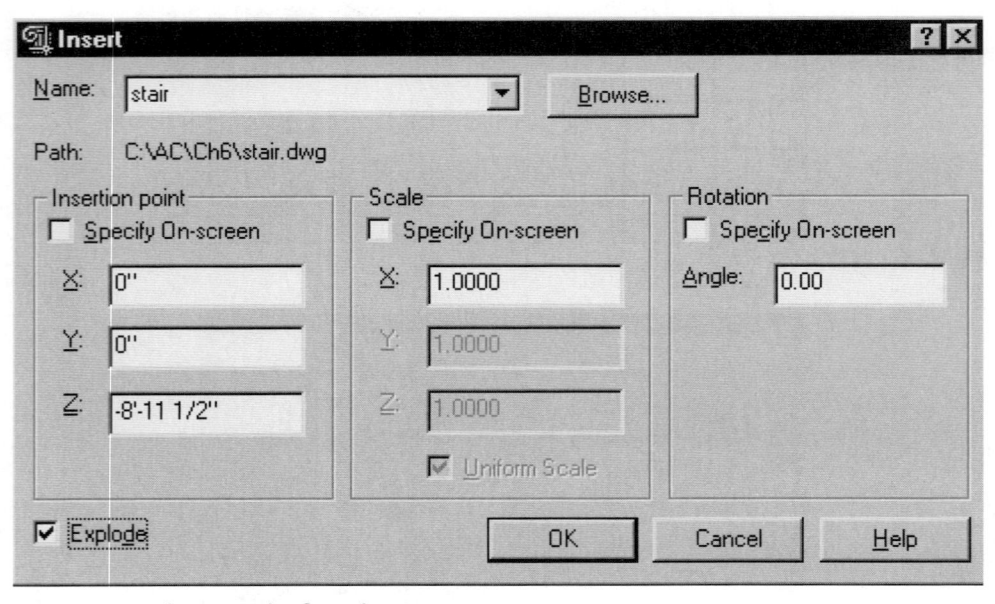

Figure 6.121 *Inserting the Stair drawing*

The stair is inserted below the floor plan as shown in Figure 6.122.

47. Select **Top View** from the **View** flyout on the **Standard** toolbar.

48. Select the stair and then right-click and choose **Edit Stair Style** from the shortcut menu.

49. Select the **Display Props** tab of the **Stair Styles** dialog box.

50. Select the Stair Style property source, select the **Attach Override** button and then select the **Display Props** button to open the **Entity Properties** dialog box, as shown in Figure 6.123.

Stairs and Railings 495

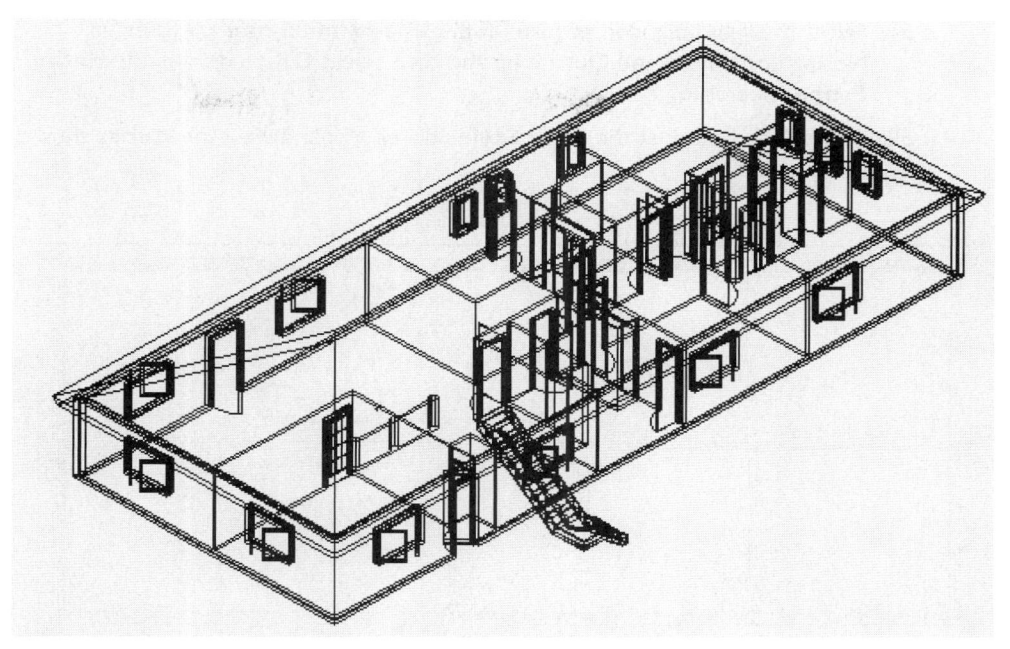

Figure 6.122 *Stair insert below the current floor plan*

Component	Visible	Layer	Color	Linetype	Lineweight	Lt Scale
Stringer up	♀	0	ByBlock	BYBLOCK	ByBlock	1.0000
Riser up	♀	0	ByBlock	HIDDEN4	ByBlock	1.0000
Nosing up	♀	0	ByBlock	BYBLOCK	ByBlock	1.0000
Path up	♀	0	ByBlock	BYBLOCK	ByBlock	1.0000
Outline up	♀	0	ByBlock	BYBLOCK	ByBlock	1.0000
Stringer down	♀	0	ByBlock	BYBLOCK	ByBlock	1.0000
Riser down	♀	0	ByBlock	HIDDEN4	ByBlock	1.0000
Nosing down	♀	0	ByBlock	BYBLOCK	ByBlock	1.0000
Path down	♀	0	ByBlock	BYBLOCK	ByBlock	1.0000
Outline down	♀	0	ByBlock	BYBLOCK	ByBlock	1.0000

Figure 6.123 *Layer/Color/Linetype tab of the Entity Properties dialog box*

51. Select the light bulb icon to turn off the visibility of Stringer up, Riser up, Nosing up, Path up, and Outline up and then select **OK** to dismiss the **Entity Properties** dialog box.

52. Select **OK** to dismiss the **Stair Styles** dialog box and view the stair as shown in Figure 6.124.

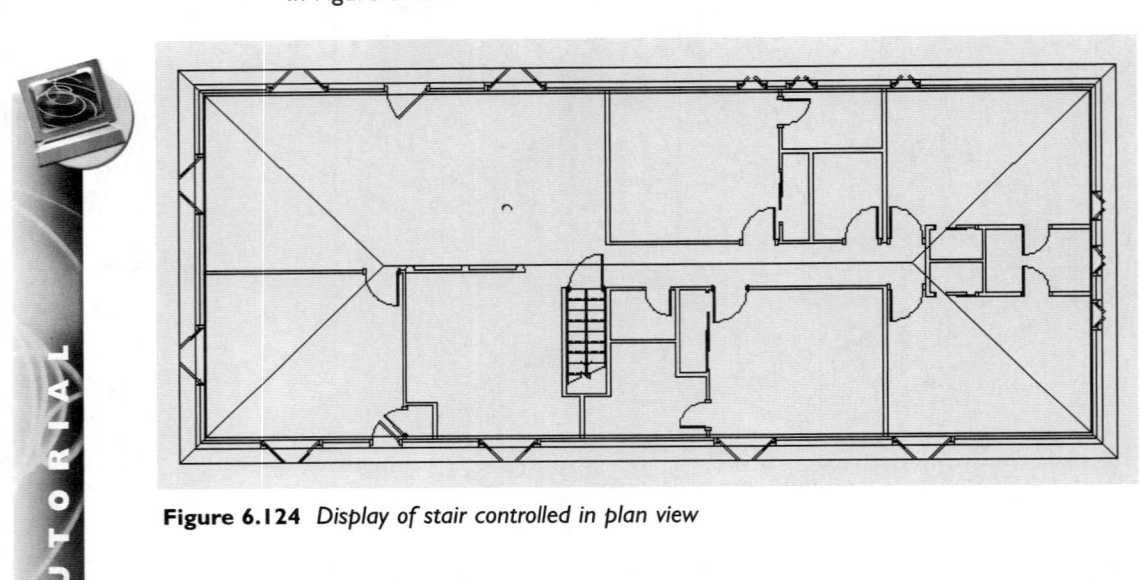

Figure 6.124 *Display of stair controlled in plan view*

53. Save the drawing.

SUMMARY

1 Stairs are created in a drawing with the **Add Stair** command.
2 Design of the number of risers, tread dimension and run of the stair is set in the **Stair Properties** dialog box.
3 Stairs are generated in the drawing from start of the flight up from the zero Z coordinate.
4 Straight stairs are created from the Multi-landing stair shape.
5 The shapes of a stair can be U Shaped, U Shaped Winder, Multi-landing and Spiral.
6 The Multi-landing shape is used to create a straight stair.
7 Code limits and railing default dimensions are set in the **AEC DwgDefaults** tab of the **Options** dialog box.
8 The start point and center of a spiral stair can be edited with grips.
9 **Modify Stair (STAIRMODIFY)** and {**STAIRPROPS**) are used to change the size, components, and style of a stair.

10. The **Stair Styles** command (**STAIRSTYLE**) is used to create, edit, import, and export stair styles and to purge them from a drawing.

11. The **Dimensions** tab of the **Stair Styles** dialog box allows you to define the number, location, and size of stringers.

12. The display of the lower or upper half of the stair is controlled in the **Display Properties** dialog box for a stair style.

13. The **Add Railing** command is used to place a railing on a stair or as a freestanding railing.

14. The **RAILINGPROPS** command allows you to edit the size, location, and profile used for the railing posts, handrail, guardrail, and balusters.

REVIEW QUESTIONS

1. The stair style that does not have open risers is _____.
2. A stair shape with two equal runs and a landing is the _____.
3. A stair with different length runs and a landing is the _____.
4. The shape used to create a straight stair is the _____.
5. The number of risers is specified in the _____ dialog box.
6. The display of the lower half of the stairs is controlled by the settings in the _____.
7. The _____ cannot be edited with the **Modify Stair** command.
8. Create Housed stairs by editing _____ dialog box.
9. Aec Profiles can be specified for the handrail in the _____ dialog box.
10. Stair default settings are established in the _____ dialog box.
11. If the **Tread** edit box is inactive, the tread has been _____ in the **Design Rules** tab of the Stair Properties dialog box.
12. The **Next Turn** options in which the treads turn in the landing are the _____ turn and _____ turn.
13. The default values for the **Code Limits** are set in the _____ dialog box.
14. The grips of the spiral stair are located at the _____ of the arc formed by the spiral stair and at the _____ of the stair.
15. All stairs are created in the _____ direction from start point.
16. The size and position of the stringer are defined in the _____ command.

17. To turn off the upper half of the stair, the _____ components are turned off in the **Display Properties** dialog box.

18. Describe the procedure for inserting a freestanding railing.

19. A handrail and guardrail can be placed in center of a stair by _____.

20. Stair styles created in a drawing can be used in other drawings by _____ the stair style.

PROJECT

EXERCISE 6.1 CREATING A MULTI-LANDING STAIR WITH RAILINGS

1. Open ADT Tutor\Ch6\Ex6-7.dwg and save the file to your directory as Proj6-7.dwg.

2. Create a multi-landing stair. Set the **Code Limits** as follows: **Upper Riser**= 7-3/4", **Lower Riser** = 6", **Upper Tread** = 12", **Lower Tread** = 10". The rise to the stair is 9'–6" and the width is 44". Use 16 risers and record below the riser height, tread size, and straight length.

3. Attach a handrail to the wall side of the stair. Use a circular Aec Profile for the handrail profile. The right side of the stair should have a railing, which consists of handrails, guardrails, balusters, posts, and dynamic posts. Set the dynamic post distance to 7'–0". Use the HR_CAP1 Aec Profile for the guardrail and a circular handrail. The completed stair and handrail should appear as shown in Figure 6.125.

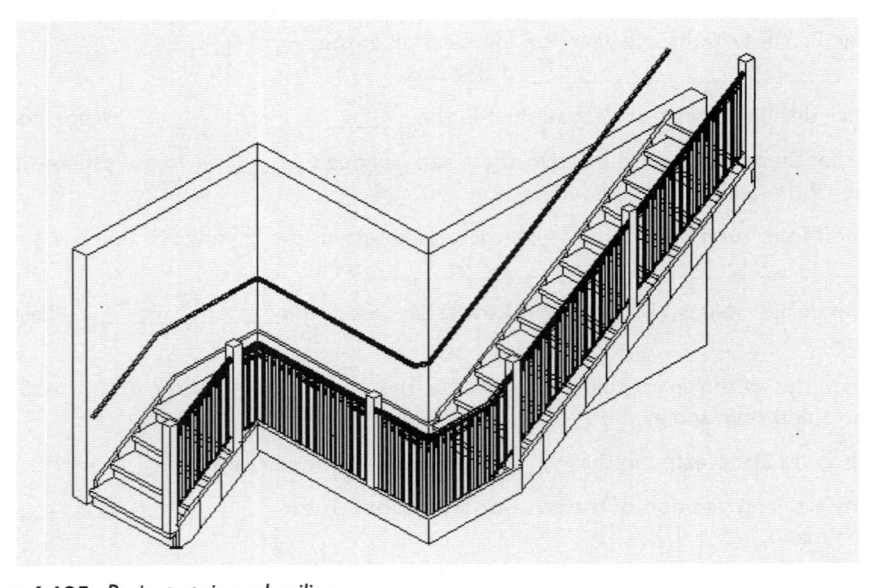

Figure 6.125 *Project stair and railing*

CHAPTER 7

Using and Creating Symbols

INTRODUCTION
Architectural Desktop includes hundreds of symbols created for two-dimensional and three-dimensional representations of building components. The two-dimensional symbols include the plan and elevation views. The symbols are selected from the AutoCAD DesignCenter and include blocks, multi-view blocks, and mask blocks. The contents of the AutoCAD DesignCenter can also be customized through the **Create AEC Content Wizard**.

OBJECTIVES
After completing this chapter, you will be able to

- Set the scale and units of a drawing
- Specify the default symbol menu
- Use the DesignCenter to insert Architectural Desktop symbols in a drawing
- Insert and modify multi-view blocks
- Define the display representation for a multi-view block
- Import and export multi-view blocks
- Create and insert masking blocks
- Attach masking blocks to Architectural Desktop objects to control display
- Create a masking block using the **Create AEC Content Wizard**

SETTING THE SCALE FOR SYMBOLS AND ANNOTATION

Prior to symbols being inserted in the drawing, the scale of the plotted drawing should be established. Some of the symbols included in the DesignCenter are created at a specific size, while others are sized according to the scale of the drawing. For example, an electrical receptacle is sized according to the scale of the drawing, and a cabinet will be inserted at its actual size. If a receptacle symbol is placed in a drawing with the scale set to 1/4"=1'–0", it is drawn with a 6" diameter circle. This symbol, if placed in a drawing with the 1/8"=1'–0" scale, would be drawn with a 12" diameter circle because the scale has changed. The scale of a drawing sets the scale factor, which is used as a multiplier for selected symbols and annotation. The scale factor is the ratio between the size of the AutoCAD entity and its display printed on paper. The typical architectural scales and the associated scale factors are listed in the **Scale** tab of the **Drawing Setup** dialog box when the **AECDWGSCALESETUP** command is selected. See Table 7.1 for **AECDWGSCALESETUP** command access.

Menu bar	Documentation>Set Drawing Scale
Command prompt	AECDWGSCALESETUP

Table 7.1 *AECDWGSCALESETUP command access*

The **Drawing Setup** dialog box opens when you choose **Documentation>Set Drawing Scale** from the menu bar. The **Drawing Setup** dialog box opens with the **Scale** tab selected, as shown in Figure 7.1.

Scale

The **Drawing Setup** dialog box consists of four tabs: **Units**, **Scale**, **Layering**, and **Display**. The settings of the **Units** tab control the drawing scales displayed in the **Scale** tab. In Figure 7.1, the scales listed are Architectural because the units are set to inches or architectural in the **Units** tab. The options of the **Scale** tab are described below:

> **Drawing Scale** – The **Drawing Scale** list box consists of a list of scales. The scale list varies according to the units selected for the drawing.
>
> **Custom Scales** – The **Custom Scales** edit field is inactive when a predefined drawing scale is selected from the **Drawing Scale** list. However, its associated scale factor is displayed in the **Custom Scales** field. In Figure 7.1, the 1/8"=1'–0" scale is selected in the **Drawing Scale** list and 96 is displayed in the **Custom Scales** edit field. If the Other scale option is selected in the **Drawing Scale** list, the **Custom Scale** field becomes active, allowing you to enter a scale factor of your choice.

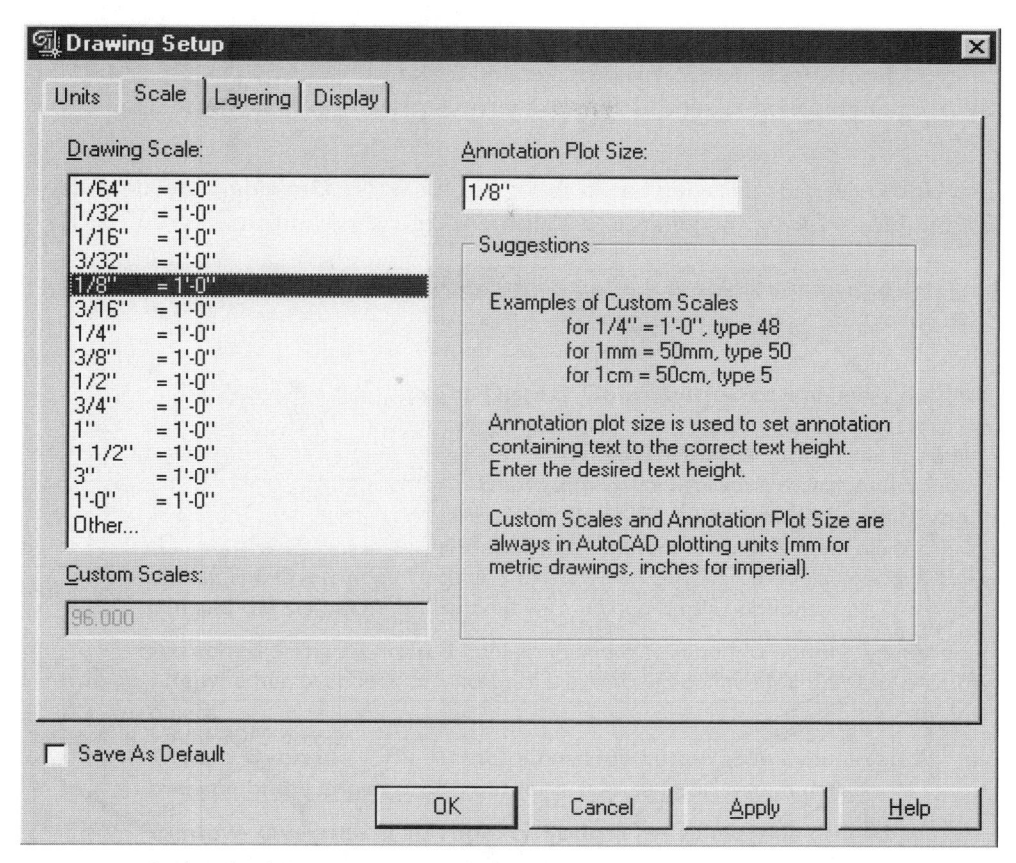

Figure 7.1 *Scale tab of the Drawing Setup dialog box*

Annotation Plot Size – The **Annotation Plot Size** edit field allows you to specify the height of text when plotted. The **Annotation Plot Size** of 1/8" as shown in Figure 7.1 will set the **DIMSCALE** dimensioning variable to 96; therefore the text of leaders and dimensions will be sized by a scale factor of 96. **Annotation Plot Size** does not set the height of **MTEXT**.

Save As Default – If the **Save As Default** check box is selected, the scale, units, and layer settings are saved and applied to new drawings if the new drawing contains no drawing setup. The Aec arch (imperial) template has a default scale of 1/8"=1'–0" and the Aec arch (metric) template has a default scale of 1=50 mm. The scale settings will be saved only to the current drawing if you clear the **Save as Default** check box.

Units

The **Units** tab allows you to set the units of the Architectural Desktop symbols. It controls the units of symbols inserted in Architectural Desktop. Therefore, if the Aec arch (imperial) template is used to create a new drawing, the units are architectural and the dimensions of a wall are displayed in feet and inches. If the **AECDWGSCALESET-UP** command is used to set the units to metric, the wall's properties are described in metric units.

The **Units** tab of the **Drawing Setup** dialog box includes a **Drawing Units** drop-down list and **Linear, Area, Angular,** and **Volume** sections, as shown in Figure 7.2.

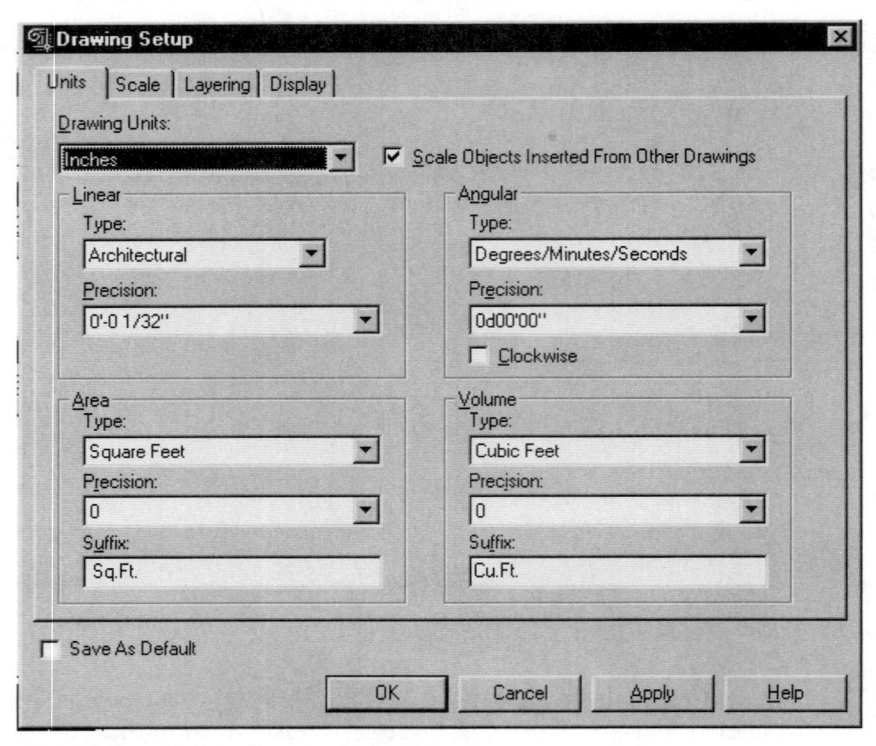

Figure 7.2 *Units tab of the Drawing Setup dialog box*

The options of the **Units** tab are described below:

> **Drawing Units** – The **Drawing Units** list contains the following options: Inches, Feet, Millimeters, Centimeters, Decimeters, and Meters. The **Drawing Units** of the Aec arch (Imperial) template is Inches. Therefore each drawing unit of the drawing is equal to an inch. The remaining sections of the **Units** dialog box vary according to the selection in the **Drawing Units** list. If Meters is selected as the **Drawing Units**, **Linear Types** are limited to Decimal and Scientific. Therefore, the **Drawing Units** should be selected before you edit the remainder of the dialog box.

Scale Objects Inserted from Other Drawings – If this check box is selected, the objects from other drawings will scaled up or down to match the new units.

Linear

Type – The **Type** list consists of **Units** options for linear measure. If **Drawing Units** is set to inches, the options of the **Type** list include Scientific, Decimal, Engineering, Architectural, and Fractional. If **Drawing Units** is set to any drawing unit other than Inches, the Decimal and Scientific Types are the only options.

Precision – The **Precision** list includes nine degrees of precision for each type of linear units. The Inch drawing unit includes precision from 0 to 1/256". The **Precision** options specify how precise information regarding Architectural Desktop is displayed. If the precision is set to 1/2, the length of a wall drawn 2'–3-5/16" will be displayed as 2'–3-1/2" in the **Wall Properties** dialog box. Therefore the precision options round off the wall length according to the precision specified.

Area

Type – The **Type** list consists of units for the area of Architectural Desktop entities. The options for area using imperial units include square inches, square feet, or square yards, while the options for area using metric units include square millimeters, square centimeters, square decimeters, and square meters.

Precision – Includes precision options for displaying area data.

Suffix – The suffix appended to the **Type** for area can be selected from the predefined list.

Angular

Type – The **Type** options for angular measure are identical to the angular options of the **UNITS** command. **Type** options include Degrees, Degrees/Minutes/Seconds, Grads, Radians, and Surveyor.

Precision – The **Precision** of the angular measure includes nine options, with precision up to one hundred of a millionth.

Clockwise – The **Clockwise** check box allows you to select clockwise measurement; the default angular measurement is in a counter-clockwise direction.

Volume

Type – The **Type** list includes options for imperial volumes as follows: cubic feet, cubic inches and cubic yards. The options for metric volumes include cubic millimeters, cubic centimeters, and cubic meters.

Precision – The **Precision** option provides options for the display of volume data.

Suffix – The **Suffix** option includes suffix options to append to the volume data.

CHANGING UNITS IN THE MIDDLE OF A DRAWING

If the **Drawing Units** are changed in the middle of a drawing, a **Drawing Setup** message box will open with a warning that you have changed the drawing units. The message box includes two check boxes that allow you to resize model or paper space objects based on the new units. Selecting the check boxes shown in the dialog box in Figure 7.3 will resize entities in paper space and model space.

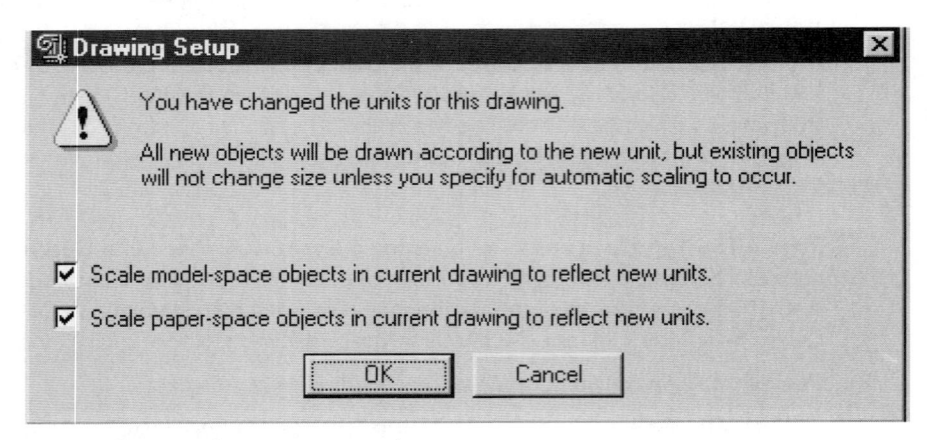

Figure 7.3 *Changing the units message box*

If **Scale model-space object in current drawing to reflect new units** is selected, previously drawn objects will be scaled to match the change in the units. This feature allows you to begin a drawing as a metric drawing and convert the entities to imperial by changing the drawing units and selecting the toggles of this dialog box.

Layering

The **Layering** tab of the **Drawing Setup** dialog box allows you to select a predefined layer standard for the drawing. It includes a **Layer Standards/Key File to Auto Import** edit field and a **Default Layer Standard Layer Key Style** list. The "Standard *current drawing" layer key is displayed in the **Layer Key Style** list shown in Figure 7.4.

The options of the **Layering** tab are described below.

Layer Standards/Key File to Auto-Import – The **Layer Standards/Key File to Auto Import** edit field allows you to select a layer standards file to

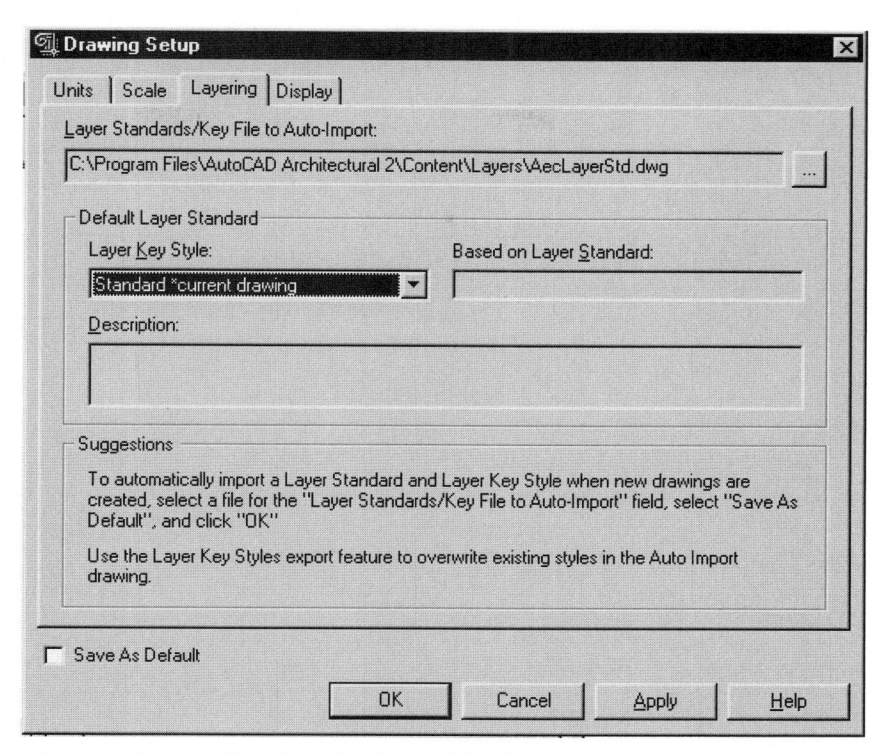

Figure 7.4 *Layering tab of the Drawing Setup dialog box*

import layer standards into a drawing. The default layer standards of the Aec arch (imperial) template are included in the \AutoCAD Architectural 2\Content\Layers\AecLayerStd.dwg file.

Layer Key Style–The **Layer Key Style** list includes the layer key styles available in a drawing.

Table 7.2 shows the layer key styles and descriptions available.

Architectural Desktop elements placed in a drawing are automatically placed on the appropriate layer according to the layer key style. Therefore, a range is inserted on the appropriate layer according to the layer key style used in the drawing. The Current layer key style places the object on the layer that is current at the time of insertion. The AIA and BS1192 layer standards place objects on the layer specified in the layer standard. The Standard layer key style places objects on the default layer. The default layer is named the same as the object's layer key. Table 7.3 shows the names of the layers for an appliance when each of the respective layer key style options is selected.

Layer Key Style	Description
Current	Non-Standard
AIA (256 Color)	American Institute of Architects Layer Guidelines
BS1192 Aug Version 2 (256 Color)	BS1192 Layer Guidelines (Aug Version 2)
BS1192 Descriptive (256 Color)	Based on BS1192 Guidelines, but using a descriptive element rather than CSI
Standard	Non-Standard

Table 7.2 *Layer key styles and descriptions*

Layer Name	Layer Key Style
Current Layer	Current
A-Flor-Appl	AIA (256 Color)
A700G4	BS1192 Aug Version 2 (256 Color)
A-Appliance-G	BS1192 Descriptive (256 Color)
Appl	Standard

Table 7.3 *Layer names for an appliance using layer key styles*

Using a layer standard enhances the drawing because the display of similar objects can be controlled as a group in the drawing. Architectural Desktop places the objects on the correct layer automatically when the object is inserted in the drawing. If the layer key style is changed in the middle of a drawing, all previous objects remain on the layer of their initial insertion. Therefore, you should determine the layer key style prior to beginning the drawing.

Display

The **Display** tab of the **Drawing Setup** dialog box allows you to set default display representations of objects displayed in the drawing. As shown in Figure 7.5, the **Display** tab consists of list boxes for the **Object Type, Display Representations**, and **Drawing Default Display Configuration** and an **Edit** button for editing entity properties of an object.

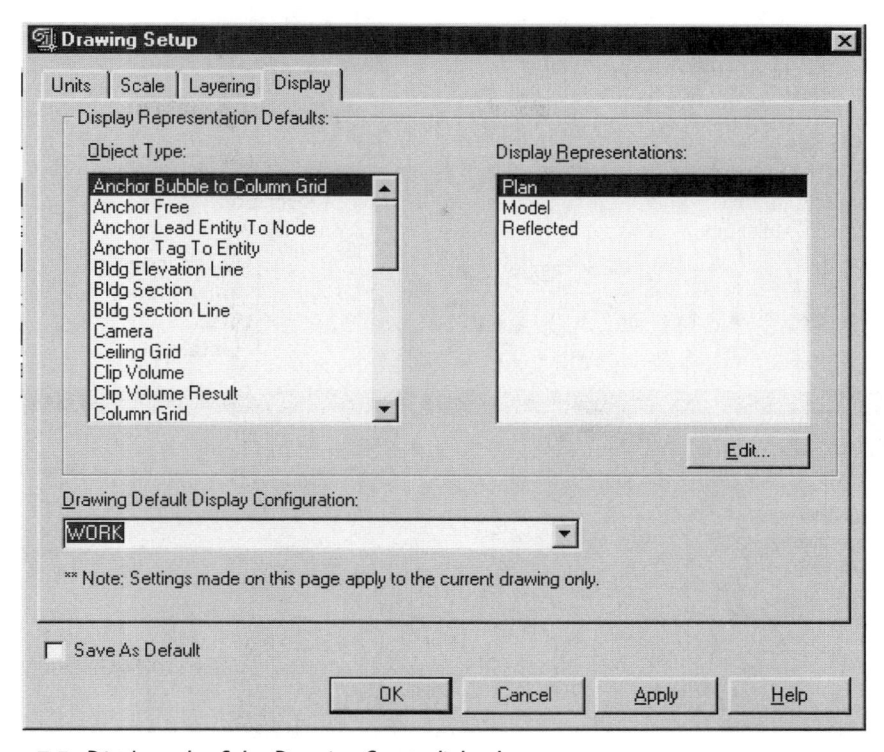

Figure 7.5 *Display tab of the Drawing Setup dialog box*

>**Object Type** – The **Object Type** list box lists all the types of objects used in Architectural Desktop.
>
>**Display Representations** – The **Display Representations** list box includes the display representations active for the object listed in the **Object Type** list.
>
>**Drawing Default Display Configuration** – The **Drawing Default Display Configuration** list shows the display configurations.
>
>**Edit** – The **Edit** button allows you to edit the entity properties of each component of an object.
>
>**Save as Default** – This check box should be deselected if you want to have your settings apply only to the current drawing for units, scale, layering, and display.

USING THE DESIGNCENTER

After specifying the units, scale, and layer standards of the drawing, you are ready to insert symbols and other documentation. Architectural symbols are inserted in a drawing through the DesignCenter. The AutoCAD DesignCenter is modified to include the display of a Custom view. Figure 7.6 shows the DesignCenter window and the commands of the DesignCenter.

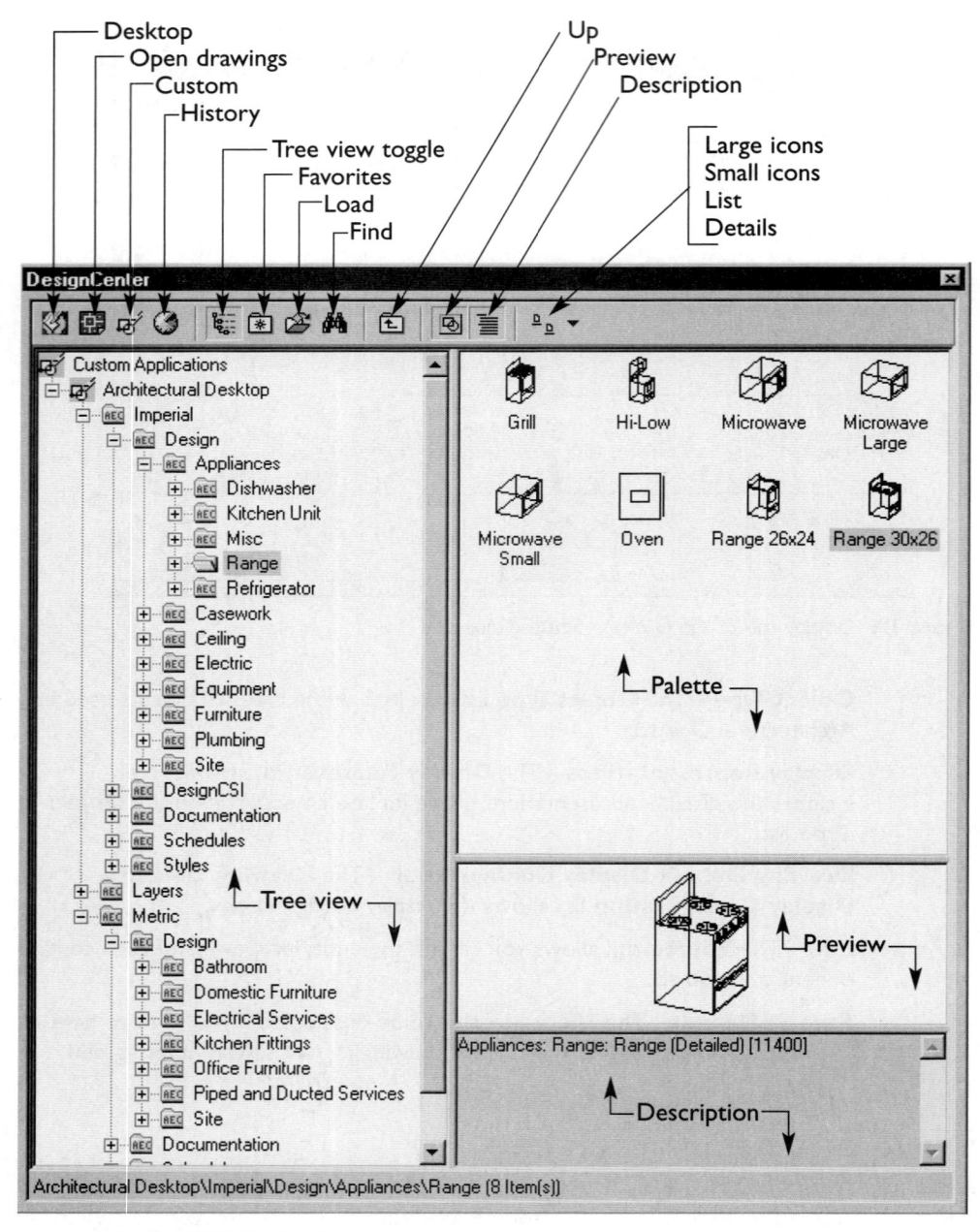

Figure 7.6 *AutoCAD DesignCenter*

The commands of the DesignCenter toolbar are described below:

Desktop – Toggles on the display of the files and folders of the computer.

Open Drawings – Displays all drawings that are open in the current Architectural Desktop session.

Custom – Displays the Architectural Desktop Design Content.

History – Displays the file path of the last 20 files accessed by the DesignCenter.

Tree View Toggle – Toggles ON/OFF the Tree View.

Favorites – Toggles to the Favorites folder of Architectural Desktop.

Load – Opens the **Load Center Palette** dialog box to load a palette from the desktop.

Find – Opens the **Find** dialog box, allowing you to find objects within other drawings.

Up – Opens the folder one level above the current container.

Preview – Toggles open/closed a preview of a selected symbol.

Description – Opens/closes a text description window at the bottom of palette.

Views – Allows you to select from a flyout Large Icon, Small Icon, List or Detail View of the block.

The Custom view includes all the symbols and documentation of Architectural Desktop. When symbols are selected from the menu bar or from a toolbar, the AutoCAD DesignCenter opens to the Architectural Desktop node in the Tree View as shown in Figure 7.6. The symbols are located in the Imperial\Design, Imperial\Design CSI, and Metric\Design directories of the Tree View. Therefore three menus are available for selecting symbols.

DEFINING THE DEFAULT SYMBOL MENU

The default symbol menu can be set to allow you to quickly select symbols. During installation, you are prompted to select either the CSI Master Format or AutoCAD Architectural Desktop 1.0 organization. Selecting the AutoCAD Architectural Desktop 1.0 option loads the symbol libraries, which include descriptive menu options. If you select the CSI Master Format, the symbol menu is organized based upon the CSI Master Format numbers, with the symbols grouped according to the CSI Master Format numbering system.

The AutoCAD Architectural Desktop 1.0 system includes the Imperial Design and the Metric Design menus. Either of these symbol menus can be set as the default. To set the Imperial Design or the Metric Design menu as the default menu, select the

AEC Content tab of the **Options** dialog box shown in Figure 7.7.

The **AEC Content** tab includes a **Content Path** edit field and a **Content Menu** toggle. The options of the **AEC Content** tab are described below.

> **Content Path** – Displays the current path to the active directory for the symbols. The path shown in Figure 7.7, *AutoCAD Architectural 2\Content*, provides access to the Metric or Imperial symbols.
>
> **Browse** – Opens the **Browse Folder** dialog box, allowing you to edit the default Content folder.
>
> **Imperial** – The **Imperial** radio button toggles the Imperial Design symbols as the default for the DesignCenter.
>
> **Metric** – The **Metric** radio button toggles the Metric Design symbols as the default for the DesignCenter.
>
> **Display Edit Schedule Data Dialog During Tag Insertion** – This check box will open the **Edit Schedule Data** dialog box when tags are inserted. The **Edit Schedule Data** dialog box is used to enter data for schedules, as shown in Figure 7.8.

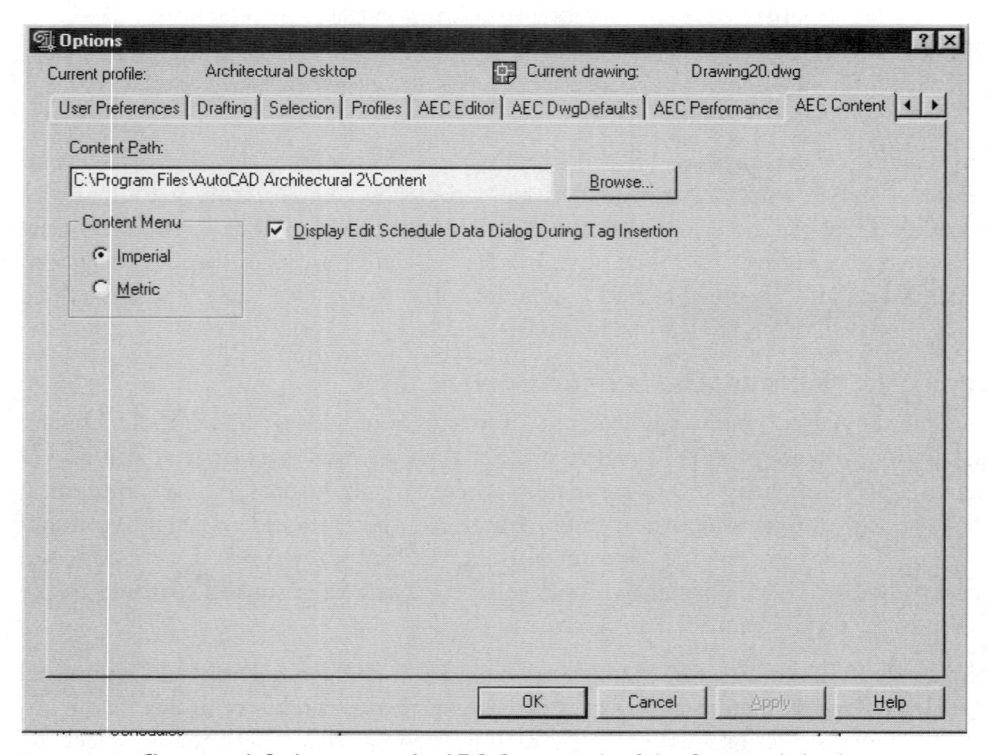

Figure 7.7 *Changing default menu in the AEC Content tab of the Options dialog box*

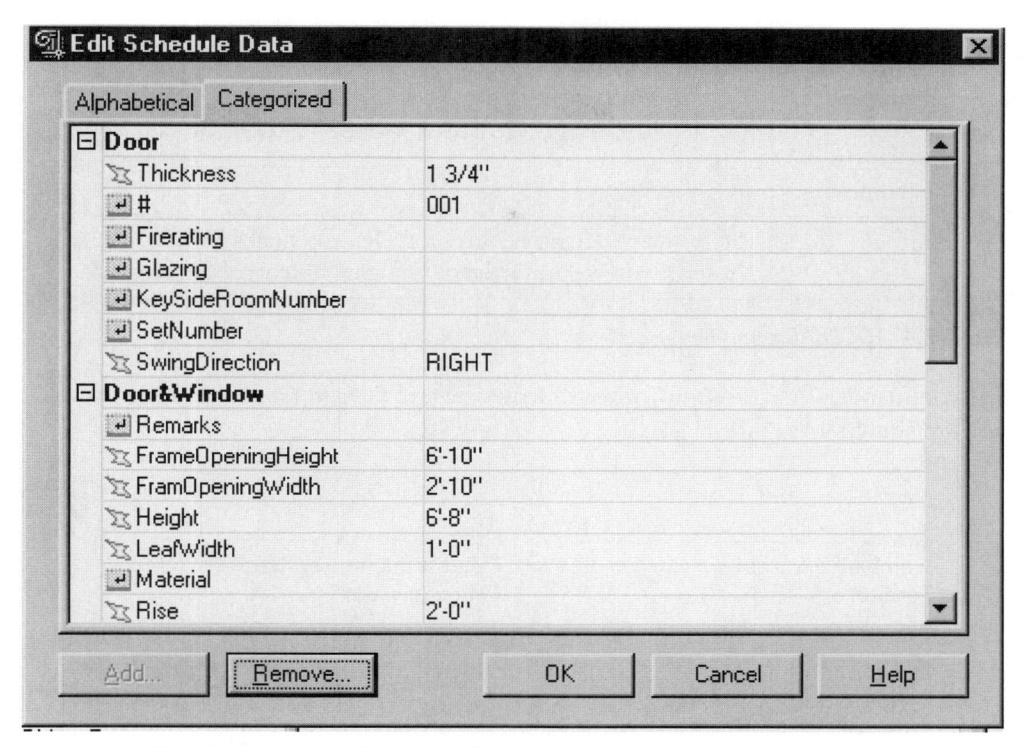

Figure 7.8 *Edit Schedule Data dialog box for entering text*

If the **Imperial** radio button in the **Content Menu** section is selected, the **Imperial** menu will be displayed. Choosing **Design> Design Content** from the menu bar will then open the Symbol menu, which includes Appliances, Casework, Ceiling Fixtures, Electrical Fixtures, Equipment, Furniture, Plumbing Fixtures, and Site. If the **Metric** option is selected, the **Design>Design Content** menu will include Bathroom Fittings, Domestic Furniture, Electrical Services, Kitchen Fittings, Office Furniture, Piped and Ducted Services, and Site. The radio buttons of the **AEC Content** tab determine whether the content menu is metric or imperial.

DISPLAYING DESIGN CONTENT TOOLBARS

Symbols can be selected from the **Design Content - Imperial** or the **Design Content - Metric** toolbar regardless of the **Metric** and **Imperial** toggles of the **AEC Content** tab. The **Design Content - Imperial** and **Design Content - Metric** toolbars are opened through the **TOOLBAR** command. See Table 7.4 for **TOOLBAR** command access.

Menu bar	View>Toolbars; select **AECARCHX** menu group, and then select Design Content—Imperial or Design Content—Metric toolbar
Command prompt	TOOLBAR
Shortcut	Right-click over an Architectural Desktop toolbar; then choose Design Content—Imperial or Design Content—Metric

Table 7.4 *TOOLBAR command access*

The commands of the **Design Content - Imperial** and **Design Content - Metric** toolbars are shown in Figure 7.9.

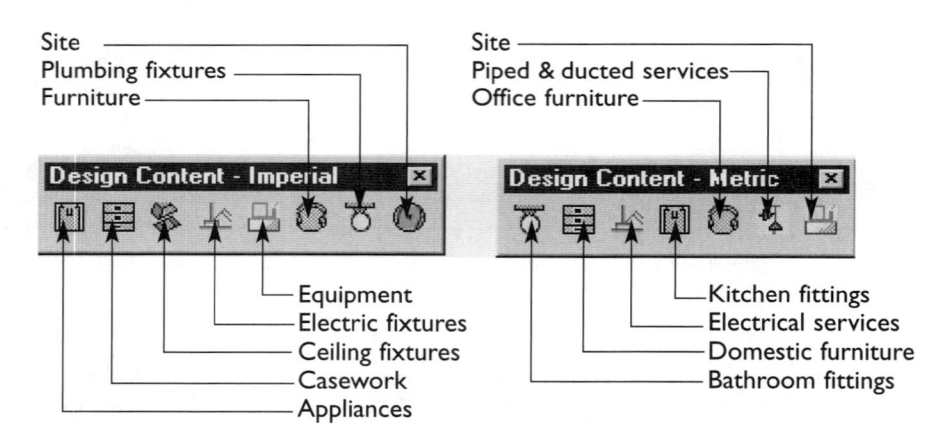

Figure 7.9 *Design Content toolbars*

If the CSI Master Format menu is selected during installation, the symbols will be organized according to the CSI Master Format numbering system; the **Design Content** menu will include Division 1 General Requirements, Division 2 Site Construction, Division 10 Specialities, Division 11 Equipment, Division 12 Furnishings, Division 13 Special Construction, Division 14 Conveying Systems, Division 15 Mechanical, and Division 16 Electrical.

INSERTING SYMBOLS

Establishing the default symbol menu allows you to quickly select symbols from the toolbar or menu bar. Selecting a symbol menu item opens the AutoCAD DesignCenter with the specified symbol container selected in the Tree View. Most symbol containers consist of several folders of symbols, which are displayed in the palette. Each folder can be opened by a double-click on the folder. You can drag the desired symbol from the palette or you can double-click on the symbol and insert the symbol using a dialog box. If you drag the symbol into the drawing, you should set Running Object Snap modes prior to beginning the drag operation. Running Object Snaps can be set while the AutoCAD DesignCenter is open; however, once a symbol is selected, object snaps cannot be altered.

To drag a symbol into the drawing, select the symbol in the palette with the left mouse button and continue to hold the button down as you drag the symbol on to the drawing area. (As you move the mouse over the DesignCenter, a slash circle appears that warns you not to release the symbol over the DesignCenter, which will void the drag operation.) The symbol will be displayed attached to the pointer as the mouse is moved over the drawing area. The default rotation of the symbol is displayed in the drawing prior to your specifying the insertion point. Symbols selected from the AutoCAD DesignCenter become a block in the drawing. Therefore, the **INSERT** command can also be used to insert additional instances of the symbol in the drawing.

STEPS TO DRAGGING SYMBOLS INTO A DRAWING

1. Right-click over the OSNAP button of the Status Bar, and choose **Settings** from the shortcut menu.
2. Select desired Running Object Snap modes from the **Drafting Settings** dialog box.
3. Select **OK** to close the **Drafting Settings** dialog box.
4. Choose **Documentation>Set Drawing Scale** from the menu bar.
5. Select a scale from the scale list of the **Scale** tab in the **Drawing Setup** dialog box as shown in Figure 7.10.
6. Select **OK** to dismiss the **Drawing Setup** dialog box.
7. Select a symbol group such as Appliances from the **Design Content - Imperial** toolbar to open the DesignCenter to Appliances as shown in Figure 7.11.
8. Select the **Preview** button of the AutoCAD DesignCenter to preview the symbol.
9. Double-click on the Refrigerator container in the palette.
10. Select a symbol such as Side-Side refrigerator and continue to hold down the left mouse button down as you move the pointer outside the palette area of the DesignCenter, as shown in Figure 7.12.
11. Release the mouse button over the drawing area to select the location to insert the symbol.

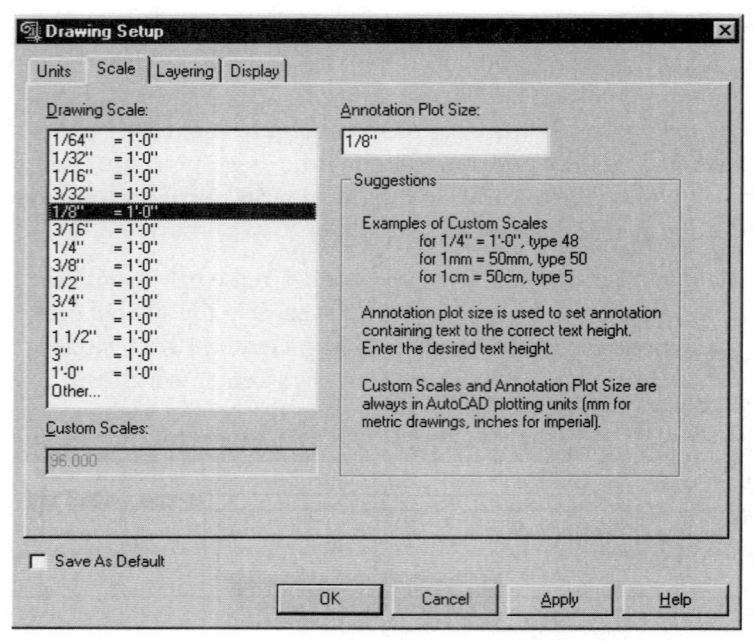

Figure 7.10 *Setting the scale in the Scale tab of the Drawing Setup dialog box*

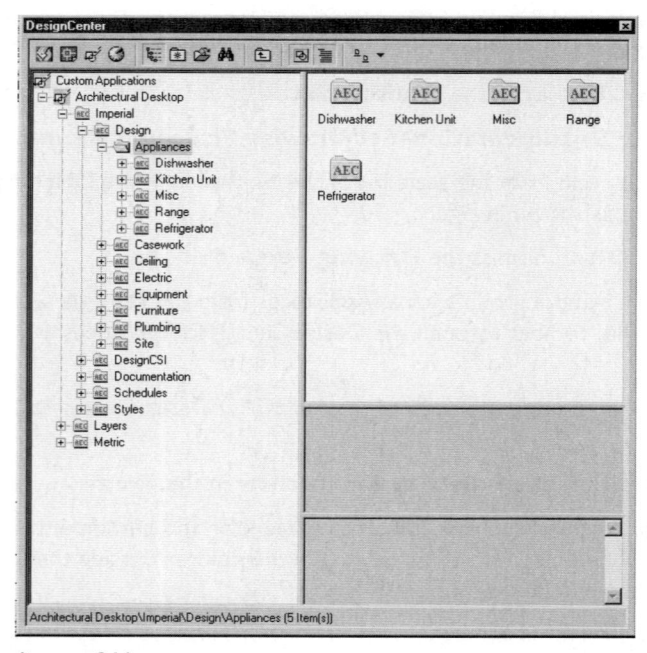

Figure 7.11 *Appliances folder*

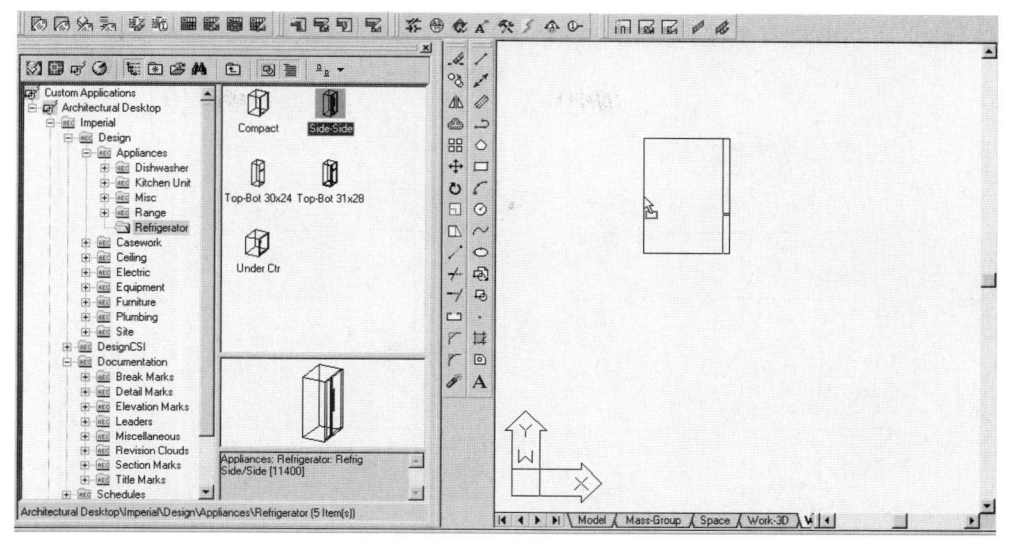

Figure 7.12 *Selecting Side-Side refrigerator from the palette*

12. Select the rotation angle with the mouse or type the angle for the rotation in the command line. The symbol is placed in the drawing.

TUTORIAL 7.1 INSERTING SYMBOLS USING DRAG AND DROP

1. Open *ADT Tutor\Ch7\Ex 7-1.dwg*.
2. Choose **File>SaveAs** from the menu bar and save the drawing as *Symbol1.dwg* in your student directory.
3. Choose **Documentation>Set Drawing Scale** from the menu bar.
4. Select the 1/8"=1'–0" scale from the **Drawing Scale** list box.
5. Verify that **Annotation Plot Size** is 1/8".
6. Select the **Units** tab and verify that the **Drawing Units** are set to Inches. Verify that the **Type** and **Precision** are as shown in Figure 7.13.
7. Select the **Layering** tab and select AIA (256 Color) as the **Layer Key Style**.
8. Select **OK** to dismiss the **Drawing Setup** dialog box.
9. Right-click over the drawing area and choose **Options** from the shortcut menu.
10. Select the **AEC Content** tab of the **Options** dialog box.
11. Select the **Imperial** radio button in the **Content Menu** section.

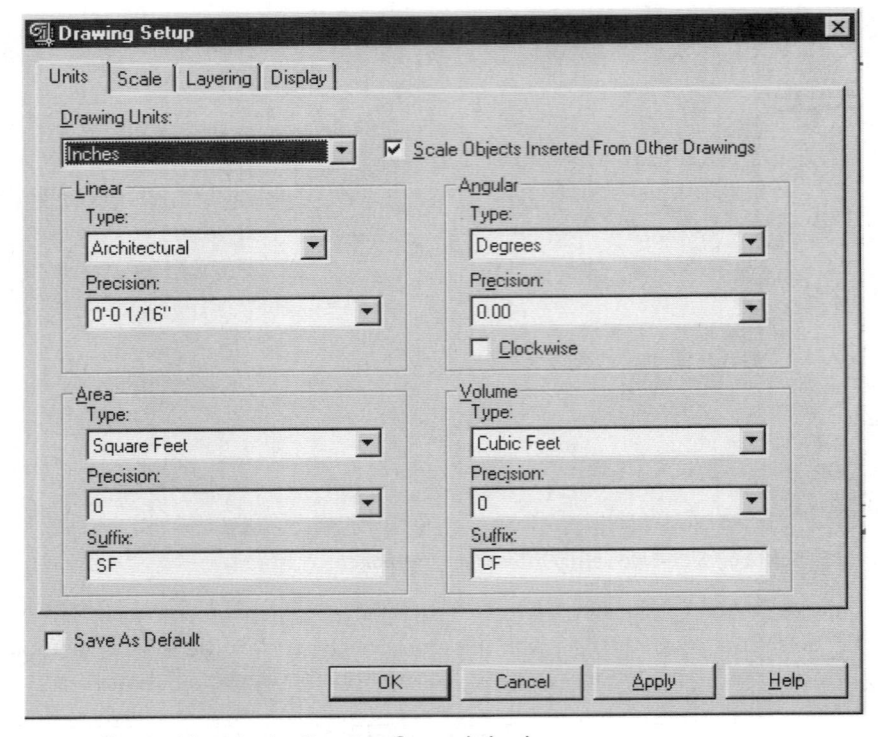

Figure 7.13 Setting Units in the Drawing Setup dialog box

12. Select **OK** to dismiss the **Options** dialog box.
13. Select **Design>Design Content>Furniture** to open the DesignCenter, as shown in Figure 7.14.
14. Double-click on the Bed folder in the palette of the DesignCenter.
15. Select the Double bed symbol and continue to hold down the left mouse button to drag the symbol to the drawing area as shown in Figure 7.15.
16. Enter a 0 rotation angle in the command line.
17. Save the drawing.

INSERTING AND MODIFYING MULTI-VIEW BLOCKS

Many of the symbols inserted in the DesignCenter of Architectural Desktop are multi-view blocks. Multi-view blocks have special display control properties, which allow the block to be displayed or not displayed when viewed from specified directions. A multi-view block of a duplex outlet can be defined to display only in the top viewing position and not in other viewing positions. Controlling display according to viewing direction is defined for each display representation.

Using and Creating Symbols 517

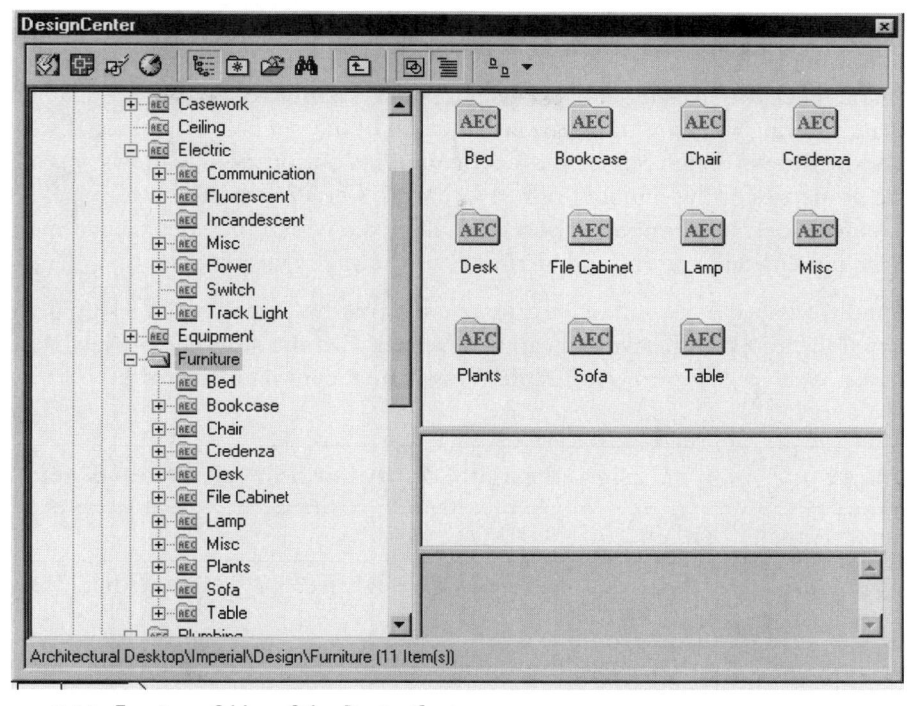

Figure 7.14 *Furniture folder of the DesignCenter*

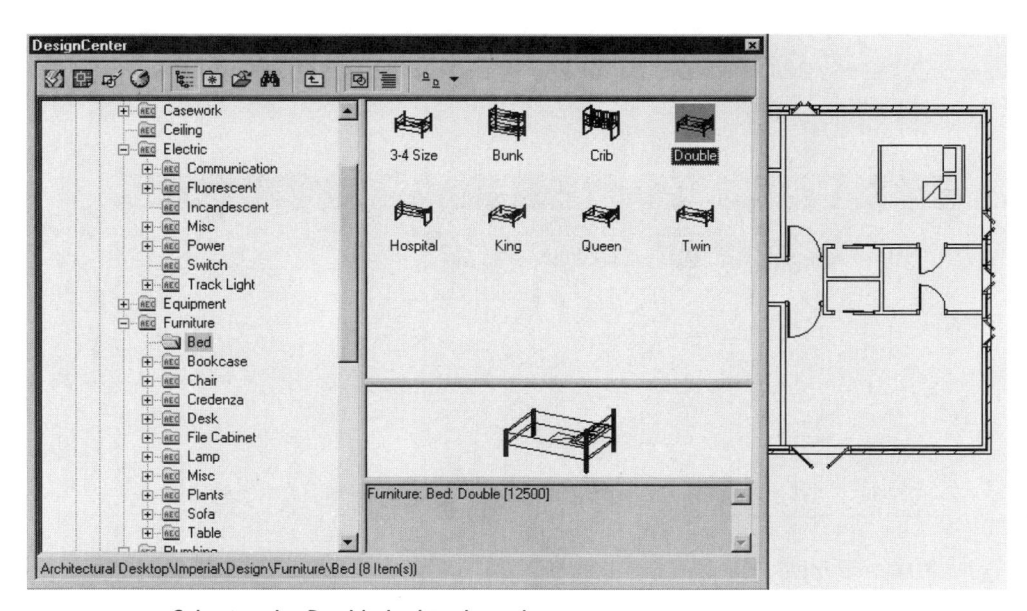

Figure 7.15 *Selecting the Double bed in the palette*

You can insert symbols from the AutoCAD DesignCenter using the **Add Multi-View Block** command (**MVBLOCKADD**) rather than the drag and drop approach. If you double-click on the symbol in the palette, the **Add Multi-View Blocks** dialog box will open, allowing you to set the scale and rotation of the symbol. In addition, the **Add Multi-View Blocks** dialog box includes a **Properties** button, which allows you to set specific properties of the symbol prior to insertion. The **Properties** button opens the **Multi-View Block Reference Properties** dialog box, which includes tabs to modify the style, dimensions, offsets, and attributes of the multi-view block.

After multi-view block symbols are inserted from the AutoCAD DesignCenter, additional instances of the symbol can be inserted with the **Add Multi-View Blocks** dialog box. See Table 7.5 for **Add Multi-View Block** command access.

Menu bar	Design>Design Content>Add Multi-View Block
Command prompt	MVBLOCKADD
Shortcut	Right-click over drawing area; then choose Utilities>Multi-View Blocks>Add
Double-click	Double-click on a symbol in the palette in the AutoCAD DesignCenter

Table 7.5 *Add Multi-View Block command access*

Selecting the **MVBLOCKADD** command opens the **Add Multi-View Blocks** dialog box, as shown in Figure 7.16.

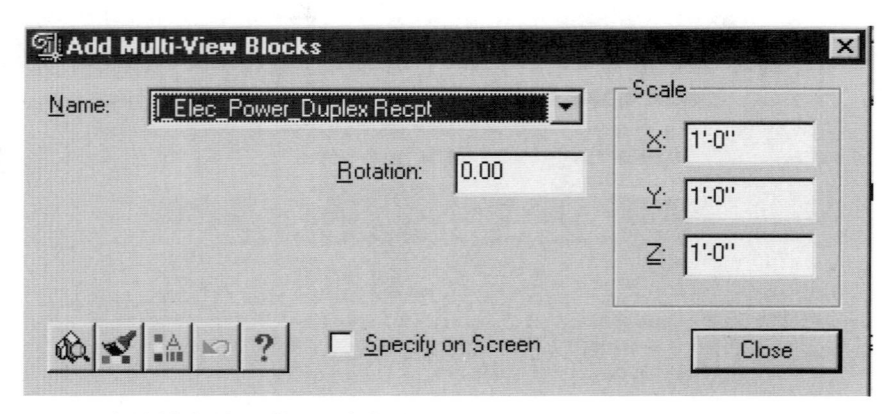

Figure 7.16 *Add Multi-View Blocks dialog box*

The **Add Multi-View Blocks** dialog box consists of a **Name** list box and **Rotation** and **Scale** edit fields. The options of the **Add Multi-View Blocks** dialog box are described below.

Name – The **Name** list shows all multi-view blocks that have been inserted in the drawing. You can select a multi-view block from the list and insert additional instances of the multi-view block in the drawing without using the AutoCAD DesignCenter.

Rotation – The **Rotation** edit field allows you to preset the rotation of the multi-view block prior to insertion. The default rotation of the selected symbol is displayed about the pointer while the **Add Multi-View Blocks** dialog box is open. You can enter a different rotation for the symbol at any time prior to specifying the insertion point.

Scale – The X, Y, and Z edit fields allow you to view or change the scale factor for the symbol prior to insertion.

Specify on Screen – The **Specify on Screen** check box, if selected, allows you to dynamically set the scale for the drawing when the multi-view block is inserted.

Floating Viewer – The **Floating Viewer** button opens the **Viewer** window, allowing you to preview the multi-view block.

Match – The **Match** button allows you to select an existing multi-view block and match its properties. When you select the **Match** button, you are prompted to select a multi-view block. Selecting a block allows you to match its name, scale, and rotation.

Properties – The **Properties** button opens the **Multi-View Block Reference Properties** dialog box. It includes the following tabs: **General**, **Style**, **Dimensions**, **Offsets**, and **Attributes**. These tabs allow you to edit the dimensions, insertion point and attributes of the multi-view block prior to insertion.

Undo – The **Undo** button allows you to undo a block insertion and remain in the **MVADDBLOCK** command.

? – The **?** button opens help on the topic Add Mulit-View Block.

If you double-click on a symbol in the AutoCAD DesignCenter, the **Add Multi-View Blocks** dialog box opens with the selected symbol listed in the **Name** list. You can edit the **Add Multi-View Blocks** dialog box, presetting the rotation and scale, or move the mouse to the drawing area to specify the insertion point. The symbol is attached to the pointer scaled and rotated as specified in the dialog box. Prior to specifying the insertion point, you can return to the **Add Multi-View Blocks** dialog box and edit the scale or rotation.

After selecting an insertion point for the multi-view block, you can place additional insertions of the block by specifying the locations in the drawing area without editing the **Add Multi-View Blocks** dialog box. Interrupt object snaps can also be specified during the insertion of the multi-view block.

USING MULTI-VIEW BLOCK PROPERTIES TO ENHANCE PLACEMENT

Selecting the **Properties** button of the **Add Multi-View Blocks** dialog box will open the **Multi-View Block Reference Properties** dialog box. This dialog box, shown in Figure 7.17, includes the following tabs: **General**, **Style**, **Dimensions**, **Offsets**, and **Attributes**. This dialog box allows you to change the scale or size of the block, change the insertion point location, and add attributes to the block.

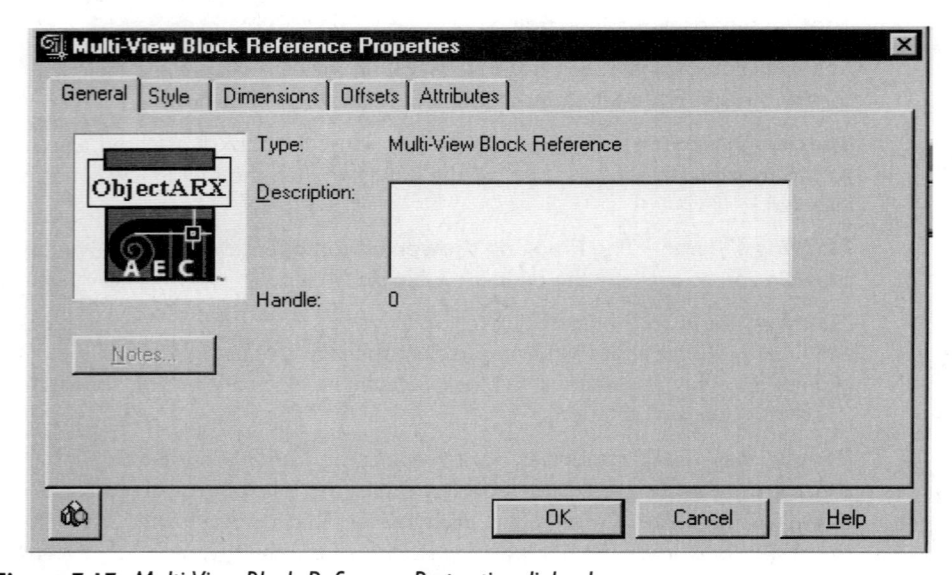

Figure 7.17 *Multi-View Block Reference Properties dialog box*

The options of each tab of the **Multi-View Block Reference Properties** dialog box are described below.

General

The **General** tab includes a **Description** edit field and a **Notes** button, as shown in Figure 7.17.

> **Description** – The **Description** edit field allows you to enter a description for the multi-view block.
>
> **Notes** – Selecting the **Notes** button opens the **Notes** dialog box. The **Notes** button is active only if a multi-view block has been inserted.

Style

The **Style** tab lists the multi-view blocks that have been inserted in the current drawing, as shown in Figure 7.18.

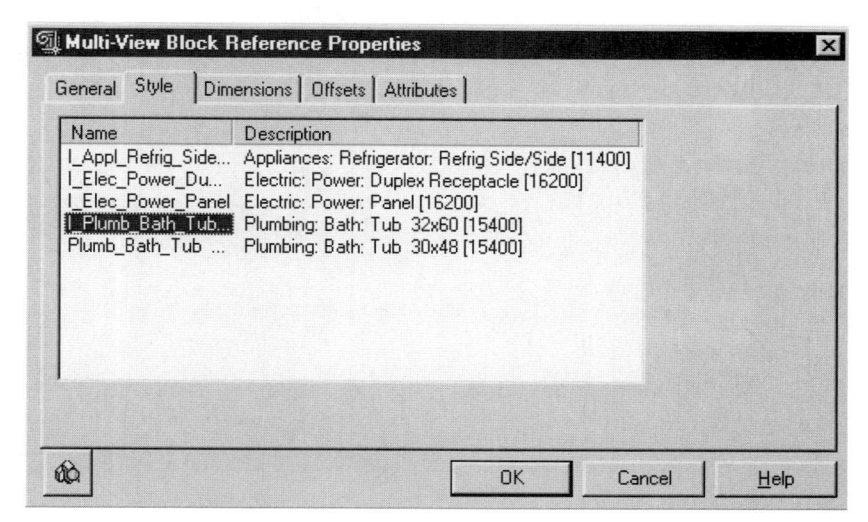

Figure 7.18 *Style list in the Style tab of the Multi-View Block Reference Properties dialog box*

The **Style** tab does not allow you to edit the name or description of the multi-view block. If you select another multi-view block from the list, the current block will change to the block selected, and the new block is substituted for the original block.

> **Note:** Attribute and offset editing of the original multi-view block will be lost when the new block is substituted.

Dimensions

The **Dimensions** tab includes edit fields for the X, Y, and Z scales as shown in Figure 7.19.

The X, Y, and Z scale factors allow you to change the size of the multi-view block. Changing the scale factors affects only the blocks inserted during the current insertion. It overrides the defaults for the multi-view block but does not change the original definition.

Offsets

The **Offsets** tab includes a **View Blocks** list box and edit fields for changing the X, Y, or Z offset for insertion, as shown in Figure 7.20.

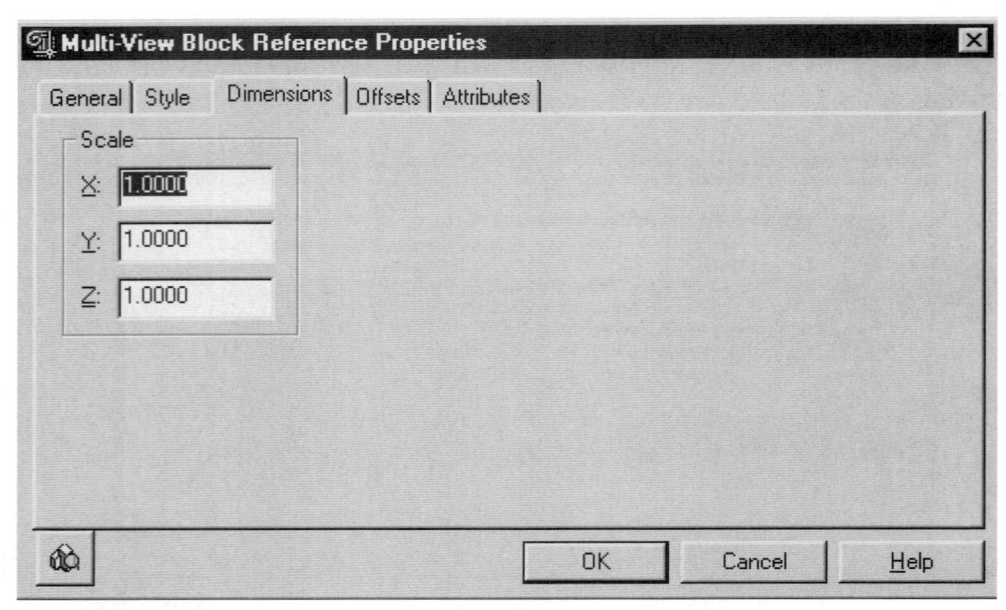

Figure 7.19 Dimensions tab of the Multi-View Block Reference Properties dialog box

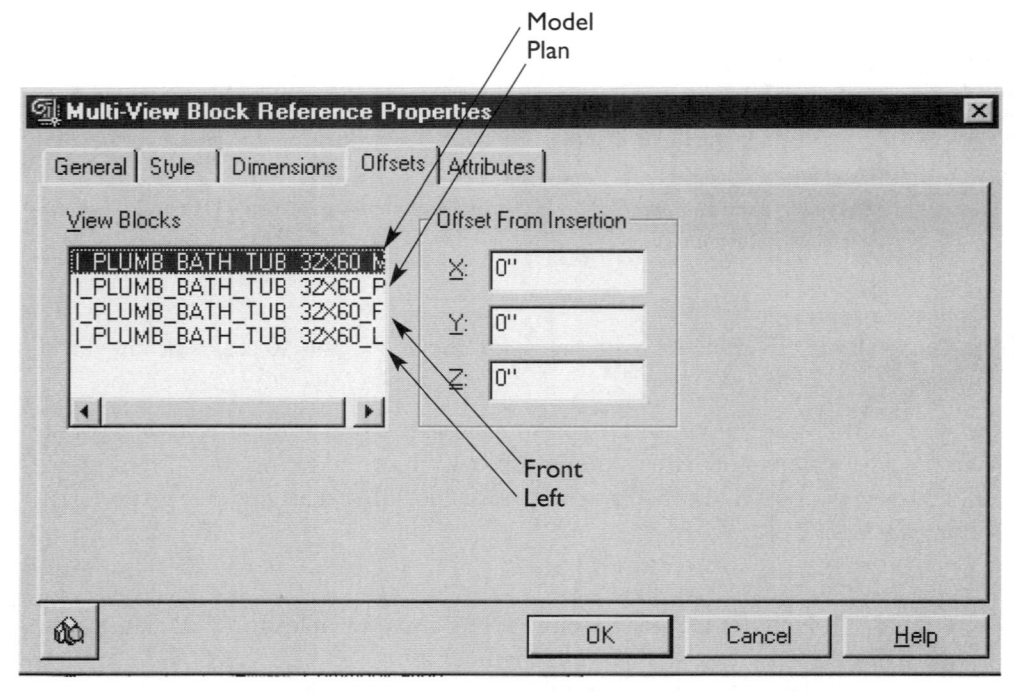

Figure 7.20 Editing the View Blocks of the Multi-View Block Reference Properties dialog box

View Blocks – The **View Blocks** list box shows all defined view controls for the multi-view block. The view blocks listed in the figure above include four multi-view block view definitions. Each of the four names in the list box represents a display method for controlling the view of a single block. The **Offset** for each of the four view blocks can be specified in the **Offset From Insertion** section.

Offset From Insertion – The **Offset From Insertion** section allows you to modify the location of the insertion point for placing the multi-view block. In Figure 7.21, the default insertion point for the top tub is in the middle of the drain end of the tub. Changing the value of the **Y** edit field to 16 will change the insertion point to the lower left corner of the tub, as shown in the bottom tub of the figure. The insertion point offset for each view of the multi-view block should be edited in the **Offsets tab**.

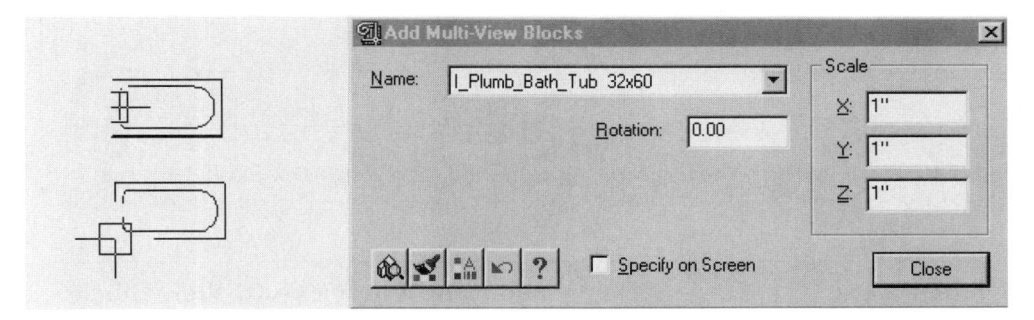

Figure 7.21 *Insertion base point changed for a block*

Note: The insertion point offsets for all views defined for the multi-view block should be edited. If both insertion point offsets are not edited, the block will be positioned differently when viewed in the plan versus the model view of the block.

Attributes

The **Attributes** tab, shown in Figure 7.22, includes a **View Blocks** list box and attributes for each of the blocks.

You can add the values for attributes for the block by entering the text in the **Value** column. The **Prompt** column cannot be edited.

EDITING A MULTI-VIEW BLOCK

After a multi-view block has been inserted in a drawing, it is modified by the **Modify Multi-View Block** command (**MVBLOCKMODIFY**) or the **MVBLOCKPROPS** command. The **MVBLOCKMODIFY** command allows you to change the scale, rotation, and to access the **MVBLOCKPROPS** command. See Table 7.6 for **Modify Multi-View Block** command access.

[Figure 7.22 screenshot: Multi-View Block Reference Properties dialog box, Attributes tab showing View Blocks list with "I_ELEC_MISC_CIRCUIT_BREAK" selected and Prompt "Symbol code" with Value "CB"]

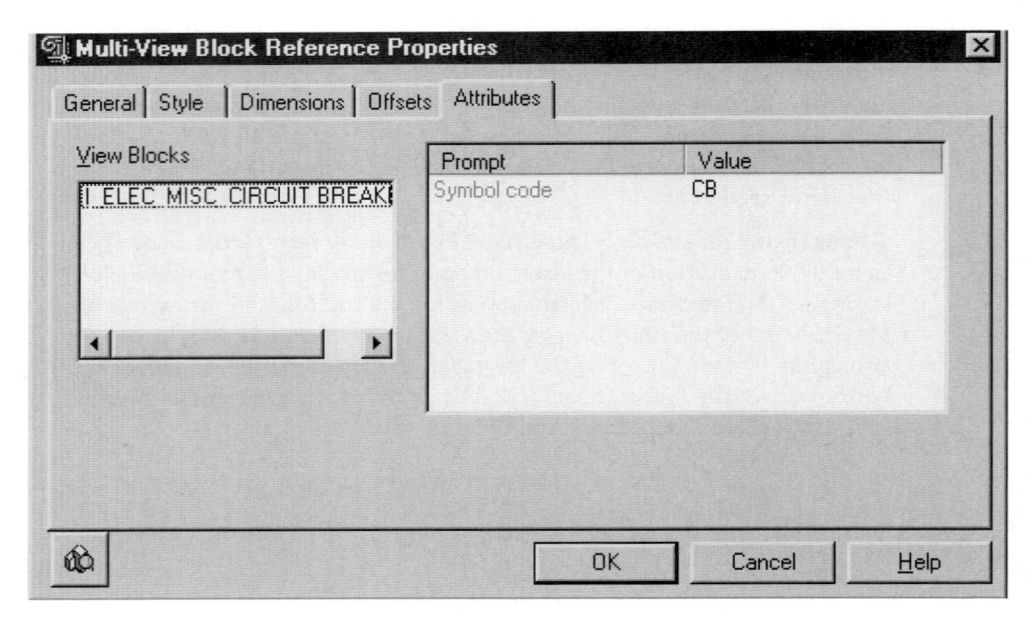

Figure 7.22 *Attribute value editing in the Attributes tab of the Multi-View Block Reference Properties dialog box*

Menu bar	Design>Design Content>Modify Multi-View Block
Command prompt	MVBLOCKMODIFY
Shortcut	Right-click over the drawing area; then choose Utilities>Multi-View Blocks>Modify
Shortcut	Select a multi-view block; then right-click and choose Multi-View Block Modify

Table 7.6 *Modify Multi-View Block command access*

After selecting the **MVBLOCKMODIFY** command from the menu bar or the command line, you will be prompted to select a multi-view block. Selecting the multi-view block opens the **Modify Multi-View Blocks** dialog box, as shown in Figure 7.23.

You can edit the **Modify Multi-View Blocks** dialog box or respond in the command line to change the multi-view block. The command line options include **Name/X scale/Y scale/Z scale/Rotation/Match**. In the command sequence shown below, the **Name** option was selected. In response to the prompt to enter the name of the multi-view block, a question mark was entered to view the multi-view blocks defined for the drawing. To select a multi-view block, you must type the full name of the multi-

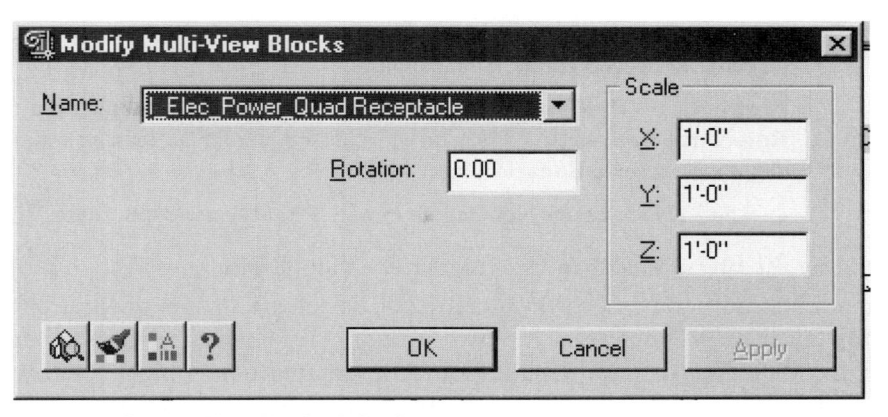

Figure 7.23 Modify Multi-View Blocks dialog box

view block in the command line. Therefore, it is much faster to select the name of the multi-view block from the dialog box.

```
Command: MvBlockModify
Select multi-view blocks: (Select a multi-view block) 1 found
Select multi-view blocks: ENTER (Press ENTER to end the selec-
tion.)
Multi-view block modify [NAme/X scale/Y scale/Z
scale/Rotation/Match]: NA ENTER
Multi-view block name or [?] <I_Elec_Power_Duplex
Recpt>: ? ENTER
Defined multi-view blocks.
  I_Elec_Power_Duplex Recpt
  I_Elec_Power_Quad Receptacle
Multi-view block name or [?] <I_Elec_Power_Duplex
Recpt>: I_Elec_Power_Quad
Receptacle
```

The options included in the **Modify Multi-View Blocks** dialog box are described below:

Name – The **Name** list includes all the multi-view blocks that have been used in the drawing. You can select a multi-view block name from the list to change the multi-view block.

Rotation – The **Rotation** edit field allows you to change the rotation of the multi-view block selected.

Scale – The **Scale** section includes edit fields for X, Y, and Z. You can enter a scale factor to change the size of the multi-view block.

Floating Viewer – The **Floating Viewer** button opens the **Viewer** and allows you to view the multi-view block in the **Viewer**.

Match – The **Match** button allows you to select an existing block and match its name, scale, and rotation properties.

Properties – The **Properties** button opens the **Multi-View Block Reference Properties** dialog box. The dialog box for an existing multi-view block includes the **Location** tab.

? – The **?** button opens Help to the Modify a Multi-View Block topic.

Changing the Properties of an Existing Multi-View Block

To modify a multi-view block in more detail, select the **Properties** button of the **Modify Multi-View Blocks** dialog box. It includes tabs for adding the attributes and changing the offset of the insertion point for the multi-view block. Selecting the **Properties** button of the **Modify Multi-View Blocks** dialog box executes the **MVBLOCKPROPS** command. See Table 7.7 for **MVBLOCKPROPS** command access.

Command prompt	MVBLOCKPROPS
Shortcut	Select multi-view block and right-click; then choose Multi-View Block Properties

Table 7.7 *MVBLOCKPROPS command access*

After selecting the **MVBLOCKPROPS** command, you are prompted to select a multi-view block. Selection of a multi-view block opens the **Multi-View Block Reference Properties** dialog box, as shown in Figure 7.24.

The **Multi-View Block Reference Properties** dialog box for an existing block includes a **Location** tab, as shown in Figure 7.25.

Options of the **Location** tab are as follows:

World Coordinate System/Current Coordinate System – One of the radio buttons is selected to indicate which system is used to display the coordinate location of the multi-view block.

Insertion Point – The **Insertion Point** section includes X, Y, and Z fields, which specify the coordinates of the insertion point of the multi-view block. You can change the coordinates of the multi-view block to move the multi-view block.

Normal – The **Normal** edit fields X, Y, and Z indicate the direction a three-dimensional block is projected. A 1.000 in the Z edit field indicates that the symbol is developed perpendicular to the XY plane.

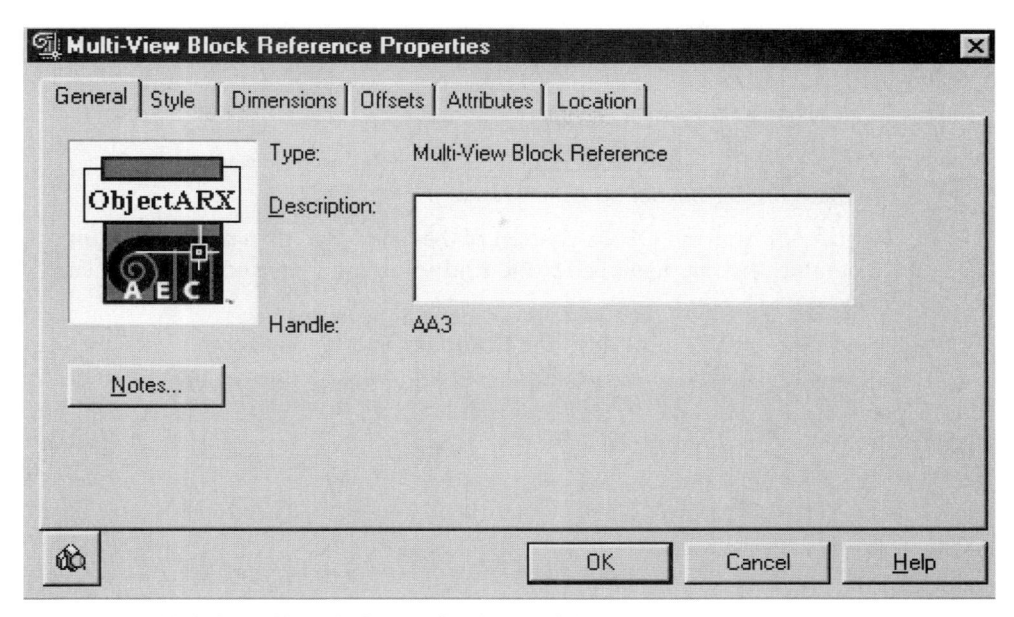

Figure 7.24 *Multi-View Block Reference Properties dialog box*

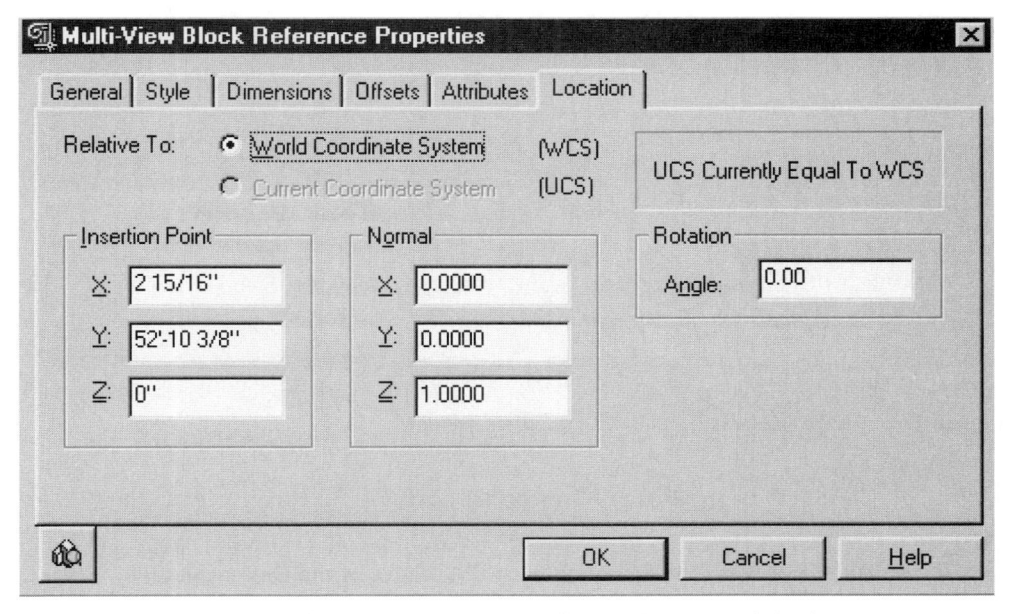

Figure 7.25 *Location tab of the Multi-View Block Reference Properties dialog box*

>**Rotation** – The **Rotation** section indicates the angular rotation of the multi-view block. You can change the rotation angle to rotate the multi-view block.

TUTORIAL 7.2 INSERTING MULTI-VIEW BLOCKS WITH PRECISION

1. Open *ADT Tutor\Ch7\EX7-2.dwg*.

2. Choose **File>SaveAs** from the menu bar and save the drawing as *MVBLOCK2.dwg* in your student directory.

3. Right-click over the OSNAP button of the Status Bar; then choose **Settings** from the shortcut menu. Select the **Endpoint** and **Intersection** object snap modes.

4. Select **Zoom Window** from the **Standard** toolbar and zoom a window around the bath area, as specified in the following command sequence.

   ```
   Command: '_zoom
   Specify corner of window, enter a scale factor (nX or nXP), or
   [All/Center/Dynamic/Extents/Previous/Scale/Window] <real time>: _w
   Specify first corner: (Select point P1 in Figure 7.26.)
   Specify opposite corner: (Select point P2 in Figure 7.26.)
   ```

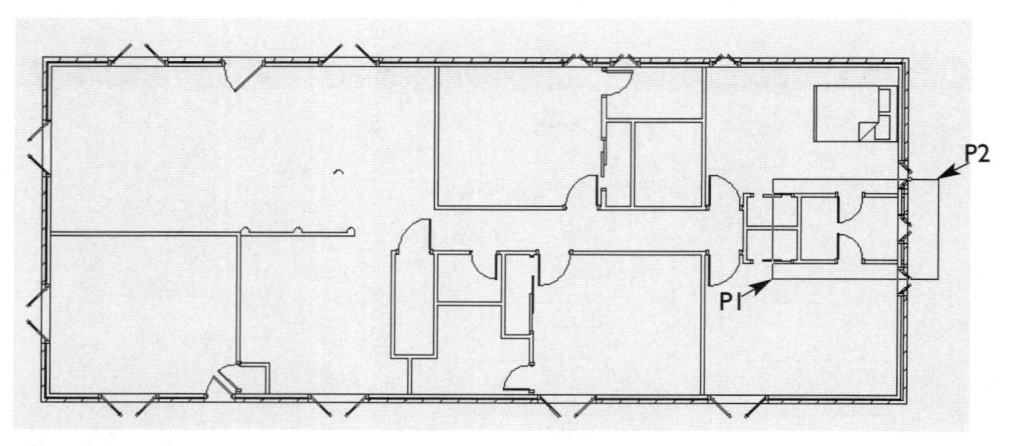

Figure 7.26 *Select locations P1 and P2 for ZOOM window*

5. Choose **Design>Design Content>Plumbing Fixtures** to open the DesignCenter.

6. Double-click on the Toilet Folder in the palette of the DesignCenter.

7. Double-click on the Tank 1 toilet symbol to open the **Add Multi-View Blocks** dialog box.

8. Move the mouse to the drawing area to determine the default symbol rotation, as shown in Figure 7.27.

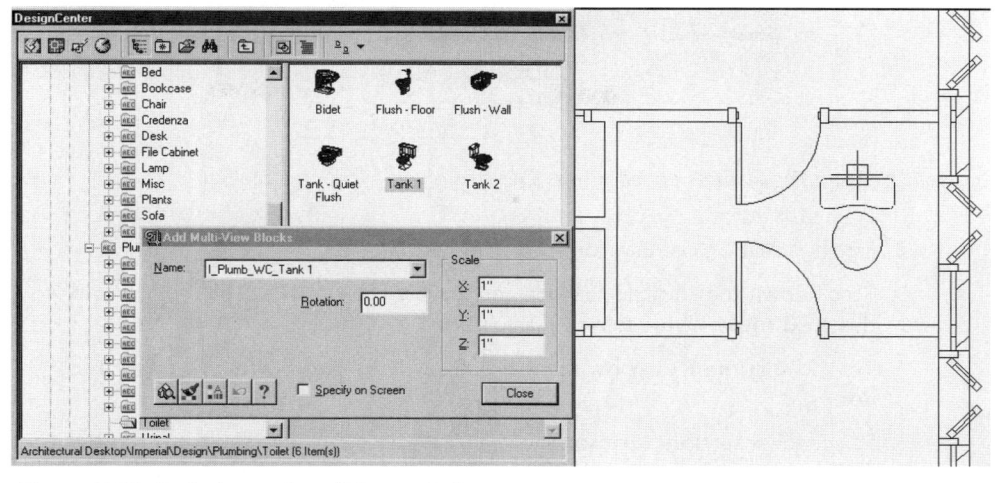

Figure 7.27 *Default rotation of the symbol*

9. Type **270** in the **Rotation** edit field of the **Add Multi-View Blocks** dialog box.

10. Select the drawing area to activate the drawing area and select the location for the toilet, as shown in Figure 7.28.

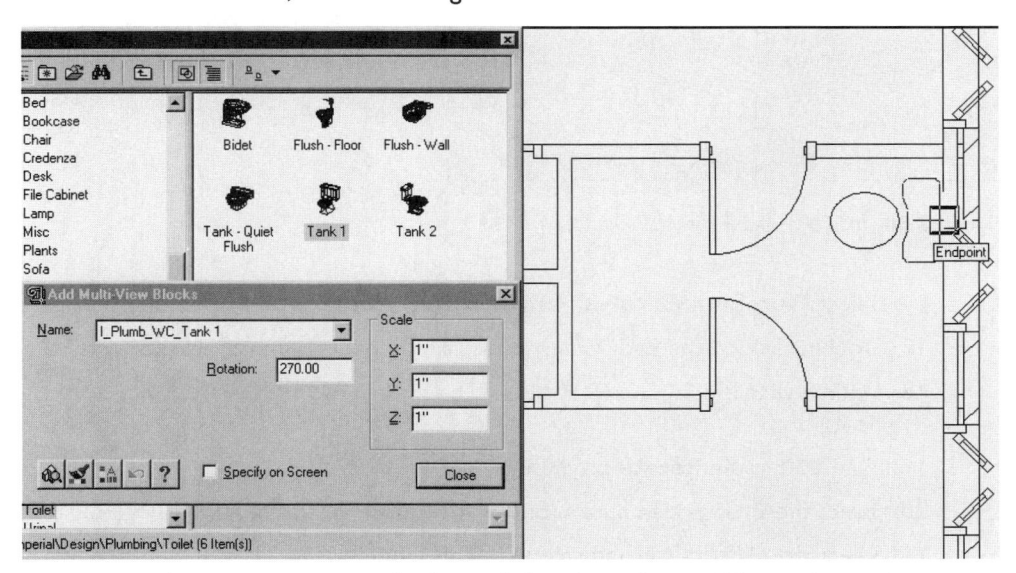

Figure 7.28 *Locate insertion point for the symbol at the endpoint of the window jamb*

11. Select **Plumbing Fixtures** from the **Design Content - Imperial** toolbar as shown in Figure 7.29.

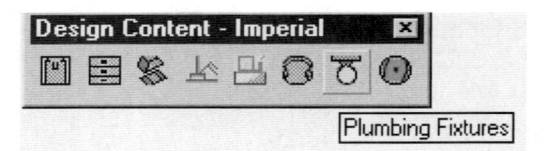

Figure 7.29 *Plumbing Fixtures option on the Design Content - Imperial toolbar*

12. Double-click on the Lavatory folder in the palette.
13. Scroll down the Lavatory palette and double-click on the Wall symbol to open the **Add Multi-View Blocks** dialog box.
14. Type **270** in the **Rotation** edit field of the **Add Multi-View Blocks** dialog box.
15. Select the insertion point for the Wall lavatory symbol as shown in Figure 7.30.

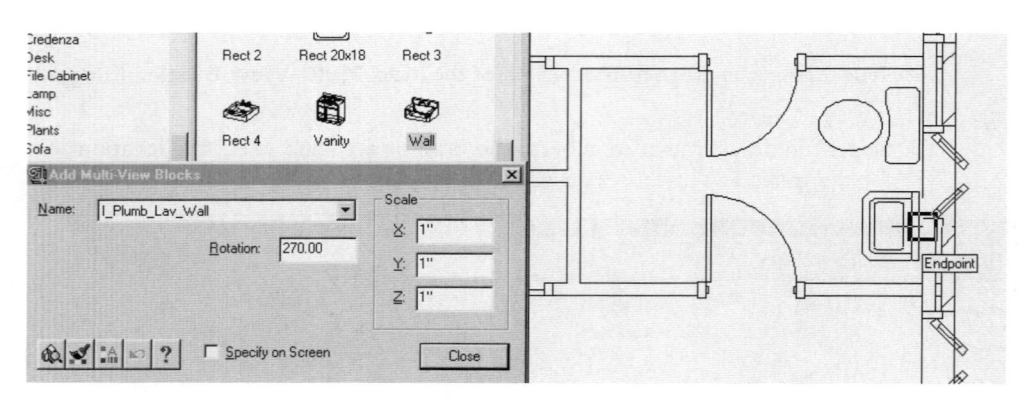

Figure 7.30 *Insertion point for Wall lavatory*

16. Select **Plumbing Fixtures** from the **Design Content - Imperial** toolbar.
17. Double-click on the Bath folder to display the Bath contents in the palette.
18. Double-click the Tub 30x60 symbol to open the **Add Multi-View Blocks** dialog box.
19. Type **270** in the **Rotation** edit field.
20. Select the **Properties** button of the **Add Multi-View Blocks** dialog box.
21. Select the **Offsets t**ab and select the I_PLUMB_BATH_TUB_30X60_P view block.
22. Change the **Y Offset From Insertion** to 15" as shown in Figure 7.31.

Figure 7.31 *Editing the offset of the insertion point for the tub symbol*

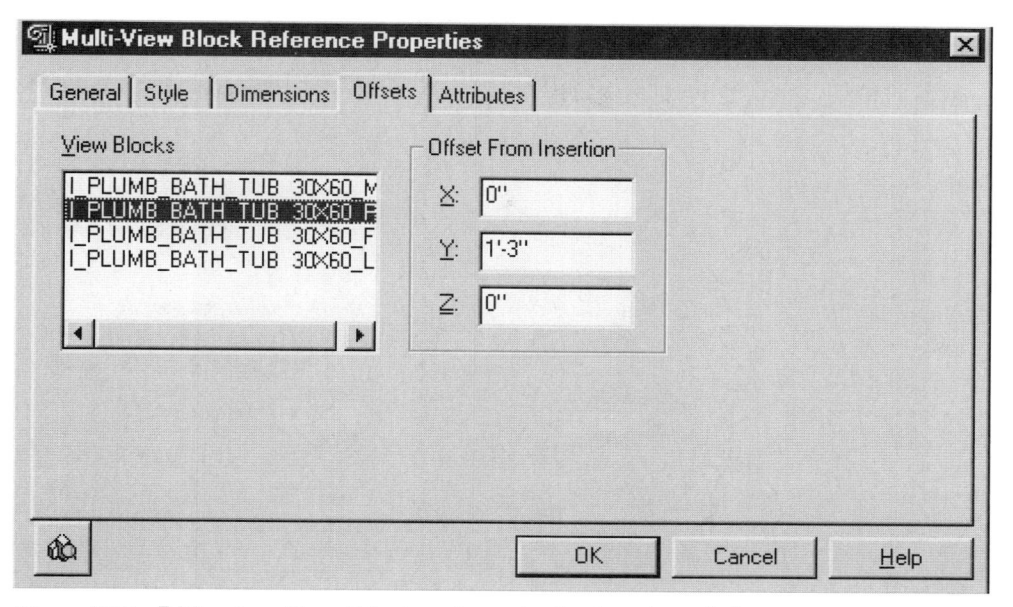

23. Select each of the view blocks in the **View Blocks** list and change the **Y Offset From Insertion** to 15".
24. Select **OK** to dismiss the **Multi-View Block Reference Properties** dialog box.
25. Select a location for the tub at the end of the wall as shown in Figure 7.32.
26. Select **Zoom Previous** from the **Standard** toolbar, and then select **Zoom Window** from the **Standard** toolbar and zoom a window around the kitchen area as specified in the following command sequence.

    ```
    Command: '_zoom
    Specify corner of window, enter a scale factor (nX
    or nXP), or
    [All/Center/Dynamic/Extents/Previous/Scale/Window]
    <real time>: _w
    Specify first corner: (Select point P1 in Figure 7.33.)
    Specify opposite corner: (Select point P2 in Figure 7.33.)
    ```

27. Select **Appliances** from the **Design Content - Imperial** toolbar.
28. Double-click on the Refrigerator folder in the palette.
29. Double-click on the Side-Side symbol to open the **Add Multi-View Blocks** dialog box.

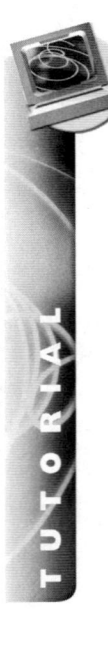

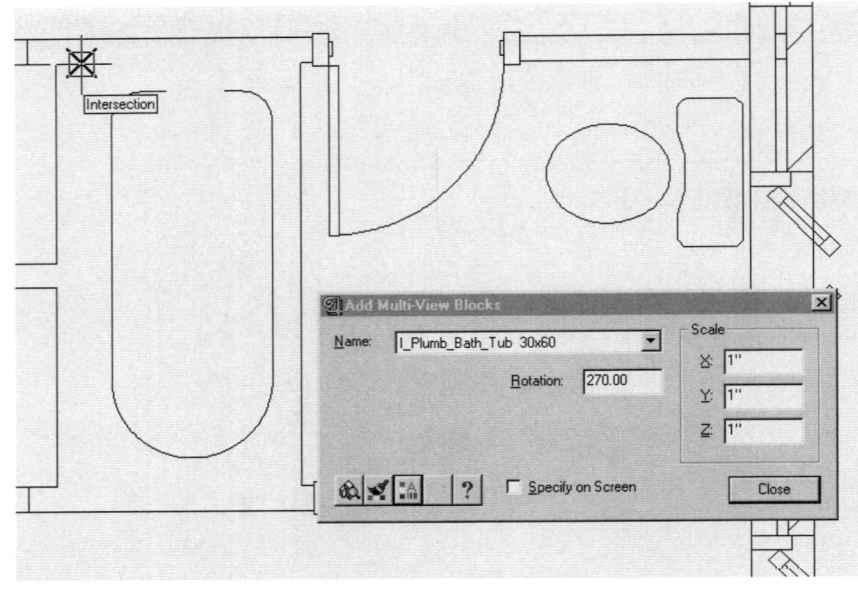

Figure 7.32 *Placing the tub with edited Offset From Insertion*

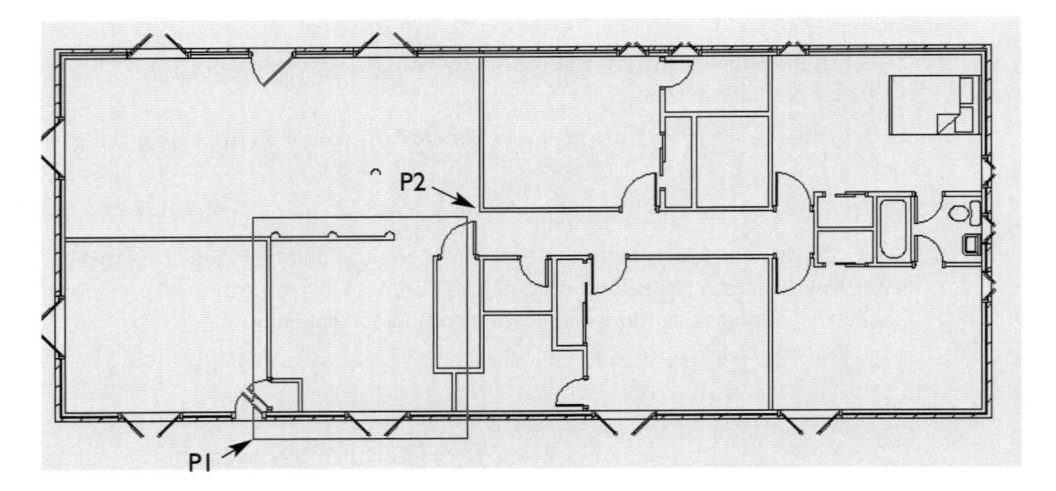

Figure 7.33 *Zoom window location*

30. Select the **Properties** button of the **Add Multi-View Blocks** dialog box.
31. Select the **Offsets** tab and select each of the four view blocks in the **View Blocks** list and change the **Offset From Insertion** for the **Y** edit field to 18" as shown in Figure 7.34. Editing the **Y** edit field will move the insertion point to the lower left corner of the refrigerator.

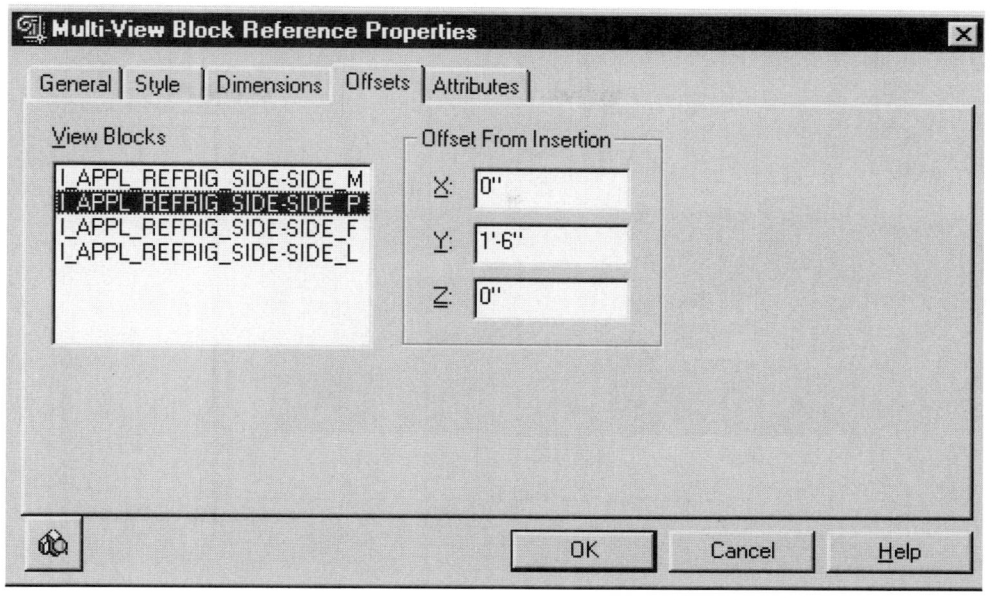

Figure 7.34 *Editing the Y Offset From Insertion*

32. Select **OK** to dismiss the **Multi-View Block Reference Properties** dialog box.
33. Select the location for the refrigerator in the corner as shown in Figure 7.35.
34. Select **Casework** from the **Design Content - Imperial** toolbar.
35. Double-click on the Base Cabinet folder in the palette.
36. Double-click on the 18in Wide base cabinet symbol to open the **Add Multi-View Blocks** dialog box.
37. Type **90** in the **Rotation** edit field of the **Add Multi-View Blocks** dialog box.
38. Select the upper left end of the refrigerator as the insertion point for the cabinet, as shown in Figure 7.36.
39. Select **Appliances** from the **Design Content - Imperial** toolbar.
40. Double-click on the Range folder in the palette.
41. Double-click on the Range 30x26 symbol to open the **Add Multi-View Blocks** dialog box.
42. Select the **Properties** button of the **Add Multi-View Blocks** dialog box and then select the **Offsets** tab. Select each of the four view blocks in the **View Blocks** list and change the **Offset From Insertion** for the **Y** edit field to 15". Editing the **Y** edit field will move the insertion point to the lower left corner for all viewing directions.

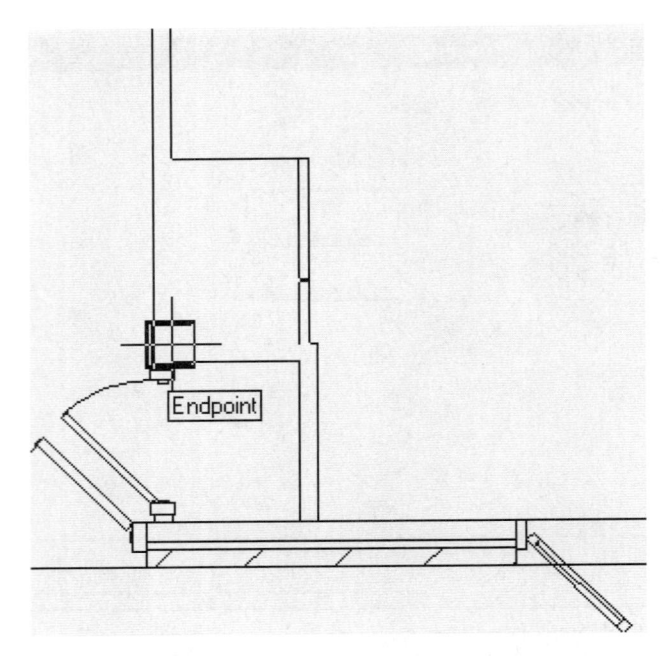

Figure 7.35 *Placing the refrigerator with edited Offset From Insertion base point*

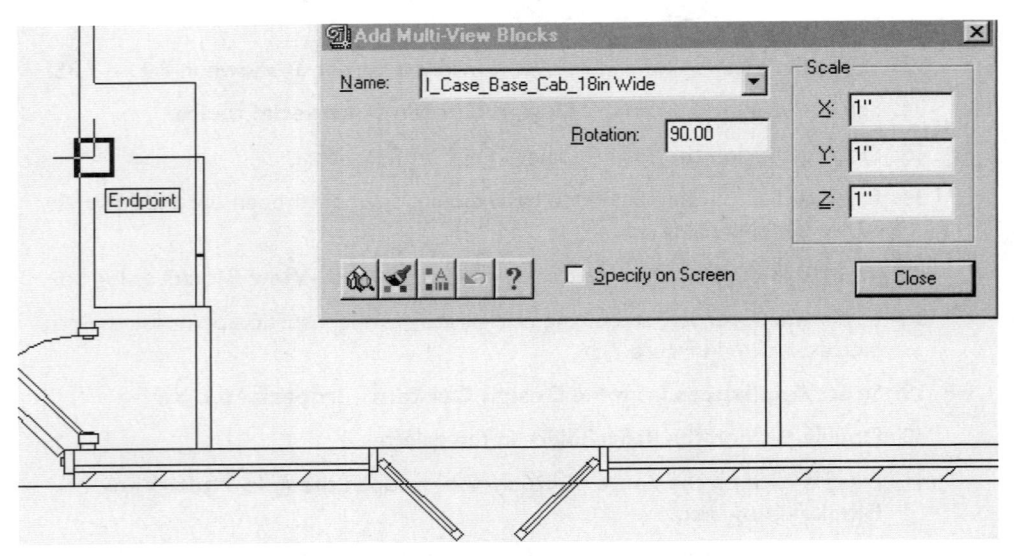

Figure 7.36 *Placing the 18in Wide base cabinet*

43. Select the upper left corner of the 18" cabinet as the insertion point for the range as shown in Figure 7.37.

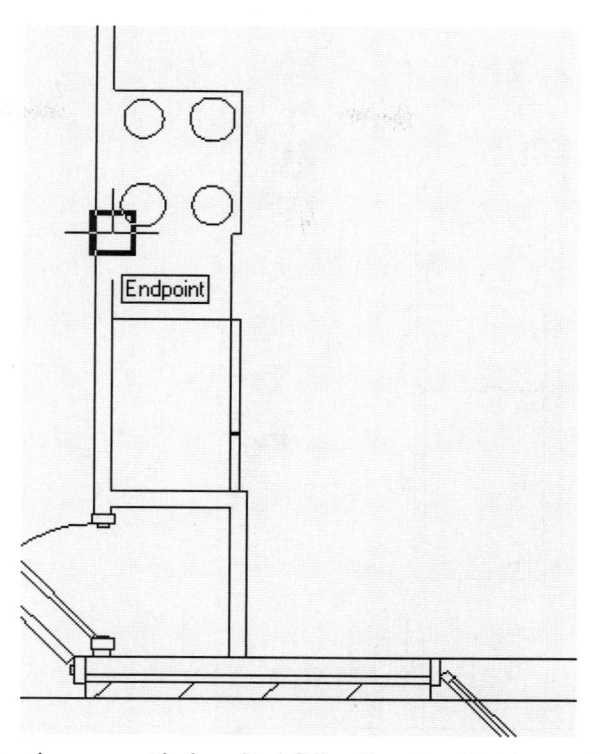

Figure 7.37 *Placing the range with the edited Offset From Insertion base point*

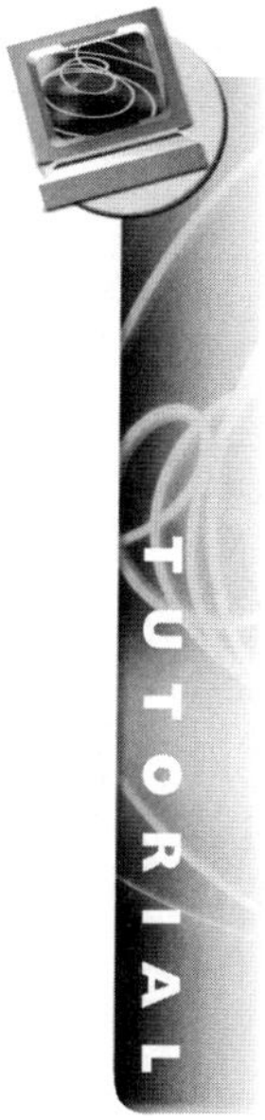

44. Select **Casework** from the **Design Content - Imperial** toolbar.
45. Double-click on the Base Cabinet folder in the palette.
46. Double-click on the 12in Wide symbol to open the **Add Multi-View Blocks** dialog box.
47. Type **90** in the **Rotation** edit field of the **Add Multi-View Blocks** dialog box.
48. Select the upper left corner of the range as the insertion point for the cabinet, as shown in Figure 7.38.
49. Select the **Close** button of the **Add Multi-View Blocks** dialog box.
50. Double-click on the 18in Wide symbol in the DesignCenter palette to open the **Add Multi-View Blocks** dialog box.
51. Select the **Properties** button of the **Add Multi-View Blocks** dialog box.
52. Select the **Offsets** tab of the **Multi-View Block Reference Properties** dialog box and select the I_CASE_BASE_CAB_18IN WIDE_P view block.

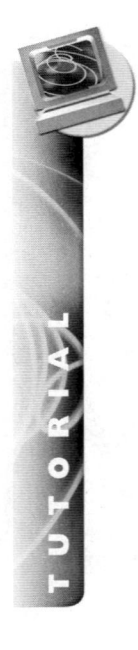

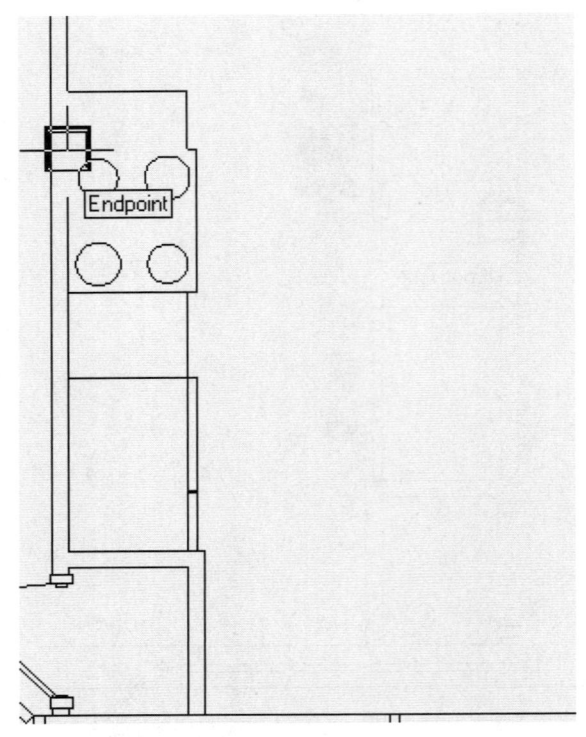

Figure 7.38 *Placing 12in Wide base cabinet*

53. Type −18" in the **X** offset edit field to move the insertion point to the upper right corner.
54. Select the remaining view blocks and type −18" in the **X** offset edit field to move the insertion point to the upper right corner.
55. Select the end of the wall as the insertion point for the base cabinet, as shown in Figure 7.39.
56. Scroll down the Base Cabinet palette and double-click on the 39in Wide base cabinet.
57. Select the **Properties** button of the **Add Multi-View Blocks** dialog box.
58. Select the **Offsets** tab, and then select each of the view blocks and set the **X** offset to −39".
59. Select **OK** to dismiss the **Multi-View Reference Properties** dialog box.
60. Select the insertion point for the cabinet at the upper left corner of the 18" base cabinet, as shown in Figure 7.40.
61. Select **Appliances** from the **Design Content - Imperial** toolbar.

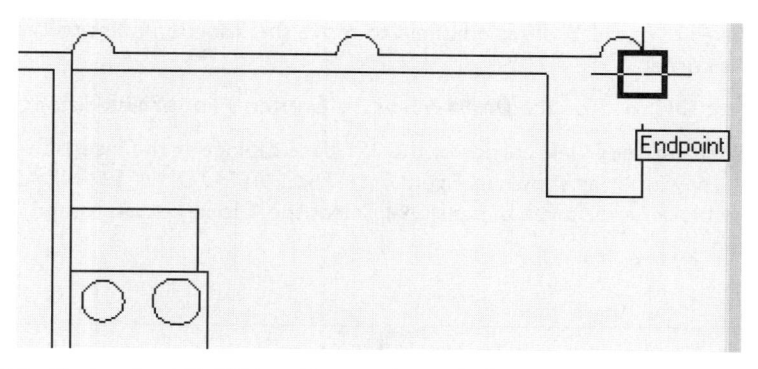

Figure 7.39 Placing the 18in Wide cabinet at the end of the wall

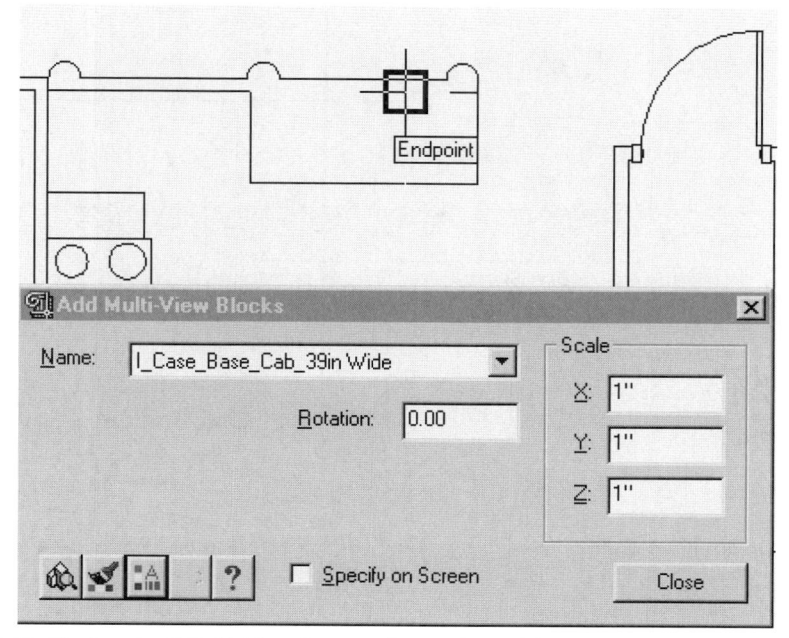

Figure 7.40 Placing the 39in Wide cabinet with an edited Offset From Insertion value

62. Double-click on the Dishwasher folder in the palette.
63. Double-click on the Standard symbol to open the **Add Multi-View Blocks** dialog box.
64. Type **270** in the **Rotation** edit field.
65. Select the **Properties** button of the **Add Multi-View Blocks** dialog box.
66. Select the **Offsets** tab and select the I_APPL_DISH_STANDARD_P view block.

67. Type –12" in the **Y** offset edit field to move the insertion point to the upper right corner.
68. Select **OK** to close the **Multi-View Reference Properties** dialog box.
69. Select the upper left corner of the 39" base cabinet as the insertion point for the dishwasher, as shown in Figure 7.41. The I_APPL_DISH_STANDARD_P view block is not three dimensional. Select the **Close** button to end the command.

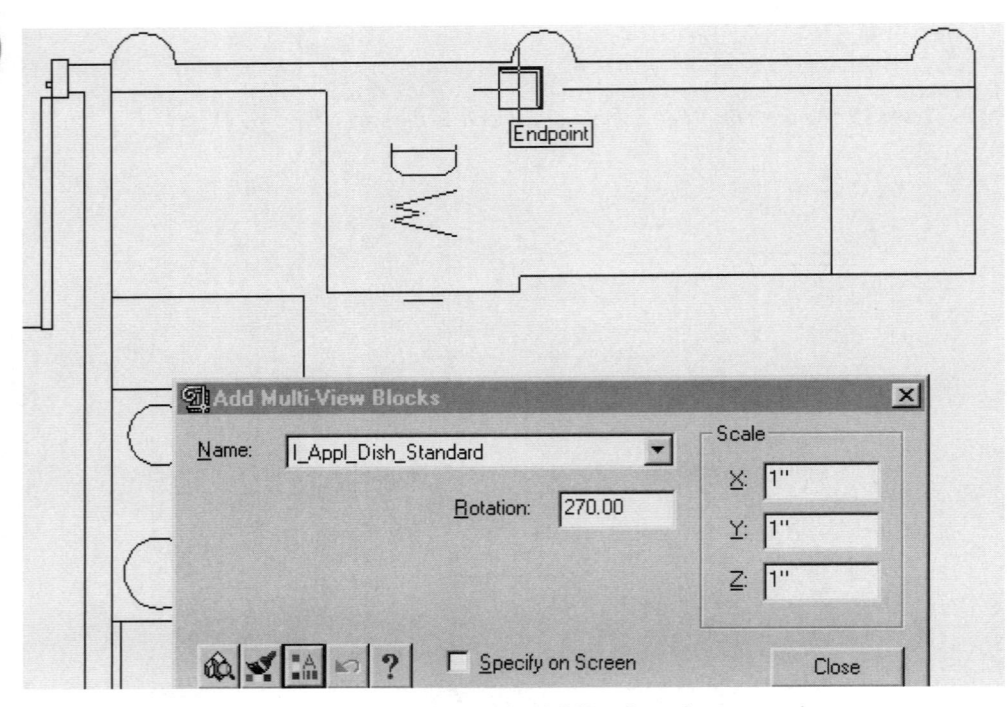

Figure 7.41 *Placing the dishwasher with an edited Offset From Insertion value*

70. Select **Layers** from the **Object Properties** toolbar; then select A-Flor-Case and select **Freeze**.
71. Select **OK** to dismiss the **Layer Properties Manager**.
72. Add the counter top by first selecting **Add Wall** from the **Walls** toolbar.
73. Edit the **Add Walls** dialog box as follows: **Style** = Casework_Counter, **Group** = Standard, **Offset** = 0, **Height** = 40, and **Justify** = Right.
74. Select the **Properties** button of the **Add Walls** dialog box.
75. Select the **Dimensions** tab of the **Wall Properties** dialog box.

76. Edit the **Wall Cleanup Radius** to 3".
77. Select **OK** to close the **Wall Properties** dialog box.
78. Refer to the following command sequence and Figure 7.42 to draw a counter top wall for the cabinets.

```
Command: _AecWallAdd
Start point or
[STyle/Group/WIdth/Height/OFfset/Justify/Match/Arc]:
(Select point P1 in the Figure 7.42.)
End point or
[STyle/Group/WIdth/Height/OFfset/Justify/Match/Arc/
Undo]: (Select point P2 in Figure 7.42.)
End point or
[STyle/Group/WIdth/Height/OFfset/Justify/Match/Arc/
Undo]: (Select point P3 in Figure 7.42.)
End point or
[STyle/Group/WIdth/Height/OFfset/Justify/Match/Arc/
Undo/Close/ORtho]: ENTER
Command: ENTER
Command: _AecWallAdd
Start point or
[STyle/Group/WIdth/Height/OFfset/Justify/Match/Arc]:
(Select point P4 in Figure 7.42.)
End point or
[STyle/Group/WIdth/Height/OFfset/Justify/Match/Arc]:
(Select point P5 in Figure 7.42.)
End point or
[STyle/Group/WIdth/Height/OFfset/Justify/Match/Arc/
Undo]: ENTER
Command: ENTER
Command: _AecWallAdd
Start point or
[STyle/Group/WIdth/Height/OFfset/Justify/Match/Arc]:
(Select point P6 in Figure 7.42.)
End point or
[STyle/Group/WIdth/Height/OFfset/Justify/Match/Arc]:
(Select point P7 in Figure 7.42.)
End point or
[STyle/Group/WIdth/Height/OFfset/Justify/Match/Arc/
Undo]: ENTER
```

79. Select **Layers** from the **Object Properties** toolbar; then select A-Flor-Case and select **Thaw**.
80. Select **OK** to dismiss the **Layer Properties Manager**.

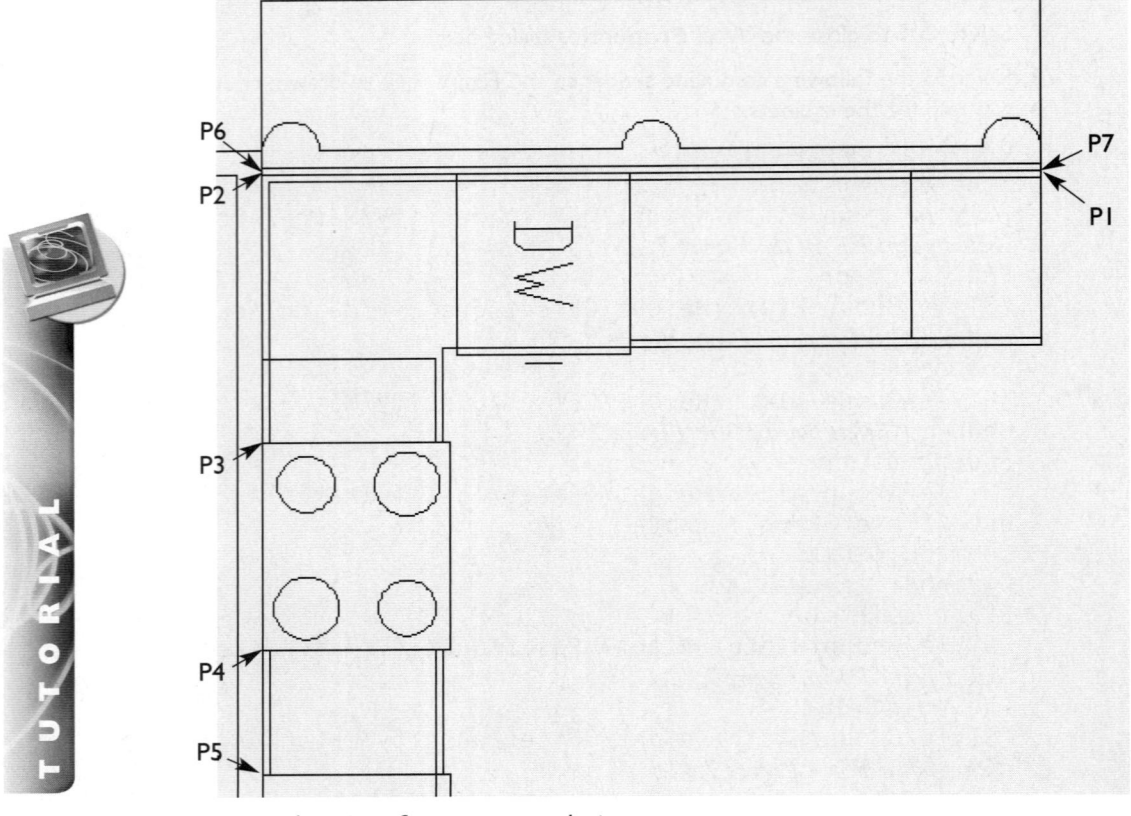

Figure 7.42 *Location of counter top endpoints*

81. Select **Plumbing Fixtures** from the **Design Content - Imperial** toolbar.
82. Double-click on the Sink folder in the palette.
83. Double-click on the Kitchen Double A symbol to open the **Add Multi-View Blocks** dialog box.
84. Type **270** in the **Rotation** edit field of the **Add Multi-View Blocks** dialog box and select the **Properties** button.
85. Select the **Offsets** tab of the **Multi-View Reference Properties** dialog box and select each view block; then set the insertion offsets for each view block to **Z** = 36 and **X** = 2.
86. Select **OK** to close the **Multi-View Reference Properties** dialog box.
87. Type **MID** and press ENTER to select the **Midpoint** object snap and select the back of the 39in Wide cabinet symbol as shown in Figure 7.43.

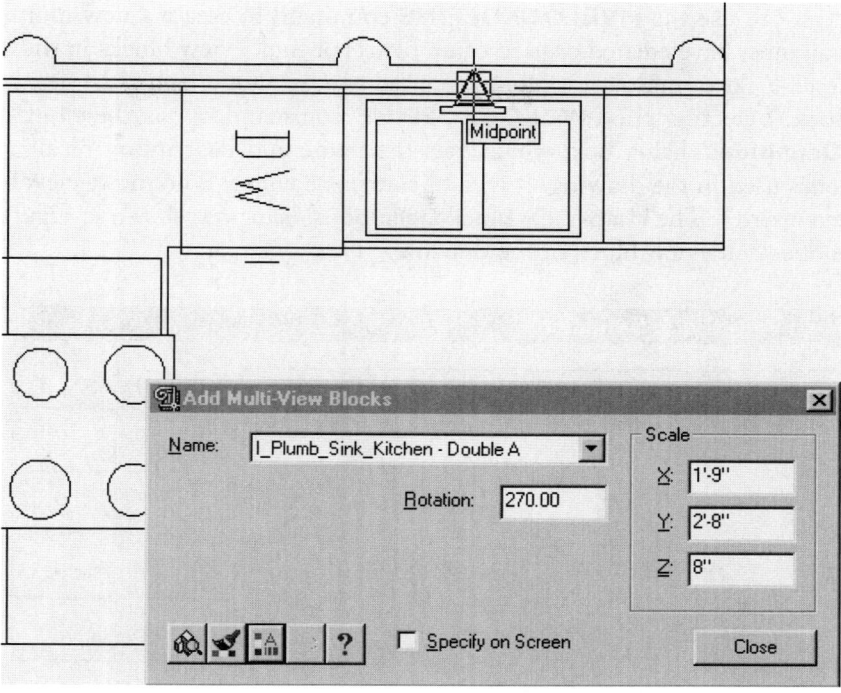

Figure 7.43 *Placing the sink at the midpoint of the 39" base cabinet*

88. Select the **Close** button of the **Add Multi-View Blocks** dialog box.
89. Save the drawing.

CREATING A MULTI-VIEW BLOCK

The **MVBLOCKDEFINE** command will create multi-view blocks from other multi-view blocks or other blocks in a drawing. This command allows you to create your own multi-view blocks, which control the visibility of the block according to the viewing direction. See Table 7.8 for **Define Multi-View Block (MVBLOCKDEFINE)** command access.

Menu bar	Design>Design Content>Define Multi-View Block
Command prompt	MVBLOCKDEFINE
Shortcut	Right-click over the drawing area; then choose Utilities>Multi-View Blocks>Table

Table 7.8 *Define Multi-View Block command access*

Before you can use the **MVBLOCKDEFINE** command to create a new multi-view block, you must have created or used other blocks or multi-view blocks in the drawing. These blocks or multi-view blocks are used as a base for creating the new multi-view block. Selecting the **MVBLOCKDEFINE** command opens the **Multi-View Block Definitions** dialog box, which lists the name and description of all multi-view blocks used in the drawing. It will, of course, be empty if no multi-view blocks have been inserted. The **Multi-View Block Definitions** dialog box shown in Figure 7.44 lists the one multi-view block of the drawing.

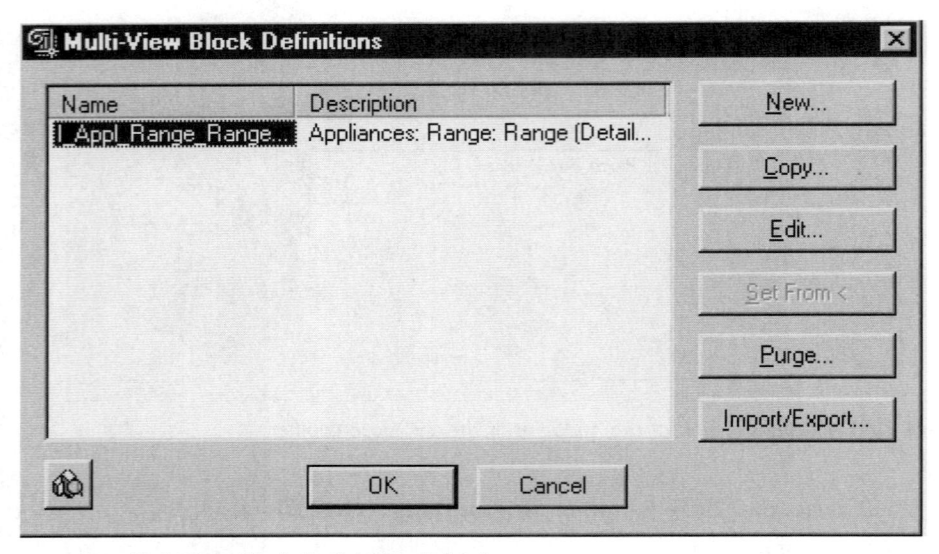

Figure 7.44 *Multi-View Block Definitions dialog box*

The buttons of the **Multi-View Block Definitions** dialog box are described below.

> **New** – Opens the **Name** dialog box, which allows you to create a new name for a multi-view block.
>
> **Copy** – Allows you to create a copy of a multi-view block. To create a copy of the multi-view block, select a multi-view block from the list box and then select **Copy**. The **Name** dialog box will open, allowing you to assign a new name for the copied multi-view block.
>
> **Edit** – Allows you to define the properties of the multi-view block. Selecting the **Edit** button opens the **Multi-View Block Definition Properties** dialog box, which includes **General** and **View Blocks** tabs, as shown in Figures 7.45 and 7.46. The **Multi-View Block Definition Properties** dialog box is used to define the specific properties of the new multi-view block.
>
> **Set From>** – Inactive for this command.

Purge – Opens the **Purge Multi-View Block Definitions** dialog box. This dialog box is used to remove multi-view blocks from the drawing. Before you purge a multi-view block from a drawing, all instances of the multi-view block must be erased from the drawing.

Import/Export – Opens the **Import/Export** dialog box, which allows you to import or export multi-view blocks to or from other drawing files.

DEFINING THE PROPERTIES OF A NEW MULTI-VIEW BLOCK

After you create a name for the multi-view block, selecting the **Edit** button allows you to define the specific properties. Selecting the **Edit** button or double-clicking on the name of the multi-view block in the list box will open the **Multi-View Block Definition Properties** dialog box, which includes the two tabs described below.

General

The **General** tab, shown in Figure 7.45, allows you to edit the name or description of the multi-view block. The **Notes** button is active if you selected an existing multi-view block. Selecting the **Notes** button displays the **Notes** dialog box that includes a **Text Notes** tab, which allows you to add text regarding the multi-view block, and a **Reference Docs** tab, which allows you to list any files relevant to the multi-view block.

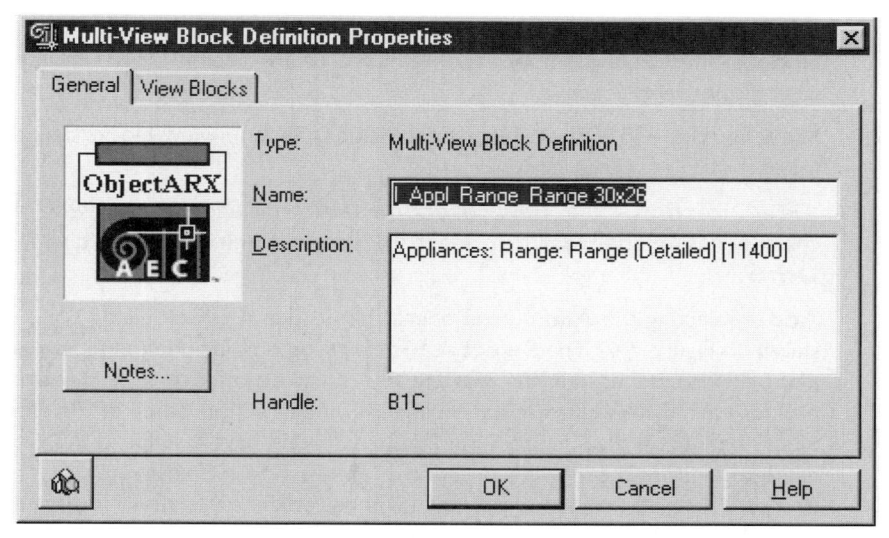

Figure 7.45 *General tab of the Multi-View Block Definitions Properties dialog box*

View Blocks

The **View Blocks** tab, shown in Figure 7.46, allows you to define the display representation and viewing directions for the display of the block. The description of its options follows.

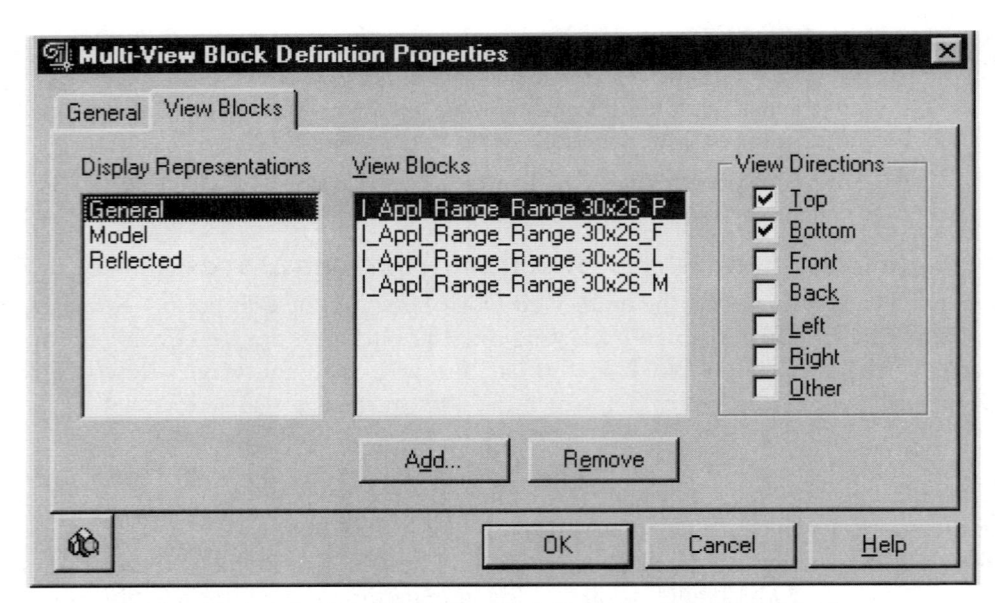

Figure 7.46 *View Blocks tab of the Multi-View Block Definition Properties dialog box*

>**Display Representations** – The display representations for multi-view blocks include General, Model, and Reflected. You can define block visibility for each display representation.
>
>**View Blocks** – The **View Blocks** list includes all multi-view blocks defined for each display representation.
>
>**View Direction** – The **View Direction** section includes seven viewing check boxes for the block: **Top**, **Bottom**, **Front**, **Back**, **Left**, **Right**, and **Other**.
>
>**Add** – Selecting the **Add** button opens the **Select A Block** dialog box shown in Figure 7.47. The **Select A Block** dialog box includes all blocks and multi-view blocks used in the drawing. You can select a block to define its display for each display representation and viewing direction. Some of the view blocks defined in the Architectural Desktop folders have a suffix to the block name to indicate the viewing direction. In Figure 7.47, suffixes were used: F (front), L (left), M (model), and P (plan). However, the same view block can be defined for each of the display representations and view.
>
>**Remove** – Selecting the **Remove** button removes the block for the display representation and viewing direction.

The properties of a multi-view block should be defined based upon the intended use of the multi-view block and which display configuration is assigned to the viewport for displaying the drawing. The General, Model and Reflected display representations

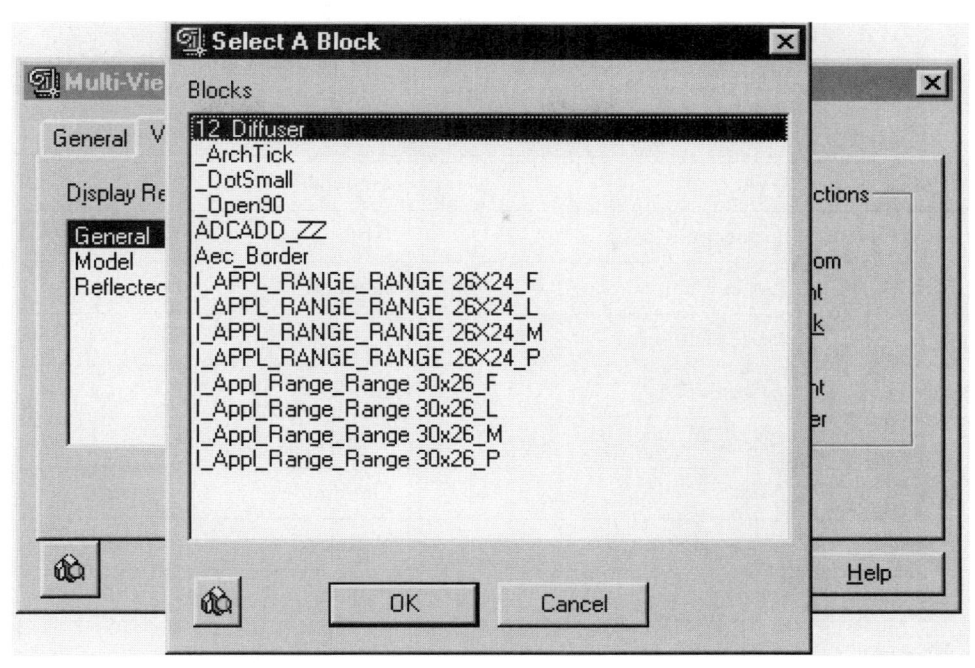

Figure 7.47 Select a Block dialog box to specify the block

can be assigned to the multi-view block. Table 7.9 shows the display representations assigned to the respective display configurations for each layout tab.

Therefore, if a multi-view block is defined for the General display representation, it will be displayed in the following layout tabs: Model, Space (Plan), Work-3D (Plan), Work-FLR, Work-SEC, Plot-FLR and Plot-SEC. The Plot-FLR layout tab is used for the plotting of floor plans, and therefore the multi-view block with a General display representation would be displayed on the floor plan. The Reflected display representation would display a multi-view block in only the Work-RCP and Plot_RCP layout tabs. Symbols created for a floor plan should be assigned the General display representation.

If you are developing an electrical plan to include receptacles, switches, and incandescent lights, all of these symbols must have the General display representation. The multi-view block definition for duplex receptacles and switches is General. However, the incandescent light symbol has the Reflected display representation. Therefore, you can edit the incandescent light symbol to a General display representation and all symbols will be visible in the Plot-FLR tab. In the following tutorial you will create a new symbol and modify existing multi-view block symbols.

Layout Tab	Display Configuration	Display Representation Active for Multi-View Block
Model	Work	General
Mass Group	Concept-Mass	None
Space	Space Plan	General
	Space Model	Model
Work-3D	Work-3D	Model
	Work-Plan	General
Work-FLR	Work	General
Work-RCP	Work-Reflected	Reflected
Work-SEC	Work	General
Plot_FLR	Plot	General
Plot_RCP	Plot Reflected	Reflected
Plot-SEC	Plot	General

Table 7.9 *Display configuration assignments*

TUTORIAL 7.3 CREATING AND MODIFYING MULTI-VIEW BLOCKS

1. Open *ADT Tutor\Ch7\EX7-3.dwg*.
2. Choose **File>SaveAs** from the menu bar and save the drawing as *ElecMVBlock3.dwg* in your student directory.
3. Choose **Documentation>Set Drawing Scale** from the menu bar.
4. Select the 1/8"=1'-0" scale from the **Drawing Scale** list box.
5. Verify that the **Annotation Plot Size** is 1/8" and then select **OK** to close the **Drawing Setup** dialog box.
6. Right-click over the OSNAP button in the Status Bar; choose **Settings**, thencheck **Nearest** object snap mode and clear all other object snap modes from the **Drafting Settings** dialog box.
7. Select **Electric Fixtures** from the **Design Content - Imperial** toolbar as shown in Figure 7.48.
8. Double-click on the Power folder in the palette.

9. Double-click on the Duplex Recpt symbol to open the **Add Multi-View Blocks** dialog box.
10. Select a location along the left wall as shown in Figure 7.48.

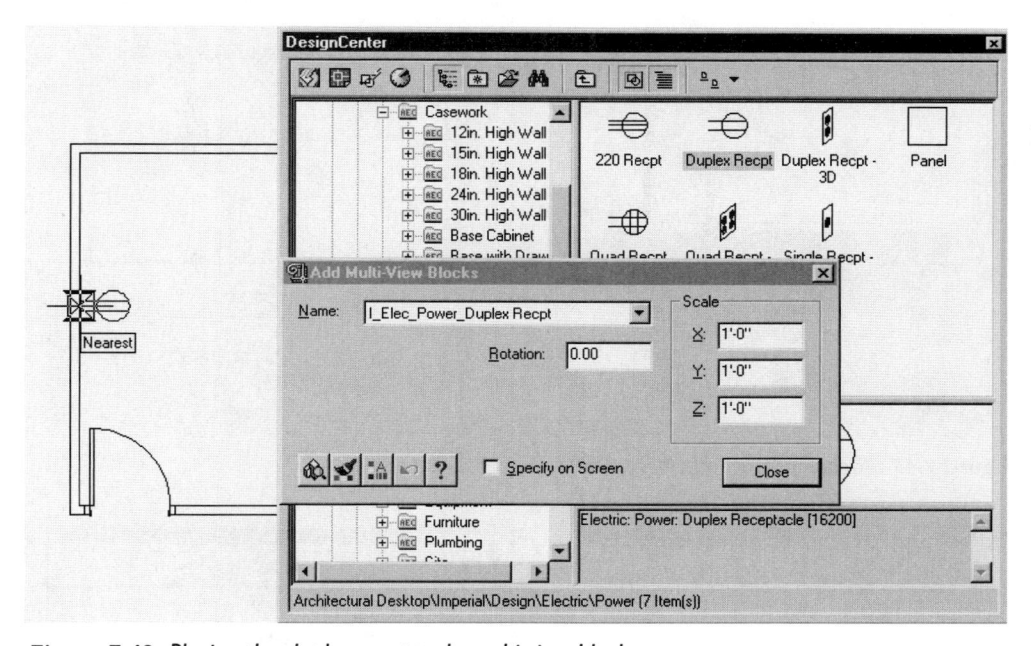

Figure 7.48 *Placing the duplex receptacle multi-view block*

11. Type **270** in the **Rotation** edit field to rotate the symbol; then select the drawing area and select the two additional locations, as shown in Figure 7.49.
12. Close the **Add Multi-View Blocks** dialog box.
13. Select the Switch folder in the DesignCenter Tree.
14. Double-click on the 3-Way symbol to open the **Add Multi-View Blocks** dialog box.
15. Type **270** in the **Rotation** edit field to rotate the switch symbol.
16. Select the drawing area and then select a location for the switch on the wall, as shown in Figure 7.50.
17. Type **90** in the **Rotation** edit field of the **Add Multi-View Blocks** dialog box.
18. Select the drawing area and then select a location on the lower wall for the switch, as shown in Figure 7.51.

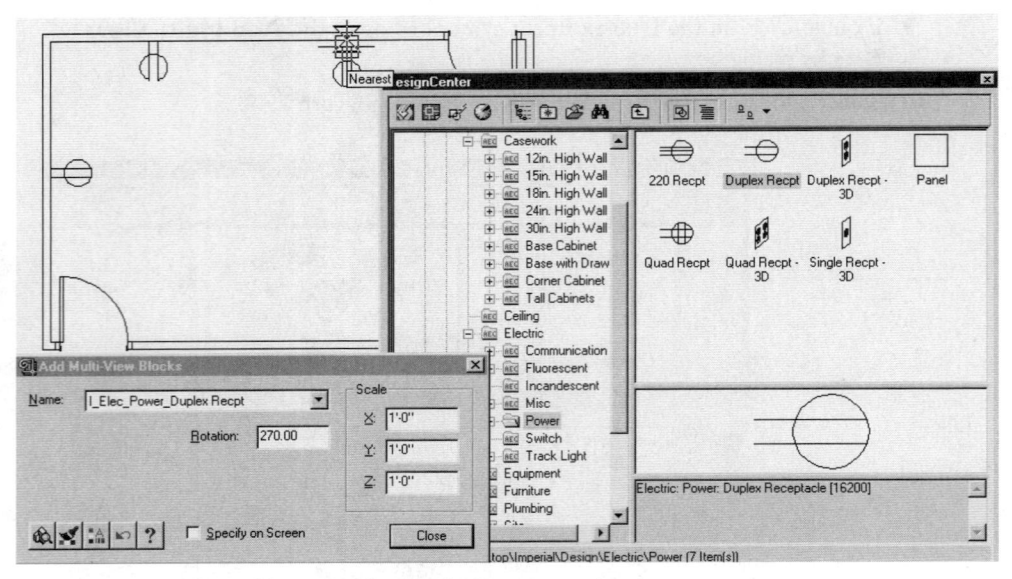

Figure 7.49 *Additional locations for the duplex receptacle*

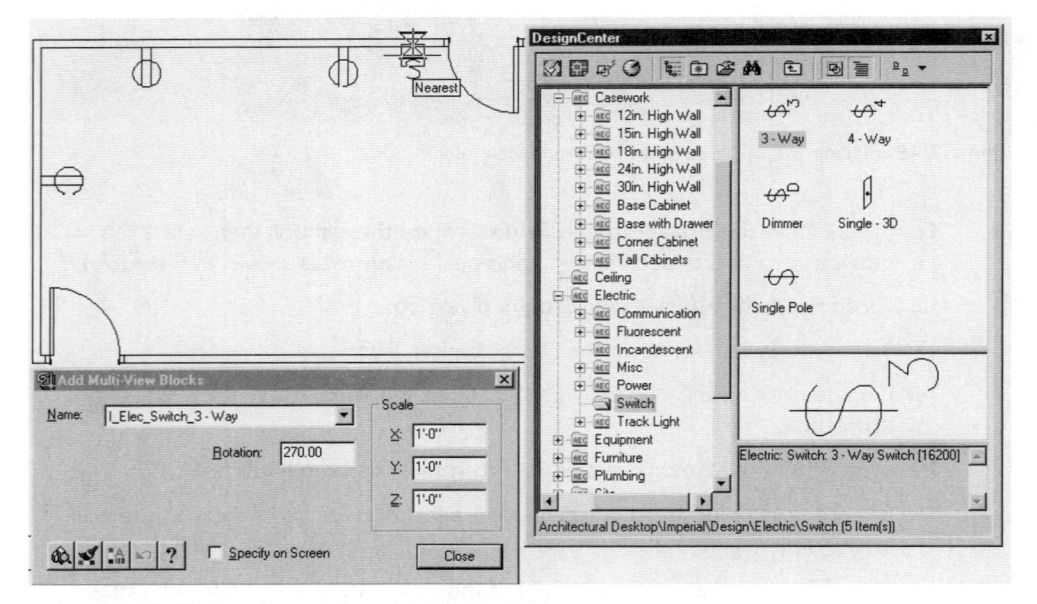

Figure 7.50 *Placing the switch multi-view block*

19. Select the **Close** button to close the **Add Multi-View Blocks** dialog box.
20. Select the Work-RCP tab. The door leaf, swings, receptacles, and switches are not shown with this display configuration.

Using and Creating Symbols 549

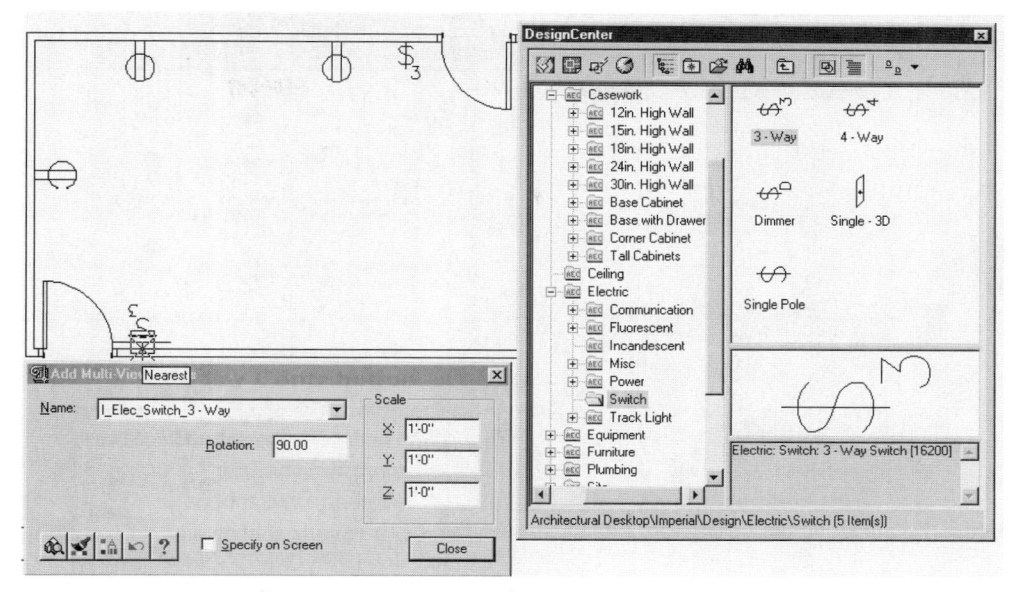

Figure 7.51 *Additional location for a switch multi-view block*

21. Select **Electric Fixtures** from the **Design Content - Imperial** toolbar.
22. Double-click on the Incandescent folder in the palette.
23. Double-click on the Ceiling symbol to open the **Add Multi-View Blocks** dialog box.
24. Select a location for the ceiling incandescent symbol near the center of the room, as shown in Figure 7.52.
25. Select the **Close** button to close the **Add Multi-View Blocks** dialog box.
26. Select the Work-FLR layout tab; the Ceiling incandescent symbol is not displayed in this viewport.
27. Select the Work-RCP layout tab.
28. Select the Ceiling incandescent symbol; then right-click and choose **Edit Multi-View Block Definition**.
29. Select the **View Blocks** tab of the **Multi-View Blocks Definition Properties** dialog box.
30. Select the General display representation of the **View Blocks** tab.
31. Select the **Add** button, and then select the I_ELEC_INC_CEILING_P from the **Select A Block** dialog box, as shown in Figure 7.53.
32. Select **OK** to dismiss the **Select A Block** dialog box.

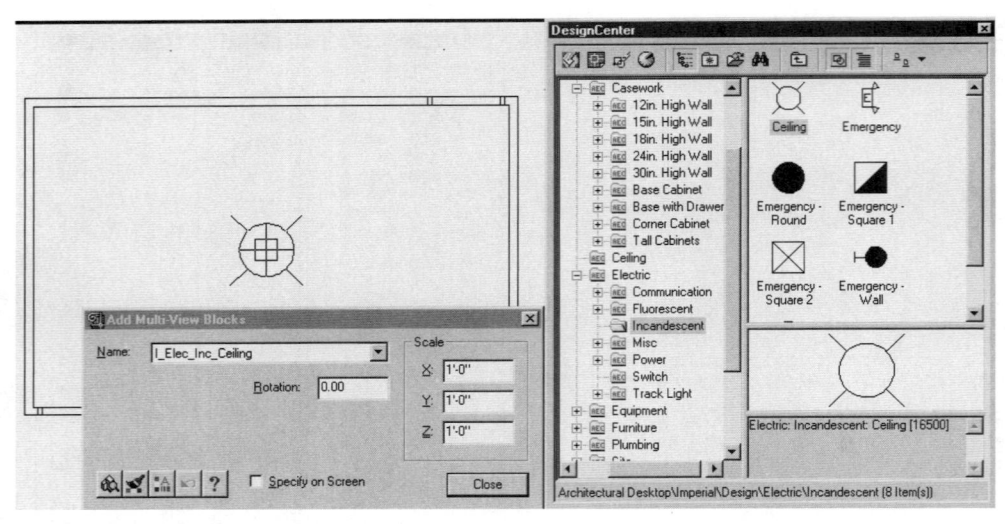

Figure 7.52 *Placing the Ceiling light symbol*

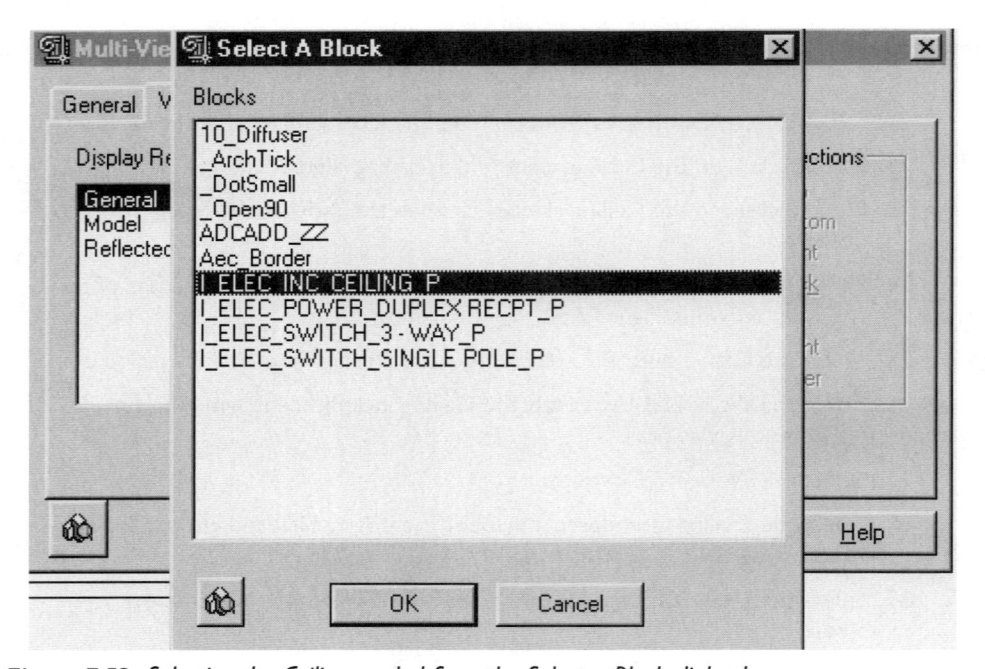

Figure 7.53 *Selecting the Ceiling symbol from the Select a Block dialog box*

33. Clear all **View Directions** except **Top**, as shown in Figure 7.54.
34. Select **OK** to dismiss the **Multi-View Blocks Definition Properties** dialog box.

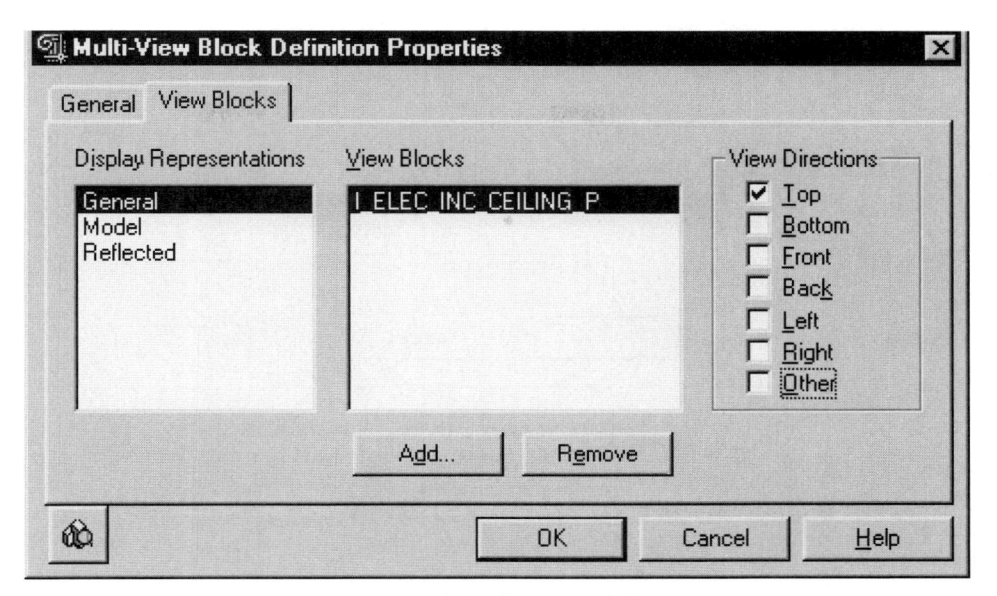

Figure 7.54 *Editing the View Directions of the Ceiling symbol*

35. Select the Work-FLR layout tab to view the Ceiling incandescent symbol and other electrical symbols.
36. Select the Plot-FLR layout tab to view all the electrical symbols.
37. To create a multi-view block from a block, choose **Design>Design Content>Define Multi-View Block**.
38. Select the **New** button and enter the name **10_DIFFUSER** in the **Name** dialog box, as shown in Figure 7.55.
39. Select **OK** to dismiss the **Name** dialog box.
40. Select the **Edit** button in the **Multi-View Block Definitions** dialog box.
41. Type the following in the **Description** edit field: **10" Ceiling Diffuser**.
42. Select the **View Blocks** tab, and then select the General display representation.
43. Select the **Add** button to open the **Select a Block** dialog box.
44. Select the 10_Diffuser block from the **Select A Block** dialog box, as shown in Figure 7.56.
45. Select **OK** to dismiss the **Select a Block** dialog box.
46. Clear all **View Directions** except **Top**.
47. Select the Reflected display representation, and then select the **Add** button.

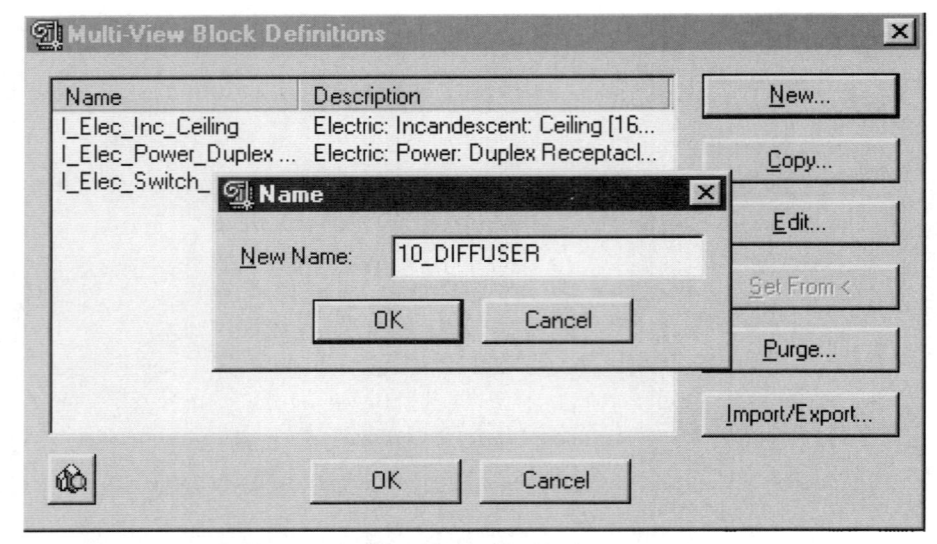

Figure 7.55 *Entering a name for the new multi-view block*

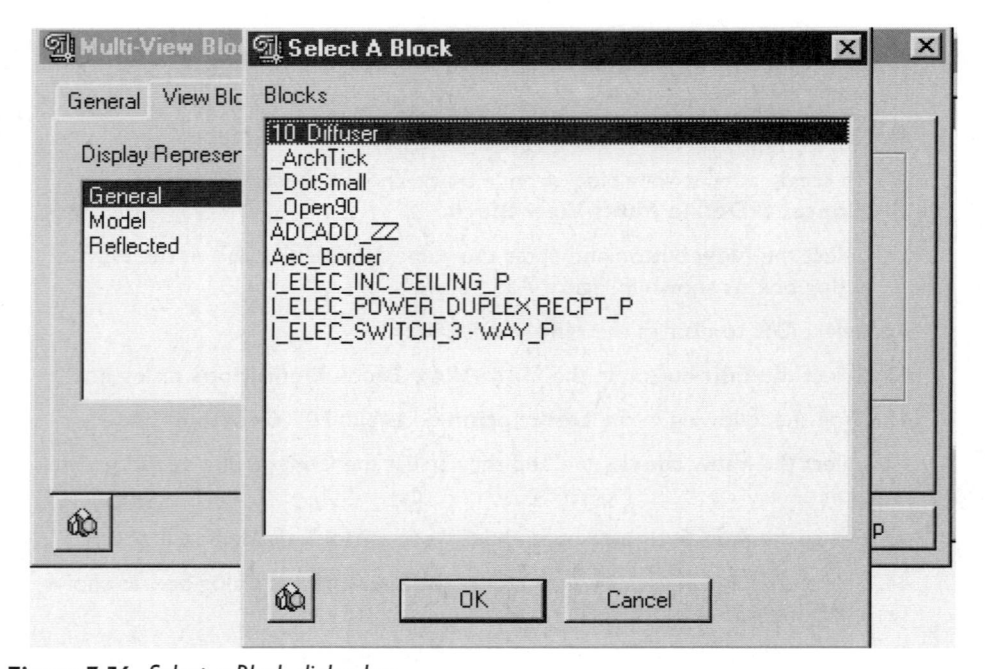

Figure 7.56 *Select a Block dialog box*

48. Select the 10_Diffuser from the **Select A Block** dialog box.
49. Select **OK** to dismiss the **Select a Block** dialog box.

50. Clear all **View Directions** except **Top**.
51. Select **OK** to dismiss the **Multi-View Block Definition Properties** dialog box.
52. Select **OK** to dismiss the **Multi-View Block Definitions** dialog box.
53. Choose **Design>Design Content>Add Multi-View Block** from the menu bar.
54. Select 10_DIFFUSER from the **Name** list and change the X, Y, Z scale factors to 1.
55. Select the drawing area and then select a location for the diffuser, as shown in Figure 7.57.

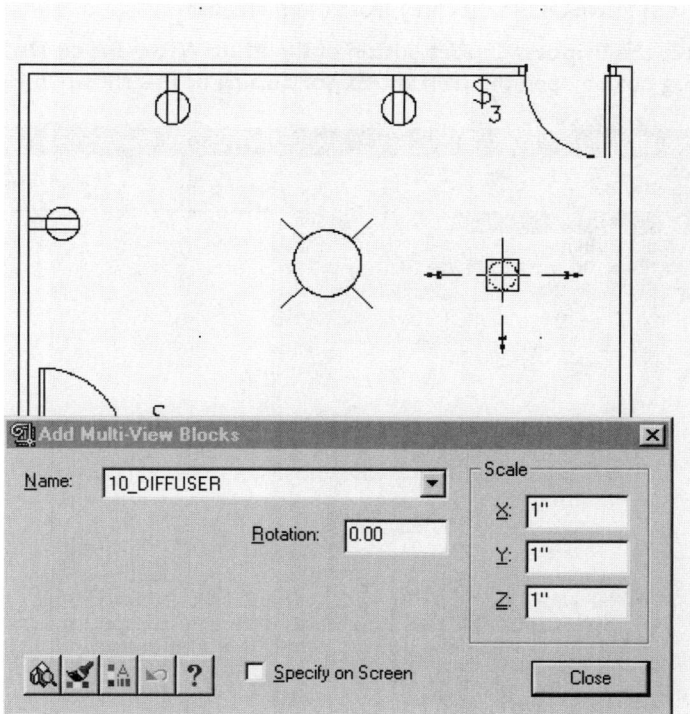

Figure 7.57 *Placing the 10_DIFFUSER in the room*

56. Select the **Close** button to close the **Add Multi-View Blocks** dialog box.
57. Select the Work-FLR layout tab to view all electrical symbols and the 10_DIFFUSER block.
58. Save the drawing.

IMPORTING AND EXPORTING MULTI-VIEW BLOCKS

The multi-view blocks allow you to insert symbols in a drawing with more display control than with blocks. You can transfer the multi-view blocks of one drawing to another drawing by exporting the multi-view block definition. Exporting multi-view blocks is similar to creating WBLOCKS in AutoCAD. However, exporting and importing multi-view blocks allow you to transfer one or all of the multi-view blocks in one operation.

STEPS TO EXPORTING MULTI-VIEW BLOCKS

1. Open a drawing that has multi-view blocks.
2. Choose the **Design>Design Content>Define Multi-View Block** command (**MVBLOCKDEFINE**) from the menu bar.
3. Select the **Import/Export** button of the **Multi-View Block Definitions** dialog box to open the **Import/Export** dialog box as shown in Figure 7.58.

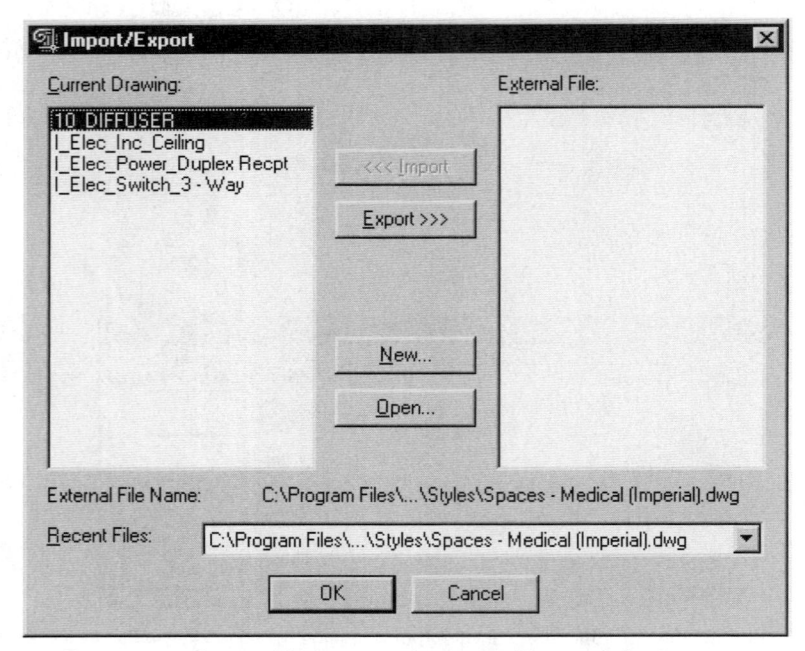

Figure 7.58 *Exporting multi-view blocks using the Import/Export dialog box*

4. Select the **Open** button of the **Import/Export** dialog box.
5. Select a target file from the **File to Import From** dialog box, as shown in Figure 7.59. The file selected must not be open when you select it.

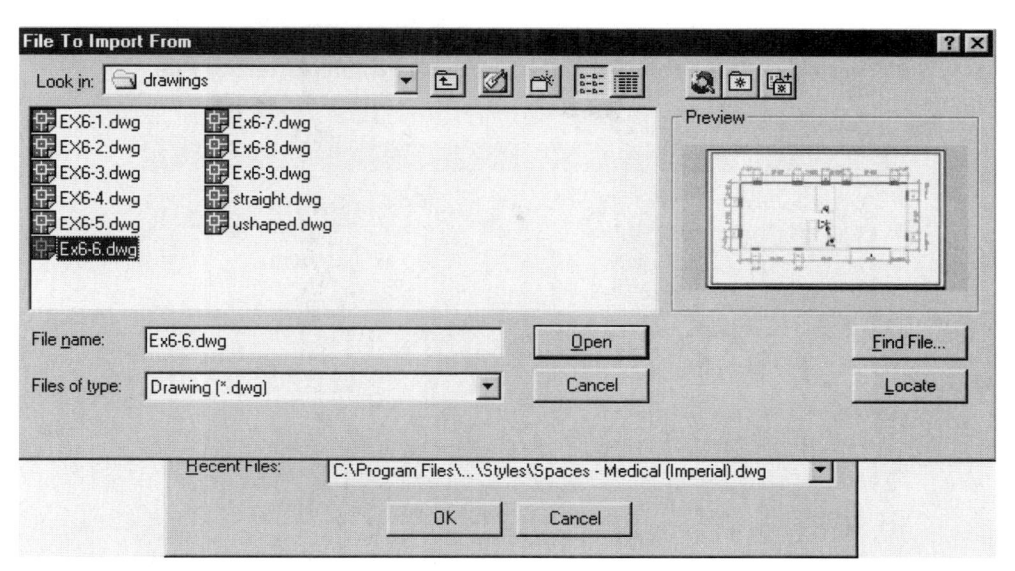

Figure 7.59 *Selecting a target file for export*

6. Select the **Open** button of the **File to Import From** dialog box.
7. Select the multi-view blocks from the **Current Drawing** list for export as shown in Figure 7.60.

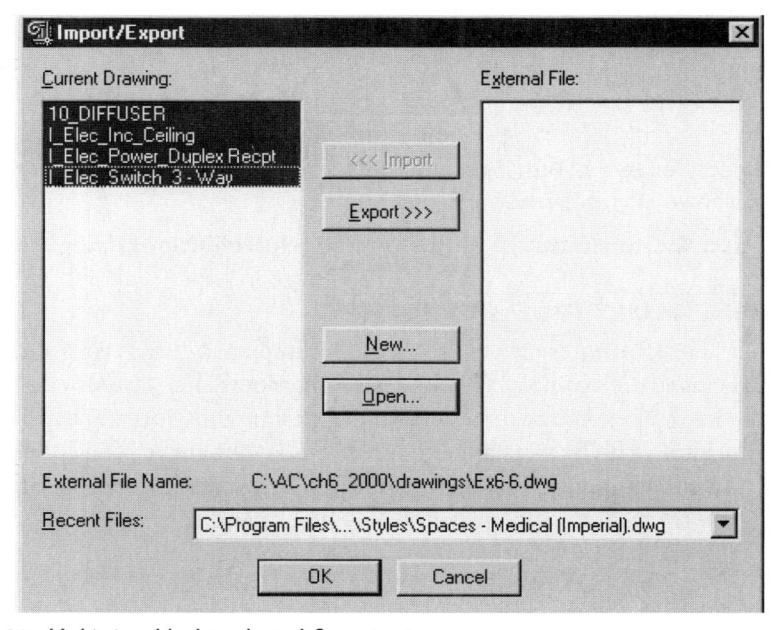

Figure 7.60 *Multi-view blocks selected for export*

8. Select the **Export** button of the **Import/Export** dialog box as shown in Figure 7.61. The multi-view blocks are copied to the **External File**.

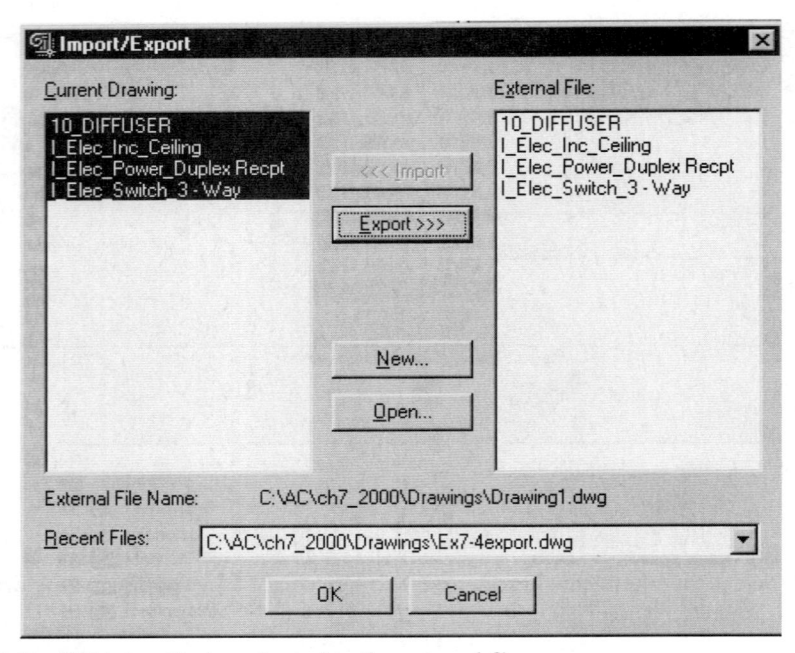

Figure 7.61 *Multi-view blocks exported to the external file*

9. Select **OK** to dismiss the **Import/Export** dialog box.
10. Select **Yes** to the AutoCAD message box to verify multi-view block transfer, as shown in Figure 7.62.
11. Select **OK** to dismiss the **Multi-View Block Definitions** dialog box.

INSERTING MULTIPLE FIXTURES

Included in the DesignCenter of Architectural Desktop are symbols consisting of several fixtures. Included are symbols of entire restroom layouts. The layouts can be inserted in the drawing and exploded, allowing you to design a restroom according to the existing room shape. The fixtures are anchored to a layout curve to evenly space the fixtures. (Layout curves are discussed in Chapter 13, Drawing Commercial Structures.) The symbols are three-dimensional AutoCAD blocks and they can be used for elevation details. See Table 7.10 for **Fixture Layout** command access.

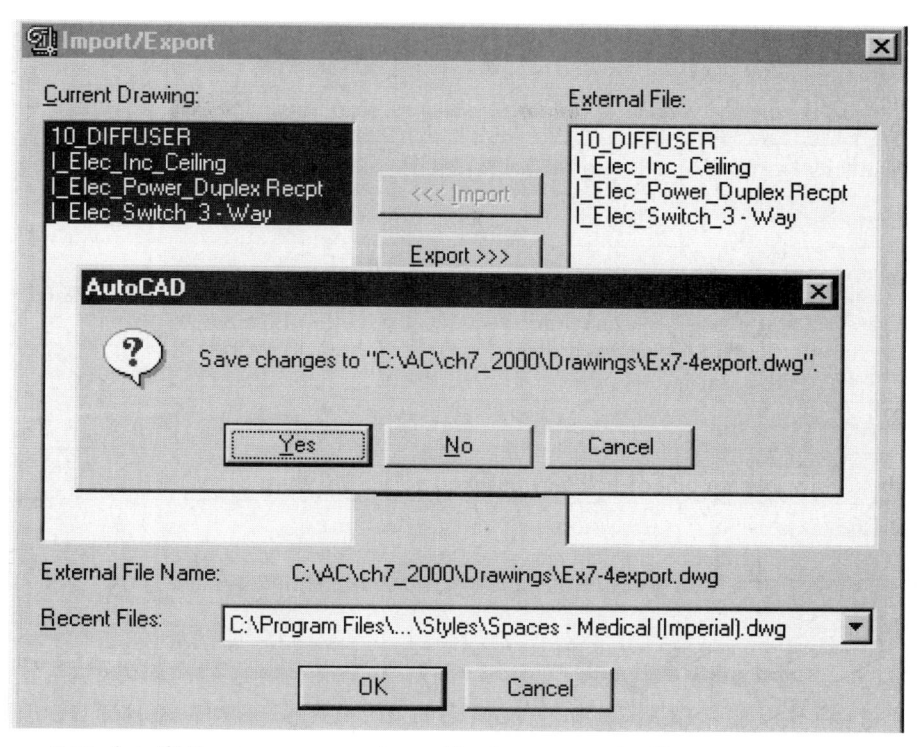

Figure 7.62 AutoCAD message requesting verification to save the file

Menu bar	Design>Design Content>Fixture Layout

Table 7.10 Fixture Layout command access

Table 7.11 lists the symbols and contents of the Fixture Layout container.

Fixture layout symbols are AutoCAD blocks and are inserted with the aid of the **Insert** dialog box. If you double-click on the fixture layout symbol, the **Insert** dialog box will open, as shown in Figure 7.63.

The **Insert** dialog box allows you to change the scale and rotation of the block prior to insertion. Selecting the **Explode** check box will explode the block upon insertion. The symbol, when exploded, can consist of blocks, multi-view block, walls, and other Architectural Desktop objects. Each component of the original symbol can then be manipulated to create a customized layout of fixtures.

Symbol	Description
1 Lavatory	1 lavatory in centered counter
2 Lavatories	2 lavatories in counter
3 Lavatories	3 lavatories in counter evenly spaced
4 Lavatories	4 lavatories in counter evenly spaced
5 Lavatories	5 lavatories in counter 3' spacing
Restroom (Men)	2 lavatories, 2 urinals, 1 toilet stall, 1 accessible toilet stall
Restroom (Women)	2 lavatories, 3 toilet stalls, 1 accessible toilet stall
Stall	Standard toilet stall
Stall End (Accessible)	Accessible stall end application
Stall (Accessible)	Accessible stall corner application
Stall (Alternate)	Standard stall
Urinal	1 urinal

Table 7.11 *Fixture layouts for restrooms*

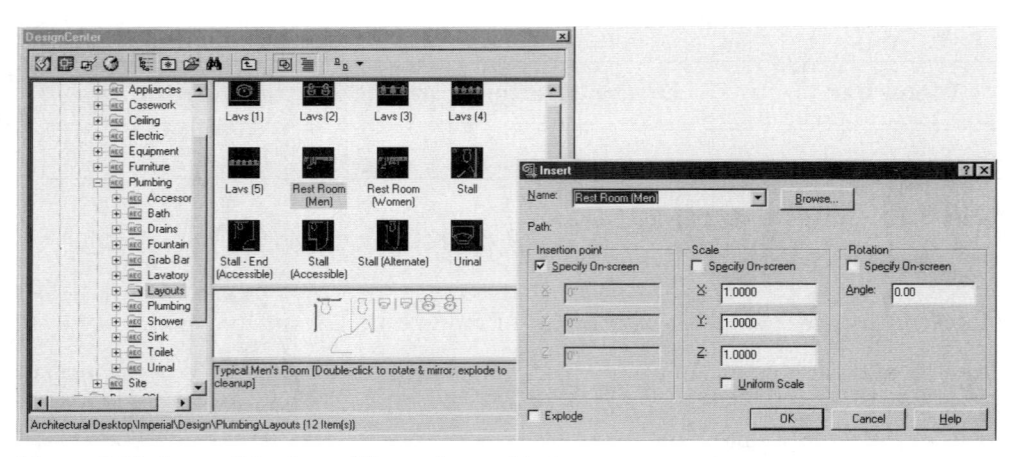

Figure 7.63 *Insert dialog box of Fixture Layout blocks*

CREATING MASKING BLOCKS

Some symbols in the DesignCenter of Architectural Desktop are masking blocks. These symbols can be used to mask or block the display of other Architectural

Desktop objects that are located within the masking block boundary. Masking blocks are often used in reflected ceiling plans or other plan views to block the display of entities. The majority of masking blocks are in the Ceiling and Electrical Fixtures folders of the DesignCenter. The masking blocks of the Ceiling Fixtures folder include Access Panel, Diffuser Return, and Diffuser Supply. These duct symbols are designed to fit in a suspended ceiling grid. All the symbols of the Electric Fixtures>Fluorescent folder are masking blocks. The symbols of the Fluorescent folder are designed for insertion on a ceiling grid. The Architectural Desktop ceiling grid includes layout nodes, which will cause the fluorescent fixture to snap in the grid. When you select a fluorescent fixture, you are prompted to `Select a Layout Node` for the insertion of the fixture. Ceiling grids and other Architectural Desktop grids will be discussed in Chapter 13. In Figure 7.64, the 2 x 4 fluorescent light in the upper right corner is masking the display of the ceiling grid.

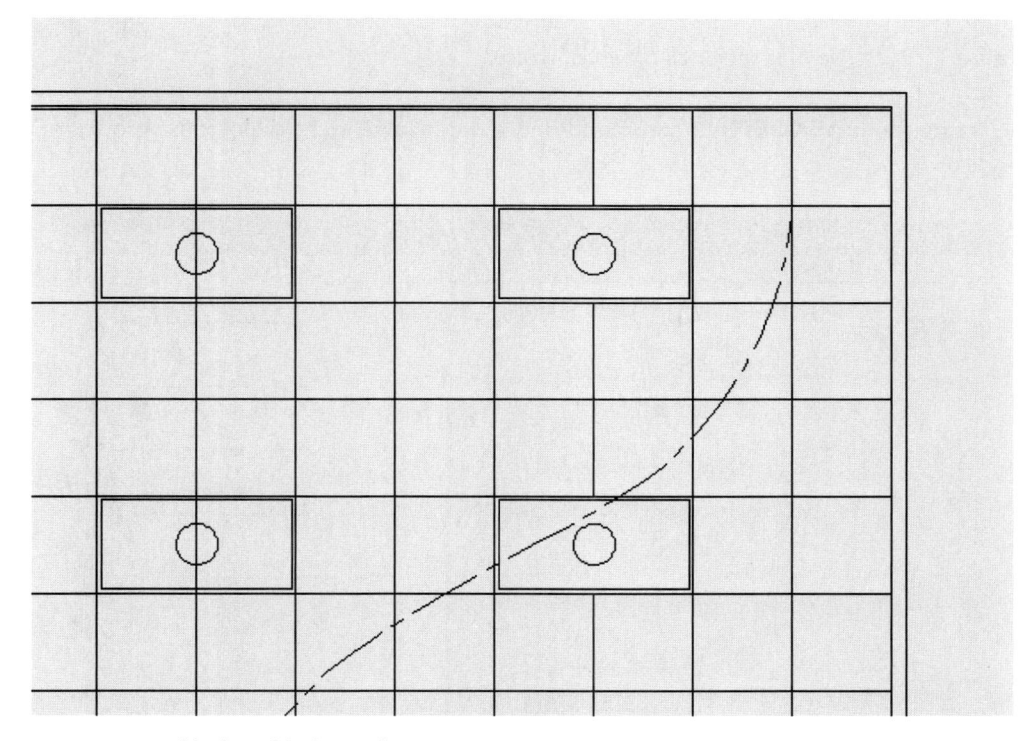

Figure 7.64 *Masking Blocks applications*

Masking blocks do not automatically block all entities that are within their boundaries. The masking block must be attached to the objects you want to block before they perform the mask function. The fluorescent fixtures on the left in Figure 7.64 are masking blocks, but the masking block has not been attached to the ceiling grid, and therefore the grid remains displayed. In the figure, the center line representing the raceway is crossing the fluorescent fixture, which is a masking block. This line is displayed because it is not an AEC object and cannot be attached to the masking block.

Several masking blocks are included in the Architectural Desktop DesignCenter; however, any closed polyline can be made into a masking block. Masking blocks are defined by the **Define Masking Block** command (**MASKDEFINE**). See Table 7.12 for **Define Masking Block** command access.

Menu bar	Desktop>AEC Masking>Define Masking Block
Command prompt	MASKDEFINE
Shortcut	Right-click over the drawing area; then choose Utilities>Masks>Table

Table 7.12 *Define Masking Block command access*

Selecting the **MASKDEFINE** command opens the **Mask Blocks** dialog box, shown in Figure 7.65.

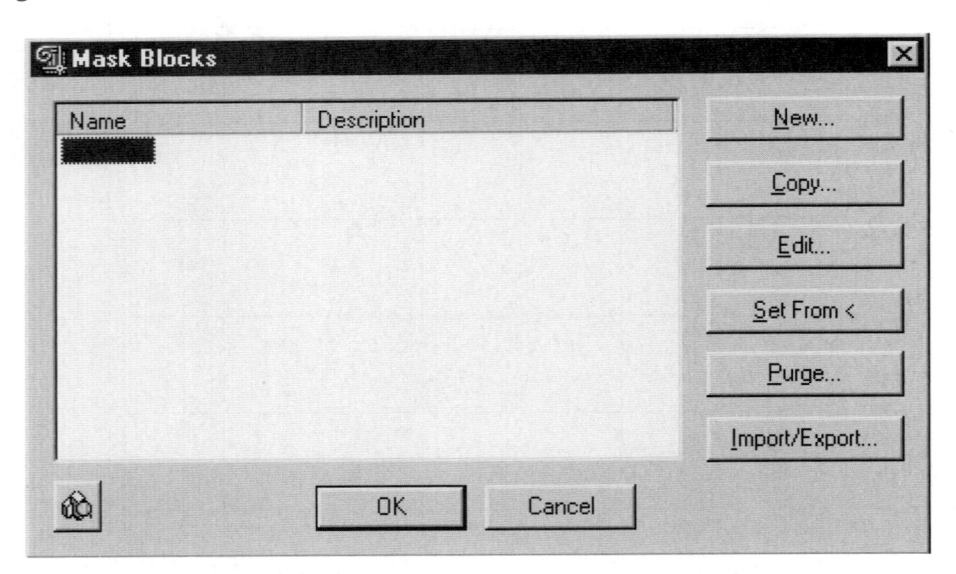

Figure 7.65 *Mask Blocks dialog box*

The **Mask Blocks** dialog box consists of a list of masking blocks used in the current drawing and six buttons on the right: **New, Copy, Set From<, Edit, Import/Export,** and **Purge**.

New – Open the **Name** dialog box and assign a new name for a masking block.

Copy – Copies an existing masking block to a new name. If you select a mask block from the list of mask blocks and then select the **Copy** button, the **Name** dialog box will open. The **Name** dialog box allows you to type a new name for the copied mask block.

Edit – Opens the **Mask Block Definition Properties** dialog, allowing you to change the description and display properties of the masking block.

Set From< – Selects the entities that define the mask block. If you select a mask block from the list of mask blocks and then select the **Set From<** button, you will be prompted to select a closed polyline. The select polyline will be defined as the mask block.

Purge – Deletes mask block names from the mask block list. Mask blocks cannot be purged if instances of the mask block remain in the drawing.

Import/Export – Transfers mask blocks to or from other drawings.

DEFINING MASK BLOCK PROPERTIES

Selecting the **Edit** button of the **Mask Blocks** dialog box opens the **Mask Block Definition Properties** dialog box, which allows you to define the specific properties of the mask block. You need to create the mask block with the **New** or **Copy** button prior to editing the mask block. Selecting the **Set From<** button allows you to select the geometry that is associated with the mask block. The tabs of the **Mask Block Definition Properties** dialog box are described below.

General

The **General** tab, shown in Figure 7.66, includes the name and description of the block, which can be edited.

The **Notes** button opens the **Notes** dialog box, which includes a **Text** tab (for adding notes regarding the symbol) and **Reference Docs** tab (for listing files that include related information).

Display Props

The **Display Props** tab, shown in Figure 7.67, includes the display representation drop-down list and a property source list, which defines how the masking block will be displayed.

The options of the **Display Props** tab are described below.

Display Representation – The **Display Representation** list displays the current display representation for the viewport. In Figure 7.67, the Reflected display representation is used for the mask block. The mask block can be displayed with the General or Reflected display representation.

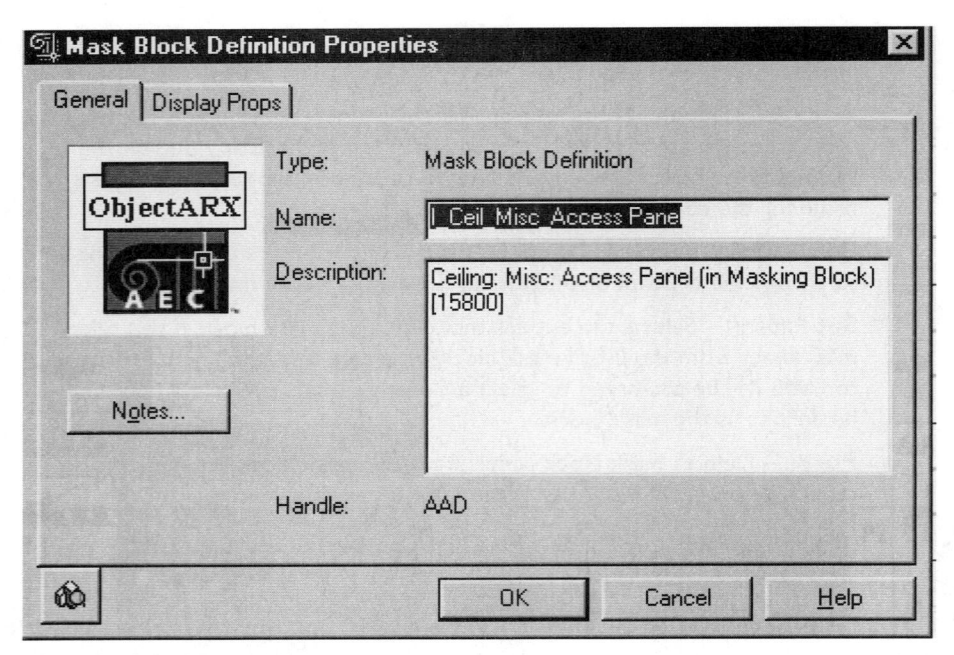

Figure 7.66 Mask Block Definition Properties dialog box

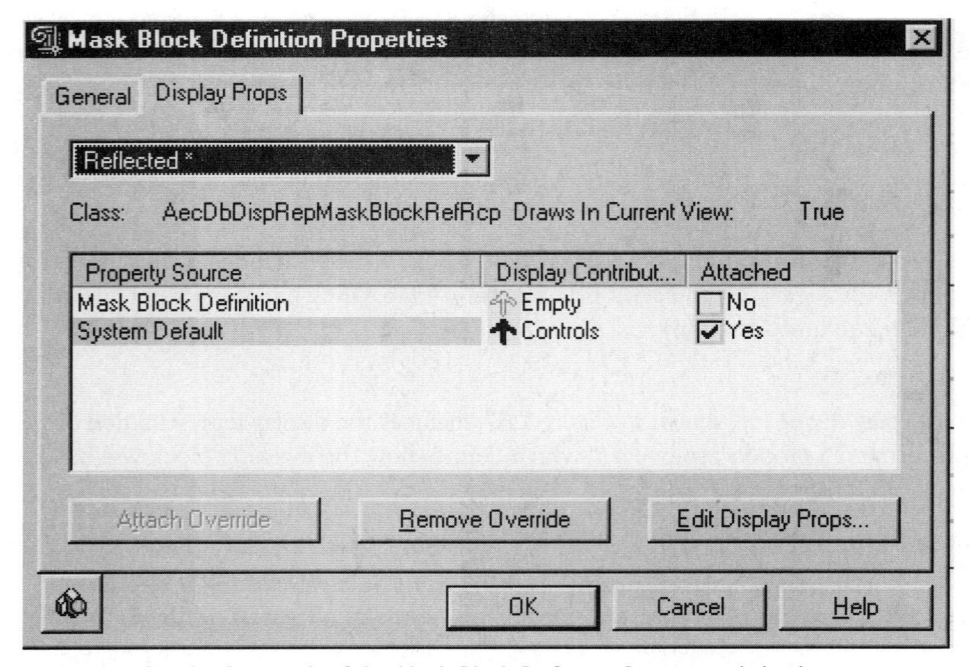

Figure 7.67 Display Props tab of the Mask Block Definition Properties dialog box

Property Source – The **Property Source** column includes the Mask Block Definition and the System Default options. You can control the display properties of the masking block specific to the masking block or globally apply them by editing the System Default. Define properties by selecting the Mask Block Definition or the System Default property and attaching an override to the display property definition.

Display Contribution – The **Display Contribution** column indicates the state of the property source. **Display Contribution** can be Empty, Controls, or Overridden.

Attached – The **Attached** column includes Yes and No options. The Yes option indicates an override or exception has been defined for the display of the object and the No option indicates that no overrides have been defined for the display of the object.

Attach Override – Defines new properties for the display of the mask block.

Remove Override – Removes the display properties that are attached as an override.

Edit Display Props – Defines display properties of the mask block. Selecting the **Edit Display Props** button opens the **Entity Properties** dialog box, as shown in Figure 7.68.

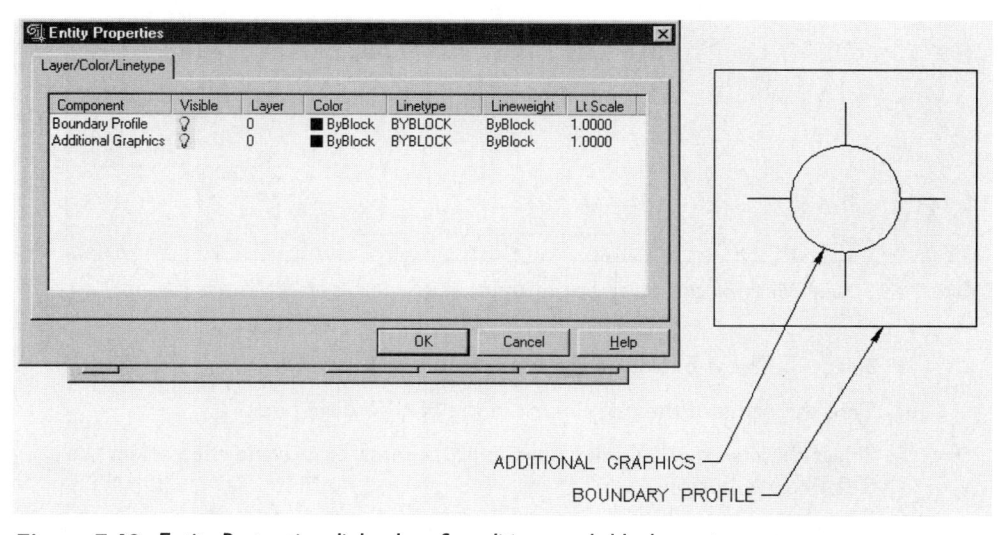

Figure 7.68 *Entity Properties dialog box for editing mask block components*

The **Entity Properties** dialog box consists of a **Layer/Color/Linetype** tab, which allows you to set the visibility, layer, color, linetype, lineweight, and linetype scale of the mask block components. The two components of a mask block are the Boundary Profile and Additional Graphics. The Boundary Profile is the geometry elements, consisting of closed polylines. These polylines can be attached to Architectural Desktop objects and block the display of part of the Architectural Desktop objects. The Additional Graphics components are other entities that are defined with the mask block but may not be polylines. Additional Graphics can include text, arcs, circles, or other AutoCAD entities. Therefore, you can turn off or on the visibility of these components.

STEPS TO CREATING A MASK BLOCK

1. Draw one or more closed polylines the size and shape of the desired symbol.
2. Choose **Desktop>AEC Masking>Define Masking Block** from the menu bar to open the **Mask Blocks** dialog box, as shown in Figure 7.69.

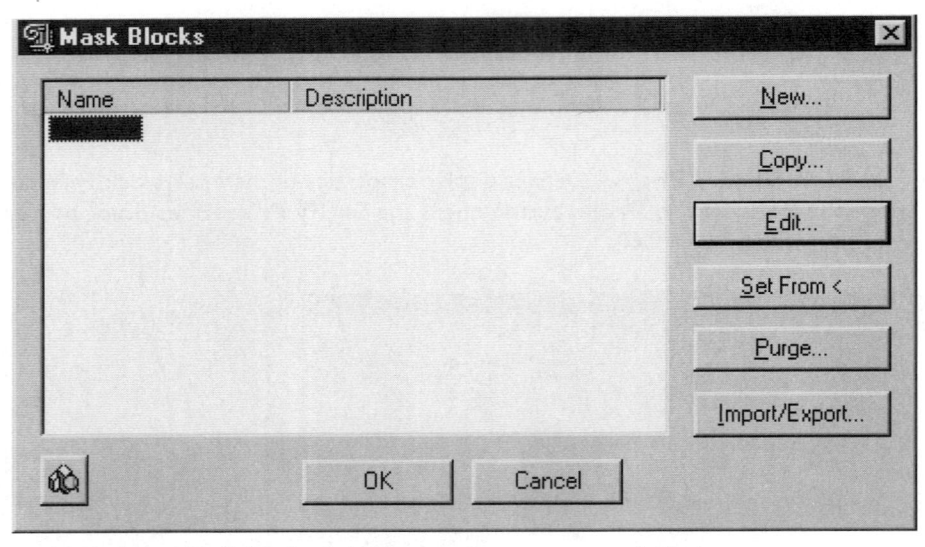

Figure 7.69 *Using Mask Blocks dialog box to create a new mask block*

3. Select the **New** button of the **Mask Blocks** dialog box.
4. Type the name of the mask block in the **Name** dialog box.
5. Select the new mask block from the list of mask blocks and then select the **Set From<** button.
6. Select a closed polyline as shown in Figure 7.70 and respond to the following command line prompts:

   Command: _AecMaskDefine
   Select a closed polyline: *(Select the closed polyline at P1 in Figure 7.70.)*

```
Add another ring? [Yes/No] <N>:
```
(Type Y to add other polyines to the selection or type N to terminate selection.)
```
Insertion base point:
```
(Select the location for the insertion base point.)
```
Select additional graphics:
```
(Select addition entities or press ENTER *to end selection.)*

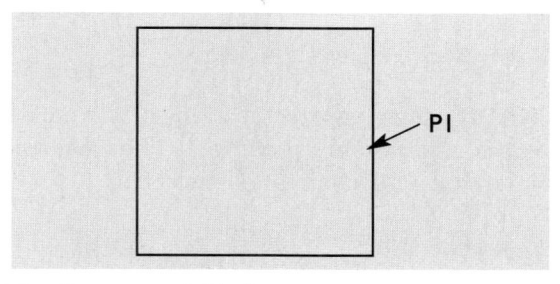

Figure 7.70 *Closed polyline for new mask block*

A mask block is created from the closed polyline.

8. Select the **Edit** button of the **Mask Block Definitions Properties** dialog box.
9. Type a description in the **Description** edit field.
10. Select the **Display Props** tab and define the display properties of the mask block.
11. Select **OK** to dismiss the **Mask Block Definition Properties** dialog box.
12. Select **OK** to dismiss the **Mask Blocks** dialog box.

In the command sequence above, you are prompted to select a closed polyline, an insertion point, and finally additional graphics to define the mask block. After selecting the first polyline you are prompted: `Add another ring?`. If you respond yes, additional polylines can be added to the mask block. The additional polylines can be defined as a void. Polylines defined as a void will be subtracted or form a hole and that boundary will not block the display of Architectural Desktop objects. The next prompt is to select an insertion point. The insertion point can be located on or off the polylines. The final prompt is to select additional graphics, which can include any AutoCAD entity. The additional graphics prompt allows you to include text as part of the mask block.

INSERTING MASKING BLOCKS

Insert masking blocks in the drawing from the AutoCAD DesignCenter or by selecting the **Add Masking Block** command (**MASKADD**). See Table 7.13 for **Add Masking Block** command access.

Menu bar	Desktop>AEC Masking>Add Masking Block
Command prompt	MASKADD
Shortcut	Right-click over the drawing area; then choose Utilities>Masks>Add.

Table 7.13 *Add Masking Block command access*

Selecting the **MASKADD** command when masking blocks are defined opens the **Add Mask Blocks** dialog box as shown in Figure 7.71. The most recent masking block is attached to the pointer when the command is selected.

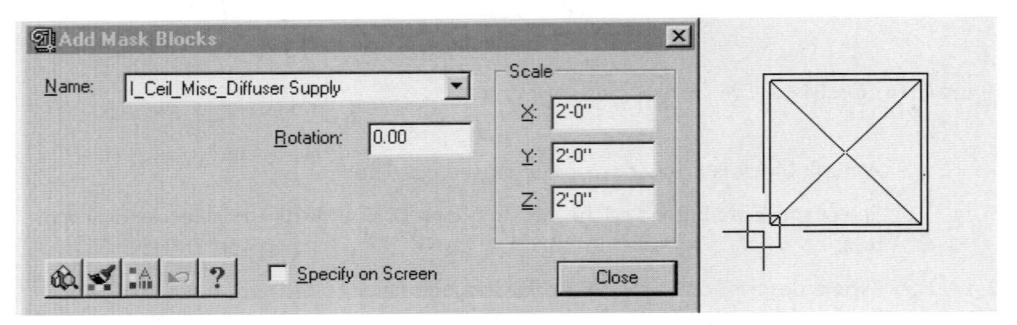

Figure 7.71 *Inserting mask blocks with the Add Mask Blocks dialog box*

If no masking blocks have been defined or inserted from the AutoCAD DesignCenter, the **Add Mask Blocks** dialog box will not open and an AutoCAD message box will open as shown in Figure 7.72.

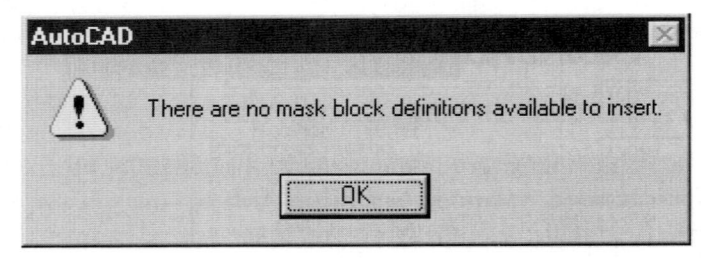

Figure 7.72 *AutoCAD message box indicating no mask blocks are defined in the drawing*

The **Add Mask Blocks** dialog box includes the edit fields for setting the scale and rotation of the masking block. The options are described below.

Name – The **Name** list includes all the masking blocks of the drawing.

Rotation – The **Rotation** edit field allows you to edit the masking block prior to insertion.

Scale – The **Scale** X, Y, Z edit fields allow you to increase or decrease the scale of the masking block.

Specify on Screen – The **Specify on Screen** check box allows you to specify the scale in the drawing area. If this check box is selected, the first selection of the mouse specifies the location of the insertion point and the next selection of the mouse specifies the diagonal corner to define the scale to fit in the masking block.

Floating Viewer – Selecting the **Floating Viewer** button will open the **Viewer** window and allow you to view the masking block prior to insertion.

Match – Selecting the **Match** button will allow you to select another masking block and copy its properties. The name, X scale, Y scale, Z scale, and rotation can be matched from other mask blocks.

Properties – Selecting the **Properties** button opens the **Mask Block Properties** dialog box, which includes **General**, **Style**, and **Dimension** tabs. The options of the **Mask Block Properties** dialog box are described in next section, Changing Mask Blocks.

Undo – Selecting the **Undo** button allows you to undo the last insertion operation of the mask block.

?–The **?** button opens Help to the Add a Masking Block help topic.

CHANGING MASKING BLOCKS

You can change masking blocks after they are inserted by selecting the **Modify Masking Block** command (**MASKMODIFY**). See Table 7.14 for **Modify Masking Block** command access.

Menu bar	Desktop>AEC Masking>Modify Masking Block
Command prompt	MASKMODIFY
Shortcut	Right-click over the drawing area; then choose Utilities>Masks>Modify.

Table 7.14 *Modify Masking Block command access*

Selecting the **MASKMODIFY** command opens the **Modify Mask Blocks** dialog box as shown in Figure 7.73.

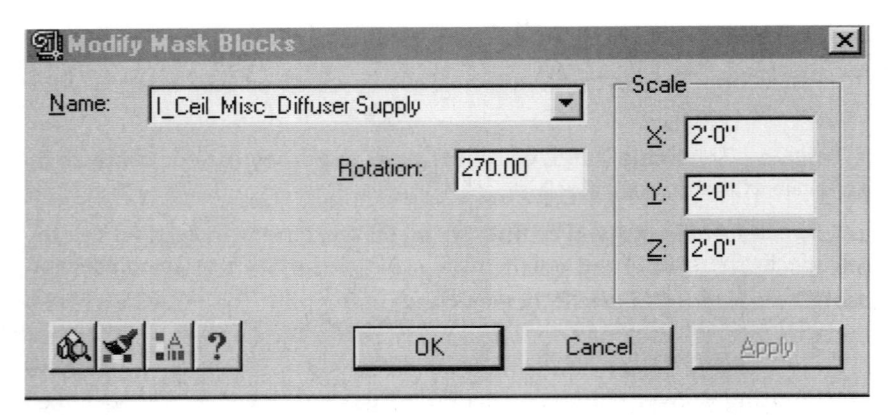

Figure 7.73 Modify Mask Blocks dialog box

The **Modify Mask Blocks** dialog box includes the same options as the **Add Mask Blocks** dialog box, except there is no **Undo** button or **Specify on Screen** check box. However, selecting the **Properties** button opens the **Mask Block Properties** dialog box, shown in Figure 7.74, which includes **General**, **Style**, **Dimensions**, and **Location** tabs.

Figure 7.74 Mask Block Properties dialog box

A description of each tab of the **Mask Block Properties** dialog box follows.

General

The **General** tab allows you to add or edit the description of the selected masking block. The **Notes** button allows you to add text notes or notes regarding files to the **Notes** dialog box.

Style

The **Style** tab, shown in Figure 7.75, includes a list of mask block styles defined in the drawing. If more than one masking block has been defined for the drawing, you can select a different masking block to change the block.

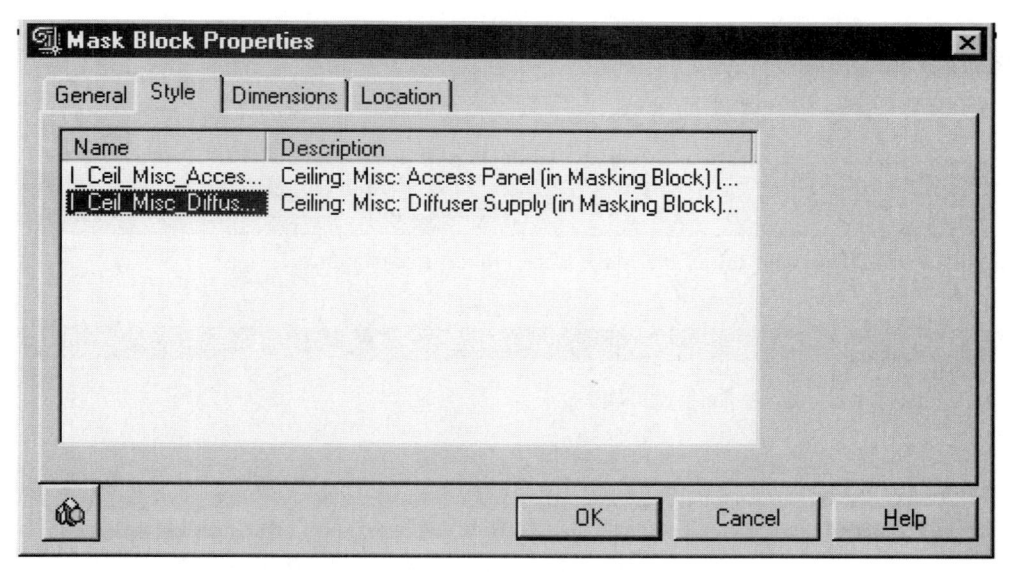

Figure 7.75 *The Style tab of the Mask Block Properties dialog box*

Dimensions

The **Dimensions** tab includes edit fields for changing the X, Y, and Z scales, as shown in Figure 7.76.

Location

The **Location** tab allows you to identify the insertion point location or edit the location of an existing masking block. The location tab shown in Figure 7.77 is for a mask block inserted in a suspended ceiling, and therefore the Z value is 8'–0".

The options of the **Location** tab are as follows:

> **World Coordinate System/Current Coordinate System** – The radio buttons indicate the location of the masking block relative to a coordinate system.

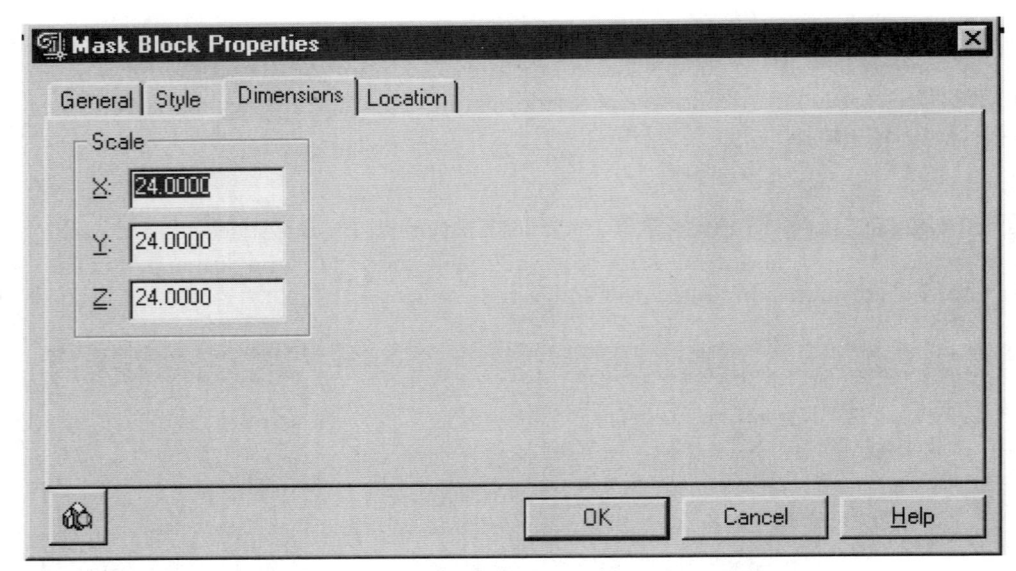

Figure 7.76 *Dimensions tab of the Mask Block Properties dialog box*

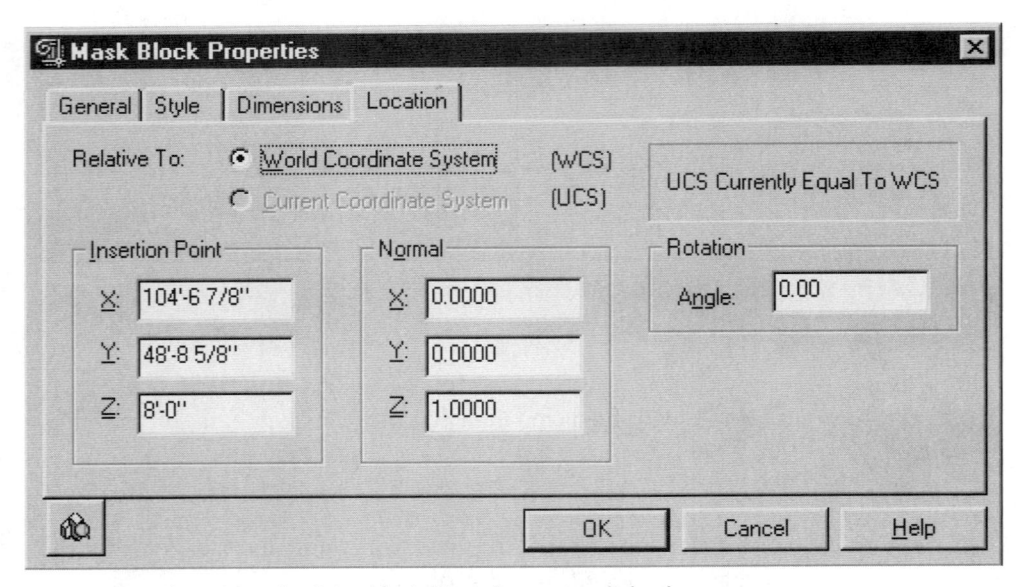

Figure 7.77 *Location tab of the Mask Block Properties dialog box*

Insertion Point – The **Insertion Point** section includes X, Y, and Z edit fields, which specify the coordinates of the insertion point of the masking block.

Normal – The **Normal** edit fields X, Y, and Z indicate the direction a three-dimensional block is projected. A 1 in the Z edit field indicates the symbol is developed perpendicular to the XY plane.

Rotation – The **Rotation** section indicates the angular rotation of the masking block.

ATTACHING OBJECTS TO MASKING BLOCKS

If the masking blocks have been inserted in the drawing, they can perform their masking function by being attached to selected Architectural Desktop objects. *Masking blocks will only block the display of selected Architectural Desktop objects when attached to the objects.* Attached objects can also be detached to return the visibility of the object. The **Attach Objects to Mask** command (**MASKATTACH**) is used to associate the masking block with the selected Architectural Desktop object. See Table 7.15 for **Attach Objects to Mask** command access.

Menu bar	Desktop>AEC Masking>Attach Objects to Mask
Command prompt	MASKATTACH
Shortcut	Right-click over the drawing area; choose select Utilities>Masks>Attach
Shortcut	Select the masking block; then right-click and choose Attach

Table 7.15 *Attach Objects to Mask command access*

If you select the **MASKATTACH** command, you will be prompted to select a masking block and then to select the objects to apply the mask. The command sequence for attaching the masking block to the wall using the **MASKATTACH** command is shown below.

```
Command: _AecMaskAttach
Select mask blocks to be attached: 1 found (Select the
```
masking block at P1 in Figure 7.78.)
```
Select mask blocks to be attached: ENTER
```
(Press ENTER to end selection.)
```
Select AEC entity to be masked:
```
(Select the wall at P2 to attach the wall to the masking block.)
```
[1] Masks attached.
```
(The Select Views dialog box opens listing the current view as shown in Figure 7.79.)

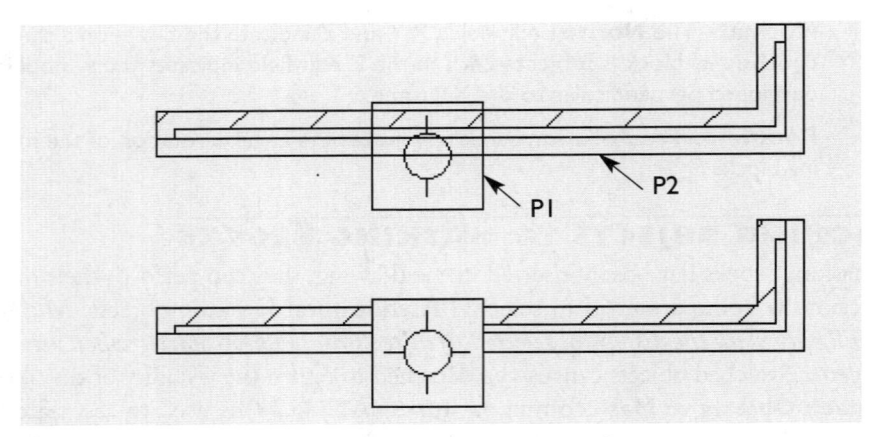

Figure 7.78 Mask Block and a wall to attach the masking block to

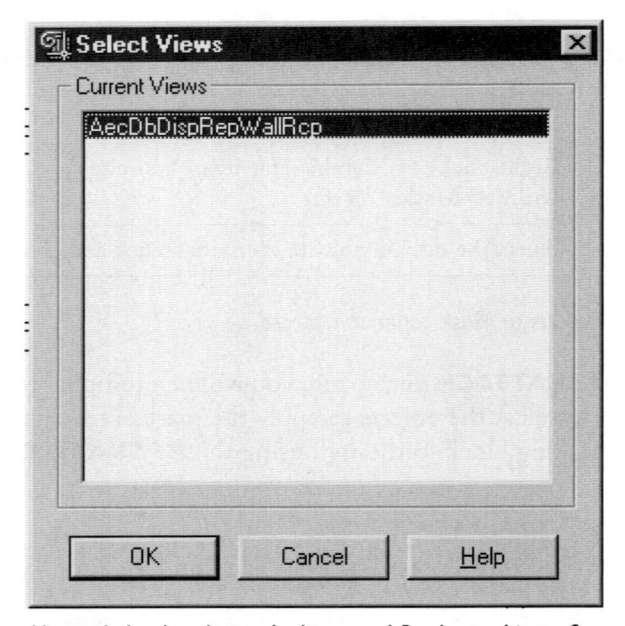

Figure 7.79 Select Views dialog box listing Architectural Desktop objects for attachment

The result of attaching the wall to the masking block is shown in the lower half of Figure 7.78. The Architectural Desktop objects that are attached to a masking block can be identified by the **LIST** command.

Masking blocks can be detached from the Architectural Desktop object by the **Detach Objects from Mask** command (**MASKDETACH**). See Table 7.16 for **Detach Objects from Mask** command access.

Menu bar	Desktop>AEC Masking>Detach Objects from Mask
Command prompt	MASKDETACH
Shortcut	Right-click over the drawing area; then choose Utilities>Masks>Detach
Shortcut	Select the masking block; then right-click and choose Detach

Table 7.16 *Detach Objects from Mask command access*

The command sequence for detaching a masking block from a wall using the MASKDETACH command is shown below.

```
Command: maskdetach
Select a mask block: (Select the masking block at P1 in Figure
7.80.)
Select AEC entity to be detached: (Select the wall at P2 in
Figure 7.80.)
1 mask(s) detached.
```

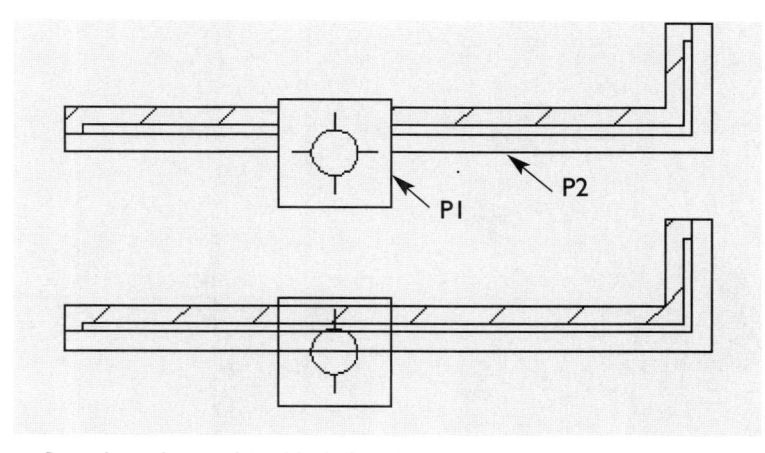

Figure 7.80 *Detaching the masking block from the the wall*

Detaching the block returns the visibility of the wall through the symbol, as shown in the lower portion of Figure 7.80.

CREATING SYMBOLS FOR THE DESIGNCENTER

New blocks, multi-view blocks, and masking blocks can be created in a drawing file; however, these symbols will not be included in the AutoCAD Architectural Desktop Content of the DesignCenter. The **Create AEC Content Wizard** can be used for adding symbols, drawings, or custom commands to the DesignCenter. The **Create AEC Content Wizard** includes options to add each block of the current drawing to the contents of the AutoCAD Architectural Desktop Content. Options of the **Create AEC Content Wizard** allow you to predefine the scale, rotation, attribute text angle, and layer. Placing the symbol on a predefined layer improves a drawing's compliance with office layer standards. See Table 7.17 for **Create AEC Content** command access.

Menu bar	Design>Design Content>Create AEC Content
Command prompt	CREATECONTENT

Table 7.17 *Create AEC Content command access*

Selecting the **CREATECONTENT** command opens the **Create AEC Content Wizard**, as shown in Figure 7.81.

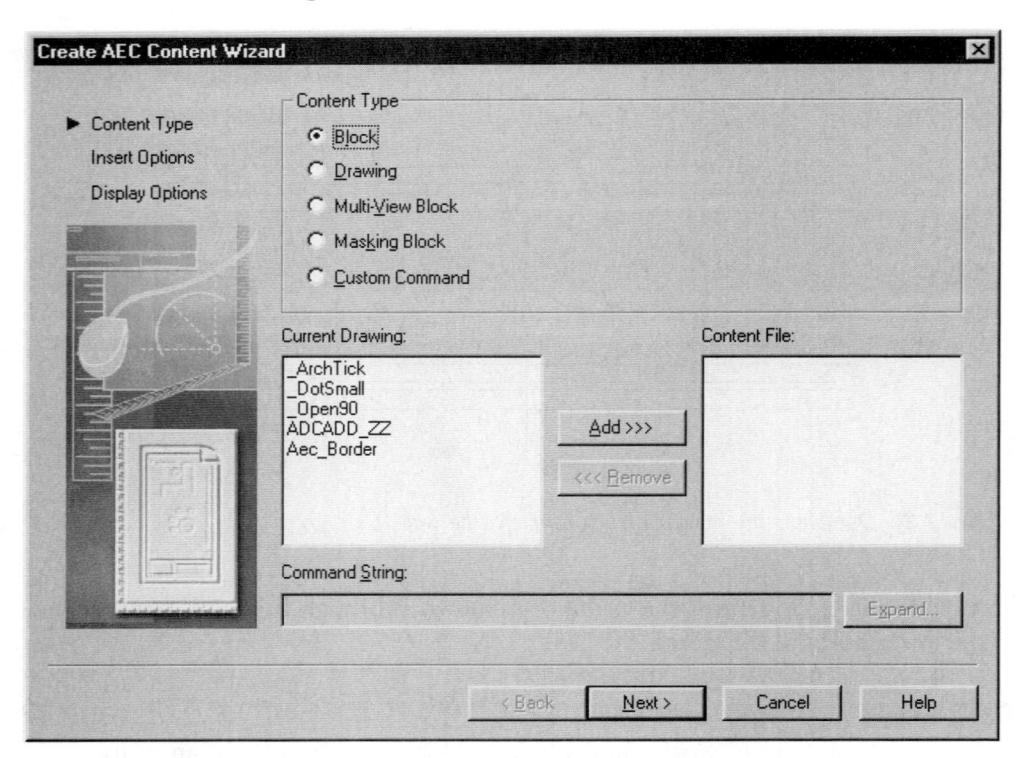

Figure 7.81 *Content Type page of the Create AEC Content Wizard*

The **Create AEC Content Wizard** consists of three pages: **Content Type, Insert Options**, and **Display Options**. You can display each page by selecting the **Next** or **Back** button at the bottom of each page. Each page is identified in the upper left corner.

CONTENT TYPE

The Content Type page consists of the following options:

> **Content Type** – The **Content Type** section allows you to specify the type of content you want to add to the DesignCenter: **Block, Drawing, Multi-View Block, Masking Block**, and **Custom Command**. When you select a content type, all named objects of that type will be displayed in the **Current Drawing** list box. In Figure 7.81, the **Block** type is selected; therefore the blocks of the drawing are displayed in the **Current Drawing** list.
>
> **Current Drawing** – The **Current Drawing** list includes objects in the drawing classified according to the content type specified in the Content Type section.
>
> **Content File** – The **Content File** list includes objects of the specified content type that have been exported to the AutoCAD Architectural Desktop Content.
>
> **Add/Remove** – The **Add** and **Remove** buttons allow you to add or delete the content from the **Current Drawing** and **Content File** lists. Content must be added to the **Current Drawing** list to complete the **Create AEC Content Wizard** operations.
>
> **Command String** – The **Command String** edit field allows you to enter a command string and command responses to execute a command in AutoCAD. Selecting the command string icon created for the DesignCenter will execute the command. The **Expand** button allows you to enter up to 255 characters in the command string.

INSERT OPTIONS

The **Insert Options** page shown in Figure 7.82 allows you to predefine the scale, elevation, layer, and attribute properties of the content.

> **Explode on Insert** – The **Explode on Insert** check box, if selected, will apply the **EXPLODE** command to the block upon insertion.
>
> **Preset Elevation** – The value entered in the **Preset Elevation** field presets the Z coordinate value of the content.
>
> **Anchor Type** – The **Anchor Type** drop-down list allows you to attach the object to one of the following: cell, curve, leader, node, space boundary, tag, volume or wall. Specific information regarding the use of anchors is presented in Chapter 13.

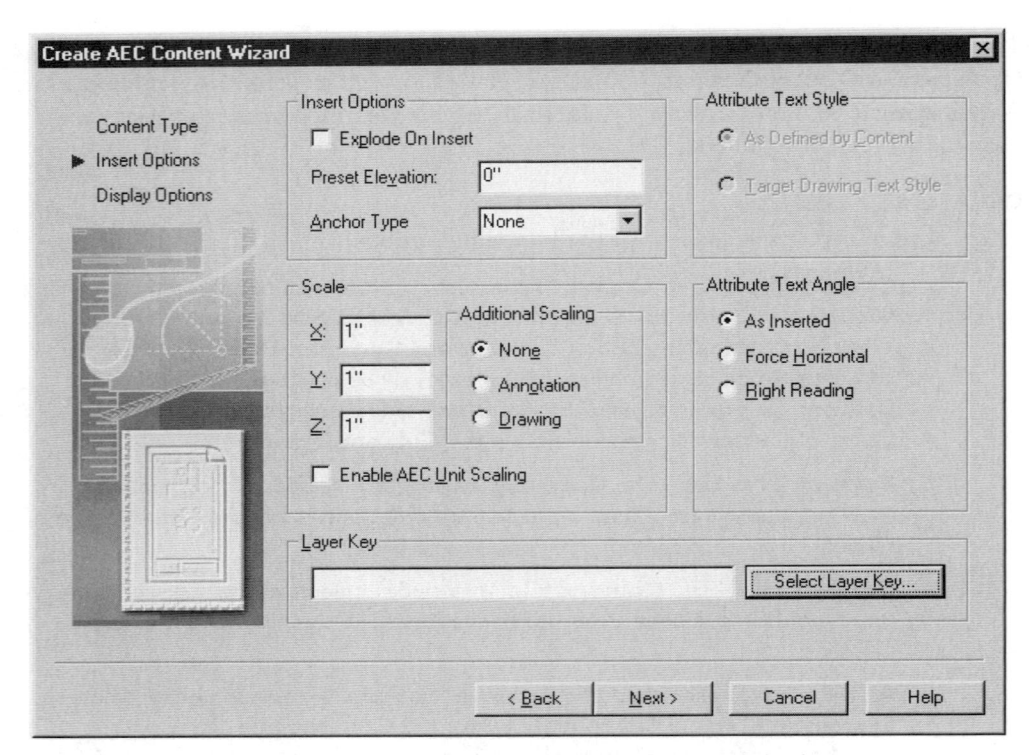

Figure 7.82 *Insert Options page of the Create AEC Content Wizard dialog box*

Scale

> **X, Y, Z** – The X, Y, and Z **Scale** edit fields allow you to preset the scale value in the X, Y, Z directions of the content upon insertion.

Additional Scaling

> **None** – The **None** option scales a multi-view block at the scale defined in the X, Y, Z edit fields. Select the **None** option if the block was drawn actual size and the units of the multi-view block match the units of target drawings.
>
> **Annotation** – The **Annotation** option applies the **Annotation Plot Size** factor to multi-view blocks. If a multi-view block was created 1 unit long, when the annotation plot size is set at 1/8" and the scale factor set at 48, the multi-view block would be inserted 6" long. The actual multi-view block dimension is multiplied by the annotation factor and the drawing scale factor.
>
> **Drawing** – The **Drawing** scale factor adjusts the scale of the multi-view block by multiplying the drawing scale factor of the drawing by the multi-view block size. Table 7.18 indicates the scaling effects of each option given the annotation plot size and the drawing scale factor.

Additional Scaling	Multi-View Block Size	Annotation Plot Size	Drawing Scale Factor	Size when inserted
None	1	1/8"	48	1
Annotation	1	1/8"	48	6 (1x1/8x48)
Drawing	1	1/8"	48	48 (1x48)

Table 7.18 *Effects of additional scaling options*

> **Enable AEC Unit Scaling** – The **Enable AEC Unit Scaling** check box applies scaling factors to blocks.
>
> **Attribute Text Style** – The **Attribute Text Style** toggles allow you to select the text style for the attribute text to be either the text style as defined in the block or the default text style of the target drawing.
>
> **Attribute Text Angle** – The **Attribute Text Angle** toggles allow you to choose the text orientation of attribute text. The attribute text angle options are **As Inserted**, **Force Horizontal** or **Right Reading**.
>
> **Select Layer Key** – The **Select Layer Key** button allows you to specify the layer for the block to be inserted on. Selecting the **Select Layer Key** will open the **Select Layer Key** dialog box shown in Figure 7.83. It includes a list of the layer keys, names, and descriptions for the layer standard used in the drawing. A layer key is a name assigned to the layer in Architectural Desktop. The name of the layer displayed in the **Object Properties** dialog box will be the name displayed in the **Layer Name** column.

DISPLAY OPTIONS

The **Display Options** page, shown in Figure 7.84, includes options to define the location of the symbol within the DesignCenter and the icon used to represent the symbol.

> **File Name** – The **File Name** edit field allows you to enter the path and file name for the symbol. Selecting the **Browse** button will open the **Save Content File** dialog box, which allows you to easily specify the name and path for the symbol to be placed. The **File Name** is required and must be specified to create the AEC Content.
>
> **Icon** – The icon for the symbol can be specified as a new icon or default icon. The default icon is the preview graphics for the file of the symbol. The **New Icon** button allows you to select a bitmap file to be used as the symbol.

Figure 7.83 Layer Key list in the Select Layer Key dialog box

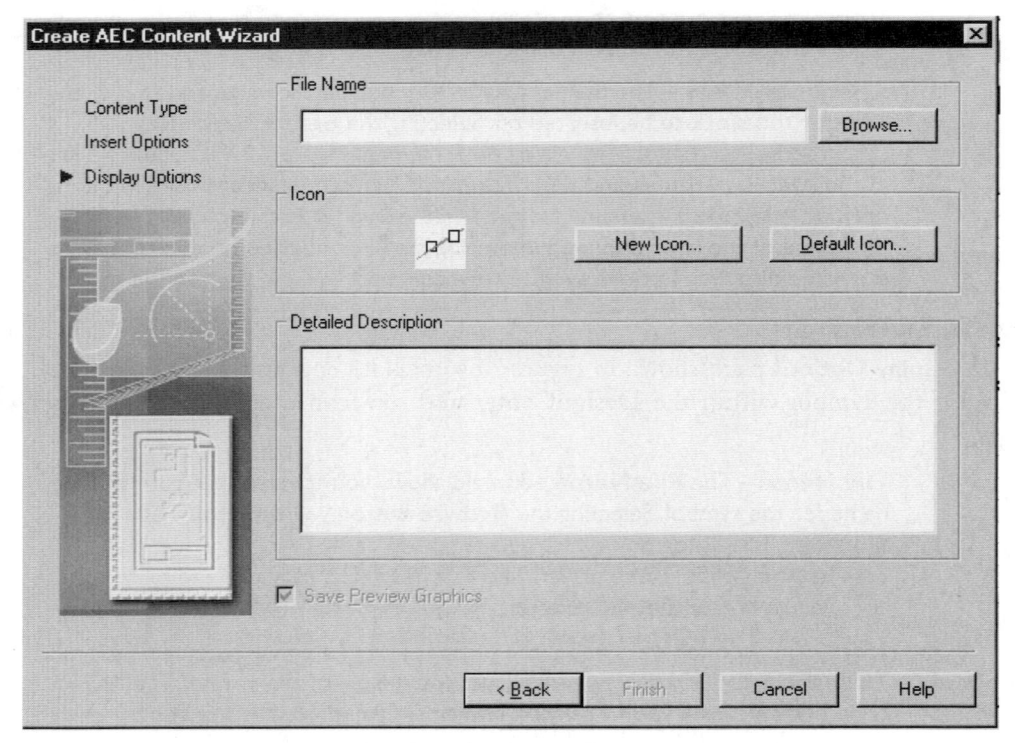

Figure 7.84 Display Options page of the Create AEC Content Wizard

Detailed Description – **Detailed Description** is an edit field for entering text that describes the block. The text entered will be displayed in the description field of the DesignCenter when the symbol is selected.

Save Preview Graphics – The **Save Preview Graphics** check box saves the last view of the symbol drawing as the preview of the symbol.

TUTORIAL 7.4 CREATING AND MODIFYING MASK BLOCKS

1. Open *ADT Tutor\Ch7\EX7-4.dwg*.
2. Choose **File>SaveAs** from the menu bar and save the drawing as *Speaker4.dwg* in your student directory.
3. Choose **Documentation>Set Drawing Scale** from the menu bar.
4. Select the 1/8"=1'-0" drawing scale from the **Drawing Scale** list box.
5. Verify that the **Annotation Plot Size** is 1/8".
6. Select **OK** to dismiss the **Drawing Setup** dialog box.
7. Right-click over the drawing area and then choose **Utilities>Masks>Table** from the shortcut menu.
8. Select the **New** button from the **Mask Blocks** dialog box.
9. Type **Speaker** in the **Name** dialog box.
10. Select **OK** in the **Name** dialog box.
11. Select the **Edit** button of the **Mask Blocks** dialog box.
12. Type **2 x 4 Speaker** in the **Description** edit field of the **General** tab, as shown in Figure 7.85.
13. Select **OK** to dismiss the **Mask Block Definition Properties** dialog box.
14. Select the **Set From<** button in the **Mask Blocks** dialog box and select the entities specified in the following command sequence and in Figure 7.86.

    ```
    Command: MaskDefine
    Select a closed polyline: (Select the rectangle at P1 in Figure 7.86.)
    Add another ring? [Yes/No] <N>: ENTER
    Insertion base point: int of (Select the lower left corner of the rectangle at P2 in Figure 7.86.)
    Select additional graphics: (Select the hexagon at P3 in Figure 7.86.) 1 found
    Select additional graphics: (Select the text at P4 in Figure 7.86.) 1 found, 2 total
    Select additional graphics: ENTER
    Command:
    ```

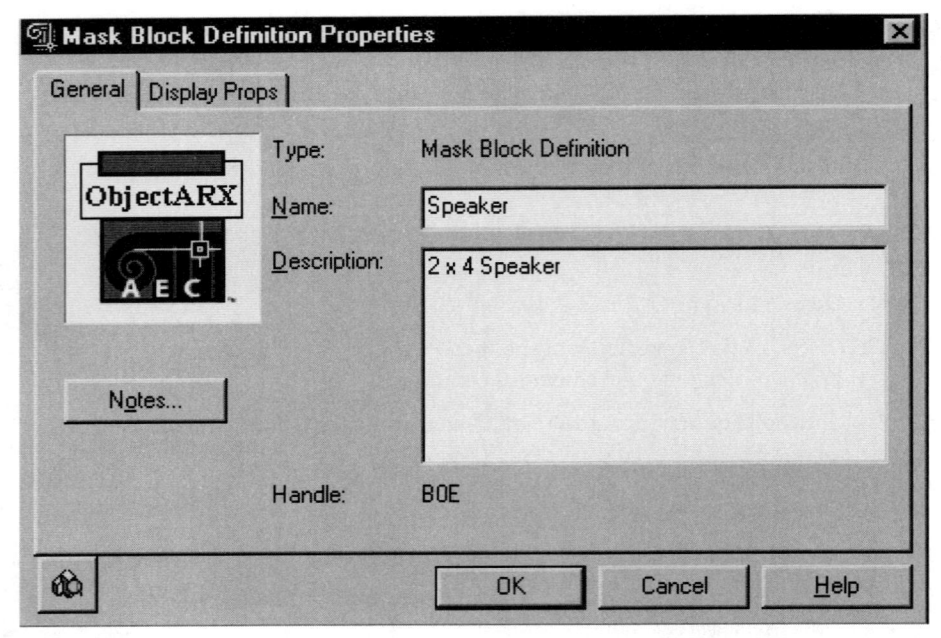

Figure 7.85 *Entering a description in the General tab of the Mask Block Definition Properties dialog box*

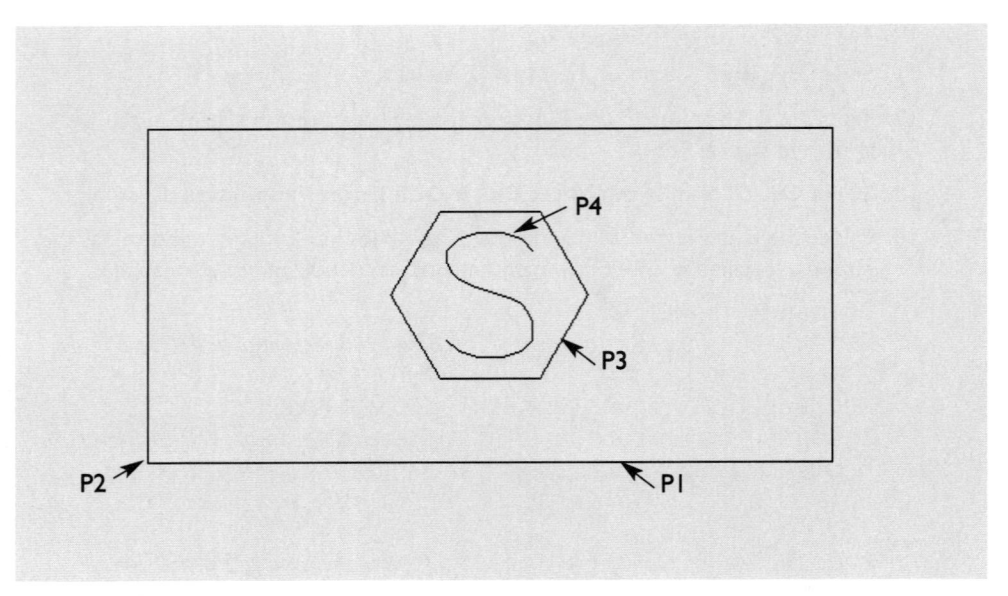

Figure 7.86 *Selecting entities for the mask block*

15. Select **OK** to dismiss the **Mask Blocks** dialog box.
16. Choose **Design>Design Content>Create AEC Content** from the menu bar.
17. Select the **Masking Block** radio button for **Content Type**; then select Speaker in the **Current Drawing** list.
18. Select the **Add** button to list the Speaker mask block in the **Content File** list box as shown in Figure 7.87.

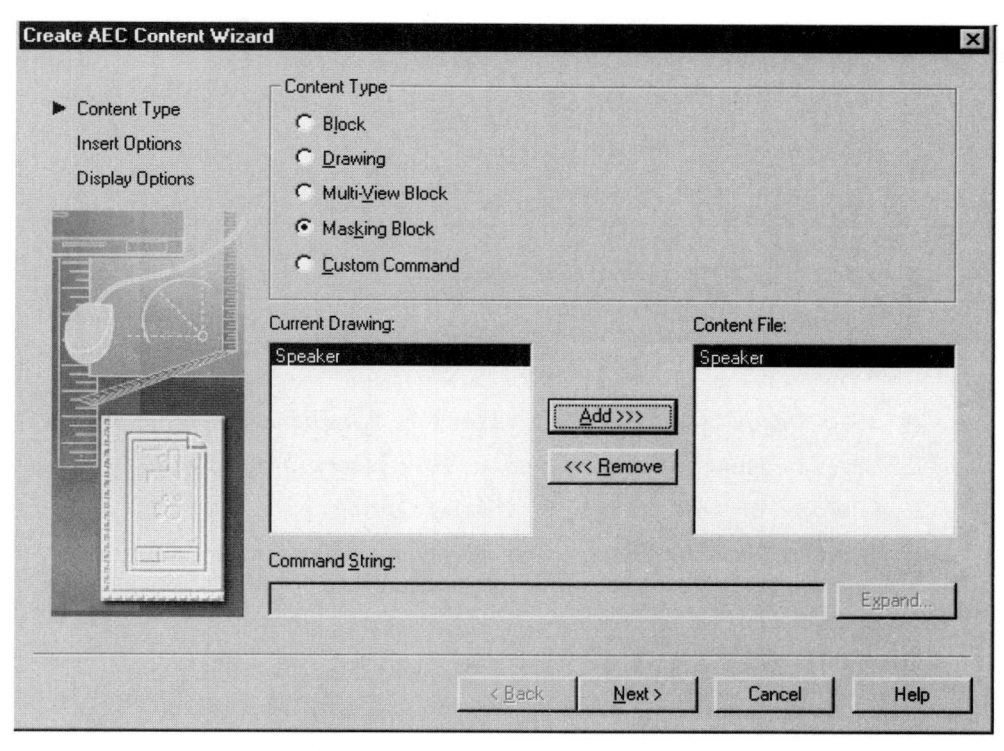

Figure 7.87 *Selecting the Speaker masking block for the content file*

19. Select the **Next** button at the bottom of the **Create AEC Content Wizard**.
20. Type **8'** in the **Preset Elevation** edit field.
21. Select **Enable AEC Unit Scaling**.
22. Select the **None** radio button in the **Additional Scaling** section.
23. Select the **Select Layer Key** button and select the CEILOBJ layer key name as shown in Figure 7.88.

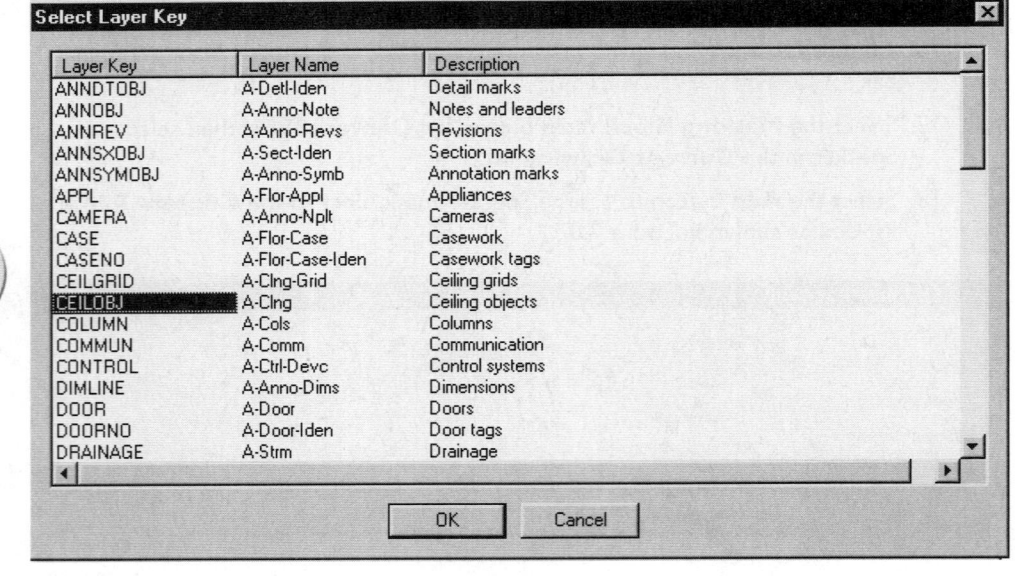

Layer Key	Layer Name	Description
ANNDTOBJ	A-Detl-Iden	Detail marks
ANNOBJ	A-Anno-Note	Notes and leaders
ANNREV	A-Anno-Revs	Revisions
ANNSXOBJ	A-Sect-Iden	Section marks
ANNSYMOBJ	A-Anno-Symb	Annotation marks
APPL	A-Flor-Appl	Appliances
CAMERA	A-Anno-Nplt	Cameras
CASE	A-Flor-Case	Casework
CASENO	A-Flor-Case-Iden	Casework tags
CEILGRID	A-Clng-Grid	Ceiling grids
CEILOBJ	A-Clng	Ceiling objects
COLUMN	A-Cols	Columns
COMMUN	A-Comm	Communication
CONTROL	A-Ctrl-Devc	Control systems
DIMLINE	A-Anno-Dims	Dimensions
DOOR	A-Door	Doors
DOORNO	A-Door-Iden	Door tags
DRAINAGE	A-Strm	Drainage

Figure 7.88 *CEILOBJ Layer Key selected for the masking block*

24. Select **OK** to dismiss the **Select Layer Key** dialog box.
25. Select the **Next** button at the bottom of the **Insert Options** page.
26. Select the **Browse** button of the **Display Options** page.
27. Select the \AutoCAD Architectural 2\Content\Imperial\Design\Ceiling directory and type **Speaker** in the **File name** edit field, as shown in Figure 7.89.
28. Select the **Save** button in the **Save Content File** dialog box.
29. Type **2 x 4 Ceiling Speaker** in the **Detailed Description** edit field.
30. Select the **Finish** button to dismiss the **Create AEC Content Wizard**.
31. Save the drawing.
32. Open *ADT Tutor\Ch7\EX7-5.dwg* and save the drawing as *AECcontent5.dwg* in your student directory.
33. Choose **Design>Design Content>Ceiling Fixtures**.
34. Double-click on the new symbol named Speaker to open the **Add Mask Blocks** dialog box as shown in Figure 7.90.
35. Select locations at points P1 and P2 for the insertion point of the speaker as shown in Figure 7.91.
36. Select each speaker, and then right-click and choose **Attach** from the shortcut menu.

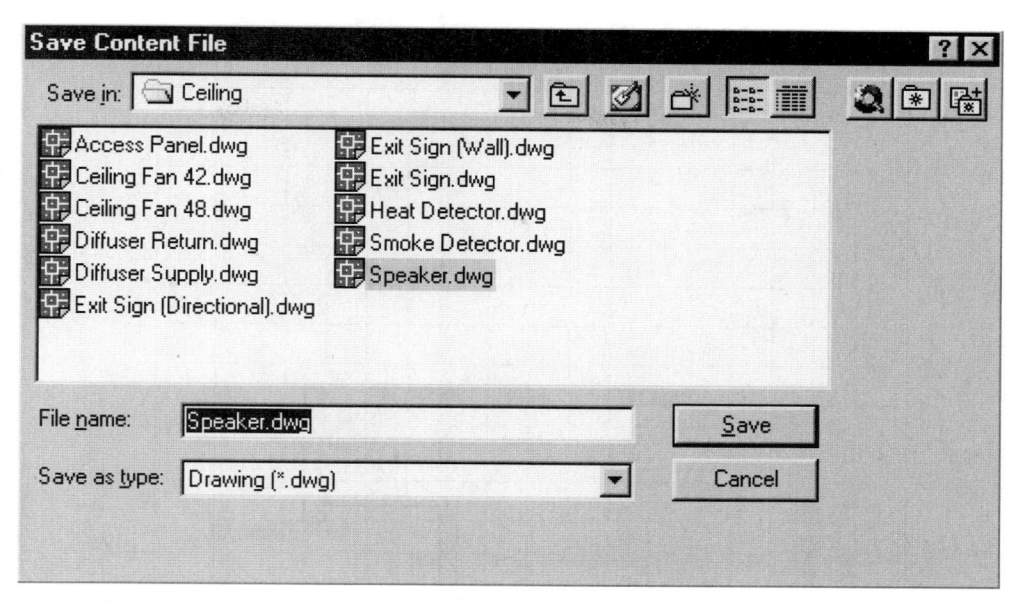

Figure 7.89 *Selecting a file for the Speaker masking block*

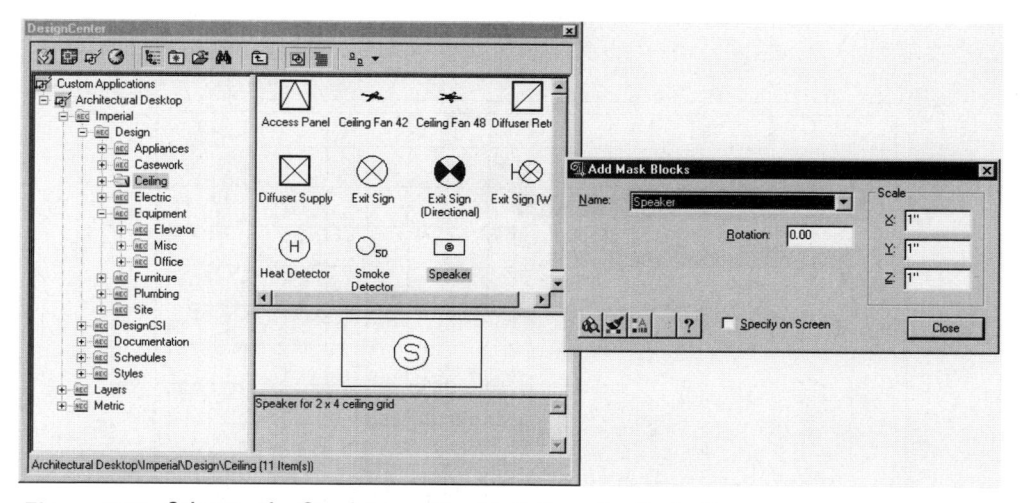

Figure 7.90 *Selecting the Speaker masking block from the DesignCenter*

37. Select the ceiling grid to display the **Select Views** dialog box, as shown in Figure 7.92.
38. Select **OK** to accept the ceiling grid object. The mask blocks block the display of the ceiling grid, as shown in Figure 7.93.

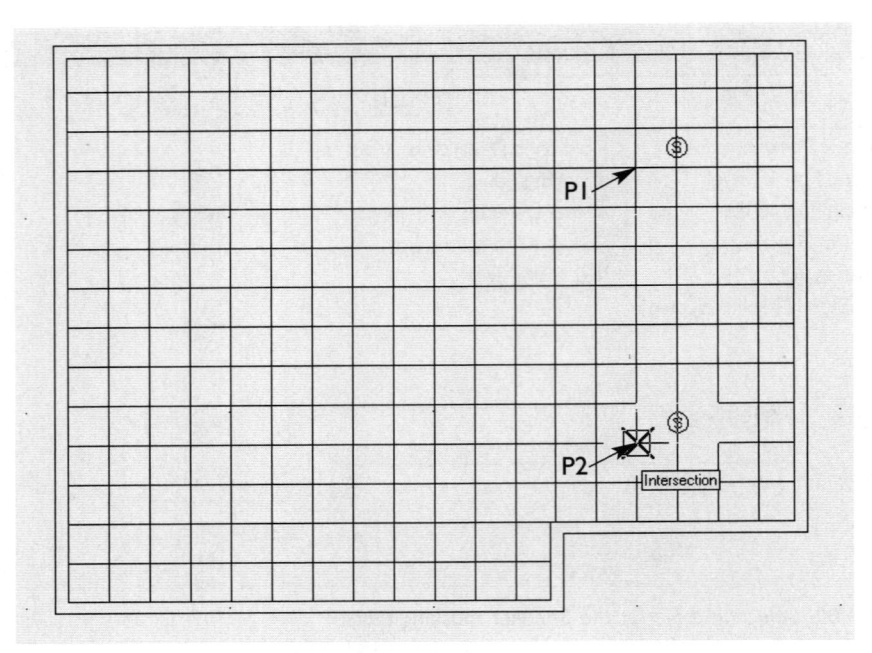

Figure 7.91 Locations for inserting the mask block

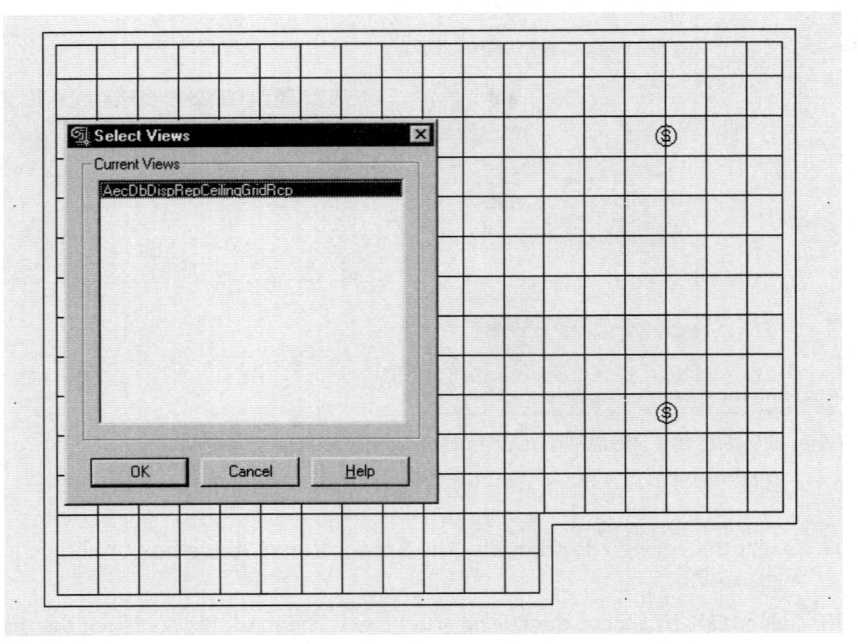

Figure 7.92 Select Views dialog box for verifying grid selection for mask block

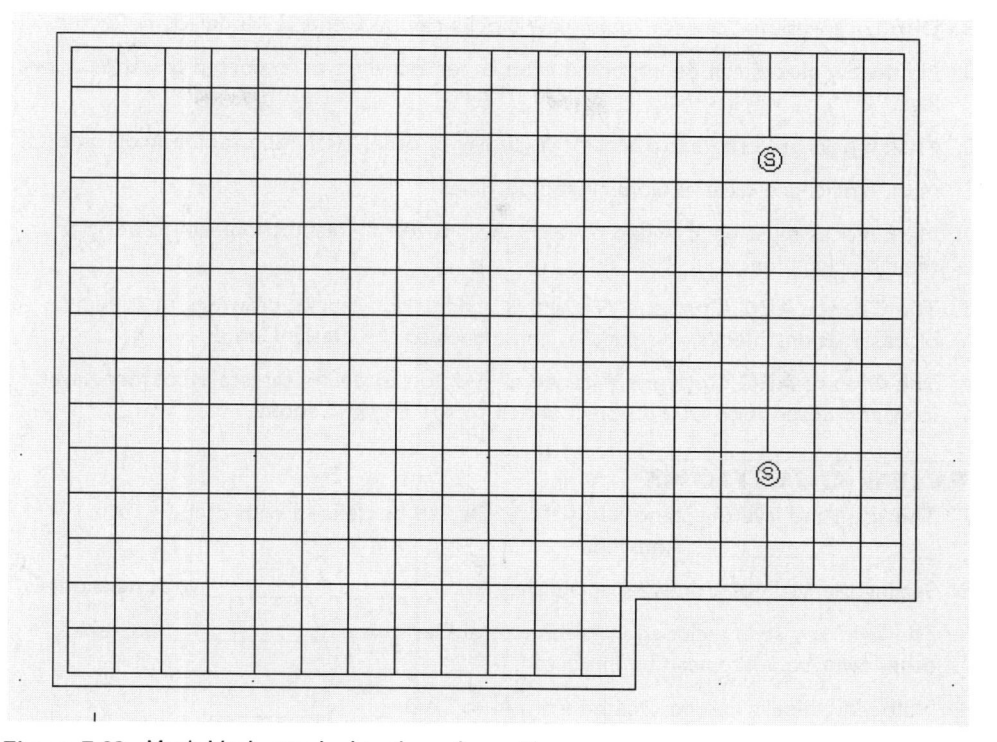

Figure 7.93 Mask block attached to the ceiling grid

39. Save the drawing.

SUMMARY

1. The scale for symbols and annotation is set with the **DWGSCALESETUP** command.

2. Define a layer standard for the drawing by selecting the **Layering** tab of the **Drawing Setup** dialog box.

3. Symbols are inserted in the drawing from the Custom View of the DesignCenter.

4. The default symbol is set to **Imperial** or **Metric** in the **AEC Content** tab of the **Options** dialog box.

5. Multi-view blocks include display control based upon the viewing direction and the display representation.

6. Multi-view blocks can be created from other multi-view blocks or blocks.

7. The **Multi-View Block Reference Properties** dialog box allows you to edit the location of the insertion point, insert attribute data, scale, rotate, and edit the style of the multi-view block

8. Display representations for multi-view blocks can be General, Model, or Reflected.
9. Multi-view blocks can be imported from other drawings or exported to other drawings.
10. Mask blocks are created to block the display of other Architectural Desktop objects.
11. Mask blocks are created from closed polylines.
12. Mask blocks must be attached to other Architectural Desktop objects to perform their blocking function.
13. The **Create AEC Content Wizard** is used to add blocks, drawings, multi-view blocks, masking blocks, and custom commands to the DesignCenter.
14. The **Create AEC Content Wizard** allows you to define the scale, rotation, layer, icon, and description of the object added to the DesignCenter.

REVIEW QUESTIONS

1. The scale and layering standard of a drawing can be defined with the _____ command.
2. Setting the scale of a drawing establishes the _____ of dimensions.
3. The layer key style included in Architectural Desktop is _____; however, other layer standards can be imported.
4. Multi-view blocks can be created from AutoCAD _____.
5. The **Add Multi-View Blocks** dialog box is opened by _____ in the DesignCenter.
6. The **Multi-View Block Reference Properties** dialog box allows you to _____.
7. If the **Offset from Insertion** value is edited for a multi-view block, it should be changed for _____.
8. Casework counter tops can be added with the _____ command.
9. The display representations for the multi-view block are _____, _____, and _____.
10. Blocks created to block the display of other Architectural Desktop objects are _____ blocks.
11. The **Create AEC Content Wizard** allows you to preset the _____, _____, and _____ of the block.
12. If the **Annotation** scale is selected in the **Create AEC Content Wizard** for a block of a circle drawn 14" in diameter, the diameter of the circle when inserted in a drawing with 1/8" **Annotation Plot Size** and **Drawing Scale Factor** 48 is_____.

PROJECT

EXERCISE 7.1 INSERTING SYMBOLS

Open *AC\Ch7\Ex7-6* and save the file to your student directory as *Proj7-6*. Insert the following symbols, as shown in Figure 7.94.

Bath 1

    1. Insert a Tub 30 x 60 from the Bath container; adjust the insertion offset distance as necessary.

    2. Insert the Tank 1 water closet from the Toilet container.

    3. Insert the Wall Lavatory from the Lavatory container.

Bath 2

    1. Insert the Vanity from the Lavatory container at the midpoint of the wall.

    2. Insert a Tub 30 x 60 from the Bath container; adjust the insertion offset distance as necessary.

    3. Insert the Tank 1 water closet from the Toilet container.

Bath 3

    1   Insert the Tank 1 water closet from the Toilet container.

    2. Insert the Wall Lavatory from the Lavatory container.

Laundry

    Insert Washer 30 x 26 and Dryer 30 x 26 from the Misc container of Appliances.

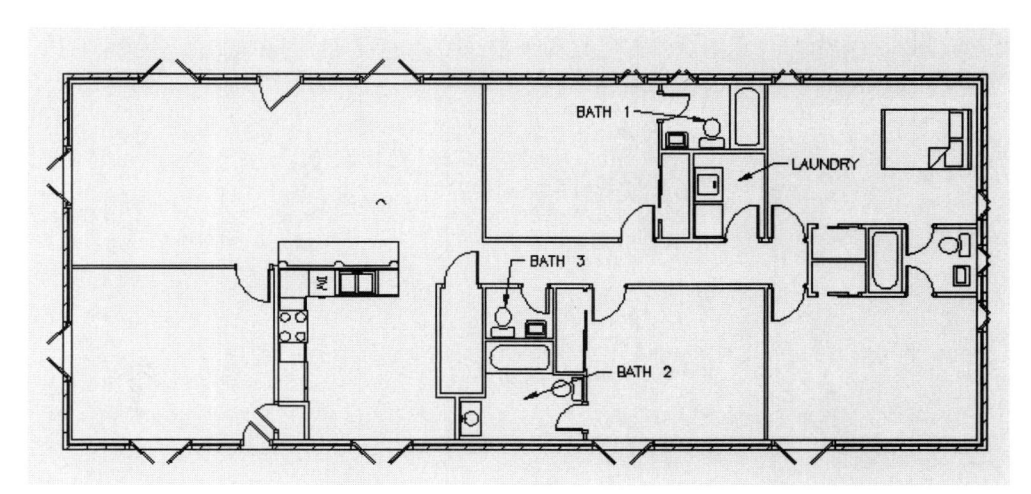

Figure 7.94 *Symbol insertion locations*

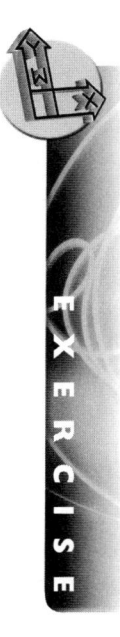

CHAPTER 8

Annotating the Drawing

INTRODUCTION

The annotation of a drawing includes the dimensions, notes, schedules, and associated symbols for specifying the size and location of architectural features. The symbols, dimensions, and leaders for annotating the drawing are selected from the AutoCAD DesignCenter. The annotation components are placed on the correct layer and are scaled according to the scale of the drawing to comply with the drafting layer standards. AutoCAD attributes are used with the annotation components to place text in the symbol.

The content of tags and schedules are defined in the schedule table style. Each schedule table style utilizes one or more property sets, which are defined for objects listed in the schedule. Schedules can be created that will be automatically updated when changes are made in the drawing. The content of a schedule can be exported to external files to enhance the use of the schedule data. Annotation components are selected from the **Annotation - Imperial**, **Annotation - Metric**, **Schedule - Imperial**, and **Schedule - Metric** toolbars.

OBJECTIVES

After completing this chapter, you will be able to

- Use **Annotation - Imperial** and **Annotation - Metric** toolbars to insert break marks, detail marks, and interior and exterior elevation marks
- Insert leaders and place dimensions on a floor plan
- Place fire rating lines on walls
- Create match lines, north arrows, and datum elevations
- Create revision clouds using various styles of revision clouds

Insert section marks for the identification and location of sections

Insert title marks and bar scales in a drawing

Create tags for doors, windows, rooms, finish, objects, and walls

Create Aec Spaces for room and finish tags

Insert schedule data in tags using **Attach/Edit Schedule Data** command

Create a schedule using the **Add Schedule Table** command

Edit and update the content of existing schedules and tags

PLACING ANNOTATION ON A DRAWING

Prior to annotation being placed, the scale of the drawing must be set with the **Set Drawing Scale** (**AECDWGSCALESETUP**) command. This command includes options for setting the annotation size and the scale factor of the drawing. Annotation commands selected from the **Annotation - Imperial** and **Annotation - Metric** toolbars are scaled according to the scale factor established in the **Set Drawing Scale** command.

You can place annotation in a drawing by selecting commands from the **Annotation - Imperial** or **Annotation - Metric** toolbar. They include the following: **Break Marks, Detail Marks, Elevation Marks, Leaders, Miscellaneous, Revision Clouds, Section Marks,** and **Title Marks**. Access the Annotation toolbars as follows (Table 8.1):

Menu bar	View>Toolbars, choose AECARCHX menu group; then choose Annotation - Imperial or Annotation - Metric
Command prompt	TOOLBAR
Shortcut	Right-click over an Architectural Desktop toolbar; then choose Annotation - Imperial or Annotation - Metric toolbar from the toolbar list

Table 8.1 *Accessing the Annotation toolbars*

The commands on the **Annotation - Imperial** and **Annotation - Metric** toolbars are shown in Figure 8.1.

When you select any of the commands on the **Annotation - Imperial** and **Annotation - Metric** toolbars, the AutoCAD DesignCenter opens. Place annotation symbols by selecting a symbol from the palette of the AutoCAD DesignCenter. The symbols can consist of polylines, text, blocks, or other entities.

Annotating the Drawing 591

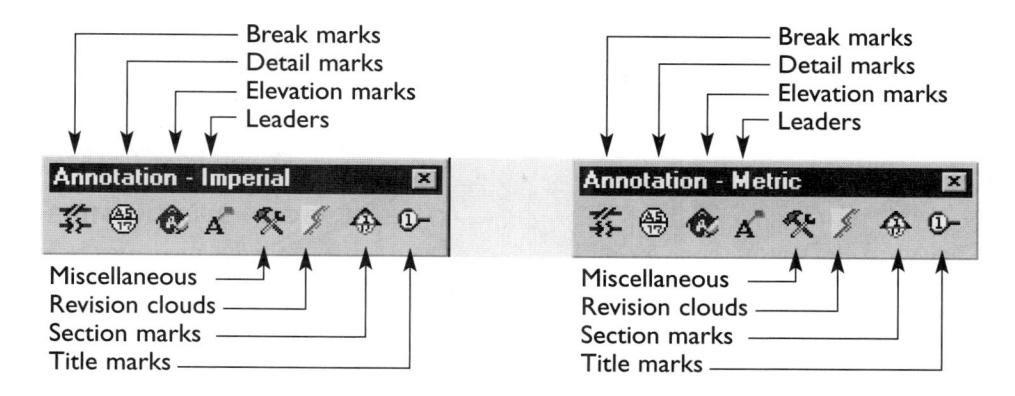

Figure 8.1 Commands on the Annotation - Imperial and Annotation - Metric toolbars

CREATING BREAK MARKS

Break marks are placed in a drawing to indicate that a feature continues beyond the drawing and is not shown. Break marks are often used in detailing of columns or wall sections. Inserting a break mark allows you to enlarge the scale of the drawing of the object by removing part of the object from the detail. Access the **Break Marks** command as follows (Table 8.2).

Menu bar	Documentation>Break Marks
Command prompt	ANNOBREAKMARKADD
Toolbar	Select Break Marks from the Annotation - Imperial or Annotation - Metric toolbar as shown in Figure 8.1

Table 8.2 Accessing the Break Marks command

Break marks are placed on the A_Anno_Symb layer. Figure 8.2 shows examples of break marks included in the Break Marks symbol folder.

Table 8.3 lists the name and description of the symbols.

When you select the **Break Mark** command from the **Annotation - Imperial** or the **Annotation - Metric** toolbar, the AutoCAD DesignCenter opens to the Break Marks folder. Selecting a specific break mark from the palette executes the **ANNOBREAKMARKADD** command. The first six lines of this command sequence require no user input because the options are predefined in the command string. Therefore, the command window from the **ANNOBREAKMARKADD** command, as shown below, will include several lines of text that do not require user input.

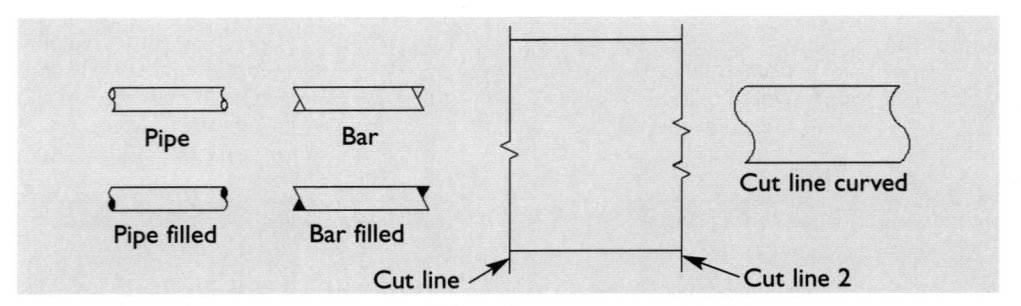

Figure 8.2 Applications of break marks available in the Break Marks folder

Name of Symbol	Symbol Option	Description
Pipe	Anno_Break_Pipe	Pipe Break Mark
Bar (Filled)	Anno_Break_Bar_Solid	Bar Break Mark (with Solid Fill)
Pipe (Filled)	Anno_Break_Pipe_Solid	Pipe Break Mark (with Solid Fill)
Bar	Anno_Break_Bar	Bar Break Mark
Cut Line 1	Anno_Break_Single	Straight Cut Line (Single Z)
Cut Line 2	Anno_Break_Double	Straight Cut Line (Double Z)
Cut Line (Curved)	Anno_Break_Curve	Curved Cut Line

Table 8.3 Symbol descriptions

To place a break mark, click on the desired symbol in the palette and continue to hold down the left mouse button as you drag the mouse into the drawing area. If you release the mouse button over the drawing area, you are prompted in the command line to specify the first point and second point of the break line. After selecting the beginning and end of the break line, you are prompted to select entities to trim that cross the break line. Shown below is the command sequence used to place the Anno_Break_Bar_Solid break line on the entity shown in Figure 8.3.

```
Command: _AecAnnoBreakMarkAdd
Adding: Bar Break Mark (w/ Solid Fill)
Specify first point of break line or [Symbol/Type]:
_SYMBOL
Specify symbol block name or
[?]<Anno_Break_Bar_Solid>: Anno_Break_Bar_Solid
```

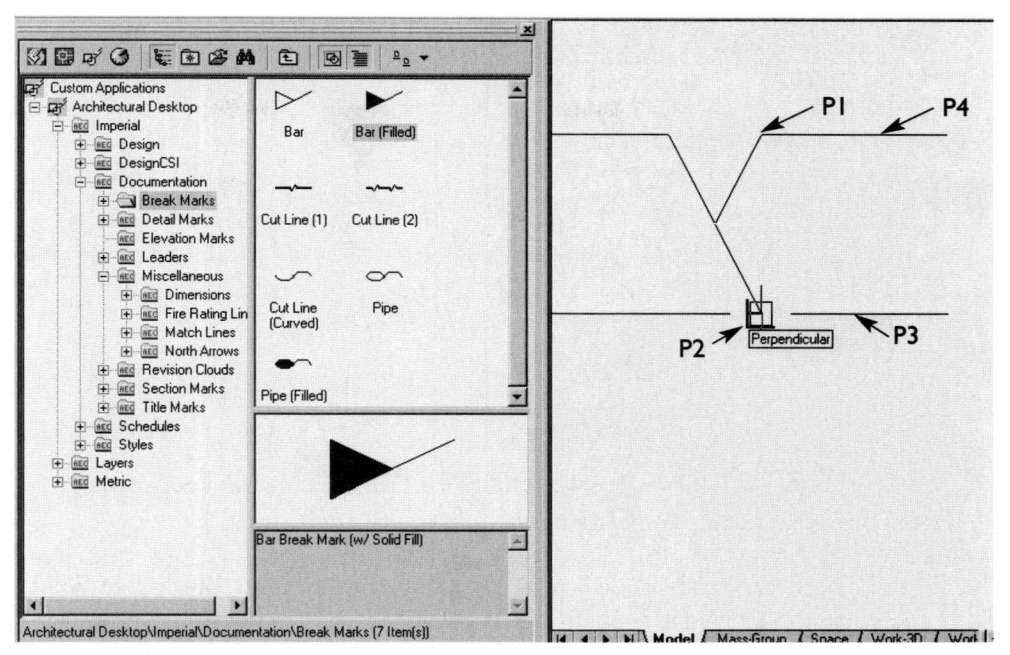

Figure 8.3 *Selecting location for placement of the Anno_Break_Bar_Solid break line*

```
Specify first point of break line or [Symbol/Type]:
_TYPE
Specify break type or [STretched/SCaled]<SCaled>:
_Scaled
Specify first point of break line or [Symbol/Type]:
_Endofparam

Specify first point of break line or [Symbol/Type]:
nea to
```
(Select the entity at point P1 in Figure 8.3.)
```
Specify second point of break line or [Symbol/Type]:
perp ENTER
to
```
(Select the entity at point P2 in Figure 8.3.)
```
Select objects to trim <None> or [Symbol/Type]:
```
(Select the entity at point P3 in Figure 8.3.)
```
Select objects to trim <None> or [Symbol/Type]:
```
(Select the entity at point P4 in Figure 8.3.)
```
Select objects to trim <None> or [Symbol/Type]:
ENTER
Command:
```
(Figure 8.4 shows the result.)

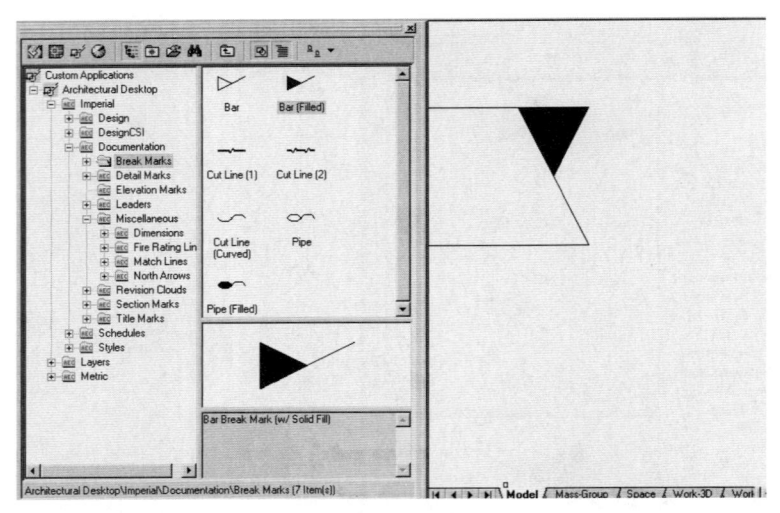

Figure 8.4 Anno_Break_Bar_Solid applied to the entities

The Nearest object snap mode is used to specify the top of the Anno_Break_Bar_Solid and Perpendicular object snap mode is used to place the bottom of the symbol perpendicular to the entity. The control of the symbol perpendicular to the horizontal or vertical entities can be also be established through the AutoCAD Ortho toggle.

The **ANNOBREAKMARKADD** command includes the **Symbol** and **Type** options, as shown in the command sequence above. The purpose of each option is described below.

> **Symbol** – The **Symbol** option specifies the break mark symbol. When a break mark is selected from the DesignCenter palette, the symbol name is predefined. However, you can type the name of another break mark symbol in the command line at this prompt. The names of the break mark symbols listed in Table 8.1 must be correctly typed in the command line to select a different break mark symbol.
>
> **Type** – The **Type** option allows you to specify the **STretched** or **SCaled** insertion method for placing the break mark. The **Type** is predefined when you select a specific break mark from the palette of the DesignCenter. The **Scaled** option is used to place all the break marks except the Single Z and Double Z break marks. The **Scaled** option creates a break proportionate to the size of the object. The **Stretched** option scales the Z portion of the break mark according to the scale of the drawing established by the **Set Drawing Scale** command and the length of the break line is determined by the stretching of the line to fit the application.

Although the **Symbol** and **Type** options are available in the command line, you can select them more easily by changing the selection from the AutoCAD DesignCenter.

PLACING STRAIGHT CUT LINES

Straight cut lines are placed in the drawing with the **Stretched** type of placement. Therefore the scale of the Cut Line (1) or Cut Line (2) is specified in the Set Drawing Scale command. When you place the Cut Line (1) or Cut Line (2), the first point of the break line specifies the location of one end of the break line and second point specifies the other end of the break line. Regardless of the length of the break line, the Z portion remains the same size. Figure 8.5 shows two Cut Line (1) symbols, which have different lengths. The Z portion of the cut line retains the same size regardless of length.

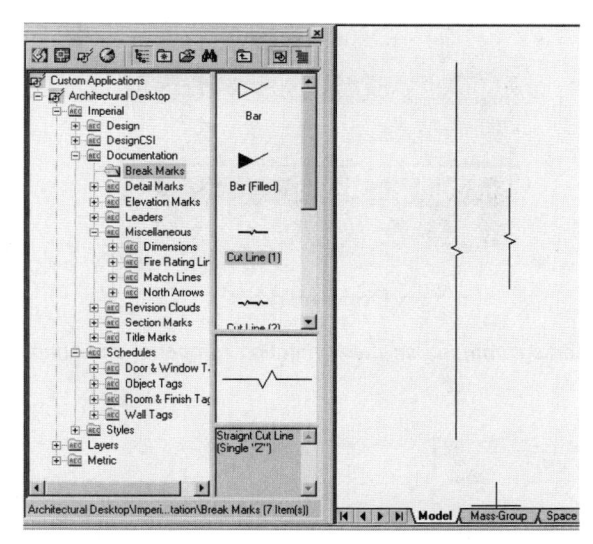

Figure 8.5 *Cut Line 1*

PLACING DETAIL MARKS

Detail marks are placed on the drawing to specify the location from which a detached detail has been developed. The detail usually provides an enlarged scale view of the construction with additional annotation or dimensions. In addition, the **Detail Marks** command includes options for placing cutting plane lines to specify the location of the cutting plane for a section. The **Detail Marks** command is accessed as follows (Table 8.4).

When you select the **Detail Marks** command, the AutoCAD DesignCenter opens to the Detail Marks folder, which includes detail symbol marks and other enhancements for placing detail boundaries and section marks. The geometry of the detail marks is placed on the A-Det-Iden layer. Figure 8.7 shows the symbols of the Detail Marks folder, and Table 8.5 gives a brief description of each.

Menu bar	Documentation>Detail Marks
Command prompt	AECANNODETAILBOUNDARYADD and AECANNODETAILMARKADD
Toolbar	Select Detail Marks from the Annotation - Imperial or Annotation - Metric toolbar as shown in Figure 8.6

Table 8.4 *Accessing the Detail Marks command*

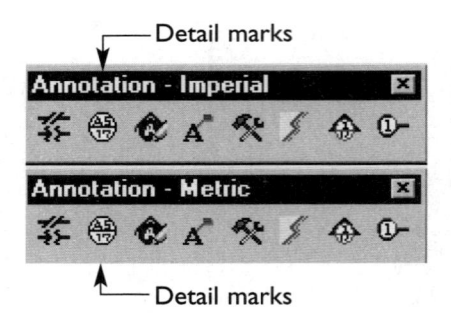

Figure 8.6 *Detail Marks command on the Annotation - Imperial and Annotation - Metric toolbars*

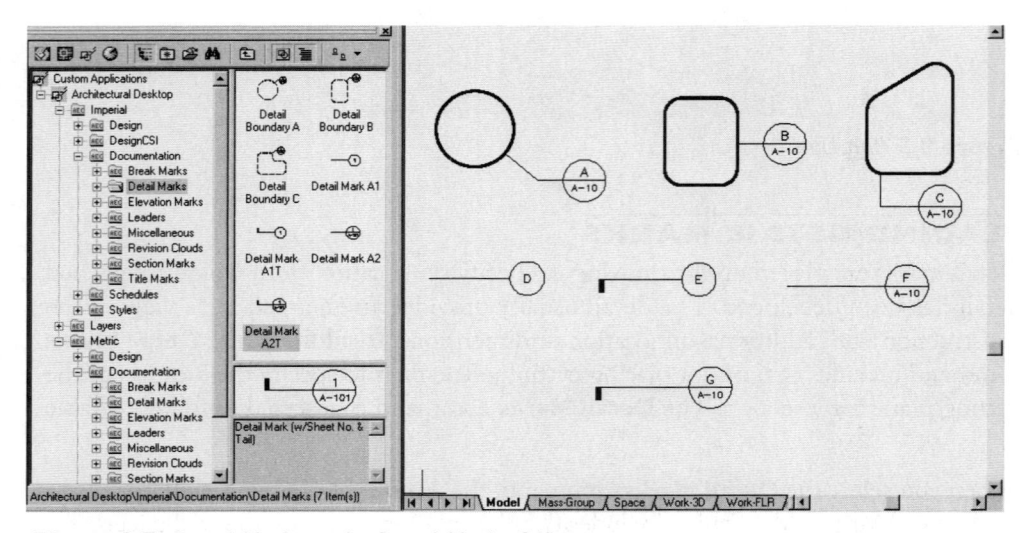

Figure 8.7 *Detail Marks in the Detail Marks folder*

The detail boundary symbols are placed with the **AECANNODETAILBOUNDARYADD** command. This command includes three options: **Circle, Rectangle,** and

Name of Symbol	Symbol	Option	Description
Detail Boundary A	Anno_Detail_A2	Circle option	Detail Boundary (Circle)
Detail Boundary B	Anno_Detail_A2	Rectangle option	Detail Boundary (Rectangle)
Detail Boundary C	Anno_Detail_A2	Freeform option	Detail Boundary (Freeform)
Detail Mark A1	Anno_Detail_A1		Detail Mark
Detail Mark A1T	Anno_Detail_A1	Tail option	Detail Mark (w/ Tail)
Detail Mark A2	Anno_Detail_A2		Detail Mark (w/ Sheet No.)
Detail Mark A2T	Anno_Detail_A2	Tail option	Detail Mark (w/ Sht. No.& tail)

Table 8.5 *Description of Detail Marks*

Freeform. Detail Boundary A allows you to create a circle with a leader, which includes the detail number. The command sequence that follows was used to place the Detail Boundary A symbol shown in Figure 8.8. The first 10 lines of the command sequence do not require user input because the responses are predefined when the command is selected from the palette of the DesignCenter.

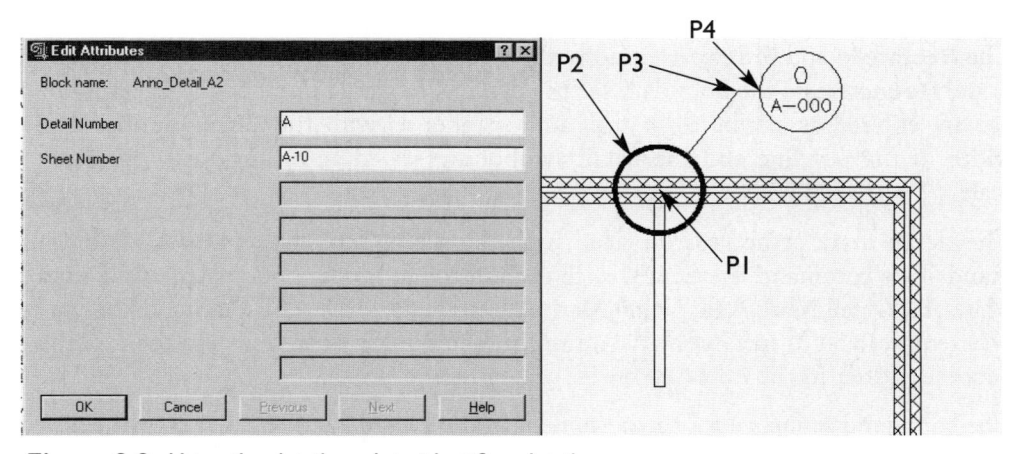

Figure 8.8 *Using the detail mark to identify a detail*

```
Command: _AecAnnoDetailBoundaryAdd
Adding: Detail Boundary (Circle)
Specify center of detail circle or
[Symbol/Linetype/pline Width]: _SYMBOL
Symbol block name or [?]<Anno_Detail_A2>:
Anno_Detail_A2
Specify center of detail circle or
[Symbol/Linetype/pline Width]: _TYPE
[Circle/Rectangle/Freeform]: _Circle
Specify center of detail circle or
[Symbol/Linetype/pline Width]: _LINETYPE
Linetype or [?]<DASHED2>: Dashed2
Specify center of detail circle or
[Symbol/Linetype/pline Width]: _WIDTH
Pline width<1/16">: .03125
Specify center of detail circle or
[Symbol/Linetype/pline Width]: _Endofparam

Specify center of detail circle or
[Symbol/Linetype/pline Width] (Select point P1 in Figure 8.8.)
Specify radius of circle: (Select point P2 in Figure 8.8.)
Specify first point of leader line outside of cir-
cle: (Select point P3 in Figure 8.8.)
Specify next point of leader line <end line>: (Select
point P4 in Figure 8.8.)
Specify next point of leader line <end line>: ENTER
(Enter attribute text in the Edit Attributes dialog box as shown on the left
in Figure 8.8, and then select OK to dismiss the Edit Attributes dialog box.)
Select block reference:
Command:
```

The **Rectangle** and **Freeform** options allow you to select the vertex of the rectangle or polygon to create the boundary for the detail. The detail number and sheet number are entered as attributes in the same manner as with the **Circle** option. The width of the polyline and the detail symbol are scaled according to the drawing scale.

The detail mark symbols are placed with the **ANNODETAILMARKADD** command. This command is executed with each of the four detail mark symbols: Detail Mark A1, Detail Mark A1T, Detail Mark A2, or Detail Mark A2T. Placing these symbols requires you to specify the beginning and end of the detail line and to enter the detail identifier in the **Edit Attributes** dialog box.

The command sequence for the placement of the Detail Mark A2T is shown below and the detail mark is shown in Figure 8.9.

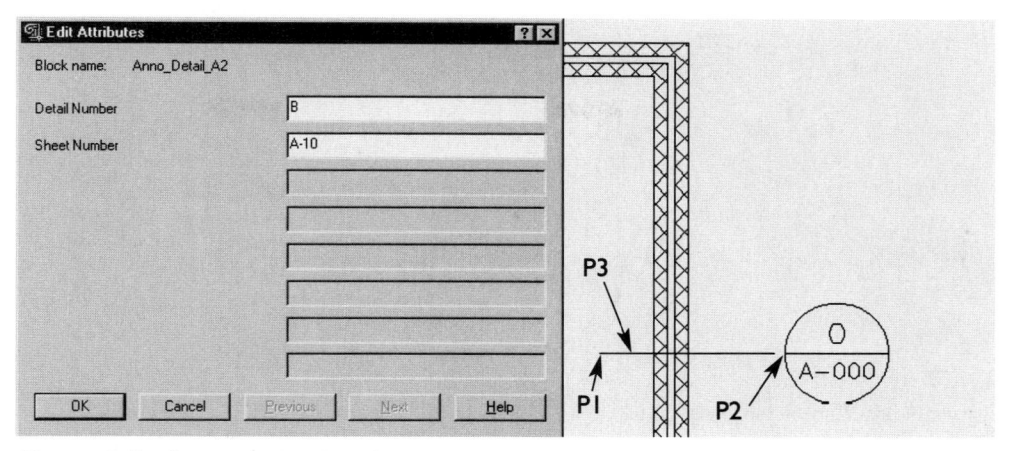

Figure 8.9 *Creating a detail mark with a tail*

```
Command: _AecAnnoDetailMarkAdd
Adding: Detail Mark (w/Sheet No. & Tail)
Specify first point of detail line or [Symbol/Tail]:
_SYMBOL
Symbol block name or [?]<NONE>: Anno_Detail_A2
Specify first point of detail line or [Symbol/Tail]:
_TAIL
Tail length,width<0.000,0.000>: .1875,.0625
Specify first point of detail line or [Symbol/Tail]:
_Endofparam

Specify first point of detail line or [Symbol/Tail]:
```
(Specify the beginning of detail line at P1 in Figure 8.9.)
```
Specify next point of line or [Symbol/Tail]:
```
(Specify the end of the detail line at P2 in Figure 8.9.)
```
Specify next point of line or [Break/Symbol/Tail]:
ENTER
```
(Enter text in the Edit Attributes dialog box.)
```
Select block reference:
Command:
Specify side for tail:
```
(Specify the direction for the tail at P3 in Figure 8.9.)
```
Command:
```
(Figure 8.10 shows the result.)

PLACING ELEVATION SYMBOLS

Elevation symbols are placed in a drawing to identify the location and orientation of elevation drawings. Elevations can be extracted from the three-dimensional model with the **Create Elevation** command (**BLDGELEVATIONLINEGENERATE**). (The

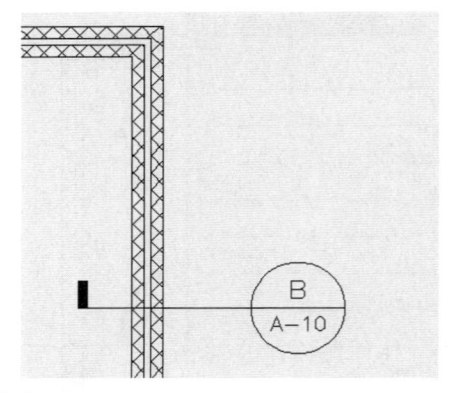

Figure 8.10 *Completed detail mark*

Create Elevation command will be discussed in detail in Chapter 9.) When an elevation is created, it is often necessary to specify the orientation of the elevation using an elevation symbol placed in the plan view. Interior or exterior elevations can be developed to describe the building construction. As shown in Figure 8.12, interior and exterior elevation symbols are included in the **Elevation Marks** folder. Access Elevation Marks as follows (Table 8.6).

Menu bar	Documentation>Elevation Marks
Command prompt	ANNOELEVATIONMARKADD; then specify symbol
Toolbar	Select Elevation Marks from the Annotation - Imperial or Annotation - Metric toolbar as shown in Figure 8.11

Table 8.6 *Accessing Elevation Marks*

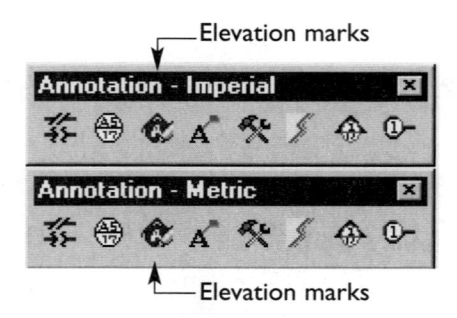

Figure 8.11 *Elevation Marks on the Annotation - Imperial and Annotation - Metric toolbars*

When you select the **Elevation Marks** command of the **Annotation - Imperial** or **Annotation - Metric** toolbar, the AutoCAD DesignCenter will open to the Elevation Marks folder, which consists of the elevation marks listed in Table 8.7 and shown in the palette of the DesignCenter in Figure 8.12.

Name of Symbol	Symbol Option	Description
Elevation Mark A1	Anno_Elevation_A1	Elevation Mark
Elevation Mark A2	Anno_Elevation_A2	Elevation Mark (with Sheet No.)
Elevation Mark B1	Anno_Elevation_B1 Interior	Elevation Mark (4 Way- 1/2/3/4)
Elevation Mark B2	Anno_Elevation_B2 Interior	Elevation Mark (4 Way – N/E/S/W)
Elevation Mark C1	Anno_Elevation_C1 Interior	Elevation Mark (Single)
Elevation Mark C2	Anno_Elevation_C2 Interior	Elevation Mark (Single – Inverted Text)

Table 8.7 *Description of Elevation Mark symbols*

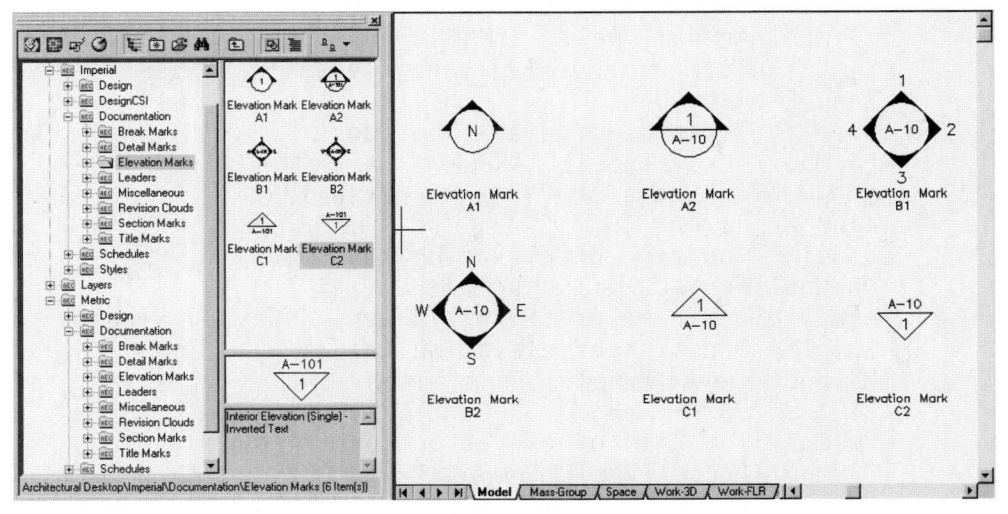

Figure 8.12 *Elevation marks in the Elevation Marks folder*

Selecting an elevation mark from the AutoCAD DesignCenter palette will execute the **ANNOELEVATIONMARKADD** command, which allows you to place one of the six elevation marks listed in Table 8.7. The following command sequence was used to place an Elevation Mark A1 in the drawing shown in Figure 8.13.

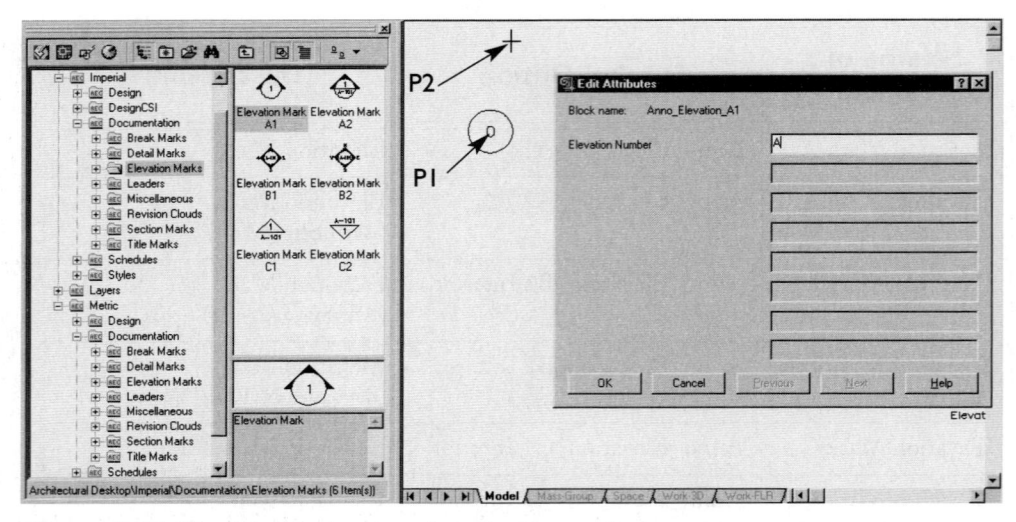

Figure 8.13 *Specifying the location of an elevation mark*

```
Command: _AecAnnoElevationMarkAdd
Adding: Elevation Mark
Specify location of elevation tag or
[Symbol/Arrow/Type]: _SYMBOL
Symbol block name or [?]<Anno_Elevation_B1>:
Anno_Elevation_A1
Specify location of elevation tag or
[Symbol/Arrow/Type]: _ARROW
Arrow block name or [?]<Anno_Arrow_1234>:
Anno_Arrow_A1
Specify location of elevation tag or
[Symbol/Arrow/Type]: _TYPE
Arrow block name or [Single/4-Way]<4-Way>: _Single
Specify location of elevation tag or
[Symbol/Arrow/Type]: _Endofparam

Specify location of elevation tag or
[Symbol/Arrow/Type]:
```
(Specify a location for the center of the elevation mark at point P1 in Figure 8.13.)

```
Specify direction for elevation or
[Symbol/Arrow/Type]:
```
(Move the mouse in the direction of viewing and select the viewing direction at point P2 in Figure 8.13; then type the elevation identification in the Edit Attribute dialog box.)
```
Select block reference:
Command:
Add AEC elevation object(s)? [Yes/No] <Y>: n
```
ENTER
(Figure 8.14 shows the result.)

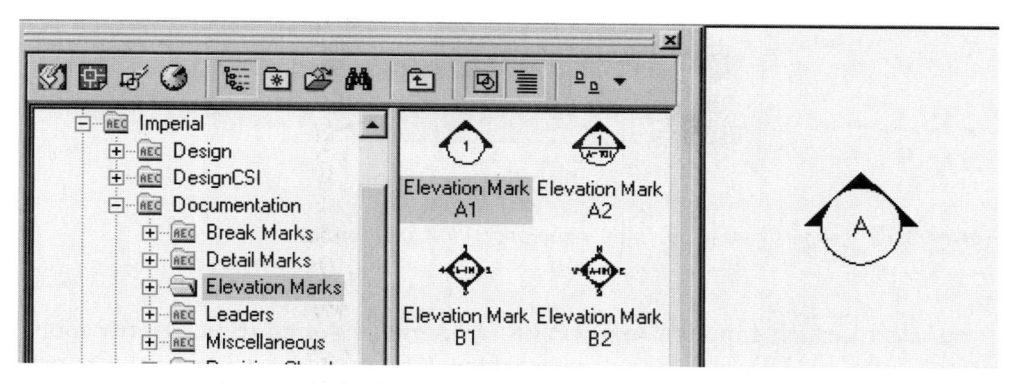

Figure 8.14 *Elevation mark inserted*

To place an elevation mark, click on the desired symbol in the palette and continue to hold down the left mouse button as you drag the mouse in the drawing area. When you release the mouse button over the drawing area, the elevation symbol is dynamically displayed on the pointer. Select a location in the drawing area to specify the insertion point for the elevation mark and select in the drawing area again to specify the viewing direction for the elevation mark. After you establish the viewing direction, the **Edit Attributes** dialog box opens, which allows you to specify the name of the elevation. Finally, you are prompted `Add AEC Elevation Line [Yes/No] <Y>`. If you place an AEC Elevation Line in the drawing, the line can be used later to develop an elevation view.

PLACING LEADERS IN THE DRAWING

Leaders are placed in a drawing to identify building components and sizes. Leaders are part of the tools used to dimension a drawing; they can be placed in a drawing with the AutoCAD dimensioning commands. However, the **Leaders** command on the **Annotation - Imperial** and **Annotation - Metric** toolbars includes enhancements: balloons, attribute text, and curved leaders. The symbols and the leaders are scaled according to the scale of the drawing and the symbol is placed on the A-Anno-Note layer. The command to place the leader is **Leaders** (**ANNOLEADERADD**). See Access the **ANNOLEADERADD** command as follows (Table 8.8).

Menu bar	Documentation>Leaders
Command prompt	ANNOLEADERADD
Toolbar	Select Leaders from the Annotation - Imperial or Annotation - Metric toolbar as shown in Figure 8.15

Table 8.8 *Accessing Leaders*

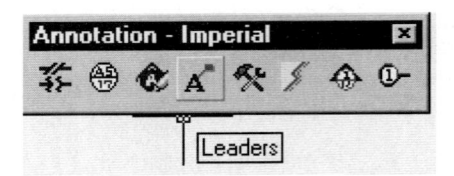

Figure 8.15 *Leaders command on the Annotation - Imperial toolbar*

If you select Leaders from the **Annotation - Imperial** or **Annotation - Metric** toolbar, the AutoCAD DesignCenter opens to the Leaders folder, which includes ten leader options as listed in Table 8.9.

Name of Symbol	Symbol Option	Description
Spline (Circle)	Anno_Circle	Spline Leader (w/ Circle Tag)
Spline (Diamond)	Anno_Diamond	Spline Leader (w/ Diamond Tag)
Spline (Hexagon)	Anno_Hexagon	Spline Leader (w/ Hexagon Tag)
Spline (Square)	Anno_Square	Spline Leader (w/ Square Tag)
Spline (Text)	Text	Spline Leader (w/ Text)
Straight (Circle)	Anno_Circle	Straight Leader (w/ Circle Tag)
Straight (Diamond)	Anno_Diamond	Straight Leader (w/ Diamond Tag)
Straight (Hexagon)	Anno_Hexagon	Straight Leader (w/ Hexagon Tag)
Straight (Square)	Anno_Square	Straight Leader (w/ Square Tag)
Straight (Text)	Text	Straight Leader (w/ Text)

Table 8.9 *Description of Leader symbols*

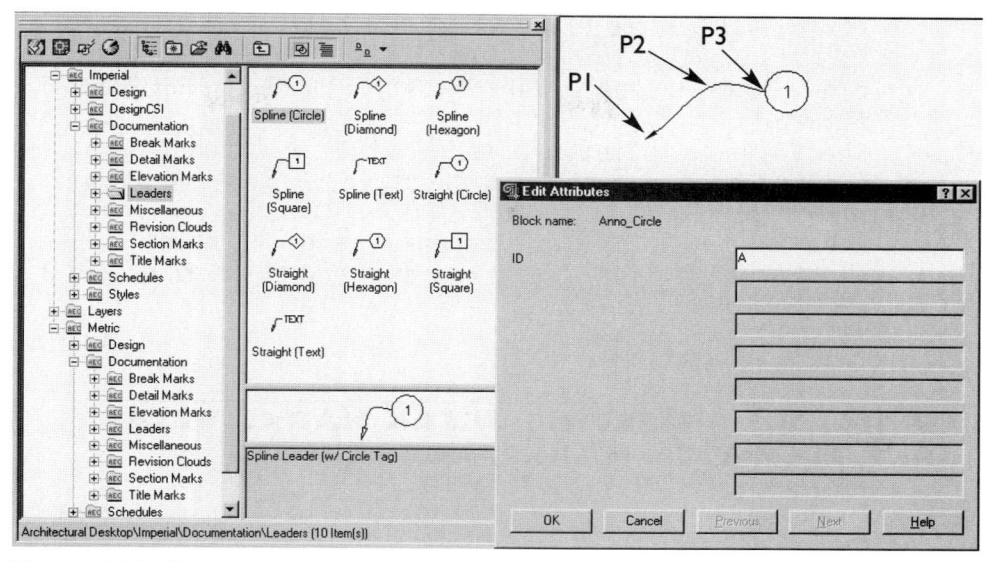

Figure 8.16 *Placing a leader in the drawing*

Place leaders in the drawing by specifying the beginning of the leader and vertex locations for the leader to pass through. Add text to the Spline (Text) and Straight (Text) symbols by typing text in the command line. The text for the tag symbols is inserted in the **Edit Attributes** dialog box.

```
Command: _AecAnnoLeaderAdd
Adding: Spline Leader (w/ Circle Tag)
Specify first point of leader line or [Symbol
block/leader Type/Dimension
style]: _SYMBOL
Symbol block or [?]<Anno_Circle>: Anno_Circle
Specify first point of leader line or [Symbol
block/leader Type/Dimension
style]: _TYPE
leader Type or [Straight/sPline]<sPline>: _Spline
Specify first point of leader line or [Symbol
block/leader Type/Dimension
style]: _DIMSTYLE
Dimension style or [?]<Aec_Arch_I$7>: Current
Specify first point of leader line or [Symbol
block/leader Type/Dimension
style]: _Endofparam

Specify first point of leader line or [Symbol
block/leader Type/Dimension
```

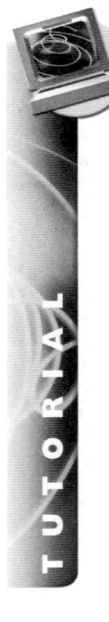

style]: *(Select point P1 in Figure 8.16 to start the leader.)*
Specify next point of leader line or [Symbol block/leader Type/Dimension style]: *(Select point P2 in Figure 8.16.)*
Specify next point of leader line or [Symbol block/leader Type/Dimension style]: *(Select point P3 in Figure 8.16.)*
Specify next point of leader line or [Symbol block/leader Type/Dimension style]: ENTER *(Insert text in the Edit Attributes dialog box.)*
Select block reference:
Command:

TUTORIAL 8.1 INSERTING BREAK MARKS, DETAIL MARKS, ELEVATION MARKS, AND LEADERS

1. Open *ADT Tutor\Ch8\Ex8-1.dwg*.
2. Choose **File>SaveAs** from the menu bar and save the drawing as *Annotation1.dwg* in your student directory.
3. Right-click over an Architectural Desktop toolbar and choose **Annotation - Imperial** from the toolbar list.
4. Select **Zoom Window** from the **View** flyout on the **Standard** toolbar and zoom a window around the gas line as shown in the following command sequence and Figure 8.17.

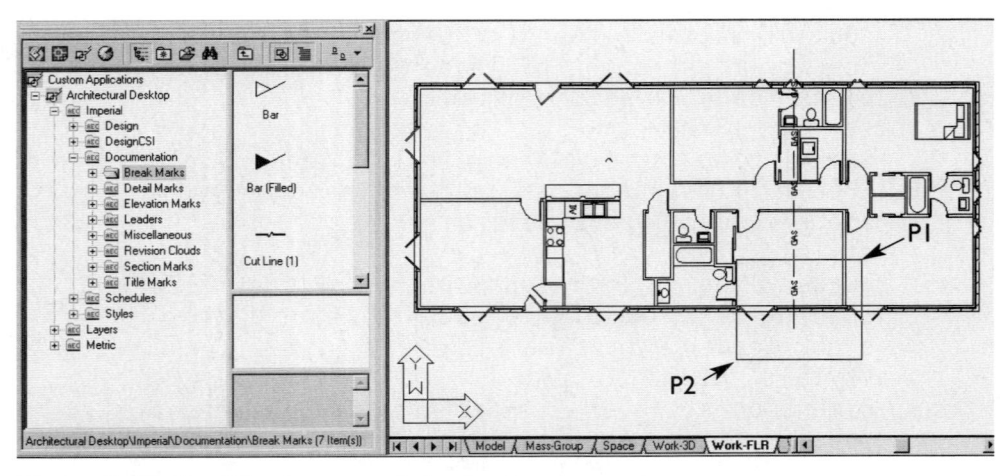

Figure 8.17 *Specifying the zoom window location*

```
Specify corner of window, enter a scale factor (nX or nXP), or
```

```
[All/Center/Dynamic/Extents/Previous/Scale/Window]
<real time>: _w
Specify first corner: (Select point P1 in Figure 8.17).
Specify opposite corner: (Select point P2 in Figure 8.17.)
```
5. Select **Break Marks** from the **Annotation - Imperial** toolbar.

6. Select the scroll bar of the DesignCenter and scroll down to Cut Line (Curved) break mark.

7. Select the Cut Line (Curved) break mark and drag the break mark to the drawing area by continuing to hold down the left mouse button as you move from the DesignCenter.

8. Release the left mouse button over the drawing area and specify the beginning of the Cut Line (Curved) break mark as shown in the command sequence and in Figure 8.18.

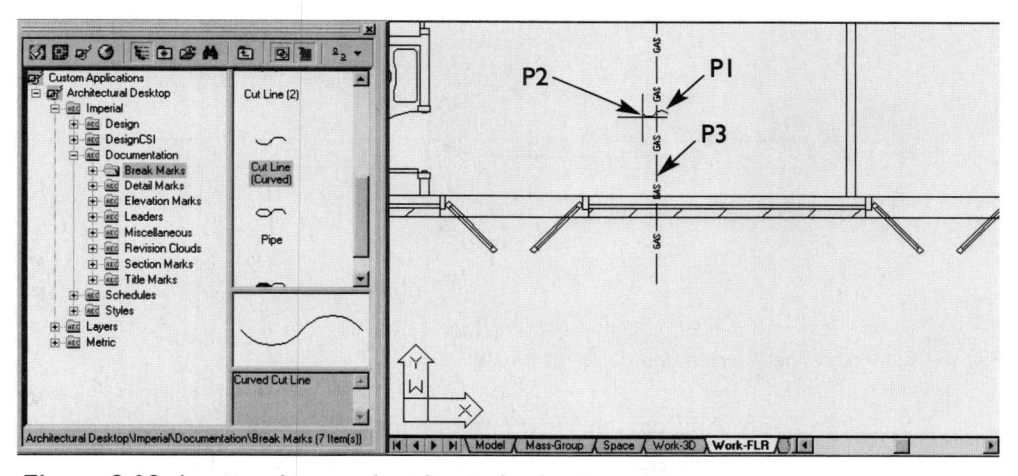

Figure 8.18 *Inserting the curved cut line in the drawing*

```
Command: _AecAnnoBreakMarkAdd
Adding: Curved Cut Line
Specify first point of break line or [Symbol/Type]:
_SYMBOL
Specify symbol block name or [?]<Anno_Break_Curve>:
Anno_Break_Curve
Specify first point of break line or [Symbol/Type]:
_TYPE
Specify break type or [STretched/SCaled]<SCaled>:
_Scaled
Specify first point of break line or [Symbol/Type]:
_Endofparam
```

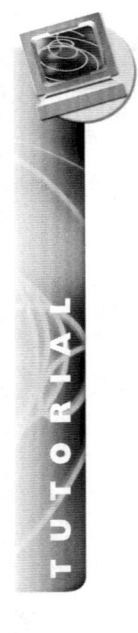

```
Specify first point of break line or [Symbol/Type]:
```
(Select at point P1 in Figure 8.18 to start the break.)
```
Specify second point of break line or [Symbol/Type]:
```
(Select at point P2 in Figure 8.18 to end the break.)
```
Select objects to trim <None> or [Symbol/Type]:
``` *(Select the gas line below the break near point P3 in Figure 8.18, to trim the lower portion of the line.)*
```
Select objects to trim <None> or [Symbol/Type]:
```
ENTER *(Pressing* ENTER *ends the command and trims the line as shown in Figure 8.19.)*

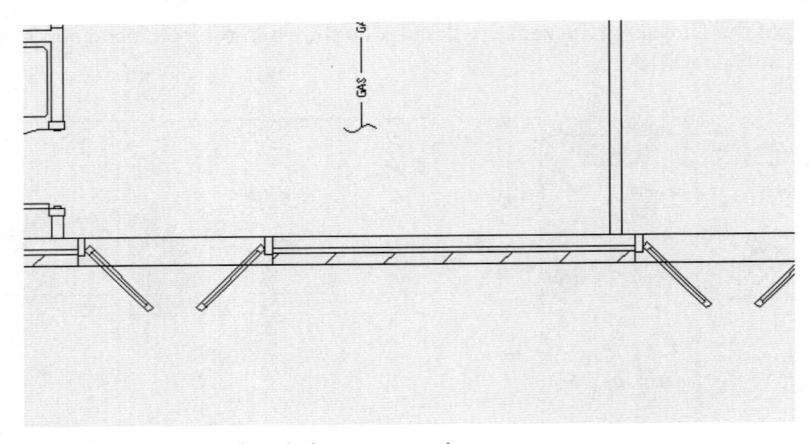

Figure 8.19 *Cut line inserted and object trimmed*

9. Select **Zoom All** from the **View** flyout on the **Standard** toolbar.

10. Select **Zoom Window** from the **View** flyout on the **Standard** toolbar and zoom a window around the kitchen area as specified in the following command sequence and Figure 8.20.

```
Command: '_zoom
Specify corner of window, enter a scale factor (nX
or nXP), or
[All/Center/Dynamic/Extents/Previous/Scale/Window]
<real time>: _w
Specify first corner:
``` *(Select point P1 in Figure 8.20.)*
```
Specify opposite corner:
``` *(Select point P2 in Figure 8.20.)*

11. Select **Detail Marks** from the **Annotation - Imperial** toolbar.

12. Select Detail Boundary B from the palette of the DesignCenter and drag the symbol to the drawing area by continuing to hold down the mouse button as you move the mouse into the drawing area.

Annotating the Drawing 609

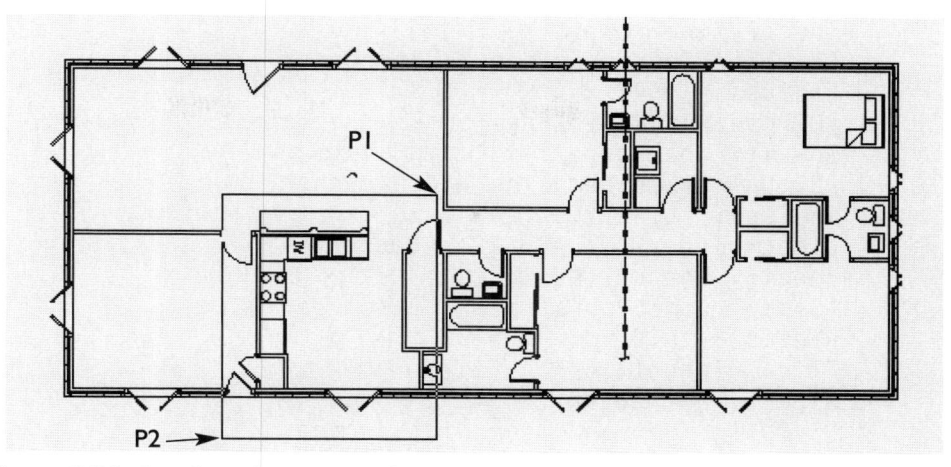

Figure 8.20 *Specifying the zoom window location*

13. Release the mouse button over the drawing area and specify the corners of the Detail Boundary B as specified in the following command sequence and Figure 8.21.

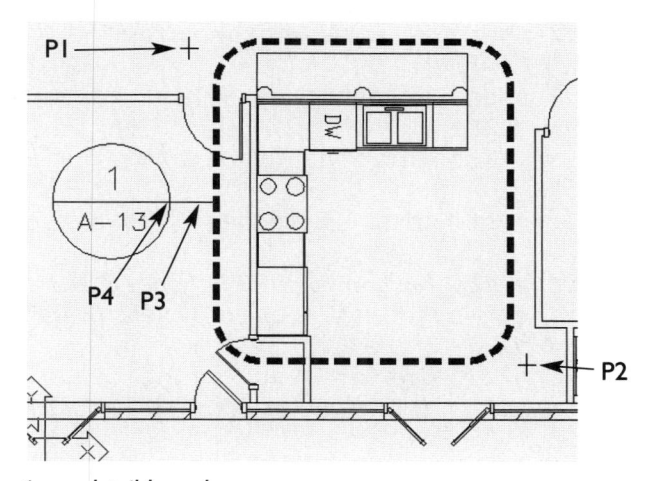

Figure 8.21 *Creating a detail boundary*

```
Command: _AecAnnoDetailBoundaryAdd
Adding: Detail Boundary (Rectangle)
Specify one corner of detail box or
[Symbol/Linetype/pline Width/corner
Radius]: _SYMBOL
Symbol block name or [?]<Anno_Detail_A2>:
Anno_Detail_A2
```

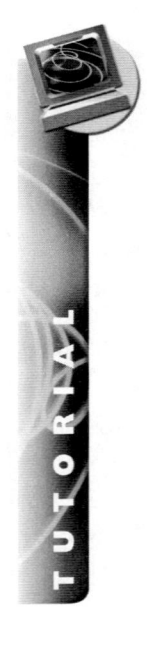

```
Specify one corner of detail box or
[Symbol/Linetype/pline Width/corner
Radius]: _TYPE
[Circle/Rectangle/Freeform]: _Rectangle
Specify one corner of detail box or
[Symbol/Linetype/pline Width/corner
Radius]: _LINETYPE
Linetype or [?]<DASHED2>: Dashed2
Specify one corner of detail box or
[Symbol/Linetype/pline Width/corner
Radius]: _WIDTH
Pline width<1/16">: .03125
Specify one corner of detail box or
[Symbol/Linetype/pline Width/corner
Radius]: _RADIUS
Corner radius<1/4">: .25
Specify one corner of detail box or
[Symbol/Linetype/pline Width/corner
Radius]: _Endofparam
```

Specify one corner of detail box or
[Symbol/Linetype/pline Width/corner Radius]: *(Select near point P1 in Figure 8.21.)*
Specify opposite corner of detail box: *(Select near point P2 in Figure 8.21.)*
Hint: Use MIDpoint osnap to start line at corner or side...
Specify first point of leader line on boundary: mid ENTER
of *(Select the detail box near point P3 in Figure 8.21.)*
Specify next point of leader line: *(Select a point near P4 in Figure 8.21.)*
Specify next point of leader line <end line>: ENTER *(Press* ENTER *to end the command and view the Detail Mark as shown in Figure 8.21.)*
Select block reference: *(Type the detail number and sheet number in the Edit Attributes dialog box as shown in Figure 8.22.)*

14. Select **Elevation Marks** from the **Annotation - Imperial** toolbar.

15. Select Elevation Mark B1 from the palette of the DesignCenter and drag the symbol to the drawing area by continuing to hold down the left mouse button as you move the mouse over the drawing area.

16. Release the mouse button over the drawing area and specify the center and rotation of the elevation mark as shown in Figure 8.23 and the following command sequence.

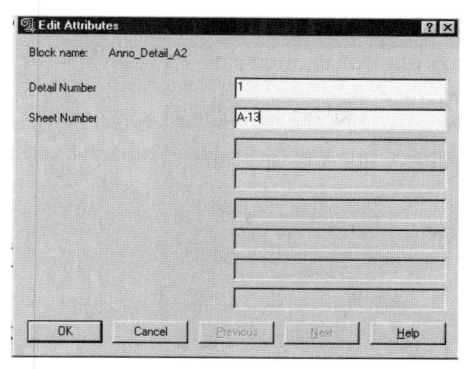

Figure 8.22 Inserting detail title in the Edit Attributes dialog box

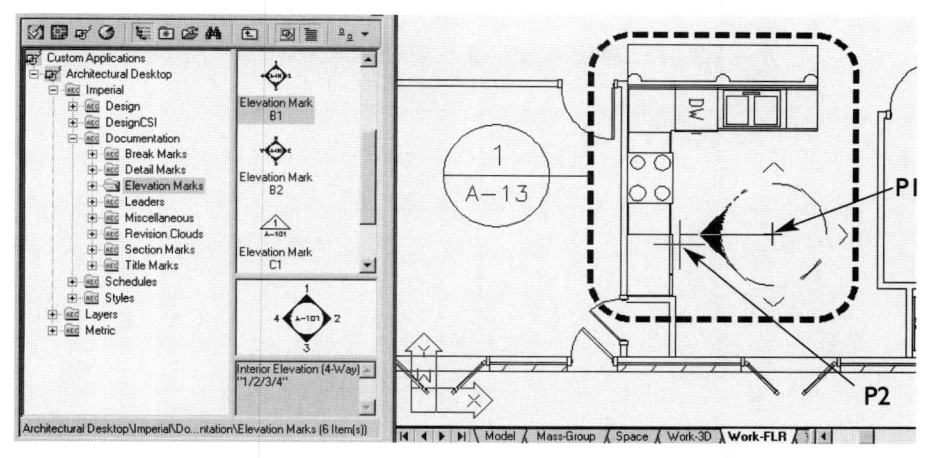

Figure 8.23 Insertion point for the interior elevation mark specified

```
Command: _AecAnnoElevationMarkAdd
Adding: Interior Elevation (4-Way) "1/2/3/4"
Specify location of elevation tag or
[Symbol/Arrow/Type]: _SYMBOL
Symbol block name or [?]<Anno_Elevation_B1>:
Anno_Elevation_B1
Specify location of elevation tag or
[Symbol/Arrow/Type]: _ARROW
Arrow block name or [?]<Anno_Arrow_1234>:
Anno_Arrow_1234
Specify location of elevation tag or
[Symbol/Arrow/Type]: _TYPE
Arrow block name or [Single/4-Way]<4-Way>: _4-Way
Specify location of elevation tag or
[Symbol/Arrow/Type]: _Endofparam
```

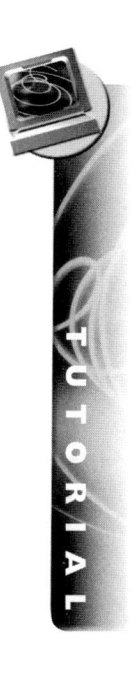

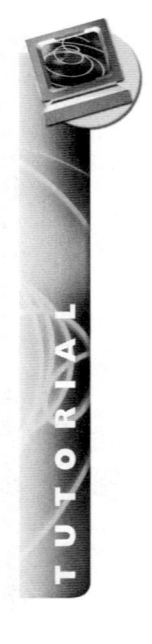

```
Specify location of elevation tag or
[Symbol/Arrow/Type]:
``` *(Select the center of the elevation mark near point P1 in Figure 8.23.)*
```
Specify direction for first elevation number or
[Symbol/Arrow/Type]:
``` *(Select the direction for the first elevation by selecting a point near P2 in Figure 8.23.)*
```
Select block reference:
``` *(Specify the sheet number in the Edit Attributes dialog box as shown in Figure 8.24.)*
```
Command:
Add AEC elevation lines [Yes/No] <N>: ENTER
``` *(Press ENTER to accept the NO default. The elevation line will not be used for extracting an elevation.)*
```
Note: Erase arrows, as required.
```

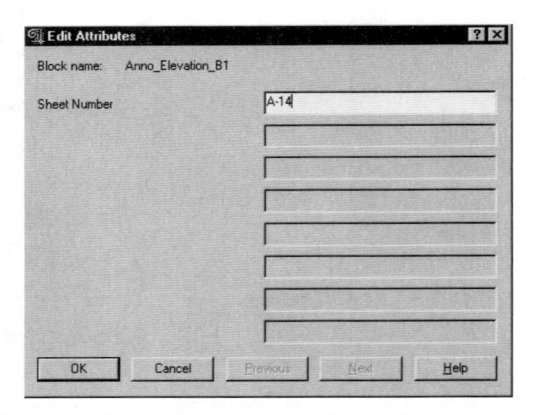

Figure 8.24 *Elevation mark defined in the Edit Attributes dialog box*

17. Select **Erase** from the **Modify** toolbar.
18. Select the text "3" and "4" to remove these components from the elevation symbol as shown in Figure 8.25.
19. Select Ortho on the Status Bar to turn OFF Ortho.
20. Select **Leaders** from the **Annotation - Imperial** toolbar.
21. Select (Spline) Text from the palette of the DesignCenter and drag the symbol to the drawing area by continuing to hold down the left mouse button as you move the mouse over the drawing area.
22. Release the mouse button over the drawing area and specify the beginning and vertices of the leader as described in the command sequence and Figure 8.26.

    ```
    Command: _AecAnnoLeaderAdd
    Adding: Spline Leader (w/ Text)
    ```

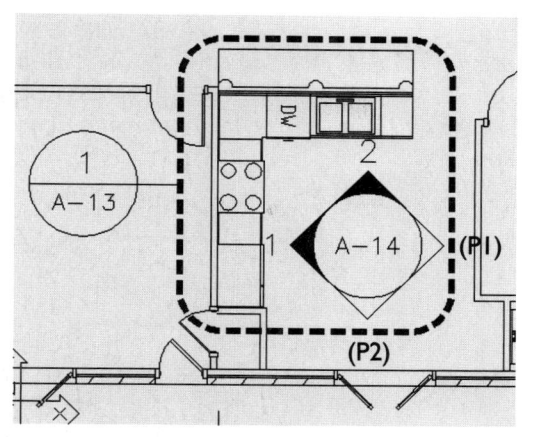

Figure 8.25 *Elevation symbol location mark edited*

```
Specify first point of leader line or [Symbol
block/leader Type/Dimension
style]: _SYMBOL
Symbol block or [?]<TEXT>: Text
Specify first point of leader line or [Symbol
block/leader Type/Dimension
style]: _TYPE
leader Type or [Straight/sPline]<sPline>: _Spline
Specify first point of leader line or [Symbol
block/leader Type/Dimension
style]: _DIMSTYLE
Dimension style or [?]<Aec_Arch_I$7>: Current
Specify first point of leader line or [Symbol
block/leader Type/Dimension
style]: _Endofparam

Specify first point of leader line or [Symbol
block/leader Type/Dimension
style]:
```
(Specify the beginning of leader at point P1 in Figure 8.26.)
```
Specify next point of leader line or [Symbol
block/leader Type/Dimension style]:
```
(Specify a vertex of the leader at point P2 in Figure 8.26.)
```
Specify next point of leader line or [Symbol
block/leader Type/Dimension style]:
```
(Specify the end of the leader location at point P3 in Figure 8.26.)
```
Specify next point of leader line or [Symbol
block/leader Type/Dimension style]: ENTER
```
(Press ENTER *to end the vertex selection for the leader.)*

Enter first line of text: SLIDE IN RANGE ENTER *(Type the desired text for the note.)*
Enter next line of text: ENTER *(Press ENTER to end text input for the leader note and complete the leader command as shown in Figure 8.26.)*

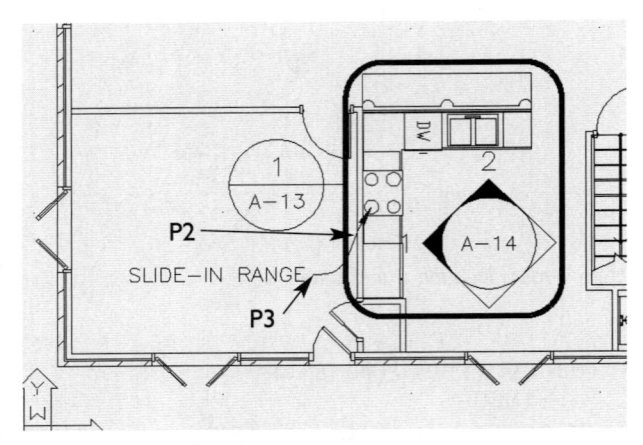

Figure 8.26 *Leader text inserted in the drawing*

23. Save the drawing.

MISCELLANEOUS

The annotation commands accessed through the **Miscellaneous** command on the **Annotation - Imperial** and **Annotation - Metric** toolbars include the datum point marker and the following folders: Dimensions, Fire Rating Lines, Match Lines, and North Arrows. These folders include commands for placing the annotation symbol and creating and placing the symbol on the correct layer. Access **Miscellaneous** as follows (Table 8.10).

Menu bar	Documentation>Miscellaneous
Command prompt	AECDCSETIMPMISCELLANEOUS
Toolbar	Select Miscellaneous from the Annotation - Imperial or Annotation - Metric toolbar as shown in Figure 8.27

Table 8.10 *Accessing the Miscellaneous comand*

CREATING DIMENSIONS

Dimensions can be placed to locate walls through the **Create Wall Dimensions** command on the **Walls** toolbar. The **Create Wall Dimensions** command, as discussed

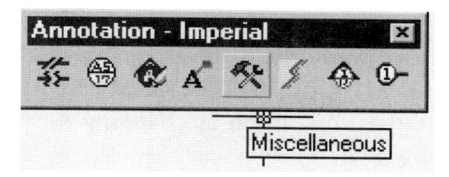

Figure 8.27 *Miscellaneous command on the Annotation - Imperial or Annotation - Metric toolbar*

in Chapter 2, can place many dimensions that are not necessary or are in the wrong position. Therefore, some of the dimensions placed with that command can be erased. Additional dimensions can be added through the **Dimension** toolbar or the **Miscellaneous** button of the **Annotation - Imperial** or **Annotation - Metric** toolbar.

> **Note:** The **Dimensions** option of the **Miscellaneous** button ensures that dimensions match the style and layer of other dimensions inserted with the **Create Wall Dimensions** command.

The content of the Dimensions folder is similar to the dimensioning commands on the **Dimension** toolbar; however, these dimensions are created according to the scale of the drawing and placed on the A-Anno-Dims layer, which has the color 221. Table 8.11 shows the dimension options included in the Dimensions folder.

Dimension Name	Dimension Option	Description
Aligned	DimAligned	Aligned dimensions (w/ layer keying)
Angular	DimAngular	Angular dimension (with layer keying)
Baseline	DimBaseline	Baseline dimension (with layer keying)
Continue	DimContinue	Continue dimension (with layer keying)
Linear	DimLinear	Linear dimension (with layer keying)
Radius	DimRadius	Radius dimension (with layer keying)

Table 8.11 *Accessing Dimensions options*

Dimensions are placed in the drawing in the same manner as if they were selected from the **Dimension** toolbar. To place a dimension in the drawing, select the dimension option from the palette of the Dimensions folder and drag the dimension into the drawing. Specify the beginning and end of the dimension and then specify the location of the dimension line. A dimension is placed with the Aec_Arch_I dimension style, included in the Aec_Arch (imperial) template. The scale of the dimension style (**DIMSCALE**) is specified when the scale of the drawing is specified through the **Set Drawing Scale** command (**AECDWGSCALESETUP**). You can edit dimensions placed through the **Dimension** option by changing the properties of the dimensioning style.

The command sequence that follows was used to place the linear dimension shown in Figure 8.28.

```
Command: _DimLinear
Specify first extension line origin or <select object>: end of ENTER (Select point P1 in Figure 8.28.)
Specify second extension line origin: end of ENTER (Select point P2 in Figure 8.28.)
Specify dimension line location or [Mtext/Text/Angle/Horizontal/Vertical/Rotated]: (Select point P3 in Figure 8.28.)
Dimension text = 18'-10 1/8"
```

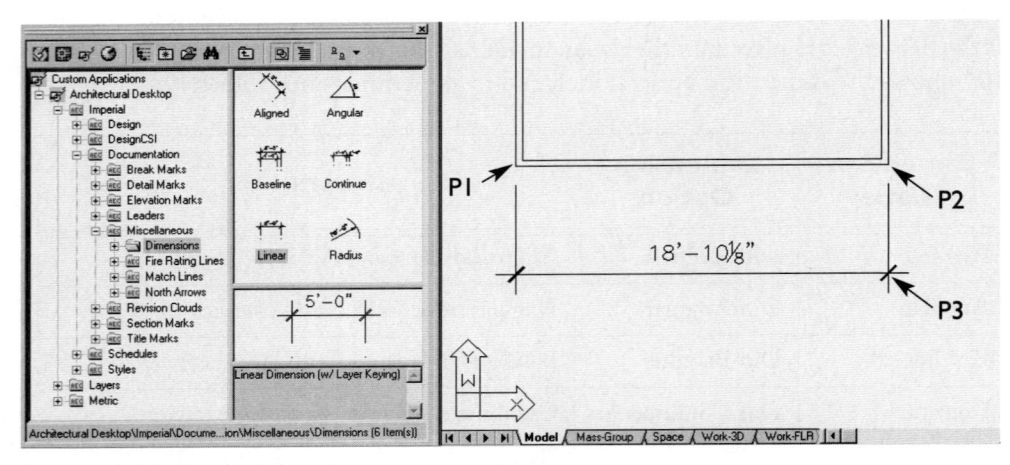

Figure 8.28 *Placing linear dimensions on a drawing*

As shown in Figure 8.28, the defaults for the Aec_Arch_I dimension style include tick marks for the arrowheads, architectural units, and a diagonal fraction bar. If a horizontal fraction bar is preferred, you can edit the fraction bar setting of the Aec_Arch_I dimension style, and all dimensions placed will be changed.

STEPS TO CHANGING THE FRACTION FORMAT OF THE AEC_ARCH_I STYLE

1. Select the **Dimension Style** button on the **Dimension** toolbar to open the **Dimension Style Manager**, as shown in Figure 8.29.

Annotating the Drawing **617**

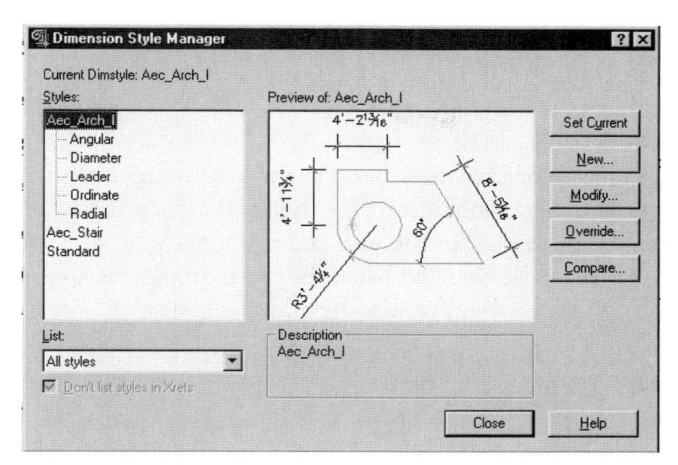

Figure 8.29 *Dimension Style Manager*

2. Select the **Modify** button of the **Dimension Style Manager** to open the **Modify Dimension Style: Aec_Arch_1** dialog box.

3. Select the **Primary Units** tab of the **Modify Dimension Style: Aec_Arch_1** dialog box as shown in Figure 8.30.

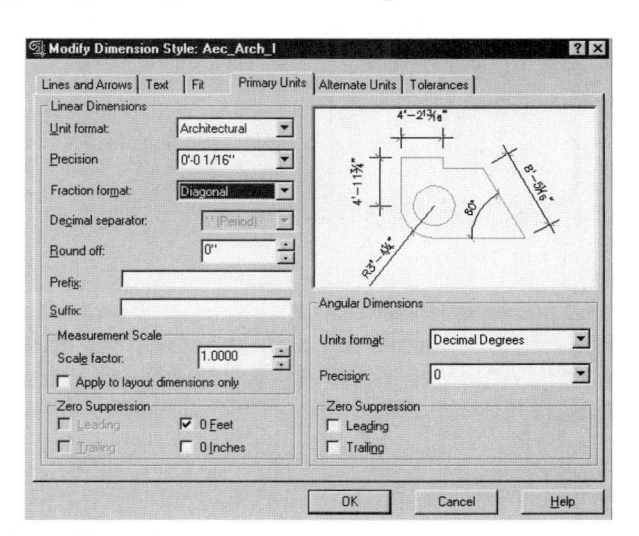

Figure 8.30 *Primary Units tab*

4. Select the **Fraction format** drop-down list and choose the Horizontal option.

5. Select **OK** to dismiss the **Modfiy Dimension Style: Aec_Arch_1** dialog box.

6. Select the **Close** button in the **Dimension Style Manager** dialog box.

All dimensions of the Aec_Arch_I dimension style should have a horizontal fraction bar.

FIRE RATING LINES

The Fire Rating Lines folder includes symbols for identifying walls according to the fire rating of the wall. The symbols create a polyline along the center of the wall with a specific linetype that distinguishes the wall as a fire-rated wall. The width, color, and linetype of the polyline can be changed when the fire rating line is applied. Figure 8.31 shows the fire rating lines of the Fire Rating Lines folder.

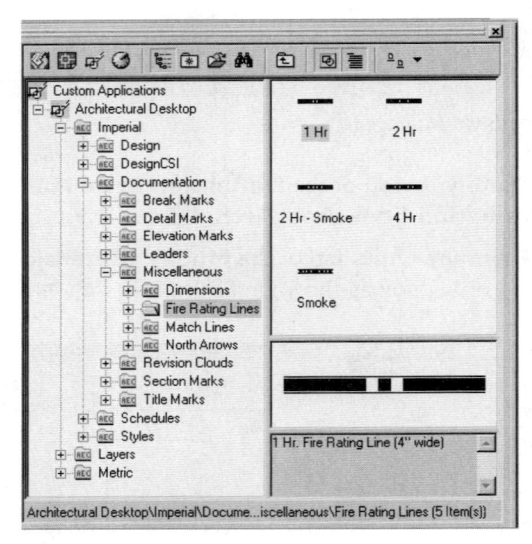

Figure 8.31 *Fire Rating Line palette*

The fire rating lines are created with the **ANNORATINGLINEADD** command. This command is executed when a fire rating line is selected from the DesignCenter. The options of this command include **Width, Color,** and **Linetype**. Each fire rating line has a default line width of at least 4" and its color is assigned by layer. The polyline is created on the A-Wall-Fire layer, which has the color 14 (brown). The linetype varies according to the hours of the fire rating. Table 8.12 lists the name and linetype option of the various wall ratings.

The fire rating line symbol cannot be inserted without the line being attached to an existing wall. When you select an option from the Fire Rating Lines folder and drag the symbol to the drawing area, you are prompted to select a wall. If you want to change the width of the polyline, you can enter a **W** to select the width option of the command and then you can enter a different width for the polyline width. The

Name of Fire Rating	Linetype Option	Description
1 Hr.	Aec_Rating_1Hr	1 Hr. Fire Rating Line (4" wide)
2 Hr.	Aec_Rating_2Hr	2 Hr. Fire Rating Line (4" wide)
2 Hr-Smoke	Aec_Rating_2Hr-Smoke	2 Hr. Fire & Smoke Rating Line (5" wide)
4 Hr	Aec_Rating_4Hr	4 Hr. Fire Rating Line (6" wide)
Smoke	Aec_Rating_Smoke	Smoke Rating Line (4" wide)

Table 8.12 *Accessing fire ratings*

width must be set prior to selecting the beginning or end of the wall. The width of existing fire rating lines can be also be changed with the **Width** option of the **PEDIT** command.

After selecting the wall, you are prompted to specify the first intersection and continuing intersections of walls to apply the fire rating symbol. The polyline will be placed centered about the width of the wall without the need to specify the beginning or end with an object snap mode. The vertices of the fire rating line will snap to wall endpoints without the use of object snap modes to specify the endpoints.

The command line sequence below was used to create the fire rating line for the wall shown in Figure 8.32.

```
Command: _AecAnnoRatingLineAdd
Adding: 1 Hr. Fire Rating Line (4" wide)
Select first end or intersection of rated wall or
[Width/Color/Linetype]: _WIDTH
Specify polyline width<4">: 4
Select first end or intersection of rated wall or
[Width/Color/Linetype]:
_COLOR Specify polyline color<BYLAYER>_ByLayer
Select first end or intersection of rated wall or
[Width/Color/Linetype]:
_LINETYPE
Specify polyline linetype or [?]<Aec_Rating_1Hr>:
Aec_Rating_1Hr
Select first end or intersection of rated wall or
```

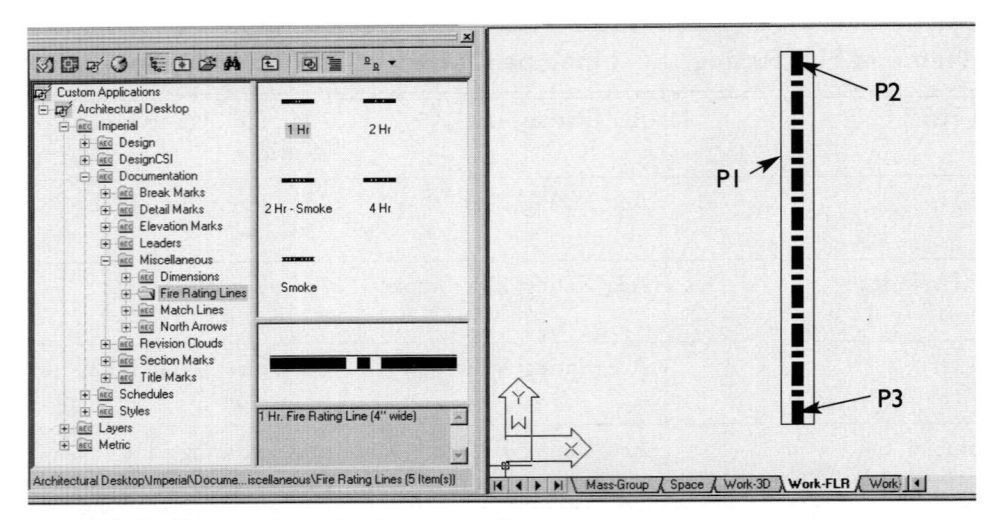

Figure 8.32 *Placing a fire rating line on a wall*

```
[Width/Color/Linetype]:
_Endofparam

Select first end or intersection of rated wall or
[Width/Color/Linetype]:
```
(Select the wall to apply the polyline at point P1 in Figure 8.32.)
```
Specify next end or intersection of rated wall or
[Width/Color/Linetype]:
```
(Select the beginning of the wall at point P2 in Figure 8.32.)
```
Specify next end or intersection of rated wall or
[Width/Color/Linetype]:
```
(Select the end of the wall near point P3 in Figure 8.32.)
```
Specify next end or intersection of rated wall or
[Width/Color/Linetype]: ENTER
```

MATCH LINES

Match lines are placed on a drawing to represent buildings that have dimensions too great for plotting the entire building on a single sheet. The match line provides a reference line in which the plotted drawings can be aligned. The Match Lines folder includes two match lines, as shown in Figure 8.33 and listed in Table 8.13.

Insert match lines in the drawing by selecting the match line option from the Match Line folder of the DesignCenter and continue holding down the left mouse button as you move the mouse to the drawing area. Releasing the mouse button over the drawing area allows you to specify the beginning and end of the match line. After you specify the length of the match line, the **Edit Attributes** dialog box opens, allowing you to insert the name of the match line. The match line created is a polyline. Match lines

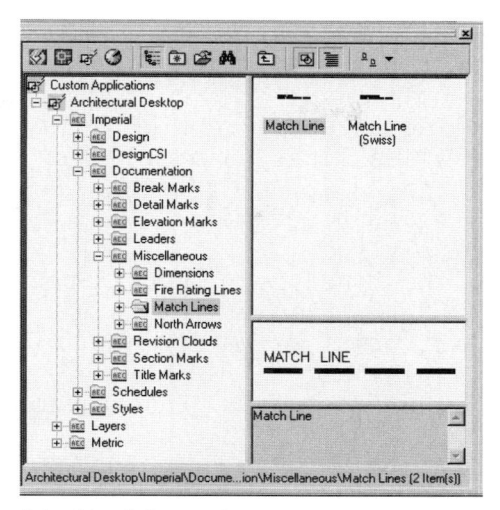

Figure 8.33 Options of the Match Line palette

Match Line	Symbol Option	Description
Match Line	Anno_Match_Line	Match Line
Match Line (Swiss)	Anno_Match_Line	Match Line (Swiss font)

Table 8.13 Match lines

are inserted with the **ANNOMATCHLINEADD** command, executed when a Match Line option is selected from the DesignCenter folder. The polyline width, color, and linetype are preset when the match line is selected from the DesignCenter. The match line symbol is placed on the A-Anno-Note layer, which has the color 211 (a hue of magenta).

The command line sequence below was used to create the match line shown in Figure 8.36.

```
Command: _AecAnnoMatchLineAdd
Adding: Match Line
Specify first point of match line or [Symbol/poly-
line Width/polyline
Color/polyline Linetype]: _SYMBOL
Specify symbol block name or [?]<Anno_Match_Line>:
Anno_Match_Line
Specify first point of match line or [Symbol/poly-
line Width/polyline
Color/polyline Linetype]: _WIDTH
```

```
Specify polyline width<1/16">: .0625
Specify first point of match line or [Symbol/poly-
line Width/polyline
Color/polyline Linetype]: _LINETYPE
Specify polyline linetype or [?]<Aec_Match_Line>:
Aec_Match_Line
Specify first point of match line or [Symbol/poly-
line Width/polyline
Color/polyline Linetype]: _COLOR Specify polyline
color<BYLAYER>_ByLayer
Specify first point of match line or [Symbol/poly-
line Width/polyline
Color/polyline Linetype]: _Endofparam

Specify first point of match line or [Symbol/poly-
line Width/polyline
Color/polyline Linetype]:
```
(Select the beginning of the match line at point P1 in Figure 8.34.)
```
Specify next point of match line or [Symbol/polyline
Width/polyline
Color/polyline Linetype]:
```
(Select the end of the match line at point P2 Figure 8.34.)

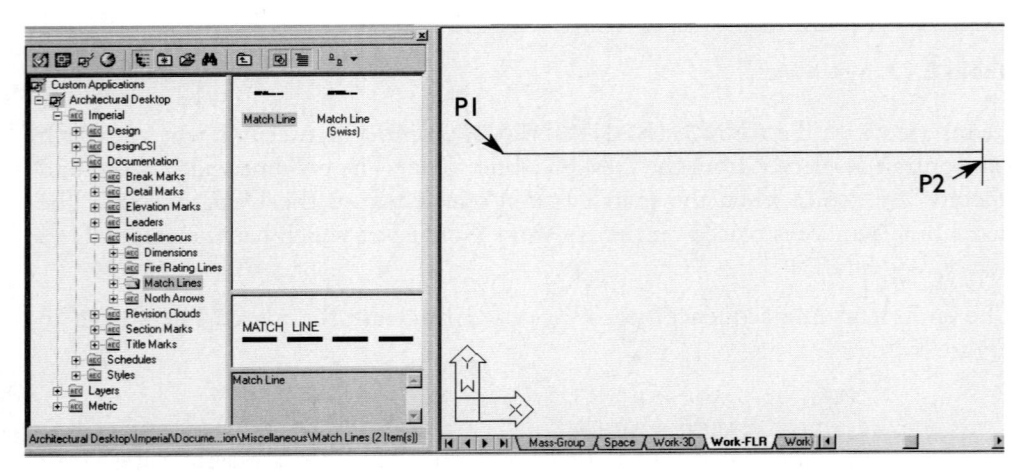

Figure 8.34 *Creating a match line*

```
Specify next point of match line <End> or
[Symbol/polyline Width/polyline
Color/polyline Linetype]: ENTER
```
(Pressing ENTER *ends selecting the location of the match line and opens the Edit Attributes dialog box as shown in Figure 8.35.)*

```
Select block reference: (Type the name of the match line in the
Edit Attributes dialog box as shown in Figure 8.35, and then select the OK
button to dismiss the Edit Attributes dialog box.)
Command:
```

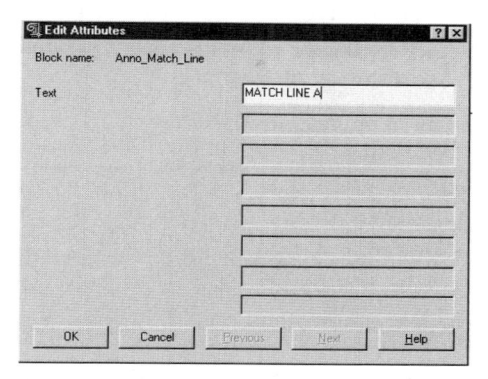

Figure 8.35 Specifying the name of the match line

The match line is inserted as shown in Figure 8.36.

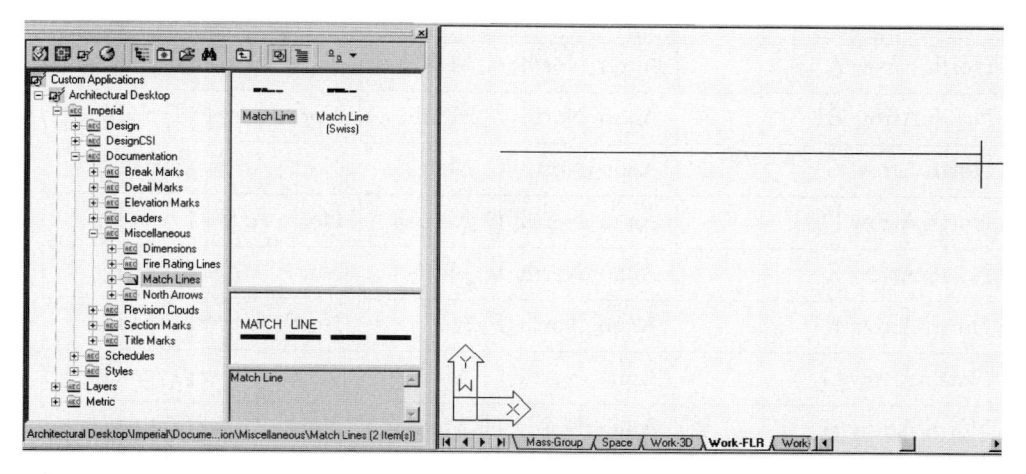

Figure 8.36 Match Line A placed in the drawing

NORTH ARROWS

The North Arrows folder includes 13 types of north arrows. When you select the north arrow from the DesignCenter, the **ANNOSYMBOLADD** command is executed to insert the north arrow symbol. All the north arrow options allow you to specify the location for the center of the symbol and the rotation. The north arrow is placed on the A-Anno-Note layer. The north arrow options shown in the DesignCenter in Figure 8.37 and listed in Table 8.8 can be selected.

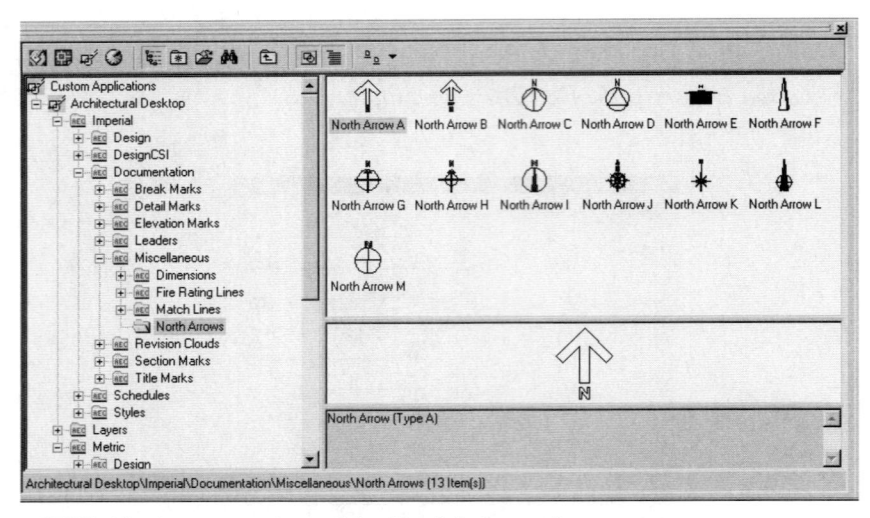

Figure 8.37 *North arrow options of the North Arrows palette*

Name of North Arrow	Symbol Option	Description
North Arrow A	Anno_North_A_Metric	North Arrow (Type A)
North Arrow B	Anno_North_B_Metric	North Arrow (Type B)
North Arrow C	Anno_North_C_Metric	North Arrow (Type C)
North Arrow D	Anno_North_D_Metric	North Arrow (Type D)
North Arrow E	Anno_North_E_Metric	North Arrow (Type E)
North Arrow F	Anno_North_F_Metric	North Arrow (Type F)
North Arrow G	Anno_North_G_Metric	North Arrow (Type G)
North Arrow H	Anno_North_H_Metric	North Arrow (Type H)
North Arrow I	Anno_North_I_Metric	North Arrow (Type I)
North Arrow J	Anno_North_J_Metric	North Arrow (Type J)
North Arrow K	Anno_North_K_Metric	North Arrow (Type K)
North Arrow L	Anno_North_L_Metric	North Arrow (Type L)
North Arrow M	Anno_North_M_Metric	North Arrow (Type M)

Table 8.14 *North Arrow options*

To place a north arrow in a drawing, select a north arrow option from the North Arrows folder and continue to hold down the left mouse button as you move the mouse over the drawing area. Release the mouse button over the drawing to specify the location of the center of the north arrow. After selecting the center location of the north arrow you can type the rotation from due east or select the angle with the mouse in the drawing area.

The command line sequence below was used to create the north arrow shown in Figure 8.38.

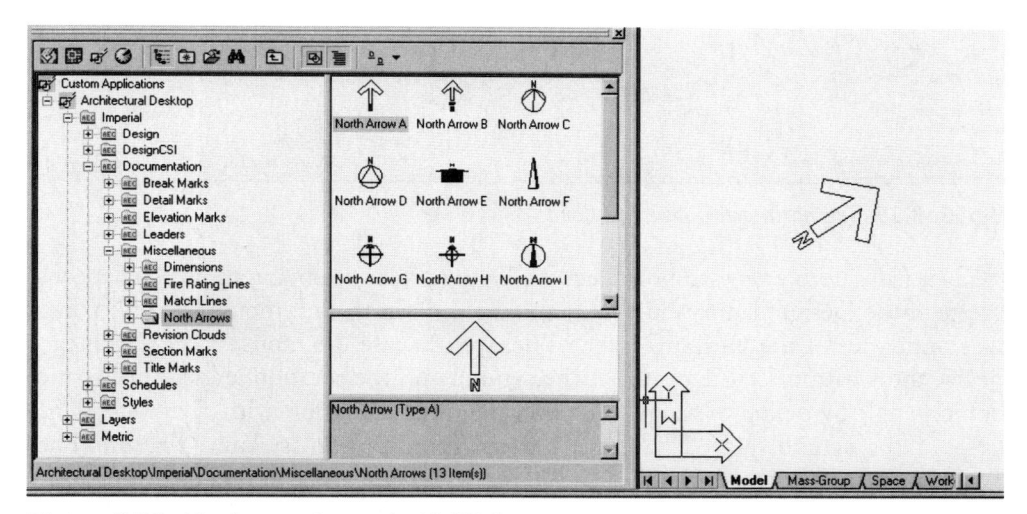

Figure 8.38 *North arrow inserted with 35-degree rotation*

```
Command: _AecAnnoSymbolAdd
Adding: North Arrow (Type A)
Specify insertion point or [Symbol]: _SYMBOL
Specify symbol block name or
[?]<Anno_North_A_Metric>: Anno_North_A_Metric
Specify insertion point or [Symbol]: _Endofparam
Specify insertion point or [Symbol]:
```
(Select the center location of the north arrow symbol shown as point P1 in Figure 8.39.)
`Specify rotation: 35` *(Type the angle of rotation from due east.)*
(The north arrow rotated 35 degrees is shown in Figure 8.39.)

DATUM ELEVATION

The datum point symbol is the only Miscellaneous symbol not included in a folder. It is used to specify the vertical distance from a reference elevation of a building component. Datum point symbols are often used on section views or plot plans. The symbol is placed in the drawing with the **ANNOSYMBOLADD** command.

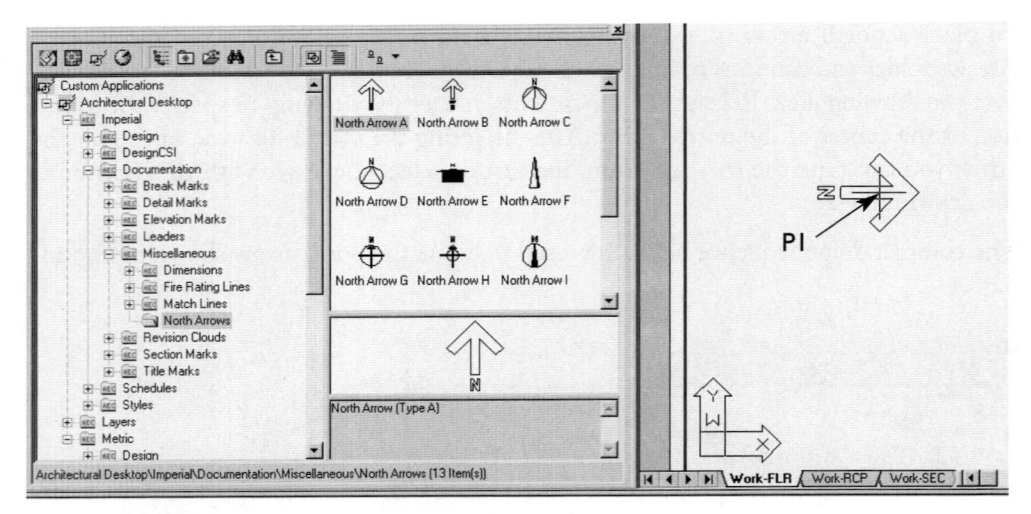

Figure 8.39 *The north arrow insertion point*

To place the datum point symbol, select the Datum Point symbol in the Miscellaneous palette of the DesignCenter and continue to hold down the left mouse button to drag the symbol out of the DesignCenter. When you release the mouse button, you can specify the location for the center of the symbol and the rotation angle. Specify the rotation angle by typing the angle in the command line or select with the mouse. The name of the symbol placed with the Datum Point option is Anno_Datum. The symbol is placed on the A-Anno-Symb layer, which has the color 111.

The command line sequence below was used to create the datum point symbol shown in Figure 8.40.

```
Command: _AecAnnoSymbolAdd
Adding: Datum Point
Specify insertion point or [Symbol]: _SYMBOL
Specify symbol block name or [?]<Anno_Datum>:
Anno_Datum
Specify insertion point or [Symbol]: _Endofparam
```

Specify insertion point or [Symbol]: *(Specify the location for the center of the symbol at point P1 in Figure 8.40.)*
Specify rotation: 90 *(Type the rotation of the datum point symbol relative to due east.)*

REVISION CLOUDS

The revision cloud allows you to identify an area of the drawing that has been revised or add a note that applies to this portion of the building. The revision cloud is created free style with a series of arcs connected to enclose an area of the drawing.

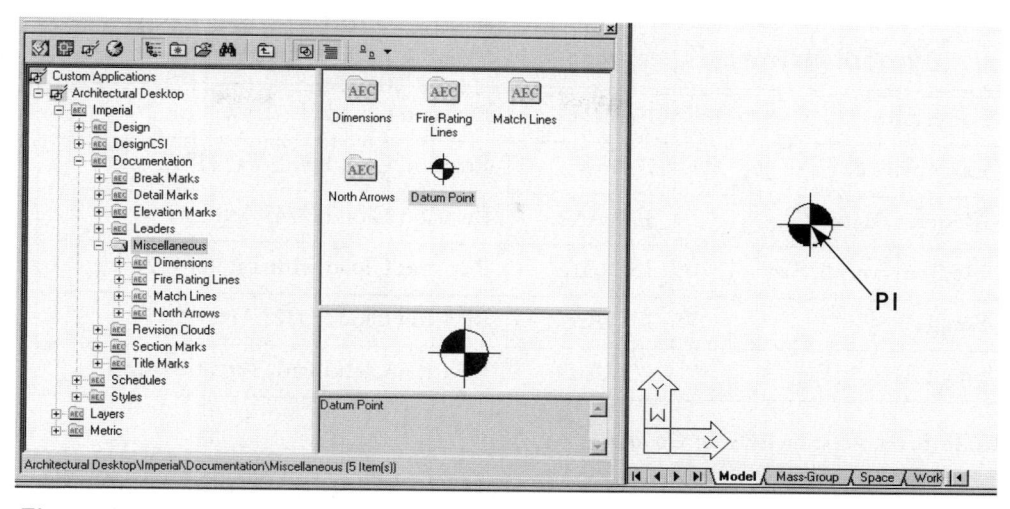

Figure 8.40 *Specifying the datum point location*

Access **Revision Cloud** command as follows (Table 8.15).

Menu bar	Documentation>Revision Cloud
Command prompt	ANNOREVISIONCLOUDADD
Toolbar	Select Revision Cloud from the Annotation - Imperial or Annotation - Metric toolbar as shown in Figure 8.41

Table 8.15 *Accessing Revision Clouds*

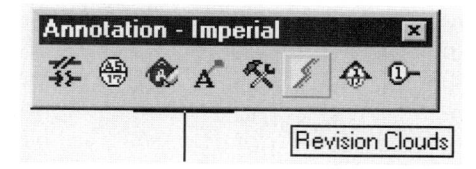

Figure 8.41 *The Revision Cloud command on the Annotation - Imperial toolbar*

When you select a revision cloud from the DesignCenter, the **ANNOREVISION-CLOUDADD** command is executed. The command allows you to draw a series of polyline arcs. The width of the arc is adjusted according to the scale factor of the drawing. Table 8.16 shows the six types of revision clouds in the Revision Clouds folder. The radius of the arc varies according to the symbol selected from the palette.

Symbol Name	Arc	Description
Large Arcs	1" Arcs	Revision Cloud (1" Arcs)
Large Arcs & Tag	1" Arcs	Revision Cloud with Tag (1" Arcs)
Medium Arcs	3/4" Arcs	Revision Cloud (3/4" Arcs)
Medium Arcs & Tag	3/4 " Arcs	Revision Cloud with Tag (3/4" Arcs)
Small Arcs	1/2" Arcs	Revision Cloud (1/2" Arcs)
Small Arcs & Tag	1/2" Arcs	Revision Cloud with Tag (1/2" Arcs)

Table 8.16 *Acessing Revision Cloud arcs*

The arc radius of the revision cloud is scaled according to the drawing scale factor defined by the **DWGSCALESETUP** command. Revision clouds are placed on the A-Anno-Revs layer, which has the color 213 (a hue of magenta).

To place a revision cloud, select a Revision Cloud symbol from the palette of the DesignCenter and continue to hold down the left mouse button to drag the symbol out of the DesignCenter. When you release the mouse button, you are prompted to specify the beginning of the cloud and move the mouse in the direction to define the cloud. The arcs of the revision cloud form a continuous chain and will close to the first arc start point.

The command line sequence below was used to create the north arrow shown in Figure 8.42.

```
Command: _AecAnnoRevisionCloudAdd
Adding: Revision Cloud (1" Arcs)
Specify cloud starting point or [Symbol block/pline
Color/Arc length/pline
Width]: _SYMBOL
Specify symbol block or [?]<NONE>: None
Specify cloud starting point or [Symbol block/pline
Color/Arc length/pline
Width]: _ARCLENGTH
Specify arc length <1/2">: 1
Specify cloud starting point or [Symbol block/pline
Color/Arc length/pline
Width]: _WIDTH
Specify cloud polyline width<1/16">: .03125
Specify cloud starting point or [Symbol block/pline
Color/Arc length/pline
```

```
Width]: _COLOR Specify cloud color<BYLAYER>_ByLayer
Specify cloud starting point or [Symbol block/pline
Color/Arc length/pline
Width]: _Endofparam

Specify cloud starting point or [Symbol block/pline
Color/Arc length/pline
Width]:
```
(Select point P1 in Figure 8.42 to start the revision cloud.)
```
Cloud will close when returned to start point...
Guide crosshairs along cloud path (counter-clock-
wise)
```
(Move the mouse in the direction to form the enclosure and close the revision cloud.)

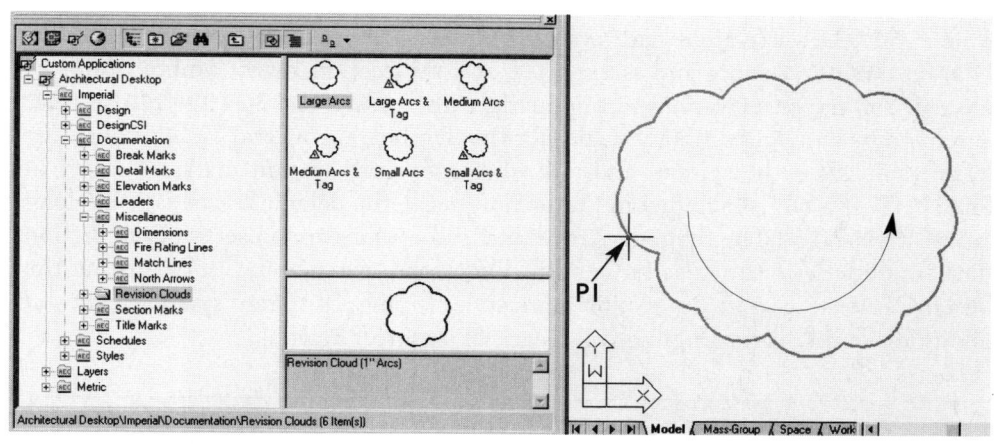

Figure 8.42 *Creating a revision cloud*

SECTION MARKS

Section marks are placed in a drawing to specify the location of the cutting plane line and the section identifier. The section identifier can include the section number and or the sheet number where the section is located. The section identifier is placed in the drawing by the attributes of the section mark block. Access **Section Marks** command as follows (Table 8.17).

Menu bar	Documentation>Section Marks
Command prompt	ANNOSECTIONMARKADD
Toolbar	Select Section Marks from the Annotation - Imperial or Annotation - Metric toolbar as shown in Figure 8.43

Table 8.17 *Accessing the Section Marks command*

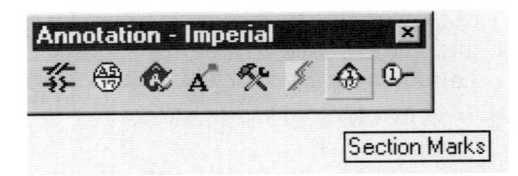

Figure 8.43 *Section Marks command on the Annotation - Imperial toolbar*

The Section Marks folder includes four section mark symbols. Each symbol option includes a cutting plane line and a section mark symbol for identifying the section. The command to place the section mark allows you to specify the length of the section line and direction. The size of the symbol and the associated text is scaled according to the scale factor specified in the **DWGSCALESETUP** command.

When you select a section mark option from the DesignCenter, the **ANNOSECTIONMARKADD** command is executed and the section mark symbol option is selected. You can create various section marks with the **ANNOSECTIONMARKADD** command by applying different values for the Symbol, Arrow and Tail options. The default values for the symbol and tail options of each section mark are listed in Table 8.18. If a tail is included in the section mark, the default length is 3/16" long and width is 1/16". The Symbol, Arrow and Tail options are preset for each section mark included in the DesignCenter. Therefore, you should select from the DesignCenter to change the section mark style. Typing a different symbol, arrow, or tail option in the command line will create an incorrect symbol.

Symbol Name	Symbol Option	Tail	Description
Section Mark A1	Anno_Section_A1	Anno_Arrow_A1	Section mark
Section Mark A1T	Anno_Section_A1	Anno_Arrow_A1	Section mark with tail
Section Mark A2	Anno_Section_A2	Anno_Arrow_A2	Section mark with sheet number
Section Mark A2T	Anno_Section_A2	Anno_Arrow_A1	Section mark with sheet number and tail

Table 8.18 *Section Mark options*

The section mark is placed on the A_Sect_Iden layer, which has the color 111 (a hue of green).

To place a section mark in the drawing, select **Section Mark** from the **Annotation - Imperial** or **Annotation - Metric** toolbar to open the AutoCAD DesignCenter. Select a section mark from the Section Marks palette and continue to hold down the left mouse button as you drag the symbol to the drawing area. When you release the mouse button over the drawing area, you can select the beginning and end of the section line for the section mark. The **Edit Attributes** dialog box is used to enter the section number and the sheet number of the drawing.

When you place the section mark, you are prompted Add AEC Section Object? A yes response to this prompt will use the section line as a section object for the development of a section using the **BLDGSECTIONLINEGENERATE** command. The use of the **BLDGSECTIONLINEGENERATE** command to generate a section drawing from the model will be discussed in Chapter 9.

The command line sequence below was used to create the Section Mark A2T shown in Figure 8.46.

```
Command: _AecAnnoSectionMarkAdd
Adding: Section Mark (w/Sheet No. & Tail)
Specify first point of section line or
[Symbol/Arrow/Tail]: _SYMBOL
Symbol block name or [?]<Anno_Section_A2>:
Anno_Section_A2
Specify first point of section line or
[Symbol/Arrow/Tail]: _ARROW
Arrow block name or [?]<Anno_Arrow_A2>:
Anno_Arrow_A2
Specify first point of section line or
[Symbol/Arrow/Tail]: _TAIL
Tail length,width<0.188,0.063>: .1875,.0625
Specify first point of section line or
[Symbol/Arrow/Tail]: _Endofparam

Specify first point of section line or
[Symbol/Arrow/Tail]:
```
(Select point P1 in Figure 8.44.)
```
Specify next point of line or [Symbol/Arrow/Tail]:
```
(Select point P2 in Figure 8.44.)
```
Specify next point of line or
[Break/Symbol/Arrow/Tail]:  ENTER
```
(Press ENTER to end selection of vertices of section line.)

```
Select block reference:
```
(Type section number and sheet number in the Edit Attributes dialog box as shown in Figure 8.45.)

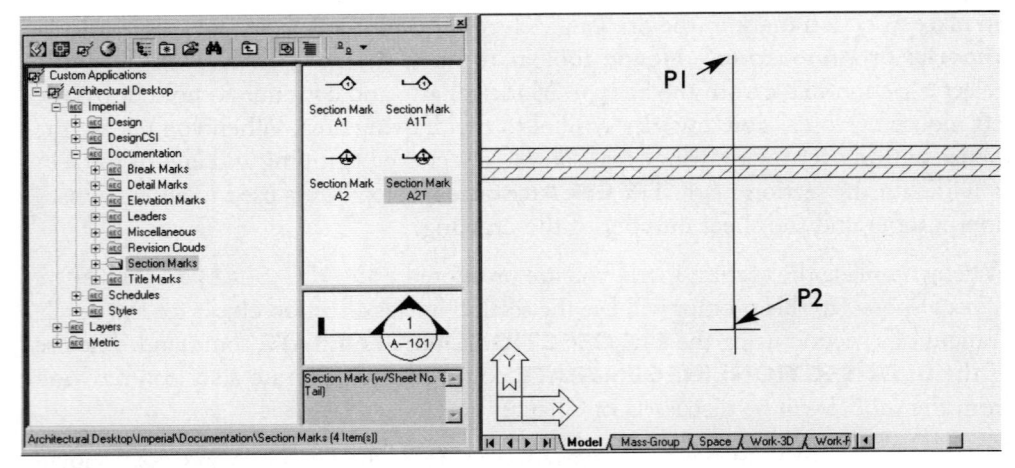

Figure 8.44 *Specifying the location of the section mark*

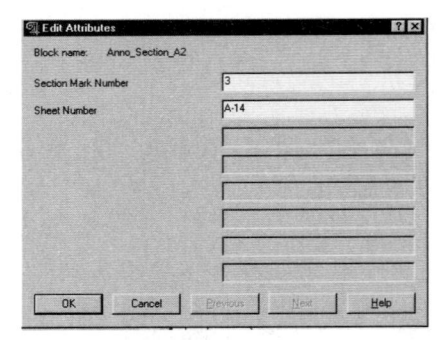

Figure 8.45 *Defining the name of the section mark in the Edit Attributes dialog box*

```
Specify side for Arrow:  (Select point P3 in Figure 8.46 to specify
viewing direction of the section.)
Add AEC section object? [Yes/No] <N>:  ENTER  (Press
ENTER to decline the use of the section mark as a section object.)
Command:
```

TITLE MARKS

The **Title Marks** command includes options for placing titles of a drawing or a bar scale. The scale of the drawing is included in the title and bar scale symbols. The scale is preset according to the scale specified in the **DWGSCALESETUP** command. See Access **Title Marks** as shown in Table 8.19.

When you select the **Title Marks** command, the AutoCAD DesignCenter opens to the Title Marks folder, which includes two title marks and a bar scale symbol as shown in Figure 8.48. The symbols are placed on the A-Anno-Ttlb layer, which has the color 111 (a hue of green).

Annotating the Drawing 633

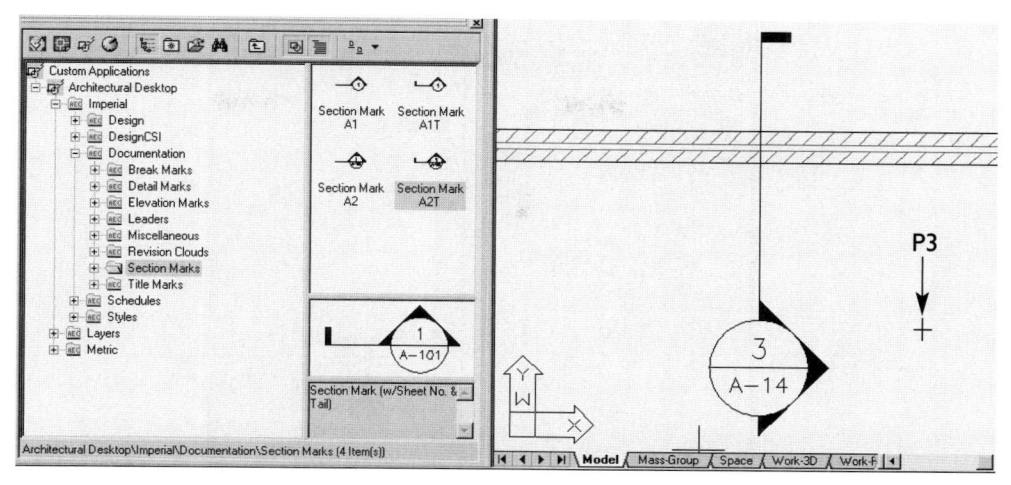

Figure 8.46 *Specifying the viewing direction of the section mark*

Menu bar	Documentation>Title Marks
Toolbar	Select Title Marks from the Annotation - Imperial or Annotation - Metric toolbar as shown in Figure 8.47

Table 8.19 *Accessing Title Marks*

Figure 8.47 *The Title Marks command on the Annotation - Imperial toolbar*

Bar Scale

Place the Bar Scale symbol in the drawing by selecting the Bar Scale (Inches) option from the Title Marks folder and dragging the mouse to the drawing area and then releasing the mouse button. After releasing the mouse button, you specify the location and rotation of the bar scale in the drawing. The Bar Scale symbol is placed with the **ANNOBARSCALEADD** command.

The command sequence that follows was used to place the bar scale shown in Figure 8.49.

```
Command: _AecAnnoBarScaleAdd
Adding: Bar Scale (Inches)
```

[Figure 8.48 screenshot]

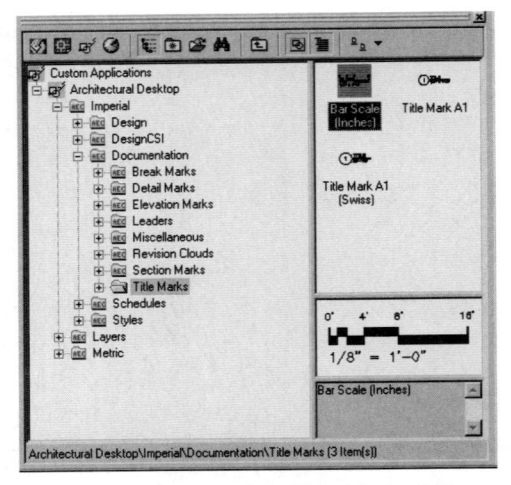

Figure 8.48 *Title marks and the bar scale of the Title Marks folder*

```
Specify insertion point or [Symbol]: _SYMBOL
Specify symbol block name or [?]<Anno_Bar_Scale_I>:
Anno_Bar_Scale_I
Specify insertion point or [Symbol]: _Endofparam

Specify insertion point or [Symbol]:
```
(Select point P1 in Figure 8.49.)
```
Specify rotation: 0
```
(Type the degrees of rotation from due east to specify the rotation of the bar scale.)

[Figure 8.49 screenshot]

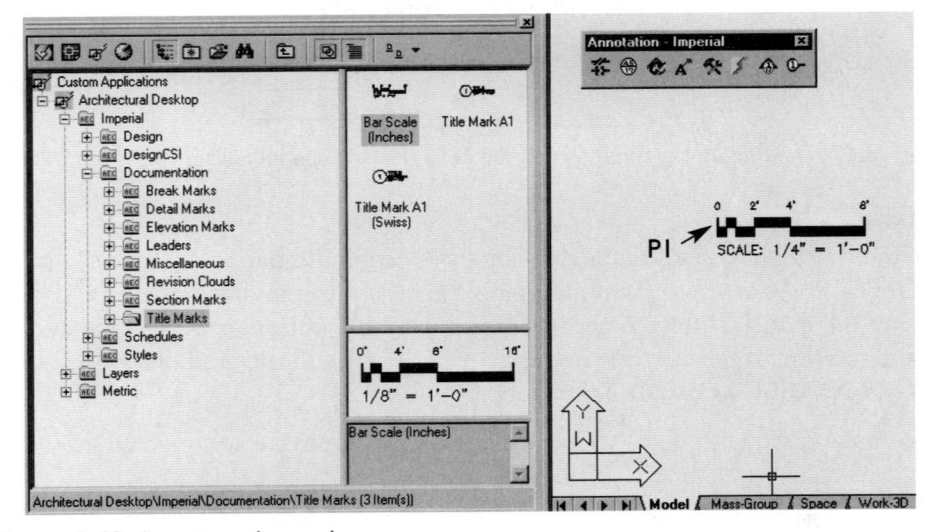

Figure 8.49 *Inserting a bar scale*

The text of the Bar Scale can be edited with the **DDATTE** command, which opens the **Edit Attributes** dialog box, allowing you to type the text changes.

Title Marks

The Title Marks palette includes two options for placing a title, as shown in Figure 8.48. The scale of the drawing is preset in the **DWGSCALESETUP** command. The two title mark symbols included in the palette are Title Mark A1 and Title Mark A1 (Swiss). Title marks are placed in the drawing with the **ANNOTITLEMARKADD** command. This command and its options are preset when a title mark is selected from the DesignCenter.

To place a title mark symbol in the drawing, select a Title Mark symbol from the DesignCenter and continue to hold down the left mouse button as you drag the symbol into the drawing area. When you release the mouse button, you can specify the location and rotation for the symbol. After you specify the location and rotation, the **Edit Attributes** dialog box opens, allowing you to enter the title number, title, and scale of the symbol.

The command sequence that follows was used to place the Title Mark A1 symbol.

```
Command: _AecAnnoTitleMarkAdd
Adding: Title Mark (w/ Scale)
Specify location of symbol or [Symbol block/TExt block/pline Width]: _SYMBOL
Specify symbol block name or [?]<Anno_Title_A1>: Anno_Title_A1
Specify location of symbol or [Symbol block/TExt block/pline Width]: _TEXT
Specify text block name or [?]<Anno_Title_T1>: Anno_Title_T1
Specify location of symbol or [Symbol block/TExt block/pline Width]: _TAG
Specify scale attribute tag<Scale>: Scale
Specify location of symbol or [Symbol block/TExt block/pline Width]: _WIDTH
Specify polyline width<0">: .025
Specify location of symbol or [Symbol block/TExt block/pline Width]: _Endofparam

Specify location of symbol or [Symbol block/TExt block/pline Width]:
```
(Select point P1 in Figure 8.52 to specify the location of the title mark.)
`Select block reference:` *(Type the title mark number in the Edit Attributes dialog box, as shown in Figure 8.50, and then select OK to dismiss the dialog box.)*

```
Command: _DDATTE
Select block reference: (Type the name and scale of the drawing
in the Edit Attributes dialog box as shown in Figure 8.51 and then select
OK to dismiss the Edit Attributes dialog box.)
```

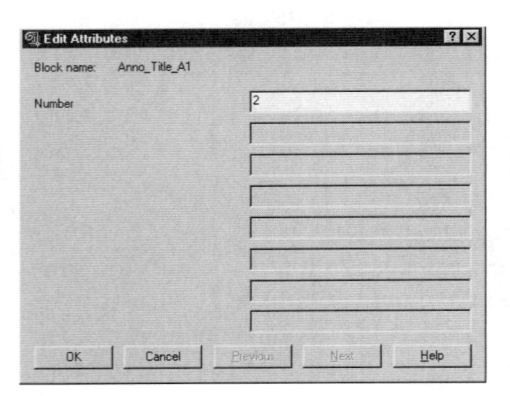

Figure 8.50 *Specifying the title number in the Edit Attributes dialog box*

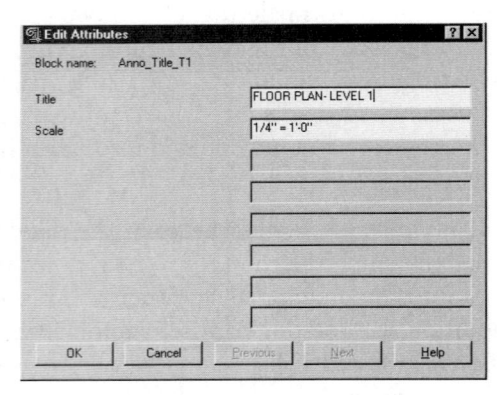

Figure 8.51 *Entering the title and scale of the drawing in the Edit Attributes dialog box*

```
Command:
Specify endpoint of line: (Select point P2 to specify the length of
the title bar as shown in Figure 8.52.)
```

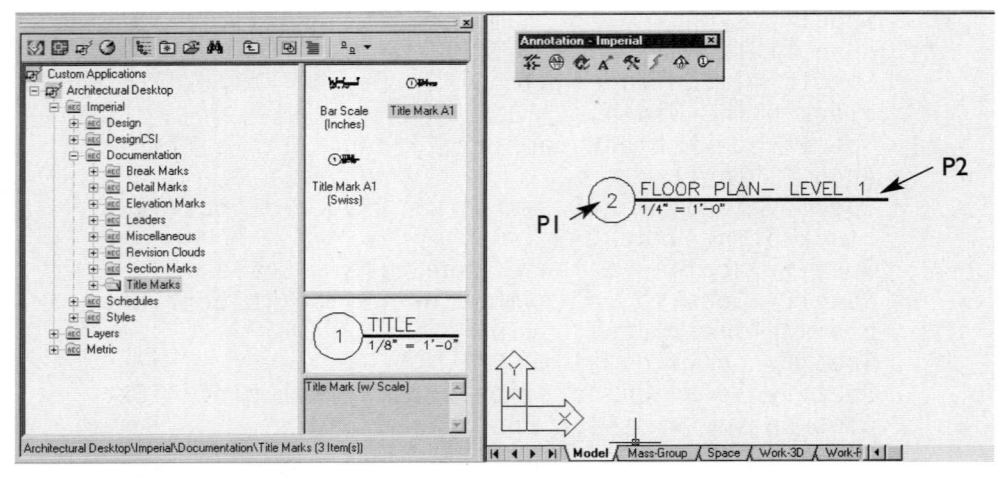

Figure 8.52 *Specifying the location of the title mark in the drawing*

TUTORIAL 8.2 PLACING DIMENSIONS

1. Open *Ex8-2.dwg*.

2. Choose **File>SaveAs** from the menu bar and save the drawing as *Annotation2.dwg* in your student directory.
3. Right-click over an Architectural Desktop toolbar and choose **Annotation - Imperial** from the toolbar list.
4. Select an exterior wall, and then right-click and choose **Edit Wall Style** to open the **Wall Style Properties** dialog box.
5. Select the **Display Props** tab, and then select the Wall Style property source and select the **Edit Display Props** button to open the **Entity Properties** dialog box, as shown in Figure 8.53.

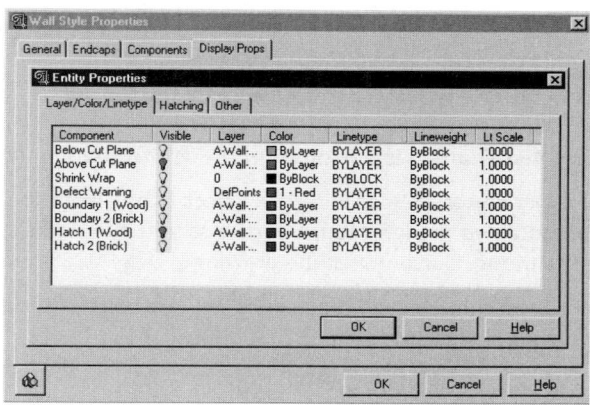

Figure 8.53 *Turning off the display of wall components to enhance dimensioning*

6. Turn off Boundary 2 (Brick) and Hatch 2 (Brick) and then select **OK** to dismiss the **Entity Properties** dialog box.
7. Select **OK** to dismiss the **Wall Style Properties** dialog box. The brick component is not displayed, which will enhance the selection of the wood stud component during dimensioning.
8. Select **Zoom Window** from the **Standard** toolbar and zoom a window around the lower left corner of house as shown in the following command sequence.

```
Command: '_zoom
Specify corner of window, enter a scale factor (nX
or nXP), or
[All/Center/Dynamic/Extents/Previous/Scale/Window]
<real time>: _w
Specify first corner: (Select point P1 in Figure 8.54.)
Specify opposite corner: (Select point P2 in Figure 8.54.)
```

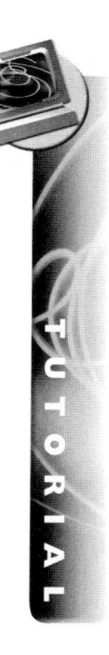

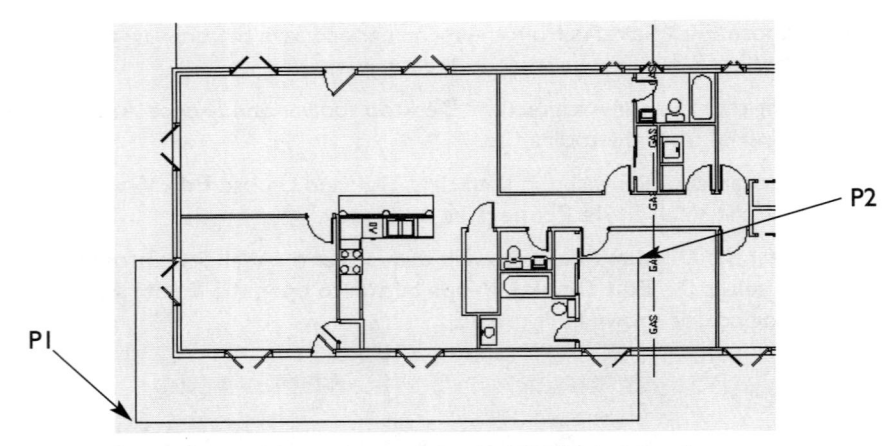

Figure 8.54 *Specifying the window locations for the ZOOM command*

9. Select the **Miscellaneous** button from the **Annotation - Imperial** toolbar.
10. Select the Dimensions folder, and then select Linear from the palette and continue to hold down the left mouse button as you move to the drawing area. Release the mouse button over the drawing area.
11. Select the extension line and dimension line locations as specified in the following command sequence.

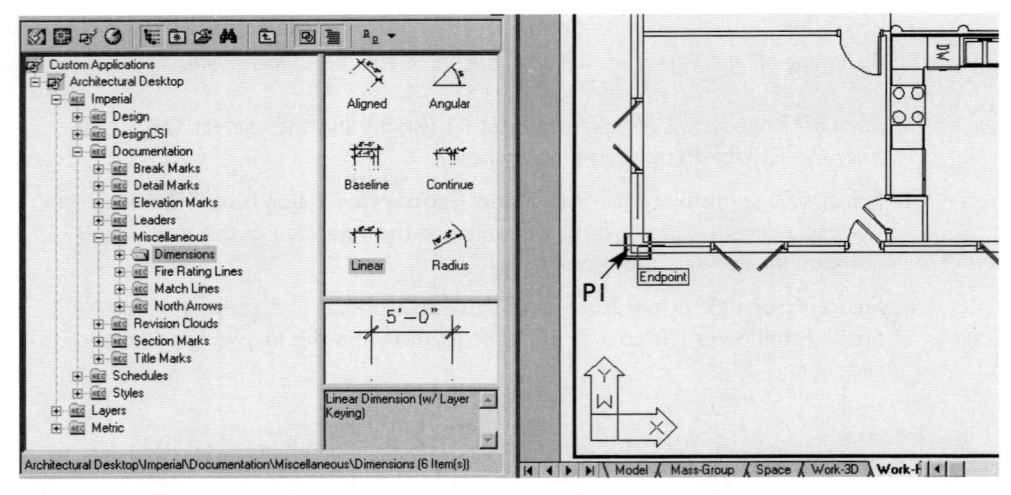

Figure 8.55 *Specifying the location of the extension line for the dimension*

```
Command: _DimLinear
Specify first extension line origin or <select
object>: end ENTER
```

of *(Select the endpoint of the exterior wall at P1 in Figure 8.55.)*
Specify second extension line origin: mid ENTER
of *(Select the midpoint of the window at P2 in Figure 8.56.)*
Specify dimension line location or
[Mtext/Text/Angle/Horizontal/Vertical/Rotated]: *(Select point P3 in Figure 8.56.)*
Specify dimension line location or
[Mtext/Text/Angle/Horizontal/Vertical/Rotated]:
Dimension text = 6'-10 13/16"

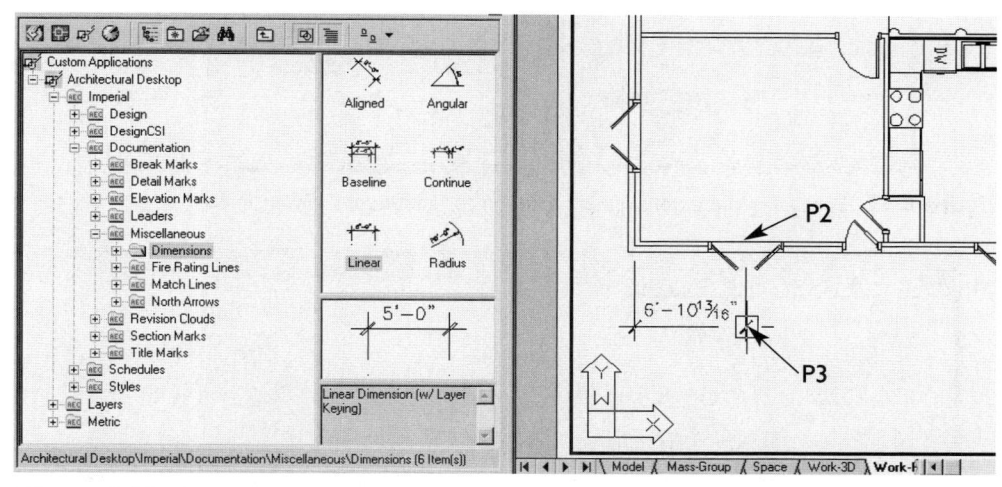

Figure 8.56 *Specifying the extension line location for a window*

12. Add additional dimensions by dragging Continue from the Dimensions palette.

13. Select the extension line for continuous dimensions as specified in the following command sequence.

    Command: _DimContinue
    Specify a second extension line origin or
    [Undo/Select] <Select>: mid ENTER
    Of *(Select the front door at point P1 in Figure 8.57.)*
    Dimension text = 6'-8 3/16"
    Specify a second extension line origin or
    [Undo/Select] <Select>: mid ENTER
    Of *(Select the window at point P2 in Figure 8.58.)*
    Dimension text = 9'-10 5/8"
    Specify a second extension line origin or
    [Undo/Select] <Select>: mid ENTER
    Of *(Select the window at point P3 in Figure 8.59.)*
    Dimension text = 18'-0 3/8"

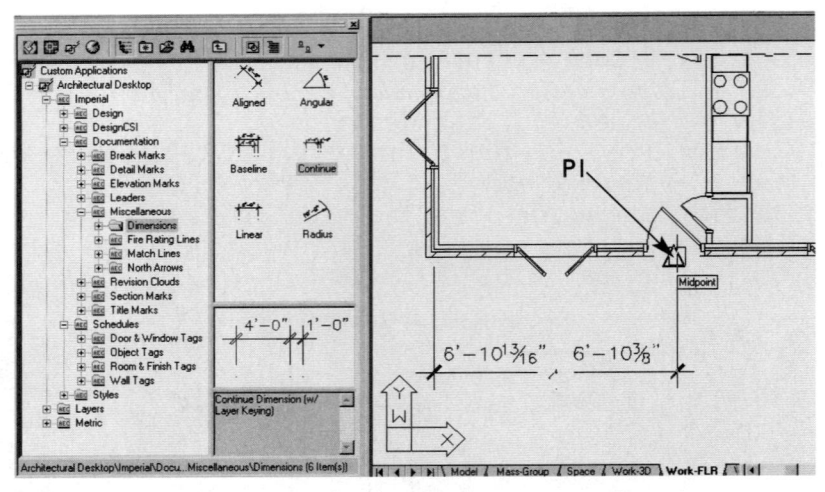

Figure 8.57 *Specifying the extension line location for continuous dimensions*

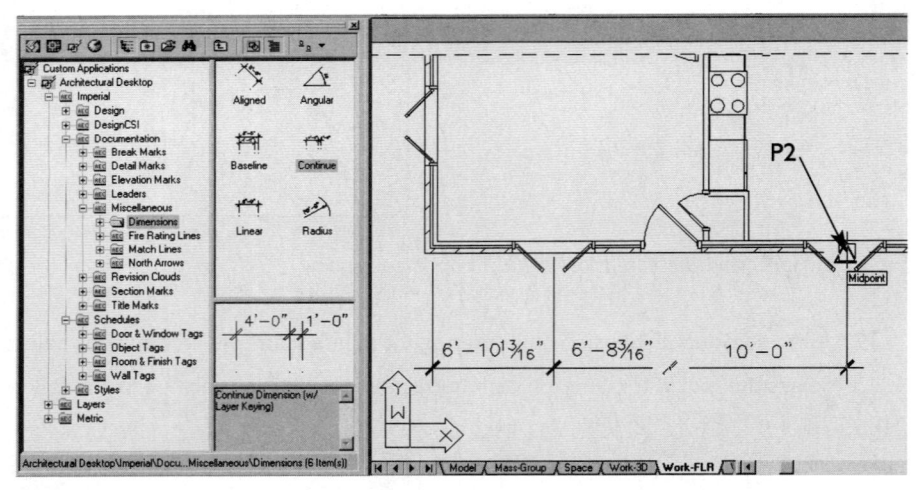

Figure 8.58 *Specifying the extension line location for dimensioning the window*

```
Specify a second extension line origin or
[Undo/Select] <Select>: mid ENTER
Of (Select the window at point P4 in Figure 8.59.)
Dimension text = 13'-6"
Specify a second extension line origin or
[Undo/Select] <Select>: end ENTER
of (Select the window at point P5 in Figure 8.59.)
Dimension text = 13'-3"
Specify a second extension line origin or
```

```
[Undo/Select] <Select>: ENTER (Press ENTER to end the dimen-
sion chain.)
Select continued dimension: ENTER (Press ENTER to end con-
tinued dimensioning.)
```

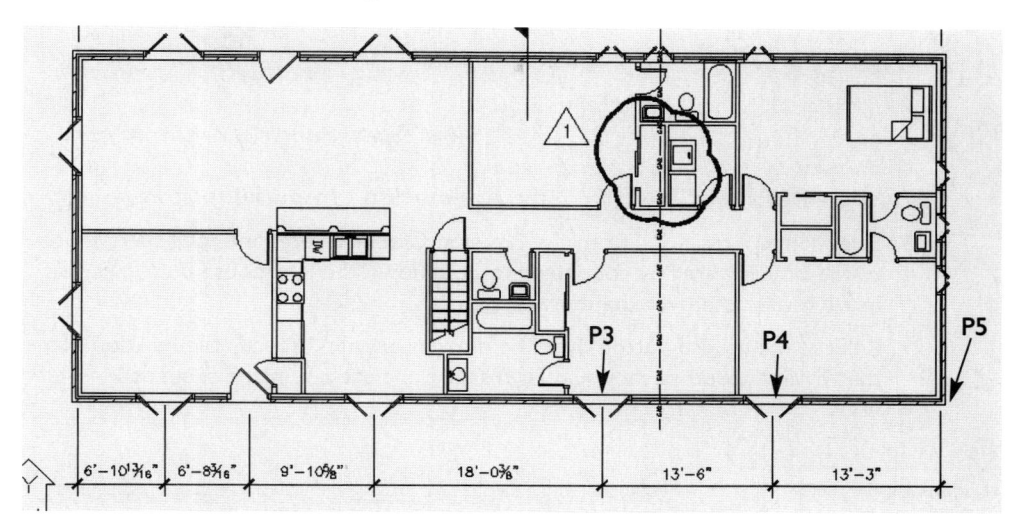

Figure 8.59 *Extension line locations to complete the chain of dimensions*

14. Select the North Arrows folder in the Tree of the DesignCenter.
15. Select North Arrow C from the palette of the DesignCenter. Continue holding down the left mouse button and release over the drawing area.
16. Specify the location and rotation of the north arrow as described in the following command sequence.

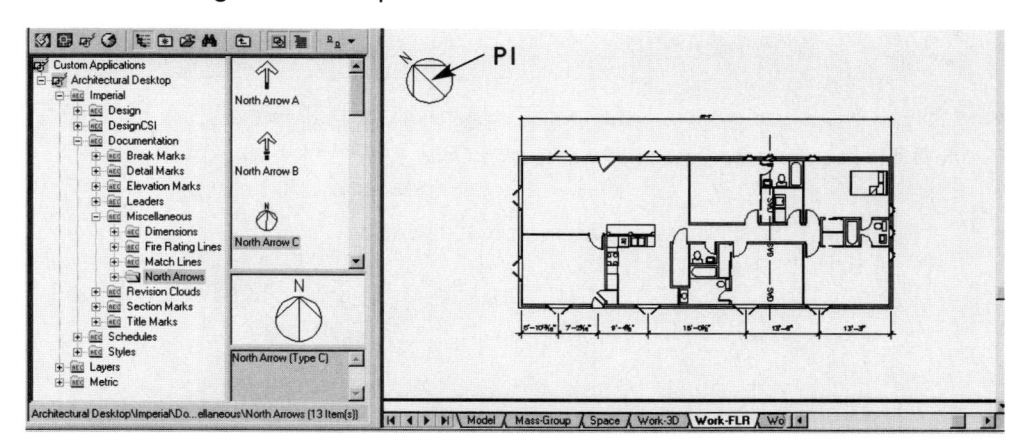

Figure 8.60 *Specifying the location of the north arrow*

```
Command: _AecAnnoSymbolAdd
Adding: North Arrow (Type C)
Specify insertion point or [Symbol]: _SYMBOL
Specify symbol block name or [?]<NONE>:
Anno_North_C_Metric
Specify insertion point or [Symbol]: _Endofparam
Specify insertion point or [Symbol]:
```
(Specify the insertion point near point P1 in Figure 8.60.)
```
Specify rotation: 135 ENTER
```
(Type the angle of deflection from due east to specify the north direction.)

17. Select **Revision Clouds** from the **Annotation - Imperial** toolbar.
18. Select Small Arcs & Tag from the Revision Clouds palette and drag the symbol to the drawing area by continuing to hold down the left mouse button as you move the mouse over the drawing area.
19. Release the mouse button over the drawing area and specify the location of the revision cloud as shown in Figure 8.61 and the command sequence.

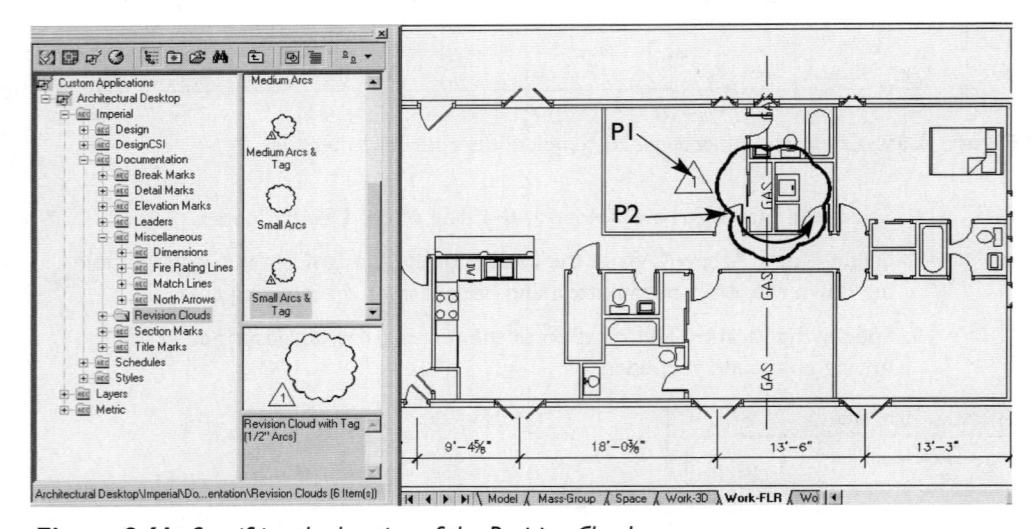

Figure 8.61 *Specifying the location of the Revision Cloud*

```
Command: _AecAnnoRevisionCloudAdd
Adding: Revision Cloud with Tag (1/2" Arcs)
Specify cloud starting point or [Symbol block/pline
Color/Arc length/pline Width]: _SYMBOL
Specify symbol block or [?]<Anno_Revision_A>:
Anno_Revision_A
Specify cloud starting point or [Symbol block/pline
```

```
Color/Arc length/pline
Width]: _ARCLENGTH
Specify arc length <1/2">: .5
Specify cloud starting point or [Symbol block/pline
Color/Arc length/pline
Width]: _WIDTH
Specify cloud polyline width<1/16">: .03125
Specify cloud starting point or [Symbol block/pline
Color/Arc length/pline
Width]: _COLOR Specify cloud color<BYLAYER>_ByLayer
Specify cloud starting point or [Symbol block/pline
Color/Arc length/pline
Width]: _Endofparam

Specify cloud starting point or [Symbol block/pline
Color/Arc length/pline
Width]:
```
(Select point P1 in Figure 8.61.)
```
Cloud will close when returned to start point...
Guide crosshairs along cloud path (counter-clock-
wise)
```
(Move the mouse from point P1 in a counter-clockwise direction to form the cloud as shown in Figure 8.61.)
```
Specify center point of revision tag <None>:
```
(Select point P2 in Figure 8.61 to specify the location for the tag.)
```
Select block reference:
```
(Type 1 in the Edit Attributes dialog box.)

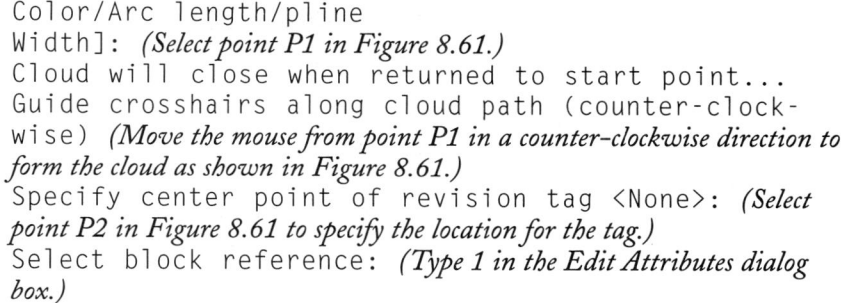

20. Select **Section Marks** from the **Annotation - Imperial** toolbar.

21. Select Section Mark A2 from the Section Marks palette and drag the symbol to the drawing area by continuing to hold down the left mouse button as you move the mouse over the drawing area.

22. Release the mouse button over the drawing area and specify the location of section line as shown in Figure **8.63** and the command sequence.

```
Command: _AecAnnoSectionMarkAdd
Adding: Section Mark (w/Sheet No.)
Specify first point of section line or
[Symbol/Arrow/Tail]: _SYMBOL
Symbol block name or [?]<NONE>: Anno_Section_A2
Specify first point of section line or
[Symbol/Arrow/Tail]: _ARROW
Arrow block name or [?]<NONE>: Anno_Arrow_A2
Specify first point of section line or
[Symbol/Arrow/Tail]: _TAIL
Tail length,width<0.000,0.000>: 0,0
Specify first point of section line or
```

```
[Symbol/Arrow/Tail]: _Endofparam

Specify first point of section line or
[Symbol/Arrow/Tail]:
```
(Select point P1 in Figure 8.63 to begin the section line.)
```
Specify next point of line or [Symbol/Arrow/Tail]:
```
(Select point P2 in Figure 8.63 to end the section line.)
```
Specify next point of line or
[Break/Symbol/Arrow/Tail]: ENTER
```
(Press ENTER to end selection of section line location.)
```
Select block reference:
```
(Type 3 in the Section Mark Number edit field and type A-12 in the Sheet Number edit field of the Edit Attributes dialog box as shown in Figure 8.62. Select OK to dismiss the Edit Attributes dialog box.)

Figure 8.62 *Inserting text in the Edit Attributes dialog box for the Section Mark symbol*

```
Command:
Specify side for Arrow:
```
(Select near point P3 in Figure 8.63 to specify the viewing direction of the section.)
```
Add AEC section object? [Yes/No] <N>: ENTER
```
(Press ENTER to decline use of the section line as a section object for the development of a section.)

23. Select **Title Marks** from the **Annotation - Imperial** toolbar.

24. Select Title Mark A1 from the Title Marks palette and drag the symbol to the drawing area by continuing to hold down the left mouse button as you move the mouse over the drawing area.

25. Release the mouse button over the drawing area and specify the location of title mark symbol as shown in Figure 8.66 and the command sequence.

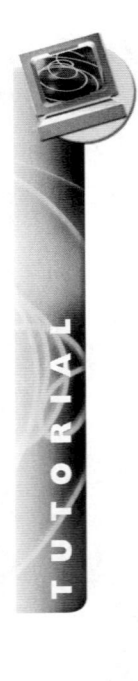

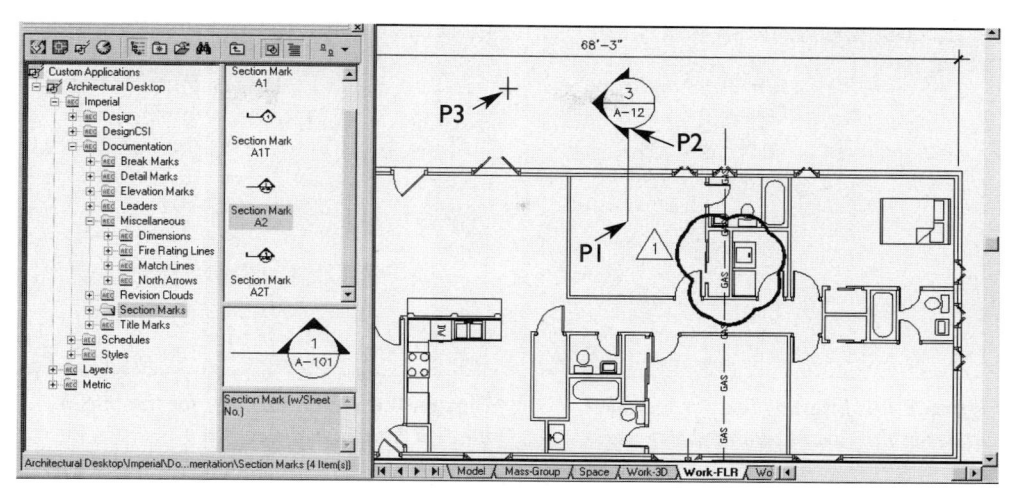

Figure 8.63 *Specify the viewing direction for the section mark*

```
Command: _AecAnnoTitleMarkAdd
Adding: Title Mark (w/ Scale)
Specify location of symbol or [Symbol block/TExt
block/pline Width]: _SYMBOL
Specify symbol block name or [?]<Anno_Title_A1>:
Anno_Title_A1
Specify location of symbol or [Symbol block/TExt
block/pline Width]: _TEXT
Specify text block name or [?]<Anno_Title_T1>:
Anno_Title_T1
Specify location of symbol or [Symbol block/TExt
block/pline Width]: _TAG
Specify scale attribute tag<Scale>: Scale
Specify location of symbol or [Symbol block/TExt
block/pline Width]: _WIDTH
Specify polyline width<0">: .025
Specify location of symbol or [Symbol block/TExt
block/pline Width]: _Endofparam
```

Specify location of symbol or [Symbol block/TExt block/pline Width]: *(Specify the location of the title symbol at point P1 in Figure 8.66.)*
Select block reference: *(Type 1 in the title number edit field of the Edit Attributes dialog box as shown in Figure 8.64.)*
(Type FLOOR PLAN in the title edit field of the Edit Attributes dialog box as shown in Figure 8.65.)

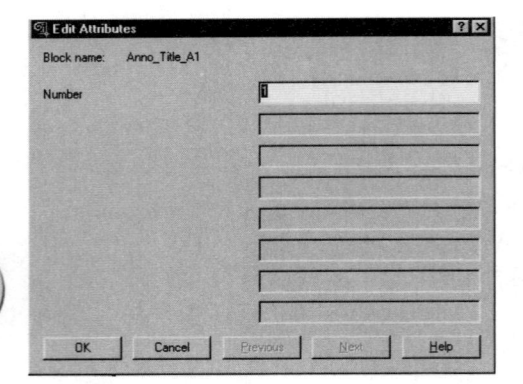

Figure 8.64 *Inserting text for the title number in the Edit Attributes dialog box*

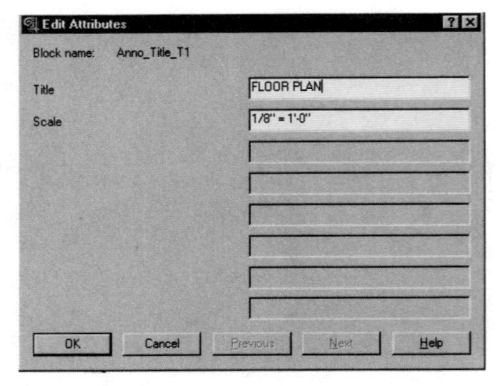

Figure 8.65 *Inserting text for the title in the Edit Attributes dialog box*

```
Command:
Specify endpoint of line: (Select point P2 as the end of the title
line as shown in Figure 8.66.)
Command:
```

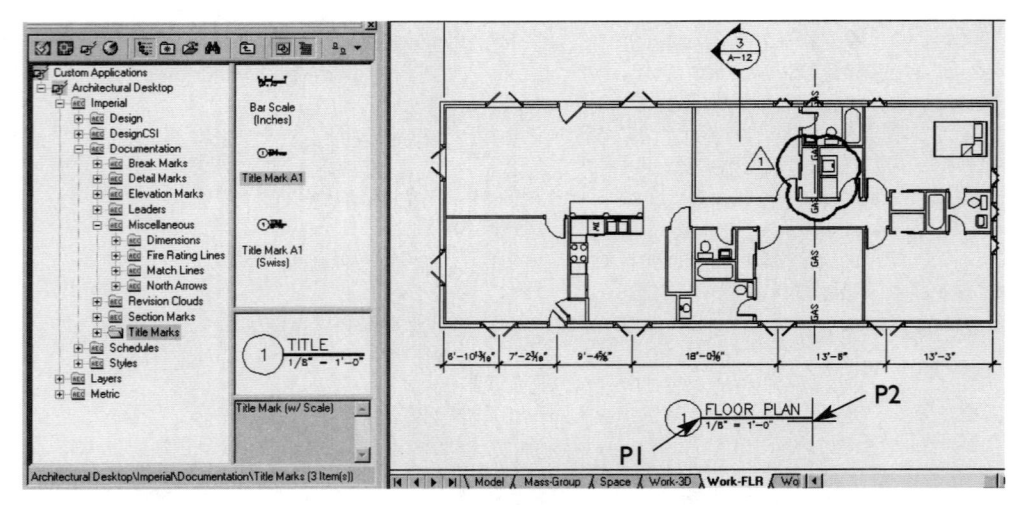

Figure 8.66 *Specifying the location and length of the title bar*

26. Select the exterior wall, and then right-click and choose **Edit Wall Style** to open the **Wall Style Properties** dialog box.

27. Select the **Display Props** tab, and then select the Wall Style property source and select **Edit Display Props** button to open the **Entity Properties** dialog box.

28. Select the light bulb icon to turn ON Boundary 2 (Brick) and Hatch 2 (Brick) and then select **OK** to dismiss the **Entity Properties** dialog box.
29. Select **OK** to dismiss the **Wall Style Properties** dialog box.
30. Save the drawing.

CREATING TAGS AND SCHEDULES FOR OBJECTS

Schedules can be created to list doors, windows, walls, or any object in the drawing. The schedule provides detail information regarding the size and construction of the object. Each object of a schedule is usually assigned a mark, which is placed as a tag near the object in the drawing. Schedules can be used to determine the quantity of a certain type of building component.

The information included in the schedule is extracted from the schedule data linked to the object. The schedule data for the object consists of property sets, as shown in Figure 8.67. Each property set consists of properties such as height, width, hardware, or materials. Some properties are automatically created when the object is inserted in the drawing. However, you can define other properties manually by editing the schedule data. When the schedule is created, the data from the property sets is extracted and included in the schedule. Architectural Desktop preferences can be set to automatically display the schedule data when a tag is attached to an object. Therefore, upon insertion of the tag you can enter the data for the schedule.

The content of the schedule is determined by the defined components for a schedule table style. The components available for the schedule table style of a window are shown in the **Edit Schedule Data** dialog box of Figure 8.70. The **Schedule Table Style Properties** dialog box allows you to define which properties of the property set to include in the schedule. Architectural Desktop includes nine schedule table styles, which can be imported and used to create a schedule. You can also modify these schedule table styles or create new ones to meet your schedule needs.

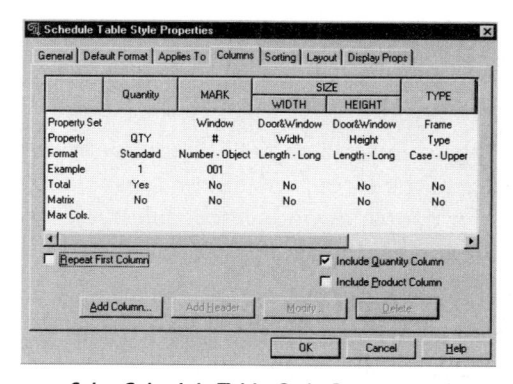

Figure 8.67 *Propety sets of the Schedule Table Style Properties dialog box*

Therefore, the schedule is created based upon the format of a schedule table style, which extracts property information from the schedule data associated with the Architectural Desktop object. The tags and properties included in the schedule can be dynamically updated as the drawing is changed.

> **Note:** The properties or components of the property set assigned to the tag and schedule style are defined by the Property Set Define (**PROPERTYSETDEFINE**) command. The use of the property set definitions (**PROPERTYSETDEFINE**) command to create customized property sets and schedules is discussed in the Enhancing Schedules and Schedule Data.pdf located on the CD.

The tags and schedule commands for Architectural Desktop objects are located on the **Schedule - Imperial** and **Schedule -Metric** toolbars. They are shown in Figure 8.68.

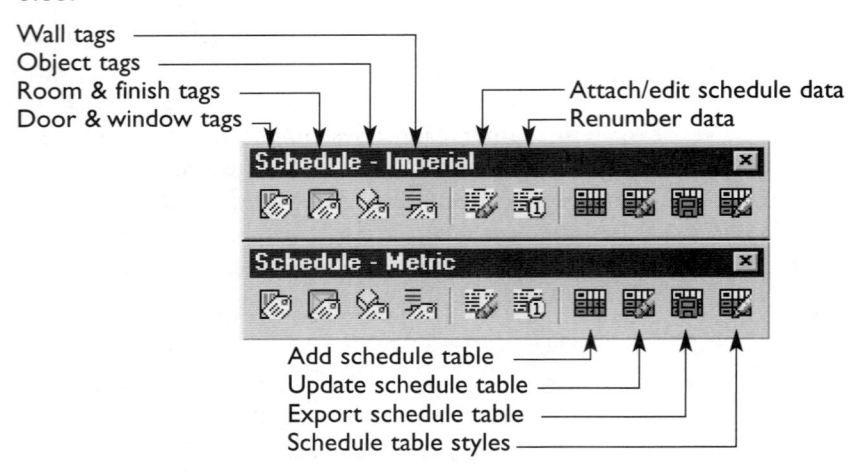

Figure 8.68 *Commands on the Schedule - Imperial and Schedule - Metric toolbars*

Tags can be applied to Architectural Desktop objects. The tag is a multi-view block with attributes attached to provide the tag mark.

DOOR AND WINDOW TAGS

The **Door & Window Tags** command on the **Schedule - Imperial** toolbar opens the AutoCAD DesignCenter to the Door & Window Tags folder. The command allows you to specify the mark for door and window schedules. Access the **Door & Window Tags** command using the toolbar in Figure 8.69.

When you select the **Door & Window Tags** command on the **Schedule - Imperial** or **Schedule - Metric** toolbar, the AutoCAD DesignCenter opens to the Door & Window Tags folder, which consists of a door and a window tag. Each of these tags uses the **AECANNOSCHEDULETAGADD** command to insert the tag. The tag

Annotating the Drawing 649

Figure 8.69 *Door & Window Tags command on the Schedule - Imperial toolbar*

for the door utilizes the Aec_Door_Tag symbol and the window tag utilizes the Aec_Window_Tag symbol. The door tag is placed on the A-Door-Iden layer, which has the color 21 (yellowish-pink). The window tag is placed on the A-Glaz-Iden layer, which has the color 91 (a hue of green).

When the tag is selected and attached to a door or window, the **Edit Schedule Data** dialog box opens, displaying the property sets information regarding the selected door or window, as shown in Figure 8.70. You can enter or review the values in the **Edit Schedule Data** dialog box prior to placing the tag.

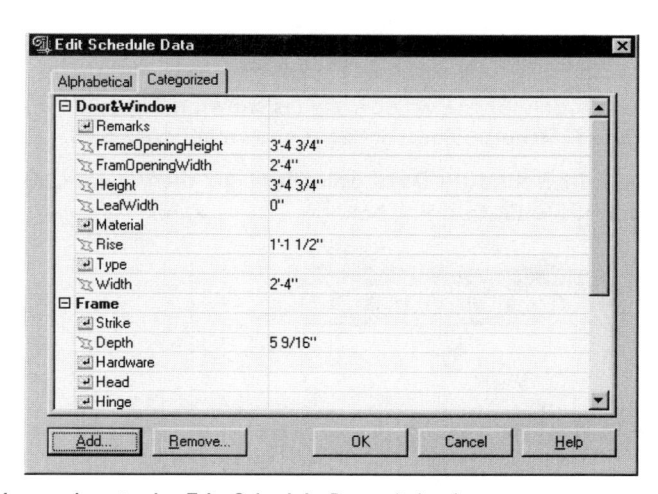

Figure 8.70 *Insert data in the Edit Schedule Data dialog box*

The **Edit Schedule Data** dialog box opens automatically only if the **Display Edit Schedule Data Dialog During Tag Insertion** check box is selected in the **AEC Content** tab of the **Options** dialog box. As shown in Figure 8.71, this check box is selected by default in drawings created from the Aec_Arch (Imperial) template.

STEPS TO PLACING A WINDOW TAG

1. Select **Door & Window Tags** from the **Schedule - Imperial** toolbar.
2. Select the Window Tag symbol from the Door & Window Tags palette;

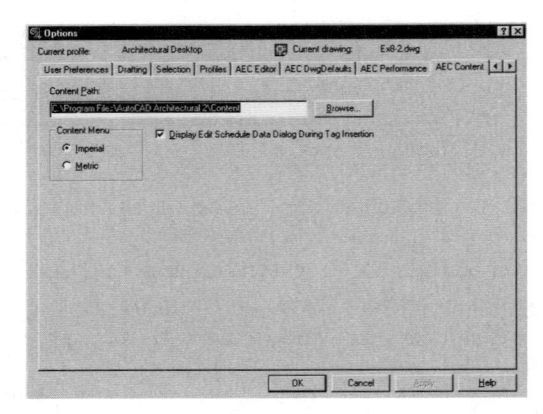

Figure 8.71 *Display Edit Schedule Data Dialog During Tag Insertion check box on the AEC Content tab*

continue holding down the left mouse button as you move the mouse to the drawing area and then release the mouse button.

3. Select a window for the placement of a tag.
4. Select a location for the window tag (as shown in Figure 8.72). The **Edit Schedule Data** dialog box opens as shown in Figure 8.70.
5. Edit the **Edit Schedule Data** dialog box, and then select **OK** to dismiss the **Edit Schedule Data** dialog box.

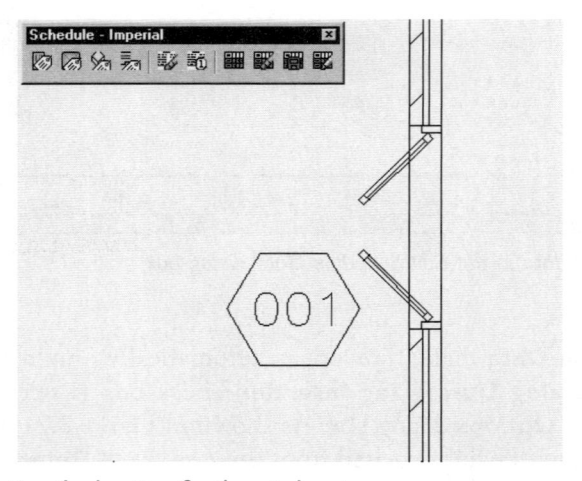

Figure 8.72 *Selecting the location for the window tag*

6. Continue to select additional windows to place additional tags or press ENTER to end the command.

ROOM AND FINISH TAGS

The **Room & Finish Tags** command on the **Schedule - Imperial** and **Schedule - Metric** toolbars opens the AutoCAD DesignCenter to the Room & Finish Tags folder. These tags allow you to specify the room number, finish, and area of a room for the development of a schedule. Access the **Room & Finish Tags** command as shown in Table 8.20.

Menu bar	Documentation>Schedule Tags> Room & Finish Tags
Command prompt	AECDCSETIMPROOMANDFINISHTAGS
Toolbar	Select Room & Finish Tags from the Schedule - Imperial or Schedule - Metric toolbar, as shown in Figure 8.73

Table 8.20 *Accessing the Room & Finish Tags*

Figure 8.73 *Room & Finish Tags on the Schedule - Imperial toolbar*

Creating Spaces

The room and finish tags are specifically designed for use with Aec Spaces, created with the **SPACEADD** command. When the tags are applied to an Aec Space, the area of the space or room is extracted from the properties of the Aec Space object. Design with Aec Spaces allows you to place space objects, which represent different rooms. Spaces can be assigned the names of the various rooms such as living, dining, and kitchen. Architectural Desktop includes predefined space styles for residential and commercial construction. These predefined spaces can be placed in the drawing to perform space planning. Detail information regarding space planning is included in Chapter 13.

However, if a floor plan is created without the use of the space planning commands, the Aec Spaces can be created from the existing wall geometry. The room tags can then be attached to the Aec Space. The creation of an Aec Space allows you to quickly determine the area of a room. You can create an Aec Space from an existing floor plan by drawing a polyline that outlines the limits of a room and then converting the polyline to a space using the **Convert to Spaces** command (**SPACECONVERT**). Access the **SPACECONVERT** command as shown in Table 8.21.

Menu bar	Concept>Space Planning>Convert to Spaces
Command prompt	SPACECONVERT
Toolbar	Select Convert to Spaces from the Space Planning toolbar as shown in Figure 8.74
Shortcut	Right-click over the drawing area and choose Design>Spaces>Convert

Table 8.21 *Accessing the Space Convert command*

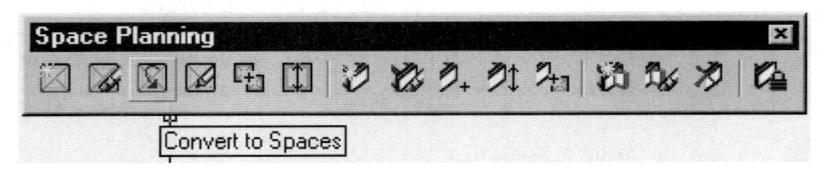

Figure 8.74 *Convert to Spaces command on the Space Planning toolbar*

When you select the **Convert to Spaces** command, you are prompted to select a polyline. You are then prompted if you wish to erase the layout geometry. If you retain the layout geometry, the polyline can be edited and used to determine new areas. After you select the polyline, the **Space Properties** dialog box opens, allowing you to select a space style. The Standard space style can be selected, because it can be used for spaces from 1 to 10,000 square feet in area. The Standard space style is the default space style and can be used to label the room with the room and finish tags.

> **Note:** You can use the BOUNDARY command to create a polyline boundary of the room. Edit the door opening property of each door to zero percent and then use the **BOUNDARY** command, **Pick Points** option to create a polyline that outlines the limits of the room. The **Convert to Space** command can be used to convert the polyline to an AEC Space.

The display of the hatch pattern representing the AEC Space may be turned off by freezing the A-Area-Spce layer or selecting the Plot-FLR layout tab. Room and finish tags are displayed in the Plot-FLR layout.

Placing Room and Finish Tags

When you select the **Room & Finish Tags** command on the **Schedule - Imperial** or **Schedule - Metric** toolbar, the AutoCAD DesignCenter opens to the Room & Finish Tags folder. Figure 8.75 shows the three tags included in the Room & Finish Tags folder.

Each of these tags uses the **AECANNOSCHEDULETAGADD** command to insert the tag. The tags for the room and finish tags utilize one of three space tag symbols. The

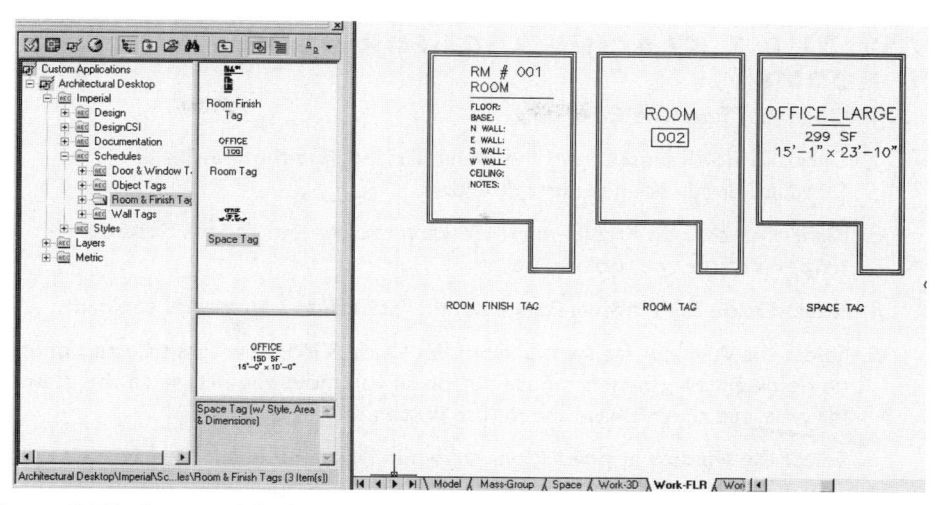

Figure 8.75 *Room and finish tags*

Room Finish Tag symbol is placed on the A-Flor-Iden layer, which has the color 11 (yellowish-pink). The Room Tag and Space Tag symbols are placed on the A-Area-Iden layer, which has the color 131 (a hue of cyan).

To place a room and finish tag, select a tag from the Room & Finish Tags palette and continue to hold down the left mouse button as you drag the mouse to the drawing area. When you release the mouse button over the drawing area, you are prompted to select a space to attach the tag. After you select a space, the **Edit Schedule Data** dialog opens, displaying the property sets information regarding the selected space. The content of the **Edit Schedule Data** dialog box varies according to tag, allowing you to insert the necessary data for the tag. Figure 8.76 shows the **Edit Schedule Data** dialog box for the Room Finish tag.

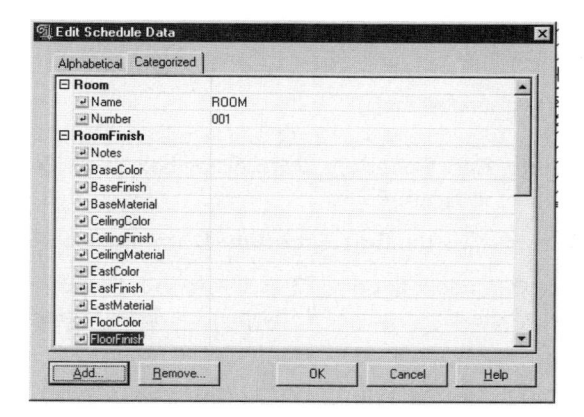

Figure 8.76 *Adding schedule data to the Room Finish tag*

TUTORIAL 8.3 PLACING TAGS FOR DOORS, WINDOWS AND ROOMS

1. Open *ADT Tutor\Ch8\Ex8-3.dwg*.
2. Choose **File>SaveAs** from the menu bar and save the drawing as *Schedule1.dwg* in your student directory.
3. Right-click over an Architectural Desktop toolbar and choose **Schedule - Imperial** from the toolbar list.
4. Select **Door & Window Tags** from the **Schedule - Imperial** toolbar.
5. Select the Window Tag symbol from the Door & Window Tags palette, continue holding down the left mouse button as you move the mouse to the drawing area, and then release the mouse button.
6. Select the window at point P1, as shown in Figure 8.77, and then select a point near P2 to locate the tag and open the **Edit Schedule Data** dialog box.

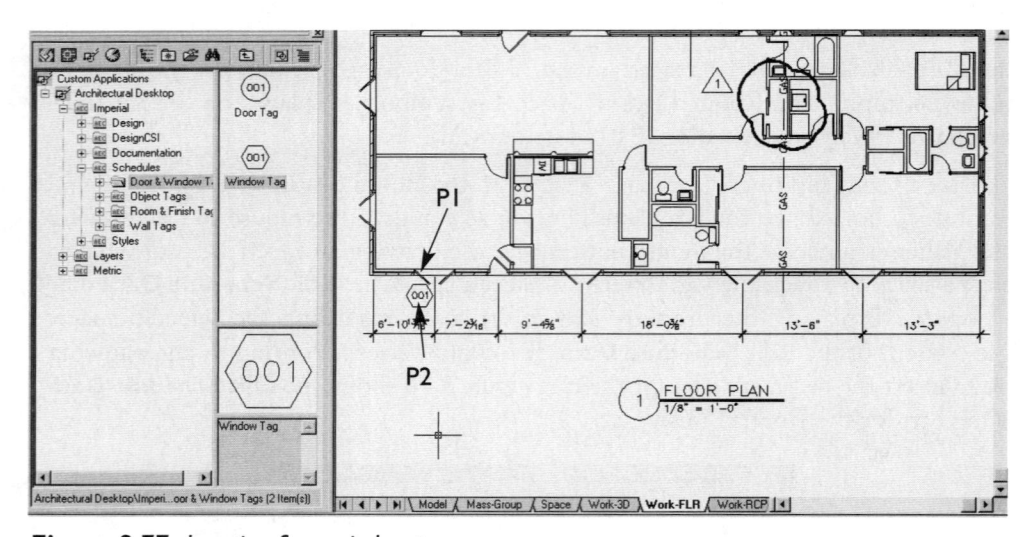

Figure 8.77 *Location for a window tag*

7. Scroll down the **Edit Schedule Data** dialog box to verify that the Window # is 001, as shown in Figure 8.78.
8. Select **OK** to dismiss the **Edit Schedule Data** dialog box.
9. Select the next window at point P3 as shown in Figure 8.79, and then specify the location of the tag at point P4 to open the **Edit Schedule Data** dialog box.
10. Scroll down the **Edit Schedule Data** dialog box to verify that the Window # is 002, as shown in Figure 8.80.

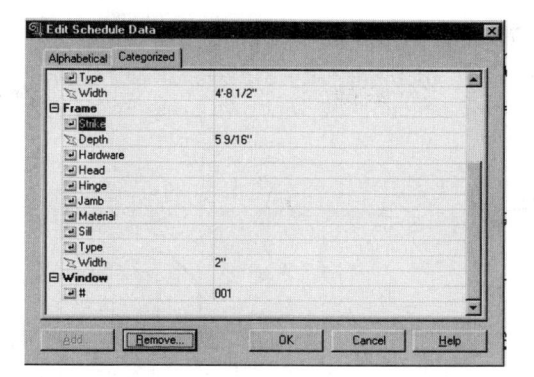

Figure 8.78 *Adding data to the Edit Schedule Data dialog box for the window tag*

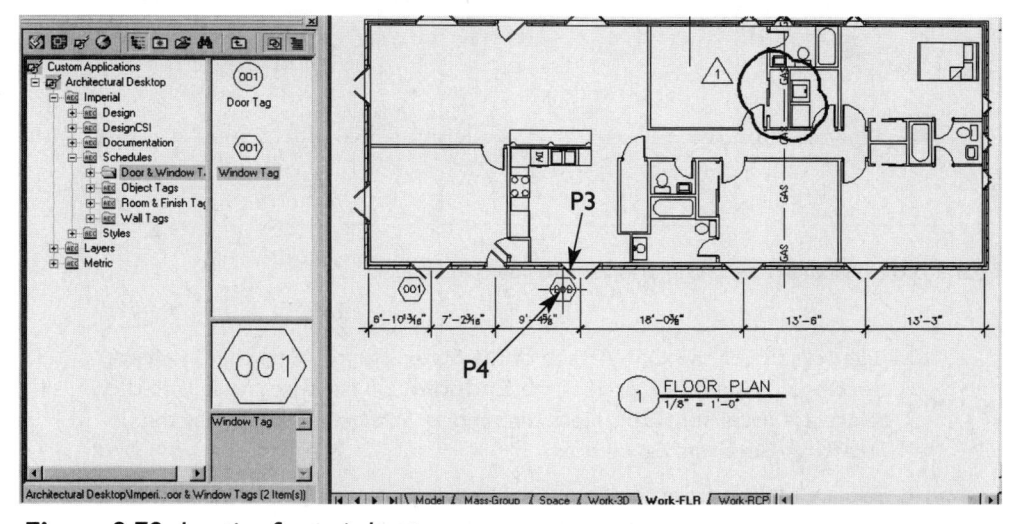

Figure 8.79 *Location for a window tag*

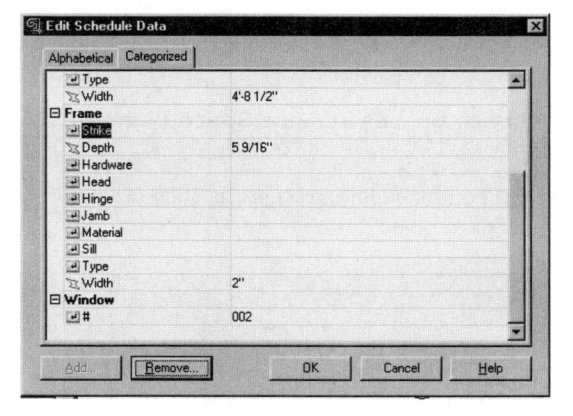

Figure 8.80 *Adding the window number to the Edit Schedule Data dialog box*

11. Select the **OK** button to dismiss the **Edit Schedule Data** dialog box.
12. Continue to place additional window tags for all 14 windows, as shown in Figure 8.81. After selecting the windows, press ENTER to end the command.

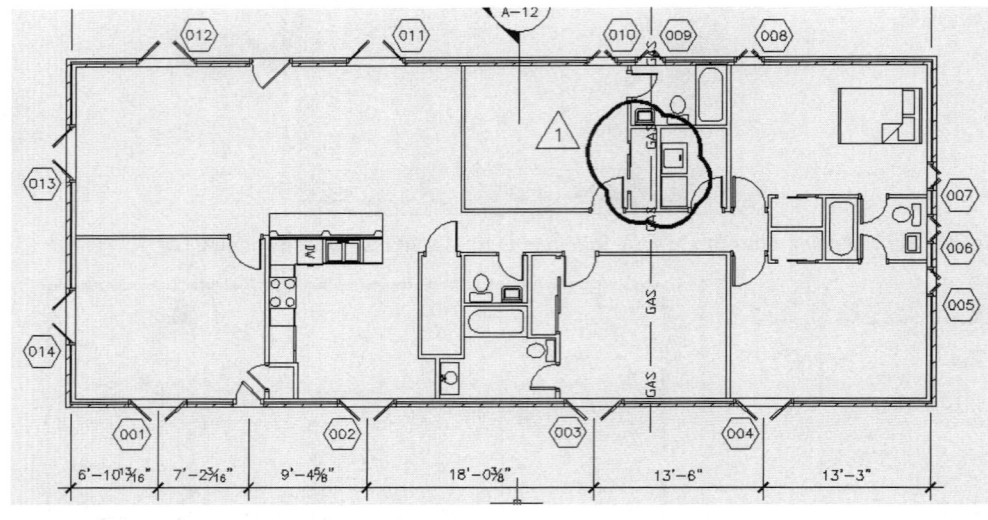

Figure 8.81 *Window tags and marks for the floor plan*

13. Right-click over the OSNAP tab of the Status Bar to open the **Drafting Settings** dialog box. Deselect the **Endpoint** object snap mode, and then select the **Intersection** object snap mode. Select **OK** to dismiss the **Drafting Settings** dialog box.
14. Select **Polyline** from the **Draw** toolbar and create a polyline that begins at P1 and continues through points P2, P3, P4 to close at P1, as shown in Figure 8.82.
15. Right-click over any Architectural Desktop toolbar and choose **Space Planning** from the toolbar list.
16. Select **Convert to Spaces** from the **Space Planning** toolbar as shown in Figure 8.83.
17. Respond to the command line prompts as shown below to convert the polyline to an Aec Space object.

```
Command: _AecSpaceConvert
Select a polyline: (Select the polyline at point P1 in Figure 8.82.)

Erase layout geometry? [Yes/No] <Y>: ENTER (Press ENTER
to accept the default to erase the polyline after the space is created.)
```

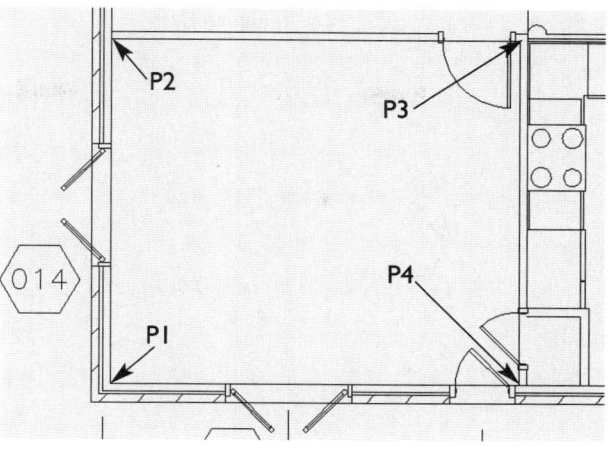

Figure 8.82 Creating a polyline for the room space

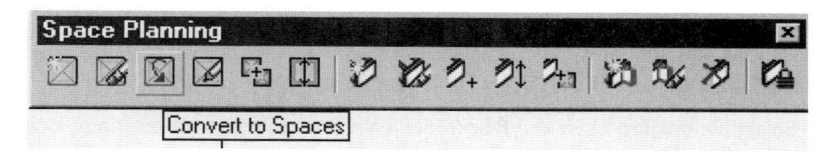

Figure 8.83 Convert to Spaces command on the Space Planning toolbar

(Edit the Space Properties dialog box shown in Figure 8.84 by selecting the Style tab to verify that the style is Standard. Select the Dimensions tab and edit B-Space Height to 8'-0" as shown in Figure 8.84 and then select OK to dismiss the Space Properties dialog box.)

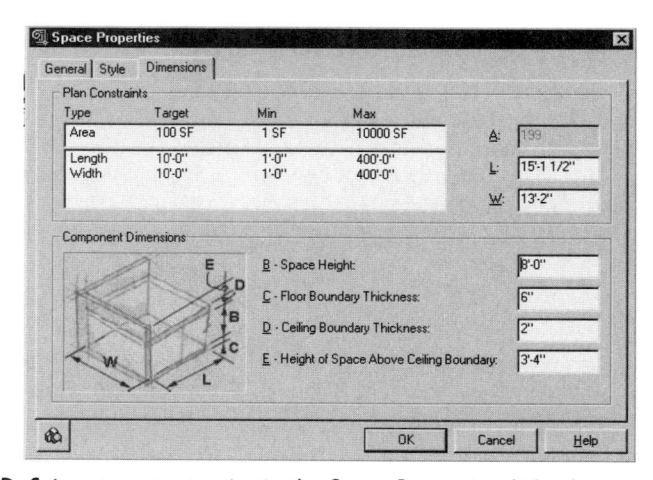

Figure 8.84 Defining space properties in the Space Properties dialog box

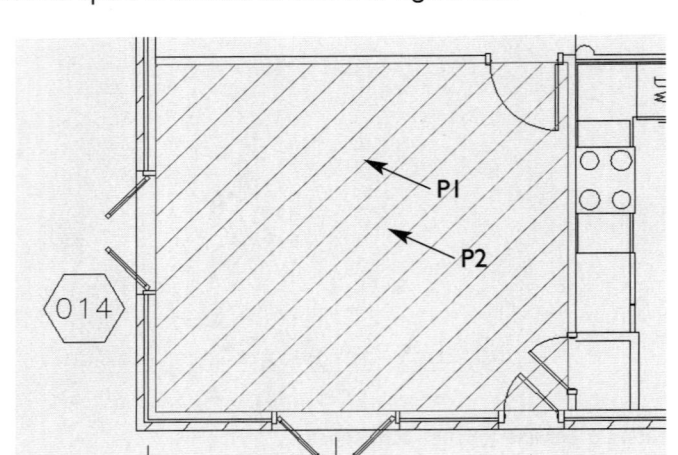

The Aec Space is created as shown in Figure 8.85.

Figure 8.85 *Aec Space*

18. Select **Room & Finish Tags** from the **Schedule - Imperial** toolbar.
19. Select Room Tag from the Room & Finish Tags palette and continue holding down the left mouse button as you drag the symbol to the drawing area; release the mouse button and select the Aec Space.
20. Respond to the command line prompts to place the Room tag as shown in the command sequence below and in Figures 8.85 and 8.86.

```
Command: _AecAnnoScheduleTagAdd
Property set definitions drawing <C:\Program
Files\AutoCAD Architectural
2\Content\Imperial\Schedules\PropertySetDefs.dwg>:
PropertySetDefs.dwg
Select object to tag [Symbol/Leader/Dimstyle]: _SYM-
BOL
Enter tag symbol name <Aec_Window_Tag>: Aec_Room_Tag
Select object to tag [Symbol/Leader/Dimstyle]:
_LEADER
Enter leader type [None/STraight/SPline] <None>:
None
Select object to tag [Symbol/Leader/Dimstyle]: _DIM-
STYLE
Enter leader dimstyle CURRENT <Aec_Arch_I$7>:
Current

Select object to tag [Symbol/Leader/Dimstyle]:
```
(Select the Aec Space at P1 in Figure 8.85.)

```
Specify location of tag [Centered]: (Select a position near
```
*point P2 in Figure 8.85. Type LIVING in the Room Name edit field of the
Edit Schedule Data dialog box as shown in Figure 8.86. Select OK to dismiss the Edit Schedule Data dialog box.)*
```
Select object to tag [Symbol/Leader/Dimstyle]: ENTER
```
(Press ENTER to end the selection of spaces for room tags.)

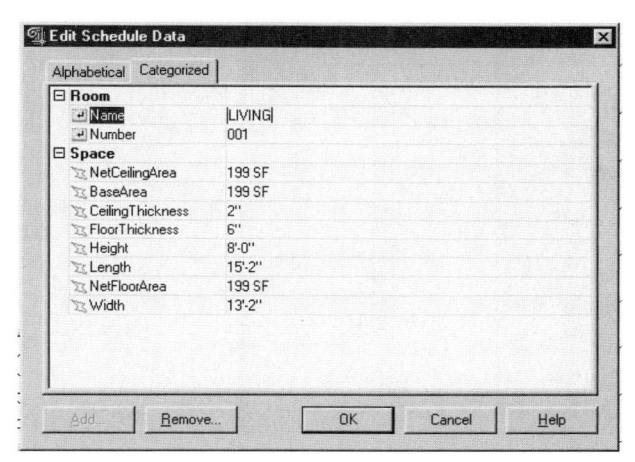

Figure 8.86 *Inserting room name and number in the Edit Schedule Data dialog box*

The Room tag is placed as shown in Figure 8.87.

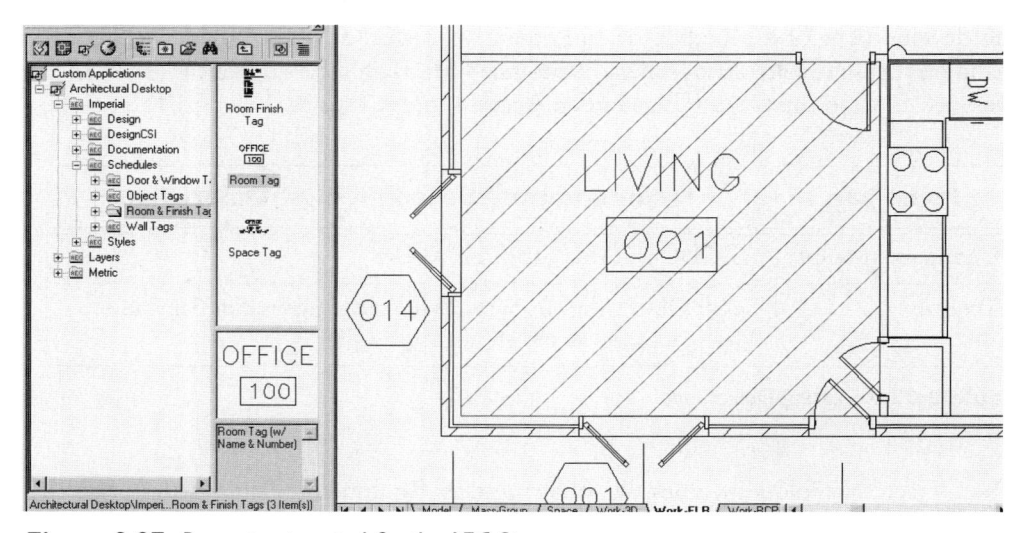

Figure 8.87 *Room tag inserted for the AEC Space*

21. Select **Layers** on the **Object Properties** toolbar and select the **Freeze** icon to freeze the A-Area-Spce layer, and then select **OK** to dismiss the **Layer Properties Manager**. The Room tag is displayed without the Aec Space, as shown in Figure 8.88.

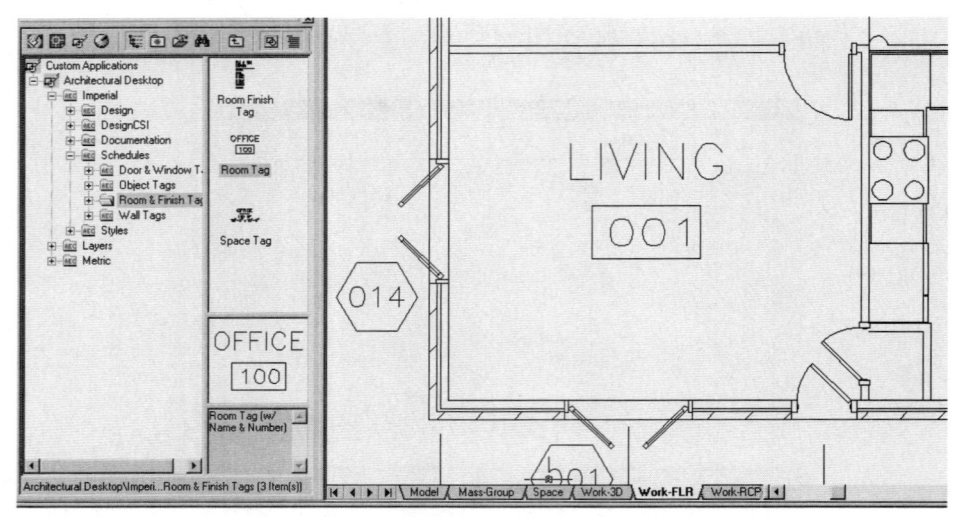

Figure 8.88 *A-Area-Spce layer frozen to hide the AEC Space*

22. Save the drawing.

OBJECT TAGS

Tags for equipment and furniture can be placed in the drawing to identify a mark for the development of a schedule of equipment or furniture. The **Object Tags** command on the **Schedule - Imperial** and **Schedule - Metric** toolbars opens the AutoCAD DesignCenter to the Object Tags folder. Access **Object Tags** as shown in Table 8.22.

Menu bar	Documentation>Schedule Tags>Object Tags
Command prompt	AECDCSETIMPOBJECTTAGS
Toolbar	Select Object Tags from the Schedule - Imperial or Schedule - Metric toolbars as shown in Figure 8.89

Table 8.22 *Accessing Object Tags*

The Object Tags folder includes the following tags: Equipment, Equipment (Leader), Furniture, and Furniture (Leader). Each of the tags can be used to tag block references or multi-view block references. Therefore most of the symbols created from the

Figure 8.89 *Object Tags command on the Schedule - Imperial toolbar*

Design Content - Imperial and **Design Content - Metric** toolbars can be tagged with object tags. However, masking blocks cannot be tagged with the **Object Tags** command.

The **AECANNOSCHEDULETAGADD** command is used to place the Equipment and Furniture tags. The Equipment and Furniture (Leader) symbol tags are placed on the A-Furn-Iden layer, which has the color 11 (yellowish-pink). The Equipment and Equipment (Leader) symbol tags are placed on the A-Eqpm-Iden layer, which has the color 90 (green).

To place an object tag in a drawing, select **Object Tags** from the **Schedule - Imperial** or **Schedule - Metric** toolbar and then select a tag symbol from the Object Tags folder of the AutoCAD DesignCenter. Then select a block to open the **Edit Schedule Data** dialog box. The tag number and other related information can be typed in the **Edit Schedule Data** dialog box.

The command line sequence below was used to create the Furniture tag as shown in Figure 8.91.

```
Command: _AecAnnoScheduleTagAdd
Property set definitions drawing <C:\Program
Files\AutoCAD Architectural
2\Content\Imperial\Schedules\PropertySetDefs.Dwg>:
PropertySetDefs.Dwg
Select object to tag [Symbol/Leader/Dimstyle]:
_SYMBOL
Enter tag symbol name <Aec_Furniture_Tag>:
Aec_Furniture_Tag
Select object to tag [Symbol/Leader/Dimstyle]:
_LEADER
Enter leader type [None/STraight/SPline] <None>:
None
Select object to tag [Symbol/Leader/Dimstyle]:
_DIMSTYLE
Enter leader dimstyle CURRENT <Aec_Arch_I$7>:
Current
```

```
Select object to tag [Symbol/Leader/Dimstyle]:
```
(Select the bed at point P1 in Figure 8.91.)
```
Specify location of tag [Centered]:
```
(Select the tag location at P2 in Figure 8.91.)
(Type 1 in the Furniture # edit field shown in the Edit Schedule Data dialog box as shown in Figure 8.90.)

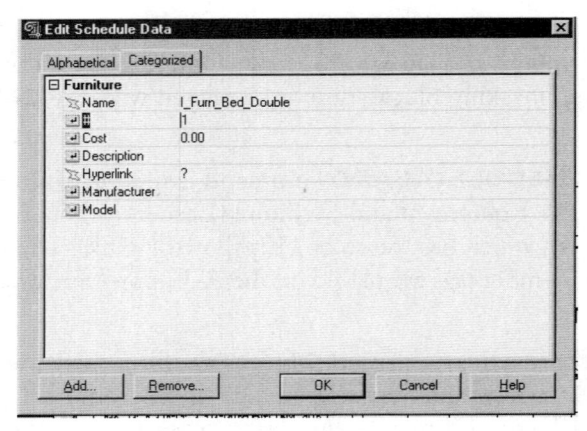

Figure 8.90 *Inserting tag mark number in the Edit Schedule Data dialog box*

```
Select object to tag [Symbol/Leader/Dimstyle]: ENTER
```
(Press ENTER to end tag insertion.)

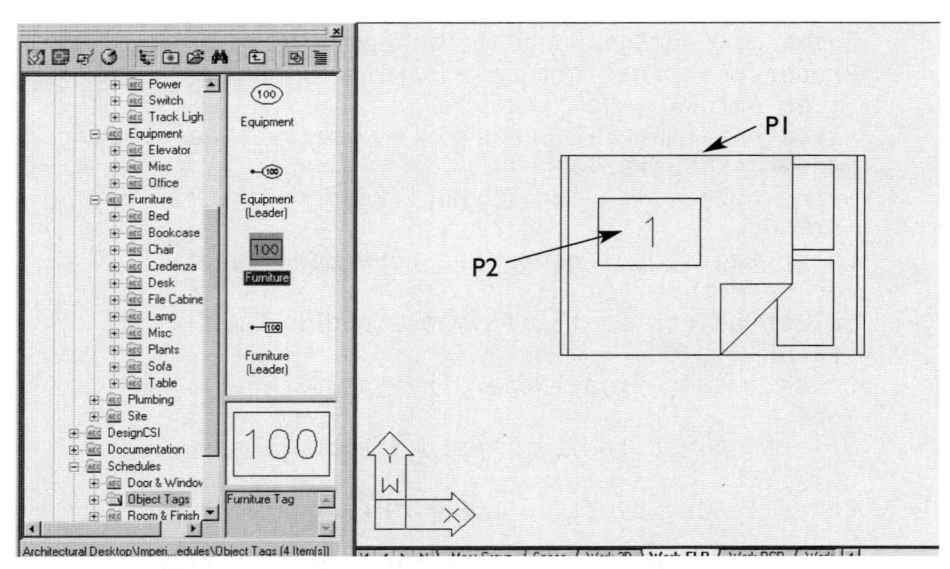

Figure 8.91 *Attaching the tag to the furniture*

WALL TAGS

Wall Tags are used to identify different types of wall construction. A wall tag identifies a wall that is detailed in a different location or in a wall schedule. The **Wall Tags** command on the **Schedule - Imperial** and the **Schedule - Metric** toolbars opens the AutoCAD DesignCenter to the Wall Tags folder. There are two wall tags included in the Wall Tags folder: Wall Tags and Wall Tags (Leader). Access **Wall Tags** as shown in Table 8.23.

Menu bar	Documentation>Schedule Tags>Wall Tags
Command prompt	AECDCSETIMPWALLTAGS
Toolbar	Select Wall Tags from the Schedule - Imperial or Schedule - Metric toolbar as shown in Figure 8.92

Table 8.23 *Accessing Wall Tags*

Figure 8.92 *Wall Tags command on the Schedule - Imperial toolbar*

Wall tags can be attached to a wall. The wall tag is placed on the A-Wall-Iden layer, which has the color 201 (magenta). The **AECANNOSCHEDULETAGADD** command is used to place the wall tags symbol.

To place a wall tag in a drawing, select **Wall Tags** from the **Schedule - Imperial** or **Schedule - Metric** toolbar; drag the symbol to the drawing area. Then select a wall to open the **Edit Schedule Data** dialog box, which allows you to enter the tag number in the Type property.

The command line sequence below was used to create the Wall Tag [Leader] as shown in Figure 8.94.

```
Command: _AecAnnoScheduleTagAdd
Property set definitions drawing <C:\Program
Files\AutoCAD Architectural
2\Content\Imperial\Schedules\PropertySetDefs.dwg>:
PropertySetDefs.dwg
Select object to tag [Symbol/Leader/Dimstyle]: _SYM-
BOL
```

```
Enter tag symbol name <Aec_Wall_Tag>: Aec_Wall_Tag
Select object to tag [Symbol/Leader/Dimstyle]:
_LEADER
Enter leader type [None/STraight/SPline] <None>:
Straight
Select object to tag [Symbol/Leader/Dimstyle]: _DIM-
STYLE
Enter leader dimstyle CURRENT <Aec_Arch_I$7>:
Aec_Wall_Tag
Duplicate definition of block Aec_Arrow_Bar ignored.

Select object to tag [Symbol/Leader/Dimstyle]:
```
(Select the wall at point P1 in Figure 8.94.)
```
Specify next point of leader line or [New] :
```
(Select a point near P2 in Figure 8.94 to end the leader.)
```
Specify next point of leader line or <End leader>:
ENTER
```
(Press ENTER *to end the leader and enter "A" in the Type property of the Edit Schedule Data dialog box as shown in Figure 8.93.)*
(Select OK to dismiss the Edit Schedule Data dialog box.)

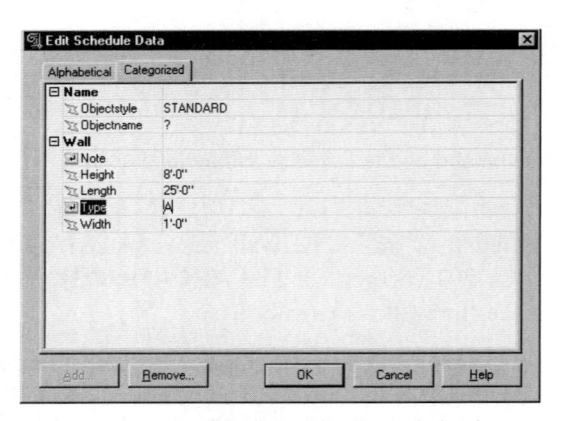

Figure 8.93 *Entering tag mark in the Edit Schedule Data dialog box*

```
Select object to tag [Symbol/Leader/Dimstyle]: ENTER
```
(Press ENTER *to end the command.)*

EDITING TAGS AND SCHEDULE DATA

When tags are inserted in the drawing, the information for the schedule is typed in the **Edit Schedule Data** dialog box. This data can be edited by the **Attach/Edit Schedule Data** command (**PROPERTYDATAEDIT**). This command allows you to select the tag and reopen the Edit Schedule Data dialog box. Access **Attach/Edit Schedule Data** command as shown in Table 8.24.

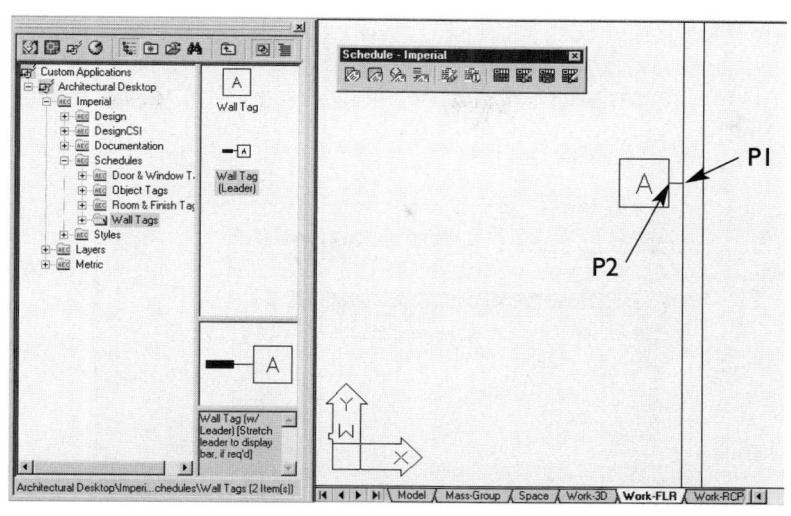

Figure 8.94 Attaching the wall tag to a wall

Menu bar	Documentation>Schedule Data> Attach/Edit Schedule Data
Command prompt	PROPERTYDATAEDIT
Toolbar	Select Attach/Edit Schedule Data from the Schedule - Imperial or Schedule - Metric toolbar as shown in Figure 8.95

Table 8.24 Accessing Attach/Edit Schedule Data

Figure 8.95 Attach/Edit Schedule Data command on the Schedule - Imperial toolbar

When you select this command, you are prompted to select an object, which must be a tag or the object of a tag. When the object is selected, the **Edit Schedule Data** dialog box will open, allowing you to edit the properties.

The command line sequence below was used to edit the tag shown in Figure 8.96.

```
Command: _AecPropertyDataEdit
Select objects: (Select the tag at point P1 in Figure 8.96.)
1 found
```

```
Select objects: ENTER (Press ENTER to end selection of tags.)
1 tag was forwarded to its owner.
(Type changes in the Edit Schedule Data dialog box as shown in Figure
8.96.)
1 object(s) selected.
```

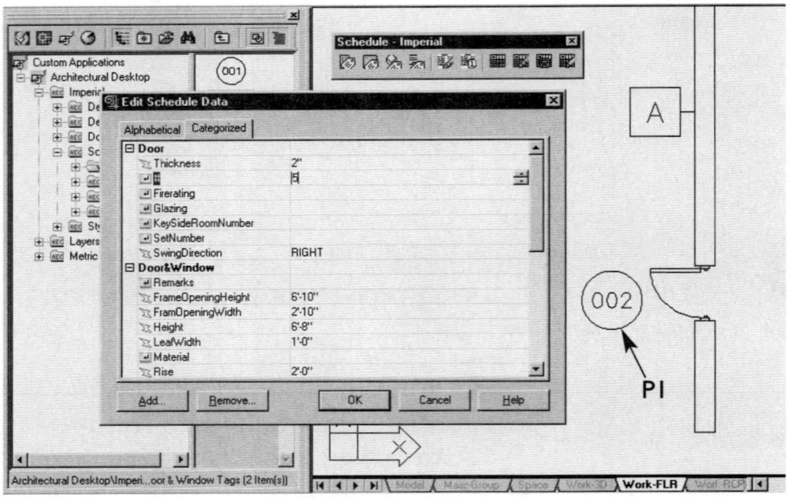

Figure 8.96 Placing data in the Edit Schedule Data dialog box

If the **Edit Schedule Data** is changed for one tag and the object has additional tags attached to it, all tags will change when one tag is changed.

> **Note:** Additional information regarding creating schedules is included in the Enhancing Schedules and Schedule Data.pdf file on the CD

USING SCHEDULE TABLE STYLES

Schedule tables are used to organize and present the data inserted in the **Edit Schedule Data** dialog box in the form of a schedule. Most schedules include a mark column, which is defined when a tag is placed in the drawing. Schedule tables can be developed to display the schedule data in the desired order. Schedule tables are included in the \AutoCAD Architectural 2\Content\Imperial\Schedules\Schedule Tables (Imperial).dwg Schedule tables can be created, copied, edited, imported, or exported by the **Schedule Table Styles** command (**TABLESTYLE**). Access the **Schedule Table Styles** command as shown in Table 8.25.

When you select the **Schedule Table Styles** command (**TABLESTYLE**), the **Schedule Table Styles** dialog box opens, which displays a list of all schedule tables that have been used in the drawing. The **Schedule Table Styles** dialog box shown in Figure 8.98 does

Menu bar	Documentation>Schedules Tables> Schedule Table Styles
Command prompt	TABLESTYLE
Toolbar	Select Schedule Table Styles from the Schedule - Imperial toolbar, shown in Figure 8.97, or Schedule - Metric toolbar

Table 8.25 *Accessing the Schedule Table Styles command*

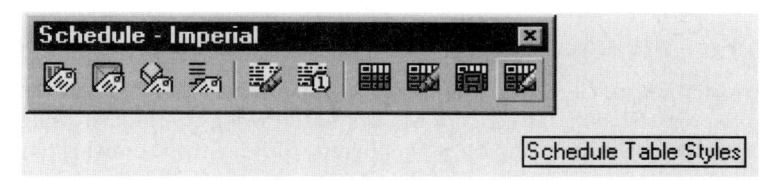

Figure 8.97 *Selecting the Work-Plan display representation set*

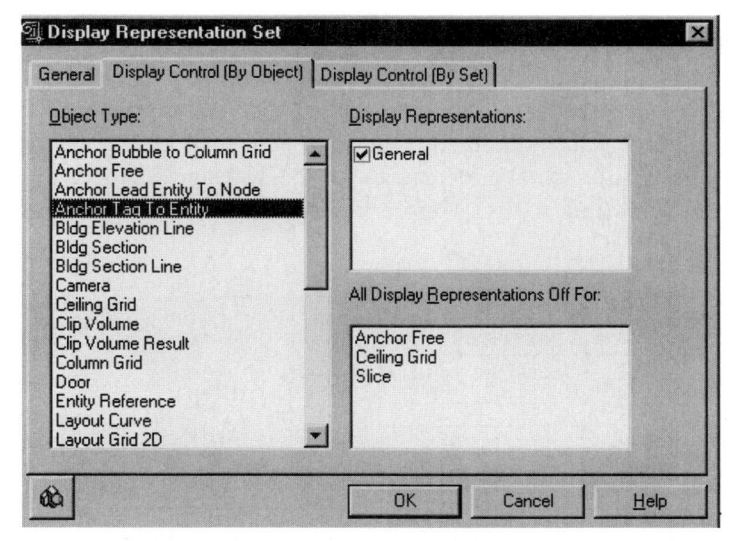

Figure 8.98 *Selecting the General display representation for the Anchor Tag to Entity object*

not include a schedule table list because no schedules have been created or imported to the drawing.

The **Schedule Table Styles** dialog box includes six buttons for creating, copying, editing, purging, importing and exporting schedule tables. The purpose of each of these buttons is described below.

New – Creates a new name for a schedule table. Selecting the **New** button opens the **Name** dialog box, which allows you to type a new name.

Copy – Allows you to select a schedule table from the schedule table list and then select the **Copy** button to enter a new name for the copied schedule table.

Edit – Allows you to change the contents of new or existing schedules. The **Edit** button opens the **Schedule Table Style Properties** dialog box, which includes tabs for defining the format, layout, sorting methods, and organization of the schedule data.

Set From< – Inactive in this dialog box.

Purge – Allows you to purge unused schedule tables from the drawing.

Import/Export – Opens the **Import/Export** dialog box, which allows you to export schedule tables from the current drawing to other drawings and files. Schedule tables can also be imported to the current drawing from other drawing files. The schedule tables included in Architectural Desktop are located in the \AutoCAD Architectural 2\Content\Imperial\Schedules\Schedule Tables (Imperial).dwg.

The **Schedule Table Styles** command is used to create schedule tables; however, there are nine schedule tables available in the *\AutoCAD Architectural 2\Content\Imperial\ Schedules\Schedule Tables (Imperial)* drawing or the *\AutoCAD Architectural 2\Content\ Metric\Schedules\Schedule Tables (Imperial)* drawing, which can be imported to the current drawing. These tables, shown in Figures 8.99, 8.100, and 8.101, can be edited to refine the format according to user preferences.

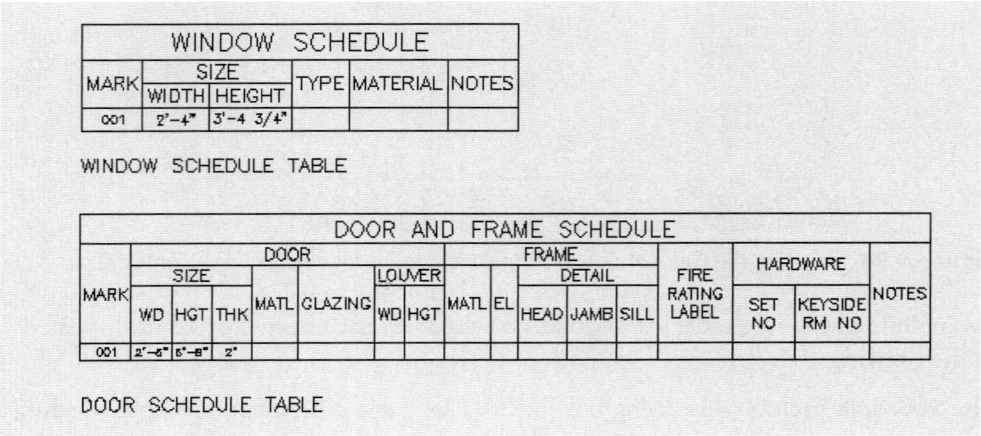

Figure 8.99 *Door and Window Schedules*

ROOM FINISH DOT SCHEDULE

NO	NAME	FLOOR MATL	FIN	COL	WALLS NORTH MATL	FIN	COL	SOUTH MATL	FIN	COL	EAST MATL	FIN	COL	WEST MATL	FIN	COL	CEILING MATL	FIN	COL	HEIGHT	NOTES
001	ROOM	●	●	●	●	●	●	●	●	●	●	●	●	●	●	●	●	●	●	9'-0"	?

ROOM FINISH MATRIX SCHEDULE TABLE

ROOM FINISH SCHEDULE

ROOM NO	ROOM NAME	FLOOR	WALLS N	S	E	W	CEILING MATL	HEIGHT	NOTES
001	ROOM	?	?	?	?	?	?	9'-0"	?

ROOM FINISH SCHEDULE TABLE

ROOM SCHEDULE

NO	NAME	LENGTH	WIDTH	HEIGHT	AREA
001	ROOM	11'-11"	8'-4"	9'-0"	100 SF
					100 SF

ROOM SCHEDULE TABLE

SPACE INVENTORY

SPACE	LOCATION SITE	BLDG	FLOOR	ZONE	DEPT	OWNER	AREA	QTY	TOTAL
?	?	?	?	?	?	?	100 SF	1	100 SF
								1	100 SF

SPACE INVENTORY SCHEDULE TABLE

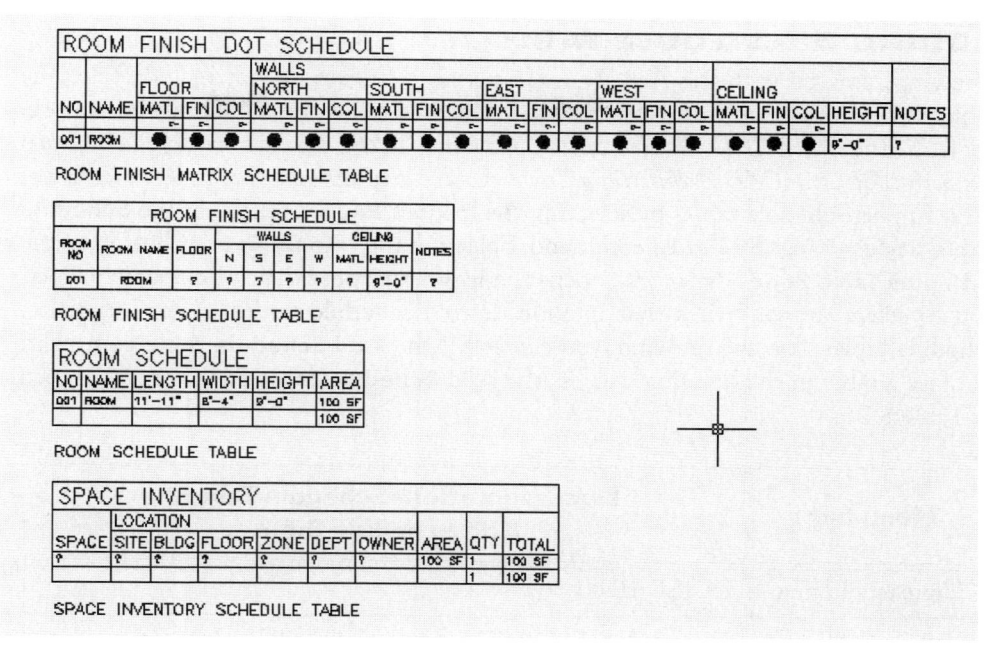

Figure 8.100 *Room and Space Schedules*

WALL TYPE SCHEDULE

TYPE	STYLE	NOTES
A	STANDARD	

WALL SCHEDULE TABLE

EQUIPMENT SCHEDULE

NO	DESCRIPTION	MANUFACTURER	MODEL	$ COST	HYPERLINK
1				0.00	
				0.00	

EQUIPMENT SCHEDULE TABLE

FURNITURE SCHEDULE

NO	DESCRIPTION	MANUFACTURER	MODEL	$ COST	HYPERLINK
A				0.00	
				0.00	

FURNITURE SCHEDULE TABLE

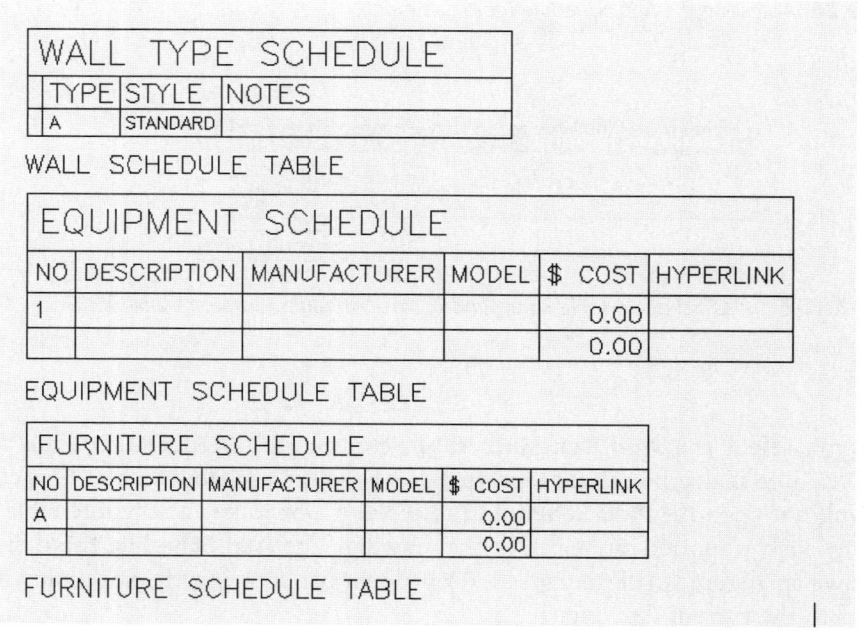

Figure 8.101 *Wall, Equipment, amd Furniture Schedules*

ADDING A SCHEDULE TABLE

Drawings created with the Aec Arch (Imperial) template do not include schedule tables. Therefore prior to the insertion of a schedule table, the table must be imported from the *AutoCAD Architectural 2\Content\Imperial\Schedules\Schedule Tables (Imperial)* or *AutoCAD Architectural 2\Metric\Schedules\Schedule Tables (Metric)* drawings. Import schedule tables by selecting the **Import/Export** option of the **Schedule Table Styles** (**TABLESTYLE**) command. Selecting the **Import/Export** button of the **Schedule Table Styles** dialog box opens the **Import/Export** dialog box, which allows you to select drawing files that include desired schedule tables. After importing schedule tables into the drawing, you can select the **Add Schedule Table** command to place a table in the drawing. Access the **Add Schedule Table** command as shown in Table 8.26.

Menu bar	Documentation>Schedule Tables> Add Schedule Table
Command prompt	TABLEADD
Toolbar	Select Add Schedule Table from the Schedule - Imperial or Schedule - Metric toolbars as shown in Figure 8.102

Table 8.26 *Accessing the Add Schedule Table command*

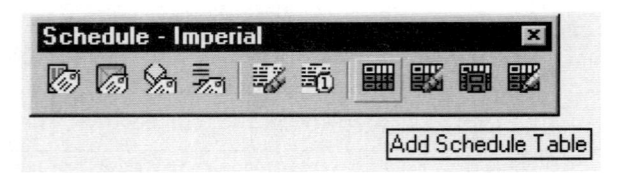

Figure 8.102 *Add Schedule Table command on the Schedule - Imperial toolbar*

When you select the **Add Schedule Table** command (**TABLEADD**), the **Add Schedule Table** dialog box opens, allowing you to select a schedule table. This command will not open the **Add Schedule Table** dialog box unless a schedule table has been previously imported or created in the drawing. The **Add Schedule Table** dialog box shown in Figure 8.103 includes a drop-down list showing the schedule tables attached to the current drawing.

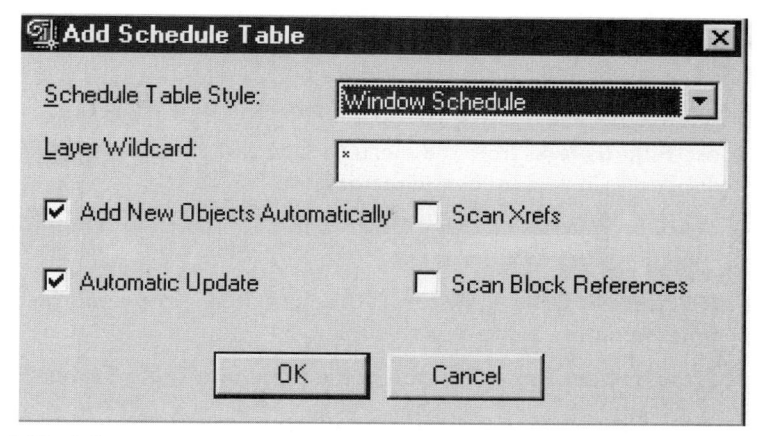

Figure 8.103 *Selecting schedule table from the Add Schedule Table dialog box*

The **Add Schedule Table** dialog box consists of the following options.

> **Schedule Table Style** – This list includes all schedule table styles that have been created or imported to the drawing.
>
> **Layer Wildcard** – The **Layer Wildcard** field allows you create a layer filter to limit the search for objects that apply to the schedule. The asterisk wildcard will include all layers of the drawing.
>
> **Add New Objects Automatically** – If the **Add New Object Automatically** check box is selected, new objects added to the drawing will automatically be added to the schedule. This check box ensures that the schedule is up to date as the drawing is being developed.
>
> **Automatic Update** – Selecting this check box will update the schedule as changes are made in the drawing.
>
> **Scan Xrefs** – The **Scan Xrefs** check box allows the search of objects to include external reference files that have been attached to the drawing. If you are developing a Window schedule, the windows of the current file and the external reference file will be included in the schedule when selected.
>
> **Scan Block References** – If drawings that include Architectural Desktop objects are inserted in the current drawing, these objects will be included in the schedule if this check box is selected. If selected, it will allow you to develop a schedule from the information included in reference drawings.

After editing the **Add Schedule Table** dialog box, you are prompted to select objects to be included in the schedule. Selecting the objects allows you to specify the location of the schedule in the drawing area. When you specify the location of the schedule, you can specify its size by selecting the location of the upper left corner and the lower right corner or you can accept the default size by pressing ENTER.

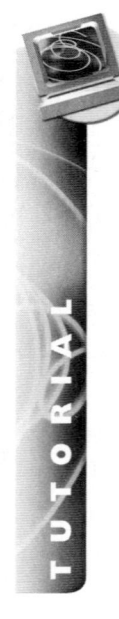

TUTORIAL 8.4 STEPS FOR CREATING A WINDOW SCHEDULE

1. Open *Ex8-4.dwg.* from the *ADT Tutor\Ch8* directory
2. Choose **File>SaveAs** from the menu bar and save the drawing as *Schedule2a.dwg* in your student directory.
3. Right-click over an Architectural Desktop toolbar and choose **Schedule - Imperial** from the toolbar list.
4. Select **Schedule Table Styles** from the **Schedule - Imperial** toolbar to open the **Schedule Table Styles** dialog box.
5. Select the **Import/Export** button of the **Schedule Table Styles** dialog box.
6. Select the **Open** button of the **Import/Export** dialog box.
7. Select the *\AutoCAD Architectural 2\Content\Imperial\Schedules\Schedule Tables (Imperial)* drawing in the **File to Import From** dialog box.
8. Select the **Open** button to select the file and close the **File to Import From** dialog box.
9. Select all the schedule tables in the **External File** list by selecting the top schedule and continue to hold down the left mouse button as you move the mouse down to the end of the list, as shown in Figure 8.104.

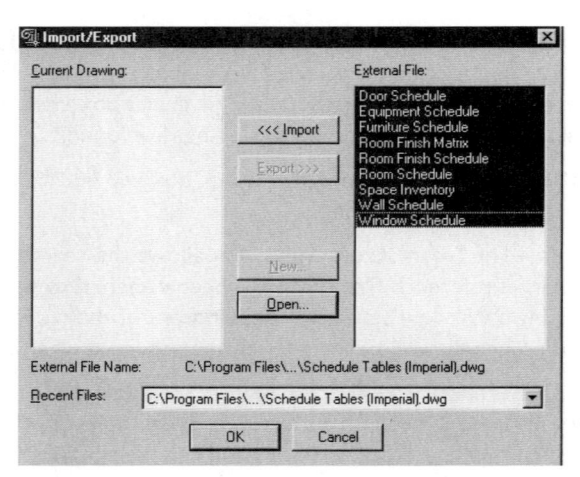

Figure 8.104 *Importing schedule tables using the Import/Export dialog box*

10. Select the **Import** button to import all the schedule tables to the current drawing.
11. Select **OK** to dismiss the **Import/Export** dialog box.

12. Select **OK** to dismiss the **Schedule Table Styles** dialog box.
13. Select **Add Schedule Table** from the **Schedule - Imperial** toolbar as shown in Figure 8.105.

Figure 8.105 *Add Schedule Table command on the Schedule - Imperial toolbar*

14. Select Window Schedule from the **Schedule Table Style** list, verify that the **Layer Wildcard** is set to an asterisk, and select **Add New Object Automatically** and **Automatic Update**, as shown in Figure 8.106.

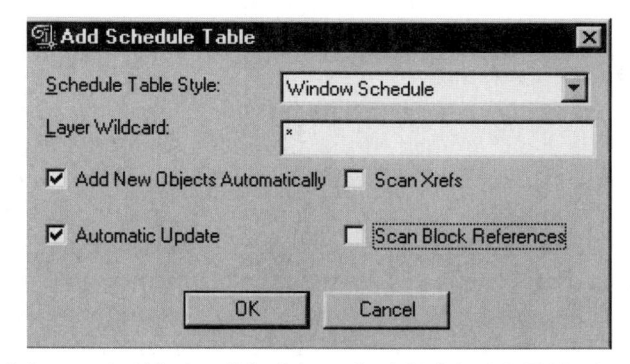

Figure 8.106 *Selecting the Window Schedule in the Add Schedule Table dialog box*

15. Select **OK** to dismiss the **Add Schedule Table** dialog box.
16. Respond to the command prompts as shown below to place the schedule.

    ```
    Command: _AecTableAdd
    Select objects: all ENTER (Selects all objects in the drawing.)
    199 found
    78 were not in current space.
    107 were filtered out.
    Select objects: ENTER (Press ENTER ends the selection.)
    Upper left corner of table: (Select a point near P1 in Figure 8.107.)
    Lower right corner (or RETURN): ENTER (Press ENTER to accept the default schedule table size.)
    ```

17. Save the drawing.

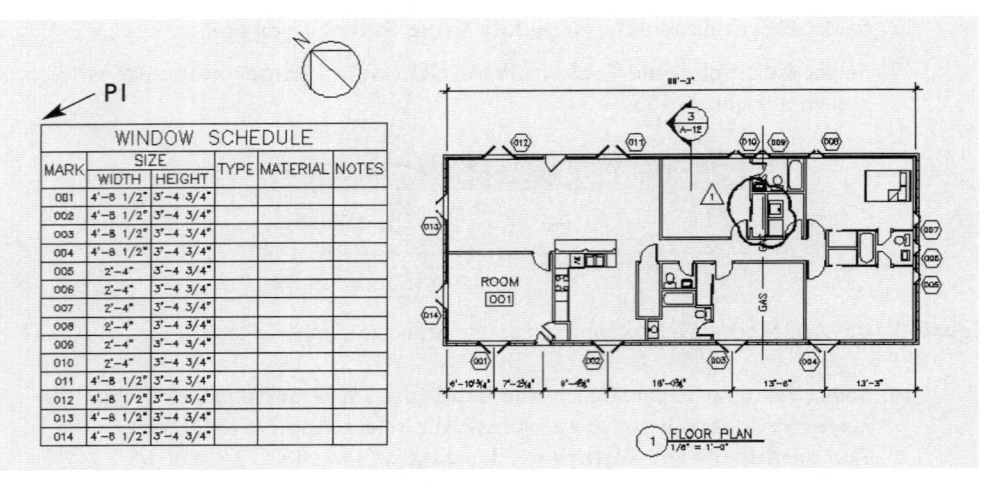

Figure 8.107 *Placing the window schedule in the drawing*

RENUMBERING TAGS

Tags are placed on a drawing and assigned a mark in a manner that allows you to predict the location of the next number. In the previous tutorial, the window tags were numbered increasing by one as you followed around the house in a counter-clockwise direction. Placing tag numbers or letters in a drawing in a systematic manner makes it easier for the craftsman to interpret the drawing. As the design develops, additional windows or doors can be inserted in a drawing, which creates disorder in the tag numbering sequence. Therefore, the **Renumber Data** command (**AECPROPERTYRENUMBERDATA**) can be used to renumber existing tags. Access the **Renumber Data** command as shown in Table 8.27.

Menu bar	Documentation>Schedule Data>Renumber Data
Command prompt	AECPROPERTYRENUMBERDATA
Toolbar	Select Renumber Data from the Schedule - Imperial or Schedule - Metric toolbar as shown in Figure 8.108

Table 8.27 *Accessing the Renumber Data command*

Figure 8.108 *Renumber Data command on the Schedule - Imperial and Schedule - Metric toobars*

When you select the **Renumber Data** command, the **Data Renumber** dialog box opens, as shown in Figure 8.109. This dialog box allows you to edit the tag number.

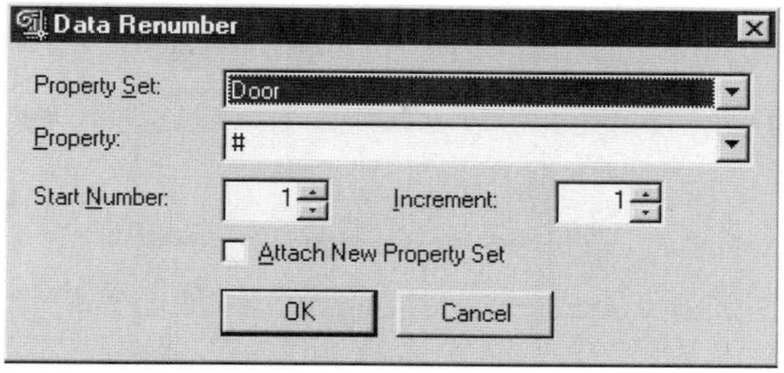

Figure 8.109 *Data Renumber dialog box*

The **Data Renumber** dialog box consists of the following options.

> **Property Set** – The **Property Set** drop-down list shows property sets defined for the drawing.
>
> **Property**–The property of the defined property set is specified for the renumbering operation.
>
> **Start Number** – The **Start Number** edit field allows you to specify the beginning number of the renumbering operation.
>
> **Increment** – The **Increment** value increases the tag number by adding the increment number to the previous tag number.
>
> **Attach New Property Set** – When you select the **Attach New Property Set** check box, the property of the objects in the drawing will change as you renumber them.

UPDATING A SCHEDULE

The content of a drawing can change after a schedule is developed; therefore, the schedule must be updated to reflect the latest condition of the drawing. When the schedule is created with the **Add Schedule Table** command, you can elect to select **Automatic Update** and **Add New Objects Automatically**, as shown in Figure 8.110, in the **Add Schedule Table** dialog box.

If these check boxes are selected, the schedule will be updated automatically when the drawing is changed. When only the **Add New Object Automatically** check box is selected, if a new object is added to the drawing, a line will be drawn across the table. This line indicates the table needs to be updated; it will be removed when the table is updated with the **Update Schedule Table** command (**TABLEUPDATENOW**). If the **Automatic Update** check box is selected, a new insertion in the drawing of an

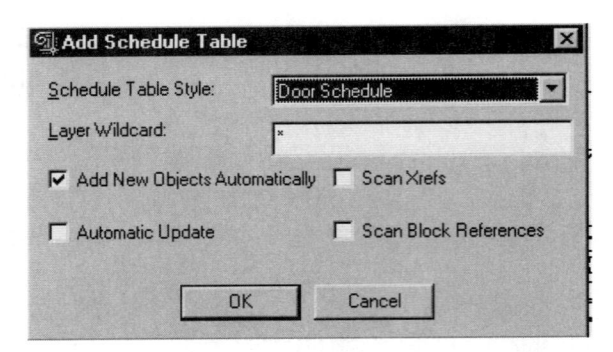

Figure 8.110 *Update check boxes in the Add Schedule Table dialog box*

object related to the schedule will cause the schedule to be updated immediately. However, these check boxes can slow down the response time in large drawings. Therefore, you can elect to deselect the check boxes when the schedule is created and update the schedule with the **Update Schedule Table** command. Access the **Update Schedule Table** (**TABLEUPDATENOW**) command as follows (Table 8.28)

Menu bar	Documentation>Schedule Tables> Update Schedule Table
Command prompt	TABLEUPDATENOW
Toolbar	Select Update Schedule Table from the Schedule - Imperial or Schedule - Metric toolbar shown in Figure 8.111

Table 8.28 *Accessing the Update Schedule Table command*

Figure 8.111 *Update Schedule Table command on the Schedule - Imperial toolbar*

When you select the **Update Schedule Table** command from the **Schedule - Imperial** or **Schedule -Metric** toolbar, you are prompted to select a schedule table. You can select one or more tables, and they will be updated as shown in the following command sequence.

```
Command: _AecTableUpdateNow
Select schedule tables: (Select a schedule table) 1 found
```

```
Select schedule tables:    ENTER (Pressing ENTER ends selection
of tables.)
Command: (Tables are updated to the current drawing.)
```

EDITING THE CELLS OF A SCHEDULE

The values in a schedule can be changed by the **Attach/Edit Schedule Data** command. This command prompts you to select the tag or the associated object and change the value of the property in the **Edit Schedule Data** dialog box. You can also edit the value of a property for an object by selecting the data in the schedule. The **Edit Table Cell** command (**TABLECELLEDIT**) is used to edit the data directly in the schedule. Access the **Edit Table Cell** command as shown in Table 8.29.

Menu bar	Documentation>Schedule Tables>Edit Table Cell
Command prompt	TABLECELLEDIT

Table 8.29 *Accessing the Edit Table Cell command*

When you select the **TABLECELLEDIT** command, you are prompted to select a cell within a schedule. Selecting the cell will open the **Edit Schedule Property** dialog box, as shown in Figure 8.112.

Figure 8.112 *Editing a cell of the schedule in the Edit Schedule Property dialog box*

The **Edit Schedule Property** dialog box allows you to enter a new value in the schedule. Included in the dialog box is the name of the property and property set that is defined for the cell. To change the value of the cell, overtype a new value and select the **OK** button; the schedule is then changed.

The **TABLECELLEDIT** command can also be used to globally edit the schedule. A global edit of the schedule would allow you to change the material of all windows to wood or steel without editing the material property value for each window. If you select the border of the schedule rather than a cell, you can edit the entire schedule, as shown in the command sequence below.

```
Command: _AecTableCellEdit
Select schedule table item (or the border for all
items):
```
(Select the border of the schedule as shown at P1 in Figure 8.113.)

Figure 8.113 *Selecting a schedule for editing*

(The manufacturer can then be changed in the Edit Property dialog box to define a different manufacturer for the entire schedule, as shown in Figure 8.114.)
```
Command: _AecTableCellEdit
Select schedule table item (or the border for all
items):
```
ENTER *(Press* ENTER *to end selection of a cell or schedule.)*

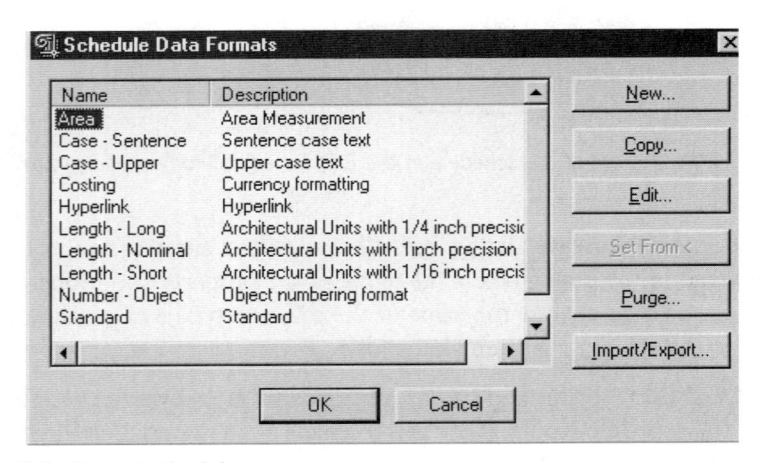

Figure 8.114 *Revised schedule*

Annotating the Drawing 679

SUMMARY

1. The scale factor for break marks, detail marks, elevation marks, and dimensions are set through the **DWGSCALESETUP** command.
2. Text for title and detail marks is placed in the drawing through attributes.
3. Leaders created from the options of the Leader folder are placed on the A-Anno-Note layer.
4. Dimensions should be placed in the drawing from the dimensioning commands of the Miscellaneous folder.
5. Fire Rating Lines to indicate fire-rated walls are selected from the Miscellaneous folder of the **Annotation - Imperial** or **Annotation - Metric** toolbar.
6. Match lines, north arrows, and datum elevations are selected from the **Miscellaneous** command on the **Annotation - Imperial** or **Annotation - Metric** toolbar.
7. Place revision clouds with revision tags in a drawing by selecting **Revision Clouds** from the **Annotation - Imperial** or **Annotation - Metric** toolbar.
8. Text for title marks, section marks, and bar scales are placed in the drawing as attributes.
9. Tags for Architectural Desktop objects can be placed in the drawing from the **Door & Window Tag**, **Room & Finish Tag**, **Object Tag**, or **Wall Tag** command on the **Schedule - Imperial** and **Schedule - Metric** toolbars.
10. Room and finish tags must be attached to an AEC Space.
11. The **Convert to Spaces** command (**SPACECONVERT**) is used to create AEC Spaces from polylines.
12. The **Attach/Edit Schedule Data** command (**PROPERTYDATAEDIT**) is used to insert or change data for objects.
13. The **Schedule Table Styles** command (**TABLESTYLE**) is used to define the components of a schedule.
14. The **Add Schedule Table** command (**TABLEADD**) is used to place a table style in the drawing.

REVIEW QUESTIONS

1. Break marks are scaled proportional to the drawing by _____.
2. The Elevation Mark B2 (Interior Elevation Mark 4 Way) can be used as _____ .
3. Place leaders in the drawing by selecting _____.
4. The precision of dimensions placed in the drawing from the **Miscellaneous** command on the **Annotation - Imperial** and **Annotation - Metric** toolbars should be changed by _____.
5. Change the width of the Fire Rating Lines by selecting the _____ command from the **Modify II** toolbar.

6. A tag is a _____ used to assign a property value to an object.
7. AEC Spaces include an area property included in the _____ tag.
8. The command used to define the association between Architectural Desktop objects and the property set is the_____.
9. A schedule can be inserted in a new drawing if _____, _____, and_____.
10. The **Attach/Edit Schedule Data** command is used to edit _____ data.
11. A schedule with a diagonal line across the schedule indicates_____.
12. Text data in a schedule is revised by the _____ command.
13. Schedule data can be exported to _____ spreadsheet.

PROJECTS

EXERCISE 8.1 CREATING DOOR AND WINDOW SCHEDULES

1. Open *ADT Tutor\Ch8\Ex8-8.dwg* and save the file to your student directory as *Proj8-8.dwg*.
2. Insert the wall tags as shown in Figure 8.115.
3. Edit the Edit Schedule Data dialog box for each window tag as shown in Figure 8.115
4. Import the Wall Schedule from the *\AutoCAD Architectural\Content\Imperial\Schedules\Schedule Tables (Imperial).dwg*.
5. Insert the Wall Schedule in the drawing and select each wall tagged for the schedule.
6. Save the drawing.

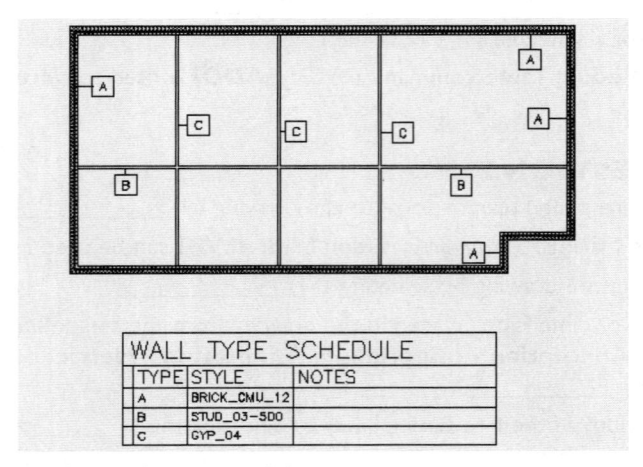

Figure 8.115 *Wall Schedule for Exercise 8.1*

CHAPTER 9

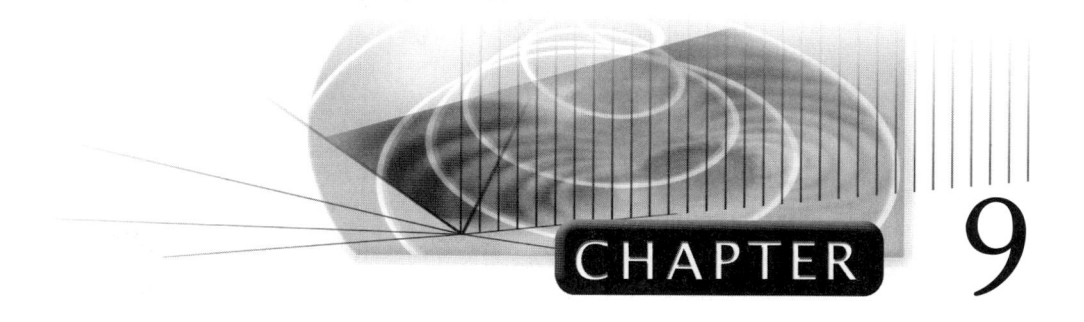

Creating Elevations and Sections

INTRODUCTION

This chapter includes the commands to draw elevations and sections of a building. The **Create Elevations** and **Create Sections** commands create skeleton 3D objects, which include the basic geometry for creating more detailed drawings. Prior to the creation of elevations and sections, the floor plans of each level are attached as external reference files to construct a model drawing. As changes occur in each level, the model will automatically be updated because it is created from external reference files. The **Update Elevation** and **Update Section** commands allow you to update the elevation or section based on changes that have been made in the model since the section or elevation was created. Elevation and section working drawings are developed from the elevation or section object through the **Hidden Line Projection** command (**CREATEHLR**), which creates a 2D block that can be exported to a separate drawing file for development as a working drawing.

OBJECTIVES

Upon completion of this chapter, you will be able to

- Use external reference files to create a 3D model of a building

- Use the **Add Elevation Mark** command (**BLDGELEVATIONLINEADD**) to define the location and viewing direction for an elevation object

- Create an elevation object using the **Create Elevation** command (**BLDGELEVATIONLINEGENERATE**)

- Edit the elevation object with the **Update Elevation** command (**BLDGSECTIONUPDATE**)

- Define the dimensions of the projection box for an elevation or section object

Use **Entity Display** command (**ENTDISLAY**) to edit the display properties of the elevation, section, and section line marks

Create blocks for the development of elevation and section working drawing with the **Hidden Line Projection** command (**CREATEHLR**)

Place the cutting plane line for a section object through the model using the **Add Section Mark** command (**BLDGSECTIONLINEADD**)

Convert existing polylines to a section line mark using the **BLDGSECTIONLINECONVERT** command

Create a section object using the **Create Section** command (**BLDGSECTIONLINEGENERATE**)

Revise a section object using the **Update Section** command (**BLDGSECTIONUPDATE**) to reflect changes made in the model

Edit the description and location properties of a section or elevation object using the **BLDGSECTIONPROPS** command

CREATING THE MODEL FOR ELEVATIONS AND SECTIONS

The elevation or section of a building should be developed from a model drawing, which consists of the floor plans for each level of the building. The model drawing allows you to put together floor plans for each level, which remain as separate drawings as the design develops. Each floor plan will be attached to the model drawing as a reference file. Therefore, as the floor plans are revised, the model will be revised when the reference file is reloaded into the model.

Prior to attaching the floor plans as reference files, freeze the layer of objects not essential to the development of the elevation. Layers for plumbing fixtures, dimensions, section marks, and other annotations should be frozen in each floor plan. Objects on the frozen layers will not be included in the elevation when the elevation is generated. The opening percent of exterior doors and windows should be set to zero because the display of these features in an elevation is usually in the closed position.

Because attaching each floor plan as a reference file creates the model drawing, the insertion point for each floor plan should be defined according to the Z coordinate elevation of its floor. Therefore, the first floor plan will be inserted at 0,0,0, and the second level will be inserted at 0,0,9' if the finish floor to finish floor dimension is 9'. After all floor plans are attached at their respective Z coordinate elevation, the building elevation or building section object can be developed from the model drawing.

If the floor plans for each level are developed from each other without moving those bearing walls or columns that align vertically, the floor plans can be inserted at X=0 and Y=0, with the Z coordinate set according to the distance between floors. In the previous tutorials, the basement floor plan was developed as a copy of the first floor. Therefore the exterior walls of each level remained aligned vertically when the second floor was inserted at 0,0,9'.

However, if the walls that should align vertically are moved relative to X,Y coordinate location, each plan must be adjusted with the **Move** command after being inserted at the appropriate elevation. The model can be used to check the vertical alignment of the structural elements of the building.

Note: If you create a layer for each level and insert the reference file on that layer, you can turn off or on the display of the entire reference file by freezing or thawing the layer of its insertion point.

TUTORIAL 9.1 CREATING THE MODEL FOR SECTIONS AND ELEVATIONS

1. Open *ADT Tutor\Ch9\Ex9-1.dwg*.
2. Choose **File>SaveAs** from the menu bar and save the drawing as *Floor1.dwg* in your student directory.
3. Close *Floor1.dwg*.
4. Open *ADT Tutor\Ch9\Ex9-2* and save the drawing as *Floor2.dwg* in your student directory.
5. Select all the exterior windows and then right-click and choose **Window Properties** from the shortcut menu.
6. Select the **Dimensions** tab of the **Window Properties** dialog box and change the **Open %** to zero, as shown in Figure 9.1.

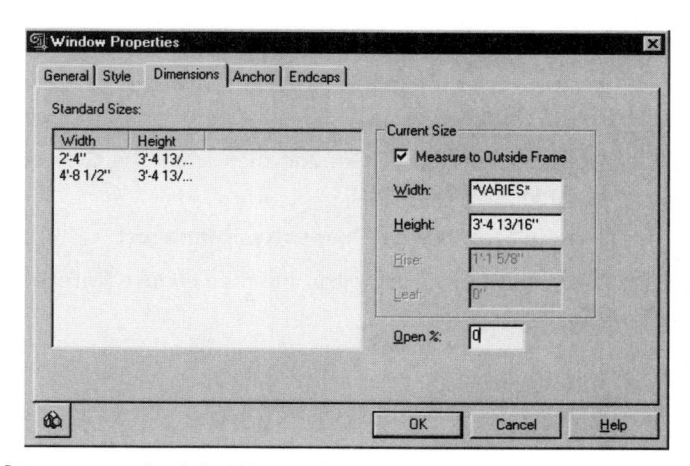

Figure 9.1 *Dimensions tab of the Window Properties dialog box*

7. Select **OK** to dismiss the **Window Properties** dialog box.
8. Select the two exterior doors; then right-click and choose **Door Properties** from the shortcut menu.
9. Select the **Dimensions** tab of the **Door Properties** dialog box and change the **Open %** to zero, as shown in Figure 9.2.

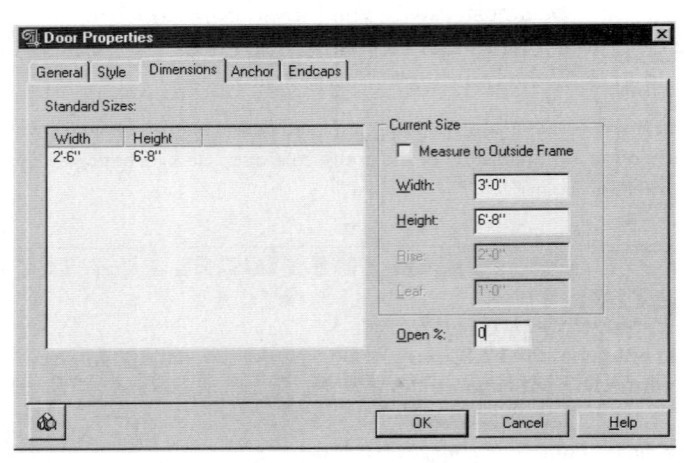

Figure 9.2 *Dimensions tab of the Door Properties dialog box*

10. Select **OK** to dismiss the **Door Properties** dialog box.
11. Save the *Floor2.dwg* drawing.
12. Close the *Floor2.dwg* drawing.
13. Create a new drawing in your student directory named *Model.dwg* using the Aec_Arch (Imperial) template.
14. Select **Layers** from the **Object Properties** toolbar, and then select the **New** button and enter **Level1** and **Level2** as the names of the new layers.
15. Select Level1 layer from the layer list, and then select the **Current** button of the **Layer Properties Manager**.
16. Select **OK** to dismiss the **Layer Properties Manager**.
17. Right-click over the **Standard** toolbar and then choose **Reference** from the toolbar list.
18. Select **External Reference Attach** from the **Reference** toolbar as shown in Figure 9.3.
19. Select *Floor1.dwg* from your student directory in the **Select Reference File** dialog box.

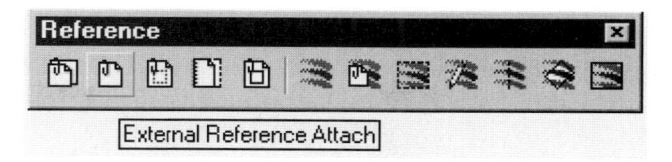

Creating Elevations and Sections   685

Figure 9.3 *External Reference Attach command on the Reference toolbar*

20. Select **Open** to open the **External Reference** dialog box.
21. Edit the **External Reference** dialog box as shown in Figure 9.4. Set the **Reference Type** to Attachment, clear the **Specify On-screen** check box and set **X** = 0, **Y** = 0, **Z** = 0, **Scale** = 1, and **Rotation** = 0. Select **OK** to dismiss the **External Reference** dialog box.

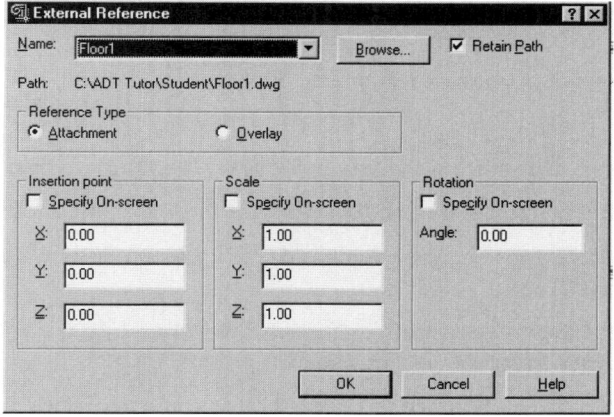

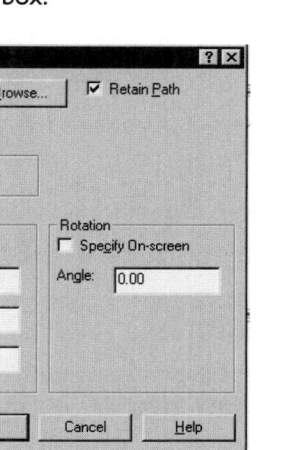

Figure 9.4 *Settings in the External Reference dialog box*

22. Select the **Layers** button on the **Object Properties** toolbar and freeze the following layers:

    Floor1|A-Anno-Dims

    Floor1|A-Anno-Legn

    Floor1|A-Area

    Floor1|A-Area-Iden

    Floor1|A-Door-Iden

    Floor1|A-Flor-Hral

    Floor1|A-Flor-Strs

    Floor1|A-Glaz-Iden

23. Select **OK** to dismiss the **Layer Properties Manager** dialog box.
24. Select the **SW Isometric** option on the **View** flyout of the **Standard** toolbar to view *Floor1.dwg* inserted at 0,0,0 as shown in Figure 9.5.

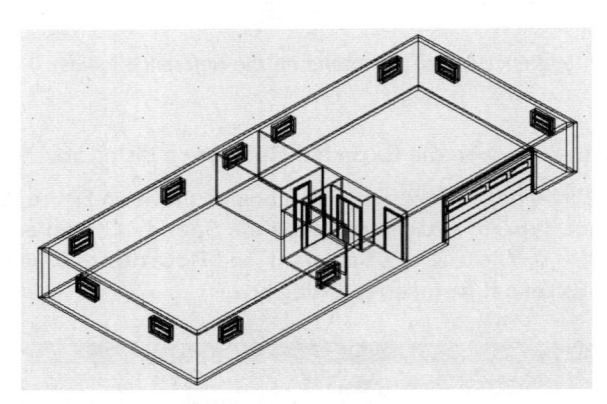

Figure 9.5 *Display of Floor1 attached as a reference file*

25. Set Level2 current from the **Layer** flyout on the **Object Properties** toolbar.
26. Select **External Reference Attach** from the **Reference** toolbar.
27. Select *Floor2.dwg* from your student directory in the **Select Reference File** dialog box.
28. Select the **Open** button of the **Select Reference File** dialog box to open the External Reference dialog box.
29. Edit the **External Reference** dialog box as shown in Figure 9.6. Set the **Reference Type** to Attachment, clear the **Specify On-screen** check box and set **X** = 0, **Y** = 0, **Z** = 8' 11", **Scale** = 1, and **Rotation** = 0.
30. Select **OK** to dismiss the **External Reference** dialog box.
31. Select the **Layers** button on the **Object Properties** toolbar and freeze the following layers:

    Floor2|A-Anno-Dims

    Floor2|A-Anno-Note

    Floor2|A-Anno-Revs

    Floor2|A-Anno-Ttlb

    Floor2|A-Area

    Floor2|A-Area-Iden

    Floor2|A-Area-Spce

Creating Elevations and Sections 687

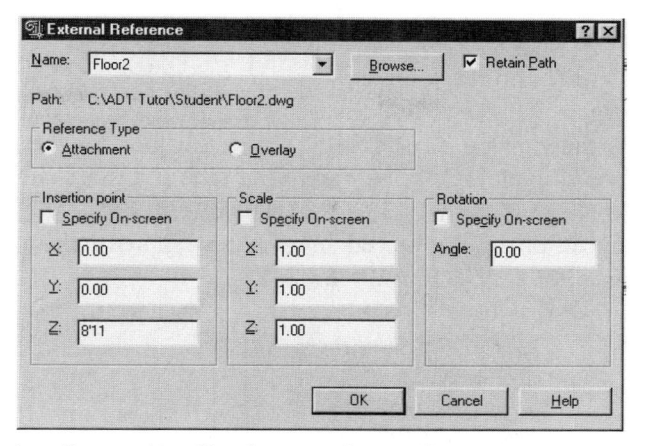

Figure 9.6 *Settings for attaching Floor2 as a reference file*

        Floor2|A-Flor-Appl
        Floor2|A-Flor-Case
        Floor2|A-Flor-Pfix
        Floor2|A-Flors-Strs
        Floor2|A-Furn
        Floor2|A-Glaz-Iden
        Floor2|A-Sect_Iden
        Floor2|Gas

32. Select **OK** to dismiss the **Layer Properties Manager**.
33. Enter **HI** in the command line to hide lines of the two external reference files inserted, as shown in Figure 9.7.
34. Save the drawing.

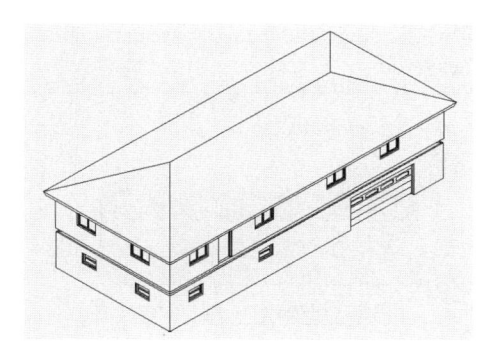

Figure 9.7 *Display of Floor1 and Floor2 reference files attached to the model*

CREATING A BUILDING ELEVATION LINE

The commands to create and revise an elevation are located on the Elevation/Section toolbar shown in Figure 9.8. Create an elevation by first drawing a building elevation line, which defines the viewing direction and projection box for the elevation. The properties of the building elevation line define the size of a three-dimensional projection box. Only those objects that fall within the box will be included in the elevation object. The elevation line is placed in the drawing with the **Add Elevation Mark** command (**BLDGELEVATIONLINEADD**).

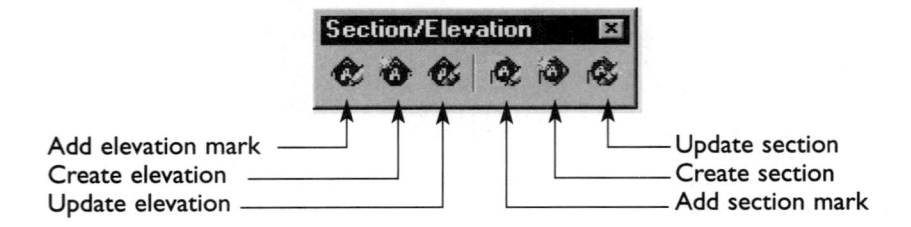

Figure 9.8 *Commands on the Section/Elevation toolbar*

The **Add Elevation Mark** command (**BLDGELEVATIONLINEADD**) is used to define the location and viewing direction of the elevation. It allows you to draw a line near the object to define the viewing plane for the elevation. See Table 9.1 for **Add Elevation Mark** command access.

Menu bar	Design>Generate Elevations>Add Elevation Mark
Command prompt	BLDGELEVATIONLINEADD
Toolbar	Select Add Elevation Mark from the Section/Elevation toolbar as shown in Figure 9.9
Shortcut	Right-click over the drawing area and choose Design>Elevation Lines>Add

Table 9.1 *Add Elevation Mark command access*

Figure 9.9 *Add Elevation Mark command on the Section/Elevation toolbar*

When the **Add Elevation Mark** command is selected, you are prompted in the command line to define the start and endpoint of the elevation line. The **Add Elevation Mark** command creates the AEC_BDG_ELEVLINE object. The start and end points selected define the viewing plane and direction of view. If a horizontal building elevation line is created from left to right, the viewing direction will be above the line. The following command line sequence was used to create the building elevation line shown in Figure 9.10.

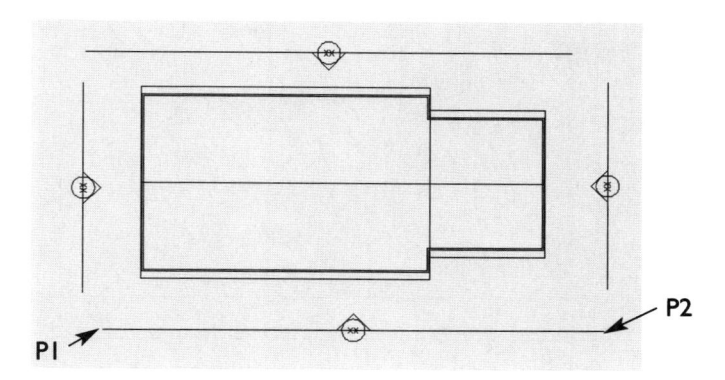

Figure 9.10 *Specify the Add Elevation Mark beginning and ending points to determine view direction*

```
Command: _AecBldgElevationLineAdd
Elevation line start point: (Select point P1 in Figure 9.10.)
Elevation line end point: (Select point P2 in Figure 9.10.)
Command:
```

The **BLDGELEVATIONLINEREVERSE** command can be used to change the viewing direction of an existing building elevation line. This command changes the viewing direction 180 degrees. See Table 9.2 for **BLDGELEVATIONLINEREVERSE** command access.

Command prompt	BLDGELEVATIONLINEREVERSE
Shortcut	Select a building elevation line; then right-click and choose Reverse

Table 9.2 *BLDGELEVATIONLINEREVERSE command access*

After placing the building elevation line, you can edit the text in the elevation bubble with the **MVBLOCKPROPS** command. This command allows you to change the attribute properties of the multi-view block. To change the default text of the building elevation line, select the circle or text of the elevation bubble, and then right-click and choose **Multi-View Block Properties** from the shortcut menu. Change the text by selecting the **Attributes** tab of the **Multi-View Block Reference Properties** dialog box, as shown in Figure 9.11.

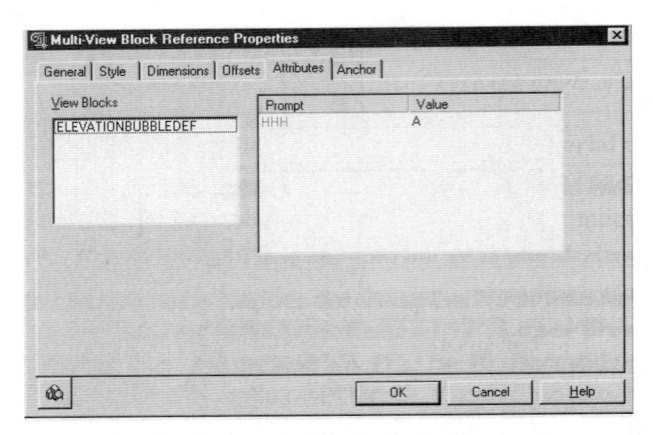

Figure 9.11 *Insert text for the elevation bubble in the Attributes tab*

The text in the attribute **Value** field can be changed to reflect the name of the elevation for the elevation bubble.

CREATING AN ELEVATION

After you place the building elevation line, you can generate an elevation from the model using the **Create Elevation** command (**BLDGELEVATIONLINEGENERATE**) on the **Section/Elevation** toolbar. This command creates a 3D-elevation object named AEC_BLDG_SECTION object. The elevation object is created on the A-Sect layer, which has the color 242 (a hue of brown). A new elevation can be created with the command or existing elevations can be updated to reflect recent changes in the model. The elevation can remain coincident with the original model or can be displaced to another location in the drawing. Therefore, after creating an elevation, you can repeat the command and reselect the model to update the elevation to reflect changes in the design model. See Table 9.3 for **Create Elevation** command access.

When you choose the **Create Elevation** command, you are prompted in the command line to select an elevation line, select the entities for the elevation, and then specify the location of the elevation, as shown in the command sequence below.

```
Command: _AecBldgElevationLineGenerate
Select elevation line: (Select a building elevation line.)
```

Creating Elevations and Sections 691

Menu bar	Design>Generate Elevations>Create Elevation
Command prompt	BLDGELEVATIONLINEGENERATE
Toolbar	Select Create Elevation from the Section/Elevation toolbar as shown in Figure 9.12
Shortcut	Right-click over the drawing area and choose Design>Elevation Lines>Generate
Shortcut	Select a building elevation line; then right-click and choose Generate Elevation from the shortcut menu

Table 9.3 *Create Elevation command access*

Figure 9.12 *Create Elevation command on the Section/Elevation toolbar*

```
Select entities for elevation: (Select the model.) Specify
opposite corner: 13 found
Select entities for elevation: ENTER (Press ENTER to end
object selection.)
```

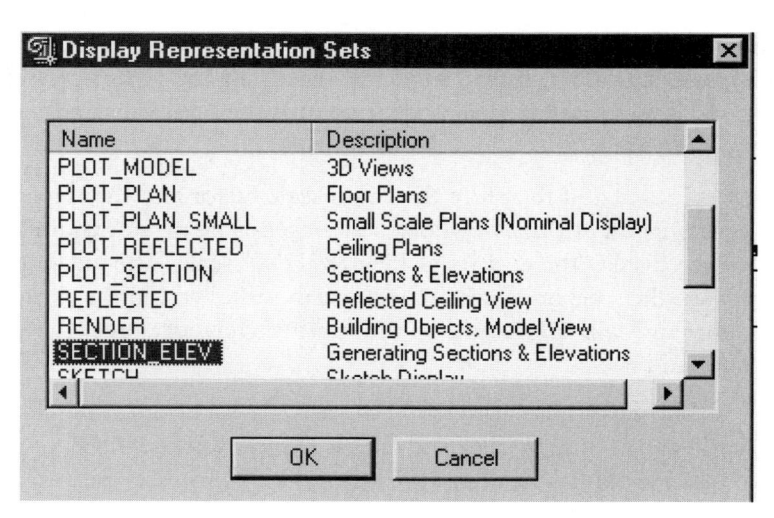

Figure 9.13 *Specify the display representation set for the elevation*

> *(Select the SECTION_ELEV display representation as shown in Figure 9.13.)*
> `Elevation to go to [Existing]<New>:` *(Specify Existing or New elevation.)*
> `Specify base point or displacement:` *(Specify the location of the elevation.)*

After selecting the building elevation line, you are prompted to select the entities for the elevation. The **Create Elevation** command will develop an elevation from all entities selected. The entities selected can be AutoCAD entities or Architectural Desktop objects. If no entities are selected, the command is terminated. After selecting the entities for the elevation, you are prompted to select the display representation for the elevation. The **Display Representation Sets** dialog box opens, allowing you to select a display representation from the list of display representations. The SECTION_ELEV display representation set should be selected for elevation views because it includes the typical display sets of architectural objects for elevation views.

After you specify the display representation set, you are asked if you want to create a new elevation or an existing elevation. Selecting the **Existing** option allows you to update a previous elevation according to the changes in the model. The **New** option allows you to create a new elevation, and you will be prompted to specify the location of the elevation.

CONTROLLING THE VIEW OF THE ELEVATION

After you create an elevation from the model, the reference files can be frozen and the elevation object can be viewed from the left, right, front or rear to obtain a preliminary elevation drawing. Obtain these views by selecting the view from the **View** toolbar or the **View** flyout on the **Standard** toolbar. The Plot-SEC layout tab, shown in Figure 9.14, is specifically designed to provide the front and right side views of an object. It consists of two viewports: the top viewport is the right view and the bottom viewport is the front view.

Each viewport is locked and therefore the zoom scale factor and the viewing direction cannot be changed until the viewport is unlocked. The default zoom scale factor is 1/96 × p, to display the entities at the 1/8"=1'-0" scale. The viewport can be unlocked to change the scale or pan the model space entities within the viewport. The **-VPORTS** command will allow you to unlock the viewport. See Table 9.4 for **-VPORTS** command access.

When the Plot-SEC layout tab is selected, paper space is current. Therefore, if you select the viewport and then right-click and choose **Display Locked>OFF**, you can turn off the lock of the viewport. When the display is unlocked, you can then use the **PAN** and **ZOOM** commands to change the display of the viewport.

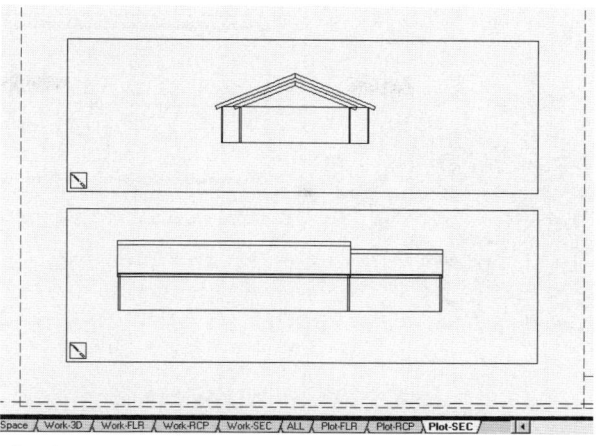

Figure 9.14 *Plot-Sec layout tab*

Command prompt	–VPORTS
Shortcut	Select the viewport; then right-click and choose Display Locked>ON/OFF

Table 9.4 *–VPORTS command access*

The default shape of viewport can be adjusted to increase the utilization of the paper. When paper space is current, you can adjust the floating viewport shape by selecting the grips of the viewport. When the grips are selected, you can stretch the viewport grips to increase the viewport size. The floating viewports are created on the Viewport layer, which is defined as a no plot layer. Therefore, the geometry of the viewport will not print or plot. You can create the preliminary elevations quickly in the Plot-SEC layout by specifying viewing direction and scale of the floating viewports.

TUTORIAL 9.2 CREATING THE ELEVATION

1. Open *Model.dwg* from your student directory.
2. Choose **File>SaveAs** from the menu bar and save the drawing as *Elev1* in your student directory.
3. Select the Work-3D layout tab and then select the right viewport to activate the plan view.
4. Select **Add Elevation Mark** from the **Section/Elevation** toolbar.
5. Respond to the following command line prompts to place the building elevation line as shown below.

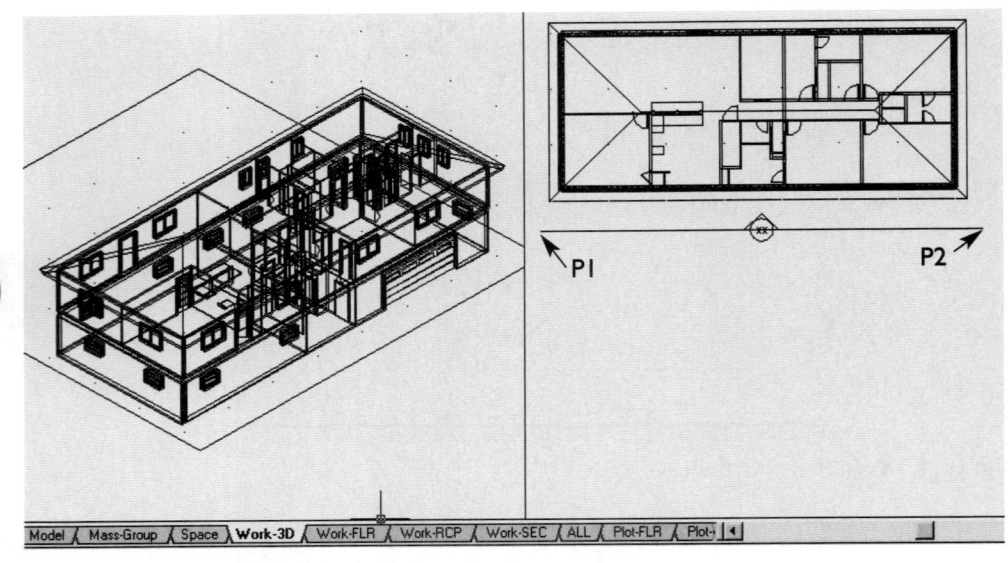

Figure 9.15 *Specify the location for the building elevation line*

(Select in the right viewport to activate the viewport.)
```
Command: _AecBldgElevationLineAdd
Elevation line start point: (Select point P1 in Figure 9.15.)
Elevation line end point: (Select point P2 in Figure 9.15.)
```

6. Select the elevation bubble of the building elevation line; then right-click and choose **Multi-View Block Properties** from the shortcut menu.
7. Select the **Attributes** tab of the **Multi-View Block Properties** dialog box as shown in Figure 9.16.

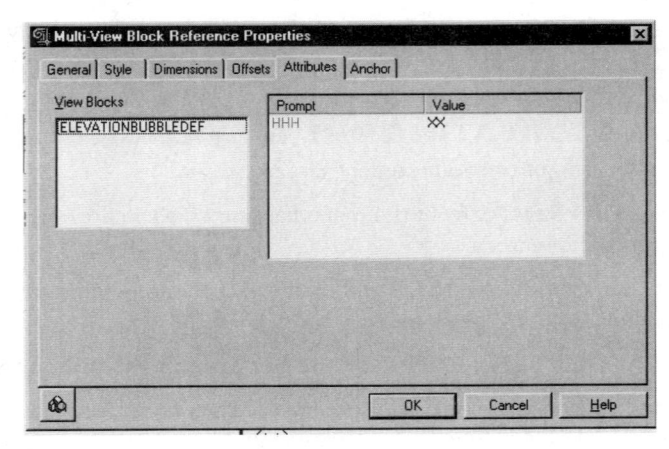

Figure 9.16 *Edit Attributes of the elevation bubble*

8. Type **A** in the **Value** edit field of the **Attributes** tab.
9. Select **OK** to dismiss the **Multi-View Block Reference Properties** dialog box.
10. Select the building elevation line; then right-click and choose **Generate Elevation** from the shortcut menu.
11. Respond to the command line prompts as shown in the command sequence below and in Figure 9.17.

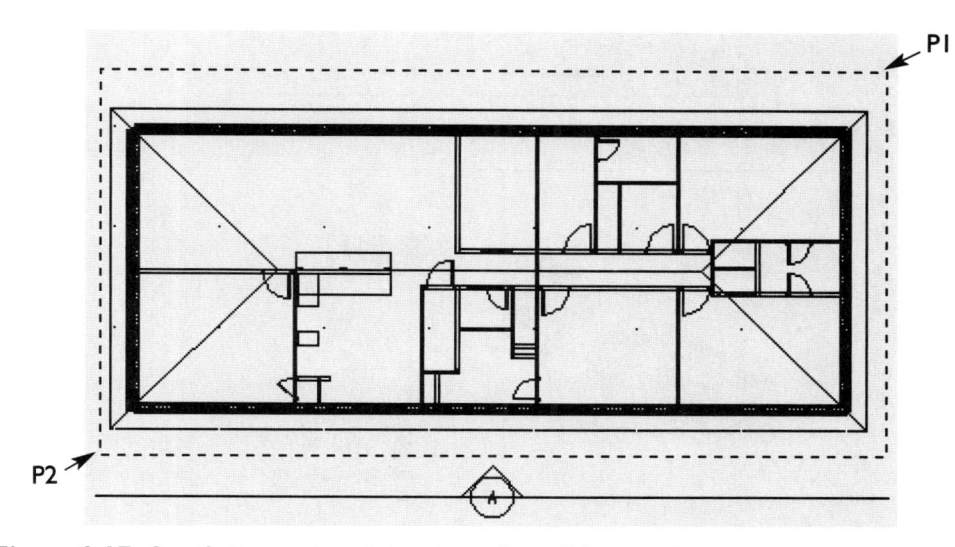

Figure 9.17 *Specify the crossing window for entities of the elevation*

```
Command: BldgElevationLineGenerate
Select entities for elevation:
```
(Specify the crossing window from point P1 to P2 in Figure 9.17.) `Specify opposite corner: 2 found`
`Select entities for elevation:` ENTER
(Select the SECTION_ELEV display representation set as shown in Figure 9.18.)
`Elevation to go to [Existing]<New>:` ENTER *(Press ENTER to create a new elevation.)*
`Specify base point or displacement: 10',0` ENTER
(Specify base point for the elevation.)
`Specify second point or <use first point as displacement>: 100',0` ENTER *(Specify the location of the elevation 100' to the right.)*

12. Select the Work-SEC layout tab to view the elevation as shown in Figure 9.19.

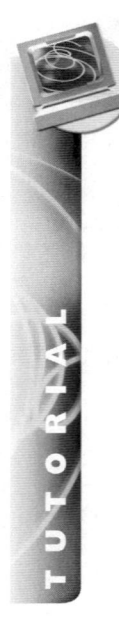

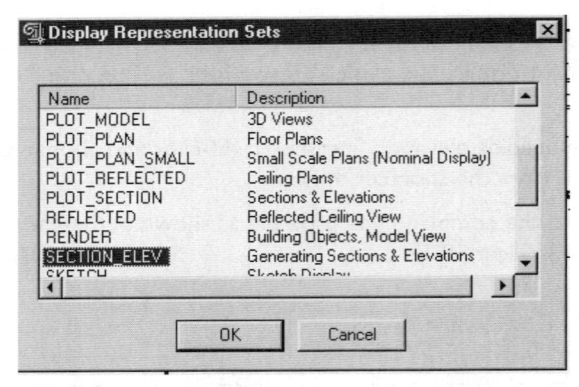

Figure 9.18 *Select the display representation for the elevation*

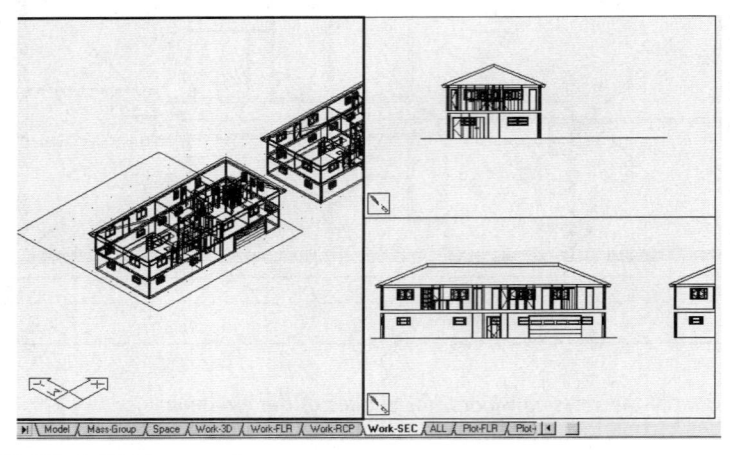

Figure 9.19 *Work-SEC layout tab*

13. Select the lower right viewport; then select the **Pan** command from the **Standard** toolbar.
14. Pan the front view to the left as shown in Figure 9.20.
15. Select the Plot-SEC layout tab.
16. Select the lower viewport and then right-click and choose **Display Locked>NO** from the shortcut menu.
17. Select the upper viewport and then right-click and choose **Display Locked>NO** from the shortcut menu.
18. Select Paper in the Status Bar to toggle ON model space.
19. Select inside the lower viewport and then select **Pan** from the **Standard** toolbar and pan the view of the elevation object to the center of the viewport window.

Creating Elevations and Sections 697

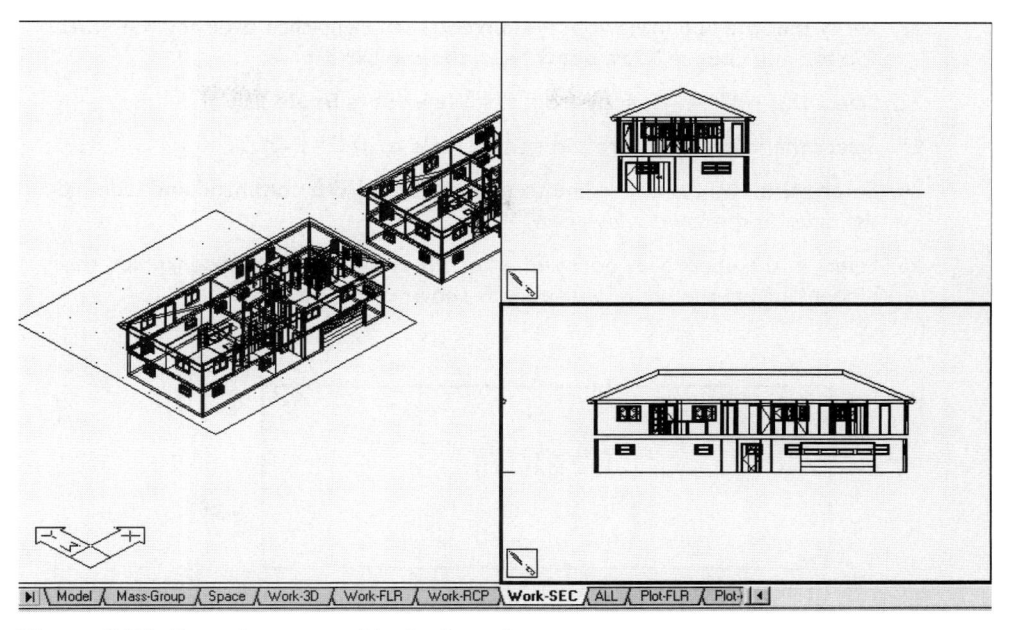

Figure 9.20 *Front view centered in the front viewport*

20. Select inside the upper viewport and then select **Pan** from the **Standard** toolbar and pan the view of the elevation object to the center of the upper viewport window, as shown in Figure 9.21.

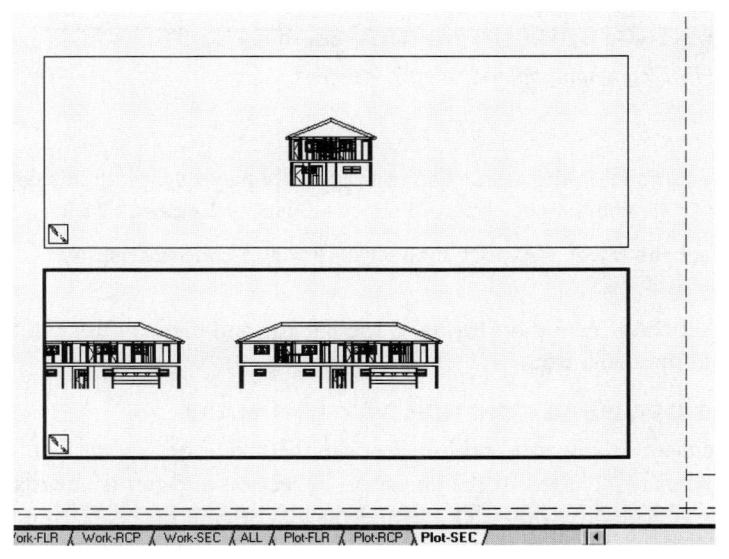

Figure 9.21 *Plot-SEC layout tab*

21. Verify that the upper viewport is current; then right-click over any AutoCAD toolbar and choose **Viewports** from the toolbar list.
22. Select the 1/4"= 1'-0" scale from the **Viewports Scale** list.
23. Select the lower viewport and edit its scale to 1/4"=1'-0".
24. Enter **HI** in the command line to execute the **HIDE** command and hide hidden lines in the lower viewport.
25. Select in the upper viewport and enter **HI** in the command line to hide the hidden lines in the upper viewport as shown in Figure 9.22.

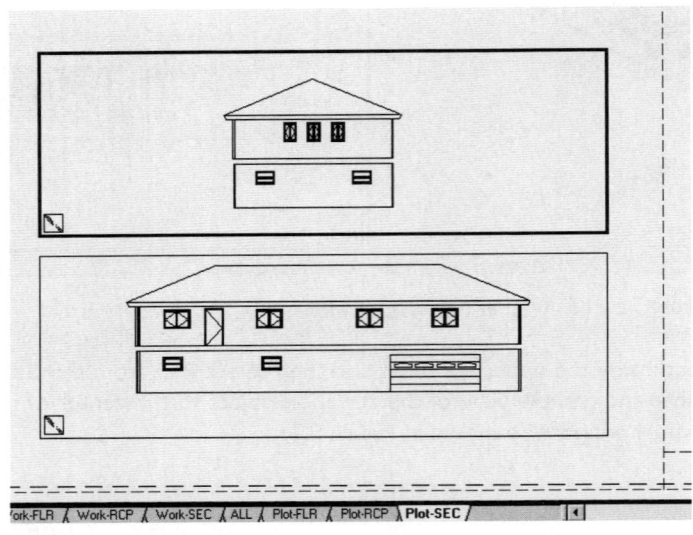

Figure 9.22 *HIDE command applied to each viewport*

26. Select Model in the Status Bar to toggle ON paper space; then select the upper viewport, right-click, and choose **Display Locked>YES**.
27. Select the lower viewport; then right-click and choose **Display Locked>YES**.
28. Select **Save** from the **Standard** toolbar bar, and then choose **File>Close** from the menu bar.

PROPERTIES OF THE SECTION/ELEVATION MARK

The building elevation line and building section line properties are similar. They allow you to define what is included in the elevation or section and refine the display of the object using graphic subdivisions. The properties of the building elevation line can be edited with the **Elevation Mark Properties** command (**BLDGELEVATIONLINE-PROPS**). See Table 9.5 for **Elevation Mark Properties** command access.

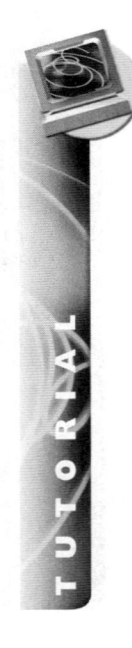

Menu bar	Design>Generate Elevations> Elevation Mark Properties
Command prompt	BLDGELEVATIONLINEPROPS
Shortcut	Right-click over the drawing area and choose Design>Elevation Lines>Properties from the shortcut menu
Shortcut	Select a building elevation line; then right-click and choose Elevation Line Properties

Table 9.5 *Elevation Mark Properties command access*

When you select a building elevation line and then right-click and choose **Elevation Line Properties** from the shortcut menu, the **Section/Elevation Line Properties** dialog box opens. Selecting the **Dimensions** tab of the **Section/Elevation Line Properties** dialog box allows you to define the properties of the building elevation line as shown in Figure 9.23.

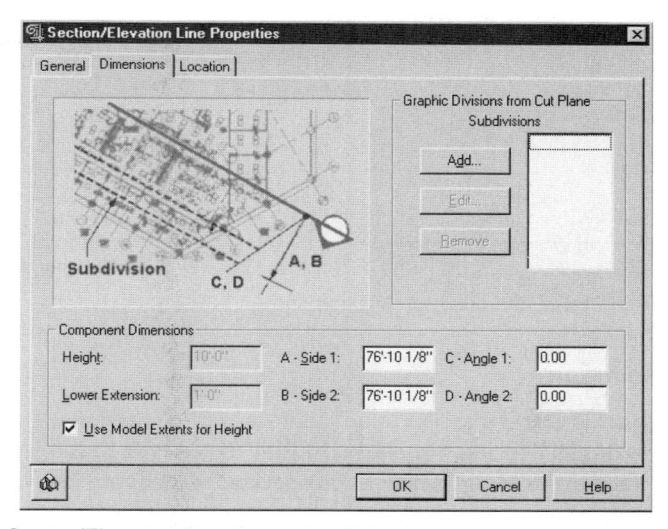

Figure 9.23 *Section/Elevation Line Properties dialog box*

The **Section/Elevation Line Properties** dialog box consists of three tabs: **General**, **Dimensions**, and **Location**.

General
The **General** tab allows you to specify a description for the elevation. The **Notes** button opens the **Notes** dialog box, which includes tabs for **Text** and **Reference Docs**.

These tabs allow you to enter text notes regarding the elevation line and list the files related to the elevation. The **Reference Docs** tab can include names of the drawing files extracted from the elevation object.

Dimensions

The **Dimensions** tab, shown in Figure 9.23, allows you to define the reach of the elevation or section. The **A-Side 1** and **B-Side 2** dimensions are relative to the first and last points selected, as shown in Figure 9.24, when the elevation line is created. The options of the **Dimensions** tab are described below.

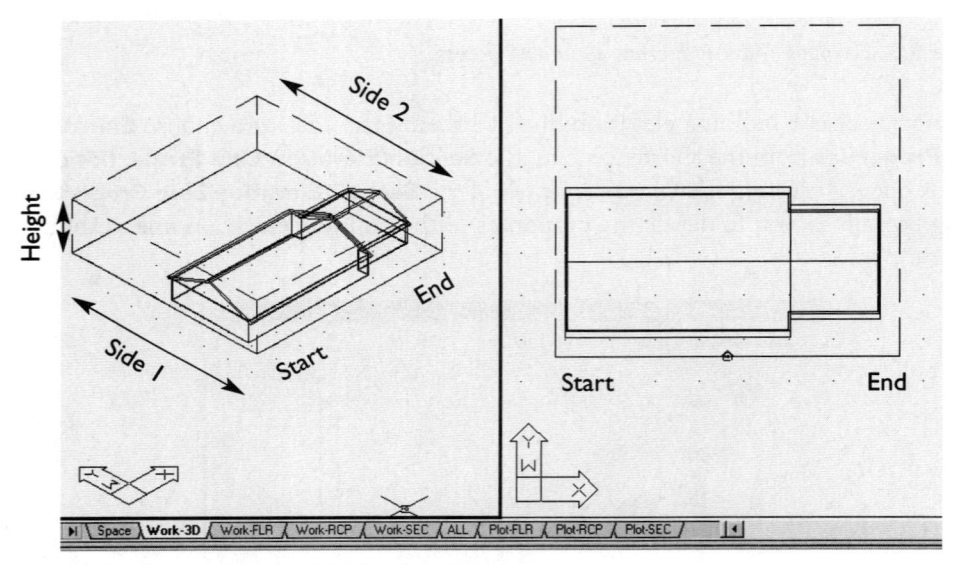

Figure 9.24 *Dimensions of the building elevation line*

Component Dimensions

Height – The **Height** dimension is the vertical dimension up from the building elevation line plane. If the **Use Model Extents for Height** check box is not selected, the **Height** dimension specified will control the vertical reach above the plane of the building elevation line.

Lower Extension – The **Lower Extension** is the vertical dimension below the building elevation line plane. It allows you to include features below the plane in the elevation.

Use Model Extents for Height – The **Use Model Extents for Height** check box will define the height of the elevation to include all model space components. This check box eliminates the need for determining the height of the building prior to creating the elevation.

A-Side 1 – The **A-Side 1** dimension is the distance from the first point of the building elevation line measured in the view direction of the building elevation line. The first point is the beginning point selected when the building elevation line is created.

B-Side 2 – The **B-Side 2** dimension is the distance from the second point of the building elevation line measured in the view direction of the building elevation line. The depth dimensions of each side can vary allowing the elevation to include only selected building components.

C-Angle 1 – The **C-Angle 1** dimension is the deflection from a line perpendicular to the building elevation line. Positive angles deflect to the left of the viewing direction, while negative angles deflect to the right. The angle of deflection is relevant to your orientation when facing toward the viewing direction for the elevation as shown in Figure 9.25.

D-Angle 2 – The **D-Angle 2** dimension is the deflection from the second point along a line perpendicular to the building elevation line. The positive angles deflect to the left and the negative angles deflect to the right, when oriented from the second point.

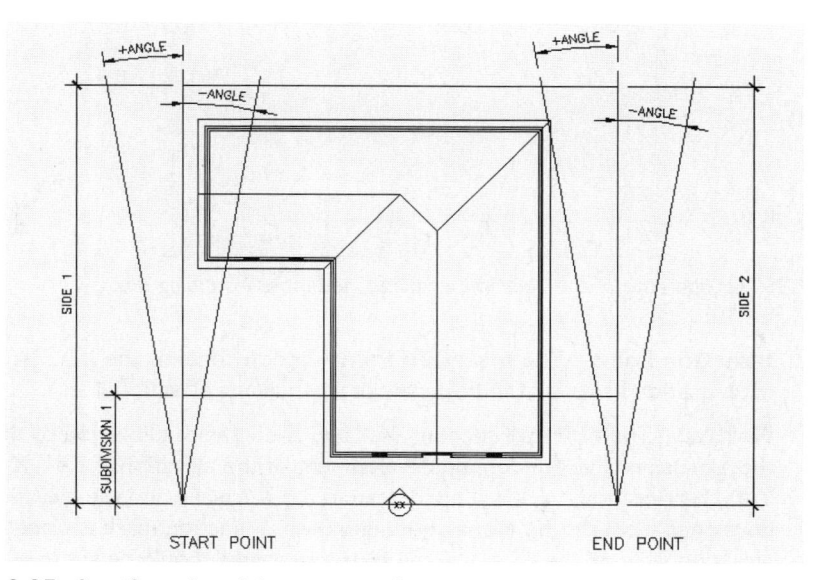

Figure 9.25 *Specify angles of the projection box*

Graphic Divisions from Cut Plane Subdivisions

Add – The **Add** button allows you to add subdivisions to the reach of the elevation.

Edit – The **Edit** button allows you to change the depth dimension of a subdivision.

Remove – The **Remove** button allows you to delete a defined subdivision.

The display of the entities within the subdivision is defined in the **Entity Display** dialog box for the AEC_BLDG_SECTION. The AEC_BLDG_SECTION object can be a building elevation or a building section. The subdivisions allow you to create elevations or sections that emphasize those planes closest to the observer by increasing the line weight or color of the entities within those subdivisions closest to the observer. The **ENTDISPLAY** command and the **Entity Display** dialog box are described in the next section.

Location

The **Location** tab, shown in Figure 9.26, allows you to move or edit the orientation of the section or elevation. The options of the **Location** tab are described below.

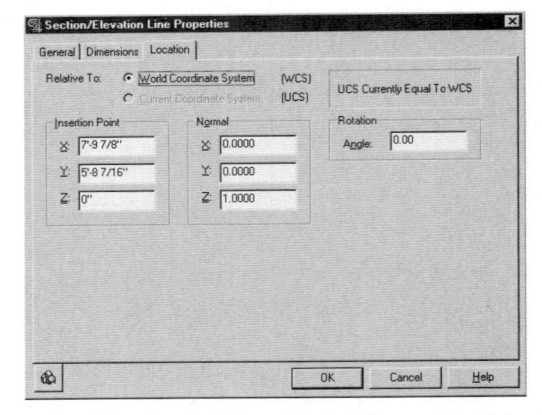

Figure 9.26 *Location tab of the Section/Elevation Line Properties dialog box*

Insertion Point – The **Insertion Point** section contains the X, Y, Z coordinate location fields for the beginning of the building elevation line.

Normal – The **Normal** section's X, Y, and Z edit fields allow you to define the extrusion direction for the development of the elevation. If Z=1, X=0, and Y=0, the elevation mark will be located in the XY plane and the elevations developed from the XY plane. If the elevation or section mark is place in an elevation view or the YZ plane, the **Normal** values should be set to X=1, Y=0 and Z=0 to develop an elevation view.

Rotation – The **Rotation Angle** edit field allows you to enter an angle of rotation for the building elevation line.

World Coordinate System / User Coordinate System – The **WCS** and **UCS** toggles indicate the coordinate system active when the building elevation line is inserted in the drawing.

ENTITY DISPLAY OF THE ELEVATION OR SECTION

The properties of the subdivisions of the elevation or section are defined in the **Section/Elevation Line Properties** dialog box. However, the display of the entities within each subdivision of the elevation is controlled by the **ENTDISPLAY** command. See Table 9.6 for **ENTDISPLAY** command access.

Command prompt	**ENTDISPLAY**
Shortcut	Select the Elevation or Section; then right-click and choose Entity Display from the shortcut menu

Table 9.6 *ENTDISPLAY command access*

The **ENTDISPLAY** command opens the **Entity Display** dialog box, which includes the **AutoCAD Props** and **Display Props** tabs. This command can be used on any Architectural Desktop object to edit its display properties. The **Entity Display** dialog box for a selected section or elevation object is shown in Figure 9.27.

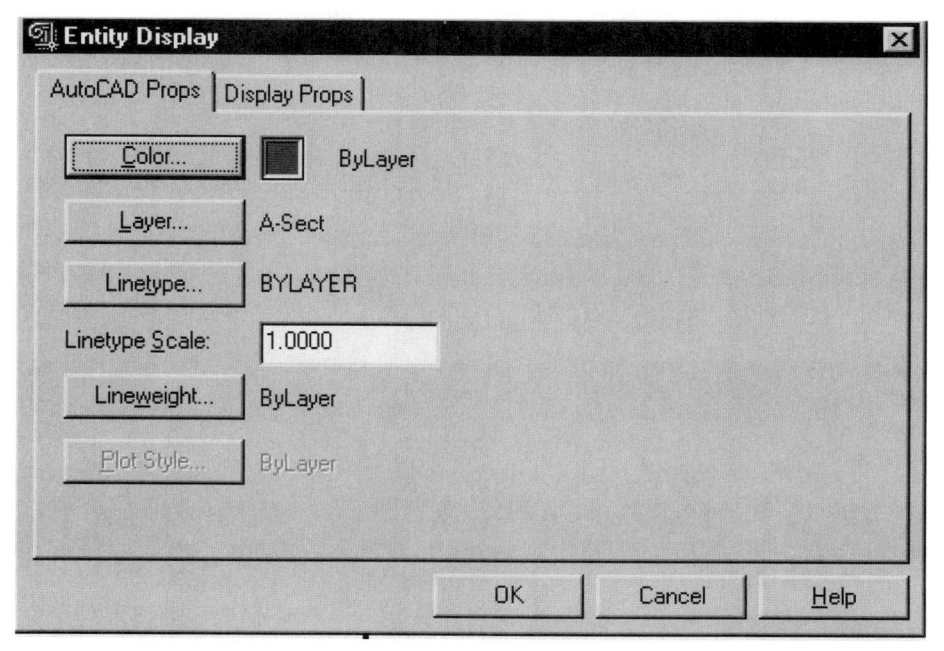

Figure 9.27 *Entity Display dialog box*

The **AutoCAD Props** and **Display Props** tabs are described below.

AutoCAD Props

The **AutoCAD Props** tab, shown in Figure 9.27, allows you to control the color, linetype, line weight, and LTSCALE of the object. You can change the color or other properties of the object by selecting the respective button and then editing the dialog box to change the specified property. You should avoid controlling the display of the elevation using AutoCAD commands because the color, layer, and linetype are defined according to the object and the AIA Layer Guidelines.

Display Props

The **Display Props** tab, shown in Figure 9.28, includes the current display representation set for the current viewport and the display settings of the **Property Source**.

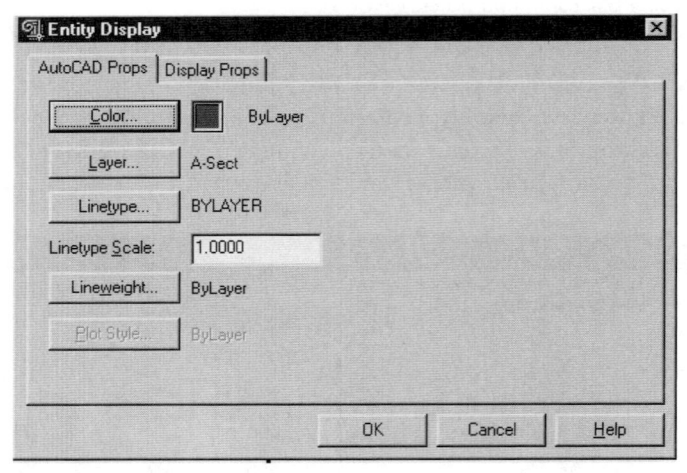

Figure 9.28 *Display Props tab of the Entity Display dialog box*

The **Display Props** tab consists of the following options:

> **Display Representation** – The drop-down list at the top of the **Display Props** tab includes the display representation sets available for the object. The one marked by an asterisk is the current display representation set used in the active viewport.
>
> **Property Source** – The **Property Source** column lists object display categories for the current viewport. The Bldg Section category controls the display of elevations and sections. The System Default category displays the section or elevation using the default display for all sections.
>
> **Display Contribution** – The **Display Contribution** column indicates the status of the current state of object display. The options of the display contribution are Controls, Empty, and Overridden. The Controls option indicates that the display of the object is being controlled by that display category.

Empty indicates that the display category has not been overridden. An Overridden display indicates that exceptions to the default display settings have been created.

Attached – The **Attached** column has Yes and No options, which indicate whether a display category has exceptions attached as overrides.

If the System Default settings are controlling display of the section or elevation, it will be listed as Controls and the **Attached** column checked as Yes. When System Default settings are overridden for a section or elevation, the Controls Display Contribution column will be marked with a red X to indicate that the System Default is overridden. The Overridden option is used to display the entities in a subdivision differently.

When System Default is overridden, you can set the display properties of each subdivision of the section or elevation. If you select the **Edit Display Props** button to edit the display properties of the Bldg Section, the **Entity Properties** dialog box opens, as shown in Figure 9.29. The **Entity Properties** dialog box allows you to edit the visibility, color, layer, linetype, lineweight, and linetype scale of the entities that are within the reach of the respective subdivision.

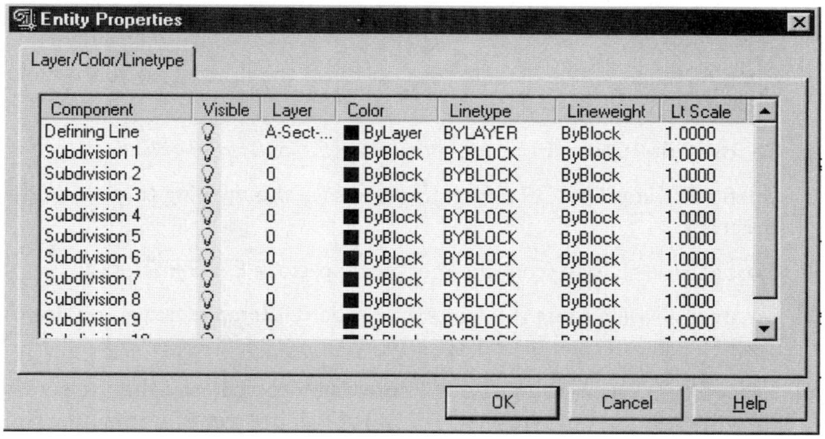

Figure 9.29 Entity Properties dialog box

A front elevation of the L-shaped building shown in Figure 9.30 could be created that emphasizes the planes closest to the observer. To control the display of these entities, you create elevation subdivisions, as shown in Figure 9.30. The depth of the first subdivision should be set to include the front projection of the L-shaped building. After the elevation is created, you can edit the entity properties of each subdivision. The line weight or color of the entities in this subdivision could be set to emphasize the subdivision.

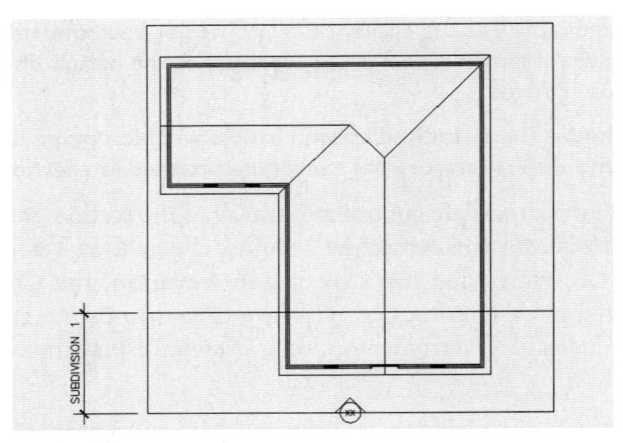

Figure 9.30 *Subdivision of the building elevation line*

TUTORIAL 9.3 REFINING THE DISPLAY OF THE ELEVATION

The following tutorial uses external references to develop the drawing. Therefore, the files copied from the CD should be saved as directed to your student directory. The model will be developed from the files located in your student directory.

1. Open *ADT Tutor\Ch9\Ex9-3-FND.dwg* and save the drawing to your student directory.
2. Choose **File>Close** from the menu bar to close *Ex9-3-FND.dwg*.
3. Open *ADT Tutor\Ch9\Ex9-3-FLR1.dwg* and save the drawing to your student directory.
4. Choose **File>Close** from the menu bar to close *Ex9-3-FLR1.dwg*.
5. Create a new file using the Aec Arch (Imperial) template; save the drawing as *Elev2* to your student directory.
6. Select **Layers** from the **Object Properties** toolbar, and then select the **New** button and type **Level1** and **Level2** as the names of the new layers.
7. Select Level1 from the layer list and then select the **Current** button.
8. Select **OK** to dismiss the **Layer Properties Manager**.
9. Select the Work-3D layout tab.
10. Right-click over the **Standard** toolbar and then choose **Reference** from the toolbar list.
11. Select **External Reference Attach** from the **Reference** toolbar.
12. Select *Ex9-3-FND.dwg* from your student directory in the **Select Reference File** dialog box.

13. Select **Open** to open the **External Reference** dialog box.
14. Edit the **External Reference** dialog box as shown in Figure 9.31. Set the **Reference Type** to Attachment, clear the **Specify On-screen** check box and set the **Insertion point** to **X** = 0, **Y** = 0, **Z** = 0, **Scale** = 1, and **Rotation** = 0.

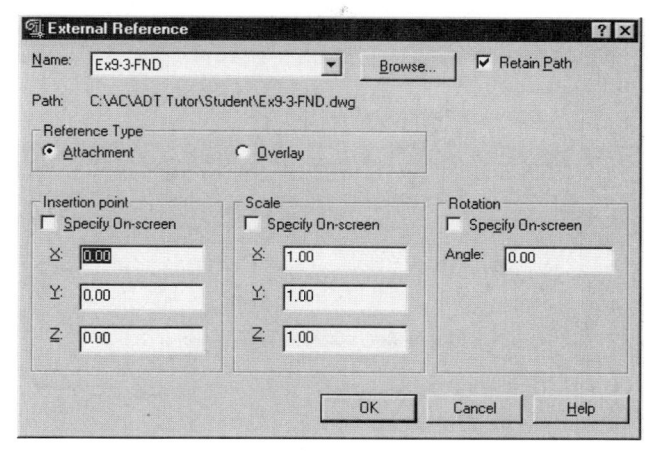

Figure 9.31 *Settings of the External Reference dialog box for inserting level1*

15. Select **OK** to dismiss the **External Reference** dialog box.
16. Select Level2 from the **Layer** flyout on the **Object Properties** toolbar to make Level2 the current layer.
17. Select **External Reference Attach** from the **Reference** toolbar.
18. Select *Ex9-3-FLR1.dwg* from your student directory in the **Select Reference File** dialog box.
19. Select **Open** to open the **External Reference** dialog box.
20. Edit the **External Reference** dialog box as shown in Figure 9.32. Set the **Reference Type** to Attachment, clear the **Specify On-Screen** check box and set the **Insertion point** to **X** = 0, **Y** = 0, **Z** = 4'–6", **Scale** = 1, and **Rotation** = 0.
21. Select **OK** to dismiss the **External Reference** dialog box.
22. Select the right viewport to activate the plan view viewport of the Work-3D layout.
23. Select **Add Elevation Mark** from the **Section/Elevation** toolbar.
24. Respond to the following command line prompts to place the building elevation line.

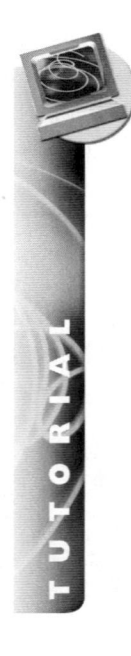

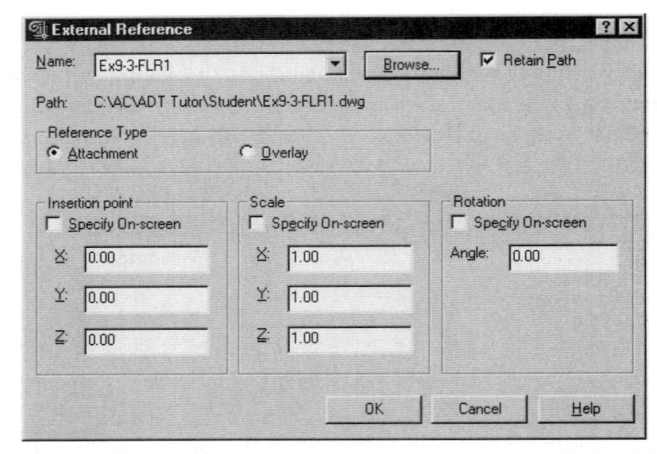

Figure 9.32 *Settings of the External Reference dialog box for inserting level 2*

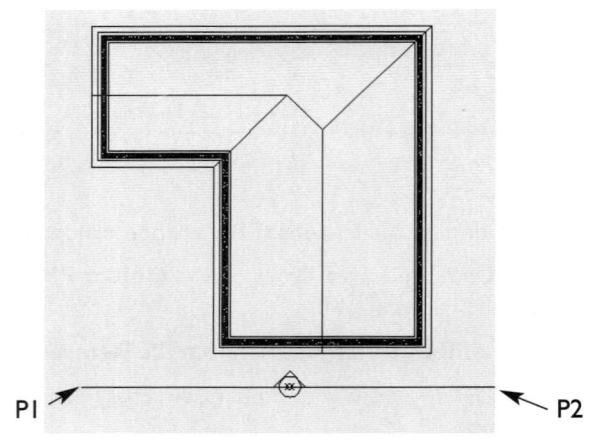

Figure 9.33 *Specify the location for the building elevation line*

```
Command: _AecBldgElevationLineAdd
Elevation line start point: (Select point P1 in Figure 9.33.)
Elevation line end point: (Select point P2 in Figure 9.33.)
```

25. Select the elevation bubble of the building elevation line; then right-click and choose **Multi-View Block Properties** from the shortcut menu.
26. Select the **Attributes** tab of the **Multi-View Block Reference Properties** dialog box as shown in Figure 9.34.
27. Type **S** in the **Value** edit field of the **Attributes** tab.
28. Select **OK** to dismiss the **Multi-View Block Reference Properties** dialog box.

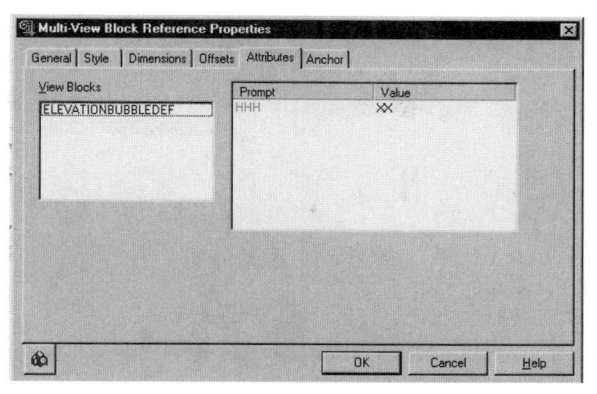

Figure 9.34 *Specify the text for the elevation bubble in Attributes tab*

29. Select the building elevation line; then right-click and choose **Elevation Line Properties** to open the **Section/Elevation Line Properties** dialog box.
30. Select the **Dimensions** tab and then select the **Add** button and type **40'** in the **Add Subdivision** dialog box, as shown in Figure 9.35.

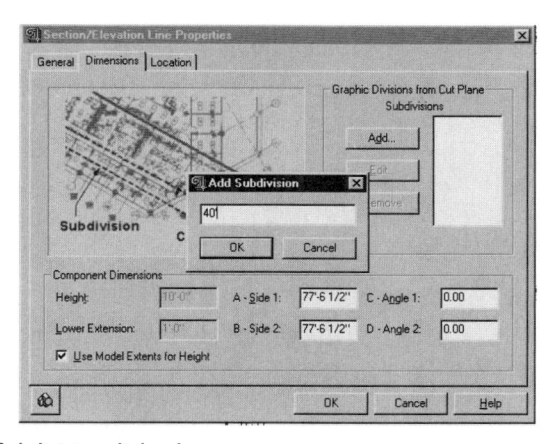

Figure 9.35 *Add Subdivision dialog box*

31. Select **OK** to dismiss the **Add Subdivision** dialog box.
32. Verify that **Use Model Extents for Height** is selected and retain the default dimensions for **A-Side 1, B-Side 2, C-Angle1**, and **D-Angle 2** as shown in Figure 9.35. The **A-Side 1** and **B-Side 2** dimensions shown here may differ from those in your drawing because the elevation line can differ in length.
33. Select **OK** to dismiss the **Section/Elevation Line Properties** dialog box.
34. Select the elevation line; then right-click and choose **Generate Elevation** from the shortcut menu.

35. Refer to Figure 9.36 and the following command sequence to create the elevation.

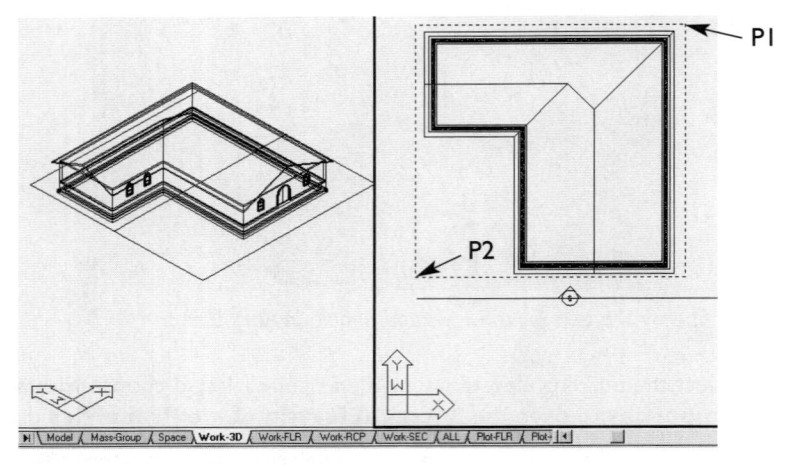

Figure 9.36 *Selecting objects for the elevation*

```
Command: BldgElevationLineGenerate
Select entities for elevation:
```
(Specify the crossing window from point P1 to P2 in Figure 9.36.) `Specify opposite corner: 2 found`
`Select entities for elevation:` ENTER
(Select the SECTION_ELEV display representation set from the Display Representation Sets dialog box, and then select OK to dismiss the dialog box.)
`Elevation to go to [Existing]<New>:` ENTER *(Press* ENTER *to create a new elevation.)*
`Specify base point or displacement: 10',0` ENTER *(Specify base point for the elevation.)*
`Specify second point or <use first point as displacement>: 100',0` ENTER *(Specify the location of the elevation 100' to the right.)*
(Elevation created as shown in Figure 9.37.)

36. Select the elevation object in the 3D view; then right-click and choose **Entity Display** from the shortcut menu.
37. Select the **Display Props** tab of the **Entity Display** dialog box.
38. Select Bldg Section from the **Property Source** list and then select the **Attach Override** button.
39. Select the **Edit Display Props** button to open the **Entity Properties** dialog box.

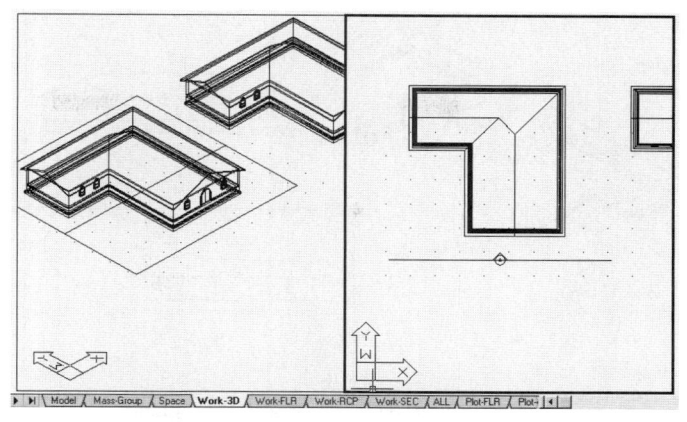

Figure 9.37 *Elevation object created from the model*

40. Select the ByBlock Color option in the **Color** column for Subdivision 1 and then select Blue from the **Select Color** dialog box as shown in Figure 9.38.

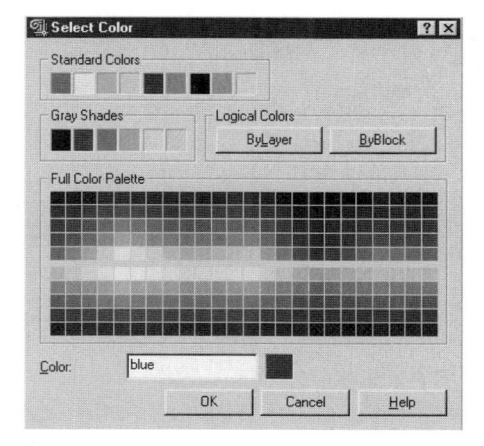

Figure 9.38 *Specifying color in the Select Color dialog box*

41. Select **OK** to dismiss the **Select Color** dialog box.
42. Select **OK** to dismiss the **Entity Properties** dialog box.
43. Select **OK** to dismiss the **Entity Display** dialog box.
44. Select the Work-SEC layout tab to view the elevation as shown in Figure 9.39.
45. Select the lower right viewport.
46. Select the **Pan** command from the **Standard** toolbar and pan the elevation to the center of the viewport as shown in Figure 9.40.

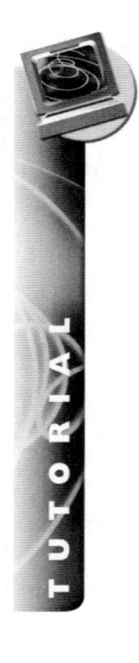

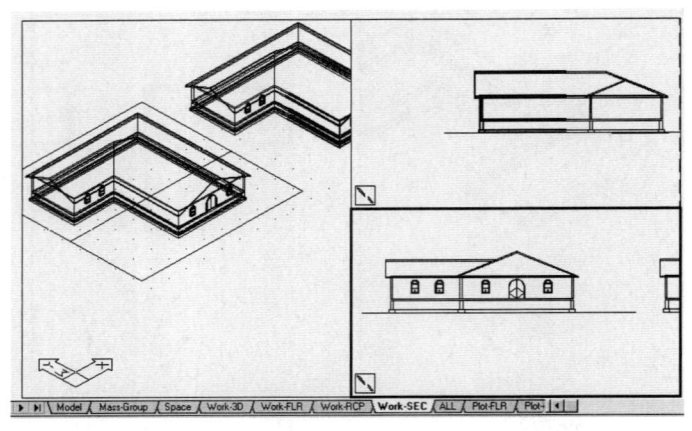

Figure 9.39 *Work-SEC layout*

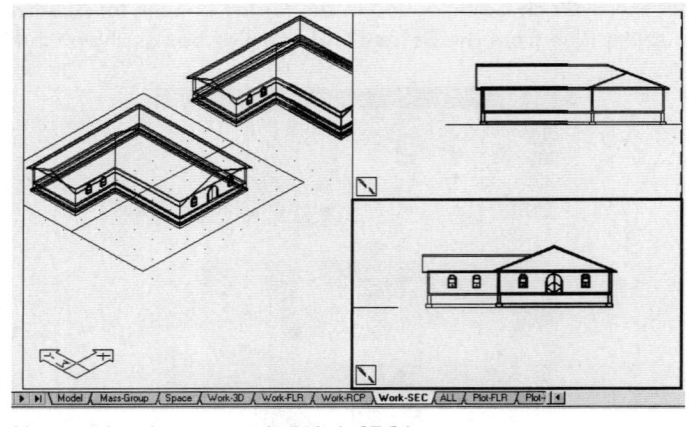

Figure 9.40 *Viewing the elevation in the Work-SEC layout*

47. Select the elevation; then right-click and choose **Entity Display** from the shortcut menu.
48. Select the **Display Props** tab, and then select Bldg Section in the **Property Source** list.
49. Select the **Edit Display Props** button to open the **Entity Properties** dialog box.
50. Select the light bulb icon to turn off the display of Subdivision 2 as shown in Figure 9.41. The display of the rear portion of the building is turned off, as shown in Figure 9.42.
51. Select **OK** to dismiss the **Entity Properties** dialog box.

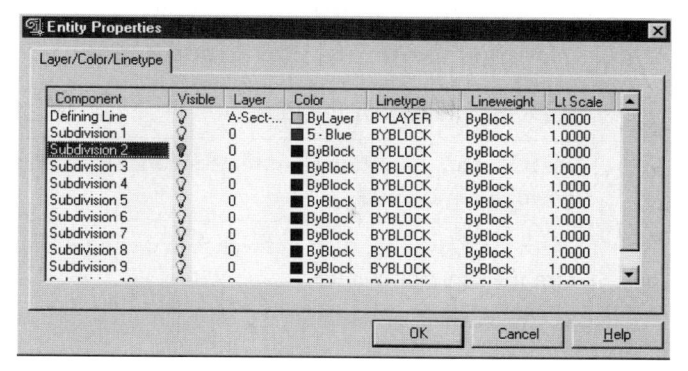

Figure 9.41 *Editing the properties of the subdivision*

52. Select **OK** to dismiss the **Entity Display** dialog box.

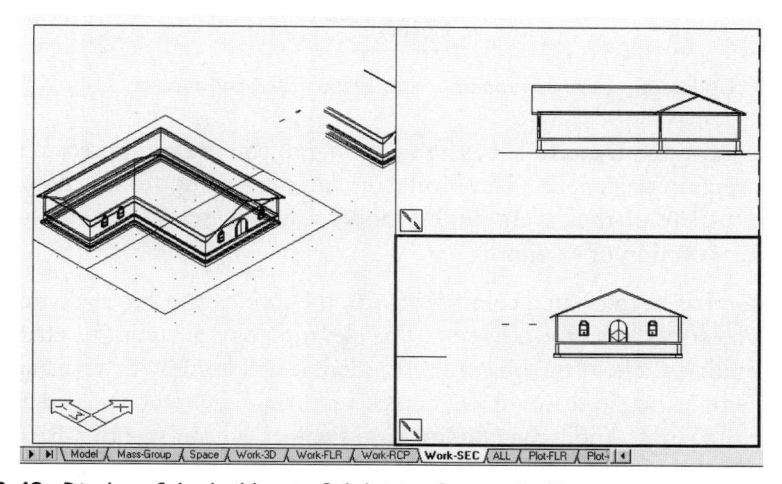

Figure 9.42 *Display of the building in Subdivision 2 turned off*

53. Save the drawing.

REVISING THE ELEVATION WITH UPDATE ELEVATION

Because elevations and sections are generated from a model that can change as the design develops, the **Update Elevation** and **Update Section** commands are provided to allow you to revise the elevation or section to the current state of the model. The elevation and section objects are not dynamically linked and the **Update Elevation** or **Update Section** command must be used to obtain a current view. The **Update Elevation** command (**BLDGSECTIONUPDATE**) will update the elevation to the latest design of the model. See Table 9.7 for **Update Elevation** command access.

Menu bar	Design>Generate Elevations>Update Elevation
Command prompt	BLDGSECTIONUPDATE
Toolbar	Select Update Elevation from the Section/Elevation toolbar as shown in Figure 9.43
Shortcut	Select the Elevation; then right-click and choose Update Section from the shortcut menu

Table 9.7 *Update Elevation command access*

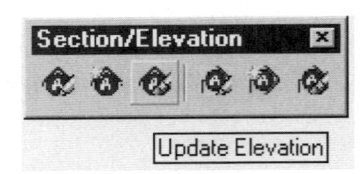

Figure 9.43 *Update Elevation command on the Section/Elevation toolbar*

When you choose the **Update Elevation** command, you are prompted to select the elevation or section object. The elevation is updated based upon the changes in the model. The display of the layers of the model must be set ON to include those objects in the elevation or section.

If the elevation has been created coincident with the model, you can cycle through the selection of objects to select the elevation. The elevation object cannot be changed with AutoCAD editing tools. If the object is exploded, it will literally fly apart and not be a useful drawing. The elevation object consists of all the geometry necessary to create elevation drawings. Apply the **Hidden Line Projection** command (**CREATEHLR**) to create a 2D block from the elevation object. The 2D block is used to create elevation working drawings.

CREATING 2D ELEVATION DRAWINGS

The **Hidden Line Projection** command (**CREATEHLR**) is used to create 2D drawings from a view of a 3D object. The **Hidden Line Projection** command creates a 2D block, which can then be used to develop elevation working drawings. Therefore, the **Hidden Line Projection** command (**CREATEHLR**) should be applied to the final elevation object to capture it as a 2D block for the development as a working elevation. See Table 9.8 for **Hidden Line Projection** command access.

Prior to selecting the Hidden Line Projection command, set the View of the elevation object according to the required elevation view. When you choose the **Hidden Line Projection** command, you will be prompted in the command line to select objects to convert to the block, as shown in the command sequence below and in Figure 9.44.

Menu bar	Desktop>AEC Utilities>Hidden Line Projection
Command prompt	CREATEHLR

Table 9.8 *Hidden Line Projection command access*

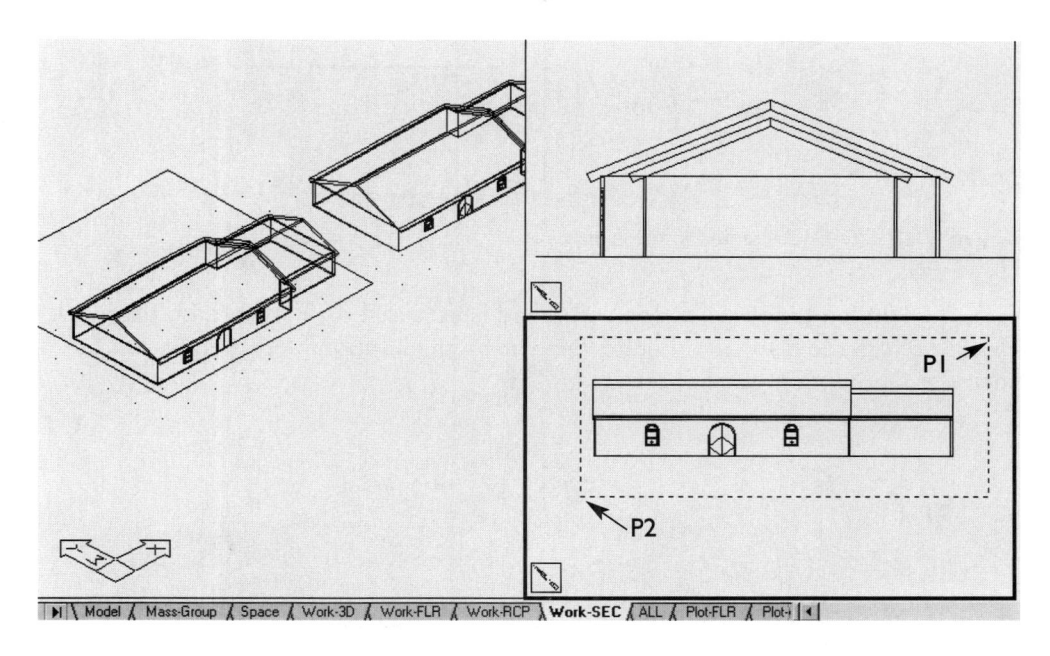

Figure 9.44 *Selecting the elevation object for the Hidden Line Projection command*

```
Command: _AecCreateHLR
Select objects:
```
(Specify a window from point P1 to P2 in Figure 9.44.) `Specify opposite corner: 4 found`
`Select objects:` ENTER *(Press ENTER to end selection of objects.)*
`Block insertion point: 0,0` *(Specify the location of the block.)*
`Insert in plan view [Yes/No] <Y>:` ENTER *(Inserts the block in plan view in the XY plane as shown in Figure 9.45.)*

After selecting the objects, you are prompted to specify the insertion point for the block. The insertion point can be at 0,0 or any X,Y location within the limits of the drawing. The location of the block is temporary because it should be moved to a separate drawing file through the **WBLOCK** command. The `Insert in plan view` prompt allows you to insert the block in the plan view or in the view that is current

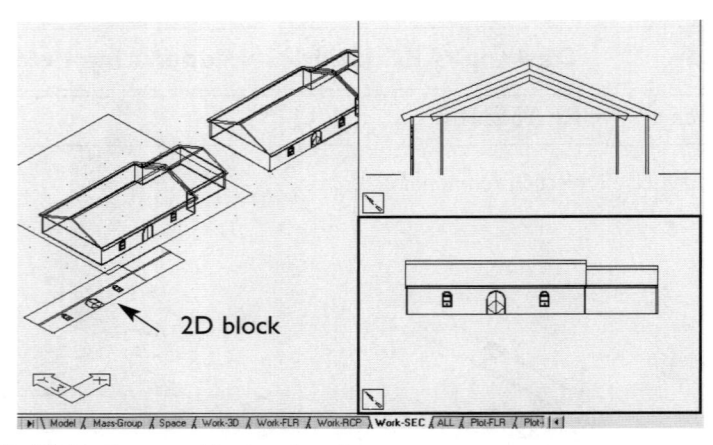

Figure 9.45 *2D block created in the plan view*

when the **Hidden Line Projection** command is executed. In Figure 9.45, the block was inserted in the plan view. Figure 9.46 shows an elevation block that was created when the **Plan** option was not selected.

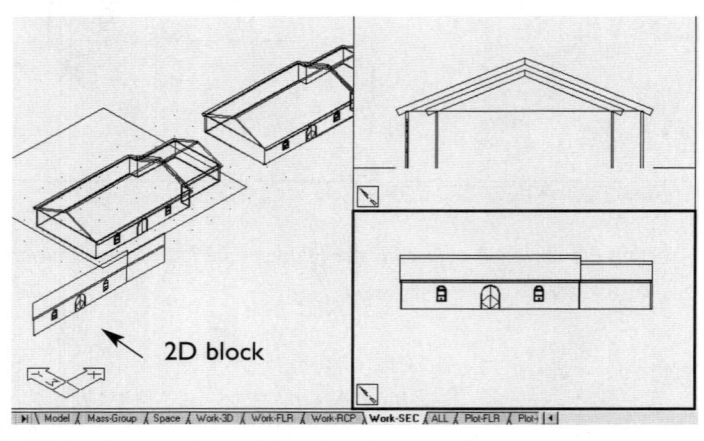

Figure 9.46 *2D block created parallel to the elevation plane*

When the **Plan** option is not selected, the block is created as a plane parallel to the view. Therefore, to dimension, hatch, or add annotation to this block, you need to rotate the UCS parallel to the view. If the **Plan** option is selected, the 2D block can be edited when viewed from the Top View.

The 2D block created by the **Hidden Line Projection** command is an anonymous block, created on the current layer. To control the display of the 2D block, you can create a layer for each elevation and then apply the **Hidden Line Projection** command when the respective layer for the elevation view is current. To assign an appropriate name to the entity, export the object to a separate file using the **WBLOCK** command. The **WBLOCK** command allows you to create a separate file in which the contents of the drawing can be exploded and developed into a working drawing. If the 2D block is exploded, the entities of the block are assigned explicit color and linetype. Each entity of the block retains its layer assignment. If the **Hidden Line Projection** command is applied to an elevation, its entities will reside on the current layer. When the 2D block is exploded, the entities will reside on layer A-Sect, with explicit color number 242 and continuous linetype.

In conclusion, for obtaining front, right, rear, and left side elevations, the elevation object or the model must be viewed from the front, right, rear and left views. Then execute the **Hidden Line Projection** command (**CREATEHLR**) while viewing the object. The **Hidden Line Projection** command creates a 2D block, which is then exported to individual drawing files for additional detail development.

TUTORIAL 9.4 CREATING ELEVATION DRAWINGS USING HIDDEN LINE PROJECTION

1. Open *Ex9-4.dwg* from the \ADT_Tutor\Ch9 directory and save the drawing to your student directory.
2. Choose **File>Close** from the menu bar to close *Ex9-4.dwg*.
3. Open *Elev3.dwg* from the \ADT_Tutor\Ch9 directory and save the drawing to your directory.
4. Select Level2 from the **Layer** flyout on the **Object Properties** toolbar.
5. Select **External Reference Attach** from the **Reference** toolbar.
6. Select *Ex9-4.dwg* from your student directory in the **Select Reference File** dialog box.
7. Select **Open** to open the **External Reference** dialog box.
8. Edit the **External Reference** dialog box: set **Reference Type** to Attachment, clear the **Specify On-screen** check box and set the **Insertion point** to **X** = 0, **Y** = 0, **Z** = 4'-6", **Scale** = 1, and **Rotation** = 0.
9. Select **OK** to dismiss the **External Reference** dialog box. Reference files are attached as shown in Figure 9.47.
10. Select the building elevation line; then right-click and choose **Generate Elevation** from the shortcut menu.
11. Refer to Figure 9.48 and the following command sequence to create the elevation.

    ```
    Command: BldgElevationLineGenerate
    Select entities for elevation: (Specify the crossing window
    ```

Figure 9.47 Reference files attached to create the model

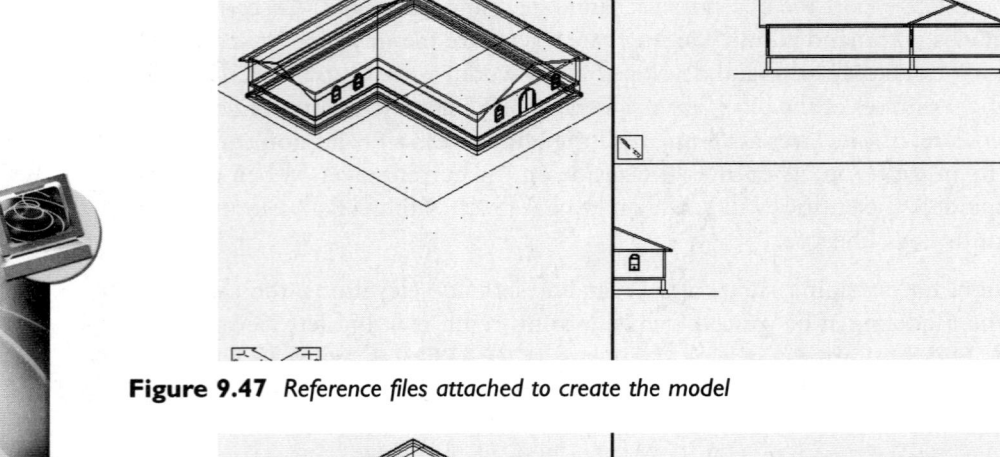

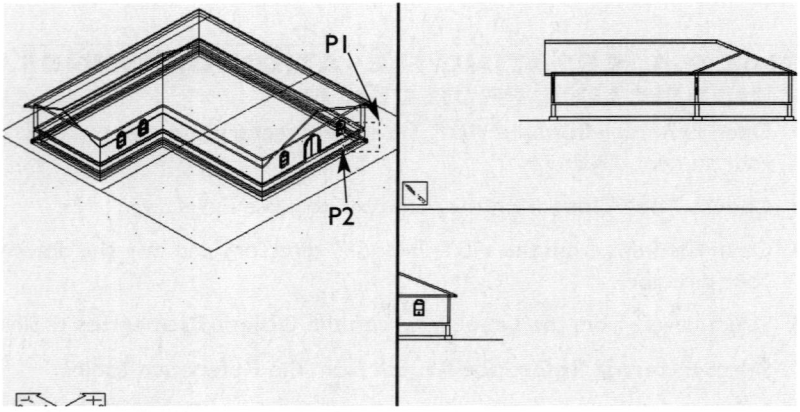

Figure 9.48 Selection of the building for the elevation

from point P1 to P2 in Figure 8.48.) `Specify opposite corner: 2 found`
`Select entities for elevation:` ENTER
(Select the SECTION_ELEV display representation set from the Display Representation Sets dialog box, and then select OK to dismiss the dialog box.)
`Elevation to go to [Existing]<New>:` ENTER *(Press* ENTER *to create a new elevation.)*
`Specify base point or displacement: 10',0` ENTER
(Specify base point for the elevation.)
`Specify second point or <use first point as displacement>: 100',0` ENTER *(Specify the location of the elevation 100' to the right.)*
(Elevation created as shown in Figure 9.49.)

Creating Elevations and Sections 719

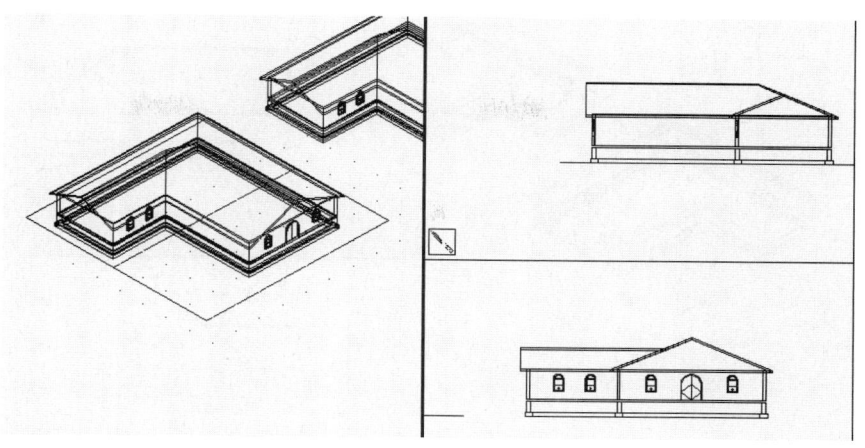

Figure 9.49 *Elevation object created from the reference files*

12. Choose **File>SaveAs** and save the file as *Elev4.dwg* in your directory; then choose **File>Close** to close the *Elev4.dwg* file.
13. Open *Ex9-4.dwg* from your student directory.
14. Select a line of the roof to display the grips of the roof; then right-click and choose **Roof Modify** from the shortcut menu.
15. Type **10** in the **Rise** edit field of the **Modify Roof** dialog box; then select **OK** to dismiss the **Modify Roof** dialog box.
16. Select **Save** from the **Standard** toolbar and then choose **File>Close** from the menu bar to close the file.
17. Open *Elev4.dwg* from your student directory.
18. Select the Work-SEC layout tab.
19. Select the lower right viewport to make it current.
20. Select **Update Elevation** from the **Section/Elevation** toolbar, and then select the elevation object. Elevation is updated as shown in Figure 9.50.
21. Choose **Desktop>AEC Utilities>Hidden Line Projection** from the menu bar and respond to the command line prompts as shown in the following command sequence and Figure 9.51.

```
Command: _AecCreateHLR
Select objects: (Select the Elevation at point P1 in Figure 9.51.) 1 found
Select objects: ENTER (Press ENTER to end object selection.)
Block insertion point: 0,0
Insert in plan view [Yes/No] <Y>: ENTER (Press ENTER to
```
insert the elevation in the plan view as shown in Figure 9.51.)

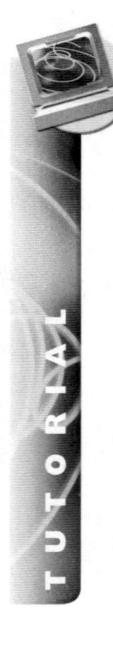

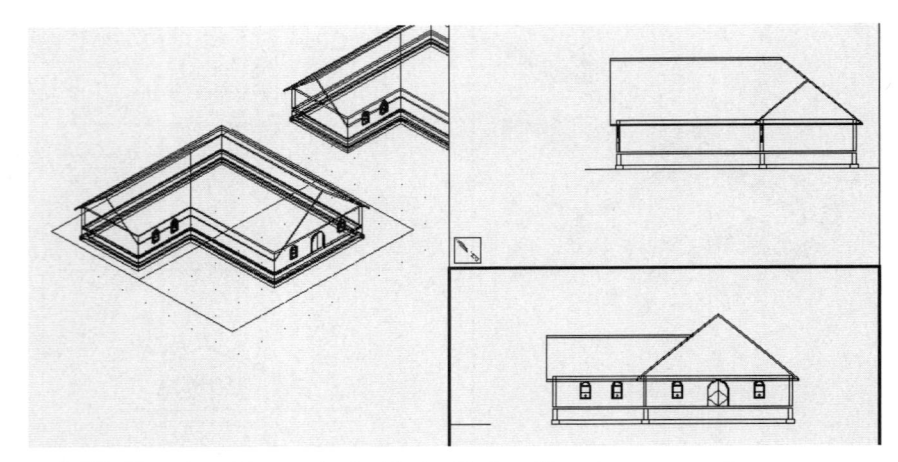

Figure 9.50 *Elevation object revised with the Update Elevation command*

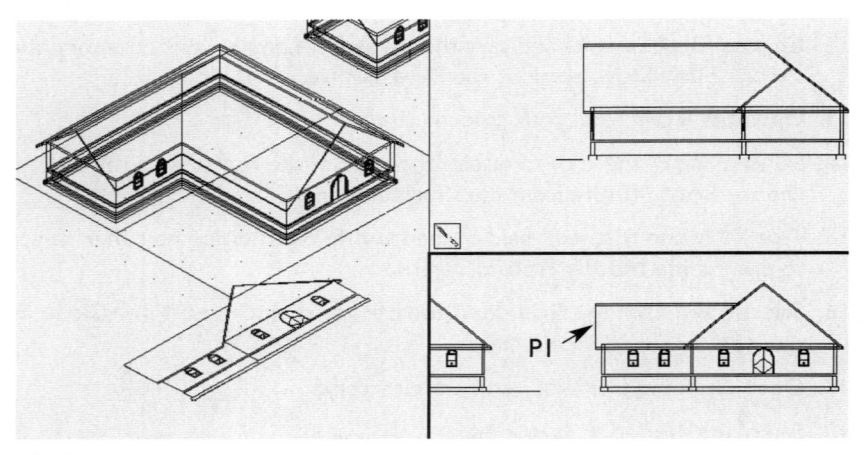

Figure 9.51 *2D block created from the elevation object*

22. To export the elevation, type **WBLOCK** in the command line and edit the following options of the **Write Block** dialog box as shown in Figure 9.52: toggle ON the **Objects** radio button, **Base point X** = 0, **Y** = 0, **Z** = 0, **Destination File Name** = Front.dwg, **Location** = your student directory, **Insert units** = Unitless.
23. Select the **Select Objects** button of the **Write Block** dialog box and select the block at point P1, as shown in Figure 9.53.
24. Select **OK** to dismiss the **Write Block** dialog box.
25. Save the drawing.
26. Open the *FRONT.dwg* drawing from your student directory.

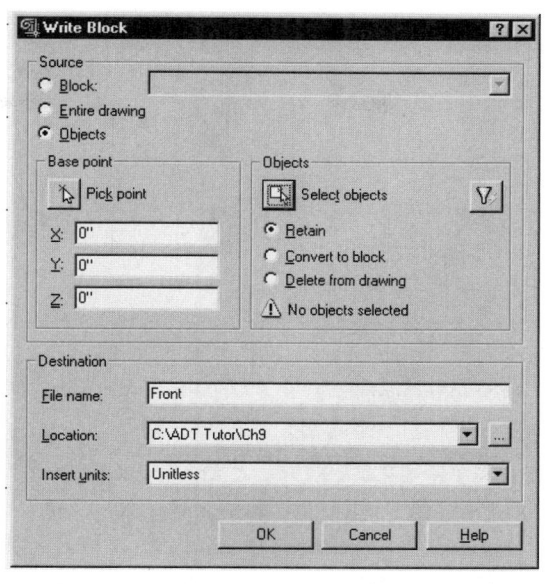

Figure 9.52 *Editing the Write Block dialog box to export the 2D block*

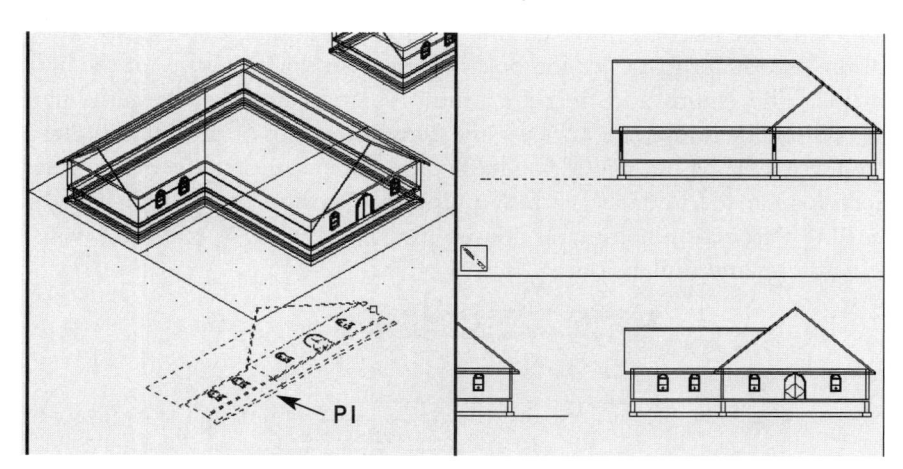

Figure 9.53 *Select the 2D block for export*

27. Select the Model layout tab.
28. Select **Zoom Extents** icon from the **Zoom** flyout on the **Standard** toolbar to view the elevation as shown in Figure 9.54.
29. Select **Explode** from the **Modify** toolbar, and then select the block. The drawing is ready for development as an elevation working drawing.
30. Save the drawing.

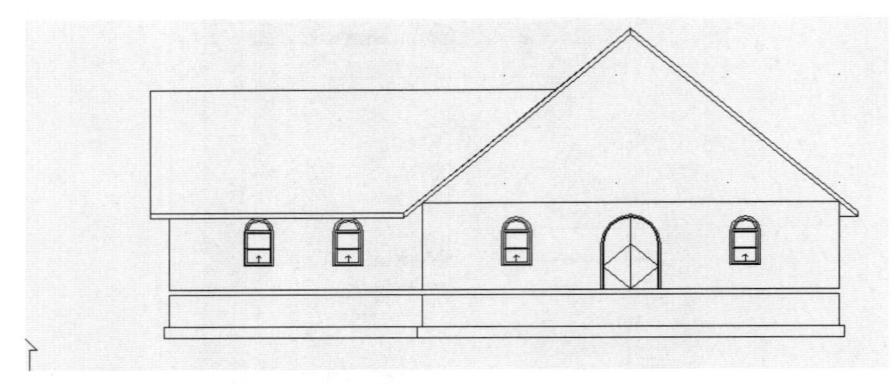

Figure 9.54 Front elevation drawing file

USING THE CREATE SECTION COMMAND

A section through the model is created by a procedure similar to that for the elevation. The commands to place the section mark and create and update the section are located on the **Section/Elevation** toolbar, as shown in Figure 9.8. The commands allow you to specify the section mark as a polyline with one or more vertices. The section mark represents the location of the cutting line for creating full or offset sections. The reach of the section, or its projection box, is defined in the properties of the building section line. The length and height of the projection box are defined when the building section line is inserted; however, the **Dimension** tabs of the **Section/Elevation Line Properties** dialog box allows you to edit the dimensions for the building section line. Sections can be updated with the **Update Section** command. Finally, the **Hidden Line Projection** command is used to create the 2D section drawings that include hatching and dimensioning.

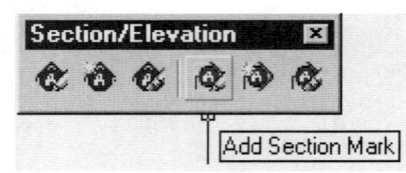

Figure 9.55 Add Section Mark command on the Section/Elevation toolbar

PLACING A SECTION MARK

The **Add Section Mark** command (**BLDGSECTIONLINEADD**) will allow you to create the location and viewing direction of a building section line. If the building section line is created with one or more vertices, an offset section will be created. The height and length of the projection box for the section is defined in the command sequence when the building section line is placed. See Table 9.9 for **Add Section Mark** command access.

Creating Elevations and Sections 723

Menu bar	Design>Generate Section>Add Section Mark
Command prompt	BLDGSECTIONLINEADD
Toolbar	Select Add Section Mark from the Section/Elevation toolbar as shown in Figure 9.55
Shortcut	Right-click over the drawing area and choose Design>Section Lines>Add

Table 9.9 *Add Section Mark command access*

When you choose **Add Section Mark**, you are prompted in the command line to identify the points for the building section line to pass through. The **Add Section Mark** command creates the AEC BDG SECTIONLINE object, which is located on the A-Sect-Iden layer. After you specify the location of the building section line, the length and height of the projection box are specified in the command line as shown in the following command sequence and in Figure 9.56.

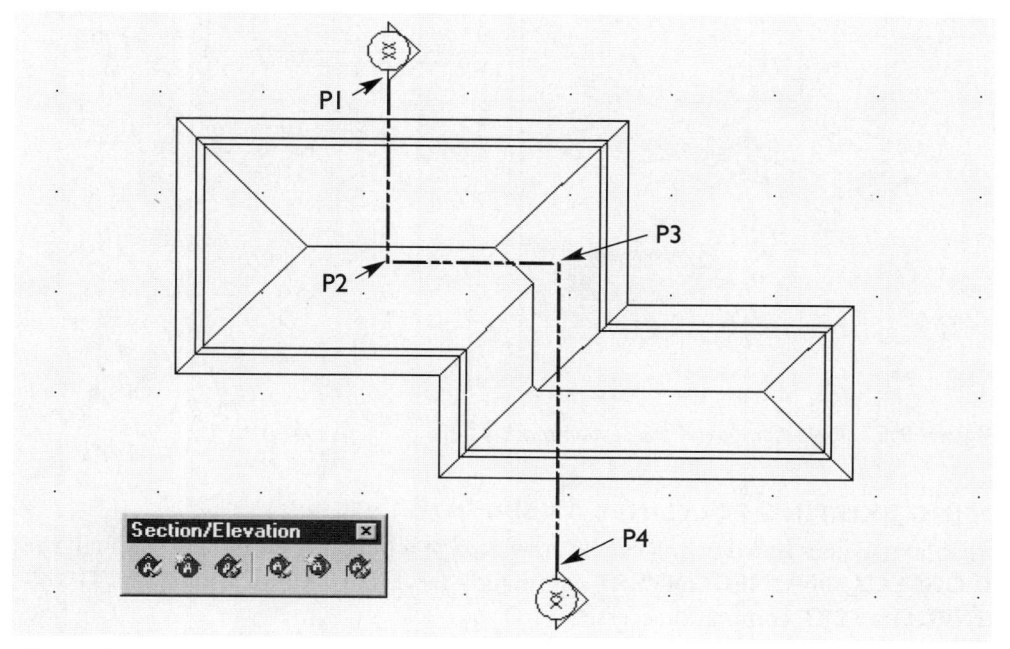

Figure 9.56 *Specifying the building section line*

```
Command: _AecBldgSectionLineAdd
Create polyline for section:
```

```
Point:  (Select point P1 in Figure 9.56.)
Point:  (Select point P2 in Figure 9.56.)
Point:  (Select point P3 in Figure 9.56.)
Point:  (Select point P4 in Figure 9.56.)
Point:  ENTER
Enter length <20'-0">: 10' ENTER
Enter height <20'-0">: 20' ENTER
```

Figure 9.57 shows the projection box defined by the building section line. Only those objects included in this projection box will be included in the section object. The size of the projection box can be changed in the **Dimensions** tab of the **Section/Elevation Line Properties** dialog box. Enter the identification of the section in the section mark bubble by editing the attribute properties of the multi-view block. The **MVBLOCKPROPS** command can be used to edit the attributes for each of the section bubbles. If two section bubbles are not required, one of the bubbles can be erased.

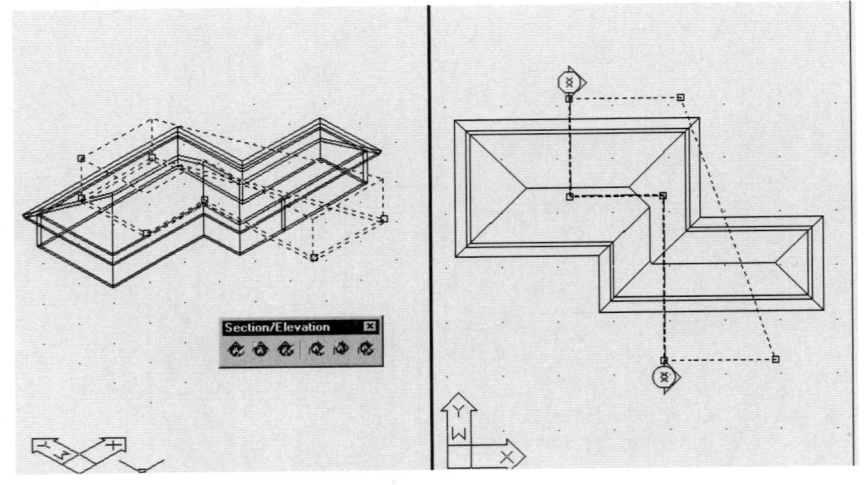

Figure 9.57 *Projection box of the section mark*

USING EXISTING POLYLINES AS SECTION LINES

Existing polylines in a drawing can be converted to a building section line through the **BLDGSECTIONLINECONVERT** command. See Table 9.10 for **BLDGSECTION-LINECONVERT** command access.

The command line sequence for the **BLDGSECTIONLINECONVERT** command is shown below. Unlike with the **Add Section Mark** command, you do not specify the vertices of the building section line; you simply select an existing polyline. When you select the existing polyline, its geometry is used as the building section line. The view-

Command prompt	**BLDGSECTIONLINECONVERT**
Shortcut	Right-click over the drawing area; then choose Design>Section Lines>Convert

Table 9.10 *BLDGSECTIONLINECONVERT command access*

ing direction of the building section line is determined from the polyline. The orientation of viewing direction is ahead, with the beginning segment of the polyline on the left and the ending polyline segments on the right. The length and height of the projection box for the section are defined in the command line when the polyline is converted. Figure 9.58 shows a polyline on the left, which was used to create the building section line shown on the right.

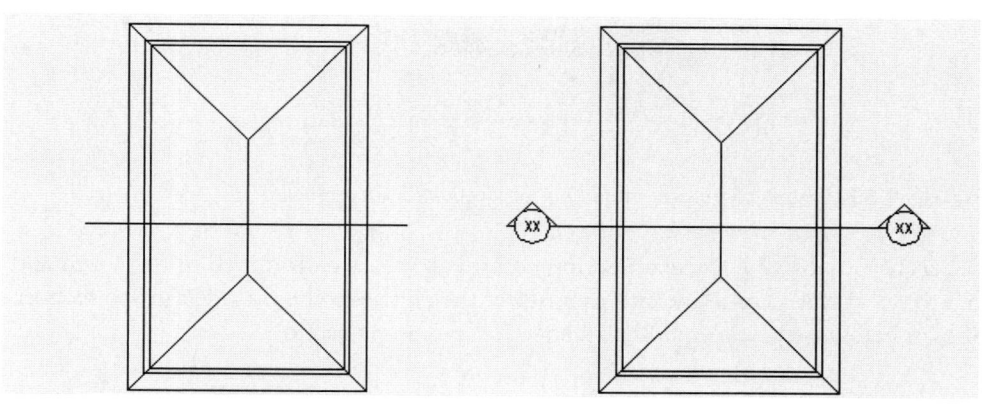

Figure 9.58 *Polyline converted to a section line*

The following command sequence was used to convert the polyline to a building section line as shown in Figure 9.58.

```
Command: BldgSectionLineConvert
Select a 2D polyline: (Select a polyline.)
Enter length <20'-0">: ENTER (Pressing ENTER accepts 20' length
of the projection box.)
Enter height <20'-0">: ENTER (Pressing ENTER accepts the 20'
height dimension of the projection box.)
```

CREATING THE SECTION

After the building section line is placed, the section is developed with the **Create Section** command (**BLDGSECTIONLINEGENERATE**). The **Create Section** command creates a section object in the same manner as an elevation object is created with the **Create Elevation** command. See Table 9.11 for **Create Section** command access.

Menu bar	Design>Generate Sections>Create Section
Command prompt	BLDGSECTIONLINEGENERATE
Toolbar	Select Create Section from the Section/Elevation toolbar as shown in Figure 9.59
Shortcut	Right-click over the drawing area and choose Design>Section Lines>Generate
Shortcut	Select the section line; then right-click and choose Generate Section

Table 9.11 *Create Section command access*

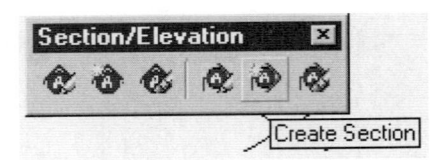

Figure 9.59 *Create Section command on the Section/Elevation toolbar*

When you choose the **Create Section** command, you are prompted in the command line to select the section line and then select the entities to be included in the section. Shown below is the command sequence for creating a section.

```
Command: _AecBldgSectionLineGenerate
Select section line: (Select a section line.)
 Select entities for section: (Select a crossing window from
P1 to P2 in Figure 9.60.) Specify opposite corner: 11 found
Select entities for section: ENTER
(Specify the SECTION_ELEV display representation set and then select OK
to dismiss the Display Representation Sets dialog box.)
Section to go to [Existing]<New>: ENTER
Specify base point or displacement: 0,0 ENTER
Specify second point or <use first point as dis-
placement>: 100',0 ENTER
```

The section consists of only those entities within the dimensions of the section projection box. After creating the section, you should freeze the display of the model and set the view of the section appropriate to the viewing direction of the section line. In the example above, a left view of the section would create the correct view of the section as shown in Figure 9.61.

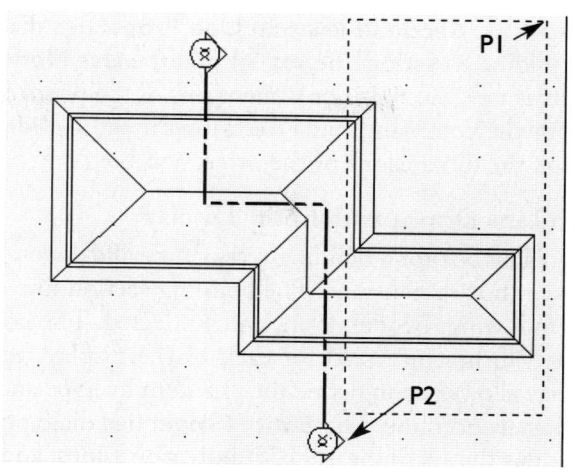

Figure 9.60 *Selecting objects for the section*

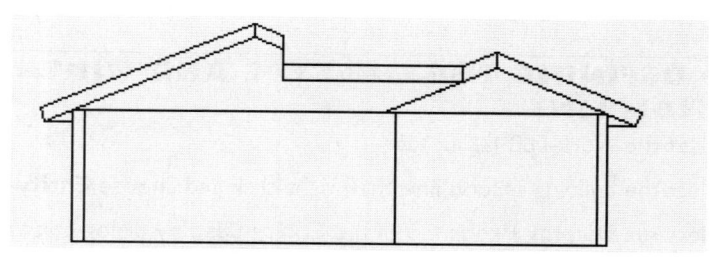

Figure 9.61 *Section object developed from the building section line*

PROPERTIES OF THE SECTION MARK

The properties of a building section line are identical to those described earlier in the Properties of the Section/Elevation Mark section of this chapter. See Table 9.12 for **Elevation/Section Line Properties** command access.

Menu bar	Design>Generate Sections> Elevation/Section Line Properties
Command prompt	BLDGSECTIONLINEPROPS
Shortcut	Right-click over the drawing area and choose Design>Section Lines>Properties from the shortcut menu
Shortcut	Select a building section line; then right-click and choose Section Line Properties

Table 9.12 *Elevation/Section Line Properties command access*

The default settings for the **Section/Elevation Line Properties** dialog box are identical to those for a building elevation line, except that the **Use Model Space Extents for Height** check box is not selected. The dimensions of the prism for projecting the section are defined in the command line. You can edit the **A-Side 1** and **B-Side 2** dimensions to change the dimensions of the projection box.

Controlling Section Line Display with Entity Display

The **Entity Display** dialog box for a building section line allows you to control its display and the projection box dimensions. The building section line display representations include the following: Model, Plan, and Reflected. The components of the building section line can be controlled for each of the display representation sets. Controlling the display allows you to define the phantom linetype and increase the line weight of the building section line. The **Entity Properties** dialog box for the building section line includes the Defining Line, Subdivision Lines, and Boundary components. Each of these components can be controlled for visibility, color, linetype, lineweight, and LTSCALE, as described in the steps below.

STEPS TO DEFINING THE LINETYPE AND WIDTH OF THE SECTION LINE

1. Select the Work-FLR layout tab.
2. Select the building section line; then right-click and choose **Entity Display**.
3. Select the **Display Props** tab of the **Entity Display** dialog box as shown in Figure 9.62.

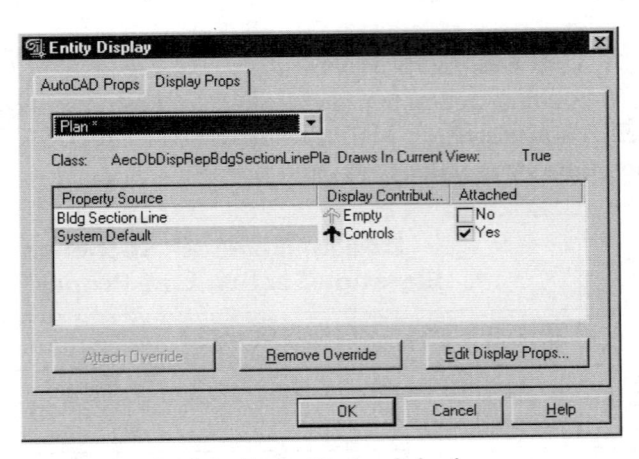

Figure 9.62 *Display Props tab of the Entity Display dialog box*

4. Verify that the Plan display representation is current in the **Display Representation** list.

5. Select the Bldg Section Line source and then select the **Attach Override** button.
6. Select the **Edit Display Props** button to open the **Entity Properties** dialog box.
7. Select the **Linetype** option for the Defining Line component, as shown in Figure 9.63.

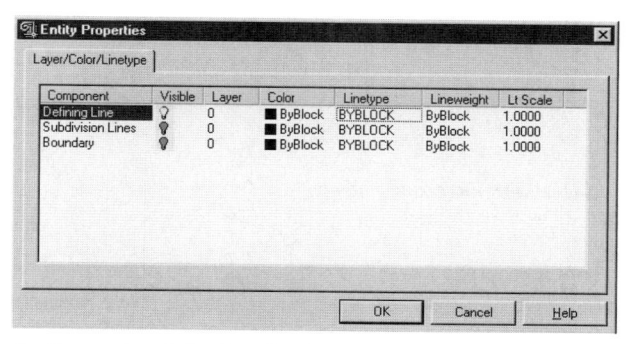

Figure 9.63 *Entity Properties dialog box for the building section line*

8. Select Phantom4 from the **Select Linetype** dialog box.
9. Select **OK** to dismiss the **Select Linetype** dialog box.
10. Select the **Lineweight** option for the **Defining Line** component (section line) to open the **Lineweight** dialog box as shown in Figure 9.64.

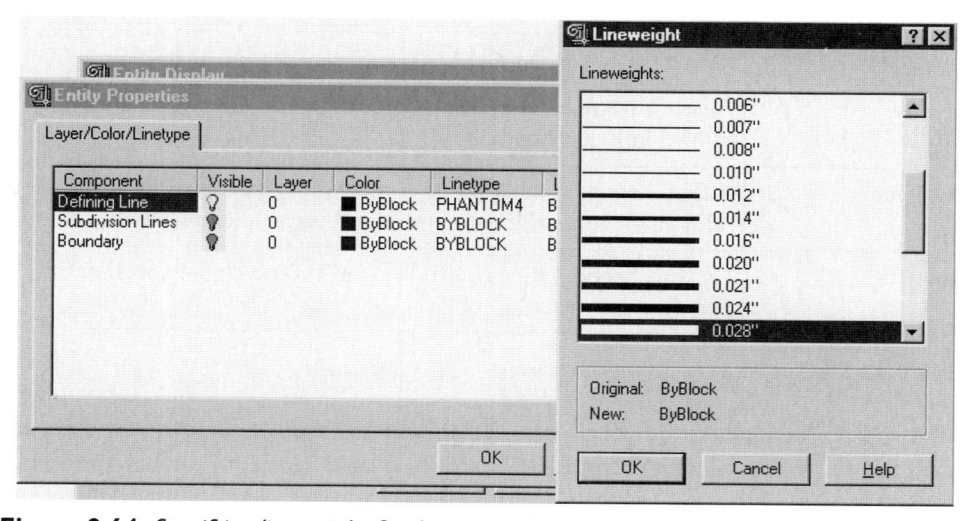

Figure 9.64 *Specifying line weight for the section line*

11. Select the 0.028 line weight; then select **OK** to dismiss the **Lineweight** dialog box.
12. Select **OK** to dismiss the **Entity Properties** dialog box.
13. Select **OK** to dismiss the **Entity Display** dialog box.
14. Toggle on LWT in Status Bar to view lineweights in a layout. All section lines will be displayed with the 0.028 lineweight and as phantom lines, as shown in Figure 9.65.

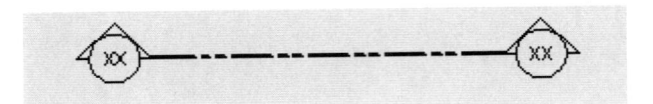

Figure 9.65 *Section Line with increased lineweight*

Note: Verify appropriate lineweight and linetype scale in the layout tab used for plotting. Lineweight should be toggled off when working in the model space layout.

REVISING THE SECTION WITH THE UPDATE SECTION COMMAND

A section can be updated as the model changes with the **Update Section** command (**BLDGSECTIONUPDATE**). You can capture the changes to the model in the section by selecting the **Update Section** command. See Table 9.13 for **Update Section** command access.

Menu bar	Design>Generate Sections>Update Section
Command prompt	BLDGSECTIONUPDATE
Toolbar	Select Update Section from the Section/Elevation toolbar as shown in Figure 9.66
Shortcut	Right-click over the drawing area; choose Design>Sections>Update from the shortcut menu
Shortcut	Select a Section; then right-click and choose Update Section for the shortcut menu

Table 9.13 *Update Section command access*

If you select the **Update Section** command from the **Section/Elevation** toolbar, you are prompted to select a section to update, as shown in the following command sequence. The entities selected for the original section are reexamined to determine if any changes have occurred.

Creating Elevations and Sections 731

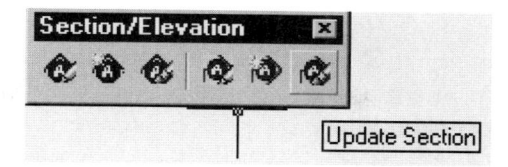

Figure 9.66 *Update Section command on the Section/Elevation toolbar*

```
_AecBldgSectionUpdate
Select section: (Select the section object.)
(The section is automatically updated.)
```

PROPERTIES OF SECTIONS AND ELEVATIONS

The properties of sections and elevations can be determined by the **BLDGSECTIONPROPS** command. Elevation objects are considered section objects when their properties are identified by this command. See Table 9.14 for **BLDGSECTIONPROPS** command access.

Menu bar	Design>Generate Sections> Section/Elevation Properties
Command prompt	BLDGSECTIONPROPS
Shortcut	Right-click over the drawing area and choose Design>Sections>Properties from the shortcut menu
Shortcut	Select a Section or Elevation; then right-click and choose Section Properties

Table 9.14 *BLDGSECTIONPROPS command access*

When you select the **BLDGSECTIONPROPS** command and select either a section or elevation, the **Section Properties** dialog box opens, as shown in Figure 9.67. It includes **General** and **Location** tabs. The contents of the **Section Properties** dialog box are described below.

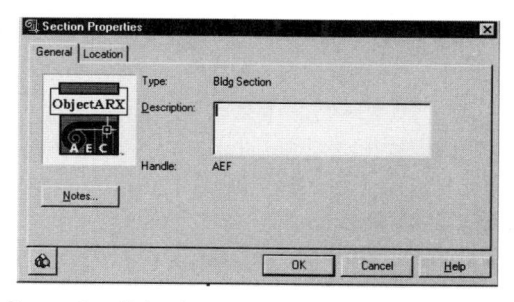

Figure 9.67 *Section Properties dialog box*

General

The **General** tab, shown in Figure 9.67, allows you to enter a description of the section or elevation. The **Notes** button opens the **Notes** dialog box, which includes **Text** and **Reference Docs** tabs, providing space for records regarding the elevation or section.

Location

The **Location** tab includes edit fields that describe the insertion point, normal, and rotation properties of the section or elevation.

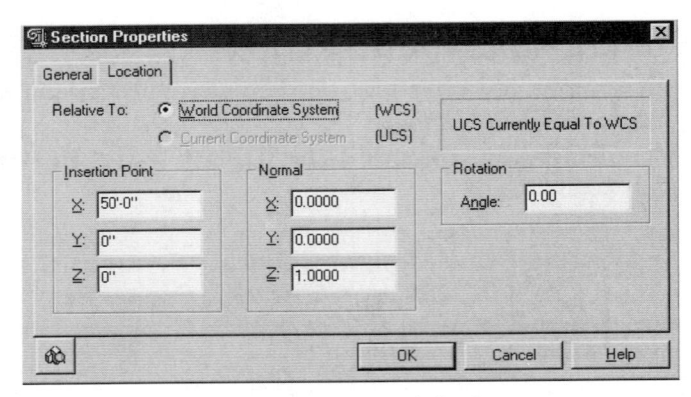

Figure 9.68 *Location tab of the Section Properties dialog box*

> **Insertion Point** – The **Insertion Point** section shows the X, Y, Z coordinate location of the section object relative to the World Coordinate System or the User Coordinate System.
>
> **Normal** – The **Normal** edit fields define the orientation of the object. The Z=1 orientation projects the object parallel to the XY plane.
>
> **Rotation** – The **Rotation** edit field allows you to rotate the section or elevation.

TUTORIAL 9.5 CREATING A SECTION USING HIDDEN LINE PROJECTION

1. Open *Elev1* from your student directory.
2. Choose **File>SaveAs** from the menu bar and save the file as *Section1.dwg* in your student directory.
3. Select the Work-3D layout tab.
4. Select the **Layer** flyout and thaw Floor2|A-Flor-Case and Floor2|A-Flor-Appl layers.

5. Select **Add Section Mark** from **Section/Elevation** toolbar and place the section line as shown in Figure 9.69 and the command sequence below.

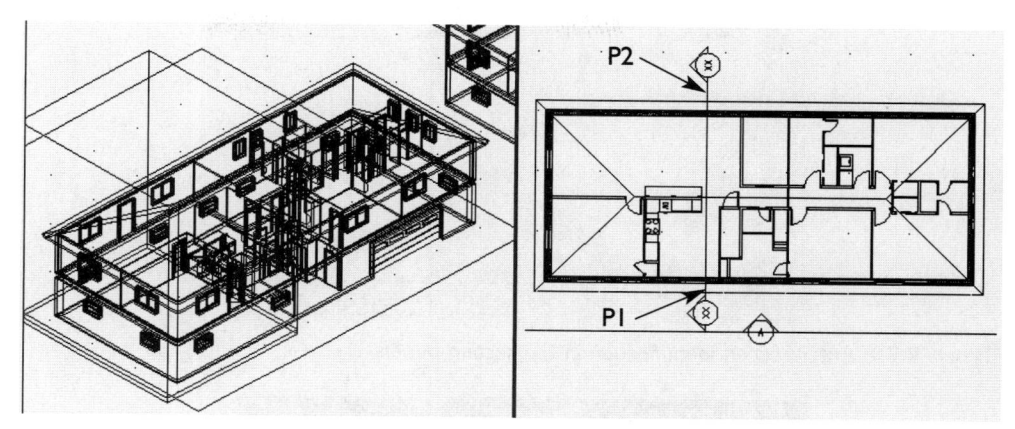

Figure 9.69 *Specify the Section Line location*

```
Command: _AecBldgSectionLineAdd
Create polyline for section:
Point: (Select point P1 in Figure 9.69.)
Point: (Select point P2 in Figure 9.69.)
Point: ENTER
```

Enter length <20'-0">: 30' ENTER *(Specify the length of the projection box for the section.)*

Enter height <20'-0">: 40' ENTER *(Specify the height of the projection box for the section.)*

6. Select the upper and lower section identification bubbles; then right-click and choose **Multi-View Block Properties** from the shortcut menu.
7. Select the **Attributes** tab of the **Multi-View Block Reference Properties** dialog box.
8. Change the **Value** of the attribute to 1 as shown in Figure 9.70.
9. Select **OK** to dismiss the **Multi-View Block Reference Properties** dialog box.
10. Select the section line mark; then right-click and choose **Entity Display** to open the **Entity Display** dialog box as shown in Figure 9.71.
11. Select Bldg Section Line in the **Property Source** list, and then select the **Attach Override** button.

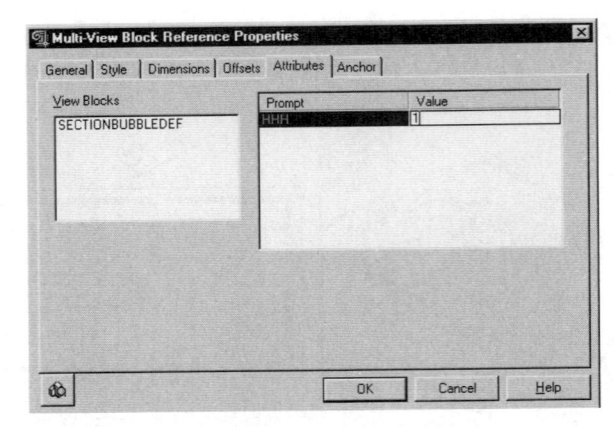

Figure 9.70 *Enter section identification in the section bubble*

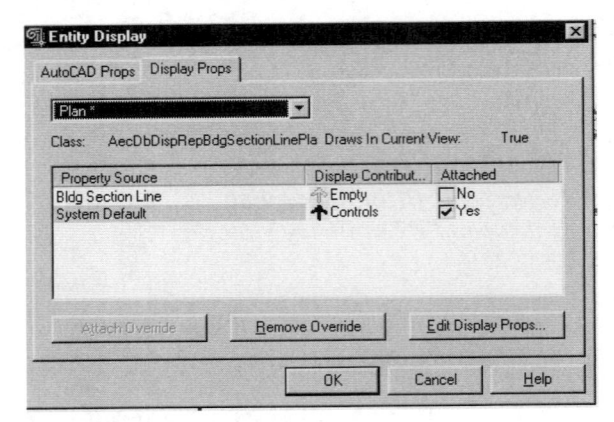

Figure 9.71 *Display Props tab of the Entity Display dialog box*

12. Select the **Edit Display Props** button of the **Entity Display** dialog box to open the **Entity Properties** dialog box.
13. Select the **Linetype** column of the Defining Line component to open the **Linetype** dialog box.
14. Select Phantom4 linetype, and then select **OK** to dismiss the **Linetype** dialog box.
15. Select the **Lineweight** column of the Defining Line component to open the **Lineweight** dialog box.
16. Scroll down the **Lineweight** dialog box and select the 0.021" line weight.
17. Select **OK** to dismiss the Lineweight dialog box.
18. Select **OK** to dismiss the **Entity Properties** dialog box.

19. Select **OK** to dismiss the **Entity Display** dialog box.
20. Select the section line, and then right-click and choose **Generate Section** from the shortcut menu.
21. Refer to the following command sequence below and Figure 9.72 to create the section.

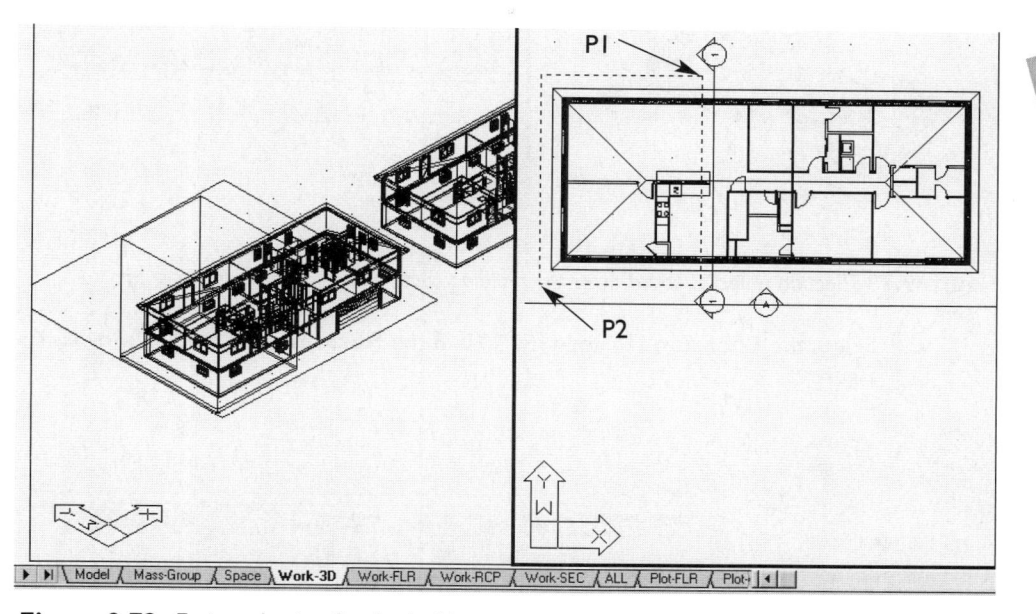

Figure 9.72 *Entity selection for the building section line*

```
Command: BldgSectionLineGenerate
Select entities for section: (Select points P1 and P2 in
Figure 9.72 to specify the corners of a crossing selection window.) Specify
opposite corner: 2 found
Select entities for section: ENTER
(Select SECTION_ELEV from the Display Representation Sets dialog box;
then select OK to dismiss the dialog box.)
Section to go to [Existing]<New>: n ENTER
Specify base point or displacement: 0,0 ENTER
Specify second point or <use first point as dis-
placement>: -70',0 ENTER
```

22. Select the left viewport; then select **Zoom Out** from the **Standard** toolbar and zoom out to display the section as shown in Figure 9.73.
23. Select the section; then right-click and choose **Section Properties** from the shortcut menu.

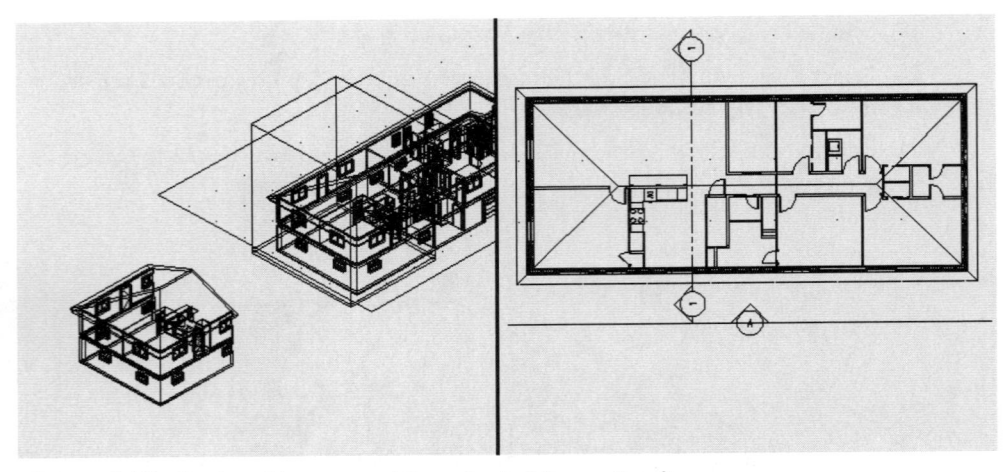

Figure 9.73 *Section object created from the building section line*

24. Select the **Location** tab and type **270** in the **Rotation** edit field as shown in Figure 9.74.

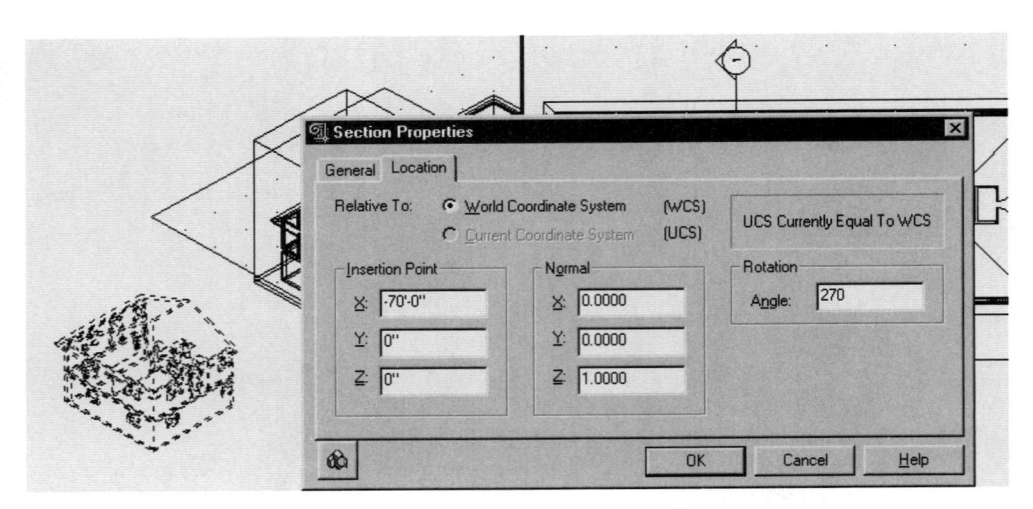

Figure 9.74 *Editing the rotation of the section in the Section Properties dialog box*

25. Select **OK** to dismiss the **Section Properties** dialog box.
26. Select the right viewport and then select the **Front** view option on the **View** flyout on the **Standard** toolbar.
27. Select the **Pan** command from the **Standard** toolbar and pan the view of the section as shown in Figure 9.75.

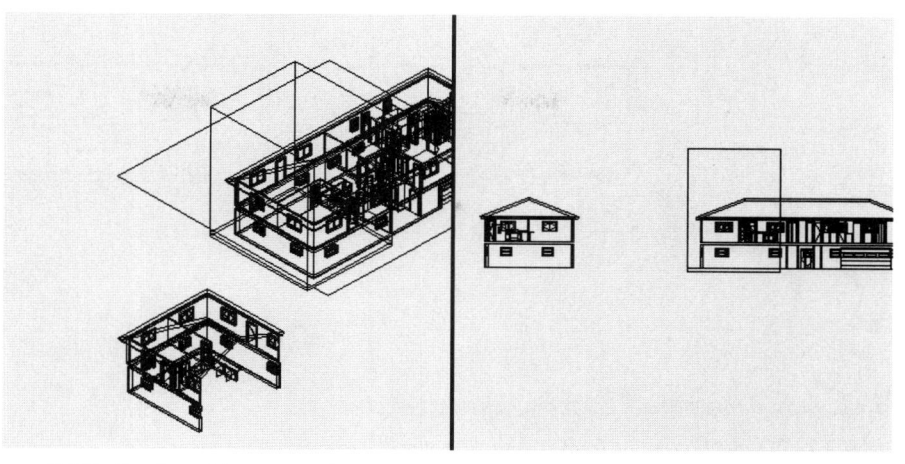

Figure 9.75 Section object rotated

28. Choose **Desktop>AEC Utilities>Hidden Line Projection** from the menu bar.
29. Refer to Figure 9.76 and the command sequence to create a 2D block of the section.

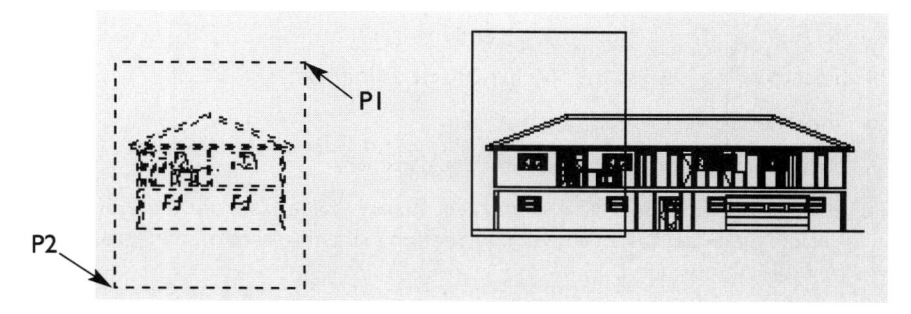

Figure 9.76 Specify the object for the Hidden Line Projection command

```
Command: AECCREATEHLR
Select objects: (Select a crossing window from point P1 to P2 in
Figure 9.76.) Specify opposite corner: 1 found
Select objects: ENTER
Block insertion point: 0,0 ENTER
Insert in plan view [Yes/No] <Y>: ENTER
```
(Select Model layout tab to view the section object as shown in Figure 9.77.)

30. Select the Model layout tab; then select **Zoom Extents** from the **Standard** toolbar.

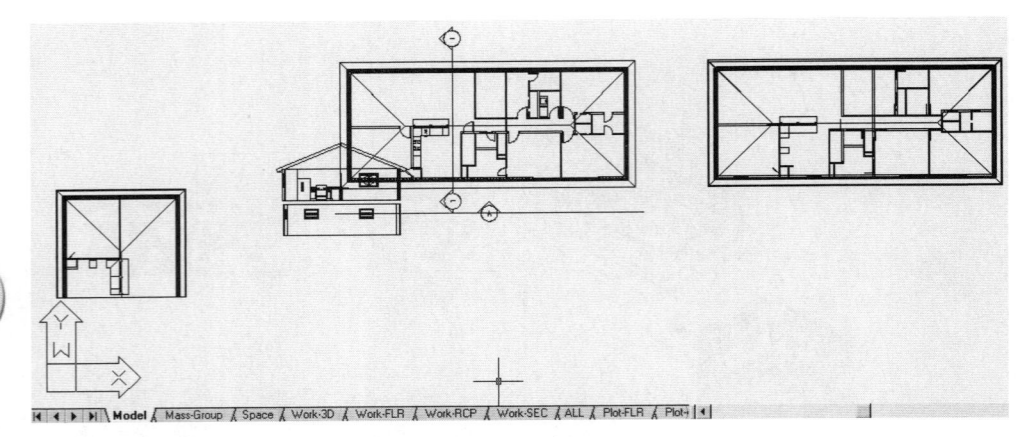

Figure 9.77 *2D block created from the section object*

31. Type **WBLOCK** in the command line to open the **Write Block** dialog box.
32. Edit the **Write Block** dialog box: toggle on the **Objects** radio button, set the **Base point X** = 0, **Y** = 0, **Z** = 0, type **Section2.dwg** in the **File Name** field, set the **Units** to Unitless, and then set the directory to your student directory.
33. Select the **Select Objects** button and select the 2D block shown in Figure 9.77.
34. Select **OK** to dismiss the **Write Block** dialog box.
35. Save the drawing.
36. Open *Section2.dwg* from your student directory.
37. Select the Model layout tab and select **Zoom Extents** from the **Zoom** fly-out on the **Standard** toolbar. The Section2 drawing, shown in Figure 9.78, can now be dimensioned.

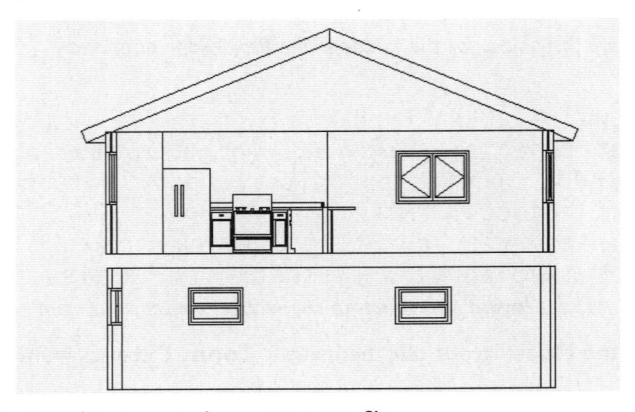

Figure 9.78 *Section object created as a separate file*

38. Explode the 2D block.
39. Save the drawing.

SUMMARY

1. Sections and elevations should be developed from a model that consists of the floor plans of the building attached as external reference files.
2. Building elevation lines are created with the **Add Elevation Mark** command (**BLDGELEVATIONLINEADD**) to define the location and direction of view of the model.
3. The elevation object is a 3D object created with the **Create Elevation** command (**BLDGELEVATIONLINEGENERATE**) for the development elevation views.
4. Changes made in the model can be reflected in the elevation and section objects with the **Update Elevation** and **Update Section** commands (**BLDGSECTIONUPDATE**).
5. Defining the display of the Defining Line component in the object's **Entity Display** dialog box can change the linetype, color, and visibility of building section lines.
6. Elevation working drawings are created from 2D blocks extracted from the 3D elevation with the **Hidden Line Projection** command (**CREATEHLR**).
7. The location and viewing direction of a section is defined in the properties of the building section line, which is placed in the drawing with the **Add Section Mark** command (**BLDGSECTIONLINECONVERT**).
8. Editing the properties of the building section line can change the size of the projection box of a section.
9. Sections can be rotated and the location of the section determined by the **BLDGSECTIONPROPS** command.

REVIEW QUESTIONS

1. The floor plans for each level of a building are inserted in a model drawing as _____.
2. Each floor plan inserted in the model should be inserted at a Z coordinate equal to the _____.
3. The floor structural system is represented by a _____ in the model.
4. How should the floor plan drawing be edited prior to attaching the floor plan to the model drawing?
5. What does the building elevation line specify when inserted in the drawing?
6. The building elevation line is placed on the _____ layer.
7. The _____ layout tabs, which consist of locked viewport displays, are preset to the viewing directions for elevations.

8. The **BLDGELEVATIONLINEREVERSE** command is used to_____.
9. The content of the elevation is specified in the _____ dimensions of the elevation line.
10. How are changes in the section made to reflect the changes in the model?
11. Building section lines should be assigned the phantom linetype by _____.
12. Subdivisions of an elevation allow you to modify the _____ of an elevation.
13. Identify the start point for the building elevation line shown in Figure 9.79.

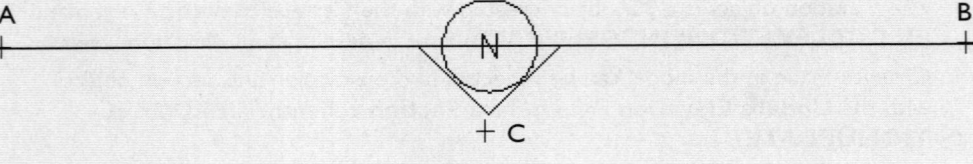

Figure 9.79 *Building elevation line*

PROJECT

EXERCISE 9.1 CREATING 3D ELEVATION AND 2D ELEVATION DRAWINGS

1. Open *Section1* from your student directory.
2. Save the file as *Proj9-6* in your student directory.
3. Create a 3D elevation object of the model and develop north and east elevations of the model displayed to 1/4"= 1'–0" Scale in the Plot-Sec layout tab (see Figure 9.80).
4. Create a longitudinal Section B-B of the model (see Figure 9.80).
5. Create 2D blocks of each elevation and the section and export the blocks to separate drawing files named: *North Elevation*, *East Elevation* and *Section B-B*.

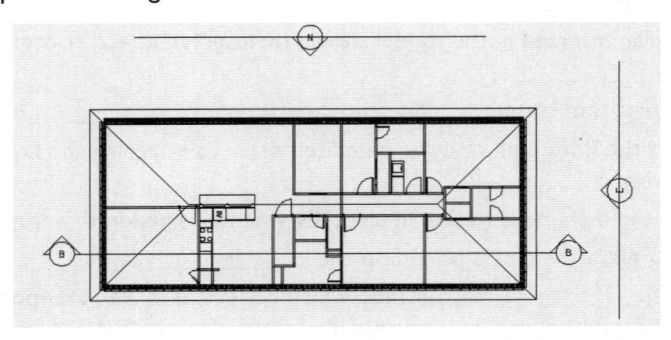

Figure 9.80 *Building elevation lines and building section lines for Exercise 9.1*

CHAPTER 10

Creating Mass Models

INTRODUCTION

Designers create physical models to refine their design and to communicate to clients the form or shape of the proposed building. Computer models can serve the same purpose as physical models, except that the computer model is easily edited to meet the intent of the designer. Computer models also allow the designer to perform space planning as the model is developed. Architectural Desktop includes the tools for creating mass models of the building. Computer models are created by the joining together of a combination of mass elements in the form of twelve possible prism shapes, extrusions of profiles, or revolutions of profiles. The mass elements can be joined together through Boolean operations of addition, subtraction or intersection. The mass elements can also be used to simulate such building components as columns, floors, or ceiling objects in other drawings.

Once a mass model is created, you can slice the model to create floorplates that can be used for the development of floor plans. After the floor plans are developed, the plans can be brought back into a model to create a more detailed model of the building.

OBJECTIVES

Upon completion of this chapter the student will be able to

Create mass elements using the **Add Mass Element** command (**MASSELE-MENTADD**)

Edit the size of a mass element using its grips

Edit the size, shape, and other properties using the **Modify Mass Element** commands (**MASSELEMENTMODIFY** and **MASSELEMENTPROPS**)

Extrude or revolve an AEC Profile to create a mass element

Create an AEC Profile and edit AEC Profiles to create custom mass elements

Create a mass element to represent a floor system for a building

CREATING MASS MODELS

Mass models are created from mass elements. The mass model should be constructed in the Mass-Group layout tab, which is included in drawings created from the Aec arch (imperial) template. The display configuration of the Mass-Group layout tab controls the display of mass elements. Mass elements are displayed in the left viewport and the defined group is displayed in the right viewport. A group consists of several mass elements combined to form one entity, as shown in Figure 10.1. Therefore, the mass elements become the building blocks for creating the mass model. Mass elements are created on layer A-Mass, which has the color 210 (a hue of magenta). The default viewing direction for both viewports is set to southwest isometric.

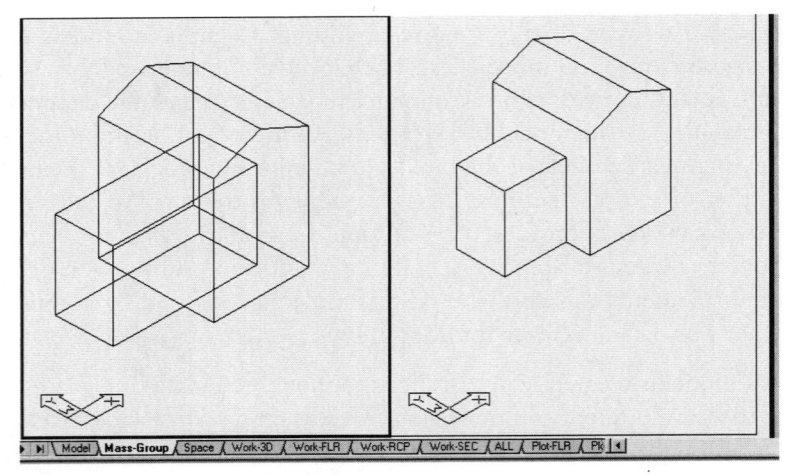

Figure 10.1 Mass-Group layout for display of mass elements

INSERTING MASS ELEMENTS

The mass element shapes included in the **Shape** list of the **Add Mass Element** dialog box are the Arch, Barrel Vault, Box, Cone, Cylinder, Dome, Doric Column, Gable, Pyramid, Isosceles Triangle, Right Triangle and Sphere. These mass element shapes (except the Doric Column) are included on the **Mass Elements** toolbar, as shown in Figure 10.2. Custom mass elements can also be created as extrusions or revolutions from a closed polyline. The **Add Mass Element** command is used to insert mass elements in the drawing. The dimensions of the mass element are defined in the

Add Mass Element dialog box. See Table 10.1 for **Add Mass Element** command access.

Menu bar	Concept>Mass Elements>Add Mass Element
Command prompt	MASSELEMENTADD
Toolbar	Select a mass element shape from the Mass Elements toolbar as shown in Figure 10.3
Shortcut	Right-click over the drawing area; then choose Utilities>Massing>Mass Element>Add

Table 10.1 *Add Mass Element command access*

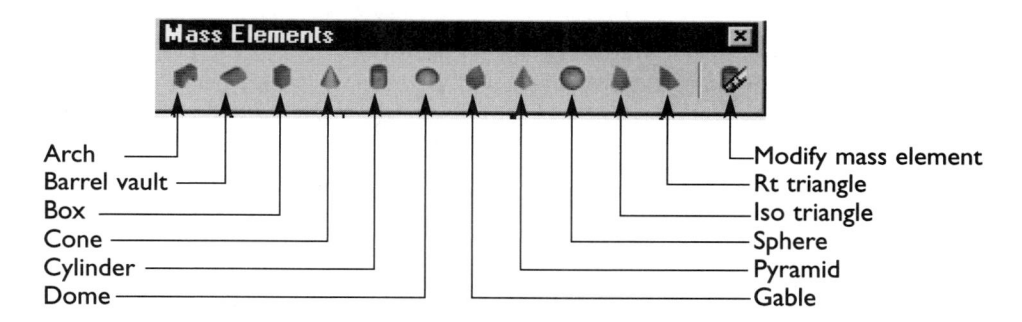

Figure 10.2 *Mass Elements toolbar*

When a mass element option is selected from the **Mass Elements** toolbar, the **Add Mass Element** dialog box opens, as shown in Figure 10.3. The **Add Mass Element** dialog box is not active until you select the dialog box with a left click of the mouse; the dialog box allows you to specify the dimensions of the mass element. After editing the **Add Mass Element** dialog box, you can select the left viewport to activate the viewport and place the mass element.

When the **Add Mass Element** dialog box opens, you can select the shape and edit the dimensions in the dialog box or select the options from the command line. Selecting the options from the **Add Mass Element** dialog box provides quick access to defining the mass element shape. The options of the **Add Mass Element** dialog box are described below.

> **Shape** – The **Shape** list includes each of the following prism shapes: Arch, Barrel Vault, Box, Cone, Cylinder, Dome, Doric Column, Gable, Pyramid, Isosceles Triangle, Right Triangle, Sphere, Extrusion, and Revolution.

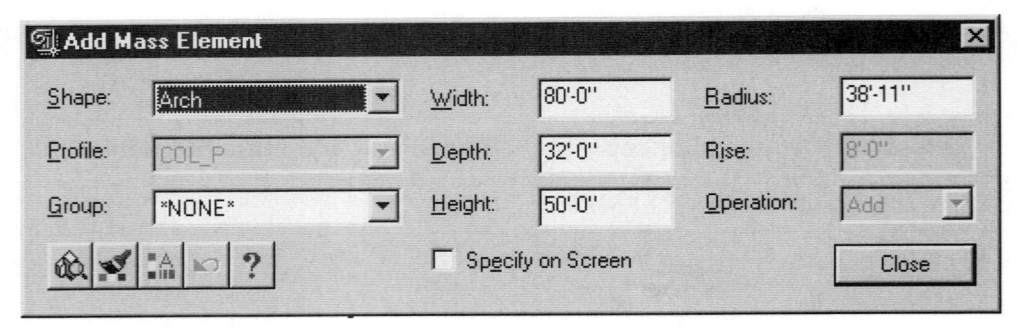

Figure 10.3 *Add Mass Element dialog box*

Profile – The **Profile** option allows you to select predefined AEC Profiles for the Extrusion or Revolution shape.

Group – The **Group** list includes group names defined in the drawing. Selecting a group name from the list will add the mass element to the group upon insertion. The group name in the **Add Mass Element** dialog box shown in Figure 10.3 displays *NONE* because no named groups have been created. Group names allow various mass elements to be associated with the group. Once a group name is defined, this edit box allows you to set a group name current and then create a mass element assigned to the group.

Width – The **Width** edit field allows you to insert the width dimension of the mass element.

Depth – The **Depth** edit field allows you to insert the depth dimension of the mass element.

Height – The **Height** edit field allows you to insert the height dimension of the mass element.

Radius – The **Radius** edit field is active if the mass element includes a radial shape; you can specify the radius of the radial shape.

Rise – The **Rise** edit field allows you to specify the rise of the Gable mass element.

Operation – The **Operation** drop-down list is active if a group is defined for the mass element. The **Operation** edit field has the options of Add, Subtract, and Intersect. Once a group has been created, mass elements can be inserted and defined to add, subtract or intersect their mass with other mass elements. If the operation is set to Subtract, the mass element being inserted will subtract its shape from the existing mass elements of the group. The default setting for operation is Add; therefore a new mass element would be added to previous mass elements of the group.

Floating Viewer – The **Floating Viewer** button opens the **Viewer**, which allows you to view the mass element prior to insertion.

Match – The **Match** button allows you to select an inserted mass element and match its properties for the current mass element.

Properties – The **Properties** button opens the **Mass Element Properties** dialog box. It includes a **General** and a **Dimensions** tab. The **General** tab, shown in Figure 10.4, allows you to add a description to the mass element.

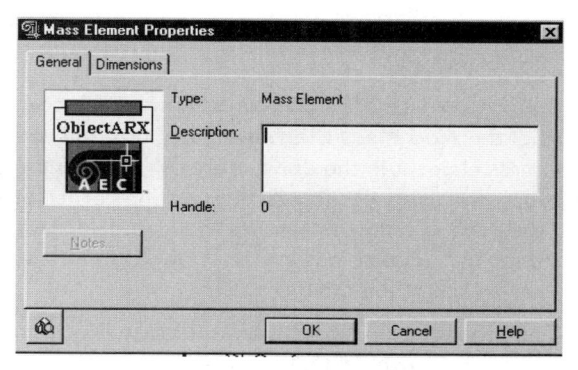

Figure 10.4 *General tab of the Mass Element Properties dialog box*

The **Dimensions** tab, shown in Figure 10.5, provides another means for defining the shape and size of the mass element.

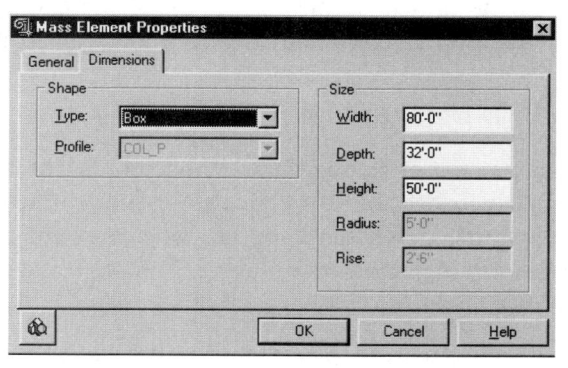

Figure 10.5 *Dimensions tab of the Mass Element Properties dialog box*

Undo – The **Undo** button allows you to undo the insertion of a mass element and select a different location for it while remaining in the **MASSELEMENTADD** command. When a mass element is inserted, you are prompted for the insert point and rotation. You can insert additional mass elements using the same settings by specifying another insert point and rotation. Once a mass element has been inserted, you can enter **U** in the command line or select the **Undo** button to withdraw the last insert point and rotation angle and remain in the **AECMASSELEMENTADD** command. This allows you to retain the shape and dimensions of the mass element within the **Add Mass Element** dialog box and remain in the **AECMASSELEMENTADD** command. Use of this option saves time because it allows you to modify shapes and dimensions without exiting the command and losing the settings.

? – The **?** button opens Architectural Desktop Help file to the Overview of Creating Mass Elements.

Specify on Screen – The **Specify on Screen** check box toggles off the dimensions of the **Add Mass Element** dialog box and you can define the size of the mass element in the drawing area. When you use the **Specify on Screen** option, the first selection establishes one corner or center of the shape and the next selection establishes the diagonal location of the base or radius. Once the base dimensions are established, the height of the object is established by a click of the mouse.

Close – The **Close** button closes the **Add Mass Element** dialog box and ends the **MASSELEMENTADD** command.

Selecting Mass Element Shapes

The **Shape** drop-down list shows the twelve mass elements and the Extrusion and Rotation options. The shapes included in the **Add Mass Element** dialog box are described below. Each shape has a specified insertion point or handle in which the mass element is inserted in the drawing. The size of the shape can be edited with its grips or the **Modify Mass Element** command, which is located on the **Mass Elements** toolbar.

Arch

Define the Arch mass element by setting the width, depth, height, and radius. It consists of a box prism, which has the semi-circular prism subtracted from the box. The radius dimension of the Arch determines the size of the semi-circular shape. The center for the Arch is located in the bottom plane. The grips and the dimensions of the Arch are shown in Figure 10.6.

There is a point at the bottom and top central grip that can be snapped to with the Node object snap. The point located in the bottom plane is used by default as the insertion point. If other mass elements are placed on top of the Arch, the top node can be snapped to with the Node object snap to align the mass elements together. The grip labeled A can be used to edit the radius of the arch. The other grips, when edited, will expand prism symmetrically, allowing the arch to remain centered in the prism as the prism dimensions change.

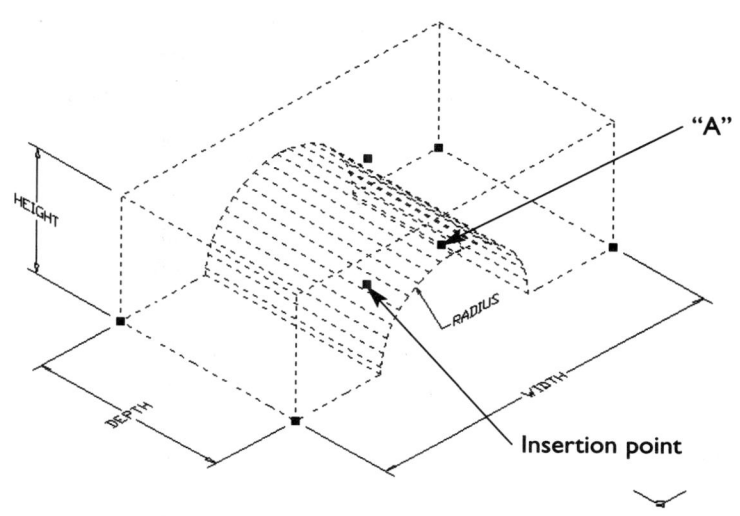

Figure 10.6 Arch mass element

Barrel Vault

The Barrel Vault is a prism consisting of a semi-circle that has been extruded. Setting the width and radial dimensions creates a prism that is one-half of a cylinder. The grips of the Barrel Vault are located at the four corners, center of the base, and at the top of the radial plane. The insertion point for placing the Barrel Vault is located at the point centered on the bottom plane. The grips of the Barrel Vault are shown in Figure 10.7.

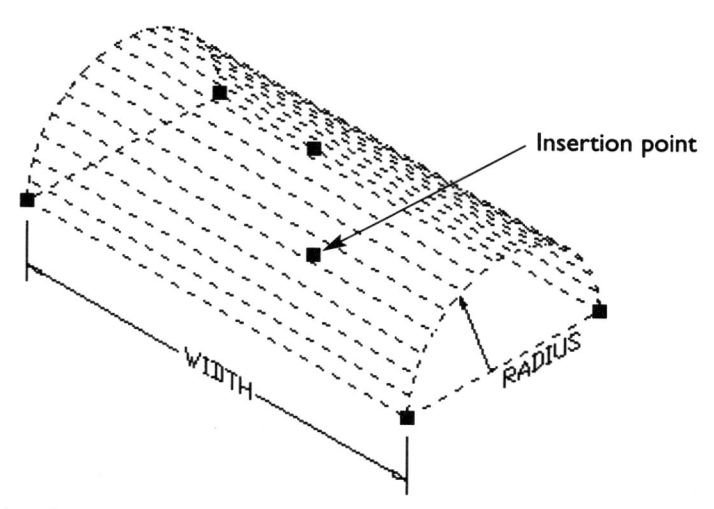

Figure 10.7 Barrel vault

If you edit the grip for the radial dimension, the radius of the semi-circle will change. To edit the width, select any of the four grips located at the corners of the base, and the Barrel Vault will stretch in the direction the mouse is moved.

Box

A Box prism is a shape frequently used in buildings, and its simplicity makes the union of multiple Boxes easy to visualize. Width, depth and height are the dimensions controlled in forming various sizes of rectangular prisms. The grip points for the Box are shown in Figure 10.8.

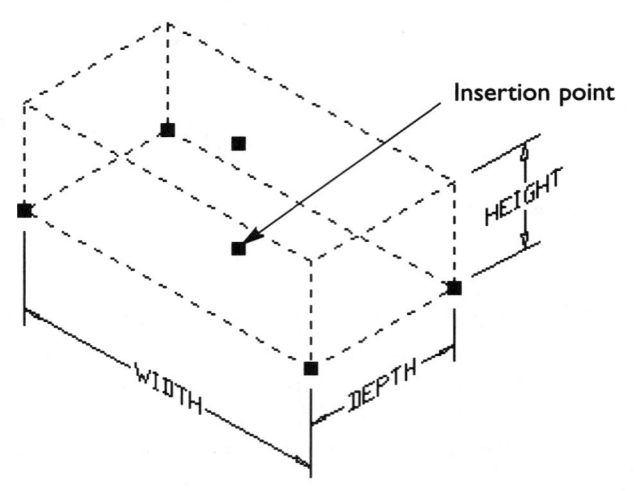

Figure 10.8 *Box mass element*

The insertion point for the Box is located in the center of the bottom plane. To edit the depth or width dimensions of the Box, select the grips located on the corners and stretch the Box. Edit the height of the Box by selecting the grip located at the top of the Box. You can snap to the grips at the top and bottom of the Box with the Node object snap mode if other mass elements are stacked together.

Cone

Create the Cone prism by establishing the radius and height dimensions. Cones are seldom used in building designs except for roof forms. The insertion point for the cone is located at its center on the bottom plane. The grips of a Cone are shown in Figure 10.9.

When locating other mass elements relative to the Cone, you can use the point at the center of the bottom plane of the Cone. Quadrant and Center object snap modes cannot be used to locate the Cone.

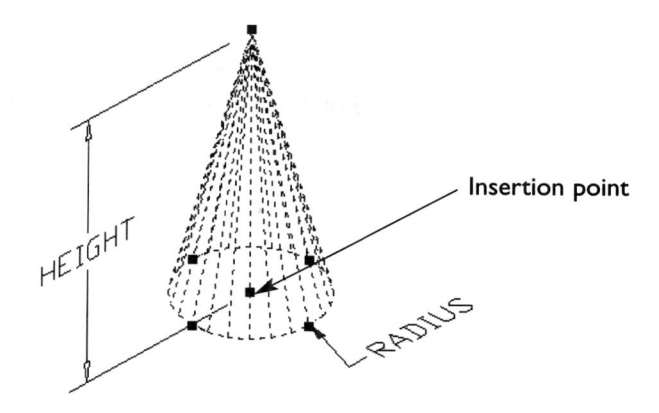

Figure 10.9 *Cone mass element*

Cylinder

Design the Cylinder prism by establishing the radius and height. When you insert the Cylinder, the insertion point is located at the center of the bottom plane. The grips of the Cylinder are shown in Figure 10.10.

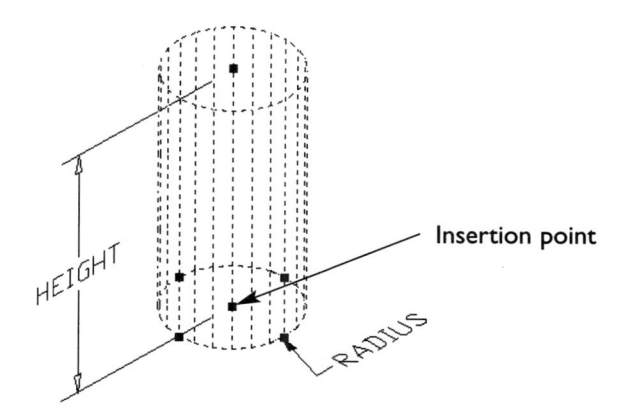

Figure 10.10 *Cylinder mass element*

Dome

The Dome mass element is one-half a sphere. Only the radius dimension is set to create the size of the Dome. The point located in the center of the bottom plane is used to place the insertion point of the Dome. The grips of a Dome are shown in Figure 10.11.

The four grips at the quadrant points of the Dome can be used to increase the radius

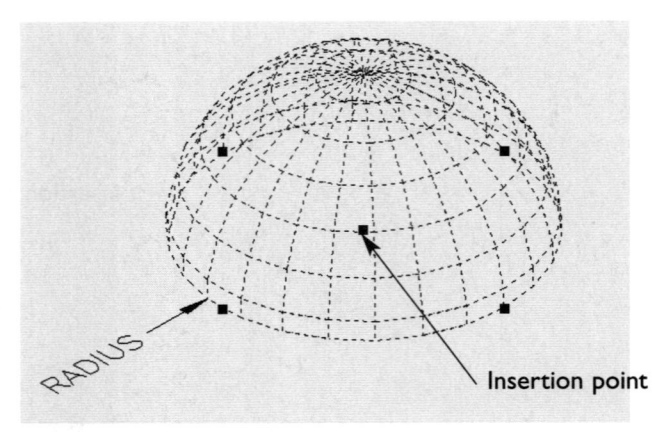

Figure 10.11 *Dome mass element*

Doric Column

The Doric Column is not a simple prism but a complex shape that can be used to create columns for a building. Other columns are available through extrusion of profiles; typical round, wide flange, and tube columns are available as profiles that can be extruded. However, the Doric Column is not extruded as a profile. To create a Doric Column, you set the height and radius. The taper of the column is fixed and increases from the radius dimension set, which is the dimension at the top of the column. The taper of the column increases the diameter at the bottom of the column. The grips of the Doric Column are shown in Figure 10.12.

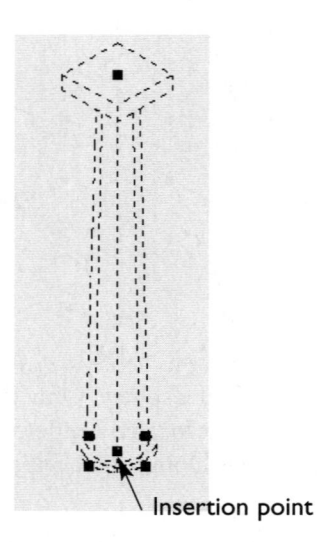

Figure 10.12 *Doric Column mass element*

Gable

The Gable shape is frequently used for residential and commercial buildings because it includes the Box shape and the gable roof. Create the Gable mass element by establishing the depth, width, height and rise as shown in Figure 10.13.

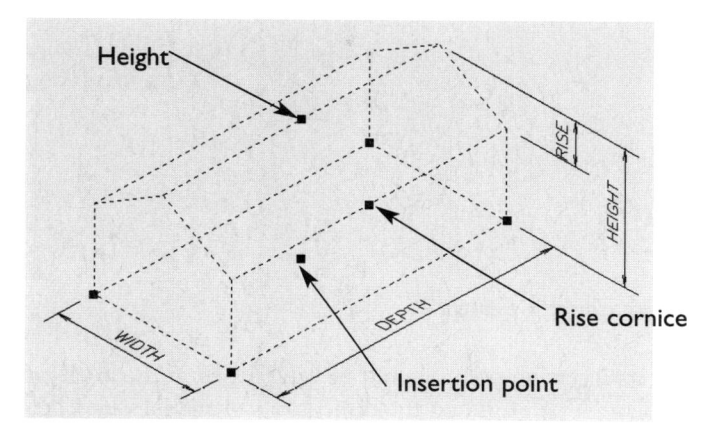

Figure 10.13 *Gable mass element*

When designing a Gable, you need to determine the distance from the top ridge down to the cornice. The four grips on the base of the Gable allow you to stretch the depth and width dimensions of the building. Gable mass elements are inserted in the drawing with the insertion point centered on the bottom of the base plane. The height is edited by the grip located at the top of the ridge of the gable while the rise can be edited with the grip at the cornice as shown in Finish 10.13. The ridge extends parallel to the width of the building.

Pyramid

Create the Pyramid mass element by setting the depth, height, and width dimensions. The insertion point for the Pyramid is located on the center of the bottom plane of the Pyramid. The grips for the Pyramid are shown in Figure 10.14.

Editing the four grips located on the corners of the Pyramid mass element will change the depth and width dimensions. The grip at the apex can be used to edit the height, while the base will remain stationary. The bottom central grip can be used to move the Pyramid rather than edit its dimensions. Pyramid mass elements are usually used for roof elements, which necessitates that the insertion point be moved up in the Z direction to the height of the base of the roof plane.

Isosceles Triangle

Create the Isosceles Triangle mass element by setting the width, depth, and height dimensions. This mass element is also a shape used for roofs, and therefore, when it

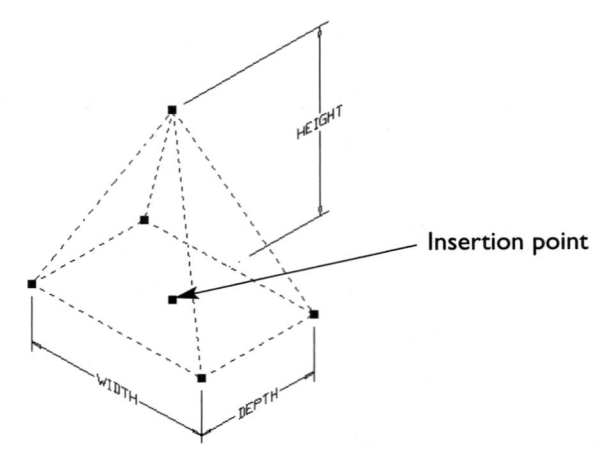

Figure 10.14 *Pyramid mass element*

is used as a roof, its insertion point should be established at the Z elevation of the bottom of the roof plane. The grips and the dimensions of the Isosceles Triangle are shown in Figure 10.15.

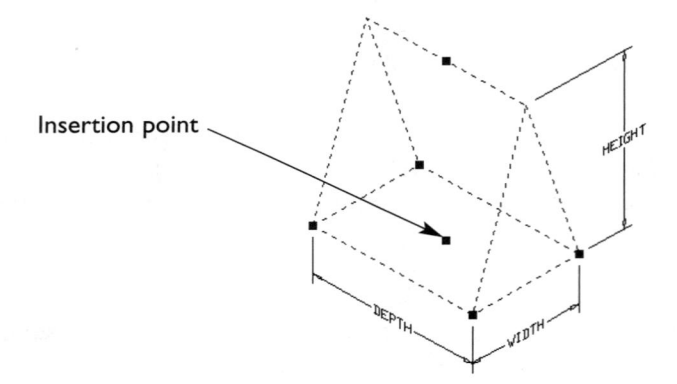

Figure 10.15 *Isosceles Triangle mass element*

The grips of the Isosceles Triangle edit the width, depth and height of the mass element similarly to the Pyramid. The insertion point for the Isosceles Triangle is located in the center of the bottom plane.

Right Triangle
Establishing the height and width of the mass element sets the dimensions of the legs of the Right Triangle. Right Triangles can be used as mass elements to form roofs or walls. The grips and dimensions of the Right Triangle are shown in Figure 10.16.

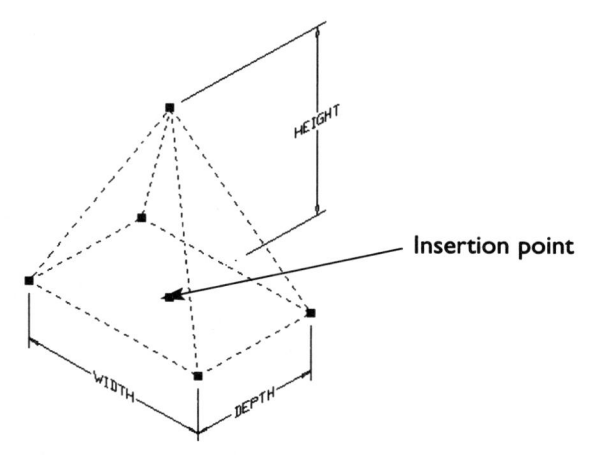

Figure 10.16 Right Triangle mass element

The insertion point for this mass element is located in the center of the bottom plane.

Sphere

The Sphere has one dimension: its radius. The grips are located at the four quadrants of the Sphere parallel to the XY plane, as shown in Figure 10.17. Editing these grips will cause the radius of the Sphere to change. The grip located at the bottom of the Sphere is at the zero Z axis location and can be used move the sphere.

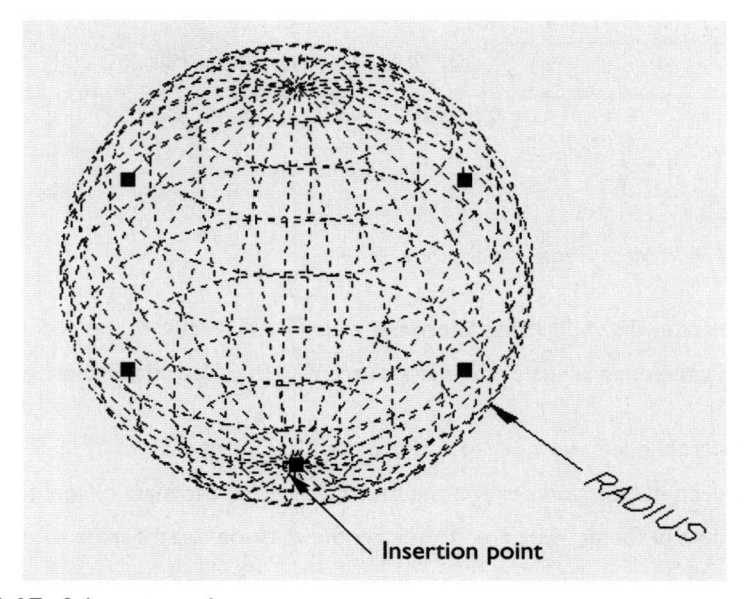

Figure 10.17 Sphere mass element

The insertion point for the Sphere is located on the bottom of the sphere. Although a Sphere is seldom used as a wall or roof element, it can be used to edit other mass elements or for ornament.

Revolution and Extrusions

Revolution and Extrusions are included in the Shape list. Although they are not mass elements, they are methods for creating a mass element by either extruding or revolving a profile. Any closed polyline can be converted to an AEC Profile and extruded or revolved to create a mass element. The AEC Profiles included in the Profiles list are column shapes such as circular, tube, and wide flange columns. Using AEC Profiles with the Extrusion and Rotation options allows you to create custom mass shapes.

STEPS TO INSERTING A MASS ELEMENT

1. Select the Mass-Group layout.
2. Right-click over an Architectural Desktop toolbar and choose **Mass Elements** from the toolbar list.
3. Select a shape option from the **Mass Elements** toolbar to open the **Add Mass Element** dialog box, as shown in Figure 10.18.

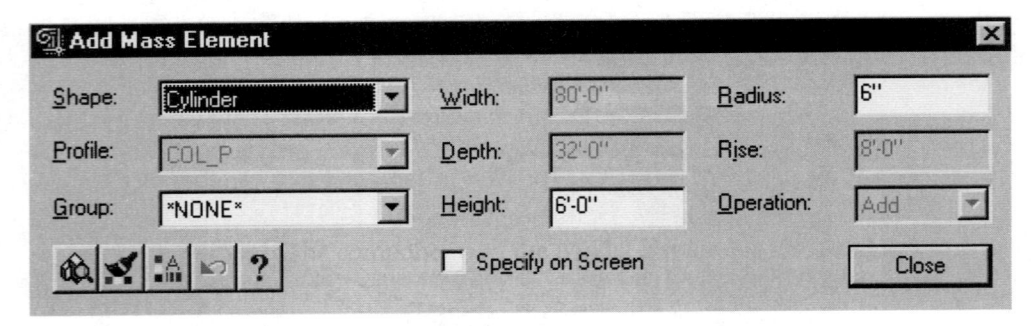

Figure 10.18 *Add Mass Element dialog box*

4. Select in the **Add Mass Element** dialog box to activate the dialog box.
5. Edit the dimensions of the mass element in the **Add Mass Element** dialog box.
6. Select the left viewport to activate the Mass viewport.
7. Select an insertion point in the drawing area for the mass element.
8. Select in the drawing area to specify the rotation for the mass element.

TUTORIAL 10.1 CREATING MASS ELEMENTS

1. Open EX10-1.dwg from the \ADT_Tutor\Ch10\ directory.
2. Choose **File>SaveAs** from the menu bar and save the drawing as *Mass1.dwg* in your student directory.
3. Select the Mass-Group layout tab.
4. Right-click over any Architectural Desktop toolbar and then select **Mass Elements** from the toolbar list.
5. Select the **Gable** command from the **Mass Elements** toolbar as shown in Figure 10.19.

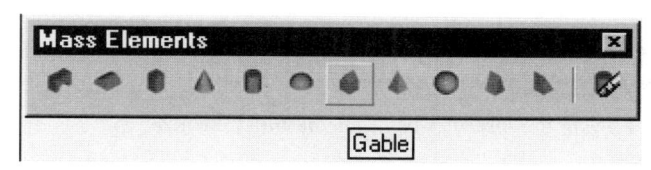

Figure 10.19 *Mass Elements toolbar*

6. Select in the **Add Mass Element** dialog box and edit **Width** to 26'; **Depth** to 50'; **Height** to 28'; **Rise** to 5', as shown in Figure 10.20. Verify that the **Shape** is Gable and **Group** is *None* in the **Add Mass Element** dialog box. Do not select the **Close** button.

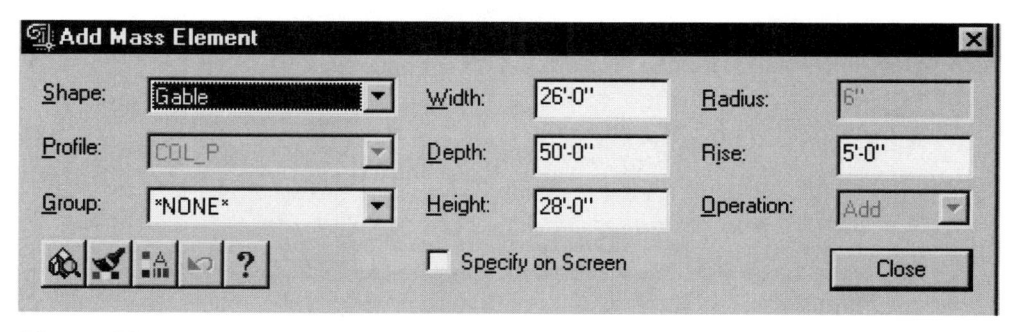

Figure 10.20 *Specify settings in the Add Mass Element dialog box*

7. Select in the left viewport of the drawing area to activate the drawing area.
8. Move the pointer and notice that the handle for the mass element is centered on the bottom plane of the mass element, as shown in Figure 10.21.

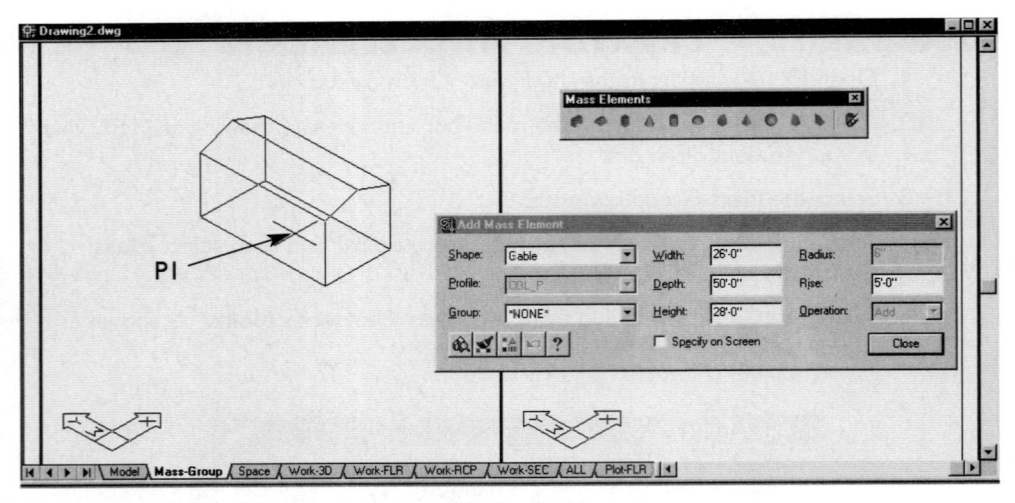

Figure 10.21 *Specifying the insertion point for the mass element*

9. Respond to the following command line prompts to insert the mass element.

   ```
   Command: _AecMassElementAdd
   Insert point or
   [SHape/WIdth/Depth/Height/RIse/Match]: _shape
   Shape [BOx/Arch/BArrel
   vault/DORic/COne/CYlinder/DOMe/Gable/Isoc
   triangle/Right
   triangle/Pyramid/Sphere]: _gable

   Insert point or
   [SHape/WIdth/Depth/Height/RIse/Match]:
   ```
   (Select a point near the center of the left viewport near point P1 in Figure 10.21.)
   ```
   Rotation or [SHape/WIdth/Depth/Height/RIse/Match]
   <0.00>: 90 ENTER
   ```
   (Mass element inserted as shown in Figure 10.22.)

10. Select in the **Add Mass Element** dialog box and edit the **Width** to 20', **Depth** to 30', **Height** to 20', and **Rise** to 5'.

11. Select in the left viewport of the drawing area to activate the Mass viewport.

12. Respond to the following command line prompts to insert the mass element as shown in the command sequence and in Figure 10.23.

    ```
    Insert point or
    [SHape/WIdth/Depth/Height/RIse/Match]:
    ```
    (Select a point near P1 in Figure 10.23.)
    ```
    Rotation or [SHape/WIdth/Depth/Height/RIse/Match]
    <0.00>: 90 ENTER
    ```

Creating Mass Models 757

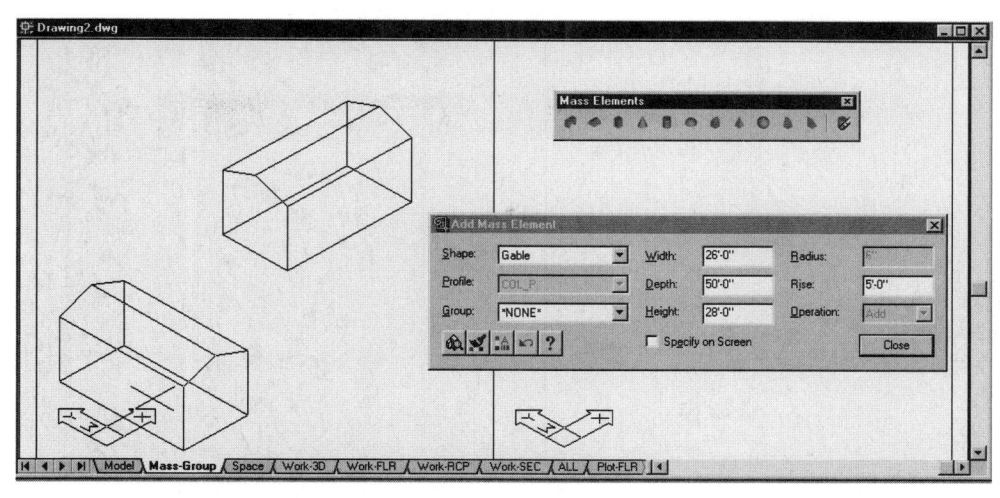

Figure 10.22 *Mass element inserted*

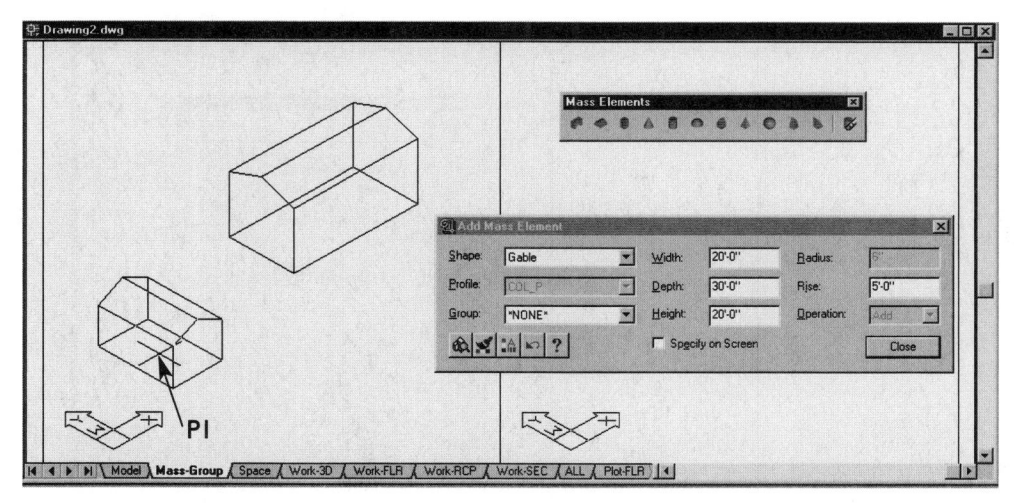

Figure 10.23 *Specifying the insertion point for the mass element*

```
Insert point or
[SHape/WIdth/Depth/Height/RIse/Match/Undo]: ENTER
```
(Press ENTER *to end the MASSELEMENTADD command.)*
(Mass element inserted as shown in Figure 10.24.)

13. Select the **Move** command from the **Modify** toolbar and move the last mass element as shown in the following command sequence and in Figure 10.25.

```
Command: m MOVE
Select objects:
```
(Select the mass element at point P1 in Figure 10.25.) 1 found

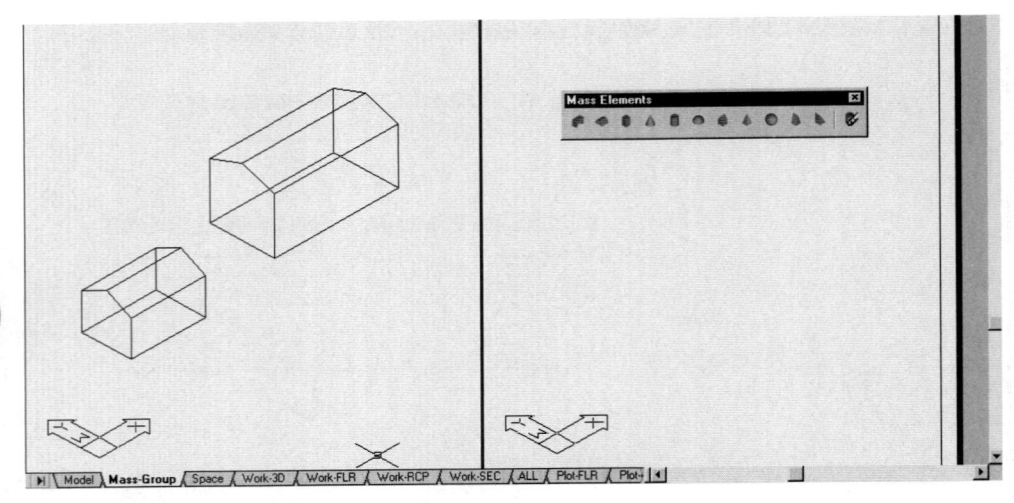

Figure 10.24 *Mass element inserted*

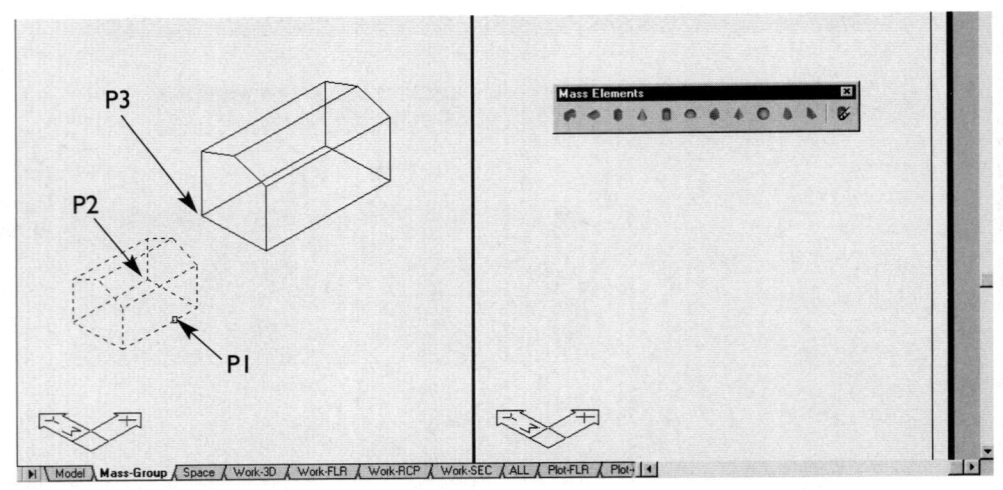

Figure 10.25 *Moving the second mass element to connect to the first mass element*

```
Select objects: ENTER (Press ENTER to end object selection.)
 Specify base point or displacement: end ENTER
of (Select the end of mass element at point P2 in Figure 10.25.)
Specify second point of displacement or <use first
point as displacement>: end ENTER
 Of (Select the end of the first mass element at point P3 in Figure 10.25.)
```

14. Enter **HI** in the command line to execute the **HIDE** command and display the mass elements as shown in Figure 10.26.

15. Save the drawing.

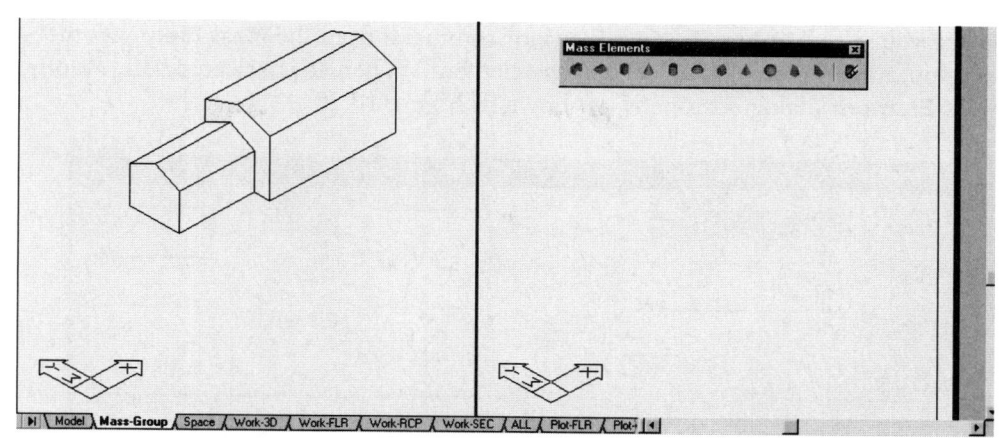

Figure 10.26 *Mass Elements connected*

MODIFYING MASS ELEMENTS

To modify mass elements, you can edit the grips of the mass element or apply the **Modify Mass Element** command. The **Modify Mass Element** command opens the **Modify Mass Element** dialog box, which allows you to change the size of the mass element in a dialog box. See Table 10.2 for **Modify Mass Element** command access.

Menu bar	Concept>Mass Elements>Modify Mass Element
Command prompt	MASSELEMENTMODIFY
Toolbar	Select Modify Mass Element from the Mass Elements toolbar as shown in Figure 10.27
Shortcut	Right-click over the drawing area and choose Utilities>Massing>Mass Element>Modify
Shortcut	Select a mass element; then right-click and choose Modify Element from the shortcut menu

Table 10.2 *Modify Mass Element command access*

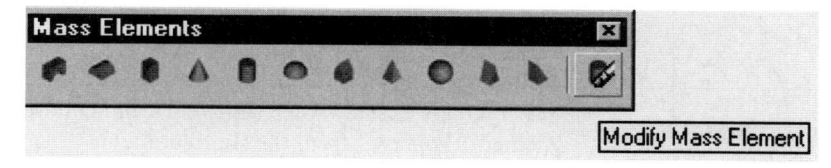

Figure 10.27 *Modify Mass Element command*

When you select the **Modify Mass Element** command from the **Mass Elements** toolbar, you are prompted to select a mass element. When it is selected, the **Modify Mass Element** dialog box opens, as shown in Figure 10.28.

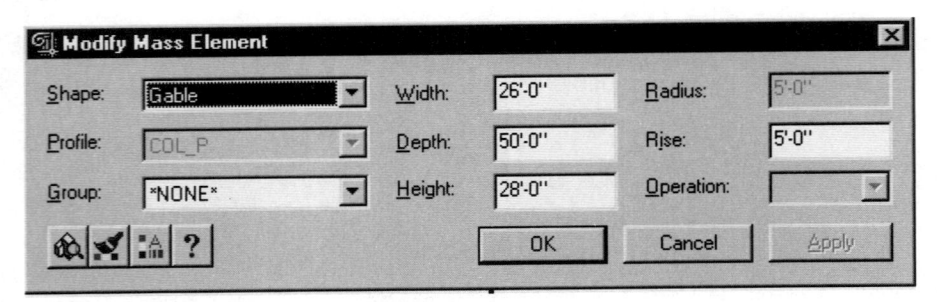

Figure 10.28 *Modify Mass Element dialog box*

The options of the **Modify Mass Element** dialog box are identical to those of the **Add Mass Element** dialog box. The **Modify Mass Element** dialog box displays the dimensions of the current mass element. To modify the mass element, you edit the dimensions shown in this dialog box. Some of the edit boxes of this dialog may be dimmed out if they are not appropriate to the mass element. Unless a mass group has been defined for the mass element, the **Operations** edit box will be dimmed out. If group operations are defined for the mass element, the following options for Boolean operations will be available in the **Operation** list: Add, Subtract, and Intersect.

Change the Properties of a Mass Element

The properties of a mass element can be changed by the **MASSELEMENTPROPS** command. This command allows you to change the dimensions, group assignment, and location of the mass element. The **Location** tab allows you to rotate the mass element after it has been inserted. See Table 10.3 for **MASSELEMENTPROPS** command access.

Command prompt	MASSELEMENTPROPS
Toolbar	Select Modify Mass Element from the Mass Elements toolbar; then select the Properties button.
Shortcut	Right-click over the drawing area and choose Utilities>Massing>Mass Element>Properties
Shortcut	Select a mass element; then right-click and choose Element Properties from the shortcut menu

Table 10.3 *MASSELEMENTPROPS command access*

When the **Properties** button of the **Modify Mass Element** dialog box is selected, the **Mass Element Properties** dialog box opens, as shown in Figure 10.29. The **Mass Element Properties** dialog box consists of the following four tabs: **General**, **Dimensions**, **Mass Group**, and **Location**.

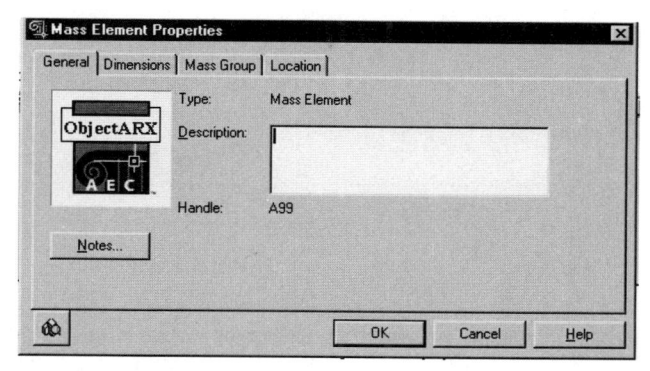

Figure 10.29 *General tab of the Mass Element Properties dialog box*

General
The **General** tab, shown in Figure 10.29, allows you to edit the **Notes** and **Description** for the mass element. The description would allow you to note the area and proposed spaces for the mass element. You can type a description in the **Description** edit field to describe the mass element. If you select the **Notes** button, the **Text Notes** and **Reference Docs** tabs can be used to enter additional information regarding the mass element.

Dimensions
The **Dimensions** tab, as shown in Figure 10.30, displays the current dimensions of the mass element. This tab can be used to change the size of the mass element.

Mass Group
The **Mass Group** tab, shown in Figure 10.31, allows you to redefine the mass element to a different group or to change the Boolean operation of the mass element. The Boolean operations of add, subtract and intersect are available only if the mass element has been defined to a group. If mass elements are defined to a group, you can then also edit the group assignment in the list.

Location
The **Location** tab, shown in Figure 10.32, allows you to view the actual X,Y,Z location of the mass element. You can change the insertion point of the mass element by editing the **X**, **Y**, and **Z** fields of the **Insertion Point** edit boxes. The **Normal** section allows you to change the plane orientation of the mass element. In essence, you

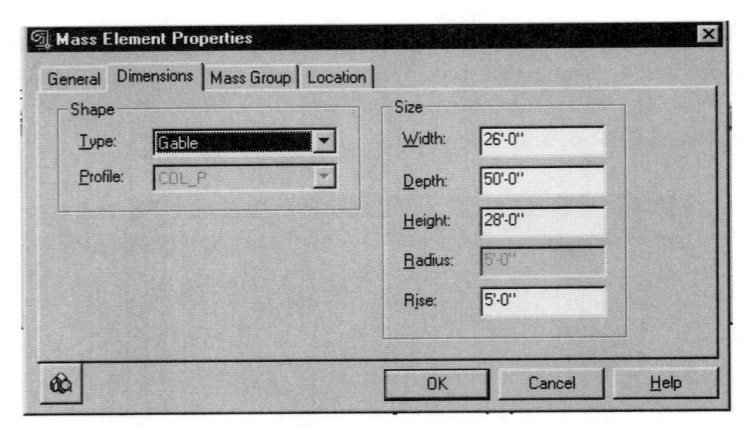

Figure 10.30 *Dimensions tab of the Mass Element Properties dialog box*

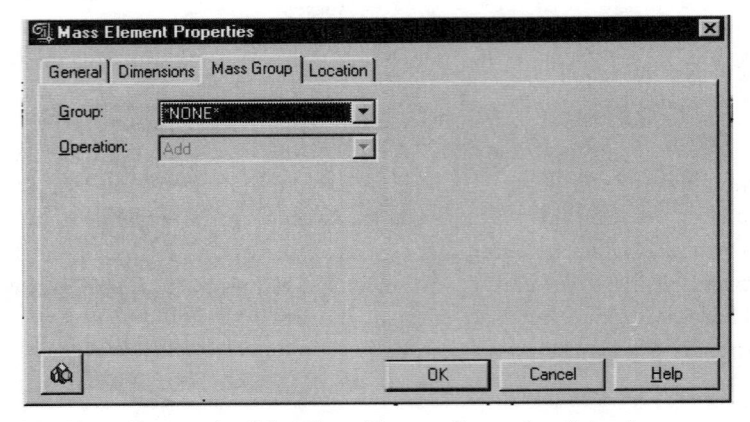

Figure 10.31 *Mass Group tab of the Mass Element Properties dialog box*

change the planar orientation for which the mass element is developed. As shown in Figure 10.32, the **Normal Z** value is set to 1 and **X** and **Y** values are set to zero. These settings position the working plane parallel to the XY plane and the mass elements are developed normal or perpendicular to the XY plane. If the **Normal Y** were set to 1 with **X** set to zero and **Z** set to zero, the working plane would be normal to the Y axis. Mass elements would then be developed perpendicular to the XZ plane. The rotation angle is measured counter-clockwise relative to the three o'clock position.

MODIFYING MASS ELEMENTS WITH GRIPS

You can modify mass elements by selecting and editing the grips of the mass element. The grips of each mass element are shown in Figures 10.6 through 10.17. A mass element can be moved, mirrored, rotated, scaled or stretched by its grips. Using the stretch operation with grips allows you to modify mass elements to fit within existing mass

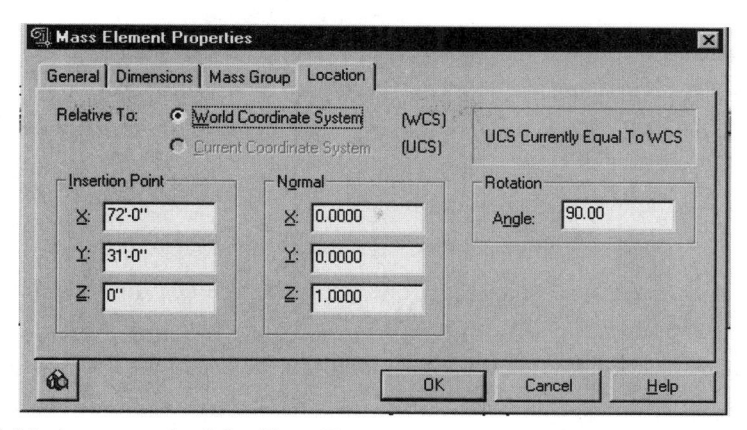

Figure 10.32 *Location tab of the Mass Element Properties dialog box*

elements. The mass element retains its shape; only the dimensions of the mass element change when the grips of the mass element are edited. If Ortho is ON, the dimensions of the object will adjust in either the X or Y direction, whereas if Ortho is OFF, both the X and Y dimensions of the object adjust. In Figure 10.33, the grip at point P1 was selected and stretched to point P2. The grips at points P3, P4, P5, and P6 can be stretched to adjust the X or Y dimensions of the box.

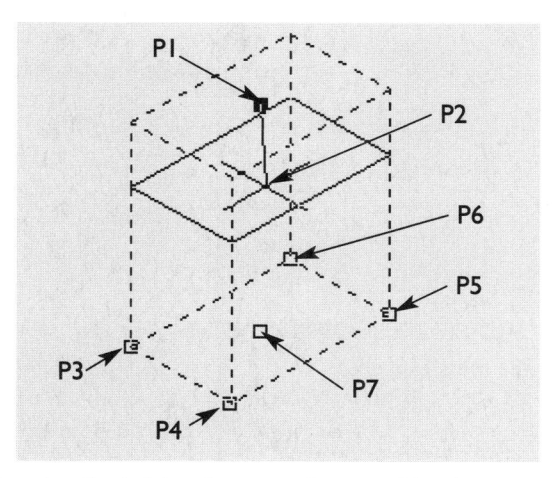

Figure 10.33 *Changing the height of a mass element with grips*

If the grip shown at P1 in Figure 10.33 is selected, the box can be stretched to change the height of the box. Each of the mass elements has a grip centered about the top plane, which can be stretched to change the height dimension of the mass element. The grip shown at P7 above can be used to move the mass element to another location.

TUTORIAL 10.2 MODIFYING MASS ELEMENTS

1. Open *EX10-2.dwg* from the \ADT_Tutor\Ch10\ directory.
2. Choose **File>SaveAs** from the menu bar and save the drawing as *Mass2* in your student directory.
3. Select the Mass-Group layout tab.
4. Select the **Box** command from the **Mass Elements** toolbar.
5. Select in the **Add Mass Element** dialog box and edit the **Width** to 80', **Depth** to 60', **Height** to 78'.
6. Select in the left viewport of the drawing area to activate the Mass viewport.
7. Move the pointer and notice the handle for the Box mass element is located in the center of the bottom plane.
8. Respond to the command line prompts to insert the mass element as shown in the following command sequence and in Figure 10.34.

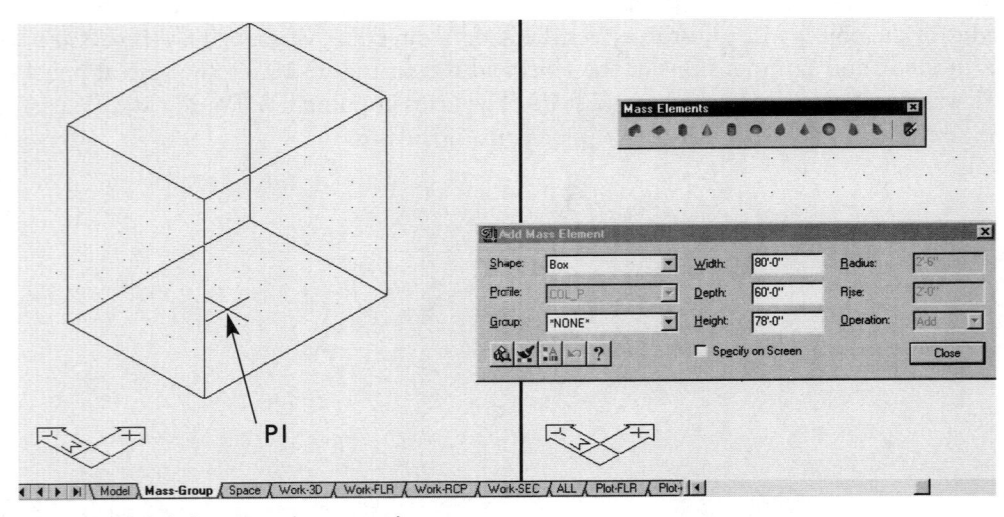

Figure 10.34 *Inserting the mass element*

```
Command: _AecMassElementAdd
First Corner or [SHape]: _shape
Shape [BOx/Arch/BArrel
vault/DORic/COne/CYlinder/DOMe/Gable/Isoc
triangle/Right
triangle/Pyramid/Sphere]: _box
First Corner or [SHape]: DBOX
```

```
Insert point or [SHape/WIdth/Depth/Height/Match]:
(Select a point near point P1 in the left viewport
as shown in Figure 10.34.)
Rotation or [SHape/WIdth/Depth/Height/Match] <0.00>:
```
ENTER *(Press* ENTER *to accept the default rotation of 0°.)*

```
Insert point or
[SHape/WIdth/Depth/Height/Match/Undo]: ENTER (Press
```
ENTER *to end the command.)*

9. Select the **Box** command from the **Mass Elements** toolbar.
10. Select in the **Add Mass Element** dialog box and edit the **Width** to 24", **Depth** to 36", and **Height** to 20'.
11. Select in the left viewport of the drawing area to activate the Mass viewport.
12. Respond to the command line prompts to insert the mass element as shown in the following command sequence and in Figure 10.35.

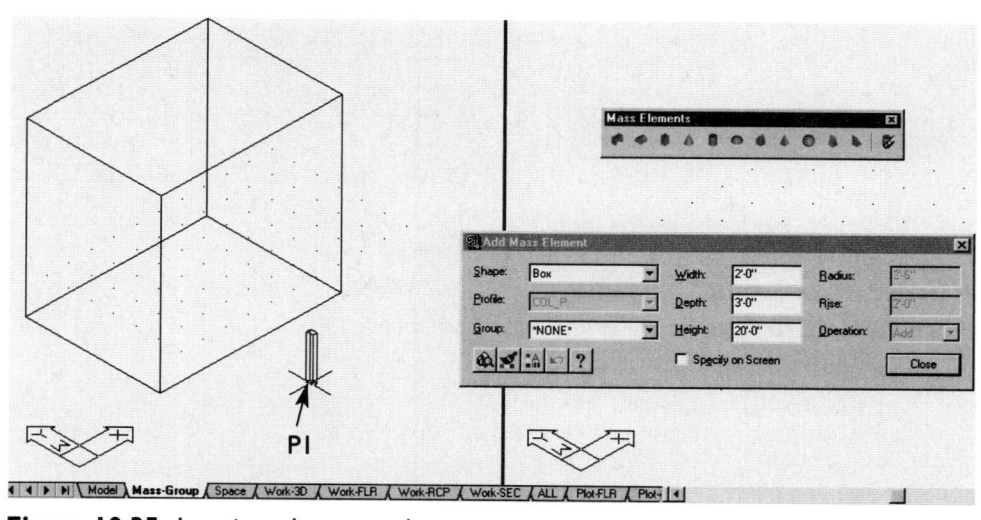

Figure 10.35 *Inserting a box mass element*

```
Command: _AecMassElementAdd
Insert point or [SHape/WIdth/Depth/Height/Match]:
_shape
Shape [BOx/Arch/BArrel
vault/DORic/COne/CYlinder/DOme/Gable/Isoc
triangle/Right
triangle/Pyramid/Sphere]: _box
```

```
Insert point or [SHape/WIdth/Depth/Height/Match]:
```
(Select a point near point P1 in Figure 10.35.)
```
Rotation or [SHape/WIdth/Depth/Height/Match] <0.00>:
```
ENTER *(Press* ENTER *to accept 0° rotation.)*
```
Insert point or
[SHape/WIdth/Depth/Height/Match/Undo]:
```
ENTER *(Press* ENTER *to end the command.)*

13. Select the **Zoom Window** command from the **Standard** toolbar and zoom a window from point P1 to P2 in Figure 10.36, as shown in the following command sequence.

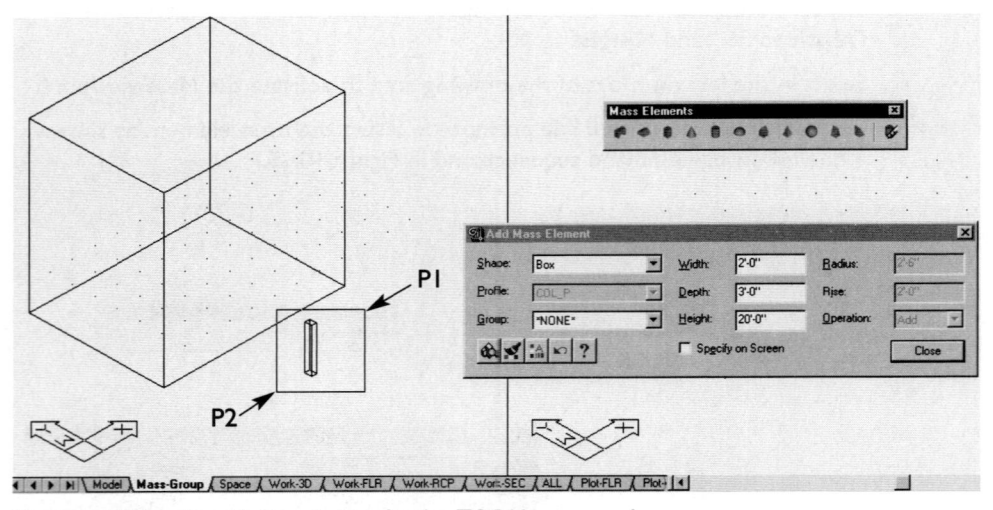

Figure 10.36 *Specify the window for the ZOOM command*

```
Command: '_zoom
Specify corner of window, enter a scale factor (nX
or nXP), or
[All/Center/Dynamic/Extents/Previous/Scale/Window]
<real time>: _w
Specify first corner:
```
(Select point P1 in Figure 10.36.)
```
Specify opposite corner:
```
(Select point P2 in Figure 10.36 to display the box mass element.)

14. Verify that Ortho is toggled ON in the Status Bar, and then select the **Copy** command from the **Modify** toolbar and respond to the following command line prompts to copy the box as shown in Figure 10.37.

```
Command: _copy
Select objects:
```
(Select the box at point P1 in Figure 10.37.) 1

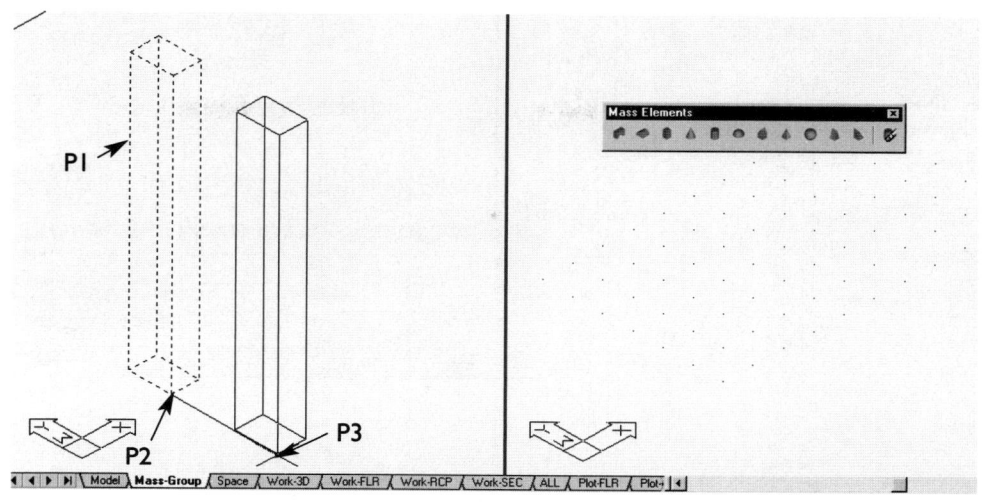

Figure 10.37 *Specifying the object and location of the COPY command*

```
found
Select objects:   ENTER  (Press ENTER to end object selection.)
Specify base point or displacement, or [Multiple]:
end ENTER
of
```
(Select the box at point P2 and move the mouse to the right toward point P3 in Figure 10.37.)
```
Specify second point of displacement or <use first
point as displacement>: 20' ENTER
```
(The box is copied using direct distance entry to the right as shown in Figure 10.38.)

15. Select the **Box** command from the **Mass Elements** toolbar.
16. Select in the **Add Mass Element** dialog box and select **Specify on Screen**.
17. Select in the left viewport to activate the Mass viewport.
18. Respond to the command line prompts as shown in the following command sequence and in Figures 10.38 and 10.39.

```
Command: _AecMassElementAdd
First Corner or [SHape]: _shape
Shape [BOx/Arch/BArrel
vault/DORic/COne/CYlinder/DOMe/Gable/Isoc
triangle/Right
triangle/Pyramid/Sphere]: _box

First Corner or [SHape]: end ENTER
Of
```
(Select the top corner of the box at P1 in Figure 10.38.)
```
Second Corner or [SHape]: end ENTER
```

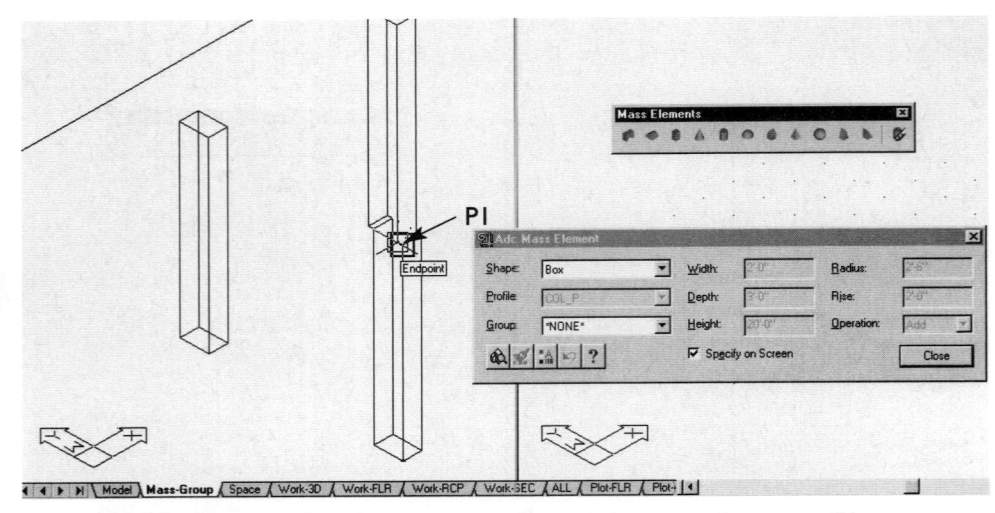

Figure 10.38 *Creating a box element on top of an existing mass elements at P1*

```
Of (Select the top corner of the box at P2 in Figure 10.39.)
Height or [SHape]: 48 ENTER
Rotation or [SHape] <0.00>: ENTER
First Corner or [Shape/Undo]: ENTER
```

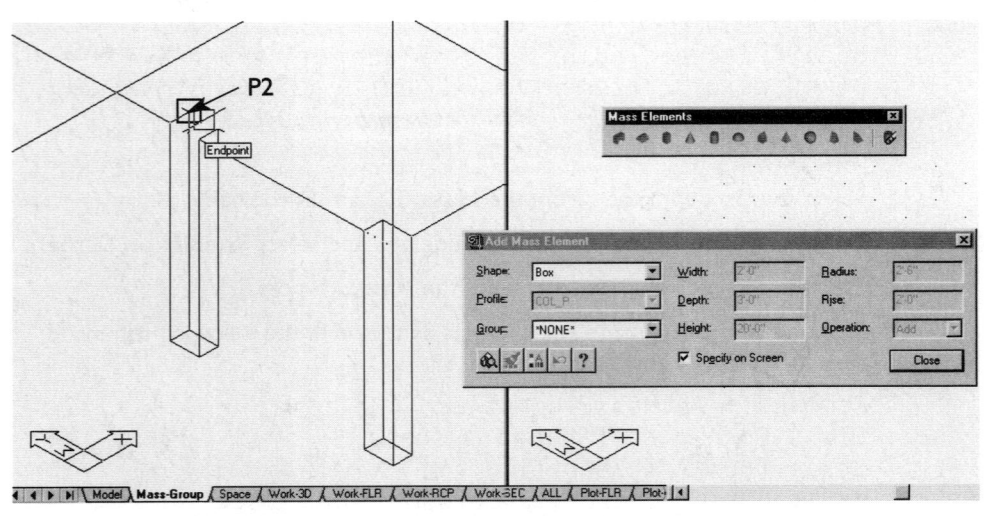

Figure 10.39 *Specifying the diagonal corner of the mass element*

19. Select **Zoom Extents** from the **Standard** toolbar.

Creating Mass Models 769

20. Select the Box mass element at point P1 as shown in Figure 10.40, and then right-click and choose **Element Modify** to open the **Modify Mass Element** dialog box.

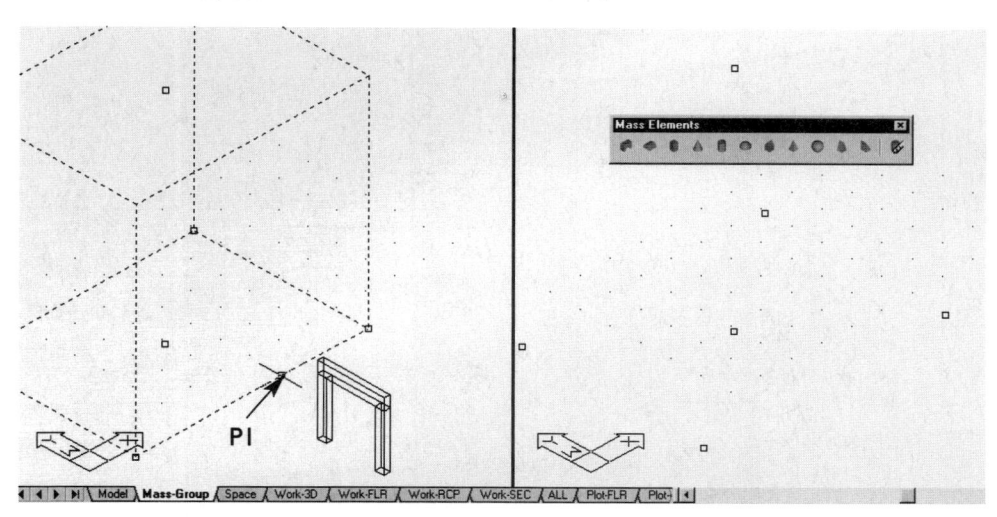

Figure 10.40 *Editing a mass element*

21. Select in the **Modify Mass Element** dialog box and then edit the **Height** dimension to 40'.
22. Select the **Apply** button and then the **OK** button to display the Box mass element as shown in Figure 10.41.

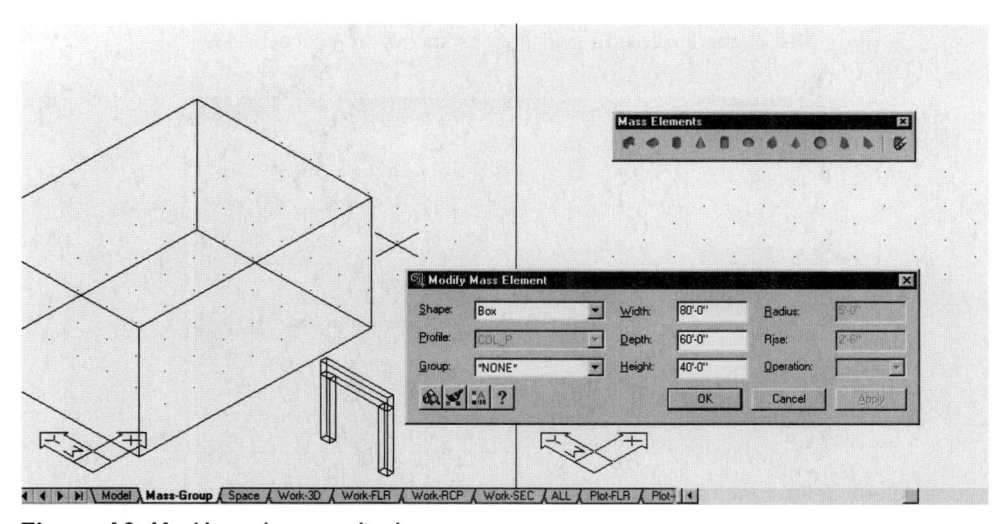

Figure 10.41 *Mass element edited*

23. Select F8 to turn OFF Ortho.
24. Select the Box mass element at point P1 as shown in Figure 10.42, and then select the grip located at point P2 and stretch the grip to a location near P3 as shown in Figure 10.42.

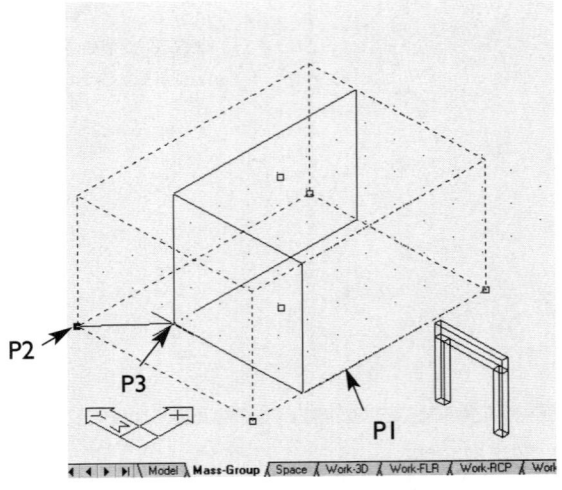

Figure 10.42 *Stretching the Grips of the Box mass element*

25. Select the Box mass element at Point P1 as shown in Figure 10.42; then right-click and choose **Element Properties** from the shortcut menu.
26. Select the **Location** tab of the **Mass Element Properties** dialog box.
27. Type **350** in the **Rotation** edit field as shown in Figure 10.43.

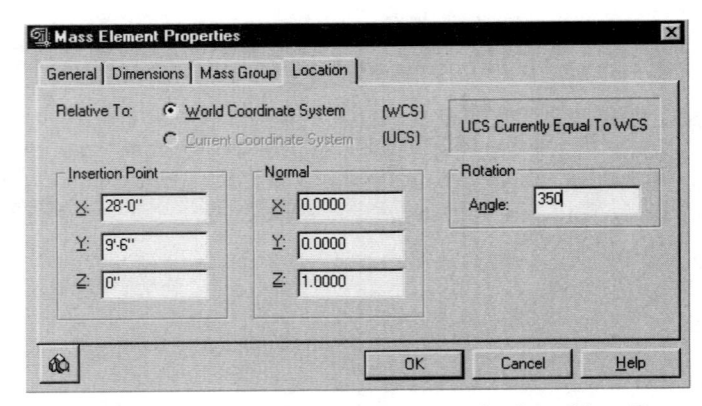

Figure 10.43 *Editing the rotation angle in the Location tab of the Mass Element Properties dialog box.*

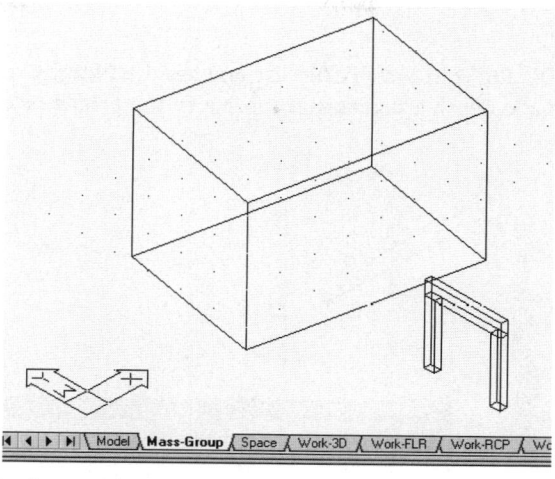

28. Select **OK** to dismiss the **Mass Element Properties** dialog box and display the revised Box mass element, as shown in Figure 10.44.

Figure 10.44 *Revised mass elements*

29. Save your drawing.

CREATING MASS ELEMENTS USING EXTRUSION AND REVOLUTION

In addition to the twelve mass elements prisms, you can also create a mass element from an extrusion or revolution of a profile. You can create your own profile or use the profiles included in the software. The profiles that are available include round columns, tubular columns, wide flange columns, handrail components and various unit geometric shapes such as the circle, hexagon, and octagon. To access a profile for extrusion, select the Extrusion or Revolution option from the **Shape** list of the **Add Mass Element** dialog box. When the Extrusion or Revolution option is selected, the **Profile** edit field of the **Add Mass Element** dialog box becomes active, allowing you to select a profile from the list. The dimensions of the mass element can then be specified in the **Add Mass Element** dialog box.

TUTORIAL 10.3 CREATING MASS ELEMENTS BY EXTRUDING PROFILES

1. Open *Ex10-3.dwg* from the *\ADT_Tutor\Ch10* directory.
2. Choose **File>SaveAs** from the menu bar and save the file to your student directory as *Mass3*.
3. Select the Mass-Group layout.
4. Select the left viewport and select **Zoom Extents** from the **Standard** toolbar to view the existing Box mass element.

5. Choose **Concept>Mass Elements>Add Mass Element** from the menu bar.
6. Select in the **Add Mass Element** dialog box and set the **Shape** to Extrusion.
7. Select COL_P6 from the **Profile** list, and set the **Height** to 7'. Notice that the width and depth are preset according to the profile, as shown in Figure 10.45.

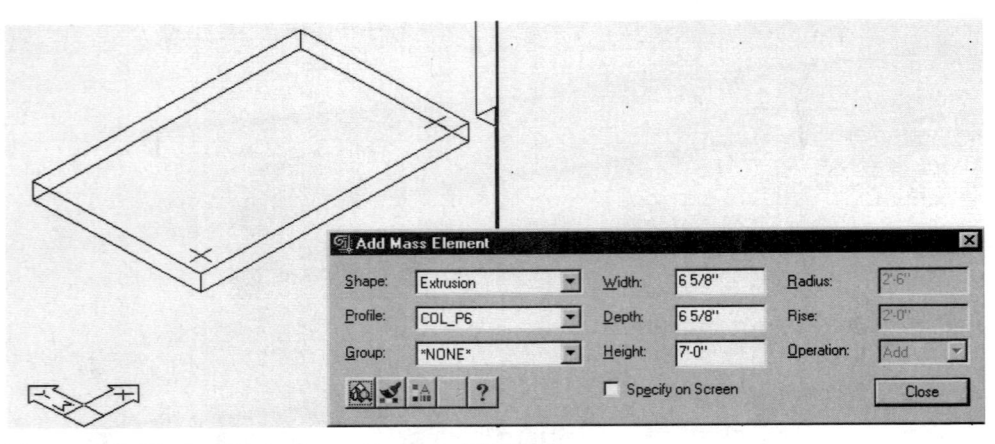

Figure 10.45 *Creating an extrusion of an AEC Profile*

8. Select in the drawing area and respond to the command line prompts to insert the column as shown in the following command sequence and in Figures 10.46 and 10.47.

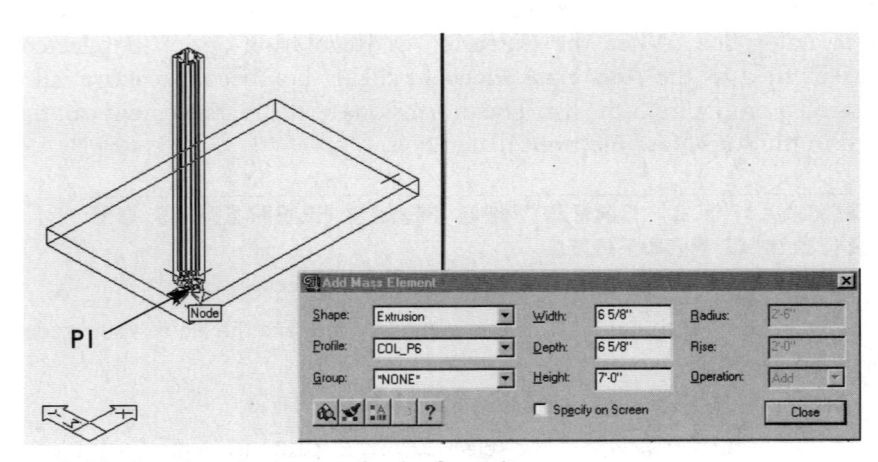

Figure 10.46 *Specifying the location for the first column*

```
Command: _AecMassElementAdd
Insert point or [SHape/WIdth/Depth/Height/Match]:
DBOX
Insert point or [SHape/WIdth/Depth/Height/Match]:
DBOX
Insert point or [SHape/WIdth/Depth/Height/Match]:
node ENTER
Of
```
(Select the point at P1 shown in Figure 10.46.)
```
Rotation or [SHape/WIdth/Depth/Height/Match] <0.00>:
ENTER
Insert point or
[SHape/WIdth/Depth/Height/Match/Undo]: node ENTER
of
```
(Select the point at P2 shown in Figure 10.47.)
```
Rotation or [SHape/WIdth/Depth/Height/Match] <0.00>:
ENTER
Insert point or
[SHape/WIdth/Depth/Height/Match/Undo]: ENTER
```
(Press ENTER to end the command.)

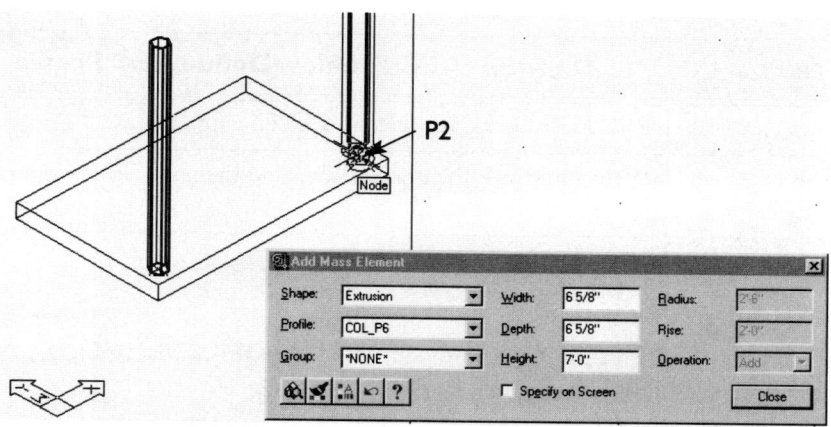

Figure 10.47 *Specifying the location for the second column*

The drawing should now appear as shown in Figure 10.48.

9. Save your drawing.

CREATING NEW PROFILES

In addition to using the profiles included in the Architectural Desktop, you can create your own profiles. Profiles are created from 2D closed polylines. They can be created for complex building components or for the footprint of a building. By creating a profile and extruding it, you can create the model more quickly than when using

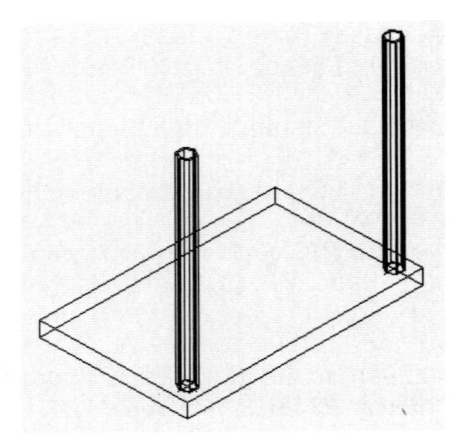

Figure 10.48 *Column mass elements inserted*

individual mass elements. The polyline can consist of straight lines or arcs. See Table 10.4 for **Define AEC Profile** command access.

Menu bar	Desktop>AEC Profiles>Define AEC Profile
Command prompt	PROFILEDEFINE; select the Define option
Shortcut	Right-click over the drawing area and then choose Utilities>Profiles>Add

Table 10.4 *Define AEC Profile command access*

When you select the **Define AEC Profile** command (**PROFILEDEFINE**), the **Profiles** dialog box opens, as shown in Figure 10.49.

The **Profiles** dialog box lists the profiles that have been created as well as any profiles that you create. The list consists of shapes used for columns, handrails, and basic polygons. The buttons to the right in the dialog box allow you create and modify profiles. A description of each button of the Profiles dialog box follows.

> **New** – Creates a new name for a profile. When you select the **New** button, the **Name** dialog box opens, allowing you to enter the name of a new profile, as shown in Figure 10.50.
>
> The name of the profile cannot exceed 31 characters and cannot include spaces. Names of profiles are displayed in alphabetical order.
>
> **Copy** – Copies the existing geometry of the profile and assigns a new name to the profile. To copy a profile, select a profile from the profile list and then

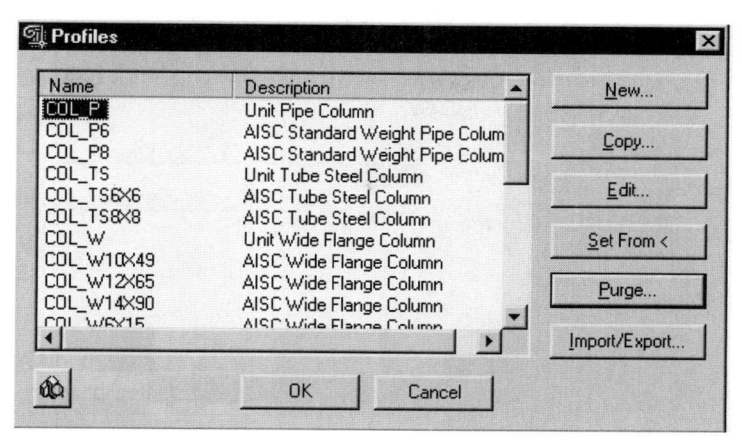

Figure 10.49 *Profiles dialog box*

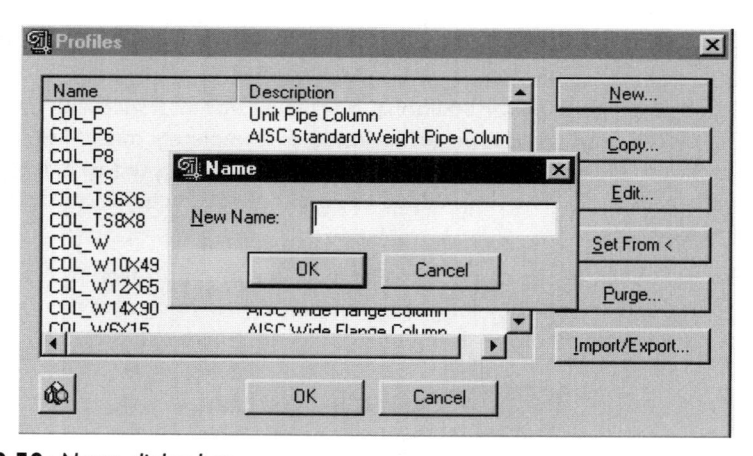

Figure 10.50 *Name dialog box*

select the **Copy** button. The **Name** dialog box opens, allowing you to assign a name to the new profile.

Edit – Allows you to edit the name and description of the profile. To edit a profile, select the profile from the profile list and then select the **Edit** button. The **Profile Definition Properties** dialog box opens, allowing you to edit the name and description of the profile, as shown in Figure 10.51.

The description for the profile is not required; however, it allows you to better identify the size and shape of the profile.

Set From< – Links the graphics for an AEC Profile to its name and description. When the **Set From<** button is selected, you are prompted to select a closed polyline. Because the polyline will be extruded or revolved, it must be closed. (A closed polyline has no gaps between line segments.) After you

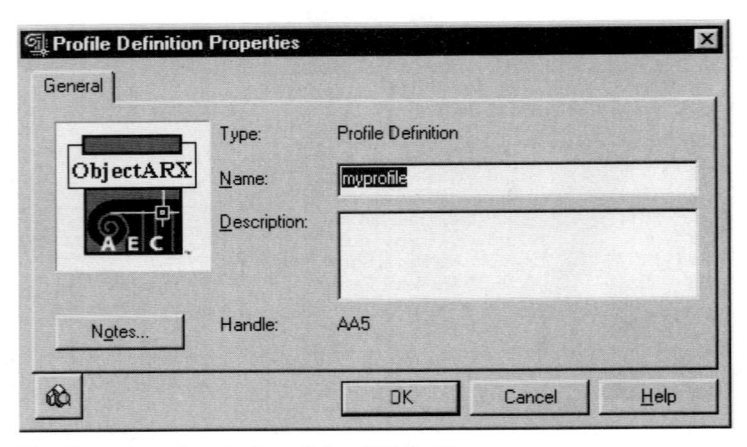

Figure 10.51 *Creating a description of the AEC Profile*

select the closed polyline, you are prompted Add another ring? If you add another ring, you are prompted to select another closed polyline. The closed polylines used for additional rings must not intersect with each other. The closed polylines can be either inside or completely outside the original polyline. If additional rings are added, you will be prompted to add them as voids or as solids. The rings added as voids are subtracted from the first shape.

Purge – Deletes the named profiles from the list of profiles. You can delete profiles that are not needed in the drawing. When you select the **Purge** button, the **Purge Profiles** dialog box opens, as shown in Figure 10.52.

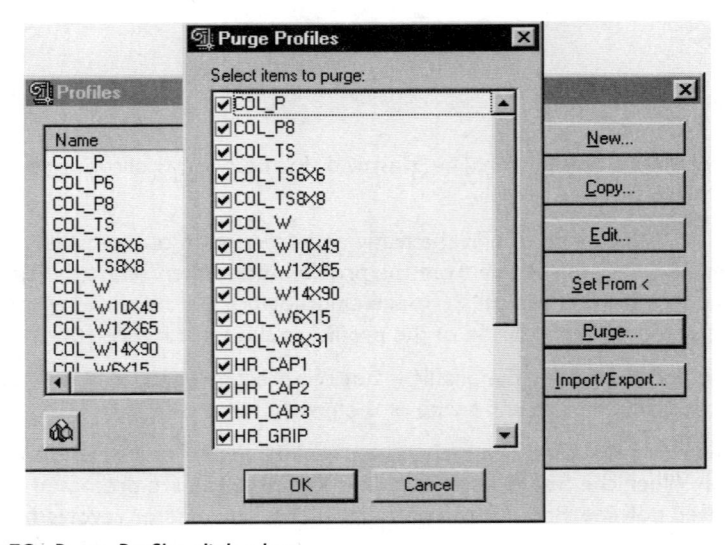

Figure 10.52 *Purge Profiles dialog box*

The dialog box lists all the profiles of the drawing with a check adjacent to the profile name. All profiles checked in the list will be removed from the file. If you clear the check from a profile in the list, that profile will not be purged. Only profiles that have not been used in the drawing will be listed in the **Purge** profile list. Profiles that are used in the drawing may not be purged; if profiles are purged at the end of the drawing, all profiles used in the drawing will be retained. You can restore the profiles of the Aec Arch (imperial) template to a drawing by importing profiles from the *\AutoCAD Architectural2\Content\Imperial\Styles\Profiles (Imperial).dwg*. The Import/Export button also allows you import profiles from other drawing files.

Import/Export – Imports or exports profiles to or from other drawing files. When you select the **Import/Export** button, the **Import/Export** dialog box opens, as shown in Figure 10.53.

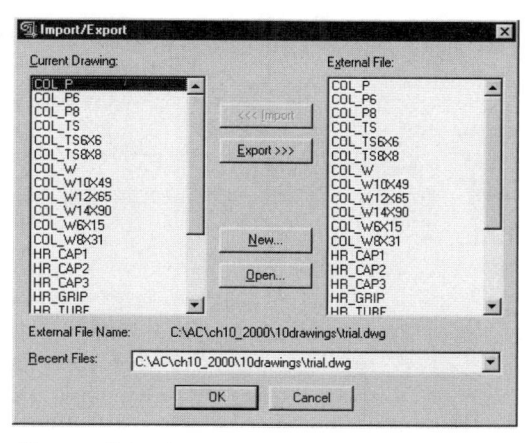

Figure 10.53 *Import/Export dialog box*

STEPS TO CREATING A NEW PROFILE

1. Draw a closed polyline.
2. Choose **Desktop>AEC Profiles>Define AEC Profile** from the menu bar to open the **Profiles** dialog box, as shown in Figure 10.54.
3. Select the **New** button to open the **Name** dialog box.
4. Type the name of the new profile in the **Name** dialog box.
5. Select **OK** to dismiss the **Name** dialog box.
6. Select the **Edit** button to open the **Profile Definition Properties** dialog box, as shown in Figure 10.55.

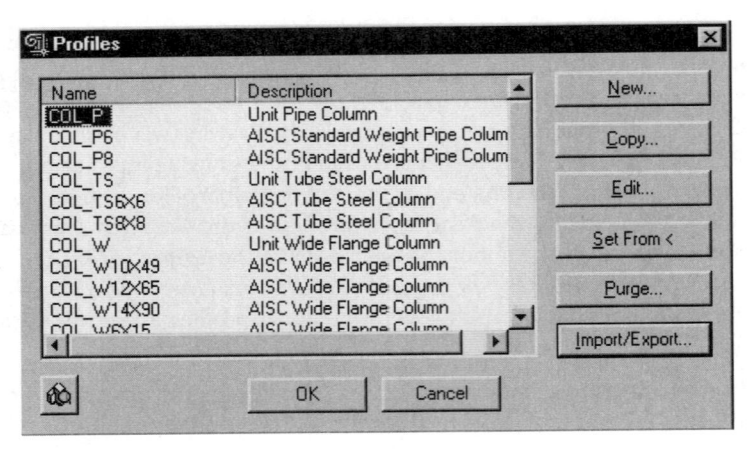

Figure 10.54 Profiles dialog box

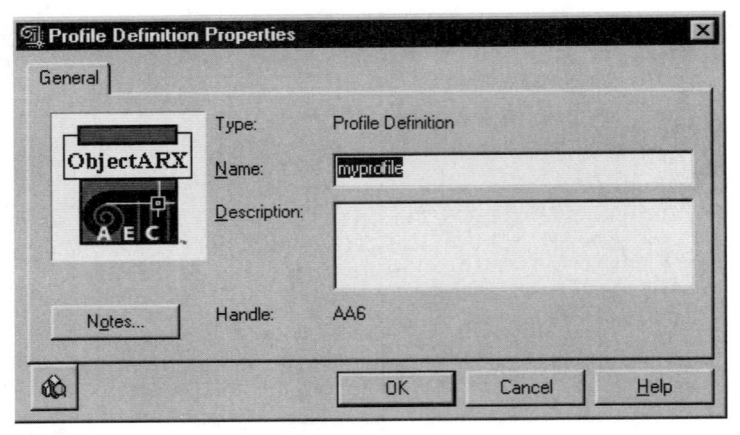

Figure 10.55 Creating a description for an AEC Profile

7. Type a description in the **Description** edit field, and then select **OK** to dismiss the **Profile Definition Properties** dialog box.

8. Select the **Set From<** button of the **Profiles** dialog box and then select the closed polyline, as shown in the following command sequence and in Figure 10.56.

```
Command: _AecProfileDefine
Select a closed polyline: 
```
(Select the polyline at P1 in Figure 10.56.)
```
Add another ring? [Yes/No] <N>: N ENTER
```
(Press ENTER to end selection and return to the Profiles dialog box.)

9. Select **OK** to dismiss the **Profiles** dialog box. The new profile can now be extruded or revolved to create a custom mass element.

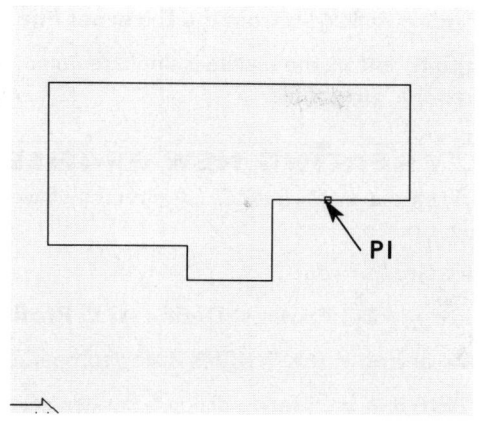

Figure 10.56 *Select a closed polyline*

Once a profile has been created for a drawing, it can be exported to other drawings, or the profiles of other drawings can be imported into the current drawing. Profiles can be used to create the outline of a floor plan or they can be individual components of a building.

STEPS TO EXPORTING A PROFILE

1. Choose **Desktop>AEC Profiles>Define AEC Profile** from the menu bar.
2. Select the **Import/Export** button of the **Profiles** dialog box.
3. Select the **Open** button of the **Import/Export** dialog box.
4. Select the *Architectural2\Content\Imperial\Styles\Profiles(Imperial).dwg* file from the **File to Import** dialog box.
5. Select **Open** to select the file and dismiss the **File to Import From** dialog box.
6. Select one or more wall styles from the **Current Drawing** list.
7. Select the **Export** button; the profile is exported from the current drawing to the external file.

STEPS TO IMPORTING A PROFILE

1. Choose **Desktop>AEC Profiles>Define AEC Profile** from the menu bar.
2. Select the **Import/Export** button of the **Profiles** dialog box.
3. Select the **Open** button and open the *Architectural2\Content\Imperial\Styles\Profiles (Imperial).dwg* file.
4. Select **Open** to select the file and dismiss the **Files to Import From** dialog box.

5. Select one or more wall styles from the **External File** list.
6. Select the **Import** button; the profile is imported from the **External Drawing** list to the current file.

TUTORIAL 10.4 CREATING NEW PROFILES

1. Open *Ex10-4* from the \ADT_Tutor\Ch10\ directory. Save the drawing as *Mass4* in your student directory.
2. Select the Mass-Group layout.
3. Choose **Desktop>AEC Profiles>Define AEC Profile** from the menu bar.
4. Select the **New** button in the **Profiles** dialog to open the **Name** dialog box.
5. Type **PILASTER_8** in the **Name** dialog box; then select **OK** to dismiss the **Name** dialog box.
6. Double-click on the name PILASTER_8 and enter the following description: **Footprint 8" Pilaster 32' by 81'** in the **Description** field.
7. Select **OK** to dismiss the **Profile Definition Properties** dialog box.
8. Select the **Set From<** button and respond as shown in the following command sequence and in Figure 10.57.

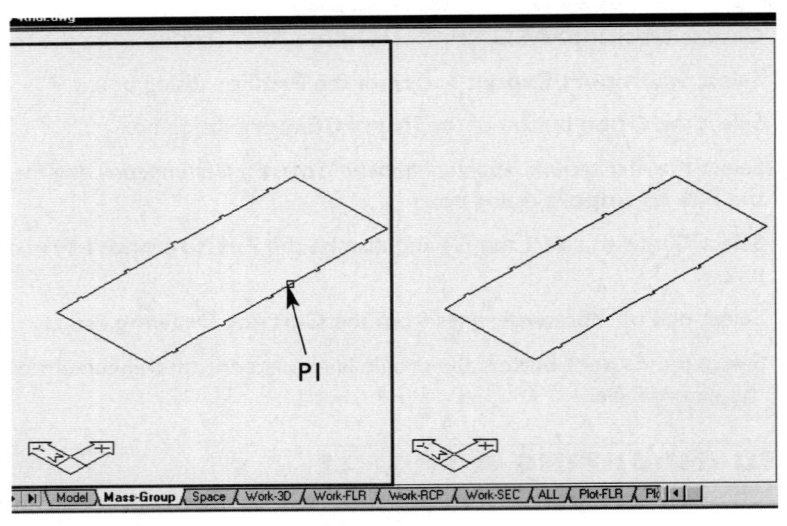

Figure 10.57 *Selecting the closed polyline*

```
Command: _AecProfileDefine
Select a closed polyline:
```
(Select the polyline at point P1 in Figure 10.57.)

Add another ring? [Yes/No] <N>: ENTER *(Press ENTER to end selection of objects and return to the Profiles dialog box.)*

9. Select **OK** to dismiss the **Profiles** dialog box.
10. Choose **Concept>Mass Elements>Add Mass Element** from the menu bar to open the **Add Mass Element** dialog box.
11. Select in the **Add Mass Element** dialog box and set the **Shape** to Extrusion and **Height** to 18'. Select PILASTER_8 from the **Profile** list.
12. Select in the left viewport to activate the drawing area.
13. Respond to the command line prompts as shown in the command sequence and in Figure 10.58 to place the Pilaster_8 mass element.

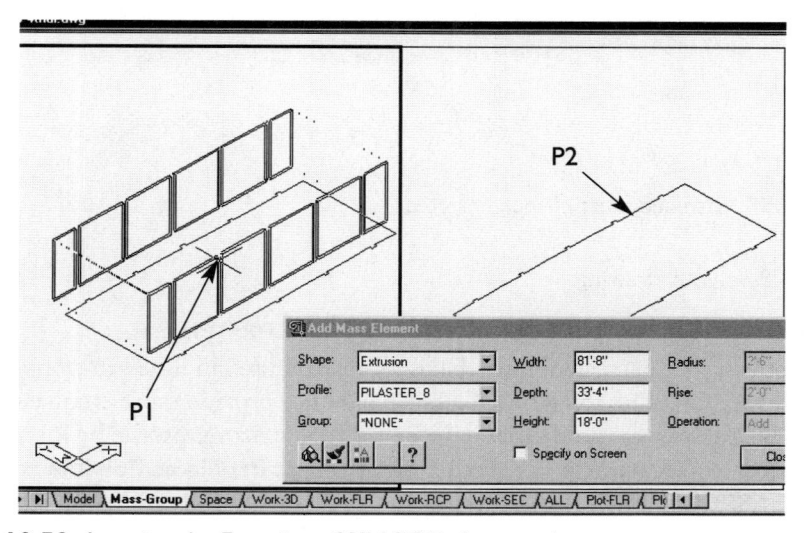

Figure 10.58 *Inserting the Extrusion of PILASTER_8 mass element*

```
Command: _AecMassElementAdd
Insert point or [SHape/WIdth/Depth/Height/Match]:
DBOX
Insert point or [SHape/WIdth/Depth/Height/Match]:
```
(Select a point near P1 in Figure 10.58.)
```
Rotation or [SHape/WIdth/Depth/Height/Match] <0.00>:
```
ENTER *(Press ENTER to accept the 0° rotation.)*
```
Insert point or
[SHape/WIdth/Depth/Height/Match/Undo]:
```
ENTER *(Press ENTER to end the command.)*

14. Select **Erase** from the **Modify** toolbar and then select the polyline used to create the profile as shown at P2 in Figure 10.58. The extruded profile is created as shown in Figure 10.59.

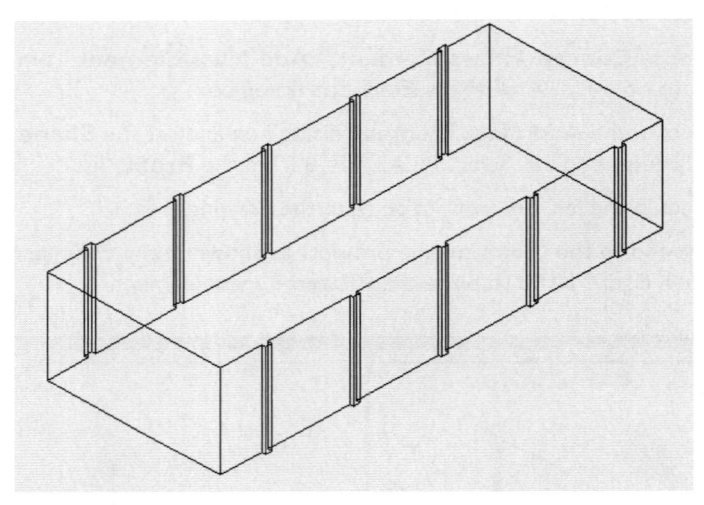

Figure 10.59 *Extruded mass element inserted*

15. Save your drawing.

CREATING NEW PROFILES FROM EXISTING PROFILES

Once a profile has been created, it can be modified and used in future drawings. A profile created for a building footprint can be changed to provide more emphasis to certain features. You can create the new profile from an existing profile by inserting the profile back in the drawing as a polyline. The **Insert Profile as Polyline** command (**PROFILEASPOLYLINE**) is used to insert a profile in a drawing as a polyline. See Table 10.5 for **Insert Profile as Polyline** command access.

Menu bar	Desktop>AEC Profiles>Insert Profile as Polyline
Command prompt	PROFILEASPOLYLINE or PROFILE; select Define option
Shortcut	Right-click over the drawing area, and choose Utilities>Profiles>Add and then select the Define option

Table 10.5 *Insert Profile as Polyline command access*

When you choose the **Desktop>AEC Profiles>Insert Profile as Polyline** command from the menu bar, the **Profile Definitions** dialog box opens, as shown in Figure 10.60.

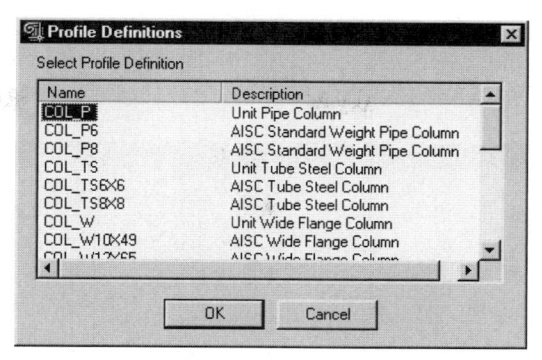

Figure 10.60 *Profile Definitions dialog box*

The **Profile Definitions** dialog box lists the profiles of the drawing. You can select a profile from the list, select **OK** to dismiss the **Profile Definitions** dialog box, and then specify the location in the drawing area. The location for the profile can be specified in the command line. The profile can then be edited and saved as a profile for future use.

IMPORTING PROFILES

Importing profiles allows you to use custom-developed profiles in future drawings. The ability to import and export profiles saves time because designs from one file can be transferred to another file. Importing and editing profiles allows designs to be quickly edited and used in future drawings.

Profiles can be imported to the *\AutoCAD Architectural\Content\Imperial\Profiles (Imperial).dwg* or to a drawing created in your directory. You can use profiles in the same manner as WBLOCKS to share designs across drawing files.

TUTORIAL 10.5 IMPORTING AND EXPORTING PROFILES

1. Open *Ex10-5* from the *\ADT_Tutor\Ch10* directory.
2. Choose **File>SaveAs** from the menu bar and save the drawing as *Mass5.dwg* in your student directory.
3. Choose **Desktop>AEC Profiles>Define AEC Profile** from the menu bar to open the **Profiles** dialog box.
4. Select the **Import/Export** button to open the **Import/Export** dialog box.
5. Select the **Open** button and select *Mass4* from your student directory in the **Files to Import From** dialog box. Select the **Open** button to select the file and close the **Files to Import From** dialog box.

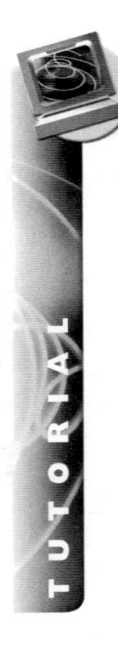

6. The **External File** list on the right should now display the profiles from the *Mass4.dwg* drawing. Scroll down the profile list and select PILASTER_8, as shown in Figure 10.61.

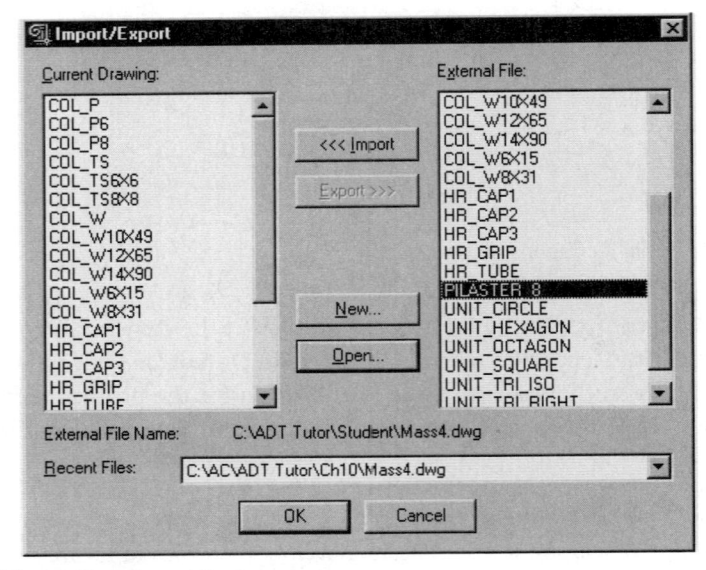

Figure 10.61 *Profile selected for import*

7. Select the **Import** button, and the PILASTER_8 profile is added to the **Current Drawing** list of profiles.
8. Select **OK** to dismiss the **Import/Export** dialog box
9. Select **OK** to dismiss the **Profiles** dialog box.
10. Choose **Desktop>AEC Profiles>Insert Profile as Polyline** from the menu bar to open the **Profile Definitions** dialog box.
11. Select PILASTER_8 from the **Profile Definitions** dialog box as shown in Figure 10.62.
12. Select **OK** to dismiss the **Profile Definitions** dialog box.
13. Select a location in the left viewport near point P1, as shown in Figure 10.62, to insert the profile as a polyline.
14. Select the Model layout tab.
15. Select Ortho in the Status bar to turn ON Ortho.
16. Select the **Stretch** command from the **Modify** toolbar and respond to the command line prompts as shown in the following command sequence and in Figures 10.63 and 10.64.

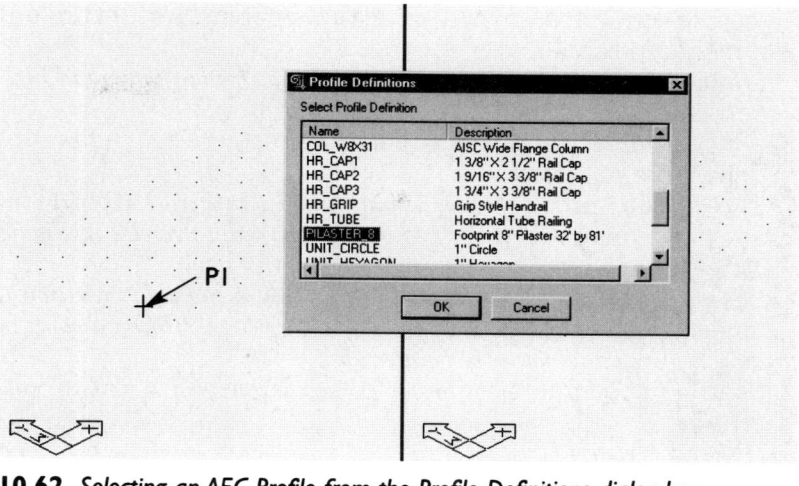

Figure 10.62 *Selecting an AEC Profile from the Profile Definitions dialog box*

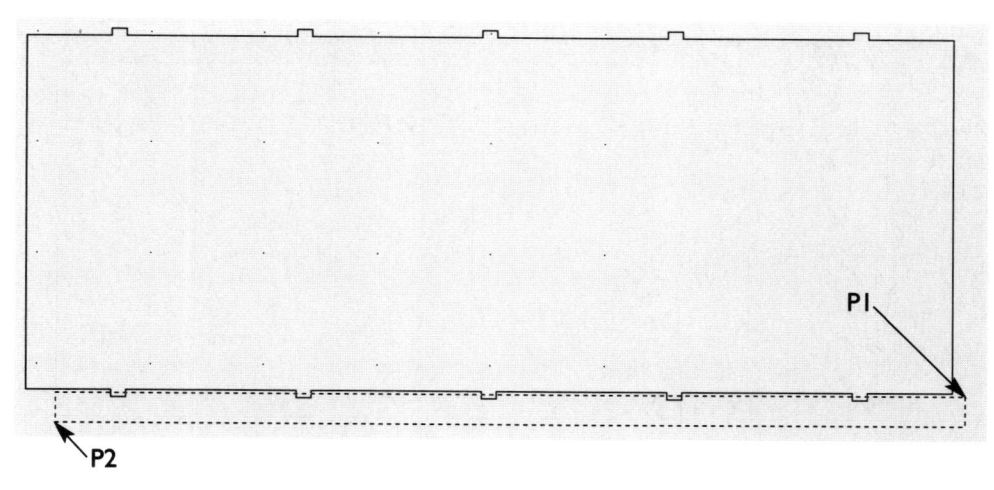

Figure 10.63 *Stretching the pilasters*

```
Command: _stretch
Select objects to stretch by crossing-window or
crossing-polygon...
Select objects: (Select point P1 in Figure 10.63.)
Specify opposite corner: (Select point P2 in Figure 10.63.) 1
found
Select objects: ENTER (Press ENTER to end object selection.)
Specify base point or displacement: (Select a point in the
drawing area and move the pointer down.)
```

```
Specify second point of displacement: <Ortho on> 8
ENTER (Pilaster stretched 8")
Command:   ENTER  (Press ENTER to repeat the last command.)
Command: _stretch
Select objects to stretch by crossing-window or
crossing-polygon...
Select objects: (Select a point near P1 in Figure 10.64.)
Specify opposite corner: (Select a point near P2 in Figure
10.64.) 1 found
Select objects: ENTER  (Press ENTER to end object selection.)
Specify base point or displacement: (Select a point in the
drawing area and move the pointer up.)
Specify second point of displacement: <Ortho on> 8
ENTER
```
(Pilaster stretched 8".)

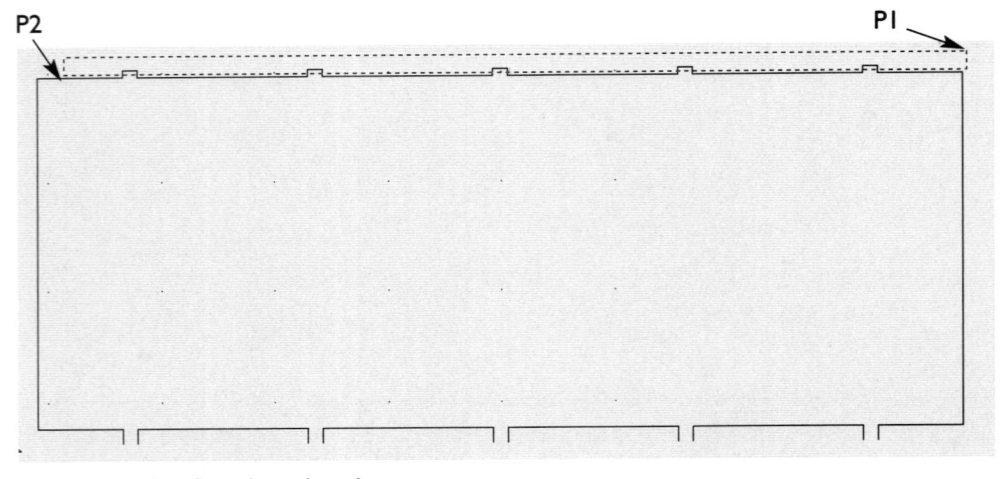

Figure 10.64 *Stretching the pilasters up*

17. Choose **Desktop>AEC Profiles>Define AEC Profile** from the menu bar to open the **Profiles** dialog box.

18. Select the **New** button and type **PILASTER_16** in the **Name** dialog box. Select **OK** to dismiss the **Name** dialog box.

19. Double-click over the PILASTER_16 profile name to open the **Profile Definition Properties** dialog box and then type **Footprint 16" Pilaster 32' by 81'** in the **Description** edit field.

20. Select **OK** to dismiss the **Profile Definition Properties** dialog box.

21. Select the **Set From<** button in the **Profiles** dialog box and respond as shown in the command sequence and in Figure 10.65.

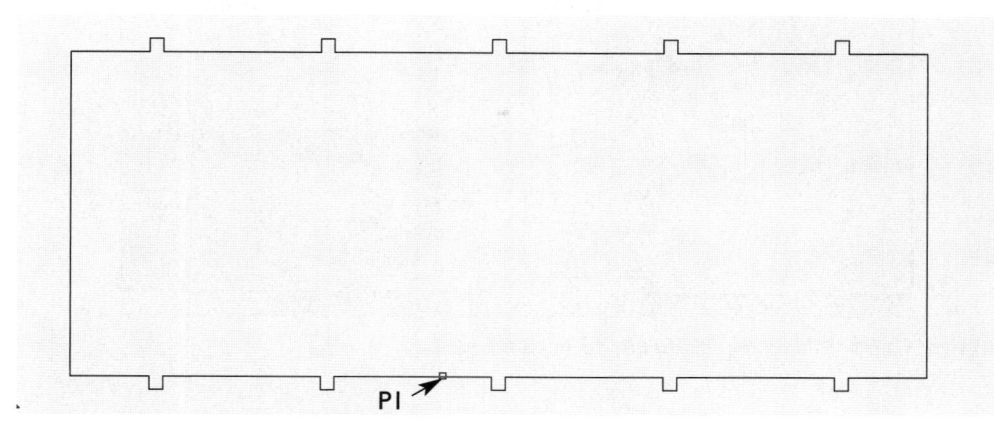

Figure 10.65 *Defining a new AEC Profile from a polyline*

```
Command: _AecProfileDefine
Select a closed polyline: (Select the polyline at point P1 in
Figure 10.65.)
Add another ring? [Yes/No] <N>: ENTER (Press ENTER to end
selection of objects.)
```

22. Select **OK** to dismiss the **Profiles** dialog box.
23. Select the Mass-Group layout.
24. Choose **Concept>Mass Elements>Add Mass Element** from the menu bar to open the **Add Mass Element** dialog box.
25. Select in the **Add Mass Element** dialog box and edit the **Shape** to Extrusion, **Profile** to PILASTER_16, **Width** to 50', **Depth** to 50', **Height** to 18' as shown in Figure 10.66.
26. Select in the left viewport to activate the drawing area and then respond to the command line prompts as shown in the following command sequence.

```
Command: _AecMassElementAdd
Insert point or
[SHape/WIdth/Depth/Height/RIse/Match]: DBOX
Insert point or [SHape/WIdth/Depth/Height/Match]:
DBOX

Insert point or [SHape/WIdth/Depth/Height/Match]:
```
(Insert a location near point P1 in Figure 10.66.)

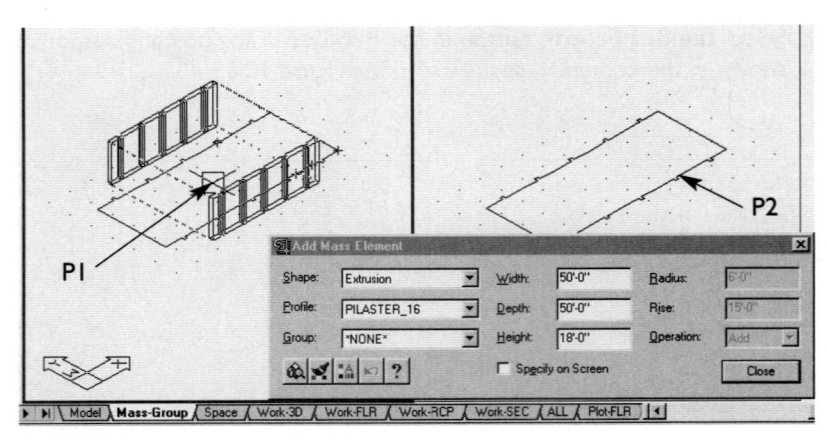

Figure 10.66 *Adding the PILASTER_16 mass element*

```
Rotation or [SHape/WIdth/Depth/Height/Match] <0.00>:
ENTER (Press ENTER to accept 0° rotation.)
Insert point or
[SHape/WIdth/Depth/Height/Match/Undo]: ENTER (Press
ENTER to end the command.)
```

27. Select the **Erase** command from the **Modify** toolbar, and then select the polyline at P2 as shown in Figure 10.66. The extruded mass element is shown in Figure 10.67.
28. Save the drawing.

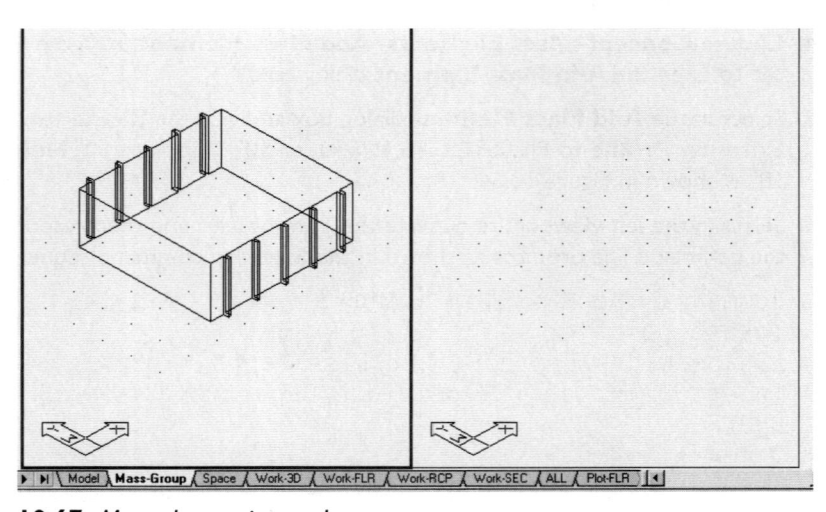

Figure 10.67 *Mass element inserted*

CREATING MASS ELEMENTS TO REPRESENT BUILDING COMPONENTS

Mass elements can be created to represent building components in working drawings. Because mass elements are three dimensional, they can be inserted in a drawing to represent building components. If you create a mass element to represent floor and ceiling systems, the mass element will enhance a floor plan when it is inserted in the model drawing. To create a mass element representing the floor or ceiling system, you need to extract the plan dimensions from a 2D floor plan and specify the height of the mass element according to the thickness of the floor or ceiling structural system. The **Hidden Line Projection** command can be used to extract the 2D drawing of the floor plan. The area representing the stairwell or plumbing chases, which extend from floor to floor, must be subtracted from the overall plan to accurately represent the floor system.

TUTORIAL 10.6 CREATING A MASS ELEMENT FOR A FLOOR SYSTEM

1. Open *Ex10-6* from the *\ADT_Tutor\Ch10* directory.
2. Choose **File>SaveAs** from the menu bar and save the file as *Mass6* to your student directory.
3. Select the Model layout tab.
4. Choose **Desktop>AEC Utilities>Hidden Line Projection** from the menu bar.
5. Respond to the command line prompts as shown in the following command sequence and in Figure 10.68.

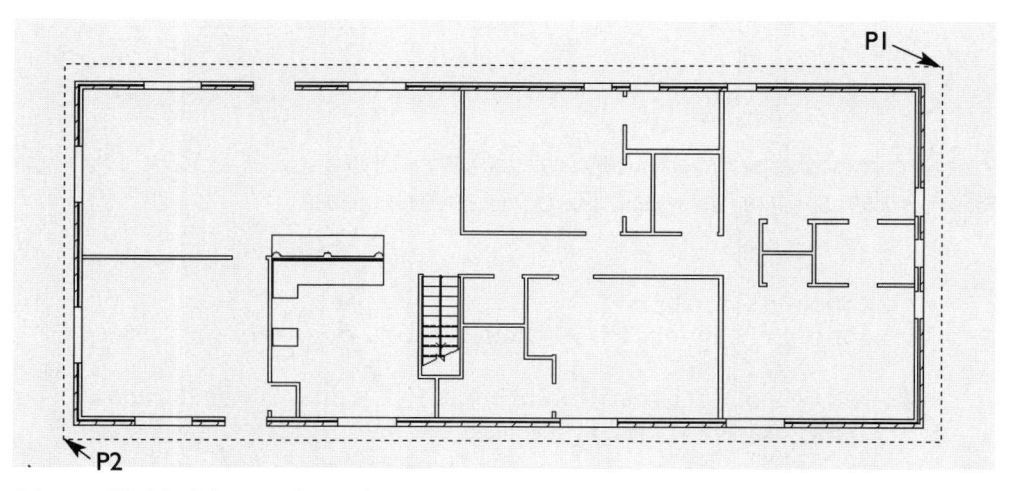

Figure 10.68 *Selecting objects for the Hidden Line Projection command*

```
Command: _AecCreateHLR
Select objects: (Select a point from P1 to P2 in Figure 10.68.)
Specify opposite corner: 35 found
Select objects: ENTER (Press ENTER to end object selection.)
Block insertion point: 125',10' ENTER (Specify the location
for the 2D block.)
Insert in plan view [Yes/No] <Y>: ENTER (Specify the 2D
block created in plan view.)
```

6. Select **Zoom Extents** from the **Standard** toolbar to view the floor plan and the 2D block, as shown in Figure 10.69.

7. Select **Layers** from the **Object Properties** toolbar to open the **Layer Properties Manager** dialog box.

8. Select the **New** button and type **Profile** as the name of the new layer, and then select the **Current** button.

9. Select **OK** to dismiss the **Layer Properties Manager** dialog box.

10. Select **Zoom Window** from the **Standard** toolbar and respond to the command line prompts as shown in the following command sequence.

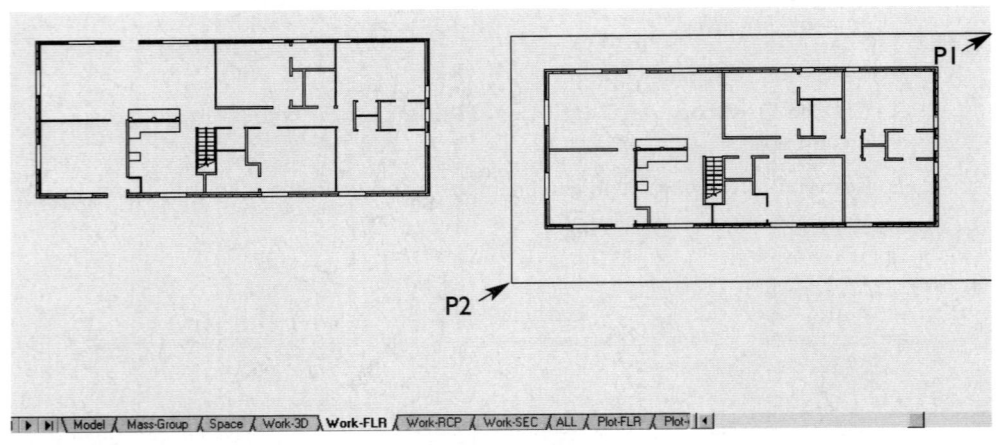

Figure 10.69 *Specifying the window size for the ZOOM command*

```
Command: '_zoom
Specify corner of window, enter a scale factor (nX
or nXP), or
[All/Center/Dynamic/Extents/Previous/Scale/Window]
<real time>: _w
Specify first corner: (Select a point near point P1 in Figure
10.69.)
```

Specify opposite corner: *(Select a point near point P2 in Figure 10.69.)*

11. Select the **Rectang** command from the **Draw** toolbar and respond to the command line prompts as shown in the following command sequence.

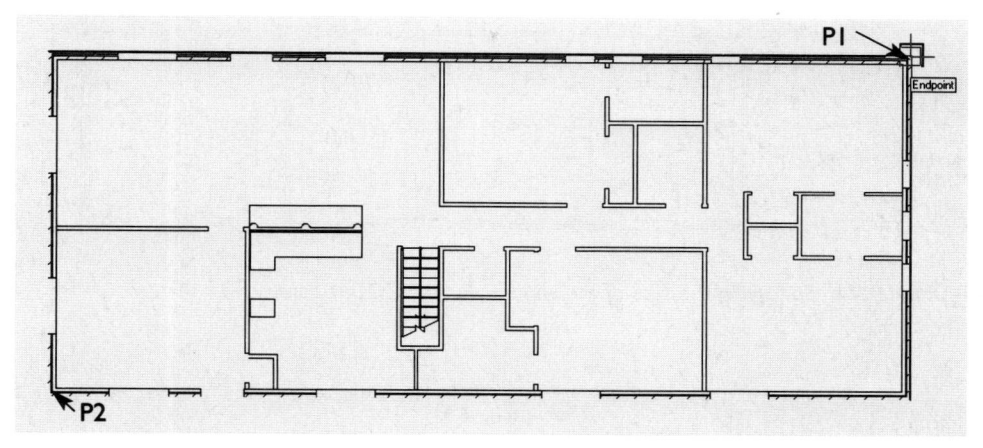

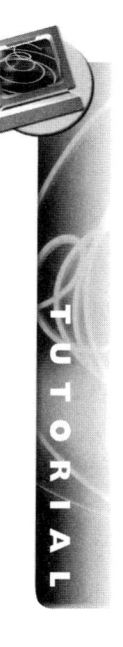

Figure 10.70 *Select the exterior surface of the corner of the building.*

```
Command: _rectang
Specify first corner point or
[Chamfer/Elevation/Fillet/Thickness/Width]: end ENTER
Of
```
(Select the outside corner of the brick veneer at point P1 in Figure 10.70.)
```
Specify other corner point: end ENTER
Of
```
(Select the diagonal corner of the brick veneer wall at point P2 in Figure 10.70.)
```
Command:
```

12. Select **Zoom Window** from the **Standard** toolbar and specify the window size as shown in the following command sequence.

```
Command: '_zoom
Specify corner of window, enter a scale factor (nX
or nXP), or
[All/Center/Dynamic/Extents/Previous/Scale/Window]
<real time>: _w
Specify first corner:
```
(Select point P1 in Figure 10.71.)
```
Specify opposite corner:
```
(Select point P2 in Figure 10.71.)

13. Select the **Rectang** command from the **Draw** toolbar and respond to the command line prompts as shown in the following command sequence.

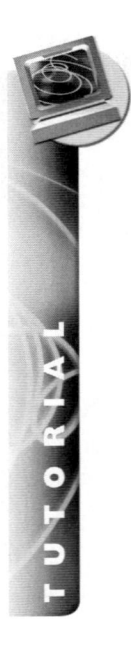

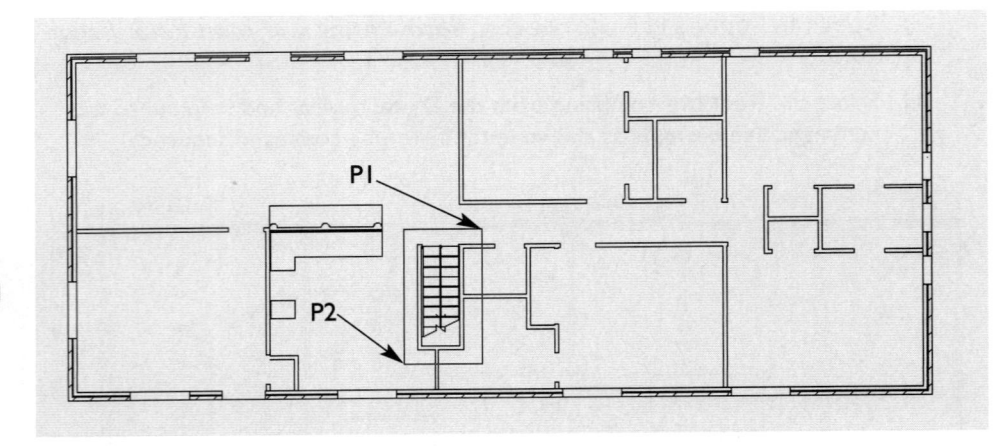

Figure 10.71 *Specifying the size of the window for the ZOOM command*

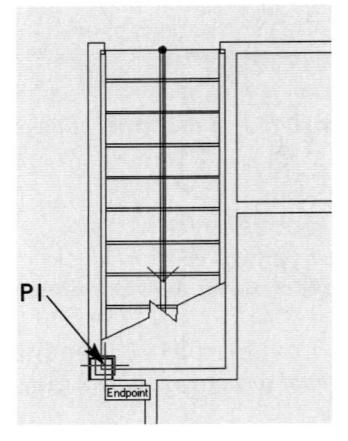

Figure 10.72 *Specifying a corner of the stairwell*

```
Command: _rectang
Specify first corner point or
[Chamfer/Elevation/Fillet/Thickness/Width]: end ENTER
Of (Select the corner of the stairwell at point P1 in Figure 10.72.)
Specify other corner point: int ENTER
Of (Select the diagonal corner of the stairwell at point P2 in Figure 10.73.)
Command:
```

14. Select **Layers** from the **Object Properties** toolbar and freeze all layers except the current layer. Select **OK** to dismiss the **Layer Properties Manager**.
15. Select **Zoom Extents** from the **Standard** toolbar.

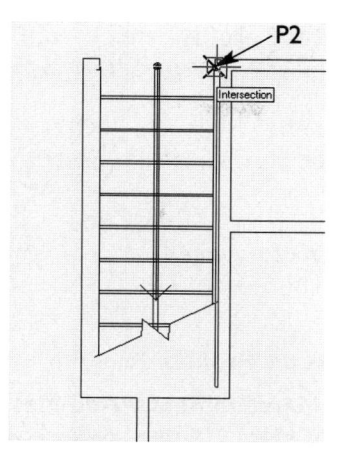

Figure 10.73 *Specifying a corner of the stairwell*

16. Choose **Desktop>AEC Profiles>Define Aec Profile** from the menu bar to open the **Profiles** dialog box.
17. Select the **New** button in the **Profiles** dialog box to open the **Name** dialog box. Type **Floor** in the **Name** dialog box.
18. Select **OK** to dismiss the **Name** dialog box.
19. Select the **Edit** button and type **Floor system for model** in the **Description** edit field.
20. Select the **OK** button to dismiss the **Profile Definition Properties** dialog box.
21. Select the **Set From<** button and respond to the command prompts as shown in the following command sequence and in Figure 10.74.

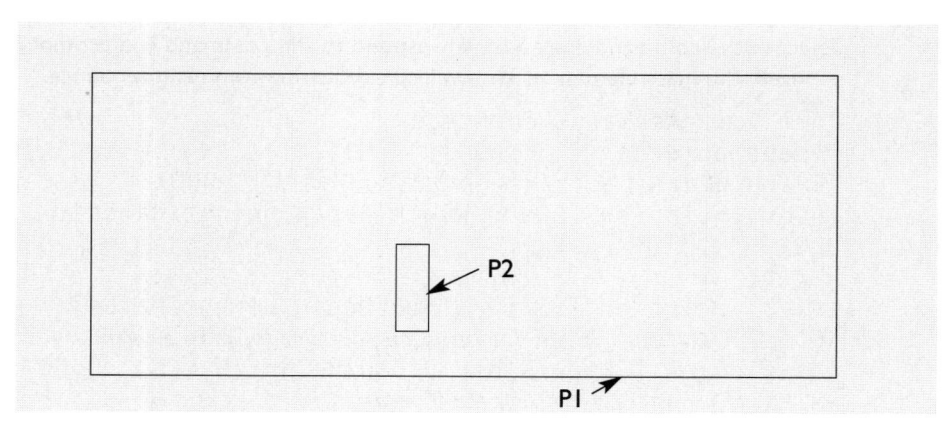

Figure 10.74 *Polylines for the AEC Profile*

```
Command: _AecProfileDefine
Select a closed polyline: (Select the polyline at point P1 in
Figure 10.74.)
Add another ring? [Yes/No] <N>: y ENTER
Select a closed polyline: (Select the rectangle at point P2 in
Figure 10.74.)
Ring is a void area? [Yes/No] <Y>: ENTER (Press ENTER to
create a void in the first polyline.)
Add another ring? [Yes/No] <N>: ENTER (Press ENTER to end
the command.)
```

22. Select **OK** to dismiss the **Profiles** dialog box.
23. Choose **Concept>Mass Elements>Add Mass Element** from the menu bar to open the **Add Mass Element** dialog box.
24. Select in the **Add Mass Element** dialog box and edit the **Shape** to Extrusion, **Profile** to Floor, and **Height** to 11" as shown in Figure 10.75. (Note: the **Height** is set to 11" because it represents the floor system for the building.)

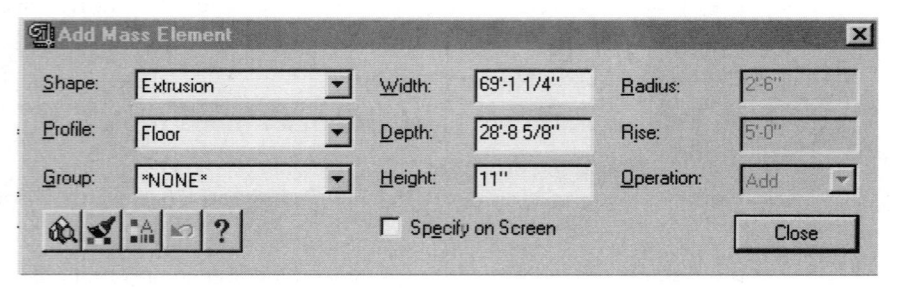

Figure 10.75 *Add Mass Element dialog box*

25. Select in the command line and then respond to the command line prompts to insert the mass element as shown in the following command sequence.

```
Command: _AecMassElementAdd
Insert point or
[SHape/WIdth/Depth/Height/RIse/Match]: DBOX
Insert point or [SHape/WIdth/Depth/Height/Match]:
DBOX

Insert point or [SHape/WIdth/Depth/Height/Match]:
50',10' ENTER (Specify the insertion point for the mass element.)
Rotation or [SHape/WIdth/Depth/Height/Match] <0.00>:
ENTER (Press ENTER to accept the 0° rotation.)
Insert point or
```

[SHape/WIdth/Depth/Height/Match/Undo]: ENTER *(Press ENTER to end the command.)*
26. Select **Layers** from the **Object Properties** toolbar, and then thaw the A-Wall layer. Select **OK** to dismiss the **Layer Properties Manager**.
27. Select **Zoom Extents** from the **Standard** toolbar.
28. Select the **SWIsometric** view from the **Standard** toolbar.
29. Select **Move** from the **Modify** toolbar and move the mass element to the exterior walls, as shown in the following command sequence and in Figure 10.76.

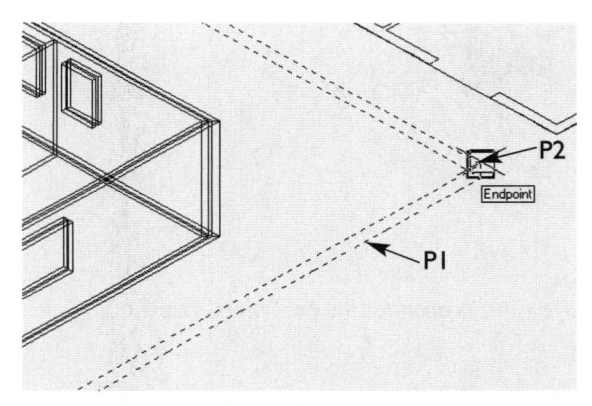

Figure 10.76 *Selecting the basepoint for moving the floor mass element*

```
Command: m MOVE
Select objects: (Select the mass element at point P1 in Figure
10.76.) 1 found
Select objects:   ENTER (Press ENTER to end the object selection.)
Specify base point or displacement: end ENTER (Specify
the End object snap mode.)
of (Select the upper corner of the mass element at point P2 in Figure
10.76.)
Specify second point of displacement or <use first
point as displacement>: end ENTER
Of (Select the lower right corner of the building at point P3 in Figure 10.77.)
```

30. Enter **HI** in the command line to view the model as shown in Figure 10.78.
31. Save the drawing.

Note: The Mass6 drawing can now be inserted as a reference drawing in the Model drawing developed in Chapter 9 to develop updated elevations and sections.

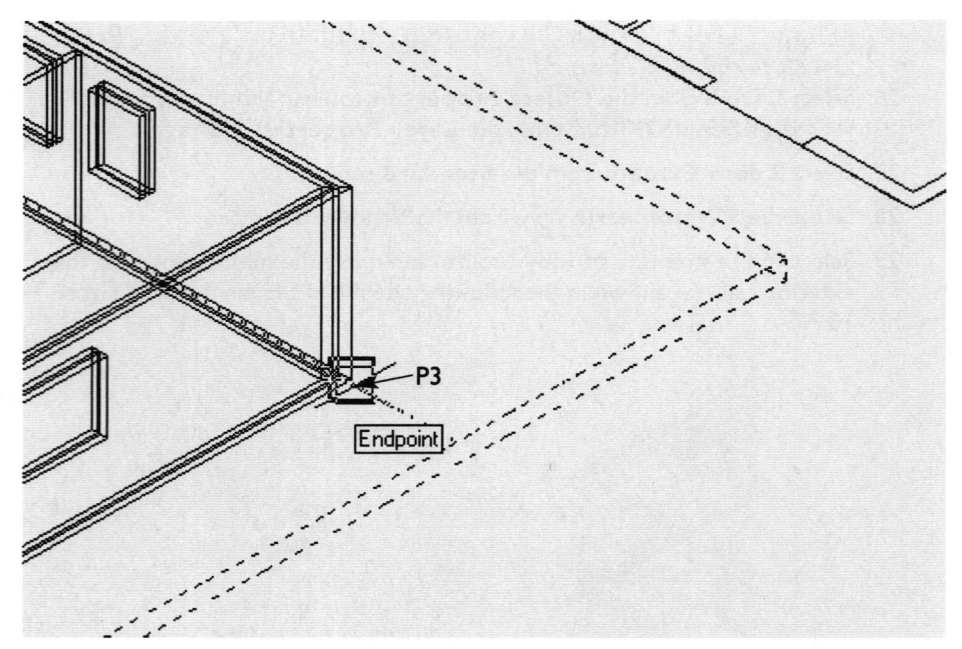

Figure 10.77 *Specifying the location for the floor mass element*

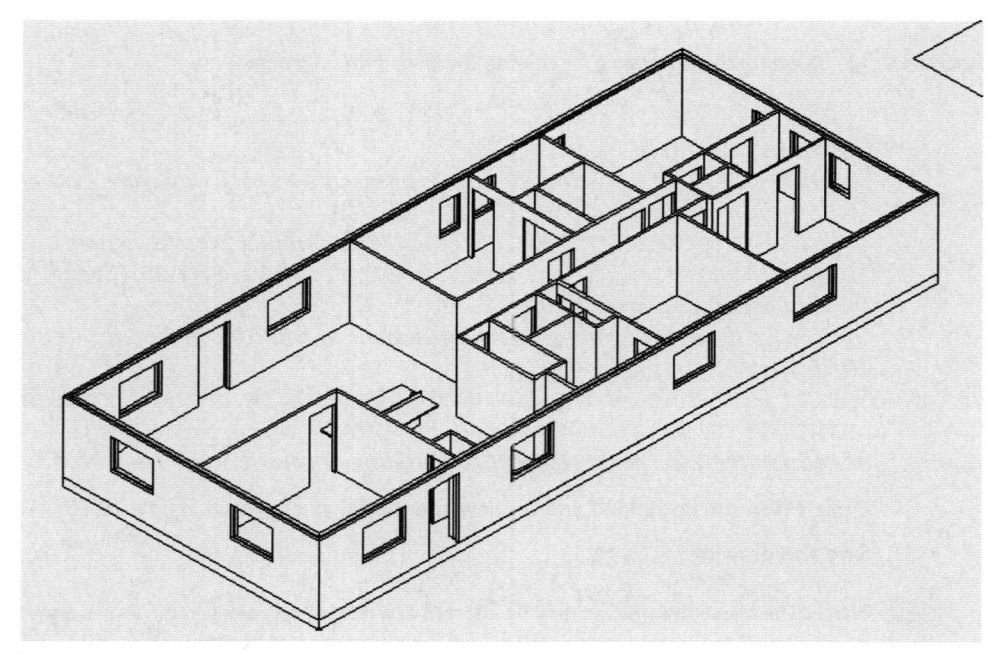

Figure 10.78 *Floor mass element moved below walls*

SUMMARY

1. The Mass-Group layout tab is used to develop mass models.
2. Mass elements are created with the **Add Mass Element** command (**MASSELEMENTADD**).
3. Mass elements are placed on the A-Mass layer.
4. AEC Profiles can be extruded or revolved to create a mass element.
5. Mass elements can be modified by grips or the **Modify Mass Element** command (**MASSELEMENTMODIFY**).
6. AEC Profiles are created from closed 2D polylines through the **Define AEC Profile** command (**PROFILEDEFINE**).
7. The **Insert Profile as Polyline** command (**PROFILEASPOLYLINE**) can be used to insert profiles in a drawing as polylines.
8. Profiles can be imported or exported across drawings.

REVIEW QUESTIONS

1. The insertion point or handle for inserting mass elements is located _____.
2. The rise of the Gable mass element is the distance from_____.
3. A Doric Column can be created by _____.
4. Three properties of a mass element changed by the **Modfy Mass Element** command are:_____,_____, and _____.
5. The rotation of a mass element is changed by the _____command.
6. The layout tab for creating mass models is _____.
7. The layout tab for viewing mass groups is _____.
8. AEC Profiles are converted to polylines by _____.
9. AEC Profiles should be transferred to other drawings _____.
10. Mass elements are placed on the _____layer.
11. What command is used to create an AEC Profile?
12. Which viewport of the Mass-Group layout tab displays the mass elements?_____
13. What is the ring of the AEC Profile?
14. How is a void created in the AEC Profile development?
15. Should a mass element be exploded to modify its size?
16. What are the Boolean operations that can be applied to mass elements that belong to a group?
17. List the mass element shapes that you can create.

PROJECT

EX10.1 CREATING A FLOOR SYSTEM USING MASS ELEMENTS

1. Open *Ex10-7* from *\ADT_Tutor\Ch10* directory.
2. Save the drawing to your student directory as *Proj10-7*.
3. Freeze the following layers: A-Anno-Dims, A-Anno-Legn, A-Area, A-Area-Iden, A-Flor-Strs, A-Glaz, A-Glaz-Iden.
4. Create a 2D block using the **Hidden Line Projection** command to represent the floor.
5. Create an AEC Profile from the 2D block named Floor1.
6. Create a mass element from the AEC Profile and extrude it 6" high to represent the basement concrete floor system.
7. Move the mass element below the walls of the first floor as shown in Figure 10.79.

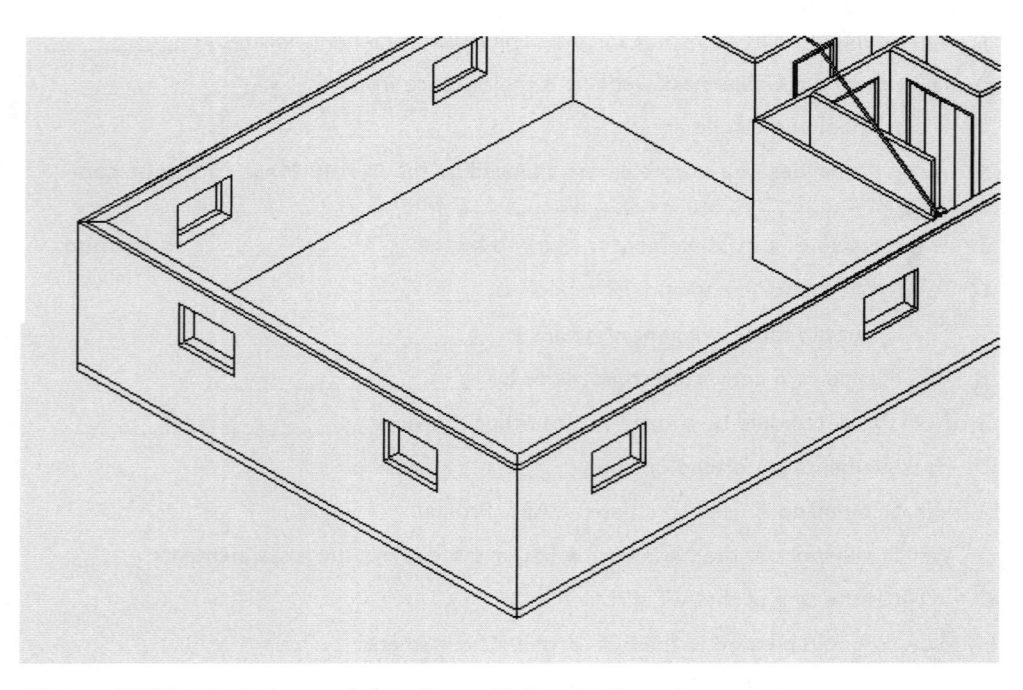

Figure 10.79 *Mass element for the floor added to the floor plan.*

CHAPTER 11

Creating Complex Models

INTRODUCTION

In the previous chapter, you created and edited mass elements. Most models usually include several mass elements that create the model when combined. In this chapter, you will create individual mass elements and use Boolean operations to add, subtract, or intersect the mass elements. The purpose of each mass element inserted is to add, subtract or intersect with other mass elements. The mass elements carve out or project from a base mass element to create the model. Each element of the model is created on the XY plane and extruded up in the positive Z direction. In this chapter, you will move mass elements up or down in the Z direction to create a more complex model. The model shown in Figure 11.1 consists of two Gable mass elements, four Box elements, one Arch, and one Barrel Vault. Most of these mass elements were moved up in the positive Z direction to create the model. Three of the mass elements were used to subtract from the remaining elements of the model.

For a model to be designed using various mass elements, a group must be defined. The use of groups allows you to explore various design options and manipulate mass elements according to their group. This chapter will also include the use of the Model Explorer to create a model. The Model Explorer includes a Tree and graphic display tools that enhance model construction. Perspective views of the model can be captured in the Model Explorer and saved as bitmap files for presentations. Finally, this chapter will include the development of floorplate slices of the model.

OBJECTIVES

After completing this chapter, you will be able to

Use the **Add Mass Group** command (**AECMASSGROUPADD**) to create a mass group

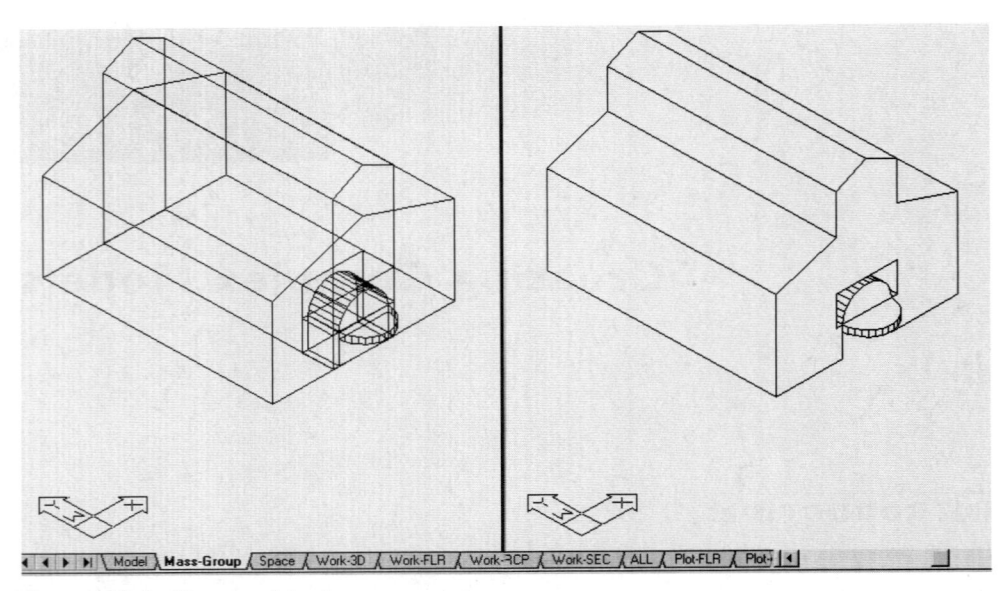

Figure 11.1 *Mass model using mass groups*

Use **Attach Element** (**MASSGROUPATTACH**) and **Detach Element** (**MASSGROUPDETACH**) to attach or detach mass elements to or from the mass group

Combine mass elements of a mass group using Boolean additive, subtractive, and intersect operations

Create and edit mass elements and mass groups using the Model Explorer

Create mass models with referenced mass elements using the **Reference Element/Group** command (**ENTREF**)

Create floorplate Slices using the **Generate Slice** command (**AECSLICECREATE**)

Attach and detach mass elements to or from a slice using the **Attach Objects** and **Detach Objects** commands (**AECSLICEATTACH** and **AECSLICEDETACH**)

Move mass elements in the X, Y, Z directions to create a model

Convert a slice to a polyline using the **AECSLICETOPLINE** command

CREATING GROUPS FOR MASS ELEMENTS

For complex models to be created, a group must be defined and the mass elements attached to that group. When the mass element is created, its group can be defined in the **Group** list of the **Add Mass Element** dialog box. However, existing mass elements can be modified and defined to a group. Defining a group and performing operations on the group are accomplished easily through the **Mass Tools** toolbar. The commands on the **Mass Tools** toolbar are shown in Figure 11.2.

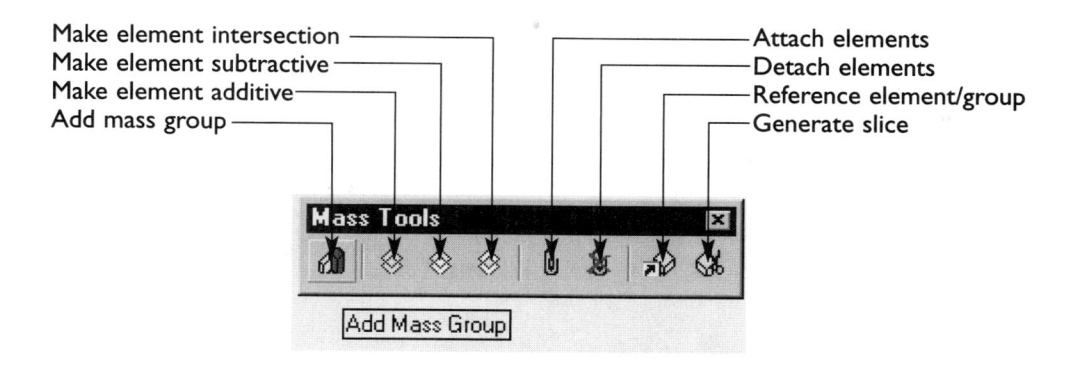

Figure 11.2 *Commands on the Mass Tools toolbar*

The **Mass Tools** toolbar includes the following commands: **Add Mass Group**, **Make Element Additive**, **Make Element Subtractive**, **Make Element Intersection**, **Attach Elements**, **Detach Elements**, **Reference Element/Group**, and **Generate Slice**.

The **Add Mass Group** command creates a named group to attach mass elements. See Table 11.1 for **Add Mass Group** command access.

Menu bar	Concept>Mass Grouping Tools>Add Mass Group
Command prompt	MASSGROUPADD
Toolbar	Select Add Mass Group from the Mass Tools toolbar as shown in Figure 11.2
Shortcut	Right-click over the drawing area; then choose Utilities>Massing>Mass Group>Add

Table 11.1 *Add Mass Group command access*

When you select the **Add Mass Group** command (**MASSGROUPADD**), you are prompted to select the location and rotation of a mass group marker, as shown in the command sequence and in Figure 11.3.

```
Command: _AecMassGroupAdd
```

```
Location: (Select a location in the right viewport of the Mass-Group
layout.)
Rotation angle <0.00>: ENTER (Press ENTER to accept 0° rota-
tion.)
```

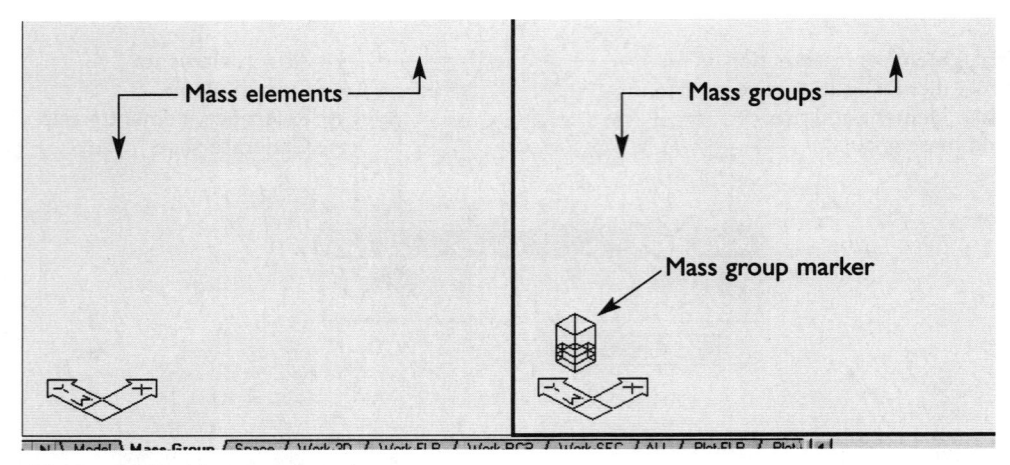

Figure 11.3 *Mass group marker*

The mass group marker in Figure 11.3 is magnified for display. The command sequence above does not prompt you to assign a name for the group. Architectural Desktop will assign a name to the mass group, such as 4A8. The assigned name is the AutoCAD handle name. AutoCAD handle names exist for all entities created in AutoCAD; they are not intended to describe the entity. To aid in modeling, names for mass elements or groups can be assigned in the Model Explorer. The procedure for assigning names to the components in the Model Explorer is discussed later in this chapter.

The mass group marker is used to graphically display the presence of a group. The exact location of the marker is not important. You can insert the marker at any location in the right viewport of the Mass-Group layout. The display configuration for the right viewport of the Mass-Group layout is defined to display mass groups and not mass elements. The group marker is placed on the A-Mass-Grps layer; the color of the layer is 130 (a hue of cyan).

ADDING MASS ELEMENTS TO A GROUP

You can assign mass elements to a group when they are created or you can assign their group definition after insertion by modifying the properties of the mass element: select the group from the **Group** list in the **Add Mass Element** dialog box, shown in Figure 11.9.

USING ATTACH ELEMENTS TO ADD MASS ELEMENTS TO A GROUP

If a mass element exists, you can assign it to a group by selecting the **Modify Mass Element** command and specifying the group from the **Group** list of the **Modify Mass Element** dialog box. You can also assign mass elements to a group by the **Attach Elements** command (**MASSGROUPATTACH**). See Table 11.2 for **Attach Elements** command access.

Menu bar	Concept>Mass Grouping Tools>Attach Elements
Command prompt	MASSGROUPATTACH
Toolbar	Select Attach Elements from the Mass Tools toolbar as shown in Figure 11.4
Shortcut	Right-click over the drawing area; then choose Utilities>Massing>Mass Group>Attach Elements
Shortcut	Select Mass Group Marker or a Mass Group; then right-click and choose Attach Elements or Add New Element

Table 11.2 *Attach Elements command access*

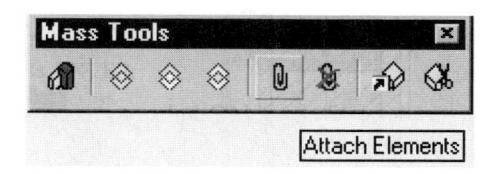

Figure 11.4 *Attach Elements command on the Mass Tools toolbar*

When you select **Attach Elements** from the **Mass Tools** toolbar, you are prompted to select a mass group and then select mass elements to be assigned to the mass group. The following command sequence and Figure 11.5 illustrate the attachment of a mass element to a group.

```
Command: _AecMassGroupAttach
Select a mass group: (Select the mass group marker at point P1 in
the right viewport in Figure 11.5.)
Select elements to attach: (Select a mass element at point P2 in
the left viewport in Figure 11.5.) 1 found
Select elements to attach: ENTER (Press ENTER to end the
command.)
Attaching elements...
1 element(s) attached.
(Element added as shown in Figure 11.6.)
```

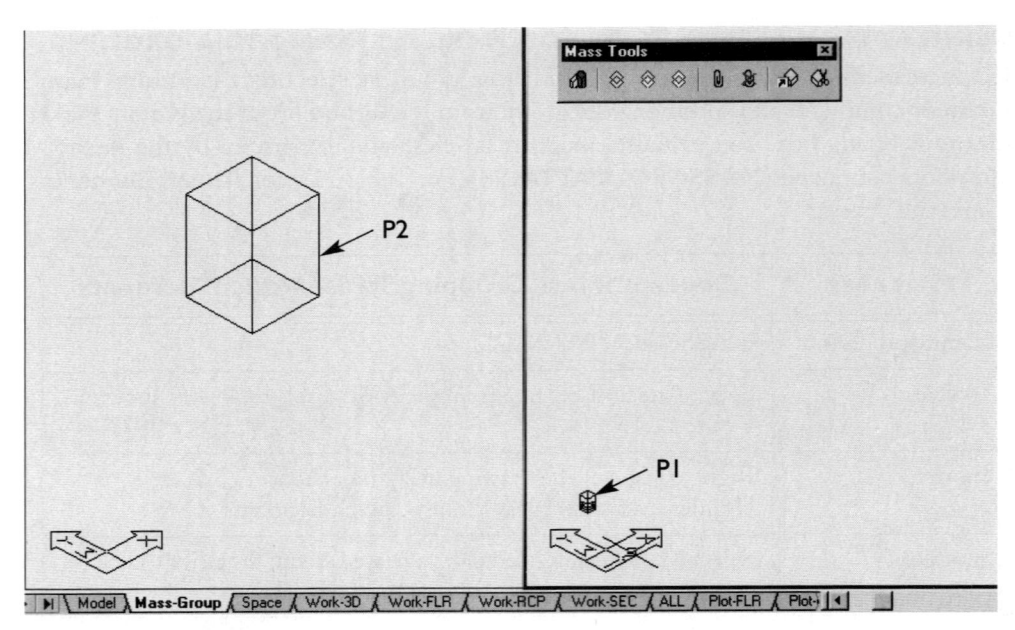

Figure 11.5 *Mass Group Marker*

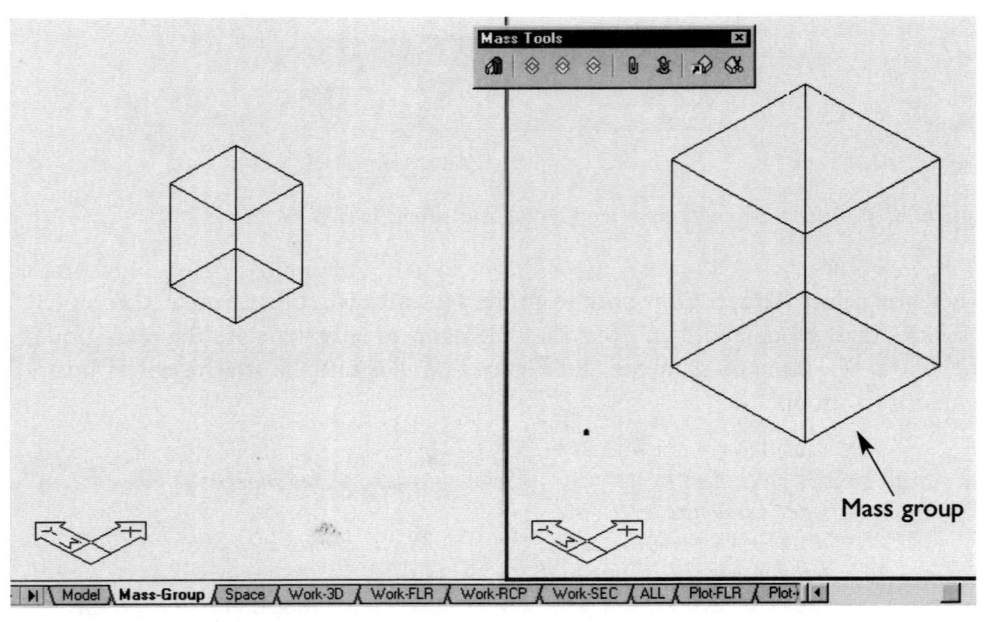

Figure 11.6 *Mass element added to the mass group*

By default, when mass elements are added to a group, they add their mass to existing mass elements of the group. The mass elements combine to form one object, adding their volume and geometry to other mass elements. Mass elements can also be defined to subtract or intersect their mass with other mass elements of a group. These Boolean operations can be changed as the model develops. Several mass groups can be defined for the drawing according to the need of the model.

DETACHING ELEMENTS FROM A MASS GROUP

Elements that have been attached to a group can be detached from the group. Detached mass elements are not erased from the drawing. The mass element no longer functions in the group to add, subtract or intersect its mass with other elements of the group; therefore, the selected mass element will not be shown in the right viewport of the Mass-Group layout. See Table 11.3 for **Detach Elements** command access.

Menu bar	Concept>Mass Grouping Tools>Detach Elements
Command prompt	MASSGROUPDETACH
Toolbar	Select Detach Elements from the Mass Tools toolbar as shown in Figure 11.7
Shortcut	Right-click over the drawing area; then choose Utilities>Massing>Mass Group>Detach Elements
Shortcut	Select a Mass Group Marker or a Mass Group; then right-click and choose Detach Elements

Table 11.3 *Detach Elements command access*

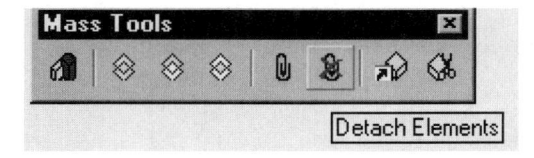

Figure 11.7 *Detach Elements command on the Mass Tools toolbar*

When you select **Detach Elements** (**MASSGROUPDETACH**) from the **Mass Tools** toolbar, you are prompted to select a group and the mass elements to detach from the group, as shown in the following command sequence.

```
Command: _AecMassGroupDetach
Select a mass group: (Select a mass group marker or mass group.)
Select elements to detach: (Select a mass element that is in the
selected group.) 1 found
```

Select elements to detach: ENTER *(Press ENTER to end the command.)*
Detaching elements...
1 element(s) removed.

TUTORIAL 11.1 CREATING MASS ELEMENTS WITH GROUPS

1. Open *Ex11-1* from the *\ADT_Tutor\Ch11* directory.
2. Choose **File>SaveAs** from the menu bar and save the file as *Group1.dwg* in your student directory.
3. Right-click over any Architectural Desktop toolbar and choose **Mass Tools** from the toolbar list.
4. Select the **Add Mass Group** command from the **Mass Tools** toolbar.
5. Respond to the command line prompts as shown in Figure 11.8 and in the command sequence below.

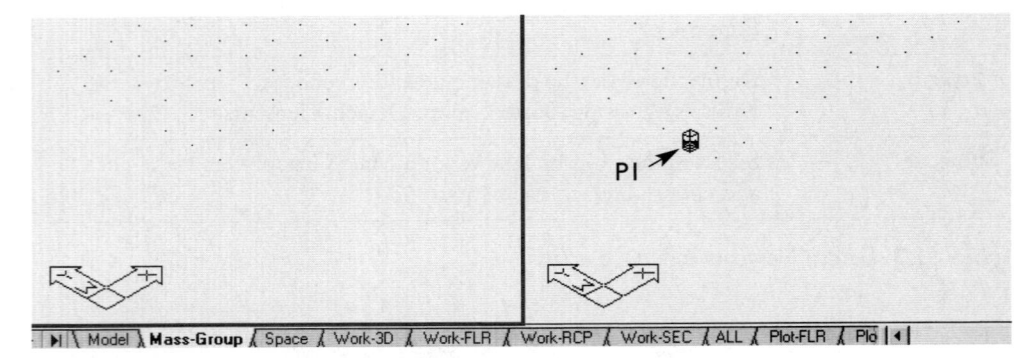

Figure 11.8 *Location for the mass group marker*

```
Command: _AecMassGroupAdd
```
(Select in the right viewport to activate the viewport.)
Location: *(Select a location near point P1 in Figure 11.8.)*
Rotation angle <0.00>: ENTER *(Press ENTER to end the command.)*

6. Select the **Gable** command from the **Mass Elements** toolbar to open the **Add Mass Element** dialog box.
7. Select in the **Add Mass Element** dialog box and edit the **Width** to 26', **Depth** to 50', **Height** to 44', and **Rise** to 10'. Set the group from None to the group name listed in the **Group** list as shown in Figure 11.9. (Note: the group name listed varies across drawings.)

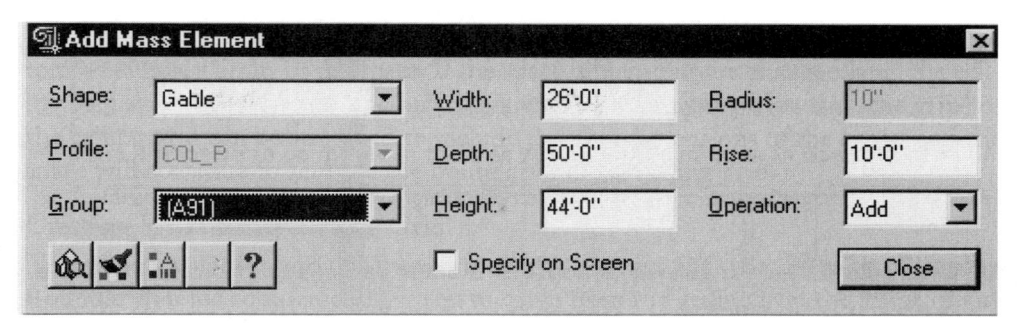

Figure 11.9 *Settings in the Add Mass Element dialog box*

8. Respond to the command line prompts as shown in Figure 11.10 and the command sequence below.

   ```
   Command: _AecMassElementAdd
   First Corner or [SHape]: _shape
   Shape [BOx/Arch/BArrel
   vault/DORic/COne/CYlinder/DOMe/Gable/Isoc
   triangle/Right
   triangle/Pyramid/Sphere]: _gable
   First Corner or [SHape]: DBOX
   ```
   (Select the left viewport to activate the viewport.)
   ```
   Insert point or
   [SHape/WIdth/Depth/Height/RIse/Match]:
   ```
   (Select a point near point P1 in Figure 11.10.)
   ```
   Rotation or [SHape/WIdth/Depth/Height/RIse/Match]
   <0.00>: ENTER
   ```
   (Press ENTER *to accept 0° rotation.)*
   ```
   Insert point or
   [SHape/WIdth/Depth/Height/RIse/Match/Undo]: ENTER
   ```
   (Press ENTER *to end the command.)*

9. Right-click over the OTRACK in the Status Bar, and then choose **Settings** from the shortcut menu to open the **Drafting Settings** dialog box. Verify that **Endpoint** and **Midpoint** object snaps are checked and that **Object Snap On (F3)** and **Object Snap Tracking On (F11)** are checked, as shown in Figure 11.11.

10. Select **OK** to dismiss the **Drafting Settings** dialog box.

11. Select the **Gable** command from the **Mass Elements** toolbar.

12. Select in the **Add Mass Element** dialog box and edit the **Width** to 4', **Depth** to 8', **Height** to 43', and **Rise** to 3'. Set the **Group** to None as shown in Figure 11.12.

13. Respond to the command line prompts to place the mass element as shown in the command sequence below and in Figure 11.13.

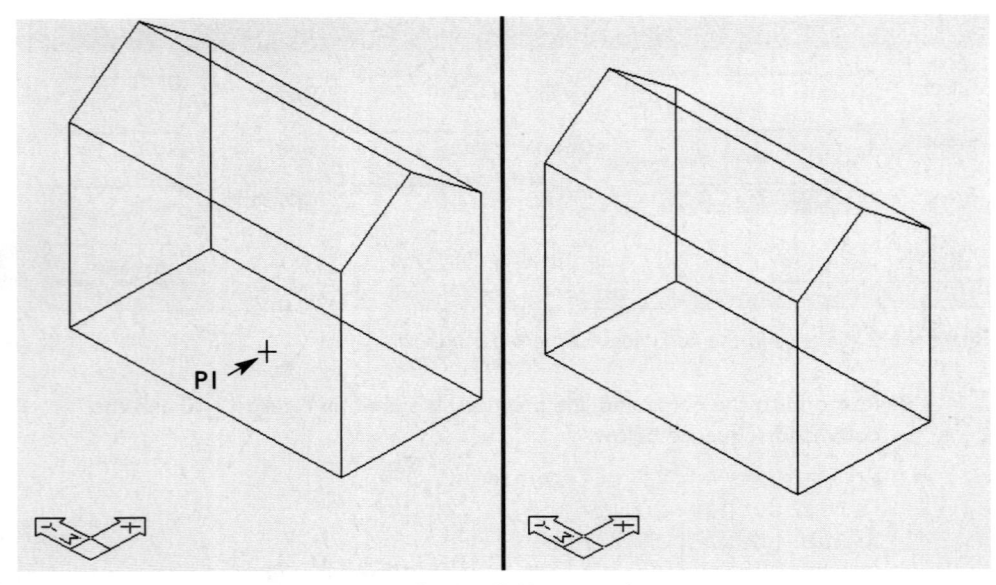

Figure 11.10 *Selecting the location for the Gable mass element*

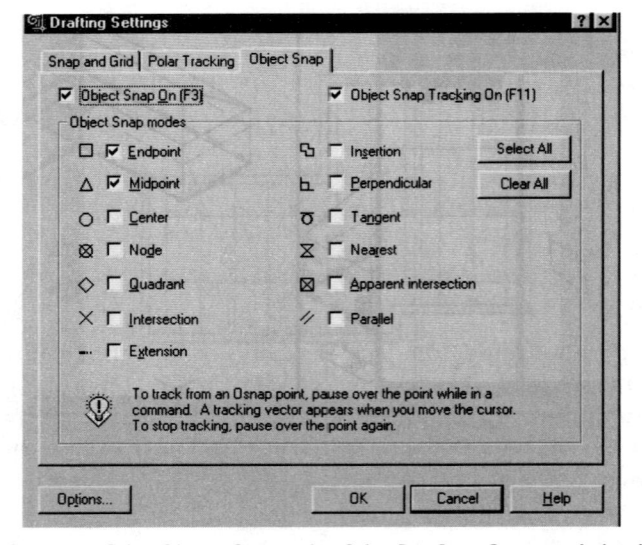

Figure 11.11 *Settings of the Object Snap tab of the Drafting Settings dialog box*

```
Command: _AecMassElementAdd
First Corner or [SHape]: _shape
Shape [BOx/Arch/BArrel
vault/DORic/COne/CYlinder/DOMe/Gable/Isoc
triangle/Right
```

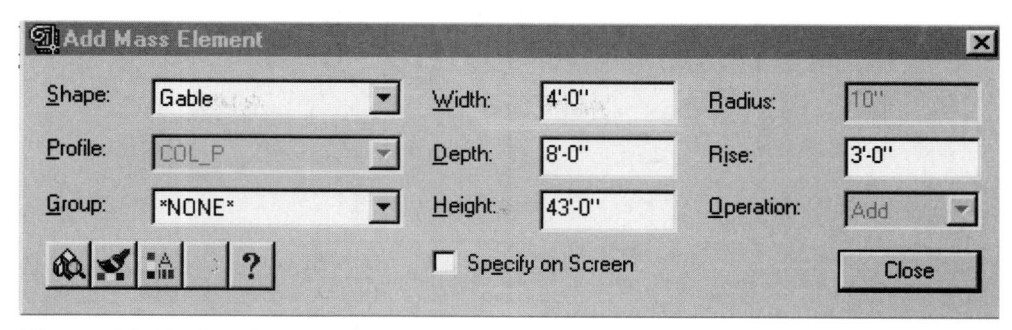

Figure 11.12 *Specifying the size of the dormer mass element*

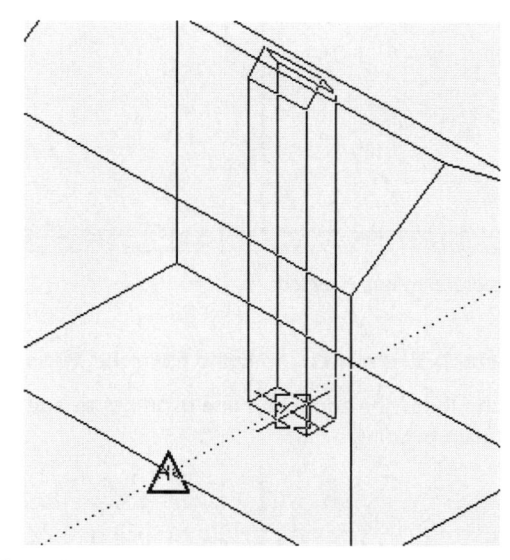

Figure 11.13 *Specifying the dormer location using object tracking*

```
                triangle/Pyramid/Sphere]: _gable
                First Corner or [SHape]: DBOX
                (Select the left viewport to activate the viewport.)
                (Move the pointer near the midpoint of the front wall of the building and
                pause until the Midpoint object is activated, and then move the pointer
                toward the rear along the extension as shown in Figure 11.13.)
                Insert point or
                [SHape/WIdth/Depth/Height/RIse/Match]: 6' ENTER
                Rotation or [SHape/WIdth/Depth/Height/RIse/Match]
                <0.00>: 90 ENTER (Press ENTER to specify a 90° rotation.)
                Insert point or
                [SHape/WIdth/Depth/Height/RIse/Match/Undo]: ENTER
                (Press ENTER to end the command.)
```

14. Notice that the dormer does not appear in the right viewport. Your screen should now appear as shown in Figure 11.14.

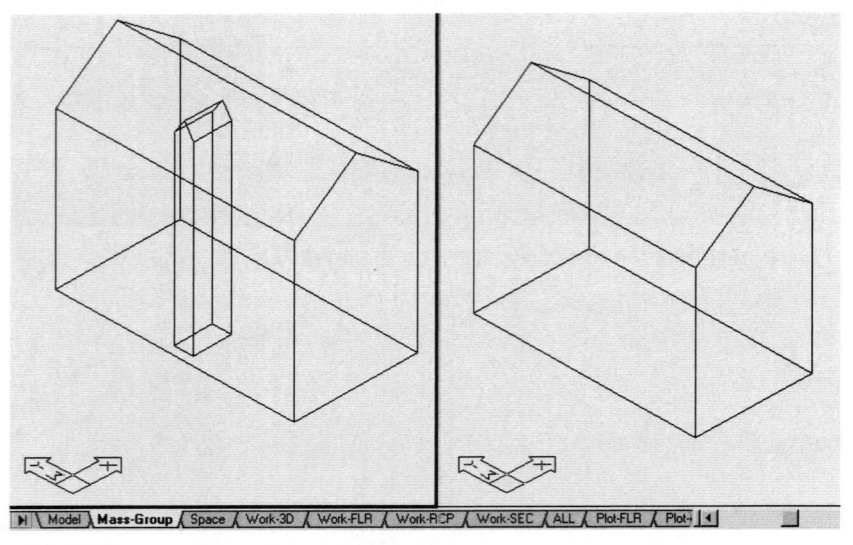

Figure 11.14 *Dormer mass element inserted*

15. Select the **Attach Elements** command from the **Mass Tools** toolbar.
16. Respond to the following command line prompts to add the dormer mass element to the main building.

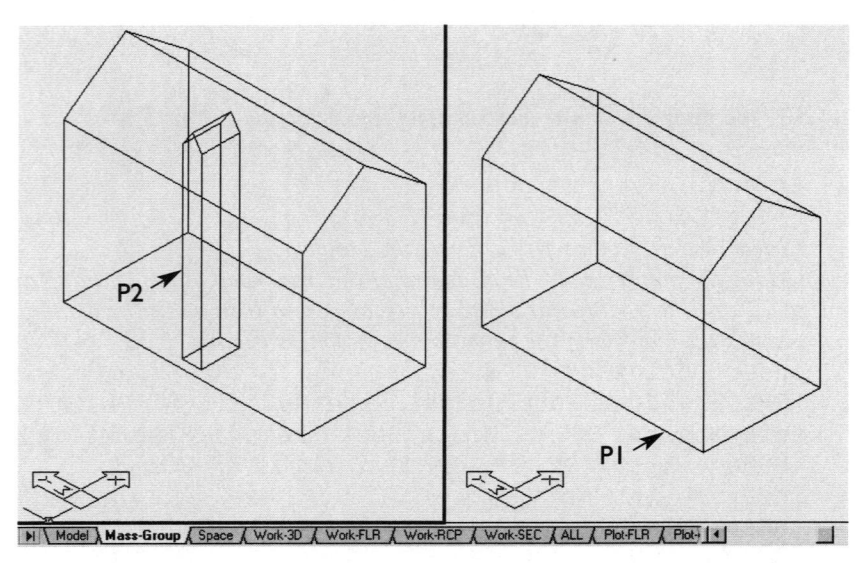

Figure 11.15 *Creating a mass group*

(Select the right viewport to activate the viewport.)
Command: _AecMassGroupAttach
Select a mass group: *(Select the building at point P1 in the right viewport in Figure 11.15.)*
(Select the left viewport to activate the left viewport.)
Select elements to attach: *(Select the dormer Gable mass element at point P2 in left viewport in Figure 11.15.)* 1 found
Select elements to attach: ENTER
Attaching elements...
1 element(s) attached.
(Gable mass element attached to the main building as shown in Figure 11.16.)

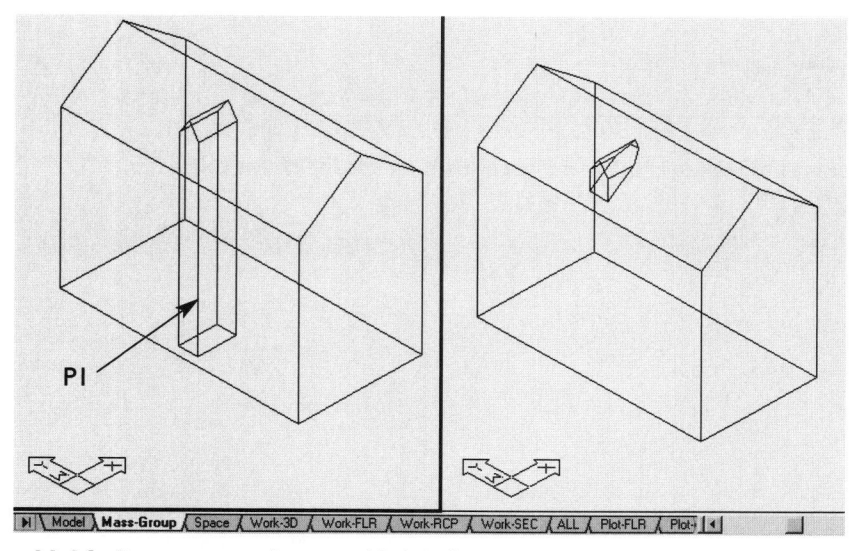

Figure 11.16 *Dormer mass element added to the mass group*

17. Verify that the left viewport is active.
18. Select the Gable mass element at point P1 in Figure 11.16; then right-click and choose **Element Modify** from the shortcut menu.
19. Select the **Modify Mass Element** dialog box and edit the **Depth** to 11'.
20. Select the **OK** button to execute the change and dismiss the **Modify Mass Element** dialog box. The edited group, shown in the right viewport, should appear as shown in Figure 11.17.
21. Select the right viewport; then enter **HI** to execute the **HIDE** command to display the group, as shown in Figure 11.18.

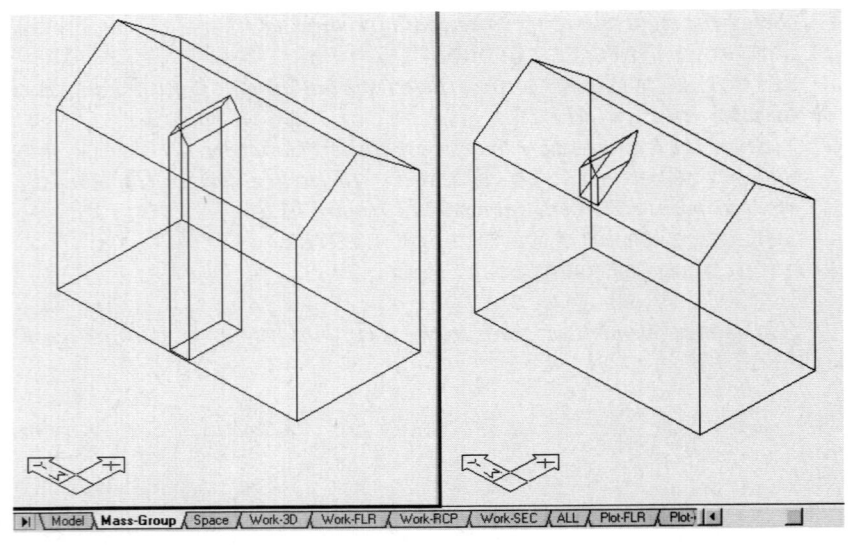

Figure 11.17 Changes in the mass group when the mass element changes

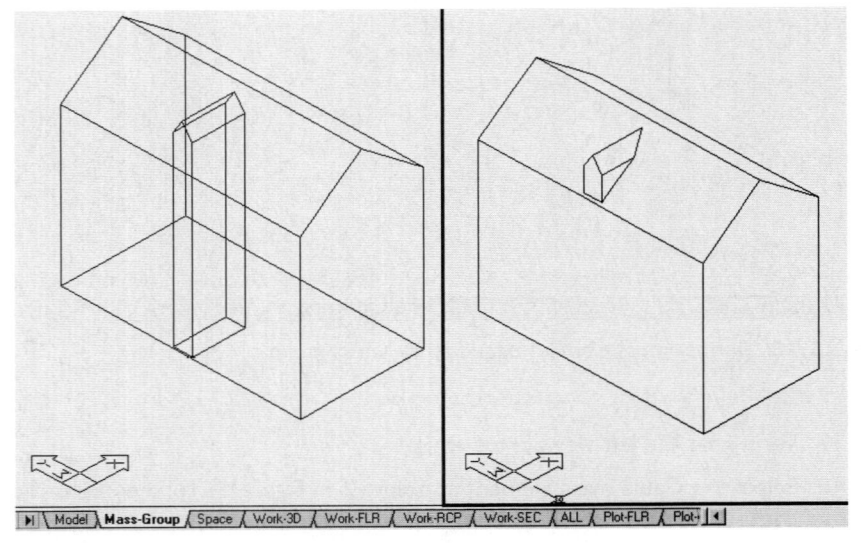

Figure 11.18 Mass group display using the HIDE command

22. Save your drawing.

BOOLEAN OPERATIONS WITH MASS ELEMENTS

When mass elements are attached to a group, the volume of the mass element can add, subtract or intersect with the other elements. The operation options of addition, subtraction and intersection are on the **Mass Tools** toolbar, as shown in Figure 11.2.

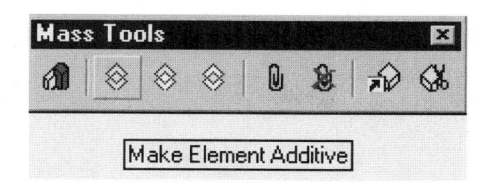

Figure 11.19 *Make Element Additive on the Mass Tools toolbar*

To insert mass elements with the add operation, you can use the **Make Element Additive** command. See Table 11.4 for **Make Element Additive** command access.

Menu bar	Concept>Mass Grouping Tools>Make Additive
Command prompt	AecMassElementOpAdd
Toolbar	Select Make Element Additive from the Mass Tools toolbar as shown in Figure 11.19
Shortcut	Select a Mass Element; then right-click and choose Operation>Additive from the shortcut menu

Table 11.4 *Make Element Additive command access*

Addition is similar to performing a union of the mass elements. Figure 11.20 shows two mass elements attached to a group with the additive operation. The Barrel Vault and the Gable are displayed in the right viewport as a group.

When mass elements are inserted with the subtractive operation, the volume of one mass element is cut from the other. The element selected for the subtraction must be created after others within the group. The Model Explorer, discussed later in this chapter, allows you to change the sequence of the mass elements. See Table 11.5 for **Make Element Subtractive** command access.

Menu bar	Concept>Mass Grouping Tools>Make Subtractive
Command prompt	MASSELEMENTOPSUBTRACT
Toolbar	Select Make Element Subtractive from the Mass Tools toolbar as shown in Figure 11.21
Shortcut	Select a Mass Element; then right-click and choose Operation>Subtractive from the shortcut menu

Table 11.5 *Make Element Subtractive command access*

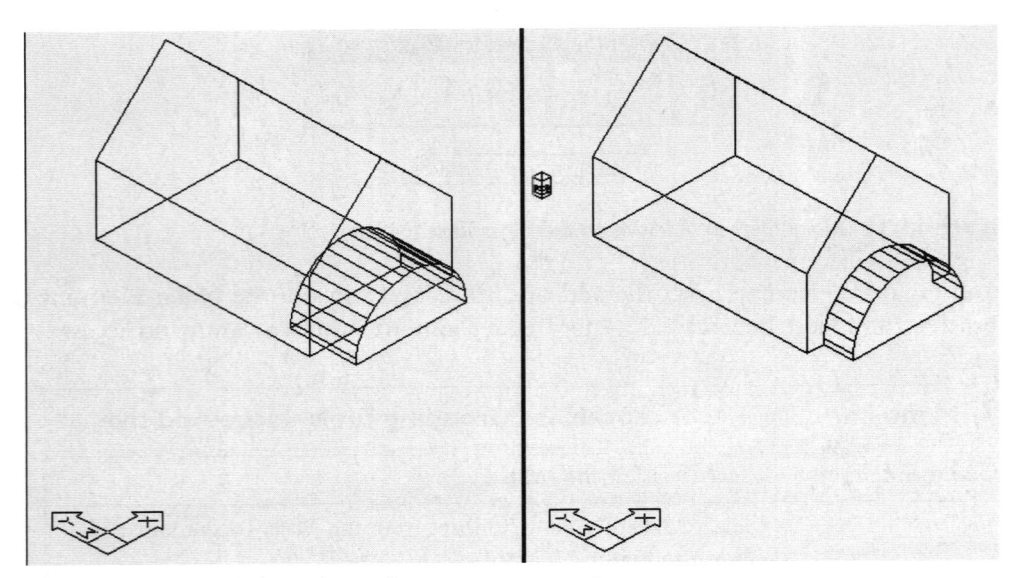

Figure 11.20 *Additive operation between two mass elements*

Figure 11.21 *Make Element Subtractive command on the Mass Tools toolbar*

In Figure 11.22, the Gable mass element was created first; then the Barrel Vault mass element was created and attached to the group. The Barrel Vault mass element is subtracted from the Gable mass element.

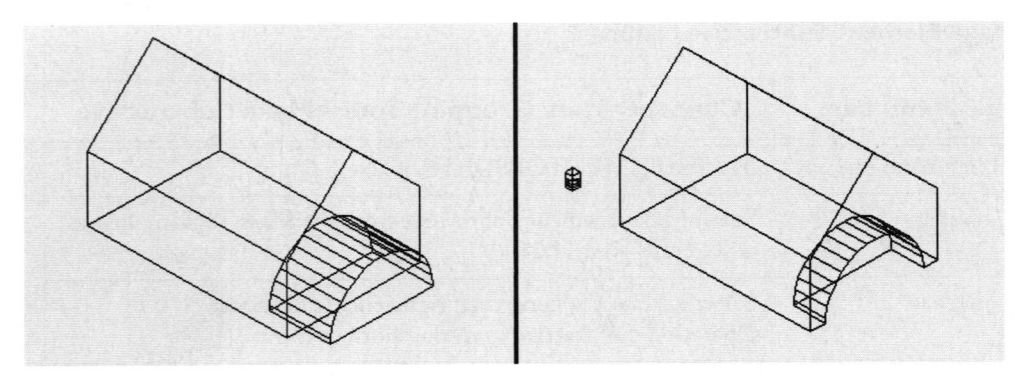

Figure 11.22 *Barrel Vault subtracted from the Gable mass element*

Recessed wall segments of a model can be created with the subtractive operation. In Figure 11.23, the model consists of one box 75' wide, 60' deep and 34' high with 12 boxes inserted with the subtractive operation to cut 2' wide by 4' high slots horizontally around the building. The ability to subtract a mass element often decreases the complexity and number of individual mass elements when you construct a model.

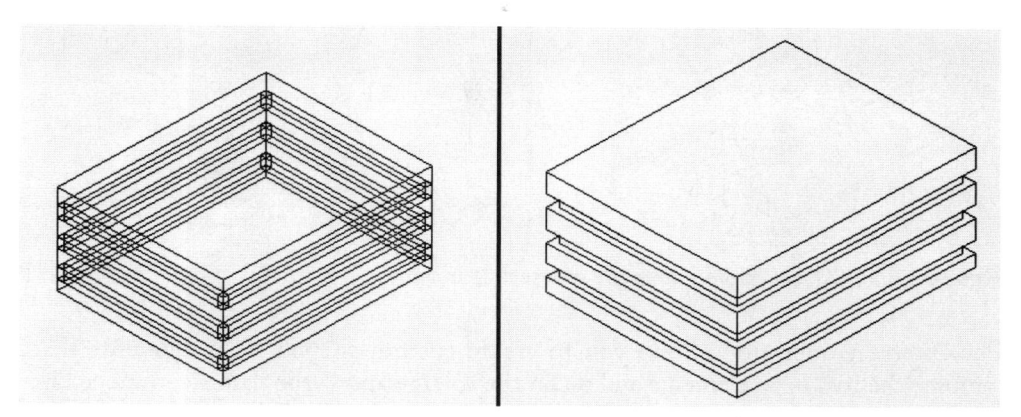

Figure 11.23 *Complex mass model using subtraction of mass elements*

The intersect operation joins the two mass elements and creates a mass group, which consists of the mass or volume that is common to the two mass elements. See Table 11.6 for **Make Element Intersection** command access.

Menu bar	Concept>Mass Grouping Tools>Make Intersection
Command prompt	MASSELEMENTOPINTERSECT
Toolbar	Select Make Element Intersection from the Mass Tools toolbar as shown in Figure 11.24
Shortcut	Select a Mass Element; then right-click and choose Operation>Intersect from the shortcut menu

Table 11.6 *Make Element Intersection command access*

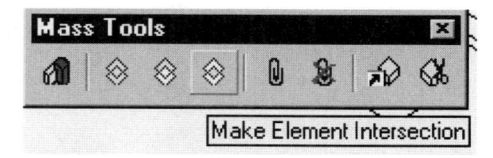

Figure 11.24 *Make Element Intersection command on the Mass Tools toolbar*

In Figure 11.25, the Barrel Vault is defined to the mass group with the intersect operation. The mass that is common to the Gable and the Barrel Vault is displayed in the right viewport.

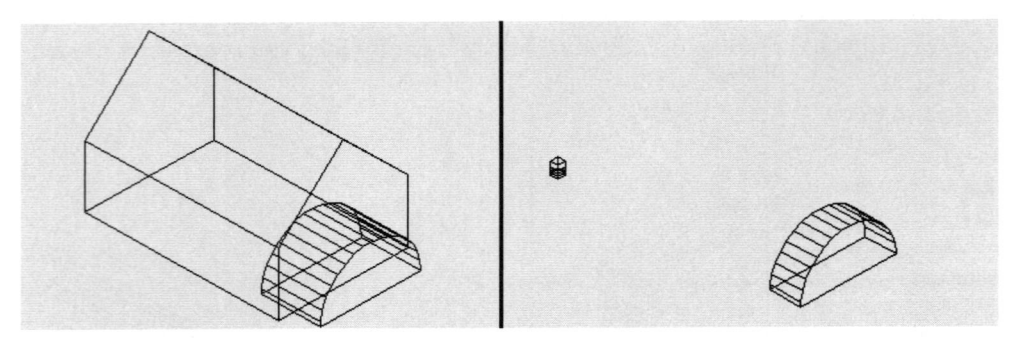

Figure 11.25 *Model created using the intersect operation*

The intersect operation allows you to create custom-shaped mass elements that combine the two mass elements and retain the volume and shape that is common. The group of mass elements shown in Figure 11.26 consists of a Gable and an Isosceles Triangle. The Isosceles Triangle is defined to intersect with the Gable to form a gambrel roof shape.

USING THE MODEL EXPLORER

The Model Explorer can be used to better visualize and edit groups of mass elements. The **Show Model Explorer** command opens the Model Explorer window, which

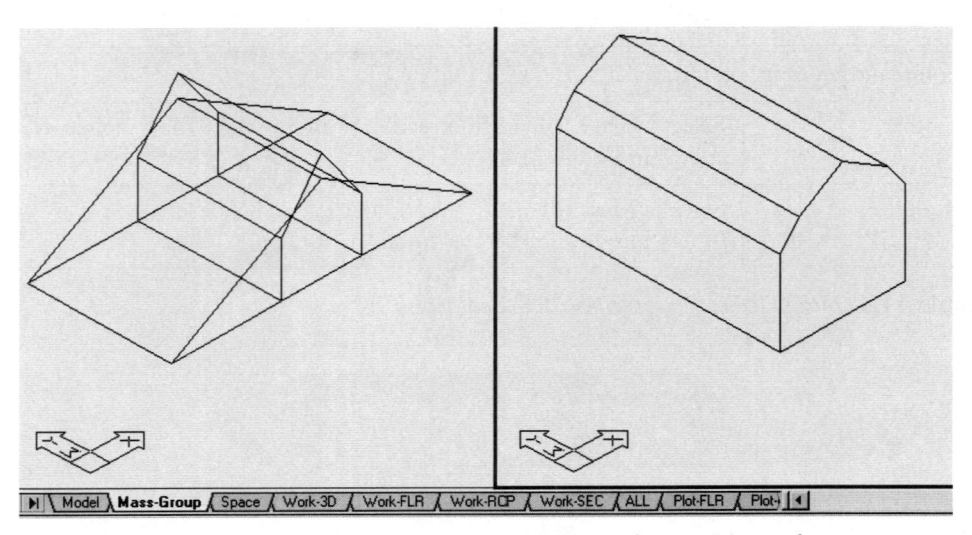

Figure 11.26 *Gambrel roof created with intersect operation between mass elements*

includes the **Object Viewer** toolbar and **Model Explorer** toolbar. These toolbars include view controls and commands for creating and editing the mass elements and groups. See Table 11.7 for **Show Model Explorer** command access.

Menu bar	Concept>Show Model Explorer
Command prompt	MODELEXPLORER
Shortcut	Right-click over the drawing area; then choose Utilities>Massing>Show Model Explorer from the shortcut menu
Shortcut	Select a Group; then right-click and choose Show Model Explorer from the shortcut menu

Table 11.7 *Show Model Explorer command access*

When the Model Explorer is selected from the menu bar, the Model Explorer window opens, as shown in Figure 11.27.

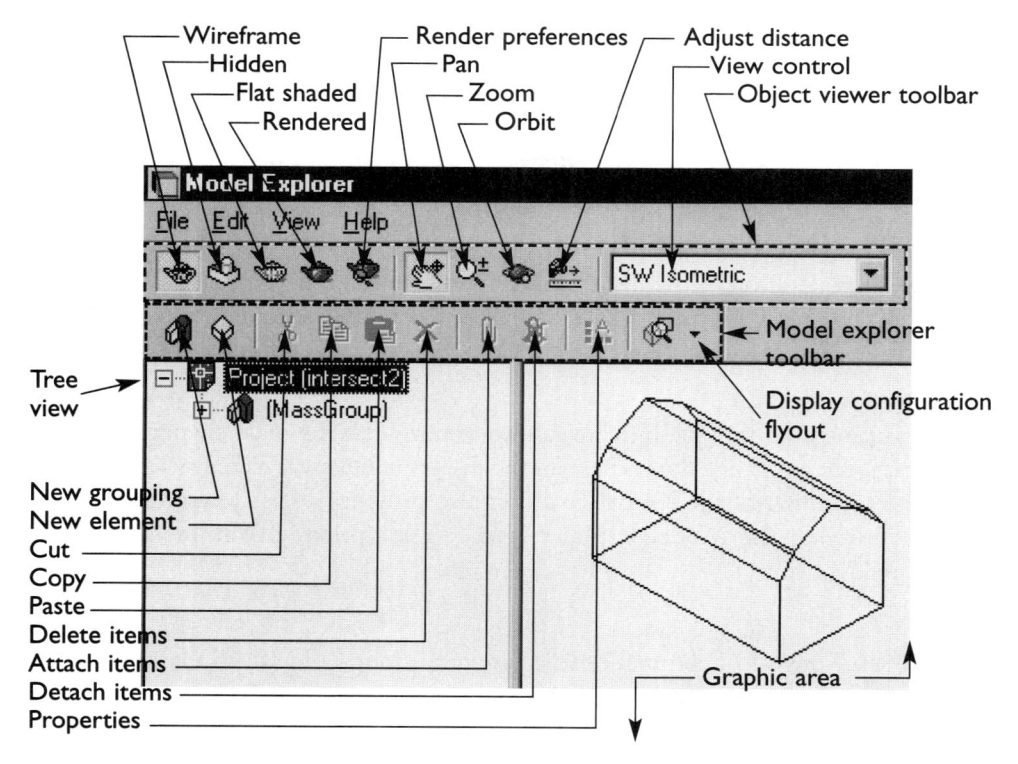

Figure 11.27 *Commands of the Model Explorer*

The Model Explorer consists of a command area at the top, a Tree View on the left and the Object Viewer on the right. You can select mass elements or groups in the Tree View and edit each component. If you select the plus sign of a mass group, the tree will expand and display the mass elements of the mass group. The graphics area shown in Figure 11.27 shows the mass group as it is displayed in the right viewport of the Mass-Group viewport. Figure 11.28 shows the Tree View of the mass group expanded. You can select each of the mass elements to display them in the graphics area. If you select a mass element and then right-click, the shortcut menu allows you to edit the mass element, as shown in Figure 11.28.

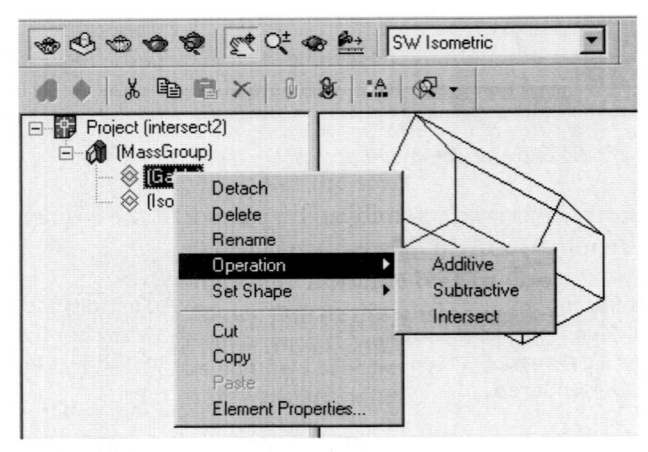

Figure 11.28 *Shortcut menu for a mass element in the Tree*

The shape and operation of the mass element can be edited. If you select a mass element within the Tree View and right-click, you can choose **Element Properties** from the shortcut menu. That option allows you to change the dimensions of the mass element.

The shortcut menu of the graphics area allows you to view the mass element or mass group in perspective, change the zoom, pan the display and copy the display as a bitmap image. The menu bar of the Model Explorer also provides access to commands for viewing and saving the Model Explorer image. The options of the menu bar are described below.

File

The **File** menu consists of commands to create new groups and attach mass elements to the group. The group operations of the **File** menu operate relative to the element or the group that is highlighted in the Tree View area. If the **New Element** command on the File menu is selected when a group is highlighted, the new mass element will be added to the group selected. The **File** menu commands are described below.

New Grouping – Creates a new group at a location specified; the group will be listed at the end of the Tree.

New Element – Creates a new mass element. The new element will be attached to the group highlighted in the Tree.

Detach – Detaches an element from the group in the Tree.

Delete – Removes the selected element or group from the Tree.

Rename – Allows you to rename or append the name of a group or mass element.

Properties – Opens the **Mass Group Properties** or **Mass Element Properties** dialog box, which allows you to edit the object.

SaveAs – Allows you to save the image in the graphics area as a bitmap file.

Close – Closes the Model Explorer window.

Edit

The **Edit** pull-down menu includes **Undo** and **Redo** for operations within the Model Explorer. The **Copy, Cut,** and **Paste** commands allow you to highlight an element and copy its definition to the clipboard and then insert it in another location in the Tree. The contents of the **Edit** pull-down are summarized below:

Undo – Performs an undo on operations within the Model Explorer.

Redo – Performs a redo; it will undo the operations of the last undo.

Cut – Removes a selected element from a group, allowing you to paste the element into a new group.

Copy – Allows you to copy a mass element to the clipboard, which can then be pasted in a new group.

Paste – Allows you to insert mass elements from the clipboard to the highlighted group.

The **Copy, Cut** and **Paste** commands are also included in the **Model Explorer** toolbar. However, the **Undo** and **Redo** commands exist only in the pull-down menu. Once you exit the Model Explorer, the AutoCAD **UNDO** and **REDO** commands will operate on the commands and elements created in the Model Explorer.

View

The **View** pull-down menu includes toggles for opening toolbars, the Object Viewer display window, and the project Tree, as well as controlling the view of the object. The view control commands of the pull-down menu are also available from the shortcut menu of the Object Viewer. Selecting options from the shortcut menu is recommended, because it provides quick access to view controls. The **View** pull-down menu commands are summarized below:

Object Viewer – Toggles on/off the display of the Object Viewer.

Object Viewer Toolbar – Toggles on/off the Model Explorer **Object Viewer** toolbar.

Tree View Toolbar – Toggles on/off the Model Explorer **Tree View** toolbar.

Status Bar – Allows you to toggle on/off the status bar at the bottom of the Model Explorer window.

Pan – Executes the real-time **PAN** command in the Object Viewer of the Model Explorer.

Zoom – Executes the real-time **ZOOM** command in the Object Viewer.

Orbit – Executes the **3DORBIT** command in the Object Viewer to view the object at virtually any rotation and tilt angle.

More – The **More** flyout provides access to **Zoom Extents**, **Zoom Center**, **Zoom In**, and **Zoom Out**.

Shading Modes – The **Shading Modes** flyout allows you to select the type of shading display of the mass element or mass group in the Object Viewer. The options available in the flyout include **Wireframe**, **Hidden**, **Flat Shaded**, and **Rendered**.

Preset Views – The **Preset Views** flyout consists of the following preset views for the Object Viewer: **Top**, **Bottom**, **Left**, **Right**, **Front**, **Back**, **South West Isometric**, **South East Isometric**, **North East Isometric**, and **North West Isometric**.

Auto Zoom Extents – A **Zoom Extents** toggle for the Model Explorer. When **Auto Zoom Extents** is toggled ON, the Object Viewer will be displayed using **Zoom Extents** when the Model Explorer is opened.

Show All Layers – Toggles the display of all layers in the Object Viewer. The layer display of objects in the Model Explorer will include those layers that are thawed or on in the AutoCAD drawing area.

Help

The **Help** menu includes **Help Topics** and **About Model Explorer** options. The **Help Topics** option opens the Help file to the Model Explorer help topic. The **About Model Explorer** option simply defines the version of the Model Explorer.

MOUSE OPERATIONS IN THE MODEL EXPLORER

Usage of the mouse in the Model Explorer provides an extension of the command menu. When the Model Explorer is opened and the mouse is positioned over the Tree View, the mouse pointer is displayed as a pointer, allowing you to select mass elements or groups from the tree. When you select a mass element or group, you can right-click and display the shortcut menu of the Tree View, as shown in Figure 11.29.

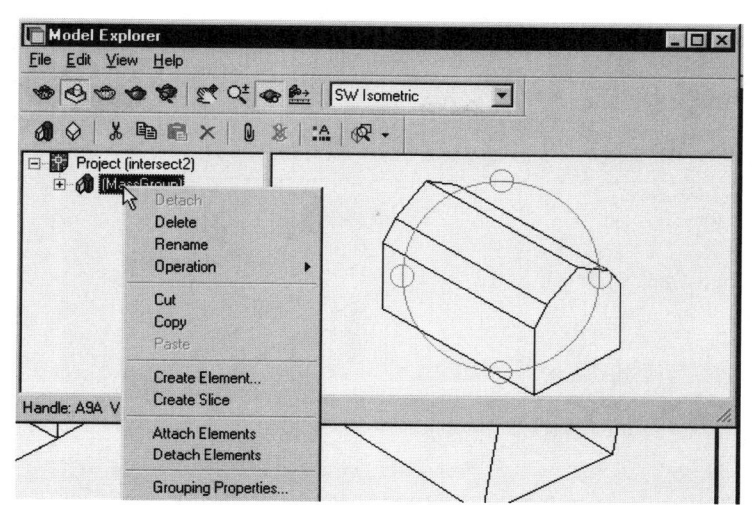

Figure 11.29 *Shortcut menu for the mass group item in the Tree*

When the mouse pointer is positioned over the Object Viewer, the pointer switches to a real time **ZOOM**, **PAN**, or **3DORBIT** cursor, as shown in Figure 11.30.

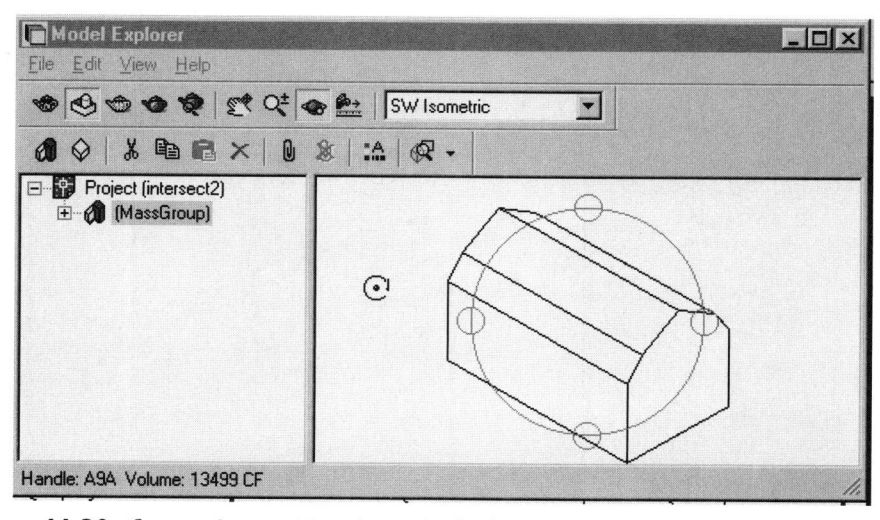

Figure 11.30 *Cursor when positioned outside the Arcball*

The cursor of the **3d Orbit** command varies according to its position relative to the Arcball. The cursor shown in Figure 11.30 is outside the arcball. If the cursor is moved inside the Arcball, it changes to move the view about the axis, as shown in Figure 11.31.

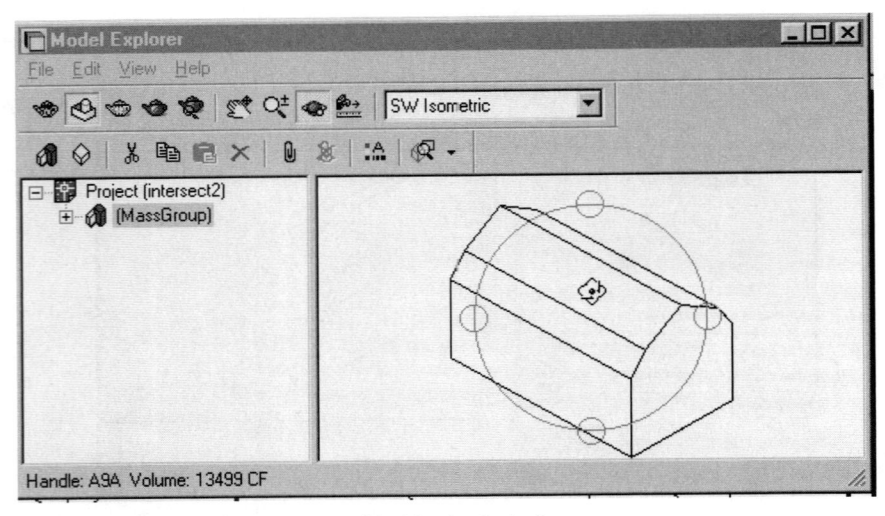

Figure 11.31 *Cursor when positioned inside the Arcball*

When the cursor is positioned over the Arcball, the cursor changes to move the view about the sphere of the Arcball, as shown in Figure 11.32.

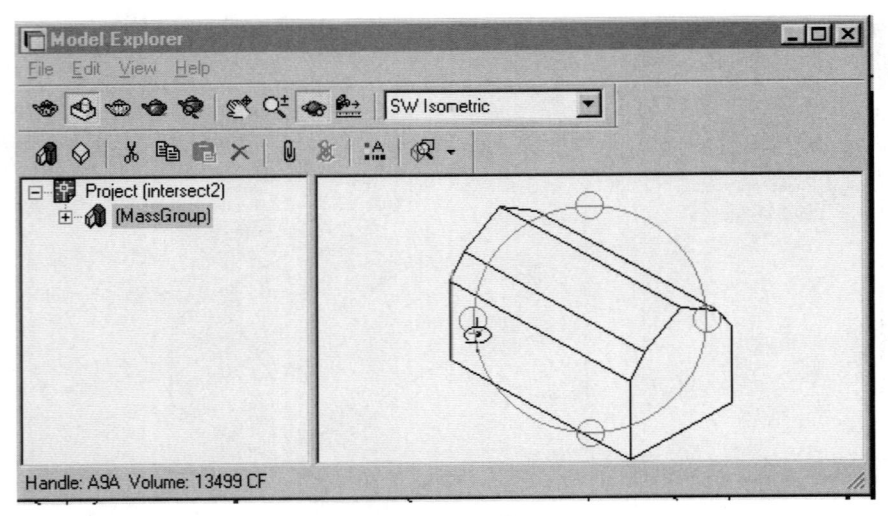

Figure 11.32 *Cursor when positioned on the Arcball*

Using the Object Viewer Shortcut Menu

The shortcut menu of the Object Viewer is shown in Figure 11.33. This shortcut menu provides quick access to commands to control the view in the Object Viewer.

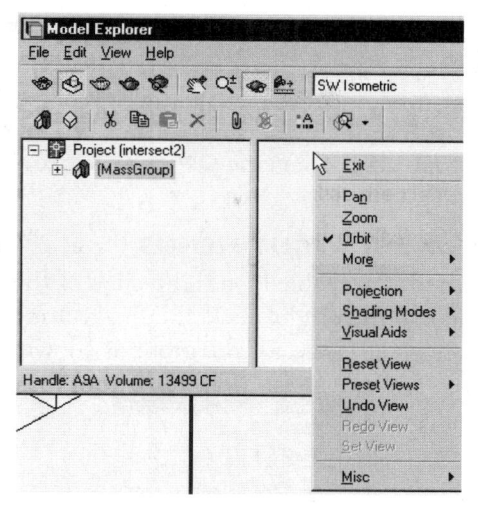

Figure 11.33 *Shortcut menu for the graphics area of the Object Viewer*

The options of this shortcut menu are described below.

Exit – Closes the Model Explorer. To retain the Model Explorer after selecting commands from the shortcut menu, select in the Tree View.

Pan – Executes the real time **PAN** command.

Zoom – Executes real time **ZOOM** command.

Orbit – Executes the **3DORBIT** command in the Object Viewer.

More – The **More** flyout includes **Zoom Window**, **Zoom Extents**, **Zoom Center**, **Zoom In**, and **Zoom Out**.

Projection – The **Projection** flyout includes **Parallel** and **Perspective** options. The **Perspective** option allows you to view the model in a perspective view.

Shading Modes – The **Shading Modes** flyout consists of **Wireframe**, **Hidden**, **Flat Shaded**, and **Rendered** modes.

Visual Aids – The **Visual Aids** flyout consists of the **Compass** and **Grids** options.

Reset View – Resets the view to the top view.

Preset View – The **Preset Views** flyout includes the **Top**, **Bottom**, **Left**, **Right**, **Front**, **Back**, **Southwest**, **Southeast**, **Northeast**, **Northwest** view directions.

Undo View – Performs an **Undo** on previous view settings; returns the view of the model to the previous settings.

Redo View – Performs a **Redo** on last view of the model, returning the view of the model to the settings prior to the undo.

Misc–The **Misc** flyout includes the following options: **Copy to Clipboard, Saveas Bitmap, Render Preferences.** The **Copy to Clipboard** option allows you to copy the image in the Object Viewer to the clipboard and paste the image in other programs.

USING THE TREE VIEW SHORTCUT MENUS

There are two shortcut menus available by a right-click of the mouse in the Project Tree. If you select a group and then right-click, the shortcut menu includes edit operations for a group. The shortcut menu for the group is shown in Figure 11.34.

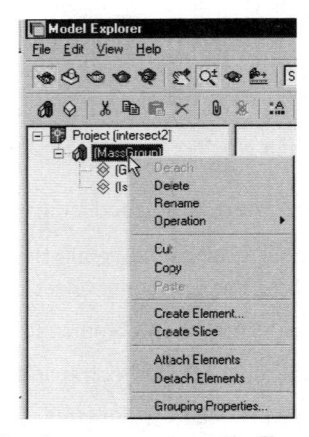

Figure 11.34 *Shortcut menu for the mass group in the Tree*

If you select a mass element and then right-click, the shortcut menu includes options to edit the selected mass element. The shortcut menu for a mass element is shown in Figure 11.35.

The shortcut menu for the mass element allows you to change the shape of the mass element or the Boolean operation assigned to it. The **Element Properties** option of this shortcut menu will open the **Mass Element Properties** dialog box, allowing you to edit of all the dimensions of the mass elements. The **Rename** option allows you to append names to the groups and each mass element. You can also append names to groups and mass elements by left-clicking twice on the name in the Tree. Naming the mass elements and groups greatly enhances model development in the Model Explorer.

CREATING MASS ELEMENTS WITH THE MODEL EXPLORER

When creating a model using the Model Explorer, you can resize the Model Explorer window to allow a view of the left viewport of the Mass-Group layout. This allows

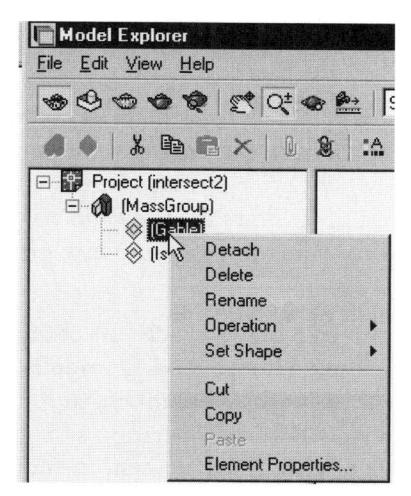

Figure 11.35 *Shortcut menu of the Gable mass element in the Tree*

you to view the mass elements regardless of the Boolean operation in the right viewport and to manipulate the model in the Model Explorer, as shown in Figure 11.36. Resizing the Model Explorer window allows you to toggle between the AutoCAD drawing area and the Model Explorer to develop the model.

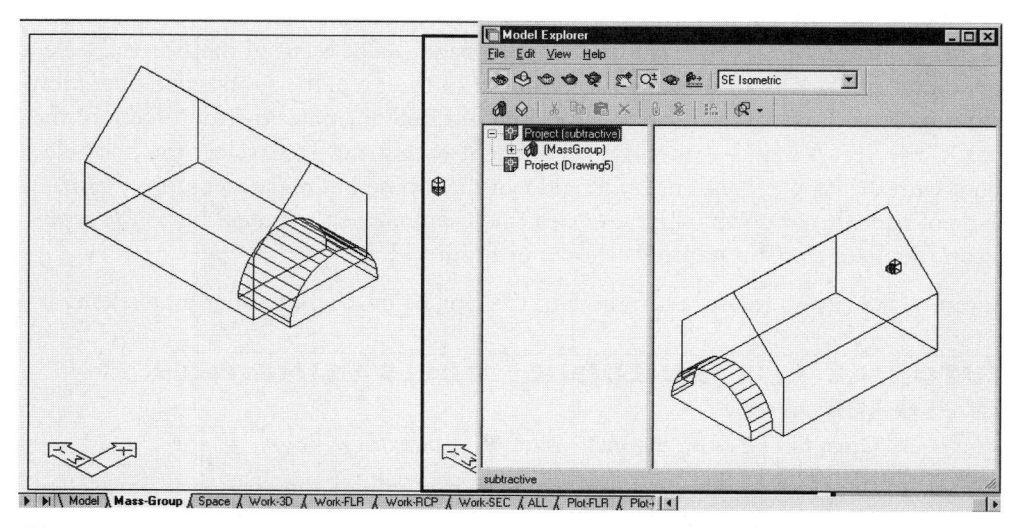

Figure 11.36 *Model Explorer resized to cover the Mass-Group viewport*

You can create a mass element for a model by selecting the **New Element** command from the **Model Explorer** toolbar. Selecting the option opens the **Add Mass Element** dialog box. You can define the size and shape of the mass element in the **Add Mass**

Element dialog box and select in the left viewport of the Mass-Group layout to specify the location and rotation of the mass element.

You can create a new mass group by selecting the **New Grouping** command from the **Model Explorer** toolbar. The location for the mass group marker should be selected in the right viewport of the Mass-Group layout. Therefore the position of the Model Explorer, as shown in Figure 11.36, allows full view and access to the left viewport for mass elements, while allowing you to place the mass group markers in the right viewport.

To add mass elements from the Tree View, you can select the mass group in the Model Explorer and then right-click and choose **Add Elements** from the shortcut menu. Using this technique places the mass element in the mass group selected. The Additive, Subtractive or Intersect Boolean operation symbols associated with a mass element will prefix the name of each mass element in the Tree, as shown in Figure 11.37.

Using the Model Explorer to Reposition Mass Elements in the Tree View

Mass elements are added to the Tree View as they are created in the drawing. The subtractive and intersect operations may require you to reposition the order of the mass elements in the Tree View. If you are subtracting mass elements to remove part of a mass element, the smaller mass element must occur below the larger mass element in the tree. Shown in Figure 11.37 in the Model Explorer is a Cylinder, which is subtracted from the Box. As a result of the subtraction, a slot is cut in the Box mass element. If the Box is moved below the Cylinder in the Tree, the subtractive operation will not create the slot. The Tree View in Figure 11.38 illustrates the movement of the Box in the Tree View. To move a mass element within the Tree, select the mass element with the mouse, continue holding down the mouse button as you move the mass element name up or down within the tree, and then release the mouse button to reposition the mass element name, as shown in Figure 11.38.

Figure 11.38 shows the result of moving the Box below the Cylinder mass element.

TUTORIAL 11.2 BUILDING MODELS USING THE EXPLORER

1. Open *Ex11-2* from the *\ADT_Tutor\Ch11* directory.
2. Choose **File>SaveAs** from the menu bar and save the drawing as *Group2.dwg* in your student directory.
3. Choose **Concept>Show Model Explorer** from the menu bar to open the Model Explorer.
4. Select the **Resize** button of the Model Explorer and drag the Model Explorer window to the size shown in Figure 11.39.

Creating Complex Models 827

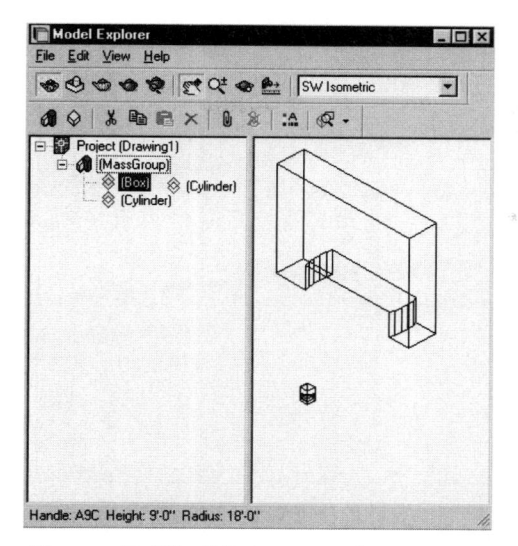

Figure 11.37 *Cylinder mass element moved in the Tree*

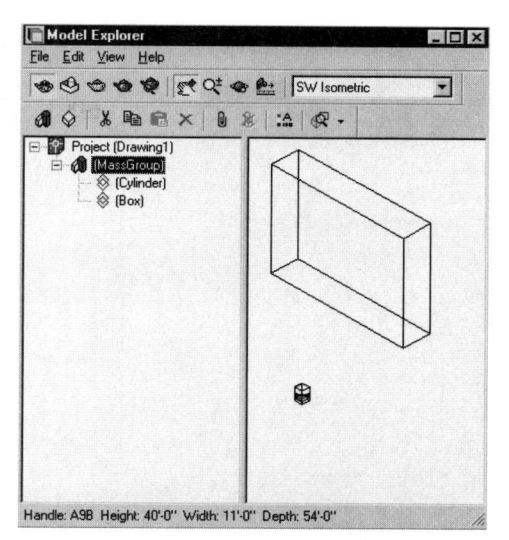

Figure 11.38 *Result of moving the Cylinder mass element above the Box mass element*

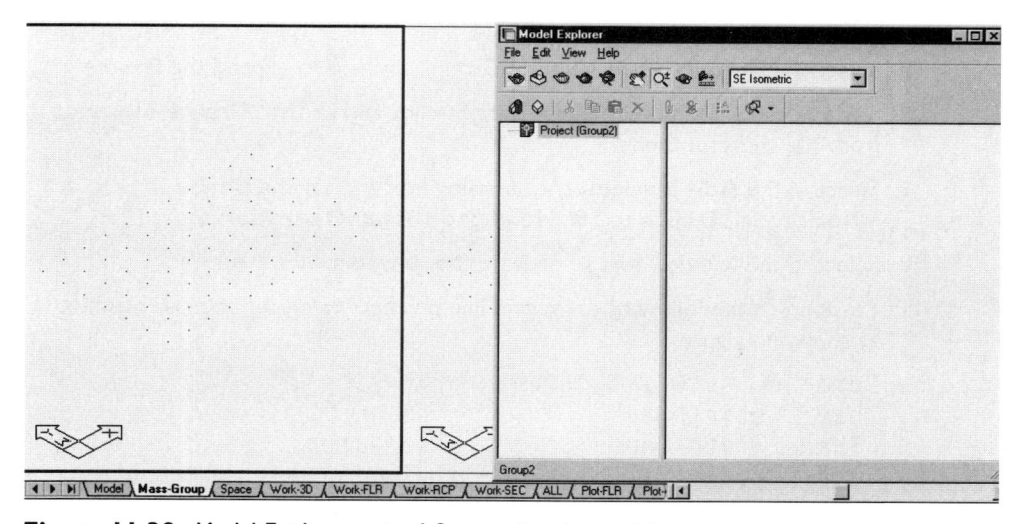

Figure 11.39 *Model Explorer resized for creating the model*

5. Select the **New Grouping** command from the **Model Explorer** toolbar.
6. Respond to the command line prompts as shown in the command sequence below and in Figure 11.40.

```
Command: AecMassingEvent
Location: (Select the right viewport to activate the viewport, then select
```

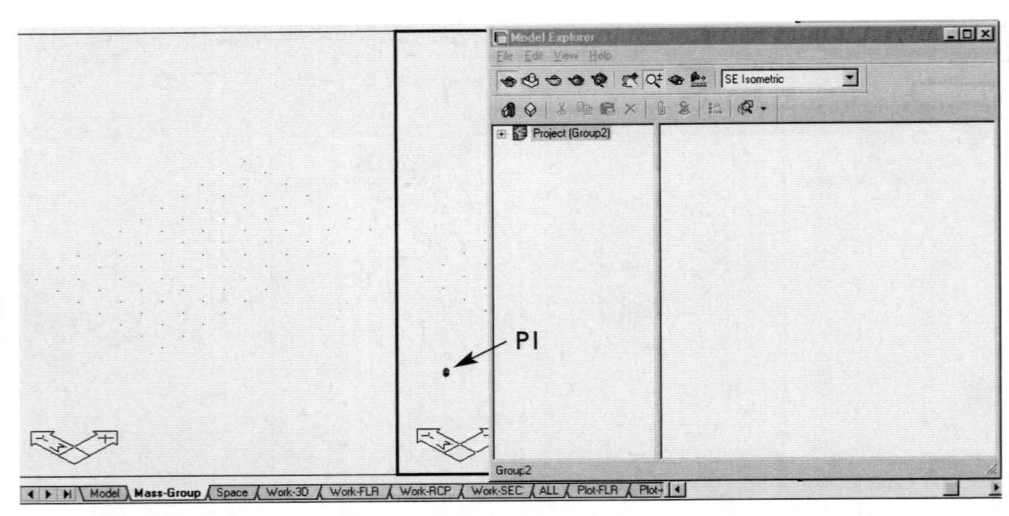

Figure 11.40 *Mass Group Marker specified in the right viewport*

    a point near P1 as shown in the Figure 11.40.)
    Rotation angle <0.00>: ENTER *(Press* ENTER *to accept 0° rotation.)*

7. Select the Plus sign of the Project (Group2) in Tree to expand the Project.

8. Select MassGroup in the Tree; then right-click and choose **Create Element** from the shortcut menu.

9. Select in the **Add Mass Element** dialog box and set the **Shape** to Box, **Width** to 50', **Depth** to 50', **Height** to 18' and **Operation** to Add.

10. Select in the left viewport to activate the viewport.

11. Respond to the following command line prompts to place the mass element as shown in Figure 11.41.

    ```
    Command: AecMassingEvent
    Insert point or
    [SHape/WIdth/Depth/Height/RIse/Match]: DBOX
    Insert point or [SHape/WIdth/Depth/Height/Match]:
    ```
    (Select a location near point P1 in Figure 11.41.)
    ```
    Rotation or [SHape/WIdth/Depth/Height/Match] <0.00>:
    ENTER
    ```
    (Press ENTER *to accept 0° rotation.)*
    ```
    Insert point or
    [SHape/WIdth/Depth/Height/Match/Undo]: ENTER
    ```
    (Press ENTER *to end the command and display the group in the Model Explorer.)*

12. Expand the Tree of the Model Explorer and select the Box in the Tree to display the mass element in the Object Viewer of the Model Explorer.

Creating Complex Models 829

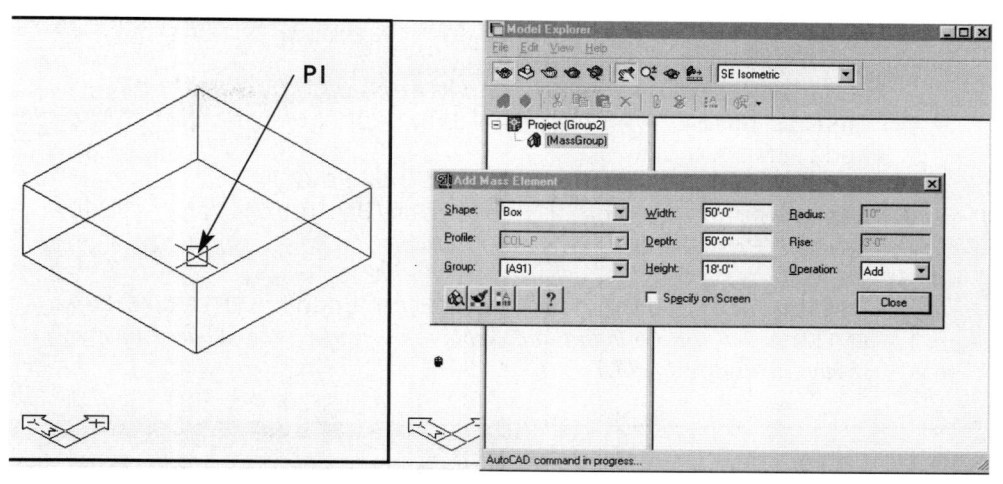

Figure 11.41 *Location for the Box mass element*

13. Select the mass group in the Tree; then right-click and choose **Create Element** from the shortcut menu.
14. Select in the **Add Mass Element** dialog box and set the **Shape** to Box, **Width** to 20', **Depth** to 20', **Height** to 25', and **Operation** to Add.
15. Select in the left viewport to activate the viewport.
16. Respond to the following command line prompts to place the mass element as shown in Figure 11.42.

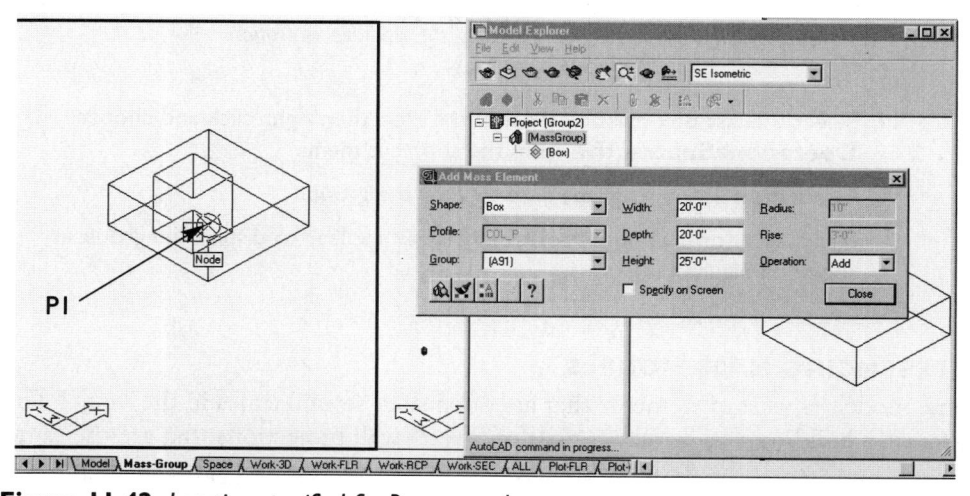

Figure 11.42 *Location specified for Box mass element*

```
Command: AecMassingEvent
Insert point or
[SHape/WIdth/Depth/Height/RIse/Match]: DBOX
Insert point or [SHape/WIdth/Depth/Height/Match]:
Node ENTER
Of
```
(Select a location near point P1 in Figure 11.42.)
```
Rotation or [SHape/WIdth/Depth/Height/Match] <0.00>:
```
ENTER *(Press ENTER to accept 0° rotation.)*
```
Insert point or
[SHape/WIdth/Depth/Height/Match/Undo]: ENTER
```
(Press ENTER to end the command and display the group in the Model Explorer, as shown in Figure 11.43.)

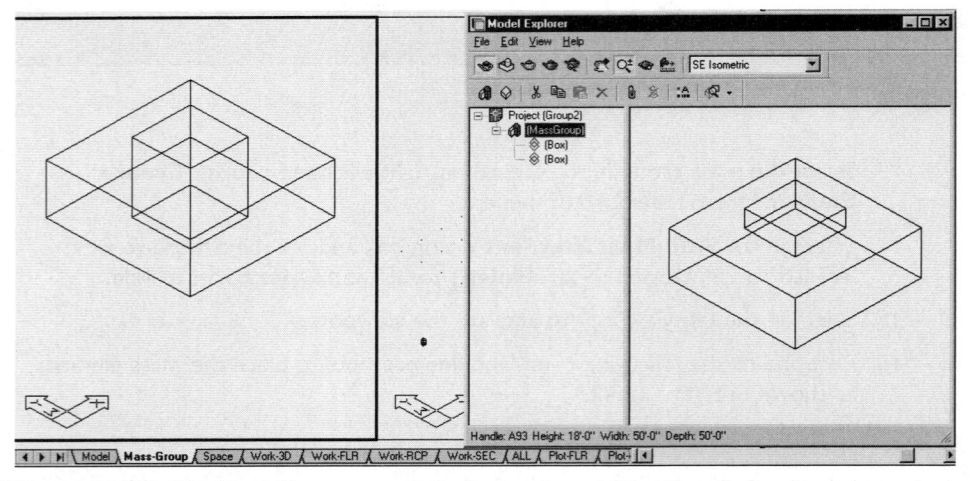

Figure 11.43 *Box Mass Elements combined using the additive operation*

17. Select the last Box mass element in the Tree; then right-click and choose **Operation>Subtractive** from the shortcut menu.
18. Select MassGroup from the Tree to view the group.
19. Select **Hidden** from the **Model Explorer** toolbar to display the group, as shown in Figure 11.44.
20. Save the drawing.

REFERENCING MASS MODELS

Some models consist of elements that are duplicated several times in the model. For instance, a building design may consist of several wall projections that are the same in shape and size. If the designer changes one of these projections, the other projections should also be changed. To avoid the need to edit several projections, you can

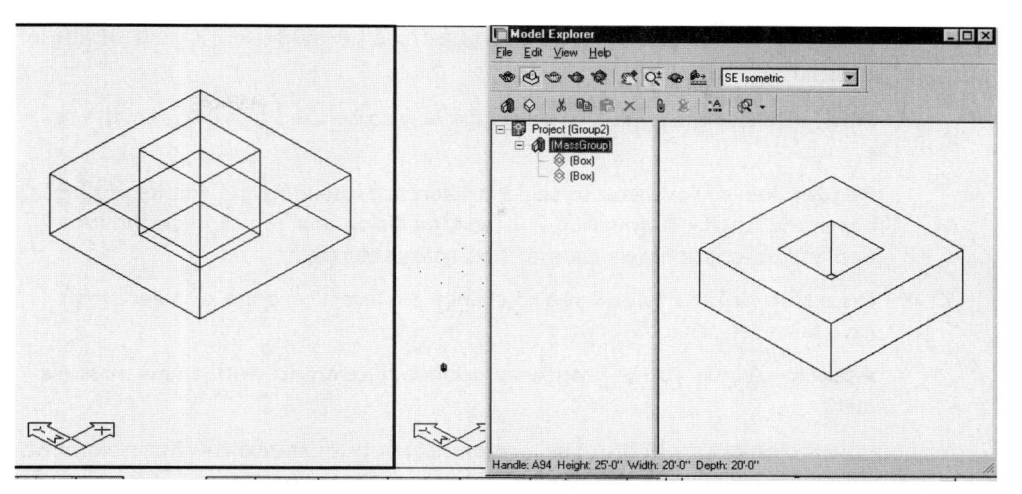

Figure 11.44 *Box mass elements combined with the subtractive operation*

insert one projection and insert additional copies by referencing the first projection. This technique of creating references allows the designer to change only the original mass element. The command to insert a reference of a mass element is the **Reference Element/Group** command (**AECENTREF**). This command can also be used to copy other AutoCAD entities such as polylines, lines, and arcs. See Table 11.8 for **Reference Element/Group** command access.

Menu bar	Concept>Mass Grouping Tools> Reference Element/Group
Command prompt	AECENTREF
Toolbar	Select Reference Element/Group from the Mass Tools toolbar as shown in Figure 11.45

Table 11.8 *Reference Element/Group command access*

When you select the **Reference Element/Group** command (**AECENTREF**) from the **Mass Tools** toolbar, you are prompted in the command with four options:

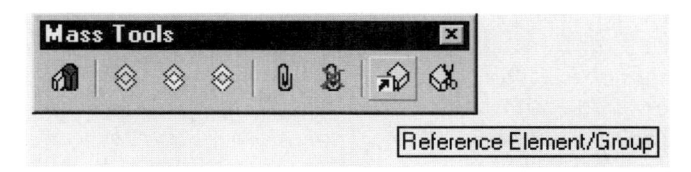

Figure 11.45 *Reference Element/Group command of the Mass Tools toolbar*

`ADd/Properties/Insertion point/Attach`. The purpose of each of these options is summarized below.

> **Add** – Defines the mass element in drawing to be used as source for the reference.
>
> **Properties** – Allows you to select a reference insertion and edit its properties in the **Entity Reference Properties** dialog box. You can edit the location and description properties of the entity reference.
>
> **Insertion point** – Allows you to change the insertion point of objects that have been referenced.
>
> **Attach** – Allows you to change an instance of reference with a new mass element.

To define a mass element to be used as a reference, you must add the mass element as a reference using the **Add** option of the **Reference Element/Group** command. A reference insertion of a Box mass element is shown in the following command sequence and in Figure 11.46.

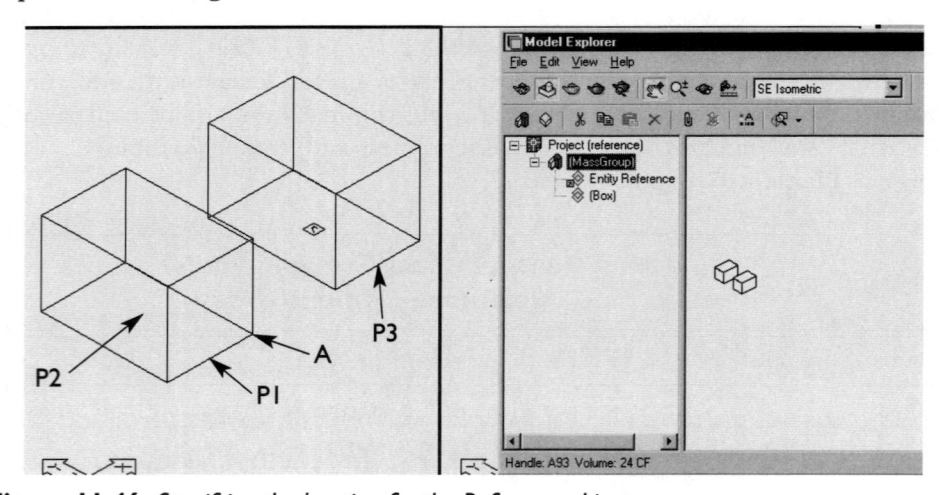

Figure 11.46 *Specifying the location for the Reference object*

```
Command: _AecEntRef
Entity reference [ADd/Properties/Insertion
point/ATtach]: AD ENTER
Select an entity to reference: (Select the Box mass element at
P1 in Figure 11.46.)
Insertion point: node ENTER (Specify the Node object snap mode.)
Of (Select a point near P2 in Figure 11.46 to select the node of the Box
mass element.)
```

Location: 42',0 ENTER *(Specify the location of the reference mass element.)*

Rotation angle <0.00>: ENTER *(Select the default 0° rotation.)*
(Mass element referenced as shown at point P3 in Figure 11.46.)

The referenced mass elements inserted in the drawing are copies of the mass element, which will change when the original mass element is changed. Referenced mass elements include a reference marker located at the insertion point of the reference object. The Model Explorer indicates only one Box mass element in the Tree. Editing this Box mass element will cause the other two insertions to change to the revised dimensions. Referenced mass elements do not have grips; therefore you can edit their size only by changing the size of the reference source.

The other options of the **Reference Element/Group** command allow you to edit the insertion point for the entity, change the entity used for the reference, and edit the properties. Changing the insertion point for the reference mass element results in the object shifting to a new position relative to the original insertion point. In Figure 11.46, the insertion point created when the reference was added was located in the center of the bottom plane. The following command sequence illustrates how the insertion point can be changed.

Command: _AecEntRef
Entity reference [ADd/Properties/Insertion point/ATtach]: I ENTER
Select reference: *(Select the Box mass element at point P3 in Figure 11.46.)*
Select new insertion point: endp ENTER
Of *(Select the corner at "A" in Figure 11.46.)*
Entity reference [ADd/Properties/Insertion point/ATtach]: ENTER *(Press ENTER to end the command.)*
(The Box mass element shifts to the insertion point as shown in Figure 11.47.)

The **Attach** option of the **Reference Element/Group** command (**AECENTREF**) allows you to change the original mass element used as a reference. When this option is used, the insertion point is retained but a different mass element can be substituted for each insertion point. Figure 11.48 shows the mass elements of Figure 11.46 with an additional cylinder mass added.

The command sequence for changing the mass element used as a reference is shown below.

Command: _AecEntRef
Entity reference [ADd/Properties/Insertion point/ATtach]: at ENTER

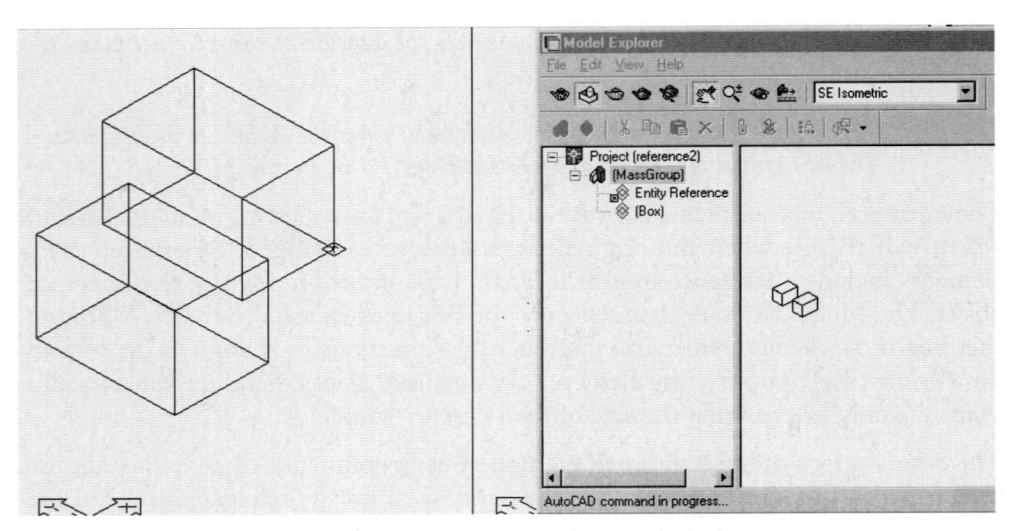

Figure 11.47 *Box mass element repositioned according to the insertion point*

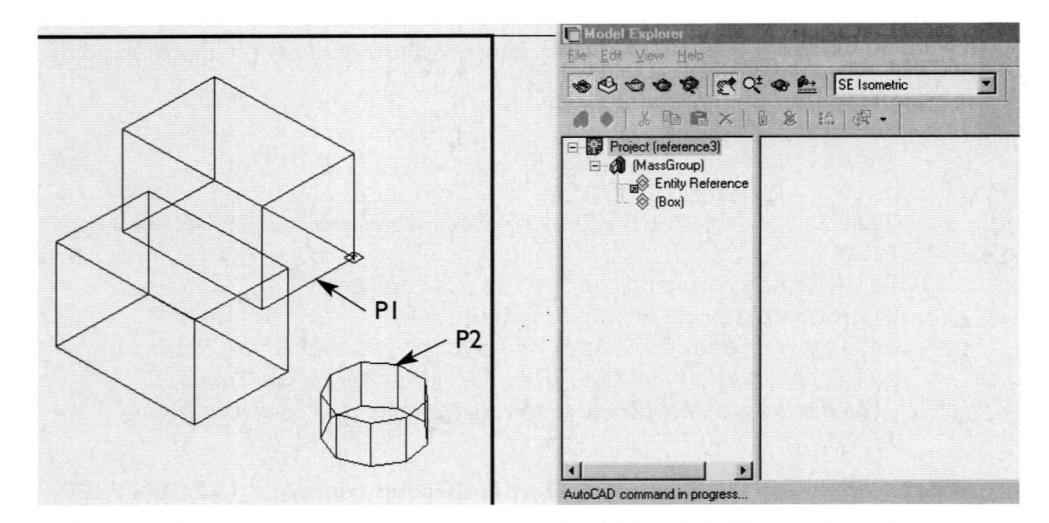

Figure 11.48 *Selection of mass elements for Reference Element/Group command*

```
Select reference: (Select reference entity at point P1 in Figure
11.48.)
Select new entity to reference: (Select Cylinder mass element
at point P2 in Figure 11.48.)
Entity reference [ADd/Properties/Insertion
point/ATtach]: ENTER (Press ENTER to end the command.)
(Figure 11.49 shows the resulting substitution.)
```

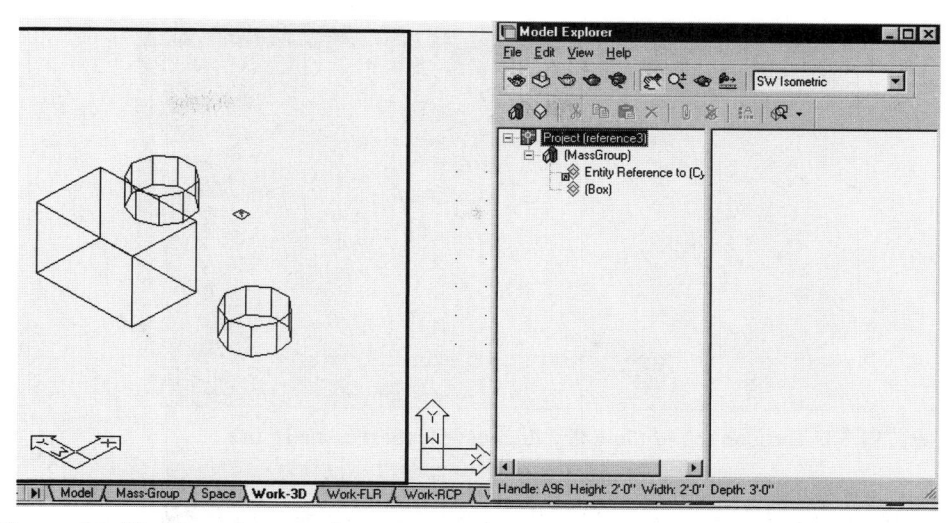

Figure 11.49 *Mass element substitution attached*

Selecting the **Properties** option of the **Reference Element/Group** command (**AECENTREF**) opens the **Properties** dialog box shown in Figure 11.50. The **Properties** dialog box consists of the **General** tab and the **Location** tab. The **General** tab allows you to type a description in the **Description** field to describe the reference.

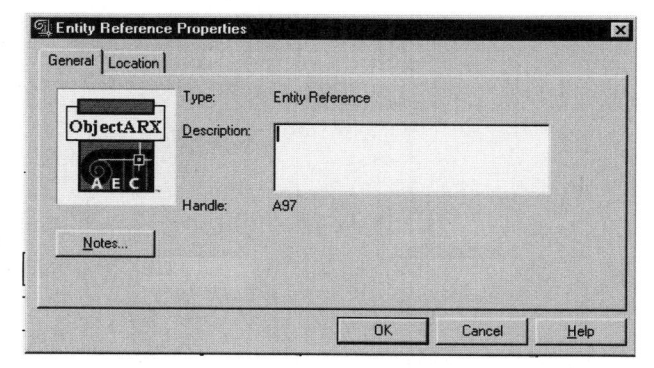

Figure 11.50 *General tab of the Entity Reference Properties dialog box*

The **Location** tab, shown in Figure 11.51, includes edit fields for the coordinate insertion point of the reference, rotation, normal and coordinate system. This dialog box allows you to change the location of the reference.

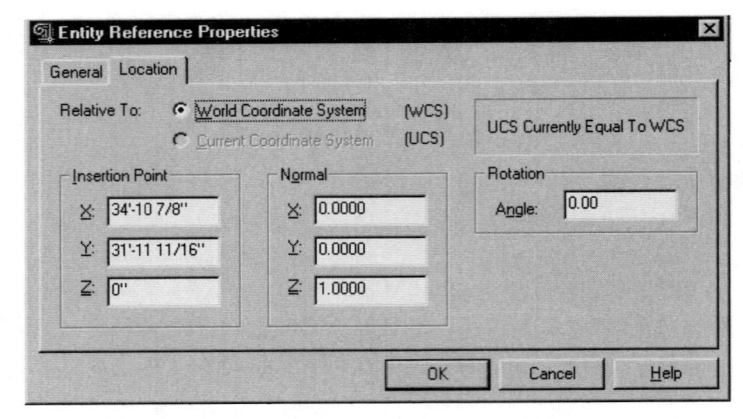

Figure 11.51 *Location tab of the Entity Reference Properties dialog box*

TUTORIAL 11.3 CREATING REFERENCE MASS ELEMENTS

1. Open *EX11-3* from the *\ADT_Tutor\Ch11* directory.

2. Choose **File>SaveAs** and save the drawing as *Group3.dwg* in your student directory.

3. Choose **Concept>Show Model Explorer** from the menu bar and resize the Model Explorer window as shown in Figure 11.52.

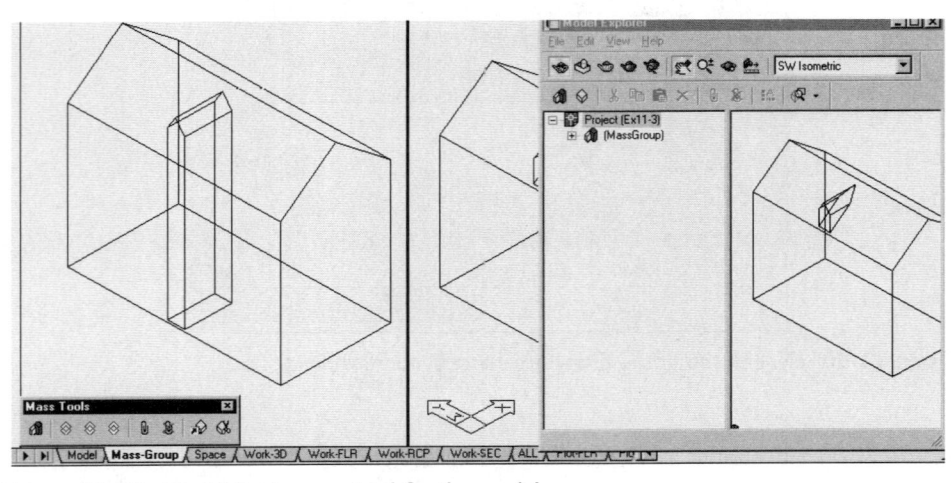

Figure 11.52 *Model Explorer resized for the model*

4. Left-click on the plus sign of the MassGroup to expand the MassGroup. The elements of this group should be displayed indicating additive and subtractive operations for each mass element.

5. Verify that Ortho is toggled ON in the Status Bar.
6. Select the **Reference Element/Group** command from the **Mass Tools** toolbar.
7. Respond to the command line prompts as shown in the following command sequence and in Figures 11.53 and 11.54.

```
Command: _AecEntRef
Entity reference [ADd/Properties/Insertion
point/ATtach]: ad ENTER
Select an entity to reference:
```
(Select the dormer at point P1 in Figure 11.53.)
```
Insertion point:
```
(Select the endpoint of the dormer at point P2 in Figure 11.53.)
(Press F3 to toggle OFF Object Snaps.)
(Move the pointer to the right to extend from the endpoint.)
```
Location: 10' ENTER
Rotation angle <90.00>: ENTER
```
(Press ENTER to accept the 90° rotation.)
(Dormer mass element reference insertion as shown in Figure 11.54.)
```
Entity reference [ADd/Properties/Insertion
point/ATtach]: ad ENTER
Select an entity to reference:
```
(Select the dormer at P1 in Figure 11.54.)
(Press F3 to toggle ON Object Snaps.)
```
Insertion point:
```
(Select the endpoint of the dormer at point P2 in Figure 11.54.)

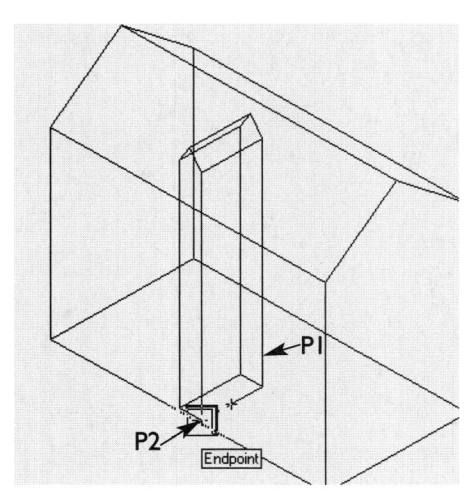

Figure 11.53 *Instance of reference mass element for a dormer*

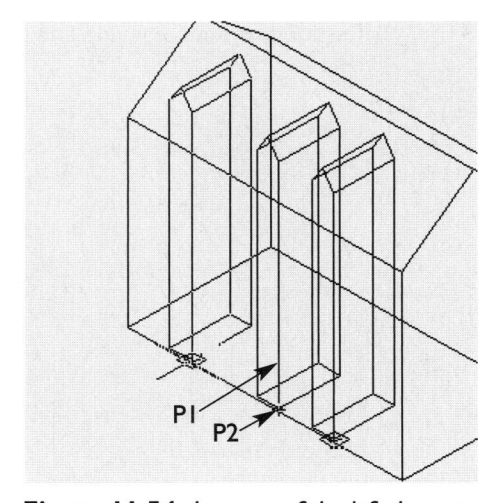

Figure 11.54 *Instance of the left dormer as a reference mass element*

(Press F3 *to toggle OFF Object Snaps.)*
(Move the pointer to the left to extend from the endpoint as shown in Figure 11.55.)
```
Location: 10' ENTER
Rotation angle <90.00>: ENTER
Entity reference [ADd/Properties/Insertion
point/ATtach]: ENTER
```
(Press ENTER *to end the command.)*

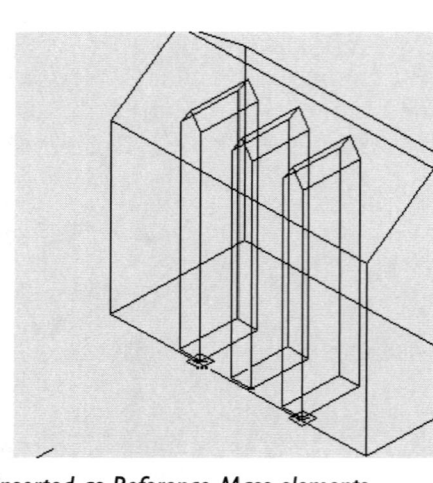

Figure 11.55 *Dormer inserted as Reference Mass elements*

8. Select the MassGroup in Tree View; then right-click and choose **Attach Elements** from the shortcut menu.
9. Respond to the command line prompts as shown in the command sequence below.

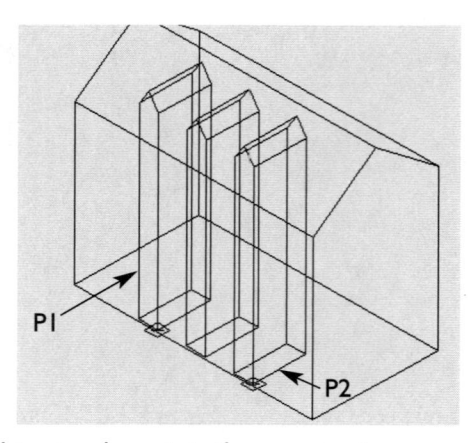

Figure 11.56 *Adding the mass elements to the mass group*

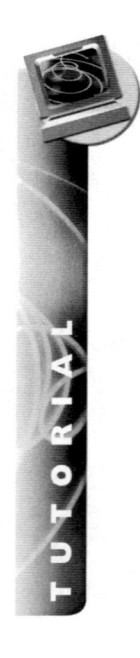

```
Command: AecMassingEvent
Select elements to attach: (Select the dormer mass element on
left at point P1 in Figure 11.56.) 1 found
Select elements to attach: (Select the dormer mass element on
right at point P2 in Figure 11.56.) 1 found, 2 total
Select elements to attach: ENTER (Press ENTER to end the
command.)
Element attached.
Element attached.
```

10. Select the dormer Gable mass element in the Model Explorer; then right-click and choose **Rename** from the shortcut menu. Type **Dormer** and press ENTER in the field of the Tree for the dormer mass element, as shown in Figure 11.57.

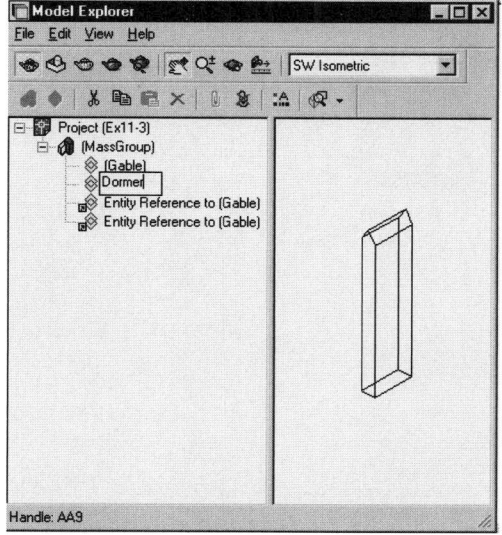

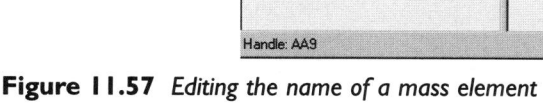

Figure 11.57 *Editing the name of a mass element*

11. Select Dormer mass element in the Model Explorer Tree; then right-click and choose **Element Properties** to open the **Mass Element Properties** dialog box.

12. Select the **Dimensions** tab of the **Mass Element Properties** dialog box. Edit the **Width** dimension to 8' as shown in Figure 11.58.

13. Select **OK** to dismiss the **Mass Element Properties** dialog box.

14. Select the MassGroup of the Tree to view the three dormers 8' wide in the building as shown in Figure 11.59. Note: Although the dormers are now 8'

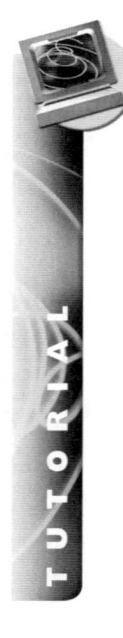

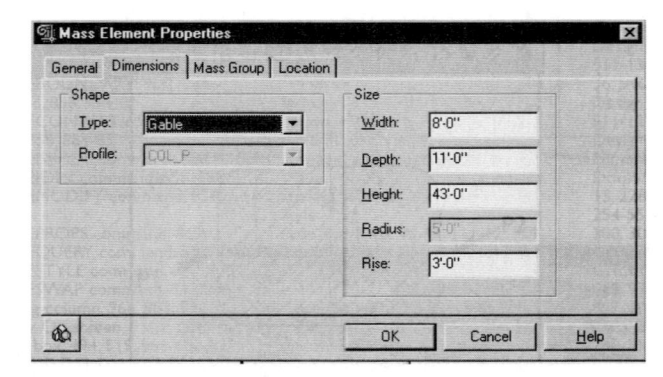

Figure 11.58 *Editing the dimensions of the mass element*

wide, the insertion point remains in the original location at the endpoint of the original 4' dormer.

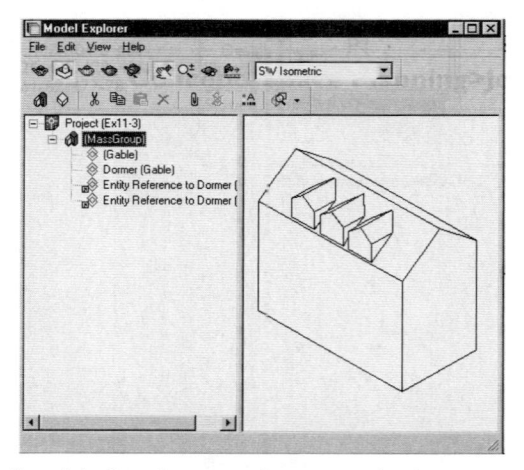

Figure 11.59 *View of model after the mass element is edited*

15. Save your drawing.

CREATING MASS ELEMENTS ABOVE THE XY PLANE

Mass models are created using the AutoCAD World Coordinate System. Because the default elevation of mass elements when inserted is at the zero elevation, most models will require some mass elements to be moved in the Z direction above the Z=0 coordinate. Mass elements can be moved and copied to create the desired mass model. The default elevation of a mass element when inserted is at the zero elevation; once inserted, it can be moved in the Z direction to create such shapes as balconies, porches, or elevated projections. You can use the AutoCAD **MOVE** command to move the mass elements by entering a displacement in the Z direction. The Z coordinate of the **Location** tab in the **Mass Element Properties** dialog box can also be edited to move the mass element in the Z direction.

Because tracking operates in the Z=0 plane, you can place the mass element using tracking at this elevation and then move the mass element. The grip at the insertion point of the mass element can also be edited to move the mass element above the Z=0 plane.

CREATING FLOORPLATE SLICES AND BOUNDARIES

For determining the shape of individual floor plans, the model is sliced to create floorplates. The slice of the model can be created at various elevations to simulate the anticipated elevations of the finish floors. The slice should be taken at the finish floor elevation. A multi-story building would be sliced at the elevations of each finish floor. The exact elevation of finish floors may have to be adjusted later in the design, and therefore the slice elevation can be edited easily. This floorplate becomes a horizontal section from which floor plans can be developed or the geometry can be used to develop space plans.

The slice is created from a slice marker, created through the **Generate Slice** command (**AECSLICECREATE**). After the slice markers are created, the mass elements or the mass groups are attached to the slice markers to create a slice of the mass elements. See Table 11.9 for Generate Slice command access.

Menu bar	Concept>Slice Floorplates>Generate Slice
Command prompt	AECSLICECREATE
Toolbar	Select Generate Slice from the Mass Tools toolbar
Shortcut	Right-click over the drawing area; then choose Utilities>Massing>Slice>Add.

Table 11.9 *Generate Slice command access*

When you select the **Generate Slice** command from the **Mass Tools** toolbar, you are prompted to specify the number of slices, the location for the slice marker, rotation,

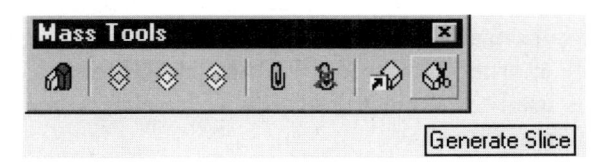

Figure 11.60 *Generate Slice command on the Mass Tools toolbar*

the elevation of the first slice, and the distance between slices. The slice marker is a graphic marker indicating the slice; the actual slice is executed when the slice marker is attached to the mass elements or mass group. The slice marker and the geometry of the slice through mass elements is created on the A-Mass_Slce layer, which has the color 240 (a hue of red) and Dashed2 linetype.

Shown below is the command line sequence that was used for creating the slice shown in Figure 11.61.

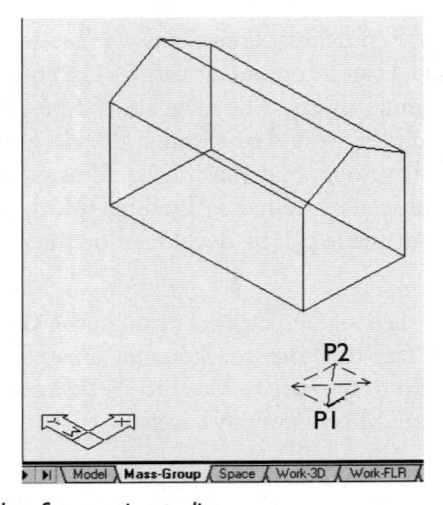

Figure 11.61 *Slice marker for creating a slice*

```
Command: AECSLICECREATE
Number of slices <1>: 2 ENTER
First corner: (Select a point near point P1 in Figure 11.61.)
Second Corner or [Width]: (Select a point near point P2 in Figure 11.61.)
Rotation <0.00>: ENTER (Press ENTER to accept the 0° rotation.)
Starting height <0">: 3' ENTER
Distance between slices <1'-0">: 9' ENTER
```
(Slice markers are created for each of the slices, as shown in Figure 11.62.)

The slice marker that is created by the **Generate Slice** command includes the elevation of each slice. The location, size, and rotation of the marker can be anywhere on the drawing. Once the marker is placed, the **Attach Objects** command (**AEC-SLICEATTACH**) is used to assign the objects to be sliced. See Table 11.10 for **Attach Objects** command access.

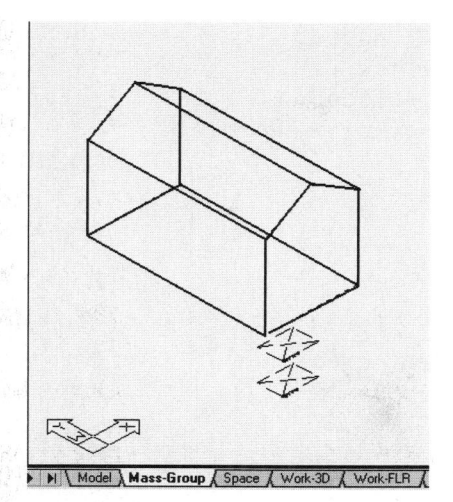

Figure 11.62 *Slice markers for each elevation*

Menu bar	Concept>Slice Floorplates>Attach Objects
Command prompt	AECSLICEATTACH
Shortcut	Select a slice marker; then right-click and choose Attach Objects
Shortcut	Right-click over the drawing area; then choose Utilities>Massing>Slice>Attach Elements

Table 11.10 *Attach Objects command access*

When you choose **Attach Objects** command (**AECSLICEATTACH**) from the menu bar, you are prompted to select the slice marker and the mass elements or mass group for the slice. The command line prompts for attaching the mass elements to a slice marker are shown in the command line sequence and in Figure 11.63.

```
Command: _AecSliceAttach
Select slices: (Select the slice marker at P1 in Figure 11.63.) 1
found
Select slices: (Select the slice marker at P2 in Figure 11.63.) 1
found, 2 total
Select slices: ENTER (Press ENTER to end slice marker selection.)
Select elements to attach: (Select the mass element at P3 in
Figure 11.63.) 1 found
Select elements to attach: ENTER (Press ENTER to end the
selection of mass elements or mass groups.)
Adding elements...
```

```
1 element(s) added.
Adding elements...
1 element(s) added.
```
(Slices created as shown in Figure 11.64.)

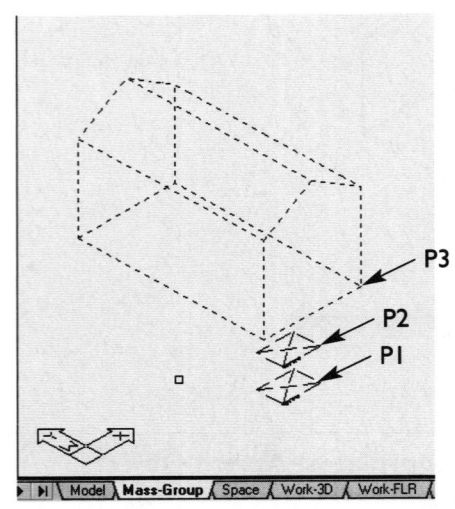

Figure 11.63 *Slice markers selected for creating a slice*

Figure 11.64 *Slices created for each elevation*

Mass elements and mass groups that have been attached to a slice can be detached from the slice through the **Detach Objects** command (**AECSLICEDETACH**), which allows you to select objects to remove from the slice. See Table 11.11 for **Detach Objects** command access.

Menu bar	Concept>Slice Floorplates>Detach Objects
Command prompt	AECSLICEDETACH
Shortcut	Select a slice marker or the slice; then right-click and choose Detach Objects
Shortcut	Right-click over the drawing area; then choose Utilities>Massing>Slice>Detach Objects

Table 11.11 *Detach Objects command access*

When you select the **Detach Objects** command, you are prompted to select the slice or slice marker and then to specify the mass elements or mass group to remove from the slice. The command line prompts for detaching a mass element from a slice are shown below.

```
Command: _AecSliceDetach
Select slices: (Select the slice marker or the slice at P1 in Figure
11.65.) 1 found
Select slices: ENTER (Press ENTER to end the selection of slices.)
Select elements to detach: 1 found (Select the mass
element at point P2 in Figure 11.65.)
Select elements to detach: (Press ENTER to end the selection of
mass elements or mass groups.)
Removing elements...
1 element(s) removed.
```
(Slice is revised as shown in Figure 11.66.)

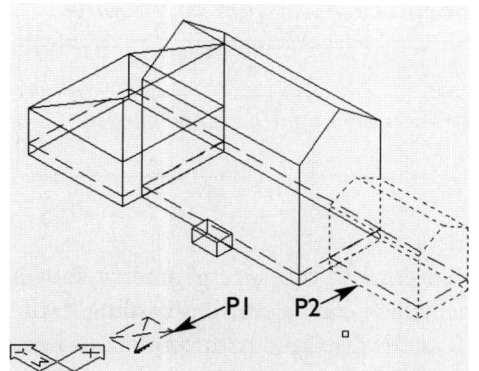

Figure 11.65 *Selection of slice and mass element for the Detach Objects command*

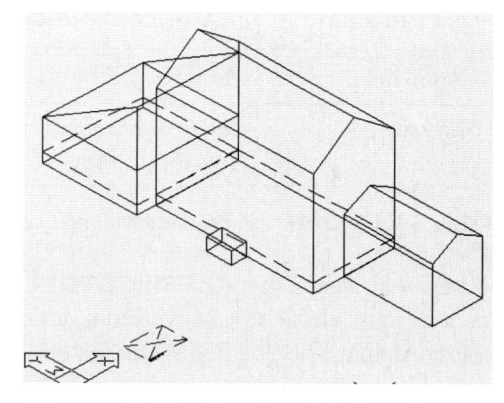

Figure 11.66 *Slice detached from the mass element*

The elevation of the slice can be revised after the **Generate Slice** command has been executed. Changing the elevation of the slice allows you to adjust the finish floor elevation as the design changes. To change the elevation of the slice, select the **Set Slice Elevation** command (**AECSLICEELEVATION**). See Table 11.12 for **Set Slice Elevation** command access.

Menu bar	Concept>Slice Floorplates>Set Slice Elevation
Command prompt	AECSLICEELEVATION
Shortcut	Select a slice marker or the slice; then right-click and choose Set Elevation

Table 11.12 *Set Slice Elevation command access*

When you select the **Set Slice Elevation** command, you are prompted in the command line to specify the elevation. If you type a different elevation in response to the command prompt, the slice will move to the specified elevation. The geometry of the slice cannot be edited. The AutoCAD **MOVE** command can be used to move the slice marker in the drawing. When the slice marker is moved, the slice through the mass element moves.

CONVERTING THE SLICE TO A POLYLINE

The slice can be converted to a polyline and the polyline edited to create the shape of the floor plan. The polyline can then be used to create a space boundary. To convert a slice to a polyline, select the **Convert to Polyline** command (**AECSLICETO-PLINE**). See Table 11.13 for **Convert to Polyline** command access.

Menu bar	Concept>Slice Floorplates>Convert to Polyline
Command prompt	AECSLICETOPLINE
Shortcut	Select a slice marker or the slice; then right-click and choose Convert to Polyline

Table 11.13 *Convert to Polyline command access*

When you select the **AECSLICETOPLINE** command, the slice geometry shown on each mass element is converted to a polyline, which can be edited according to the desired shape. The polyline is created on layer 0 and is displayed with continuous linetype. The command line prompts for the **Convert to Polyline** command are shown below.

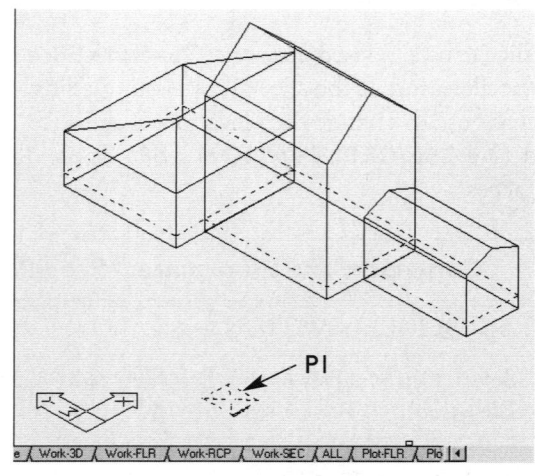

Figure 11.67 *Slice selected for conversion to polyline*

```
Command: _AecSliceToPline
Select slices: (Select the slice marker or slice at P1 in Figure 11.67.)
1 found
Select slices: ENTER (Press ENTER to end selection of slice geome-
try.)
1 polyline(s) created.
```
(Polyline created from the slice as shown in Figure 11.68.)

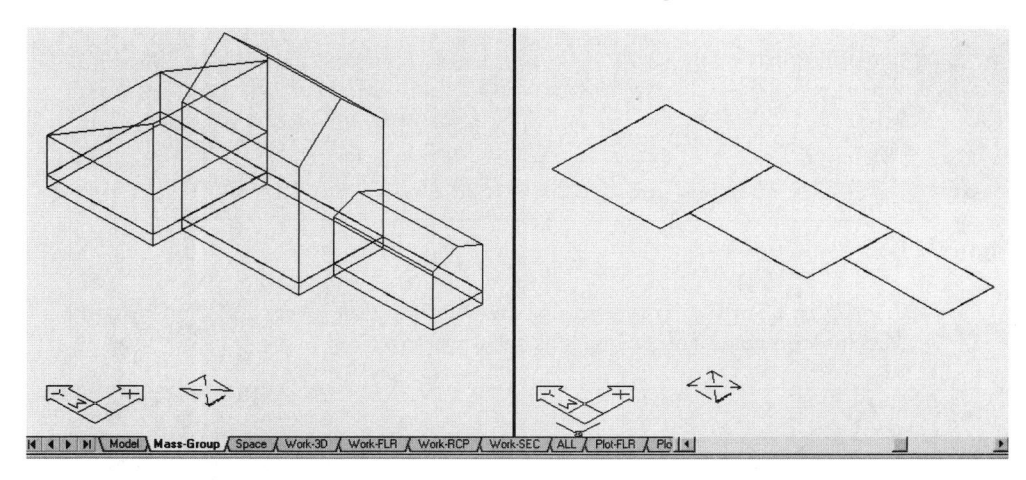

Figure 11.68 *Polyline created from the slice as shown with continuous line*

TUTORIAL 11.4 CREATING A FLOORPLATE SLICE AND BOUNDARY

1. Open *Ex11-4* from the \ADT_Tutor\Ch11 directory.
2. Choose **File>SaveAs** from the menu bar and save the drawing as *Group4* in your student directory.
3. Choose **Concept>Show Model Explorer** from the menu bar.
4. Select the right viewport to activate the viewport.
5. Choose **Desktop>AEC Profiles>Define AEC Profile** from the menu bar to open the **Profiles** dialog box, as shown in Figure 11.69.
6. Select the **New** button of the **Profiles** dialog box and type **Canopy** in the **Name** dialog box.
7. Select **OK** to dismiss the **Name** dialog box.
8. Select the **Edit** button to open the **Profiles Definition Properties** dialog box, and then type **Circular Canopy** in the **Description** field of the dialog box.
9. Select **OK** to dismiss the **Profiles Definition Properties** dialog box.

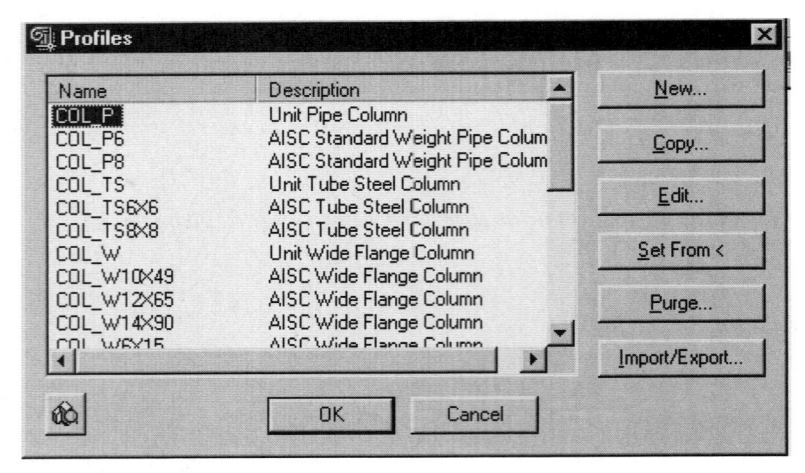

Figure 11.69 *Profiles dialog box*

10. Select the **Set From<** button of the **Profiles** dialog box, and then respond to the command line prompts as shown below.

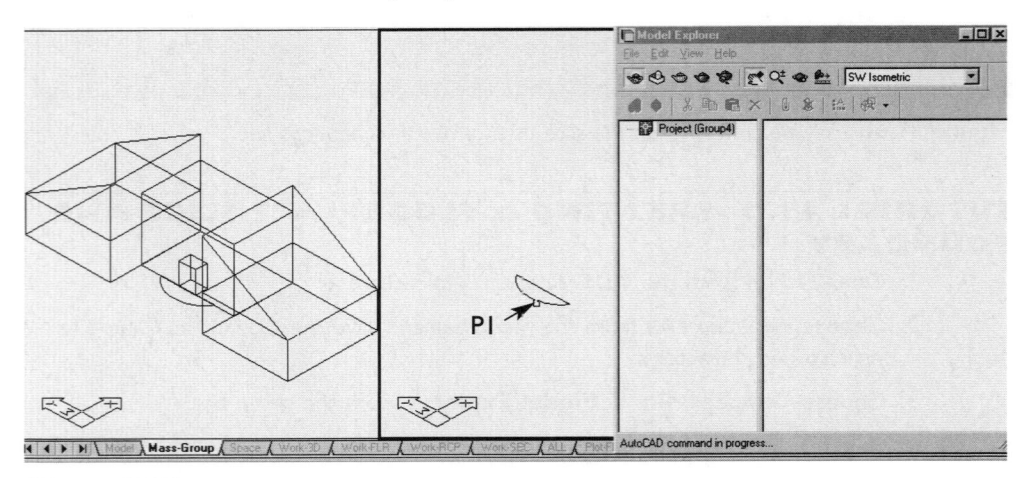

Figure 11.70

```
Command: _AecProfileDefine
Select a closed polyline: (Select the closed polyline as shown at
P1 in Figure 11.70.)
Add another ring? [Yes/No] <N>: ENTER (Press ENTER to end
selection of closed polylines and end the command.)
```

11. Select **OK** to dismiss the **Profiles** dialog box.

12. Select **Erase** from the **Modify** toolbar and erase the original polylines used for the AEC Profile as shown at P1 in Figure 11.70.
13. Select the **New Element** command from the **Model Explorer** toolbar.
14. Select in the **Add Mass Element** dialog box and set the **Shape** to Extrusion. Select Canopy from the **Profile** list and set the **Height** to 16".
15. Select the left viewport to activate the viewport and then respond to the command prompts as shown in the following command sequence.

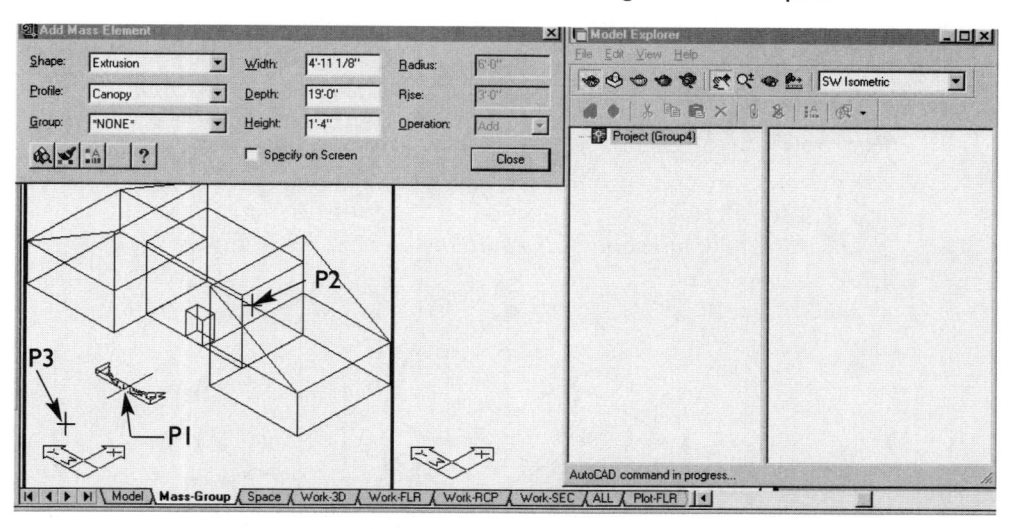

Figure 11.71 *Location for insertion of the Canopy mass element*

```
Command: AecMassingEvent
Insert point or [SHape/WIdth/Depth/Height/Match]:
```
(Select a point near point P1 in Figure 11.71.)
```
Rotation or [SHape/WIdth/Depth/Height/Match] <0.00>:
```
ENTER *(Press* ENTER *to accept 0° rotation.)*
```
Insert point or
[SHape/WIdth/Depth/Height/Match/Undo]:
```
ENTER *(Press* ENTER *to end the command.)*

16. Select **Zoom Window** from the **Standard** toolbar; then select near points P2 and P3 to specify the window location as shown in Figure 11.71.
17. Select **Move** from the **Modify** toolbar and move the Canopy to the midpoint of the building as described in the following command sequence.

```
Command: _move
Select objects:
```
(Select the Canopy at point P1 in Figure 11.72.) 1 found

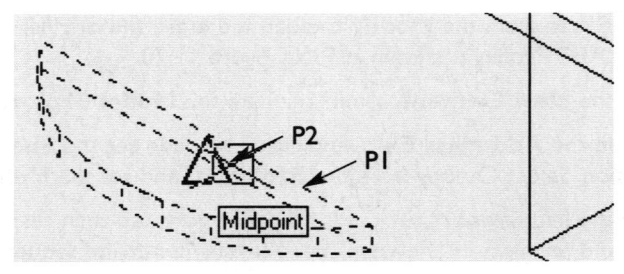

Figure 11.72 *Selecting midpoint of the base of canopy*

```
Select objects: ENTER (Press ENTER to end object selection.)
Specify base point or displacement: mid ENTER
of (Select the midpoint of the base of the Canopy at P2 in Figure 11.72.)
Specify second point of displacement
or <use first point as displacement>: mid ENTER
   of (Select the midpoint of the building at point P3 in Figure 11.73.)
```

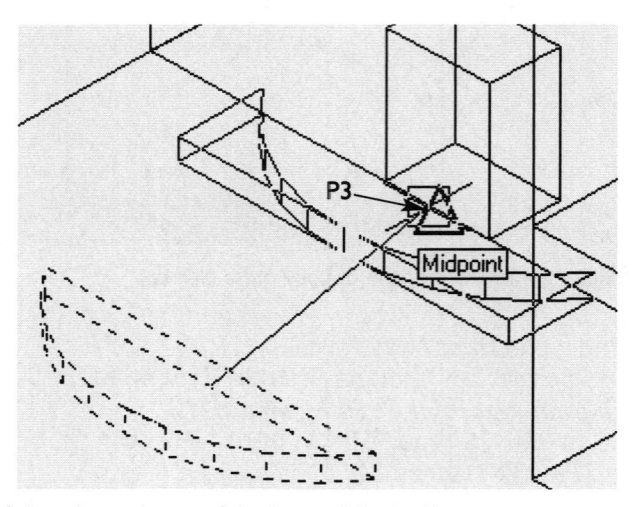

Figure 11.73 *Select the midpoint of the base of the building*

18. Select **Zoom Extents** from the **Standard** toolbar.
19. Select in the right viewport and then select the **New Grouping** command from the **Model Explorer** toolbar.
20. Respond to the following command line prompts to create the new group.

    ```
    Command: AecMassingEvent
    Location: (Select a point near point P1 in Figure 11.74.)
    Rotation angle <0.00>: ENTER (Press ENTER to accept a rotation of 0°.)
    ```

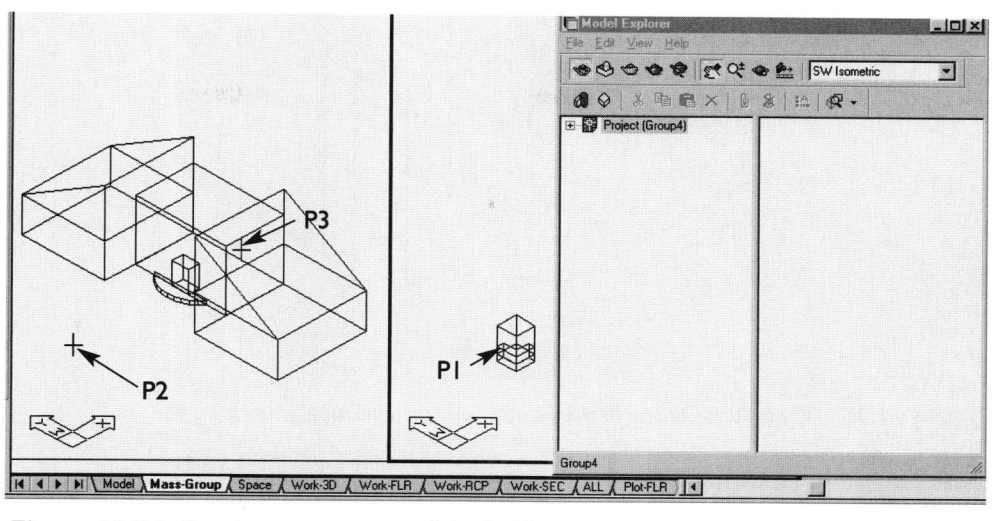

Figure 11.74 *Creating a mass group of the building components*

21. Expand the Project and select the MassGroup of the Tree View in the Model Explorer; then right-click and choose **Attach Elements** from the shortcut menu.
22. Select the left viewport and select all mass elements. When finished selecting objects, press ENTER.
23. Verify that Ortho is toggled ON in the Status bar.
24. Select **Zoom Window** from the **Standard** toolbar and select a window in the left viewport around the canopy from P2 to P3 as shown in Figure 11.74.
25. Select the left viewport, and then select the Canopy mass element to display the grips of the Canopy. Select the grip at P1, as shown in Fig. 11.75. Respond to the following command line prompts to move the Canopy 10' in the vertical direction, as shown in Figure 11.75.

    ```
    Command:
    ** STRETCH **
    Specify stretch point or [Base
    point/Copy/Undo/eXit]: <Ortho on> @0,0,10' ENTER
    ```
    (Canopy is moved 10' above the base of the building.)

26. Expand the MassGroup and select the last Box mass element listed in the Tree View of the Model Explorer; then right-click and choose **Operation>Subtractive** from the shortcut menu.
27. Select in the right viewport to display the subtraction of the Box mass element from the building, as shown in Figure 11.76.

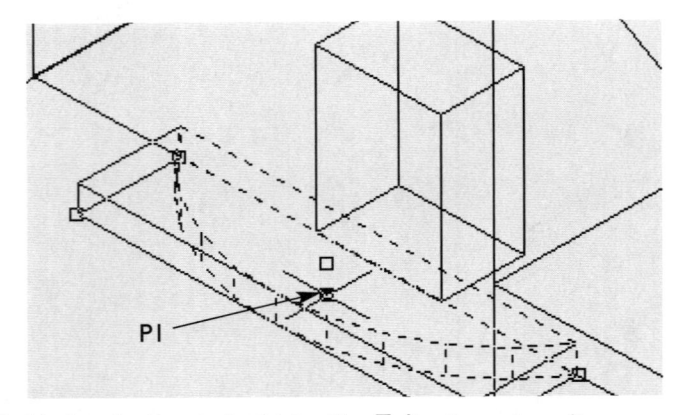

Figure 11.75 *Moving the Canopy in the positive Z direction using grips*

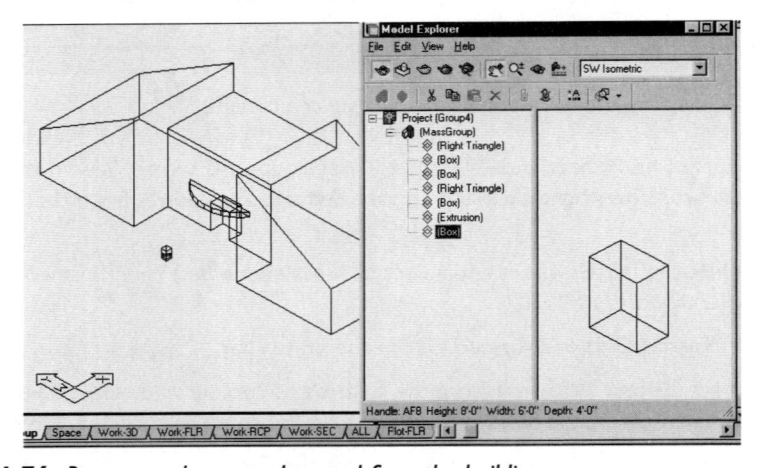

Figure 11.76 *Box mass element subtracted from the building*

28. Select the **Generate Slice** command from the **Mass Tools** toolbar. Respond to the following command sequence to place the slice marker in the right viewport.

    ```
    Command: _AecSliceCreate
    Number of slices <1>: ENTER (Press ENTER to accept the default
    of one slice.)
    First corner: (Select a point near point P1 in Figure 11.77.)
    Second Corner or [Width]: (Select a point near P2 in Figure
    11.77.)
    Rotation <0.00>: ENTER (Press ENTER to accept the 0° rotation of
    the Slice Marker.)
    Starting height <0">: 48 ENTER (Specify the height of the slice.)
    ```

29. Verify that the right viewport is active.

30. Select the slice marker; then right-click and choose **Attach Objects** from the shortcut menu. Respond to the command line prompts as shown in the following command sequence.

   ```
   Command: AecSliceAttach
   Select elements to attach: (Select a point near P3 in Figure 11.77.)
   Specify opposite corner: (Select a point near P4 in Figure 11.77.) 3 found
   1 was filtered out.
   Select elements to attach: ENTER (Press ENTER to end object selection.)
   Adding elements...
   1 element(s) added.
   ```
   (Slice attached to the building as shown in Figure 11.77.)

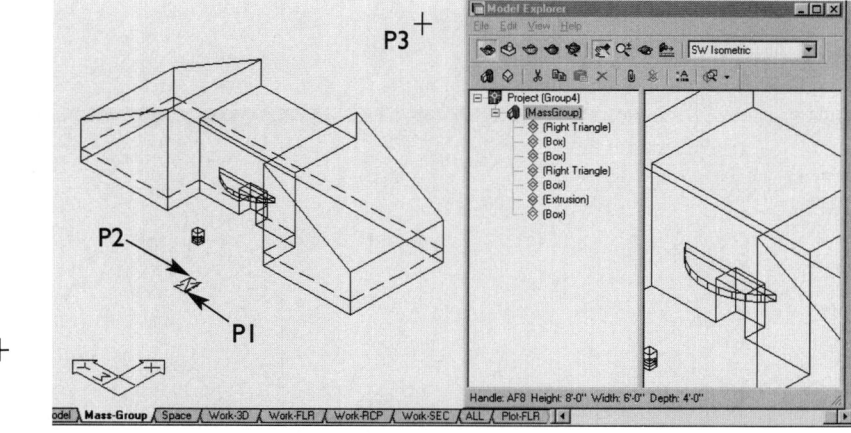

Figure 11.77 *Creating a slice of the mass group*

31. Select the **Maximize Window** button in the Model Explorer window. Select MassGroup in the Tree view; then position the pointer over the Graphics Area and right-click and choose **More>Zoom Extents** from the shortcut menu to display the building, as shown in Figure 11.78.

32. Position the pointer over the Graphics Area; then right-click and choose **Projection>Perspective** from the shortcut menu to display the building in perspective.

33. Position the pointer over the Graphics Area; then right-click and choose **Shading Modes>Hidden** from the shortcut menu to display the building as shown in Figure 11.79.

34. Choose **Close** from the menu bar of the Model Explorer.

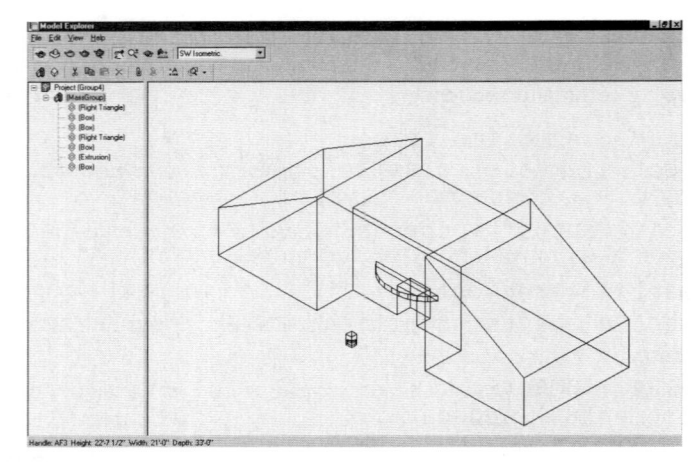

Figure 11.78 *Hidden display of the model in the Model Explorer*

35. Save the drawing.

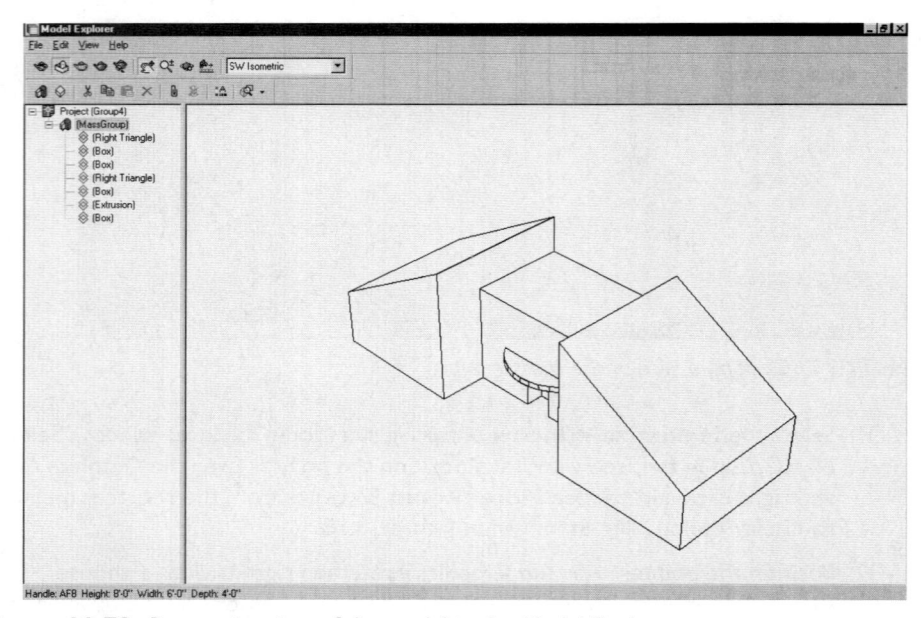

Figure 11.79 *Perspective view of the model in the Model Explorer*

SUMMARY

1. Creating groups for mass elements allows you to conduct Boolean operations of additive, subtractive and intersection on various mass elements to create complex shapes.

2. The Model Explorer allows you to view mass groups and the Tree View of the components.
3. The Model Explorer allows you to name mass elements and mass groups.
4. Mass elements can be moved in the Tree View to perform Boolean operations.
5. Referenced instances of the original mass shapes will be changed automatically when the original mass is edited.
6. Mass elements can be moved in the Z direction to develop complex mass shapes.
7. Slice markers are placed with the **Slice Generate** command (**AECSLICECREATE**).
8. Mass elements and mass groups can be attached to a slice marker to create a slice of the mass element or mass group.
9. Slices can be converted to polylines with the **AECSLICETOPLINE** command.

REVIEW QUESTIONS

1. The **Add Mass Group** command is used to create a _____.
2. The mass elements are attached to a group by the _____ command.
3. Elements attached to a mass group marker are by default attached with the _____ Boolean operation.
4. How is the Boolean operation of a mass element specified in the **Add Mass Element** dialog box?
5. The _____ _____ can be used to assign names to mass elements.
6. The _____ _____ _____ command can be used to create copies of mass elements that will change if the original mass element changes.
7. You can move mass elements above the Z=0 plane by selecting the _____ tab of the **Modify Mass Element** dialog box.
8. The _____ command is used to create a slice through a mass element or mass group.
9. Describe the procedure to change the elevation of a slice through a mass element using the shortcut menu.
10. What is the purpose of the **Convert to Polyline** command (**AECSLICETOPLINE**)?
11. What is the name of the layer created for mass elements and mass groups?
12. What are the advantages of the Model Explorer?

PROJECT

EX 11.1 CREATING A MASS GROUP AND SLICING THE MASS MODEL

1. Open *Ex11-5* from the *\ADT_Tutor\CH11* directory.
2. Choose **File>SaveAs** from the menu bar and save the file as *Proj11-5* in your student directory.
3. Move each of the mass elements in the drawing together to create a model as shown in Figure 11.80.
4. Create slice markers at 1', and 12' elevations.
5. Create a mass group of all the mass elements.
6. Slice the mass model at each of the elevations.
7. Create a perspective view of the model in the Model Explorer, as shown in Figure 11.80.
8. Save the drawing.

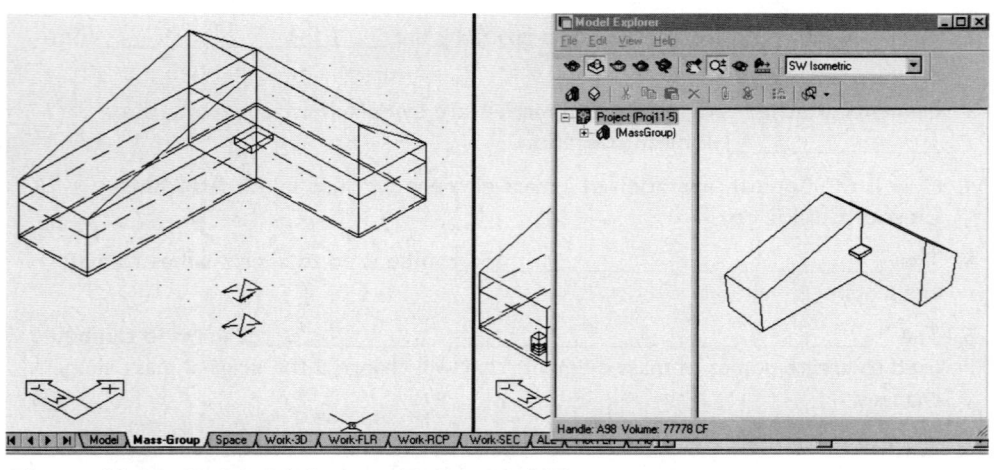

Figure 11.80 *Model and slices used in Exercise 11.1*

CHAPTER 12

Creating Spaces and Boundaries

INTRODUCTION

In the previous chapter, a mass model was created and sliced to create a floorplate. This chapter explains how to convert the slice to a space or to a space boundary. A space is an object that graphically represents the function of an area in the building. A space can be created with floor or ceiling components to represent the actual building space. Spaces can be created to represent such rooms or areas as offices, restrooms, or corridors. These predefined spaces are placed in the drawing to determine the size of the building and desired location of the spaces. Spaces can also be imported into the drawing to assist in space planning.

Space boundaries provide a representation of walls, which contain the space. The space and space boundaries combine to create a three-dimensional object that represents the walls, ceilings, and floors of the building. This space boundary can be converted to walls later in the design. Spaces and space boundaries are temporary objects created to aid in visualization, modeling, and presentation of a design.

Spaces and space boundaries can be created with or without the use of a model or floorplate. Design with spaces and space boundaries allows you to design from the inside out to create the building shape according to the desired function.

OBJECTIVES

Upon completion of this chapter the student will be able to

Insert a space in the drawing using the **Add Spaces** command (**SPACEADD**)

Modify the shape, size, and insertion point of a space

Use the **Space Styles** command (**SPACESTYLE**) to create spaces with specific area and dimensions

Modify a space using the **Modify Spaces** command (**SPACEMODIFY**) to change the space style, floor boundary, and ceiling boundary

Create a space from a polyline using the **Convert to Spaces** command (**SPACECONVERT**)

Divide and combine spaces using the **Divide Spaces** and **Join Spaces** commands (**SPACEDIVIDE** and **SPACEJOIN**)

Create space boundaries with specific height, width, and justification properties using the **Add Boundary** command (**SPACEBOUNDARYADD**)

Combine independent boundaries with the **Merge Boundaries** command (**SPACEBOUNDARYMODIFY**)

Attach a space to a boundary using the **Attach Spaces to Boundary** command (**SPACEBOUNDARYMERGESPACE**) and detach a space from a boundary using the **Split Boundary** command (**SPACEBOUNDARYSPLIT**)

Convert spaces, slices, and sketches to space boundaries using the **SPACEBOUNDARYCONVERT** command

Insert a space boundary in an existing space boundary with the **Add Boundary Edges** command (**SPACEBOUNDARYADDEDGES**)

Modify each space boundary segment using the **Edit Boundary Edges** command (**SPACEBOUNDARYEDGE**)

Delete a space boundary segement using the **Remove Boundary Edges** command (**SPACEBOUNDARYREMOVEEDGES**)

Release or attach objects to the anchor of the space boundary using the **Anchor to Boundary** command (**SPACEBOUNDARYANCHOR**)

Summarize the spaces of the drawing with the **Space Inquiry** command (**SPACEQUERY**)

Generate walls from space boundaries using the **Generate Walls** command (**SPACEBOUNDARYGENERATEWALLS**)

CREATING SPACES FOR SPACE PLANNING

A space is an Architectural Desktop object that assists you in planning. Spaces and space boundaries are created and edited in the space layout. Figure 12.1 shows a space on the left and a space boundary on the right. The space is represented as a hatched object when viewed in plan view, as shown in the right viewport. The left viewport displays the model view of the space without the hatching. The floor and ceiling components of the space are displayed in the left viewport, while the space boundary shown on the right wraps around the space to represent the walls of a building. Spaces and space boundaries can be created independently of each other or a space boundary can be generated from the space object.

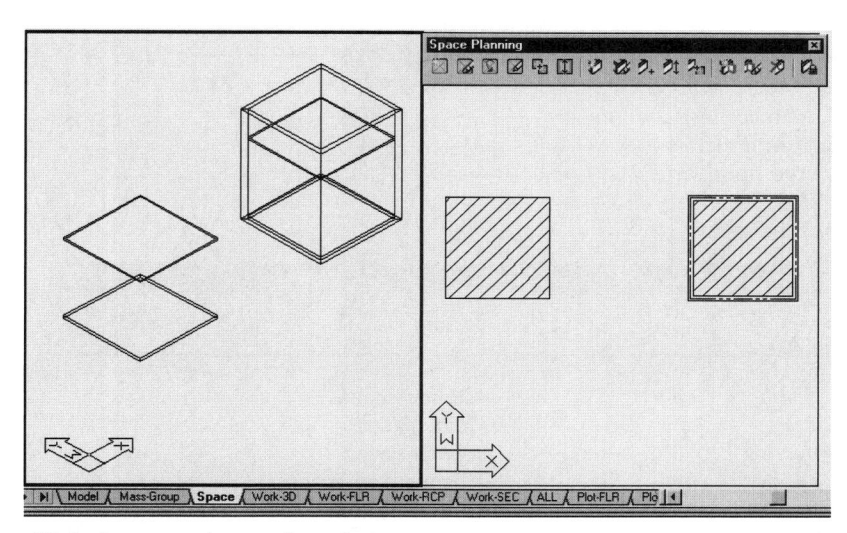

Figure 12.1 *Spaces and space boundaries*

The display configuration for the Space layout tab assists you in viewing the spaces and space boundaries. The left viewport displays the model view of the space with the ceiling and floor components. The hatching is turned OFF in the model viewport. The right viewport displays the space as hatched when viewed in the plan direction. When you have finished with space planning, you can convert the space boundary to walls. The walls are not displayed in the Space layout.

The commands for creating and editing spaces and space boundaries are located on the **Space Planning** toolbar, as shown in Figure 12.2. The left half of the toolbar includes commands for spaces and the right half includes commands for space boundaries.

The **Add Spaces** command is used to create a space. It allows you to insert predefined spaces in the drawing. See Table 12.1 for **Add Spaces** command access.

When you choose the **Add Spaces** command, the **Add Space** dialog box opens and you are prompted to specify the insertion point for the space. The space object is attached to the pointer when the command is executed. However, you select in the **Add Space** dialog box and edit the style, size, and properties of the space before specifying the insertion point, as shown in Figure 12.3.

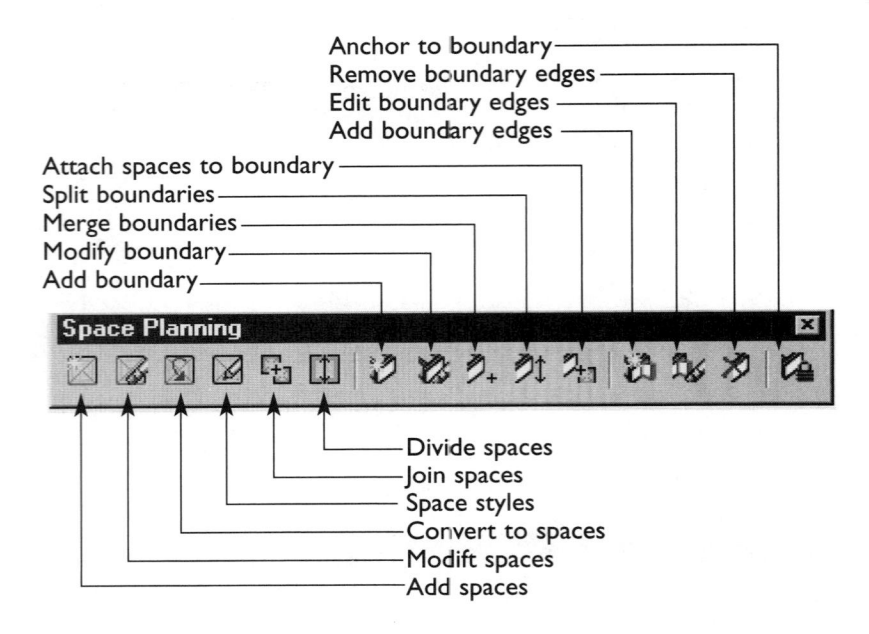

Figure 12.2 *Space Planning toolbar*

Callouts (top): Anchor to boundary; Remove boundary edges; Edit boundary edges; Add boundary edges

Callouts (left): Attach spaces to boundary; Split boundaries; Merge boundaries; Modify boundary; Add boundary

Callouts (bottom): Divide spaces; Join spaces; Space styles; Convert to spaces; Modift spaces; Add spaces

Menu bar	Concept>Space Planning>Add Spaces
Command prompt	SPACE or SPACEADD
Toolbar	Select Add Spaces from the Space Planning toolbar shown in Figure 12.2
Shortcut	Right-click over the drawing area and choose Design>Spaces>Add from the shortcut menu

Table 12.1 *Add Spaces command access*

The options of the **Add Space** dialog box are described below.

Style – The **Style** list includes the predefined space styles. The Standard space style is the default space style in drawings created from the Aec_arch (Imperial) template. Additional space styles can be listed if they are imported or created in the drawing.

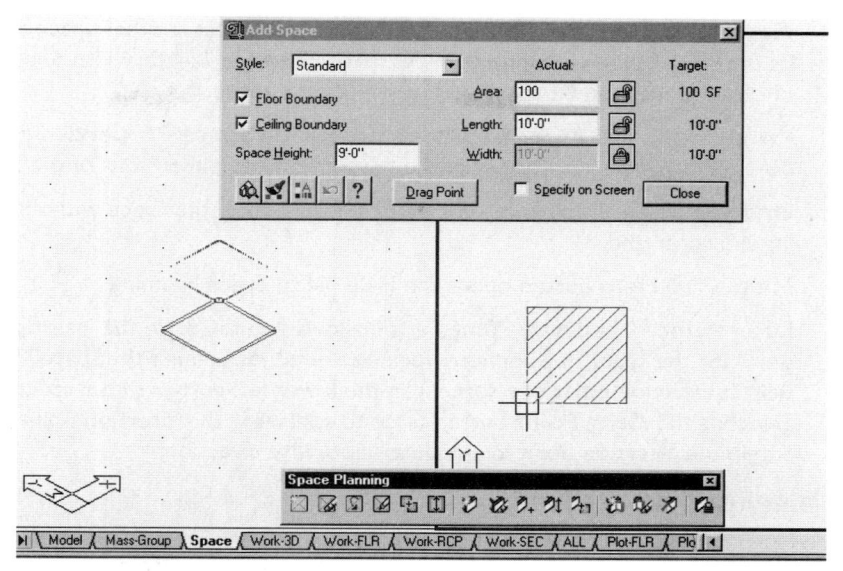

Figure 12.3 *Add Space dialog box*

>**Floor Boundary** – The **Floor Boundary** check box allows you to create a space with a floor boundary. The floor boundary simulates the existence of a floor system.
>
>**Ceiling Boundary** – The **Ceiling Boundary** check box allows you to create a space with a representation of the ceiling for the space.
>
>**Space Height** – The **Space Height** edit box defines the finish floor to finish ceiling distance between the **Floor Boundary** and the **Ceiling Boundary**.
>
>**Area** – The **Area** edit field list the area of active space. The area of the active space can be specified on screen or set to a value in the **Area** edit field. If the **Area** edit field is specified and locked, the length and width dimension of the space will vary as necessary to provide the area.
>
>**Length** – The **Length** edit field allows you to preset the length dimension of the space.
>
>**Width**–The **Width** edit field allows you to specify the width dimension of the space.
>
>**Specify on Screen** – When **Specify on Screen** is checked, you specify the length and width of the space by selecting locations in the drawing area to create the total area.
>
>**Floating Viewer** – The **Floating Viewer** button opens the **Viewer**, allowing you to see the space prior to insertion.

Match – Selecting the **Match** button allows you select another space and use its properties when creating the new space. The style, length, width, and height of other spaces can be specified for the current space.

Properties – The **Properties** button opens the **Space Properties** dialog box, which allows you to specify the description and dimensions of the space.

Undo – The **Undo** button will delete an insertion of the space without exiting the command.

Help – The **Help** button opens the Help file to Space Planning.

Drag Point – The **Drag Point** button allows you to toggle the insertion point for the space in a counter-clockwise direction around the space. The default insertion point of a space is in the lower left corner of the space. Selecting the **Drag Point** button prior to specifying the insertion point will toggle the insertion point in a counter-clockwise direction.

EDITING SPACE PROPERTIES

Each space inserted in the drawing has properties defined according to its space style. The space style of a space includes the dimensions or target area of the space and its name and description. These features can be edited in the **Space Properties** dialog box. Selecting the **Properties** button of the **Add Space** dialog box executes the **SPACEPROPS** command. See Table 12.2 for **SPACEPROPS** command access.

Command prompt	SPACEPROPS
Shortcut	Select a space; then right-click and choose Space Properties from the shortcut menu

Table 12.2 *SPACEPROPS command access*

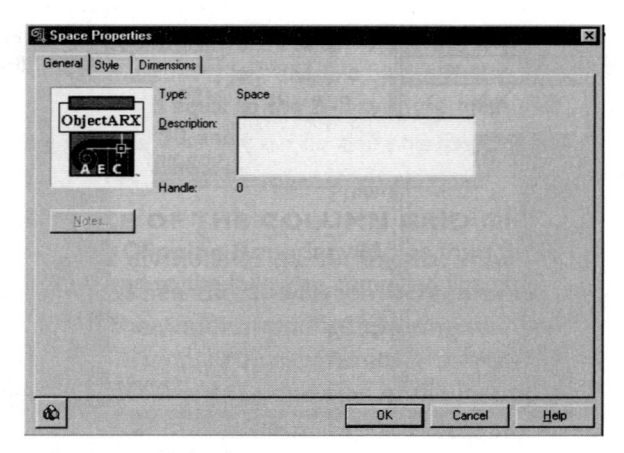

Figure 12.4 *Space Properties dialog box*

When you select the **Properties** button of the **Add Space** dialog box, the **Space Properties** dialog box opens; it includes three tabs, as shown in Figure 12.4. If you are editing an existing space, a **Location** tab is added, which specifies the location of the space.

General

The **General** tab, shown in Figure 12.4, includes a **Description** edit field, which allows you to type a description for the space. If you are editing an existing space, the **Notes** button is active and will open a **Notes** dialog box, which includes the **Text Notes** tab (allowing you to type additional notes regarding the space) and the **Reference Docs** tab (allowing you to enter the location of associated files).

Style

The **Style** tab allows you to select from the list of space styles that are in the current drawing. If no space styles have been created or imported, only the Standard space style will be included in the list, as shown in Figure 12.5.

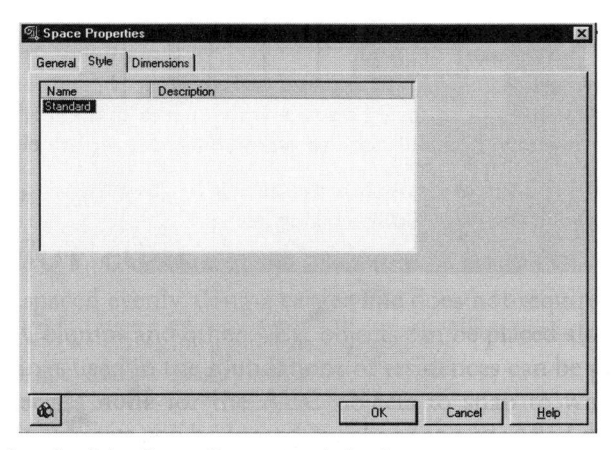

Figure 12.5 *Style tab of the Space Properties dialog box*

Dimensions

The **Dimensions** tab, shown in Figure 12.6, allows you to adjust the target dimensions of the space. Dimensions are initially specified when the space style is created. The **Dimensions** tab shown in Figure 12.6 defines the properties of the Standard space style. The Standard space style highlighted in the figure specifies a maximum area of 10,000 SF to allow the space style to be used for a various space style applications. The **Target, Minimum,** and **Maximum** values for the **Area, Length,** and **Width** are not edited in this dialog box; these options are defined through the **Space Styles** command. The options of the **Dimensions** tab are described below.

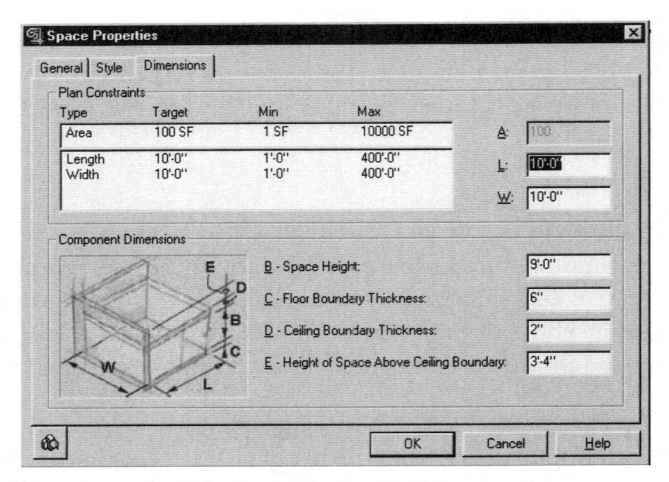

Figure 12.6 *Dimensions tab of the Space Properties dialog box*

Plan Constraints

>   **Area** – The **Area** row includes the target, minimum, and maximum areas for the space style. The area values listed define the area parameters for the space style. If you attempt to insert a space with an area that exceeds the maximum, an error message box will open to warn you of the violation.
>
>   **Length** – The target **Length** is the default length dimension of the space that is displayed in the **Add Space** dialog box.
>
>   **Width** – The target **Width** is the default width dimension of the space that is displayed in the **Add Space** dialog box.
>
>   **A:** – The **A:** value is the area as specified in the **Area** edit field of the **Add Space** dialog box. If the area in the **Add Space** dialog box is locked, the area in the **Space Properties** dialog box will be inactive. Changes in the size of the space will alter the area value shown in the A: field
>
>   **L:** – The **L:** value is the Length value as specified in the **Add Space** dialog box. This value can be edited in the **Space Properties** dialog box, and the change will be reflected in the **Add Space** dialog box.
>
>   **W:** – The **W:** value is the width of the space as specified in the **Add Space** dialog box.

Component Dimensions

The **Component Dimensions** section of the **Dimensions** tab describes the size and location of the floor and ceiling boundaries.

>   **B-Space Height** – The **B-Space Height** is the distance from finish floor to finish ceiling of the space.

C-Floor Boundary Thickness – The **C-Floor Boundary Thickness** is the thickness of the floor system as represented in the space. If a floor system is specified in the **Add Space** dialog box, a rectangular prism is created to represent the floor system. The **C-Floor Boundary Thickness** option controls the height of the rectangular prism.

D-Ceiling Boundary Thickness – The **D-Ceiling Boundary Thickness** option controls the thickness of the ceiling boundary object created to represent the ceiling.

E-Height of Space Above Ceiling Boundary – The **E-Height of Space Above Ceiling Boundary** option defines the vertical dimension of the ceiling cavity.

The default values for the **C-Floor Boundary Thickness, D-Ceiling Boundary Thickness,** and **E-Height of Space Above Ceiling Boundary** are set in the **AEC Dwg Defaults** tab of the **Options** dialog box, as shown in Figure 12.7.

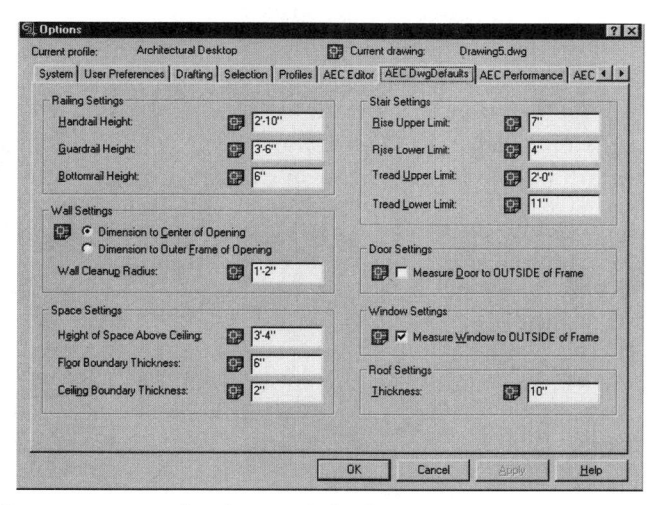

Figure 12.7 *Space settings in the Options dialog box*

The entity display of the Space layout includes the model display representation in the left viewport and the plan display representation in the right viewport. The plan display representation in the right viewport consists of the Net Boundary, Gross Boundary, and Hatching components. The display of the Gross Boundary is turned off by default. The Model viewport of the Space layout displays the space with Floor and Ceiling components. The **ENTDISPLAY** command can be used to edit the display of the space in a viewport. The default display properties have been overridden for the space shown in Figure 12.8. The Gross Boundary has been turned on in each viewport.

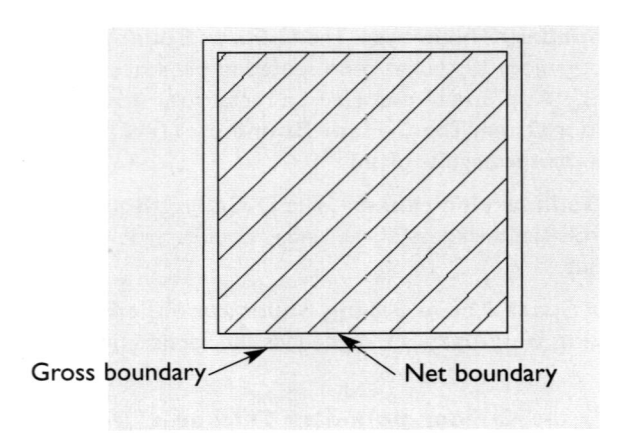

Figure 12.8 *Space geometry*

The **Net to Gross Offset** dimension of the space style defines the distance between the Net Boundary and Gross Boundary. That dimension can be set to zero or the anticipated wall width. When the boundary is created from the space, the boundary width can be determined automatically from the **Net to Gross Offset** dimension of the space style. If the **Net to Gross Offset** dimension is set to zero, you can set the Space Boundary width to a specific dimension to simulate the wall width.

Points are located on the Gross Boundary, which can be snapped to with the Node object snap mode to place the space. These points act as bumpers separating the spaces according to the **Net to Gross Offset** dimension. The points for the Node object snap are shown in Figure 12.9.

The Standard space style shown in Figure 12.10 has a **Net to Gross Offset** dimension equal to zero; therefore the Gross Boundary and Net Boundary are coincident. The points at the corners can be snapped to with the Node object snap. Therefore, you

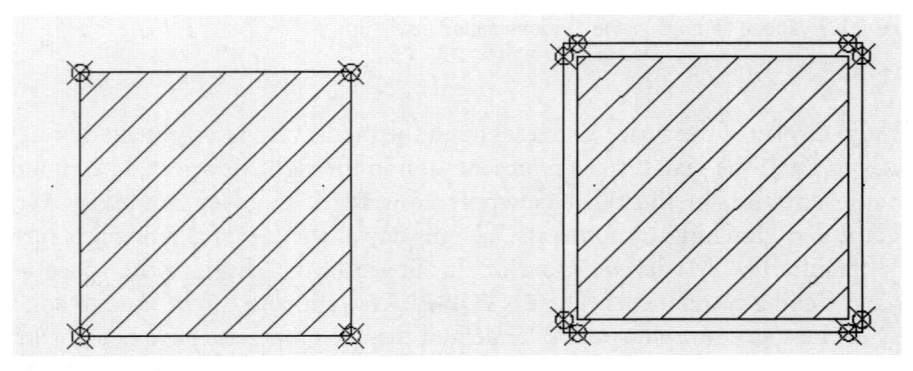

Figure 12.9 *Points of the space object*

THOMSON LEARNING
DISTRIBUTION CENTER
7625 EMPIRE DRIVE
FLORENCE, KY 41042

The enclosed materials are sent to you for your review by
MICHELLE SCARINGE

SALES SUPPORT

Date	Account	Contact
08/16/00	77855	356

SHIP TO: Kelly Liegikis
Broome Community College
901 Front Street
Binghamton NY 139021017

WAREHOUSE INSTRUCTIONS

SLA: 7   BOX: Staple

LOCATION	QTY	ISBN	AUTHOR/TITLE
K-ASY-034-01	1	0-7668-1262-6	WYATT ACCESS ACAD ARCH DESKTOP R2

INV# 27352692SM
PO# 16505062
DATE: / /
CARTON: 1 of 1
ID# 7612923

ASSEMBLY
-SLSB

VIA: UP

PAGE 1 OF 1

BATCH: 0991521

050/100

can place spaces adjacent to existing spaces snapping to the Node of the space, as defined in the **Net to Gross Offset** dimension.

Figure 12.10 shows the grips of each space on the Gross Boundary line. The grips are located at the corners and midpoint of each side of the space.

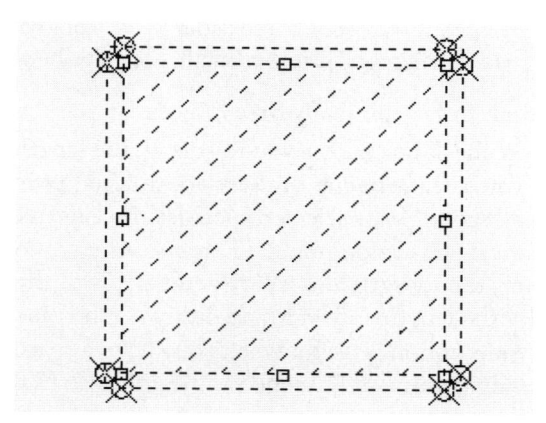

Figure 12.10 *Location of grips on a space*

POSITIONING AND SIZING OF SPACES

Spaces are placed in the drawing with the Node object snap. The points located at each of the four corners of a space allow the spaces to snap together. When you insert a space, the default handle for inserting the space is in the lower left corner of the space. Selecting the **Drag Point** button of the **Add Space** dialog box can change the position of the insertion handle. Selecting the **Drag Point** button will toggle the handle around the corners of the space in a counter-clockwise direction. This allows the positioning of spaces adjacent to others with precision from any direction.

The target area, length, and width are defined with the space style. However, you can set these dimensions dynamically upon insertion by selecting the **Specify on Screen** button of the **Add Space** dialog box. You can edit the size of the space prior to selecting the insertion point by selecting one or more of the following options in the command line.

    Style – Allows you to type the name of another space style.

    Area – Enter a new area for the space style.

    Length – Enter a new distance for the length or select two points on screen to define length.

    WIdth – Enter a new width distance or select on screen points to define the width.

Height – Enter a new height.

Move – Allows you to move the insertion point once it has been placed.

Size – Resets the size based on length and width.

Drag Point – Point-toggles the handle for the space.

MAtch – Prompts you to select an existing style; from which you can match the entire style definition or just the length, width, or height.

Using Specify on Screen to Dynamically Size Spaces

The **Specify on Screen** check box allows you to specify the length of the space in the drawing area. When you select **Specify on Screen**, the first point selected becomes the insertion point and second point selected specifies the length of the space according to the distance from the insertion point. Therefore, as you move the pointer after the first selection point, the **Length** and **Width** edit fields of the **Add Space** dialog box dynamically display the length and width according to the movement of the pointer. The **Specify on Screen** option is useful when you create a space between two existing spaces. The following command sequence and Figure 12.11 demonstrate the use of the **Specify on Screen** option.

```
( Select Add Spaces from the Space Planning toolbar and then select the
Specify on Screen check box of the Add Space dialog box.)
Command: AECSPACEADD
Insertion point or
[STyle/Area/Length/WIdth/Height/MOve/SIze/Drag
point/MAtch]: (Select point P1 in Figure 12.11.)
New size or
[STyle/Area/Length/WIdth/Height/MOve/SIze/Drag
point/MAtch]
<10'-0">: (Move the pointer to the right of P1 and select point P2 in
Figure 12.11 to specify the length.)
Rotation or
[STyle/Area/Length/WIdth/Height/MOve/SIze/Drag
point/MAtch] <0.00>: ENTER (Press ENTER to accept 0° rotation.)
```

The **Specify on Screen** check box is used to select the length of a space by selecting a point relative to the insertion point. However, if you deselect **Specify on Screen**, you can specify the length and width of a new space by selecting **WIdth** and **Length** options in the command line. Selecting the **Width** or **Length** option *before* you select the insertion point allows you to select two points on existing spaces to specify the length or width. This feature allows you to create a space to fit between existing spaces. Shown below is the command sequence in which the length and width were preset by the selection of points on existing geometry.

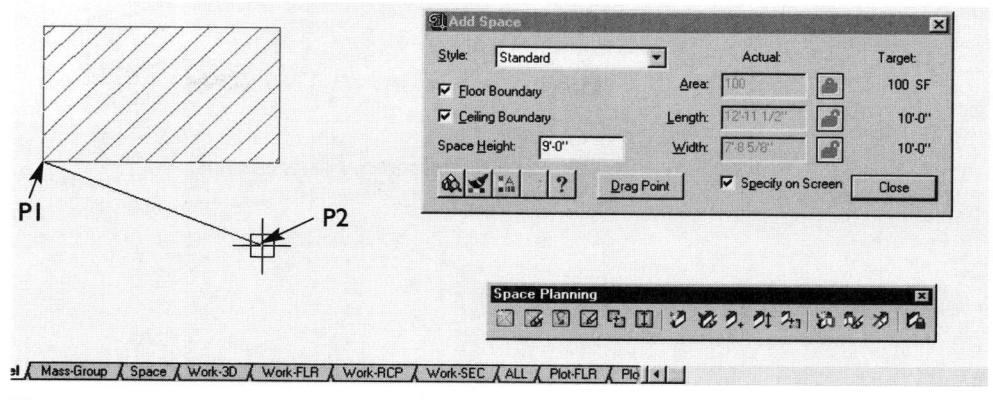

Figure 12.11 *Using Specify on Screen option*

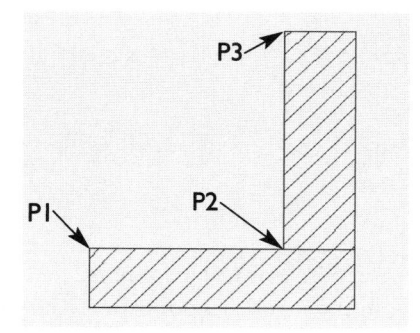

Figure 12.12 *Sizing a new space based upon existing spaces*

```
Command:
Command: _AecSpaceAdd
```
(Deselect the Specify on Screen check box.)
```
Insertion point or
[STyle/Area/Length/WIdth/Height/MOve/SIze/Drag
point/MAtch]: L
Length <10'-0">:
```
(Select point P1 in Figure 12.12.)
```
Specify second point:
```
(Select point P2 in Figure 12.12.)
```
Insertion point or
[STyle/Area/Length/WIdth/Height/MOve/SIze/Drag
point/MAtch]: WI
Width <10'-0">:
```
(Select point P2 as shown above.)
```
Specify second point:
```
(Select point P3 in Figure 12.12.)
```
Insertion point or
[STyle/Area/Length/WIdth/Height/MOve/SIze/Drag
point/MAtch]:
```
(Select point P4 in Figure 12.13.)

```
Rotation or
[STyle/Area/Length/WIdth/Height/MOve/SIze/Drag
point/MAtch] <0.00>:  ENTER  (Press ENTER to accept the 0° rota-
tion.)
```

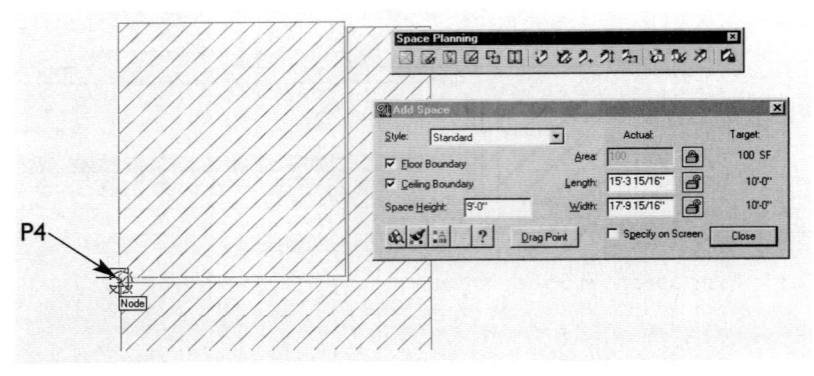

Figure 12.13 *Insertion point for the space*

The width and length of the space can be locked, allowing you to continue to insert spaces with the dimensions locked.

Dragging the Insertion Point

The default location for the insertion point of a space is located at the lower left corner of the space. You can toggle the insertion point in a counter-clockwise direction to the other corners of the space by selecting the **Drag Point** button in the **Add Space** dialog box. To view the location of the insertion point, you may need to move the **Add Space** dialog box. Figure 12.14 shows the insertion point position after the **Drag Point** button was selected twice. Each successive selection of the **Drag Point** button continues to move the insertion point around the space in a counter-clockwise direction.

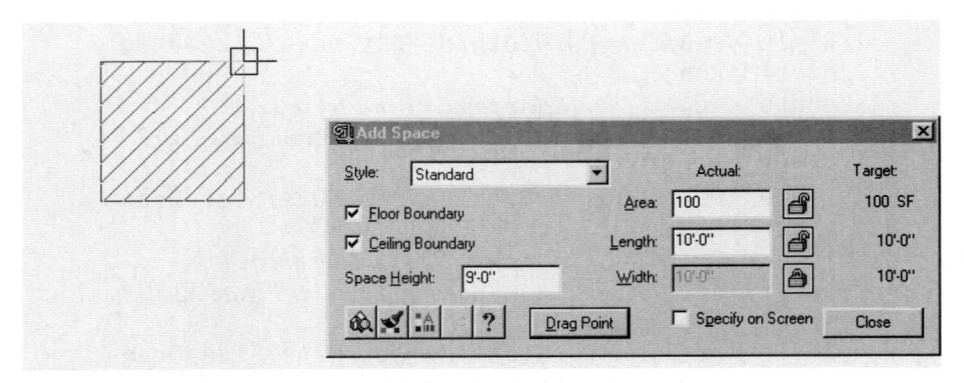

Figure 12.14 *Using Drag Point to toggle the insertion point*

TUTORIAL 12.1 INSERTING SPACES

1. Open *Ex12-1* from the *ADT Tutor\Ch12* directory.
2. Choose **File>SaveAs** from the menu bar and save the drawing as *Space1* in your student directory.
3. Select the Space layout, and then verify that the right viewport is active.
4. Right-click over an Architectural Desktop toolbar and choose **Space Planning** from the shortcut menu.
5. Select the **Add Spaces** command from the **Space Planning** toolbar.
6. Verify the following settings in the **Add Space** dialog box: set **Style** to Standard, select **Floor Boundary**, select **Ceiling Boundary**, set **Height** to 9', deselect **Specify on Screen**, toggle **Area Lock** ON, and set **Length** to 10', and **Width** to 10', as shown in Figure 12.15.

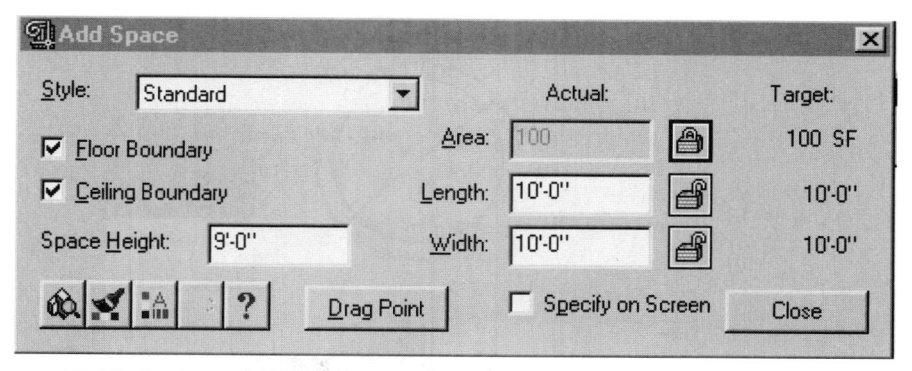

Figure 12.15 *Settings of the Add Space dialog box*

7. Respond to the command line prompts as shown below.

```
Command: _AecSpaceAdd
Insertion point or
[STyle/Area/Length/WIdth/Height/MOve/SIze/Drag
point/MAtch]: (Select a location in the drawing area of the right view-
port.)
Rotation or
[STyle/Area/Length/WIdth/Height/MOve/SIze/Drag
point/MAtch] <0.00>: ENTER (Press ENTER to accept the 0° rota-
tion.)
Insertion point or
[STyle/Area/Length/WIdth/Height/MOve/SIze/Drag
point/MAtch/Undo]: ENTER (Press ENTER to end the command.)
```

8. Right-click over the Osnap toggle of the Status Bar, choose **Settings** from the shortcut menu, and then clear all object snap modes and select the **Node** object snap mode.
9. Select the **Add Spaces** command from the **Space Planning** toolbar.
10. Select in the **Add Space** dialog box and change the **Length** to 5'. Retain the **Area Lock** and verify the following settings in the **Add Space** dialog box: set **Style** to Standard, select **Floor Boundary**, select **Ceiling Boundary**, set **Space Height** to 9' and **Width** to 20'.
11. Respond to the following command line prompts to specify the location as shown in Figure 12.16.

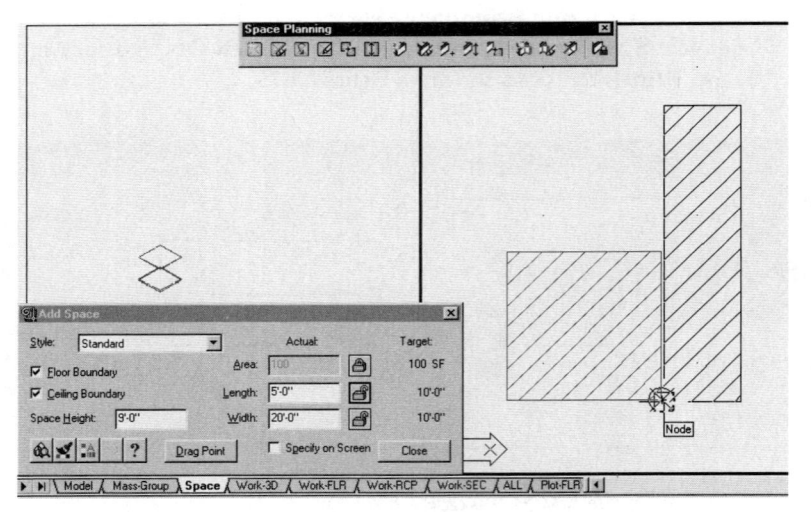

Figure 12.16 *Positioning the new space*

```
(Select in the right viewport to activate the drawing area.)
Command:
AECSPACEADD
Insertion point or
[STyle/Area/Length/WIdth/Height/MOve/SIze/Drag
point/MAtch]: (Move the pointer to the lower right corner of the previous space and select the Node of the space as shown in Figure 12.16.)
Rotation or
[STyle/Area/Length/WIdth/Height/MOve/SIze/Drag
point/MAtch] <0.00>: ENTER (Press ENTER to accept the 0° rotation.)
Insertion point or
[STyle/Area/Length/WIdth/Height/MOve/SIze/Drag
point/MAtch]: ENTER (Press ENTER to end the Add Space command.)
```

12. Select in the **Add Space** dialog box to return the focus to the **Add Space** dialog box; then verify that **Area** is locked and change the **Width** to 6'. Verify the remaining settings in the **Add Space** dialog box: set **Style** to Standard, select **Floor Boundary**, select **Ceiling Boundary**, and set **Space Height** to 9' and **Length** to 16'-8".

13. Select in the right viewport; then select the **Drag Point** button of the **Add Space** dialog box, continuing to select it until the insertion point is toggled to the upper right corner of the space.

14. Respond to the following command line prompts to place the space as shown in Figure 12.17.

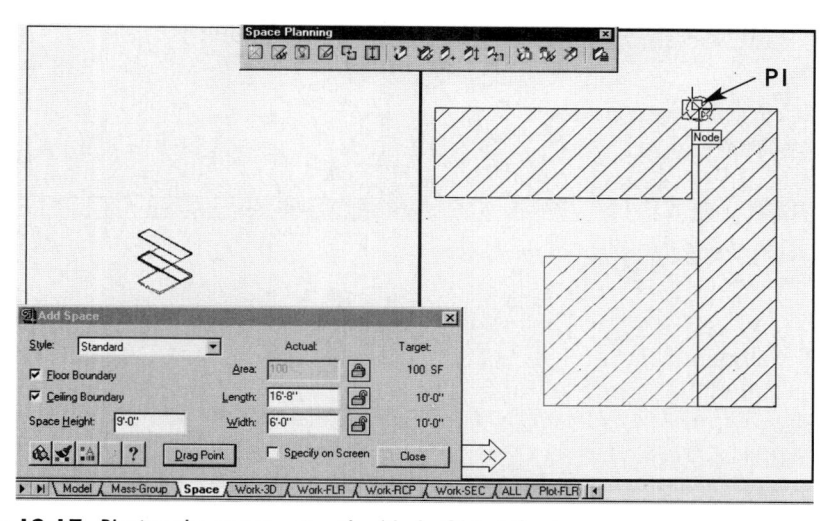

Figure 12.17 *Placing the space using the Node Object Snap*

```
Insertion point or
[STyle/Area/Length/WIdth/Height/MOve/SIze/Drag
point/MAtch]: (Move the pointer to point P1 in Figure 12.17 and
select the corner using the Node object snap.)
Rotation or
[STyle/Area/Length/WIdth/Height/MOve/SIze/Drag
point/MAtch] <0.00>: ENTER (Press ENTER to accept the 0° rota-
tion.)
Insertion point or
[STyle/Area/Length/WIdth/Height/MOve/SIze/Drag
point/MAtch/Undo]: ENTER (Press ENTER to end the command.)
```

15. Select in the drawing area of the right viewport to activate the viewport.

16. Select **Add Spaces** from the **Space Planning** toolbar and respond to the following command line prompts.

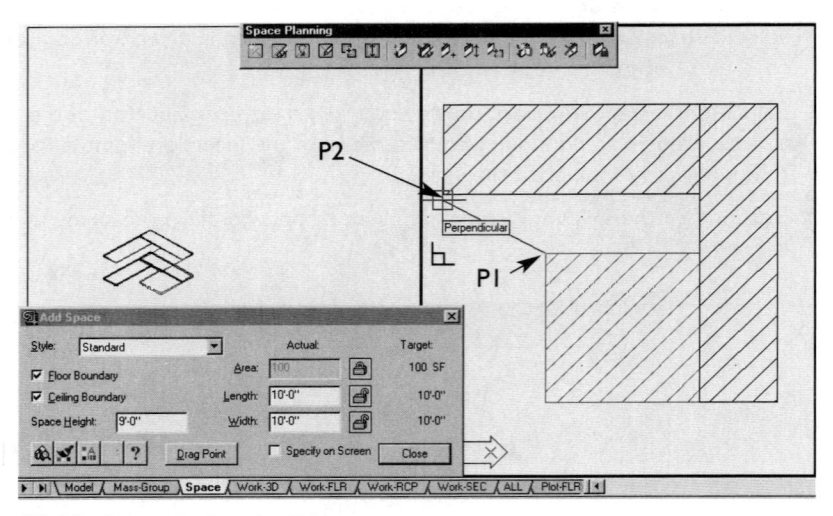

Figure 12.18 *Sizing the length of the space*

```
Command: _AecSpaceAdd
Insertion point or
[STyle/Area/Length/WIdth/Height/MOve/SIze/Drag
point/MAtch]:
L ENTER
```
(Select the Length option to define the length of the space.)
```
Length <10'-0">:
```
(Use the Node object snap to select the corner at point P1 in Figure 12.18.)
```
Specify second point: perp ENTER
To
```
(Select near P2 to specify a point perpendicular to the end of the space, as shown in Figure 12.18.)

```
Insertion point or
[STyle/Area/Length/WIdth/Height/MOve/SIze/Drag
point/MAtch]:
Wi ENTER
```
(Select the Width option to preset the width of the new space.)
```
Width <10'-0">:
```
(Use the Node object snap to select the corner at P1 in Figure 12.19.)
```
Specify second point: perp ENTER
To
```
(Select the side of the top Space at point P2 in Figure 12.19 to obtain the Width dimension.)

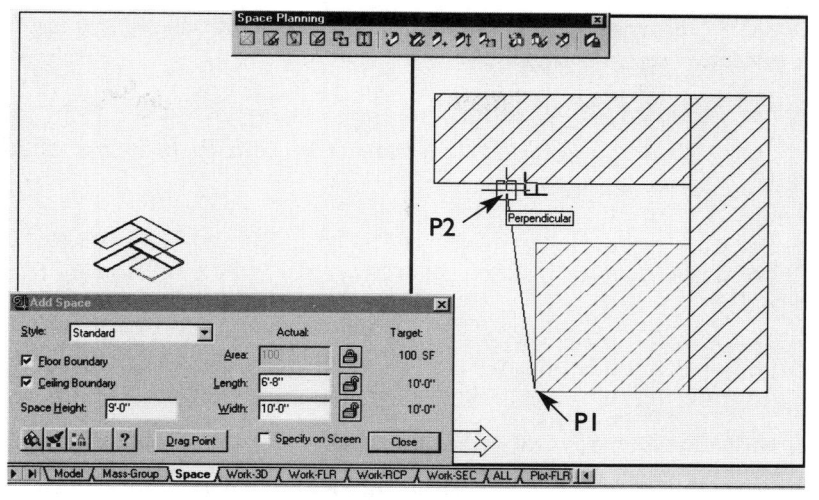

Figure 12.19 *Sizing the width of the space*

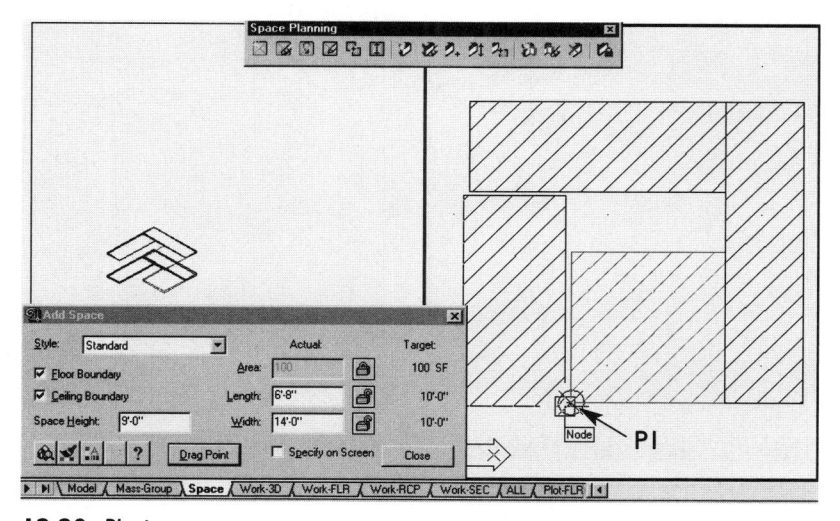

Figure 12.20 *Placing a space*

```
Insertion point or
[STyle/Area/Length/WIdth/Height/MOve/SIze/Drag
point/MAtch]:
DBOX  (Select in the Add Space dialog box to activate the dialog box.)
Insertion point or
[STyle/Area/Length/WIdth/Height/MOve/SIze/Drag
point/MAtch]:
```

DBOX *(Select the Drag Point button until the insertion point is toggled to the lower right corner.)*
Insertion point or
[STyle/Area/Length/WIdth/Height/MOve/SIze/Drag point/MAtch]: *(Select the lower left corner at P1 in Figure 12.20 using the Node object snap.)*
Rotation or
[STyle/Area/Length/WIdth/Height/MOve/SIze/Drag point/MAtch] <0.00>: ENTER *(Press ENTER to accept the 0° rotation.)*
Insertion point or
[STyle/Area/Length/WIdth/Height/MOve/SIze/Drag point/MAtch/Undo]: ENTER *(Press ENTER to end the command.)*

17. Verify that Ortho is ON in the Status Bar.
18. Select the space as shown in Figure 12.21 to display its grips.

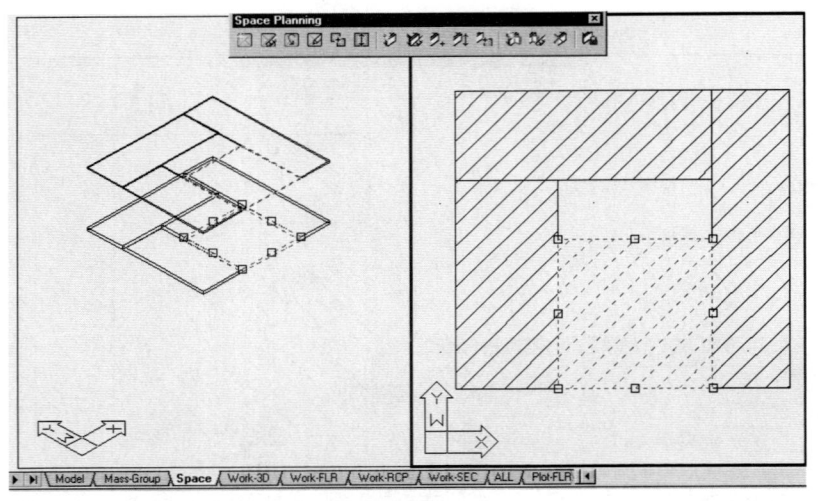

Figure 12.21 *Selecting a space for editing with grips*

19. Using grips, stretch the space up to the top space by selecting the grip at the midpoint and respond to the following command line prompts.

Command:
** STRETCH **
Specify stretch point or [Base point/Copy/Undo/eXit]: perp ENTER
To *(Select the top space near point P1 in Figure 12.22 when the Perpendicular object snap is displayed.)*

```
Command:    (Press ESC to clear the hot grip.)
Command:    (Press ESC to clear the display of grips.)
```

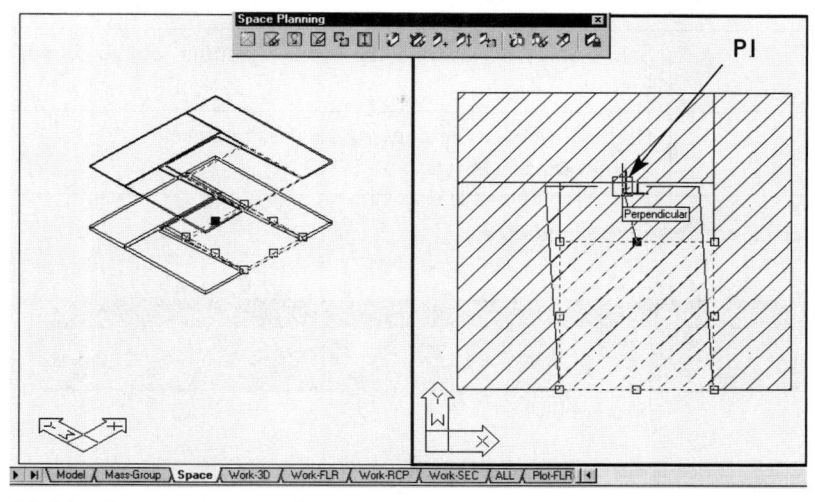

Figure 12.22 *Changing the size of a space using grips*

20. Save the drawing.

USING SPACE STYLES

Space styles can be created with preset dimensions and attributes. You can create your own space styles according to the function of the space with specific length, width, and area dimensions. Space styles are assigned a name according to the function of the space, such as office, entry, hotel room, cafeteria, or restroom. A collection of space styles is included in Architectural Desktop. They are located in the *Styles* directory of the *AutoCAD Architectural2\Content\Imperial* directory. The space styles drawings include Spaces-Commercial (Imperial), Spaces-Educational (Imperial), Spaces-Medical (Imperial), and Spaces-Residential (Imperial). You can transfer space styles from other files to the active drawing using the **Import** option of the **Space Styles** command. The **Export** option allows you to transfer the files from the active file to other files.

Space styles are created, edited, imported, or exported with the **Space Styles** command. See Table 12.3 for **Space Styles** command access.

When you select the **Space Styles** command from the **Space Planning** toolbar, the **Space Styles** dialog box opens, as shown in Figure 12.24. The **Space Styles** dialog box includes a list of space styles loaded in the drawing and six buttons to create, copy, edit, set from, purge and import or export space styles.

Menu bar	Concept>Space Planning>Space Styles
Command prompt	SPACESTYLE
Toolbar	Select Space Styles from the Space Planning toolbar shown in Figure 12.23
Shortcut	Right-click over the drawing area and choose Design>Spaces>Styles

Table 12.3 *Space Styles command access*

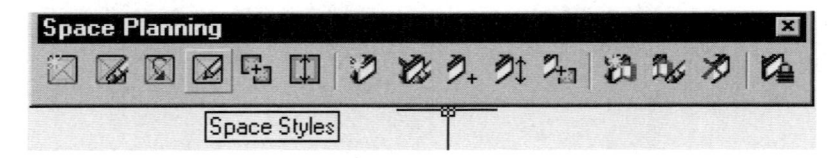

Figure 12.23 *Space Styles command on the Space Planning toolbar*

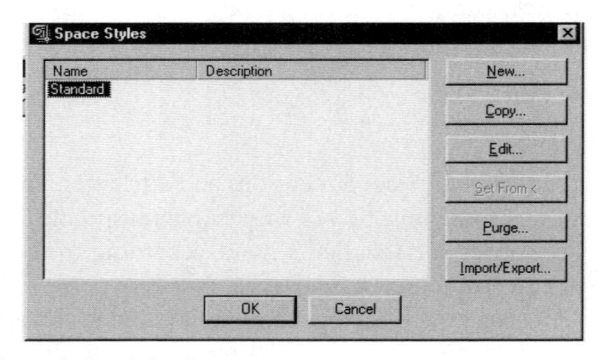

Figure 12.24 *Space Styles dialog box*

The buttons on the right of the **Space Styles** dialog box are described below.

> **New** – Creates a new name for a space style. The properties of the new space style are identical to the Standard space style.
>
> **Copy** – Copies an existing space style to a new name. To create a copy of a space style, select a space style from the space style list, and then select the **Copy** button to create a new name for the space style.
>
> **Edit** – Defines the area, dimensions, and other properties of the space style. Edit a space style by selecting the space style from the space style list and then selecting the **Edit** button, which opens the **Space Style Properties** dialog box, allowing you to define specific dimensions of the space style.

Set from< – Inactive for the **Space Styles** command.

Purge – Deletes unused space styles from the drawing.

Import/Export – Transfers saved space styles to other files or copies space styles from other files to the current drawing.

CREATING A SPACE STYLE

A new space style is created from the Standard space style. When you select the **New** button of the **Space Styles** dialog box, the **Name** dialog box opens, allowing you to type a name for the new style. After specifying the name, you define the properties of the style by selecting the **Edit** button, which opens the **Space Style Properties** dialog box, as shown in Figure 12.25. There are three tabs in the **Space Style Properties** dialog box: **General**, **Dimensions**, and **Display Props**; they allow you to define the properties of the space style. The tabs of the **Space Style Properties** dialog box are described below.

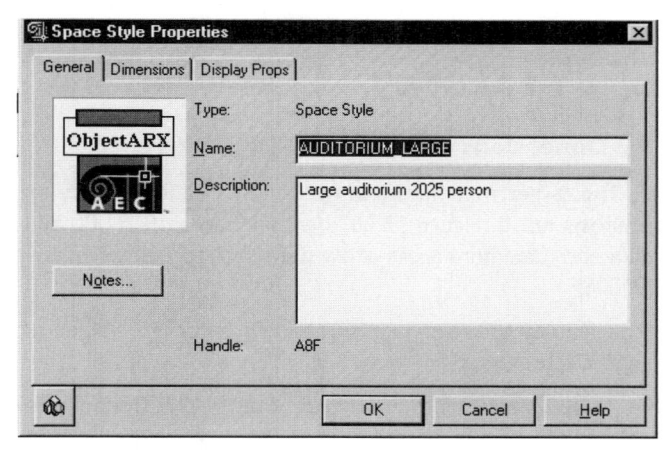

Figure 12.25 *General tab of the Space Style Properties dialog box*

General

The **General** tab, shown in Figure 12.25, allows you to add a description to the space style or edit the name. The name and description can be edited at any time during the drawing. The **Notes** button will open the **Notes** dialog box, which includes **Text Docs** and **Reference Docs** tabs. The **Text Docs** tab allows you to enter notes regarding style. The **Reference Docs** tab allows you to list the files and directories that include related information.

Dimensions

The **Dimension** tab, shown in Figure 12.26, allows you to define a target length, width, and area for the space. These dimensions form the upper and lower limits for the dynamic sizing of the space. The **Target Area, Target Length,** and **Target Width** are the values displayed in the **Add Space** dialog box.

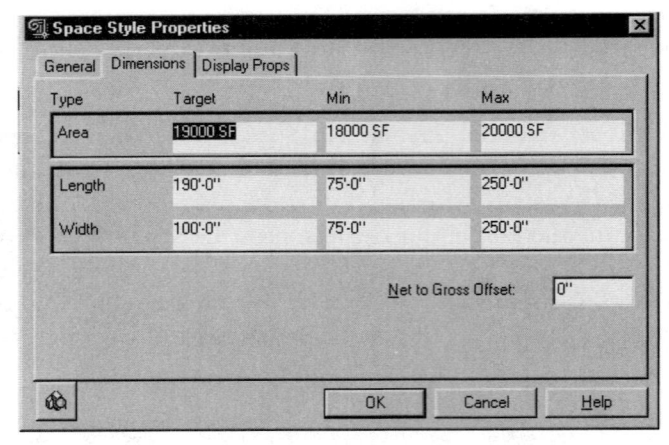

Figure 12.26 *Dimensions tab of the Space Style Properties dialog box*

The options of the **Dimensions** tab are described below.

> **Area** – The target, minimum, and maximum area for a space are defined in the **Dimensions** tab. In Figure 12.26, the target area of 19,000 SF is specified. The minimum and maximum areas allow the space to be sized between 18000SF and 20000SF.
>
> **Length** – The target, minimum, and maximum Length values allow you define the length of the space.
>
> **Width** – The target, minimum, and maximum Width values allow you to define the width of the space.
>
> **Net to Gross Offset** – The **Net to Gross Offset** is the distance each space is offset from another. The **Net to Gross Offset** can be set to zero, allowing the spaces to touch each other, or it can be set to the wall thickness.

The space style dimensions set the parameters for the space inserted in the drawing. Therefore, the minimum and maximum values must be within the range of minimum and maximum areas. The **Dimensions** tab for the AUDITORIUM_LARGE space style shown in Figure 12.26 has a target area of 19,000 square feet. The target length and width, if multiplied together, should create a target area of 19,000 square feet. The minimum and maximum values control the minimum and maximum size of the space. The length and width dimensions of the AUDITORIUM_LARGE space may not exceed values that create an area over 20,000 square feet. When the minimum and maximum values are created, the dimensions must conform to the target values. If the length and width values of the AUDITORIUM_LARGE space are set in the **Add**

Space dialog box to exceed the maximum and minimum values of the style, a warning box will be displayed.

Display Props

The **Display Props** tab allows you to define how the space style is displayed in the drawing. The display representations listed at the top of the tab include Model, Plan, Reflected, and Volume. The display representation marked with an asterisk is current for the viewport as shown in Figure 12.27. The **Display Props** tab allows you to edit the display of the space style. The **Property Source** includes Space Style and System Default properties. The buttons below the **Property Source** table—**Attach Override, Remove Override,** and **Edit Display Props**—are used to set the display of the space style.

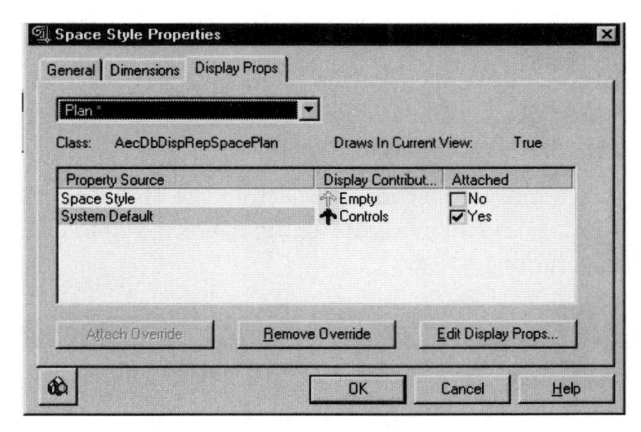

Figure 12.27 *Display Props tab of the Space Style Properties dialog box*

> **Attach Override** – Creates an override for the selected property source.
>
> **Remove Override** – Deletes the override or changes made in the display of the property source.
>
> **Edit Display Props** – Used when an override has been attached to edit the display of the space style.

Entity Properties

Selecting the **Edit Display Props** button opens the **Entity Properties** dialog box to control the visibility, layer, color, linetype, lineweight, and linetype scale of the components of the space style. Figure 12.28 shows the **Entity Properties** dialog box of the Plan display representation for the space style. The Plan display representation includes Net Boundary, Gross Boundary, and Hatch components. You can toggle the visibility of each of these components on or off by selecting the light bulb icon for the component.

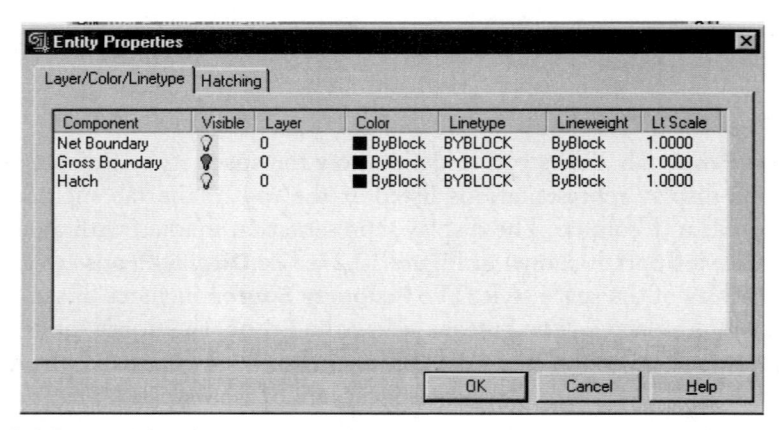

Figure 12.28 *Layer/Color/Linetype tab of the Entity Properties dialog box*

The **Entity Properties** dialog box for the Model display representation includes Floor and Ceiling components. The visibility of the Floor or Ceiling component can be toggled OFF in the **Layer/Color/Linetype** tab.

You can adjust the hatching properties by selecting the **Hatching** tab of the **Entity Properties** dialog box, as shown in Figure 12.29; it allows you to change the pattern, scale, angle, and orientation.

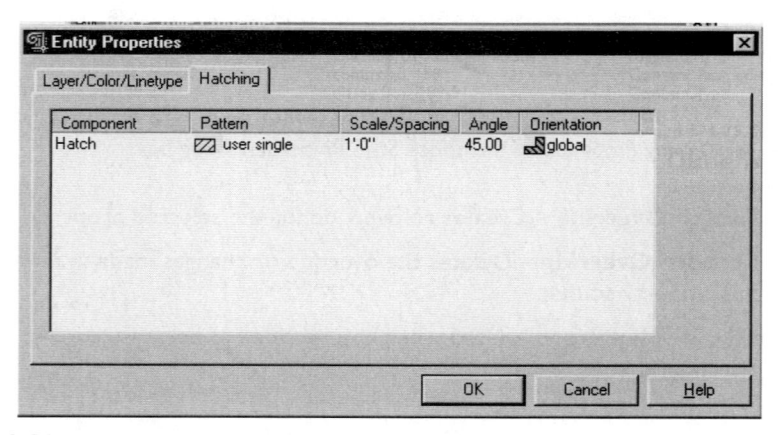

Figure 12.29 *Hatching tab of the Entity Properties dialog box*

MODIFYING SPACES

Existing spaces can be changed to different space styles or their dimensions can be changed through the **Modify Spaces** command. The **Modify Spaces** command allows you to change the length and width dimensions. See Table 12.4 for **Modify Spaces** command access.

Creating Spaces and Boundaries

Menu bar	Concept>Space Planning>Modify Spaces
Command prompt	SPACEMODIFY
Toolbar	Select Modify Spaces from the Space Planning toolbar as shown in Figure 12.30
Shortcut	Right-click over the drawing area and choose Design>Spaces>Modify
Shortcut	Select a space; then right-click and choose Space Modify

Table 12.4 *Modify Spaces command access*

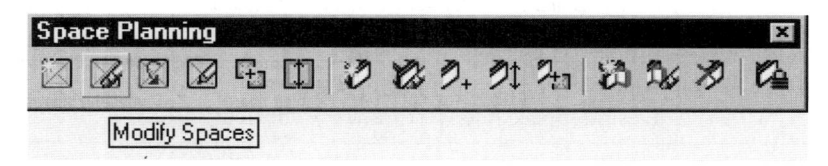

Figure 12.30 *Modify Spaces command on the Space Planning toolbar*

When you select **Modify Spaces** from the **Space Planning** toolbar, you are prompted to select a space. Selecting a space opens the **Modify Space** dialog box, as shown in Figure 12.31.

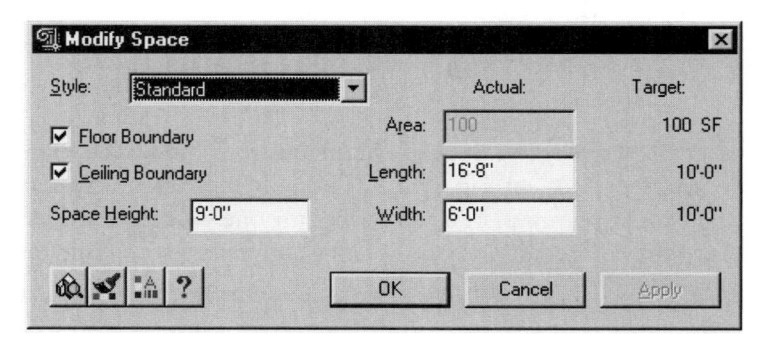

Figure 12.31 *Modify Space dialog box*

The **Modify Space** dialog box is identical to the **Add Space** dialog box, except that it does not include the locks for the sizes and the **Specify on Screen** check box. You can change a space by editing the dialog box and then selecting the **Apply** button to execute the change. Selecting the **OK** button will execute the change and close the **Modify Space** dialog box.

CONVERTING POLYLINES AND SLICES TO A SPACE

Drawing polylines and then converting the polylines to spaces can create spaces for space planning. The spaces can then be changed to a specific space name with space styles. Therefore, you can create a sketch that consists of polylines and convert the polylines to a space. Slices for floorplates can also be converted to polylines, which can then be converted to spaces. The **Convert to Spaces** command (**SPACECONVERT**) will convert a polyline to a space. See Table 12.5 for **Convert to Spaces** command access.

Menu bar	Concept>Space Planning>Convert to Spaces
Command prompt	SPACECONVERT
Toolbar	Select Convert to Spaces from the Space Planning toolbar as shown in Figure 12.32
Shortcut	Right-click over the drawing area and choose Design>Space>Convert from the shortcut menu

Table 12.5 *Convert to Spaces command access*

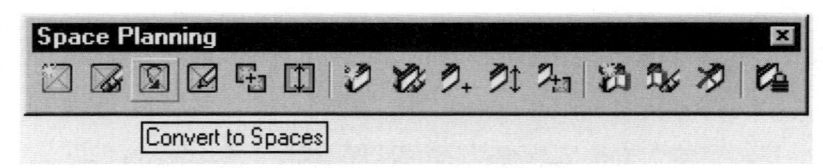

Figure 12.32 *Convert to Spaces command on the Space Planning toolbar*

When you select the **Convert to Spaces** command from the **Space Planning** toolbar, you are prompted to select a polyline. The polyline can be created from the lines, arcs, rectangles, or polygons. You can select only one polyline for the conversion. Shown below are the command line prompts for converting a polyline to a space.

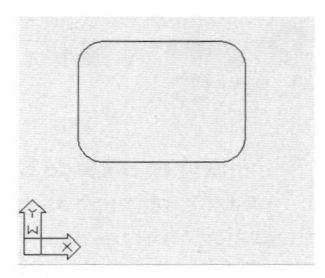

Figure 12.33 *Polyline geometry*

```
Command: _AecSpaceConvert
Select a polyline: (Select a shape as shown in Figure 12.33.)
Erase layout geometry? [Yes/No] <Y>: N ENTER (Press
```
ENTER *to retain the geometry in the drawing and open the Space Properties dialog box, as shown in Figure 12.34.)*

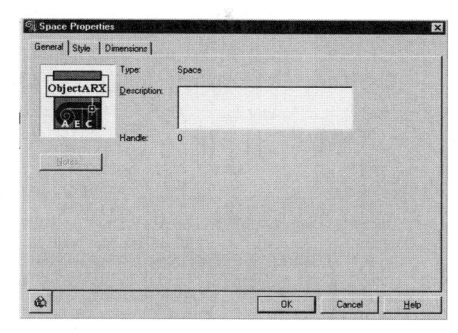

Figure 12.34 *Space Properties dialog box*

(Edit the Space Properties dialog box to select a space style, and then select OK to dismiss the dialog box and display the space, as shown in Figure 12.35.)

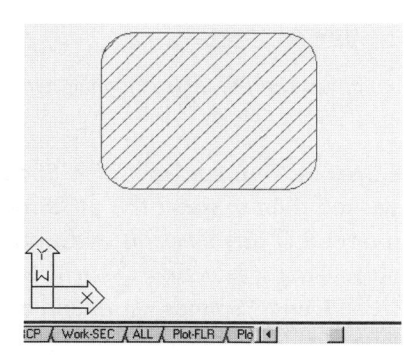

Figure 12.35 *Space created from the polyline*

CONVERTING A SLICE TO A SPACE

A slice created from a floorplate can be converted to a space. To convert the slice to a space, you must first convert the slice to a polyline using the **Convert to Polyline** command (**AECSLICETOPLINE**), as discussed in Chapter 11. The **Convert to Polyline** command allows you to select the slice and then convert it to a polyline. You can then convert the polyline to a space using the **Convert to Spaces** command (**SPACECONVERT**).

DIVIDING AND COMBINING SPACES

When a slice is converted to a space, the entire floorplate is created as one space. This space can be divided and manipulated with the **Divide Spaces** and **Join Spaces** commands. These commands allow you to cut the space into smaller spaces and join smaller spaces. The **Modify Space** command is then used to change the space style of the smaller spaces. See Table 12.6 for **Divide Spaces** command access.

Menu bar	Concept>Space Planning>Divide Spaces
Command prompt	SPACEDIVIDE
Toolbar	Select Divide Spaces from the Space Planning toolbar, shown in Figure 12.36
Shortcut	Select the space; then right-click and choose Divide from the shortcut menu

Table 12.6 *Divide Spaces command access*

Figure 12.36 *Divide Spaces command on the Space Planning toolbar*

When you select the **Divide Spaces** command from the **Space Planning** toolbar, you are prompted to select a space and then to select the dividing line start and endpoint. The dividing line can begin or end at any location within the space. A space can be divided without the aid of object snaps; selecting near the boundary of the space will select the edge of the space. The **Divide Spaces** command can be used to divide each space into a number of smaller spaces. The command line prompts for the **Divide Spaces** command are shown below.

```
Command: _AecSpaceDivide
Select a space: (Select the space at point P1 in Figure 12.37.)
Divide line start point: (Select a point near P2 in Figure 12.37.)
Divide line end point: (Select a point near P3 in Figure 12.37.)
(Space is divided as shown in Figure 12.38.)
```

The **Divide Spaces** command will divide only one space in the operation; if the dividing line crosses two spaces, only the first space encountered will be divided. In

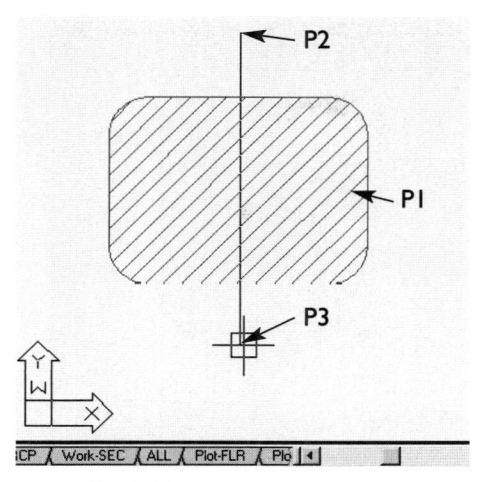

Figure 12.37 *Selection points for dividing a space*

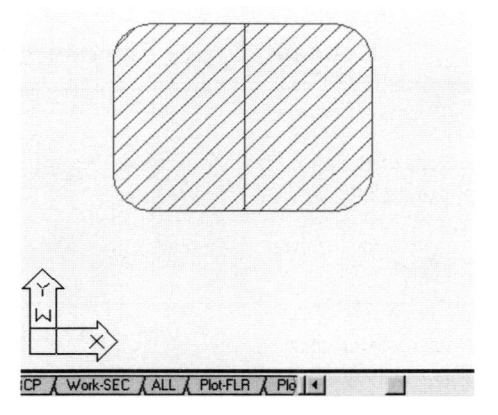

Figure 12.38 *Space divided with the Divide Spaces command*

Figure 12.39, the space was divided by a horizontal dividing line from P1 to P2. Although the dividing line crossed two spaces, only the space on the left was divided because its boundaries were encountered first.

While the **Divide Spaces** command allows you to divide a space into two components, the **Join Spaces** command allows the two spaces to be combined into one space. The **Join Spaces** command can be used to create custom shapes for a space by combining spaces. See Table 12.7 for **Join Spaces** command access.

When you select the **Join Spaces** command from the **Space Planning** toolbar, you are prompted to select a space and then to select another space to join with the first. The spaces must be adjacent to each other. The style of the new space created is determined

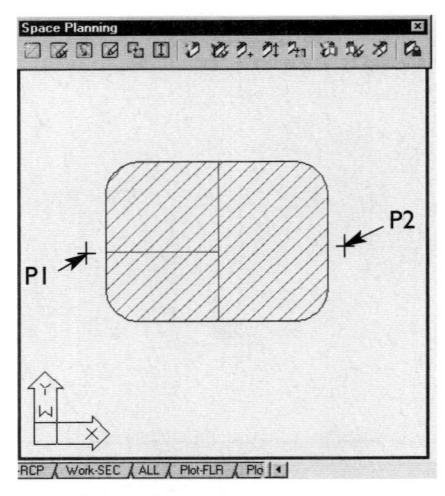

Figure 12.39 Selection points for dividing a space

Menu bar	Concept>Space Planning>Join Spaces
Command prompt	SPACEJOIN
Toolbar	Select Join Spaces from the Space Planning toolbar as shown in Figure 12.40
Shortcut	Select a space; then right-click and choose Join from the shortcut menu

Table 12.7 Join Spaces command access

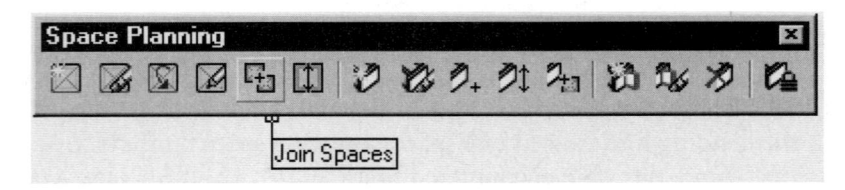

Figure 12.40 Join Spaces command on the Space Planning toolbar

from the space style of the first space selected. Therefore, if you join an office space with a corridor space and you select the office as the first space, the new space will be an office space. The command line prompts for the **Join Spaces** command are shown below.

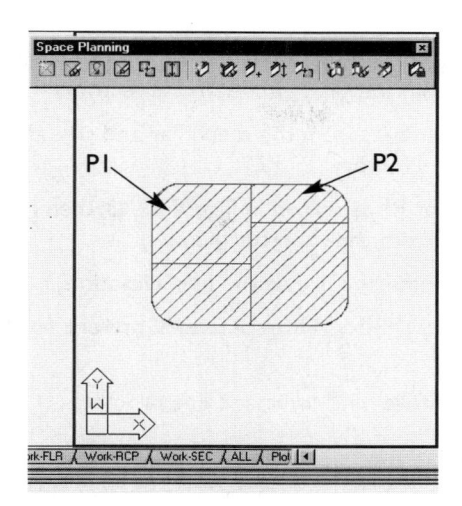

Figure 12.41 *Selection points for joining two spaces*

```
Command: _AecSpaceJoin
Select a space: 
```
(Select a space at point P1 in Figure 12.41.)
```
Select neighboring space to join with: 
```
(Select the space at point P2 in Figure 12.41.)
```
Command:
```
(Spaces are merged as shown in Figure 12.42.)

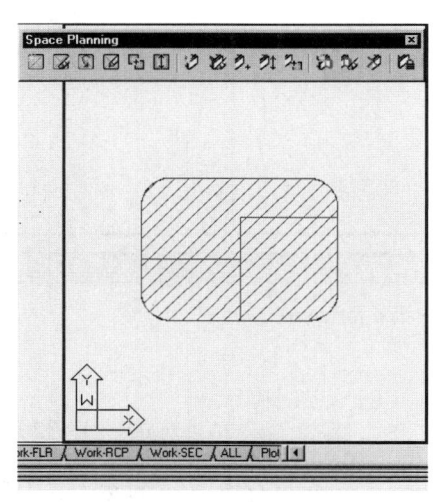

Figure 12.42 *Spaces merged with the Join Spaces command*

TUTORIAL 12.2 CREATING A SPACE STYLE

1. Open *Ex12-2* from the *\ADT_Tutor\Ch12* directory.
2. Choose **File>SaveAs** from the menu bar and save the drawing as *Space2.dwg* in your student directory.
3. Select the slice at P1 as shown in Figure 12.43; then right-click and choose **Set Elevation** from the shortcut menu.
4. Type **1'** in the command line as the new elevation.
5. Choose **Concept>Slice Floorplates>Convert to Polyline** from the menu bar.
6. Respond to the following command line prompts.

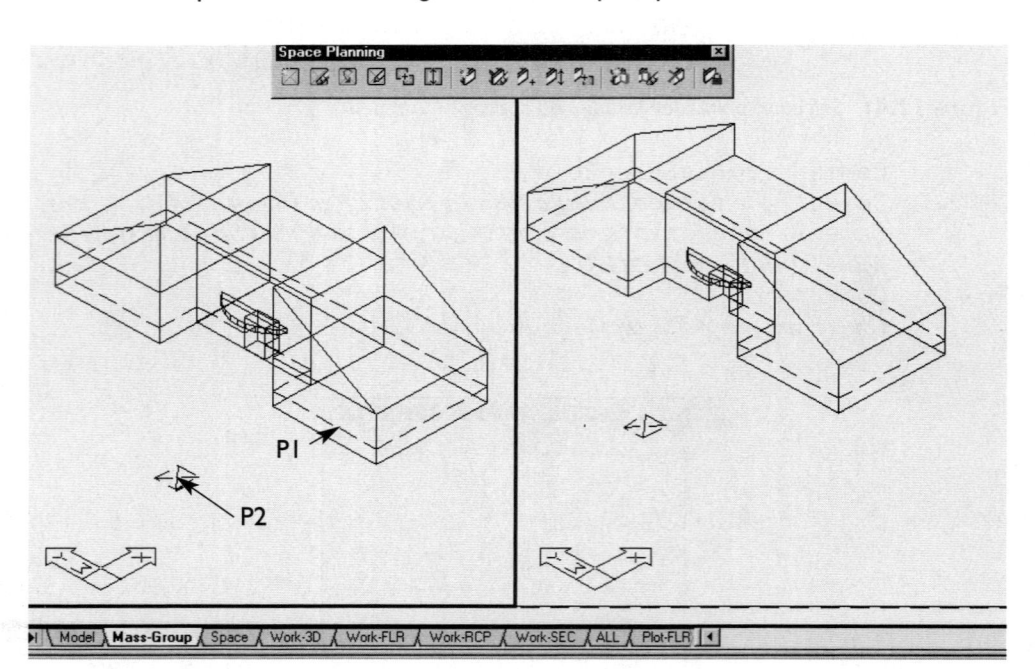

Figure 12.43 *Selecting the slice for conversion*

```
Command: _AecSliceToPline
Select slices: (Select the slice marker at P2 in Figure 12.43.) 1 found
Select slices: ENTER (Press ENTER to end slice selection.)
1 polyline(s) created.
```
(Polyline created as shown in Figure 12.44.)

Creating Spaces and Boundaries 891

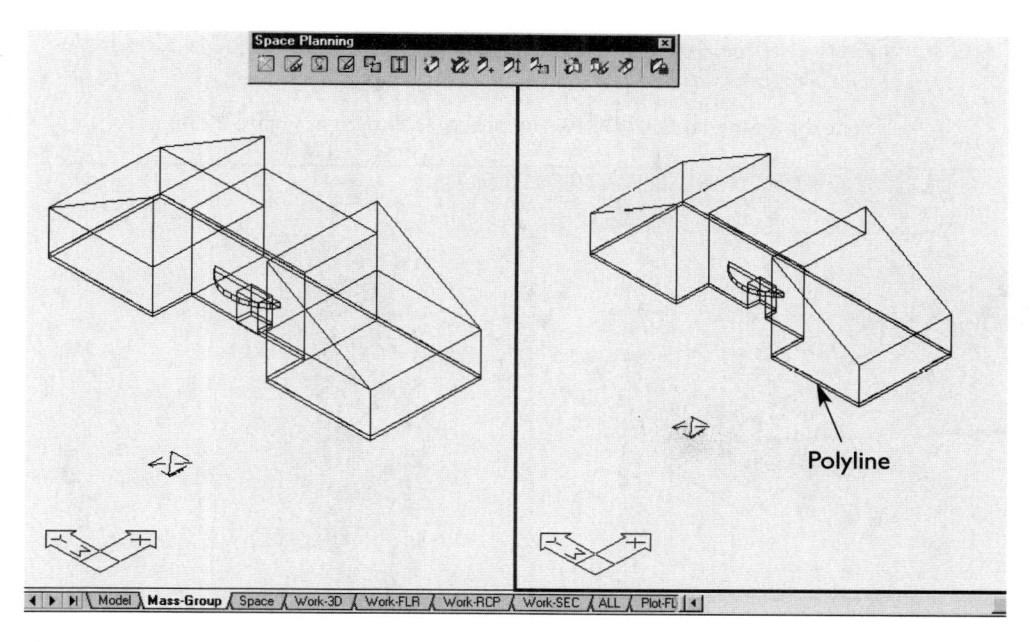

Figure 12.44 *Polyline created from the slice*

7. Select the Space layout and verify that the right viewport is active.

8. Select **Convert to Spaces** from the **Space Planning** toolbar and respond to the following command line prompts.

```
Command: _AecSpaceConvert
Select a polyline: (Select the polyline shown in the right viewport.)
Erase layout geometry? [Yes/No] <Y>: n ENTER (Press
```
ENTER *to retain the geometry and open the Space Properties dialog box.) (Select the Style tab of the Space Properties dialog box to verify the Standard Space Style, then select the Dimensions tab as shown in Figure 12.45.)*

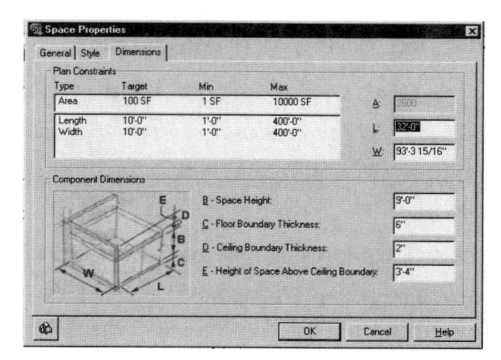

Figure 12.45 *Space Properties dialog box*

(Verify that the Area of the Space is 2600, and then select OK to dismiss the Space Properties dialog box.)

9. Select the Space layout to view the space, as shown in Figure 12.46.

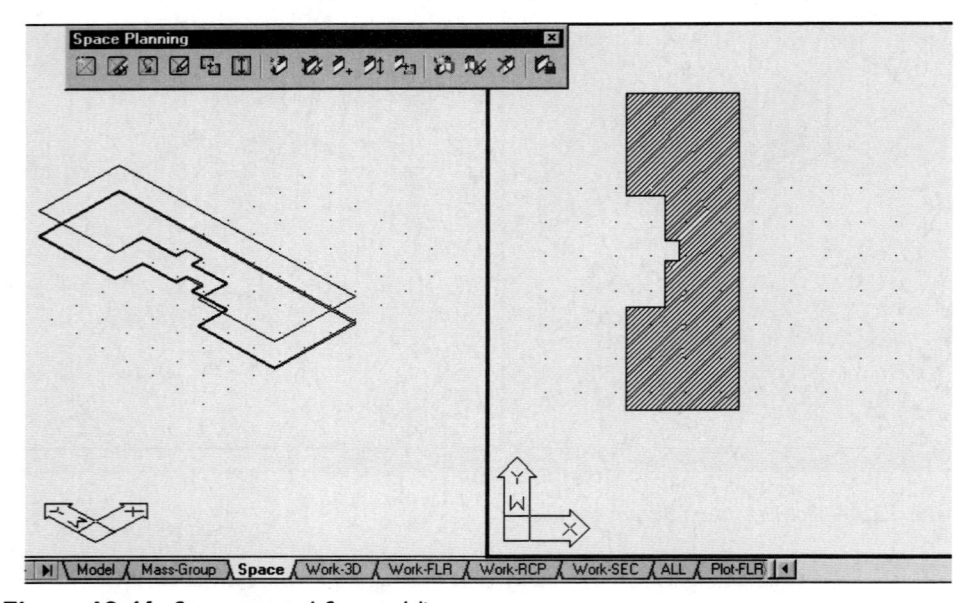

Figure 12.46 *Space created from polyline*

10. Verify that Ortho is toggled ON.
11. Select the **Divide Spaces** command from the **Space Planning** toolbar.
12. Select in the right viewport to activate the viewport and then respond to the following command line prompts to divide the space.

    ```
    Command: _AecSpaceDivide
    Select a space: (Select the space at point P1 in Figure 12.47.)
    Divide line start point: (Select point P2 with the Node object snap as shown in Figure 12.47.)
    Divide line end point: (Select a point near P3 in Figure 12.47.)
    (Space divided with the Space Divide command as shown in Figure 12.48.)
    ```

13. Select **Space Styles** from the **Space Planning** toolbar to open the **Space Styles** dialog box.
14. Select the **Import/Export** button of the **Space Styles** dialog box.
15. Select the **Open** button of the **Import/Export** dialog box, and then select the *Spaces-Commercial (Imperial).dwg* from the *AutoCAD Architectural2\Content\Imperial\Styles* directory, as shown in Figure 12.49.

Creating Spaces and Boundaries 893

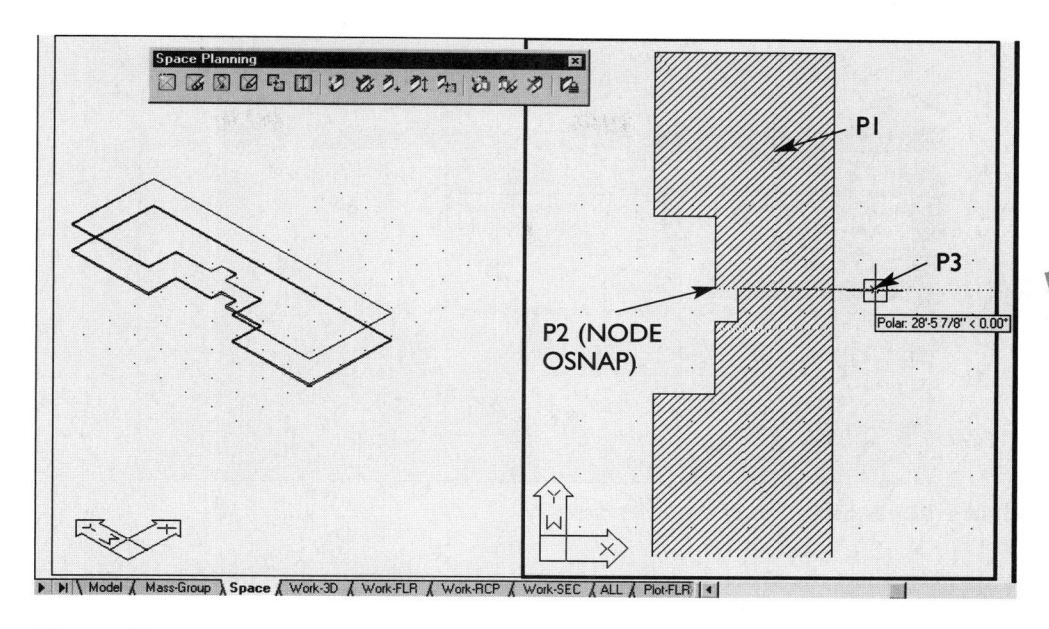

Figure 12.47 *Selecting points for dividing a space*

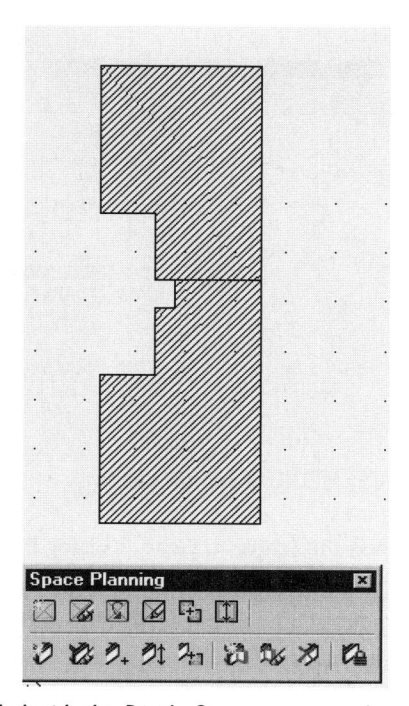

Figure 12.48 *Space divided with the Divide Spaces command*

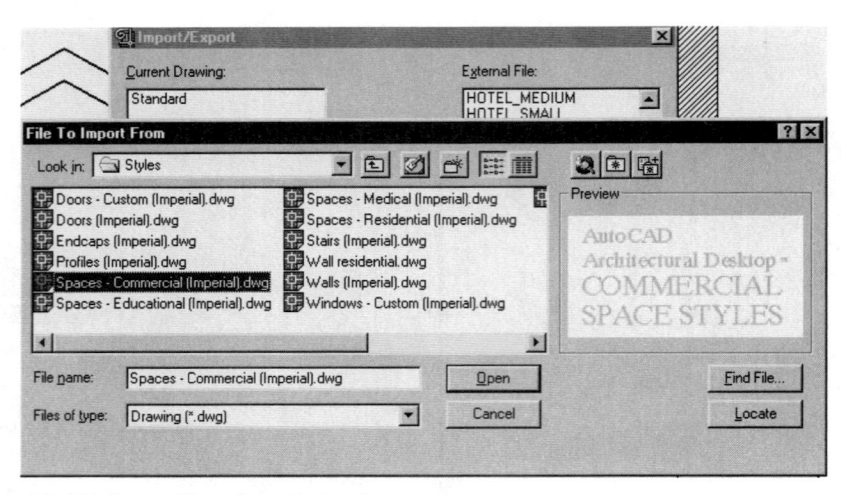

Figure 12.49 Space files of the Styles directory

16. Select the **Open** button to select the file and dismiss the **File to Import From** dialog box.
17. Select OFFICE_MEDIUM from the **External File** list and then select the **Import** button, as shown in Figure 12.50.

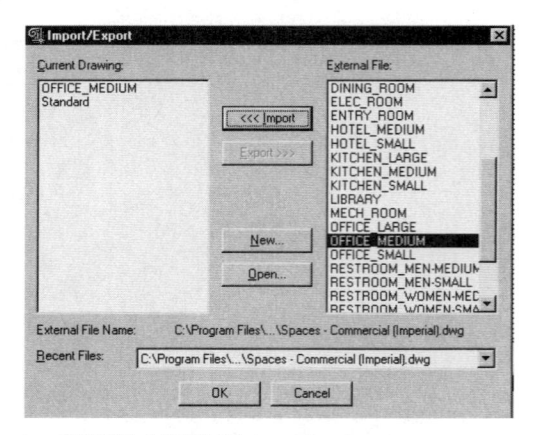

Figure 12.50 Importing OFFICE_MEDIUM into the drawing

18. Select **OK** to dismiss the **Import/Export** dialog box.
19. Select the OFFICE_MEDIUM space style, and then select the **Copy** button of the **Space Styles** dialog box to open the **Name** dialog box.
20. Type **OFFICE_CAD** in the **Name** dialog box, as shown in Figure 12.51.
21. Select **OK** to dismiss the Name dialog box.

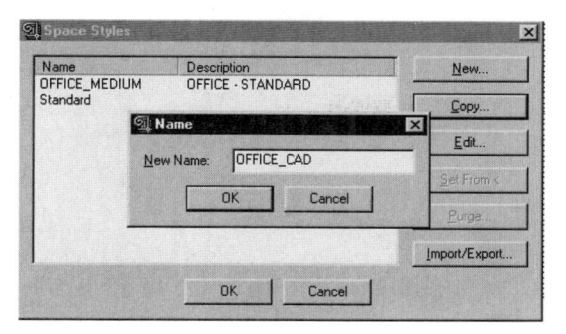

Figure 12.51 *Typing a new name for the space style*

22. Select the OFFICE_CAD space style and then select the **Edit** button to open the **Space Style Properties** dialog box.

23. Select the **Dimensions** tab and then edit the **Target Area** to 200 SF. Note that the warning dialog box will open because the area exceeds the maximum area, as shown in Figure 12.52.

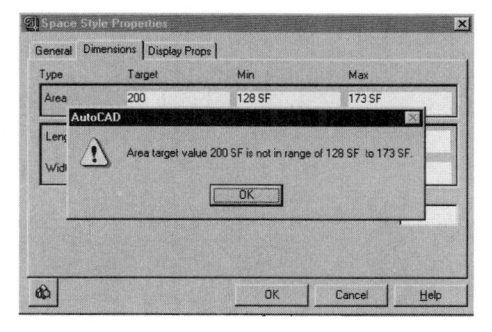

Figure 12.52 *AutoCAD warning dialog box*

24. Select **OK** to dismiss the warning box and continue to edit the dimensions as follows: **Target Length** = 15', **Target Width** = 12', **Minimum Area** = 150 SF, **Minimum Length** = 10', **Minimum Width** = 10', **Maximum Area** = 300 SF, **Maximum Length** = 25', **Maximum Width** = 25', and **Net to Gross Offset** = 0. The completed **Dimensions** tab is shown in Figure 12.53.

25. Select **OK** to dismiss the **Dimensions** tab and the **Space Style Properties** dialog box.

26. Select **OK** to dismiss the **Space Styles** dialog box.

27. Select the top space as shown in Figure 12.54; then right-click and choose **Space Modify** from the shortcut menu.

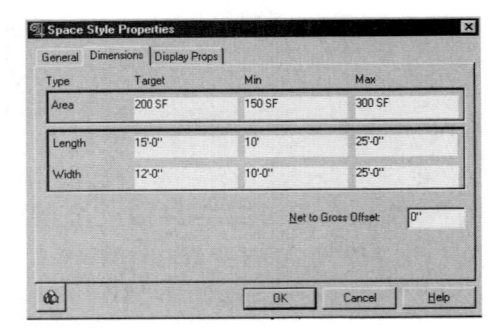

Figure 12.53 *Space style properties*

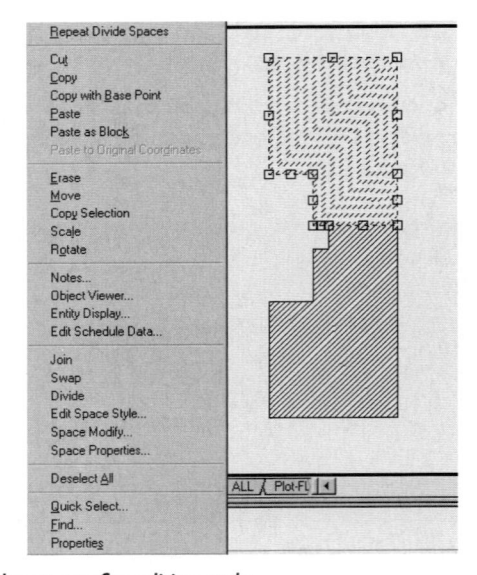

Figure 12.54 *Select the space for editing style*

    28. Select OFFICE_CAD from the **Style** list of the **Modify Space** dialog box and select the **Apply** button.

        The AutoCAD warning dialog box will open because the maximum area, maximum length, and maximum width are too small for the selected space, as shown in Figure 12.55.

    29. Press ESC to dismiss the AutoCAD Warning dialog box and the **Modify Space** dialog box.

    30. Select the **Divide Spaces** command from the **Space Planning** toolbar.

    31. Respond to the following command line prompts to divide the space.

```
Command: _AecSpaceDivide
```

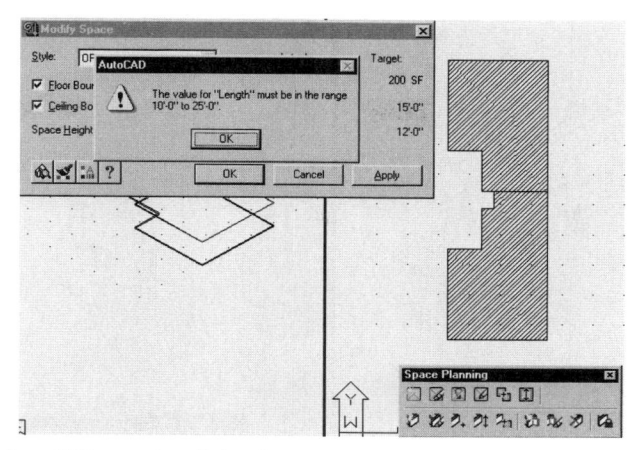

Figure 12.55 *AutoCAD warning dialog box*

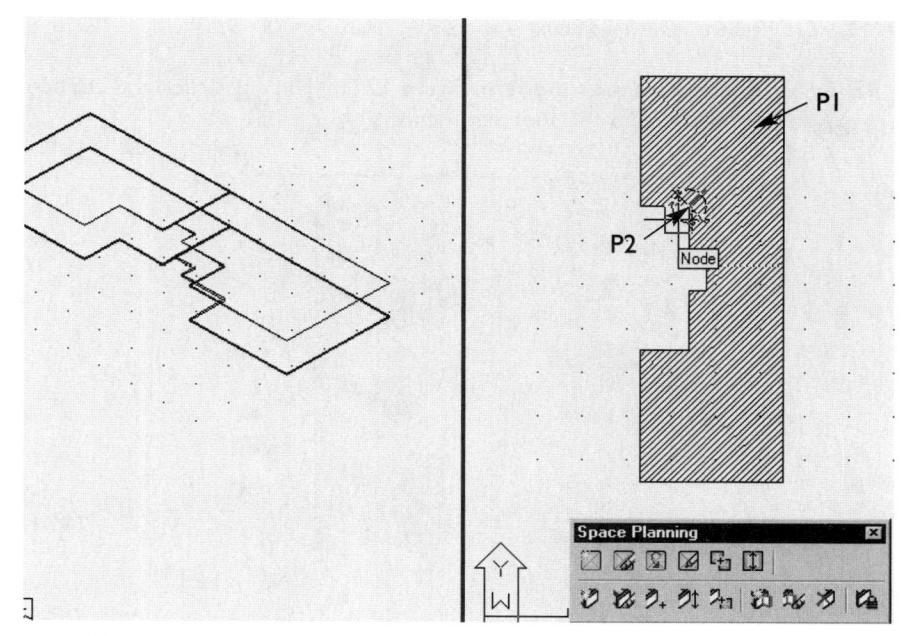

Figure 12.56 *Selection points for dividing a space*

```
Select a space:
```
(Select the space at point P1 in Figure 12.56.)
```
Divide line start point:
```
(Select point P2 with the Node object snap as shown in Figure 12.56.)
```
Divide line end point:
```
(Select a point near P3 in Figure 12.57.) (Space is divided with the Space Divide command as shown in Figure 12.57.)

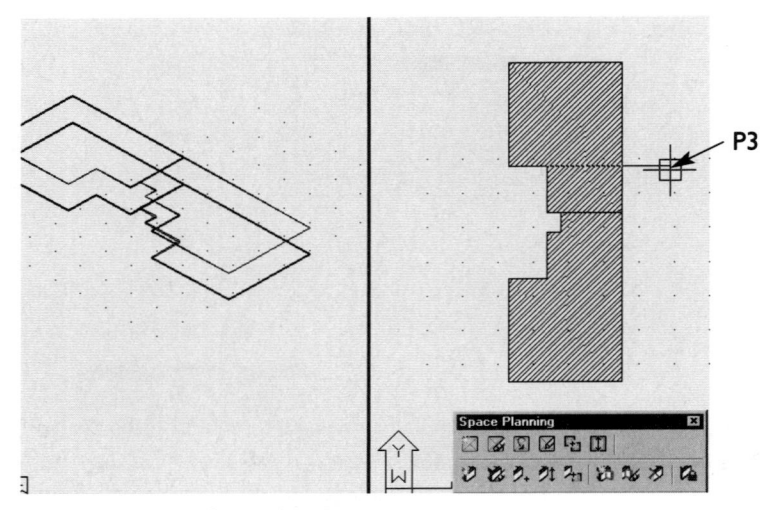

Figure 12.57 *Selection point for dividing a space*

32. Select the new space as shown in Figure 12.58; then right-click and choose **Space Modify** from the shortcut menu.

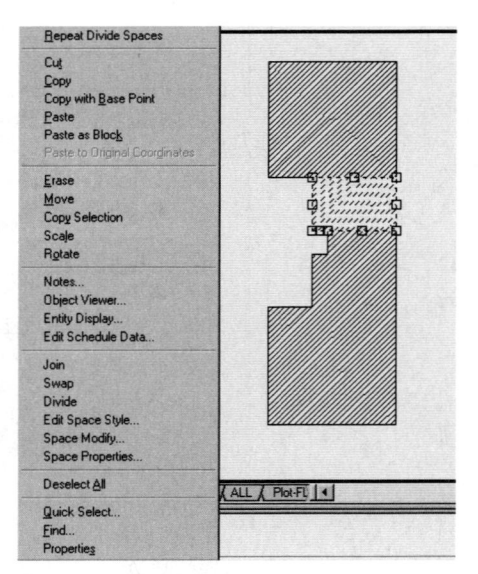

Figure 12.58 *Selecting the space for editing*

33. Select OFFICE_CAD from the **Style** list of the **Modify Space** dialog box; then select **OK** to dismiss the **Modify Space** dialog box.
34. Save the drawing.

USING SPACE BOUNDARIES

Space boundaries create a representation of the walls for a proposed building. Adding space boundaries to the space creates geometry for a total system to represent walls, floors and ceiling of each proposed space. Space boundaries can be created with solid form or as areas of separation. The solid form space boundary creates geometry similar to walls, while the space boundary as area of separation has zero wall thickness and the boundary separates the spaces without a wall. One advantage of the solid form space boundary is it can be converted to wall object.

The commands for creating and editing space boundaries are included on the **Space Planning** toolbar. The first five commands on the toolbar are used to create space boundaries and edit the relationship with spaces and space boundaries. The remaining commands on the toolbar are used to edit existing space boundaries with such commands as **Add Boundary Edges**, **Edit Boundary Edges**, and **Remove Boundary Edges**. Space boundaries can be created relative to the space they contain. Therefore, if you delete the boundary, the space will also be deleted.

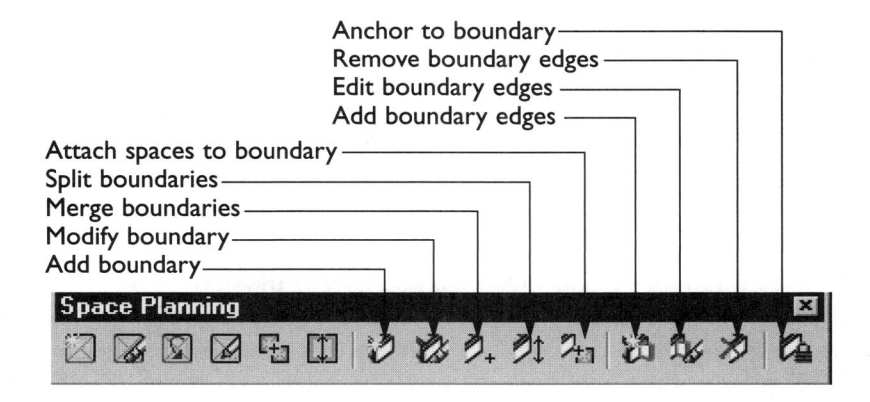

Figure 12.59 *Space boundary commands on the Space Planning toolbar*

ADDING SPACE BOUNDARIES

Space boundaries can be created with or without a model, floorplate slice, or existing space. The space boundary can be created and its associated space defined to interact together. The **Add Boundary** command is used to create a space boundary. The **Add Boundary** command and the **Add Boundary** dialog box are similar to the **Add Wall** command and its dialog box. Space boundary segments can be drawn with justification, width, and height. Space boundaries are drawn on the A-Area-Bdry layer, which has the color 140 (a hue of cyan). See Table 12.8 for **Add Boundary** command access.

Menu bar	Concept>Blocking and Boundaries>Add Boundary
Command prompt	SPACEBOUNDARYADD
Toolbar	Select Add Boundary from the Space Planning toolbar as shown in Figure 12.59
Shortcut	Right-click over the drawing area and choose Design>Space Boundaries>Add from the shortcut menu

Table 12.8 *Add Boundary command access*

When you select the **Add Boundary** command from the **Space Planning** toolbar, the **Add Space Boundary** dialog box opens, as shown in Figure 12.60.

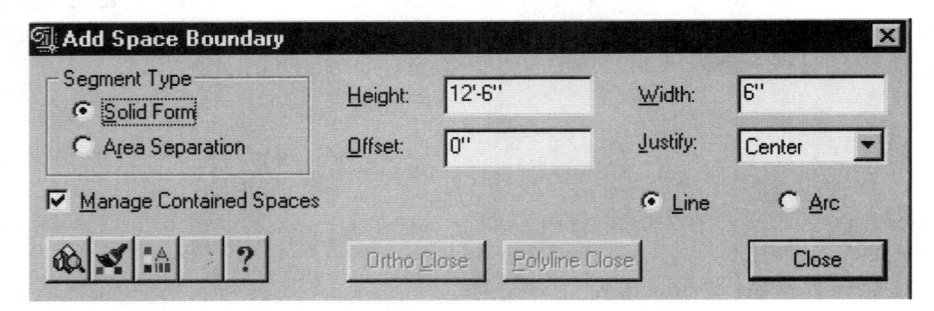

Figure 12.60 *Add Space Boundary dialog box*

The **Add Space Boundary** dialog box includes many of the options included in the **Add Wall** dialog box. However, the **Manage Contained Spaces** check box and the **Solid Form** and **Area Separation** toggles are unique to the **Add Space Boundary** dialog box. The options of the **Add Space Boundary** dialog box are described below.

> **Solid Form** – **Solid Form** space boundaries have width and represent the walls of a space; they can be converted to walls.
>
> **Area Separation** – **Area Separation** space boundaries have zero width and do not represent walls; they represent only the containment of spaces.
>
> **Manage Contained Spaces** – The **Manage Contained Spaces** check box, when selected, will create a space that is linked to the space boundary. If **Manage Contained Spaces** is not selected, no space is automatically created when a closed space boundary is created.
>
> **Height** – The **Height** edit field specifies the height of the space boundary. This field allows you to create a space boundary that extends above the ceiling and/or below the floor.

Offset – The **Offset** option allows the handle for the space boundary to be offset from the space boundary.

Width – The **Width** edit field allows you to specify the width of the space boundary.

Justify – The **Justify** option includes left, center, and right justifications. The justification of the space boundary shifts the handle for inserting the wall. A left-justified space boundary has its handle on the left side when a horizontal space boundary is drawn from left to right, as shown in Figure 12.61. The grips of the space boundary are located on its justification line.

Figure 12.61 *Justifications of the space boundaries*

Line – The **Line** toggle allows you to draw straight space boundary segments after you draw arc segments.

Arc – The **Arc** toggle allows you to draw curved space boundaries. Arc space boundaries are created in a counter-clockwise direction through the selected points.

Ortho Close – The **Ortho Close** option closes the last space boundary segment at the beginning segment, creating a segment that forms a 90-degree angle with the beginning segment.

Polyline Close – The **Polyline Close** option closes the last space boundary segment to the beginning segment.

Floating Viewer button – The **Floating Viewer** button opens the **Viewer**, allowing you to view the space boundary.

Match button – The **Match** button allows you to select an existing space boundary and determine its properties for the new space boundary. The height, width, and justification properties of an existing space boundary can be specified for the new space boundary.

Properties button – The **Properties** button opens the **Space Boundary Properties** dialog box, which allows you to specify the dimensions of the space boundary.

Undo button – The **Undo** button allows you to undo the insertion points specified for the space boundary.

Help button – The **Help** button opens the Help file on Blocking and Boundaries Overview.

SPECIFYING THE PROPERTIES OF THE SPACE BOUNDARY

Selecting the **Properties** button of the **Add Space Boundary** dialog box will open the **Space Boundary Properties** dialog box, which allows you to define the dimensions and function of the space boundary. The space and the space boundary will function together to simulate the walls, ceiling, and floor of the building. Therefore, when you edit the space boundary properties, the interaction with the space should be considered. Selecting the **Properties** button of the **Add Space Boundary** dialog box executes the **SPACEBOUNDARYPROPS** command from within the **Add Space Boundary** command. See Table 12.9 for **SPACEBOUNDARYPROPS** command access.

Command prompt	SPACEBOUNDARYPROPS
Shortcut	Right-click over the drawing area and choose Design>Space Boundaries>Properties
Shortcut	Select a space boundary; then right-click and choose Properties from the shortcut menu

Table 12.9 *SPACEBOUNDARYPROPS command access*

The **Space Boundary Properties** dialog box, shown in Figure 12.62, includes three tabs: **General**, **Dimensions**, and **Design Rules**.

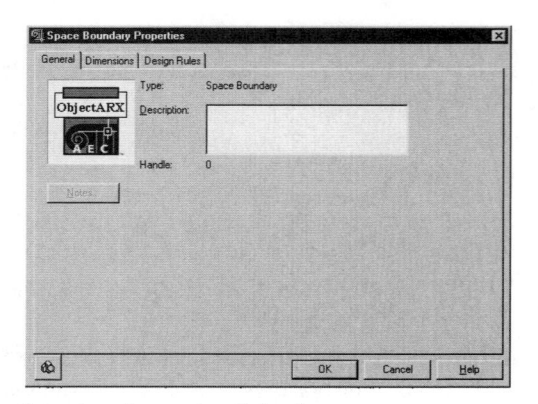

Figure 12.62 *Space Boundary Properties dialog box*

General

The **General** tab, shown in Figure 12.62, allows you to add a description to the space boundary. If the properties of an existing space boundary are edited, the **Notes** button is active. The **Notes** button opens a **Notes** dialog box, which includes **Text Notes** and **Reference Docs** tabs.

Dimensions

The **Dimensions** tab, shown in Figure 12.63, includes **Segment Type** and **Size and Placement** sections, and the **Manage Contained Spaces** check box. The **Segment Type** section includes toggles that allow you to create either **Solid Form** or **Area Separation** space boundaries. The **Size and Placement** edit fields allow you to specify the width of the boundary and the justification.

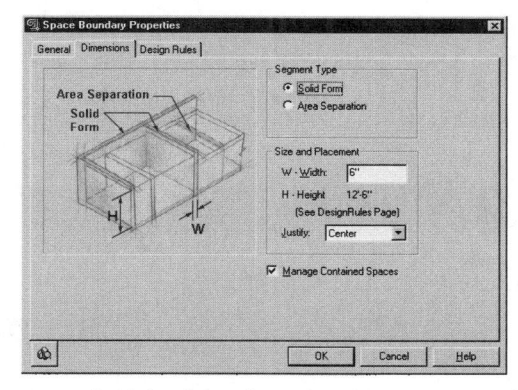

Figure 12.63 *Dimensions tab of the Space Boundary Properties dialog box*

The **Manage Contained Spaces** check box will create a space within the space boundary if selected. This check box allows you to create the space as you create the space boundary. The height of 12'-6" shown above is derived from the **Design Rules** tab, which allows you to specify an explicit height for the space boundary or to define the height of the space boundary based on the floor to ceiling height, ceiling cavity, and floor cavity, as specified in the **Space Properties** dialog box.

Design Rules

The **Design Rules** tab defines the interaction of the space with the space boundary. In the **Design Rules** tab shown in Figure 12.64, **Automatically Determine from Spaces** is checked for the boundary conditions at both the ceiling and floor. When these check boxes are selected, the space boundary height is determined from the floor to ceiling height as specified in the **Space Properties** dialog box. The **Space Properties** dialog box defines the ceiling cavity and floor thickness, which ultimately control the total space boundary height. Therefore, the space boundary height is the total of the **Space Height, Floor Boundary Thickness, Ceiling Boundary Thickness,** and **Height of Space above Ceiling Boundary** values.

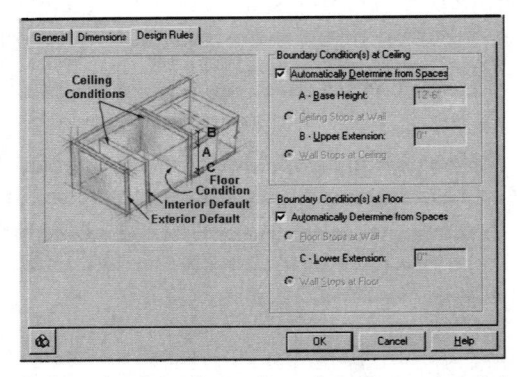

Figure 12.64 *Design Rules tab of the Space Boundary Properties dialog box*

When the **Automatically Determine from Spaces** check box is cleared, the dimensions of floor to ceiling, upper extensions of the wall, and lower extensions of the wall can be specified. The options of the **Design Rules** tab when **Automatically Determine from Spaces** is cleared are shown in Figure 12.65 and described below.

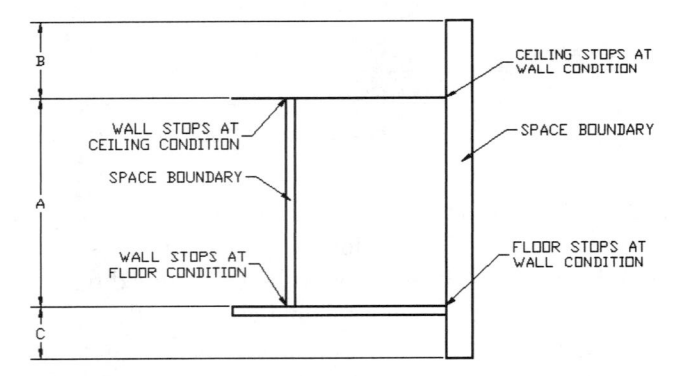

Figure 12.65 *Ceiling and Floor conditions for wall intersections*

Boundary Condition(s) at Ceiling

A-Base Height – The **A-Base Height** is the finish floor to finish ceiling distance.

Ceiling Stops at Wall – The **Ceiling Stops at Wall** toggle allows you to simulate a suspended ceiling with the walls extending above the elevation of the ceiling. This condition would be typical if the boundary is intended to contain the spread of fire or to reduce sound transmission.

B-Upper Extension – When the **Ceiling Stops at Wall** toggle is selected, this edit field is active and you can specify how far the wall extends above the ceiling. The upper extension is the vertical dimension of the ceiling cavity.

Wall Stops at Ceiling – Selecting this radio button terminates the wall at the ceiling height.

Boundary Condition(s) at Floor

Floor Stops at Wall – The **Floor Stops at Wall** option is checked if the wall extends below the floor. Space boundaries for exterior or bearing walls would extend below the floor; therefore this option should be toggled on for the boundary.

C-Lower Extension – The **C-Lower Extension** edit field specifies the distance below the floor that the space boundary extends. When the floor stops at the wall, this edit field is inactive.

Wall Stops at Floor – The **Wall Stops at Floor** toggle allows the space boundary to begin at the floor elevation. The **Wall Stops at Floor** condition would be typical of interior /non-bearing walls.

When **Automatically Determine from Spaces** is checked, the space properties control the finish ceiling, ceiling cavity, and floor boundary. Therefore, if you create a 16' space boundary, the floor and ceiling components of the space are positioned within the space boundary according to the dimension specified in the **Space Properties** dialog box. If the **B-Space Height** (Ceiling to Floor dimension) were increased, the space boundary would automatically increase to comply with the new ceiling height. The **Height** of the space boundary is altered to comply with dimensions specified in the **Space Properties** dialog box. Figure 12.66 shows the elevation view of a 16' space boundary and the **Dimensions** tab of the **Space Properties** dialog for the space contained.

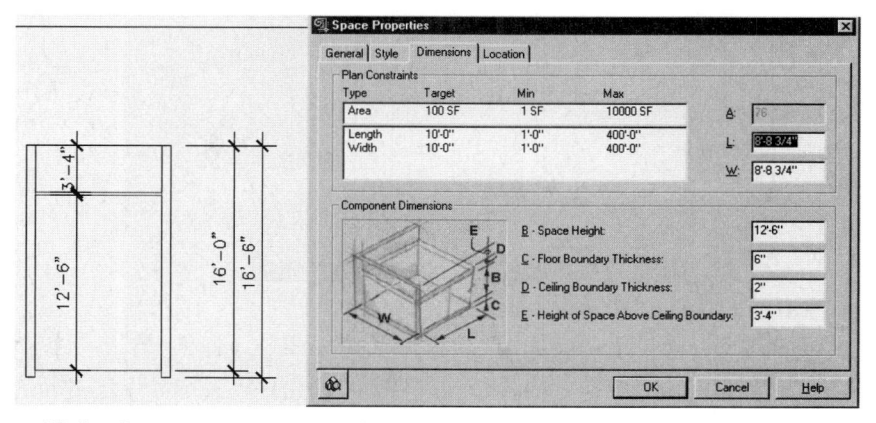

Figure 12.66 *Dimension properties of a space boundary*

If the **Space Properties** dialog box were changed as shown in Figure 12.67, the height of the space boundary would automatically change to comply with the revision.

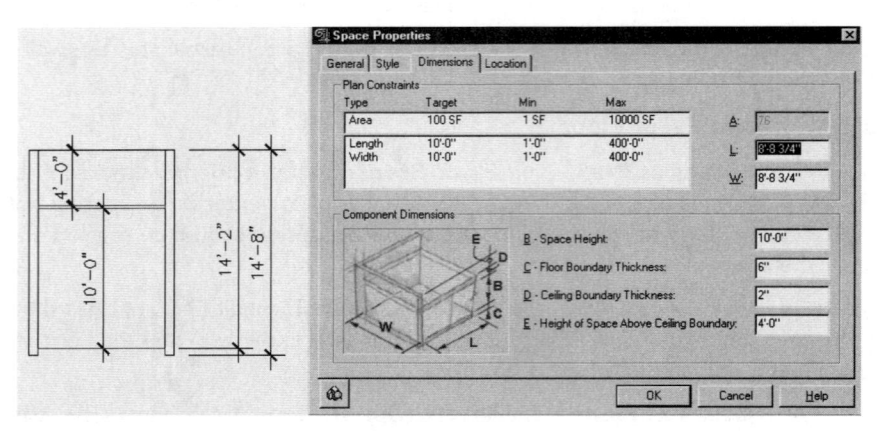

Figure 12.67 *Space boundary adjusted to ceiling dimensions*

Space boundaries have grips located at the endpoints and midpoint on the justification line. Selecting the endpoint grips allows you to stretch a space boundary in alignment with an adjoining space boundary. In Figure 12.68, Ortho is toggled ON and the endpoint grip of the vertical space boundary is selected and stretched along the justification line of the horizontal space boundary. The lengths of the justification lines are displayed dynamically as the grip is stretched.

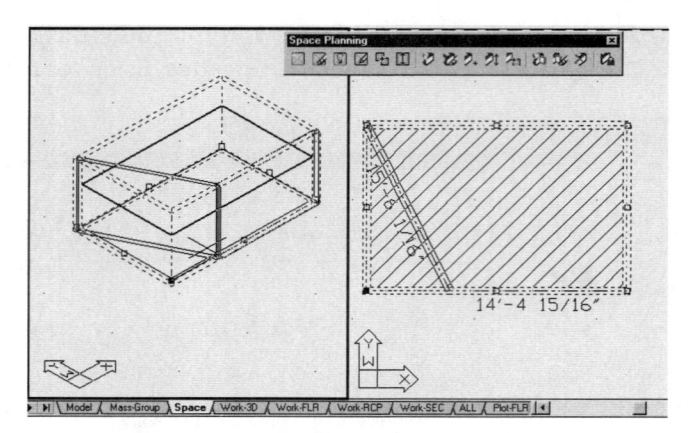

Figure 12.68 *Grip editing of a space boundary*

If you select the midpoint grip with Ortho ON, you can resize the space boundary while retaining the rectangular shape. Space boundaries added to an existing shape must intersect the justification line to close the boundary. You can select the grip of the space boundary segment and stretch the segment to intersect with the justification line.

TUTORIAL 12.3 CREATING SPACE BOUNDARIES

1. Open *Ex12-3* from the \ADT_Tutor\Ch12 directory
2. Choose **File>SaveAs** from the menu bar and save the file as *Boundary1.dwg* in your student directory.
3. Select the Space layout and verify that the right viewport is active and that Ortho is ON in the Status Bar.
4. Select **Add Boundary** from the **Space Planning** toolbar.
5. Select in the **Add Space Boundary** dialog box and then edit the values as follows: set **Segment Type** to **Solid Form**, select **Manage Contained Spaces**, set **Height** to 14', **Offset** to 0, **Width** to 7 5/8", and **Justify** to Right, as shown in Figure 12.69.

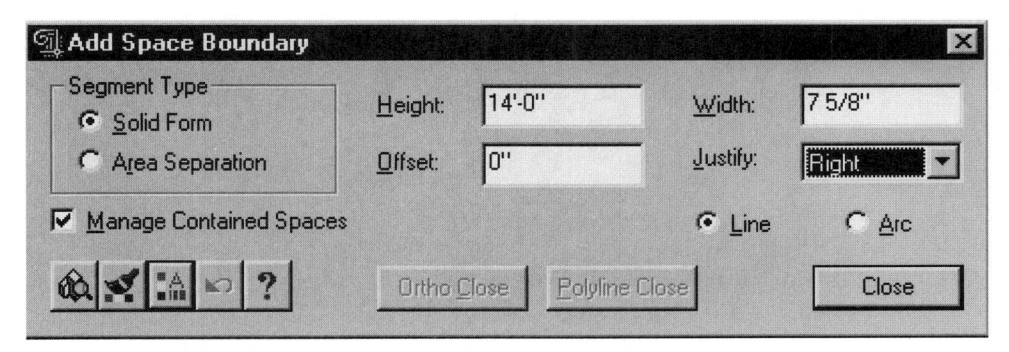

Figure 12.69 *Settings in the Add Space Boundary dialog box*

6. Select the **Properties** button to open the **Space Boundary Properties** dialog box, and then select the **Design Rules** tab. Verify that **Automatically Determine from Spaces** is selected for the **Boundary Conditions at Ceiling** and **Boundary Conditions at Floor**. Select **OK** to dismiss the **Space Boundary Properties** dialog box.
7. Select in the right viewport to activate the viewport; then respond to the following command line prompts.

```
Command: _AecSpaceBoundaryAdd
Start point or [SOlid
form/SEparation/MANage/Width/Height/OFfset/Justify/
MATch/Arc]: (Select a point near point P1 in Figure 12.70.)
(Move the mouse to the right.)
```

```
End point or [SOlid
form/SEparation/MANage/Width/Height/OFfset/Justify/
MATch/Arc]: 50' ENTER
```
(Point P2 created in Figure 12.70.)
(Move the mouse up.)
```
End point or [SOlid
form/SEparation/MANage/Width/Height/OFfset/Justify/
MATch/Arc/Undo]: 32' ENTER
```
(Point P3 created as shown in Figure 12.70.)
(Move the pointer to the left.)
```
End point or [SOlid
form/SEparation/MANage/Width/Height/OFfset/Justify/
MATch/Arc/Undo/Close/ORtho]:
DBOX
```
(Select the Ortho Close button of the Add Space Boundary dialog box.)
```
Point on segment in direction of close:
```
(Select a point near P4 as shown in Figure 12.70 to create the rectangular space boundary. The completed space boundary is shown in Figure 12.71.)

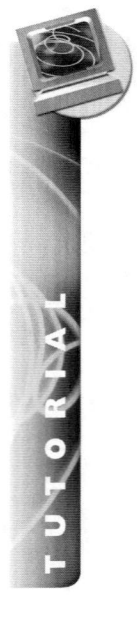

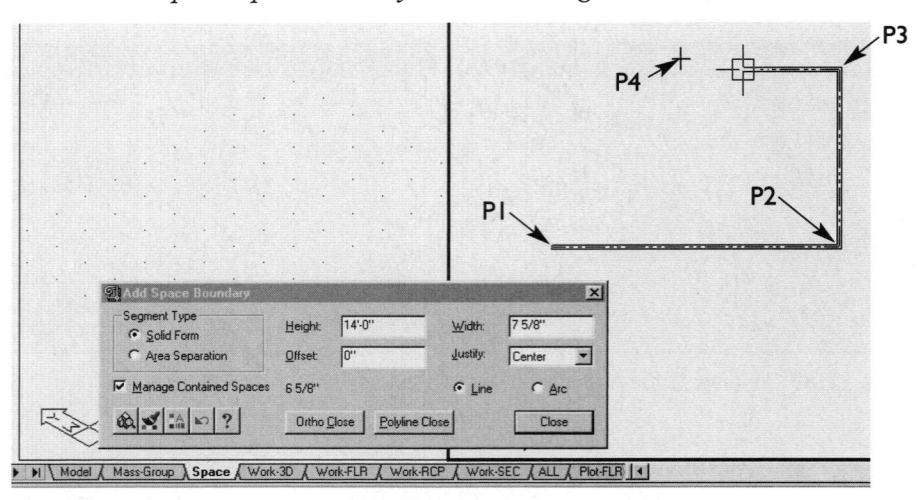

Figure 12.70 *Using Ortho Close to close space boundary*

8. Select the space; then right-click and choose **Space Properties** from the shortcut menu.

9. Select the **Dimensions** tab and change the **B-Space Height** to 9'.

10. Select **OK** to dismiss the **Space Properties** dialog box and view the space boundary reduced in height, as shown in Figure 12.71.

11. Save the drawing.

Figure 12.71 *Space and space boundary created*

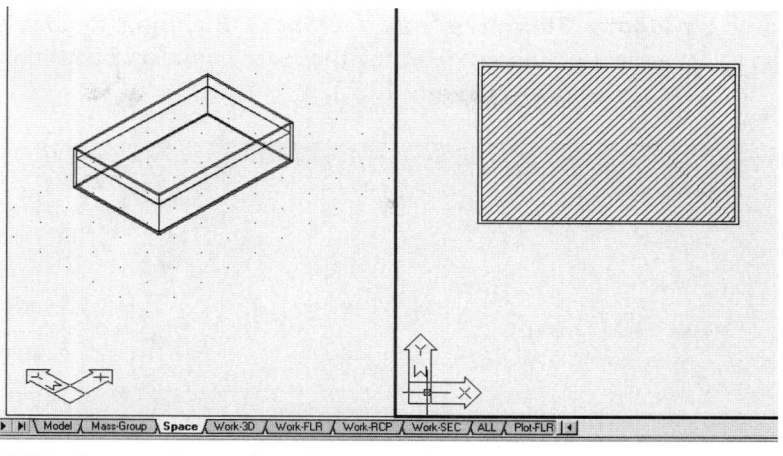

MODIFYING SPACE BOUNDARIES

The space boundaries created in the previous tutorial created a space when the space boundaries closed to the start point. When **Manage Contained Spaces** is selected in the **Add Space Boundary** dialog box, space boundaries that close will create a space using the Standard space style. The remaining space boundary commands on the **Space Planning** toolbar will modify space boundaries or change the association with the space. The **Modify Boundary** command allows you to change the properties of the space boundary. See Table 12.10 for **Modify Boundary** command access.

Menu bar	Concept>Blocking and Boundaries> Modify Boundary
Command prompt	SPACEBOUNDARYMODIFY
Toolbar	Select Modify Boundary from the Space Planning toolbar as shown in Figure 12.72
Shortcut	Design>Space Boundaries>Modify

Table 12.10 *Modify Boundary command access*

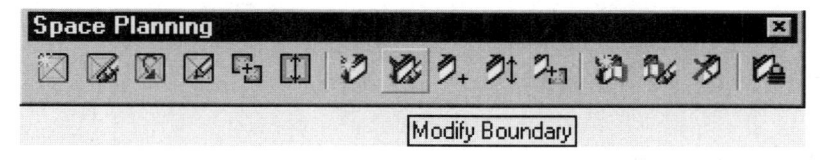

Figure 12.72 *Modify Boundary command on the Space Planning toolbar*

When you select **Modify Boundary** from the **Space Planning** toolbar, you are prompted to select a space boundary. Selecting the space boundary opens the **Modify Boundary** dialog box, as shown in Figure 12.73.

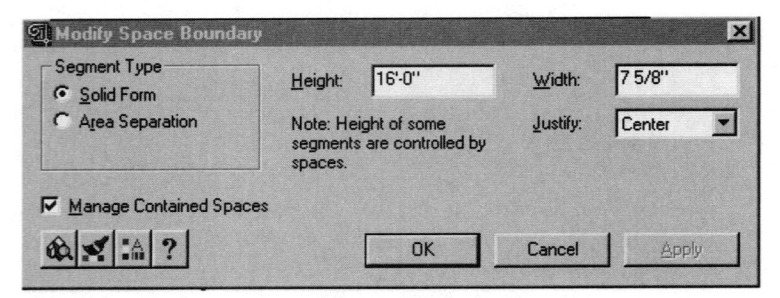

Figure 12.73 *Modify Space Boundary dialog box*

After selecting a space boundary, you can edit the height, width, and justification of the space boundary, and then select **OK** to execute the change. The **Modify Space Boundary** dialog box includes a **Properties** button that provides access to the **Space Boundary Properties** dialog box, which includes the **Design Rules** and **Dimensions** tabs for specifying the relationship with ceiling and floor components.

Using Merge Boundaries

Space boundaries created in separate operations can be merged together with the **Merge Boundaries** command. In Figure 12.74, the horizontal space boundary was created after the vertical space boundary. Therefore the space boundaries are independent of each other. The **Merge Boundaries** command will clean up the intersection of space boundaries and merge them together.

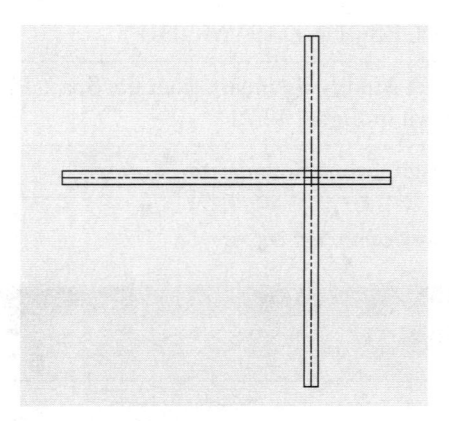

Figure 12.74 *Independent space boundaries*

Space boundaries that are merged together act as one space boundary when edited. See Table 12.11 for **Merge Boundaries** command access.

Menu bar	Concept>Blocking and Boundaries> Merge Boundaries
Command prompt	SPACEBOUNDARYMERGE
Toolbar	Select Merge Boundaries from the Space Planning toolbar as shown in Figure 12.75
Shortcut	Select a space boundary; then right-click and choose Merge Boundary

Table 12.11 *Merge Boundaries command access*

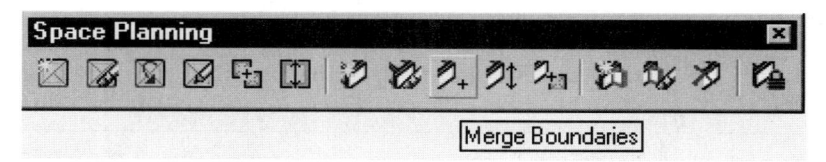

Figure 12.75 *Merge Boundaries command on the Space Planning toolbar*

When you select the **Merge Boundaries** command from the **Space Planning** toolbar, you are prompted to select the two boundaries for merging. The space boundaries can be disconnected from each other or they can cross, as shown in Figure 12.76. The command line prompts for the **Merge Boundaries** command are shown below.

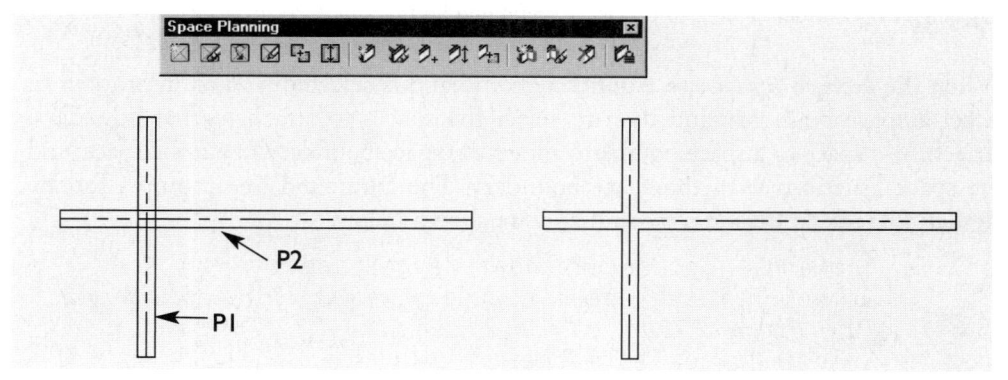

Figure 12.76 *Selection points for merging space boundaries*

```
Command: _AecSpaceBoundaryMerge
Select gaining space boundary:
```
(Select the space boundary at

point P1 in Figure 12.76.)
`Select space boundary to merge:` *(Select the space boundary at point P2 in Figure 12.76.)*
(Space boundaries merged as shown at right in Figure 12.76.)

Attaching Spaces to Boundaries

Space boundaries that are added to an existing space boundary are disconnected, as shown on the left in Figure 12.78. The **Attach Spaces to Boundary** command connects the space to a new space boundary. When a space is connected to a space boundary, the changes in the space properties can alter the space boundary. See Table 12.12 for **Attach Spaces to Boundary** command access.

Menu bar	Concept>Blocking and Boundaries> Attach Spaces to Boundary
Command prompt	SPACEBOUNDARYMERGESPACE
Toolbar	Select Attach Spaces to Boundary from the Space Planning toolbar as shown in Figure 12.77

Table 12.12 *Attach Spaces to Boundary command access*

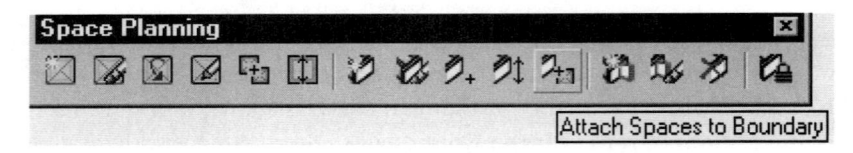

Figure 12.77 *Attach Spaces to Boundary command on the Space Planning toolbar*

When the **Attach Spaces to Boundary** command is selected, you are prompted to select a space boundary and then to select the spaces to attach to that boundary. Attaching a space to a space boundary allows the space boundary to cut the space, and the space functions with the space boundary. The command line prompts for the **Attach Spaces to Boundary** command are shown below.

```
Command: _AecSpaceBoundaryMergeSpace
Select gaining space boundary:
```
(Select the space boundary at P1 in Figure 12.78.)
```
Select spaces to merge.
Select spaces:
```
(Select the space at P2 in Figure 12.78.) `1 found`
`Select spaces:` ENTER *(Press* ENTER *to end space selection.)*
`Merged 1 space(s).`
(Space attached to the new space boundary as shown at right in Figure 12.78.)

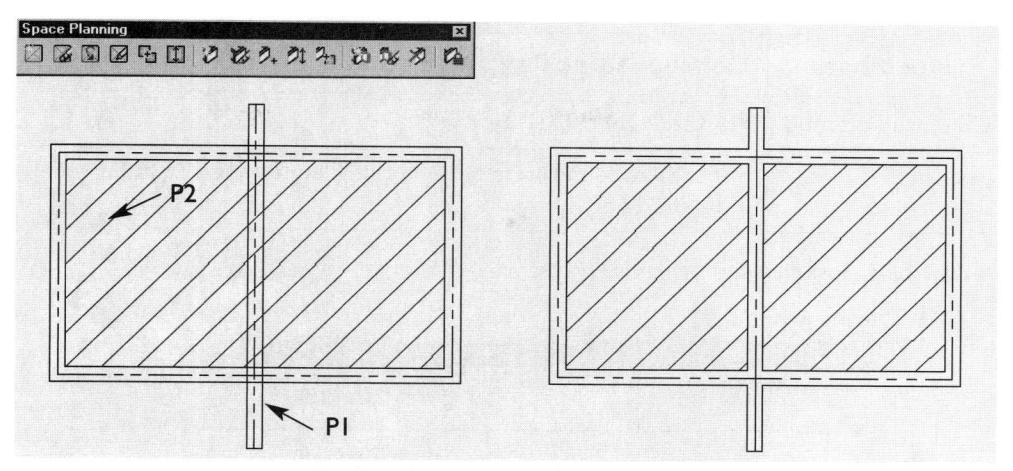

Figure 12.78 *Selection point for attaching a space to a space boundary*

When the space boundary was attached to the space, as shown above, the space boundary was modified as shown in Figure 12.79.

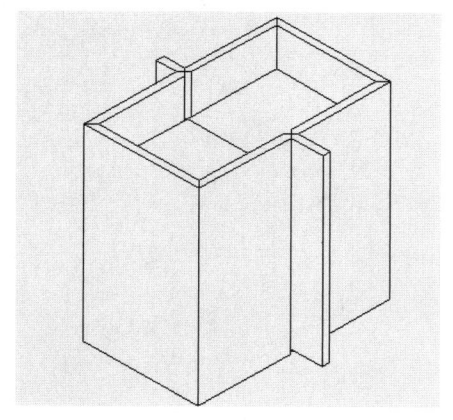

Figure 12.79 *Modified space boundary*

Splitting the Space from the Boundary

The **Split Boundary** command can be used to detach a space boundary from a space. When the **Split Boundary** command is applied to a space and its space boundary, the association is dropped. The space boundary can then act independently of the space. See Table 12.13 for **Split Boundary** command access.

When you select the **Split Boundary** command from the **Space Planning** toolbar, you are prompted to select a space to convert to a new space boundary. Selecting the space splits the association between the space and space boundary. The command line prompts for the **Split Boundary** command are shown below.

Menu bar	Concept>Blocking and Boundaries>Split Boundary
Command prompt	SPACEBOUNDARYSPLIT
Toolbar	Select Split Boundary from the Space Planning toolbar, as shown in Figure 12.80

Table 12.13 *Split Boundary command access*

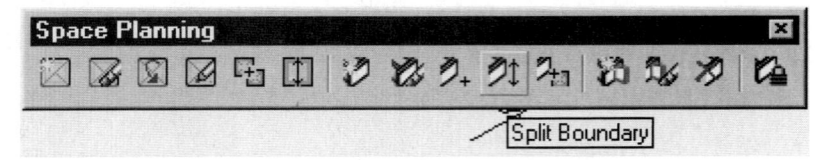

Figure 12.80 *Split Boundary command on the Space Planning toolbar*

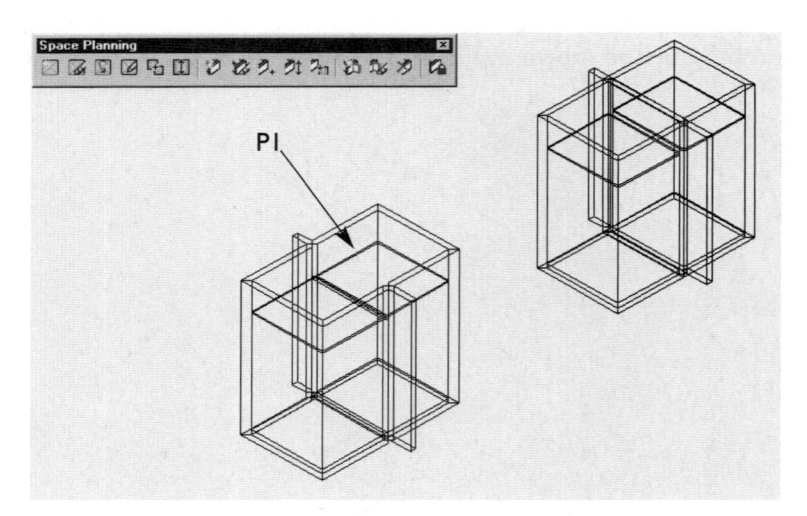

Figure 12.81 *Selecting the space to split from the space boundary*

```
Command: _AecSpaceBoundarySplit
Select spaces to convert to new space boundary.
Select spaces: (Select a space at point P1 in Figure 12.81.) 1
found
Select spaces: ENTER (Press ENTER to end the selection of spaces.)
Transferred 1 space(s) to new space boundary.
(Space boundary modified as shown at right in Figure 12.81.)
```

TUTORIAL 12.4 WORKING WITH BOUNDARIES

1. Open *Ex12-4* from the *\ADT_Tutor\Ch12* directory.

2. Choose **File>SaveA**s from the menu bar and save the file in your student directory as *Boundary2.dwg*.

3. Select the exterior space boundary to verify that the two interior space boundaries are independent of the exterior space boundary, as shown in Figure 12.82.

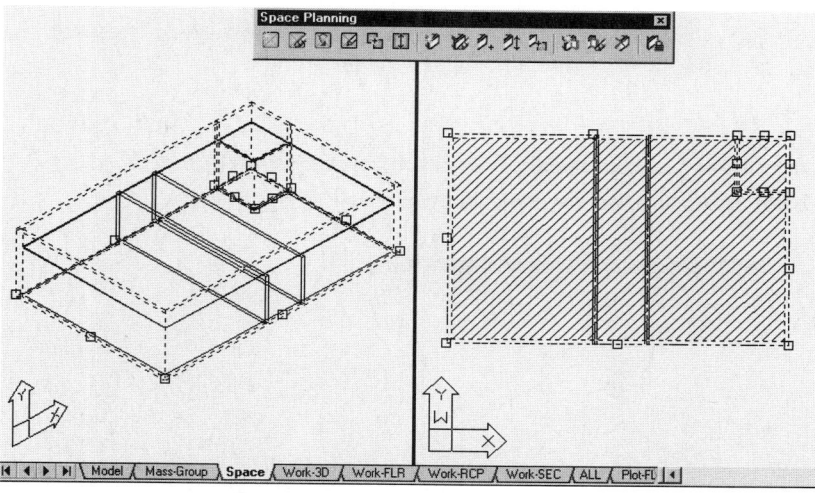

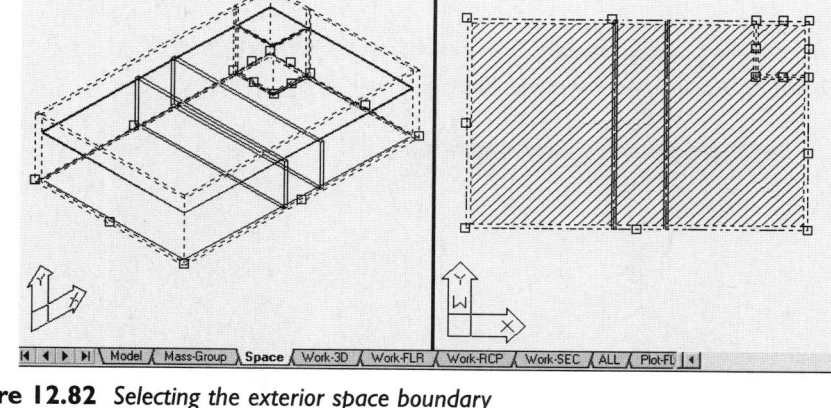

Figure 12.82 *Selecting the exterior space boundary*

4. Press ESC to clear the selection, and then press ESC again to clear the display of the grips.

5. Select **Merge Boundaries** from the **Space Planning** toolbar; then respond to the following command line prompts.

   ```
   Command: _AecSpaceBoundaryMerge
   Select gaining space boundary:
   ```
   (Select the exterior space boundary at P1 in Figure 12.83.)
   ```
   Select space boundary to merge:
   ```
   (Select the interior space boundary at P2 in Figure 12.83.)
   (Space boundaries merged as shown in Figure 12.84.)

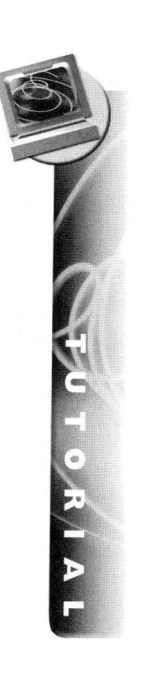

6. Select **Merge Boundaries** from the **Space Planning** toolbar, and then respond to the following command line prompts.

   ```
   Command: _AecSpaceBoundaryMerge
   Select gaining space boundary:
   ```
   (Select the exterior space boundary at P1 in Figure 12.85.)

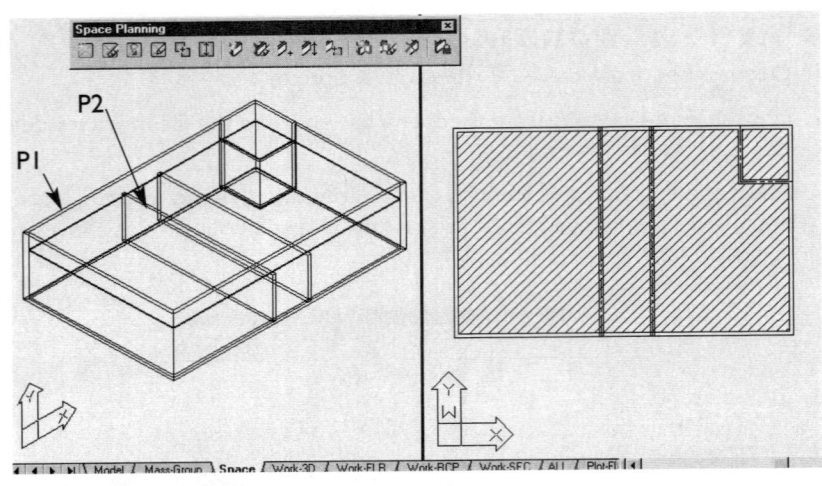

Figure 12.83 *Selection points for merging boundaries*

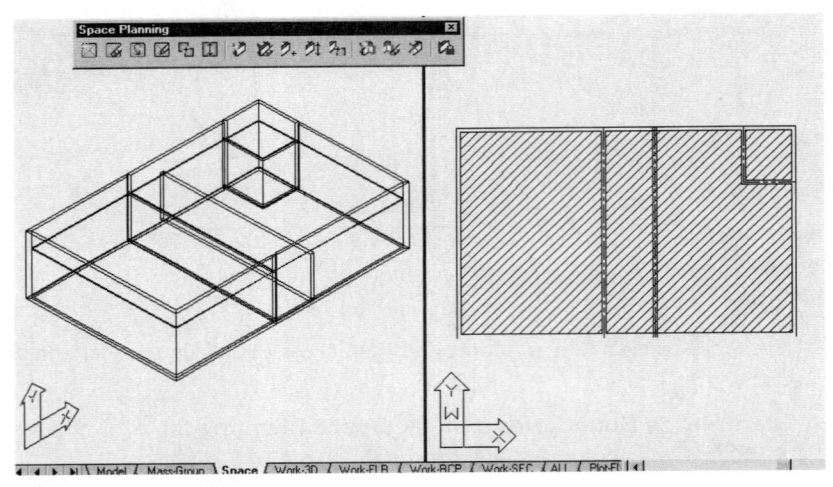

Figure 12.84 *Space boudaries merged*

> Select space boundary to merge: *(Select the interior space boundary at P2 in Figure 12.85.)*
> *(Space boundaries merged as shown in Figure 12.86.)*

7. Select the middle space at P1 in Figure 12.86; then right-click and choose **Space Properties** from the shortcut menu.
8. Select the **Dimensions** tab of the **Space Properties** dialog box.
9. Change the **B-Space Height** to 8' and the **Height of Space Above Ceiling Boundary** to 4'-4", as shown in Figure 12.87.

Creating Spaces and Boundaries 917

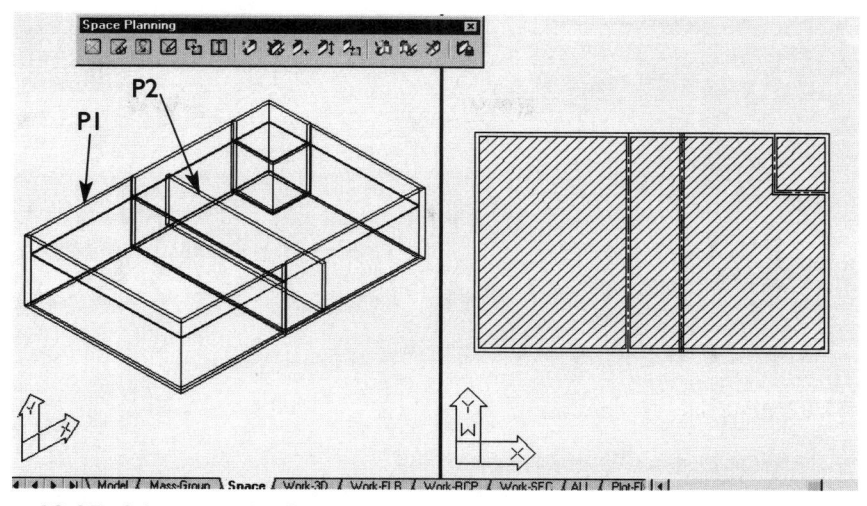

Figure 12.85 *Selection points for merging boundaries*

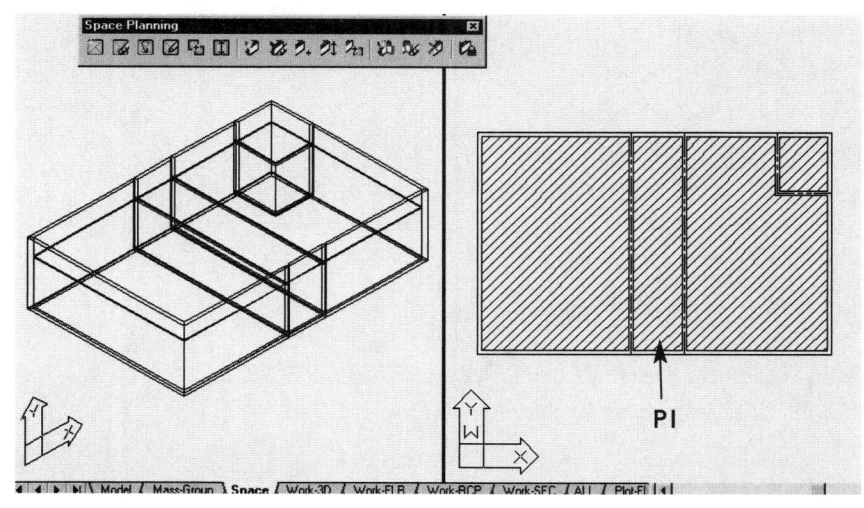

Figure 12.86 *Selecting space for editing*

10. Select **OK** to dismiss the **Space Properties** dialog box.

11. Select **Zoom Window** from the **Standard** toolbar and respond to the following command line prompts.

    ```
    Command: '_zoom
    Specify corner of window, enter a scale factor (nX
    or nXP), or
    [All/Center/Dynamic/Extents/Previous/Scale/Window]
    <real time>: _w
    ```

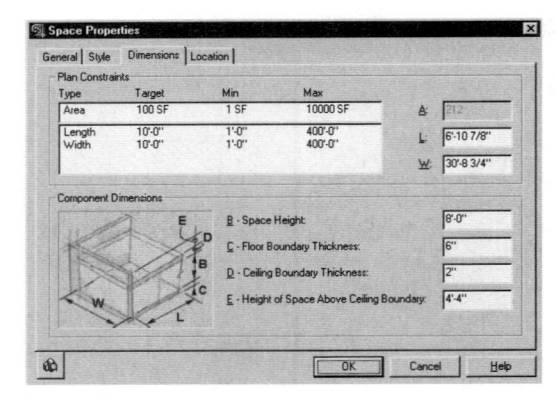

Figure 12.87 *Modifying ceiling cavity*

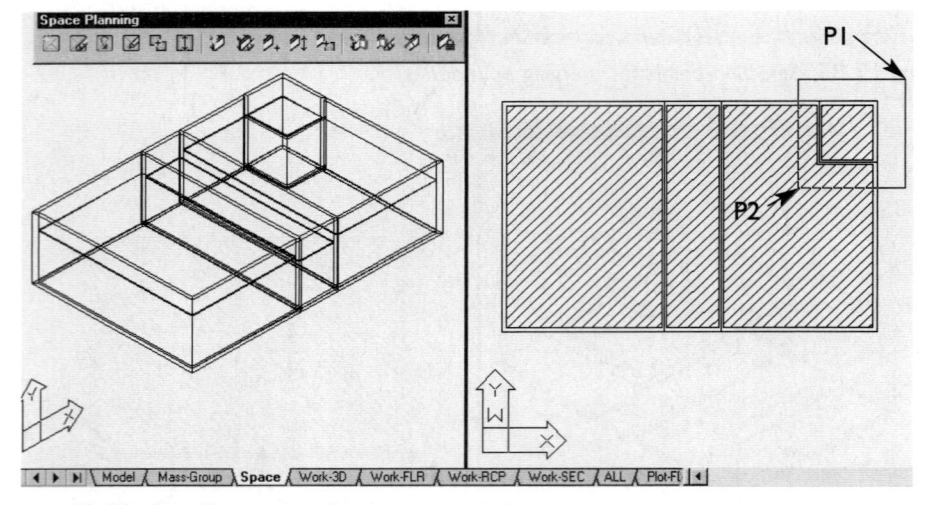

Figure 12.88 *Specifying points for the zoom window*

```
Specify first corner: (Select point P1 in Figure 12.88.)
Specify opposite corner: (Select point P2 in Figure 12.88.)
```

12. Select **Split Boundary** from the **Space Planning** toolbar; then respond to the following command line prompts.

```
Command: _AecSpaceBoundarySplit
Select spaces to convert to new space boundary.
Select spaces: (Select the space at point P1 in Figure 12.89.) 1 found
Select spaces: ENTER
Transferred 1 space(s) to new space boundary.
```
(The space is split from the space boundary as shown in Figure 12.90.)

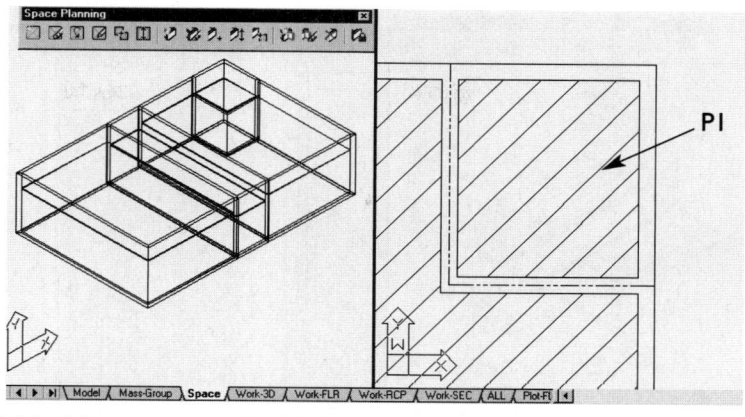

Figure 12.89 *Selecting a space for the Split Boundary command*

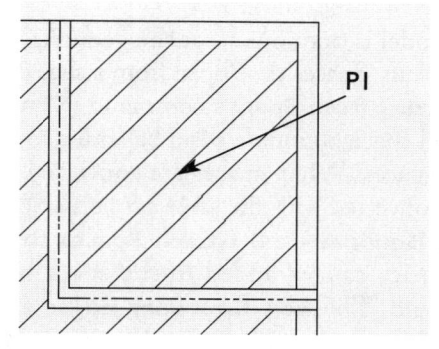

Figure 12.90 *Space split from the space boundary*

13. Select the **Erase** command from the **Modify** toolbar; then select the small space at point P1 in Figure 12.90 to delete the space and the boundary, as shown in the following command line prompts.

    ```
    Command: _erase
    Select objects: (Select the space at point P1 in Figure 12.90.) 1 found
    Select objects: ENTER (Press ENTER to end the command.)
    (The space is removed as shown in Figure 12.91.)
    ```

14. Save the drawing.

CREATING SPACES BOUNDARIES FROM SPACES, SLICES, AND SKETCHES

Space boundaries can be created from spaces, slices, and sketches. The slice taken through the mass model can be converted to a polyline, which can be used to create

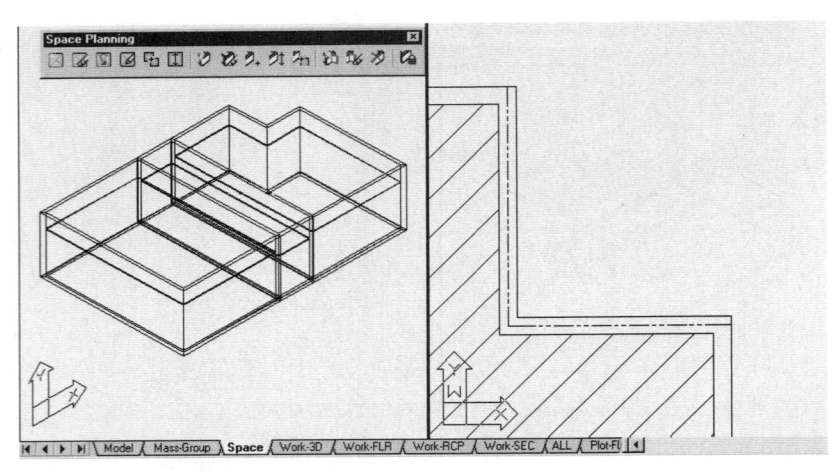

Figure 12.91 *Selected space removed*

a space. Therefore, the model is one source for the geometry to create space boundaries and ultimately floor plans. Spaces developed from a slice can be converted to space boundaries with the **Convert from Spaces** command (**SPACEBOUNDARYCONVERTSPACE**). Additional space boundaries can be added to create a space boundary plan that includes room spaces. When spaces are converted to space boundaries, all the space boundaries are converted with the same property. Space boundaries can then be edited with the **Split Boundary** and **Attach Spaces to Boundary** commands. Additional space boundaries can be added to the space boundary with the **Add Boundary Edges** command. The **Edit Boundary Edges** and **Remove Boundary Edges** commands are used to edit the added space boundaries. Converting a space that has been developed from a slice allows you to convert the floorplate slice to a floor plan. See Table 12.14 for **Convert from Spaces** command access.

Menu bar	Concept>Blocking and Boundaries> Convert to Boundary>Convert from Spaces
Command prompt	SPACEBOUNDARYCONVERTSPACE
Shortcut	Right-click over the drawing area and choose Design>Space Boundaries>Convert from the shortcut menu

Table 12.14 *Convert from Spaces command access*

When you choose the **Convert from Spaces** command from the menu bar, you are prompted to select spaces to convert. The space boundary that is created has area separation segments; they have no height or width. Therefore, to convert the area separation space boundaries to solid form, you use the **Modify Boundary** command. Shown below are the command line prompts for converting a space to a space boundary.

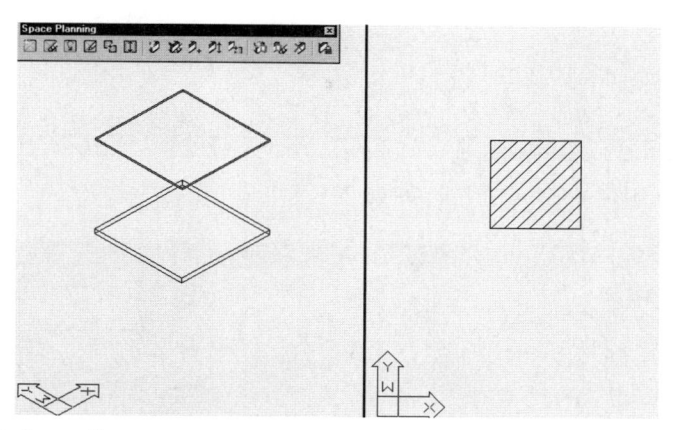

Figure 12.92 *Space for conversion to space boundaries*

```
Command: _AecSpaceBoundaryConvertSpace
Select spaces to convert to new space boundary.
Select spaces: (Select the space shown in the right viewport in Figure
12.92.) 1 found
Select spaces: ENTER (Press ENTER to end space selection.)
Transferred 1 space(s) to new space boundary.
```
(Space boundary created as an area separation.)

(Select Modify Boundary from the Space Planning toolbar.)
```
Command: _AecSpaceBoundaryModify
Select space boundaries: L ENTER (Select the space boundary
using the Last selection option.)
1 found
Select space boundaries: ENTER
Space boundary modify [SOLid
form/SEparation/Width/Height/Justify/MATch/MANage]:
DBOX
```
(Space Boundary Properties dialog box opens as shown in Figure 12.93.)

(Select the Solid Form radio button as shown in Figure 12.94.)

(Select OK to dismiss the Space Boundary Properties dialog box and view the space boundary as shown in Figure 12.95.)

Space boundaries can also be converted from a slice. The **Convert from Slice** command (**SPACEBOUNDARYCONVERTSLICE**) is used to convert a slice to a space

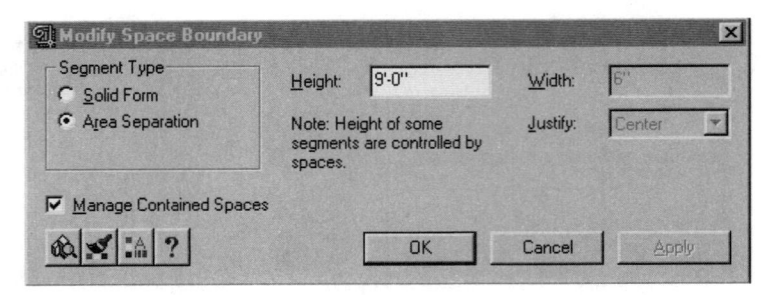

Figure 12.93 Modify Space Boundary dialog box

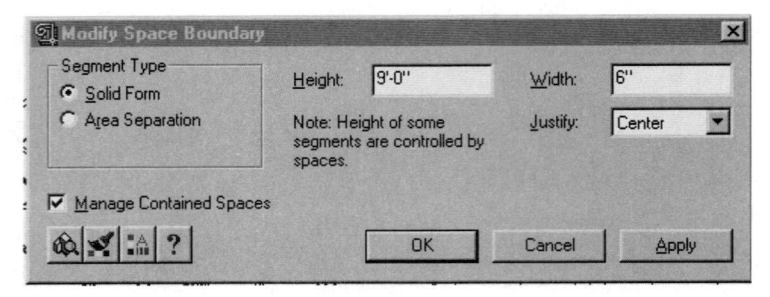

Figure 12.94 Modifying space to the solid form

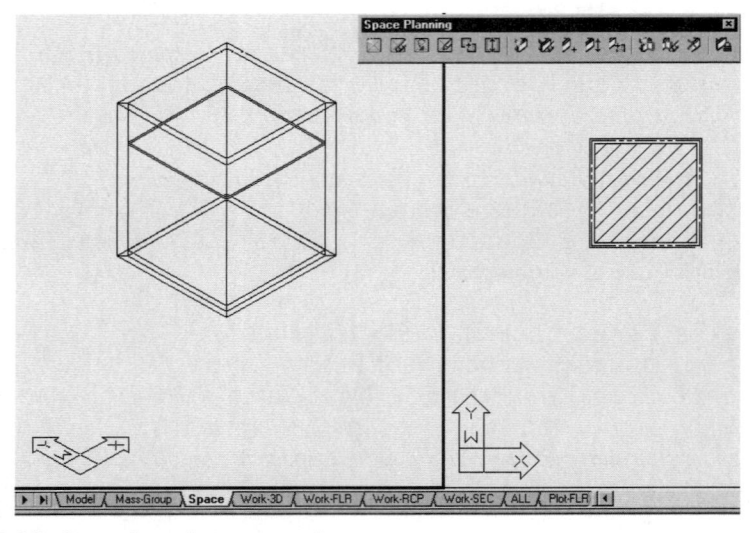

Figure 12.95 Space boundaries created

boundary. The space boundary will include a Standard space. See Table 12.15 for **Convert from Slice** command access.

Menu bar	Concept>Blocking and Boundaries> Convert to Boundary>Convert from Slice
Command prompt	SPACEBOUNDARYCONVERTSLICE
Shortcut	Right-click over the drawing area and choose Design>Space Boundaries>Convert from the shortcut menu

Table 12.15 *Convert from Slice command access*

When you choose the **Convert from Slice** command (**SPACEBOUNDARYCONVERTSLICE**), you are prompted to select a slice. The slice is displayed in the Mass-Group layout; therefore, you must select the Mass-Group layout and then select the slice. When you select the slice, the **Space Boundary Properties** dialog box opens, allowing you to define the properties of the space boundary, which is displayed in the Space layout.

STEPS TO CREATING SPACE BOUNDARY FROM A SLICE

1. Select the Mass-Group layout.
2. Choose **Concept>Blocking and Boundaries>Convert to Boundary>Convert from Slice** from the menu bar.
3. Select the slice marker; the **Space Boundary Properties** dialog box opens.
4. Edit the **General**, **Dimensions**, and **Design Rules** tabs, and then select **OK** to dismiss the **Space Boundary Properties** dialog box.
5. Select the Space layout to view the space boundary.

A sketch that includes lines, arcs, circles, or polylines can be converted to space boundaries. The command to convert AutoCAD geometry to space boundaries is the **Convert from Sketching** command (**SPACEBOUNDARYCONVERTEDGE**). See Table 12.16 for **Convert from Sketching** command access.

Menu bar	Concept>Blocking and Boundaries> Convert to Boundary>Convert from Sketching
Command prompt	SPACEBOUNDARYCONVERTEDGE
Shortcut	Right-click over the drawing area and choose Design>Space Boundaries>Convert from the shortcut menu

Table 12.16 *Convert from Sketching command access*

When you choose the **Convert from Sketching** command, you are prompted to select lines, arcs, circles, or polylines to convert. After you select the entities, the **Space Boundary Properties** dialog box opens, allowing you to set the properties of the space boundary.

ADDING BOUNDARY EDGES

The **Convert from Space**, **Convert from Slice**, and **Convert from Edge** commands allow you to quickly convert the floorplate of a model or sketch to a space boundary. Additional boundaries can then be added to the space boundary with the **Add Boundary Edges** command, which allows you to insert space boundaries or partitions within the larger space boundary. These boundaries are attached to the larger space boundary. If the space boundary were inserted with the **Add Boundary** command, the space boundary would have to be merged with the larger space boundary. See Table 12.17 for **Add Boundary Edges** command access.

Menu bar	Concept>Blocking and Boundaries> Add Boundary Edges
Command prompt	SPACEBOUNDARYADDEDGES
Toolbar	Select Add Boundary Edges from the Space Planning toolbar as shown in Figure 12.96
Shortcut	Select a Space Boundary then select Add from the shortcut menu

Table 12.17 *Add Boundary Edges command access*

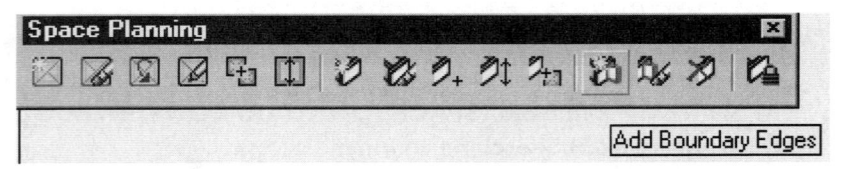

Figure 12.96 *Add Boundary Edges command on the Space Planning toolbar*

The **Add Boundary Edges** command allows you to add space boundary partitions within the overall space boundary plan. If the new space boundary totally encloses or connects to the justification line, the space will also be divided. When you select **Add Boundary Edges** from the **Space Planning** toolbar and you are prompted to select a space boundary, the **Add Space Boundary** dialog box opens. Boundary edges that connect to the justification lines of an existing space will divide the space. If the boundary edges do not intersect the justification line, the boundary edge is not merged. The justification line is most obvious when an existing space boundary is center justified.

In Figure 12.97, the vertical space boundaries were added with the **Add Boundary Edges** command. The space boundary on the left does not intersect with the justification line, and therefore the space is not divided. When creating boundary edges, you can extend the boundary beyond the justification line to ensure an intersection, as shown on the right. The short extension can then be easily trimmed with the **Remove Boundary Edges** command, discussed later in this chapter. You can determine if the space was divided by the boundary by selecting the space on one side of the boundary; if the space on the other side highlights, the boundary did not divide the space.

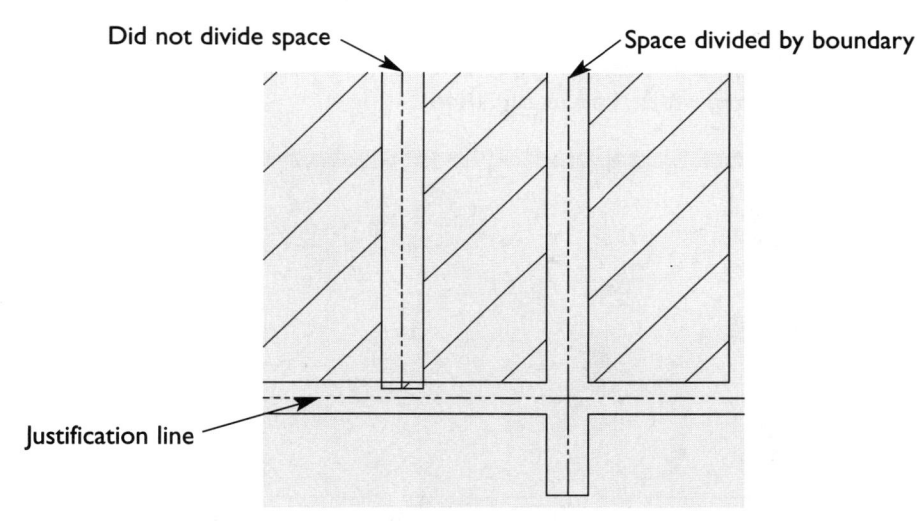

Figure 12.97 *Intersection of space boundaries with the justification line*

EDITING BOUNDARY EDGES

The edges of existing space boundaries can be changed with the **Edit Boundary Edges** command. **Edit Boundary Edges** allows you to change the dimensions and design rules of the space boundary. See Table 12.18 for **Edit Boundary Edges** command access.

Menu bar	Concept>Blocking and Boundaries> Edit Boundary Edges
Command prompt	SPACEBOUNDARYEDGE
Toolbar	Select Edit Boundary Edges from the Space Planning toolbar as shown in Figure 12.98
Shortcut	Select a Space Boundary; then choose Modify Edges from the shortcut menu

Table 12.18 *Edit Boundary Edges command access*

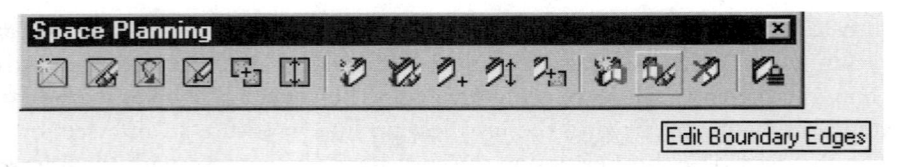

Figure 12.98 *Edit Boundary Edges command on the Space Planning toolbar*

When you select **Edit Boundary Edges** from the **Space Planning** toolbar, you are prompted to select the edge of a space boundary. Selecting the space boundary opens the **Boundary Edge Properties** dialog box, which consists of two tabs: **Dimensions** and **Design Rules**. The **Dimensions** tab, shown in Figure 12.99, allows you to change the segment type, width, and justification.

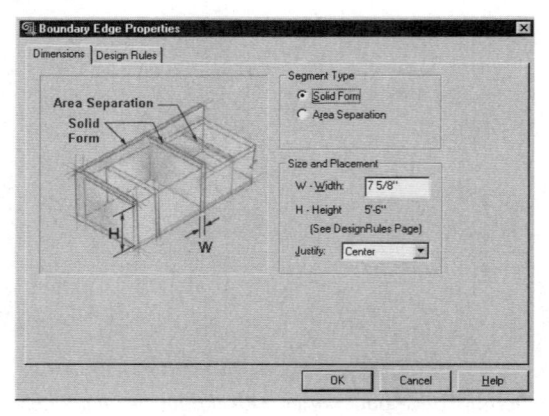

Figure 12.99 *Dimensions tab of the Boundary Edge Properties dialog box*

The **Design Rules** tab, shown in Figure 12.100, allows you to change how the ceiling and floor components will fit the space boundary.

DELETING A BOUNDARY USING REMOVE BOUNDARY EDGES

Space boundaries that are added to existing space boundaries cannot be selected individually. If you select one component of the space boundary, the entire boundary is selected. Therefore, to remove a component of the space boundary, you can use the **Remove Boundary Edges** command. See Table 12.19 for **Remove Boundary Edges** command access.

When you select **Remove Boundary Edges** from the **Space Planning** toolbar, you are prompted to select a space boundary edge to remove. Selecting a space boundary removes the space boundary, and any intersections are mended. Shown below is the command line prompt for removing space boundaries.

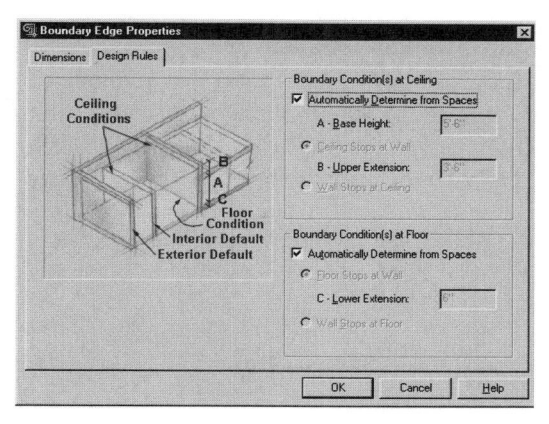

Figure 12.100 *Design Rules tab of the Boundary Edge Properties dialog box*

Menu bar	Concept>Blocking and Boundaries> Remove Boundary Edges
Command prompt	SPACEBOUNDARYREMOVEEDGES
Toolbar	Select Remove Boundary Edges from the Space Planning toolbar, as shown in Figure 12.101
Shortcut	Select a space boundary; then select Remove Edges from the shortcut menu

Table 12.19 *Remove Boundary Edges command access*

Figure 12.101 *Remove Boundary Edges command on the Space Planning toolbar*

```
Command: _AecSpaceBoundaryRemoveEdges
Select edges of a space boundary:
```
(Select space boundary at P1 in Figure 12.102.)
```
Select edges of a space boundary:
```
(Select space boundary at P2 in Figure 12.102.)
```
Select edges of a space boundary: ENTER
```

Note: The **Remove Boundary Edges** command can be used as a **TRIM** command for trimming space boundaries.

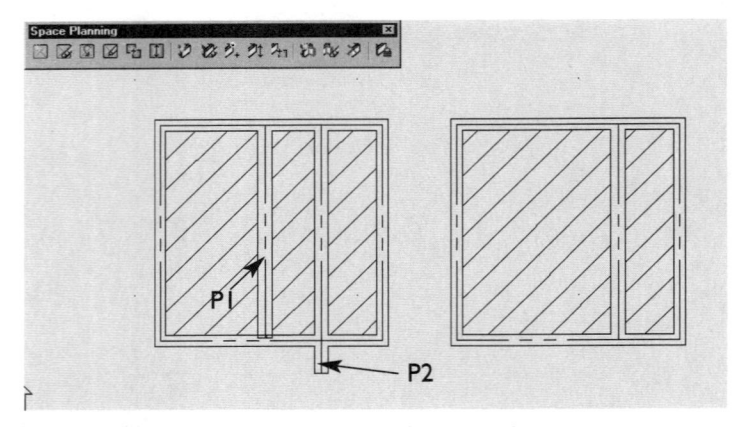

Figure 12.102 *Selecting space boundary edges for removal*

USING THE ANCHOR TO BOUNDARY COMMAND

The **Anchor to Boundary** command allows you to attach objects to the space boundary or free objects from it. The anchor of doors and windows to the space boundary can be released or reattached with the **Anchor to Boundary** command. Freeing the door or window of the anchor allows you to move it out of the space boundary. See Table 12.20 for **Anchor to Boundary** command access.

Menu bar	Concept>Blocking and Boundaries> Anchor to Boundary
Command prompt	SPACEBOUNDARYANCHOR
Toolbar	Select Anchor to Boundary from the Space Planning toolbar, as shown in Figure 12.103

Table 12.20 *Anchor to Boundary command access*

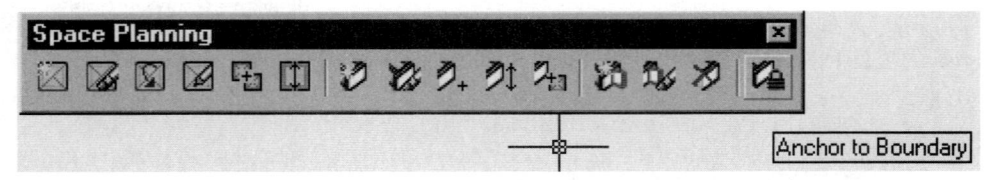

Figure 12.103 *Anchor to Boundary command on the Space Planning toolbar*

When you select the **Anchor to Boundary** command, you are prompted to select the **Free** or **Attach** option and then to select the object with an anchor. Objects that are freed of their anchor can be removed from the wall. Shown below is the command sequence of the **Anchor to Boundary** command to free a door from a space boundary.

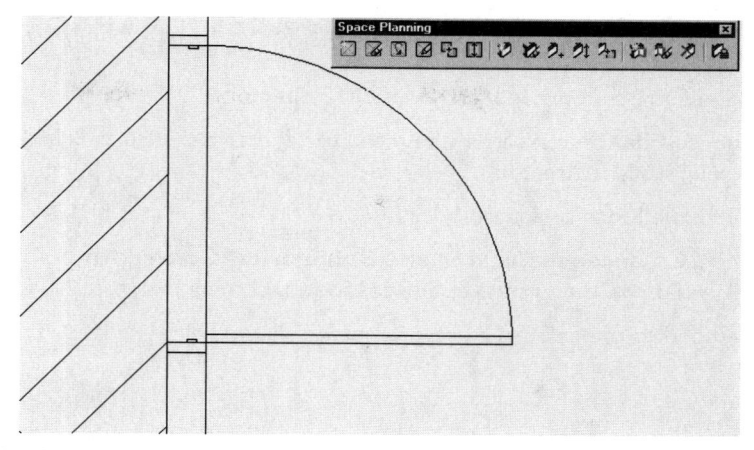

Figure 12.104 Select a door to release its anchor to the space boundary

```
Command: _AecSpaceBoundaryAnchor
Space boundary anchor [Attach objects/Free objects]:
F ENTER
Command: AnchorRelease
Select objects with anchors: (Select the door in Figure
12.104.) 1 found
Select objects with anchors: ENTER (Press ENTER to end
selection of anchored objects.)
```
   (The door frame is released from the space boundary. The door frame no longer adjusts to the width of the wall and the door can be moved out of the wall, as shown in Figure 12.105.)

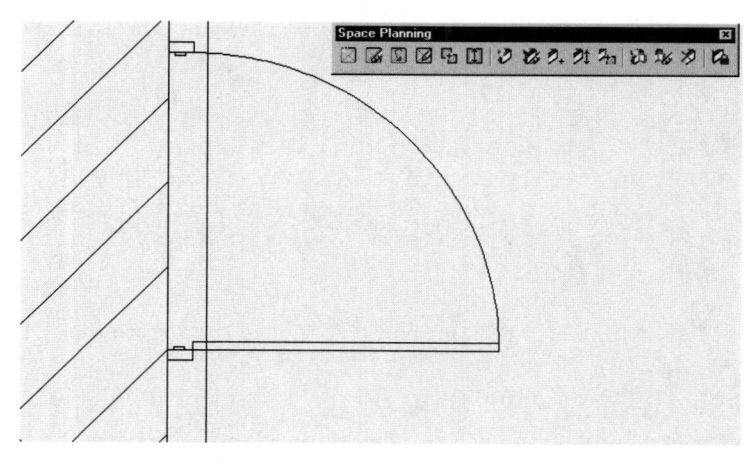

Figure 12.105 Door object anchor released from the space boundary

TUTORIAL 12.5 CONVERTING SPACES TO SPACE BOUNDARIES

1. Open *EX12-5* from the \ADT_Tutor\Ch12 directory.
2. Choose **File>SaveAs** from the menu bar and save the file as *Boundary3.dwg* in your student directory.
3. Select the Space layout tab.
4. Choose **Concept>Blocking and Boundaries>Convert to Boundary>Convert from Spaces** from the menu bar.
5. Respond to the command line prompts as shown below.

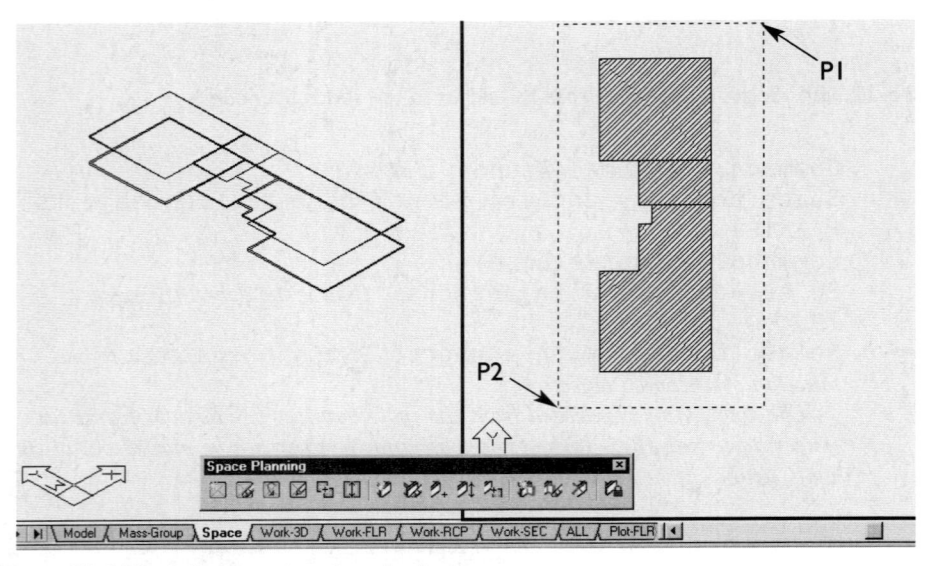

Figure 12.106 *Selection points for selecting a space*

```
Command: _AecSpaceBoundaryConvertSpace
Select spaces to convert to new space boundary.
Select spaces: (Select a point near P1 in Figure 12.106.)
Specify opposite corner: (Select a point near P2 in Figure
12.106.) 4 found
1 was filtered out.
Select spaces: ENTER (Press ENTER to end the command.)

Transferred 3 space(s) to new space boundary.
```

6. Select **Modify Boundary** from the **Space Planning** toolbar and then respond as shown in the following command line prompts.

```
Command: _AecSpaceBoundaryModify
Select space boundaries: L ENTER (Select the Last option to
select the space boundary.)
1 found
Select space boundaries: ENTER (Press ENTER to end selection.)
Space boundary modify [SOLid
form/SEparation/Width/Height/Justify/MATch/MANage]:
DBOX (Select the Properties button of the Modify Space Boundary dialog
box as shown in Figure 12.107.)
```

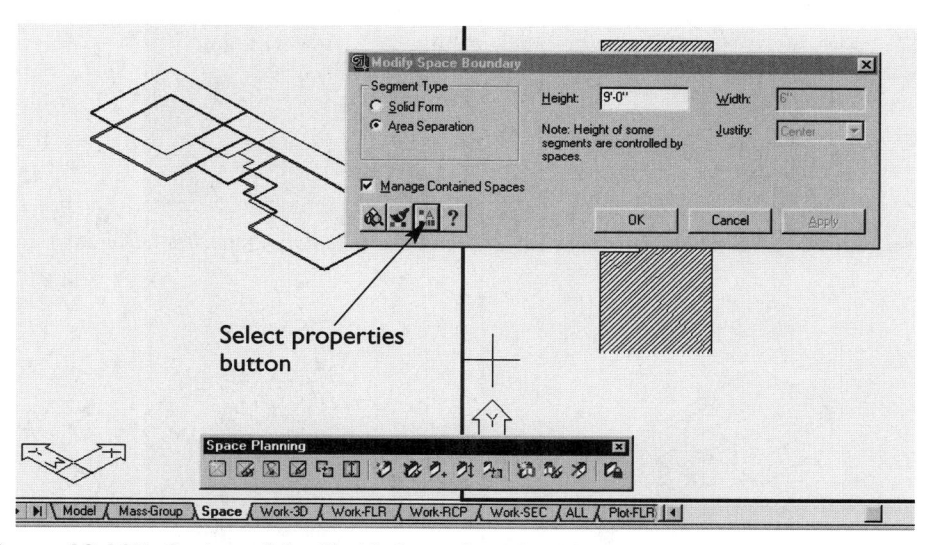

Figure 12.107 Settings of the Modify Space Boundary dialog box

(Select the Dimensions tab of the Space Boundary Properties dialog box; then change the segment type to Solid Form and Width to 7 5/8", verify that Manage Contained Spaces is selected, and Justify set to Center, as shown in Figure 12.108.)

(Select OK to dismiss the Space Boundary Properties dialog box and select OK to dismiss the Modify Space Boundary dialog box.)
(The space boundaries now have width and height as shown in Figure 12.109.)

7. Select **Zoom Window** from the **Standard** toolbar and respond to the following command line prompts.

```
Command: z ZOOM
Specify corner of window, enter a scale factor (nX
or nXP), or
```

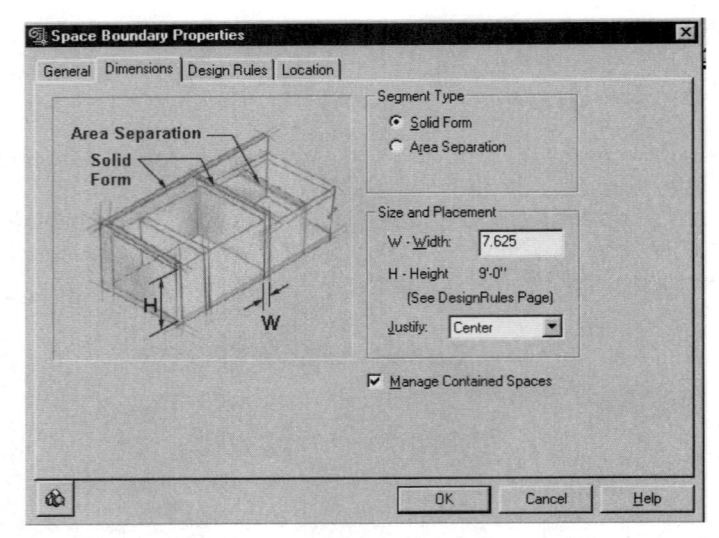

Figure 12.108 *Editing the Dimensions tab of the Space Boundary Properties dialog box*

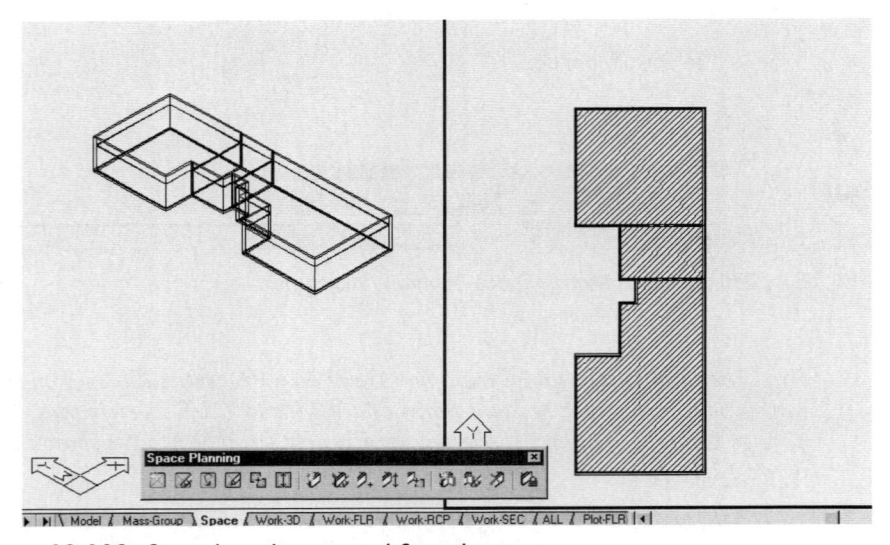

Figure 12.109 *Space boundary created from the space*

```
[All/Center/Dynamic/Extents/Previous/Scale/Window]
<real time>:
```
(Select point P1 in Figure 12.110.)
```
Specify opposite corner:
```
(Select point P2 in Figure 12.110.)

8. Right-click over the Osnap button of the Status bar, select **Settings** and verify that **Node** object snap mode is active and **Object Snap** is toggled ON.

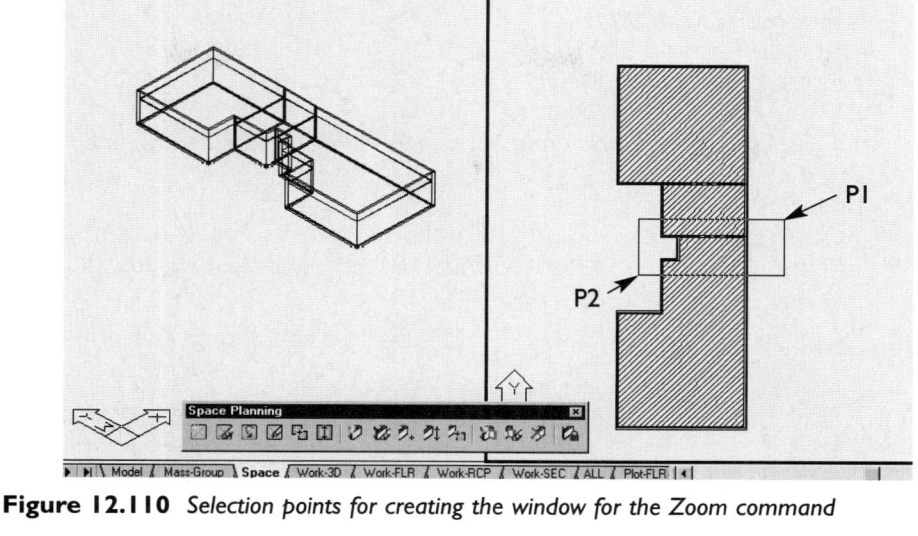

Figure 12.110 *Selection points for creating the window for the Zoom command*

9. Select **Add Boundary Edges** from the **Space Planning** toolbar. Set **Height** to 9', **Solid Form** toggled ON, **Offset** to 0, **Width** to 7 5/8, and **Justify** to Center. Respond to the following command line prompts to place the boundary.

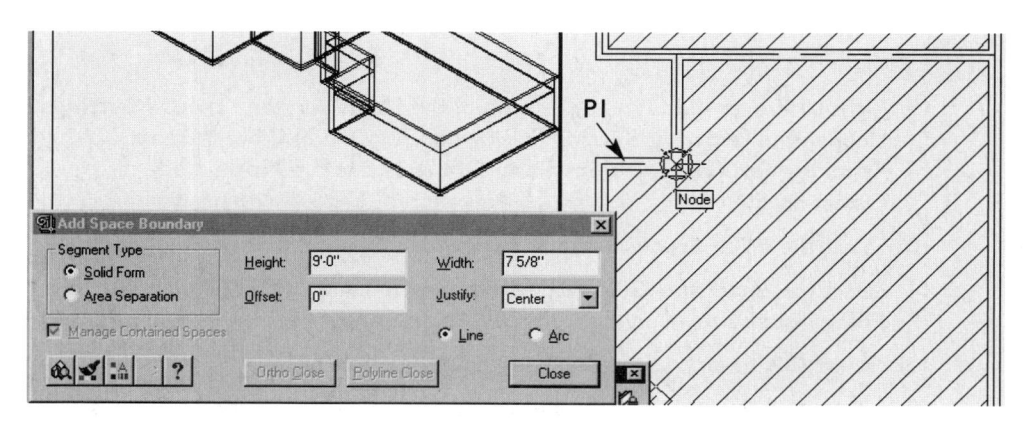

Figure 12.111 *Beginning the space boundary*

```
Command: AECSPACEBOUNDARYADDEDGES
Select a space boundary: (Select the space boundary at P1 in
Figure 12.111.)
Start point or [SOlid
form/SEparation/MANage/Width/Height/OFfset/Justify/
```

MATch/Arc]: *(Using the Node object snap, select the justification line as shown in Figure 12.111.)*
End point or [SOlid form/SEparation/MANage/Width/Height/OFfset/Justify/ MATch/Arc]: PERP ENTER
to *(Select the justification line of the space boundary at P2 in Figure 12.112.)*
End point or [SOlid form/SEparation/MANage/Width/Height/OFfset/Justify/ MATch/Arc/Undo]: ENTER *(Press ENTER to end the command.)*

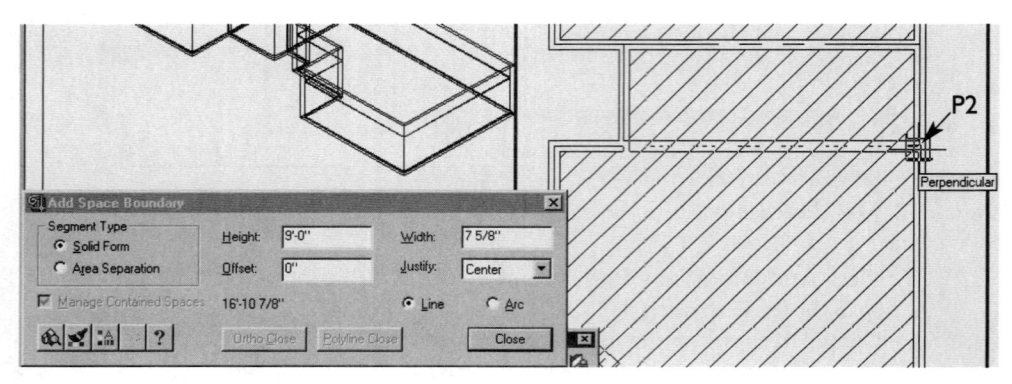

Figure 12.112 *Endpoint of the space boundary*

9. Right-click over the OTRACK button of the Status Bar; then choose **Settings** to open the **Drafting Settings** dialog box. Toggle ON **Object Snap Tracking On (F11)**, **Object Snap ON (F3)**, and the **Node** mode of Object Snap.
10. Select the **Polar Tracking** tab and verify that **Polar Tracking** is toggled ON.
11. Select the **Snap and Grid** tab; then toggle ON **Snap ON (F9)**, set the **Polar** distance to 16" and toggle ON **Polar Snap**, as shown in Figure 12.113.
12. Select **OK** to dismiss the **Drafting Settings** dialog box.
13. Select **Add Boundary Edges** from the **Space Planning** toolbar; then select the space boundary at P1, as shown in Figure 12.114.
14. Edit the **Add Space Boundary** dialog box as follows: set **Solid Form** ON, **Height** to 9', **Offset** to 0, **Width** to 7 5/8", and **Justify** to Right.
15. Respond to the following command line prompts.

Command: AECSPACEBOUNDARYADDEDGES
Select a space boundary: *(Space boundary selected at P1 in*

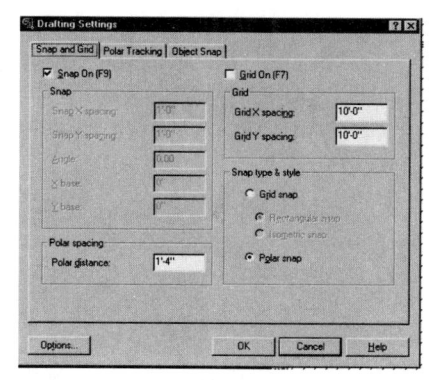

Figure 12.113 Snap and Grid settings of the Drafting Settings dialog box

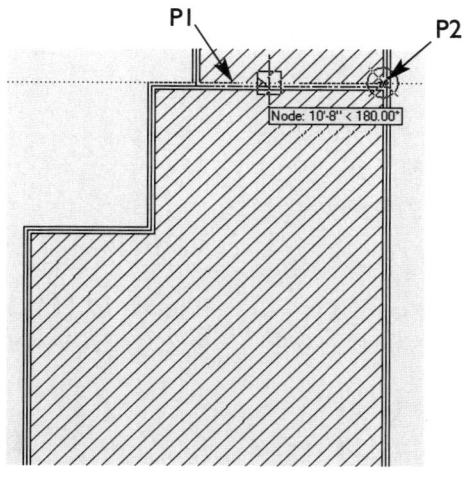

Figure 12.114 Using polar tracking to place the space boundary

Figure 12.114.)
```
1 was filtered out.
Start point or [SOlid
form/SEparation/MANage/Width/Height/OFfset/Justify/
MATch/Arc]:
```
(Move the pointer to display the Node object snap over the corner at point P2 on the space in Figure 12.114; then move to the left using polar tracking until 10'-8" is displayed in the polar tracking tip as shown above, and left-click to start the boundary.)
```
End point or [SOlid
form/SEparation/MANage/Width/Height/OFfset/Justify/
MATch/Arc]:
```
(Move the pointer down in a vertical direction until 30'8 is displayed in the polar tracking tip, as shown in Figure 12.115, and then left-click to specify the endpoint.)

```
End point or [SOlid
form/SEparation/MANage/Width/Height/OFfset/Justify/
MATch/Arc/Undo]: PERP ENTER
to
```
(Select the justification line of the exterior wall on the left as shown in Figure 12.116.)

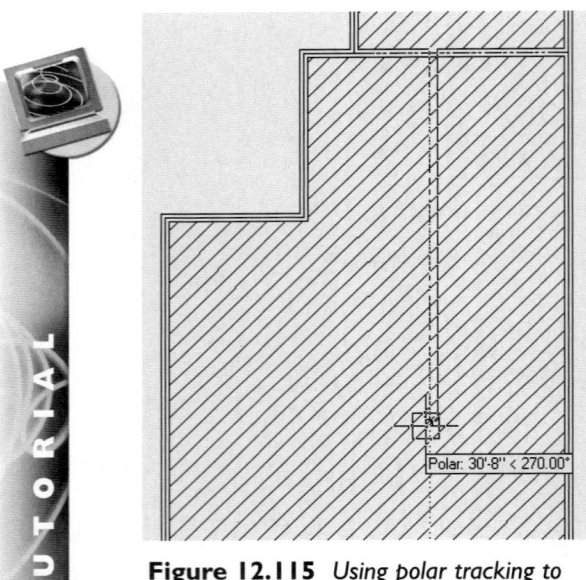

Figure 12.115 *Using polar tracking to place the second vertex of the space boundary*

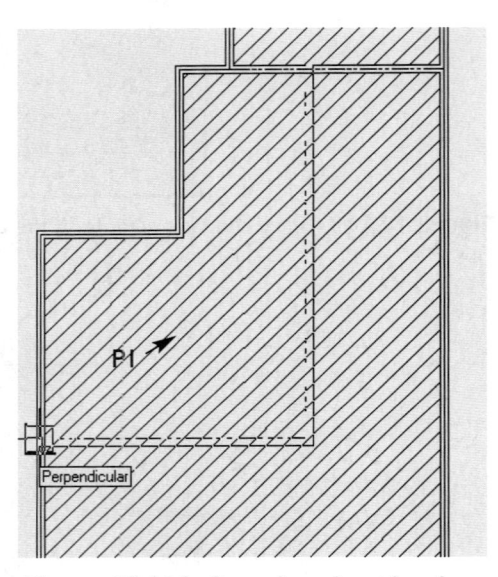

Figure 12.116 *Space boundary placed with precision using polar tracking*

```
End point or [SOlid
form/SEparation/MANage/Width/Height/OFfset/Justify/
MATch/Arc/Undo/Close/ORtho]: ENTER
```
(Press ENTER to end the command.)

16. Select the space at P1 shown in Figure 12.116. The space should highlight, indicating that the space is independent of the remainder of the building, as shown in Figure 12.117.

17. Select **Add Boundary Edges** from the **Space Planning** toolbar; then select the space boundary at P1 in Figure 12.118. Edit the **Add Space Boundary** dialog box as follows: set **Solid** Form ON, **Height** to 8', **Offset** to 0, **Width** to 3.5", and **Justify** to Right.

18. Respond to the following command line prompts.

```
Command: AECSPACEBOUNDARYADDEDGES
Select a space boundary:
```
(Select space boundary at P1 in Figure 12.118.)

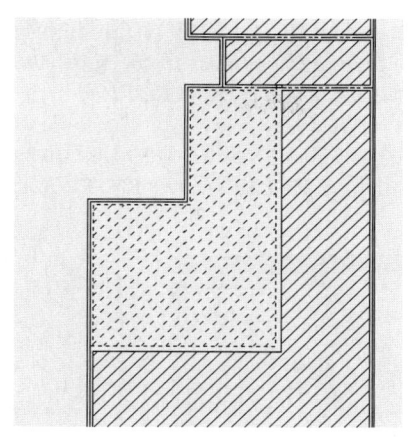

Figure 12.117 *Space selected*

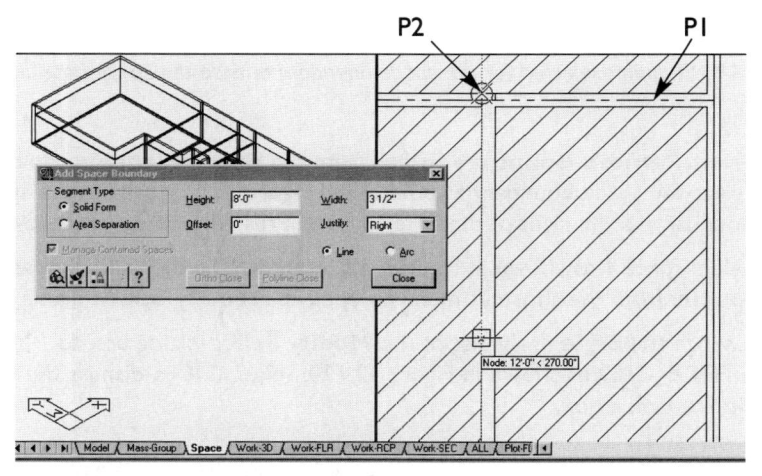

Figure 12.118 *Using polar tracking to place a space boundary*

```
1 was filtered out.
```
(Select the right viewport to activate the drawing area.)
```
Start point or [SOlid
form/SEparation/MANage/Width/Height/OFfset/Justify/
MATch/Arc]:
```
*(Move the pointer to display the Node object snap **above the justification line** at point P2 in Figure 12.119. Then move the pointer down using polar tracking until 12' is displayed in the polar tracking tip and left-click, as shown in Figure 12.118.)*

```
End point or [SOlid
```

```
form/SEparation/MANage/Width/Height/OFfset/Justify/
MATch/Arc]:
```
(Move the pointer to the right then select to the right of the exterior wall P3 as shown in Figure 12.119.)
```
End point or [SOlid
form/SEparation/MANage/Width/Height/OFfset/Justify/
MATch/Arc/Undo]: ENTER
```
(Press ENTER to end the command.)

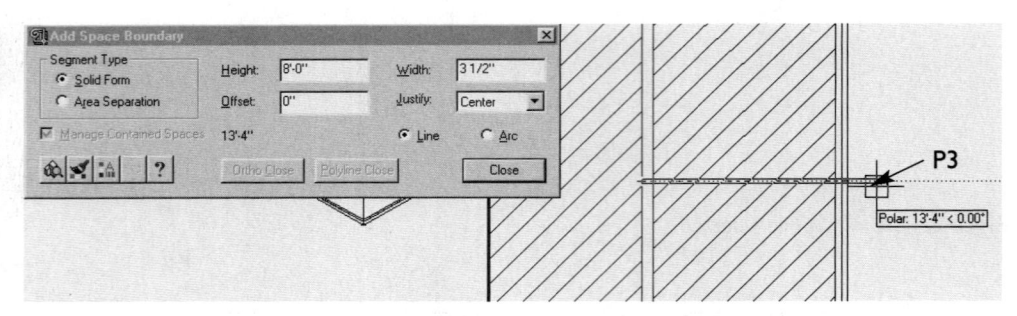

Figure 12.119 *Endpoint selected for the space boundary outside the space boundary*

19. Select **Remove Boundary Edges** from the **Space Planning** toolbar; then select the space boundary at point P3 in Figure 12.119 to trim the space boundary to the right of the exterior wall. Press ENTER to end the command.

20. Select space 1 as shown in Figure 12.120; then right-click and choose **Space Modify** from the shortcut menu to open the **Modify Space** dialog box.

21. Change the **Style** in the list of the **Modify Space** dialog box to OFFICE_CAD, as shown in Figure 12.120. Select **OK** to dismiss the **Modify Space** dialog box.

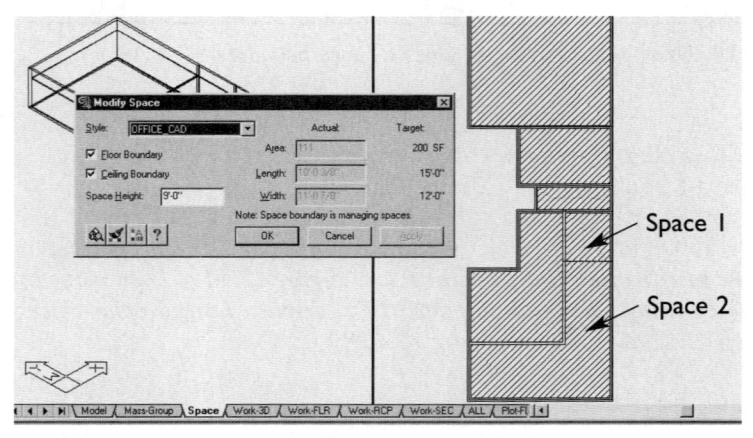

Figure 12.120 *Spaces identified for the Modify Space command*

Creating Spaces and Boundaries 939

22. Select space 2 as shown in Figure 12.120; then right-click and choose **Space Modify** from the shortcut menu to open the **Modify Space** dialog box.

23. Change the **Style** in the list of the **Modify Space** dialog box to OFFICE_MEDIUM. Select **OK** to dismiss the **Modify Space** dialog box.

24. Select **Space Styles** from the **Space Planning** toolbar to open the **Space Styles** dialog box.

25. Select the **Import/Export** button of the **Space Styles** dialog box to open the **Import/Export** dialog box.

26. Select the **Open** button to open the **File to Import From** dialog box.

27. Select the *Spaces Commercial (Imperial)* from the *\AutoCAD Architectural2\Content\Imperial\Styles* directory; then select **Open** to complete the selection and close the dialog box.

28. Select all the styles from the **External File** list, and then select the **Import** button.

29. Verify **Leave Existing** is toggled ON, and then select **OK** in the **Import/Export Duplicate Names Found** dialog box, as shown in Figure 12.121.

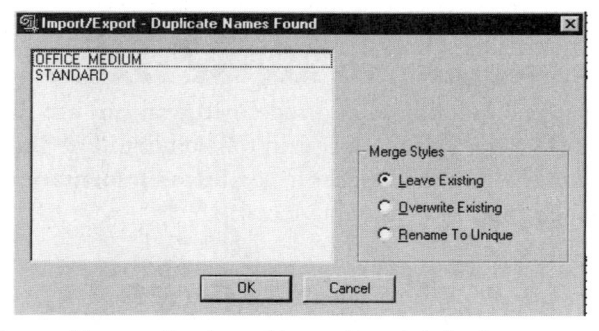

Figure 12.121 *Import/Export –Duplicate Names Found dialog box*

30. Select **OK** to dismiss the **Import/Export** dialog box.
31. Select **OK** to dismiss the **Space Styles** dialog box.
32. Select space 3 as shown in Figure 12.122; then right-click and choose **Space Modify**. Change the space style in the **Style** list of the **Modify Space** dialog box to CORRIDOR. Select **OK** to dismiss the **Modify Space** dialog box.
33. Select space 4 as shown in Figure 12.122; then right-click and choose **Space Modify**. Change the space style in the **Style** list of the **Modify Space** dialog box to CONFERENCE_LARGE. Select **OK** to dismiss the **Modify Space** dialog box.

34. Select space 5 as shown in Figure 12.122; then right-click and choose **Space Modify**. Change the space style in the **Style** list of the **Modify Space** dialog box to LIBRARY. Select **OK** to dismiss the **Modify Space** dialog box.

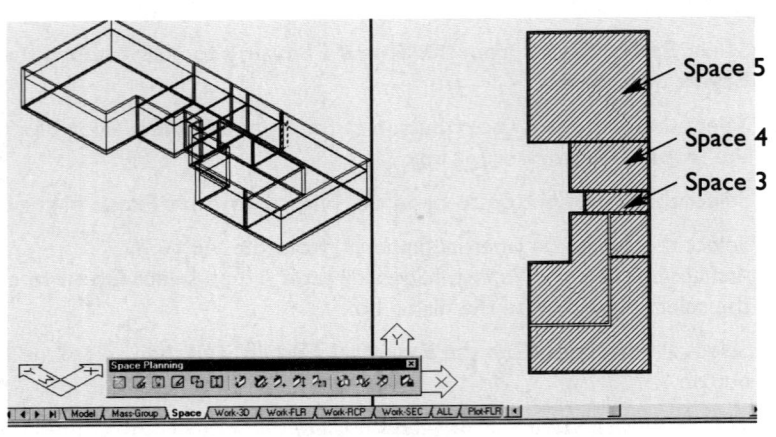

Figure 12.122 *Spaces identified for the Modify Space command*

35. Save the drawing.

USING SPACE INQUIRY TO ADJUST SPACES

Once all spaces have been identified and redefined, you can use the **Space Inquiry** command to create a summary of the spaces in the building. The **Space Inquiry** command creates a summary list of the spaces in the **Space Information** dialog box. See Table 12.21 for **Space Inquiry** command access.

Menu bar	Concept>Space Inquiry
Command prompt	SPACEQUERY
Shortcut	Right-click over the drawing area and choose Design>Spaces>Query

Table 12.21 *Space Inquiry command access*

When you choose the **Space Inquiry** command from the menu bar, the **Space Information** dialog box opens; it includes a **Space Info Total** tab and a **Space Information** tab. The **Space Info Total** tab for a building, shown in Figure 12.123, summarizes the spaces of the building.

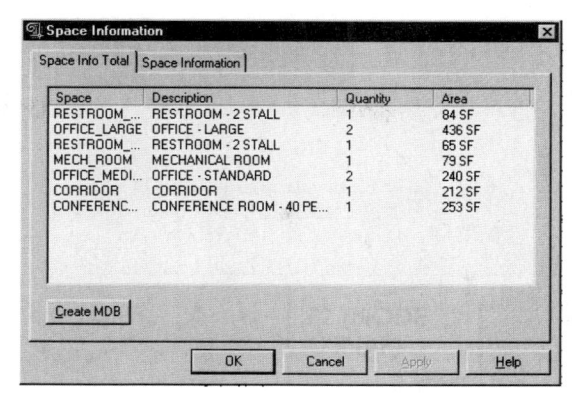

Figure 12.123 *Space Information dialog box*

The **Space Information** tab provides a list of the spaces with the actual area and the minimum and maximum areas defined in the space styles. If a space is changed in the drawing area, it will be updated in the **Space Information** dialog box. As shown in Figure 12.124, the RESTROOM_WOMEN-SMALL space is smaller than the minimum defined for that space style. Selecting RESTROOM_WOMEN-SMALL in the **Space Information** dialog box will highlight the space in the drawing. Therefore the designer can return to the drawing and adjust sizes according to the space style definitions.

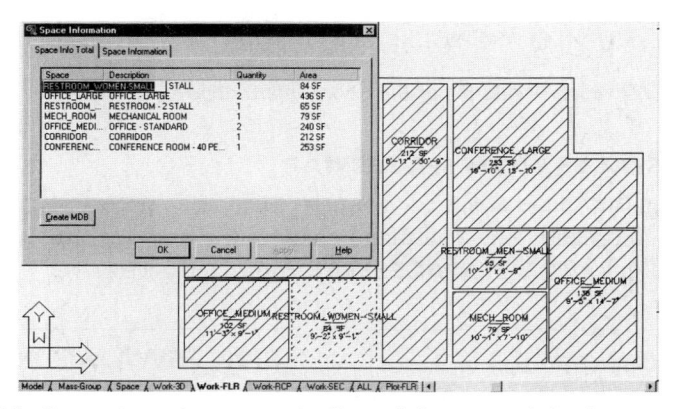

Figure 12.124 *Comparing spaces using the Space Information dialog box*

The **Create MDB** button of the **Space Information** dialog box creates a Microsoft Access file that contains the space and area data. When you select the **Create MDB** button, the **Database File** dialog box opens, allowing you to specify the directory and file name for the database file, as shown in Figure 12.125.

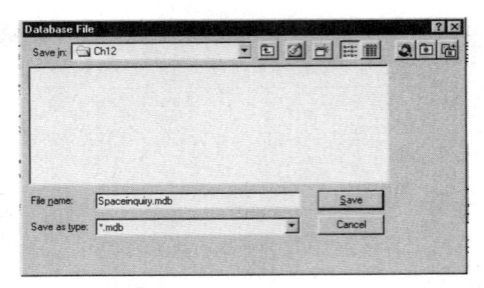

Figure 12.125 *Database File dialog box*

The space information file format is readable by Microsoft Access. Figure 12.126 shows a database file created for the drawing shown in Figure 12.124, opened from the Microsoft Access program.

Figure 12.126 *MicroSoft Access file of the Space Information*

CHANGING SPACES WITH SPACESWAP

After reviewing the space assigned, you can swap the spaces assigned using the **SPACESWAP** command. This command changes the name of the style assigned to the space. See Table 12.22 for **SPACESWAP** command access.

Command prompt	SPACESWAP
Shortcut	Select the space; then right-click and choose Swap from the shortcut menu

Table 12.22 *SPACESWAP command access*

When you choose the **Swap** command from the shortcut menu, you are prompted to select another space to swap with the previously selected space. The space name assigned to each space is swapped and the information totals of the **Space Information** dialog box will automatically update to the new space assignment. However, the room title tag does not automatically update; the Space Tag must be deleted and re-inserted to correctly title and summarize the space for the schedules.

GENERATING ROOM LABELS

You can generate room labels for the respective spaces by selecting the **Room & Finish Tags** command on the **Schedule - Imperial** toolbar. Selecting the **Space Tag** option in the Room & Finish Tags palette allows you to create room labels according to the space style name. The **Space Tag** option also provides the area data in the tag for the room. The A-Area-Spce layer can be frozen to hide the hatch of the space and provide a floor plan with space tags, as shown in Figure 12.127.

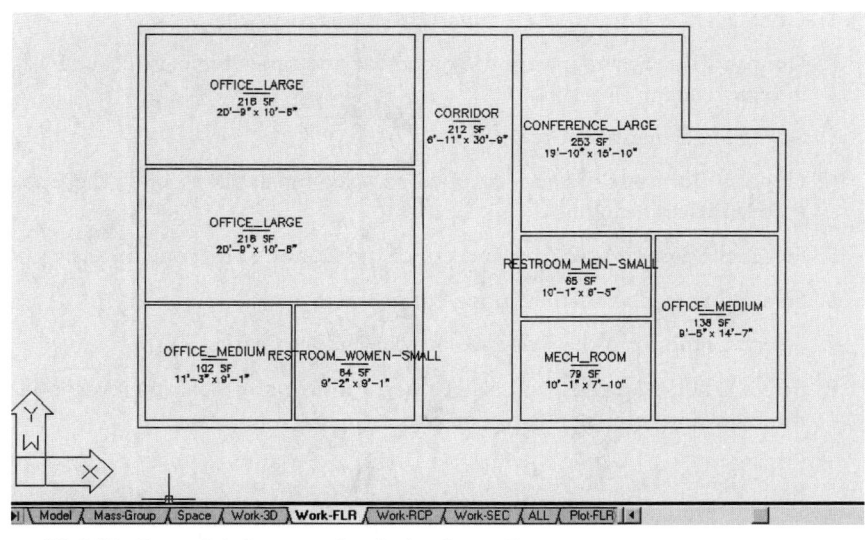

Figure 12.127 *Room labels created with the Space Tag*

GENERATING WALLS FROM SPACE BOUNDARIES

Walls can be created from the space boundaries by the **Generate Walls** command (**SPACEBOUNDARYGENERATEWALLS**), which copies the geometry of the space boundary and creates walls on the A-Wall layer. Therefore, you can freeze the A-Area-Bdry layer to turn off the display of the space boundary and view only the wall entities of the floor plan when working in the Work-FLR layout. However, selecting the Plot-FLR layout will display the walls and space tags without the display of the space boundaries. Changes to the walls or boundaries are independent of each other after the conversion; therefore the **Generate Walls** command should be executed when the room sizes are finalized. See Table 12.23 for **Generate Walls** command access.

Menu bar	Concept>Generate Walls
Command prompt	SPACEBOUNDARYGENERATEWALLS
Shortcut	Select a space boundary; then right-click and choose Generate Walls from the shortcut menu

Table 12.23 *Generate Walls command access*

When you select the Generate Walls command, you are prompted to select space boundaries, which are immediately turned into walls.

TUTORIAL 12.6 ADJUSTING SPACES

1. Open *EX12-6* from the *ADT_Tutor\Ch12* directory.
2. Choose **File>SaveAs** from the menu bar and save the file as *Boundary4.dwg* in your student directory.
3. Select the Space layout tab.
4. Choose **Concept>Space Inquiry** from the menu bar to open the **Space Information** dialog box.
5. Select the **Space Information** tab of the **Space Information** dialog box.
6. Select **OK** to dismiss the **Space Information** dialog box.
7. Select Ortho from the Status Bar to toggle ON Ortho.
8. Select **Divide Spaces** from the **Space Planning** toolbar; then respond to the following command line prompts.

    ```
    Command: AECSPACEDIVIDE
    Select a space: (Select the space at point P1 in Figure 12.128.)
    Divide line start point: mid ENTER
    of (Select the midpoint of the exterior space boundary P2 in Figure 12.128.)
    Divide line end point: (Select a point near P3 in Figure 12.129.)
    ```

9. Select **Edit Boundary Edges** from the **Space Planning** toolbar, then select the space boundary at P1 as shown in Figure 12.130.
10. Edit the **Width** of the space boundary to 3 1/2 and the justification to Right as shown in Figure 12.130.
11. Select **OK** to dismiss the **Boundary Edge Properties** dialog box.
12. Choose **Concept>Space Inquiry** from the menu bar to open the **Space Information** dialog box.
13. Select the **Space Information** tab.

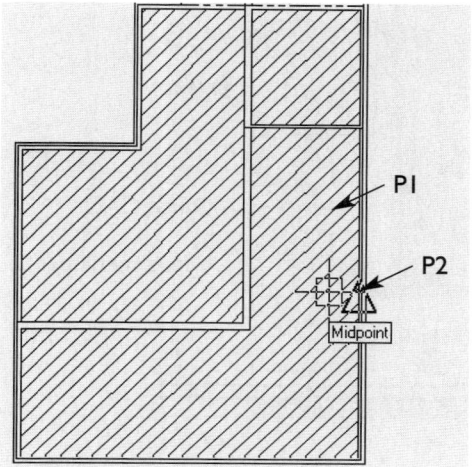

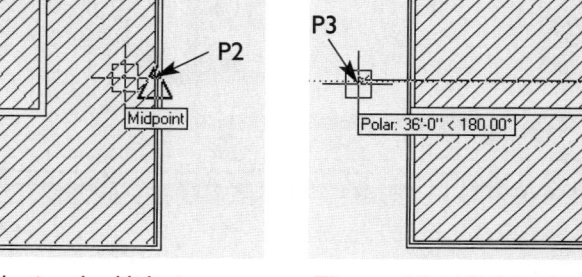

Figure 12.128 *Selecting the Midpoint object snap for beginning the space boundary*

Figure 12.129 *Selecting the endpoint of the space boundary*

14. Select OFFICE_CAD in the **Space Information** dialog box to verify the location of the OFFICE_CAD space in the drawing area and then select OFFICE_MEDIUM to identify its location in the drawing area, as shown in Figure 12.131.

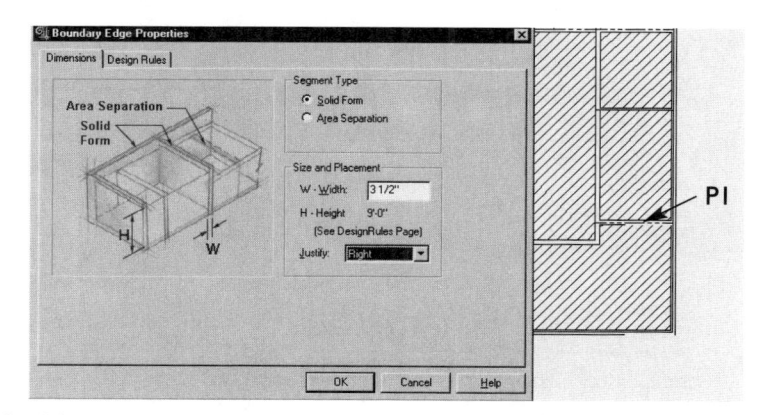

Figure 12.130 *Editing the space boundary width and justification*

Based upon the minimum and maximum areas, it would be more appropriate if the OFFICE_MEDIUM, which is highlighted in Figure 12.131, were swapped with the OFFICE_CAD space.

15. Select **OK** to dismiss the **Space Information** dialog box.

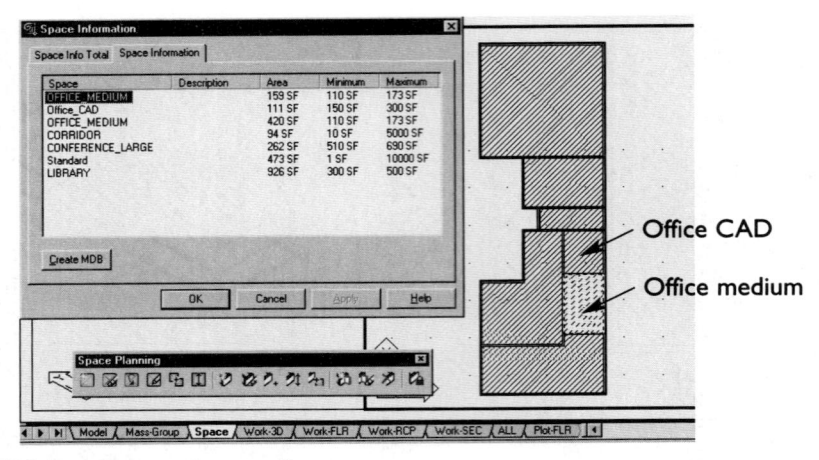

Figure 12.131 *Reviewing total areas of spaces*

16. Select OFFICE_MEDIUM as shown above; then right-click and choose **Swap** from the shortcut menu.
17. Select the OFFICE_CAD space, as shown in Figure 12.131, when prompted to select the other space.
18. Choose **Concept>Space Inquiry** from the menu bar to open the **Space Information** dialog box.
19. Select the **Space Information** tab to verify the space swap, as shown in Figure 12.132.

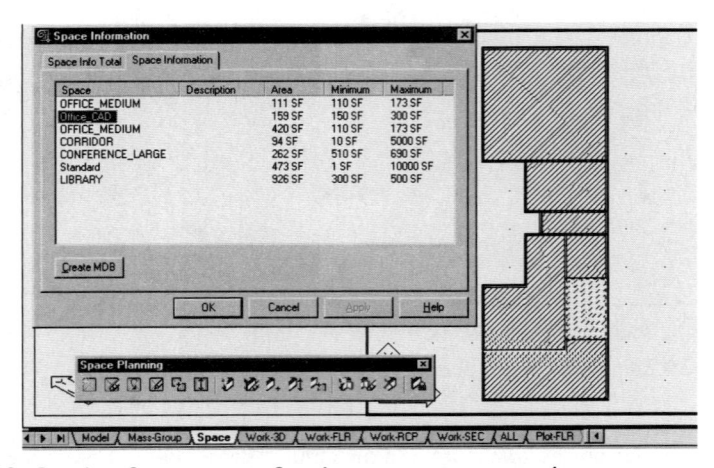

Figure 12.132 *Results of space area after the spaces are swapped*

20. Select **OK** to dismiss the **Space Information** dialog box.

21. Press F3 to toggle OFF object snaps.
22. Choose **Documentation>Schedule Tags>Room & Finish Tags** from the menu bar to open the DesignCenter to the Room & Finish Tags palette.
23. Select **Space Tag** by double-clicking it in the Room & Finish Tags palette; then select the center of the top space. Review the content of the **Edit Schedule Data** dialog box, then select **OK** to dismiss the **Edit Schedule Data** dialog box.
24. Select the **Space Tag** command from the Room & Finish Tags palette and continue to insert space tags for each space, as shown in Figure 12.133.

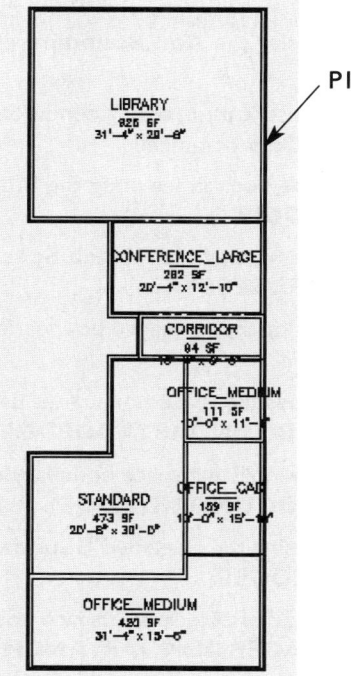

Figure 12.133 *A-Area-Spce layer frozen*

25. Select **Layers** from the **Object Properties** toolbar and freeze the A-Area-Spce layer.
26. Choose **Concept>Generate Walls** from the menu bar and then select the space boundary at P1 in Figure 12.133.
27. Select the Work-FLR layout to view the walls.
28. Save the drawing.

SUMMARY

1. Use the **Add Spaces** command (**SPACEADD**) to create and modify spaces.
2. Use the **Space Styles** command (**SPACESTYLE**) to create and modify space style definitions.
3. To change the style, floor boundary, and ceiling boundary of a space, select the **Space Modify** command (**SPACEMODIFY**).
4. Convert polylines to spaces using the **Convert to Spaces** command (**SPACECONVERT**).
5. Divide spaces with the **Divide Spaces** command (**SPACEDIVIDE**).
6. Combine adjacent spaces with the **Join Spaces** command (**SPACEJOIN**).
7. Create space boundaries using the **Add Boundary** command (**SPACEBOUNDARYADD**).
8. Change the height, width, justification, ceiling condition, and floor condition using the **SPACEBOUNDARYPROPS** command.
9. Combine independent space boundaries using the **Merge Boundaries** command (**SPACEBOUNDARYMODIFY**).
10. Attach a space to a boundary using the **Attach Spaces to Boundary** command (**SPACEBOUNDARYMERGESPACE**).
11. Convert space, slices, and sketches to space boundaries with the **SPACEBOUNDARYCONVERT** command.
12. Add space boundaries to existing space boundaries using the **Add Boundary Edges** command (**SPACEBOUNDARYADDEDGES**).
13. Change the properties of an existing space boundary using the **Edit Boundary Edge**s command (**SPACEBOUNDARYEDGE**).
14. Delete a space boundary using the **Remove Boundary Edges** command (**SPACEBOUNDARYREMOVEEDGES**).
15. Release or attach the anchor of objects in a space boundary using the **Anchor to Boundary** command (**SPACEBOUNDARYANCHOR**).
16. Determine the areas of the spaces in a plan using the **Space Inquiry** command (**SPACEQUERY**).
17. Create walls from space boundaries using the **Generate Walls** command (**SPACEBOUNDARYGENERATEWALLS**).

REVIEW QUESTIONS

1. Use the Node object snap to place spaces by snapping to points on the _____ boundary.
2. Selecting the **Drag Point** button toggles the insertion point of a space in a _____ from the default insertion point.

3. The default insertion point of a space is located in the _____ of a space.
4. The _____ check box of the **Add Space** dialog box requires that you specify the length and width options in the command line and then select the distance with the mouse for each dimension.
5. The distance between the Net and Gross space boundaries is specified in the _____ option of the **Space Style Properties** dialog box.
6. The space style definition includes _____, _____ length and _____ length.
7. The **Divide Spaces** command requires that the points specified be _____ on the space.
8. The **Join Spaces** command can be used to _____.
9. Select the _____ command to change the width of a space boundary.
10. The grips are displayed on the _____ of the space boundary.
11. Space boundaries with a width dimension greater than zero are _____.
12. Specify the location of the floor and ceiling of a space boundary by selecting the **Modify Space** command and editing the _____ of the contained space.
13. The _____ command can be used to trim a space boundary that extends beyond other space boundaries.
14. To detach a space from its boundary, use the _____ command.
15. A polyline is changed to a space boundary by the _____ command.
16. Space boundaries can be converted to walls by the _____ command.
17. Summarize the area per space by selecting the _____ command.
18. The summary of the areas in the building can be exported to the _____ database.
19. Insert additional space boundaries attached to the adjacent spaces by selecting the _____ command.
20. Space boundaries are displayed in the _____ layout tab.

PROJECT

EX12.1 POSITIONING SPACES TO CREATE SPACE BOUNDARIES AND WALLS

1. Open *Ex12-7* from the *ADT Tutor\Ch12* directory and save the drawing in your student directory as *Proj12-7*.

2. Select each of the spaces and position them as shown in Figure 12.134. Use the Node object snap to position the spaces.
3. Use grip editing to stretch the spaces to form the building shape as shown in Figure 12.135.

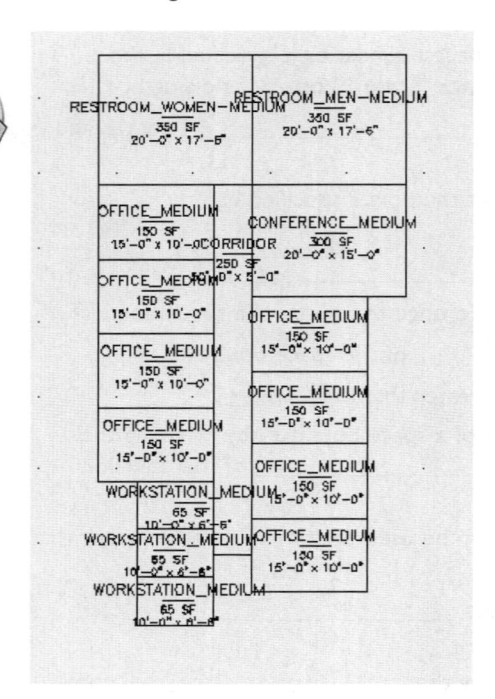

Figure 12.134 *Position of spaces to create a plan*

Figure 12.135 *Spaces stretched using grips*

4. Convert the spaces to space boundaries and set the width to 7 5/8", height to 9' and center justified, as shown in Figure 12.136.
5. Edit the walls at the end of the corridor, as shown in Figure 12.137, using the **Add Boundary Edges** and **Remove Boundary Edges** commands.
6. Set the drawing scale to 1/4"=1'-0" and insert space tags for each space. Create a list of spaces using **Space Inquiry** as shown in Figure 12.138.
7. Turn off hatching of the spaces by editing the System Default for spaces.
8. Convert the space boundaries to walls.

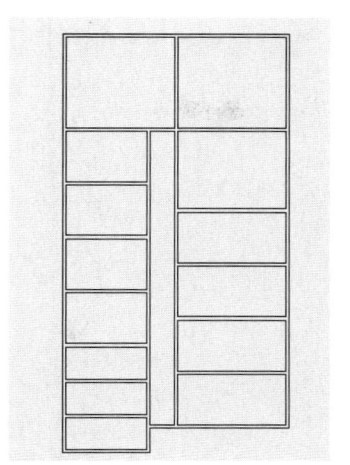

Creating Spaces and Boundaries 951

Figure 12.136 *Space boundaries created from spaces*

Figure 12.137 *Editing space boundaries for the corridor*

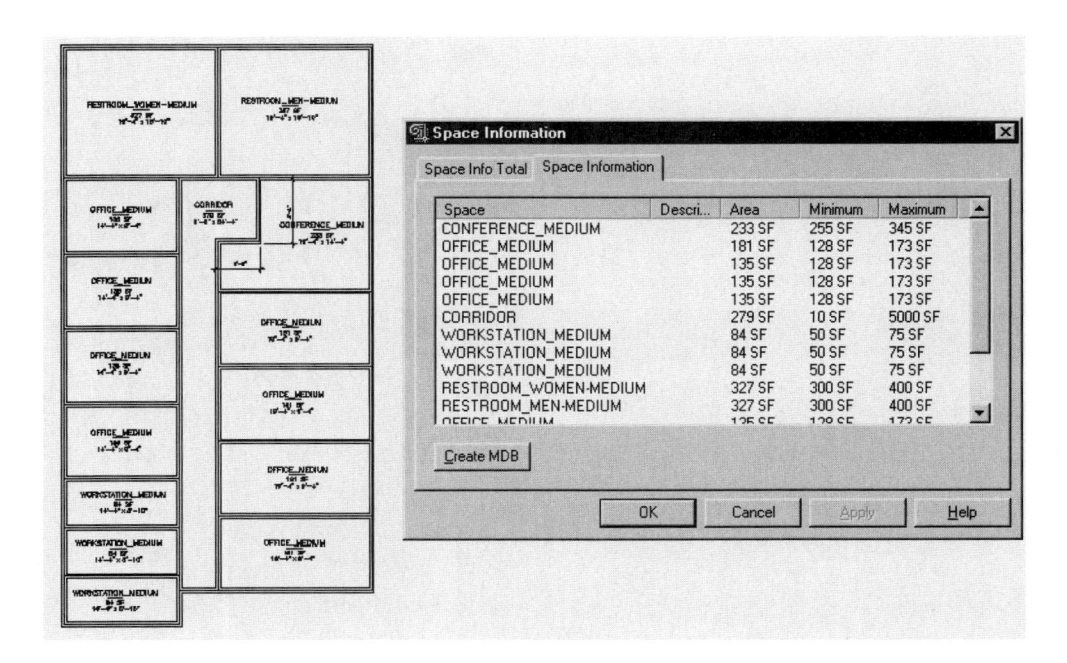

Figure 12.138 *Space summary for the building*

CHAPTER 13

Drawing Commercial Structures

INTRODUCTION

The development of drawings for commercial building requires the insertion of structural grids and ceiling grids. Layout curves can be used to insert columns or other AEC objects spaced uniformly along the layout curve. Structural grids, layout curves, and ceiling grids are presented in this chapter. In addition, the development of views and walkthroughs using a camera and various paths are presented.

OBJECTIVES

Upon completion of this chapter the student will be able to

Create a structural column grid using the **Add Column Grid** command (**COLUMNGRIDADD**)

Dynamically size a structural grid using the **Specify on Screen** option of the **Add Column Grid** command

Use the **Layout Mode** command (**COLUMNGRIDXMODE / COLUMN-GRIDYMODE**) to add or remove column grid lines

Insert mass elements in the column grid using AEC Profiles

Adjust the location of the column grid using **Start Offset** and **End Offset** options

Change the column grid line location using the **Column Grid Properties** command (**COLUMNGRIDPROPS**)

Change the size and dimension of the column grid using the **Column Grid Modify** command (**COLUMNGRIDMODIFY**)

Label the column grid using the **Column Grid Labeling** command (**COLUMNGRIDLABEL**)

Create dimensions for the column grid using the **Column Grid Dimension** command (**COLUMNGRIDDIM**)

Insert columns independent of a column grid with the **Add Column** command (**COLUMNCOVERADD**)

Create a ceiling grid with boundaries using the **Add Ceiling Grid** command (**CEILINGGRIDADD**)

Adjust the size and dimension of the ceiling grid components using the **Ceiling Grid Properties** command (**CEILINGGRIDPROPS**) and **Modify Ceiling Grid** command (**CEILINGGRIDMODIFY**)

Add and remove ceiling grid lines using the **CEILINGGRIDXADD / CEILINGGRIDYADD** and **CEILINGGRIDXREMOVE / CEILINGGRIDYREMOVE** commands

Use the **Attach Clipping Boundary/Holes** command (**CEILINGGRIDCLIP**) to trim the ceiling grid

Use the **Add Layout Curves** command (**LAYOUTCURVEADD**) to place columns and other AEC objects symmetrically along a layout curve

Modify the intersection of columns to walls using the **Add Interference Condition** command (**WALLINTERFERENCEADD**)

Place a camera in the drawing using the **Add Camera** command (**AECCAMERAADD**)

Create a video of a walkthrough using the **Create Video** command (**AECCAMERAVIDEO**)

CREATING STRUCTURAL GRIDS

Column grids are developed to uniformly place structural columns in the plan view. The size and orientation of the structural grid are dependent upon the floor plan and maximum span of the structural components. The column grid is created on the A-Grid layer, which has the color 173 (a hue of blue) with Center2 linetype.

The column grid can be created with mass elements attached to the grid. The mass elements created in the grid to represent the columns are placed on the A-Cols layer, which has the color 110 (a hue of green). The column shape assigned to the grid is assigned to each grid intersection.

The commands for creating, editing, and labeling the structural grids are located on the **Grids/Columns** toolbar as shown in Figure 13.1.

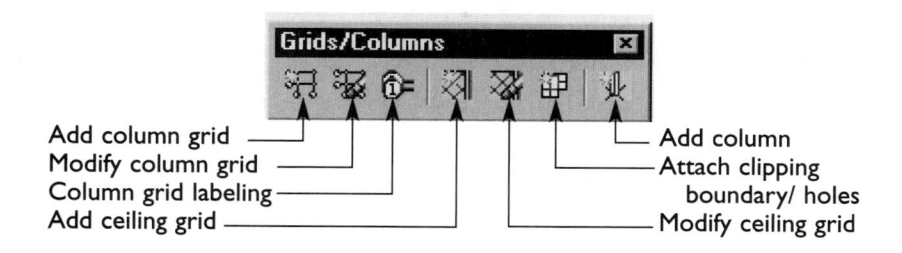

Figure 13.1 *Commands on the Grids/Columns toolbar*

The **Add Column Grid** command (**COLUMNGRIDADD**) is used to create a column grid, which can be rectangular or radial in shape. See Table 13.1 for **Add Column Grid** command access.

Menu bar	Design>Grids>Add Column Grid
Command prompt	COLUMNGRIDADD
Toolbar	Select Add Column Grid from the Grids/Columns toolbar as shown in Figure 13.1
Shortcut	Right-click over the drawing area; then choose Design>Column Grids>Add

Table 13.1 *Add Column Grid command access*

When you select the **Add Column Grid** command, the **Add Column Grid** dialog box opens, allowing you to specify the parameters of the column grid, as shown in Figure 13.2.

Figure 13.2 shows the **Add Column Grid** dialog box for a rectangular column grid. The rectangular column grid allows you to specify the overall size of the grid according to the **X-Width** and **Y-Depth** dimensions. The **X-Width** and **Y-Depth** dimensions

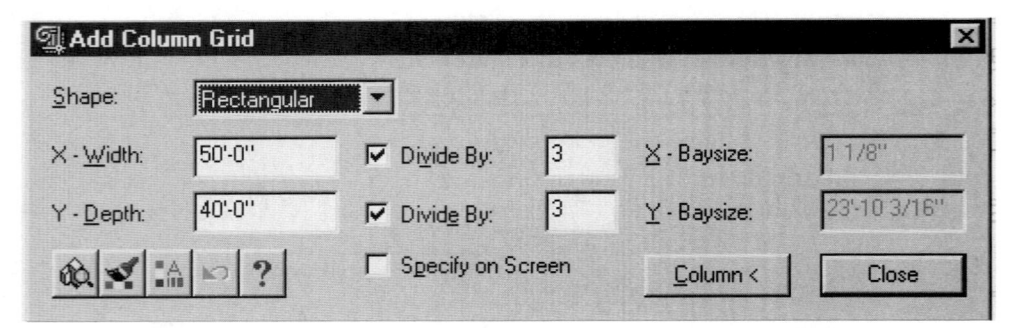

Figure 13.2 *Add Column Grid dialog box*

can be divided by the number of bays to determine the bay dimension, or the **X-Baysize** and **Y-Baysize** dimensions can be specified. The column grid is developed relative to its insertion point, which is in the lower left corner, as shown in Figure 13.3. The options of the **Add Column Grid** dialog box for a rectangular column grid are described below.

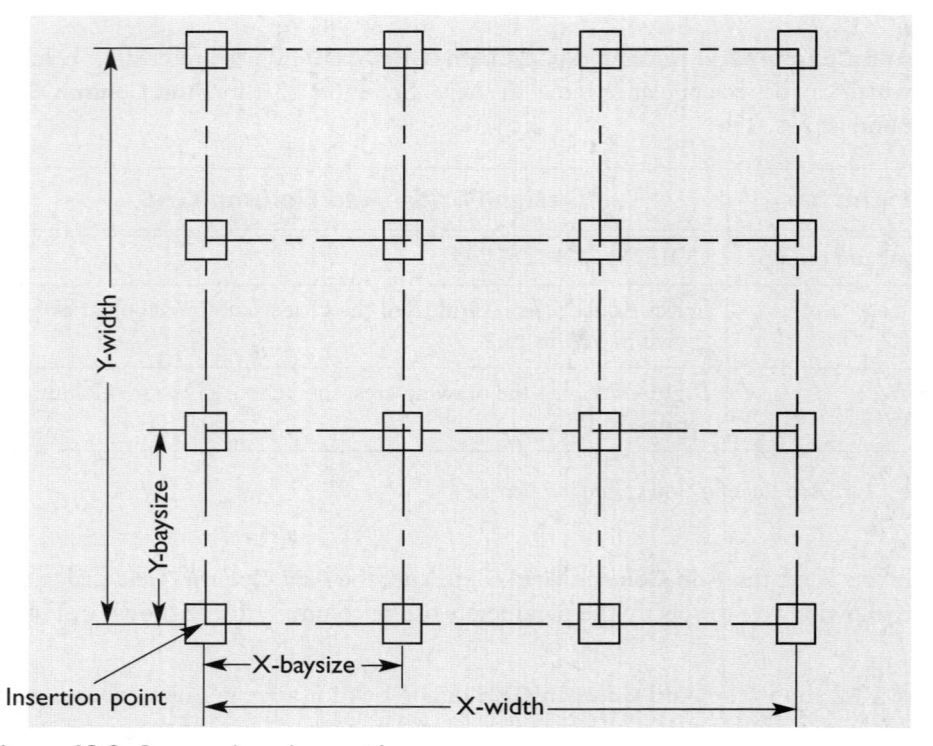

Figure 13.3 *Rectangular column grid*

Shape – The **Shape** list includes radial and rectangular grid shapes. The options of the dialog box change according to the shape selected.

X-Width – The **X-Width** dimension of the grid is the overall horizontal dimension of the grid.

Y-Depth – The **Y-Depth** dimension of the grid is the vertical dimension of the grid.

Divide by – The **Divide by** check boxes allow you to specify the number of bays to divide the overall **X-Width** or the **Y-Depth** to create a bay.

Specify on Screen – The **Specify on Screen** check box allows you to specify a bay dimension in the **X-Width** and **Y-Depth** directions by selecting points in the drawing area.

X-Baysize – The **X-Baysize** edit field specifies an absolute value for the horizontal dimension of the bay.

Y-Baysize – The **Y-Baysize** edit field specifies a specific vertical dimension for the bay.

Column – The Column button opens the **Add Mass Element to Column** dialog box, which includes options to specify a mass element at each column grid intersection.

Floating Viewer – The **Floating Viewer** button opens the **Viewer**, allowing you to view the column grid.

Match – The **Match** button allows you to select an existing grid and match the parameters of the grid. The **X-Width, Y-Depth, X-Baysize**, and **Y-Baysize** dimensions can be matched from an existing column grid.

Properties – The **Properties** button opens the **Column Grid Properties** dialog box; it includes options to adjust the dimensions of the column grid.

Undo – The **Undo** button will reverse the insertion of the grid, and you will remain in the **Add Column Grid** command.

Help – The **Help** button opens the Help file to the Column Grid Overview topic.

Create a radial column grid by selecting the Radial shape from the **Shape** list of the **Add Column Grid** dialog box. The **Add Column Grid** dialog box is converted for the radial shape to include edit fields for the radius and angle of the column grid. The radius of the column grid specifies the overall size and the angle specifies the included angle of the grid, as shown in Figure 13.4. The insertion point of the grid is at the center of the radius. The **Divide by** edit fields allow you to specify the number of bays of the column grid. In Figure 13.4, the radial dimension has been divided into four equal spaces, while the angle has been divided into three equal spaces.

The options of the **Add Column Grid** dialog box for a radial column grid are described below.

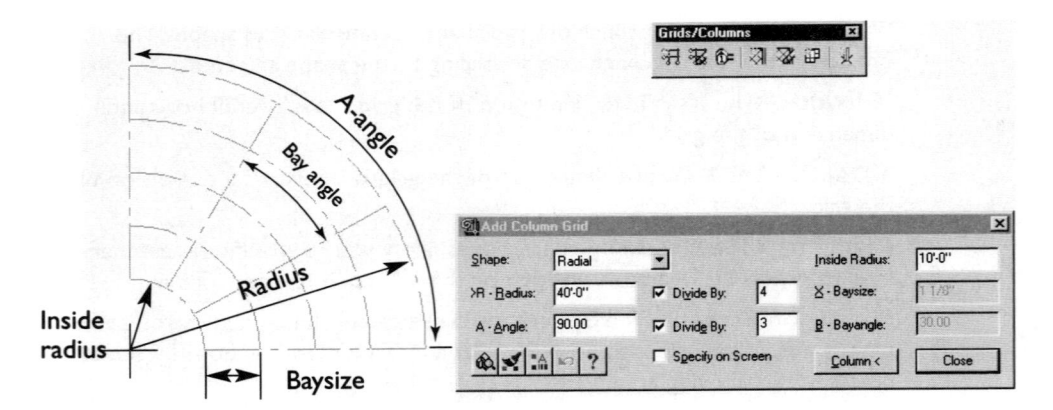

Figure 13.4 *Radial column grid*

>**R-Radius** – The **>R-Radius** edit field specifies the distance from the insertion point to the furthermost column grid.

A-Angle – The **A-Angle** edit field specifies the included angle of the column grid. The column grid is developed in counter-clockwise direction from the three o'clock direction.

Inside Radius – The **Inside Radius** specifies the distance from the insertion point to the inner most radial column grid.

X-Baysize – The **X-Baysize** is the distance between curved column grid lines. The **X-Baysize** specifies the grid size when the **Radius Divide by** check box is not selected.

B-Bayangle – The **B-Bayangle** is the angular measure between the radial column grid lines. The **B-Bayangle** specifies the angle between radial column grids when the **Divide by** check box is not selected. Three bays would be created if the **A-Angle** were set to 90 and the **B-Bayangle** were set to 30. If the **A-Angle** measure is not divisible by the **B-Bayangle** measure, the column grid is created equal to the **B-Bayangle** measure. In Figure 13.4, if the **A-Angle** is set to 90, and the **B-Bayangle** is set to 50, then a 50-degree column grid would be created.

SPECIFYING THE SIZE OF THE COLUMN GRID ON SCREEN

The size and number of divisions of the rectangular column grid can be specified dynamically on screen. When the **Specify on Screen** option is selected, you can specify the overall size of the column grid by selecting two diagonal points. Moving the pointer from the last diagonal point dynamically sets the number of divisions. As you move the pointer back toward the direction of the insertion point, the number of bays increases. Movement of the pointer in a horizontal direction toward the insertion point increases the number of divisions along the horizontal grid line. Vertical movement of the pointer in the direction toward the insertion point will increase the number of divisions along the vertical grid line. If the **Divide by** check box is selected, the number of bays is displayed in the **Add Column Grid** dialog box as the pointer is moved. When **Divide by** is cleared, the size of the bay in the X and Y directions is displayed in the **Add Column Grid** dialog box.

STEPS TO DYNAMICALLY SIZING A RECTANGULAR COLUMN GRID

1. Select the **Add Column Grid** command from the **Grids/Columns** toolbar.
2. Select the Rectangular shape from the **Shape** list.
3. Select the **Specify on Screen** check box.
4. Select **Divide by** for both the **X-Width** and **Y-Depth** dimensions.
5. Select the insertion point at P1 as shown in Figure 13.5.
6. Select the diagonal corner of the column grid at P2 in Figure 13.5.

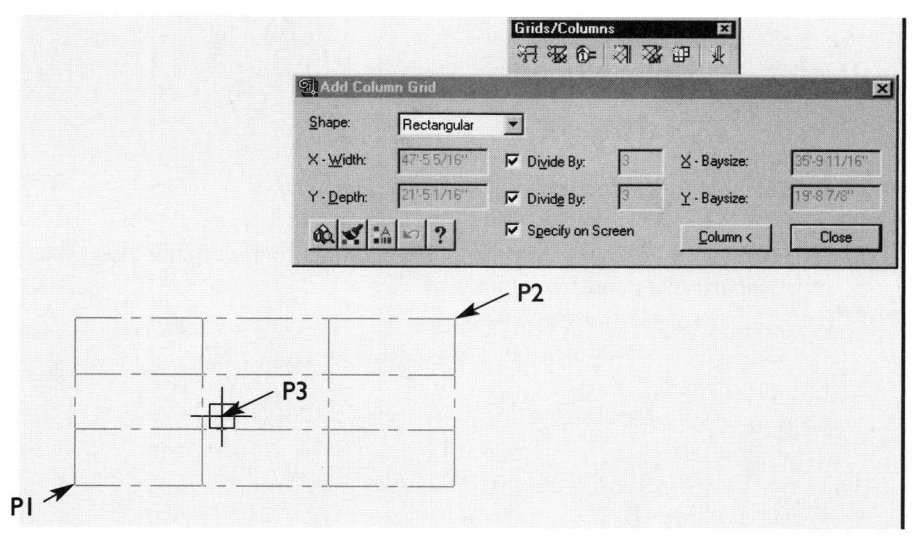

Figure 13.5 *Selecting points for the Specify on Screen option*

7. Move the pointer horizontally toward P1 to point P3 in Figure 13.5 to specify the number of **X-Width** divisions and **Y-Depth** divisions.
8. Press ENTER twice to end the command.

Radial column grids can also be specified on screen. Selecting points in the drawing area specifies the insertion point, outer radius and inner radius. Select the number of divisions by moving the pointer toward the inside radius and down.

STEPS TO DYNAMICALLY SIZING A RADIAL COLUMN GRID

1. Select the **Add Column Grid** command from the **Grids/Columns** toolbar.
2. Select the Radial shape from the **Shape** list.
3. Select the **Specify on Screen** check box.
4. Select **Divide by** for both the **X-Width** and **Y-Depth** dimensions.

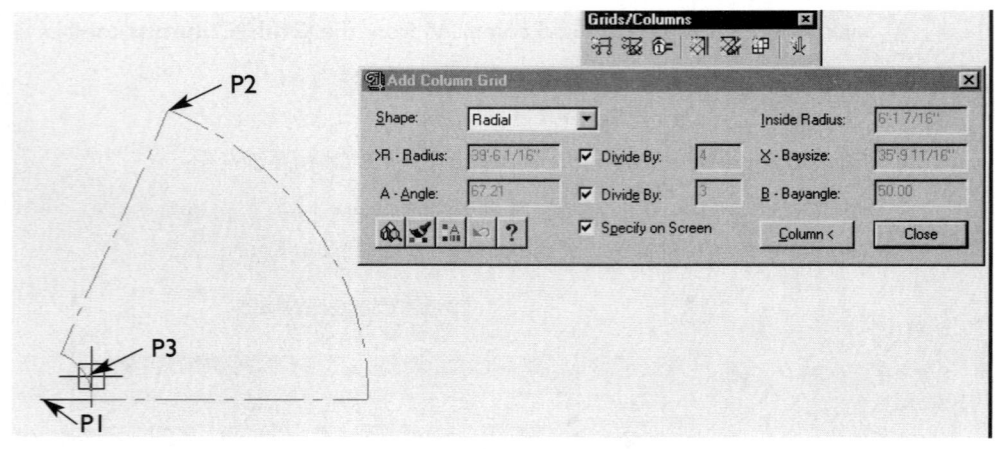

Figure 13.6 *Specifying the insertion point of a radial column grid*

5. Respond to the following command line prompts to specify the insertion point, inner radius, and the number of bays.

```
Command: AECCOLUMNGRIDADD
Insertion point or [WIdth/Depth/Xbay/Ybay/XDivide by
toggle/YDivide by
toggle/Match]: DBOX

Insertion point or [Radius/Angle/Inside
radius/Xbay/Bay angle/XDivide by
toggle/YDivide by toggle/Match]:
```
(Select point P1 in Figure 13.6 to specify the insertion point.)

```
New size or [Radius/Angle/Inside radius/Xbay/Bay
angle/XDivide by
toggle/YDivide by toggle/Match]:
```
(Select point P2 in Figure 13.6 to specify the radial dimension.)

```
New inside radius or [Radius/Angle/Inside
radius/Xbay/Bay angle/XDivide by
```

```
toggle/YDivide by toggle/Match] <0">:
```
(Select point P3 in Figure 13.6 to specify the inner radius.)

```
New size for cell or [Radius/Angle/Inside
radius/Xbay/Bay angle/XDivide by
toggle/YDivide by toggle/Match]:
```
(Move the mouse to point P4 as shown in Figure 13.7 to specify the number of bays.)

```
Rotation or [Radius/Angle/Inside radius/Xbay/Bay
angle/XDivide by
toggle/YDivide by toggle/Match] <0.00>:
```
ENTER *(Press* ENTER *to accept 0° rotation.)*

```
Insertion point or [Radius/Angle/Inside
radius/Xbay/Bay angle/XDivide by
toggle/YDivide by toggle/Match/Undo]:
```
ENTER *(Press* ENTER *to end the command.)*

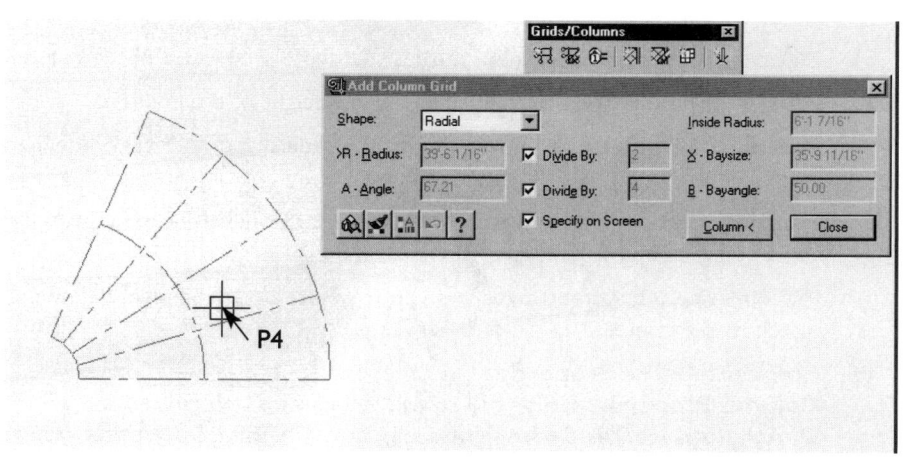

Figure 13.7 *Specifying on screen the number of divisions*

DEFINING THE COLUMN FOR THE GRID

Selecting the **Column** button of the **Add Column Grid** dialog box opens the **Add Mass Element as Column** dialog box, which can then be used to specify the column for each grid intersection. Predefined shapes or custom shapes can be specified for the column grid. The **Predefined Shape** option includes only the Box and Cylinder shapes. The **Custom Shape** option allows you to select from the AEC Profiles of the drawing. They are extruded according to the height dimension to create the column. The AEC Profiles include pipe, tube, and wide flange shapes. The size of the column is specified in the width, depth, height, and radius edit fields, as shown in Figure 13.8.

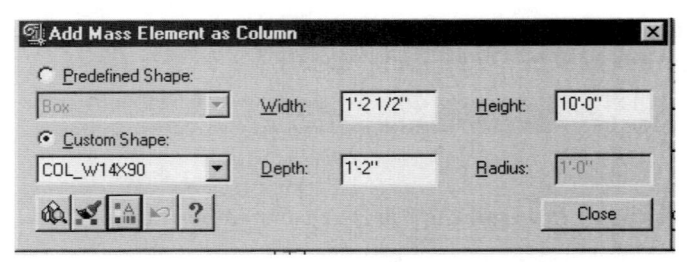

Figure 13.8 *Specifying the size of column in the Add Mass Element as Column*

The options of the **Add Mass Element as Column** dialog box are described below.

Predefined Shape – The **Predefined Shape** radio button and list include the Box and Cylinder mass elements for the column.

Custom Shape – The **Custom Shape** radio button activates the **Custom Shape** list, which shows the AEC Profiles of the drawing; they include the tube column, pipe, and wide flange column shapes.

Width – The **Width** edit field specifies the width dimension of the column.

Depth – The **Depth** edit field specifies the depth dimension of the column.

Height – The **Height** edit field specifies the height of the column.

Radius – The **Radius** edit field specifies the radius of cylindrical predefined shapes.

Floating Viewer – The **Floating Viewer** button opens the **Viewer**, allowing you to view the mass element.

Match – The **Match** button allows you to select an existing mass element to match its parameters.

Properties – The **Properties** button opens the **Mass Element as Column Properties** dialog box, which includes a **General** tab and a **Dimensions** tab. The **Dimensions** tab, shown in Figure 13.9, allows you to specify the dimensions of the mass element.

Undo – The Undo button is inactive when you place the column in a grid.

Help – The **Help** button opens the Help file to the Column Overview help topic

If you select the **Add Column Grid** command and do not select the **Column<** button, no column will be placed in the drawing. If you select the **Column<** button, a default column shape will be inserted when you select the **Close** button of the **Add Mass Element as Column** dialog box. However, if the **Add Mass Element as Column** dialog box is open and you decide not to insert a column, you can close the dialog box without inserting the default column shape by select the **X** in the upper right corner of the dialog box.

Drawing Commercial Structures 963

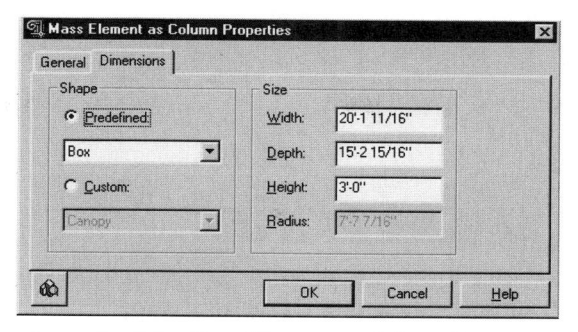

Figure 13.9 *Dimensions tab of the Mass Element as Column Properties dialog box*

REFINING THE COLUMN GRID WITH PROPERTIES

The **Properties** button of the **Add Column Grid** dialog box executes the **Column Grid Properties** command (**COLUMNGRIDPROPS**) and opens the **Column Grid Properties** dialog box. It consists of four tabs, which allow you to modify the grid to create offsets. Each grid line can be altered in the **Dimensions** tab. The properties of an existing column grid can also be edited. See Table 13.2 for **Column Grid Properties** command access.

Command prompt	COLUMNGRIDPROPS
Shortcut	Select the column grid; then right-click and choose Column Grid Properties

Table 13.2 *Column Grid Properties command access*

The **Properties** button of both the **Add Column Grid** and **Modify Column Grid** dialog boxes executes the **Column Grid Properties** command from within the **Add Column Grid** or **Modify Column Grid** command. When you choose the **Column Grid Properties** command from the shortcut menu, the **Column Grid Properties** dialog box opens, as shown in Figure 13.10.

The **Column Grid Properties** dialog box of an existing column grid includes a **Location** tab, which allows you to adjust the position of the grid. The **Dimensions** tab lists each of the grid lines in the X and Y directions. The location of each grid line listed can be edited in this tab. The tabs of the **Column Grid Properties** dialog box are described below.

General

The **General** tab, as shown in Figure 13.10, includes a **Description** edit field that allows you to add a description to the column grid.

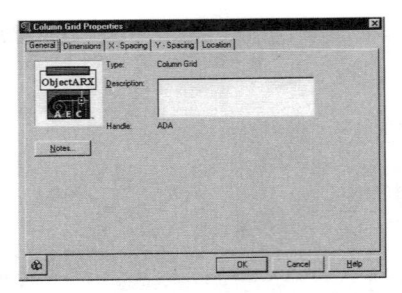

Figure 13.10 *General tab of the Column Grid Properties dialog box*

Dimensions

The **Dimensions** tab, shown in Figure 13.11, includes three sections to adjust the overall size and the specific offsets for the horizontal and vertical grid lines. The **Dimensions** tab and its options are described below.

Figure 13.11 *Dimensions tab of the Column Grid Properties dialog box*

> **X-Width** – The **X-Width** is the overall width of the column grid in the horizontal direction.
>
> **Y-Depth** – The **Y-Depth** is the overall depth of the column grid in the vertical direction.
>
> **Automatic Spacing** – The **Automatic Spacing** check box turns ON spacing the grid automatically according to a bay size or a specific number of bays. If the **Automatic Spacing** check box is not selected, you can edit the column grid location using grips. Figure 13.12 shows two column grids: the column grid on the left was created with **Automatic Spacing** selected, while the **Properties** command was used to clear the **Automatic Spacing** check box for the column grid on the right. The grips of the column grid shown on the right can be selected and stretched to a new location.

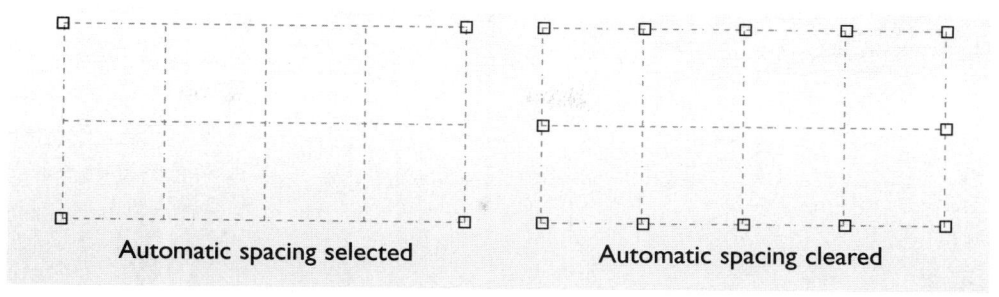

Figure 13.12 *Display of grips with Automatic Spacing selected and cleared*

Space Lines Evenly – The **Space Lines Evenly** toggle creates a column grid with the specified number of bays evenly spaced within the overall **X-Width** or **Y-Depth**.

Repeat Bay Size – The **Repeat Bay Size** toggle creates column grid lines spaced as specified by the bay size dimension.

Start Offset – The **Start Offset** edit field allows you to specify the distance from the insertion point to start the column grid. The **Start Offset** for the **X-Width** dimension offsets the grid in a horizontal direction; The offset distance must be positive. The **Start Offset** for the **X-Axis** of the column grid shown in Figure 13.13 is 2'-0" from the insertion point and its **Y-Axis** offset is 5'-0". **Repeat Bay Size** is toggled ON to create the column grid.

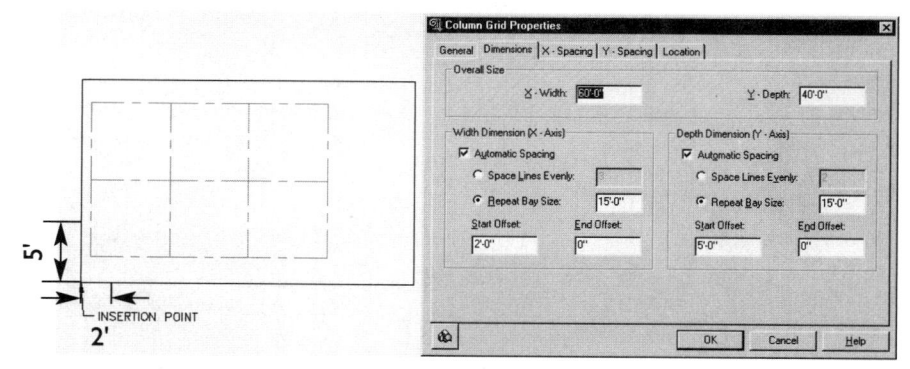

Figure 13.13 *Start offsets specified in the Dimensions tab of the Column Grid Properties dialog box*

End Offset – The **End Offset** edit field allows you to specify the distance from the diagonal corner of the overall column grid to end the grid. The end offset can be specified for the width and depth of the grid. If **Space Lines Evenly** is toggled ON, the column grid can be moved down when the end offsets are specified as 2'-0", as shown in Figure 13.14.

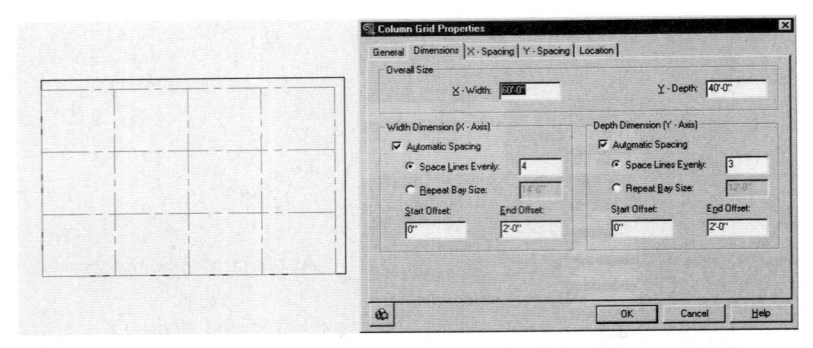

Figure 13.14 *End offset specified in the Dimensions tab of the Column Grid Properties dialog box*

X-Spacing

The **X-Spacing** tab lists the grid lines that run in the vertical direction parallel to the Y axis. The **X-Spacing** tab of the column grid shown in Figure 13.15 includes four vertical grid lines; however, only three lines are listed in the table. The column grid lines are described in terms of their distance from the insertion point. If the start offset is zero, the first column grid line passes through the insertion point. The **Distance to Line** column in the table lists the distance from each column grid line to the insertion point. The **Spacing** column lists the distance between column grid lines.

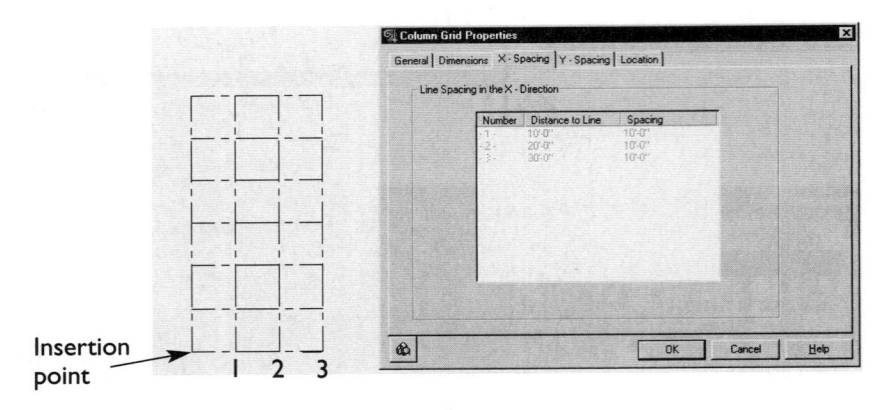

Figure 13.15 *X-Direction column grid lines*

The column grid locations can be edited in the **X-Spacing** tab if **Automatic Spacing** is cleared in the **Dimensions** tab. Change a column grid location by selecting the column line number and then typing a different value in the **Distance to Line** column, as shown in Figure 13.16. If **Automatic Spacing** is selected, the **Distance to Line** column cannot be edited.

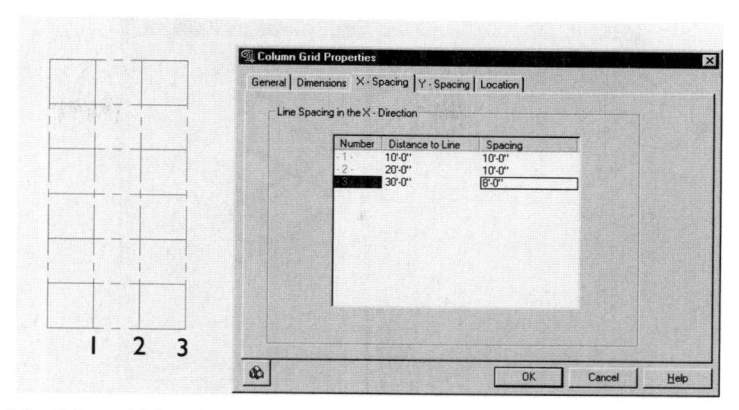

Figure 13.16 *Editing X-Direction column grid lines*

Y-Spacing

The **Y-Spacing** tab lists the grid lines that run in the horizontal direction, or parallel to the X axis. The **Y-Spacing** tab shown in Figure 13.17 includes six column grid lines in the table. The distance from the insertion point to each column grid line is recorded in the **Distance to Line** column. The column grid line, which passes through the insertion point of the grid, is not listed in the table.

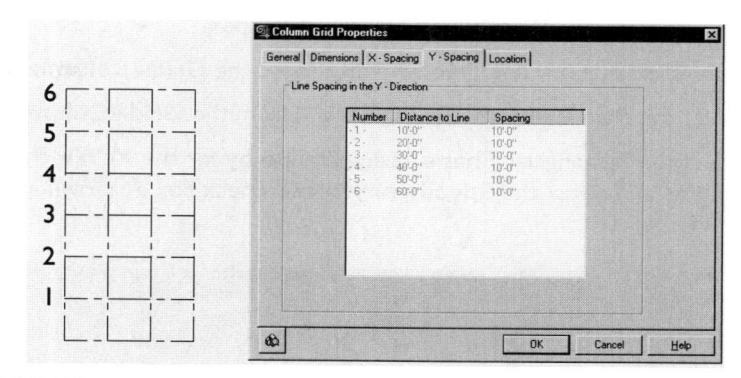

Figure 13.17 *Y-Spacing tab*

> **Note:** The **Distance to Line** and **Spacing** values of the column grid line can be adjusted in the **Y-Spacing** tab of the **Column Grid Properties** dialog box when **Automatic Spacing** is cleared in the **Dimensions** tab.

Location

The **Location** tab lists the insertion point, normal values, and rotation of the column grid and allows you to edit its location. The **Location** tab shown in Figure 13.18 describes the column grid relative to the World Coordinate System.

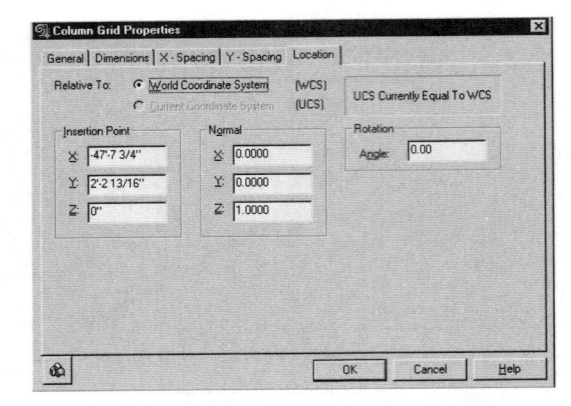

Figure 13.18 *Location tab of the Column Grid Properties dialog box*

TUTORIAL 13.1 CREATING A COLUMN GRID

1. Open *Ex13-1* from the \ADT_Tutor\Ch13 directory.
2. Save the drawing to your student directory as *Column1.dwg*.
3. Right-click over the Osnap button of the Status bar and then choose **Settings**.
4. Turn **Object Snap** On and then deselect **Node** and select **Endpoint**. Select **OK** to dismiss the **Drafting Settings** dialog box.
5. Right-click over any ADT toolbar and choose the **Grids/Columns** toolbar.
6. Select the **Add Column Grid** command from the **Grids/Columns** toolbar.
7. Select the Rectangular **Shape**, check **Divide by** for the **X-Width** and **Y-Depth**, and select the **Specify on Screen** check box as shown in Figure 13.19.

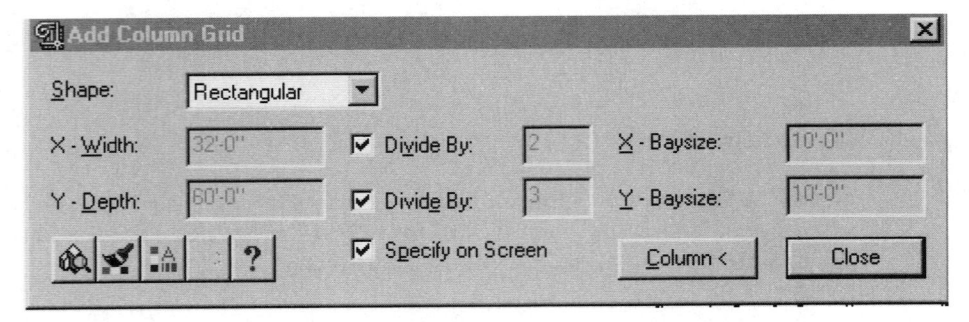

Figure 13.19 *Add Column Grid dialog box*

8. Select the **Column<** button to open the **Add Mass Element as Column** dialog box.

9. Select the **Custom Shape** radio button and then select Col_TS6x6 from the shape list and set **Width** to 6", **Depth** to 6", and **Height** to 12'-0", as shown in Figure 13.20.

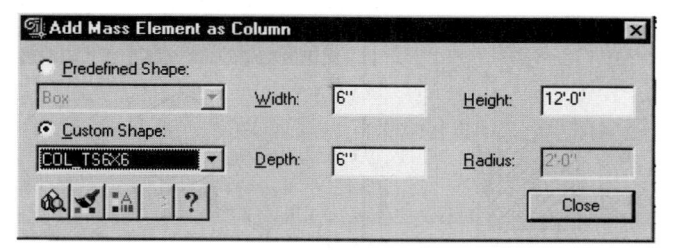

Figure 13.20 *Add Mass Element as Column dialog box*

10. Select **Close** to dismiss the **Add Mass Element as Column** dialog box.
11. Respond to the following command sequence to place the column grid.

    (Select in the left floating viewport to activate the viewport.)
    ```
    Command: AECCOLUMNGRIDADD
    Insertion point or [WIdth/Depth/Xbay/Ybay/XDivide by
    toggle/YDivide by
    toggle/Match]: DBOX
    [SHape/WIdth/Depth/Height/Match]: DBOX
    [SHape/WIdth/Depth/Height/Match]: DBOX
    [SHape/WIdth/Depth/Height/Match]:
    Insertion point or [WIdth/Depth/Xbay/Ybay/XDivide by
    toggle/YDivide by
    toggle/Match]:
    ```
    (Select the lower left corner of the building using the Endpoint object snap, as shown in Figure 13.21.)

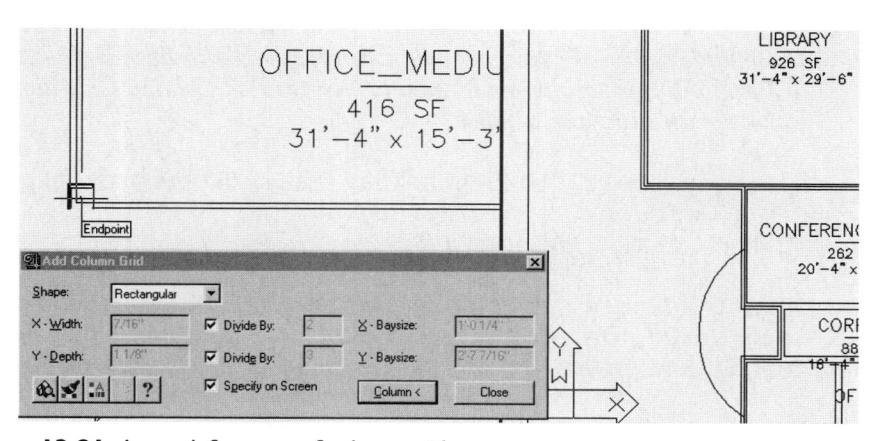

Figure 13.21 *Lower left corner of column grid*

(Select the right floating viewport to activate the viewport.)

`New size or [WIdth/Depth/Xbay/Ybay/XDivide by toggle/YDivide by toggle/Match]:` *(Select the upper right corner of the building as shown in Figure 13.22.)*

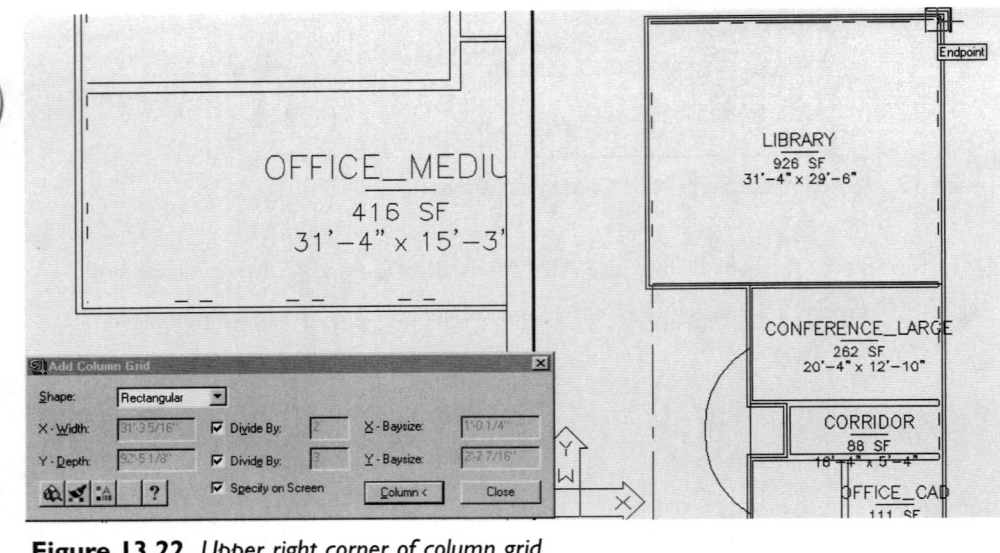

Figure 13.22 *Upper right corner of column grid*

(Press F3 *to toggle OFF Object Snaps)*
`New size for cell or [WIdth/Depth/Xbay/Ybay/XDivide by toggle/YDivide by toggle/Match]:` *(Drag the pointer left and down to near P1 in Figure 13.23 and select in the drawing area when the X-Width Divide by: is equal to 2 and Y-Depth Divide by: is equal to 3 in the Add Column Grid dialog box as shown in Figure 13.23.)*

`Rotation or [WIdth/Depth/Xbay/Ybay/XDivide by toggle/YDivide by toggle/Match]`
`<0.00>:` ENTER *(Press* ENTER *to accept 0° rotation.)*

`Insertion point or [WIdth/Depth/Xbay/Ybay/XDivide by toggle/YDivide by toggle/Match/Undo]:` ENTER *(Press* ENTER *to end the command.)*

12. Save the drawing.

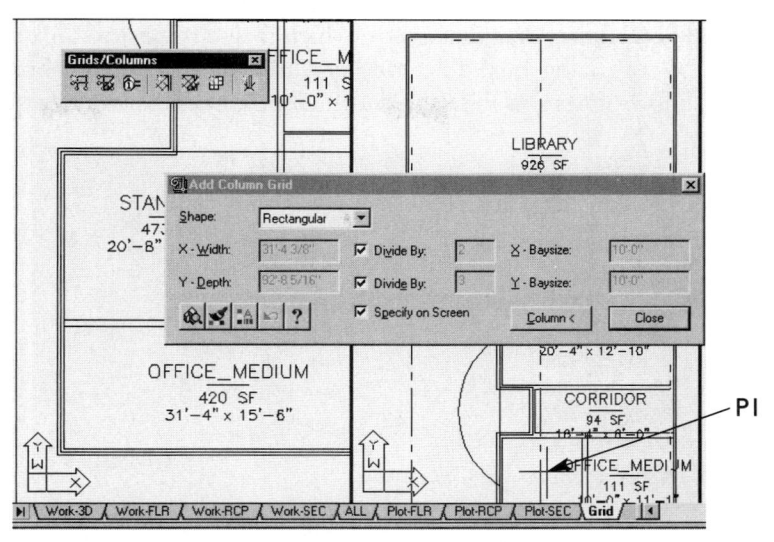

Figure 13.23 *Dragging pointer to specify X and Y bay quantity*

MODIFYING THE COLUMN GRID

The column grid can be modified by the **Column Grid Properties** command (**COLUMNGRIDPROPS**) or the **Column Grid Modify** command (**COLUMNGRIDMODIFY**). The **Column Grid Properties** command, discussed earlier, allows you to edit the position of each column grid line. The **Column Grid Modify** command includes options for changing the overall size of the grid, **X-Baysize**, **Y-Baysize**, or the number of bays. See Table 13.3 for **Column Grid Modify** command access.

Menu bar	Design>Grids>Modify Column Grid
Command prompt	COLUMNGRIDMODIFY
Toolbar	Select Modify Column Grid from the Grids/Columns toolbar as shown in Figure 13.24
Shortcut	Right-click over the drawing area; then choose Design>Column Grids>Modify
Shortcut	Select a column grid; then right-click and choose Column Grid Modify

Table 13.3 *Column Grid Modify command access*

When you select the **Column Grid Modify** command from the **Grids/Columns** toolbar, you are prompted to select a column grid. The selection of the column grid opens the **Modify Column Grid** dialog box, which includes most of the options of the **Add Column Grid** dialog box. The **Modify Column Grid** dialog box, shown in Figure 13.24, allows you to modify the width, depth, bay size and bay quantity.

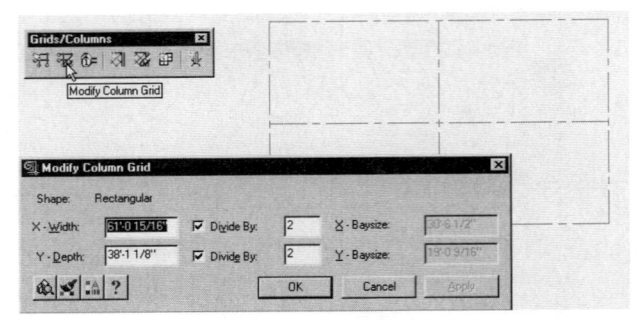

Figure 13.24 *Modify Column Grid dialog box*

The **Properties** button of the **Modify Column Grid** dialog box executes the **Column Grid Properties** command from within the **Modify Column Grid** command.

ADDING AND REMOVING COLUMN GRID LINES

Additional column grid lines can be added to the grid if the column grid is set to the **Manual** mode. A column grid is changed to **Manual** through the **Layout Mode** command, which is included in the **X-Axis** and **Y-Axis** options of the column grid shortcut menu. The cascade menu options of the **X-Axis** and **Y-Axis** shortcut menus include **Add Grid Line**, **Remove Grid Line**, and **Layout Mode**. These options allow you to quickly change the column grid mode and add or remove column grid lines in either the X axis or the Y axis.

Using the Layout Mode

The **Layout Mode** command on the shortcut menu allows you to define the spacing system used in the column grid. See Table 13.4 for **Layout Mode** command access.

Choosing the **Layout Mode** command from the shortcut menu allows you to select one of the following modes for the column grid: **Manual**, **Repeat**, or **Space Evenly**.

Command prompt	COLUMNGRIDXMODE or COLUMNGRIDYMODE
Shortcut	Select the column grid; then right-click and choose X-Axis>Layout Mode or Y-Axis>Layout Mode from the shortcut menu

Table 13.4 *Layout Mode command access*

These options are equivalent to the options of the **Width Dimension (X-Axis)** or **Depth Dimension (Y-Axis)** of the **Column Grid Properties** dialog box. Selecting the **Manual** option allows you to add or remove individual column grid lines. The **Manual** option is equivalent to clearing the **Automatic Spacing** check box in the **Dimensions** tab of the **Column Grid Properties** dialog box. The **Repeat** option allows you to specify in the command line the bay size to be repeated in the column grid. The **Space Evenly** option allows you to specify the number of equal spaces in the column grid. These options allow you to edit the column grid by entering values in the command line. The **Layout Mode** command allows you to quickly toggle the column grid to **Manual** and add or remove column grids.

Adding Column Grid Lines

You can add column grid lines for the X axis or the Y axis by choosing the **Add Grid Line** command (**COLUMNGRIDXADD** or **COLUMNGRIDYADD**) from the shortcut menu. See Table 13.5 for **Add Grid Line** command access.

Command prompt	COLUMNGRIDXADD or COLUMNGRIDYADD
Shortcut	Select the column grid; then right-click and choose X-Axis>Add Grid Line or Y-Axis>Add Grid Line from the shortcut menu

Table 13.5 *Add Grid Line command access*

When you select **COLUMNGRIDXADD** or **COLUMNGRIDYADD**, you are prompted to specify the location for the new line. The new line can be specified in the command line or selected with the pointer. The distance inserted in the command line is relative to the insertion point of the column grid. The new column grid line is combined with the remainder of the column grid. The procedure for adding a new column grid line is shown in the following command line sequence.

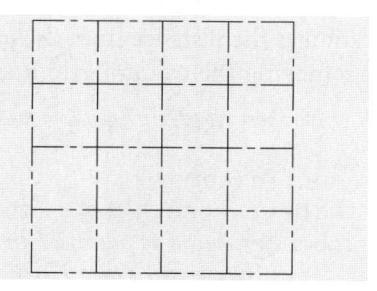

Figure 13.25 *Column grid prior to adding a grid line*

(Select the column grid as shown in Figure 13.25; then right-click and choose X-Axis>Add Grid Line.)
```
Command: ColumnGridXAdd
Enter X length <10'-0">: 15'
```
ENTER *(Column grid line added 15' from the insertion point along the x axis as shown in Figure 13.26.)*

Insertion point

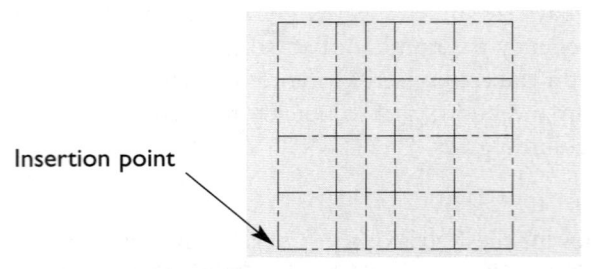

Figure 13.26 *Column grid line added at 15' from the insertion point*

Removing Column Grid Lines

You can remove column grid lines by choosing the **Remove Grid Line** command (**COLUMNGRIDXREMOVE** or **COLUMNGRIDYREMOVE**) from the shortcut menu. See Table 13.6 for **Remove Grid Line** command access.

Command prompt	**COLUMNGRIDXREMOVE or COLUMNGRIDYREMOVE**
Shortcut	Select the column grid; then right-click and choose X-Axis>Remove Grid Line or Y-Axis>Remove Grid Line from the shortcut menu

Table 13.6 *Remove Grid Line command access*

When you choose the **Remove Grid Line** command, you are prompted in the command line to specify the location of the grid line to remove. Selecting near the line with the pointer or typing the location as the distance from the insertion point specifies the location. The procedure for removing the column grid line is shown below.

(Select the column grid; then right-click and choose X-Axis>Remove Grid Line.)
```
Command: ColumnGridXRemove
Enter approximate X length to remove <10'-0">:
```
(Select the location of the column grid line at point P1 in Figure 13.27.)
(Column grid removed as shown in Figure 13.28.)

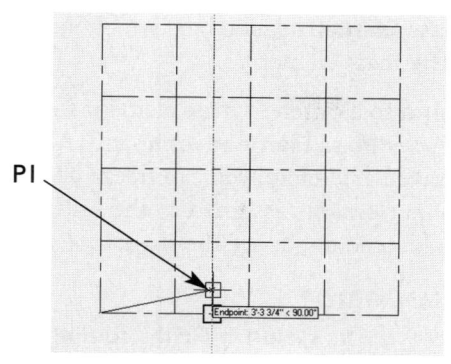

Figure 13.27 *Column grid line selected*

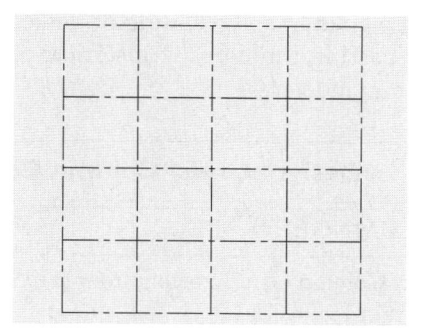

Figure 13.28 *Selected column grid line removed*

EDITING THE COLUMN MASS ELEMENTS

You can change the mass element inserted to represent a column by selecting the **Modify Column** command or the **Modify Mass Element** command on the **Mass Elements** toolbar. See Table 13.7 for **Modify Column** command access.

Menu bar	Design>Columns>Modify Column
Command prompt	MASSELEMENTMODIFY
Toolbar	Select Modify Mass Element from the Mass Elements toolbar
Shortcut	Select the column; then right-click and choose Element Modify from the shortcut menu

Table 13.7 *Modify Column command access*

When you choose the **Modify Column** command from the menu bar, the **Modify Mass Element** command is used to edit the column. Selecting the **Modify Column** command opens the **Modify Mass Element** dialog box, which includes options for changing the size and shape of the mass element.

To edit all columns of the grid to a different mass element shape, select mass elements and edit them using the **Modify Mass Element** command. All Column mass elements are located on the A-Cols layer. Therefore, you can freeze all layers except the A-Cols layer and then select all mass elements assigned to the grid. The AutoCAD **PROPERTIES** command can also be used to change the shape and size of the mass elements.

LABELING THE COLUMN GRID

The column grid is labeled with the **Column Grid Labeling** command. The column grid can be labeled automatically with numbers or letters or each column grid line can be assigned a unique label. If the column grid labels are automatically generated, the numbers or letters can be assigned in an ascending or descending pattern generated from the column grid line passing through the insertion point. See Table 13.8 for **Column Grid Labeling** command access.

Menu bar	Design>Grids>Column Grid Labeling
Command prompt	COLUMNGRIDLABEL
Toolbar	Select Column Grid Labeling from the Grids/Columns toolbar shown in Figure 13.29
Shortcut	Select a column grid; then right-click and choose Column Grid Labeling

Table 13.8 *Column Grid Labeling command access*

Figure 13.29 *Column Grid Labeling command on the Grids/Columns toolbar*

When you select the **Column Grid Labeling** command from the **Grids/Columns** toolbar, you are prompted to select a column grid. Selecting the column grid opens the **Column Grid Labeling** dialog box, as shown in Figure 13.30.

The **Column Grid Labeling** dialog box consists of the **X-Labeling** and **Y-Labeling** tabs. The column grid lines for each grid direction are labeled in the tab. The **Automatically Calculate Values for Labels** check box allows the numbering of the column grid lines

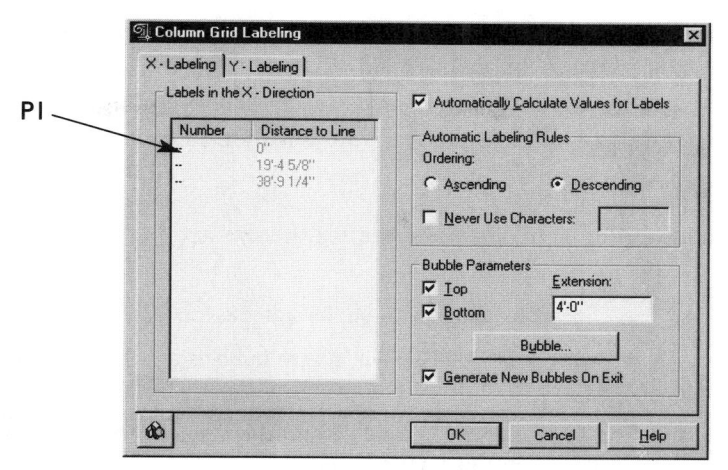

Figure 13.30 *X-Labeling tab of the Column Grid Labeling dialog box*

to be assigned in ascending or descending order from the first column grid line. The number assigned to the first column grid line as shown at P1 in Figure 13.30 defines the beginning character of the sequence. In the figure, if 1 is assigned at P1 and **Ascending** order is toggled ON, the columns will be labeled 1, 2, 3. The options of the X-Labeling and Y-Labeling tabs are described below.

Labels in the X- Direction

The **Labels in the X-Direction** list box shows each grid label and the distance from the insertion point.

> **Number** – The **Number** column allows you to assign an alphanumeric character for each column grid line. If **Automatically Calculate Values for Labels** has been selected, the labels are generated from the label specified in the top field of the **Number** column. You can override the column grid labels generated with the **Automatically Calculate Values for Labels** check box by selecting the grid label and typing the character for the column grid in the **Number** column.
>
> **Distance to Line** – The **Distance to Line** column lists the distance for each column grid line from the insertion point of the column grid.
>
> **Automatically Calculate Values for Labels** – The **Automatically Calculate Values for Labels** check box will automatically assign the labels to the column grid from the beginning column grid in ascending or descending order.

Automatic Labeling Rules

> **Ascending** – The **Ascending** toggle assigns the column grid label in increasing numeric or alphabetic order.

Descending – The **Descending** toggle assigns the column grid label in decreasing numeric or alphabetic order.

Never Use Characters – The **Never Use Characters** toggle allows you to identify characters to exclude when the labels are automatically generated.

Bubble Parameters

The **Bubble Parameter** section allows you to define the size and location of the bubbles used to identify the column grid.

Top – The **Top** check box will create bubbles at the top of the column grid.

Bottom – The **Bottom** check box will create bubbles at the bottom of the column grid.

Extension – The **Extension** edit field defines the distance from the grid to extend the grid centerline and place the bubble.

Bubble – The **Bubble** button opens the **MvBlockDef Select** dialog box, which allows you to select the multi-view block for the bubble.

Generate New Bubbles On Exit – The **Generate New Bubbles On Exit** check box redisplays the position and size of the bubbles according to the changes. The size of the column bubbles is determined by the **AECD-WGSCALESETUP** command, accessed through **Documentation>Set Drawing Scale** in the menu bar.

Note: the **Y-Labeling** tab includes similar options for specifying the labeling of the column grid in the Y direction.

DIMENSIONING THE COLUMN GRID

The column grid can be automatically dimensioned through the **Dimensions** command (**COLUMNGRIDDIM**). The entire column grid is dimensioned on each side using the current dimension style. The dimensions are placed on the A-Anno-Dims layer. See Table 13.9 for **Dimensions** command access.

Command prompt	COLUMNGRIDDIM
Shortcut	Select the column grid; then right-click and choose Dimensions from the shortcut menu

Table 13.9 *Dimensions command access*

When you enter the **COLUMNGRIDDIM** command in the command line, you are prompted to select a column grid. Selecting the column grid allows you to dimension the column grid on each side. Shown below is the command sequence for dimensioning a column grid.

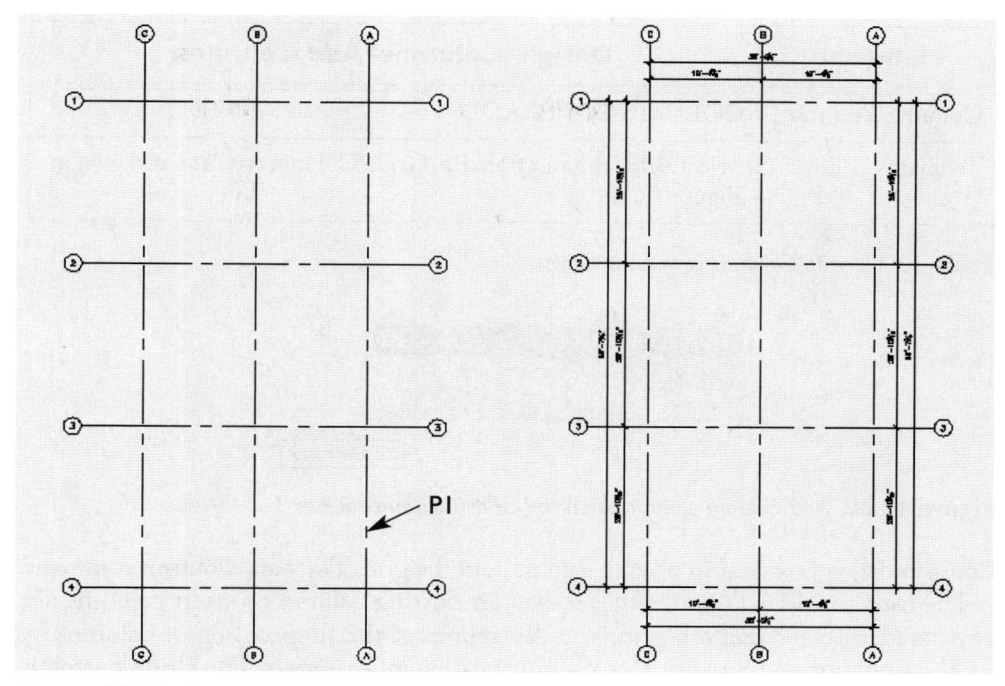

Figure 13.31 *Dimensioned column grid*

(Select the column grid; then right-click and choose Dimensions from the shortcut menu.)
Command: COLUMNGRIDDIM
Select column layout grids: *(Select the column grid at P1 in Figure 13.31.)* 1 found
Select column layout grids: ENTER *(Press* ENTER *to end column grid selection.)*
Enter offset distance <4'-0">: ENTER *(Press* ENTER *to accept the dimension line offset distance.)*
(Column grid is dimensioned as shown on right in Figure 13.31.)

ADDING A COLUMN WITHOUT A GRID

Columns can be added to the drawings that are not associated with a grid. The **Add Columns** command (**COLUMNCOVERADD**) can be used to create an isolated column or to cover an existing column. The **Add Columns** command adds a mass element as a representation of the column. See Table 13.10 for **Add Columns** command access.

When you select the **Add Column** command from the **Grids/Columns** toolbar, the **Add Mass Element as Column** dialog box opens, as shown in Figure 13.33. The **Add Mass Element as Column** dialog box is identical to the **Add Mass Element as**

Menu bar	Design>Columns>Add Columns
Command prompt	COLUMNCOVERADD
Toolbar	Select Add Column from the Grids/Columns toolbar as shown in Figure 13.32

Table 13.10 *Add Columns command access*

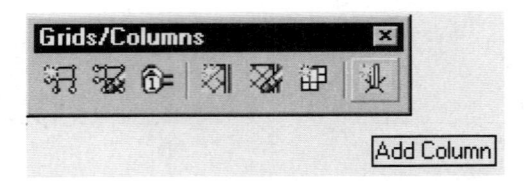

Figure 13.32 *Add Column command on the Grids/Columns toolbar*

Column dialog box used in placing a column in the grid. The **Add Column** command can be used to add a column as cover over an existing column or insert columns not associated with the column grid. You can represent the fireproofing of columns by adding a column to represent the column encasement materials. The **Undo** button is inactive when you insert isolated columns; however, after you insert a column, the **Undo** button can be used to remove the insertion while the dialog box remains open.

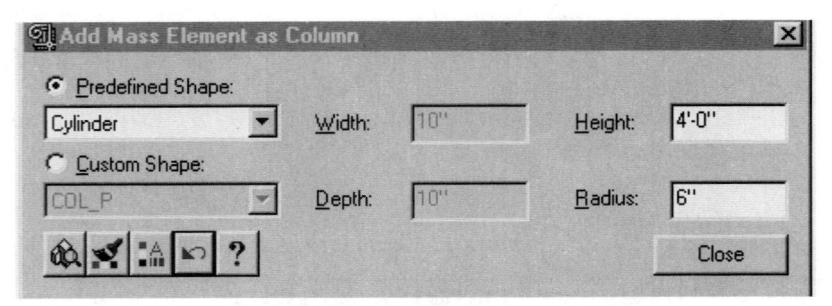

Figure 13.33 *Add Mass Element as Column dialog box*

Columns can be created from one or more mass elements. Figure 13.34 shows a column created from the wide flange mass element shape for the full height of the column and cylinder column to cover the lower portion of the column. The two mass elements are attached to a mass group; therefore, they are joined through the additive operation.

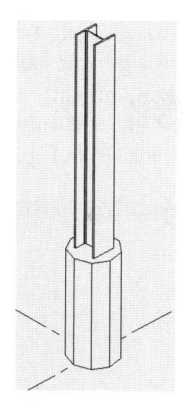

Figure 13.34 *Mass elements combined as a mass group to create a composite column*

TUTORIAL 13.2 LABELING A COLUMN GRID

1. Open *Ex13-2* from the *\ADT_Tutor\Ch13* directory.

2. Save the drawing to your student directory as *Column2.dwg*.

3. Choose **Documentation>Set Drawing Scale** from the menu bar and set the **Drawing Scale** to 1/8"= 1'-0".

4. Select **OK** to dismiss the **Drawing Setup** dialog box.

5. Select **Column Grid Labeling** from the **Grids/Columns** toolbar; then select the column grid to open the **Column Grid Labeling** dialog box.

6. Select the **X-Labeling** tab; then select the **Automatically Calculate Values for Labels** check box, select the **Descending** radio button, verify that **Bubble Parameter Top** is checked, and clear the **Bottom** bubble check box, set the extension to 12'-0" and select **Generate New Bubbles On Exit**. Type **C** in the top row of the **Number** column, as shown in Figure 13.35.

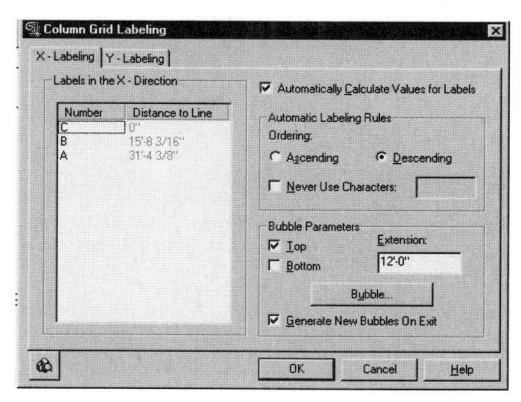

Figure 13.35 *X-Labeling tab of the Column Grid Labeling dialog box*

7. Select the **Y-Labeling** tab; then select **Automatically Calculate Values for Labels**, select the **Descending** radio button, verify that the **Bubble Parameter Right** is checked and clear the **Left** bubble, set the extension to 12'-0" and select **Generate New Bubbles On Exit**. Type **4** in the top row of the **Number** column as shown in Figure 13.36.

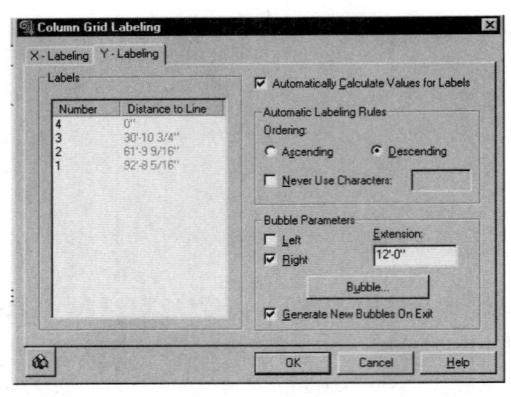

Figure 13.36 *Y-Labeling tab of the Column Grid Labeling dialog box*

8. Select **OK** to dismiss the **Column Grid Labeling** dialog box and view the column grid as shown in Figure 13.37.

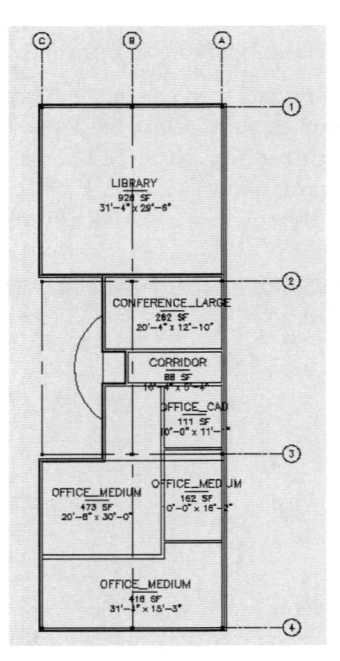

Figure 13.37 *Labeling applied to the column grid*

9. Select the column grid; then right-click and choose **Y-Axis>Layout Mode** and respond as follows in the command line.

   ```
   Command: ColumnGridYMode
   Node layout mode Y axis [Manual/Repeat/Space evenly]
   <Manual>: ENTER
   Command:
   ```

10. Select the column grid; then right-click and choose **Column Grid Properties** from the shortcut menu.

11. Select the **Y-Spacing** tab and type **29'-6"** in the **Spacing** column for number 1 and type **33'-8"** in the **Spacing** column for number 2, as shown in Figure 13.38.

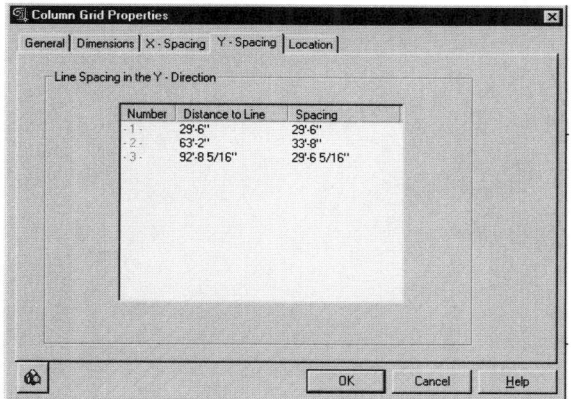

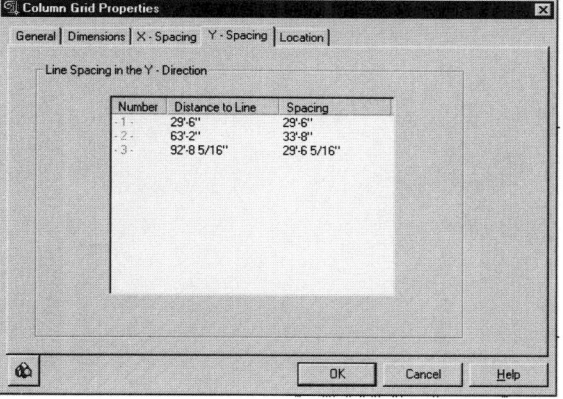

Figure 13.38 *Y-Spacing tab of the Column Grid Properties dialog box*

12. Select **OK** to dismiss the **Column Grid Properties** dialog box.

13. Select **Zoom Window** from the **Standard** toolbar and zoom a window around column B3 as shown in Figure 13.39.

14. Select the column at B3; then right-click and choose **Element Modify** from the shortcut menu to open the **Modify Mass Element** dialog box as shown in Figure 13.39.

15. Edit the **Shape** to Doric, **Height** to 9', and **Radius** to 4 in the **Modify Mass Element** dialog box as shown in Figure 13.39.

16. Select **OK** to dismiss the **Modify Mass Element** dialog box.

17. Verify that Ortho is ON in the Status bar.

18. Select **Zoom Previous** from the **Standard** toolbar; then select **Zoom Window** and zoom a window around column B4 as shown in Figure 13.40.

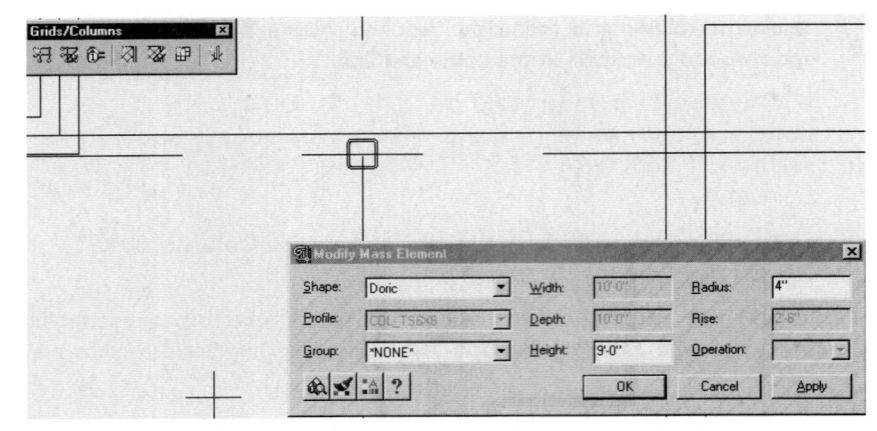

Figure 13.39 *Column B3*

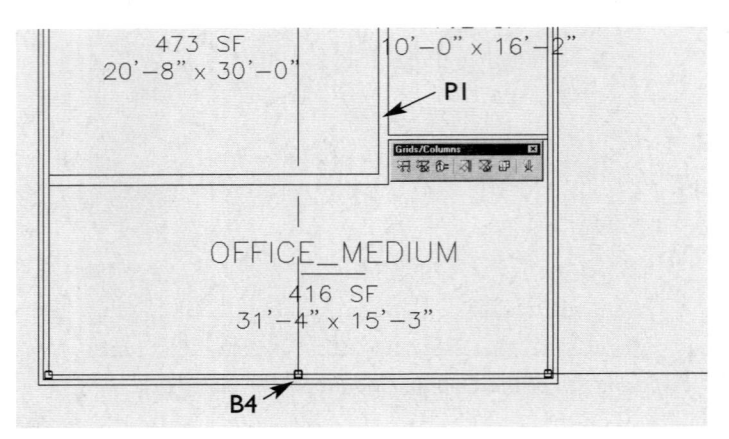

Figure 13.40 *Column B4*

19. Select the column grid; then right-click and choose **X-Axis>Layout Mode** and respond to the command line prompt as follows.

    ```
    Command: ColumnGridXMode
    Node layout mode X axis [Manual/Repeat/Space evenly]
    <Manual>: ENTER (Press ENTER to edit the X axis of the column grid.)
    ```

20. Press F3 to turn OFF Object Snaps and press F10 to turn OFF Polar Tracking.
21. Select the column grid; then right-click and choose **X-Axis>Add Grid Line** from the shortcut menu.
22. Select near P1 in the center of the wall as shown in Figure 13.40.
23. Select the column grid; then right-click and choose **Column Grid Labeling** from the shortcut menu.

24. Select the **X-Labeling** tab of the **Column Grid Labeling** dialog box.
25. Type **D** in the top row in the **Number** column of the **X-Labeling** tab as shown in Figure 13.41.

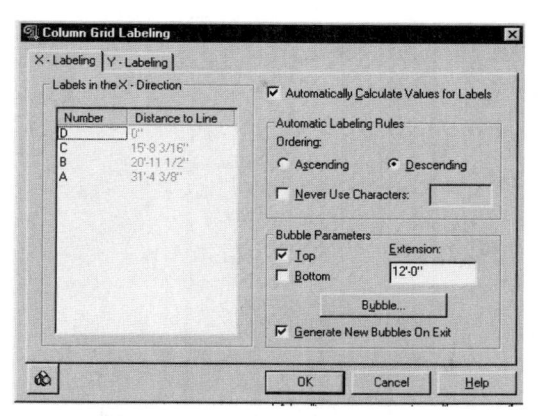

Figure 13.41 *X-Labeling of the Column Grid Labeling dialog box*

26. Select the **Y-Labeling** tab and renumber them by placing a **4** in the top value. Note: this is needed because after you add a column line in step 22, the **Y-Column** labels disappeared.
27. Select **OK** to dismiss the **Column Grid Labeling** dialog box.
28. Select **Zoom Previous** to view the overall column grid plan as shown in Figure 13.42.
29. Select the column grid; then right-click and choose **Dimensions** from the shortcut menu and respond to the command line prompts as shown below.

```
Command: ColumnGridDim
Enter offset distance <4'-0">: ENTER
```
(Press ENTER *to accept the 4'-0" offset of the dimension line and view the dimensions as shown in Figure 13.42.)*

30. Save the drawing.

CREATING CEILING GRIDS

Ceiling grids represent the building components of the suspended ceiling. The ceiling grid is created as an object in a rectangular shape; when it is inserted, its size should exceed the maximum dimensions of the room. If the ceiling grid is attached to a space or closed polyline as a boundary, only that part of the ceiling grid inside the room will be displayed. The ceiling grids are displayed in the Work-RCP, Work-SEC, and Plot-RCP layouts. The ceiling grid should be anchored to a space or a polyline; however, it can be placed free standing. Ceiling grids attached to a space are inserted at the ele-

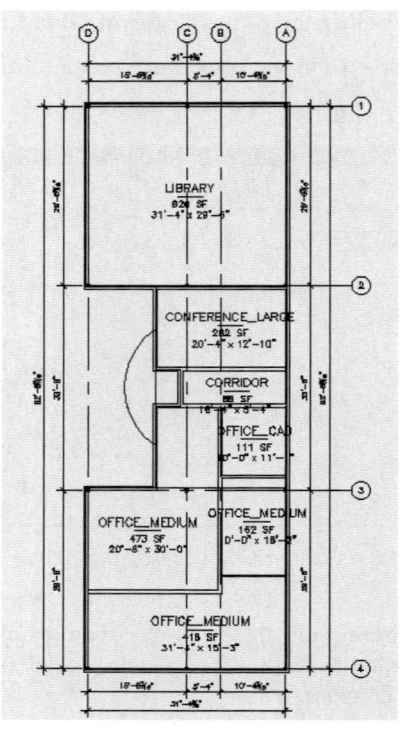

Figure 13.42 *Dimensions added to the column grid*

vation of the ceiling boundary in the space. The command to place a ceiling grid is the **Add Ceiling Grid** (**CEILINGGRIDADD**). See Table 13.11 for **Add Ceiling Grid** command access.

Menu bar	Design>Grids>Add Ceiling Grid
Command prompt	CEILINGGRIDADD
Toolbar	Select Add Ceiling Grid from the Grids/Columns toolbar as shown in Figure 13.43
Shortcut	Right-click in the drawing area; then choose Design>Ceiling Grid>Add

Table 13.11 *Add Ceiling Grid command access*

Ceiling grids are displayed in Work-RCP layout; the display configuration of this layout allows display of the walls, space boundary, and reflected ceiling. Therefore,

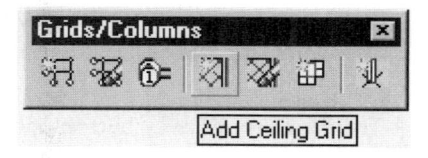

Figure 13.43 *Add Ceiling Grid command on the Grids/Columns toolbar*

select the Work-RCP layout prior to selecting the **Add Ceiling Grid** command. When you select the **Add Ceiling Grid** command, the **Add Ceiling Grid** dialog box opens, allowing you to specify the size and properties of the ceiling grid, as shown in Figure 13.44.

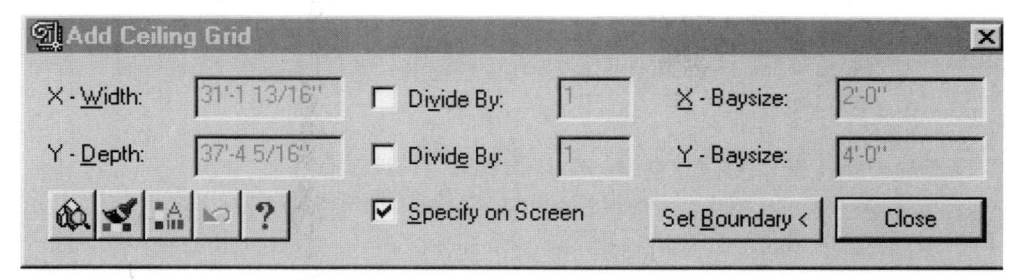

Figure 13.44 *Add Ceiling Grid dialog box*

The **X-Width** and **Y-Depth** of the ceiling grid determine the overall size of the grid. The size of the divisions of the ceiling grid can be specified in the baysize edit fields or the number of divisions can be specified in the **Divide By** edit field. The options of the **Add Ceiling Grid** dialog box are described below.

> **X-Width** – The **X-Width** value is the dimension parallel to the X axis of the ceiling grid.
>
> **Y-Depth** – The **Y-Depth** value is the dimension parallel to the Y axis of the ceiling grid.
>
> **Divide By** – The **Divide By** check box and value determine the number of divisions to divide the **X-Width** or **Y-Depth** to create the ceiling grid unit width and unit depth.
>
> **X-Baysize** – If the **Divide By** check box is cleared, the X axis dimension of the bay can be specified in this edit field.
>
> **Y-Baysize** – If the **Divide By** check box is cleared, the Y axis dimension of the bay is specified in this edit field.
>
> **Specify on Screen** – The **Specify on Screen** check box allows you to select the diagonal corners of the ceiling to specify the width and depth.

Set Boundary< – The **Set Boundary<** button allows you to select a space or closed polyline as the boundary. The ceiling grid is displayed only inside the boundary.

Floating Viewer – The **Floating Viewer** button opens the **Viewer**, allowing you to view the ceiling grid.

Match – The **Match** button allows you to select a ceiling grid to match its properties of **X-Width**, **Y-Depth**, **X-Baysize**, and **Y-Baysize**.

Properties – The **Properties** button opens the **Ceiling Grid Properties** dialog box, which allows you to adjust the size and location of the ceiling grid.

Undo – The **Undo** button deletes the insertion of the active ceiling grid and allows the ceiling grid to be placed in a different location while you remain in the **Add Ceiling Grid** command.

Help – The **Help** button opens the Help file to the Ceiling Grid Overview help topic.

STEPS FOR PLACING A CEILING GRID

1. Select one of the following layouts: Work-RCP, Work-SEC, or Plot-RCP.
2. Select the **Add Ceiling Grid** command from the **Grids/Columns** toolbar.
3. Respond to the following command line prompts:

   ```
   Command: _AecCeilingGridAdd
   Insertion point or [WIdth/Depth/Xbay/Ybay/XDivide by toggle/YDivide by toggle/Match]:
   ```
   (Select a location at point P1 in the drawing area as shown in Figure 13.45.)
   ```
   Rotation or [WIdth/Depth/Xbay/Ybay/XDivide by toggle/YDivide by toggle/Match]
   <0.00>:
   ``` ENTER *(Press ENTER to accept 0° rotation.)*
   ```
   Insertion point or [WIdth/Depth/Xbay/Ybay/XDivide by toggle/YDivide by toggle/Match]:
   ``` ENTER *(Press ENTER to end the command).*
   *(A free standing ceiling grid is placed as shown in Figure 13.45.)*

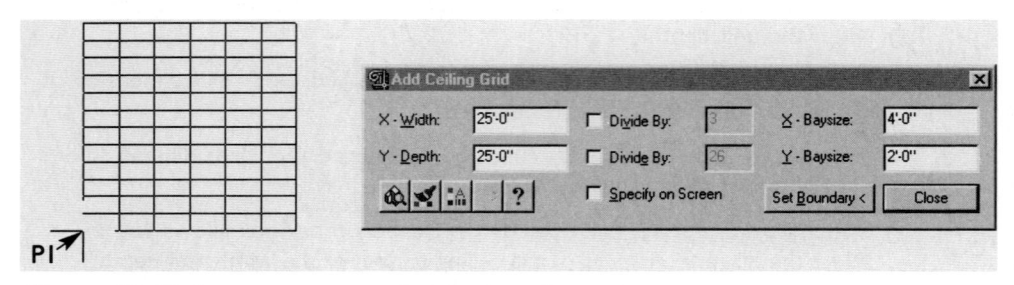

**Figure 13.45** *Insertion point specified for the ceiling grid*

The **Set Boundary<** button of the **Add Ceiling Grid** dialog box allows you to select a closed polyline or space to attach the ceiling grid. The closed polyline or space creates a boundary for the ceiling grid, which turns off the display of the ceiling grid outside the boundary. The ceiling grid is created at elevation Z=0 unless it is attached to a boundary that has elevation. Attaching the ceiling grid to an AEC Space will place the ceiling at the elevation of the ceiling boundary. Ceiling grids attached to any closed polyline will be bound by the polyline and placed at the elevation of the polyline. Using polylines and spaces as the boundaries for the ceiling grid creates ceiling grids that fit irregularly shaped spaces. You can center the ceiling grid in the room using the **MOVE** command after the grid is inserted. The ceiling grid has grips at each of the four corners of the grid, as shown in Figure 13.46. The grips can be used to stretch or adjust the grid to the size of the room.

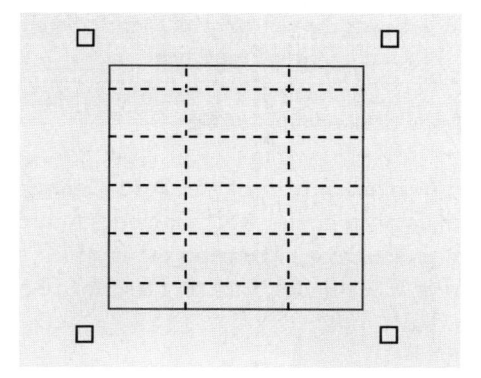

**Figure 13.46** *Grips of the ceiling grid*

The **Specify on Screen** check box allows you to specify the ceiling grid size by selecting points in the drawing area. The **Specify on Screen** check box and the **Set Boundary<** button options allow you to create a ceiling grid that fits the room and that can be centered in the room. When you select the **Specify on Screen** check box, you are prompted to select the insertion point and the diagonal corner to specify the new size. Selecting two points equal to or greater than the room size creates a ceiling grid that will cover the room. When the **Add Ceiling Grid** command is selected, the default width and depth are retained from the last ceiling grid insertion. Therefore, if the ceiling grid size fits the room, it is easier to specify an insertion point that can center the ceiling about the room.

> **Note:** You can use the AutoCAD **BOUNDARY** and **SPACEBOUNDARYCONVERTSPACE** commands to create the boundaries for ceiling grids of floor plans that were developed without space planning.

## ADJUSTING THE PROPERTIES OF THE CEILING GRID

The **Properties** button of the **Add Ceiling Grid** dialog box executes the **Ceiling Grid Properties** command from within the **Add Ceiling Grid** command. The **Ceiling Grid Properties** command (**CEILINGGRIDPROPS**) can be used to specify the properties of ceiling grids before or after they are inserted. See Table 13.12 for **Ceiling Grid Properties** command access.

| Command prompt | CEILINGGRIDPROPS |
|---|---|
| Toolbar | Select the Add Ceiling Grid or Modify Ceiling Grid on the Grids/Columns toolbar; then select the Properties button of the dialog box |
| Shortcut | Select the ceiling grid; then right-click and choose Ceiling Grid Properties |

**Table 13.12** *Ceiling Grid Properties command access*

When you select the **Properties** button of the **Add Ceiling Grid** dialog box, the **Ceiling Grid Properties** dialog box opens, which includes four tabs as shown in Figure 13.47. If an existing ceiling is selected, a **Location** tab is added that describes the position of the grid in the World Coordinate System. Described below are the tabs of the **Ceiling Grid Properties** dialog box.

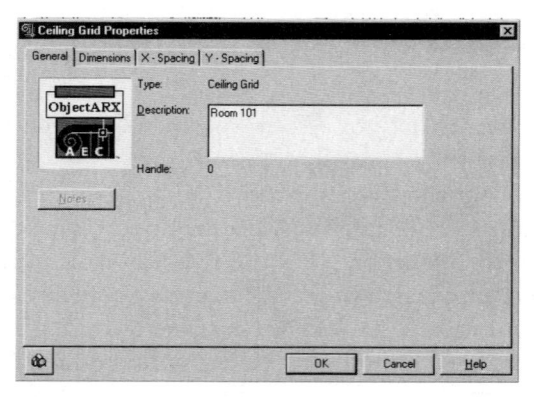

**Figure 13.47** *CeilingGrid Properties dialog box*

### General

The **General** tab of the **Ceiling Grid Properties** dialog box allows you to add a description to the ceiling grid. The **Description** field can be used to identify the properties of the ceiling grid or the room identification, as shown in Figure 13.47. The

**Ceiling Grid Properties** dialog box for existing ceiling grids includes an active **Notes** button. Selecting the **Notes** button opens the **Notes** dialog box, which includes the **Text Notes** and **Reference Docs** tabs.

*Dimensions*

The **Dimensions** tab, shown in Figure 13.48, includes three sections: **Overall Size**, **Width Dimension (X-Axis)** and **Depth Dimension (Y-Axis)**. The edit fields of these sections allow you to specify the size and location of the ceiling grid. All the options of this tab are available in the **Add Ceiling Grid** dialog box except for **Start Offset** and **End Offset**, which allow you to shift the ceiling grid to start or end a specified distance from the insertion point. Described below are the options of the **Dimensions** tab.

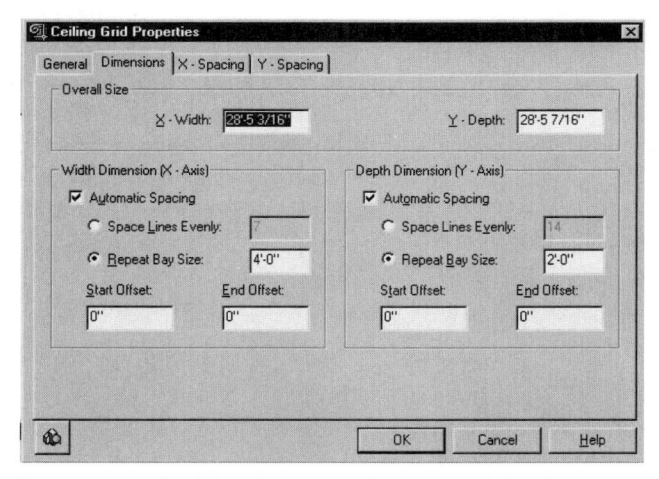

**Figure 13.48** *Dimensions tab of the Ceiling Grid Properties dialog box*

Overall Size

> **X-Width** – The **X-Width** is the dimension of the ceiling grid parallel to the X axis.
>
> **Y-Depth** – The **Y-Depth** is the dimension of the ceiling grid parallel to the Y axis.

Width Dimension (X-Axis) and Depth Dimension (Y-Axis)

> **Automatic Spacing** – The **Automatic Spacing** check box creates a uniform ceiling grid according to the bay size or the number of divisions specified. If **Automatic Spacing** is cleared, each ceiling grid line can be moved or removed, and additional lines can be added. Ceiling grids created with **Automatic Spacing** cleared have grips located at the endpoints of each ceiling grid line of the specified axis.

**Space Lines Evenly** – The **Space Lines Evenly** radio button allows you to specify the number of divisions to divide the overall grid along the specified axis to create the grid.

**Repeat Bay Size** – The **Repeat Bay Size** radio button allows you to specify the width of the ceiling grid along the specified axis. This option can be used to create a ceiling grid to fit modular ceiling tile sizes of 2' x 2' or 2' x 4'.

**Start Offset** – The **Start Offset** option allows you to specify a distance along the axis to start the ceiling grid.

**End Offset** – The **End Offset** option allows you to specify a distance to limit the end of the ceiling grid.

*X-Spacing*

The **X-Spacing** tab includes a list of the grid lines in the X direction created in the ceiling grid. The **Number** column, as shown in Figure 13.49, lists each grid line, and its distance from the insertion point is shown in the **Distance to Line** column. If **Automatic Spacing** is not selected in the **Dimensions** tab, the **Distance to Line** and **Spacing** can be edited to adjust the position of a ceiling grid line.

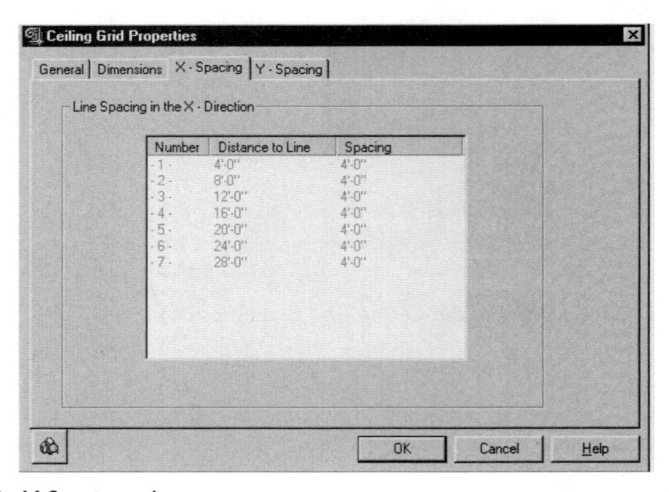

**Figure 13.49** *X-Spacing tab*

*Y-Spacing*

The **Y-Spacing** tab includes a list of the grid lines in the Y direction created in the ceiling grid. The list includes **Number, Distance to Line**, and **Spacing** columns, as shown in Figure 13.50. The **Distance to Line** column is the distance the each grid line is placed from the insertion point. If **Automatic Spacing** is not selected in the **Dimensions** tab, the **Distance to Line** and **Spacing** can be edited to adjust the position of the ceiling grid line.

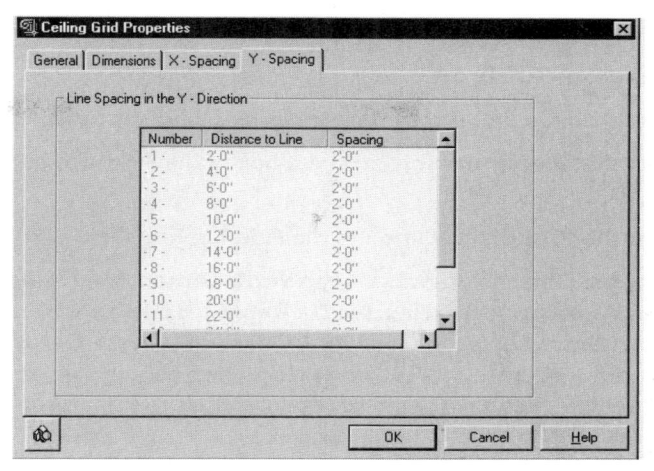

**Figure 13.50** *Y-Spacing tab of the Ceiling Grid Properties dialog box*

*Location*

A **Location** tab is included in the **Ceiling Grid Properties** dialog box if an existing ceiling grid is selected. The **Location** tab, shown in Figure 13.51, allows you to edit the location of the grid. You can move ceiling grids created without boundaries attached to spaces to a ceiling elevation by editing the Z coordinate of the insertion point to the Z elevation of the ceiling.

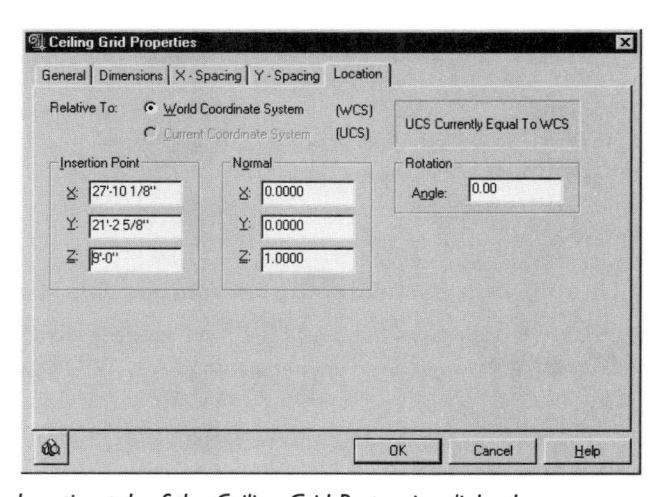

**Figure 13.51** *Location tab of the Ceiling Grid Properties dialog box*

## TUTORIAL 13.3  CREATING A CEILING GRID

1. Open *Ex13-3.dwg* from the *\ADT_Tutor\Ch13* directory.

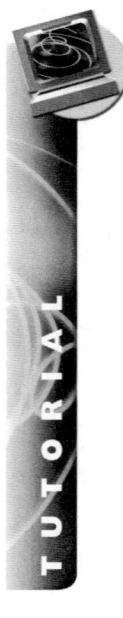

2. Save the file to your student directory as *Ceiling1.dwg*.
3. Select Work-RCP layout.
4. Select **Add Ceiling Grid** from the **Grids/Columns** toolbar.
5. Select the **Specify on Screen** check box of the **Add Ceiling Grid** dialog box.
6. Select the **Properties** button of the **Add Ceiling Grid** dialog box.
7. Select the **Dimensions** tab; in the **Width Dimension (X-Axis)** section, select **Automatic Spacing**, set the **Repeat Bay Size** ON, bay size = 2'-0", and set **Start Offset** to 0 and **End Offset** to 0. In the **Depth Dimension (Y-Axis)** section, select **Automatic Spacing**, set the **Repeat Bay Size** ON, set bay size to 4'-0", and set **Start Offset** to 0 and **End Offset** to 0, as shown in Figure 13.52.

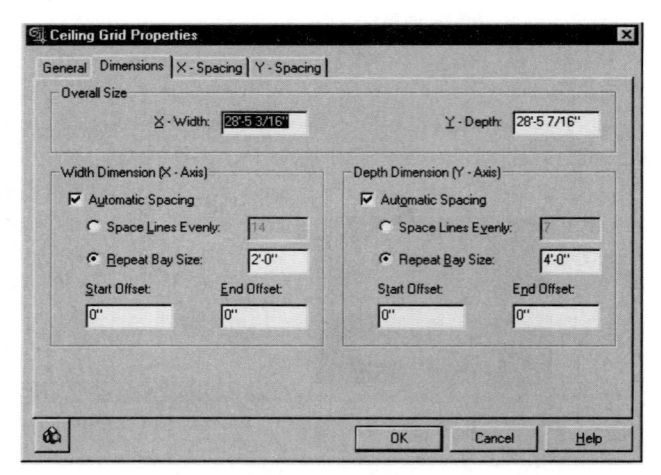

**Figure 13.52** *Setting up the Dimensions tab of Ceiling Grid Properties dialog box*

8. Select **OK** to dismiss the **Ceiling Grid Properties** dialog box.
9. Select the **Set Boundary<** button of the **Add Ceiling Grid** dialog box and respond to the following command line prompts.

```
Command: _AecCeilingGridAdd
Insertion point or [WIdth/Depth/Xbay/Ybay/XDivide by toggle/YDivide by toggle/Match]: DBOX
```
*(Select in the drawing area to activate the viewport.)*
```
Select a space or closed pline:
```
*(Select the space at P1 in Figure 13.53.)*
```
Insertion point or [WIdth/Depth/Xbay/Ybay/XDivide by
```

```
toggle/YDivide by
toggle/Match]: (Select a position outside of the room near P2 in Figure
13.53.)
New size or [WIdth/Depth/Xbay/Ybay/XDivide by tog-
gle/YDivide by toggle/Match]: (Select a point outside of the
room near P3 in Figure 13.53.)
Rotation or [WIdth/Depth/Xbay/Ybay/XDivide by tog-
gle/YDivide by toggle/Match]
<0.00>: ENTER (Press ENTER to accept the 0° rotation.)
Insertion point or [WIdth/Depth/Xbay/Ybay/XDivide by
toggle/YDivide by
toggle/Match/Undo]: ENTER (Press ENTER to end the command.)
```

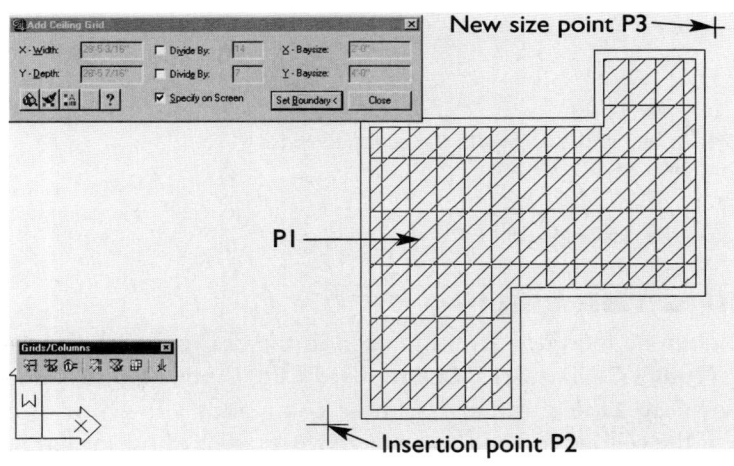

Figure 13.53 *Specifying the location of the ceiling grid*

10. Toggle OFF Ortho in the Status Bar.
11. Select the **Move** command from the **Modify** toolbar and shift the ceiling grid to the center of the room as described in the following command line sequence.

```
Command: MOVE
Select objects: (Select the ceiling grid.) 1 found
Select objects: ENTER (Press ENTER to end selection.)
Specify base point or displacement: (Select a point in the
drawing area.)
Specify second point of
displacement or <use first point as displacement>:
(Move the pointer to position the ceiling grid centered in the room as shown
in Figure 13.53.)
```

12. Select **SW Isometric** from the **View** flyout of the **Standard** toolbar to view the ceiling grid in the ceiling boundary, as shown in Figure 13.54.

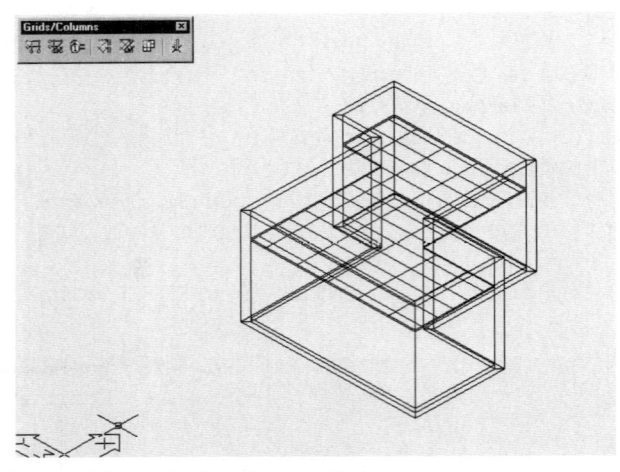

Figure 13.54 *Ceiling grid attached to the specified space*

13. Save the drawing.

CHANGING THE CEILING GRID

Ceiling grids can be modified after insertion in the drawing with the **Ceiling Grid Properties, Modify Ceiling Grid, Ceiling Grid Clip, Layout Mode, Add Grid Line,** and **Remove Grid Line** commands. These commands allow you to change the dimensions of the ceiling grid and add or remove sections of the grid as needed. The **Modify Ceiling Grid** command is used to change the dimensions and orientation of the ceiling grid. See Table 13.13 for **Modify Ceiling Grid** command access.

Menu bar	Design>Grids>Modify Ceiling Grid
Command prompt	CEILINGGRIDMODIFY
Toolbar	Select Modify Ceiling Grid from the Grids/Columns toolbar as shown in Figure 13.55
Shortcut	Right-click over the drawing area; then choose Design>Ceiling Grids>Modify from the shortcut menu
Shortcut	Select the ceiling grid; then right-click and choose Ceiling Grid Modify from the shortcut menu

Table 13.13 *Modify Ceiling Grid command access*

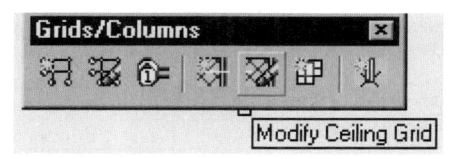

Figure 13.55 *Modify Ceiling Grid command on the Grids/Columns toolbar*

When you select the **Modify Ceiling Grid** command from the **Grids/Columns** toolbar, you are prompted to select a ceiling grid. Selecting a ceiling grid opens the **Modify Ceiling Grid** dialog box, as shown in Figure 13.56. Change the ceiling grid by editing the options of the dialog box, and then select **OK** or **Apply** to execute the changes.

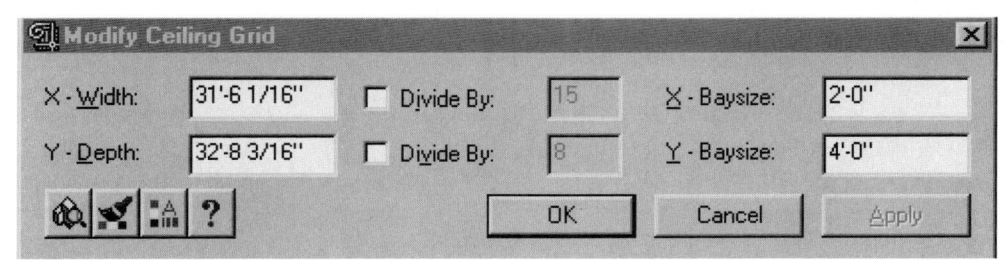

Figure 13.56 *Modify Ceiling Grid dialog box*

ADDING BOUNDARIES AND HOLES

Boundaries and holes can be added to the ceiling grid with the **Attach Clipping Boundary/Holes** command (**CEILINGGRIDCLIP**). This command has three options: **Set a Boundary, Add a Hole,** and **Remove a Hole**. The command can be used to create a boundary for a ceiling that has not been bounded to a space or polyline. The **Add a Hole** and **Remove a Hole** options allow you to carve out an area from the ceiling grid for penetrations through the ceiling, such as columns, shafts, and plumbing chases. See Table 13.14 for **Attach Clipping Boundary/Holes** command access.

Menu bar	Design>Grids>Attach Clipping Boundary/Holes
Command prompt	CEILINGGRIDCLIP
Toolbar	Select Attach Clipping Boundary/Holes from the Grids/Columns toolbar as shown in Figure 13.57

Table 13.14 *Attach Clipping Boundary/Holes command access*

Figure 13.57 *Attach Clipping Boundary/Holes command on the Grids/Columns toolbar*

When you select the **Attach Clipping Boundary/Holes** command from the toolbar, you are prompted in the command line to select one of three options: Set boundary/Add hole/Remove hole. If a grid has been placed without being attached to a boundary, the **Attaching Clipping Boundary/Holes** command can used to edit the ceiling grid and assign a boundary, which can be a space or closed polyline. The command sequence for assigning a boundary to an existing ceiling grid is shown below.

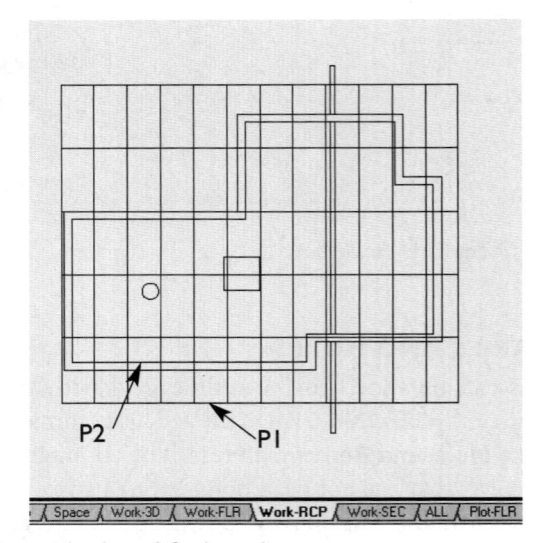

Figure 13.58 *Ceiling grid selected for boundary*

```
Command: _AecCeilingGridClip
Ceiling grid clip [Set boundary/Add hole/Remove
hole]: S ENTER
Select ceiling grids: (Select a ceiling grid as shown at P1 in
Figure 13.58.) 1 found
Select ceiling grids: ENTER (Press ENTER to end the selection.)
Select a closed polyline or space entity for bound-
ary: (Select a wall as shown at P2 in Figure 13.58.)
Ceiling grid clip [Set boundary/Add hole/Remove
hole]: ENTER (Press ENTER to end the command and display the grid as
shown in Figure 13.59.)
```

If a space is selected as the boundary of the ceiling grid, the ceiling grid is restricted to the boundary limits of the space. If the space dimensions are changed later, the ceiling grid does not stretch to the new dimensions of the space. However, the grips of the ceiling grid can be stretched to include the new dimensions of the space, and the space will continue to serve as the boundary for the ceiling grid.

Creating Holes in the Ceiling Grid

The **Add a Hole** and **Remove a Hole** options of the **Attach Clipping Boundary/Holes** command allow you to carve from the ceiling grid the shape of any closed polyline or AEC enitity. The polyline does not have to have the same elevation or intersect with the plane of the ceiling grid. AEC entities such as mass elements inserted as columns or walls can also be used as the cutting edge to create the hole. Using the **Add a Hole** option allows the ceiling grid to be placed continuously over interior walls and then trimmed by the interior walls. This technique allows the pattern or alignment of the ceiling grid to continue across the wall. The **Remove a Hole** option allows you to edit the grid and remove any holes that have been created. The following command sequence was used to edit the ceiling grid as shown in Figure 13.60.

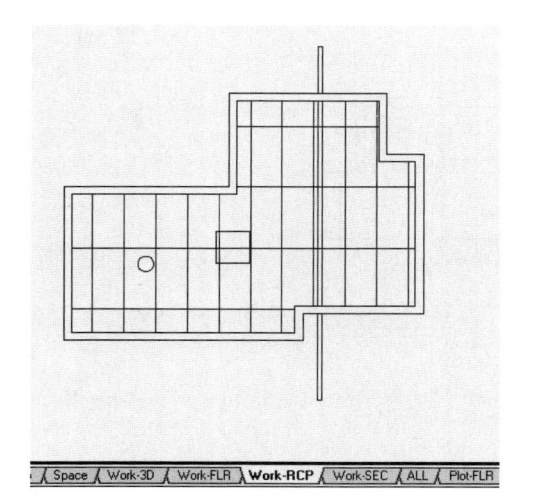

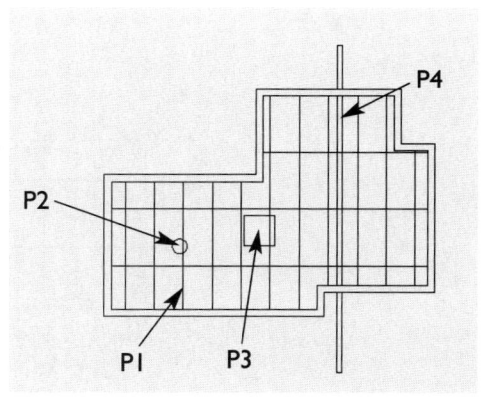

Figure 13.59 Ceiling grid bound to the space

Figure 13.60 Selecting objects to trim the ceiling grid

```
Command: _AecCeilingGridClip
Ceiling grid clip [Set boundary/Add hole/Remove
hole]: A ENTER (Selects the Add Hole option.)
Select ceiling grids: (Select the ceiling grid at P1 in Figure
13.60.) 1 found
```

```
Select ceiling grids: ENTER (Press ENTER to end selection.)
Select a closed polyline or AEC entity for hole:
```
(Select the column at P2 in Figure 13.60.)
```
Ceiling grid clip [Set boundary/Add hole/Remove
hole]: A ENTER
Select ceiling grids:
```
 (Select the ceiling grid at P1 in Figure 13.60.) `1 found`
```
Select ceiling grids: ENTER (Press ENTER to end selection.)
Select a closed polyline or AEC entity for hole:
```
(Select the rectangle polyline at P3 in Figure 13.60.)
```
Ceiling grid clip [Set boundary/Add hole/Remove
hole]: A ENTER
Select ceiling grids:
``` *(Select the ceiling grid at P1 in Figure 13.60.)* `1 found`
```
Select ceiling grids: ENTER (Press ENTER to end selection.)
Select a closed polyline or AEC entity for hole:
```
(Select the wall at P4 in Figure 13.60.)
```
Ceiling grid clip [Set boundary/Add hole/Remove
hole]: ENTER (Press ENTER to end the command.)
```
(Figure 13.61 shows the result.)

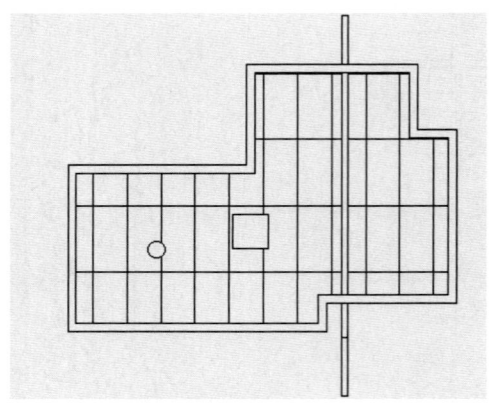

Figure 13.61 *Ceiling grid trimmed by selected objects*

ADDING AND REMOVING CEILING GRID LINES

The **Layout Mode** command allows the ceiling grid to be set to **Manual,** and ceiling grid lines can then be added or removed from the grid. The **Layout Mode** command (**CEILINGGRIDXMODE** or **CEILINGGRIDYMODE**) allows editing of each ceiling grid line. You can edit grid lines in the X axis direction by selecting the **Layout Mode** command for the X axis (**CEILINGGRIDXMODE**). Edit the ceiling grid lines in the Y direction by selecting the **Layout Mode** command for the Y axis (**CEILINGGRIDYMODE**). See Table 13.15 for **Layout Mode** command access.

Command prompt	**CEILINGGRIDXMODE OR CEILINGGRIDYMODE**
Shortcut	Select the ceiling grid; then right-click and choose X-Axis>Layout Mode or Y-Axis>Layout Mode from the shortcut menu

Table 13.15 *Layout Mode command access*

When you select the **Layout Mode** command for either the X or Y axis, the ceiling grid is released from the automatic mode and additional ceiling grid lines can be added or removed. Selecting the **Manual** option is equivalent to deselecting **Automatic Spacing** in the **Dimensions** tab of the **Ceiling Grid Properties** dialog box.

When the layout mode has been toggled to **Manual**, grips are displayed for each of the ceiling grid lines. Figure 13.62 shows the grips of a ceiling grid after the X-Axis layout mode has been toggled to **Manual**.

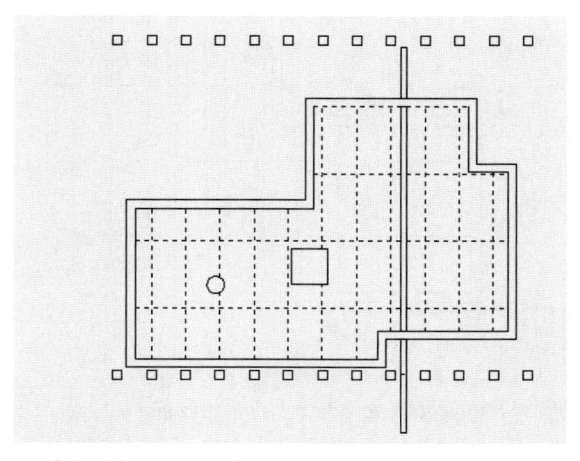

Figure 13.62 *Grips of the X axis using X-Axis manual layout mode*

Adding Ceiling Grid Lines

After the layout mode has been toggled to **Manual,** you can add ceiling grid lines with the **Add Grid Line** command (**CEILINGGRIDXADD** or **CEILINGGRIDYADD**). See Table 13.16 for **Add Grid Line** command access.

When you select the **Add Grid Line** command (**CEILINGGRIDXADD** or **CEIL-INGGRIDYADD**), you are prompted to specify the location for the new line. You can specify the location by typing a distance in the command line or you can left-click with the mouse to select the location. If a distance is entered in the command line, it is mea-

Command prompt	**CEILINGGRIDXADD or CEILINGGRIDYADD**
Shortcut	Select the ceiling grid; then right-click and choose X-Axis>Add Grid Line or Y-Axis>Add Grid Line from the shortcut menu

Table 13.16 *Add Grid Line command access*

sured relative to the insertion point of the grid. The procedure for adding a ceiling grid line is shown in the following command line sequence.

Note: Layout mode must be set to **Manual** prior to grid lines being added.

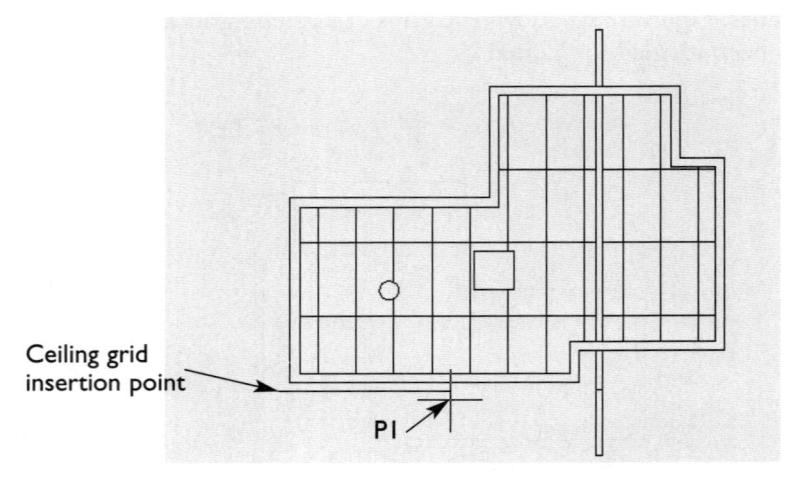

Figure 13.63 *Specifying the location to add a ceiling grid line*

(Verify that Ortho is toggled ON in the Status Bar and grid layout mode is Manual.)
(Select the ceiling grid; then right-click and choose X-Axis>Add Grid Line)
```
Command: CeilingGridXAdd
Enter X length <10'-0">:
```
(Select a point near P1 in Figure 13.63.)
(Ceiling grid line added as shown in Figure 13.64.)

Removing Ceiling Grid Lines

Ceiling grid lines can be removed if the layout mode has been set to **Manual** for the axis of the ceiling grid and you select the **Remove Grid Line** command (**CEILING-**

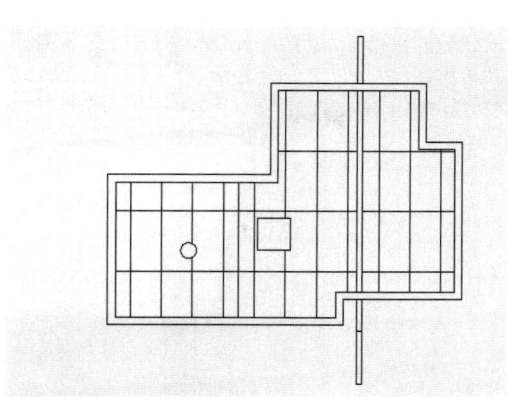

Figure 13.64 *Ceiling grid line added with the CEILINGGRIDXADD command*

GRIDXREMOVE or **CEILINGGRIDYREMOVE**). See Table 13.17 for **Remove Grid Line** command access.

Command prompt	CEILINGGRIDXREMOVE or CEILINGGRIDYREMOVE
Shortcut	Select the ceiling grid; then right-click and choose X-Axis>Remove Grid Line or Y-Axis>Remove Grid Line from the shortcut menu

Table 13.17 *Remove Grid Line command access*

When the layout mode of the ceiling grid is set to Manual, you can remove selected ceiling grid lines. The following command line sequence removes a ceiling grid line.

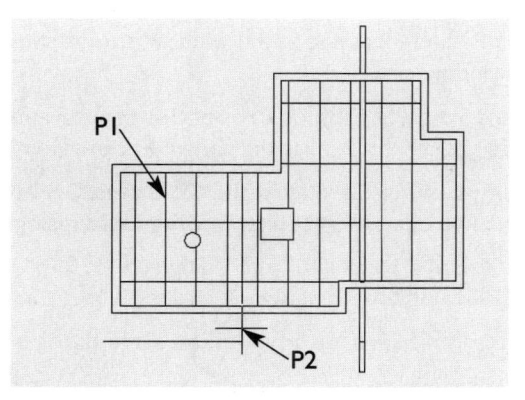

Figure 13.65 *Specifying the location to remove a ceiling grid line*

(Verify that Ortho is toggled ON in the Status Bar.)
(Select the ceiling grid at P1 in Figure 13.65; then right-click and choose X-Axis>Remove Grid Line from the shortcut menu.)
```
Command: CeilingGridXRemove
```

```
Enter approximate X length to remove <10'-0">:
```
(Select near the grid line at P2 in Figure 13.65 to remove the grid line. Grid line is removed as shown in Figure 13.66.)

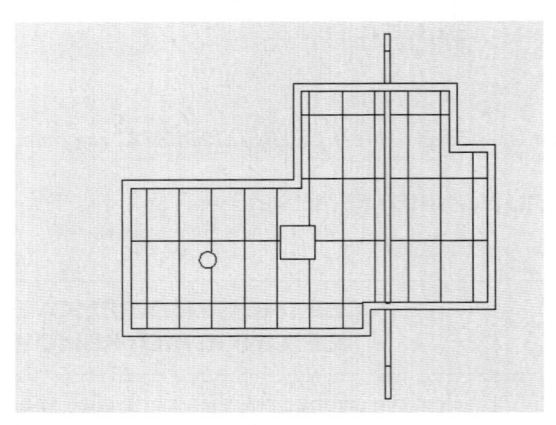

Figure 13.66 *Ceiling grid line removed from the ceiling grid*

TUTORIAL 13.4 EDITING THE CEILING GRID

1. Open *Ex13-4.dwg* from the \ADT_Tutor\Ch13 directory.
2. Save the drawing as *Ceiling2.dwg* in your student directory.
3. Select the Work-RCP layout.
4. Select **Zoom Window** from the **View** flyout on the **Standard** toolbar and select points P1 and P2 as the zoom window size, as shown in Figure 13.67, to specify the window size.
5. Select **Layers** from the **Object Properties** toolbar and freeze the A-Area-Spce layer. Select **OK** to dismiss the **Layer Properties Manager** dialog box.
6. Select **Modify Ceiling Grid** from the **Columns/Grids** toolbar; then select the ceiling grid to open the **Modify Ceiling Grid** dialog box.
7. Change the **X-Baysize** to 4'-0" and **Y-Baysize** to 2'-0" in the **Modify Ceiling Grid** dialog box, as shown in Figure 13.68.
8. Select **OK** to dismiss the **Modify Ceiling Grid** dialog box.
9. Select the ceiling grid to display its grips; then select the grip in the lower right corner and drag the grip to the right near P2, as shown in Figure 13.69.

Drawing Commercial Structures 1005

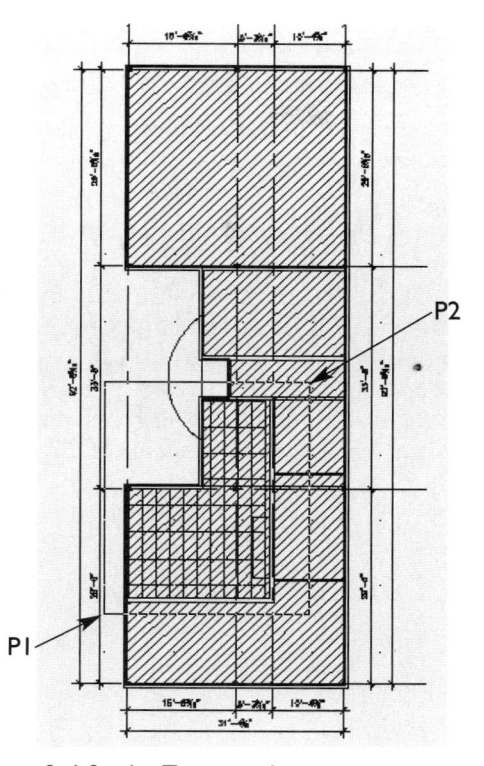

Figure 13.67 *Points specified for the Zoom window*

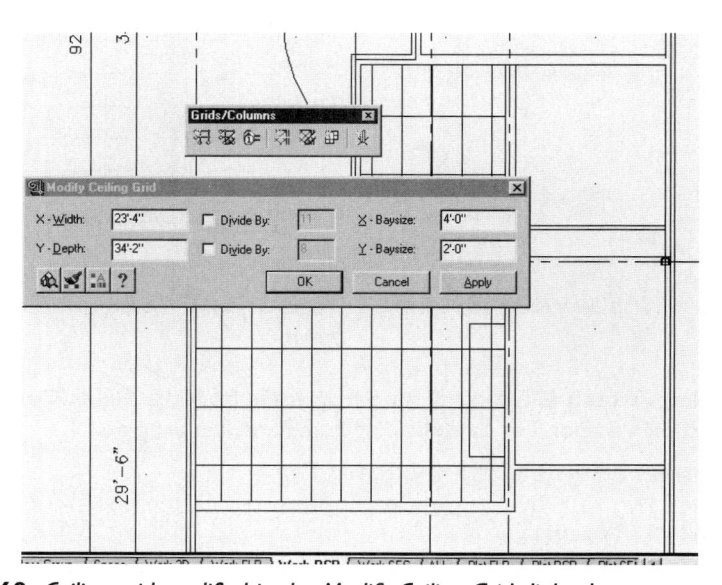

Figure 13.68 *Ceiling grid modified in the Modify Ceiling Grid dialog box*

10. Press ESC twice to clear the grips.
11. Select Ortho in the Status Bar to toggle OFF Ortho.
12. Select the **Move** command from the **Modify** toolbar and respond to the following command line prompts.

    ```
    Command: _move
    Select objects: (Select the ceiling grid.) 1 found

    Select objects: ENTER (Press ENTER to end selection.)
    Specify base point or displacement: (Select a point near P1 in the drawing area in Figure 13.69.)
    Specify second point of
    displacement or <use first point as displacement>:
    ```
    (Move the mouse to center the grid as shown in Figure 13.70; then left-click.)

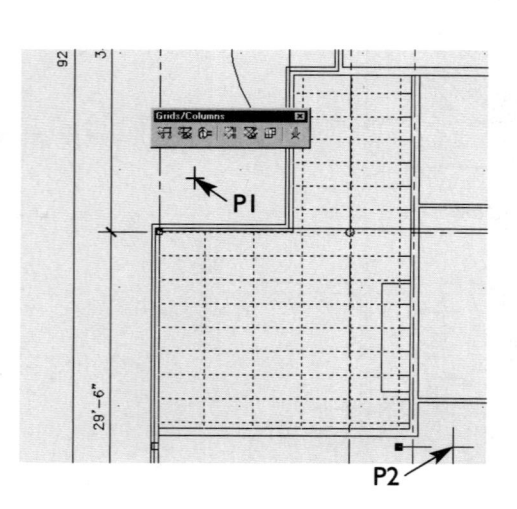

Figure 13.69 Stretch the ceiling grid using grips

Figure 13.70 Ceiling grid centered in the room

13. Select **Attach Clipping Boundary/Holes** from the **Grids/Columns** toolbar; then respond to the following command line prompts.

    ```
    Command: _AecCeilingGridClip
    Ceiling grid clip [Set boundary/Add hole/Remove hole]: A ENTER
    Select ceiling grids: (Select the ceiling grid at point P1 in Figure 13.70.) 1 found
    ```

Select ceiling grids: ENTER
Select a closed polyline or AEC entity for hole:
(Select the polyline at P2 in Figure 13.70.)
Ceiling grid clip [Set boundary/Add hole/Remove hole]: ENTER *(Press ENTER to end the command and display the ceiling grid as shown in Figure 13.71.)*

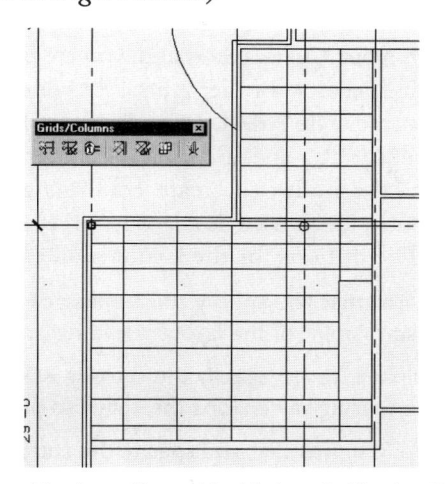

Figure 13.71 *Hole created in the ceiling grid with Attach Clipping Boundary/Holes command*

14. Save the drawing.

USING LAYOUT CURVES TO PLACE COLUMNS

Placing columns spaced evenly along a center line does not require the creation of a column grid line. Columns and other AEC objects can be placed along a layout curve. Isolated pier footings used in the foundations of residences can be placed on a layout curve, which inserts a node for the AEC objects to snap to when placed in the drawing. The AEC objects can be spaced evenly or at a specified distance apart. See Table 13.18 for **Add Layout Curve** command access.

Menu bar	Desktop>Layout Tools>Add Layout Curve
Command prompt	LAYOUTCURVEADD
Toolbar	Select Add Layout Curve from the AEC Layout Tools toolbar as shown in Figure 13.72
Shortcut	Right-click over the drawing area; then choose Utilities>Layout Tool>Layout Curve from the shortcut menu

Table 13.18 *Add Layout Curve command access*

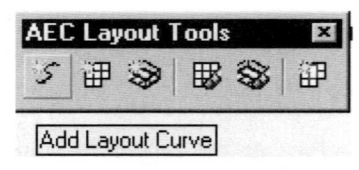

Figure 13.72 *Add Layout Curve command on the AEC Layout Tools toolbar*

When you select the **Add Layout Curve** command, you are prompted to select a curve. The curve can be a straight line, arc, or other AutoCAD entity. Walls and boundaries can be used as the layout curve. The **Add Layout Curve** command allows you to use layout nodes equally spaced or at a specified distance apart. Architectural Desktop objects can be inserted at the nodes to create columns along an arc or irregular curve. After selecting the entity to be used as a layout curve, you are prompted to select the mode for the layout. The options for the layout mode are described below.

> **Manual** – Requires that you specify the number of nodes and the specific distance from the start point of the layout curve for each of the nodes.
>
> **Repeat** – Requires that you specify the starting offset distance, ending offset distance, and the spacing value along the curve to place the nodes.
>
> **Space Evenly** – Requires that you specify the starting offset distance, ending offset distance, and the number of nodes you want placed along the curve.

The **Add Layout Curve** command is used in the following command sequence to place nodes along a curved wall. The nodes can then be used to locate the columns, as shown in Figure 13.74.

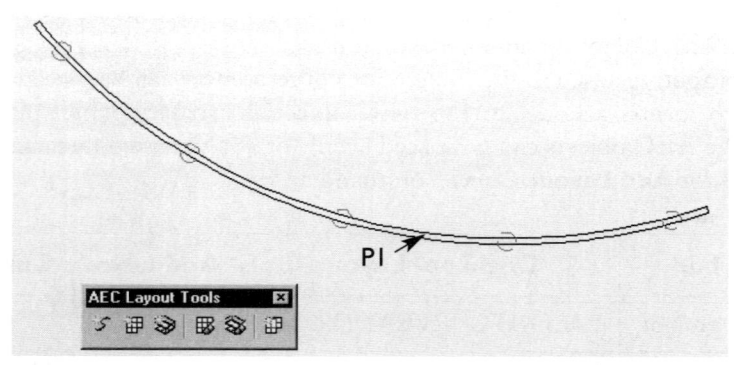

Figure 13.73 *Nodes placed on the arc with the Add Layout Curve command*

```
Command: _AecLayoutCurveAdd
Select a curve: (Select the wall at P1 in Figure 13.73.)
Select node layout mode [Manual/Repeat/Space evenly]
```

```
<Manual>: S ENTER
Start offset <0">: 2' ENTER
End offset <0">: 2' ENTER
Number of nodes <3>: 5 ENTER
```
(The Add Column command can then be used to place columns along the wall as shown in Figure 13.74.)

Figure 13.74 *Columns added to the layout curve at each node*

REFINING THE INTERSECTION OF COLUMNS AND WALLS

A column inserted along a wall can project from the wall, with the two displayed as separate entities, or it can be merged with the wall. The **Add Interference Condition** command (**WALLINTERFERENCEADD**) is used to merge the two entities or to adjust the intersection. It inserts a shrink-wrap wall component to trace the intersection of the column and wall. See Table 13.19 for **Add Interference Condition** command access.

Menu bar	Design>Walls>Wall Tools>Interference Condition
Command prompt	WALLINTERFERENCEADD
Toolbar	Select Add Interference Condition from the Walls toolbar as shown in Figure 13.75

Table 13.19 *Add Interference Condition command access*

Figure 13.75 *Add Interference Condition command on the Walls toolbar*

The shrink-wrap effect can be **Additive, Subtractive** or **Ignore**, as described below.

Additive – Merges the column to the wall and outlines the perimeter.

Subtractive – Subtracts the column from the wall and outlines the boundary.

Ignore – Allows the wall to pass through the column without shrink-wrapping the intersection.

You can increase the lineweight of the shrink-wrap wall component by editing the display properties of the wall to emphasize the intersection of the wall. The **Entity Display** command (**ENTDISPLAY**) is used to change the lineweight of the shrink-wrap wall component to emphasize the intersection. Shown below is the command sequence using the **Add Interference Condition** command selected from the **Walls** toolbar to apply shrink-wrap to the walls and columns.

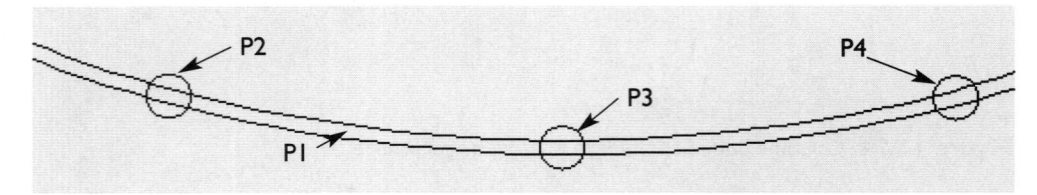

Figure 13.76 *Interference specified for each column using Add Interference Condition command*

```
Command: _AecWallInterferenceAdd
Select walls: (Select the wall at P1 in Figure 13.76.) 1 found
Select walls: ENTER (Press ENTER to end selection.)
Select AEC entities to add: (Select the column at P2 in Figure
13.76.) 1 found
Select AEC entities to add: ENTER (Press ENTER to end selection.)
Enter shrinkwrap effect
[Additive/Subtractive/Ignore]: A
1 object(s) added to wall A97.
```

(Additive shrink-wrap applied to column A as shown in Figure 13.77.)

```
Command: ENTER (Press ENTER to repeat the command.)
Command: _AecWallInterferenceAdd
Select walls: (Select the wall at P1 in Figure 13.76.) 1 found
Select walls: ENTER (Press ENTER to end wall selection.)
Select AEC entities to add: (Select the column at point P3 in
Figure 13.76.) 1 found
```

```
Select AEC entities to add: ENTER (Press ENTER to end selec-
tion.)
Enter shrinkwrap effect
[Additive/Subtractive/Ignore]: S ENTER
1 object(s) added to wall A97.
```

(Subtractive shrink-wrap applied to column B as shown in Figure 13.77.)

```
Command: ENTER (Press ENTER to repeat the command.)
Command: AECWALLINTERFERENCEADD
Select walls: (Select the wall at P1 in Figure 13.76.) 1 found
Select walls: ENTER (Press ENTER to end selection.)
Select AEC entities to add: (Select the column at P4 in Figure
13.76.) 1 found
Select AEC entities to add: ENTER (Press ENTER to end AEC
add selection.)
Enter shrinkwrap effect
[Additive/Subtractive/Ignore]: I ENTER
1 object(s) added to wall A97.
```
(Ignore shrink-wrap applied to column C as shown in Figure 13.77.)

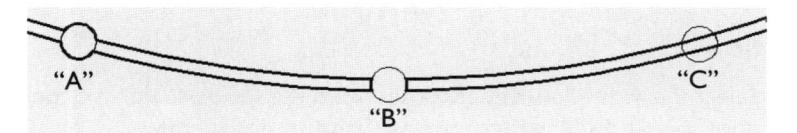

Figure 13.77 *Shrink-wrap applied with lineweights to columns with the Add Interference Condition command*

TUTORIAL 13.5 USING LAYOUT CURVES

1. Open *Ex13-5* from the *\ADT_Tutor\Ch13* directory.

2. Save the file as *Curves* in your student directory.

3. Right-click over an Architectural Desktop toolbar and choose **AEC Layout Tools** from the toolbar list.

4. Select **Add Layout Curve** from the **AEC Layout Tools** toolbar and respond to the command line prompts as shown in the following command sequence.

```
Command: _AecLayoutCurveAdd
Select a curve: (Select the curve at P1 in Figure 13.78.)
Select node layout mode [Manual/Repeat/Space evenly]
<Manual>: S ENTER
Start offset <0">: 6" ENTER
```

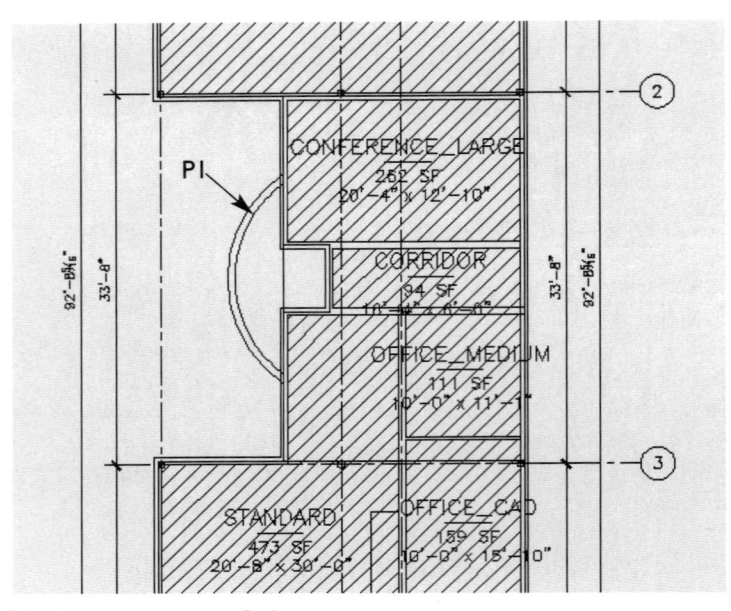

Figure 13.78 *Layout curve specified*

```
End offset <0">: 6" ENTER
Number of nodes <3>: 4 ENTER
```

5. Select the **Add Column** command from the **Grids/Columns** toolbar to open the **Add Mass Element as Column** dialog box.

6. Edit the **Add Mass Element as Column** dialog box as follows: toggle on **Predefined Shape**, set **Shape** to Cylinder; **Height** to 9', and **Radius** to 5", and respond to the following command line prompts to insert the column on the nodes of the layout curve.

```
Command: _AecColumnCoverAdd
Insert point or [SHape/Height/Radius/Match]: DBOX
Insert point or [SHape/Height/Radius/Match]: node
ENTER
Of (Select the node at P1 in Figure 13.79.)
Rotation or [SHape/Height/Radius/Match] <0.00>:
ENTER
Insert point or [SHape/Height/Radius/Match/Undo]:
node ENTER
Of (Select the node at P2 in Figure 13.79.)
Rotation or [SHape/Height/Radius/Match/Undo] <0.00>:
ENTER
Insert point or [SHape/Height/Radius/Match/Undo]:
node ENTER
```

Drawing Commercial Structures 1013

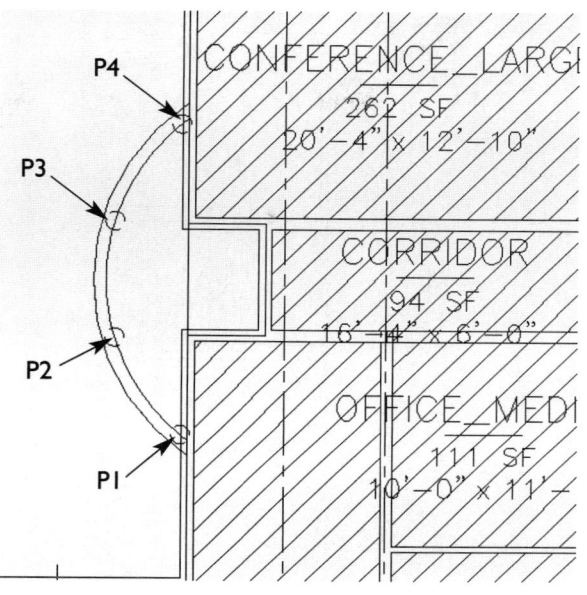

Figure 13.79 *Specifying points to insert columns*

```
Of (Select the node at P3 in Figure 13.79.)
Rotation or [SHape/Height/Radius/Match/Undo] <0.00>:
ENTER
Insert point or [SHape/Height/Radius/Match/Undo]:
node ENTER
Of (Select the node at P4 in Figure 13.79.)
Rotation or [SHape/Height/Radius/Match/Undo] <0.00>:
ENTER
Insert point or [SHape/Height/Radius/Match/Undo]:
ENTER (Press ENTER to end the command.)
```

7. Select **Zoom Window** from the Standard toolbar and select points near P1 and P2 to specify the Zoom window as shown in Figure 13.80.

8. Select **Add Interference Condition** from the **Walls** toolbar; then respond to the following command line prompts as shown below.

```
Command: _AecWallInterferenceAdd
Select walls: Specify opposite corner: (Select the wall
using a crossing window from P3 to P4 in Figure 13.80.) 2 found, 1
total
1 was filtered out.

Select walls: ENTER (Press ENTER to end wall selection.)
Select AEC entities to add: (Select the column at P5 in Figure
13.80.) 1 found
```

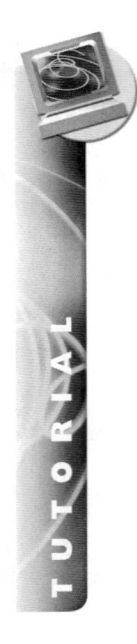

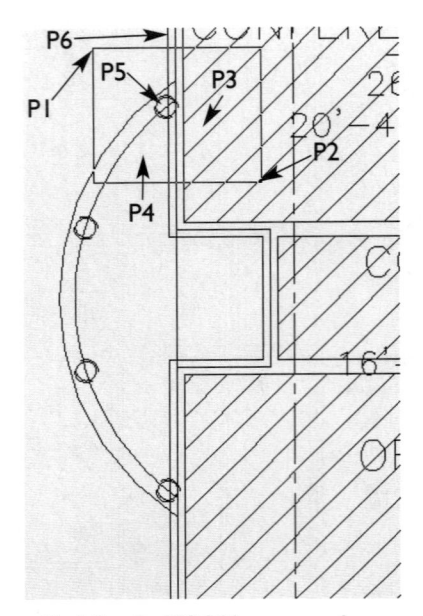

Figure 13.80 *Locations specified for the ZOOM command*

```
Select AEC entities to add: ENTER (Press ENTER to end selec-
tion.)
Enter shrinkwrap effect
[Additive/Subtractive/Ignore]: A ENTER
1 object(s) added to wall BB1.
```
(Shrink-wrap additive applied to the column and wall.)

9. Select A-Area-Bdry from the **Layer** flyout and freeze the layer.
10. Select the wall at P6 as shown in Figure 13.80; then right-click and choose **Entity Display** from the shortcut menu.
11. Select the **Display Props** tab of the **Entity Display** dialog box.
12. Select the Wall **Property Source**; then select the **Attach Override** button and select the **Edit Display Props** button to open the **Entity Properties** dialog box.
13. Select the **Lineweight** column of the Shrink Wrap row and select 0.031" from the **Lineweight** dialog box as shown in Figure 13.81. Select **OK** to dismiss the **Lineweight** dialog box.
14. Select **OK** to dismiss the **Entity Properties** dialog box.
15. Select **OK** to dismiss the **Entity Display** dialog box.
16. Toggle on Lineweights in the Status Bar to display the shrink-wrap as shown in Figure 13.82.

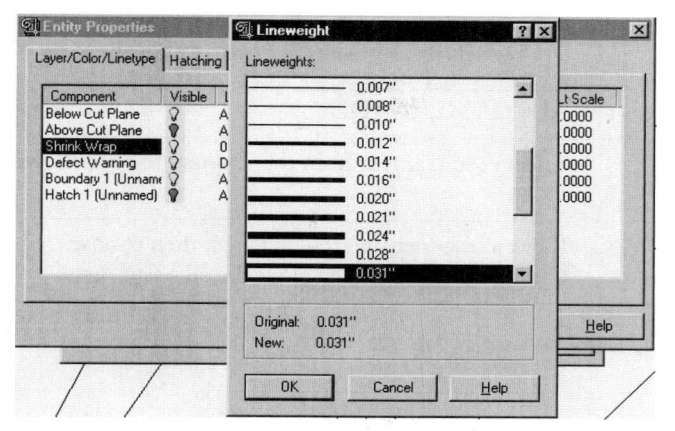

Figure 13.81 *Lineweight selected for the Shrink Wrap object*

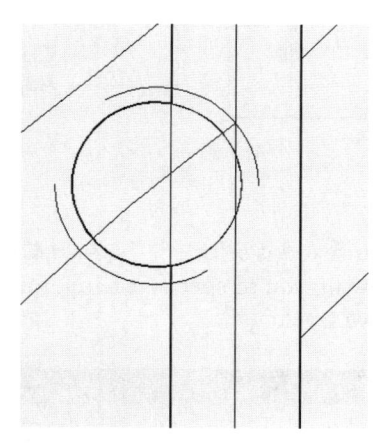

Figure 13.82 *Lineweights changed to emphasize the shrink-wrap*

17. Save the drawing.

VIEWING THE MODEL

Most frequently, the designs of commercial buildings are presented to clients in form of perspective drawings or a walkthrough. Architectural Desktop includes a camera to facilitate the view of the building model. The camera provides a quick visualization tool for presenting the model. The model can be viewed through one or more cameras placed in the plan view of the drawing. The still view or walkthrough presentation can be created with the camera; the camera is inserted in the floor plan and pointed toward objects of interest. The **Add Camera** and **Create Video** commands are included on the **Camera** toolbar shown in Figure 13.83. See Table 13.20 for **Add Camera** command access.

Menu bar	Design>Generate Perspectives>Add Camera
Command prompt	AECCAMERAADD
Toolbar	Select Add Camera from the Camera toolbar shown in Figure 13.83
Shortcut	Right-click over the drawing area; then choose Utilities>Cameras>Add from the shortcut menu

Table 13.20 *Add Camera command access*

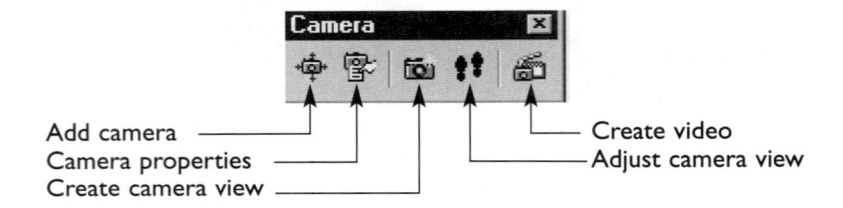

Figure 13.83 *Camera toolbar*

When the **Add Camera** command is selected, the **Add Camera** dialog box opens, as shown in Figure 13.84, allowing you to specify a name for the camera, the zoom factor, and the height of the eye level.

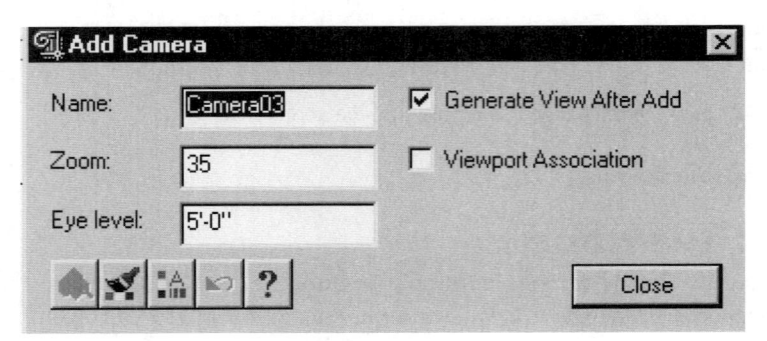

Figure 13.84 *Add Camera dialog box*

After setting the eye level and zoom of the camera, you can select the insertion point and target of the camera. A camera symbol and the camera name are placed in the drawing. The target can be adjusted with grips after you insert it in the drawing.

You can obtain the view of the camera by selecting **Create Camera View** from the **Camera** toolbar. When you select the **Create Camera View** command, you are prompted to select a camera. Selecting a camera provides a view through the camera of the building. See Table 13.21 for **Create Camera View** command access.

Menu bar	Design>Generate Perspectives> Create Camera View
Command prompt	AECCAMERAVIEW
Toolbar	Select Create Camera View from the Camera toolbar shown in Figure 13.83
Shortcut	Right-click over the drawing area and choose Utilities>Camera>Create View

Table 13.21 *Create Camera View command access*

CREATING A WALKTHROUGH

Create a walkthrough by inserting a camera and defining the path and target for the camera to follow. The camera follows the path looking toward the target to create the walkthrough, which is recorded in video. The video is played when the **Create Video** command is executed. See Table 13.22 for **Create Video** command access.

Menu bar	Select Design>Generate Perspectives> Create Video
Command prompt	AECCAMERAVIDEO
Toolbar	Select Create Video from the Camera toolbar shown in Figure 13.83
Shortcut	Right-click over the drawing area; then choose Utilities>Cameras>Create Video

Table 13.22 *Create Video command access*

When you select **Create Video**, you are prompted to select a camera, and the **Create Video** dialog box opens as shown in Figure 13.85.

The **Create Video** dialog box allows you to name the path and target. The path and target can be assigned to a polyline path or you can select a point for the camera to follow. The options of the **Create Video** dialog box are as follows:

> **Camera Path** – The **Camera Path** list shows the paths defined for the cameras in the drawing.

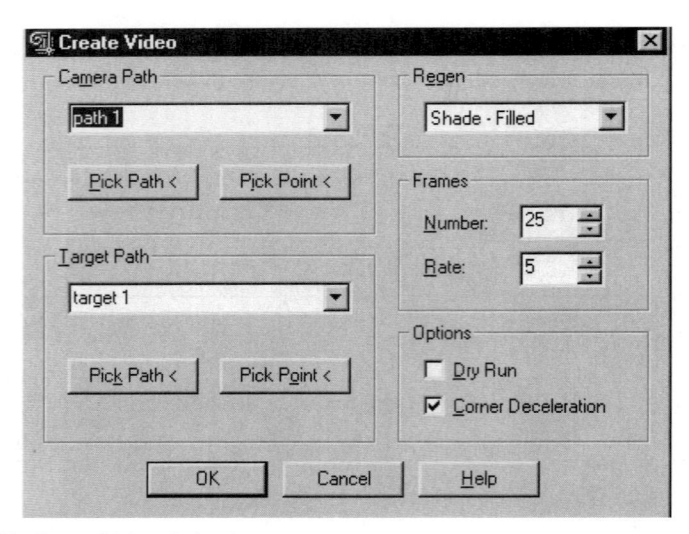

Figure 13.85 *Create Video dialog box*

Pick Path< – The **Pick Path<** button of the **Camera Path** section prompts you to select a polyline to use as the path for the camera. The **Path Name** dialog box opens if the selected polyline has not previously been defined as a path. Create a new name by typing the name in the **Path Name** dialog box.

Pick Point< – The **Pick Point<** button of the **Camera Path** section allows you to pick a location in the drawing for the camera to point.

Target Path – The **Target Path** list shows the targets defined for the camera.

Pick Path< – The **Pick Path<** button of the **Target Path** section allows you to select a polyline as the target for the camera to follow.

Pick Point< – The **Pick Point<** button of the **Target Path** section allows you to select a point in the drawing for the camera to point to as it travels the path.

Regen – The **Regen** list includes the following options: None, Hide, Shade 256 Color, Shade 256 Edge, and Shade Filled.

Frames – The **Frames** section allows you to specify the number of frames and the rate. Increasing the number of frames creates a smoother video, while decreasing the rate slows the pace of the walkthrough.

Dry Run – The **Dry Run** check box provides a dry run or animation of the camera following its path.

Corner Deceleration – The **Corner Deceleration** check box adjusts the speed of the camera as it turns the corner of the path.

TUTORIAL 13.6 CREATING A WALKTHROUGH

1. Open *Ex13-6* from the \ADT_Tutor\Ch13 directory.
2. Save the drawing in your student directory as *Video1.dwg*.
3. Select **Add Camera** from the **Camera** toolbar.
4. Edit the **Add Camera** dialog box as follows: set **Name** to Camera01, **Zoom** 35, **Eye level** 5', and select **Generate View After Add**.
5. Select the insertion point for the camera at P1 as shown in Figure 13.86.
6. Move the pointer to rotate and point the target of the camera to P2 as shown in Figure 13.86.

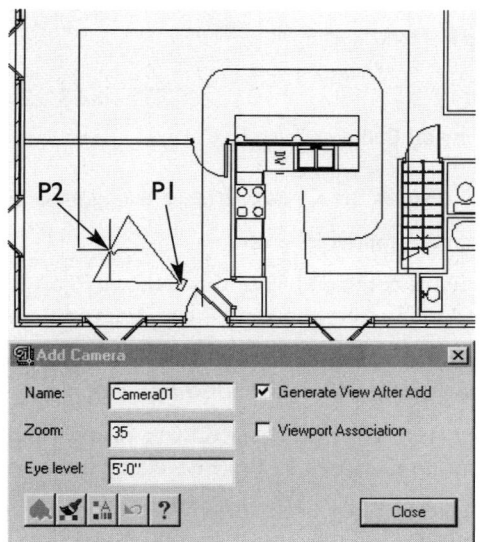

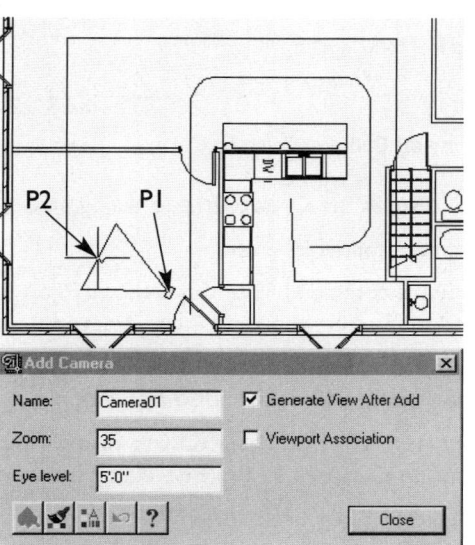

Figure 13.86 *Specifying the settings of the camera for the walkthrough*

7. Select **Top View** from the **View** flyout on the **Standard** toolbar to exit the preview and return to the drawing area.
8. Select **Create Video** from the **Camera** toolbar.
9. Select Camera01 to open the **Create Video** dialog box.
10. Edit the **Create Video** dialog box as follows: select the **Pick Path<** button and select the path shown in Figure 13.87. In the **Target Path** section, select the **Pick Path<** button and select the **Target** polyline as shown in Figure 13.87. Select Shade 256 Color from the **Regen** list, edit **Frames Number** to 50 and **Frames Rate** to 5, and select **Dry Run** and **Corner Deceleration**, as shown in Figure 13.87.

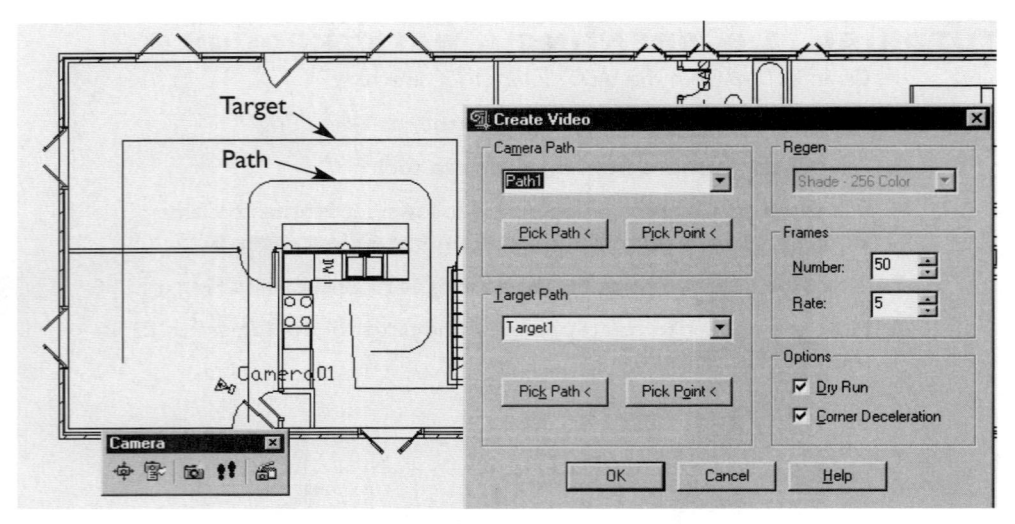

Figure 13.87 *Editing the Add Camera dialog box*

11. Select **OK** to dismiss the **Create Video** dialog box.
12. Watch the camera follow its path.
13. Select the **Layers** command from the **Object Properties** toolbar, select the Path layer, and then select the sunshine to freeze for the Path layer.
14. Select the **Create Video** command from the **Camera** toolbar; then select Camera01 to open the **Create Video** dialog box.
15. Deselect **Dry Run**, and then select **OK** to dismiss the **Create Video** dialog box and open the **Camera Video File** dialog box, shown in Figure 13.88.
16. Type **Video1** in the **Camera Video File** dialog box, select your student directory in the **Save in** list, and select the **Save** button to dismiss the **Camera Video File** dialog box.

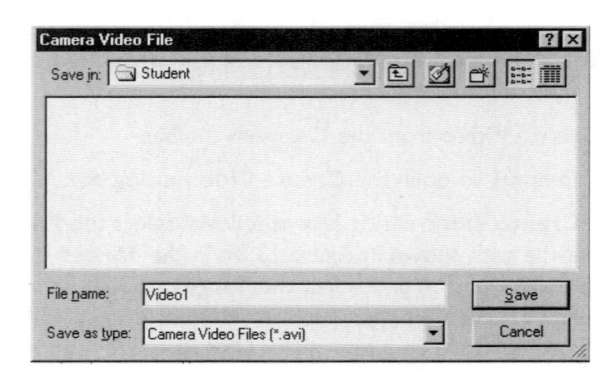

Figure 13.88 *Camera Video File dialog box*

17. When the **Video Compression** dialog box opens, select **Full Frames Uncompressed**, and then select **OK** to dismiss the **Video Compression** dialog box.
18. Select **Yes** when prompted to play the video in the **AEC Camera** message box.
19. View the video.
20. Select the **Close** button of the Video1.AVI window.
21. Select **Top View** on the **Standard** toolbar to return to the drawing.
22. Save the drawing.

SUMMARY

1. Column grids are placed in the drawing with the **Add Column Grid** command (**COLUMNGRIDADD**).
2. Set the **Layout Mode** to **Manual** for an axis to edit the column grid lines of that axis.
3. The start offset and end offset distances of the column grid are relative to the insertion point of the column grid.
4. The **Column Grid Modify** command (**COLUMNGRIDMODIFY**) is used to change the size and dimensions of the column grid.
5. The column grid can be dimensioned with the **Column Grid Dimension** command (**COLUMNGRIDDIM**).
6. Ceiling grids are added to the drawing with the **Add Ceiling Grid** command (**CEILINGGRIDADD**).
7. The Modify Ceiling Grid command (**CEILINGGRIDMODIFY**) and the **Ceiling Grid Properties** command (**CEILINGGRIDPROPS**) are used to change a ceiling grid.
8. The **Attach Clipping Boundary/Holes** command (**CEILINGGRIDCLIP**) can be used to trim a ceiling grid by a polyline or AEC object.
9. The **Add Layout Curve** command (**LAYOUTCURVEADD**) can be used to create nodes that are evenly spaced along a line, arc, or polyline.
10. The intersection of columns and walls can be added, subtracted, or ignored through the **Add Interference Condition** command (**WALLINTERFERENCEADD**).
11. Obtain views of the building by placing a camera with the **Add Camera** command (**AECCAMERAADD**).
12. A video of a walkthrough in the building is developed with the **Create Video** command (**AECCAMERAVIDEO**).

REVIEW QUESTIONS

1. The columns of a column grid are placed on the _____ layer and the column grid lines are placed on the _____ layer.
2. The distance measured between column grids horizontally is the ___ baysize.
3. The angle of the radial column grid is measured relative to the _____ position.
4. Column grid lines can be added automatically if **Automatic Spacing** is selected in the _____ tab.
5. Columns added to a column grid are edited by the _____ command.
6. The size of the text used in the column label bubble is specified in the _____ dialog box.
7. Ceiling grids can be trimmed by wall elements with the _____ command.
8. Ceiling grids only have four grips when _____ is toggled _____ in the **Dimensions** tab of the **Ceiling Grid Properties** dialog box.
9. A _____ can be used to place columns spaced evenly without the column grid.
10. The _____ command is used to modify the wall to column intersection.
11. You can combine two columns to create one column by creating a _____.
12. Ceiling grids should always be sized _____ than the room used to bound the ceiling.

PROJECTS

EXERCISE 13.1 CREATING COLUMNS AND COLUMN GRIDS

1. Open *EX13-7* from the \ADT_Tutor\Ch13 directory.
2. Save the drawing as *Proj13-1* in your student directory.
3. Create a column grid using the **Specify on Screen** option. Specify the column grid at the inner surface of the diagonal corners of the building. Specify ColTS_8x8 columns that are 9' high.
4. Specify the drawing scale as 1/8"=1'-0".
5. Use **Column Grid Labeling** to label the columns descending from the upper right corner. Specify the upper right corner column as A1 and label the remaining columns using the automatic labeling value option.
6. Edit column A4 and B4 to a Doric mass element 9' high and 8" in diameter.

7. Dimension the column grid as shown in Figure 13.89.
8. Save the drawing.

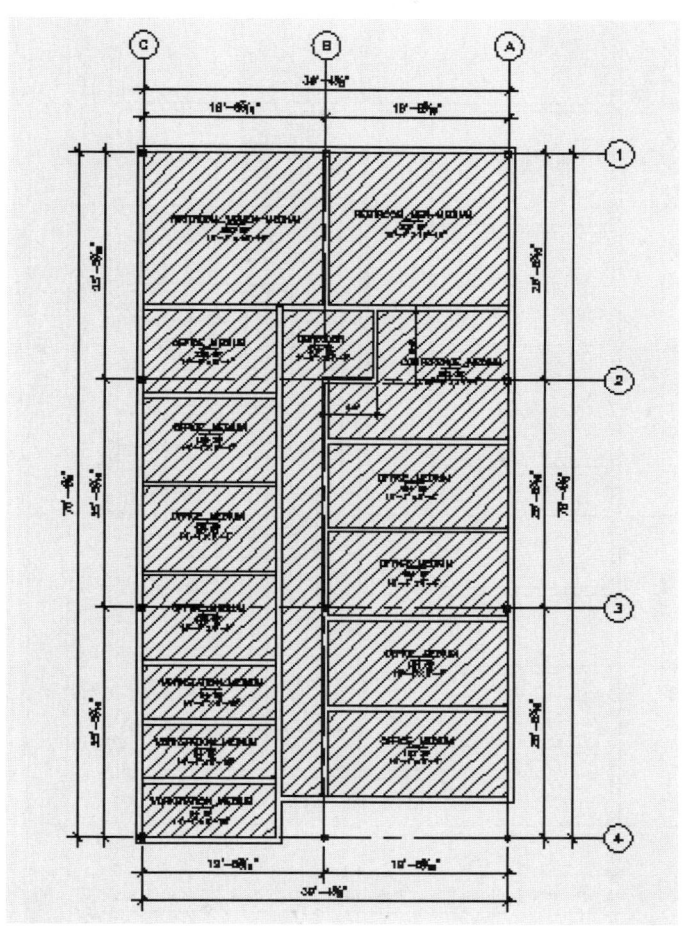

Figure 13.89 *Column grid placed in the drawing*

EXERCISE 13.2 CREATING A CEILING GRID

1. Open *EX13-8* from the *\ADT_Tutor\Ch13* directory.
2. Save the drawing as *Proj13-2* in your student directory.
3. Create suspended ceilings in the OFFICE_MEDIUM rooms on the right side of the building using the space as the boundary.
4. Specify ceiling panels equal to 4' x 2'.

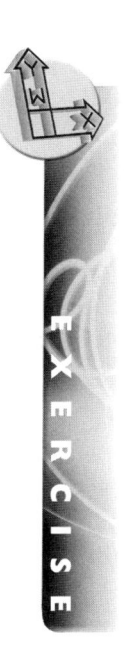

5. Trim the ceiling grid by the chase in the corner as shown at P1 in Figure 13.90.

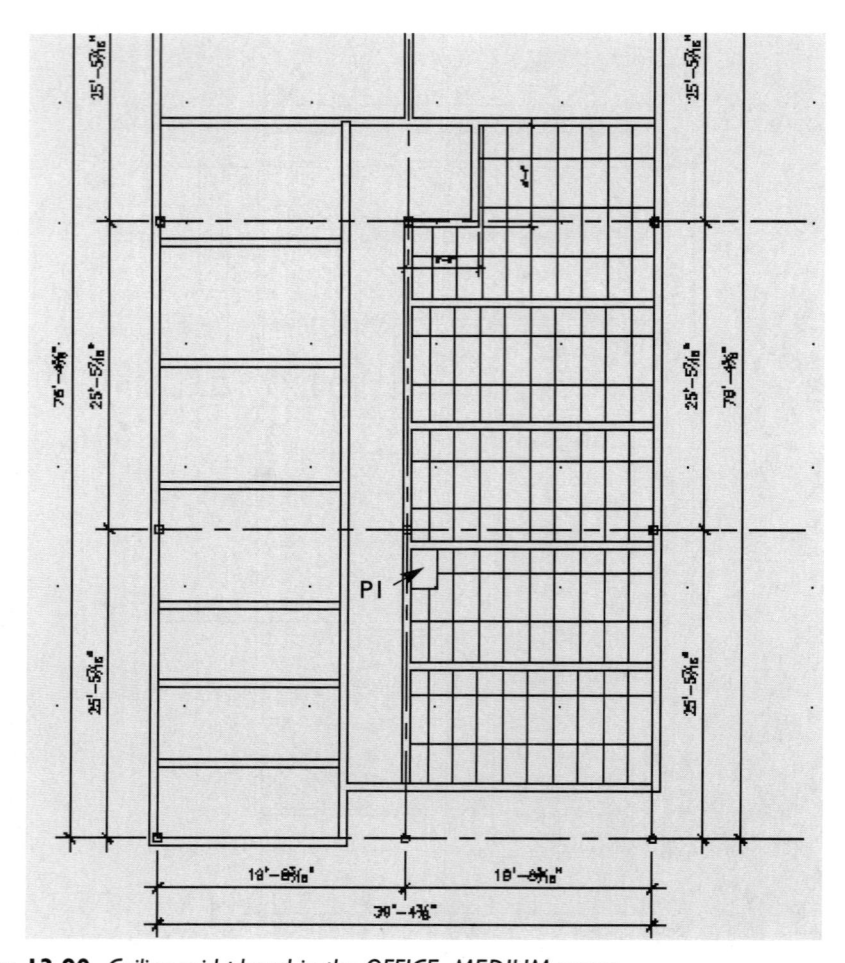

Figure 13.90 *Ceiling grid placed in the OFFICE_MEDIUM rooms*

6. Freeze the A-Area-Spce layer.
7. Save the drawing.

INDEX

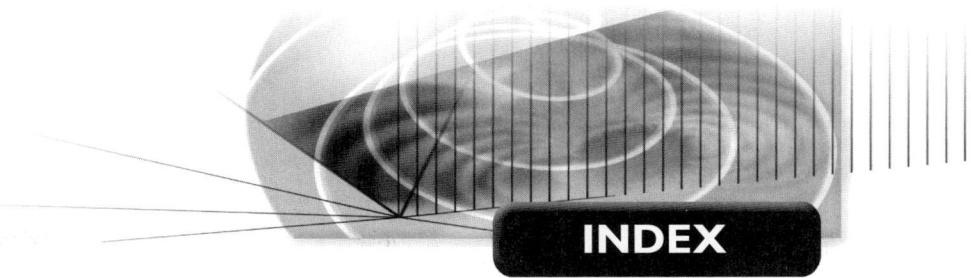

? button, 44, 323, 401, 403, 469, 519, 566-67, 746

3d Orbit command, 821
3DORBIT cursor, 821

A (Height) dimension, 332
A-Angle edit field, 958
A-Base Height, 904
About Model Explorer option, 820
Above Cut Plane, 138-39
ACAD menu group, 10
Add
  button, 117-18, 121-22, 136, 176-77, 186-87, 197, 208, 248, 254, 256, 289, 293, 544, 549, 551, 574-75, 581, 699, 701, 709
  option, 832
Add a Hole option, 997, 999
Add Boundary
  command, 858, 899-900, 907, 924
  dialog box, 899
Add Boundary Edges command, 858, 899, 920, 924-25, 933-34, 936
Add Camera
  command, 954, 1015-16, 1019
  dialog box, 1016, 1019
Add Ceiling Grid
  command, 954, 986-90, 994
  dialog box, 987-91, 994
Add Column command, 954, 979-80, 1012
Add Column Grid
  command, 953, 955, 957, 959-61, 968
  dialog box, 955-59, 961, 963, 972
Add Door command, 217-21, 230, 232, 238-39, 254, 269
Add Doors dialog box, 31-32, 219-21, 225-26, 230-32, 254, 309
Add Elements shortcut, 826
Add Elevation Mark command, 681, 688-89, 693, 707
Add Gable button, 95
Add Grid Line command, 972-73, 984, 996, 1001-2
Add Interference Condition command, 954, 1009-11, 1013
Add Layout Curves command, 954, 1007-9, 1011
Add Mask Blocks dialog box, 566-68, 582
Add Masking Block command, 565-66
Add Mass Element
  command, 741-43, 772, 781, 787, 794
  dialog box, 742-46, 754-56, 760, 764-65, 767, 771-72, 781, 787, 794, 801-2, 806-7, 825-26, 828, 849
Add Mass Element as Column dialog box, 957, 961-62, 968-69, 979-80, 1012

Add Mass Group command, 361-62, 364, 376, 799, 801, 806
Add Multi-View Blocks
  command, 518
  dialog box, 518-20, 528-37, 540-, 547-49, 553
Add New Objects Automatically check box, 671, 673, 675-76
Add Opening
  command, 298-300
  dialog box, 298-300
Add Override button, 166
Add Railing
  command, 399-400, 466, 485-88
  dialog box, 466-69, 472, 485-88
Add Roof
  command, 319-20, 327, 333, 338, 343, 351, 360, 368, 373, 386-87
  dialog box, 320-21, 330, 337-40, 358
Add Schedule Table
  command, 590, 668, 673, 675
  dialog box, 670-71, 673, 675
Add Section Mark command, 682, 722-24, 733
Add Space dialog box, 859-63, 865, 867-68, 870-73, 880-81, 883
Add Space Boundary
  command, 902
  dialog box, 900-2, 907, 909, 924, 934, 936
Add Spaces command, 857, 859-60, 871-72, 874
Add Stair command, 399-401, 411, 414, 416, 420, 425, 429, 433, 436, 482
Add Stairs dialog box, 400-5, 411, 414, 416, 420-21, 423-27, 429-30, 433-34, 439-440, 482-85
Add Standard Size dialog box, 248, 252-54, 256, 289-90, 293, 296
Add Subdivision dialog box, 709
Add to List button, 12-13, 15. (See also Profiles tab)
Add Tread Depth check box, 408
Add Wall
  button, 84
  command, 41, 43, 45, 47-49, 51, 53, 57, 70, 84, 112, 211, 899
  dialog box, 900
Add Walls dialog box, 43-45, 47-52, 54-55, 57, 70, 84, 111-12, 114, 124-25, 127, 168-69, 177, 190, 538
Add Window command, 269-71
Add Windows dialog box, 270-74, 277, 280, 283
Additional Scaling section, 581-82
Additive shrink-wrap effect, 1010
AEC Anchors toolbar. (See Toolbars, AEC Anchors)
AEC arch. (See Templates)
AEC Camera message box, 1021
AEC Content tab, 416, 510-11, 515, 649

AEC DwgDefaults tab, 63, 416, 419, 427, 469, 472-73, 480, 865
AEC Editor, 416
AEC Layer Utilities toolbar. (See Toolbars, AEC Layer Utilities)
AEC Layout Tools
  command, 1011
  toolbar. (See Toolbars, AEC Layout Tools)
AEC Performance, 416
AEC Plan Rotation toolbar. (See Toolbars, AEC Plan Rotation)
AECADDWALL command, 41, 45, 51, 53, 57
AECANNODETAILBOUND-ARYADD command, 596-98
AECANNODETAILMARKADD command, 596, 598-99
AECANNOSCHEDULETAGADD command, 648, 652, 661, 663
AECARCHX menu group, 10-11, 14, 42, 50
AECCAMERAADD command, 954, 1016
AECCAMERAVIDEO command, 954, 1017
AECCAMERAVIEW command, 1017
AECDCSETIMPMISCELLANEOUS command, 614
AECDCSETIMPROOMANDFIN-ISHTAGS command, 651
AECDCSETIMPWALLTAGS command, 663
AECDISPLAYCONFIGATTACH command, 28
AECDISPLAYCONFIGDEFINE command, 29
AECDISPLAYSETDEFINE command, 26
AECDWGSCALESETUP command, 500, 502, 590, 616, 978
AECENTREF command, 831, 835
AECMASSELEMENTADD command, 746
AECMASSELEMENTOPADD command, 813
AECMASSGROUPADD command, 799
AECMODIFYWALL command, 41
AECPROFILEDEFINE command, 305
AECPROPERTYRENUMBERDATA command, 674
AECSLICEATTACH command, 800, 842-43
AECSLICECREATE command, 800, 841
AECSLICEDETACH command, 800, 844
AECSLICEELEVATION command, 845
AECSLICETOPLINE command, 800, 846, 885
AECWALLADD command, 42

AECWALLCONVERT, 41, 100-2
AECWALLDIM, 102-3, 105
AECWALLMODIFY, 89, 141, 190
AECWALLSTYLE, 113, 152
AECWINDOWADD, 281
A-Height edit box, 472, 485-86
AIA layer standard, 16
Align Right radio button, 459
Alignment option, 451-52, 456
All
  default option, 469
  tab, 18-19, 34-35, 50, 57
Anchor
  tab, 260-62, 277, 302, 476-78
  Type drop-down, 575-76
Anchor to Boundary command, 858, 928-29
Angle
  column, 199
  edit field, 454-55
Angular section, 502
ANNOBARSCALEADD command, 633-34
ANNOBREAKMARKADD command, 591-94
ANNOELEVATIONMARKADD command, 600, 602
ANNOLEADERADD command, 603-4
ANNOMATCHLINEADD command, 621-23
ANNORATINGLINEADD command, 618-20
ANNOREVISIONCLOUDADD command, 627-29
ANNOSECTIONMARKADD command, 629-32
ANNOSYMBOLADD command, 623-26
Annotating documents. (See Documentation menu)
Annotation components
  Imperial toolbar. (See Toolbars, Annotation - Imperial)
  Metric toolbar. (See Toolbars, Annotation - Metric)
  option, 576
Plot Size edit field, 501, 515, 546, 576, 579
ANNOTITLEMARKADD command, 635-36
Apply and Close button, 13, 15
Apply button, 88, 90, 229, 254, 337, 441, 461, 769, 883, 896, 997
Apply to drop-down list, 176-77, 186-87
Arc radio button, 901
Arch mass element, 746-47
Architectural Desktop Imperial Content submenu, 3. (See also Design Content)
Area
  edit field, 861, 873
  Lock, 871-72
  option, 867, 880

1025

Separation radio button, 900, 903
Areas section, 502
ARRAY command, 272
Arrow section, 454-55. (See also Other tab)
As Inserted option, 576-77
as Pline option, 161, 180-81
Ascending radio button, 977
A-Side 1 dimension, 699-701, 709, 728
A-Stairway Width dimension, 406-7, 420-21
A-Start Endcap button, 163-65
A-Stringer End Offset option, 450-51, 456-58
Attach Clipping Boundary/Holes command, 954, 997-99, 1006
Attach Element command, 800
Attach Elements command, 362-64, 376, 801, 803, 810, 838, 851
Attach Objects command, 800, 842-43, 853
Attach Objects to Mask command, 571
Attach option, 832-33, 928
Attach Override button, 129-30, 144, 184, 197, 249, 256, 291, 453, 494, 562-63, 710, 729, 733, 881, 1014
Attach Spaces to Boundary command, 858, 912-13, 920
Attach/Edit Schedule Data command, 590, 664-65
Attached column, 130, 453, 562-63, 705
Attached to drop-down list, 467, 485-89
Attribute Text Angle section, 576-77
Attribute Text Style section, 576-77
Attributes tab, 519-20, 523-24, 690, 694, 708, 733
Auto project option, 380, 382
Auto Track tooltip, 45
Auto Zoom Extents, 820
Auto-Adjust to Width of Wall check box, 244-45, 255, 287, 293
AutoCAD 2000, 2
  editing commands, 83-84
AutoCAD Props tab, 703-4
Automatic Offset/Center check box, 219-20, 230, 254, 269, 271-72, 280, 299
Automatic Placement check box, 467, 485-89
Automatic Spacing check box, 964-67, 973, 991-92, 994, 1001
Automatic Update check box, 671, 673, 675-76
Automatically Calculate Values for Labels check box, 976-77, 981-82
Automatically Determine from Spaces check box, 903-5, 907
A-Width edit field, 244, 252, 255, 287, 293

B (Overhang) dimension, 332
Back button, 574-75
Balusters
check box, 466, 472-73, 479, 485-89
option, 469
Bar
menu, 6
Scale symbol, 633-35

Barrel Vault, 747-48
Base Point section, 492-93, 720-21, 738
Base Value edit field, 125, 135, 196-97
Base Width, 125, 196
Baseline Justification line, 124
B-Bayangle, 958
B-Depth edit field, 244-45, 252, 255, 287, 293
Below Cut Plane, 138-39, 208
B-End Endcap button, 163-65
B-Height edit field, 472-73
Bitmap image file, 2
BLDGELEVATIONLINEADD command, 681, 688-89
BLDGELEVATIONLINEGENERATE command, 599-600, 681, 690-91
BLDGELEVATIONLINEPROPS command, 698-
BLDGELEVATIONLINEREVERSE command, 689-90
BLDGSECTIONLINEADD command, 682, 722—24
BLDGSECTIONLINECONVERT command, 682, 724-25
BLDGSECTIONLINEGENERATE command, 631, 682, 725-26
BLDGSECTIONLINEPROPS command, 682, 727
BLDGSECTIONPROPS command, 682, 731
BLDGSECTIONUPDATE command, 681-82, 713-14, 730-31
Block
drop-down list, 455
radio button, 574-75
Bottom check box, 978, 981
Bottom Elevation Offset, 127-28, 207-8
Bottomrail
checkbox, 467, 472-73, 479, 485-89
default height, 469-70
Boundary, 138-39
command, 652, 989
Boundary Edge Properties dialog box, 926, 944
Box, 748
command, 764-65, 767
Break Mark section, 454-55. (See also Other tab)
Break Marks command, 490-91, 607
Browse
button, 494, 577-78, 582
option, 510
Browse Folder dialog box, 510
B-Side 2 dimension, 699-701, 709, 728
B-Space Height, 864, 905, 908, 916
B-Straight Length dimension, 406-7, 412-15, 420, 435-36, 441, 482-85
B-Stringer Width option, 450-51, 456-58
Bubble
button, 978, 981
Parameter section, 978, 981-82
B-Upper Extension, 905

C (Eave) dimension, 332
Camera
Path list, 1017-18
toolbar. (See Toolbars, Camera)
Camera Video File dialog box, 1020

C-Angle 1 dimension, 699, 701, 709
Ceiling Boundary check box, 861, 871-73
Ceiling Boundary Thickness, 904
Ceiling Grid Clip command, 996
Ceiling Grid Properties
command, 954, 990, 996
dialog box, 988, 990-94, 1001
Ceiling Stops at Wall, 904-5
CEILINGGRIDADD command, 954, 986
CEILINGGRIDCLIP command, 954, 997
CEILINGGRIDMODIFY command, 954, 996
CEILINGGRIDPROPS command, 954, 990
CEILINGGRIDXADD command, 954, 1001
CEILINGGRIDXMODE command, 1000-1
CEILINGGRIDXREMOVE command, 954, 1002-3
CEILINGGRIDYADD command, 954, 1001
CEILINGGRIDYMODE command, 1000-1
CEILINGGRIDYREMOVE command, 954, 1002-3
Center Mark, 436
Center option, 263-64
C-Floor Boundary Thickness, 865
CHAMFER command, 84
Change Swing Opening command, 229
C-Height edit field, 472-73
Circle option, 596-98
Cleanup
  groups, creating, 188-93
  radius, 41, 69
  settings, 94
Cleanup Group Definitions command, 189
dialog box, 189-91
Cleanup Group Definition Properties dialog box, 191-92
Cleanup Settings section, 190
Clockwise check box, 503
Close
button, 14, 44, 50, 323, 535, 538, 541, 548-49, 553, 746, 755, 962, 969, 1021
command, 819, 853
option, 50
C-Lower Extension, 905
Code Limits section, 407-8, 412-15
Cold grips, 79
Color
column, 711
option, 618
Column button, 957, 961
Column Grid Dimension command, 954
Column< button, 962, 968
Column Grid Labeling command, 954, 976, 981, 984
dialog box, 976-78, 981-82, 985
Column Grid Modify command, 971
Column Grid Properties
command, 953, 957, 963, 971-72
dialog box, 963-68, 973, 983
COLUMNCOVERADD command, 954, 979-80
COLUMNGRIDADD command, 953, 955

COLUMNGRIDDIM command, 954, 978-79
COLUMNGRIDLABEL command, 954, 976
COLUMNGRIDMODIFY command, 953, 971
COLUMNGRIDPROPS command, 953, 963, 971
COLUMNGRIDXADD command, 973
COLUMNGRIDXMODE command, 953
COLUMNGRIDXREMOVE command, 974
COLUMNGRIDYADD command, 973
COLUMNGRIDYMODE command, 953
COLUMNGRIDYREMOVE command, 974
Command String edit field, 574-75
Compass. (See Visual Aids option)
Component
Dimensions section, 864-85
list, 121-22
Name drop-down list, 166, 176-77, 186-87
Offset dialog box, 124, 135-36, 196-97, 211
Relationships section, 407-8
Width dialog box, 125, 135-36, 196-97
Components tab, 116, 121-28, 135, 161, 166, 183, 196, 207, 210-11, 386-87
Concept menu, 6-7
Cone, 748-49
Configuration tab, 29-31, 67, 74
Configurations, plot, 16
Content File list, 574-75, 581
Content Menu
section, 511, 515
toggle, 510
Content Path edit field, 510
Content Type
page, 574-75
section, 574-75, 581
Controls Display Contribution column, 130
Convert from Edge command, 924
Convert from Sketching command, 923-24
Convert from Slice command, 921-24
Convert from Spaces command, 920, 924, 930
Convert to Polyline command, 846, 885, 890
Convert to Roof command, 319, 358-60, 368, 373
Convert to Spaces command, 651-52, 656, 891
Convert to Spaces command, 858, 884-85
Convert to Walls command, 41-42, 100-2
Converting boundaries to walls. (See Concept menu)
Copy
button, 114, 158, 174, 189, 241, 285, 306, 448, 542, 560-61, 668, 774-75, 878, 894
command, 9, 78, 766, 819
operation, 117
shortcut command, 80, 283
COPY-STRETCH operation, 79-80, 279

Index

Corner Deceleration check box, 1018-19
Create AEC Content command, 574, 581
Create AEC Content Wizard, 499, 573-79, 581-82
Create Camera View command, 1016-17
Create Element shortcut, 828-29
Create Elevation command, 599-600, 681, 690-92, 725
Create MDB button, 941
Create New Drawing dialog box, 50, 53, 57, 112, 115, 310
Create Section, 681-82, 725-26
Create Video
  command, 954, 1015, 1017, 1019-20
  dialog box, 1017-20
Create Wall Dimensions command, 102-5, 107, 615
CREATECONTENT command, 574
CREATEHLR command, 681-82, 714-15, 717
Creating mass models. (See Concept menu)
C-Riser Count dimension, 406-7, 412-15, 429, 438, 461-62, 482-85
CSI
  Imperial Content submenu, 3. (See also Design Content)
  Masterformat, 3
C-Stringer Waist/Slab Depth option, 450-51, 456-58
Current button, 684, 790
Current Coordinate System radio button, 526, 569-70, 702
Current Drawing list, 151, 153-54, 313, 464, 555, 574-75, 581, 779, 784
Current Size section, 226-27, 274
Curved segment toggle option, 47-50
Custom
  list, 245
    option, 245
    radio button, 289, 308
  Custom Command, 509, 574-75
  Custom Scales edit field, 500
  Custom Shape
    list, 962
    radio button, 961-62, 969
Custom viewport, 16. (See also Templates)
Customize option, 10, 14, 42, 50. (See also Menu, shortcut)
Cut command, 9, 819
Cut operation, 117
Cut Plane Elevation edit field, 459-60
Cut Plane section, 454-55. (See also Other tab)
C-Width edit field, 244-45, 252, 255, 288, 293
Cylinder, 748

D-Angle 2 dimension, 699, 701, 709
Data Renumber dialog box, 675
Database File dialog box, 941-42
Datum Point symbol, 625-26
D-Ceiling Boundary Thickness, 865
DDATTE command, 635
D-Depth edit field, 244-45, 252, 254, 288, 293

DDVIEW command, 19
Default
  drawing name, 5
  Layer Standard Layer Key Style list, 504
  profile, 13
  symbol menu, defining, 509-11
Defect Warning, 65, 138-39
Define AEC Profile command, 305-7, 774, 777, 779-80, 783, 786, 793, 847
Define Masking Block command, 560, 564
Define Multi-View Block command, 541, 554
Define option, 161, 180-81
DefPoints layer, 65
Delete
  button, 12, 117-18
  command, 819
  operation, 117
Depth Dimension (Y-Axis) option, 973, 991, 994
Depth edit field, 178-79, 473-75, 486, 744, 755-56, 764-65, 787, 806-7, 811, 828-29, 962, 969
Descending radio button, 978, 981-82
Description
  command 509
  edit field, 15, 116-17, 133, 207, 210, 255, 293, 456, 520, 551, 565, 579, 761, 778, 780, 786, 793, 835, 847, 863, 963, 990-91
Design Content
  menu, 511-12
  submenu
    Architectural Desktop Imperial Content, 3
    CSI Imperial Content, 3
    Metric Content, 3
  toolbars
    Imperial. (See Toolbars, Design Content - Imperial)
    Metric. (See Toolbars, Design Content - Metric)
Design menu, 6-9
Design Rules tab, 242, 245-46, 252, 256, 287-89, 293, 308, 402, 405, 407-8, 411-16, 420-21, 425, 429, 438, 461, 469, 471, 473-75, 482, 486, 902-7, 910, 923, 926-27
DesignCenter, 3, 507-12
Desktop, 509
Desktop
  menu, 6-7, 25-31
  Preferences, 63
Destination File Name edit field, 720-21
Detach command, 819
Detach Element command, 800
Detach Elements command, 801, 805
Detach Objects command, 800, 844-45
Detach Objects from Mask command, 572-73
Detail Marks
  command, 590, 595, 608
  placing, 595-99
Detailed Description edit field, 578-79, 582
Developing documents. (See Documentation menu)
Dim Style drop-down list, 454-55

Dimension
  properties, setting, 105-7
  option, 616
  Style button, 616
  Style Manager dialog box, 106-7, 616-18
  Styles, 106-7
  toolbar. (See Toolbars, Dimension)
Dimensions
  command, 978, 985
  option, 615
  section, 472-75
  tab, 63-64, 69, 94-96, 101, 225-28, 231, 242-44, 252, 254, 260, 273-75, 277, 280, 287-89, 293, 299-300, 302, 330, 333-34, 345-46, 348, 350, 405-6, 411-15, 420, 429, 436, 438, 449-51, 456, 461, 469, 471-73, 482, 485-89, 519-21, 538, 567-69, 683-84, 699-702, 709, 722, 724, 745, 761, 839, 863-67, 879-81, 895, 902-3, 905, 908, 910, 916, 923, 926, 962-67, 973, 991-92, 994, 1001
Dimensions Measured to Inside Frame system, 274-75
DIMSCALE dimensioning variable, 501, 616
Display
  Configuration dialog box, 29-30, 37, 67, 73-74
  configuration, 16, 25, 28-31
  Configurations dialog box, 29-30, 37, 66-67
  Configurations, 36
  Contribution, 129-30, 453, 562-63, 704-5
  Control (By Set) dialog box, 28
  control. (See Desktop menu)
  Control tab
    (by Object), 37, 68, 75-76
    (by Set), 28
  Edit Schedule
    Data Dialog During Tag Insertion check box, 510, 649
    dialog box, 510-11
      Locked>OFF, 692
      Locked>NO, 696
      Locked>YES, 698
      Options page, 575, 577-79, 582
  Properties tab, 116, 129-30
  Props
    button, 494
    tab, 129-30, 144, 148, 184, 197, 208, 235, 239, 242, 248-51, 256, 287, 291-92, 326, 387, 449-53, 459, 490, 494, 561-63, 565, 637, 646, 703-5, 710, 712, 728, 879, 881, 1014
  Representation
    list, 68, 75, 453, 561-62, 704, 728
    sets, 25-28
  Representation Set
    dialog box, 26-27, 37-38, 68, 75-76
    drop-down list, 129, 506-7
  Representation Sets dialog box, 68, 74-76, 691-92
  representations, 25-26, 544
  tab, 500, 506-7
DISPLAYCONFIGATTACH command, 65-66, 72
DISPLAYCONFIGDEFINE command, 66-67, 73
DISPLAYSETDEFINE command, 68, 74
Distance edit field, 454-55

Distance to Line column, 966-67, 977, 992
Divide By
  check boxes, 957-60, 968, 987
  edit fields, 957, 987
Divide Spaces command, 858, 886-87, 892, 896, 944
Documentation menu, 6-7
Dome, 749-50
Door
  editing, 260-64
  inserting, 7. (See also Design menu)
  insertion point, 222-25
  object, 1
  Properties option, 9-10
  shifting within a wall, 263-64
  swing, setting, 221-22
  thickness, 245
Door & Window Tags command, 648-49, 654
Door Properties
  command, 260, 684
  dialog box, 225-27, 231, 260-63, 684
Door style category, 249
Door Style Properties dialog box, 242-44, 246-47, 252, 254-56, 258, 287, 308
Door Styles
  command, 240-42, 251, 255
  dialog box, 241-43, 251-52, 254-55, 258, 308-10, 312-14
Door Type section, 245-47, 252, 256, 308
DOORADD, 218-19, 239
DOORMODIFY command, 228
DOORPROPS command, 260
Doors/Windows/Openings toolbar. (See Toolbars, Doors/Windows/Openings)
DOORSTYLE command, 240
Doric Column, 750
Dormers, creating, 366-79
DOUBLE_JOIST wall style, 205
D-Overall Height dimension, 406-7, 420-21, 429, 438, 482-85
Drafting Settings dialog box, 70, 325-26, 338, 342, 350, 368, 384, 387, 513, 546, 656, 807, 934, 968
Drafting tab, 45
Drag Point
  button, 862, 867, 870, 873
  option, 868
Draw toolbar. (See Toolbars, Draw)
Drawing
  Default Display Configuration drop-down list, 506-7
  Defaults tab, 416-17
  radio button, 574-75
  scale factor, 576, 981
  Scale list, 500-1, 515, 546, 579
  setup. (See Desktop menu)
  Setup
    dialog box, 500-2, 504-7, 513-16, 546, 579, 981
    message box, 504
  Units drop-down list, 502-3, 515
  walls, 41-50
DRAWINGSETUP command, 105-6
Dry Run check box, 1018-20
D-Stringer Total Depth option, 450-51

DWGSCALESETUP command, 628, 630, 632, 635
Dynamic Posts check box, 472-73, 485-89

Edge Offset
button, 121, 124-25, 196-97
value, 136, 208, 211, 332
Edit
button, 26, 29, 36-37, 67-68, 73-74, 76, 114, 116-18, 130, 133, 158, 166, 168, 174, 183, 189, 196, 207, 210, 241-42, 248, 252, 255, 285, 290, 293, 306, 308, 448, 456, 506-7, 542, 551, 560-61, 579, 668, 699, 701, 775, 777, 793, 847, 878-79, 895
Door Style option, 9-10
Wall Style shortcut, 144, 148, 267
Edit Attributes dialog box, 598-99, 603, 605, 620, 631, 635
Edit Boundary Edges command, 858, 899, 920, 925-26, 944
Edit Display Props button, 129-30, 144, 184, 208, 237, 239, 249, 256, 291-92, 326, 387, 453, 459, 490, 562-63, 637, 646, 705, 710, 712, 729, 734, 881, 1014
Edit Multi-View Block Definition, 549
Edit Override button, 166
Edit pull-down, 819
Edit Roof Edges dialog box, 346-48, 375-77
Edit Roof Edges/Faces, 319
command, 346-47, 376
dialog box, 346-48
Edit Schedule Data dialog box, 647, 649-50, 653-56, 661-666, 677-78, 947
Edit Schedule Property dialog box, 677
Edit Stair Style shortcut, 490, 494
Edit Standard Size dialog box, 248, 290
Edit Table Cell command, 677-78
Edit Vertex button, 95
Edit Wall Style shortcut, 235, 239, 326, 386, 637, 646
Editing operations, 78-84
E-Door Thickness edit field, 245, 252, 255
E-Glass Thickness edit field, 293
E-Height of Space Above Ceiling Boundary, 865
Electric groups, 3
Element Modify, 769, 811, 983
Element Properties, 770, 818, 824, 839
Elevation edit field, 454-55
Elevation Line Properties, 699, 709
Elevation Mark Properties command, 698-99
Elevation Marks
command, 590, 601, 610
folder, 600
Elevation/Section Line Properties command, 727
dialog box, 728
Elevations, generating, 7. (See also Design menu)
Enable AEC Unit Scaling check box, 576-77, 581
End Elevation Offset edit box, 176-78, 186-87
END object snap mode, 56

End Offset
edit field, 477-78, 965, 991-92, 994
option, 953
Endcap Overrides, 163
Endcap Styles, 156-58, 167, 172
dialog box, 157-58, 161, 168-69, 172-73
Endcaps
creating, 154-58
drop=down list, 262-63
editing, 163-65
tab, 116-17, 119-20, 158-59, 168, 260, 262, 268, 277, 302
Endpoint object snap mode, 528, 656, 807, 968
E-Nosing Depth option, 450-51, 456-58
ENTDISPLAY command, 682, 702-3, 865, 1010
Entity Display
command, 682, 710, 712, 728, 733, 1010, 1014
dialog box, 702-4, 710-11, 728, 730, 733-35, 1014
Entity Properties dialog box, 130-31, 138-40, 144, 184, 197, 199, 208, 210, 237-39, 249-50, 256, 258, 326, 387, 453-55, 459-60, 490, 496, 563-64, 637, 646-47, 705, 710-12, 728-30, 733, 881-82, 1014
Entity Reference Properties dialog box, 832
ENTREF command, 800
Erase, 206
command, 202-3, 213, 612, 782, 788, 849, 919
E-Top Offset dimension, 406-7, 482-85
Existing option, 692
Exit option, 33, 823. (See also Viewer)
Expand button, 574-75
Explode, 494, 721
checkbox, 557
EXPLODE command, 84, 575
Explode on Insert check box, 575-76
Export
button, 12, 150-52, 154, 313, 556, 779
option, 877
Exporting wall styles, 148-52
EXPRESS menu group, 10
Extend, 199-200, 203-4
Extension edit field, 978
External Drawing list, 152, 780
External File list, 151, 154-55, 207, 311, 464-65, 480, 672, 780, 784, 894, 939
External Reference Attach, 684-86, 706-7, 717
External Reference dialog box, 685-87, 707-8, 717
Eye Level edit field, 1019

Face column, 332
Favorites, 509
File
menu, 5, 14, 50, 53, 57
Name edit field, 14, 313, 492-93, 577-78, 582, 738
to Import From dialog box, 150-51, 206-7, 310-11, 464, 480, 554-55, 672, 779, 783, 894, 939
FILLET command, 84, 170

Find
command, 509
dialog box, 509
Finish button, 582
Fire Rating Lines, 518-20
Fixture Layout command, 556-58
Flat Shaded. (See Shading Mode option)
Flip Hinge command, 224
Flip Swing command, 224-25
Floating Viewer button, 31, 44, 90, 94, 117, 119, 219-20, 271, 274, 323, 401, 403, 449, 468, 519, 525, 566-67, 744, 861, 901, 957, 962, 988
Floor Boundary check box, 861, 871-73
Floor Boundary Thickness, 904
Floor radio button, 333, 350
Floor Stops at Wall, 905
FLOORLINE command, 380, 382-84
Floors, creating additional, 193-205
Force Horizontal option, 576-77
Formats, schedule, 16
Foundation plan, creating a, 205-13
Frame section, 244-45, 287, 293
Frames Number, 1019-20
Frames Rate, 1019-20
Frames section, 1018
Free option, 928
Freeform option, 596-99
Freeze button, 326, 365, 660
F-Top Depth dimension, 406-7, 482-85
Full Frames Uncompressed, 1021
F-Width size, 247, 289

Gable
command, 755, 806-7
check box, 320-21, 337, 342, 350
roof, creating, 337-41
shape, 751
G-Bottom Offset dimension, 406-7, 482-85
General tab, 29-30, 94-95, 116-19, 210, 225-26, 242-44, 252, 255, 260, 273, 277, 287, 293, 299, 302, 330, 333, 405, 449, 469, 471, 519-20, 542-43, 561-62, 567-69, 579-80, 699-700, 731-32, 745, 761, 835, 863-67, 879, 902-3, 923, 962-64, 990-91
Generate
Elevation, 695, 709, 717
New Bubbles On Exit check box, 978, 981-82
polyline option, 380, 382
Section, 735
Slice command, 801, 841-42, 845, 852
View After Add, 1019
Walls command, 858, 943-44, 947
Generating
elevations, 7. (See also Design menu)
roofs, 7. (See also Design menu)
sections, 7. (See also Design menu)
G-Height size, 247, 289
Glass Thickness edit field, 288
Graft display representation, 65-66, 76
Grid, 41, 61-62. (See also Visual Aids option)

Grids/Colums toolbar. (See Toolbars, Grids/Columns)
Grips, 41, 60-61, 78-83, 223-24, 277-80, 348-50, 442-46
modifying mass elements with, 762-63
Group drop-down list, 43-44, 190, 538, 744, 755, 801-3, 806-7
Groups
Electric, 3
Plumbing Fixtures, 3
menu, 10-11, 50
symbol, 3-4
Guardrail
check box, 466, 472-73, 479, 485-89
default height, 469
option, 469

Half Landing radio button, 402-3. (See also Stairs, multi-landing)
Half Turn radio button, 402-3. (See also Stairs, multi-landing)
Handrail
check box, 466, 472-73, 479, 485-89
default height, 469
option, 469
Handrail Extensions at Landing options, 477
Hatch Pattern dialog box, 139-40, 197, 199
Hatch, 138-39
Hatching tab, 130-32, 139-40, 144, 197, 882
H-Bottom Depth dimension, 406-7, 482-85
Head Height edit field, 271, 280
Height
changing, 94
edit field, 43-44, 127, 210-11, 219, 230, 248, 271, 280, 332, 402, 411, 414, 434, 440-41, 489, 538, 699-700, 744, 755-56, 764-65, 772, 781, 787, 794, 806-7, 828-29, 849, 871, 900, 907, 933-34, 936, 962, 969, 983, 1012
of space boundary, 905
option, 868
Height of Space above Ceiling Boundary, 904, 916
Help
button, 44, 90, 219-20, 271, 320, 862, 900, 902, 957, 962, 988
menu, 8, 820
Topics option, 820
HI command, 309, 329, 341, 350, 364-66, 376, 379, 687, 698, 758, 795, 811
Hidden. (See Shading Mode option)
Hidden Line Projection command, 681-82, 714-17, 719, 722, 737, 789
HIDE command, 309, 329, 341, 350, 364-66, 376, 379, 698, 758, 811
Hinge location, changing, 224
History, 509
Horizontal Components section, 471-72
Hot grips, 79
H-Rise size, 247, 289

Ignore shrink-wrap effect, 1010
Image file, bitmap, 2

Images, 2
Imperial
  menu, 511
  radio button, 510-11, 515
Import
  button, 12, 150, 152, 464-65, 480, 672, 780, 784, 894, 939
  option, 877
Import/Export
  button, 114, 150-53, 158, 175, 190, 206, 242, 285, 306, 310, 312-13, 448, 463-64, 480, 543, 554, 560-61, 668, 670, 672, 777, 779, 783, 878-79, 892, 939
  dialog box, 150-52, 154-55, 158, 206-7, 310-12, 314, 463-65, 480, 543, 554, 556, 668, 670, 672, 777, 779, 783-84, 892, 894, 939
Import/Export - Duplicate Names Found dialog box, 151, 312-14, 939
Importing wall styles, 148-52
Index, 121
Insert Block, 494
INSERT command, 513
Insert dialog box, 557
Insert Options page, 575-77, 582
Insert Profile as Polyline command, 782, 784
Insert Units drop-down list, 493-94, 720-21
Insert Vertex, 95
Inserting
  doors, 7. (See also Design menu)
  symbols, 7. (See also Design menu)
  windows, 7. (See also Design menu)
Insertion Point
  edit fields, 761-63
  location, 98, 475-76
  option, 832
  section, 494, 526, 570, 702, 707, 717, 732
Inside Radius edit field, 958
Intersection object snap mode, 528, 656
Isometric view, 2
Isosceles Triangle, 751-52

J-Leaf size, 247, 289
Join Spaces command, 858, 886-89
J-Tr. Depth dimension, 407-8, 482-85
Justification, 211
  option, 402
  section, 467, 478
  setting, 45-47, 61-62, 440-41, 489
Justify
  changing, 94, 142
  drop-down list, 901, 907, 933-34, 936
  edit field, 43-44, 482-85, 538

K-R. Depth dimension, 407-8, 482-85

Labels in the X-Direction list box, 977
Landings. (See Stairs, multi-landing)
Layer
  flyout, 707, 717, 732
  Key Style list, 504-5, 515
  management. (See Desktop menu)
  standards, 16

Standards/Key File to Auto-Import edit field, 504-5
Wildcard edit field, 671, 673
Layer/Color/Linetype tab, 138, 256, 258, 453-54, 563, 882
Layer Name column, 576-77
Layer Properties Manager dialog box, 206, 326, 329, 365-66, 538-39, 660, 684, 686-87, 706, 790, 792, 795, 1004
Layering tab, 500, 504-6, 515
Layers
  button, 326, 328, 660, 684-86, 706, 790, 792, 795
  option, 181, 206, 947, 1004, 1020
Layout Mode command, 972-73, 983-84, 996, 1000-1
LAYOUTCURVEADD command, 954, 1007
Layouts
  selection, 21-25
  tabs, 18-21
  using, 17-21
Leaders command, 590, 603-4, 612
Leaf edit field, 248
Leave Existing option, 151, 939
Length
  changing the, 94
  edit field, 176, 178, 861, 871-73
  option, 867-68, 880
Lengthen command, 84
Libraries, symbol, 16
Line
  command, 50
  radio button, 901
Linear
  section, 502-3
  Types, 502
Linetype
  column, 208, 734
  dialog box, 734
  option, 618-20, 729
Lineweight
  column, 1014
  dialog box, 729-30, 734, 1014
  option, 729, 734
List, 1
LIST command, 572
List of Components section, 473-74
L-Nosing dimension, 407-8
Load, 509
Load Center Palette dialog box, 509
Location
  list, 492-94, 720-21
  tab, 97-98, 337, 409, 471, 475-76, 526, 568-70, 702, 731-32, 736, 760-62, 770, 835-36, 841, 963, 967-68, 990, 993
Lower Extension edit field, 699-700
Lower rails option, 469

Make Element Additive command, 801, 813
Make Element Intersection command, 801, 815-16
Make Element Subtractive command, 801, 813-15
Manage Contained Spaces check box, 900, 903, 907, 909
Management, layer. (See Desktop menu)
Manual mode, 972-73, 1000-2, 1008

Mask Block Definition Properties dialog box, 561-62, 565, 579-80
Mask Block Properties dialog box, 567-70
Mask Blocks dialog box, 560-61, 564-65, 579, 581
MASKADD command, 565-66
MASKATTACH command, 571
MASKDEFINE command, 560
MASKDETACH command, 572-73
Masking Block radio button, 574-75, 581
Masking blocks
  attaching objects to, 571-73
  changing, 567-71
  creating, 558-65
  inserting, 565-67
MASKMODIFY command, 567
Mass Element as Column Properties dialog box, 962
Mass Element Properties dialog box, 745, 761-63, 770-71, 819, 824, 839, 841
Mass Elements
  creating with Model Explorer, 824-26
  toolbar. (See Toolbars, Mass Elements)
Mass Group Properties dialog box, 819
Mass Group tab, 761-62
Mass Tools toolbar. (See Toolbars, Mass Tools)
MASSELEMENTADD command, 741, 745-46
MASSELEMENTMODIFY command, 741, 759, 975
MASSELEMENTOPINTERSECT command, 815
MASSELEMENTOPSUBTRACT command, 813
MASSELEMENTPROPS command, 741, 760
Mass-Group layout tab, 17-18, 34-35
MASSGROUPADD command, 361, 801
MASSGROUPATTACH command, 362, 800, 803
MASSGROUPDETACH command, 800, 805
Match
  button, 44, 90-91, 219-20, 271, 319, 323, 401, 403, 468, 519, 526, 566-67, 745, 862, 901, 957, 962, 988
  command line option, 524, 868
  lines, 620-23
Maximize Window button, 853
Maximum
  Area, 895
  Length, 895
  Width, 895
Measure to Center check box, 176, 179
Measure to Inside of Frame option, 287
Measure to Outside of Frame check box, 227-28, 231, 274, 280 system, 275
Measured Normal to: Roof or Floor, 333-34, 354, 356
Menu
  bar, 6
  Concept, 6-7
  Design, 6-9
  Desktop, 6-7
  Documentation, 6-7

File, 818-19
File, SaveAs, 5, 14, 50, 53, 57, 70, 79, 81, 84, 92, 100, 103, 133, 181, 195, 206, 230, 251, 264, 280, 292, 307, 313, 325, 358, 363, 419, 427, 443, 455, 480, 494, 515, 528, 546, 579, 606, 637, 654, 672, 683, 693, 719, 732, 755, 764, 771, 783, 789, 806, 826, 836, 847, 871, 890, 907, 915, 930, 944
groups, 10-11
Help, 8
shortcut, 8-10, 14
Utilities, 8-9
Walls, 8
Merge Boundaries command, 858, 910-12, 915
Metric
Content submenu, 3. (See also Design Content)
  option, 511
  radio button, 510
Midpoint object snap, 807
Minimum
  Area, 895
  Length, 895
  Width, 895
MIRROR
  command, 443
  editing operation, 78-79
Mirror X check box, 176, 179
Mirror Y check box, 176, 179
Misc option, 34, 824. (See also Viewer)
Miscellaneous
  button, 638
  command, 590, 614
Model Explorer
  toolbar. (See Toolbars, Model Explorer)
  window, 816-26
Model layout tab, 18
Modifier Style drop-down list, 176-77, 186-87
Modify Boundary
  command, 909-10, 921, 930
  dialog box, 910
Modify
  button, 617
  shortcut, 938
Modify Ceiling Grid
  command, 954, 996-97, 1004
  dialog box, 997, 1004
Modify Column command, 975-76
Modify Column Grid
  command, 963
  dialog box, 963
Modify Dimension Style dialog box, 106-7
Modify Dimension Style: Aec_Arch_I dialog box, 617
Modify Door command, 9-10, 228-29, 239
Modify Doors dialog box, 228-29, 239, 254-55, 258
Modify Mask Blocks dialog box, 567-67
Modify Masking Block command, 567
Modify Mass Element
  command, 741, 746, 759-60, 803, 975-76
  dialog box, 759-61, 769, 803, 811, 976, 983
Modify Multi-View Block command, 523-24

Modify Multi-View Blocks dialog box, 524-26
Modify Opening dialog box, 302
Modify Railing
　command, 399-400, 478
　dialog box, 479
Modify Roof
　command, 319, 336
　dialog box, 336-37, 365, 719
Modify Space dialog box, 883, 896, 898, 938-40
Modify Space Boundary dialog box, 910
Modify Spaces command, 858, 882-83, 886
Modify Stair
　command, 399, 440-41
　dialog box, 440-41, 461
Modify toolbar. (See Toolbars, Modify)
Modify Wall command, 41, 88-94, 195
Modify Window
　command, 276
　dialog box, 296
Modify Windows dialog box, 277
More option, 33, 38, 820, 823. (See also Viewer)
Move command, 429, 432, 445, 683, 757, 795, 840, 846, 849, 868, 989, 995, 1006
MOVE editing operation, 78-79
M-Riser dimension, 407-8, 412-15, 420-422, 429, 438, 461
MTEXT, 501
Multi-landing, 401-2. (See also Add Stairs dialog box)
Multi-View Block Definition Properties dialog box, 542-45, 549-51, 553
Multi-View Block Definitions dialog box, 542-43, 553-54, 556
Multi-View Block Properties, 690, 694, 708, 733
Multi-View Block radio button, 574-75
Multi-View Block Reference Properties dialog box, 518-23, 526-27, 531, 533, 535-36, 538, 540, 690, 694-95, 708-9, 733
Multi-view blocks
　editing, 523-27
　inserting and modifying, 516-20
　placement, 520-23
Multiple Document Environment. (See AutoCAD 2000)
MVBLOCKADD command, 518-19
MVBLOCKDEFINE command, 541-42, 554
MVBLOCKMODIFY commannd, 523-24
MVBLOCKPROPS command, 523, 526, 689-90, 724

Name, default drawing, 5
Name
　column, 451-52, 456
　command line option, 524
　dialog box, 115-16, 133, 167, 182-83, 196, 207, 210, 242, 251, 255, 285-86, 292-93, 307-8, 448-49, 455-56, 542, 551, 561, 564, 579, 668, 774-75, 777, 780, 786, 793, 847, 879, 894, 1019
　edit field, 116-17, 122, 135, 196, 208

list, 68, 130, 519, 525, 553, 567
value, 136
Net to Gross Offset option, 866-67, 880, 895
Never Use Characters radio button, 978
New. (See Standard toolbar)
New
　button, 113, 115, 133, 150, 153, 158, 167, 174, 182, 182-83, 189, 241-42, 251, 255, 285-86, 306-8, 313, 448-49, 455, 464, 542, 551, 560-61, 564, 579, 668, 684, 706, 774, 777, 780, 786, 790, 793, 847, 878-79
　option, 692
New Drawing File dialog box, 150, 153, 313, 464
New Element command, 818-19, 825, 849
New Grouping command, 819, 826-27, 850
New Icon button, 577-78
Next button, 574-75, 581, 582
Next Turn
　button, radio, 402-3
　section, 401-3, 420, 423-24, 426, 429, 433-35, 482-85
Node object snap mode, 59-60, 872, 932, 934, 968
None
　option, 576
　radio button, 581
Normal
　edit fields, 409, 475-76, 526, 570-71, 702, 732, 761-63
　value, 98
Z column, 409
North Arrows, 623-25
Note: Dimensions Measured to Inside Frame, 231
Notes
　button, 94, 116-18, 244, 287, 330, 405, 449, 471, 520, 543, 561, 569, 699, 732, 761, 863, 879, 903, 991
　dialog box, 116-18, 244, 287, 330, 405, 449, 471, 520, 543, 561, 569, 699, 732, 863, 879, 903, 991
N-Tread dimension, 408, 412-15, 420, 422, 429, 438, 461, 482-85
Number column, 977, 981-82, 985, 992

ObjectARX technology, 1-2
Object
　schedules, creating, 647-49
　Snap tab, 41, 85-86, 338
　tags, creating, 647-49
Type list, 37-38, 68, 75-76, 506-7
Viewer button, 2, 226-27
Object Properties
　dialog box, 577
　toolbar. (See Toolbars, Object Properties)
Object Snap
　check box, 325-26, 932
　toolbar. (See Toolbars, Object Snap)
Object Snap On (F3) check box, 338, 342, 350, 368, 384, 387, 807, 934, 968
Object Snap Tracking On (F11) check box, 807, 934
Object Tags command, 660-62
Object Viewer
　shortcut menu, 822-24
　toggle, 820

toolbar. (See Toolbars, Object Viewer)
Objects, 1
　radio box, 492-93, 720-21, 738
Offset
　column, 454-56, 459
　edit field, 43-44, 55-56, 472, 477, 479, 485-89, 522-23, 538, 901, 907, 933-34, 936
　option, 41, 45, 263, 380, 382
OFFSET command, 76-78, 86
Offset From Insertion section, 522-23, 532-33
Offsets list, 519-23, 530, 532-33, 535-337, 540
Open
　a Drawing button, 3
　button, 150, 152, 207, 310-11, 464, 480, 554-55, 685-86, 717, 779, 783, 892, 894, 939
　check box, 401
　dimension, 408
　Drawings, 509
Open % edit field, 275, 280, 683-84
Open Risers check box, 441, 482-85
Open Tread check box, 440-41
Opening Endcap Style, 119-20 button, 119-20, 268
Opening (percent) edit field, 219-20, 230, 239
Opening Percent command, 229-30
Opening Properties dialog box, 299-300, 302-4
OPENINGADD command, 298-300
OPENINGFLIPHINGE command, 224
OPENINGFLIPSWING command, 224-25
OPENINGPERCENT command, 229-30
Openings, creating, 298-309
Operand setting, 124
Operation drop-down list, 744, 760, 828-29
Operations edit field, 760
Operator setting, 124
OPTIONS command, 12, 16, 416, 419, 427
Options
　dialog box, 14, 45, 63, 399, 416-17, 419, 429-28, 469-70, 472-73, 480, 515-16, 865
　shortcut, 515
OR command, 52
Orbit option, 33, 820, 823. (See also Viewer)
Ortho Close, 41, 44, 52-56, 901
Orthographic view, 2
Other tab, 130-32, 139, 141-42, 184, 453-55, 459
Overall Size section, 991
Overhang
　check box, 319, 323, 327, 337-38, 343
　edit field, 323
Overhang toggle. (See Overhang check box)
Override Ending Priority button, 166
Override Starting Priority button, 166
Overwrite Existing option, 151

Pan
　command, 9, 324, 692, 696-97, 711, 736, 820, 823
　cursor, 821
　option, 33. (See also Viewer)
Parallel option, 823
Paste command, 9
Paste
　command, 819
　operation, 117
Path Name dialog box, 1018
Pattern column, 139-40, 197, 199
PEDIT command, 182, 619
Perpendicular object snap, 86
Perspective
　option, 823, 853
　view, 2
Pick Path< button, 1018-19
Pick Point< button, 1018
Pick Points option, 652
Plan
　Constraints, 864
　option, 716
Plate Height edit field, 321, 327, 338, 343, 368
Plot configurations, 16
Plot-FLR layout tab, 18, 38
Plot-RCP layout tab, 18
Plot-SEC layout tab, 18
Plumbing Fixtures group, 3
Polar distance edit field, 934-35
Polar Snap, 934-35
Polar Tracking
　tab, 934
　toggle, 934
POLYGON command, 100
Polyline
　command, 656
　Close, 41, 44, 50, 52, 901
Position Within (Y) section, 262
Post Extrusion Above Top Railing edit field, 472-73, 485-89
Posts
　check box, 466, 472-73, 479, 485-89
　option, 469
Precision drop-down list, 503-4, 515
Predefined Shape radio button, 961-62, 1012
Preset Elevation edit field, 575-76
Preset Views option, 33, 820, 823. (See also Viewer)
Preview button, 509, 513
Primary Units tab, 617
Priority
　edit field, 121-22, 166, 208
　number, 96, 135
Overrides list, 163, 165-66
　value, 136
Priority Override dialog box, 166
PRise option, 322
Profile
　creating a, 12-34
　drop-down list, 772, 781, 787, 794, 849
　edit field, 771
　option, 744
　shapes, 16
Profile Definitions dialog box, 782-84
Profile Definition Properties dialog box, 775-78, 780, 786, 793, 847
Profile Name
　drop-down list, 473-75, 486, 489
　section, 473

Index 1031

PROFILEASPOLYLINE command, 782
PROFILEDEFINE command, 774
Profiles
  creating new, 773-79
  dialog box, 305-8, 473-74, 774-78, 780-81, 784, 786-87, 793-94, 847-48
  tab, 12-14
Project option, 380, 382
Projection option, 33, 823. (See also Viewer)
Prompt column, 523
Properties
  button, 44, 90, 219-20, 225, 231, 271, 277, 280, 299, 319, 323, 330, 333, 337, 401, 403-4, 411, 414, 429, 436, 441, 461, 469, 482, 485, 489, 518-20, 532-33, 535-38, 540, 567-68, 745, 862-63, 901-2, 907, 910, 957, 962-64, 988, 990, 994
  command, 819, 976
  dialog box, 835
  option, 832, 835
  Property Source, 129-30, 144, 184, 208, 249, 256, 453, 526, 530, 562-63, 704, 710, 712, 733, 881, 1014
PROPERTYDATAEDIT command, 664-65
PROPERTYSETDEFINE command, 648
PSlope option, 322
Purge button, 114, 152, 158, 175, 189, 241, 285, 306, 448, 465, 543, 560-61, 668, 776, 878-79
Purge Multi-View Block Definitions dialog box, 543
Purge Profiles dialog box, 306, 776
Purge Stair Styles dialog box, 465
Purge Wall Styles dialog box, 152-53
Pyramid, 751

Quarter Landing radio button, 401-2. (See also Stairs, multi-landing)
Quarter Turn radio button, 401-2. (See also Stairs, multi-landing)

>R-Radius edit field, 958
Radius
  changing the, 94
  cleanup, 41, 62-66
  edit field, 744, 962, 983, 1012
  property, 332
Radius Divide by check box, 958
RAILING command, 466
Railing Properties dialog box, 471-72, 475-77, 485-89
Railing Rules tab, 405-6, 408-9, 477-78
Railing Settings section, 416-17, 469
RAILINGADD command, 399, 466
RAILINGMODIFY command, 399, 478
RAILINGPROPS command, 470-71, 475
RE command, 69, 364
RECTANG command, 100, 791
Rectangle option, 596-99
Redistribute Posts check box, 477
Redo command, 819
Redo View option, 34, 824. (See also Viewer)

Reference Doc tab, 94, 117-18
Reference Docs tab, 244, 287, 330, 449, 471, 543, 561, 699-700, 732, 761, 863, 879, 903, 991
Reference Element/Group command, 800-1, 831-35, 837
Reference File tab, 405
Reference Type radio button, 685-86, 707, 717
Regen list, 1018-19
  REGENALL command, 75
  REGENERATION command, 364
Relative to-Current Coordinate System, 475-76
Relative to-World Coordinate System option, 475-76
Remove a Hole option, 997, 999
Remove Boundary Edges command, 858, 899, 920, 925-27, 938
Remove button, 121-22, 176-77, 210, 248, 291, 544, 699, 702
Remove Grid Line command, 972, 974, 996, 1002-4
Remove Override button, 129, 166, 249, 291-92, 453, 562-63, 881
Rename
  button, 12
  command, 819, 839
  option, 824
Rename to Unique option, 151-52
Render Preferences dialog box, 34
Rendered images, 2
Rendered. (See Shading Mode option)
Renumber Data command, 674-75
Repeat Bay Size radio button, 965, 992, 994
Repeat mode, 972-73, 1008
Reposition Along Wall command, 268-69
Reposition Within command, 263
Reposition Within Wall command, 263-66, 269
REPOSITIONALONG command, 268-69
REPOSITIONWITHIN command, 263
Representation Sets, 37
Reset
  button, 12
  View option, 33, 823. (See also Viewer)
Resize button, 826
Reverse Wall command, 99
Reverse Wall Start/End command, 99
Revision Clouds, 626-29
  command, 590, 627, 642
Revolution and Extrusions, 754
Right Reading option, 576-77
Right Triangle, 752-53
Rise
  edit field, 248, 275, 321-22, 325, 327, 332, 338, 343, 365, 719, 744, 755-56, 806-7
  Lower edit field, 412-15, 419, 422, 427, 480
  Upper edit field, 412-15, 419, 422, 427, 480
Roof,
  editing with grips, 348-50
  flat, creating a, 386-95
  gamble, 350-57
Roof Edit Edges/Faces command, 346-47, 374

Roof Faces (by Edge)
  list box, 332-34
  section, 330, 332-35, 345-46, 375-76
Roof Modify, 719
Roof object, 1
Roof Properties
  command, 333-34, 345-46, 348
  dialog box, 319, 330, 333-37, 345-48, 354-56
Roof radio button, 333
Roof Thickness (all Faces) edit box, 333-34, 354
Roof/Floor Line tab, 95-96
ROOFADD command, 319-20, 323
ROOFCONVERT command, 358
ROOFEDITEDGES command, 346-47
ROOFMODIFY command, 336
ROOFPROPS command, 333-34
Roofs toolbar. (See Toolbars, Roofs)
Roofs, generating, 7. (See also Design menu)
Room & Finish Tags command, 651-53, 658, 943, 947
Room labels, generating, 943
ROTATE editing operation, 78-79
Rotation
  Angle edit filed, 702
  command line option, 524-25
  edit field, 97-98, 334, 473-76, 519, 525, 529, 530, 533, 535, 537, 540, 547, 566-68, 685-86, 707, 717, 732, 736, 770
  section, 527, 570-71
Run dimension, 333-34

Sash section, 288
Save As Default check box, 501, 507
Save button, 14, 153, 313, 582, 719
Save Content File dialog box, 577, 582
Save Drawing As dialog box, 14
Save In list, 14, 313, 1020
Save Preview Graphics check box, 578-79
SaveAs command, 819, 836
Scale
  edit field, 519, 566-68, 576, 685-86, 707, 717
  editing operation, 78-79
  model-spcae object in current drawing to reflect new units check box, 504
  Objects Inserted from Other Drawings check box, 503
  section, 525, 576
  tab, 500-1
Scale/Spacing edit field, 199
SCaled insertion method, 594
Scaled option, 594
Scan Block References check box, 671
Scan Xrefs check box, 671
Schedule - Imperial toolbar. (See Toolbars, Schedule - Imperial)
Schedule - Metric toolbar. (See Toolbars, Schedule - Metric)
Schedule formats, 16
Schedule Table Style
  list, 671, 673
  Properties dialog box, 647, 668

Schedule Table Styles
  command, 666-68, 670, 672
  dialog box, 666-68, 670, 672-73
Schedules, 1
  creating, 647-49
Section/Elevation
  Line Properties dialog box, 699, 702-3, 709, 722, 724
  toolbar. (See Toolbars, Section/Elevation)
Section Marks, 629-32
  command, 590, 629, 631, 643
Section Properties
  command, 735
  dialog box, 731-32, 736
Sections, generating, 7. (See also Design menu)
Segment option
  curved, 44
  straight, 44, 47
Segment Type section, 903, 907
Segments, 332
Select A Block dialog box, 544-45, 549, 551-52
Select a Template list, 50
Select All button, 116-18
Select an Endcap Style dialog box, 121, 160, 165, 168, 173, 268
Select Color dialog box, 711
Select Layer Key
  button, 576-77, 581
  dialog box, 577-78, 581
Select Linetype dialog box, 208, 210, 729
Select Objects button, 720-21, 738
Select Reference Document File dialog box, 117
Select Reference File dialog box, 684, 686, 706-7, 717
Select Views dialog box, 583-84
Selected Roof Edges
  list box, 332
  sections, 330, 334-35, 346, 354
Set a Boundary option, 997
Set Boundary< button, 988-89, 994
Set Current button, 12, 15, 19, 50, 57. (See also Profiles tab)
Set Drawing Scale command, 594, 616, 978, 981
Set From< button, 114, 158, 167, 172, 174, 176, 183, 189, 241, 285, 306-7, 448, 542, 560-61, 564, 579, 668, 775-76, 778, 780, 793, 848, 878-79
Set Slice Elevation command, 845-46
Set View option, 34. (See also Viewer)
Shaded images, 2
Shading Mode option, 33, 820, 823. (See also Viewer)
Shape
  drop-down list, 320-21, 420, 425, 429, 434, 439, 440, 482-85, 742-44, 746, 755, 771-72, 781, 787, 794, 828-29, 849, 957, 959-60, 968, 983, 1012
  section, 245-46, 252-53, 256, 288, 293
SingleSlope option, 325, 327, 338, 343
Shapes, profile, 16
Shift to Add, 79
Shortcut menu, 8-10, 14
Show All Layers, 820

Show Model Explorer command, 816-17, 826, 836, 847
Shrink Wrap, 138-39
Size
  edit field, 454-55
  list, 219, 270
  option, 868
Size and Placement section, 903
Slicing floorplates. (See Concept menu)
Slope value, 322, 332, 346, 375
Snap and Grid tab, 934
Snap, 61-62
Solid Form radio button, 900, 903, 907, 933-34, 936
Solid Slab option, 450-51
Source section, 492-93
Space boundary height, 903-4
Space Boundary Properties dialog box, 901-3, 907, 910, 913, 923
Space Evenly mode, 972-73, 1008
Space Height
  edit box, 861, 872-73
  value, 904
Space Info Total tab, 940-41
Space Information
  dialog box, 940-41, 943-46
  tab, 940-41, 944, 946
Space Inquiry command, 858, 940, 944
Space Lines Evenly radio button, 965, 992
Space Modify shortcut, 895, 898, 938-40
Space planning. (See Concept menu)
Space layout tab, 18, 34-35
Space Planning
  shortcut, 871
  toolbar. (See Toolbars, Space Planning)
Space Properties dialog box, 652, 862-64, 905-6, 908, 916-17
Space Style Properties dialog box, 878-81, 895
Space Styles
  command, 857, 863, 877-79, 892, 939
  dialog box, 877-79, 892, 894-95, 939
Space Tag option, 943, 947
SPACEADD command, 651, 857, 860
SPACEBOUNDARYADD command, 858, 900
SPACEBOUNDARYADDEDGES command, 858, 924
SPACEBOUNDARYANCHOR command, 858, 928-29
SPACEBOUNDARYCONVERT command, 858
SPACEBOUNDARYCONVERTEDGE command, 923
SPACEBOUNDARYCONVERTSLICE command, 921-23
SPACEBOUNDARYCONVERTSPACE command, 920, 989
SPACEBOUNDARYEDGE command, 858, 925
SPACEBOUNDARYGENERATEWALLS command, 858, 943-44
SPACEBOUNDARYMERGE command, 911
SPACEBOUNDARYMERGESPACE command, 858, 912
SPACEBOUNDARYMODIFY command, 858, 909

SPACEBOUNDARYPROPS command, 902
SPACEBOUNDARYREMOVEEDGES command, 858, 927
SPACEBOUNDARYSPLIT command, 858, 914
SPACECONVERT command, 651-52, 858, 884-85
SPACEDIVIDE command, 858, 886
SPACEJOIN command, 858, 888
SPACEMODIFY command, 858, 883
SPACEPROPS command, 862
SPACEQUERY command, 857, 940
SPACESTYLE command, 857, 878
SPACESWAP command, 942
Spacing column, 966, 983, 992
Specify On-screen
  check box, 494, 519, 566-68, 685-86, 707, 717, 746, 767, 861, 868, 871, 883, 957-60, 968, 987, 989, 994
  option, 867-68, 953
Sphere, 753-54
Spiral, 401-2. (See also Add Stairs dialog box)
Split Boundary command, 858, 913-14, 918, 920
Stair Properties
  command, 399
  dialog box, 400, 402, 404-6, 408-16, 420-22, 425, 429, 429, 435-36, 438-39, 461-62, 477-78, 482-83
Stair Settings section, 416-17, 427
Stair Styles
  command, 399-400, 446-48, 464
  dialog box, 447-52, 455-59, 463-65, 480, 490, 494, 496
Stair Values section, 406-7
Stairs
  multi-landing, 433-35
  object, 1
STAIRADD command, 399-400
STAIRMODIFY command, 399, 441
STAIRPROPS command, 399, 405, 409
Stairs Modify command, 461
Stairs/Railings toolbar. (See Toolbars, Stairs/Railings)
STAIRSTYLE command, 399, 447
Standard Sizes section, 226-27, 254, 274
Standard Sizes tab, 242, 247-48, 252, 256, 287, 289-91, 293
Standard toolbar. (See Toolbars, Standard)
Standards, layer, 16
Start
  button, 3, 13
  from Scratch button, 3
Start Elevation Offset edit box, 176-77, 186-87
Start Offset
  edit field, 477-78, 965, 991, 994
  option, 953, 992
Start Position Offset edit box, 176-77, 186-87
Startup dialog box, 3-5, 13
Status Bar toggle, 820
Stop, 245
Straight
  segment toggle option, 47
  walls, creating, 47-48
Stretch command, 490, 784

STRETCH editing operation, 78-79
STretched insertion method, 594-95
Stretched option, 594
Stringer Alignment dialog box, 456, 458-59
Stringer
  option, 450-51
  radio button, 456, 458-59
Stringer Values section, 450-51, 456, 458-59
Structural grids, creating, 954-58
Style
  drop-down list, 43-44, 111-12, 114, 142, 168, 205, 210-11, 219, 239, 254, 270, 280, 283, 296, 308, 401, 405, 420-21, 429, 433, 439-41, 461, 860, 872-73, 896, 898, 938-40
  option, 867
  tab, 94-95, 225-27, 262, 273-74, 277, 405, 519-21, 567, 569, 863
Styles, 16
Subtractive shrink-wrap effect, 1010
Suffix edit box, 503-4
Sun icon. (See Thaw button)
Swap command, 942-43, 946
Symbol
  groups, 3-4
  inserting, 7, 512-16. (See also Design menu)
  libraries, 16
  option, 594
System Default category, 129-31, 144, 249

TABLEADD command, 670
TABLECELLEDIT command, 677-78
TABLESTYLE command, 666-67, 670
TABLEUPDATENOW command, 675-76
Tabs, layout, 18-21, 34
Tags, renumbering, 674-75
Target
  Area, 880, 895
  Length, 880, 895
  Path list, 1018-19
  Width, 880, 895
Technology, ObjectARX, 1-2
Templates, 4-5, 16-25, 57, 59
Temporary Tracking Point, 69-76, 85-86
Text Docs tab, 244, 287, 330, 449, 699, 879
Text Notes tab, 116-18, 405, 471, 543, 761, 863, 903
Text tab, 94, 732
Thaw button, 328, 366, 539
TILEMODE, 17
Title Marks, 632-36
  command, 590, 632-33, 644
  pallete, 635-36
TOOLBAR
  command, 10, 13-14, 42, 50, 511-12, 590
  list, 656
Toolbars
  AEC Anchors, 11
  AEC Layer Utilities, 11
  AEC Layout Tools, 11, 1011
  AEC Plan Rotation, 11
  Annotation - Imperial, 11, 589-91, 600-1, 603-4, 606-8, 610, 612, 614-15, 631, 637-38, 642-44

Annotation - Metric, 11, 589-91, 600-1, 603-4, 614-15, 631
Camera, 11, 1015-16, 1019-20
Design Content - Imperial, 11, 511-13, 529, 530-31, 533, 535-36, 546, 549, 661
Design Content - Metric, 11, 511-12, 661
DesignCenter, 509
Dimension, 106, 436, 615-16
Doors/Windows/Openings, 11-15, 228, 230, 238-39, 242, 251, 254-55, 258, 271, 285, 292, 296, 300, 308-10, 313
Draw, 10, 13, 494, 656, 791
Grids/Colums, 11, 955, 959-60, 968, 971, 976, 979, 981, 988, 994, 997, 1004, 1006, 1012
Mass Elements, 11, 13, 742-43, 746, 754-55, 764-65, 767, 806-7, 975
Mass Tools, 11, 13, 364, 376, 801, 803, 805-6, 810, 812, 837, 841, 852
Model Explorer, 817, 819, 825-27, 830, 849-50
Modify, 10, 13, 76, 83-84, 86, 170, 199, 201-3, 206, 213, 429, 490, 612, 721, 757, 766, 782, 784, 788, 795, 849, 919, 995, 1006
Object Properties, 10, 181, 206, 326, 328, 365, 538-39, 660, 684-86, 706-7, 717, 790, 792, 795, 947, 1004, 1020
Object Snap, 69-70
Object Viewer, 817, 820
Reference, 684-86, 706, 717
Roofs, 11, 327, 359, 374, 376
Schedule - Imperial, 11, 589, 648-49, 651, 658, 660-1, 663, 672-73, 676, 943
Schedule - Metric, 11, 589, 648, 651-52, 660-1, 663, 676
Section/Elevation, 11, 690-91, 693, 707, 719, 722, 731, 733
Space Planning, 11-12, 656, 859, 871-72, 874, 877-78, 883-84, 886-87, 891-92, 896, 899-900, 907, 909-11, 913, 915, 918, 924, 926, 930, 933-34, 936, 938-39, 944
Stairs/Railings, 11, 13-15, 400, 411, 414, 425, 433, 436, 447-48, 459, 463-64, 480, 482, 485
Standard, 10, 38, 50, 169, 173, 230, 300, 309-10, 324, 327, 329, 341, 354, 363-64, 368, 376, 379, 394, 429, 443, 445, 494, 528, 531, 606, 608, 637, 684, 686, 692, 696-98, 706, 719, 721, 735-38, 766, 768, 771, 790-92, 795, 849-51, 917, 930, 983, 996, 1004, 1019, 1021
Tree View, 820
View, 692
Walls, 11-15, 42-43, 50-51, 53, 57, 84, 91-92, 100, 103, 105, 112, 115, 130, 133, 152-52, 158, 166-68, 168, 172, 184, 196, 206, 210-11, 230, 539, 614, 1010, 1013
Toolbars dialog box, 10-11, 14, 42-43, 50
Top check box, 978
Top Elevation Offset, 127-28, 208, 211
Tread
  edit box, 402, 440-41

Index 1033

Lower edit box, 412-15, 419, 422, 427, 480
Upper edit box, 412-15, 419, 422, 427, 480
Tree View
  shortcut menus, 824
Toggle, 509
toolbar. (See Toolbars, Tree View)
TRIM command, 87-88, 201-2, 927
TT command, 69
Type
  drop-down list, 199, 455, 503, 515
  option, 451-52, 594

U Shaped, 401-2. (See also Add Stairs dialog box)
U Shaped Winder, 401-2. (See also Add Stairs dialog box)
Uheight option, 322
Undo
  button, 44, 219-20, 271, 320, 323, 325, 401, 403, 469, 566-68, 745-46, 862, 900, 902, 957, 962, 980, 988
  command, 819
  operation, 117
  option, 48
View option, 34, 823. (See also Viewer)
Units, 738
  tab, 500-4, 515
Up, 509
Update Elevation command, 681, 713-14, 719
Update Schedule Table command, 675-76
Update Section command, 681-82, 713, 722, 730
Upper Height edit field, 322
URise option, 322
Use a
  Template button, 3-5, 13, 50, 53, 57, 112, 115, 310
  Wizard button, 3
Use Base Width check box, 124-25, 135, 196-97
Use Drawn Size check box, 176, 178, 186-87
Use Model Extents for Height check box, 699-700, 728
USlope option, 322
Utilities menu, 8-9

V command, 19, 50
Value
  column, 523
  edit field, 690, 695, 708, 733
Vertex button, 95
Vertical Components section, 471-72
Video COmpression dialog box, 1021
View

Blocks
  list, 521-23, 531-33, 544
  tab, 542-44, 549
  command, 50
  dialog box, 19
Direction section, 544
flyout, 38, 169, 173, 251, 300, 309, 329, 341, 343, 354, 363-64, 368, 376, 379, 394, 443, 445, 494, 606, 608, 686, 692, 736, 996, 1004, 1019
pull-down menu, 819-20
Viewer, 31-34, 117, 119, 226-27, 271, 274, 320, 323, 449, 468, 519, 525, 567, 744, 901, 957, 988
Viewport Display Configuration dialog box, 28-29, 36, 72
Viewports Scale list, 698
Viewports, 17-21, 698
Views, 2, 509
Visual Aids option, 33, 823. (See also Viewer)
Volume section, 502-4
VPORTS command, 17
-VPORTS command, 19, 692-93

Wall Cleanup Radius, 63, 69, 539
WALL command, 189
Wall Endcap Style, 119-20
  button, 119-20, 160, 168
Wall error marker, 64-66, 69
Wall Interference command, 139
Wall Modifier Style Properties dialog box, 183
Wall modifier styles, 174-88
Wall Modifier Styles
  command, 174, 180
  dialog box, 174-76, 180, 182-84
Wall Modifiers
  dialog box, 186-87
  tab, 96-97, 176, 184
Wall Modify
  command, 90, 141, 190, 210
  dialog box, 89-91, 94, 141-43, 190, 192, 205, 210
Wall object, 1
Wall Properties
  button, 44
  command, 94, 161, 163
  dialog box, 44, 63-64, 69, 100, 173, 175-79, 184, 187, 190, 503, 538-39
  setting, 94-100
  shortcut menu, 63-64
  tabs, 94-98
Wall Stops at
  Ceiling, 905
  Floor, 905
Wall style name, creating, 114
Wall Style
  category, 129
  dialog box, 183-84
Wall Style Overrides, 163
  dialog box, 96-97
  tab, 96, 163-65, 173

Wall Style Properties dialog box, 116-17, 122, 127, 130, 133, 136, 144-48, 160-61, 166-68, 184, 199, 207, 210-11, 235, 238-39, 268, 386-87, 637, 646-47
Wall Style Property Source, 197
Wall Styles, 42, 115, 130, 133, 196, 206
  command, 113, 152-53, 158, 166, 168, 210
  dialog box, 113-16, 130, 133, 136, 150, 152-53, 158, 160, 167-68, 183-84, 196, 199, 206-7, 210-11
  editing, 116-30
  importing and exporting, 148-52
  purging, 152-53
Wall Tags command, 663-64
WALLENDCAP command, 160-61
WALLINTERFERENCEADD command, 954, 1009
WALLMODIFIER command, 179-81
WALLMODIFIERSTYLE command, 174
WALLPROPS command, 94, 176, 190
Walls
  check box, 42-43
  justifying, 45-47
  menu, 8
  straight, 47-48
  toolbar. (See Toolbars, Walls)
Warm grips, 79
Warning: Calculation Limits Check Failure message box, 410-11
WBLOCK command, 492, 715, 717, 720, 738
Width, 121
  button, 135, 196-97
  changing the, 94
  Dimension (X-Axis), 973, 991, 994
  dimension for a frame, 245
  edit field, 43-44, 112, 124-25, 136, 208, 219, 230, 232, 248, 270, 280, 402, 429, 434, 439-41, 473-75, 482-86, 489, 744, 755-56, 764-65, 787, 806-7, 828-29, 839, 861, 871-73, 901, 907, 933-34, 936, 944, 962, 969
  option, 618-20, 867-68, 880
Window object, 1
Window Properties
  button, 273, 683
  dialog box, 271, 273-75, 277-78, 280, 287, 683-84
Window Style Properties dialog box, 286-91, 293-96
Window Styles
  command, 283-85
  dialog box, 284-86, 292-93, 296
  list, 292
Window Type list box, 288-89, 293
WINDOWADD command, 270
WINDOWMODIFY command, 276

Windows
  Custom Imperial Style file, 3
  editing with grips, 277-80
  inserting, 7. (See also Design menu)
  Tile Vertically, 2
Wireframe. (See Shading Mode option)
Work Reflected Ceiling Viewport, 34-35
Work_Model display representation set, 36-37
Work-3D layout tab, 18
Work-FLR layout tab, 18, 53, 59, 70, 112, 115
Work-RCP layout tab, 18, 34
Work-SEC layout tab, 18
World Coordinate System radio button, 526, 569-70, 702
Write Block dialog box, 492-93, 720-21, 738

X Scale command line option, 524-25
X-Axis shortcut, 972
X-Baysize edit field, 956-58, 971, 987-88, 1004
X-Labeling tab, 976-77, 981, 985
X-Spacing tab, 966, 992
X-Width edit field, 955-57, 959-60, 964-65, 987-88, 991

Y Offset From Insertion edit box, 530-31
Y Scale command line option, 524-25
Y-Axis shortcut, 972
Y-Baysize edit field, 956-57, 971, 987-88, 1004
Y-Depth field, 955-57, 959-60, 964-65, 987-88, 991
Y-Labeling tab, 976-78, 982, 985
Y-Spacing tab, 966-67, 983, 992

Z Scale command line option, 524-25
Zoom Center. (See More option)
Zoom
  command, 9, 324, 692, 820, 823
  edit field, 1019
  cursor, 821
Zoom All, 608
Zoom Extents, 721, 737-38, 768, 771, 790, 792, 795, 850, 853`. (See also More option)
Zoom flyout, 721, 738
Zoom In. (See More option)
Zoom option, 33. (See also Viewer)
Zoom Out, 735. (See also More option)
Zoom Previous, 531, 983, 985
Zoom Window, 230, 251, 324, 327, 528, 531, 606, 608, 637, 766, 790-91, 849, 851, 917, 931, 983, 1004, 1013. (See also More option)

License Agreement for Autodesk Press, Thomson Learning™

Educational Software/Data

You the customer, and Autodesk Press incur certain benefits, rights, and obligations to each other when you open this package and use the software/data it contains. BE SURE YOU READ THE LICENSE AGREEMENT CAREFULLY, SINCE BY USING THE SOFTWARE/DATA YOU INDICATE YOU HAVE READ, UNDERSTOOD, AND ACCEPTED THE TERMS OF THIS AGREEMENT.

Your rights:

1. You enjoy a non-exclusive license to use the enclosed software/data on a single microcomputer that is not part of a network or multi-machine system in consideration for payment of the required license fee, (which may be included in the purchase price of an accompanying print component), or receipt of this software/data, and your acceptance of the terms and conditions of this agreement.

2. You own the media on which the software/data is recorded, but you acknowledge that you do not own the software/data recorded on them. You also acknowledge that the software/data is furnished "as is," and contains copyrighted and/or proprietary and confidential information of Autodesk Press or its licensors.

3. If you do not accept the terms of this license agreement you may return the media within 30 days. However, you may not use the software during this period.

There are limitations on your rights:

1. You may not copy or print the software/data for any reason whatsoever, except to install it on a hard drive on a single microcomputer and to make one archival copy, unless copying or printing is expressly permitted in writing or statements recorded on the diskette(s).

2. You may not revise, translate, convert, disassemble or otherwise reverse engineer the software/data except that you may add to or rearrange any data recorded on the media as part of the normal use of the software/data.

3. You may not sell, license, lease, rent, loan, or otherwise distribute or network the software/data except that you may give the software/data to a student or and instructor for use at school or, temporarily at home.

Should you fail to abide by the Copyright Law of the United States as it applies to this software/data your license to use it will become invalid. You agree to erase or otherwise destroy the software/data immediately after receiving note of Autodesk Press' termination of this agreement for violation of its provisions.

Autodesk Press gives you a LIMITED WARRANTY covering the enclosed software/data. The LIMITED WARRANTY can be found in this product and/or the instructor's manual that accompanies it.

This license is the entire agreement between you and Autodesk Press interpreted and enforced under New York law.

Limited Warranty

Autodesk Press warrants to the original licensee/ purchaser of this copy of microcomputer software/ data and the media on which it is recorded that the media will be free from defects in material and workmanship for ninety (90) days from the date of original purchase. All implied warranties are limited in duration to this ninety (90) day period. THEREAFTER, ANY IMPLIED WARRANTIES, INCLUDING IMPLIED WARRANTIES OF MERCHANTABILITY AND FITNESS FOR A PARTICULAR PURPOSE ARE EXCLUDED. THIS WARRANTY IS IN LIEU OF ALL OTHER WARRANTIES, WHETHER ORAL OR WRITTEN, EXPRESSED OR IMPLIED.

If you believe the media is defective, please return it during the ninety day period to the address shown below. A defective diskette will be replaced without charge provided that it has not been subjected to misuse or damage.

This warranty does not extend to the software or information recorded on the media. The software and information are provided "AS IS." Any statements made about the utility of the software or information are not to be considered as express or implied warranties. Autodesk Press will not be liable for incidental or consequential damages of any kind incurred by you, the consumer, or any other user.

Some states do not allow the exclusion or limitation of incidental or consequential damages, or limitations on the duration of implied warranties, so the above limitation or exclusion may not apply to you. This warranty gives you specific legal rights, and you may also have other rights which vary from state to state. Address all correspondence to:

Autodesk Press
3 Columbia Circle
P. O. Box 15015
Albany, NY 12212-5015